Mathematical Tools for Data Mining

Advanced Information and Knowledge Processing

Series editors
Professor Lakhmi Jain
lakhmi.jain@unisa.edu.au

Professor Xindong Wu
xwu@cs.uvm.edu

For further volumes:
http://www.springer.com/series/4738

Dan A. Simovici · Chabane Djeraba

Mathematical Tools for Data Mining

Set Theory, Partial Orders, Combinatorics

Second Edition

Springer

Dan A. Simovici MS, MS, Ph.D.
University of Massachusetts
Boston
USA

Chabane Djeraba BSc, MSc, Ph.D.
University of Sciences and Technologies
of Lille
Villeneuve d'Ascq
France

ISSN 1610-3947
ISBN 978-1-4471-7134-8 ISBN 978-1-4471-6407-4 (eBook)
DOI 10.1007/978-1-4471-6407-4
Springer London Heidelberg New York Dordrecht

Printed on acid-free paper

Springer is part of Springer Science+Business Media (www.springer.com)

Preface

The data mining literature contains many excellent titles that address the needs of users with a variety of interests ranging from decision making to pattern investigation in biological data. However, these books do not deal with the mathematical tools that are currently needed by data mining researchers and doctoral students and we felt that it is timely to produce a new version of our book that integrates the mathematics of data mining with its applications. We emphasize that this book is about mathematical tools for data mining and *not* about data mining itself; despite this, many substantial applications of mathematical concepts in data mining are included. The book is intended as a reference for the working data miner.

We present several areas of mathematics that, in our opinion are vital for data mining: *set theory*, including partially ordered sets and combinatorics; *linear algebra*, with its many applications in linear algorithms; *topology* that is used in understanding and structuring data, and *graph theory* that provides a powerful tool for constructing data models.

Our set theory chapter begins with a study of functions and relations. Applications of these fundamental concepts to such issues as equivalences and partitions are discussed. We have also included a précis of universal algebra that covers the needs of subsequent chapters.

Partially ordered sets are important on their own and serve in the study of certain algebraic structures, namely lattices, and Boolean algebras. This is continued with a combinatorics chapter that includes such topics as the inclusion–exclusion principle, combinatorics of partitions, counting problems related to collections of sets, and the Vapnik–Chervonenkis dimension of collections of sets.

An introduction to topology and measure theory is followed by a study of the topology of metric spaces, and of various types of generalizations and specializations of the notion of metric. The dimension theory of metric spaces is essential for recent preoccupations of data mining researchers with the applications of fractal theory to data mining.

A variety of applications in data mining are discussed, such as the notion of entropy, presented in a new algebraic framework related to partitions rather than random distributions, level-wise algorithms that generalize the Apriori technique, and generalized measures and their use in the study of frequent item sets.

Linear algebra is present in this new edition with three chapters that treat linear spaces, norms and inner products, and spectral theory. The inclusion of these chapters allowed us to expand our treatment of graph theory and include many new applications.

A final chapter is dedicated to clustering that includes basic types of clustering algorithms, techniques for evaluating cluster quality, and spectral clustering.

The text of this second edition, which appears 7 years after the publication of the first edition, was reorganized, corrected, and substantially amplified. Each chapter ends with suggestions for further reading. Over 700 exercises and supplements are included; they form an integral part of the material. Some of the exercises are in reality supplemental material. For these, we include solutions. The mathematics required for making the best use of our book is a typical three-semester sequence in calculus.

Boston, January 2014 Dan A. Simovici
Villeneuve d'Ascq Chabane Djeraba

Contents

Chapter 1
Sets, Relations, and Functions

1.1 Introduction

In this chapter, dedicated to set-theoretical bases of data mining, we assume that the reader is familiar with the notion of a set, membership of an element in a set, and elementary set theory. After a brief review of set-theoretical operations we discuss collections of sets, ordered pairs, and set products.

Countable and uncountable sets are presented in Sect. 1.4. An introductory section on elementary combinatorics is expanded in Chap. 3.

We present succinctly several algebraic structures to the extent that they are necessary for the material presented in the subsequent chapters. We emphasize notions like operations, morphisms, and congruences that are of interest for the study of any algebraic structure. Finally, we discuss closure and interior systems, topics that have multiple applications in topology, algebra, and data mining.

1.2 Sets and Collections

The membership of x in a set S is denoted by $x \in S$; if x is not a member of the set S, we write $x \notin S$.

Throughout this book, we use standardized notations for certain important sets of numbers:

$\mathbb{C}$	the set of complex numbers	$\mathbb{R}$	the set of real numbers
$\mathbb{R}_{\geqslant 0}$	the set of nonnegative real numbers	$\mathbb{R}_{>0}$	the set of positive real numbers
$\hat{\mathbb{R}}_{\geqslant 0}$	the set $\mathbb{R}_{\geqslant 0} \cup \{+\infty\}$	$\hat{\mathbb{R}}$	the set $\mathbb{R} \cup \{-\infty, +\infty\}$
$\mathbb{Q}$	the set of rational numbers	$\mathbb{I}$	the set of irrational numbers
$\mathbb{Z}$	the set of integers	$\mathbb{N}$	the set of natural numbers

D. A. Simovici and C. Djeraba, *Mathematical Tools for Data Mining*,
Advanced Information and Knowledge Processing, DOI: 10.1007/978-1-4471-6407-4_1,

The usual order of real numbers is extended to the set $\hat{\mathbb{R}}$ by $-\infty < x < +\infty$ for every $x \in \mathbb{R}$. In addition, we assume that

$$x + \infty = \infty + x = +\infty, \text{ and } x - \infty = -\infty + x = -\infty,$$

for every $x \in \mathbb{R}$. Also,

$$x \cdot \infty = \infty \cdot x = \begin{cases} +\infty & \text{if } x > 0 \\ -\infty & \text{if } x < 0, \end{cases}$$

and

$$x \cdot (-\infty) = (-\infty) \cdot x = \begin{cases} -\infty & \text{if } x > 0 \\ \infty & \text{if } x < 0. \end{cases}$$

Note that the product of 0 with either $+\infty$ or $-\infty$ is not defined. Division is extended by $x/+\infty = x/-\infty = 0$ for every $x \in \mathbb{R}$.

If S is a finite set, we denote by $|S|$ the number of elements of S.

Sets may contain other sets as elements. For example, the set

$$\mathcal{C} = \{\emptyset, \{0\}, \{0, 1\}, \{0, 2\}, \{1, 2, 3\}\}$$

contains the empty set $\emptyset$ and $\{0\}$, $\{0, 1\}$,$\{0, 2\}$,$\{1, 2, 3\}$ as its elements. We refer to such sets as *collections of sets* or simply *collections*. In general, we use calligraphic letters $\mathcal{C}, \mathcal{D}, \ldots$ to denote collections of sets.

If $\mathcal{C}$ and $\mathcal{D}$ are two collections, we say that $\mathcal{C}$ is *included* in $\mathcal{D}$, or that $\mathcal{C}$ is a *subcollection* of $\mathcal{D}$, if every member of $\mathcal{C}$ is a member of $\mathcal{D}$. This is denoted by $\mathcal{C} \subseteq \mathcal{D}$.

Two collections $\mathcal{C}$ and $\mathcal{D}$ are equal if we have both $\mathcal{C} \subseteq \mathcal{D}$ and $\mathcal{D} \subseteq \mathcal{C}$. This is denoted by $\mathcal{C} = \mathcal{D}$.

Definition 1.1 *Let $\mathcal{C}$ be a collection of sets. The* union of $\mathcal{C}$, *denoted by $\bigcup \mathcal{C}$, is the set defined by*

$$\bigcup \mathcal{C} = \{x \mid x \in S \textit{ for some } S \in \mathcal{C}\}.$$

If $\mathcal{C}$ is a nonempty *collection, its* intersection *is the set $\bigcap \mathcal{C}$ given by*

$$\bigcap \mathcal{C} = \{x \mid x \in S \textit{ for every } S \in \mathcal{C}\}.$$

If $\mathcal{C} = \{S, T\}$, we have $x \in \bigcup \mathcal{C}$ if and only if $x \in S$ or $x \in T$ and $x \in \bigcup \mathcal{C}$ if and only if $x \in S$ and $y \in T$. The union and the intersection of this two-set collection are denoted by $S \cup T$ and $S \cap T$ and are referred to as the union and the intersection of S and T, respectively.

We give, without proof, several properties of union and intersection of sets:

1. $S \cup (T \cup U) = (S \cup T) \cup U$ (*associativity of union*),
2. $S \cup T = T \cup S$ (*commutativity of union*),
3. $S \cup S = S$ (*idempotency of union*),
4. $S \cup \emptyset = S$,
5. $S \cap (T \cap U) = (S \cap T) \cap U$ (*associativity of intersection*),
6. $S \cap T = T \cap S$ (*commutativity of intersection*),
7. $S \cap S = S$ (*idempotency of intersection*),
8. $S \cap \emptyset = \emptyset$,

for all sets S, T, U.

The associativity of union and intersection allows us to denote unambiguously the union of three sets S, T, U by $S \cup T \cup U$ and the intersection of three sets S, T, U by $S \cap T \cap U$.

Definition 1.2 *The sets S and T are* disjoint *if $S \cap T = \emptyset$.*

A collection of sets $\mathcal{C}$ is said to be a collection of pairwise disjoint sets *if for every distinct sets S and T in $\mathcal{C}$, S and T are disjoint.*

Definition 1.3 *Let S and T be two sets. The* difference *of S and T is the set $S - T$ defined by $S - T = \{x \in S \mid x \notin T\}$.*

When the set S is understood from the context, we write $\overline{T}$ for $S - T$, and we refer to the set $\overline{T}$ as the *complement* of T with respect to S or simply the *complement of* T.

The relationship between set difference and set union and intersection is given in the following theorem.

Theorem 1.4 *For every set S and nonempty collection $\mathcal{C}$ of sets, we have*

$$S - \bigcup \mathcal{C} = \bigcap \{S - C \mid C \in \mathcal{C}\} \text{ and } S - \bigcap \mathcal{C} = \bigcup \{S - C \mid C \in \mathcal{C}\}.$$

Proof We leave the proof of these equalities to the reader.

Corollary 1.5 *For any sets S, T, U, we have*

$$S - (T \cup U) = (S - T) \cap (S - U) \text{ and } S - (T \cap U) = (S - T) \cup (S - U).$$

Proof Apply Theorem 1.4 to $\mathcal{C} = \{T, U\}$.

With the notation previously introduced for the complement of a set, the equalities of Corollary 1.5 become

$$\overline{T \cup U} = \overline{T} \cap \overline{U} \text{ and } \overline{T \cap U} = \overline{T} \cup \overline{U}.$$

The link between union and intersection is given by the *distributivity properties* contained in the following theorem.

Theorem 1.6 *For any collection of sets* $\mathcal{C}$ *and set* T*, we have*

$$\left(\bigcup \mathcal{C}\right) \cap T = \bigcup \{C \cap T \mid C \in \mathcal{C}\}.$$

If $\mathcal{C}$ *is nonempty, we also have*

$$\left(\bigcap \mathcal{C}\right) \cup T = \bigcap \{C \cup T \mid C \in \mathcal{C}\}.$$

Proof We prove only the first equality; the proof of the second one is left as an exercise for the reader.

Let $x \in (\bigcup \mathcal{C}) \cap T$. This means that $x \in \bigcup \mathcal{C}$ and $x \in T$. There is a set $C \in \mathcal{C}$ such that $x \in C$; hence, $x \in C \cap T$, which implies $x \in \bigcup \{C \cap T \mid C \in \mathcal{C}\}$.

Conversely, if $x \in \bigcup\{C \cap T \mid C \in \mathcal{C}\}$, there exists a member $C \cap T$ of this collection such that $x \in C \cap T$, so $x \in C$ and $x \in T$. It follows that $x \in \bigcup \mathcal{C}$, and this, in turn, gives $x \in (\bigcup \mathcal{C}) \cap T$.

Corollary 1.7 *For any sets* T, U, V*, we have*

$$(U \cup V) \cap T = (U \cap T) \cup (V \cap T) \text{ and } (U \cap V) \cup T = (U \cup T) \cap (V \cup T).$$

Proof The corollary follows immediately by choosing $\mathcal{C} = \{U, V\}$ in Theorem 1.6.

Note that if $\mathcal{C}$ and $\mathcal{D}$ are two collections such that $\mathcal{C} \subseteq \mathcal{D}$, then

$$\bigcup \mathcal{C} \subseteq \bigcup \mathcal{D} \text{ and } \bigcap \mathcal{D} \subseteq \bigcap \mathcal{C}.$$

We initially excluded the empty collection from the definition of the intersection of a collection. However, within the framework of collections of subsets of a given set S, we will extend the previous definition by taking $\bigcap \emptyset = S$ for the empty collection of subsets of S. This is consistent with the fact that $\emptyset \subseteq \mathcal{C}$ implies $\bigcap \mathcal{C} \subseteq S$.

The *symmetric difference* of sets denoted by $\oplus$ is defined by $U \oplus V = (U - V) \cup (V - U)$ for all sets U, V.

Theorem 1.8 *For all sets* U, V, T*, we have*

(i) $U \oplus U = \emptyset$;
(ii) $U \oplus V = V \oplus T$;
(iii) $(U \oplus V) \oplus T = U \oplus (V \oplus T)$.

Proof The first two parts of the theorem are direct applications of the definition of $\oplus$. We leave to the reader the proof of the third part (the associativity of $\oplus$).

The next theorem allows us to introduce a type of set collection of fundamental importance.

Theorem 1.9 *Let* $\{\{x, y\}, \{x\}\}$ *and* $\{\{u, v\}, \{u\}\}$ *be two collections such that* $\{\{x, y\}, \{x\}\} = \{\{u, v\}, \{u\}\}$*. Then, we have* $x = u$ *and* $y = v$*.*

Proof Suppose that $\{\{x, y\}, \{x\}\} = \{\{u, v\}, \{u\}\}$.

If $x = y$, the collection $\{\{x, y\}, \{x\}\}$ consists of a single set, $\{x\}$, so the collection $\{\{u, v\}, \{u\}\}$ will also consist of a single set. This means that $\{u, v\} = \{u\}$, which implies $u = v$. Therefore, $x = u$, which gives the desired conclusion because we also have $y = v$.

If $x \neq y$, then neither (x, y) nor (u, v) are singletons. However, they both contain exactly one singleton, namely $\{x\}$ and $\{u\}$, respectively, so $x = u$. They also contain the equal sets $\{x, y\}$ and $\{u, v\}$, which must be equal. Since $v \in \{x, y\}$ and $v \neq u = x$, we conclude that $v = y$.

Definition 1.10 *An* ordered pair *is a collection of sets* $\{\{x, y\}, \{x\}\}$.

Theorem 1.9 implies that for an ordered pair $\{\{x, y\}, \{x\}\}$, x and y are uniquely determined. This justifies the following definition.

Definition 1.11 *Let* $\{\{x, y\}, \{x\}\}$ *be an ordered pair. Then* x *is* the first component of p *and* y *is* the second component of p.

From now on, an ordered pair $\{\{x, y\}, \{x\}\}$ will be denoted by (x, y). If both $x, y \in S$, we refer to (x, y) as an *ordered pair on the set* S.

Definition 1.12 *Let* $\mathcal{C}$ *and* $\mathcal{D}$ *be two collections of sets such that* $\bigcup \mathcal{C} = \bigcup \mathcal{D}$. $\mathcal{D}$ *is* a refinement *of* $\mathcal{C}$ *if, for every* $D \in \mathcal{D}$, *there exists* $C \in \mathcal{C}$ *such that* $D \subseteq C$.

This is denoted by $\mathcal{C} \sqsubseteq \mathcal{D}$.

Example 1.13 Consider the collection $\mathcal{C} = \{(a, \infty) \mid a \in \mathbb{R}\}$ and $\mathcal{D} = \{(a, b) \mid a, b \in \mathbb{R}, a < b\}$. It is clear that $\bigcup \mathcal{C} = \bigcup \mathcal{D} = \mathbb{R}$.

Since we have $(a, b) \subseteq (a, \infty)$ for every $a, b \in \mathbb{R}$ such that $a < b$, it follows that $\mathcal{D}$ is a refinement of $\mathcal{C}$.

Definition 1.14 *A collection of sets* $\mathcal{C}$ *is* hereditary *if* $U \in \mathcal{C}$ *and* $W \subseteq U$ *implies* $W \in \mathcal{C}$.

Example 1.15 Let S be a set. The collection of subsets of S, denoted by $\mathcal{P}(S)$, is a hereditary collection of sets since a subset of a subset T of S is itself a subset of S.

The set of subsets of S that contain k elements is denoted by $\mathcal{P}_k(S)$. Clearly, for every set S, we have $\mathcal{P}_0(S) = \{\emptyset\}$ because there is only one subset of S that contains 0 elements, namely the empty set. The set of all finite subsets of a set S is denoted by $\mathcal{P}_{\text{fin}}(S)$. It is clear that $\mathcal{P}_{\text{fin}}(S) = \bigcup_{k \in \mathbb{N}} \mathcal{P}_k(S)$.

Example 1.16 If $S = \{a, b, c\}$, then $\mathcal{P}(S)$ consists of the following eight sets:

$$\emptyset, \{a\}, \{b\}, \{c\}, \{a, b\}, \{a, c\}, \{b, c\}, \{a, b, c\}.$$

For the empty set, we have $\mathcal{P}(\emptyset) = \{\emptyset\}$.

Definition 1.17 *Let* $\mathcal{C}$ *be a collection of sets and let* U *be a set. The* trace of the collection $\mathcal{C}$ on the set U *is the collection* $\mathcal{C}_U = \{U \cap C \mid C \in \mathcal{C}\}$.

We conclude this presentation of collections of sets with two more operations on collections of sets.

Definition 1.18 *Let $\mathcal{C}$ and $\mathcal{D}$ be two collections of sets. The collections $\mathcal{C} \vee \mathcal{D}$, $\mathcal{C} \wedge \mathcal{D}$, and $\mathcal{C} - \mathcal{D}$ are given by*

$$\mathcal{C} \vee \mathcal{D} = \{C \cup D \mid C \in \mathcal{C} \text{ and } D \in \mathcal{D}\},$$
$$\mathcal{C} \wedge \mathcal{D} = \{C \cap D \mid C \in \mathcal{C} \text{ and } D \in \mathcal{D}\},$$
$$\mathcal{C} - \mathcal{D} = \{C - D \mid C \in \mathcal{C} \text{ and } D \in \mathcal{D}\}.$$

Example 1.19 Let $\mathcal{C}$ and $\mathcal{D}$ be the collections of sets defined by

$$\mathcal{C} = \{\{x\}, \{y, z\}, \{x, y\}, \{x, y, z\}\},$$
$$\mathcal{D} = \{\{y\}, \{x, y\}, \{u, y, z\}\}.$$

We have

$$\mathcal{C} \vee \mathcal{D} = \{\{x, y\}, \{y, z\}, \{x, y, z\}, \{u, y, z\}, \{u, x, y, z\}\},$$
$$\mathcal{C} \wedge \mathcal{D} = \{\emptyset, \{x\}, \{y\}, \{x, y\}, \{y, z\}\},$$
$$\mathcal{C} - \mathcal{D} = \{\emptyset, \{x\}, \{z\}, \{x, z\}\},$$
$$\mathcal{D} - \mathcal{C} = \{\emptyset, \{u\}, \{x\}, \{y\}, \{u, z\}, \{u, y, z\}\}.$$

Unlike "$\cup$" and "$\cap$", the operations "$\vee$" and "$\wedge$" between collections of sets are not idempotent. Indeed, we have, for example,

$$\mathcal{D} \vee \mathcal{D} = \{\{y\}, \{x, y\}, \{u, y, z\}, \{u, x, y, z\}\} \neq \mathcal{D}.$$

The trace $\mathcal{C}_K$ of a collection $\mathcal{C}$ on K can be written as $\mathcal{C}_K = \mathcal{C} \wedge \{K\}$.

1.3 Relations and Functions

This section covers a number of topics that are derived from the notion of relation.

1.3.1 Cartesian Products of Sets

Definition 1.20 *Let X and Y be two sets. The* Cartesian product *of X and Y is the set $X \times Y$, which consists of all pairs (x, y) such that $x \in X$ and $y \in Y$.*

If either $X = \emptyset$ or $Y = \emptyset$, then $X \times Y = \emptyset$.

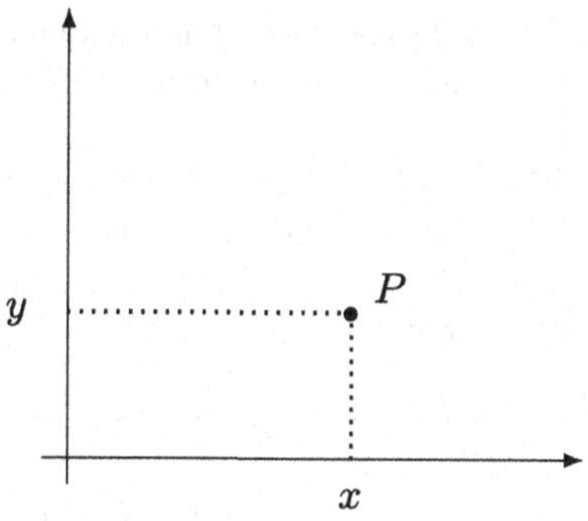

Fig. 1.1 Cartesian representation of the pair (x, y)

Example 1.21 Consider the sets $X = \{a, b, c\}$ and $Y = \{0, 1\}$. Their Cartesian product is the set $X \times Y = \{(x, 0), (y, 0), (z, 0), (x, 1), (y, 1), (z, 1)\}$.

Example 1.22 The Cartesian product $\mathbb{R} \times \mathbb{R}$ consists of all ordered pairs of real numbers (x, y). Geometrically, each such ordered pair corresponds to a point in a plane equipped with a system of coordinates. Namely, the pair $(u, v) \in \mathbb{R} \times \mathbb{R}$ is represented by the point P whose x-coordinate is u and y-coordinate is v (see Fig. 1.1)

The Cartesian product is distributive over union, intersection, and difference of sets.

Theorem 1.23 *If* $\star$ *is one of* $\cup, \cap$*, or* $-$*, then for any sets* R*,* S*, and* T*, we have*

$$(R \star S) \times T = (R \times T) \star (S \times T) \text{ and } T \times (R \star S) = (T \times R) \star (T \times S).$$

Proof We prove only that $(R - S) \times T = (R \times T) - (S \times T)$. Let $(x, y) \in (R - S) \times T$. We have $x \in R - S$ and $y \in T$. Therefore, $(x, y) \in R \times T$ and $(x, y) \notin S \times T$, which show that $(x, y) \in (R \times T) - (S \times T)$.

Conversely, $(x, y) \in (R \times T) - (S \times T)$ implies $x \in R$ and $y \in T$ and also $(x, y) \notin S \times T$. Thus, we have $x \notin S$, so $(x, y) \in (R - S) \times T$.

It is not difficult to see that if $R \subseteq R'$ and $S \subseteq S'$, then $R \times S \subseteq R' \times S'$. We refer to this property as the *monotonicity of the Cartesian product with respect to set inclusion.*

1.3.2 Relations

Definition 1.24 *A* relation *is a set of ordered pairs.*

If S *and* T *are sets and* ρ *is a relation such that* $\rho \subseteq S \times T$*, then we refer to* ρ *as a* relation from S to T.

A relation from S *to* S *is called a* relation on S.

$\mathcal{P}(S \times T)$ is the set of all relations from S to T.

Among the relations from S to T, we distinguish the *empty relation* $\emptyset$ and the *full relation* $S \times T$.

The *identity relation* of a set S is the relation $\iota_S \subseteq S \times S$ defined by $\iota_S = \{(x, x) \mid x \in S\}$. The *full relation on* S is $\theta_S = S \times S$.

If $(x, y) \in \rho$, we sometimes denote this fact by $x \,\rho\, y$, and we write $x \not\rho\, y$ instead of $(x, y) \notin \rho$.

Example 1.25 Let $S \subseteq \mathbb{R}$. The relation "less than" on S is given by

$$\{(x, y) \mid x, y \in S \text{ and } y = x + z \text{ for some } z \in \mathbb{R}_{\geqslant 0}\}.$$

Example 1.26 Consider the relation $\nu \subseteq \mathbb{Z} \times \mathbb{Q}$ given by

$$\nu = \{(n, q) \mid n \in \mathbb{Z},\ q \in \mathbb{Q},\ \text{and } n \leqslant q < n + 1\}.$$

We have $(-3, -2.3) \in \nu$ and $(2, 2.3) \in \nu$. Clearly, $(n, q) \in \nu$ if and only if n is the integral part of the rational number q.

Example 1.27 The relation δ is defined by

$$\delta = \{(m, n) \in \mathbb{N} \times \mathbb{N} \mid n = km \text{ for some } k \in \mathbb{N}\}.$$

We have $(m, n) \in \delta$ if m divides n evenly.

Note that if $S \subseteq T$, then $\iota_S \subseteq \iota_T$ and $\theta_S \subseteq \theta_T$.

Definition 1.28 *The* domain *of a relation ρ from S to T is the set*

$$\mathrm{Dom}(\rho) = \{x \in S \mid (x, y) \in \rho \text{ for some } y \in T\}.$$

The range *of ρ from S to T is the set*

$$\mathsf{Ran}(\rho) = \{y \in T \mid (x, y) \in \rho \text{ for some } x \in S\}.$$

If ρ is a relation and S and T are sets, then ρ *is a relation from S to T* if and only if $\mathrm{Dom}(\rho) \subseteq S$ and $\mathsf{Ran}(\rho) \subseteq T$. Clearly, ρ is always a relation from $\mathrm{Dom}(\rho)$ to $\mathsf{Ran}(\rho)$.

If ρ and σ are relations and $\rho \subseteq \sigma$, then $\mathrm{Dom}(\rho) \subseteq \mathrm{Dom}(\sigma)$ and $\mathsf{Ran}(\rho) \subseteq \mathsf{Ran}(\sigma)$.

If ρ and σ are relations, then so are $\rho \cup \sigma$, $\rho \cap \sigma$, and $\rho - \sigma$, and in fact if ρ and σ are both relations from S to T, then these relations are also relations from S to T.

Definition 1.29 *Let ρ be a relation. The* inverse *of ρ is the relation ρ^{-1} given by*

$$\rho^{-1} = \{(y, x) \mid (x, y) \in \rho\}.$$

The proofs of the following simple properties are left to the reader:

(i) $\text{Dom}(\rho^{-1}) = \mathsf{Ran}(\rho)$,
(ii) $\mathsf{Ran}(\rho^{-1}) = \text{Dom}(\rho)$,
(iii) if ρ is a relation from A to B, then ρ^{-1} is a relation from B to A, and
(iv) $(\rho^{-1})^{-1} = \rho$

for every relation ρ. Furthermore, if ρ and σ are two relations such that $\rho \subseteq \sigma$, then $\rho^{-1} \subseteq \sigma^{-1}$ (monotonicity of the inverse).

Definition 1.30 *Let ρ and σ be relations. The* product *of ρ and σ is the relation $\rho\sigma$, where $\rho\sigma = \{(x, z) \mid$ for some y, $(x, y) \in \rho$, and $(y, z) \in \sigma\}$.*

It is easy to see that $\text{Dom}(\rho\sigma) \subseteq \text{Dom}(\rho)$ and $\mathsf{Ran}(\rho\sigma) \subseteq \mathsf{Ran}(\sigma)$. Further, if ρ is a relation from A to B and σ is a relation from B to C, then $\rho\sigma$ is a relation from A to C.

Several properties of the relation product are given in the following theorem.

Theorem 1.31 *Let ρ_1, ρ_2, and ρ_3 be relations. We have*

(i) $\rho_1(\rho_2\rho_3) = (\rho_1\rho_2)\rho_3$ *(associativity of relation product).*
(ii) $\rho_1(\rho_2 \cup \rho_3) = (\rho_1\rho_2) \cup (\rho_1\rho_3)$ *and* $(\rho_1 \cup \rho_2)\rho_3 = (\rho_1\rho_3) \cup (\rho_2\rho_3)$ *(distributivity of relation product over union).*
(iii) $(\rho_1\rho_2)^{-1} = \rho_2^{-1}\rho_1^{-1}$.
(iv) *If* $\rho_2 \subseteq \rho_3$, *then* $\rho_1\rho_2 \subseteq \rho_1\rho_3$ *and* $\rho_2\rho_1 \subseteq \rho_3\rho_1$ *(monotonicity of relation product).*
(v) *If S and T are any sets, then $\iota_S\rho_1 \subseteq \rho_1$ and $\rho_1\iota_T \subseteq \rho_1$. Further, $\iota_S\rho_1 = \rho_1$ if and only if $\text{Dom}(\rho_1) \subseteq S$, and $\rho_1\iota_T = \rho_1$ if and only if $\mathsf{Ran}(\rho_1) \subseteq T$. (Thus, ρ_1 is a relation from S to T if and only if $\iota_S\rho_1 = \rho_1 = \rho_1\iota_T$.)*

Proof We prove (i), (ii), and (iv) and leave the other parts as exercises.

To prove Part (i), let $(a, d) \in \rho_1(\rho_2\rho_3)$. There is a b such that $(a, b) \in \rho_1$ and $(b, d) \in \rho_2\rho_3$. This means that there exists c such that $(b, c) \in \rho_2$ and $(c, d) \in \rho_3$. Therefore, we have $(a, c) \in \rho_1\rho_2$, which implies $(a, d) \in (\rho_1\rho_2)\rho_3$. This shows that $\rho_1(\rho_2\rho_3) \subseteq (\rho_1\rho_2)\rho_3$.

Conversely, let $(a, d) \in (\rho_1\rho_2)\rho_3$. There is a c such that $(a, c) \in \rho_1\rho_2$ and $(c, d) \in \rho_3$. This implies the existence of a b for which $(a, b) \in \rho_1$ and $(b, c) \in \rho_3$. For this b, we have $(b, d) \in \rho_2\rho_3$, which gives $(a, d) \in \rho_1(\rho_2\rho_3)$. We have proven the reverse inclusion, $(\rho_1\rho_2)\rho_3 \subseteq \rho_1(\rho_2\rho_3)$, which gives the associativity of relation product.

For Part (ii), let $(a, c) \in \rho_1(\rho_2 \cup \rho_3)$. Then, there is a b such that $(a, b) \in \rho_1$ and $(b, c) \in \rho_2$ or $(b, c) \in \rho_3$. In the first case, we have $(a, c) \in \rho_1\rho_2$; in the second, $(a, c) \in \rho_1\rho_3$. Therefore, we have $(a, c) \in (\rho_1\rho_2) \cup (\rho_1\rho_3)$ in either case, so $\rho_1(\rho_2 \cup \rho_3) \subseteq (\rho_1\rho_2) \cup (\rho_1\rho_3)$.

Let $(a, c) \in (\rho_1\rho_2) \cup (\rho_1\rho_3)$. We have either $(a, c) \in \rho_1\rho_2$ or $(a, c) \in \rho_1\rho_3$. In the first case, there is a b such that $(a, b) \in \rho_1$ and $(b, c) \in \rho_2 \subseteq \rho_2 \cup \rho_3$. Therefore, $(a, c) \in \rho_1(\rho_2 \cup \rho_3)$. The second case is handled similarly. This establishes $(\rho_1\rho_2) \cup (\rho_1\rho_3) \subseteq \rho_1(\rho_2 \cup \rho_3)$.

The other distributivity property has a similar argument.

Finally, for Part (iv), let ρ_2 and ρ_3 be such that $\rho_2 \subseteq \rho_3$. Since $\rho_2 \cup \rho_3 = \rho_3$, we obtain from (ii) that

$$\rho_1\rho_3 = (\rho_1\rho_2) \cup (\rho_1\rho_3),$$

which shows that $\rho_1\rho_2 \subseteq \rho_1\rho_3$. The second inclusion is proven similarly.

Definition 1.32 *The* n-power *of a relation* $\rho \subseteq S \times S$ *is defined inductively by* $\rho^0 = \iota_S$ *and* $\rho^{n+1} = \rho^n\rho$ *for* $n \in \mathbb{N}$.

Note that $\rho^1 = \rho^0\rho = \iota_S\rho = \rho$ for any relation ρ.

Example 1.33 Let $\rho \subseteq \mathbb{R} \times \mathbb{R}$ be the relation defined by

$$\rho = \{(x, x+1) \mid x \in \mathbb{R}\}.$$

The zero-th power of ρ is the relation $\iota_{\mathbb{R}}$. The second power of ρ is

$$\begin{aligned}\rho^2 = \rho \cdot \rho &= \{(x, y) \in \mathbb{R} \times \mathbb{R} \mid (x, z) \in \rho \text{ and } (z, y) \in \rho \text{ for some } z \in \mathbb{R}\}\\ &= \{(x, x+2) \mid x \in \mathbb{R}\}.\end{aligned}$$

In general, $\rho^n = \{(x, x+n) \mid x \in \mathbb{R}\}$.

Definition 1.34 *A relation* ρ *is a* function *if for all* x, y, z, $(x, y) \in \rho$ *and* $(x, z) \in \rho$ *imply* $y = z$; ρ *is a* one-to-one relation *if, for all* x, x', *and* y, $(x, y) \in \rho$ *and* $(x', y) \in \rho$ *imply* $x = x'$.

Observe that $\emptyset$ is a function (referred to in this context as the *empty function*) because $\emptyset$ satisfies vacuously the defining condition for being a function.

Example 1.35 Let S be a set. The relation ρ on $S \times \mathcal{P}(S)$ given by $\rho = \{(x, \{x\}) \mid x \in S\}$ is a function.

Example 1.36 For every set S, the relation ι_S is both a function and a one-to-one relation. The relation ν from Example 1.26 is a one-to-one relation, but it is not a function.

Theorem 1.37 *For any relation* ρ, ρ *is a function if and only if* ρ^{-1} *is a one-to-one relation.*

Proof Let ρ be a function, and let $(y_1, x), (y_2, x) \in \rho^{-1}$. Definition 1.29 implies that $(x, y_1), (x, y_2) \in \rho$ so $y_1 = y_2$, so ρ^{-1} is one-to-one.

Conversely, assume that ρ^{-1} is one-to-one and let $(x, y_1), (x, y_2) \in \rho$. Applying Definition 1.29, we obtain $(y_1, x), (y_2, x) \in \rho^{-1}$ and, since ρ^{-1} is one-to-one, we have $y_1 = y_2$. This shows that ρ is a function.

Example 1.38 We observed that the relation ν introduced in Example 1.26 is one-to-one. Therefore, its inverse $\nu^{-1} \subseteq \mathbb{Q} \times \mathbb{Z}$ is a function. In fact, ν^{-1} associates to each rational number q its integer part $\lfloor q \rfloor$.

Definition 1.39 *A relation ρ from S to T is* total *if* $\mathrm{Dom}(\rho) = S$ *and is* onto *if* $\mathsf{Ran}(\rho) = T$.

Any relation ρ is a total and onto relation from $\mathrm{Dom}(\rho)$ to $\mathsf{Ran}(\rho)$. If both S and T are nonempty, then $S \times T$ is a total and onto relation from S to T.

It is easy to prove that a relation ρ from S to T is a total relation from S to T if and only if ρ^{-1} is an onto relation from T to S.

If ρ is a relation, then one can determine whether or not ρ is a function or is one-to-one just by looking at the ordered pairs of ρ. Whether ρ is a total or onto relation from A to B depends on what A and B are.

Theorem 1.40 *Let ρ and σ be relations.*

(i) *if ρ and σ are functions, then $\rho\sigma$ is also a function;*
(ii) *if ρ and σ are one-to-one relations, then $\rho\sigma$ is also a one-to-one relation;*
(iii) *if ρ is a total relation from R to S and σ is a total relation from S to T, then $\rho\sigma$ is a total relation from R to T;*
(iv) *if ρ is an onto relation from R to S and σ is an onto relation from S to T, then $\rho\sigma$ is an onto relation from R to T;*

Proof To show Part (i), suppose that ρ and σ are both functions and that (x, z_1) and (x, z_2) both belong to $\rho\sigma$. Then, there exists a y_1 such that $(x, y_1) \in \rho$ and $(y_1, z_1) \in \sigma$, and there exists a y_2 such that $(x, y_2) \in \rho$ and $(y_2, z_2) \in \sigma$. Since ρ is a function, $y_1 = y_2$, and hence, since σ is a function, $z_1 = z_2$, as desired.

Part (ii) follows easily from Part (i). Suppose that relations ρ and σ are one-to-one (and hence that ρ^{-1} and σ^{-1} are both functions). To show that $\rho\sigma$ is one-to-one, it suffices to show that $(\rho\sigma)^{-1} = \sigma^{-1}\rho^{-1}$ is a function. This follows immediately from Part (i).

We leave the proofs for the last two parts of the theorem to the reader.

The properties of relations defined next allow us to define important classes of relations.

Definition 1.41 *Let S be a set and let $\rho \subseteq S \times S$ be a relation. The relation ρ is:*

(i) reflexive *if $(s, s) \in \rho$ for every $s \in S$;*
(ii) irreflexive *if $(s, s) \notin \rho$ for every $s \in S$;*
(iii) symmetric *if $(s, s') \in \rho$ implies $(s', s) \in \rho$ for $s, s' \in S$;*
(iv) antisymmetric *if $(s, s'), (s', s) \in \rho$ implies $s = s'$ for $s, s' \in S$;*
(v) asymmetric *if $(s, s') \in \rho$ implies $(s', s) \notin \rho$; and*
(vi) transitive *if $(s, s'), (s', s'') \in \rho$ implies $(s, s'') \in \rho$.*

Example 1.42 The relation ι_S is reflexive, symmetric, antisymmetric, and transitive for any set S.

Example 1.43 The relation δ introduced in Example 1.27 is reflexive since $n \cdot 1 = n$ for any $n \in \mathbb{N}$.

Suppose that $(m, n), (n, m) \in \delta$. There are $p, q \in \mathbb{N}$ such that $mp = n$ and $nq = m$. If $n = 0$, then this also implies $m = 0$; hence, $m = n$. Let us assume that $n \neq 0$. The previous equalities imply $nqp = n$, and since $n \neq 0$, we have $qp = 1$. In view of the fact that both p and q belong to $\mathbb{N}$, we have $p = q = 1$; hence, $m = n$, which proves the antisymmetry of ρ.

Let $(m, n), (n, r) \in \delta$. We can write $n = mp$ and $r = nq$ for some $p, q \in \mathbb{N}$, which gives $r = mpq$. This means that $(m, r) \in \delta$, which shows that δ is also transitive.

Definition 1.44 *Let S and T be two sets and let $\rho \subseteq S \times T$ be a relation.*

The image of an element $s \in S$ under the relation ρ *is the set* $\rho(s) = \{t \in T \mid (s, t) \in \rho\}$.

The preimage of an element $t \in T$ under ρ *is the set* $\{s \in S \mid (s, t) \in \rho\}$, *which equals* $\rho^{-1}(t)$, *using the previous notation.*

The collection of images of S under ρ *is*

$$IM_\rho = \{\rho(s) \mid s \in S\},$$

while the collection of preimages of T *is*

$$PIM_\rho = IM_{\rho^{-1}} = \{\rho^{-1}(t) \mid t \in T\}.$$

If $\mathcal{C}$ and $\mathcal{C}'$ are two collections of subsets of S and T, respectively, and $\mathcal{C}' = IM_\rho$ and $\mathcal{C} = PIM_\rho$ for some relation $\rho \subseteq S \times T$, we refer to $\mathcal{C}'$ as the dual class relative to ρ *of $\mathcal{C}$.*

Example 1.45 Any collection $\mathcal{D}$ of subsets of S can be regarded as the collection of images under a suitable relation. Indeed, let $\mathcal{C}$ be such a collection. Define the relation $\rho \subseteq S \times \mathcal{C}$ as $\rho = \{(s, C) \mid s \in S, C \in \mathcal{C}$ and $c \in C\}$. Then, IM_ρ consists of all subsets of $\mathcal{P}(\mathcal{C})$ of the form $\rho(s) = \{C \in \mathcal{C} \mid s \in C\}$ for $s \in S$. It is easy to see that $\mathsf{PIM}_\rho(\mathcal{C}) = \mathcal{C}$.

The collection IM_ρ defined in this example is referred to as the *bi-dual collection of* $\mathcal{C}$.

1.3.3 Functions

We saw that a function is a relation ρ such that, for every x in $\text{Dom}(\rho)$, there is only one y such that $(x, y) \in \rho$. In other words, a function assigns a unique value to each member of its domain.

From now on, we will use the letters f, g, h, and k to denote functions, and we will denote the identity relation ι_S, which we have already remarked is a function, by 1_S.

If f is a function, then, for each x in $\text{Dom}(f)$, we let $f(x)$ denote the unique y with $(x, y) \in f$, and we refer to $f(x)$ as the *image of x under f*.

Definition 1.46 *Let S and T be sets. A* partial function from S to T *is a relation from S to T that is a function.*

A total function from S to T *(also called a* function from S to T *or a* mapping from S to T*) is a partial function from S to T that is a total relation from S to T.*

The set of all partial functions from S to T is denoted by $S \rightsquigarrow T$ and the set of all total functions from S to T by $S \longrightarrow T$. We have $S \longrightarrow T \subseteq S \rightsquigarrow T$ for all sets S and T.

The fact that f is a partial function from S to T is indicated by writing $f : S \rightsquigarrow T$ rather than $f \in S \rightsquigarrow T$. Similarly, instead of writing $f \in S \longrightarrow T$, we use the notation $f : S \longrightarrow T$.

For any sets S and T, we have $\emptyset \in S \rightsquigarrow T$. If either S or T is empty, then $\emptyset$ is the only partial function from S to T. If $S = \emptyset$, then the empty function is a total function from S to any T. Thus, for any sets S and T, we have

$$S \rightsquigarrow \emptyset = \{\emptyset\}, \emptyset \rightsquigarrow T = \{\emptyset\}, \text{ and } \emptyset \longrightarrow T = \{\emptyset\}.$$

Furthermore, if S is nonempty, then there can be no (total) function from S to the empty set, so we have $S \longrightarrow \emptyset = \emptyset$ if $S \neq \emptyset$.

Definition 1.47 *A one-to-one function is called an* injection.

A function $f : S \rightsquigarrow T$ is called a surjection (from S to T) *iff f is an onto relation from S to T, and it is called a* bijection (from S to T) *or a* one-to-one correspondence between S and T *if it is total, an injection, and a surjection.*

Using our notation for functions, we can restate the definition of injection as follows: f is an injection if for all $s, s' \in \text{Dom}(f)$, $f(s) = f(s')$ implies $s = s'$. Likewise, $f : S \rightsquigarrow T$ is a surjection if for every $t \in T$ there is an $s \in S$ with $f(s) = t$.

Example 1.48 Let S and T be two sets and assume that $S \subseteq T$. The *containment mapping* $c : S \longrightarrow T$ defined by $c(s) = s$ for $s \in S$ is an injection. We denote such a containment by $c : S \hookrightarrow T$.

Example 1.49 Let $m \in \mathbb{N}$ be a natural number, $m \geqslant 2$. Consider the function $r_m : \mathbb{N} \longrightarrow \{0, \ldots, m-1\}$, where $r_m(n)$ is the remainder when n is divided by m. Obviously, r_m is well-defined since the remainder p when a natural number is divided by m satisfies $0 \leqslant p \leqslant m - 1$. The function r_m is onto because of the fact that, for any $p \in \{0, \ldots, m-1\}$, we have $r_m(km + p) = p$ for any $k \in \mathbb{N}$.

For instance, if $m = 4$, we have $r_4(0) = r_4(4) = r_4(8) = \cdots = 0$, $r_4(1) = r_4(5) = r_4(9) = \cdots = 1$, $r_4(2) = r_4(6) = r_4(10) = \cdots = 2$ and $r_4(3) = r_4(7) = r_4(11) = \cdots = 3$.

Example 1.50 Let $\mathcal{P}_{\text{fin}}(\mathbb{N})$ be the set of finite subsets of $\mathbb{N}$. Define the function $\phi : \mathcal{P}_{\text{fin}}(\mathbb{N}) \longrightarrow \mathbb{N}$ as

$$\phi(K) = \begin{cases} 0 & \text{if } K = \emptyset, \\ \sum_{i=1}^{p} 2^{n_i} & \text{if } K = \{n_1, \ldots, n_p\}. \end{cases}$$

It is easy to see that ϕ is a bijection.

Since a function is a relation, the ideas introduced in the previous section for relations in general can equally well be applied to functions. In particular, we can consider the inverse of a function and the product of two functions.

If f is a function, then, by Theorem 1.37, f^{-1} is a one-to-one relation; however, f^{-1} is not necessarily a function. In fact, by the same theorem, if f is a function, then f^{-1} is a function if and only if f is an injection.

Suppose now that $f : S \rightsquigarrow T$ is an injection. Then, $f^{-1} : T \rightsquigarrow S$ is also an injection. Further, $f^{-1} : T \rightsquigarrow S$ is total if and only if $f : S \rightsquigarrow T$ is a surjection, and $f^{-1} : T \rightsquigarrow S$ is a surjection if and only if $f : S \rightsquigarrow T$ is total. It follows that $f : S \rightsquigarrow T$ is a bijection if and only if $f^{-1} : T \rightsquigarrow S$ is a bijection.

If f and g are functions, then we will always use the alternative notation gf instead of the notation fg used for the relation product. We will refer to gf as the *composition* of f and g rather than the product.

By Theorem 1.40, the composition of two functions is a function. In fact, it follows from the definition of composition that

$$\text{Dom}(gf) = \{s \in \text{Dom}(f) \mid f(s) \in \text{Dom}(g)\}$$

and, for all $s \in \text{Dom}(gf)$,

$$gf(s) = g(f(s)).$$

This explains why we use gf rather than fg. If we used the other notation, the previous equation would become $fg(s) = g(f(s))$, which is rather confusing.

Definition 1.51 *Let* $f : S \longrightarrow T$. *A* left inverse (relative to S and T) *for* f *is a function* $g : T \longrightarrow S$ *such that* $gf = 1_S$. *A* right inverse (relative to S and T) *for* f *is a function* $g : T \longrightarrow S$ *such that* $fg = 1_T$.

Theorem 1.52 *A function* $f : S \longrightarrow T$ *is a surjection if and only if* f *has a right inverse (relative to* S *and* T*).*

If S *is nonempty, then* f *is an injection if and only if* f *has a left inverse (relative to* S *and* T*).*

Proof Suppose that $f : S \longrightarrow T$ is a surjection. Define a function $g : T \longrightarrow S$ as follows: For each $y \in T$, let $g(y)$ be some arbitrarily chosen element $x \in S$ such that $f(x) = y$. (Such an x exists because f is surjective.) Then, by definition, $f(g(y)) = y$ for all $y \in T$, so g is a right inverse for f. Conversely, suppose that f has a right inverse g. Let $y \in T$ and let $x = g(y)$. Then, we have $f(x) = f(g(y)) = 1_T(y) = y$. Thus, f is surjective.

To prove the second part, suppose that $f : S \longrightarrow T$ is an injection and that S is nonempty. Let x_0 be some fixed element of S. Define a function $g : T \longrightarrow S$ as follows: If $y \in \mathsf{Ran}(f)$, then, since f is an injection, there is a unique element $x \in S$ such that $f(x) = y$. Define $g(y)$ to be this x. If $y \in T - \mathsf{Ran}(f)$, define $g(y) = x_0$. Then, it is immediate from the definition of g that, for all $x \in S$, $g(f(x)) = x$, so g is a left inverse for f. Conversely, suppose that f has a left inverse g. For all $x_1, x_2 \in S$, if $f(x_1) = f(x_2)$, we have $x_1 = 1_S(x_1) = g(f(x_1)) = g(f(x_2)) = 1_S(x_2) = x_2$. Hence, f is an injection.

Theorem 1.53 *Let $f : S \longrightarrow T$. Then, the following statements are equivalent:*

(i) *f is a bijection;*
(ii) *there is a function $g : T \longrightarrow S$ that is both a left and a right inverse for f;*
(iii) *f has both a left inverse and a right inverse.*

Furthermore, if f is a bijection, then f^{-1} is the only left inverse that f has, and it is the only right inverse that f has.

Proof (i) implies (ii): If $f : S \longrightarrow B$ is a bijection, then $f^{-1} : T \longrightarrow S$ is both a left and a right inverse for f.

(ii) implies (iii): This implication is obvious.

(iii) implies (i): If f has both a left inverse and a right inverse and $S \neq \emptyset$, then it follows immediately from Theorem 1.52 that f is both injective and surjective, so f is a bijection. If $S = \emptyset$, then the existence of a left inverse function from T to S implies that T is also empty; this means that f is the empty function, which is a bijection from the empty set to itself.

Finally, suppose that $f : S \longrightarrow T$ is a bijection and that $g : T \longrightarrow S$ is a left inverse for f. Then, we have

$$f^{-1} = 1_S f^{-1} = (gf)f^{-1} = g(ff^{-1}) = g1_T = g.$$

Thus, f^{-1} is the unique left inverse for f. A similar proof shows that f^{-1} is the unique right inverse for f.

To prove that $f : S \longrightarrow T$ is a bijection one could prove directly that f is both one-to-one and onto. Theorem 1.53 provides an alternative way. If we can define a function $g : T \longrightarrow S$ and show that g is both a left and a right inverse for f, then f is a bijection and $g = f^{-1}$.

The next definition provides another way of viewing a subset of a set S.

Definition 1.54 *Let S be a set. An* indicator function *over S is a function $I : S \longrightarrow \{0, 1\}$.*

If P is a subset of S, then the indicator function *of P (as a subset of S) is the function $I_P : S \longrightarrow \{0, 1\}$ given by*

$$I_P(x) = \begin{cases} 1 & \text{if } x \in P \\ 0 & \text{otherwise,} \end{cases}$$

for every $x \in S$.

It is easy to see that

$$I_{P\cap Q}(x) = I_P(x) \cdot I_Q(x),$$
$$I_{P\cup Q}(x) = I_P(x) + I_Q(x) - I_P(x) \cdot I_Q(x),$$
$$I_{\bar{P}}(x) = 1 - I_P(x),$$

for every $P, Q \subseteq S$ and $x \in S$.

The relationship between the subsets of a set and indicator functions defined on that set is discussed next.

Theorem 1.55 *There is a bijection* $\Psi : \mathcal{P}(S) \longrightarrow (S \longrightarrow \{0, 1\})$ *between the set of subsets of S and the set of indicator functions defined on S.*

Proof For $P \in \mathcal{P}(S)$, define $\Psi(P) = I_P$. The mapping Ψ is one-to-one. Indeed, assume that $I_P = I_Q$, where $P, Q \in \mathcal{P}(S)$. We have $x \in P$ if and only if $I_P(x) = 1$, which is equivalent to $I_Q(x) = 1$. This happens if and only if $x \in Q$; hence, $P = Q$ so Ψ is one-to-one.

Let $f : S \longrightarrow \{0, 1\}$ be an arbitrary function. Define the set $T_f = \{x \in S \mid f(x) = 1\}$. It is easy to see that f is the indicator function of the set T_f. Hence, $\Psi(T_f) = f$, which shows that the mapping Ψ is also onto and hence it is a bijection.

Definition 1.56 *A* simple function *on a set S is a function* $f : S \longrightarrow \mathbb{R}$ *that has a finite range.*

Simple functions are linear combinations of indicator functions, as we show next.

Theorem 1.57 *Let* $f : S \longrightarrow \mathbb{R}$ *be a simple function such that* $\mathsf{Ran}(f) = \{y_1, \ldots, y_n\} \subseteq \mathbb{R}$. *Then,*

$$f = \sum_{i=1}^{n} y_i I_{f^{-1}(y_i)}.$$

Proof Let $x \in \mathbb{R}$. If $f(x) = y_j$, then

$$I_{f^{-1}(y_\ell)}(x) = \begin{cases} 1 & \text{if } \ell = j, \\ 0 & \text{otherwise.} \end{cases}$$

Thus,

$$\left(\sum_{i=1}^{n} y_i I_{f^{-1}(y_i)}\right)(x) = y_j,$$

which shows that $f(x) = \left(\sum_{i=1}^{n} y_i I_{f^{-1}(y_i)}\right)(x)$.

Theorem 1.58 *Let* $f_1, \ldots, f_k$ *be k simple functions defined on a set S. If* $g : \mathbb{R}^k \longrightarrow \mathbb{R}$ *is an arbitrary function, then* $g(f_1, \ldots, f_k)$ *is a simple function on S and we have*

$$g(f_1,\ldots,f_k)(x) = \sum_{p_1=1}^{m_1} \cdots \sum_{p_k=1}^{m_k} g(y_{1p_1},\ldots,y_{kp_k}) I_{f_1^{-1}(y_{1p_1})\cap\cdots\cap f_k^{-1}(y_{kp_k})}(x)$$

for every $x \in S$, where $\mathsf{Ran}(f_i) = \{y_{i1},\ldots,y_{im_i}\}$ for $1 \leqslant i \leqslant k$.

Proof It is clear that the function $g(f_1,\ldots,f_k)$ is a simple function because it has a finite range. Moreover, if $\mathsf{Ran}(f_i) = \{y_{i1},\ldots,y_{im_i}\}$, then the values of $g(f_1,\ldots,f_k)$ have the form $g(y_{1p_1},\ldots,y_{kp_k})$, and $g(f_1,\ldots,f_k)$ can be written as

$$\begin{aligned} & g(f_1,\ldots,f_k)(x) \\ &= \sum_{p_1=1}^{m_1} \cdots \sum_{p_k=1}^{m_k} g(y_{1p_1},\ldots,y_{kp_k}) I_{f_1^{-1}(y_{1p_1})}(x) \cdots I_{f_k^{-1}(y_{kp_k})}(x) \\ &= \sum_{p_1=1}^{m_1} \cdots \sum p_k = 1^{m_k} g(y_{1p_1},\ldots,y_{kp_k}) I_{f_1^{-1}(y_{1p_1})\cap\cdots\cap f_k^{-1}(y_{kp_k})}(x) \end{aligned}$$

for $x \in S$.

Theorem 1.58 justifies the following statement.

Theorem 1.59 *If $f_1,\ldots,f_k$ are simple functions on a set S, then*

$$\begin{aligned} &\max\{f_1(x),\ldots,f_k(x)\}, \\ &\min\{f_1(x),\ldots,f_k(x)\}, \\ &f_1(x) + \cdots + f_k(x), \\ &f_1(x) \cdot \cdots \cdot f_k(x) \end{aligned}$$

are simple functions on S.

Proof The statement follows immediately from Theorem 1.58.

1.3.3.1 Functions and Sets

Let $f : S \longrightarrow T$ be a function. If $L \subseteq S$, the *the image of L under f* is the set $f(L) = \{f(s) \mid s \in L\}$.

If $H \subseteq T$, the *inverse image of H under f* is the set $f^{-1}(H) = \{s \in S \mid f(s) \in H\}$.

It is easy to verify that $L \subseteq L'$ implies $f(L) \subseteq f(L')$ (*monotonicity of set images*) and $H \subseteq H'$ implies $f^{-1}(H) \subseteq f^{-1}(H')$ for every $L, L' \in \mathcal{P}(S)$ and $H, H' \in \mathcal{P}(T)$ (*monotonicity of set inverse images*).

Next, we discuss the behavior of images and inverse images of sets with respect to union and intersection.

Theorem 1.60 *Let $f : S \longrightarrow T$ be a function. If $\mathcal{C}$ is a collection of subsets of S, then we have*

(i) $f(\bigcup \mathcal{C}) = \bigcup\{f(L) \mid L \in \mathcal{C}\}$ *and*
(ii) $f(\bigcap \mathcal{C}) \subseteq \bigcap\{f(L) \mid L \in \mathcal{C}\}$.

Proof Note that $L \subseteq \bigcup \mathcal{C}$ for every $L \in \mathcal{C}$. The monotonicity of set images implies $f(L) \subseteq f(\bigcup \mathcal{C})$. Therefore, $\bigcup\{f(L) \mid L \in \mathcal{C}\} \subseteq f(\bigcup \mathcal{C})$.

Conversely, let $t \in f(\bigcup \mathcal{C})$. There is $s \in \bigcup \mathcal{C}$ such that $t = f(s)$. Further, since $s \in \bigcup \mathcal{C}$ we have $s \in L$, for some $L \in \mathcal{C}$, which shows that $f \in f(L) \subseteq \bigcup\{f(L) \mid L \in \mathcal{C}\}$, which implies the reverse inclusion $f(\bigcup \mathcal{C}) \subseteq \bigcup\{f(L) \mid L \in \mathcal{C}\}$.

We leave to the reader the second part of the theorem.

Theorem 1.61 *Let $f : S \longrightarrow T$ and $g : T \longrightarrow U$ be two functions. We have $f^{-1}(g^{-1}(X)) = (gf)^{-1}(X)$ for every subset X of U.*

Proof We have $s \in f^{-1}(g^{-1}(X))$ if and only if $f(s) \in g^{-1}(X)$, which is equivalent to $g(f(s)) \in X$, that is, with $s \in (gf)^{-1}(X)$. The equality of the theorem follows immediately.

Theorem 1.62 *If $f : S \longrightarrow T$ is an injective function, then $f(\bigcap \mathcal{C}) = \bigcap\{f(L) \mid L \in \mathcal{C}\}$ for every collection $\mathcal{C}$ of subsets of S.*

Proof By Theorem 1.60, it suffices to show that for an injection f we have $\bigcap\{f(L) \mid L \in \mathcal{C}\} \subseteq f(\bigcap \mathcal{C})$.

Let $y \in \bigcap\{f(L) \mid L \in \mathcal{C}\}$. For each set $L \in \mathcal{C}$ there exists $x_L \in L$ such that $f(x_L) = y$. Since f is an injection, it follows that there exists $x \in S$ such that $x_L = x$ for every $L \in \mathcal{C}$. Thus, $x \in \bigcap \mathcal{C}$, which implies that $y = f(x) \in f(\bigcap \mathcal{C})$. This allows us to obtain the desired inclusion.

Theorem 1.63 *Let $f : S \longrightarrow T$ be a function. If $\mathcal{D}$ is a collection of subsets of T, then we have*

(i) $f^{-1}(\bigcup \mathcal{D}) = \bigcup\{f^{-1}(H) \mid H \in \mathcal{D}\}$ *and*
(ii) $f^{-1}(\bigcap \mathcal{D}) = \bigcap\{f^{-1}(H) \mid H \in \mathcal{D}\}$.

Proof We prove only the second part of the theorem and leave the first part to the reader.

Since $\bigcap \mathcal{D} \subseteq H$ for every $H \in \mathcal{D}$, we have $f^{-1}(\bigcap \mathcal{D}) \subseteq f^{-1}(H)$ due to the monotonicity of set inverse images. Therefore, $f^{-1}(\bigcap \mathcal{D}) \subseteq \bigcap\{f^{-1}(H) \mid H \in \mathcal{D}\}$.

To prove the reverse inclusion, let $s \in \bigcap\{f^{-1}(H) \mid H \in \mathcal{D}\}$. This means that $s \in f^{-1}(H)$ and therefore $f(s) \in H$ for every $H \in \mathcal{D}$. This implies $f(s) \in \bigcap \mathcal{D}$, so $s \in f^{-1}(\bigcap \mathcal{D})$, which yields the reverse inclusion $\bigcap\{f^{-1}(H) \mid H \in \mathcal{D}\} \subseteq f^{-1}(\bigcap \mathcal{D})$.

Note that images and inverse images behave differently with respect to intersection. The inclusion contained by the second part of Theorem 1.60 may be strict, as the following example shows.

Example 1.64 Let $S = \{s_0, s_1, s_2\}$, $T = \{t_0, t_1\}$, and $f : S \longrightarrow T$ be the function defined by $f(s_0) = f(s_1) = t_0$ and $f(s_2) = t_1$. Consider the collection $\mathcal{C} = \{\{s_0\}, \{s_1, s_2\}\}$. Clearly, $\bigcap \mathcal{C} = \emptyset$, so $f(\bigcap \mathcal{C}) = \emptyset$. However, $f(\{s_0\}) = \{t_0\}$ and $f(\{s_1, s_2\}) = \{t_0, t_1\}$, which shows that $\bigcap\{f(L) \mid L \in \mathcal{C}\} = \{t_0\}$.

Theorem 1.65 *Let $f : S \longrightarrow T$ be a function and let U and V be two subsets of T. Then, $f^{-1}(U - V) = f^{-1}(U) - f^{-1}(V)$.*

Proof Let $s \in f^{-1}(U - V)$. We have $f(s) \in U - V$, so $f(s) \in U$ and $f(s) \notin V$. This implies $s \in f^{-1}(U)$ and $s \notin f^{-1}(V)$, so $s \in f^{-1}(U) - f^{-1}(V)$, which yields the inclusion

$$f^{-1}(U - V) \subseteq f^{-1}(U) - f^{-1}(V).$$

Conversely, let $s \in f^{-1}(U) - f^{-1}(V)$. We have $s \in f^{-1}(U)$ and $s \notin f^{-1}(V)$ which amount to $f(s) \in U$ and $f(s) \notin V$, respectively. Therefore, $f(s) \in U - V$, which implies $s \in f^{-1}(U - V)$. This proves the inclusion:

$$f^{-1}(U) - f^{-1}(V) \subseteq f^{-1}(U - V),$$

which concludes the argument.

Corollary 1.66 *Let $f : S \longrightarrow T$ be a function and let V be a subset of T. We have $f^{-1}(\bar{V}) = \overline{f^{-1}(V)}$.*

Proof Note that $f^{-1}(T) = S$ for any function $f : S \longrightarrow T$. Therefore, by choosing $U = T$ in the equality of Theorem 1.65, we have

$$S - f^{-1}(V) = f^{-1}(T - V),$$

which is precisely the statement of this corollary.

1.3.4 Finite and Infinite Sets

Functions allow us to compare sizes of sets. This idea is formalized next.

Definition 1.67 *Two sets S and T are* equinumerous *if there is a bijection $f : S \longrightarrow T$.*

The notion of equinumerous sets allows us to introduce formally the notions of finite and infinite sets.

Definition 1.68 *A set S is* finite *if there exists a natural number $n \in \mathbb{N}$ such that S is equinumerous with the set $\{0, \ldots, n - 1\}$. Otherwise, the set S is said to be* infinite.

If S is an infinite set and T is a subset of S such that $S - T$ is finite, then we refer to T as a *cofinite set*.

Theorem 1.69 *If $n \in \mathbb{N}$ and $f : \{0, \ldots, n-1\} \longrightarrow \{0, \ldots, n-1\}$ is an injection, then f is also a surjection.*

Proof Let $f : \{0, \ldots, n-1\} \longrightarrow \{0, \ldots, n-1\}$ be an injection. Suppose that f is not a surjection, that is, there is k such that $0 \leqslant k \leqslant n-1$ and $k \notin \mathsf{Ran}(f)$. Since f is injective, the elements $f(0), f(1), f(n-1)$ are distinct; this leads to a contradiction because k is not one of them. Thus, f is a surjection.

Theorem 1.70 *For any natural numbers $m, n \in \mathbb{N}$, the following statements hold:*

(i) *there exists an injection from $\{0, \ldots, n-1\}$ to $\{0, \ldots, m-1\}$ if and only if $n \leqslant m$;*
(ii) *there exists a surjection from $\{0, \ldots, n-1\}$ to $\{0, \ldots, m-1\}$ if and only if $n \geqslant m > 0$ or if $n = m = 0$;*
(iii) *there exists a bijection between $\{0, \ldots, n-1\}$ and $\{0, \ldots, m-1\}$ if and only if $n = m$.*

Proof For the first part of the theorem, if $n \leqslant m$, then the mapping $f : \{0, \ldots, n-1\} \longrightarrow \{0, \ldots, m-1\}$ given by $f(k) = k$ is the desired injection. Conversely, if $f : \{0, \ldots, n-1\} \longrightarrow \{0, \ldots, m-1\}$ is an injection, the list $(f(0), \ldots, f(n-1))$ consists of n distinct elements and is a subset of the set $\{0, \ldots, m-1\}$. Therefore, $n \leqslant m$.

For the second part, if $n = m = 0$, then the empty function is a surjection from $\{0, \ldots, n-1\}$ to $\{0, \ldots, m-1\}$. If $n \geq m > 0$, then we can define a surjection $f : \{0, \ldots, n-1\} \longrightarrow \{0, \ldots, m-1\}$ by defining

$$f(r) = \begin{cases} r & \text{if } 0 \leqslant r \leqslant m-1, \\ 0 & \text{if } m \leqslant r \leqslant n-1. \end{cases}$$

Conversely, suppose that $f : \{0, \ldots, n-1\} \longrightarrow \{0, \ldots, m-1\}$ is a surjection. Define $g : \{0, \ldots, m-1\} \longrightarrow \{0, \ldots, n-1\}$ by defining $g(r)$, for $0 \leqslant r \leqslant m-1$, to be the least t, $0 \leqslant t \leqslant n-1$, for which $f(t) = r$. (Such t exists since f is a surjection.) Then, g is an injection, and hence, by the first part, $m \leqslant n$. In addition, if $m = 0$, then we must also have $n = 0$ or else the function f could not exist.

For Part (iii), if $n = m$, then the identity function is the desired bijection. Conversely, if there is a bijection from $\{0, \ldots, n-1\}$ to $\{0, \ldots, m-1\}$, then by the first part, $n \leqslant m$, while by the second part, $n \geqslant m$, so $n = m$.

Corollary 1.71 *If S is a finite set, then there is a unique natural number n for which there exists a bijection from $\{0, \ldots, n-1\}$ to S.*

Proof Suppose that $f : \{0, \ldots, n-1\} \longrightarrow S$ and $g : \{0, \ldots, m-1\} \longrightarrow S$ are both bijections. Then, $g^{-1}f : \{0, \ldots, n-1\} \longrightarrow \{0, \ldots, m-1\}$ is a bijection, so $n = m$.

If S is a finite set, we denote by $|S|$ the unique natural number that exists for S according to Corollary 1.71. We refer to $|S|$ as the *cardinality* of S.

Corollary 1.72 *Let S and T be finite sets.*

(i) *There is an injection from S to T if and only if* $|S| \leqslant |T|$.
(ii) *There is a surjection from S to T if and only if* $|S| \geqslant |T|$.
(iii) *There is a bijection from S to B if and only if* $|S| = |T|$.

Proof Let $|S| = n$ and $|T| = m$ and let $f : \{0, \ldots, n-1\} \longrightarrow S$ and $g : \{0, \ldots, m-1\} \longrightarrow T$ be bijections. If $h : S \longrightarrow T$ is an injection, then $g^{-1}hf : \{0, \ldots, n-1\} \longrightarrow \{0, \ldots, m-1\}$ is an injection, so by Theorem 1.70, Part (i), $n \leqslant m$, i.e., $|S| \leqslant |T|$. Conversely, if $n \leqslant m$, then there is an injection $k : \{0, \ldots, n-1\} \longrightarrow \{0, \ldots, m-1\}$, namely the inclusion, and $gkf^{-1} : S \longrightarrow T$ is an injection.

The other parts are proven similarly.

1.3.5 Generalized Set Products and Sequences

The Cartesian product of two sets was introduced as the set of ordered pairs of elements of these sets. Here we present a definition of an equivalent notion that can be generalized to an arbitrary family of sets.

Definition 1.73 *Let S and T be two sets. The* set product *of S and T is the set of functions of the form* $p : \{0, 1\} \longrightarrow S \cup T$ *such that* $f(0) \in S$ *and* $f(1) \in T$.

Note that the function $\Phi : P \longrightarrow S \times T$ given by $\Phi(p) = (p(0), p(1))$ is a bijection between the set product P of the sets S and T and the Cartesian product $S \times T$. Thus, we can regard a function p in the set product of S and T as an alternate representation of an ordered pair.

Definition 1.74 *Let* $\mathcal{C} = \{S_i \mid i \in I\}$ *be a collection of sets indexed by a set I. The* set product of $\mathcal{C}$ *is the set* $\prod \mathcal{C}$ *of all functions* $f : I \longrightarrow \bigcup \mathcal{C}$ *such that* $f(i) \in S_i$ *for every* $i \in I$.

Example 1.75 Let $\mathcal{C} = \{\{0, \ldots, i\} \mid i \in \mathbb{N}\}$ be a family of sets indexed by the set of natural numbers. Clearly, we have $\bigcup \mathcal{C} = \mathbb{N}$. The set $\prod \mathcal{C}$ consists of those functions f such that $f(i) \in \{0, \ldots, i\}$ for $i \in \mathbb{N}$, that is, of those functions such that $f(i) \leqslant i$ for every $i \in I$.

Definition 1.76 *Let* $\mathcal{C} = \{S_i \mid i \in I\}$ *be a collection of sets indexed by a set I and let i be an element of I. The ith* projection *is the function* $p_i : \prod \mathcal{C} \longrightarrow S_i$ *defined by* $p_i(f) = f(i)$ *for every* $f \in \prod \mathcal{C}$.

Theorem 1.77 *Let* $\mathcal{C} = \{S_i \mid i \in I\}$ *be a collection of sets indexed by a set I and let T be a set such that, for every* $i \in I$ *there exists a function* $g_i : T \longrightarrow S_i$. *Then, there exists a unique function* $h : T \longrightarrow \prod \mathcal{C}$ *such that* $g_i = p_i h$ *for every* $i \in I$.

Proof For $t \in T$, define $h(t) = f$, where $f(i) = g_i(t)$ for every $i \in I$. We have $p_i(h(t)) = p_i(f) = g_i(t)$ for every $t \in T$, so h is a function that satisfies the conditions of the statement.

Suppose now that h_1 is another function, $h_1 : T \longrightarrow \prod \mathcal{C}$, such that $g_i = p_i h_1$ and $h_1(t) = f_1$. We have $g_i(t) = p_i(h_1(t)) = p_i(f_1) = p_i(f)$, so $f(i) = f_1(i)$ for every $i \in I$. Thus, $f = f_1$ and $h(t) = h_1(t)$ for every $t \in T$, which shows that h is unique with the property of the statement.

Let $\mathcal{C} = \{S_0, \ldots, S_{n-1}\}$ be a collection of n sets indexed by the set $\{0, \ldots, n-1\}$. By Definition 1.74 the *set product* $\prod \mathcal{C}$ consists of those functions $f : \{0, \ldots, n-1\} \longrightarrow \bigcup_{i=0}^{n-1} S_i$ such that $f(i) \in S_i$ for $0 \leqslant i \leqslant n-1$.

For set products of this type, we use the alternative notation $S_0 \times \cdots \times S_{n-1}$.

If $S_0 = \cdots = S_{n-1} = S$, we denote the set product $S_0 \times \cdots \times S_{n-1}$ by S^n.

Definition 1.78 *A* sequence on S *of length* n *is a member of this set product. If the set* S *is clear from the context, then we refer to* $\mathbf{s}$ *as a* sequence.

The set of finite sequences of length n *on the set* S *is denoted by* $\mathbf{Seq}_n(S)$.

If $\mathbf{s} \in \mathbf{Seq}_n(S)$, we refer to the number n as the *length of the sequence* $\mathbf{s}$ and it is denoted by $|\mathbf{s}|$. The *set of finite sequences on a set* S is the set $\bigcup\{\mathbf{Seq}_n(S) \mid n \in \mathbb{N}\}$, which is denoted by $\mathbf{Seq}(S)$.

For a sequence $\mathbf{s}$ of length n on the set S such that $\mathbf{s}(i) = s_i$ for $0 \leqslant i \leqslant n-1$, we denote $\mathbf{s}$ as

$$\mathbf{s} = (s_0, s_1, \ldots, s_{n-1}).$$

The elements $s_0, \ldots, s_{n-1}$ are referred to as the *components* of $\mathbf{s}$.

For a sequence $\mathbf{r} \in \mathbf{Seq}(S)$, we denote the set of elements of S that occur in $\mathbf{s}$ by *set*($\mathbf{r}$).

In certain contexts, such as the study of formal languages, sequences over a nonempty, finite set I are referred to as *words*. The set I itself is called an *alphabet*. We use special notation for words. If $I = \{a_0, \ldots, a_{n-1}\}$ is an alphabet and $\mathbf{s} = (a_{i_0}, a_{i_1}, \ldots, a_{i_{p-1}})$ is a word over the alphabet I, then we write $\mathbf{s} = a_{i_0} a_{i_1} \cdots a_{i_{p-1}}$.

The notion of a relation can also be generalized.

Definition 1.79 *Let* $\mathcal{C} = \{C_i \mid i \in I\}$ *be a collection of sets. A* $\mathcal{C}$-relation *is a subset* ρ *of the generalized Cartesian product* $\prod \mathcal{C}$. *If* I *is a finite set and* $|I| = n$, *then we say that* ρ *is an* n-ary relation.

For small values of n, *we use specific terms such as* binary relation *for* $n = 2$ *or* ternary relation *for* $n = 3$.

The number n *is the* arity of the relation ρ.

Example 1.80 Let $I = \{0, 1, 2\}$ and $C_0 = C_1 = C_2 = \mathbb{R}$. Define the ternary relation ρ on the collection $\{C_0, C_1, C_2\}$ by

$$\rho = \{(x, y, z) \in \mathbb{R}^3 \mid x < y < z\}.$$

In other words, we have $(x, y, z) \in \rho$ if and only if $y \in (x, z)$.

Definition 1.81 *Let* $\boldsymbol{p}$ *and* $\boldsymbol{q}$ *be two finite sequences in* $\mathbf{Seq}(S)$ *such that* $|\boldsymbol{p}| = m$ *and* $|\boldsymbol{q}| = n$. *The* concatenation *or the* product *of* $\boldsymbol{p}$ *and* $\boldsymbol{q}$ *is the sequence* $\boldsymbol{r}$ *given by*

$$\boldsymbol{r}(i) = \begin{cases} \boldsymbol{p}(i) & \text{if } 0 \leqslant i \leqslant m-1 \\ \boldsymbol{q}(i-m) & \text{if } m \leqslant i \leqslant m+n-1. \end{cases}$$

The concatenation of $\boldsymbol{p}$ *and* $\boldsymbol{q}$ *is denoted by* $\boldsymbol{pq}$.

Example 1.82 Let $S = \{0, 1\}$ and let $\mathbf{p}$ and $\mathbf{q}$ be the sequences

$$\mathbf{p} = (0, 1, 0, 0, 1, 1), \mathbf{q} = (1, 1, 1, 0).$$

By Definition 1.81, we have

$$\mathbf{pq} = (0, 1, 0, 0, 1, 1, 1, 1, 1, 0),$$
$$\mathbf{qp} = (1, 1, 1, 0, 0, 1, 0, 0, 1, 1).$$

The example above shows that, in general, $\mathbf{pq} \neq \mathbf{qp}$.

It follows immediately from Definition 1.81 that

$$\boldsymbol{\lambda}\mathbf{p} = \mathbf{p}\boldsymbol{\lambda} = \mathbf{p}$$

for every sequence $\mathbf{p} \in \mathbf{Seq}(S)$.

Definition 1.83 *Let* $\boldsymbol{x}$ *be a sequence,* $\boldsymbol{x} \in \mathbf{Seq}(S)$. *A sequence* $\boldsymbol{y} \in \mathbf{Seq}(S)$ *is:*

(i) *a* prefix *of* $\boldsymbol{x}$ *if* $\boldsymbol{x} = \boldsymbol{yv}$ *for some* $\boldsymbol{v} \in \mathbf{Seq}(S)$;
(ii) *a* suffix *of* $\boldsymbol{x}$ *if* $\boldsymbol{x} = \boldsymbol{uy}$ *for some* $\boldsymbol{v} \in \mathbf{Seq}(S)$; *and*
(iii) *an* infix *of* $\boldsymbol{x}$ *if* $\boldsymbol{x} = \boldsymbol{uyv}$ *for some* $\boldsymbol{u}, \boldsymbol{v} \in \mathbf{Seq}(S)$.

A sequence $\mathbf{y}$ *is a* proper prefix *(a* proper suffix, *a* proper infix*) of* $\boldsymbol{x}$ *if* $\mathbf{y}$ *is a prefix (suffix, infix) and* $\mathbf{y} \notin \{\boldsymbol{\lambda}, \boldsymbol{x}\}$.

Example 1.84 Let $S = \{a, b, c, d\}$ and $\mathbf{x} = (b, a, b, a, c, a)$. The sequence $\mathbf{y} = (b, a, b, a)$ is a prefix of $\mathbf{x}$, $\mathbf{z} = (a, c, a)$ is a suffix of $\mathbf{x}$, and $\mathbf{t} = (b, a)$ is an infix of the same sequence.

For a sequence $\mathbf{x} = (x_0, \ldots, x_{n-1})$, we denote by $\mathbf{x}_{ij}$ the infix $(x_i, \ldots, x_j)$ for $0 \leqslant i \leqslant j \leqslant n-1$. If $j < i$, $\mathbf{x}_{i,j} = \boldsymbol{\lambda}$.

Definition 1.85 *Let* S *be a set and let* $\boldsymbol{r}, \boldsymbol{s} \in \mathbf{Seq}(S)$ *such that* $|\boldsymbol{r}| \leqslant |\boldsymbol{s}|$. *The sequence* $\boldsymbol{r}$ *is a* subsequence *of* $\boldsymbol{s}$, *denoted* $\boldsymbol{r} \sqsubseteq \boldsymbol{s}$, *if there is a function* $f : \{0, \ldots, m-1\} \longrightarrow \{0, \ldots, n-1\}$ *such that* $f(0) < f(1) < \cdots < f(m-1)$ *and* $\boldsymbol{r} = \boldsymbol{s}f$.

Note that the mapping f mentioned above is necessarily injective.

If $\mathbf{r} \sqsubseteq \mathbf{s}$, as in Definition 1.85, we have $r_i = s_{f(i)}$ for $0 \leqslant i \leqslant m-1$. In other words, we can write $\mathbf{r} = (s_{i_0}, \ldots, s_{i_{m-1}})$, where $i_p = f(p)$ for $0 \leqslant p \leqslant m-1$.

The set of subsequences of a sequence $\mathbf{s}$ is denoted by *SUBSEQ*($\mathbf{s}$). There is only one subsequence of $\mathbf{s}$ of length 0, namely $\boldsymbol{\lambda}$.

Example 1.86 For $S = \{a, b, c, d\}$ and $\mathbf{x} = (b, a, b, a, c, a)$ we have $\mathbf{y} = (b, b, c) \sqsubseteq \mathbf{x}$ because $\mathbf{y} = \mathbf{x}f$, where $f : \{0, 1, 2\} \longrightarrow \{0, 1, 2, 3, 4\}$ is defined by $f(0) = 0, f(1) = 2$, and $f(2) = 4$. Note that $set(\mathbf{y}) = \{b, c\} \subseteq set(\mathbf{x}) = \{a, b, c\}$.

Definition 1.87 *Let T be a set. An* infinite sequence *on T is a function of the form $s : \mathbb{N} \longrightarrow T$.*

The set of infinite sequences on T is denoted by $\mathbf{Seq}_\infty(T)$. If $s \in \mathbf{Seq}_\infty(T)$, we write $|s| = \infty$.

For $\mathbf{s} \in \mathbf{Seq}_\infty(T)$ such that $\mathbf{s}(n) = s_n$ for $n \in \mathbb{N}$, we also use the notation $\mathbf{s} = (s_0, \ldots, s_n, \ldots)$.

The notion of a subsequence for infinite sequences has a definition that is similar to the case of finite sequences. Let $\mathbf{s} \in \mathbf{Seq}_\infty(T)$ and let $\mathbf{r} : D \longrightarrow T$ be a function, where D is either a set of the form $\{0, \ldots, m-1\}$ or the set $\mathbb{N}$. Then, $\mathbf{r}$ is a subsequence of $\mathbf{s}$ if there exists a function $f : D \longrightarrow \mathbb{N}$ such that $f(0) < f(1) < \cdots < f(k-1) < \cdots$ such that $\mathbf{r} = \mathbf{s}f$. In other words, a subsequence of an infinite sequence can be a finite sequence (when D is finite) or an infinite sequence. Observe that $\mathbf{r}(k) = \mathbf{s}(f(k)) = s_{f(k)}$ for $k \in D$. Thus, as was the case for finite sequences, the members of the sequence $\mathbf{r}$ are extracted among the members of the sequence $\mathbf{s}$. We denote this by $\mathbf{r} \sqsubseteq \mathbf{s}$, as we did for the similar notion for finite sequences.

Example 1.88 Let $\mathbf{s} \in \mathbf{Seq}_\infty(\mathbb{R})$ be the sequence defined by $\mathbf{s}(n) = (-1)^n$ for $n \in \mathbb{N}$, $\mathbf{s} = (1, -1, 1, -1, \ldots)$. If $f : \mathbb{N} \longrightarrow \mathbb{N}$ is the function given by $f(n) = 2n$ for $n \in \mathbb{N}$, then $\mathbf{r} = \mathbf{s}f$ is defined by $r_k = \mathbf{r}(k) = \mathbf{s}(f(k)) = (-1)^{2k} = 1$ for $k \in \mathbb{N}$.

1.3.5.1 Occurrences in Sequences

Let $\mathbf{x}, \mathbf{y} \in \mathbf{Seq}(S)$. An *occurrence* of $\mathbf{y}$ in $\mathbf{x}$ is a pair $(\mathbf{y}, i)$ such that $0 \leqslant i \leqslant |\mathbf{x}| - |\mathbf{y}|$ and $\mathbf{y}(k) = \mathbf{x}(i + k)$ for every k, $0 \leqslant k \leqslant |\mathbf{y}| - 1$.

The set of all occurrences of $\mathbf{y}$ in $\mathbf{x}$ is denoted by $\mathsf{OCC}_{\mathbf{y}}(\mathbf{x})$.

There is an occurrence $(\mathbf{y}, i)$ of $\mathbf{y}$ in $\mathbf{x}$ if and only if $\mathbf{y}$ is an infix of $\mathbf{x}$. If $|\mathbf{y}| = 1$, then an occurrence of $\mathbf{y}$ in $\mathbf{x}$ is called an *occurrence of the symbol* $\mathbf{y}(0)$ in $\mathbf{x}$.

$|\mathsf{OCC}_{(s)}(\mathbf{x})|$ will be referred to as the number of occurrences of a symbol s in a finite sequence $\mathbf{x}$ and be denoted by $|\mathbf{x}|_s$.

Observe that there are $|\mathbf{x}|+1$ occurrences of the null sequence λ in any sequence $\mathbf{x}$.

Let $\mathbf{x} \in \mathbf{Seq}(S)$ and let $(\mathbf{y}, i)$ and $(\mathbf{y}', j)$ be occurrences of $\mathbf{y}$ and $\mathbf{y}'$ in $\mathbf{x}$. The occurrence $(\mathbf{y}', j)$ is a *part of the occurrence* $(\mathbf{y}, i)$ if $0 \leqslant j - i \leqslant |\mathbf{y}| - |\mathbf{y}'|$.

Example 1.89 Let $S = \{a, b, c\}$ and let $\mathbf{x} \in \mathbf{Seq}(S)$ be defined by $\mathbf{x} = (a, a, b, a, b, a, c)$. The occurrences $((a, b), 1)$, $((b, a), 2)$, and $((a, b), 3)$ are parts of the occurrence $((a, b, a, b), 1)$.

Theorem 1.90 *If $(y, j) \in \mathsf{OCC}_y(x)$ and $(z, i) \in \mathsf{OCC}_z(y)$, then $(z, i + j) \in \mathsf{OCC}_z(x)$.*

Proof The argument is left to the reader.

Definition 1.91 *Let* $\mathbf{x}$ *be a finite sequence and let* $(\mathbf{y}, i)$ *be an occurrence of* $\mathbf{y}$ *in* $\mathbf{x}$. *If* $\mathbf{x} = \mathbf{x}_0\mathbf{y}\mathbf{x}_1$, *where* $|\mathbf{x}_0| = i$, *then the sequence which results from the* replacement *of the occurrence* $(\mathbf{y}, i)$ *in* $\mathbf{x}$ *by the finite sequence* $\mathbf{y}'$ *is the sequence* $\mathbf{x}_0\mathbf{y}'\mathbf{x}_1$, *denoted by* $\mathbf{replace}(\mathbf{x}(\mathbf{y}, i)\mathbf{y}')$.

Example 1.92 For the occurrences $((a, b), 1), ((a, b), 3)$ of the sequence (a, b) in the sequence $\mathbf{x} = (a, a, b, a, b, a, c)$, we have

$$\mathbf{replace}\,(\mathbf{x}, ((a, b), 1), (c, a, c)) = (a, c, a, c, a, b, a, c)$$
$$\mathbf{replace}\,(\mathbf{x}, ((a, b), 3), (c, a, c)) = (a, a, b, c, a, c, a, c).$$

1.3.5.2 Sequences of Sets

Next we examine sets defined by sequences of sets.

Let $\mathbf{s}$ be a sequence of sets. The intersection of $\mathbf{s}$ is denoted by $\bigcap_{i=0}^{n-1} S_i$ if $\mathbf{s}$ is a sequence of length n and by $\bigcap_{i=0}^{\infty} S_i$ if $\mathbf{s}$ is an infinite sequence. Similarly, the union of $\mathbf{s}$ is denoted by $\bigcup_{i=0}^{n-1} S_i$ if $\mathbf{s}$ is a sequence of length n and by $\bigcup_{i=0}^{\infty} S_i$ if $\mathbf{s}$ is an infinite sequence.

Definition 1.93 *A sequence of sets* $\mathbf{s} = (S_0, S_1, \ldots)$ *is* expanding *if* $i < j$ *implies* $S_i \subseteq S_j$ *for every* i, j *in the domain of* $\mathbf{s}$.

If $i < j$ *implies* $S_j \subseteq S_i$ *for every* i, j *in the domain of* $\mathbf{s}$, *then we say that* $\mathbf{s}$ *is a* contracting *sequence of sets.*

A sequence of sets is monotonic *if it is expanding or contracting.*

Definition 1.94 *Let* $\mathbf{s}$ *be an infinite sequence of subsets of a set* S, *where* $\mathbf{s}(i) = S_i$ *for* $i \in \mathbb{N}$.

The set $\bigcup_{i=0}^{\infty} \bigcap_{j=i}^{\infty} S_j$ *is referred to as the* lower limit *of* $\mathbf{s}$; *the set* $\bigcap_{i=0}^{\infty} \bigcup_{j=i}^{\infty} S_j$ *is the* upper limit *of* $\mathbf{s}$. *These two sets will be denoted by* $\liminf \mathbf{s}$ *and* $\limsup \mathbf{s}$, *respectively.*

If $x \in \liminf \mathbf{s}$, then there exists i such that $x \in \bigcap_{j=i}^{\infty} S_j$; in other words, x belongs to almost all sets S_i.

If $x \in \limsup \mathbf{s}$, then for every i there exists $j \geqslant i$ such that $x \in S_j$; in this case, x belongs to infinitely many sets of the sequence.

Clearly, we have $\liminf \mathbf{s} \subseteq \limsup \mathbf{s}$.

Definition 1.95 *A sequence of sets* $\mathbf{s}$ *is* convergent *if* $\liminf \mathbf{s} = \limsup \mathbf{s}$. *In this case, the set* $L = \liminf \mathbf{s} = \limsup \mathbf{s}$ *is said to be the* limit *of the sequence* $\mathbf{s}$.

The limit of $\mathbf{s}$ *will be denoted by* $\lim \mathbf{s}$.

Example 1.96 Every expanding sequence of sets is convergent. Indeed, since $\mathbf{s}$ is expanding, we have $\bigcap_{j=i}^{\infty} S_j = S_i$. Therefore, $\liminf \mathbf{s} = \bigcup_{i=0}^{\infty} S_i$. On the other hand,

$\bigcup_{j=i}^{\infty} S_j \subseteq \bigcup_{i=0}^{\infty} S_i$ and therefore lim sup $\mathbf{s} \subseteq$ lim inf $\mathbf{s}$. This shows that lim inf $\mathbf{s} =$ lim sup $\mathbf{s}$, that is, $\mathbf{s}$ is convergent.

A similar argument can be used to show that $\mathbf{s}$ is convergent when $\mathbf{s}$ is contracting.

Let $\mathcal{C}$ be a collection of subsets of a set S. Denote by $\mathcal{C}_\sigma$ the collection of all unions of subcollections of $\mathcal{C}$ indexed by $\mathbb{N}$ and by $\mathcal{C}_\delta$ the collection of all intersections of such subcollections of $\mathcal{C}$,

$$\mathcal{C}_\sigma = \left\{ \bigcup_{n \leqslant 0} C_n \mid C_n \in \mathcal{C} \right\},$$

$$\mathcal{C}_\delta = \left\{ \bigcap_{n \leqslant 0} C_n \mid C_n \in \mathcal{C} \right\}.$$

Observe that by taking $C_n = C \in \mathcal{C}$ for $n \geqslant 0$, it follows that $\mathcal{C} \subseteq \mathcal{C}_\sigma$ and $\mathcal{C} \subseteq \mathcal{C}_\delta$.

Theorem 1.97 *For any collection of subsets $\mathcal{C}$ of a set S, we have $(\mathcal{C}_\sigma)_\sigma = \mathcal{C}_\sigma$ and $(\mathcal{C}_\delta)_\delta = \mathcal{C}_\delta$.*

Proof The argument is left to the reader.

The operations σ and δ can be applied iteratively. We denote sequences of applications of these operations by subscripts adorning the affected collection. The order of application coincides with the order of these symbols in the subscript. For example, $(\mathcal{C})_{\sigma\delta\sigma}$ means $((\mathcal{C}_\sigma)_\delta)_\sigma$. Thus, Theorem 1.97 can be restated as the equalities $\mathcal{C}_{\sigma\sigma} = \mathcal{C}_\sigma$ and $\mathcal{C}_{\delta\delta} = \mathcal{C}_\delta$.

Observe that if $\mathbf{c} = (C_0, C_1, \ldots)$ is a sequence of sets, then lim sup $\mathbf{c} = \bigcap_{i=0}^{\infty} \bigcup_{j=i}^{\infty} C_j \in \mathcal{C}_{\sigma\delta}$ and lim inf $\mathbf{c} = \bigcup_{i=0}^{\infty} \bigcap_{j=i}^{\infty} C_j$ belongs to $\mathcal{C}_{\delta\sigma}$, where $\mathcal{C} = \{C_n \mid n \in \mathbb{N}\}$.

1.3.6 Equivalence Relations

Equivalence relations occur in many data mining problems and are closely related to the notion of partition, which we discuss in Sect. 1.3.7.

Definition 1.98 *An* equivalence relation *on a set S is a relation that is reflexive, symmetric, and transitive.*

The set of equivalences on A is denoted by $EQ(S)$.

An important example of an equivalence relation is presented next.

Definition 1.99 *Let U and V be two sets, and consider a function $f : U \longrightarrow V$. The relation **ker**$(f) \subseteq U \times U$, called the* kernel *of f, is given by*

$$\mathbf{ker}(f) = \{(u, u') \in U \times U \mid f(u) = f(u')\}.$$

In other words, $(u, u') \in \mathbf{ker}(f)$ if f maps both u and u' into *the same* element of V.

It is easy to verify that the relation introduced above is an equivalence. Indeed, it is clear that $(u, u) \in \mathbf{ker}(f)$ for any $u \in U$, which shows that $\iota_U \subseteq \mathbf{ker}(f)$.

The relation $\mathbf{ker}(f)$ is symmetric since $(u, u') \in \mathbf{ker}(f)$ means that $f(u) = f(u')$; hence, $f(u') = f(u)$, which implies $(u', u) \in \mathbf{ker}(f)$.

Suppose that $(u, u'), (u', u'') \in \mathbf{ker}(f)$. Then, we have $f(u) = f(u')$ and $f(u') = f(u'')$, which gives $f(u) = f(u'')$. This shows that $(u, u'') \in \mathbf{ker}(f)$; hence, $\mathbf{ker}(f)$ is transitive.

Example 1.100 Let $m \in \mathbb{N}$ be a positive natural number. Define the function $f_m : \mathbb{Z} \longrightarrow \mathbb{N}$ by $f_m(n) = r$ if r is the remainder of the division of n by m. The range of the function f_m is the set $\{0, \ldots, m-1\}$.

The relation $\mathbf{ker}(f_m)$ is usually denoted by $\equiv_m$. We have $(p, q) \in \equiv_m$ if and only if $p - q$ is divisible by m; if $(p, q) \in \equiv_m$, we also write $p \equiv q (\text{mod } m)$.

Definition 1.101 *Let ρ be an equivalence on a set U and let $u \in U$.*

The equivalence class *of u is the set $[u]_\rho$, given by*

$$[u]_\rho = \{y \in U \mid (u, y) \in \rho\}.$$

When there is no risk of confusion, we write simply $[u]$ instead of $[u]_\rho$.

Note that an equivalence class $[u]$ of an element u is never empty since $u \in [u]$ because of the reflexivity of ρ.

Theorem 1.102 *Let ρ be an equivalence on a set U and let $u, v \in U$. The following three statements are equivalent:*

(i) $(u, v) \in \rho$;
(ii) $[u] = [v]$;
(iii) $[u] \cap [v] \neq \emptyset$.

Proof The argument is immediate and we omit it.

Definition 1.103 *Let S be a set and let $\rho \in EQ(S)$. A subset U of S is* ρ-saturated *if it equals a union of equivalence classes of ρ.*

It is easy to see that U is a ρ-saturated set if and only if $x \in U$ and $(x, y) \in \rho$ imply $y \in U$. It is clear that both $\emptyset$ and S are ρ-saturated sets.

The following statement is immediate.

Theorem 1.104 *Let S be a set, $\rho \in EQ(S)$, and $\mathcal{C} = \{U_i \mid i \in I\}$ be a collection of ρ-saturated sets. Then, both $\bigcup \mathcal{C}$ and $\bigcap \mathcal{C}$ are ρ-saturated sets. Also, the complement of every ρ-saturated set is a ρ-saturated set.*

Proof We leave the argument to the reader.

A more general class of relations that generalizes equivalence relations is introduced next.

Definition 1.105 *A* tolerance relation *(or, for short, a* tolerance *on a set S is a relation that is reflexive and symmetric.*

The set of tolerances on A is denoted by $TOL(S)$.

Example 1.106 Let a be a nonnegative number and let $\rho_a \subseteq \mathbb{R} \times \mathbb{R}$ be the relation defined by

$$\rho_a = \{(x, y) \in S \times S \mid |x - y| \leqslant a\}.$$

It is clear that ρ_a is reflexive and symmetric; however, ρ_a is not transitive in general. For example, we have $(3, 5) \in \rho_2$ and $(5, 6) \in \rho_2$, but $(3, 6) \notin \rho_2$. Thus, ρ_2 is a tolerance but is not an equivalence.

1.3.7 Partitions and Covers

Next, we introduce the notion of partition of a set, a special collection of subsets of a set.

Definition 1.107 *Let S be a nonempty set. A* partition *of S is a nonempty collection $\pi = \{B_i \mid i \in I\}$ of nonempty subsets of S, such that $\bigcup\{B_i \mid i \in I\} = S$, and $B_i \cap B_j = \emptyset$ for every $i, j \in I$ such that $i \neq j$.*

Each set B_i of π is a block of the partition π.

The set of partitions of a set S is denoted by PART(S). The partition of S that consists of all singletons of the form $\{s\}$ with $s \in S$ will be denoted by α_S; the partition that consists of the set S itself will be denoted by ω_S.

Example 1.108 For the two-element set $S = \{a, b\}$, there are two partitions: the partition $\alpha_S = \{\{a\}, \{b\}\}$ and the partition $\omega_S = \{\{a, b\}\}$.

For the one-element set $T = \{c\}$, there exists only one partition, $\alpha_T = \omega_T = \{\{t\}\}$.

Example 1.109 A complete list of partitions of a set $S = \{a, b, c\}$ consists of

$$\begin{aligned}
&\pi_0 = \{\{a\}, \{b\}, \{c\}\}, \ \pi_1 = \{\{a, b\}, \{c\}\},\\
&\pi_2 = \{\{a\}, \{b, c\}\}, \quad \pi_3 = \{\{a, c\}, \{b\}\},\\
&\pi_4 = \{\{a, b, c\}\}.
\end{aligned}$$

Clearly, $\pi_0 = \alpha_S$ and $\pi_4 = \omega_S$.

Definition 1.110 *Let S be a set and let $\pi, \sigma \in PART(S)$. The partition π is* finer *than the partition σ if every block C of σ is a union of blocks of π. This is denoted by $\pi \leqslant \sigma$.*

Theorem 1.111 *Let $\pi = \{B_i \mid i \in I\}$ and $\sigma = \{C_j \mid j \in J\}$ be two partitions of a set S.*

For $\pi, \sigma \in PART(S)$, we have $\pi \leqslant \sigma$ if and only if for every block $B_i \in \pi$ there exists a block $C_j \in \sigma$ such that $B_i \subseteq C_j$.

Proof If $\pi \leqslant \sigma$, then it is clear for every block $B_i \in \pi$ there exists a block $C_j \in \sigma$ such that $B_i \subseteq C_j$.

Conversely, suppose that for every block $B_i \in \pi$ there exists a block $C_j \in \sigma$ such that $B_i \subseteq C_j$. Since two distinct blocks of σ are disjoint, it follows that for any block B_i of π, the block C_j of σ that contains B_i is unique. Therefore, if a block B of π intersects a block C of σ, then $B \subseteq C$.

Let $Q = \bigcup\{B_i \in \pi \mid B_i \subseteq C_j\}$. Clearly, $Q \subseteq C_j$. Suppose that there exists $x \in C_j - Q$. Then, there is a block $B_\ell \in \pi$ such that $x \in B_\ell \cap C_j$, which implies that $B_\ell \subseteq C_j$. This means that $x \in B_\ell \subseteq C$, which contradicts the assumption we made about x. Consequently, $C_j = Q$, which concludes the argument.

Note that $\alpha_S \leqslant \pi \leqslant \omega_S$ for every $\pi \in PART(S)$.

Two equivalence classes either coincide or are disjoint. Therefore, starting from an equivalence $\rho \in EQ(U)$, we can build a partition of the set U.

Definition 1.112 *The* quotient set *of the set U with respect to the equivalence ρ is the partition U/ρ, where*

$$U/\rho = \{[u]_\rho \mid u \in U\}.$$

An alternative notation for the partition U/ρ is π_ρ.

Moreover, we can prove that any partition defines an equivalence.

Theorem 1.113 *Let $\pi = \{B_i \mid i \in I\}$ be a partition of the set U. Define the relation ρ_π by $(x, y) \in \rho_\pi$ if there is a set $B_i \in \pi$ such that $\{x, y\} \subseteq B_i$. The relation ρ_π is an equivalence.*

Proof Let B_i be the block of the partition that contains u. Since $\{u\} \subseteq B_i$, we have $(u, u) \in \rho_\pi$ for any $u \in U$, which shows that ρ_π is reflexive.

The relation ρ_π is clearly symmetric. To prove the transitivity of ρ_π, consider $(u, v), (v, w) \in \rho_\pi$. We have the blocks B_i and B_j such that $\{u, v\} \subseteq B_i$ and $\{v, w\} \subseteq B_j$. Since $v \in B_i \cap B_j$, we obtain $B_i = B_j$ by the definition of partitions; hence, $(u, w) \in \rho_\pi$.

Corollary 1.114 *For any equivalence $\rho \in EQ(U)$, we have $\rho = \rho_{\pi_\rho}$. For any partition $\pi \in PART(U)$, we have $\pi = \pi_{\rho_\pi}$.*

Proof The argument is left to the reader.

The previous corollary amounts to the fact that there is a bijection $\phi : EQ(U) \longrightarrow PART(U)$, where $\phi(\rho) = \pi_\rho$. The inverse of this mapping, $\Psi : PART(U) \longrightarrow EQ(U)$, is given by $\psi(\pi) = \rho_\pi$.

Also, note that, for $\pi, \pi' \in PART(S)$, we have $\pi \leqslant \pi'$ if and only if $\rho_\pi \subseteq \rho_{\pi'}$.

We say that a subset T of a set S is π-*saturated* if it is a ρ_π-saturated set.

Theorem 1.115 *For any mapping $f : U \longrightarrow V$, there is a bijection $h : U/\textbf{ker}(f) \longrightarrow f(U)$.*

Proof Consider the $\mathbf{ker}(f)$ class $[u]$ of an element $u \in U$, and define $h([x]) = f(x)$. The mapping h is well-defined for if $u' \in [u]$, then $(u, u') \in \mathbf{ker}(f)$, which gives $f(u) = f(u')$.

Further, h is onto since if $y \in f(U)$, then there is $u \in U$ such that $f(u) = y$, and this gives $y = h([u])$.

To prove the injectivity of h, assume that $h([u]) = h([v])$. This means that $f(u) = f(v)$; hence, $(u, v) \in \mathbf{ker}(f)$, which means, of course, that $[u] = [v]$.

An important consequence of the previous proposition is the following decomposition theorem for mappings.

Theorem 1.116 *Every mapping $f : U \longrightarrow V$ can be decomposed as a composition of three mappings: a surjection $g : U \longrightarrow U/\mathbf{ker}(f)$, a bijection $h : U/\mathbf{ker}(f) \longrightarrow f(U)$, and an injection $k : f(U) \longrightarrow V$.*

Proof The mapping $g : U \longrightarrow U/\mathbf{ker}(f)$ is defined by $g(u) = [u]$ for $u \in U$, while $k : f(A) \longrightarrow B$ is the inclusion mapping given by $k(v) = v$ for all $v \in f(U)$. Therefore, $k(h(g(u))) = k(h([u])) = k(f(u)) = f(u)$ for all $u \in U$.

A generalization of the notion of partition is introduced next.

Definition 1.117 *Let S be a set. A* cover *of S is a nonempty collection $\mathcal{C}$ of nonempty subsets of S, $\mathcal{C} = \{B_i \mid i \in I\}$, such that $\bigcup\{B_i \mid i \in I\} = S$.*

The set of covers of a set S is denoted by $COVERS(S)$.

Example 1.118 Let S be a set. The collection $\mathcal{P}_k(S)$ of subsets of S that contain k elements is a cover of S for every $k \geqslant 1$. For $k = 1$, $\mathcal{P}_1(S)$ is actually the partition α_S.

The notion of collection refinement introduced in Definition 1.12 is clearly applicable to covers and will be used in Sect. 15.5.

Definition 1.119 *A* Sperner collection *of subsets of a set S is a collection $\mathcal{C}$ such that $\mathcal{C} \subseteq \mathcal{P}(S)$ and for any $C, C' \in \mathcal{C}$, $C \neq C'$ implies that $C \not\subseteq C'$ and $C' \not\subseteq C$.*

The set of Sperner collections on S is denoted by $\mathsf{SPER}(S)$.

Let $\mathcal{C}$ and $\mathcal{D}$ be two Sperner covers of S. Define $\mathcal{C} \leqslant \mathcal{D}$ if for every $C \in \mathcal{C}$ there exists $D \in \mathcal{D}$ such that $C \subseteq D$.

The relation $\leqslant$ is a partial order on the collection $\mathsf{SPER}(S)$. Indeed, the relation $\leqslant$ is clearly reflexive and transitive. So we need to verify only that it is antisymmetric.

Suppose that $\mathcal{C}, \mathcal{D} \in \mathsf{SPER}(S)$, $\mathcal{C} \leqslant \mathcal{D}$,and $\mathcal{D} \leqslant \mathcal{C}$. If $C \in \mathcal{C}$, there exists $D \in \mathcal{D}$ such that $C \subseteq D$. On the other hand, since $\mathcal{D} \leqslant \mathcal{C}$, there exists $C' \in \mathcal{C}$ such that $D \subseteq C'$, so $C \subseteq C'$. Since $\mathcal{C}$ is a Sperner collection this is possible only if $C = C'$, so $D = C$, which implies $\mathcal{C} \subseteq \mathcal{D}$. Applying a similar argument to an arbitrary $D \in \mathcal{D}$ yields the conclusion that $\mathcal{D} \subseteq \mathcal{C}$, so $\mathcal{C} = \mathcal{D}$, which allows us to conclude that "$\leqslant$" is antisymmetric and, therefore, a partial order on $\mathsf{SPER}(S)$.

1.4 Countable Sets

A set is called *countable* if it is either empty or the range of a sequence. A set that is not countable is called *uncountable*.

Note that if S is a countable set and $f : S \longrightarrow T$ is a surjection, then T is also countable.

Example 1.120 Every finite set is countable. Let S be a finite set. If $S = \emptyset$, then S is countable. Otherwise, suppose that $S = \{a_0, \ldots, a_{n-1}\}$, where $n \geqslant 1$. Define the sequence $\mathbf{s}$ as

$$\mathbf{s}(i) = \begin{cases} a_i & \text{if } 0 \leqslant i \leqslant n-1, \\ a_{n-1} & \text{otherwise.} \end{cases}$$

It is immediate that $\mathsf{Ran}(\mathbf{s}) = S$.

Example 1.121 The set $\mathbb{N}$ is countable because $\mathbb{N} = \mathsf{Ran}\mathbf{s}$, where $\mathbf{s}$ is the sequence $\mathbf{s}(n) = n$ for $n \in \mathbb{N}$. A similar argument can be used to show that the set $\mathbb{Z}$ is countable. Indeed, let $\mathbf{t}$ be the sequence defined by

$$\mathbf{t}(n) = \begin{cases} \frac{n-1}{2} & \text{if } n \text{ is odd} \\ -\frac{n}{2} & \text{if } n \text{ is even.} \end{cases}$$

Let m be an integer. If $m > 0$, then $m = \mathbf{t}(2m - 1)$; otherwise (that is, if $m \leqslant 0$), $m = \mathbf{t}(-2m)$, so $\mathbf{z} = \mathsf{Ran}(\mathbf{t})$.

Example 1.122 We prove now that the set $N \times N$ is countable. To this end, consider the representation of pairs of natural numbers shown in Fig. 1.2. The pairs of the set $\mathbb{N} \times \mathbb{N}$ are scanned in the order suggested by the dotted arrows. The 0th pair is $(0, 0)$, followed by $(0, 1)$, $(1, 0)$, $(0, 2)$, $(1, 1)$, $(2, 0)$, etc. We define the bijection $\beta : \mathbb{N} \times \mathbb{N} \longrightarrow \mathbb{N}$ as $\beta(p, q) = n$, where n is the place occupied by the pair (p, q) in the previous list. Thus, $\beta(0, 0) = 0$, $\beta(0, 1) = 1$, $\beta(2, 0) = 5$, and so on.

In general, bijections of the form $h : \mathbb{N} \times \mathbb{N} \longrightarrow \mathbb{N}$ are referred to as *pairing functions*, so β is an example of a pairing function.

The existence of the inverse bijection $\beta^{-1} : \mathbb{N} \longrightarrow \mathbb{N} \times \mathbb{N}$ shows that $\mathbb{N} \times \mathbb{N}$ is indeed a countable set because $\mathbb{N} \times \mathbb{N} = \mathsf{Ran}(\beta^{-1})$.

Another example of a bijection between $\mathbb{N} \times \mathbb{N}$ and $\mathbb{P}$ can be found in Exercise 22.

Starting from countable sets, it is possible to construct uncountable sets, as we see in the next example.

Example 1.123 Let F be the set of all functions of the form $f : \mathbb{N} \longrightarrow \{0, 1\}$. We claim that F is not countable.

If F were countable, we could write $F = \{f_0, f_1, \ldots, f_n, \ldots\}$. Define the function $g : \mathbb{N} \longrightarrow \{0, 1\}$ by $g(n) = \overline{f_n(n)}$ for $n \in \mathbb{N}$, where $\bar{0} = 1$ and $\bar{1} = 0$. Note that $g \neq f_n$ for every f_n in F because $g(n) = \overline{f_n(n)} \neq f_n(n)$, that is, g is different from f_n

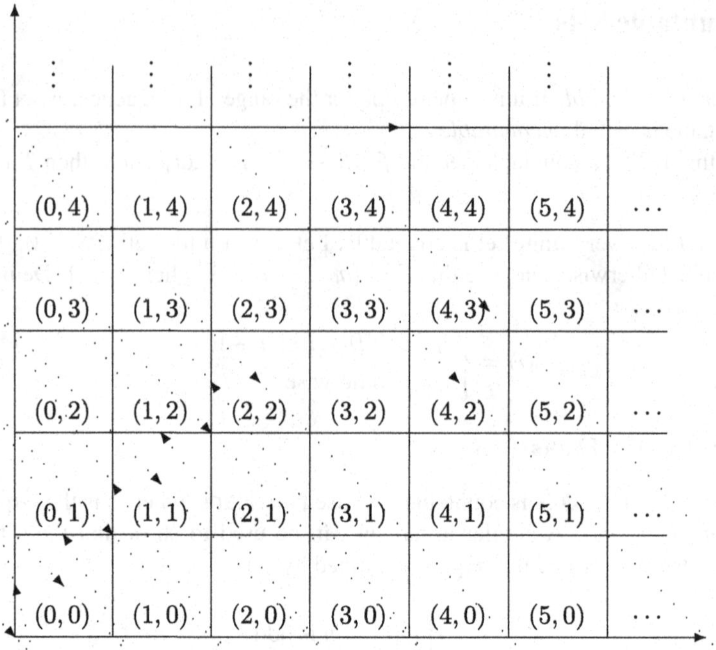

Fig. 1.2 Representation of $\mathbb{N} \times \mathbb{N}$.

at least on n for every $n \in \mathbb{N}$. But g is a function defined on $\mathbb{N}$ with values in $\{0, 1\}$, so it must equal some function f_m from F. This contradiction implies that F is not countable.

Theorem 1.124 *A subset T of a finite set S is countable.*

Proof If either S or T are empty, the statement is immediate. Suppose therefore that neither S nor T are empty and that $S = \{a_0, \ldots, a_{n-1}\}$. Since T is a subset of S, we can write $T = \{a_{i_0}, \ldots, a_{i_{m-1}}\}$, so T is the range of the sequence $\mathbf{t} : \{0, \ldots, m-1\} \longrightarrow S$ given by $\mathbf{t}(j) = a_{i_j}$ for $0 \leqslant j \leqslant m - 1$. Thus, T is countable.

Theorem 1.125 *Let $\mathcal{C} = \{C_i \mid i \in I\}$ be a collection of sets such that each set C_i is countable and the indexing set is countable. Then, $\bigcup \mathcal{C}$ is a countable set.*

Proof Without loss of generality, we can assume that none of the sets C_i is empty. Also, if $I = \emptyset$, then $\bigcup \mathcal{C}$ is empty and therefore countable.

Suppose, therefore, that C_i is the range of the sequence $\mathbf{s}_i$ for $i \in I$ and that $I \neq \emptyset$. Since I is a countable set, we can assume that I is the range of a sequence $\mathbf{z}$.

Define the function $f : \mathbb{N} \times \mathbb{N} \longrightarrow \bigcup \mathcal{C}$ by $f(p, q) = s_{z(p)}(q)$ for $p, q \in \mathbb{N}$. It is easy to verify that f is a surjection. Indeed, if $c \in \bigcup \mathcal{C}$, there exists a set C_i such that $c \in C_i$ and, since C_i is the range of the sequence $\mathbf{s}_i$, it follows that $c = \mathbf{s}_i(q)$ for some $q \in \mathbb{N}$.

Suppose that $i = \mathbf{z}(p)$. Then, we can write $c = \mathbf{s}_{\mathbf{z}(p)}(q) = f(p, q)$, which allows us to conclude that $\bigcup \mathcal{C} = \mathsf{Ran}(f)$. To obtain the enumerability of $\bigcup \mathcal{C}$, observe that this set can now be regarded as the range of the sequence $\mathbf{u}$ given by $\mathbf{u}(n) = f(\beta^{-1}(n))$, where β is the pairing function introduced in Example 1.122.

Theorem 1.126 *The set $\mathbb{Q}$ of rational numbers is countable.*

Proof We show first that the set of positive rational numbers $\mathbb{Q}_{\geqslant 0}$ is countable. Indeed, note that the function $f : \mathbb{N} \times \mathbb{N} \longrightarrow \mathbb{Q}_{\geqslant 0}$ defined by $f(m, n) = \frac{m}{n+1}$ is a surjection. A similar argument shows that the set of negative rational numbers is countable. By Theorem 1.125, the countability of $\mathbb{Q}$ follows immediately.

1.5 Multisets

Multisets generalize the notion of a set by allowing multiple copies of an element. Formally, we have the following definition.

Definition 1.127 *A* multiset *on a set S is a function $M : S \longrightarrow \mathbb{N}$. Its* carrier *is the set $carr(M) = \{x \in S \mid M(x) > 0\}$. The* multiplicity *of an element x of S in the multiset M is the number $M(x)$.*

The set of all multisets on S is denoted by $\mathcal{M}(S)$.

Example 1.128 Let PRIMES be the set of prime numbers:

$$\mathsf{PRIMES} = \{2, 3, 5, 7, 11, \ldots\}. \tag{1.1}$$

A number is determined by the multiset of its prime divisors in the following sense. If $n \in \mathbb{N}, n \geqslant 1$, can be factored as a product of prime numbers, $n = p_{i_1}^{k_1} \cdots p_{i_\ell}^{k_\ell}$, where p_i is the ith prime number and $k_1, \ldots, k_\ell$ are positive numbers, then the multiset of its prime divisors is the multiset $M_n : \mathsf{PRIMES} \longrightarrow \mathbb{N}$, where $M_n(p)$ is the exponent of the prime number p in the product (1.1).

For example, M_{1960} is given by

$$M_{1960}(p) = \begin{cases} 3 & \text{if } p = 2, \\ 1 & \text{if } p = 5, \\ 2 & \text{if } p = 7. \end{cases}$$

Thus, $carr(M_{1960}) = \{2, 5, 7\}$.

Note that if $m, n \in \mathbb{N}$, we have $M_m = M_n$ if and only if $m = n$.

We denote a multiset by using square brackets instead of braces. If x has the multiplicity n in a multiset M, we write x a number of times n inside the square brackets. For example, the multiset of Example 1.128 can be written as [2, 2, 2, 5, 7, 7].

Note that while multiplicity counts in a multiset, order does not matter; therefore, the multiset [2, 2, 2, 5, 7, 7] could also be denoted by [5, 2, 7, 2, 2, 7] or [7, 5, 2, 7, 2, 2]. We also use the abbreviation $n * x$ in a multiset to mean that x has the multiplicity n in M. For example, the multiset M_{1960} can be written as $M_{1960} = [3 * 2, 1 * 5, 2 * 7]$.

The multiset M on the set S defined by $M(x) = 0$ for $x \in S$ is the *empty multiset*.

Multisets can be combined to construct new multisets. Common set-theoretical operations such as union and intersection have natural generalizations to multisets.

Definition 1.129 *Let M and N be two multisets on a set S.*

The union of M and N *is the multiset* $M \cup N$ *defined by*

$$(M \cup N)(x) = \max\{M(x), N(x)\}$$

for $x \in S$.

The intersection of M and N *is the multiset* $M \cap N$ *defined by*

$$(M \cap N)(x) = \min\{M(x), N(x)\}$$

for $x \in S$.

The sum of M and N *is the multiset* $M + N$ *given by*

$$(M + N)(x) = M(x) + N(x)$$

for $x \in S$.

Example 1.130 Let $m, n \in \mathbb{N}$ be two numbers that have the prime factorizations

$$m = p_{i_1}^{k_1} \cdots p_{i_r}^{k_r},$$
$$n = p_{j_1}^{h_1} \cdots p_{h_s}^{h_s},$$

and let M_m, M_n be the multisets of their prime divisors, as defined in Example 1.128. Denote by $\gcd(m, n)$ the greatest common divisor of m and n, and by $\text{lcm}(m, n)$ the least common multiple of these numbers.

We have

$$M_{\gcd(m,n)} = M_m \cap M_n,$$
$$M_{\text{lcm}(m,n)} = M_m \cup M_n,$$
$$M_{mn} = M_m + M_n,$$

as the reader can easily verify.

A multiset on the set $\mathcal{P}(S)$ is referred to as a *multicollection* of sets on S.

1.6 Operations and Algebras

The notion of operation on a set is needed for introducing various algebraic structures on sets.

Definition 1.131 *Let* $n \in \mathbb{N}$. *An* n-ary operation *on a set* S *is a function* $f : S^n \longrightarrow S$. *The number* n *is the* arity *of the operation* f.

If $n = 0$, we have the special case of *zero-ary operations*. A zero-ary operation is a function $f : S^0 = \{\emptyset\} \longrightarrow S$, which is essentially a constant element of $S, f()$. Operations of arity 1 are referred to as *unary operations*.

Binary operations (of arity 2) are frequently used. For example, the union, intersection, and difference of subsets of a set S are binary operations on the set $\mathcal{P}(S)$.

If f is a binary operation on a set, we denote the result $f(x, y)$ of the application of f to x, y by xfy rather than $f(x, y)$.

We now introduce certain important types of binary operations.

Definition 1.132 *An operation* f *on a set* S *is*

(i) associative *if* $(xfy)fz = xf(yfz)$ *for every* $x, y, z \in S$,
(ii) commutative *if* $xfy = yfx$ *for every* $x, y, \in S$, *and*
(iii) idempotent *if* $xfx = x$ *for every* $x \in S$.

Example 1.133 Set union and intersection are both associative, commutative, and idempotent operations on every set of the form $\mathcal{P}(S)$.

The addition of real numbers "+" is an associative and commutative operation on $\mathbb{R}$; however, "+" is not idempotent.

The binary operation $g : \mathbb{R}^2 \longrightarrow \mathbb{R}$ given by $g(x, y) = \frac{x+y}{2}$ for $x, y \in \mathbb{R}$ is a commutative and idempotent operation of $\mathbb{R}$ that is not associative. Indeed, we have $(xgy)gz = \frac{x+y+2z}{4}$ and $xg(ygz) = \frac{2x+y+z}{4}$.

Example 1.134 The binary operations $\max\{x, y\}$ and $\min\{x, y\}$ are associative, commutative, and idempotent operations on the set $\mathbb{R}$.

Next, we introduce special elements relative to a binary operation on a set.

Definition 1.135 *Let* f *be a binary operation on a set* S.

(i) *An element* u *is a* unit *for* f *if* $xfu = ufx = x$ *for every* $x \in S$.
(ii) *An element* z *is a* zero *for* f *if* $zfu = ufz = z$ *for every* $x \in S$.

Note that if an operation f has a unit, then this unit is unique. Indeed, suppose that u and u' were two units of the operation f. According to Definition 1.135, we would have $ufx = xfu = x$ and, in particular, $ufu' = u'fu = u'$. Applying the same definition to u' yields $u'fx = xfu' = x$ and, in particular, $u'fu = ufu' = u$. Thus, $u = u'$.

Similarly, if an operation f has a zero, then this zero is unique. Suppose that z and z' were two zeros for f. Since z is a zero, we have $zfx = xfz = z$ for every $x \in S$; in particular, for $x = z'$, we have $zfz' = z'fz = z$. Since z' is zero, we also have $z'fx = xfz' = z'$ for every $x \in S$; in particular, for $x = z$, we have $z'fz = zfz' = z'$, and this implies $z = z'$.

Definition 1.136 *Let f be a binary associative operation on S such that f has the unit u. An element x has an* inverse *relative to f if there exists $y \in S$ such that $xfy = yfx = u$.*

An element x of S has at most one inverse relative to f. Indeed, suppose that both y and y' are inverses of x. Then, we have

$$y = yfu = yf(xfy') = (yfx)fy' = ufy' = y',$$

which shows that y coincides with y'.

If the operation f is denoted by "+", then we will refer to the inverse of x as the *additive inverse* of x, or the *opposite element* of x; similarly, when f is denoted by "·", we refer to the inverse of x as the *multiplicative inverse of* x. The additive inverse of x is usually denoted by $-x$, while the multiplicative inverse of x is denoted by x^{-1}.

Definition 1.137 *Let $\mathfrak{I} = \{f_i | i \in I\}$ be a set of operations on a set S indexed by a set I. An* algebra type *is a mapping $\theta : I \longrightarrow \mathbb{N}$.*

An algebra of type θ *is a pair $\mathcal{A} = (A, \mathfrak{I})$ such that*

(i) *A is a set, and*
(ii) *the operation f_i has arity $\theta(i)$ for every $i \in I$. If the type is clear from context we refer to an algebra of type θ simply as an* algebra.

The algebra $\mathcal{A} = (A, \mathfrak{I})$ is finite *if the set A is finite. The set A will be referred to as the* carrier *of the algebra $\mathcal{A}$.*

If the indexing set I is finite, we say that the type θ is a finite type *and refer to $\mathcal{A}$ as an* algebra of finite type.

If $\theta : I \longrightarrow \mathbb{N}$ is a finite algebra type, we assume, in general, that the indexing set I has the form $(0, 1, \ldots, n-1)$. In this case, we denote θ by the sequence $(\theta(0), \theta(1), \ldots, \theta(n-1))$.

Next, we discuss several algebra types.

Definition 1.138 *A* groupoid *is an algebra of type (2), $\mathcal{A} = (A, \{f\})$. If f is an associative operation, then we refer to this algebra as a* semigroup.

In other words, a groupoid is a set equipped with a binary operation f.

Example 1.139 The algebra $(\mathbb{R}, \{f\})$, where $f(x, y) = \frac{x+y}{2}$ is a groupoid. However, it is not a semigroup because f is not an associative operation.

Example 1.140 Define the binary operation g on $\mathbb{R}$ by $xgy = \ln(e^x + e^y)$ for $x, y \in \mathbb{R}$. Since

$$\begin{aligned}(xgy)gz &= \ln(e^{xgy+e^z}) = \ln(e^x + e^y + e^z)\\ xg(ygz) &= \ln(x + e^{ygz}) = \ln(e^x + e^y + e^z),\end{aligned}$$

for every $x, y, z \in \mathbb{R}$ it follows that g is an associative operation. Thus, $(\mathbb{R}, g)$ is a semigroup. It is easy to verify that this semigroup has no unit element.

Definition 1.141 A monoid *is an algebra of type* $(0, 2)$, $\mathcal{A} = (A, \{e, f\})$, *where* e *is a zero-ary operation,* f *is a binary operation, and* e *is the unit element for* f.

Example 1.142 The algebras $(\mathbb{N}, \{1, \cdot\})$ and $(\mathbb{N}, \{0, \text{gcd}\})$ are monoids. In the first case, the binary operation is the multiplication of natural numbers, the unit element is 1, and the algebra is clearly a monoid. In the second case, the binary operation $\text{gcd}(m, n)$ yields the greatest common divisor of the numbers m and n and the unit element is 0.

We claim that gcd is an associative operation. Let $m, n, p \in \mathbb{N}$. We need to verify that $\text{gcd}(m, \text{gcd}(n, p)) = \text{gcd}(\text{gcd}(m, n), p)$.

Let $k = \text{gcd}(m, \text{gcd}(n, p))$. Then, $(k, m) \in \delta$ and $(k, \text{gcd}(n, p)) \in \delta$, where δ is the divisibility relation introduced in Example 1.27. Since $\text{gcd}(n, p)$ divides evenly both n and p, it follows that $(k, n) \in \delta$ and $(k, p) \in \delta$. Thus, k divides $\text{gcd}(m, n)$, and therefore k divides $h = \text{gcd}(\text{gcd}(m, n), p)$.

Conversely, h being $\text{gcd}(\text{gcd}(m, n), p)$, it divides both $\text{gcd}(m, n)$ and p. Since h divides $\text{gcd}(m, n)$, it follows that it divides both m and p. Consequently, h divides $\text{gcd}(n, p)$ and therefore divides $k = \text{gcd}(m, \text{gcd}(n, p))$. Since k and h are both natural numbers that divide each other evenly, it follows that $k = h$, which allows us to conclude that gcd is an associative operation. Since n divides 0 evenly, for any $n \in \mathbb{N}$, it follows that $\text{gcd}(0, n) = \text{gcd}(n, 0) = n$, which shows that 0 is the unit for gcd.

Definition 1.143 A group *is an algebra of type* $(0, 2, 1)$, $\mathcal{A} = (A, \{e, f, h\})$, *where* e *is a zero-ary operation,* f *is a binary operation, and* h *is a unary operation such that the following conditions are satisfied:*

(i) e *is the unit element for* f;
(ii) $f(h(x), x) = f(x, h(x)) = e$ *for every* $x \in A$.

Note that if we have $xfy = yfx = e$, then $y = h(x)$. Indeed, we can write

$$h(x) = h(x)fe = h(x)f(xfy) = (h(x)fx)fy = efy = y.$$

We refer to the unique element $h(x)$ as the *inverse* of x. The usual notation for $h(x)$ is x^{-1}.

A special class of groups are the *Abelian groups*, also known as *commutative groups*. A group $\mathcal{A} = (A, \{e, f, h\})$ is Abelian if $xfy = yfx$ for all $x, y \in A$.

Example 1.144 The algebra $(\mathbb{Z}, \{0, +, -\})$ is an Abelian group, where "+" is the usual addition of integers, and the additive inverse of an integer n is $-n$.

Usually the binary operation of an Abelian group is denoted by "+".

Definition 1.145 A ring *is an algebra of type* $(0, 2, 1, 2)$, $\mathcal{A} = (A, \{e, f, h, g\})$, *such that* $\mathcal{A} = (A, \{e, f, h\})$ *is an Abelian group and* g *is a binary associative operation such that*

$$xg(ufv) = (xgu)f(xgv),$$
$$(ufv)gx = (ugx)f(vgx),$$

for every x, u, v ∈ A. These equalities are known as left *and* right distributivity laws, *respectively.*

The operation f is known as the ring addition, *while · is known as the* ring multiplication. *Frequently, these operations are denoted by "+" and "·", respectively.*

Example 1.146 The algebra $(\mathbb{Z}, \{0, +, -, \cdot\})$ is a ring. The distributivity laws amount to the well-known distributivity properties

$$p \cdot (q + r) = (p \cdot q) + (p \cdot r),$$
$$(q + r) \cdot p = (q \cdot p) + (r \cdot p),$$

for $p, q, r \in \mathbb{Z}$, of integer addition and multiplication.

Example 1.147 A more interesting type of ring can be defined on the set of numbers of the form $m + n\sqrt{2}$, where m and n are integers. The ring operations are given by

$$(m + n\sqrt{2}) + (p + q\sqrt{2}) = m + p + (n + q)\sqrt{2},$$
$$(m + n\sqrt{2}) \cdot (p + q\sqrt{2}) = m \cdot p + 2 \cdot n \cdot q + (m \cdot q + n \cdot p)\sqrt{2}.$$

If the multiplicative operation of a ring has a unit element 1, then we say that the ring is a *unitary ring*. We consider a unitary ring as an algebra of type (0, 0, 2, 1, 2) by regarding the multiplicative unit as another zero-ary operation.

Observe, for example, that the ring $(\mathbb{Z}, \{0, 1, +, -, \cdot\})$ is a unitary ring. Also, note that the set of even numbers also generates a ring $(\{2k \mid k \in \mathbb{Z}\}, \{0, +, -, \cdot\})$. However, no multiplicative unit exists in this ring.

Rings with commutative multiplicative operations are known as *commutative rings*. All examples of rings considered so far are commutative rings. In Sect. 5.3, we shall see an important example of a noncommutative ring.

Definition 1.148 *A* field *is a pair $\mathcal{F} = (F, \{e, f, h, g, u\})$ such that $(F, \{e, f, h, g\})$ is a commutative and unitary ring and u is a unit for the binary operation g such that every element $x \neq e$ has an inverse relative to the operation g.*

Example 1.149 The pair $\mathcal{R} = (\mathbb{R}, \{0, +, -, \cdot, 1\})$ is a field. Indeed, the multiplication "·" is a commutative operation and 1 is a multiplicative unit. In addition, each element $x \neq 0$ has the inverse $\frac{1}{x}$.

Similarly, the algebra $\mathcal{C} = (\mathbb{C}, \{0, +, -, \cdot, 1\})$ is a field referred to as the *complex field*.

If "·" is a binary operation on a set A, this operation can be extended to subsets of A. Namely, if $H, K \in \mathcal{P}(A)$, the set $H \cdot K$ is

$$H \cdot K = \{x \cdot y \mid x \in H, y \in K\}.$$

1.7 Morphisms, Congruences, and Subalgebras

Morphisms are mappings between algebras of the same type that satisfy certain compatibility conditions with the operations of the type.

Let $\theta : I \longrightarrow \mathbb{N}$ be a type. To simplify notation, we denote the operations that correspond to the same element i with the same symbol in every algebra of this type.

Definition 1.150 *Let $\theta : I \longrightarrow \mathbb{N}$ be a finite algebra type and let $\mathcal{A} = (A, \mathcal{I})$ and $\mathcal{B} = (B, \mathcal{I})$ be two algebras of the type θ. A* morphism *is a function $h : A \longrightarrow B$ such that, for every operation $f_i \in \mathcal{I}$, we have*

$$h(f_i(x_1, \ldots, x_{n_i})) = f_i(h(x_1), \ldots, h(x_{n_i}))$$

for $(x_1, \ldots, x_{n_i}) \in A^{n_i}$, where $n_i = \theta(i)$.

If the algebras $\mathcal{A}$ and $\mathcal{B}$ are the same, then we refer to f as an endomorphism *of the algebra $\mathcal{A}$.*

The set of morphisms between $\mathcal{A}$ and $\mathcal{B}$ is denoted by $\mathsf{MOR}(\mathcal{A}, \mathcal{B})$.

Example 1.151 A morphism between the groupoids $\mathcal{A} = (A, \{f\})$ and $\mathcal{B} = (B, \{f\})$ is a mapping $h : A \longrightarrow B$ such that

$$h(f(x_1, x_2)) = f(h(x_1), h(x_2)) \tag{1.2}$$

for every x_1 and x_2 in A. Exactly the same definition is valid for semigroup morphisms.

If $\mathcal{A} = (A, \{e, f\})$ and $\mathcal{B} = (B, \{e, f\}\})$ are two monoids, where e is a zero-ary operation and f is a binary operation, then a morphism of monoids must satisfy the equalities $h(e) = e$ and $h(f(x_1, x_2)) = f(h(x_1), h(x_2))$ for $x_1, x_2 \in A$.

Example 1.152 Let $(\mathbb{N}, \{0, \text{gcd}\})$ be the monoid introduced in Example 1.142. The function $h : \mathbb{N} \longrightarrow \mathbb{N}$ defined by $h(n) = n^2$ for $n \in \mathbb{N}$ is an endomorphism of this monoid because $\text{gcd}(p, q)^2 = \text{gcd}(p^2, q^2)$ for $p, q \in \mathbb{N}$.

Example 1.153 A morphism between two groups $\mathcal{A} = (A, \{e, \cdot, ^{-1}\})$ and $\mathcal{B} = (B, \{e, \cdot, ^{-1}\})$ satisfies the conditions

(i) $h(e) = e$,
(ii) $h(x_1 \cdot x_2) = h(x_1) \cdot h(x_2)$,
(iii) $h(x_1^{-1}) = (h(x_1))^{-1}$,

for $x_1, x_2 \in A$.

It is interesting to observe that, in the case of groups, the first and last conditions are consequences of the second condition, so they are superfluous. Indeed, choose $x_2 = e$ in the equality $h(x_1 \cdot x_2) = h(x_1) \cdot h(x_2)$; this yields $h(x_1) = h(x_1)h(e)$. By multiplying both sides with $h(x_1)^{-1}$ at the left and applying the associativity of the binary operation, we obtain $h(e) = e$. On the other hand, by choosing $x_2 = x_1^{-1}$, we have $e = h(e) = h(x_1)h(x_1^{-1})$, which implies $h(x_1^{-1}) = (h(x_1))^{-1}$.

Example 1.154 Let $\mathcal{A} = (A, \{0, +, ^{-1}, \cdot\})$ and $\mathcal{B} = (B, \{0, +, ^{-1}, \cdot\})$ be two rings. Then, $h : A \longrightarrow B$ is a ring morphism if $h(0) = 0$, $h(x_1 + x_2) = h(x_1) + h(x_2)$, and $h(x_1 \cdot x_2) = h(x_1) \cdot h(x_2)$.

Definition 1.155 *Let $\mathcal{A} = (A, \mathcal{I})$ be an algebra. An equivalence $\rho \in EQ(A)$ is a* congruence *if, for every operation f of the algebra, $f : A^n \longrightarrow A$, $(x_i, y_i) \in \rho$ for $1 \leqslant i \leqslant n$ implies $(f(x_1, \ldots, x_n), f(y_1, \ldots, y_n)) \in \rho$ for $x_1, \ldots, x_n, y_1, \ldots, y_n \in A$.*

Recall that we introduce the kernel of a mapping in Definition 1.99. When the mapping f is a morphism, we have further properties of $\mathbf{ker}(f)$.

Theorem 1.156 *Let $\mathcal{A} = (A, \mathcal{I})$ and $\mathcal{B} = (B, \mathcal{I})$ be two algebras of the type θ and let $h : A \longrightarrow B$ be a morphism. The relation $\mathbf{ker}(h)$ is a congruence of the algebra $\mathcal{A}$.*

Proof Let $x_1, \ldots, x_n, y_1, \ldots, y_n \in A$ such that $(x_i, y_i) \in \mathbf{ker}(h)$ for $1 \leqslant i \leqslant n$; that is, $h(x_i) = h(y_i)$ for $1 \leqslant i \leqslant n$. By applying the definition of morphism, we can write for every n-ary operation in $\mathcal{I}$

$$\begin{aligned} h(f(x_1, \ldots, x_n)) &= f(h(x_1), \ldots, h(x_n)) \\ &= f(h(y_1), \ldots, h(y_n)) \\ &= h(f(y_1, \ldots, y_n)), \end{aligned}$$

which means that $(f(x_1, \ldots, x_n), f(y_1, \ldots, y_n)) \in \mathbf{ker}(f)$. Thus, $\mathbf{ker}(f)$ is a congruence.

If $\mathcal{A} = (A, \mathcal{I})$ is an algebra and ρ is a congruence of $\mathcal{A}$, then the quotient set A/ρ (see Definition 1.112) can be naturally equipped with operations derived from the operations of $\mathcal{A}$ by

$$f([x_1]_\rho, \ldots, [x_n]_\rho) = [f(x_1, \ldots, x_n)]_\rho \tag{1.3}$$

for $x_1, \ldots, x_n \in A$. Observe first that the definition of the operation that acts on the A/ρ is a correct one for if $y_i \in [x_i]_\rho$ for $1 \leqslant i \leqslant n$, then $[f(y_1, \ldots, y_n)]_\rho = [f(x_1, \ldots, x_n)]_\rho$.

Definition 1.157 *The* quotient algebra *of an algebra $\mathcal{A} = (A, \mathcal{I})$ and a congruence ρ is the algebra $\mathcal{A}/\rho = (A/\rho, \mathcal{I})$, where each operation in $\mathcal{I}$ is defined starting from the corresponding operation f in $\mathcal{A}$ by Equality (1.3).*

Example 1.158 Let $\mathcal{A} = (A, \{e, \cdot, ^{-1}\})$ be a group. An equivalence ρ is a congruence if $(x_1, x_2), (y_1, y_2) \in \rho$ imply $(x_1^{-1}, x_2^{-1}) \in \rho$ and $(x_1 \cdot y_1, x_2 \cdot y_2) \in \rho$.

Definition 1.159 *Let $\mathcal{A} = (A, \mathcal{I})$ be an algebra. A subset B of A is* closed *if for every n-ary operation $f \in \mathcal{I}$, $x_1, \ldots, x_n \in B$ implies $f(x_1, \ldots, x_n) \in B$.*

Note that if the set B is closed, then for every zero-ary operation e of $\mathcal{I}$ we have $e \in B$.

Let $\mathcal{A} = (A, \mathcal{I})$ be an algebra and let B be a closed subset of A. The pair $(B, \mathcal{I}')$, where $\mathcal{I}' = \{g_i = f_i \restriction_B \mid i \in I\}$, is an algebra of the same type as $\mathcal{A}$. We refer to it as a *subalgebra* of $\mathcal{A}$. Often we will refer to the set B itself as a subalgebra of the algebra $\mathcal{A}$.

It is clear that the empty set is closed in an algebra $\mathcal{A} = (A, \mathcal{I})$ if and only if there is no zero-ary operation in $\mathcal{I}$.

We refer to subalgebras of particular algebras with more specific terms. For example, subalgebras of monoids or groups are referred to as *submonoids* or *subgroups*, respectively.

Theorem 1.160 *Let $\mathcal{A} = (A, \{e, \cdot, {}^{-1}\})$ be a group. A nonempty subset B of A is a subgroup if and only if $x \cdot y^{-1} \in B$ for every $x, y \in B$.*

Proof The necessity of the condition is immediate. To prove that the condition is sufficient, observe that since $B \neq \emptyset$ there is $x \in B$, so $x \cdot x^{-1} = e \in B$.

Next, let $x \in B$. Since $e \cdot x^{-1} = x^{-1}$ it follows that $x^{-1} \in B$. Finally, if $x, y \in B$, then $x \cdot y = x \cdot (y^{-1})^{-1} \in B$, which shows that B is indeed a subgroup.

Example 1.161 Let $\mathcal{A} = (A, \{e, \cdot, {}^{-1}\})$ be a group. For $u \in A$, define the set $C_u = \{x \in G \mid xu = ux\}$. If $x \in C_u$, then $xu = ux$, which implies $xux^{-1} = u$, so $ux^{-1} = x^{-1}u$. Thus, $x^{-1} \in C_u$. It is easy to see that $e \in C_u$ and $x, y \in C_u$ implies $xy \in C_u$. Thus C_u is a subgroup.

Definition 1.162 *Let $\mathcal{A} = (A, \mathcal{I})$ and $\mathcal{B} = (B, \mathcal{I})$ be two algebras of the type θ. An* isomorphism *between $\mathcal{A}$ and $\mathcal{B}$ is a bijective morphism $h : A \longrightarrow B$ between $\mathcal{A}$ and $\mathcal{B}$. An* automorphism *of $\mathcal{A}$ is an isomorphism between $\mathcal{A}$ and itself.*

Definition 1.163 *Let $\mathfrak{A} = \{\mathcal{A}_i \mid i \in I\}$ be a collection of algebras of the same type indexed by the set I, where $\mathcal{A}_i = (A_i, \mathcal{I})$.*

The product *of the collection $\mathfrak{A}$ is the algebra $\prod_{i\in I} \mathcal{A}_i = (\prod_{i\in I} A_i, \mathcal{I})$, whose operations are defined componentwise, as follows. If $f \in \mathcal{I}$ is an n-ary operation and $t_1, \ldots, t_n \in \prod_{i\in I} A_i$, where $t_k = (t_{ki})_{i\in I}$ for $1 \leqslant k \leqslant n$, then $f(t_1, \ldots, t_n) = s$, where $s = (s_i)_{i\in I}$ and $s_i = f(t_{1i}, \ldots, t_{ni})$ for $i \in I$.*

Example 1.164 Let $(A_1, \{e_1, *, {}^{-1}\})$ and $(A_2, \{e_2, *, {}^{-1}\})$ be two groups. Their product is the group $(A_1 \times A_2, (e_1, e_2), *, {}^{-1})$, where $(x_1, x_2) * (y_1, y_2) = (x_1 * y_1, x_2 * y_2)$ and $(x, y)^{-1} = (x^{-1}, y^{-1})$.

Theorem 1.165 *Let $\mathfrak{A} = \{\mathcal{A}_i \mid i \in I\}$ be a collection of algebras of the same type θ indexed by the set I, where $\mathcal{A}_i = (A_i, \mathcal{I})$, and let $\mathcal{A} = \prod_{i\in I} \mathcal{A}_i$ be their product.*

Each projection $p_j : \prod_{i\in I} A_i \longrightarrow A_j$ belongs to $\mathsf{MOR}(\mathcal{A}, \mathcal{A}_j)$ for $j \in I$. Furthermore, if $\mathcal{B}$ is an algebra of type θ and $h_i \in \mathsf{MOR}(\mathcal{B}, \mathcal{A}_i)$ for every $i \in I$, then there exists a morphism $h \in \mathsf{MOR}(\mathcal{B}, \mathcal{A})$ such that $h_i = p_i h$ for every $i \in I$.

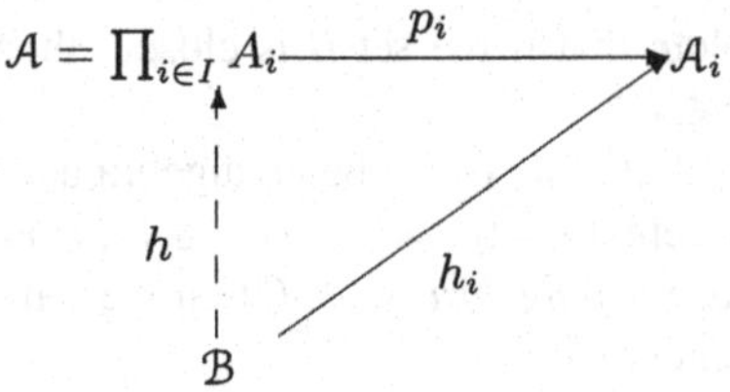

Fig. 1.3 Morphisms involved in the proof of Theorem 1.165

Proof To prove that each projection p_j is a morphism we need to show that for every n-ary operation f we have

$$p_j(f(t_1, \ldots, t_n)) = f(p_j(t_1), \ldots, p_j(t_n))$$

for every $t_1, \ldots, t_n \in \prod_{i \in I} A_i$. If $t_k = (t_{ki})_{i \in I}$ for $1 \leqslant k \leqslant n$, then $p_j(t_k) = t_{kj}$ and the previous equality follows from the definition of the operation f of $\prod_{i \in I}$.

For the second part of the theorem, let $w \in \mathcal{P}$. Since $a_i = h_i(w) \in A_i$, we have $a \in \prod_{i \in I} A_i$, where $p_i(a) = a_i$ for $i \in I$. For $h(w) = a$, we clearly have $h_i(w) = p_i(h(w))$ for $i \in I$.

1.8 Closure and Interior Systems

The notions of closure system and interior system introduced in this section are significant in algebra and topology and have applications in the study of frequent item sets in data mining.

Definition 1.166 *Let S be a set. A* closure system *on S is a collection $\mathcal{C}$ of subsets of S that satisfies the following conditions:*

(i) *$S \in \mathcal{C}$ and*
(ii) *for every collection $\mathcal{D} \subseteq \mathcal{C}$, we have $\bigcap \mathcal{D} \in \mathcal{C}$.*

Example 1.167 Let $\mathcal{C}$ be the collection of all intervals $[a, b] = \{x \in \mathbb{R} \mid a \leqslant x \leqslant b\}$ with $a, b \in \mathbb{R}$ and $a \leqslant b$ together with the empty set and the set $\mathbb{R}$. Note that $\bigcup \mathcal{C} = \mathbb{R} \in \mathcal{C}$, so the first condition of Definition 1.166 is satisfied.

Let $\mathcal{D}$ be a nonempty subcollection of $\mathcal{C}$. If $\emptyset \in \mathcal{D}$, then $\bigcap \mathcal{D} = \emptyset \in \mathcal{C}$. If $\mathcal{D} = \{\mathbb{R}\}$, then $\bigcap \mathcal{D} = \mathbb{R} \in \mathcal{C}$. Therefore, we need to consider only the case when $\mathcal{D} = \{[a_i, b_i] \mid i \in I\}$. Then, $\bigcap \mathcal{D} = \emptyset$ unless $a = \sup\{a_i \mid i \in I\}$ and $b = \inf\{b_i \mid i \in I\}$ both exist and $a \leqslant b$, in which case $\bigcap \mathcal{D} = [a, b]$. Thus, $\mathcal{C}$ is a closure system.

Example 1.168 Let $\mathcal{A} = (A, \mathcal{I})$ be an algebra and let $\mathsf{S}(\mathcal{A})$ be the collection of subalgebras of $\mathcal{A}$, $\mathsf{S}(\mathcal{A}) = \{(A_i, \mathcal{I}) \mid i \in I\}$. The collection $\mathcal{S} = \{A_i \mid i \in I\}$ is a closure system. It is clear that we have $S \in \mathcal{S}$. Also, if $\{A_i \mid i \in J\}$ is a family of subalgebras, then $\bigcap_{i \in J} A_i$ is a subalgebra of $\mathcal{A}$.

Many classes of relations define useful closure systems.

Theorem 1.169 *Let S be a set and let $REFL(S)$, $\mathsf{SYMM}(S)$ and $\mathsf{TRAN}(S)$ be the sets of reflexive relations, the set of symmetric relations, and the set of transitive relations on S, respectively. Then, $REFL(S)$, $\mathsf{SYMM}(S)$ and $\mathsf{TRAN}(S)$ are closure systems on S.*

Proof Note that $S \times S$ is a reflexive, symmetric, and transitive relation on S. Therefore, $\bigcup REFL(S) = S \times S \in REFL(S)$, $\bigcup \mathsf{SYMM}(S) = S \times S \in \mathsf{SYMM}(S)$, and $\bigcup \mathsf{TRAN}(S) = S \times S \in \mathsf{TRAN}(S)$.

Now let $\mathcal{C} = \{\rho_i \mid i \in I\}$ be a collection of transitive relations and let $\rho = \bigcap\{\rho_i \mid i \in I\}$. Suppose that $(x, y), (y, z) \in \rho$. Then $(x, y), (y, z) \in \rho_i$ for every $i \in I$, so $(x, z) \in \rho_i$ for $i \in I$ because each of the relations ρ_i is transitive. Thus, $(x, z) \in \rho$, which shows that $\bigcap \mathcal{C} \in \mathsf{TRAN}(S)$. This allows us to conclude that $\mathsf{TRAN}(S)$ is indeed a closure system. We leave it to the reader to prove that $REFL(S)$ and $\mathsf{SYMM}(S)$ are also closure systems.

Theorem 1.170 *The set of equivalences on S, $EQ(S)$, is a closure system.*

Proof The relation $\theta_S = S \times S$, is clearly an equivalence relation as we have seen in the proof of Theorem 1.169. Thus, $\bigcup EQ(S) = \theta_S \in EQ(S)$.

Now let $\mathcal{C} = \{\rho_i \mid i \in I\}$ be a collection of transitive relations and let $\rho = \bigcap\{\rho_i \mid i \in I\}$. It is immediate that ρ is an equivalence on S, so $EQ(S)$ is a closure system.

Definition 1.171 *A mapping $\mathbf{K} : \mathcal{P}(S) \longrightarrow \mathcal{P}(S)$ is a* closure operator *on a set S if it satisfies the conditions*

(i) $U \subseteq \mathbf{K}(U)$ (expansiveness),
(ii) $U \subseteq V$ *implies* $\mathbf{K}(U) \subseteq \mathbf{K}(V)$ (monotonicity), *and*
(iii) $\mathbf{K}(\mathbf{K}(U)) = \mathbf{K}(U)$ (idempotency)

for $U, V \in \mathcal{P}(S)$.

Example 1.172 Let $\mathbf{K} : \mathcal{P}(\mathbb{R}) \longrightarrow \mathcal{P}(\mathbb{R})$ be defined by

$$\mathbf{K(U)} = \begin{cases} \emptyset & \text{if} U = \emptyset, \\ [a, b] & \text{if both } a = \inf U \text{ and } b = \sup U \text{exist}, \\ \mathbb{R} & \text{otherwise}, \end{cases}$$

for $U \in \mathcal{P}(\mathbb{R})$. We leave to the reader the verification that $\mathbf{K}$ is a closure operator.

Closure operators induce closure systems, as shown by the next lemma.

Lemma 1.173 *Let $\mathbf{K} : \mathcal{P}(S) \longrightarrow \mathcal{P}(S)$ be a closure operator. Define the family of sets $\mathcal{C}_{\mathbf{K}} = \{H \in \mathcal{P}(S) \mid H = \mathbf{K}(H)\}$. Then, $\mathcal{C}_{\mathbf{K}}$ is a closure system on S.*

Proof Since $S \subseteq \mathbf{K}(S) \subseteq S$, we have $S \in \mathcal{C}_{\mathbf{K}}$, so $\bigcup \mathcal{C}_{\mathbf{K}} = S \in \mathcal{C}_{\mathbf{K}}$.

Let $\mathcal{D} = \{D_i \mid i \in I\}$ be a collection of subsets of S such that $D_i = \mathbf{K}(D_i)$ for $i \in I$. Since $\bigcap \mathcal{D} \subseteq D_i$, we have $\mathbf{K}(\bigcap \mathcal{D}) \subseteq \mathbf{K}(D_i) = D_i$ for every $i \in I$. Therefore, $\mathbf{K}(\bigcap \mathcal{D}) \subseteq \bigcap \mathcal{D}$, which implies $\mathbf{K}(\bigcap \mathcal{D}) = \bigcap \mathcal{D}$. This proves our claim.

Note that $\mathcal{C}_{\mathbf{K}}$, as defined in Lemma 1.173, equals the range of $\mathbf{K}$. Indeed, if $L \in \mathsf{Ran}(\mathbf{K})$, then $L = \mathbf{K}(H)$ for some $H \in \mathcal{P}(S)$, so $\mathbf{K}(L) = \mathbf{K}(\mathbf{K}(H)) = \mathbf{K}(H) = L$, which shows that $L \in \mathcal{C}_{\mathbf{K}}$. The reverse inclusion is obvious.

We refer to the sets in $\mathcal{C}_{\mathbf{K}}$ as the $\mathbf{K}$-*closed subsets of* S.

In the reverse direction from Lemma 1.173, we show that every closure system generates a closure operator.

Lemma 1.174 *Let $\mathcal{C}$ be a closure system on the set S. Define the mapping $\mathbf{K}_{\mathcal{C}} : \mathcal{P}(S) \longrightarrow \mathcal{P}(S)$ by $\mathbf{K}_{\mathcal{C}}(H) = \bigcap\{L \in \mathcal{C} \mid H \subseteq L\}$. Then, $\mathbf{K}_{\mathcal{C}}$ is a closure operator on the set S.*

Proof Note that the collection $\{L \in \mathcal{C} \mid H \subseteq L\}$ is not empty since it contains at least S, so $\mathbf{K}_{\mathcal{C}}(H)$ is defined and is clearly the smallest element of $\mathcal{C}$ that contains H. Also, by the definition of $\mathbf{K}_{\mathcal{C}}(H)$, it follows immediately that $H \subseteq \mathbf{K}_{\mathcal{C}}(H)$ for every $H \in \mathcal{P}(S)$.

Suppose that $H_1, H_2 \in \mathcal{P}(S)$ are such that $H_1 \subseteq H_2$. Since

$$\{L \in \mathcal{C} \mid H_2 \subseteq L\} \subseteq \{L \in \mathcal{C} \mid H_1 \subseteq L\},$$

we have

$$\bigcap\{L \in \mathcal{C} \mid H_1 \subseteq L\} \subseteq \bigcap\{L \in \mathcal{C} \mid H_2 \subseteq L\},$$

so $\mathbf{K}_{\mathcal{C}}(H_1) \subseteq \mathbf{K}_{\mathcal{C}}(H_2)$.

We have $\mathbf{K}_{\mathcal{C}}(H) \in \mathcal{C}$ for every $H \in \mathcal{P}(S)$ because $\mathcal{C}$ is a closure system. Therefore, $\mathbf{K}_{\mathcal{C}}(H) \in \{L \in \mathcal{C} \mid \mathbf{K}_{\mathcal{C}}(H) \subseteq L\}$, so $\mathbf{K}_{\mathcal{C}}(\mathbf{K}_{\mathcal{C}}(H)) \subseteq \mathbf{K}_{\mathcal{C}}(H)$. Since the reverse inclusion clearly holds, we obtain $\mathbf{K}_{\mathcal{C}}(\mathbf{K}_{\mathcal{C}}(H)) = \mathbf{K}_{\mathcal{C}}(H)$.

Definition 1.175 *Let $\mathcal{C}$ be a closure system on a set S and let T be a subset of S. The* $\mathcal{C}$ -set generated by T *is the set $\mathbf{K}_{\mathcal{C}}(T)$.*

Note that $\mathbf{K}_{\mathcal{C}}(T)$ is the least set in $\mathcal{C}$ that includes T.

Theorem 1.176 *Let S be a set. For every closure system $\mathcal{C}$ on S, we have $\mathcal{C} = \mathcal{C}_{\mathbf{K}_{\mathcal{C}}}$. For every closure operator $\mathbf{K}$ on S, we have $\mathbf{K} = \mathbf{K}_{\mathcal{C}_{\mathbf{K}}}$.*

Proof Let $\mathcal{C}$ be a closure system on S and let $H \subseteq M$. Then, we have the following equivalent statements:

1. $H \in \mathcal{C}_{\mathbf{K}_{\mathcal{C}}}$.
2. $\mathbf{K}_{\mathcal{C}}(H) = H$.
3. $H \in \mathcal{C}$.

The equivalence between (2) and (3) follows from the fact that $\mathbf{K}_{\mathcal{C}}(H)$ is the smallest element of $\mathcal{C}$ that contains H.

Conversely, let $\mathbf{K}$ be a closure operator on S. To prove the equality of $\mathbf{K}$ and $\mathbf{K}_{\mathcal{C}_{\mathbf{K}}}$, consider the following list of equal sets, where $H \subseteq S$:

1. $\mathbf{K}_{\mathcal{C}_{\mathbf{K}}}(H)$.
2. $\bigcap\{L \in \mathcal{C}_{\mathbf{K}} \mid H \subseteq L\}$.
3. $\bigcap\{L \in \mathcal{P}(S) \mid H \subseteq L = \mathbf{K}(L)\}$.
4. $\mathbf{K}(H)$.

We need to justify only the equality of the last two members of the list. Since $H \subseteq \mathbf{K}(H) = \mathbf{K}(\mathbf{K}(H))$, we have $\mathbf{K}(H) \in \{L \in \mathcal{P}(S) \mid H \subseteq L = \mathbf{K}(L)\}$. Thus, $\bigcap\{L \in \mathcal{P}(S) \mid H \subseteq L = \mathbf{K}(L)\} \subseteq \mathbf{K}(H)$. To prove the reverse inclusion, note that for every $L \in \{L \in \mathcal{P}(S) \mid H \subseteq L = \mathbf{K}(L)\}$, we have $H \subseteq L$, so $\mathbf{K}(H) \subseteq \mathbf{K}(L) = L$. Therefore, $\mathbf{K}(H) \subseteq \bigcap\{L \in \mathcal{P}(S) \mid H \subseteq L = \mathbf{K}(L)\}$.

Theorem 1.176 shows the existence of a natural bijection between the set of closure operators on a set S and the set of closure systems on S.

Definition 1.177 *Let $\mathcal{C}$ be a closure system on a set S and let T be a subset of S. The $\mathcal{C}$-closure of the set T is the set $\mathbf{K}_{\mathcal{C}}(T)$.*

As we observed before, $\mathbf{K}_{\mathcal{C}}(T)$ is the smallest element of $\mathcal{C}$ that contains T.

Example 1.178 Let $\mathbf{K}$ be the closure operator given in Example 1.172. Since the closure system $\mathcal{C}_{\mathbf{K}}$ equals the range of $\mathbf{K}$, it follows that the members of $\mathcal{C}_{\mathbf{K}}$, the $\mathbf{K}$-closed sets, are $\emptyset$, $\mathbb{R}$, and all closed intervals $[a, b]$ with $a \leqslant b$. Thus, $\mathcal{C}_{\mathbf{K}}$ is the closure system $\mathcal{C}$ introduced in Example 1.167. Therefore, $\mathbf{K}$ and $\mathcal{C}$ correspond to each other under the bijection of Theorem 1.176.

For a relation ρ, on S define ρ^+ as $\mathbf{K}_{\mathit{TRAN}(S)}(\rho)$. The relation ρ^+ is called the *transitive closure* of ρ and is the least transitive relation containing ρ.

Theorem 1.179 *Let ρ be a relation on a set S. We have*

$$\rho^+ = \bigcup\{\rho^n \mid n \in \mathbb{N} \text{ and } n \geqslant 1\}.$$

Proof Let τ be the relation $\bigcup\{\rho^n \mid n \in \mathbb{N} \text{ and } n \geqslant 1\}$. We claim that τ is transitive. Indeed, let $(x, z), (z, y) \in \tau$. There exist $p, q \in \mathbb{N}$, $p, q \geqslant 1$ such that $(x, z) \in \rho^p$ and $(z, y) \in \rho^q$. Therefore, $(x, y) \in \rho^p \rho^q = \rho^{p+q} \subseteq \rho^+$, which shows that ρ^+ is transitive. The definition of ρ^+ implies that if σ is a transitive relation such that $\rho \subseteq \sigma$, then $\rho^+ \subseteq \sigma$. Therefore, $\rho^+ \subseteq \tau$.

Conversely, since $\rho \subseteq \rho^+$ we have $\rho^n \subseteq (\rho^+)^n$ for every $n \in \mathbb{N}$. The transitivity of ρ^+ implies that $(\rho^+)^n \subseteq \rho^+$, which implies $\rho^n \subseteq \rho^+$ for every $n \geqslant 1$. Consequently, $\tau = \bigcup\{\rho^n \mid n \in \mathbb{N} \text{ and } n \geqslant 1\} \subseteq \rho^+$. This proves the equality of the theorem.

It is easy to see that the set of all reflexive and transitive relations on a set S, $\mathit{REFTRAN}(S)$, is also a closure system on the set of relations on S.

For a relation ρ on S, define ρ^* as $\mathbf{K}_{\mathit{REFTRAN}(S)}(\rho)$. The relation ρ^* is called the *transitive-reflexive closure* of ρ and is *the least transitive and reflexive relation* containing ρ. We have the following analog of Theorem 1.179.

Theorem 1.180 *Let ρ be a relation on a set S. We have*

$$\rho^* = \bigcup\{\rho^n \mid n \in \mathbb{N}\}.$$

Proof The argument is very similar to the proof of Theorem 1.179; we leave it to the reader.

Definition 1.181 *Let S be a set and let F be a set of operations on S. A subset P of S is* closed under *F, or F*-closed, *if P is closed under f for every $f \in F$; that is, for every operation $f \in F$, if f is n-ary and $p_0, \ldots, p_{n-1} \in P$, then $f(p_0, \ldots, p_{n-1}) \in P$.*

Note that S itself is closed under F. Further, if $\mathcal{C}$ is a nonempty collection of F-closed subsets of S, then $\bigcap \mathcal{C}$ is also F-closed.

Example 1.182 Let F be a set of operations on a set S. The collection of all F-closed subsets of a set S is a closure system.

Definition 1.183 *An* interior operator *on a set S is a mapping $\mathbf{I} : \mathcal{P}(S) \longrightarrow \mathcal{P}(S)$ that satisfies the following conditions:*

(i) $U \supseteq \mathbf{I}(U)$ (contraction),
(ii) $U \supseteq V$ *implies* $\mathbf{I}(U) \supseteq \mathbf{I}(V)$ (monotonicity), *and*
(iii) $\mathbf{I}(\mathbf{I}(U)) = \mathbf{I}(U)$ (idempotency),

for $U, V \in \mathcal{P}(S)$. Such a mapping is known as an interior operator on the set *S*.

Interior operators define certain collections of sets.

Definition 1.184 *An* interior system on a set *S is a collection $\mathcal{I}$ of subsets of S such that*

(i) $\emptyset \in \mathcal{I}$ *and,*
(ii) *for every subcollection $\mathcal{D}$ of $\mathcal{I}$ we have $\bigcup \mathcal{D} \in \mathcal{I}$.*

Theorem 1.185 *Let $\mathbf{I} : \mathcal{P}(S) \longrightarrow \mathcal{P}(S)$ be an interior operator. Define the family of sets $\mathcal{I}_{\mathbf{I}} = \{U \in \mathcal{P}(S) \mid U = \mathbf{I}(U)\}$. Then, $\mathcal{I}_{\mathbf{I}}$ is an interior system on S.*

Conversely, if $\mathcal{I}$ is an interior system on the set S, define the mapping $\mathbf{I}_{\mathcal{I}} : \mathcal{P}(S) \longrightarrow \mathcal{P}(S)$ by $\mathbf{I}_{\mathcal{I}}(U) = \bigcup\{V \in \mathcal{I} \mid V \subseteq U\}$. Then, $\mathbf{I}_{\mathcal{I}}$ is an interior operator on the set S.

Moreover, for every interior system $\mathcal{I}$ on S, we have $\mathcal{I} = \mathcal{I}_{\mathbf{I}_{\mathcal{I}}}$. For every interior operator $\mathbf{I}$ on S, we have $\mathbf{I} = \mathbf{I}_{\mathcal{I}_{\mathbf{I}}}$.

Proof This statement follows by duality from Lemmas 1.173 and 1.174 and from Theorem 1.176.

We refer to the sets in $\mathcal{I}_{\mathbf{I}}$ as the **I**-*open subsets of S*.

Theorem 1.186 *Let $\mathbf{K} : \mathcal{P}(S) \longrightarrow \mathcal{P}(S)$ be a closure operator on the set S. Then, the mapping $\mathbf{L} : \mathcal{P}(S) \longrightarrow \mathcal{P}(S)$ given by $\mathbf{L}(U) = S - \mathbf{K}(S - U)$ for $U \in \mathcal{P}(S)$ is an interior operator on S.*

Proof Since $S - U \subseteq \mathbf{K}(S - U)$, it follows that $\mathbf{L}(U) \subseteq S - (S - U) = U$, which proves property (i) of Definition 1.184.

Suppose that $U \subseteq V$, where $U, V \in \mathcal{P}(S)$. Then, we have $S - V \subseteq S - U$, so $\mathbf{K}(S - V) \subseteq \mathbf{K}(S - U)$ by the monotonicity of closure operators. Therefore,

$$\mathbf{L}(U) = S - \mathbf{K}(S - U) \subseteq S - \mathbf{K}(S - V) = \mathbf{L}(V),$$

which proves the monotonicity of $\mathbf{L}$.

Finally, observe that we have $\mathbf{L}(\mathbf{L}(U)) \subseteq \mathbf{L}(U)$ because of the contraction property already proven for $\mathbf{L}$. Thus, we need only show that $\mathbf{L}(U) \subseteq \mathbf{L}(\mathbf{L}(U))$ to prove the idempotency of $\mathbf{L}$. This inclusion follows immediately from

$$\mathbf{L}(\mathbf{L}(U)) = \mathbf{L}(S - \mathbf{K}(S - U)) \supseteq \mathbf{L}(S - (S - U)) = \mathbf{L}(U).$$

We can prove that if $\mathbf{L}$ is an interior operator on a set S, then $\mathbf{K} : \mathcal{P}(S) \longrightarrow \mathcal{P}(S)$ defined as $\mathbf{K}(U) = S - \mathbf{L}(S - U)$ for $U \in \mathcal{P}(S)$ is a closure operator on the same set.

In Chap. 4, we extensively use closure and interior operators.

1.9 Dissimilarities and Metrics

The notion of a metric was introduced in mathematics by the French mathematician Maurice René Fréchet in [1] as an abstraction of the notion of distance between two points. In this chapter, we explore the notion of metric and the related notion of metric space, as well as a number of generalizations and specializations of these notions.

Dissimilarities are functions that allow us to evaluate the extent to which data objects are different.

Definition 1.187 *A* dissimilarity on a set *S is a function* $d : S^2 \longrightarrow \mathbb{R}_{\geqslant 0}$ *satisfying the following conditions:*

(i) $d(x, x) = 0$ *for all* $x \in S$;
(ii) $d(x, y) = d(y, x)$ *for all* $x, y \in S$.

The pair (S, d) *is a* dissimilarity space.

The set of dissimilarities defined on a set S is denoted by $\mathcal{D}_S$.

Let (S, d) be a dissimilarity space and let $S(x, y)$ be the set of all nonnull sequences $\mathbf{s} = (s_1, \ldots, s_n) \in \mathbf{Seq}(S)$ such that $s_1 = x$ and $s_n = y$. The *d-amplitude of* $\mathbf{s}$ is the number $\&_d(\mathbf{s}) = \max\{d(s_i, s_{i+1}) \mid 1 \leqslant i \leqslant n - 1\}$.

Next we introduce the notion of extended dissimilarity by allowing ∞ as a value of a dissimilarity.

Definition 1.188 *Let S be a set. An* extended dissimilarity on *S is a function* $d : S^2 \longrightarrow \hat{\mathbb{R}}_{\geqslant 0}$ *that satisfies the conditions (DISS1) and (DISS2) of Definition 1.187.*

The pair (S, d) *is an* extended dissimilarity space.

Additional properties may be satisfied by dissimilarities. A nonexhaustive list is given next.

1. $d(x, y) = 0$ implies $d(x, z) = d(y, z)$ for every $x, y, z \in S$ (*evenness*).
2. $d(x, y) = 0$ implies $x = y$ for every x, y (*definiteness*).
3. $d(x, y) \leqslant d(x, z) + d(z, y)$ for every x, y, z (*triangular inequality*).
4. $d(x, y) \leqslant \max\{d(x, z), d(z, y)\}$ for every x, y, z (*the ultrametric inequality*).
5. $d(x, y) + d(u, v) \leqslant \max\{d(x, u) + d(y, v), d(x, v) + d(y, u)\}$ for every x, y, u, v (*Buneman's inequality*, also known as the *four-point condition*).

If $d : S^2 \longrightarrow \mathbb{R}$ is a function that satisfies the properties of dissimilarities and the triangular inequality, then the values of d are nonnegative numbers. Indeed, by taking $x = y$ in the triangular inequality, we have

$$0 = d(x, x) \leqslant d(x, z) + d(z, x) = 2d(x, z),$$

for every $z \in S$.

Various connections exist among these properties. As an example, we can show the following statement.

Theorem 1.189 *Both the triangular inequality and definiteness imply evenness.*

Proof Suppose that d is a dissimilarity that satisfies the triangular inequality, and let $x, y \in S$ be such that $d(x, y) = 0$. By the triangular inequality, we have both $d(x, z) \leqslant d(x, y) + d(y, z) = d(y, z)$ and $d(y, z) \leqslant d(y, x) + d(x, z) = d(x, z)$ because $d(y, x) = d(x, y) = 0$. Thus, $d(x, z) = d(y, z)$ for every $z \in S$.

We leave it to the reader to prove the second part of the statement.

We denote the set of definite dissimilarities on a set S by $\mathcal{D}'_S$. Further notations will be introduced shortly for other types of dissimilarities.

Definition 1.190 *A dissimilarity* $d \in \mathcal{D}_S$ *is*

(i) *a* metric *if it satisfies the definiteness property and the triangular inequality,*
(ii) *a* tree metric *if it satisfies the definiteness property and Buneman's inequality, and*
(iii) *an* ultrametric *if it satisfies the definiteness property and the ultrametric inequality.*

The set of metrics on a set S *is denoted by* $\mathcal{M}_S$. *The sets of tree metrics and ultrametrics on a set* S *are denoted by* $\mathcal{T}_S$ *and* $\mathcal{U}_S$, *respectively.*

If d *is a metric or an ultrametric on a set* S, *then* (S, d) *is a* metric space *or an* ultrametric space, *respectively.*

If d is a metric defined on a set S and $x, y \in S$, we refer to the number $d(x, y)$ as the *d-distance* between x and y or simply the *distance* between x and y whenever d is clearly understood from context.

Thus, a function $d : S^2 \longrightarrow \mathbb{R}_{\geqslant 0}$ is a metric if it has the following properties:

(i) $d(x, y) = 0$ if and only if $x = y$ for $x, y \in S$;
(ii) $d(x, y) = d(y, x)$ for $x, y \in S$;
(iii) $d(x, y) \leqslant d(x, z) + d(z, y)$ for $x, y, z \in S$.

If the first property is replaced by the weaker requirement that $d(x, x) = 0$ for $x \in S$, then we refer to d as a *semimetric* on S. Thus, if d is a semimetric $d(x, y) = 0$ does not necessarily imply $x = y$ and we can have for two distinct elements x, y of S, $d(x, y) = 0$.

The notions of extended metric and extended ultrametric are defined starting from the notion of extended dissimilarity using the same process as in the definitions of metrics and ultrametrics.

A collection of semimetrics on a set S is said to be a *gauge* on S.

Example 1.191 Let S be a nonempty set. Define the mapping $d : S^2 \longrightarrow \mathbb{R}_{\geqslant 0}$ by

$$d(u, v) = \begin{cases} 1 & \text{if } u \neq v, \\ 0 & \text{otherwise,} \end{cases}$$

for $x, y \in S$. It is clear that d satisfies the definiteness property. The triangular inequality, $d(x, y) \leqslant d(x, z) + d(z, y)$ is satisfied if $x = y$. Therefore, suppose that $x \neq y$, so $d(x, y) = 1$. Then, for every $z \in S$, we have at least one of the inequalities $x \neq z$ or $z \neq y$, so at least one of the numbers $d(x, z)$ or $d(z, y)$ equals 1. Thus d satisfies the triangular inequality. The metric d introduced here is the *discrete metric* on S.

Example 1.192 Consider the mapping $d_h : (\mathbf{Seq}_n(S))^2 \longrightarrow \mathbb{R}_{\geqslant 0}$ defined by

$$d_h(\mathbf{p}, \mathbf{q}) = |\{i \mid 0 \leqslant i \leqslant n-1 \text{ and } \mathbf{p}(i) \neq \mathbf{q}(i)\}|$$

for all sequences $\mathbf{p}, \mathbf{q}$ of length n on the set S.

Clearly, d_h is a dissimilarity that is both even and definite. Moreover, it satisfies the triangular inequality. Indeed, let $\mathbf{p}, \mathbf{q}, \mathbf{r}$ be three sequences of length n on the set S. If $\mathbf{p}(i) \neq \mathbf{q}(i)$, then $\mathbf{r}(i)$ must be distinct from at least one of $\mathbf{p}(i)$ and $\mathbf{q}(i)$. Therefore,

$$\begin{aligned} &\{i \mid 0 \leqslant i \leqslant n-1 \text{ and } \mathbf{p}(i) \neq \mathbf{q}(i)\} \\ &\subseteq \{i \mid 0 \leqslant i \leqslant n-1 \text{ and } \mathbf{p}(i) \neq \mathbf{r}(i)\} \cup \{i \mid 0 \leqslant i \leqslant n-1 \text{ and } \mathbf{r}(i) \neq \mathbf{q}(i)\}, \end{aligned}$$

which implies the triangular inequality. This is a rather rudimentary distance known as the *Hamming distance* on $\mathbf{Seq}_n(S)$. If we need to compare sequences of unequal length, we can use an extended metric d'_h defined by

$$d'_h(\mathbf{x}, \mathbf{y}) = \begin{cases} |\{i \mid 0 \leqslant i \leqslant |\mathbf{x}| - 1, x_i \neq y_i\} & \text{if } |\mathbf{x}| = |\mathbf{y}|, \\ \infty & \text{if } |\mathbf{x}| \neq |\mathbf{y}|. \end{cases}$$

We use frequently use the notions of closed sphere and open sphere.

Definition 1.193 *Let (S, d) be a metric space. The* closed sphere *centered in $x \in S$ of radius r is the set*

$$B_d(x, r) = \{y \in S | d(x, y) \leqslant r\}.$$

The open sphere *centered in $x \in S$ of radius r is the set*

$$C_d(x, r) = \{y \in S | d(x, y) < r\}.$$

Definition 1.194 *Let (S, d) be a metric space. The* diameter *of a subset U of S is the number $\mathsf{diam}_{S,d}(U) = \sup\{d(x, y) \mid x, y \in U\}$. The set U is* bounded *if $\mathsf{diam}_{S,d}(U)$ is finite.*

The diameter *of the metric space (S, d) is the number*

$$\mathsf{diam}_{S,d} = \sup\{d(x, y) \mid x, y \in S\}.$$

If the metric space is clear from the context, then we denote the diameter of a subset U just by $\mathsf{diam}(U)$.

If (S, d) is a finite metric space, then $\mathsf{diam}_{S,d} = \max\{d(x, y) \mid x, y \in S\}$.

A notion close to the notion of dissimilarity is given next.

Definition 1.195 *A* similarity on a set *S is a function $s : S^2 \longrightarrow [0, 1]$ satisfying the following conditions:*

(i) $s(x, x) = 1$ *for all $x \in S$;*
(ii) $s(x, y) = s(y, x)$ *for all $x, y \in S$.*

If $s(x, y) = 1$ implies $x = y$, then s is a definite similarity.

In other words, the similarity between an object x and itself is the largest; also, the similarity is symmetric.

Example 1.196 Let $d : S^2 \longrightarrow \mathbb{R}_{\geqslant 0}$ be a dissimilarity on S. The function $s : S^2 \longrightarrow [0, 1]$ defined by $s(x, y) = e^{-\frac{d^2(x,y)}{2\sigma^2}}$ for $x, y \in S$ and $\sigma \in \mathbb{R}$ is easily seen to be a similarity. Note that s is definite if and only if d is definite.

1.10 Rough Sets

Rough sets are approximative descriptions of sets that can be achieved using equivalences (or partitions). This fertile idea was introduced by the Polish mathematician Z. Pawlak and has generated a large research effort in mathematics and computer science due to its applications.

Unless stated otherwise, all sets in this chapter are finite.

Definition 1.197 *Let S be a set. An* approximation space on S *is a pair (S, ρ), where ρ is an equivalence relation defined on the set S.*

If S is clear from the context, we will refer to (S, ρ) just as an approximation space.

If (S, ρ) is an approximation space defined on S and U is a subset of S, then the *ρ-degree of membership of an element x of S in U* is the number

$$m_\rho(x, U) = \frac{|U \cap [x_\rho]|}{|[x]_\rho|}.$$

Clearly, we have $0 \leqslant m_\rho(x, U) \leqslant 1$.

Example 1.198 Let S be the set of natural numbers $\{0, 1, \ldots, 12\}$ and let ρ be the equivalence $\equiv_5 \cap (S \times S)$. The equivalence classes of ρ are $\{0, 5, 10\}$, $\{1, 6, 11\}$, $\{2, 7, 12\}$, $\{3, 8\}$, and $\{4, 9\}$.

If E is the subset of even members of S, then we have $m_\rho(2, E) = \frac{|E \cap \{2,7,12\}|}{|\{2,7,12\}|} = \frac{2}{3}$ and $m_\rho(4, E) = \frac{|E \cap \{4,9\}|}{|\{4,9\}|} = \frac{1}{2}$.

Definition 1.199 *Let (S, ρ) be an approximation space and let U be a subset of S. The* ρ-lower approximation *of U is the set obtained by taking the union of all ρ-equivalence classes included in the set U:*

$$\mathsf{lap}_\rho(U) = \bigcup \{[x]_\rho \in S/\rho \mid [x]_\rho \subseteq U\}.$$

The ρ-upper approximation *of U is the set obtained by taking the union of ρ-equivalence classes that have a nonempty intersection with the set U:*

$$\mathsf{uap}_\rho(U) = \bigcup \{[x]_\rho \in S/\rho \mid [x]_\rho \cap U \neq \emptyset\}.$$

The ρ-boundary *of U is the set*

$$\mathsf{bd}_\rho(U) = \mathsf{uap}_\rho(U) - \mathsf{lap}_\rho(U).$$

If $x \in \mathsf{lap}_\rho(U)$, then $x \in U$ and $m_\rho(x, U) = 1$. Thus, $\mathsf{lap}_\rho(U)$ is a strong approximation of the set U that consists of those objects of S that can be identified as members of U. This set is also known as the *ρ-positive region of U* and denoted alternatively by $\mathsf{POS}_\rho(U)$.

On the other hand, if $x \in \mathsf{uap}_\rho(U)$, then x may or may not belong to U. Thus, $\mathsf{uap}_\rho(U)$ contains those objects of S that may be members of U and we have $0 \leqslant m_\rho(x, U) \leqslant 1$. For $x \in S - \mathsf{uap}_\rho(U)$ we have $x \notin U$. This justifies naming the set $S - \mathsf{uap}_\rho(U)$ the *ρ-negative region of U*.

Note that, in general, $\mathsf{lap}_\rho(U) \subseteq \mathsf{uap}_\rho(U)$ for any set U.

The equivalence ρ is used interchangeably with the partition π_ρ in the notations introduced in Definition 1.199. For example, we can write

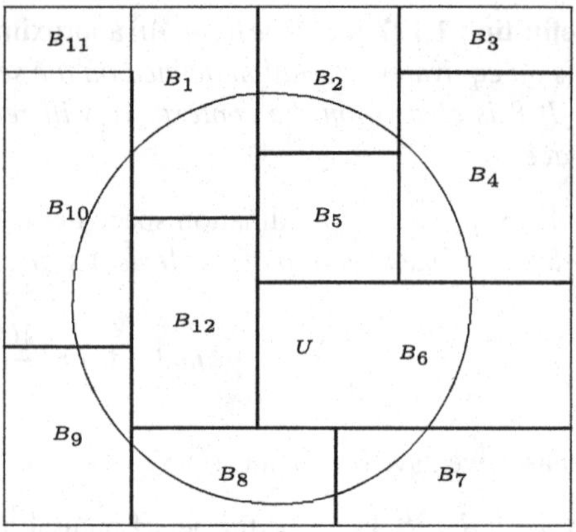

Fig. 1.4 Lower and upper approximations of set U

$$\mathsf{lap}_\rho(U) = \bigcup\{B \in \pi_\rho \mid B \subseteq U\}$$

and

$$\mathsf{uap}_\rho(U) = \bigcup\{B \in \pi_\rho \mid B \cap U \neq \emptyset\}.$$

Definition 1.200 *Let (S, ρ) be an approximation space. A set U, $U \subseteq S$ is* ρ-rough *if $\mathsf{bd}_\rho(U) \neq \emptyset$ and is* ρ-crisp *otherwise.*

Example 1.201 Let S be a set and ρ be an equivalence such that the corresponding partition π consists of 12 blocks, $B_1, \ldots, B_{12}$ (see Fig. 1.4). For the set U shown in this figure, we have

$$\begin{aligned}
\mathsf{lap}_\rho(U) &= \{B_5, B_{12}\},\\
\mathsf{uap}_\rho(U) &= \{B_1, B_2, B_4, B_5, B_6, B_7, B_8, B_9, B_{10}, B_{12}\},\\
\mathsf{bd}_\rho(U) &= \{B_1, B_2, B_4, B_6, B_7, B_8, B_9, B_{10}\}.
\end{aligned}$$

Thus, U is a ρ-rough set.

The next statement links ρ-saturated sets to ρ-crisp sets.

Theorem 1.202 *Let (S, ρ) be an approximation space. A subset U of S is ρ-crisp if and only if ρ is a π_ρ-saturated set.*

Proof Let U be a ρ-crisp set. Since $\mathsf{bd}_\rho(U) = \mathsf{uap}_\rho(U) - \mathsf{lap}_\rho(U) = \emptyset$, it follows that $\mathsf{uap}_\rho(U) = \mathsf{lap}_\rho(U)$. Thus, $[x]_\rho \cap U \neq \emptyset$ implies $[x]_\rho \subseteq U$. Clearly, if $u \in U$, then $u \in [u]_\rho \cap U$ and therefore $[u]_\rho \subseteq U$, which implies $\bigcup_{u \in U}[u]_\rho \subseteq U$. The reverse inclusion is obvious, so $\bigcup_{u \in U}[u]_\rho = U$, which means that U is ρ-saturated.

Conversely, suppose that U is ρ-saturated; that is, $\bigcup_{u \in U}[u]_\rho = U$. If $x \in \mathsf{uap}_\rho(U)$, then $[x]_\rho \cap U \neq \emptyset$, which means that $[x]_\rho \cap [u]_\rho \neq \emptyset$ for some $u \in U$. Since two equivalence classes that have a nonempty intersection must be equal, it follows that $[x]_\rho = [u]_\rho \subseteq U$, so $x \in \mathsf{lap}_\rho(U)$.

Theorem 1.203 *The following statements hold in an approximation space (S, ρ):*

(i) $\mathsf{lap}_\rho(\emptyset) = \mathsf{uap}_\rho(\emptyset) = \emptyset$ *and* $\mathsf{lap}_\rho(S) = \mathsf{uap}_\rho(S) = S$,
(ii) $\mathsf{lap}_\rho(U \cap V) = \mathsf{lap}_\rho(U) \cap \mathsf{lap}_\rho(V)$,
(iii) $\mathsf{uap}_\rho(U \cup V) = \mathsf{uap}_\rho(U) \cup \mathsf{uap}_\rho(V)$,
(iv) $\mathsf{lap}_\rho(U \cup V) \supseteq \mathsf{lap}_\rho(U) \cup \mathsf{lap}_\rho(V)$,
(v) $\mathsf{uap}_\rho(U \cap V) \subseteq \mathsf{uap}_\rho(U) \cap \mathsf{uap}_\rho(V)$,
(vi) $\mathsf{lap}_\rho(U^c) = \left(\mathsf{uap}_\rho(U)\right)^c$ *and* $\mathsf{uap}_\rho(U^c) = \left(\mathsf{lap}_\rho(U)\right)^c$,
(vii) $\mathsf{lap}_\rho(\mathsf{lap}_\rho(U)) = \mathsf{uap}_\rho(\mathsf{lap}_\rho(U)) = \mathsf{lap}_\rho(U)$, *and*
(viii) $\mathsf{uap}_\rho(\mathsf{uap}_\rho(U)) = \mathsf{lap}_\rho(\mathsf{uap}_\rho(U)) = \mathsf{uap}_\rho(U)$

for every $U, V \in \mathcal{P}(S)$.

Proof We leave the verification of these statements to the reader. □

Corollary 1.204 *Let (S, ρ) be an approximation space and let U and V be two subsets of S. If $U \subseteq V$, then $\mathsf{lap}_\rho(U) \subseteq \mathsf{lap}_\rho(V)$ and $\mathsf{uap}_\rho(U) \subseteq \mathsf{uap}_\rho(V)$.*

Proof If $U \subseteq V$, we have $U = U \cap V$, so by Part (iii) of Theorem 1.203, we have $\mathsf{lap}_\rho(U) = \mathsf{lap}_\rho(U) \cap \mathsf{lap}_\rho(V)$, which implies $\mathsf{lap}_\rho(U) \subseteq \mathsf{lap}_\rho(V)$. The second part of this statement follows from Part (iv) of the same theorem.

Definition 1.205 *Let (S, ρ) be an approximation space. A subset U of S is* ρ-definable *if $\mathsf{lap}_\rho(U) \neq \emptyset$ and $\mathsf{uap}_\rho(U) \neq S$.*

If U is not ρ-definable, then we say that U is *ρ-undefinable*. In this case, three cases may occur:

1. If $\mathsf{lap}_\rho(U) = \emptyset$ and $\mathsf{uap}_\rho(U) \neq S$, then we say that U is *internally ρ-undefinable*.
2. If $\mathsf{lap}_\rho(U) \neq \emptyset$ and $\mathsf{uap}_\rho(U) = S$, then U as an *externally ρ-undefinable set*.
3. If $\mathsf{lap}_\rho(U) = \emptyset$ and $\mathsf{uap}_\rho(U) = S$, then U is a *totally ρ-undefinable set*.

Definition 1.206 *Let (S, ρ) be a finite approximation space on S. The* accuracy of the ρ-approximation *of U is the number*

$$\mathsf{acc}\rho(U) = \frac{|\mathsf{lap}_\rho(U)|}{|\mathsf{uap}_\rho(U)|}.$$

It is clear that $0 \leqslant \mathsf{acc}\rho(U) \leqslant 1$. If $\mathsf{acc}\rho(U) = 1$, U is a ρ-crisp set; otherwise (that is, if $\mathsf{acc}\rho(U) < 1$), U is *ρ-rough*.

Example 1.207 Let S be the set of natural numbers $\{0, 1, \ldots, 12\}$ and let ρ be the equivalence $\equiv_5 \cap (S \times S)$ considered in Example 1.198. For the set E of even members of S, we have $\mathsf{uap}_\rho(E) = \emptyset$ and $\mathsf{lap}_\rho(E) = S$, so the set E is a totally ρ-undefinable set.

On the other hand, for the subset of perfect squares in S, $P = \{1, 4, 9\}$, we have

$$\begin{aligned} lap_\rho(P) &= \{4, 9\}, \\ uap_\rho(P) &= \{1, 4, 9, 6, 11\}. \end{aligned}$$

Thus, the accuracy of the ρ-approximation of P is $acc_\rho(P) = 0.4$.

The notion of a ρ-positive subset of a set is extended to equivalences as follows.

Definition 1.208 *Let S be a set, ρ and ρ' be two equivalences on S, and $\pi = \{B_1, \ldots, B_m\}$, $\sigma = \{C_1, \ldots, C_n\}$ the partitions that correspond to ρ and ρ', respectively. The* positive set of ρ' relative to ρ *is the subset of S defined by*

$$POS_\rho(\rho') = \bigcup_{j=1}^{n} lap_\rho(C_j).$$

Theorem 1.209 *Let S be a set, and ρ and ρ' two equivalences on S. We have $\rho \leqslant \rho'$ if and only if $POS_\rho(\rho') = S$.*

Proof Let $\pi = \{B_1, \ldots, B_m\}$ and $\sigma = \{C_1, \ldots, C_n\}$ be the partitions that correspond to ρ and ρ', respectively.

Suppose that $\rho \leqslant \rho'$. Then, each block C_j of σ is a union of blocks of π, so $lap_\rho(C_j) = C_j$. Therefore, we have

$$POS_\rho(\rho') = \bigcup_{j=1}^{n} lap_\rho(C_j) = \bigcup_{j=1}^{n} C_j = S.$$

Conversely, suppose that $POS_\rho(\rho') = S$, that is, $\bigcup_{j=1}^{n} lap_\rho(C_j) = S$. Since we have $lap_\rho(C_j) \subseteq C_j$ for $1 \leqslant j \leqslant n$, we claim that we must have $lap_\rho(C_j) = C_j$ for every j, $1 \leqslant j \leqslant n$. Indeed, if we have a strict inclusion $lap_\rho(C_{j_0}) \subset C_{j_0}$ for some j_0, this implies $\bigcup_{j=1}^{n} lap_\rho(C_j) \subset \bigcup_{j=1}^{n} C_j = S$, which would contradict the equality $\bigcup_{j=1}^{n} lap_\rho(C_j) = S$. Therefore, we have $lap_\rho(C_j) = C_j$ for every j, which shows that each block of σ is a union of blocks of π. Consequently, $\rho \leqslant \rho'$.

1.11 Closure Operators and Rough Sets

In Exercise 77, the reader is asked to prove that the lower approximation operator lap_ρ defined by an equivalence on a set S is an interior operator on S and the upper approximation operator uap_ρ is a closure operator on the same set. In addition, it is easy to see (Exercise 75) that

$$\begin{aligned}
\mathsf{uap}_\rho(\emptyset) &= \emptyset,\\
\mathsf{uap}_\rho(U \cup V) &= \mathsf{uap}_\rho(U) \cup \mathsf{uap}_\rho(V),\\
U &\subseteq S - \mathsf{uap}_\rho(S - \mathsf{uap}_\rho(U)),\\
\mathsf{lap}_\rho(S) &= S,\\
\mathsf{lap}_\rho(U \cap V) &= \mathsf{lap}_\rho(U) \cap \mathsf{lap}_\rho(V),\\
U &\supseteq S - \mathsf{lap}_\rho(S - \mathsf{lap}_\rho(U)),
\end{aligned}$$

for every $U, V \in \mathcal{P}(S)$.

It has been shown (see [2]) that approximation spaces can be defined starting from certain closure operators or interior operators.

Theorem 1.210 *Let S be a set and let $\mathbf{K}$ be a closure operator on S such that the following conditions are satisfied:*

(i) $\mathbf{K}(\emptyset) = \emptyset$,
(ii) $\mathbf{K}(U \cup V) = \mathbf{K}(U) \cup \mathbf{K}(V)$, *and*
(iii) $U \subseteq S - \mathbf{K}(S - \mathbf{K}(U))$,

for every $U, V \in \mathcal{P}(S)$. Then, the mapping $\mathbf{I} : \mathcal{P}(S) \longrightarrow \mathcal{P}(S)$ defined by $\mathbf{I}(U) = S - \mathbf{K}(S - U)$ for $U \in \mathcal{P}(S)$ is an interior operator on S and there exists an approximation space (S, ρ) such that $\mathsf{lap}_\rho(U) = \mathbf{I}(U)$ and $\mathsf{uap}_\rho(U) = \mathbf{K}(U)$ for every $U \in \mathcal{P}(S)$.

Proof Define ρ as $\rho = \{(x, y) \in S \times S \mid x \in \mathbf{K}(\{y\})\}$ and let $r(u)$ be the set $r(u) = \{v \in S \mid u \in \mathbf{K}(\{v\})\}$ for $u \in S$. Observe that $\mathbf{K}(\{y\}) = \{x \in S \mid y \in r(x)\}$ for $x, y \in S$. We begin the argument by proving that ρ is an equivalence.

The reflexivity of ρ follows from $x \in \mathbf{K}(\{x\})$ for $x \in S$.

We claim that $y \notin \mathbf{K}(W)$ if and only if $r(y) \cap W = \emptyset$. Since $\mathbf{K}(W) = \bigcup\{\mathbf{K}(\{x\}) \mid x \in W\}$, it follows that the statements

(i) $y \notin \mathbf{K}(W)$,
(ii) $y \notin \mathbf{K}(\{x\})$ for every $x \in W$,
(iii) $x \notin r(y)$ for every $x \in W$, and
(iv) $r(y) \cap W = \emptyset$

are equivalent, which justifies our claim.

Property (iii) of Theorem 1.210 implies that $\{y\} \subseteq S - \mathbf{K}(S - \mathbf{K}(\{y\}))$; that is, $y \notin \mathbf{K}(S - \mathbf{K}(\{y\}))$. Therefore, by the argument of the previous paragraph, we have $r(y) \cap (S - \mathbf{K}(\{y\})) = \emptyset$, which implies $r(y) \subseteq \mathbf{K}(\{y\})$. Thus, if $x \in r(y)$, it follows that $x \in \mathbf{K}(\{y\})$; that is, $y \in r(x)$. In terms of the relation ρ, this means that $(y, x) \in \rho$ implies $(x, y) \in \rho$, so ρ is a symmetric relation.

To prove the transitivity of ρ, suppose that $(x, y), (y, z) \in \rho$. We have $x \in \mathbf{K}(\{y\})$ and $y \in \mathbf{K}(\{z\})$. By the idempotency of $\mathbf{K}$, we have $\mathbf{K}(\{y\}) \subseteq \mathbf{K}(\mathbf{K}(\{z\})) = \mathbf{K}(\{z\})$, so $x \in \mathbf{K}(\{z\})$. Consequently, $(x, z) \in \rho$. This allows us to conclude that ρ is indeed an equivalence. Moreover, we realize now that $r(x)$ is exacly the equivalence class $[x]_\rho$ for $x \in S$.

An immediate consequence of this fact is that we have $x \in r(y)$ if and only if $y \in r(x)$. Therefore,

$$\mathbf{K}(\{y\}) = \{x \in S \mid y \in r(x)\} = \{x \in S \mid x \in r(y)\} = r(y),$$

and by the second property of **K**, we have

$$\mathbf{K}(U) = \bigcup\{r(u) \mid u \in U\}$$

for $U \in \mathcal{P}(S)$. Consequently, we can write

$$\begin{aligned} \mathsf{uap}_\rho(U) &= \bigcup\{r(x) \mid r(x) \cap U \neq \emptyset\} \\ &= \bigcup\{r(u) \mid u \in U\} \\ &= \mathbf{K}(U). \end{aligned}$$

Also, we have $\mathbf{K}(\{z\}) = r(z)$ for every $z \in S$.

The definition of **I** implies immediately that this function is an interior operator on S that enjoys two additional properties, namely $\mathbf{I}(S) = S$ and $\mathbf{I}(U \cap V) = \mathbf{I}(U) \cap \mathbf{I}(V)$.

By applying the definition of **I**, we have

$$\begin{aligned} \mathbf{I}(U) &= S - \mathbf{K}(S - U) \\ &= S - \bigcup\{\mathbf{K}(\{z\}) \mid z \in S - U\} \\ &= S - \bigcup\{r(z) \mid z \in S - U\} \\ &= \bigcup\{r(z) \mid z \in U\} \\ &= \mathsf{lap}_\rho(U). \end{aligned}$$

Exercises and Supplements

1. Prove that for any set S we have $\bigcup \mathcal{P}(S) = S$.
2. A set is *transitive* if $X \subseteq \mathcal{P}(X)$. Prove that $\{\emptyset, \{\emptyset\}, \{\{\emptyset\}\}\}$ is transitive.
3. Let $\mathcal{C}$ and $\mathcal{D}$ be two collections of sets such that $\mathcal{C} \subseteq \mathcal{D}$. Prove that $\bigcup \mathcal{C} \subseteq \bigcup \mathcal{D}$; also, if $\mathcal{C} \neq \emptyset$, then show that $\bigcap \mathcal{C} \supseteq \bigcap \mathcal{D}$.
4. Let $\{\mathcal{C}_i \mid i \in I\}$ be a family of hereditary collections of sets. Prove that $\bigcap_{i \in I} \mathcal{C}_i$ is also a hereditary collection of sets.
5. Let $\mathcal{C}$ be a nonempty collection of nonempty subsets of a set S. Prove that $\mathcal{C}$ is a partition of S if and only if every element $a \in S$ belongs to exactly one member of the collection $\mathcal{C}$.
6. Let S be a set and let U be a subset of S. For $a \in \{0, 1\}$, define the set

$$U^a = \begin{cases} U & \text{if } a = 1, \\ S - U & \text{if } a = 0. \end{cases}$$

(a) Prove that if $\mathcal{D} = \{D_1, \ldots, D_r\}$ is a finite collection of subsets of S, then the collection $\pi_{\mathcal{D}}$ defined by

$$\pi_{\mathcal{D}} = \{D_1^{a_1} \cap D_2^{a_2} \cap \cdots \cap D_r^{a_r} \neq \emptyset \mid (a_1, a_2, \ldots, a_r) \in \{0, 1\}^r\},$$

is a partition S.

(b) Prove that each set of $\mathcal{D}$ is a $\pi_{\mathcal{D}}$-saturated set.

7. Prove that if $\pi, \pi' \in PART(S)$ and $\pi' \leqslant \pi$, then every π-saturated set is a π'-saturated set.
8. Let $\mathcal{C}$ and $\mathcal{D}$ be two collections of subsets of a set S. Prove that if T is a subset of S, then $(\mathcal{C} \cup \mathcal{D})_T = \mathcal{C}_T \cup \mathcal{D}_T$ and $(\mathcal{C} \cap \mathcal{D})_T \subseteq \mathcal{C}_T \cap \mathcal{D}_T$.
9. Let S be a set and let $\mathcal{C}$ be a collection of subsets of S. The elements x and y of S are *separated by* $\mathcal{C}$ if there exists $C \in \mathcal{C}$ such that either $x \in C$ and $y \notin C$ or $x \notin C$ and $y \in C$. Let $\rho \subseteq S \times \mathcal{C}$ be the relation defined in Example 1.45. Prove that x and y are separated by $\mathcal{C}$ if and only if $\rho(x) \neq \rho(y)$.
10. Prove that for all sets R, S, T we have

    (a) $(R \cup S) \oplus (R \cap S) = R \oplus S$,
    (b) $R \cap (S \oplus T) = (R \cap S) \oplus (R \cap T)$.

11. Let P and Q be two subsets of a set S.

    (a) Prove that $P \cup Q = S$ if and only if $S - P \subseteq Q$.
    (b) Prove that $P \cap Q = \emptyset$ if and only if $Q \subseteq S - P$.

12. Let S be a nonempty set and let s_0 be a fixed element of S. Define the collection $[x, y] = \{\{x, s_0\}, \{y, \{s_0\}\}\}$. Prove that if $[x, x'] = [y, y']$, then $x = y$ and $x' = y'$.
13. Let S and T be two sets. Suppose that the function $p : S \times T \longrightarrow T$ given by $p(x, y) = y$ for $x \in S$ and $y \in T$ is a bijection. What can be said about the set S?
14. Let S and T be two sets. The functions $p_1 : S \times T \longrightarrow S$ and $p_2 : S \times T \longrightarrow T$ defined by $p_1(x, y) = x$ and $p_2(x, y) = y$ for $x \in S$ and $y \in T$ are the *projections* of the Cartesian product $S \times T$ on S and T, respectively. Let U be a set such that $f : U \longrightarrow S$ and $g : U \longrightarrow T$ are two functions. Prove that there is a unique function $h : U \longrightarrow S \times T$ such that $p_1 h = f$ and $p_2 h = g$.
15. Let $\mathcal{C}$ be a collection of subsets of a set S and let $\rho_{\mathcal{C}}$ and $\sigma_{\mathcal{C}}$ be the relations defined by

$$\sigma_{\mathcal{C}} = \{(x, y) \in S \times S \mid \mid x \in C \text{ if and only if } y \in C \text{ for every } C \in \mathcal{C}\}$$

and

$$\rho_{\mathcal{C}} = \{(x, y) \in S \times S \mid \mid x \in C \text{ implies } y \in C \text{ for every } C \in \mathcal{C}\}.$$

Prove that, for every collection $\mathcal{C}$, the relation $\sigma_{\mathcal{C}}$ is an equivalence and that $\rho_{\mathcal{C}}$ is a reflexive and transitive relation.

16. Prove that a relation ρ is a function if and only if $\rho^{-1}\rho \subseteq \iota_{\mathsf{Ran}(\rho)}$.
17. Prove that ρ is a one-to-one relation if and only if $\rho\rho^{-1} \subseteq \iota_{\mathsf{Dom}(\rho)}$.
18. Prove that ρ is a total relation from A to B if and only if $\iota_A \subseteq \rho\rho^{-1}$.
19. Prove that ρ is an onto relation from A to B if and only if $\iota_B \subseteq \rho^{-1}\rho$.
20. Prove that the composition of two injections (surjections, bijections) is an injection (a surjection, a bijection, respectively).
21. Let $f : S_1 \longrightarrow S_2$ be a function. Prove that, for every set $L \in \mathcal{P}(S_2)$, we have

$$S_1 - f^{-1}(S_2 - L) = f^{-1}(L).$$

22. Prove that the function $\gamma : \mathbb{N} \times \mathbb{N} \longrightarrow \mathbb{P}$ defined by $\gamma(p, q) = 2^p(2q + 1)$ for $p, q \in \mathbb{N}$ is a bijection.
23. Let $f : S \longrightarrow T$ be a function. Prove that f is an injection if and only if $f(U \cap V) = f(U) \cap f(V)$ for every $U, V \in \mathcal{P}(S)$.
24. Let T be a set and let $f_i : T \longrightarrow T$ be m injective mappings for $1 \leqslant i \leqslant m$. For $\mathbf{s} = (i_1, i_2, \ldots, i_k) \in \mathbf{Seq}(\{1, \ldots, m\})$ define $f_{\mathbf{s}}$ as the injection $f_{i_1} f_{i_2} \cdots f_{i_k}$. Let $T_{\mathbf{s}}$ be the set $f_{i_1}(f_{i_2}(\cdots(f_{i_k}(T)\cdots)))$.

    (a) Prove that if $(i_1, i_2, \ldots) \in \mathbf{Seq}_\infty(\{1, \ldots, m\})$, then $T \supseteq T_{i_1} \supseteq T_{i_1 i_2} \supseteq \cdots$.
    (b) Prove that if $\{T_{i_1}, T_{i_2}, \ldots, T_{i_m}\}$ is a partition of the set T, then $\{T_{\mathbf{u}} \mid \mathbf{u} \in \mathbf{Seq}_p(\{1, \ldots, n\})\}$ is a partition of T for every $p \geqslant 1$.

25. Let $f : S \longrightarrow T$ be a function. Prove that for every $U \in \mathcal{P}(S)$ we have $U \subseteq f^{-1}(f(U))$ and for every $V \in \mathcal{P}(T)$ we have $f(f^{-1}(V)) = V$.
26. Let S be a finite set and let $\mathcal{C}$ be a collection of subsets of S. For $x \in S$, define the mapping $\phi_x : \mathcal{C} \longrightarrow \mathcal{P}(S)$ by

$$\phi_x(C) = \begin{cases} C - \{x\} & \text{if } x \in C \text{ and } C - \{x\} \notin \mathcal{C}, \\ C & \text{otherwise,} \end{cases}$$

    for $C \in \mathcal{C}$. Prove that $|\mathcal{C}| = |\{\phi_x(C) \mid C \in \mathcal{C}\}|$.

27. Let S and T be two finite sets with the same cardinality. If $h : S \longrightarrow T$, prove that the following statements are equivalent:

    (a) h is an injection;
    (b) h is a surjection;
    (c) h is a bijection.

28. Let $\mathcal{C} = \{C_i \mid i \in I\}$ and $\mathcal{D} = \{D_i \mid i \in I\}$ be two collections of sets indexed by the same set I. Define the collections

$$\mathcal{C} \vee_I \mathcal{D} = \{C_i \vee D_i \mid i \in I\},$$
$$\mathcal{C} \wedge_I \mathcal{D} = \{C_i \wedge D_i \mid i \in I\}.$$

Prove that

$$\left(\prod \mathcal{C}\right) \cap \left(\prod \mathcal{D}\right) = \prod (\mathcal{C} \wedge_I \mathcal{D}),$$
$$\left(\prod \mathcal{C}\right) \cup \left(\prod \mathcal{D}\right) \subseteq \prod (\mathcal{C} \vee_I \mathcal{D}).$$

29. Prove that the relation $\rho \subseteq S \times S$ is

    (a) reflexive if $\iota_S \subseteq \rho$,
    (b) irreflexive if $\iota_S \cap \rho = \emptyset$,
    (c) symmetric if $\rho^{-1} = \rho$,
    (d) antisymmetric if $\rho \cap \rho^{-1} \subseteq \iota_S$,
    (e) asymmetric if $\rho^{-1} \cap \rho = \emptyset$,
    (f) transitive if $\rho^2 \subseteq \rho$.

30. Let S be a set and let ρ be a relation on S. Prove that ρ is an equivalence on S if and only if there exists a collection $\mathcal{C}$ of pairwise disjoint subsets of S such that $S = \bigcup \mathcal{C}$ and $\rho = \bigcup\{C \times C \mid C \in \mathcal{C}\}$.
31. Let ρ and ρ' be two equivalence relations on the set S. Prove that $\rho \cup \rho'$ is an equivalence on S if and only if $\rho\rho' \cup \rho'\rho \subseteq \rho \cup \rho'$.
32. Let $\mathcal{E} = \{\rho_i \mid i \in I\}$ be a collection of equivalence relations on a set S such that, for $\rho_i, \rho_j \in \mathcal{E}$, we have $\rho_i \subseteq \rho_j$ or $\rho_j \subseteq \rho_i$, where $i, j \in I$. Prove that $\bigcup \mathcal{E}$ is an equivalence on S.
33. Let ρ be a relation on a set S. Prove that the relation $\sigma = \bigcup_{n\in\mathbb{N}}(\rho \cup \rho^{-1} \cup \iota_S)^n$ is the least equivalence on S that includes ρ.
34. Let $p_1, p_2, p_3, \ldots$ be the sequence of prime numbers $2, 3, 5, \cdots$. Define the function $f : \mathbf{Seq}(\mathbb{N}) \longrightarrow \mathbb{N}$ by $f(n_1, \ldots, n_k) = p_1^{n_1} \cdots p_k^{n_k}$. Prove that $f(n_1, \ldots, n_k) = f(m_1, \ldots, m_k)$ implies $(n_1, \ldots, n_k) = (m_1, \ldots, m_k)$.
35. Let $f : S \longrightarrow T$ be a function such that $f^{-1}(t)$ is a countable set for every $t \in T$. Prove that the set S is countable.
36. Prove that $\mathcal{P}(\mathbb{N})$ is not countable.
37. Let $\mathbf{S} = (S_0, S_1, \ldots)$ be a sequence of countable sets. Prove that $\liminf \mathbf{S}$ and $\limsup \mathbf{S}$ are both countable sets.
38. Let S be a countable set. Prove that $\mathbf{Seq}(S)$ is countable. How about $\mathbf{Seq}_\infty(S)$?
39. Let S and T be two finite sets such that $|S| = m$ and $|T| = n$.

    (a) Prove that the set of functions $S \longrightarrow T$ contains n^m elements.
    (b) Prove that the set of partial functions $S \leadsto T$ contains $(n+1)^m$ elements.

40. Let I be a finite set. A *system of distinct representatives* for a collection of sets $\mathcal{C} = \{C_i \mid i \in I\}$ is an injection $r : I \to \bigcup \mathcal{C}$ such that $r(i) \in C_i$ for $i \in I$. Define the mapping $\Phi_\mathcal{C} : \mathcal{P}(I) \longrightarrow \mathcal{P}(\bigcup \mathcal{C})$ by $\Phi_\mathcal{C}(L) = \bigcup_{i\in L} C_i$ for $L \subseteq I$.

    (a) Show that if $\mathcal{C}$ has a system of distinct representatives, then $|\Phi_\mathcal{C}(L)| \geqslant |L|$ for every L such that $L \subseteq \{1, \ldots, n\}$.

(b) A subset L of I is *Φ-critical* if $|\Phi_{\mathcal{C}}(L)| = |L|$. Let $x \in \bigcup \mathcal{C}$. Define $\Phi' : \mathcal{P}(I) \longrightarrow \mathcal{P}(\bigcup \mathcal{C})$ by $\Phi'(L) = \Phi_{\mathcal{C}}(L) - \{x\}$. Prove that if no nonempty set L is Φ-critical, then $|\Phi'(L)| \geqslant |L|$ for every L.

(c) Let L be a nonempty minimal $\Phi_{\mathcal{C}}$-critical set such that $L \subset I$. Define the collection $\mathcal{D} = \{C_i - \Phi(L) \mid i \in I - L\}$. Prove that $|\Phi_{\mathcal{D}}(H)| \geqslant |H|$ for every $H \subseteq I - L$.

(d) Prove, by induction on the number $n = |I|$, the converse of the first statement: If $|\Phi_{\mathcal{C}}(L)| \geqslant |L|$ for every L in $\mathcal{P}(I)$, then a system of distinct representatives exists for the collection $\mathcal{C}$ (*Hall's matching theorem*).

41. Let M_n and M_p be the multisets of prime divisors of the numbers n and p, respectively, where $n, p \in \mathbb{N}$. Prove that $M_n + M_p = M_{np}$.
42. For two multisets M and P on a set S, denote by $M \leqslant P$ the fact that $M(x) \leqslant P(x)$ for every $x \in S$. Prove that $M \leqslant P$ implies $M \cup Q \leqslant P \cup Q$ and $M \cap Q \leqslant P \cap Q$ for every multiset Q on S.
43. Let M and P be two multisets on a set S. Define the *multiset difference* $M - P$ by $(M - P)(x) = \max\{0, M(x) - P(x)\}$ for $x \in S$.

(a) Prove that $P \leqslant Q$ implies $M - P \geqslant M - Q$ and $P - M \leqslant Q - M$ for all multisets M, P, Q on S.

(b) Prove that

$$M - (P \cup Q) = (M - P) \cap (M - Q),$$
$$M - (P \cap Q) = (M - P) \cup (M - Q),$$

for all multisets M, P, Q on S.

44. Define the symmetric difference of two multisets M and P as $(M \oplus P) = (M \cup P) - (M \cap P)$. Determine which properties of the symmetrical difference of sets can be extended to the symmetric difference of multisets.
45. Let a, b, c, d be four real numbers and let $f : \mathbb{R}^2 \longrightarrow \mathbb{R}$ be the binary operation on $\mathbb{R}$ defined by

$$f(x, y) = axy + bx + cy + d$$

for $x, y \in \mathbb{R}$.

(a) Prove that f is a commutative operation if and only if $b = c$.

(b) Prove that f is an idempotent operation if and only if $a = d = 0$ and $b + c = 1$.

(c) Prove that f is an associative operation if and only if $b = c$ and $b^2 - b - ad = 0$.

46. Let $*$ be an operation defined on a set T and let $f : S \longrightarrow T$ be a bijection. Define the operation $\circ$ on S by $x \circ y = f^{-1}(f(x) * f(y))$ for $x, y \in T$. Prove that:

(a) The operation $\circ$ is commutative (associative) if and only if $*$ is commutative (associative).
(b) If u is a unit element for $*$, then $v = f^{-1}(u)$ is a unit element for $\circ$.

47. Prove that the algebra $(\mathbb{R}_{>0}, \{*\})$, where $*$ is a binary operation defined by $x * y = x^{\log y}$, is a commutative semigroup.
48. Define the binary operation $\circ$ on $\mathbb{Z} \times \mathbb{Z}$ by $(x_1, y_1) \circ (x_2, y_2) = (x_1x_2 + 2y_1y_2, x_1y_2 + x_2y_1)$. Prove that the algebra $(\mathbb{Z} \times \mathbb{Z}, \{\circ\})$ is a commutative monoid.
49. Let $\mathcal{A} = (A, \{e, \cdot, {}^{-1}\})$ be a group and let ρ be a congruence of $\mathcal{A}$. Prove that the set $\{x \in A \mid (x, e) \in \rho\}$ is a subalgebra of $\mathcal{A}$, that is, a subgroup.
50. Let $(G, \{e, \cdot, {}^{-1}\})$ be a finite group. Then, a nonempty subset H of G is a subgroup of G if and only if $x, y \in H$ implies $x \cdot y \in H$.
Solution: The necessity of the condition is immediate. To prove that the condition is sufficient let $u \in H$ and let $f_u : G \longrightarrow G$ be the mapping defined by $f_u(x) = xu$. This mapping is injective for, if $x_1 = x_2$, we have $x_1u = x_2u$, and a left multiplication by u^{-1} yields $x_1 = x_2$. If $x \in H$, then $f_u(x) \in H$ by the condition of the theorem. Thus, the restriction of f_u to H is also a surjection because H is a finite set. Since $u \in H$, there exists $x \in H$ such that $f_u(x) = u$, that is $xu = u$. This is possible only if $x = e$, so $e \in H$.
Since $e \in H$, there exists $z \in H$ such that $zu = f_u(z) = e$ and therefore $z = u^{-1} \in H$. This argument can be applied to every $u \in H$, which allows us to reach the desired conclusion.

Let S be a set and let $(G, \{u, \cdot, {}^{-1}\})$ be a group. A *group action on S* is a binary function $f : G \times S \longrightarrow S$ that satisfies the following conditions:

(i) $f(x \cdot y, s) = f(x, f(y, s))$ for all $x, y \in G$ and $s \in S$;
(ii) $f(u, s) = s$ for every $s \in S$.
The element $f(x, s)$ is denoted by xs. If an action of a group on a set S is defined, we say that the group is acting on the set S.

The *orbit* of an element $s \in S$ is the subset O_s of S defined by $O_s = \{xs \mid x \in G\}$. The *stabilizer* of s is the subset of G given by $T_x = \{x \in G \mid xs = s\}$.

51. Let $(G, \{u, \cdot, {}^{-1}\})$ be a group acting on a set S. Prove that if $O_s \cap O_z \neq \emptyset$, then $O_s = O_z$.
52. Let $(G, \{u, \cdot, {}^{-1}\})$ be a group acting on a set S. Prove that:

(a) If $O_s \cap O_z \neq \emptyset$, then $O_s = O_z$ for every $s, z \in S$.
(b) For every $s \in S$, the stabilizer of s is a subgroup of G.

53. Let B be a subgroup of a group $\mathcal{A} = (A, \{e, \cdot, {}^{-1}\})$. Prove that:

(a) The relations ρ_B and σ_B defined by

$$\rho_B = \{(x, y) \in A \times A \mid x \cdot y^{-1} \in B\},$$
$$\sigma_B = \{(x, y) \in A \times A \mid x^{-1} \cdot y \in B\},$$

are equivalence relations on A.

(b) $(x, y) \in \rho_B$ implies $(x \cdot z, y \cdot z) \in \rho_B$ and $(x, y) \in \sigma_B$ implies $(z \cdot x, z \cdot y) \in \sigma_B$ for every $z \in A$.
(c) If A is a finite set, then $|[u]_{\rho_B}| = |[u]_{\sigma_B} = |B|$ for every $u \in G$.
(d) If A is finite and B is a subgroup of $\mathcal{A}$, then $|B|$ divides $|A|$.

54. If B is a subgroup of group $\mathcal{A} = (A, \{e, \cdot, {}^{-1}\})$, let $xB = \{x \cdot y \mid y \in B\}$ and $Bx = \{y \cdot x \mid y \in B\}$. Prove that $\rho_B = \sigma_B$ if and only if $xB = Bx$ for every $x \in A$; also, show that in this case ρ_B is a congruence of $\mathcal{A}$.
55. Let $\mathcal{A}$ and $\mathcal{B}$ be two algebras of the same type. Prove that $f : A \longrightarrow B$ belongs to $\mathsf{MOR}(\mathcal{A}, \mathcal{B})$ if and only if the set $\{(x, h(x)) \mid x \in A\}$ is a subalgebra of the product algebra $\mathcal{A} \times \mathcal{B}$.

Let $\mathcal{A} = (A, \mathcal{I})$ be an algebra. The set $\mathsf{Pol}_n(\mathcal{A})$ of *n-ary polynomials of the algebra* $\mathcal{A}$ consists of the following functions:

(i) Every projection $p_i : A^n \longrightarrow A$ is an n-ary polynomial.
(ii) If f is an m-ary operation and $g_1, \ldots, g_m$ are n-ary polynomials, then $f(g_1, \ldots, g_n)$ is an n-ary polynomial.

The set of polynomials of the algebra $\mathcal{A}$ is the set $\mathsf{Pol}(\mathcal{A}) = \bigcup_{n \in \mathbb{N}} \mathsf{Pol}_n(\mathcal{A})$.

A *k-ary algebraic function* of $\mathcal{A}$ is a function $h : A^k \longrightarrow A$ for which there exists a polynomial $p \in \mathsf{Pol}_n(\mathcal{A})$ and $n - k$ elements $a_{i_1}, \ldots, a_{i_{n-k}}$ of A such that $h(x_1, \ldots, x_k) = p(x_1, \ldots, a_{i_1}, \ldots, a_{i_{n-k}}, \ldots, x_k)$ for $x_1, \ldots, x_k \in A$.

56. Let ρ be a congruence of an algebra $\mathcal{A} = (A, \mathcal{I})$ and let $f \in \mathsf{Pol}_n(\mathcal{A})$. Prove that if $(x_i, y_i) \in \rho$ for $1 \leqslant i \leqslant n$, then $(f(x_1, \ldots, x_n), f(y_1, \ldots, y_n)) \in \rho$.
57. Let $\mathcal{A} = (A, \mathcal{I})$ be an algebra and let S be a subset of A. Define the sequence of sets $\mathbf{S} = (S_0, S_1, \ldots)$ as $S_0 = S$ and $S_{n+1} = S_n \cup \{f(a_1, \ldots, a_m) \mid f$ is an m-ary operation, $a_1, \ldots, a_m \in S_n\}$.

    (a) Prove that the least subalgebra of $\mathcal{A}$ that contains S is $\bigcup_{n \in \mathbb{N}} S_n$.
    (b) Prove by induction on n that if $a \in S_n$, then there is a finite subset U of A such that a belongs to the least subalgebra that contains U.

58. Let $\mathcal{A} = (A, \mathcal{I})$ be an algebra, S be a subset of A, and a be an element in the least subalgebra of $\mathcal{A}$ that contains S. Prove that there is a finite subset T of S such that a belongs to the least subalgebra of $\mathcal{A}$ that contains T.
59. Let $\mathbf{K}$ be a closure operator on a set S. Prove the following statements:

    (a) if $U \in \mathcal{C}_{\mathbf{K}}$ and $X \subseteq U \subseteq \mathbf{K}(X)$, then $\mathbf{K}(X) = U$;
    (b) $\mathbf{K}(X) \cap \mathbf{K}(Y) \supseteq \mathbf{K}(X \cap Y)$;
    (c) $\mathbf{K}(X) \cap \mathbf{K}(Y) \in \mathcal{C}_{\mathbf{K}}$,

    for $X, Y \in \mathcal{P}(S)$.
60. Let S and T be two sets and let $f : S \longrightarrow T$ be a function. Suppose that $\mathbf{K}$ and $\mathbf{L}$ are two closure operators on S and T, respectively, such that if $V \in \mathcal{C}_{\mathbf{L}}$, then $f^{-1}(V) \in \mathcal{C}_{\mathbf{K}}$. Prove that, for every $W \in \mathcal{C}_{\mathbf{L}}$ we have $S - f^{-1}(T - W) \in \mathcal{C}_{\mathbf{K}}$.
61. Let $\mathbf{K}$ be a closure operator on a set S. For $U \in \mathcal{P}(S)$, define the $\mathbf{K}$*-border of the set* U as $\partial_{\mathbf{K}}(U) = \mathbf{K}(U) \cap \mathbf{K}(S - U)$. Let S and T be two sets and let $\mathbf{K}, \mathbf{L}$ be two closure operators on S and T, respectively.

(a) Prove that if $f^{-1}(\mathbf{K}(V)) = \mathbf{L}(f^{-1}(V))$ for every $V \in \mathcal{P}(T)$, then $f^{-1}(\partial_{\mathbf{L}}(V)) = \partial_{\mathbf{K}}(f^{-1}(V))$.
(b) Now let $f : S \longrightarrow T$ be a bijection such that both $f^{-1}(\mathbf{K}(V)) = \mathbf{L}(f^{-1}(V))$ for every $V \in \mathcal{P}(T)$ and $f(\mathbf{K}(U)) = \mathbf{L}(f(U))$ for every $U \in \mathcal{P}(S)$. Prove that $\partial_{\mathbf{K}}(f^{-1}(V)) = f^{-1}(\partial_{\mathbf{L}}(V))$ and $\partial_{\mathbf{L}}(f(U)) = f(\partial_{\mathbf{K}}(U))$ for $U \in \mathcal{P}(S)$ and $V \in \mathcal{P}(T)$.

62. Let (S, d) be a metric space. Prove that $d(x, y) \geqslant |d(x, z) - d(y, z)|$ for all $x, y, z \in S$.
63. Let (S, d) be a metric space. Prove that $\mathit{diam}(\{x\}) = 0$ for every $x \in S$.
64. Let $B = \{x_1, \ldots, x_n\}$ be a finite subset of a metric space (S, d). Prove that

$$(n-1)\sum_{i=1}^{n} d(x, x_i) \geqslant \sum \{d(x_i, x_j) \mid 1 \leqslant i < j \leqslant n\}$$

for every $x \in S$.
Explain why this inequality can be seen as a generalization of the triangular inequality.
65. Let (S, d) be a metric space and let T be a finite subset of S. Define the mapping $D_T : S^2 \longrightarrow \mathbb{R}_{\geqslant 0}$ by

$$D_T(x, y) = \max\{|d(t, x) - d(t, y)| \mid t \in P\}$$

for $x, y \in S$.

(a) Prove that D_T is a semimetric on S and that $d(x, y) \geq D_T(x, y)$ for $x, y \in S$.
b) Prove that if $T \subseteq T'$, then $D_T(x, y) \leqslant D_{T'}(x, y)$ for every $x, y \in S$.

66. Let d be a semimetric on a set S. Prove that for $x, y, u, v \in S$ we have $|d(x, y) - d(u, v)| \leqslant d(x, u) + d(y, v)$.
67. Let (S, d) be a metric space and let $p, q, x \in S$. Prove that $d(p, x) > kd(p, q)$ for some $k > 1$, then $d(q, x) > (k-1)d(p, q)$.
68. Let (S, d) be a metric space and let $x, y \in S$. Prove that if r is a positive number and $y \in C_d(x, \frac{r}{2})$, then $C_d(x, \frac{r}{2}) \subseteq C_d(y, r)$.
69. Let (S, d) be a metric space and $p \in S$. Define the function $d_u : S^2 \longrightarrow \mathbb{R}_{\geqslant 0}$ by

$$d_u(x, y) = \begin{cases} 0 & \text{if } x = y, \\ d(x, u) + d(u, y) & \text{otherwise,} \end{cases}$$

for $x, y \in S$. Prove that d is a metric on S.
70. Let (S, d) be a metric space. Prove that $\sqrt{d}$ and $\frac{d}{1+d}$ are also metrics on S. What can be said about d^2?
71. Let (S_1, d_1) and (S_2, d_2) be two metric spaces. Define $d, e : S_1 \times S_2 \longrightarrow \mathbb{R}_{\geqslant 0}$ by $d((x_1, y_1), (x_2, y_2)) = \max\{d_1(x_1, y_1), d_2(x_2, y_2)\}$ and $e((x_1, y_1), (x_2, y_2)) = d_1(x_1, y_1) + d_2(x_2, y_2)$. Prove that both d and e are metrics on $S_1 \times S_2$.

For $p, q \in \mathbb{N}$ define the finite set $\mathbb{N}_{p,q} = \{p, p+1, \ldots, q\}$. Note that the binary equivalent of a number $n \in \mathbb{N}_{0,2^m-1}$ can be represented as a sequence of m bits, $(b_1, \ldots, b_m)$. We denote this sequence by $\beta_m(n)$. For example, $\beta_4(6) = (0, 1, 1, 0)$ and $\beta_4(13) = (1, 1, 0, 1)$.

Define $d_{h,m}(r, s)$ as $d_h(\beta_m(r), \beta_m(s))$, where d_h is the Hamming distance introduced in Example 1.192.

72. Prove that $d_{h,m}$ is a metric on $\mathbb{N}_{0,2^m-1}$.
73. Define the function $d_{\text{bin}} : \mathbb{N}^2 \longrightarrow \mathbb{N}$ as $d_{\text{bin}}(p, q) = k$ if k is the least number such that there exist $2k$ numbers $i_1, \ldots, i_k, j_1, \ldots, j_k$ in $\mathbb{N}$ where $p = q + \sum_{r=1}^{k} (-1)^{i_r} 2^{j_r}$. For example, $d_{\text{bin}}(64, 6) = 3$ because $58 = 2^6 - 2^2 - 2^1$ and we have $i_1 = 2, i_2 = i_3 = 1$, and $j_1 = 6, j_2 = 2, j_3 = 1$.
Prove that:

    (a) d_{bin} is a metric on $\mathbb{N}$;
    (b) for every numbers $p, q \in \mathbb{N}_{0,2^m-1}$ we have $d_{\text{bin}}(p, q) \leqslant d_{h,m}$.

74. Let ρ_1 and ρ_2 be two equivalences on a set S such that $\rho_1 \subseteq \rho_2$. Prove that $\mathsf{lap}_{\rho_1}(U) \supseteq \mathsf{lap}_{\rho_2}(U)$ and $\mathsf{lap}_{\rho_1}(U) \subseteq \mathsf{lap}_{\rho_2}(U)$. Conclude that $\mathsf{bd}_{\rho_1}(U) \subseteq \mathsf{bd}_{\rho_2}(U)$ for every $U \in \mathcal{P}(S)$.
75. Let (S, ρ) be an approximation space. Prove that

    (a) $\mathsf{uap}_\rho(\emptyset) = \emptyset$,
    (b) $\mathsf{uap}_\rho(U \cup V) = \mathsf{uap}_\rho(U) \cup \mathsf{uap}_\rho(V)$,
    (c) $U \subseteq S - \mathsf{uap}_\rho(S - \mathsf{uap}_\rho(U))$,
    (d) $\mathsf{lap}_\rho(S) = S$,
    (e) $\mathsf{lap}_\rho(U \cap V) = \mathsf{lap}_\rho(U) \cap \mathsf{lap}_\rho(V)$, and
    (f) $U \supseteq S - \mathsf{lap}_\rho(S - \mathsf{lap}_\rho(U))$

    for every $U, V \in \mathcal{P}(S)$.
76. Let (S, ρ) be an approximation space. A *lower (upper) sample* of a subset U of S is a subset Y of S such that $Y \subseteq U$ and $\mathsf{uap}_\rho(Y) = \mathsf{lap}_\rho(U)$ ($\mathsf{uap}_\rho(Y) = \mathsf{uap}_\rho(U)$, respectively). A lower (upper) sample of U is *minimal* if there no lower (upper) sample of U with fewer elements.
Prove that every nonempty lower (upper) sample Y of a set U has a nonempty intersection with each ρ-equivalence class included in $\mathsf{lap}_\rho(U)$ ($\mathsf{lap}_\rho(U)$, respectively). Prove that if Y is a lower (upper) minimal sample, then its intersection with each ρ-equivalence class included in $\mathsf{lap}_\rho(U)$ ($\mathsf{lap}_\rho(U)$, respectively) consists of exactly one element.
77. Let S be a set and let ρ be an equivalence on S.

    (a) Prove that lap_ρ is an interior operator on S.
    (b) Prove that uap_ρ is a closure operator on S.
    (c) Prove that the lap_ρ-open subsets of S coincide with the uap_ρ-closed subsets of S.

A *generalized approximation space* is a pair (S, ρ), where ρ is an arbitrary relation on S. Denote the set $\{y \in S \mid (x, y) \in \rho\}$ by $\rho(x)$. The lower and upper

ρ-approximations of a set $U \in \mathcal{P}(S)$ generalize the corresponding notions from approximation spaces and are defined by

$$\mathit{lap}_\rho(U) = \bigcup\{\rho(x) \mid \rho(x) \subseteq U\},$$
$$\mathit{uap}_\rho(U) = \bigcup\{\rho(x) \mid \rho(x) \cap U \neq \emptyset\},$$

for $U \in \mathcal{P}(S)$.

78. Let (S, ρ) be a generalized approximation space. Prove that

    (a) $\mathit{lap}_\rho(U) = S - \mathit{uap}_\rho(S - U)$,
    (b) $\mathit{lap}_\rho(S) = S$,
    (c) $\mathit{lap}_\rho(U \cap V) = \mathit{lap}_\rho(U) \cap \mathit{lap}_\rho(V)$,
    (d) $\mathit{lap}_\rho(U \cup V) \supseteq \mathit{lap}_\rho(U) \cup \mathit{lap}_\rho(V)$,
    (e) $U \subseteq V$ implies $\mathit{lap}_\rho(U) \subseteq \mathit{lap}_\rho(V)$,
    (f) $\mathit{uap}_\rho(U) = S - \mathit{lap}_\rho(S - U)$,
    (g) $\mathit{uap}_\rho(\emptyset) = \emptyset$,
    (h) $\mathit{uap}_\rho(U \cap V) \subseteq \mathit{uap}_\rho(U) \cap \mathit{uap}_\rho(V)$,
    (i) $\mathit{uap}_\rho(U \cup V) = \mathit{uap}_\rho(U) \cup \mathit{uap}_\rho(V)$,
    (j) $U \subseteq V$ implies $\mathit{uap}_\rho(U) \subseteq \mathit{uap}_\rho(V)$,
    (k) $\mathit{lap}_\rho((S - U) \cup V) \subseteq (S - \mathit{lap}_\rho(U)) \cup \mathit{lap}_\rho(V)$

    for $U, V \in \mathcal{P}(S)$.
79. Let (S, ρ) be a generalized approximation space, where ρ is a tolerance relation. Prove that

    (a) $\mathit{lap}_\rho(\emptyset) = \emptyset$,
    (b) $\mathit{lap}_\rho(U) \subseteq U$,
    (c) $U \subseteq \mathit{lap}_\rho(\mathit{uap}_\rho(U))$,
    (d) $\mathit{uap}_\rho(S) = S$,
    (e) $U \subseteq \mathit{uap}_\rho(U)$, and
    (f) $\mathit{uap}_\rho(\mathit{lap}_\rho(U)) \subseteq U$

    for $U, V \in \mathcal{P}(S)$.

Bibliographical Comments

Readers may find [3] a useful reference for a detailed presentation of many aspects discussed in this chapter and especially for various variants of mathematical induction. Suggested introductory references to set theory are [4, 5]. For a deeper study of algebras readers should consult the vast mathematical literature concerning general and universal algebra [6–9].

Rough sets were introduced by Z. Pawlak (see [10]). Excellent surveys supplemented by large bibliographies are [11] and [12]. The notions of lower and upper samples discussed in Exercise 76 are introduced in [13]. Various generalizations of the notion of approximation space are presented in [2].

References

1. M. Fréchet, Sur quelques points du calcul fonctionnel. Rendiconti del Circolo Matematico di Palermo **22**, 1–47 (1906)
2. Y.Y. Yao, Two views of the theory of rough sets in finite universes. Int. J. Approximate Reasoning **15**, 291–317 (1996)
3. P.A. Fejer, D.A. Simovici, *Mathematical Foundations of Computer Science*, vol. 1 (Springer-Verlag, New York, 1991)
4. P.R. Halmos, *Naive Set Theory* (Springer, New York, 1974)
5. P. Suppes, *Axiomatic Set Theory* (Dover, New York, 1972)
6. P.M. Cohn, *Universal Algebra* (D. Reidel, Dordrecht, 1981)
7. G. Birkhoff, *Lattice Theory*, 3rd edn. (American Mathematical Society, Providence, 1973)
8. J.B. Fraleigh, *A First Course in Abstract Algebra* (Addison-Wesley, Reading, 1982)
9. C. Reischer, D.A. Simovici, M. Lambert, *Introduction aux Structures Algébriques* (Éditions du Renouveau Pédagogique, Montréal, 1992)
10. Z. Pawlak, *Rough Sets: Theoretical Aspects of Reasoning About Data* (Kluwer Academic Publishing, Dordrecht, 1991)
11. J. Komorowski, Z. Pawlak, L. Polkowski, A. Skowron, in *Rough sets: A tutorial*, eds. by S. K. Pal, A. Skowron. Rough Sets Hybridization: A New Trend in Decision Making (Springer, New York, 1999), pp. 3–98
12. I. Düntsch, G. Gediga, *Rough Sets Analysis—A Road to Non-invasive Knowledge Discovery* (Methoδos, Bangor, 2000)
13. Z. Bonikowski, A certain conception of the calculus of rough sets. Notre Dame J. Formal Logic **33**, 412–421 (1992)

Chapter 2
Partially Ordered Sets

2.1 Introduction

We introduce the notion of a partially ordered set (poset) we and define several types of special elements associated with partial orders. Two partially ordered sets receive special attention: the poset of real numbers and the poset of partitions of a finite set. Partially ordered sets serve as the starting point for the study of several algebraic structures in Chap. 11.

2.2 Partial Orders

The fundamental notion of this chapter is introduced next.

Definition 2.1 *A* partial order on a set *S is a relation* $\rho \subseteq S \times S$ *that is reflexive, antisymmetric, and transitive. The pair* (S, ρ) *is referred to as a* partially ordered set *or, for short, a* poset.

When $|S|$ is finite, we refer to poset (S, ρ) as a *finite poset*.

A *strict partial order*, or more simply, *a strict order* on S, is a relation $\rho \subseteq S \times S$ that is irreflexive and transitive.

Example 2.2 The identity relation on a set S, ι_S, is a partial order; this is often referred to as the *discrete partial order* on S. Also, the relation $\theta_S = S \times S$ is a partial order on S.

Example 2.3 The relation "$\leqslant$" on the set of partitions of a set $\mathsf{PART}(S)$ introduced in Definition 1.110 is a partial order on the set $\mathsf{PART}(S)$.

Example 2.4 The divisibility relation δ introduced in Example 1.27 is a partial order on $\mathbb{N}$ since, as we have shown in Example 1.43, δ is reflexive, antisymmetric, and transitive.

D. A. Simovici and C. Djeraba, *Mathematical Tools for Data Mining*,
Advanced Information and Knowledge Processing, DOI: 10.1007/978-1-4471-6407-4_2,

For a poset (S, ρ), we prefer to use the *infix notation*; that is, write $s\rho t$ instead of $(s, t) \in \rho$. Moreover, various partial orders have their traditional notations, which we favor. For example, the relation δ introduced in Example 1.27 is usually denoted by $\mid$. Therefore, we write $m \mid n$ to denote that $(m, n) \in \delta$. Whenever practical, for generic partially ordered sets, we denote their partial order relation by $\leqslant$. Generic strict partial orders will be denoted by $<$.

Example 2.5 The inclusion relation $\subseteq$ is a partial order on the set of subsets $\mathcal{P}(S)$ of a set S.

Example 2.6 For $\mathbf{u}, \mathbf{v} \in \mathbf{Seq}(S)$ define $\mathbf{u} \leqslant_{\text{pref}} \mathbf{v}$ if $\mathbf{u}$ is a prefix of $\mathbf{v}$. Clearly, $\mathbf{u} \leqslant_{\text{pref}} \mathbf{v}$ if and only if there exists $\mathbf{t} \in \mathbf{Seq}(S)$ such that $\mathbf{v} = \mathbf{ut}$. It is immediate that "$\leqslant_{\text{pref}}$" is a reflexive relation.

If $\mathbf{u} \leqslant_{\text{pref}} \mathbf{v}$ and $\mathbf{v} \leqslant_{\text{pref}} \mathbf{u}$ there exist $\mathbf{t}, \mathbf{t}' \in \mathbf{Seq}(S)$ such that $\mathbf{v} = \mathbf{ut}$ and $\mathbf{u} = \mathbf{vt}'$. This implies $\mathbf{u} = \mathbf{utt}'$. Thus, $\mathbf{tt}' = \boldsymbol{\lambda}$, so $\mathbf{t} = \mathbf{t}' = \boldsymbol{\lambda}$, which allows us to infer that $\mathbf{u} = \mathbf{v}$. This shows that "$\leqslant_{\text{pref}}$" is antisymmetric.

Finally, suppose that $\mathbf{u} \leqslant_{\text{pref}} \mathbf{v}$ and $\mathbf{v} \leqslant_{\text{pref}} \mathbf{w}$. We have $\mathbf{v} = \mathbf{ut}$ and $\mathbf{w} = \mathbf{vs}$ for some $\mathbf{s}, \mathbf{t} \in \mathbf{Seq}(S)$. This implies $\mathbf{w} = \mathbf{uts}$, which shows that $\mathbf{u} \leqslant_{\text{pref}} \mathbf{w}$. Thus, "$\leqslant_{\text{pref}}$" is indeed a partial order on $\mathbf{Seq}(S)$.

In a similar manner, it is possible to show that the relations

$$\leqslant_{\text{suff}} = \{(\mathbf{u}, \mathbf{v}) \in (\mathbf{Seq}(S))^2 \mid \mathbf{v} = \mathbf{tu} \text{ for some } \mathbf{t} \in \mathbf{Seq}(S)\},$$
$$\leqslant_{\text{infix}} = \{(\mathbf{u}, \mathbf{v}) \in (\mathbf{Seq}(S))^2 \mid \mathbf{v} = \mathbf{tut}' \text{ for some } \mathbf{t}, \mathbf{t}' \in \mathbf{Seq}(S)\},$$

are partial orders on $\mathbf{Seq}(S)$ (exercise!).

If $(S, \leqslant)$ is a poset and $T \subseteq S$, then $(T, \leqslant_T)$ is also a poset, where $\leqslant_T = \leqslant \cap (T \times T)$ is the *trace of* $\leqslant$ *on* T.

Every strict partial order is also asymmetric. Indeed, let $<$ be a strict partial order on S and assume that $x < y$. If $y < x$, then $x < x$ due to the transitivity of $<$, which contradicts the irreflexivity of $<$. This shows that $<$ is indeed asymmetric.

A strict partial order is not, in general, a partial order since strict partial orders are irreflexive, while partial orders are reflexive. The link between partial orders and strict partial orders is given next.

Theorem 2.7 *Let $\leqslant$ be a partial order on a set S and let $<$ be the relation $\leqslant - \iota_S$. The relation $<$ is a strict partial order on S.*

If $<$ is a strict partial order on S, then the relation $\leqslant$ defined as the union $< \cup\ \iota_S$ is a partial order on S.

Proof Since $\iota_S \cap <= \emptyset$, the relation $<$ is irreflexive.

To prove the transitivity of $<$, let $x, y, z \in S$ be such that $x < y$, $y < z$. Because of the transitivity of $\leqslant$, we have $x \leqslant z$. On the other hand, we also have $x \neq z$. Indeed, if we assume that $x = z$, then we would have both $z < y$ and $y < z$, which is impossible by the asymmetry of $<$. Therefore, $(x, z) \in \leq - \iota_S = <$, which implies the transitivity of $<$.

Let $<$ be a strict partial order and let $\leqslant$ be the relation $< \cup \iota_S$. The reflexivity of $\leqslant$ is immediate.

To show that $\leqslant$ is antisymmetric, assume that $x \leqslant y$ and $y \leqslant x$. Because of the definition of $\leqslant$, we may have $x < y$ or $(x, y) \in \iota_S$ (that is, $x = y$). In the first case, we have a contradiction. Indeed, if $y < x$, this contradicts the asymmetry of $<$; if $(y, x) \in \iota_S$, we also have $(x, y) \in \iota_S$, and this contradicts the irreflexivity of $<$. Consequently, we must have $x = y$.

Let $x \leqslant y$ and $y \leqslant x$. We need to consider the following four cases.

(i) If $x < y$, $y < z$, we have $x < z$ because of the transitivity of $<$. This implies $x \leqslant z$.
(ii) If $(x, y) \in \iota_S$ and $y < z$, we have $x = y$; hence, $x < z$ and therefore $x \leqslant z$.
(iii) If $x < y$ and $(y, z) \in \iota_S$, we follow an argument similar to the one used in the previous case.
(iv) If $(x, y), (y, z) \in \iota_S$, we have $(x, z) \in \iota_S$ because of the transitivity of ι_S; hence, $x \leqslant z$.

We proved that $\leqslant$ is also transitive, and this concludes our argument.

Example 2.8 Consider the relation "$\leqslant$" on $\mathbb{R}$, which is a partial order. The strict partial order attached to it by the previous proposition is the relation "$<$".

A relation $\rho \subseteq S \times S$ is *acyclic* if $\rho^n \cap \iota_S = \emptyset$ for every $n \geqslant 1$.

Theorem 2.9 *Every strict partial order is acyclic.*

Proof Let ρ be a strict partial order relation on S. Its transitivity implies the existence of the descending sequence $\rho \supseteq \rho^2 \supseteq \cdots \supseteq \rho^n \supseteq \cdots$. Since ρ is irreflexive, we have $\rho \cap \iota_S = \emptyset$, and this implies $\rho^n \cap \iota_S = \emptyset$.

Next we introduce a graphical representation of partial orders.

Definition 2.10 *Let $(S, \leqslant)$ be a poset. The* Hasse diagram *of $(S, \leqslant)$ is the digraph of the relation $< - (<)^2$, where "$<$" is the strict partial order corresponding to $\leqslant$.*

In view of the properties of acyclic relations discussed above, it is clear that the relation $< - (<)^2$ is acyclic; therefore, the Hasse diagram is always an acyclic directed graph. We will denote this relation by "$\prec$".

Observe that $x \prec y$ if $x \neq y$, $x \leqslant y$, and there is no $u \in S$ such that $x \leqslant u$ and $u \leqslant y$. In other words, if $x \prec y$, then y *covers* x directly, without any intermediate elements.

The use of Hasse diagrams in representing posets is justified by the following statement.

Theorem 2.11 *If $\leqslant$ is a partial order on a finite set S, $<$ is the strict partial order corresponding to $\leqslant$, and $\theta = < - (<)^2$, then $\theta^* = \leqslant$.*

Proof Let $x, y \in S$ such that $x \leqslant y$. If $x = y$, then we have $(x, y) \in \iota_S \subseteq \theta^*$.

Assume now that $x \leqslant y$ and $x \neq y$, which means that $x < y$. Consider the collection $\mathcal{C}_{xy}$ of all sequences of elements of A that can be "interpolated" between x and y:

$$\begin{aligned}\mathcal{C}_{xy} = \{(\mathbf{s}(0), \ldots, \mathbf{s}(n-1)) \mid x = \mathbf{s}(0), \mathbf{s}(n-1) = y, \text{ and}\\ \mathbf{s}(i) < \mathbf{s}(i+1) \text{ for } 0 \leqslant i \leqslant n-2, n \geqslant 2\}.\end{aligned}$$

We have $\mathfrak{C}_{xy} \neq \emptyset$ since the sequence (x, y) belongs to $\mathfrak{C}_{xy}$. Furthermore, no sequence from $\mathfrak{C}_{xy}$ may contain a repetition. Since S is finite, $\mathfrak{C}_{xy}$ contains a finite number of sequences.

Let $(\mathbf{s}(0), \mathbf{s}(1), \ldots, \mathbf{s}(m-1))$ be a sequence of maximal length from $\mathfrak{C}_{xy}$, where $x = \mathbf{s}(0)$ and $y = \mathbf{s}(m-1)$.

Observe that for no pair $(\mathbf{s}(i), \mathbf{s}(i+1))$ can we have $\mathbf{s}(i) < z < \mathbf{s}(i+1)$ because, otherwise, the maximality of m would be contradicted. Therefore, $(\mathbf{s}(i), \mathbf{s}(i+1)) \in (< -(<)^2) = \theta$, and this shows that $(x, y) \in \theta^{m-1} \subseteq \theta^*$.

Conversely, if $(x, y) \in \theta^*$, there is $k \in \mathbb{N}$ such that $(x, y) \in \theta^k$, which means that there exists a sequence $(\mathbf{z}(0), \ldots, \mathbf{z}(k))$ such that

$$x = \mathbf{z}(0), (\mathbf{z}(i), \mathbf{z}(i+1)) \in \theta \text{ for } 0 \leqslant i \leqslant k-1 \text{ and } y = \mathbf{z}(k).$$

This implies $\mathbf{z}(i) \leqslant \mathbf{z}(i+1)$; hence, $(x, y) \in (\leqslant)^k \subseteq \leqslant$ because of the transitivity of $\leqslant$.

The relation θ introduced in Theorem 2.11 is called the *transitive reduction* of the partial order ρ.

Example 2.12 The Hasse diagram of the poset $(\mathcal{P}(S), \subseteq)$, where $S = \{a, b, c\}$, is given in Fig. 2.1a.

Example 2.13 Consider the poset $(\{1, 2, 3, 4, 5, 6, 7, 8\}, \delta)$, where δ is the divisibility relation introduced in Example 1.27. Its Hasse diagram is shown in Fig. 2.1b.

Definition 2.14 *Let $(S, \leqslant)$ be a poset and let $K \subseteq S$. The* set of upper bounds *of the set K is the set $K^s = \{y \in S \mid x \leqslant y$ for every $x \in K\}$.*

The set of lower bounds *of the set K is the set $K^i = \{y \in S \mid y \leqslant x$ for every $x \in K\}$.*

If $K^s \neq \emptyset$, we say that the set K is bounded above. *Similarly, if $K^i \neq \emptyset$, we say that K is* bounded below. *If K is both bounded above and bounded below we will refer to K as a* bounded set.

If $K^s = \emptyset$ ($K^i = \emptyset$), then K is said to be unbounded above (below).

Theorem 2.15 *Let $(S, \leqslant)$ be a poset and let U and V be two subsets of S. If $U \subseteq V$, then we have $V^i \subseteq U^i$ and $V^s \subseteq U^s$.*

Also, for every subset T of S, we have $T \subseteq (T^s)^i$ and $T \subseteq (T^i)^s$.

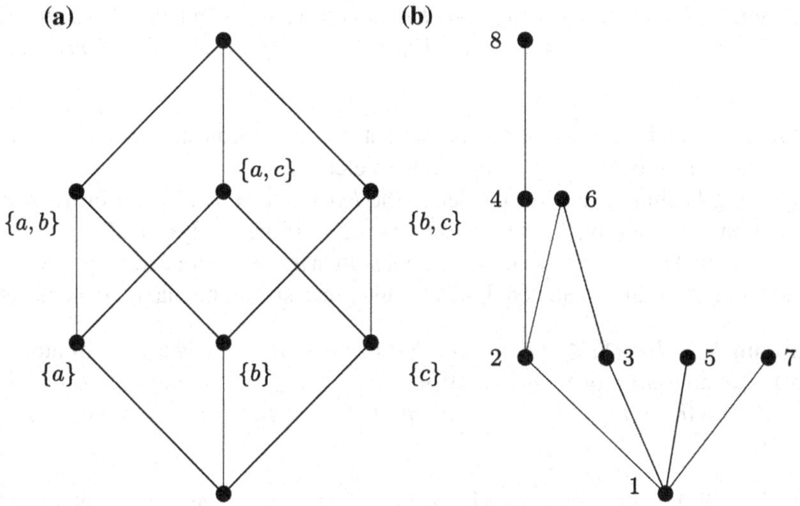

Fig. 2.1 Hasse diagrams. **a** The poset ($\mathcal{P}(S)$,$\subseteq$), and **b** The poset ({1, ..., 8}, δ)

Proof The argument for both statements of the theorem amounts to a direct application of Definition 2.14.

Note that for every subset T of a poset S, we have both

$$T^i = ((T^i)^s)^i \tag{2.1}$$

and

$$T^s = ((T^s)^i)^s. \tag{2.2}$$

Indeed, since $T \subseteq (T^i)^s$, by the first part of Theorem 2.15, we have $((T^s)^i)^s \subseteq T^s$. By the second part of the same theorem applied to T^s, we have the reverse inclusion $T^s \subseteq ((T^s)^i)^s$, which yields $T^s = ((T^s)^i)^s$.

Theorem 2.16 *For any subset K of a poset (S, ρ), the sets $K \cap K^s$ and $K \cap K^i$ contain at most one element.*

Proof Suppose that $y_1, y_2 \in K \cap K^s$. Since $y_1 \in K$ and $y_2 \in K^s$, we have $(y_1, y_2) \in \rho$. Reversing the roles of y_1 and y_2 (that is, considering now that $y_2 \in K$ and $y_1 \in K^s$), we obtain $(y_2, y_1) \in \rho$. Therefore, we may conclude that $y_1 = y_2$ because of the antisymmetry of the relation ρ, which shows that $K \cap K^s$ contains at most one element.

A similar argument can be used for the second part of the proposition; we leave it to the reader.

Definition 2.17 *Let* $(S, \leqslant)$ *be a poset. The* least (greatest) element *of the subset* K *of* S *is the unique element of the set* $K \cap K^i$ ($K \cap K^s$, *respectively) if such an element exists.*

If K is unbounded above, then it is clear that K has no greatest element. Similarly, if K is unbounded below, then K has no least element.

Applying Definition 2.17 to the set S, the *least (greatest) element* of the poset $(S, \leqslant)$ is an element a of S such that $a \leqslant x$ ($x \leqslant a$, respectively) for all $x \in S$.

It is clear that if a poset has a least element u, then u is the unique minimal element of that poset. A similar statement holds for the greatest and the maximal elements.

Definition 2.18 *Let* $(S, \leqslant)$ *be a poset that has* 0 *as its least element. An* atom *of* $(S, \leqslant)$ *is an element* x *of* S *such that* $0 \prec x$.

If $(S, \leqslant)$ *is a poset that has* 1 *as its greatest element, then* y *is a* co-atom *of* $(S, \leqslant)$ *if* $y \lessdot 1$.

Example 2.19 For the poset introduced in Example 2.12, the greatest element is $\{a, b, c\}$, while the least element is $\emptyset$.

The atoms of this poset are $\{a\}, \{b\}, \{c\}$; its co-atoms are $\{a, b\}$, $\{b, c\}$, and $\{a, c\}$.

Definition 2.20 *The subset* K *of the poset* $(S, \leqslant)$ *has a* least upper bound u *if* $K^s \cap (K^s)^i = \{u\}$.

K *has the* greatest lower bound v *if* $K^i \cap (K^i)^s = \{v\}$.

We note that a set can have at most one least upper bound and at most one greatest lower bound. Indeed, we have seen above that for any set U the set $U \cap U^i$ may contain an element or be empty. Applying this remark to the set K^s, it follows that the set $K^s \cap (K^s)^i$ may contain at most one element, which shows that K may have at most one least upper bound. A similar argument can be made for the greatest lower bound.

If the set K has a least upper bound, we denote it by $\sup K$. The greatest lower bound of a set will be denoted by $\inf K$. These notations come from the terms *supremum* and *infimum* used alternatively for the least upper bound and the greatest lower bound, respectively.

Example 2.21 A two-element subset $\{m, n\}$ of $(\mathbb{N}, \delta)$ has both an infimum and a supremum. Indeed, let p be the least common multiple of m and n. Since $(n, p), (m, p) \in \delta$, it is clear that p is an upper bound of the set $\{m, n\}$. On the other hand, if k is an upper bound of $\{m, n\}$, then k is a multiple of both m and n. In this case, k must also be a multiple of p because otherwise we could write $k = pq + r$ with $0 < r < p$ by dividing k by p. This would imply $r = k - pq$; hence, r would be a multiple of both m and n because both k and p have this property. However, this would contradict the fact that p is the least multiple that m and n share! This shows that the least common multiple of m and n coincides with the supremum of the set $\{m, n\}$. Similarly, $\inf\{m, n\}$ equals the greatest common divisor m and n.

Example 2.22 Consider a set M and the poset $(\mathcal{P}(M), \subseteq)$. Let K and H be two subsets of M. The set $\{K, H\}$ has an infimum and a supremum. Indeed, let $L = K \cap H$. Clearly, $L \subseteq K$ and $L \subseteq H$, so L is a lower bound of the set $\{K, H\}$. Furthermore, if $J \subseteq K$ and $J \subseteq H$, then $J \subseteq L$ by the definition of the intersection. This proves that the infimum of $\{K, H\}$ is the intersection $K \cap H$. A similar argument shows that $K \cup H$ is the supremum of $\{K, H\}$.

In the previous two examples, any two-element subset of the poset has both a supremum and an infimum.

For a one-element subset $\{x\}$ of a poset (S, ρ), we have $\sup\{x\} = \inf\{x\} = x$.

Definition 2.23 *A* minimal element *of a poset* $(S, \leqslant)$ *is an element* $x \in S$ *such that* $\{x\}^i = \{x\}$. *A* maximal element *of* $(S, \leqslant)$ *is an element* $y \in S$ *such that* $\{y\}^s = \{y\}$.

In other words, x is a minimal element of the poset $(S, \leqslant)$ if there is no element less than or equal to x other than itself; similarly, x is maximal if there is no element greater than or equal to x other than itself.

The set of minimal elements of a poset $(S, \leqslant)$ is denoted by $\mathsf{MIN}(S, \leqslant)$; the set of maximal elements of this poset is denoted by $\mathsf{MAX}(S, \leqslant)$.

Example 2.24 Not every subset of a poset has a least or a greatest element. Indeed, let $(\{2, 3, 4, 5, 6, 7, 8, \}, \delta)$ be a poset whose Hasse diagram is shown in Fig.2.1b. It is easy to see that

$$\mathsf{MIN}(\{2, 3, 4, 5, 6, 7, 8, \}, \delta) = \{2, 3, 5, 7\},$$
$$\mathsf{MAX}(\{2, 3, 4, 5, 6, 7, 8, \}, \delta) = \{5, 6, 7, 8\}.$$

There is no least element and there is no largest element in this poset.

Theorem 2.25 *Every finite nonempty subset* K *of a poset* $(S, \leqslant)$ *has a minimal element and a maximal element.*

Proof Suppose that $K = \{x_0, \ldots, x_{n-1}\}$ for $n \geqslant 1$. Define the element $u_0 = x_0$ and

$$u_k = \begin{cases} x_k & \text{if } x_k < u_{k-1}, \\ u_{k-1} & \text{otherwise.} \end{cases}$$

Then, u_{n-1} is a minimal element. The proof of the existence of a maximal element of K is similar.

Next, we discuss a simple property of partially ordered sets that will allow us to obtain half of some of the arguments related to the properties of partial orders for free.

Theorem 2.26 *Let* ρ *be a partial order on a set* S. *The inverse* ρ^{-1} *is also a partial order on the same set.*

Proof Since $(x, x) \in \rho$ for every $x \in S$, it follows that $(x, x) \in \rho^{-1}$ for every $x \in S$, so ρ^{-1} is reflexive.

The antisymmetry of ρ^{-1} follows from $(\rho^{-1})^{-1} = \rho$ and because of the antisymmetry of ρ.

To prove the transitivity of ρ^{-1}, assume that $(x, y) \in \rho^{-1}$ and $(y, z) \in \rho^{-1}$. This means that $(y, x), (z, y) \in \rho$, and because of the transitivity of ρ, we obtain $(z, x) \in \rho$, so $(x, z) \in \rho^{-1}$, which proves that ρ^{-1} is transitive.

Definition 2.27 *The* dual *of the poset* (S, ρ) *is the poset* (S, ρ^{-1}).

Concepts valid for a poset have a counterpart for their dual poset. For instance, x is an upper bound for the set K in the poset (S, ρ) if and only if x is a lower bound for K in the dual poset. Similarly, $t = \sup K$ in the poset (S, ρ) if and only if $t = \inf K$ in the dual poset. Similar pairs are minimal element and maximal element, infimum and supremum, etc.

If all concepts occurring in a statement about posets are replaced by their duals, we obtain the *dual statement*; the method of proving statements about posets is known as *dualization*. Furthermore, if a statement holds for a poset (S, ρ), its dual holds for the dual poset (S, ρ^{-1}). This allows us to formulate the following principle.

The Duality Principle for Posets: If a statement is true for all posets, then its dual is also true for all posets.

The validity of this principle follows from the fact that any poset can be regarded as the dual of some other poset. The duality principle allows us to simplify proofs of certain statements that concern posets. For statements involving both a concept and its dual we need to prove only half of the statement; the other half follows by applying the duality principle. For instance, once we prove the statement "any subset of a poset can have at most one least upper bound," the dual statement "any subset of a poset can have at most one greatest lower bound" follows.

2.3 The Poset of Real Numbers

For the poset $(\mathbb{R}, \leqslant)$, it is possible to give more specific descriptions of the supremum and infimum of a subset when they exist.

Theorem 2.28 *If* $T \subseteq \mathbb{R}$, *then* $u = \sup T$ *if and only if* u *is an upper bound of* T *and, for every* $\epsilon > 0$, *there is* $t \in T$ *such that* $u - \epsilon < t \leqslant u$.

The number v *is* $\inf T$ *if and only if* v *is a lower bound of* T *and, for every* $\epsilon > 0$, *there is* $t \in T$ *such that* $v \leqslant t < v + \epsilon$.

Proof We prove only the first part of the theorem; the argument for the second part is similar and is left to the reader.

Suppose that $u = \sup T$; that is, $\{u\} = T^s \cup (T^s)^i$. Since $u \in T^s$, it is clear that u is an upper bound for T. Suppose that there is $\epsilon > 0$ such that no $t \in T$ exists such that $u - \epsilon < t \leqslant u$. This means that $u - \epsilon$ is also an upper bound for T, and in this case u cannot be a lower bound for the set of upper bounds of T. Therefore, no such ϵ may exist.

Conversely, suppose that u is an upper bound of T and for every $\epsilon > 0$, there is $t \in T$ such that $u - \epsilon < t \leqslant u$. Suppose that u does not belong to $(K^s)^i$. This means that there is another upper bound u' of T such that $u' < u$. Choosing $\epsilon = u - u'$, we would have no $t \in T$ such that $u - \epsilon = u' < t \leqslant u$ because this would prevent u' from being an upper bound of T. This implies $u \in (K^s)^i$, so $u = \sup T$.

A very important axiom for the set $\mathbb{R}$ is given next.

The Completeness Axiom for $\mathbb{R}$: If T is a nonempty subset of $\mathbb{R}$ that is bounded above, then T has a supremum.

A statement equivalent to the Completeness Axiom for $\mathbb{R}$ follows.

Theorem 2.29 *If T is a nonempty subset of $\mathbb{R}$ that is bounded below, then T has an infimum.*

Proof Note that the set T^i is not empty. If $s \in T^i$ and $t \in T$, we have $s \leqslant t$, so the set T^i is bounded above. By the Completeness axiom $v = \sup T^i$ exists and $\{v\} = (T^i)^s \cap ((T^i)^s)^i = (T^i)^s \cap T^i$ by Equality (2.1). Thus, $v = \inf T$.

We leave to the reader to prove that Theorem 2.29 implies the Completeness Axiom for $\mathbb{R}$.

Another statement equivalent to the Completeness Axiom is the following.

Theorem 2.30 (Dedekind's Theorem) *Let U and V be nonempty subsets of $\mathbb{R}$ such that $U \cup V = \mathbb{R}$ and $x \in U$, $y \in V$ imply $x < y$. Then, there exists $a \in \mathbb{R}$ such that if $x > a$, then $x \in V$, and if $x < a$, then $x \in U$.*

Proof Observe that $U \neq \emptyset$ and $V \subseteq U^s$. Since $V \neq \emptyset$, it means that U is bounded above, so by the Completeness Axiom $\sup U$ exists. Let $a = \sup U$. Clearly, $u \leq a$ for every $u \in U$. Since $V \subseteq U^s$, it also follows that $a \leqslant v$ for every $v \in V$.

If $x > a$, then $x \in V$ because otherwise we would have $x \in U$ since $U \cup V = \mathbb{R}$ and this would imply $x \leqslant a$. Similarly, if $x < a$, then $x \in U$.

Using the previously introduced notations, Dedekind's theorem can be stated as follows: if U and V are nonempty subsets of $\mathbb{R}$ such that $U \cup V = \mathbb{R}$, $U^s \subseteq V$, $V^i \subseteq U$, then there exists a such that $\{a\}^s \subseteq V$ and $\{a\}^i \subseteq U$.

One can prove that Dedekind's theorem implies the Completeness Axiom. Indeed, let T be a nonempty subset of $\mathbb{R}$ that is bounded above. Therefore $V = T^s \neq \emptyset$. Note that $U = (T^s)^i \neq \emptyset$ and $U \cup V = \mathbb{R}$. Moreover, $U^s = ((T^s)^i)^s = T^s = V$ and $V^i = (T^s)^i = U$. Therefore, by Dedekind's theorem, there is $a \in \mathbb{R}$ such that $\{a\}^s \subseteq V = T^s$ and $\{a\}^i \subseteq U = (T^s)^i$. Note that $a \in \{a\}^s \cap \{a\}^i \subseteq T^s \cap (T^s)^i$, which proves that $a = \sup T$.

By adding the symbols $+\infty$ and $-\infty$ to the set $\mathbb{R}$, one obtains the set $\hat{\mathbb{R}}$. The partial order $\leqslant$ defined on $\mathbb{R}$ can now be extended to $\hat{\mathbb{R}}$ by $-\infty \leqslant x$ and $x \leqslant +\infty$ for every $x \in \mathbb{R}$.

We also extend the addition and multiplication of reals to $\hat{\mathbb{R}}$ by

$$
\begin{aligned}
x + \infty = +\infty + x &= +\infty \text{ for } -\infty < x \leqslant +\infty,\\
x - \infty = -\infty + x &= -\infty \text{ for } -\infty \leqslant x < +\infty,\\
x \cdot \infty = \infty \cdot x &= \begin{cases} -\infty & \text{if } -\infty \leqslant x < 0,\\ 0 & \text{if } x = 0\\ \infty & \text{if } 0 < x \leq +\infty \end{cases},\\
x \cdot (-\infty) = -\infty \cdot x &= \begin{cases} \infty & \text{if } -\infty \leqslant x < 0,\\ 0 & \text{if } x = 0,\\ -\infty & \text{if } 0 < x \leqslant +\infty \end{cases},\\
\frac{x}{+\infty} = \frac{x}{-\infty} &= 0 \text{ for } x \in \mathbb{R}.
\end{aligned}
$$

The operations $+\infty - \infty$ and $-\infty + \infty$ are undefined.

Note that, in the poset $(\hat{\mathbb{R}}, \leqslant)$, the sets T^i and T^s are nonempty for every $T \in \mathcal{P}(\hat{\mathbb{R}})$ because $-\infty \in T^i$ and $+\infty \in T^s$ for any subset T of $\hat{\mathbb{R}}$.

Theorem 2.31 *For every set $T \subseteq \hat{\mathbb{R}}$, both $\sup T$ and $\inf T$ exist in the poset $(\hat{\mathbb{R}}, \leqslant)$.*

Proof We present the argument for $\sup T$. If $\sup T$ exists in $(\mathbb{R}, \leqslant)$, then it is clear that the same number is $\sup T$ in $(\hat{\mathbb{R}}, \leqslant)$.

Assume now that $\sup T$ does not exist in $(\mathbb{R}, \leqslant)$. By the Completeness Axiom for $\mathbb{R}$, this means that the set T does not have an upper bound in $(\mathbb{R}, \leqslant)$. Therefore, the set of upper bounds of T in $(\hat{T}, \leqslant)$ is $T^{\hat{s}} = \{+\infty\}$. It follows immediately that in this case $\sup T = +\infty$ in $(\hat{\mathbb{R}}, \leqslant)$.

2.4 Chains and Antichains

The main notions of this section are introduced next.

Definition 2.32 *Let $(S, \leqslant)$ be a poset. A* chain *of $(S, \leqslant)$ is a subset T of S such that for every $x, y \in T$ such that $x \neq y$ we have either $x < y$ or $y < x$. If the set S is a chain, we say that $(S, \leqslant)$ is a* totally ordered set *and the relation $\leqslant$ is a* total order.

If $s \in \mathbf{Seq}(S)$ (or $s \in \mathbf{Seq}_\infty(S)$) and for every $i, j \in \mathbb{N}$ we have $s(i) < s(j)$ or $s(j) < s(i)$, we refer to the sequence s as a chain in S; if $s(i) \leqslant s(j)$ or $s(j) \leqslant s(i)$ for every $i, j \in \mathbb{N}$, then we say that s is a multichain *in $(S, \leqslant)$.*

If $S = \{x_1, \ldots, x_n\}$, the total order whose diagram is given in Fig. 2.2 is denoted by $\mathsf{TO}(x_1, \ldots, x_n)$.

Let $(S, \leqslant)$ be a poset. The elements x, y of S are *incomparable* if we have neither $x \leqslant y$ nor $y \leqslant x$. This is denoted by $x \parallel y$. It is easy to see that "$\parallel$" is a symmetric and irreflexive relation. The set of pairs of incomparable elements of a poset $(S, \leqslant)$ is

$$\mathsf{INC}(S, \leqslant) = \{(x, y) \in S \times S \mid x \not\leqslant y \text{ and } y \not\leqslant x\}.$$

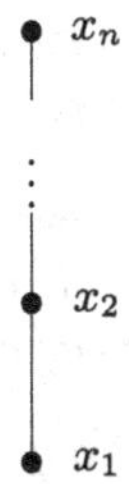

Fig. 2.2 Hasse diagram of a total order on $S = \{x_1, x_2, \ldots, x_n\}$

Definition 2.33 *An* antichain *of* $(S, \leqslant)$ *is a subset* U *of* S *such that, for every two distinct elements* $x, y \in U$, *we have* $x \parallel y$.

Example 2.34 The set of real numbers equipped with the usual partial order $(\mathbb{R}, \leqslant)$ is a chain since, for every $x, y \in \mathbb{R}$, we have either $x \leqslant y$ or $y \leqslant x$.

Example 2.35 In the poset $(\mathbb{N}, \delta)$, the set of all prime numbers is an antichain since if p and q are two distinct primes, we have neither $(p, q) \in \delta$ nor $(q, p) \in \delta$.

Example 2.36 If S is a finite set such that $|S| = n$, the set of subsets of S that contain k elements (for a fixed k, $k \leqslant |S|$) is an antichain in the poset $(\mathcal{P}(S), \subseteq)$ that contains $\binom{n}{k}$ elements.

Example 2.37 If $(S, \leqslant)$ is a poset, then both $\mathsf{MIN}(S, \leqslant)$ and $\mathsf{MAX}(S, \leqslant)$ are maximal antichains of $(S, \leqslant)$ (with respect to set inclusion).

Every finite chain of a poset has a least element and a greatest element. Indeed, by Theorem 2.25, a finite chain has a minimal element and a maximal element. Since the notions of minimal and maximal elements in a chain coincide with the notions of least element and largest element, respectively, it statement follows.

Definition 2.38 *Let* u, v *be two elements of a poset* $(S, \leqslant)$ *such that* $u \leqslant v$. *The interval determined by* u *and* v *is the set*

$$[u, v] = \{x \in S \mid u \leqslant x \leqslant v\}.$$

Example 2.39 In the poset $(\mathbb{N}, \delta)$ we have $(3, 24) \in \delta$. The interval $[3, 24]$ is

$$[3, 24] = \{3, 6, 12, 24\}.$$

Not every poset is a chain, as shown in the next example.

Example 2.40 The poset $(\mathcal{P}(S), \subseteq)$ considered in Example 2.12 is not a chain; elements of $\mathcal{P}(S)$ such as $\{a, b\}$ and $\{b, c\}$ are incomparable.

The poset from Example 2.13 is not a chain since it contains incomparable elements (for instance, $4 \parallel 6$). However, the subset $\{1, 2, 4, 8\}$ is a chain, as can be easily seen. Thus, a poset $(S, \leqslant)$ that is not a chain itself may very well contain subsets that are chains with respect to the trace of the partial order of the set itself.

Denote by *CHAINS*(S) the set of chains of a poset $(S, \leqslant)$. We use the poset $(\textit{CHAINS}(S), \subseteq)$, where the partial order relation is the set inclusion.

Theorem 2.41 *If* $\{U_i \mid i \in I\}$ *is a chain of the poset* $(\textit{CHAINS}(S), \subseteq)$ *(that is, a chain of chains of* $(S, \leqslant)$*), then* $\bigcup\{U_i \mid i \in I\}$ *is itself a chain of* $(S, \leqslant)$ *(that is, a member of* $(\textit{CHAINS}(S), \subseteq)$*).*

Proof Let $x, y \in \bigcup\{U_i \mid i \in I\}$. There are $i, j \in I$ such that $x \in U_i$ and $y \in U_j$ and we have either $U_i \subseteq U_j$ or $U_j \subseteq U_i$. In the first case, we have either $x_i \leqslant x_j$ or $x_j \leqslant x_i$ because both x and y belong to the chain U_j. The same conclusion can be reached in the second case when both x and y belong to the chain U_i. So, in any case, x and y are comparable, which proves that $\bigcup\{U_i \mid i \in I\}$ is a chain of $(S, \leqslant)$.

Definition 2.42 A well-ordered *poset is a poset for which every nonempty subset has a least element.*

A well-ordered set is necessarily a chain. Indeed, consider the well-ordered set $(S, \leqslant)$ and $x, y \in S$. Since the set $\{x, y\}$ must have a least element, we have either $x \leqslant y$ or $y \leqslant x$.

Example 2.43 The set of natural numbers is well-ordered. This property of natural numbers is known as the *well-ordering principle*.

(Well-Ordering Axiom) Given any set S, there is a binary relation ρ such that (S, ρ) is a well-ordered set.

The set $(\mathbb{R}, \leqslant)$ is not well-ordered, despite the fact that it is a chain, since it contains subsets such as $(0, 1) = \{x \mid x \in R, 0 < x < 1\}$ that do not have a least element.

Definition 2.44 *Let "<" be the strict partial order of the poset* $(S, \leqslant)$*. An* infinite descending sequence *in a poset* (S, ρ) *is an infinite sequence* $\mathbf{s} \in \mathbf{Seq}_\infty(S)$ *such that* $\mathbf{s}(n + 1) < \mathbf{s}(n)$ *for all* $n \in \mathbb{N}$.

An infinite ascending sequence *in a poset* (S, ρ) *is an infinite sequence* $\mathbf{s} \in \mathbf{Seq}_\infty(S)$ *such that* $\mathbf{s}(n) < \mathbf{s}(n + 1)$ *for all* $n \in \mathbb{N}$.

A poset with no infinite descending sequences is called Artinian. *A poset with no infinite ascending sequences is called* Noetherian.

Clearly, the range of every infinite ascending or descending sequence is a chain.

Example 2.45 The poset $(\mathbb{N}, \delta)$ is Artinian. Indeed, suppose that $\mathbf{s}$ is an infinite descending sequence of natural numbers. If $\mathbf{s}(0) \neq 0$, then the natural number $\mathbf{s}(0)$ has an infinite set of divisors $\{\mathbf{s}(0), \mathbf{s}(1), \ldots\}$. If $\mathbf{s}(0) = 0$, in view of the fact that any natural number is a divisor of 0, we obtain the impossibility of an infinite descending sequence by applying the same argument to $\mathbf{s}(1)$. However, this poset is not Noetherian. For instance, the sequence $\mathbf{z} : \mathbb{N} \longrightarrow \mathbb{N}$ defined by $\mathbf{z}(n) = 2^n$ for $n \in \mathbb{N}$ is an infinite ascending sequence.

A generalization of well-ordered posets is considered in the next definition.

Definition 2.46 A well-founded poset *is a partially ordered set where every nonempty subset has a minimal element.*

Since the least element of a subset is also a minimal element, it is clear that a well-ordered set is also well-founded. However, the inverse is not true; for instance, not every finite set is well-ordered.

Theorem 2.47 *A poset (S, ρ) is well-founded if and only if it is Artinian.*

Proof Let (S, ρ) be a well-founded poset, and suppose that $\mathbf{s}$ is an infinite descending sequence in this poset. The set $T = \{\mathbf{s}(n) \mid n \in \mathbb{N}\}$ has no minimal element since, for every $\mathbf{s}(k) \in T$, we have $(\mathbf{s}(k+1), \mathbf{s}(k)) \in \rho_1$, which contradicts the well-foundedness of (S, ρ).

Conversely, assume that (S, ρ) is Artinian; that is, there is no infinite descending sequence in (S, ρ). Suppose that K is a nonempty subset of S without minimal elements. Let x_0 be an arbitrary element of K. Such an element exists since K is not empty. Since x_0 is not minimal, there is $x_1 \in K$ such that $(x_1, x_0) \in \rho$. Since x_1 is not minimal, there is $x_2 \in K$ such that $(x_2, x_1) \in \rho$, etc., and this construction can continue indefinitely. In this way, we can build an infinite descending sequence $\mathbf{s} : \mathbb{N} \longrightarrow S$, where $\mathbf{s}(n) = x_n$ for $n \in \mathbb{N}$.

Theorem 2.47 implies immediately that any finite poset is well-founded.

Example 2.48 We will show that the poset $(\mathbb{N} \times \mathbb{N}, \preceq)$ is well-founded.

If $(m, n_0) \succ (m, n_1) \succ \ldots$ is a descending chain of pairs having the same first component, then $n_0 > n_1 > \ldots$ is a descending chain of natural numbers and such a chain is finite. Therefore, $(m, n_0) \succ (m, n_1) \succ \ldots$ must be a finite chain.

Consider now an arbitrary descending chain,

$$(p_0, q_0) \succ (p_1, q_1) \succ \ldots,$$

in $(\mathbb{N} \times \mathbb{N}, \preceq)$. We have $p_0 \geqslant p_1 \geqslant \ldots$, and in this sequence we may have only finite "constant" fragments $p_k = p_{k+1} = \cdots = p_{k+l}$. Therefore, the chain of the first components of the pairs of the sequence $(p_0, q_0) \succ (p_1, q_1) \succ \ldots$ is ultimately decreasing, and this shows that the chain is finite. Thus, this poset is Artinian and therefore, by Theorem 2.47, it is well-founded.

Definition 2.49 *A* graded poset *is a triple $(S, \leqslant, h)$, where $(S, \leqslant)$ is a poset and $h : S \longrightarrow \mathbb{N}$ is a function that satisfies the conditions:*

(i) *$x < y$ implies $h(x) < h(x)$ and*
(ii) *y covers x implies $h(y) = h(x) + 1$,*

for every $x, y \in S$. The function h is referred to as the grading function.

*The set $L_k = \{x \in S \mid h(x) = k\}$ is called the k-*th level of the poset $(S, \leqslant, h)$.

Example 2.50 Define the function $h : M_5 \longrightarrow \mathbb{N}$ by $h(0) = 0$, $h(a) = h(b) = h(c) = 1$, and $h(1) = 2$. The triple $(M_5, \leqslant, h)$ is a graded poset. Its levels are

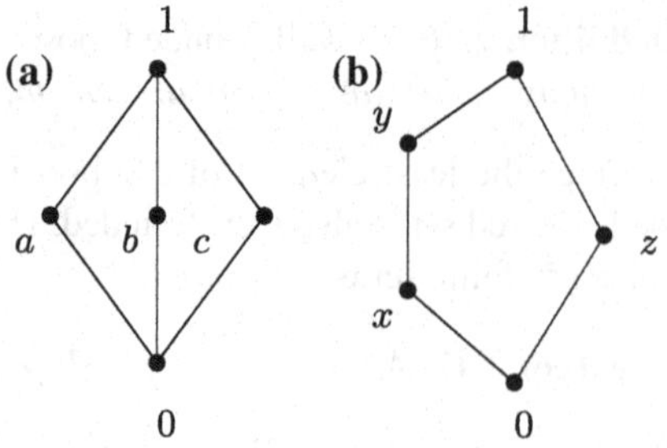

Fig. 2.3 Hasse diagrams of posets **a** $(M_5, \leqslant)$ and **b** $(N_5, \leqslant)$

$$L_0 = \{0\},$$
$$L_1 = \{a, b, c\},$$
$$L_2 = \{1\}.$$

Definition 2.51 *Let $(S, \leqslant)$ be a finite poset. The* height *of $(S, \leqslant)$, denoted by* $\mathsf{height}(S, \leqslant)$*, is the maximal number of elements of a chain.*

The width *of $(S, \leqslant)$,* $\mathsf{width}(S, \leqslant)$*, is the maximal number of elements of an antichain.*

Example 2.52 Let $S = \{s_1, \ldots, s_n\}$ be a finite set such that $|S| = n$. The poset $(\mathcal{P}(S), \subseteq)$ has height $n+1$ since a maximal chain has the form $(\emptyset, \{s_{i_1}\}, \{s_{i_1}, s_{i_2}\}, \ldots, S)$, where $(s_{i_1}, s_{i_2}, \ldots, s_{i_n})$ is a list of the elements of S. Its width is $\binom{n}{[n/2]}$.

Definition 2.53 *Let $(S, \leqslant)$ be a poset that has a least element denoted by 0.*

The height of an element *$x \in S$ (denoted by* $\mathsf{height}(x)$*) is the least upper bound of the lengths of the chains of the form $0 < x_1 < \cdots < x_k = x$.*

If x is an atom of a poset that has the least element 0, then $\mathsf{height}(x) = 1$.

Definition 2.54 *A poset $(S, \leqslant)$ satisfies the* Jordan-Dedekind *condition if all maximal chains between the same elements have the same finite length.*

Example 2.55 The poset $(M_5, \leqslant)$ whose Hasse diagram is shown in Fig. 2.3a satisfies the Jordan-Dedekind condition; the poset $(N_5, \leqslant)$ shown in Fig. 2.3b fails this condition because it contains two maximal chains $0 < x < y < 1$ and $0 < z < 1$ of different lengths between 0 and 1.

The next theorem shows that for a poset that has finite chains the Jordan-Dedekind conditions is equivalent to the fact that the poset is graded by its height function.

Theorem 2.56 *Let $(S, \leqslant)$ be a poset that has finite chains and has the least element 0. $(S, \leqslant)$ satisfies the Jordan-Dedekind condition if and only if the following conditions are satisfied:*

(i) *$x < y$ implies* $\mathsf{height}(x) < \mathsf{height}(x)$*, and*

(ii) *y covers x implies* $\mathsf{height}(y) = \mathsf{height}(x) + 1$
for every $x, y \in S$.

Proof If the height function satisfies the conditions of the theorem, then any maximal chain between the elements x and y has length $\mathsf{height}(y) - \mathsf{height}(x)$, so the Jordan-Dedekind condition is satisfied. Conversely, if the Jordan-Dedekind condition holds, then $\mathsf{height}(x)$ is the length of any maximal chain between 0 and x and the conditions of the theorem follow immediately.

It is clear that if a finite poset $(S, \leqslant)$ contains an antichain U such that $|U| = m$, then S is the union of at least m chains since no two elements of an antichain may belong to the same chain.

Theorem 2.57 (Dilworth's Theorem) *If $(S, \leqslant)$ is a finite nonempty poset such that $\mathsf{width}(S, \leqslant) = m$, then there is a partition of S into m chains.*

Proof The argument is by strong induction on $n = |S|$. If $n = 1$, then the statement holds trivially.

Suppose that the statement holds for sets with fewer than n elements, and let $(S, \leqslant)$ be a poset with $|S| = n$.

Let C be a maximal chain in $(S, \leqslant)$. Two cases may occur:

(i) If no antichain of $(S - C, \leqslant)$ has m elements, then, by the induction hypothesis, there exists a partition of $S - C$ into $m - 1$ chains, so there is a partition of S into m chains.
(ii) If $S - C$ has an antichain $U = \{u_1, \ldots, u_m\}$, define the sets UP_U and DOWN_U as

$$\begin{aligned}\mathsf{UP}_U &= \{x \in S | x \geqslant u_i \text{ for some } u_i \in U\},\\ \mathsf{DOWN}_U &= \{x \in S | x \leqslant u_i \text{ for some } u_i \in U\}.\end{aligned}$$

Note that $S = \mathsf{UP}_U \cup \mathsf{DOWN}_U$ since otherwise S would contain an antichain with more than m elements. Since $(S, \leqslant)$ is a finite poset, the chain C has a largest element t_1 and a smallest element t_0. We have the strict inclusions $\mathsf{UP}_U \subset S$ and $\mathsf{DOWN}_U \subset S$ because $t_1 \notin \mathsf{DOWN}_U$ and $t_0 \notin \mathsf{UP}_U$. Thus, both DOWN_U and UP_U have fewer than n elements.

By the induction hypothesis, we can decompose both UP_U and DOWN_U as partitions of chains, $\mathsf{UP}_U = \bigcup_{i=1}^m C^i_{\geq}$ and $\mathsf{DOWN}_U = \bigcup_{i=1}^m C^i_{\leq}$, where $u_i \in C^i_{\geq} \cap C^i_{\leq}$. Note that u_i is the least element of $C^i_{\geq}$ and the greatest element of $C^i_{\leq}$. Therefore, $C^i_{\geq} \cup C^i_{\leq}$ is a chain, which gives the desired result.

Next, we state a related statement using antichains.

Theorem 2.58 *If $(S, \leqslant)$ is a finite nonempty poset such that* $\mathsf{height}(S, \leqslant) = m$, *then there is a partition of S into m antichains.*

Proof We construct a sequence of finite posets $(S_i, \leqslant_i)$ for $0 \leqslant i \leqslant k-1$. The first poset is $(S_0, \leqslant_0) = (S, \leqslant)$.

Suppose that we defined the nonempty poset $(S_i, \leqslant_i)$. Consider the antichain $U_{i+1} = \mathsf{MAX}(S_i, \leqslant_i)$ and the poset $(S_{i+1}, \leqslant_{i+1})$, where $S_{i+1} = S_i - U_{i+1}$ and $\leqslant_{i+1} = (\leqslant_i)_{S_{i+1}}$. The process halts when $S_k = S_{k-1} - U_k = \emptyset$. It is clear that the $U_1, \ldots, U_k$ are k pairwise disjoint antichains in $(S, \leqslant)$ and that $S = \bigcup_{i=1}^k U_i$.

Since no two members of an antichain may belong to the same chain and S contains a chain having m elements, it follows that any partition of S into antichains requires at least m antichains. Therefore, we have $m \leqslant k$, which means that we need to show only that $k \leqslant m$.

To prove that $k \leqslant m$, we construct a chain $x_1 < x_2 < \cdots < x_k$ in the poset $(S, \leqslant)$ beginning with x_k. Choose x_k to be an arbitrary element of U_k. If $x_j \in U_j$ for $i \leqslant j \leqslant k$, then choose $x_{i-1} \in U_{i-1}$ such that $x_i < x_{i-1}$. This choice is possible because otherwise $x_i \in U_{i-1} = \mathsf{MAX}(S_{i-1}, \leqslant_{i-1})$, which is contradictory because $x_i \in U_i$. This proves that $\{x_1, \ldots, x_k\}$ is a chain, so $\mathsf{height}(S, \leqslant) = m \geqslant k$.

2.5 Poset Product

Let I be a set, and (S, ρ) be a poset. A partial order ρ is defined on the set of functions $I \longrightarrow S$ as $(f, g) \in \rho$ if $(f(i), g(i)) \in \rho$ for every $i \in I$ for $f, g : I \longrightarrow S$.

The relation ρ on $I \longrightarrow S$. We verify only the antisymmetry and leave for the reader the proofs of the reflexivity and transitivity. Assume that $(f, g), (g, f) \in \rho$ for $f, g : I \longrightarrow S$. We have $(f(i), g(i)) \in \rho$ and $(g(i), f(i)) \in \rho$ for every $i \in I$. Therefore, taking into account the antisymmetry of ρ, we obtain $f(i) = g(i)$ for all $i \in I$; hence, $f = g$, which proves the antisymmetry of ρ.

For a set of functions $F \subseteq I \longrightarrow S$, define the subset $F(i)$ of S as $S(i) = \{f(i) \mid f \in F\}$ for $i \in I$.

Theorem 2.59 *The subset F of the poset $(I \longrightarrow S, \rho)$ has a supremum if and only if* $\sup F(i)$ *exists for every $i \in I$ in the poset (S, ρ).*

Proof Suppose that $\sup F(i)$ exists for every $i \in I$ in the poset (S, ρ). Define the mapping $g : I \longrightarrow S$ by $g(i) = \sup F(i)$ for every $i \in I$. We claim that g is $\sup F$.

If $f \in F$, then $(f(i), g(i)) \in \rho$ for every $i \in I$ because of the definition of g. This shows that $(f, g) \in \rho$; hence, g is an upper bound of F. Let h be an upper bound of F. For every $f \in F$, we have $(f(i), h(i)) \in \rho$ for $i \in I$. The definition of g implies $(g(i), h(i)) \in \rho$ for $i \in I$; hence, $g = \sup F$.

Conversely, assume that $k = \sup F$ exists in the poset $(I \longrightarrow S, \rho)$. We prove that $k(i)$ is $\sup F(i)$ for every $i \in I$ in the poset (S, ρ).

The definition of k implies that, for every $f \in F$, we have $(f, k) \in \rho$; that is, $(f(i), k(i)) \in \rho$ for every $i \in I$. Therefore, $k(i)$ is an upper bound of the set $F(i)$ for every $i \in I$.

Let l_i be an upper bound for $F(i)$ for $i \in I$. Define the function $l : I \longrightarrow S$ as $l(i) = l_i$ for $i \in I$. Clearly, l is an upper bound of the set F in the poset $(I \longrightarrow S, \rho)$, and therefore $(k, l) \in \rho$. This, in turn, means that $(k(i), l(i)) = (k(i), l_i) \in \rho$, which shows that $\sup F(i)$ exists and is equal to $k(i)$.

Definition 2.60 *The* product of the posets $\{(S_i, \leqslant_i) \mid i \in I\}$ *is the poset* $(D, \leqslant)$, *where* $D = \prod_{i \in I} S_i$ *and "*$\leqslant$*" is the partial order introduced above on* D. *When* $I = \{1, \ldots, n\}$, *the product will be denoted by*

$$(S_1, \leqslant_1) \times \cdots \times (S_n, \leqslant_n)$$

or by $\prod_{i \in I}(S_i, \leqslant_i)$.

Theorem 2.61 *Let* $\{(S_i, \leqslant_i) \mid i \in I\}$ *be a family of partially ordered sets. If* $H \subseteq \prod_{i \in I} S_i$, *then in the product poset,* $\sup H$ *(*$\inf H$*) exists if and only if* $\sup p_i(H)$ *(*$\inf p_i(H)$*, respectively) exists for every* $i \in I$. *Moreover, if* $y = \sup H$ *(*$y = \inf H$*), then* $p_i(y) = \sup p_i(H)$ *(*$p_i(y) = \inf p_i(H)$*) for every* $i \in I$.

Proof Assume that $y_i = \sup p_i(H)$ exists for every $i \in I$. We need to prove that the element y of $\prod_{i \in I} S_i$ defined by $p_i(y) = y_i$ is $\sup H$.

Consider an arbitrary element $z \in H$. Since $p_i(z) \in p_i(H)$, we have $p_i(z) \leqslant_i y_i$, that is, $p_i(z) \leqslant_i p_i(y)$ for every $i \in I$. This means that $z \leqslant y$, which shows that y is an upper bound of H.

Suppose now that v is an arbitrary upper bound of H. To show that y is $\sup H$, we need to prove that y is the least upper bound of H; that is, $y \leqslant v$ or, equivalently, $p_i(y) \leqslant_i p_i(v)$ for every $i \in I$.

If v is an upper bound of H, then $p_i(v)$ is an upper bound of $p_i(H)$. Since $p_i(y) = y_i = \sup p_i(H)$, we obtain immediately $p_i(y) \leqslant_i p_i(v)$ for every $i \in I$.

Conversely, suppose that $\sup H$ exists. Let $y = \sup H$ and let $y_i = p_i(y)$ for every $i \in I$. We have $x_i \in p_i(H)$ if there is $x \in H$ such that $p_i(x) = x_i$. Since $x \leqslant y$, it follows that $x_i \leqslant_i p_i(y)$, which shows that $p_i(y)$ is an upper bound for $p_i(H)$.

Let w_i be an arbitrary upper bound of $p_i(H)$ for every $i \in I$. There is $w \in \prod_{i \in I} S_i$ such that $p_i(w) = w_i$, and we have $y \leqslant w$ because w is an upper bound for H. Consequently, $p_i(y) \leqslant_i p_i(w)$, and this means that $y_i = \sup p_i(H)$ for every $i \in I$.

The statement for inf follows by dualization.

Another kind of partial order that can be introduced on $S_1 \times \cdot \times S_n$ is defined next.

Theorem 2.62 *For* $f, g \in S_1 \times \cdots \times S_n$, *define* $f \preceq g$ *if* $f = g$ *or if there is* k, $1 \leqslant k \leqslant n$, *such that* $f(k) \neq g(k)$, $f(i) = g(i)$ *for* $1 \leqslant i < k$ *and* $f(k) <_k g(k)$.

The relation $\preceq$ *is a partial order on* $S_1 \times \cdots \times S_n$.

Proof The relation $\preceq$ is obviously reflexive. Suppose now that $f \preceq g$ and $g \preceq f$ and that $f \neq g$. There are $k, h \in \mathbb{N}$ such that $f(i) = g(i)$ for $1 \leqslant i < k$, $f(k) <_k g(k)$, and $f(i) = g(i)$ for $1 \leqslant i < h$, $f(h) <_h g(h)$. If $k < h$, this leads to a contradiction since we cannot have $f(k) <_k g(k)$ and $f(k) = g(k)$. The case $h < k$ also results

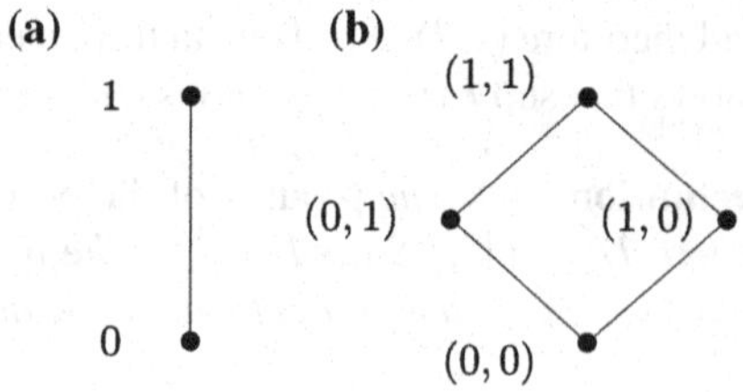

Fig. 2.4 Hasse diagrams of **a** $(\{0, 1\}, \leqslant)$ and **b** $(\{0, 1\}^2, \leqslant)$

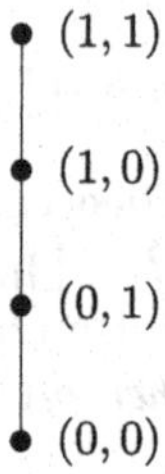

Fig. 2.5 Hasse diagram of $(\{0, 1\}^2, \preceq)$

in a contradiction. For $k = h$, the previous supposition implies $f(k) <_k g(k)$ and $g(k) <_k f(k)$, which is contradictory because "$<_k$" is a strict partial order.

Assume that $f \preceq g$ and $g \preceq l$ and that $f \neq g$, $g \neq l$. There are $k, h \in \mathbb{N}$ such that $f(i) = g(i)$ for $1 \leqslant i < k$, $f(k) <_k g(k)$, and $g(i) = l(i)$ for $1 \leqslant i < h$, $g(h) <_h l(h)$. Define p as being the least of the numbers k, h. For $1 \leqslant i < p$, we have $f(i) = g(i) = l(i)$. In addition, we have $f(p) \leqslant_p l(p)$. Three cases may occur:

1. $f(p) = g(p)$ and $g(p) <_p l(p)$ (when $k > h$),
2. $f(p) <_p g(p)$ and $g(p) = l(p)$ (when $k < h$), and
3. $f(p) <_p g(p)$ and $g(p) <_p l(p)$ (when $k = h$).

If $f = l$, then we have $f \preceq l$. Therefore, we can assume that $f \neq l$. In the first two cases mentioned above, this would imply immediately $f \preceq l$ because of the fact that $f(p) <_p l(p)$. The same conclusion can be reached in the third case because of the transitivity of the strict partial order $<_p$.

We refer to the partial order "$\preceq$" as the *lexicographic partial order* on $S_1 \times \cdots \times S_n$.

Let $\{(S_i, \leqslant_i) \mid 1 \leqslant i \leqslant n\}$ be a family of totally ordered posets. The product poset $\prod_{i=1}^{n}(S_i, \leqslant_i)$ is not necessarily a total order; however, the lexicographic product $(S_1 \times \cdots \times S_n, \preceq)$ is a total order (see Exercise 24).

Example 2.63 Consider the totally ordered set $(\{0, 1\}, \leqslant)$, whose Hasse diagram is given in Fig. 2.4a. The Hasse diagram of the poset $(S \times S, \leqslant)$ is shown in Fig. 2.4b.

On the other hand, the Hasse diagram of the poset $(\{0, 1\}^2, \preceq)$ given in Fig. 2.5 shows that "$\preceq$" is a total order on $\{0, 1\}^2$.

If $S_1 = \cdots S_n = S$, then we obtain the poset $(\mathbf{Seq}_n(S), \preceq)$.

2.6 Functions and Posets

Let $(S, \leqslant)$ and $(T, \leqslant)$ be two posets.

Definition 2.64 *A* morphism *between $(S, \leqslant)$ and $(T, \leqslant)$ or a* monotonic mapping *between $(S, \leqslant)$ and $(T, \leqslant)$ is a mapping $f : S \longrightarrow T$ such that $u, v \in S$ and $u \leqslant v$ imply $f(u) \leqslant f(v)$.*

A mapping $g : S \longrightarrow T$ is antimonotonic *if $u, v \in S$ and $u \leqslant v$ imply $g(u) \geqslant g(v)$.*

The mapping f is strictly monotonic *if $u < v$ implies $f(u) < f(v)$, where "$<$" is the strict partial order associated with the partial order "$\leqslant$".*

Note that $g : S \longrightarrow T$ is antimonotonic if and only if g is a monotonic mapping between the poset $(S, \leqslant)$ and the dual $(T, \geqslant)$ of the poset $(T, \leqslant)$.

Example 2.65 Consider a set M, the poset $(\mathcal{P}(M), \subseteq)$, and the functions $f, g : (\mathcal{P}(M))^2 \longrightarrow \mathcal{P}$, defined by $f(K, H) = K \cup H$ and $g(K, H) = K \cap H$, for $K, H \in \mathcal{P}(M)$. If the Cartesian product is equipped with the product partial order, then both f and g are monotonic. Indeed, if $(K_1, H_1) \subseteq (K_2, H_2)$, we have $K_1 \subseteq K_2$ and $H_1 \subseteq H_2$, which implies that

$$f(K_1, H_1) = K_1 \cup H_1 \subseteq K_2 \cup H_2 = f(K_2, H_2).$$

The argument for g is similar, and it is left to the reader.

Example 2.66 Let $\{(S_i, \rho_i) \mid i \in I\}$ be a collection of posets and let

$$\left(\prod_{i \in I} S_i, \rho\right)$$

be the product of these posets. The projections $p_i : \prod_{i \in I} S_i \longrightarrow S_i$ are monotonic mappings, as the reader will easily verify.

Example 2.67 Let (M, ρ) be an arbitrary poset. Any function $f : S \longrightarrow M$ is monotonic when considered between the posets (S, ι_S) and (M, ρ).

Theorem 2.68 *Let $(P, \leqslant), (R, \leqslant), (S, \leqslant)$ be three posets and let $f : P \longrightarrow R$, $g : R \longrightarrow S$ be two monotonic mappings. The mapping $gf : P \longrightarrow S$ is also monotonic.*

Proof Let $x, y \in P$ be such that $x \leqslant y$. In view of the monotonicity of f, we have $f(x) \leqslant f(y)$, and this implies $(g(f(x)) \leq g(f(y))$ because of the monotonicity of g. Therefore, gf is monotonic.

Let $(P, \leqslant)$ and $(R, \leqslant)$ be two posets. For a monotonic function $f : P \longrightarrow R$, the quotient set, $P/\mathbf{ker}(f)$ can also be organized as a poset. Indeed, if $[x], [y] \in$

$P/\mathbf{ker}(f)$, then we define $[x] \leqslant [y]$ if $f(x) \leqslant f(y)$. This partial order o $P/\mathbf{ker}(f)$ is well-defined because if $x' \in [x]$ and $y' \in [y]$, we have $(f(x'), f(y')) = (f(x), f(y))$.

Theorem 2.69 *The mapping $g : P \longrightarrow P/\mathbf{ker}(f)$ defined by $g(x) = [x]$ for $x \in P$ is a monotonic mapping between the posets $(P, \leqslant)$ and $(P/\mathbf{ker}(f), \leqslant)$.*

Proof The argument is straightforward, and it is left to the reader as an exercise.

Let $f : S \longrightarrow T$ be a monotonic bijection between the posets $(S, \leqslant)$ and $(T, \leqslant)$. As we have seen in Chapter 1, the inverse f^{-1} is also a bijection. Nevertheless, the inverse *is not* necessarily monotonic, as follows from the next example.

Example 2.70 Let $(M_5, \leqslant)$ and $(N_5, \leqslant)$ be the posets whose Hasse diagrams are given in Fig. 2.3, and consider the mapping $f : M_5 \longrightarrow N_5$ defined by $f(0) = 0$, $f(a) = y$, $f(b) = x$, $f(c) = z$, and $f(1) = 1$. The inverse bijection f^{-1} is not monotonic because we have $x \leqslant y$ in $(N_5, \leqslant)$ and $(f^{-1}(x), f^{-1}(y)) = (b, a)$ and $b \not\leqslant a$ in $(M_5, \leqslant)$.

Let $(R, \leqslant)$ and $(S, \leqslant)$ be two posets. The previous considerations justify the following definition.

Definition 2.71 *A* poset isomorphism *between the posets $(R, \leqslant)$ and $(S, \leqslant)$ is a monotonic bijective mapping $f : R \longrightarrow S$ for which the inverse mapping f^{-1} is also monotonic.*

If a poset isomorphism exists between the posets $(P, \leqslant)$ and $(S, \leqslant)$, then we refer to these posets as isomorphic.

Example 2.72 Let $\{p_1, p_2, \ldots, p_n\}$ be the first n primes, $p_1 = 2$, $p_2 = 3$, $p_3 = 5$, etc. Let $m = p_1 \cdots p_n$ be their product and let D_m be the set of all divisors of m. Consider an arbitrary set $A = \{a_1, \ldots, a_n\}$ having n elements.

The posets $(\mathcal{P}(A), \subseteq)$ and (D_m, δ) are isomorphic. Indeed, define the mapping $f : \mathcal{P}(A) \longrightarrow D_m$ by $f(\emptyset) = 1$ and $f(\{a_{i_1}, \ldots, a_{i_k}\}) = p_{i_1} \cdots p_{i_k}$.

The mapping f is bijective. Indeed, for any divisor h of m, we have $h = p_{i_1} \cdots p_{i_k}$ and therefore $h = f(\{a_{i_1}, \ldots, a_{i_k}\})$, which shows that f is surjective.

If $f(\{a_{i_1}, \ldots, a_{i_k}\}) = f(\{a_{j_1}, \ldots, a_{j_l}\})$, then $p_{i_1} \cdots p_{i_k} = p_{j_1} \cdots p_{j_l}$. This gives $k = l$ and $i_1 = j_1, \ldots, i_k = j_k$; hence, $\{a_{i_1}, \ldots, a_{i_k}\} = \{a_{j_1}, \ldots, a_{j_l}\}$, which proves that f is injective.

The mapping f is monotonic because if $\{a_{i_1}, \ldots, a_{i_k}\} \subseteq \{a_{j_1}, \ldots, a_{j_l}\}$,

$$\{i_1, \ldots, i_k\} \subseteq \{j_1, \ldots, j_l\},$$

and this means that the number $p_{i_1} \cdots p_{i_k}$ divides $p_{j_1} \cdots p_{j_l}$.

The inverse mapping $g : D_m \longrightarrow \mathcal{P}(A)$ is also monotonic; we leave the argument to the reader.

Monotonic functions map chains to chains, as we show next.

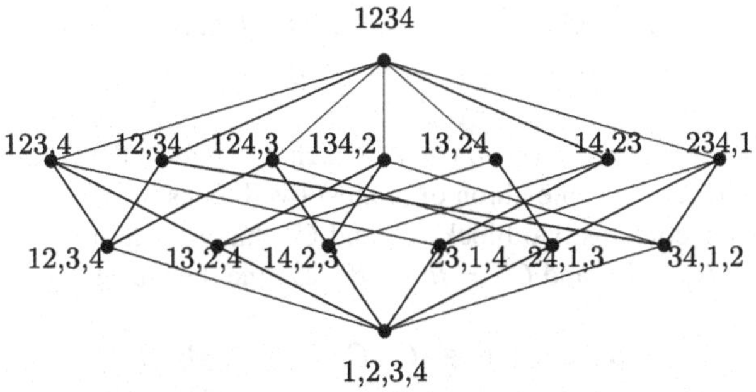

Fig. 2.6 The Hasse diagram of $(\mathit{PART}(\{1,2,3,4\}), \leqslant)$

Theorem 2.73 *Let $(P, \leqslant)$ and $(R, \leqslant)$ be two posets and $f : P \longrightarrow R$ be a monotonic function. If $L \subseteq P$ is a chain in $(P, \leqslant)$, then $f(L)$ is a chain in $(R, \leqslant)$.*

Proof Let $u, v \in f(L)$ be two elements of $f(L)$. There exist $x, y \in L$ such that $f(x) = u$ and $f(y) = v$. Since L is a chain, we have either $x \leqslant y$ or $y \leqslant x$. In the former case, the monotonicity of f implies $u \leqslant v$; in the latter situation, we have $v \leqslant u$.

2.7 The Poset of Equivalences and the Poset of Partitions

In Definition 1.110 we introduced the relation "$\leqslant$" on $\mathit{PART}(S)$ and we examined the relationships that exists between equivalences and partitions on a set. It is easy to verify that this is a partial order relation on $\mathit{PART}(S)$. Thus, the pair $(\mathit{PART}(S), \leqslant)$ is a poset.

Example 2.74 The Hasse diagram of $(\mathit{PART}(\{1,2,3,4\}), \leqslant)$ is given in Fig. 2.6.

To simplify this figure, we represent each nonempty subset of $\{1, 2, 3, 4\}$ as an increasing set of its elements and omit the outer braces; for instance, instead of $\{1, 2, 3\}$, we write 123.

The poset $(\mathit{PART}(S), \leqslant)$ has α_S as its first element and ω_S as its largest.

Theorem 2.75 *Let $\pi, \sigma \in \mathit{PART}(S)$ such that $\pi \leqslant \sigma$. The partition σ* covers the partition π *if and only if there exists a block C of σ that is the union of two blocks B and B' of π and every block of σ that is distinct of C is a block of π.*

Proof Suppose that σ is a partition that covers the partition π. Since $\pi \leqslant \sigma$, every block of σ is a union of blocks of π. Suppose that there exists a block E of σ that is the union of more than two blocks of π; that is, $E = \bigcup\{B_i \mid i \in I\}$, where $|I| \geqslant 3$, and let $B_{i_1}, B_{i_2}, B_{i_3}$ be three blocks of π included in E. Consider the partitions

$$\sigma_1 = \{C \in \sigma \mid C \neq E\} \cup \{B_{i_1}, B_{i_2}, B_{i_3}\},$$
$$\sigma_2 = \{C \in \sigma \mid C \neq E\} \cup \{B_{i_1} \cup B_{i_2}, B_{i_3}\}.$$

It is easy to see that $\pi \leqslant \sigma_1 < \sigma_2 < \sigma$, which contradicts the fact that σ covers π. Thus, each block of σ is the union of at most two blocks of π.

Suppose that σ contains two blocks C' and C'' that are unions of two blocks of π, namely $C' = B_{i_0} \cup B_{i_1}$ and $C'' = B_{i_2} \cup B_{i_3}$. Define the partitions

$$\sigma' = \{C \in \sigma \mid C \notin \{C', C''\}\} \cup \{C', B_{i_2}, B_{i_3}\},$$
$$\sigma'' = \{C \in \sigma \mid C \notin \{C', C''\}\} \cup \{B_{i_1}, B_{i_2}, C''\}.$$

Since $\pi < \sigma', \sigma'' < \sigma$, this contradicts the fact that σ covers π. Thus, we obtain the conclusion of the theorem.

We introduced the equivalence ρ_π that can be built from a partition π and the partition π_ρ that consists of the equivalence classes of ρ. Furthermore, in Corollary 1.114 we noted that $\rho = \rho_{\pi_\rho}$ and $\pi = \pi_{\rho_\pi}$. These observations can be strengthened in the framework of posets.

Theorem 2.76 *The posets* $(\mathsf{EQ}(S), \subseteq)$ *and* $(\mathsf{PART}(S), \leqslant)$ *are isomorphic.*

Proof Let $f : \mathsf{EQ}(S) \longrightarrow \mathsf{PART}(S)$ be the mapping defined by $f(\rho) = S/\rho$. We need to show that f is a monotonic bijective mapping and that its inverse mapping f^{-1} is also monotonic.

The bijectivity of f follows immediately from the remarks that precede the theorem. Let ρ_0, ρ_1 be two equivalences such that $\rho_0 \subseteq \rho_1$ and let $S/rho_0 = \{B_i \mid i \in I\}$, $S/rho_1 = \{C_j \mid j \in J\}$. Let B_i be a block in S/rho_0 and assume that $B_i = [x]_{\rho_0}$. We have $y \in B_i$ if and only if $(x, y) \in \rho_0$, so $(x, y) \in \rho_1$. Therefore, $y \in [x]_{\rho_1}$, which shows that every block $B \in S/rho_0$ is included in a block $C \in \rho_1$. This shows that $f(\rho_0) \leqslant f(\rho_1)$, so f is indeed monotonic. We leave to the reader the proof of monotonicity for f^{-1}.

Theorem 2.77 *Let* $\{\rho_i \mid i \in I\} \subseteq \mathsf{EQ}(S)$ *be a collection of equivalences. Then,* $\inf\{\rho_i \mid i \in I\} = \bigcap_{i \in I} \rho_i$.

Proof By Theorem 1.170, $\rho = \bigcap_{i \in I} \rho_i$ is the closure of the family of equivalences $\{\rho_i \mid i \in I\}$. It is clear that if $\xi \in \mathsf{EQ}(S)$ and $\rho_i \subseteq \xi$ for $i \in I$, then $\rho \subseteq \xi$.

Definition 2.78 *Let S be a set and let* $\rho, \tau \in \mathsf{EQ}(S)$. *A* (ρ, τ)-alternating sequence that joins *x to y is a sequence* $(s_0, s_1, \ldots, s_n)$ *such that* $x = s_0$, $y = s_n$, $(s_i, s_{i+1}) \in \rho$ *for every even i and* $(s_i, s_{i+1}) \in \tau$ *for every odd i, where* $0 \leqslant i \leqslant n - 1$.

lemma 2.79 *Let S be a set and let* $\rho, \tau \in \mathsf{EQ}(S)$. *If s and z are two* (ρ, τ)*-alternating sequences joining x to y and y to z, respectively, then there exists a* (ρ, τ)*-alternating sequence that joins x to z.*

Proof Let $(s_0, \ldots, s_n)$ be a (ρ, τ)-alternating sequences joining x to y and $(w_0, \ldots, w_m)$ a (ρ, τ)-alternating sequences joining y to z, where $x = s_0$, $s_n = w_0 = y$ and $w_m = z$. If $(s_{n-1}, s_n) \in \tau$, then the sequence $(s_0, \ldots, s_n, w_1, \ldots, w_m)$ is a (ρ, τ)-alternating sequence joining x to z. Otherwise, that is, if $(s_{n-1}, s_n) \in \rho$, then taking into account the reflexivity of τ we have $(s_n, w_0) = (s_n, s_n) \in \tau$. In this case, $(s_0, \ldots, s_n, s_n, w_1, \ldots, w_m)$ is a (ρ, τ)-alternating sequence joining x to z.

Theorem 2.80 *Let S be a set and let $\rho, \tau \in \mathsf{EQ}(S)$. If ξ is the relation that consists of all pairs $(x, y) \in S \times S$ that can be joined by a (ρ, τ)-alternating sequence, then $\xi = \sup\{\rho, \tau\}$.*

Proof It is easy to verify that ξ is indeed an equivalence relation. Note that we have both $\rho \subseteq \xi$ and $\tau \subseteq \xi$. Indeed, if $(x, y) \in \rho$, then (x, y, y) is a (ρ, τ)-alternating sequence joining x to y. If $(x, y) \in \tau$, then (x, x, y) is the needed alternating sequence.

Let $\zeta \in \mathsf{EQ}(S)$ such that $\rho \subseteq \zeta$ and $\tau \subseteq \zeta$. If $(x, y) \in \xi$, and $(s_0, s_1, \ldots, s_n)$ is a (ρ, τ)-alternating sequence such that $x = s_0$, $y = s_n$, then each pair (s_i, s_{i+1}) belongs to ζ. By the transitivity property, $(x, y) \in \zeta$, so $\xi \subseteq \zeta$. This implies that $\xi = \sup\{\rho, \tau\}$.

By Theorem 2.76, if $\pi, \sigma \in \mathsf{PART}(S)$ both the infimum and the supremum of the set $\{\pi, \sigma\}$ exist and their description follows from the corresponding results that refer to the equivalence relations. Namely, if $\pi, \sigma \in \mathsf{PART}(S)$, where $\pi = \{B_i \mid i \in I\}$ and $\sigma = \{C_j \mid j \in J\}$, the partition $\inf\{\pi, \sigma\}$ exists and is given by

$$\inf\{\pi, \sigma\} = \{B_i \cap C_j \mid i \in I, j \in J \text{ and } B_i \cap C_j \neq \emptyset\}.$$

The partition $\inf\{\pi, \sigma\}$ will be denoted by $\pi \wedge \sigma$.

A block of the partition $\sup\{\pi, \sigma\}$, denoted by $\pi \vee \sigma$, is an equivalence class of the equivalence $\theta = \sup\{\rho_\pi \wedge \rho_\sigma\}$. We have $y \in [x]_\theta$ if there exists a sequence $(s_0, \ldots, s_n) \in \mathbf{Seq}(S)$ such that $x = s_0$, $s_n = y$ and successive sets $\{s_i, s_{i+1}\}$ are included, alternatively, in a block of π or in a block of σ. More intuitive descriptions of $\sup\{\pi, \sigma\}$ and $\inf\{\pi, \sigma\}$ is given in Sect. 10.4 of Chap. 10.

2.8 Posets and Zorn's Lemma

A statement equivalent to a fundamental principle of set theory known as the Axiom of Choice is Zorn's lemma stated below.

> **Zorn's Lemma:** If every chain of a poset $(S, \leqslant)$ has an upper bound, then S has a maximal element.

Theorem 2.81 *The following three statements are equivalent for a poset $(S, \leqslant)$:*

(i) *If every chain of $(S, \leqslant)$ has an upper bound, then S has a maximal element (Zorn's Lemma).*

(ii) *If every chain of $(S, \leqslant)$ has a least upper bound, then S has a maximal element.*
(iii) *S contains a chain that is maximal with respect to set inclusion (Hausdorff maximality principle).*

Proof (i) implies (ii) is immediate.

(ii) implies (iii): Let $(CHAINS(S), \subseteq)$ be the poset of chains of S ordered by set inclusion. By Theorem 2.41, every chain $\{U_i \mid i \in I\}$ of the poset $(CHAINS(S), \subseteq)$ has a least upper bound $\bigcup\{U_i \mid i \in I\}$ in the poset $(CHAINS(S), \subseteq)$. Therefore, by (ii), $(CHAINS(S), \subseteq)$ has a maximal element that is a chain of $(S, \leqslant)$ that is maximal with respect to set inclusion.

(iii) implies (i): Suppose that S contains a chain W that is maximal with respect to set inclusion and that every chain of $(S, \leqslant)$ has an upper bound. Let w be an upper bound of W.

If $w \in W$, then w is a maximal element of S. Indeed, if this were not the case, then S would contain an element t such that $w < t$ and $W \cup \{t\}$ would be a chain that would strictly include W.

If $w \notin W$, then $W \cup \{w\}$ would be a chain strictly including W, which, again, would contradict the maximality of W. Thus, w is a maximal element of $(S, \leqslant)$.

Denote by $PORD(S)$ the collection of partial order relations on the set S.

Definition 2.82 *Let $\rho, \rho' \in PORD(S)$. The partial order ρ' is an* extension *of ρ if $(x, y) \in \rho$ implies $(x, y) \in \rho'$. Equivalently, we shall say that ρ'* extends *ρ.*

An important consequence of Zorn's lemma is the next statement, which shows that any partial order defined on a set can be extended to a total order on the same set.

Theorem 2.83 (Szpilrajn's Theorem) *Let $(S, \leqslant)$ be a poset. There is a total order $\leqslant'$ on S that is an extension of $\leqslant$.*

Proof Let $PORD(S, \leqslant)$ be the set of partial order relations that can be defined on the set S and contain the relation "$\leqslant$"; clearly, the relation "$\leqslant$" itself is a member of $PORD(S, \leqslant)$. We will apply Zorn's lemma to the poset $(PORD(S, \leqslant), \subseteq)$.

Let $\mathcal{R} = \{\rho_i \mid i \in I\}$ be a chain of $(PORD(S, \leqslant), \subseteq)$; that is, a chain of partial orders ρ_i relative to set inclusion such that $x \leqslant y$ implies $(x, y) \in \rho_i$ for every $i \in I$ and all $x, y \in S$. We claim that the relation $\rho = \bigcup \mathcal{R}$ is a partial order on S.

Indeed, since $\iota_S \subseteq \leqslant \subseteq \rho_i$ for $i \in I$ we have $\iota_S \subseteq \rho$, so ρ is a reflexive relation. To prove that ρ is antisymmetric let $x, y \in S$ be two elements such that $(x, y) \in \rho$ and $(y, x) \in \rho$. By the definition of ρ, there exist $i, j \in I$ such that $(x, y) \in \rho_i$ and $(y, x) \in \rho_j$. Since $\mathcal{R}$ is a chain, we have either $\rho_i \subseteq \rho_j$ or $\rho_j \subseteq \rho_i$. In the first case, both (x, y) and (y, x) belong to ρ_j, so $x = y$ because of the antisymmetry of ρ_j; in the second case, the same conclusion follows because (x, y) and (y, x) belong to ρ_i. Thus, ρ is indeed antisymmetric.

We leave it to the reader to prove the transitivity of ρ. Thus, ρ is a partial order that includes "$\leqslant$", and the arbitrary chain $\mathcal{R}$ has an upper bound. By Zorn's lemma

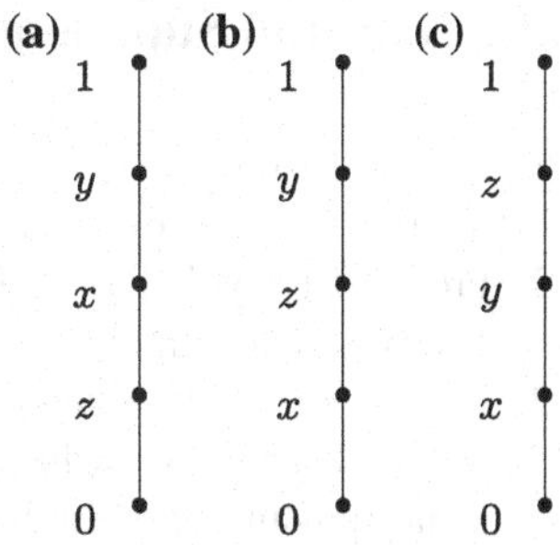

Fig. 2.7 Hasse diagrams of three total orders on the set $\{0, x, y, z, 1\}$

the poset $(PORD(S, \leqslant), \subseteq)$ has a maximal element $\leqslant'$. We now prove that $\leqslant'$ is a total order.

Suppose that (u, v) and (v, u) are two distinct ordered pairs of elements of S such that $u \not\leqslant' v$ and $v \not\leqslant' u$. We show that this supposition leads to a contradiction.

Let $\leqslant_1$ be the relation on S given by

$$\begin{aligned}\leqslant_1 = &\{(x, y) \in S \times S \mid x \leqslant' y\} \cup \{(u, v)\}\\ &\cup\{(z, v) \in S \times \{v\} \mid z \leqslant' v\} \cup \{(u, t) \in \{u\} \times S \mid u \leqslant' t\}.\end{aligned}$$

Since $\iota_S \subseteq \leqslant' \subseteq \leqslant_1$, it follows that $\leqslant_1$ is reflexive.

To prove the antisymmetry of $\leqslant_1$, suppose that $p \leqslant_1 q$ and $q \leqslant_1 p$. Since $v \not\leqslant' u$, it follows that $(p, q) \neq (u, v)$. Thus, the following cases may occur:

(i) If $p \leqslant' q$ and $q \leqslant' p$, then $p = q$ by the antisymmetry of $\leqslant'$.
(ii) If $p = u$, we have $u \leqslant_1 q$ and $q \leqslant_1 u$. By the definition of $\leqslant_1$, this implies $u \leqslant' q$ and $q \leqslant' u$, respectively, so $q = u = p$.
(iii) If $q = v$, we have $p \leqslant_1 v$ and $v \leqslant_1 p$, which imply $p \leqslant' v$ and $v \leqslant' p$, respectively. Thus, $p = v = q$.

We leave the proof of transitivity for "$\leqslant_1$" to the reader.

Note that $\leqslant'$ is strictly included in $\leqslant_1$ because $u \not\leqslant' v$. This contradicts the maximality of the partial order $\leqslant'$, so $\leqslant'$ must be a total order.

Example 2.84 Consider the poset $(N_5, \leqslant)$ introduced in Example 2.55. The posets $(N_5, \leqslant_i)$, where $1 \leq i \leqslant 3$ whose Hasse diagrams are shown in Fig. 2.7a–c are such that $\leqslant \subset \leqslant_i$ and $\leqslant_i$ is a total order for $1 \leqslant i \leqslant 3$. Also, it is easy to see that we have actually $\leqslant = \leqslant_1 \cap \leqslant_3$.

Corollary 2.85 *Let $(S, \leqslant)$ be a poset and let x and y be two incomparable elements in $(S, \leqslant)$. There exists a total order $\leqslant'$ on S that extends $\leqslant$ such that $x \leqslant' y$ and a total order $\leqslant''$ that extends $\leqslant$ such that $y \leqslant'' x$.*

Proof This statement follows immediately from Szpilrajn's theorem.

Exercises and Supplements

1. Define the relation $\leqslant$ on the set $\mathbb{N}^n$ by $(p_1, \ldots, p_n) \leq (q_1, \ldots, q_n)$ if $p_i \leqslant q_i$ for $1 \leqslant i \leqslant n$. Prove that $(\mathbb{N}^n, \leqslant)$ is a partially ordered set.
2. Prove that acyclicity is a hereditary property; this means that if a relation $\sigma \subseteq S \times S$ is acyclic and $\theta \subseteq \sigma$, then θ is also acyclic.
3. Let $f : \mathbb{R} \longrightarrow \mathbb{R}_{>0}$ and $g : \mathbb{R}_{>0} \longrightarrow \mathbb{R}$ be the functions defined by $f(x) = e^x$ for $x \in \mathbb{R}$ and $g(x) = \ln x$ for $X \in \mathbb{R}_{>0}$. Prove that f and g are mutually inverse isomorphisms between the posets $(\mathbb{R}, \geqslant)$ and $(\mathbb{R}_{>0}, \geqslant)$.
4. Let S and T be two sets and let $\sqsubseteq$ be the relation on $S \rightsquigarrow T$ defined by $f \sqsubseteq g$ if $\text{Dom}(f) \subseteq \text{Dom}(g)$ and $f(s) = g(s)$ for every $s \in \text{Dom}(f)$. Prove that $\sqsubseteq$ is a partial order on $S \rightsquigarrow T$.
5. Prove that a binary relation ρ on a set S is a strict partial order on S if and only if it is irreflexive, transitive, and antisymmetric.
6. Let $(S, \leqslant)$ be a poset. An *order ideal* is a subset I of S such that $x \in I$ and $y \leqslant x$ implies $y \in I$. If $\mathcal{I}(S, \leqslant)$ is the collection of order ideals of $(S, \leqslant)$, prove that $\mathcal{K} \subseteq \mathcal{I}(S, \leqslant)$ implies $\bigcap \mathcal{K} \in \mathcal{I}(S, \leqslant)$. Further, argue that $S \in \mathcal{I}(S, \leqslant)$.
7. Let $(S, \leqslant)$ be a poset. An *order filter* is a subset F of S such that $x \in F$ and $y \geqslant x$ implies $y \in F$. If $\mathcal{F}(S, \leqslant)$ is the collection of order filters of $(S, \leqslant)$, prove that $\mathcal{K} \subseteq \mathcal{F}(S, \leqslant)$ implies $\bigcap \mathcal{K} \in \mathcal{F}(S, \leqslant)$. Further, show that $S \in \mathcal{I}(S, \leqslant)$.
8. Let $(S, \leqslant)$ be a finite poset. Prove that S contains at least one maximal and at least one minimal element.
9. Let $(S, \leqslant)$ be a finite poset, where $S = \{x_1, \ldots, x_n\}$. Construct the sequence of posets $((S_1, \leqslant_1), (S_2, \leqslant_2), \ldots)$ as follows. Let $(S_1, \leqslant_1) = (S, \leqslant)$. For $1 \leqslant i \leqslant n$, choose x_{p_i} to be the first element of S_i in the sequence $\mathbf{s} = (x_1, \ldots, x_n)$ that is minimal in $(S_i, \leqslant)$. Define $S_{i+1} = S_i - \{x_{p_i}\}$ and $\leqslant_{i+1} = \leqslant_i \cap (S_{i+1} \times S_{i+1})$. Prove that the sequence $(x_{p_1}, \ldots, x_{p_n})$ is a total order on S that extends the partial order $\leqslant$.
10. Let S be an infinite set and let $(\mathcal{C}, \subseteq)$ be the partially ordered set of its cofinite sets. Prove that for every $U, V \in \mathcal{C}$ both $\sup\{U, V\}$ and $\inf\{U, V\}$ exist.
11. Does the poset of partial functions $(S \rightsquigarrow T, \sqsubseteq)$ introduced in Exercise 4 have a least element?
12. Let $(S, \leqslant)$ be a poset and let U and V be two subsets of S such that $U \subseteq V$. Prove that if both $\sup U$ and $\sup V$ exist, then $\sup U \leqslant \sup V$. Prove that if both $\inf U$ and $\inf V$ exist, then $\inf V \leqslant \inf U$.
13. Prove that the Completeness Axiom of $\mathbb{R}$ implies that for any positive real numbers x, y there exists $n \in \mathbb{N}$ such that $nx > y$ (Archimedes' property of $\mathbb{R}$).
14. Suppose that S and T are subsets of $\mathbb{R}$ that are bounded above. Prove that $S \cup T$ is bounded above and $\sup S \cup T = \max\{\sup S, \sup T\}$.
15. Let π and σ be two partitions of a finite set S. Prove that $|\pi| + |\sigma| \leqslant |\pi \wedge \sigma| + |\pi \vee \sigma|$.
16. Prove that if π is a partition of a set S and $|\pi| = k$, then there are $\binom{k}{2}$ partitions that cover π.

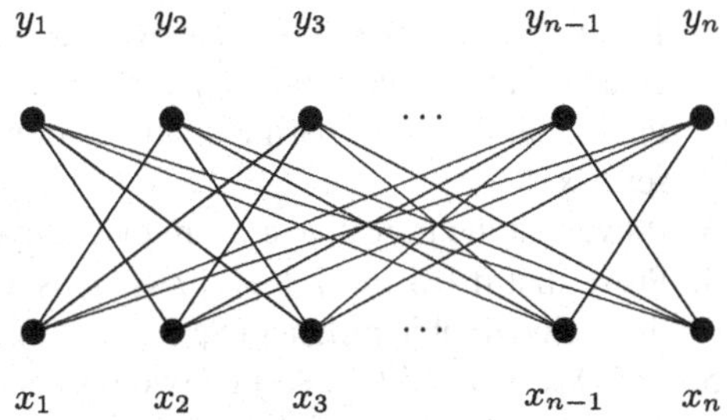

Fig. 2.8 The Hasse diagram of the standard example

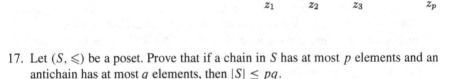

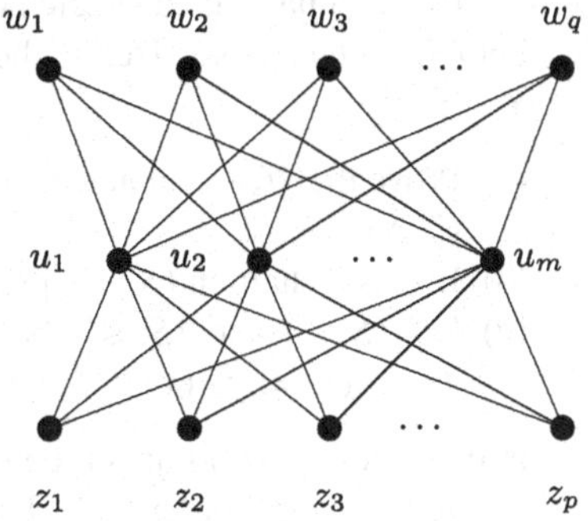

Fig. 2.9 Hasse diagram of the poset $T_{m,p,q}$

17. Let $(S, \leqslant)$ be a poset. Prove that if a chain in S has at most p elements and an antichain has at most q elements, then $|S| \leq pq$.
18. Let $(S, \leqslant)$ be a poset. Prove that $(S, \leqslant)$ is a chain if and only if for every subset T of S both $\sup T$ and $\inf T$ exist and $\{\sup T, \inf T\} \subseteq T$.

Let $(S, \leqslant)$ be a poset. A *realizer* of $(S, \leqslant)$ is a family of total orders on S, $\mathcal{R} = \{\leqslant_i \mid i \in I\}$ such that

$$\leqslant = \bigcap \{\leqslant_i \mid i \in I\}.$$

If $(S, \leqslant)$ is a finite poset, the *dimension* of $(S, \leqslant)$ is the smallest size d of a realizer of $(S, \leqslant)$. The dimension of a finite poset $(S, \leqslant)$ is denoted by $\mathit{dim}(S, \leqslant)$.

19. Let $S = \{x_1, \ldots, x_n\}$ be a finite set. Prove that the discrete partial order ι_S on S has dimension 2.

    Solution: Consider the total order $\leqslant_1 = \mathit{TO}(x_1, \ldots, x_n)$ and its dual $\leqslant_2 = \mathit{TO}(x_n, \ldots, x_2, x_1)$. Note that $(x, x') \in \leqslant_1 \cap \leqslant_2$ if and only if $x = x'$; that is, if and only if $(x, x') \in \iota_S$.
20. Let $(S_n, \leqslant)$ be the poset whose Hasse diagram is given in Fig. 2.8, where $S_n = \{x_1, \ldots, x_n, y_1, \ldots, x_n\}$. This poset was introduced in [1] and is known as the *standard example*. Prove that $\mathit{dim}(S_n, \leqslant) = n$.
21. Consider the poset $(T_{p,m,q}, \leqslant)$, whose Hasse diagram is given in Fig. 2.9. The set $T_{p,m,q}$ consists of three sets of pairwise incomparable elements $\{z_1, \ldots, z_p\}$, $\{u_1, \ldots, u_m\}$, and $\{w_1, \ldots, w_q\}$ such that $z_i < u_j < w_k$ for every $1 \leqslant i \leqslant p$,

$1 \leqslant j \leqslant m$, and $1 \leqslant k \leqslant q$. Prove that if at least one of the numbers p, m, q is greater than 1, then $\mathit{dim}(T_{p,m,q}, \leqslant) = 2$.

22. Prove that the set of partial order relations on a set S is a closure system on the set $S \times S$.
23. Prove that the transitive closure of an acyclic relation is a strict partial order.
24. Prove that if $\{(S_i, \leqslant_i) \mid 1 \leqslant i \leqslant n\}$ is a family of totally ordered posets, then the lexicographic product $(S_1 \times \cdots \times S_n, \preceq)$ is a total order.
25. Let $(S_1, \leqslant_1)$ and $(S_2, \leqslant_2)$ be two posets and let $f : S_1 \longrightarrow S_2$ be a monotonic mapping. Prove that if S_2 has a least element 0, then $f^{-1}(0)$ is an order filter of S_1, and if S_2 has a greatest element 1, then $f^{-1}(1)$ is an order ideal of S_1.
26. Let $(S, \leqslant)$ be a poset. Define the mapping $f_{\leqslant} : S \longrightarrow \mathcal{P}(S)$ by $f_{\leqslant}(x) = \{y \in S \mid x < y\}$.

    (a) Prove that $f_{\leqslant}$ is an antimonotonic mapping between the posets $(S, \leqslant)$ and $(\mathcal{P}(S), \subseteq)$.
    (b) If C is a chain in $(S, \leqslant)$, prove that $f_{\leqslant}(C)$ is a chain in $(\mathcal{P}(S), \subseteq)$.
    (c) Let $(S, \leqslant)$ and $(S, \leqslant')$ be two posets defined on the set S. Prove that $f_{\leqslant \cap \leqslant'}(x) = f_{\leqslant}(x) \cap f_{\leqslant'}(x)$ for every $x \in S$.

27. In the proof of Szpilrajn's theorem, we introduced the set of partial orders that extend the partial order "$\leqslant$". The inclusion between relations defines a partial order on $\mathit{PORD}(S, \leqslant)$. We saw that the maximal elements of $\mathit{PORD}(S, \leqslant)$ are total orders on S and that the least element of $\mathit{PORD}(S, \leqslant)$ is the relation $\leqslant$ itself.
    Let $(S, \leqslant)$ be a poset. Prove that there exists a collection of total orders $\{\leqslant_i \mid i \in I\}$ on S such that $\leqslant = \bigcap_{i \in I} \leqslant_i$.
    Solution: If $\leqslant$ is itself a total order, then the desired collection of total orders consists of $\leqslant$ itself. Suppose therefore that $\leqslant$ is not total, and let $\mathit{INC}(S, \leqslant)$ be the set of all pairs of incomparable elements of $(S, \leqslant)$.
    For each pair $(x, y) \in \mathit{INC}(S, \leqslant)$, consider the total orders $\leqslant'_{xy}$ and $\leqslant''_{xy}$ that extend $\leqslant$ such that $x \leqslant'_{xy} y$ and $y \leqslant'_{xy} x$. Clearly,

$$\leqslant \subseteq \bigcap \{\leqslant'_{xy} \cap \leqslant''_{xy} \mid (x, y) \in \mathit{INC}(S, \leqslant)\}.$$

    Suppose that $\bigcap \{\leqslant'_{xy} \cap \leqslant''_{xy} \mid (x, y) \in \mathit{INC}(S, \leqslant)\}$ contains a pair of elements $(r, s)\mathit{INC}(S, \leqslant)$. Then, we have both $r \leqslant'_{rs} s$ and $r \leqslant''_{rs} s$. Since $s \leqslant''_{rs} r$, this would imply $r = s$ by the antisymmetry of $\leqslant''_{rs}$. This, however, contradicts the incomparability of (r, s) in $(S, \leqslant)$. Thus, for any pair $(u, v) \in \bigcap \{\leqslant'_{xy} \cap \leqslant''_{xy} \mid (x, y) \in \mathit{INC}(S, \leqslant)\}$, we have $u \leqslant v$ or $v \leqslant u$, which shows that

$$\leqslant = \bigcap \{\leqslant'_{xy} \cap \leqslant''_{xy} \mid (x, y) \in \mathit{INC}(S, \leqslant)\}.$$

A poset $(S, \leqslant)$ is *locally finite* if every interval $[x, y]$ of S is a finite set.

28. Prove that the poset $(\mathbb{N}, \leqslant)$ is locally finite.

29. Let S be a finite set. Prove that the poset $(\mathbf{Seq}(S), \leqslant_{inf})$, where $\leqslant_{inf}$ is the partial order introduced in Example 2.6, is locally finite.

Let $(P, \leqslant)$, and $(Q, \leqslant)$ be two posets. Their product is the poset $(P \times Q, \leqslant)$ where $(x, y) \leqslant (x', y')$ if $x \leqslant x'$ and $y \leqslant y'$.

30. Let $(P, \leqslant)$, and $(Q, \leqslant)$ be two posets. Prove that $(P \times Q, \leqslant)$ is locally finite if and only if both $(P, \leqslant)$ and $(Q, \leqslant)$ are locally finite.
31. Prove that if $(P, \leqslant)$ and $(Q, \leqslant)$ are graded posets by the grading functions h and g, respectively, then $(P \times Q, \leqslant)$ is graded by the function f defined by $f(p, q) = h(p)g(q)$ for $(p, q) \in P \times Q$.
32. Let $\zeta : S \times S \longrightarrow \mathbb{R}$ be the Riemann function of a locally finite poset $(S, \leqslant)$, and let ζ^k be the product $\zeta * \zeta * \cdots * \zeta$, which contains k ζ factors, where $k \in \mathbb{N}$. Prove that:

    (a) $\zeta^2(x, y) = |[x, y]|$ if $x \leqslant y$.
    (b) $\zeta^k(x, y)$ gives the number of multichains of length k that can be interpolated between x and y.

Bibliographical Comments

There is a vast body of literature dealing with posets and their applications and a substantial number of references that focus on combinatorial study of posets. Among these we mention [2–5].

Two very useful referrences are [6] and [7].

References

1. B. Dushnik, E.W. Miller, Partially ordered sets. Am. J. Math. **63**, 600–610 (1941)
2. W.M. Trotter, *Combinatorics and Partially Ordered Sets* (The Johns Hopkins University Press, Baltimore, 1992)
3. W.T. Trotter, Partially ordered sets, in *Handbook of Combinatorics*, ed. by R.L. Graham, M. Grötschel, L. Lovász (The MIT Press, Cambridge, 1995), pp. 433–480
4. R.P. Stanley, *Enumerative Combinatorics*, vol. 1 (Cambridge University Press, Cambridge, 1997)
5. J.H. van Lint, R.M. Wilson, *A Course in Combinatorics* (Cambridge University Press, Cambridge, second edition, 2002)
6. M. Barbut, B. Montjardet, *Ordre ar Classification - Algèbre et Combinatoire* (Hachette Université, Paris, 1970)
7. N. Caspard, B. Leclerc, B. Montjardet, *Finite Ordered Sets* (Cambridge University Press, Cambridge, 2012)

Chapter 3
Combinatorics

3.1 Introduction

Combinatorics is the area of mathematics concerned with counting collections of mathematical objects. We begin by discussing several elementary combinatorial issues such as permutations, the power set of a finite sets, the inclusion-exclusion principle, and continue with more involved combinatorial techniques that are relevant for data mining, such as the combinatorics of locally finite posets, Ramsey's Theorem, various combinatorial properties of collection of sets. The chapter concludes with two sections dedicated to the Vapnik-Chervonenkis dimension of a collection and to the Sauer–Shelah theorem.

3.2 Permutations

Definition 3.1 *A* permutation *of a set S is a bijection $f : S \longrightarrow S$.*

A permutation f of a finite set $S = \{s_0, \ldots, s_{n-1}\}$ is completely described by the sequence $(f(s_0), \ldots, f(s_{n-1}))$. No two distinct components of such a sequence may be equal because of the injectivity of f, and all elements of the set S appear in this sequence because f is surjective. Therefore, the number of permutations equals the number of such sequences, which allows us to conclude that there are $n(n-1)\cdots 2\cdot 1$ permutations of a finite set S with $|S| = n$.

The number $n(n-1)\cdots 2\cdot 1$ is denoted by $n!$. This notation is extended by defining $0! = 1$ to capture the fact that there exists exactly one bijection of $\emptyset$, namely the empty mapping.

The *set of permutations* of the set $S = \{1, \ldots, n\}$ is denoted by $PERM_n$. If $f \in PERM_n$ is such a permutation, we write

$$f : \begin{pmatrix} 1 & \cdots & i & \cdots & n \\ a_1 & \cdots & a_i & \cdots & a_n \end{pmatrix},$$

D. A. Simovici and C. Djeraba, *Mathematical Tools for Data Mining*,
Advanced Information and Knowledge Processing, DOI: 10.1007/978-1-4471-6407-4_3,

where $a_i = f(i)$ for $1 \leq i \leq n$. To simplify the notation, we specify f just by the sequence $(a_1, \ldots, a_i, \ldots, a_n)$.

Since $\{1, \ldots, n\}$ is a finite set, for every $x \in \{1, \ldots, n\}$ and $f \in PERM_n$ there exists $k \in \mathbb{N}$ such that $x = f^k(x)$. If k is the least number with this property, the set $\{x, f(x), \ldots, f^{k-1}(x)\}$ is the *cycle* of x and is denoted by $C_{f,x}$. The number $|C_{f,x}|$ is the *length of the cycle*.

Cycles of length 1 are said to be *trivial*.

Note that each pair of elements $f^i(x)$ and $f^j(x)$ of $C_{f,x}$ are distinct for $0 \leq i, j \leq |C_{f,x}| - 1$.

If $z \in C_{f,x}$ and $|C_{f,x}| = k$, then $z = f^j(x)$ for some j, $0 \leq j \leq k - 1$. Since $x = f^k(x)$, it follows that $x = f^{k-j}(z)$, which shows that $x \in C_{f,z}$. Consequently, $C_{f,x} = C_{f,z}$.

Thus, the cycles of a permutation $f \in PERM_n$ form a partition π_f of $\{1, \ldots, n\}$.

Definition 3.2 *A* k-cyclic permutation *of* $\{1, \ldots, n\}$ *is a permutation such that* π_f *consists of a cycle of length k,* $(j_1, \ldots, j_k)$ *and a number of* $n - k$ *cycles of length* 1.

A transposition *of* $\{1, \ldots, n\}$ *is a* 2*-cyclic permutation.*

Note that if f is a transposition of $\{1, \ldots, n\}$, then $f^2 = 1_S$.

Theorem 3.3 *Let* f *be a permutation in* $PERM_n$*, and* $\pi_f = \{C_{f,x_1}, \ldots, C_{f,x_m}\}$ *be the cycle partition associated to* f*. Define the cyclic permutations* $g_1, \ldots, g_m$ *of* $\{1, \ldots, n\}$ *as*

$$g_p(t) = \begin{cases} f(t) & \text{if } t \in C_{f,x_p}, \\ t & \text{otherwise.} \end{cases}$$

Then, $g_p g_q = g_q g_p$ *for every* p, q *such that* $1 \leq p, q \leq m$.

Proof Observe first that $u \in C_{f,x}$ if and only if $f(u) \in C_{f,x}$ for any cycle $C_{f,x}$.

We can assume that $p \neq q$. Then, the cycles C_{f,x_p} and C_{f,x_q} are disjoint. If $u \notin C_{f,x_p} \cup C_{f,x_q}$, then we can write $g_p(g_q(u)) = g_p(u) = u$ and $g_q(g_p(u)) = g_q(u) = u$.

Suppose now that $u \in C_{f,x_p} - C_{f,x_q}$. We have $g_p(g_q(u)) = g_p(u) = f(u)$. On the other hand, $g_q(g_p(u)) = g_q(f(u)) = f(u)$ because $f(u) \notin C_{f,x_q}$. Thus, $g_p(g_q(u)) = g_q(g_p(u))$. The case where $u \in C_{f,x_q} - C_{f,x_p}$ is treated similarly. Also, note that $C_{f,x_p} \cap C_{f,x_q} = \emptyset$, so, in all cases, we have $g_p(g_q(x)) = g_q(g_p(u))$.

The set of cycles $\{g_1, \ldots, g_m\}$ is the *cyclic decomposition of the permutation* f.

Definition 3.4 *A* standard transposition *is a transposition that changes the places of two adjacent elements.*

Example 3.5 The permutation $f \in PERM_5$ given by

$$f : \begin{pmatrix} 1\,2\,3\,4\,5 \\ 1\,3\,2\,4\,5 \end{pmatrix}$$

is a standard transposition of the set $\{1, 2, 3, 4, 5\}$.

On the other hand, the permutation

$$g : \begin{pmatrix} 1\,2\,3\,4\,5 \\ 1\,5\,3\,4\,2 \end{pmatrix}$$

is a transposition but not a standard transposition of the same set because the pair of elements involved is not consecutive.

If $f \in PERM_n$ is specified by the sequence $(a_1, \ldots, a_n)$, we refer to each pair (a_i, a_j) such that $i < j$ and $a_i > a_j$ as an *inversion* of the permutation f. The set of all such inversions is denoted by $INV(f)$. The number of elements of $INV(f)$ is denoted by $inv(f)$.

A *descent* of a permutation $f \in PERM_n$ is a number j such that $1 \leq j \leq n-1$ and $a_j > a_{j+1}$. The set of descents of f is denoted by $D(f)$.

Example 3.6 Let $f \in PERM_6$ be:

$$f : \begin{pmatrix} 1\,2\,3\,4\,5\,6 \\ 4\,2\,5\,1\,6\,3 \end{pmatrix}.$$

We have $INV(f) = \{(4,2), (4,1), (4,3), (2,1), (5,1), (5,3), (6,3)\}$ and $inv(f) = 7$. Furthermore, $D(f) = \{1, 3, 5\}$.

It is easy to see that the following conditions are equivalent for a permutation $f \in PERM_n$:

(i) $f = 1_S$;
(ii) $inv(f) = 0$;
(iii) $D(f) = \emptyset$.

Theorem 3.7 *Every permutation $f \in PERM_n$ can be written as a composition of transpositions.*

Proof If $D(f) = \emptyset$, then $f = 1_S$ and the statement is vacuous. Suppose therefore that $D(f) \neq \emptyset$, and let $j \in D(f)$, which means that (a_j, a_{j+1}) is an inversion f. Let g be the standard transposition that exchanges a_j and a_{j+1}. It is clear that $inv(gf) = inv(f) - 1$. Thus, if g_i are the transpositions that correspond to all standard inversions of f for $1 \leq i \leq p = inv(f)$, it follows that $g_p \cdots g_1 f$ has 0 inversions and, as observed above, $g_p \cdots g_1 f = 1_S$. Since $g^2 = 1_S$ for every transposition g, we have $f = g_p \cdots g_1$, which gives the desired conclusion.

Theorem 3.8 *If $f \in PERM_n$, then $inv(f)$ equals the least number of standard transpositions, and the number of standard transpositions involved in any other factorization of f as a product of standard transposition differs from $inv(f)$ by an even number.*

Proof Let $f = h_q \cdots h_1$ be a factorization of f as a product of standard transpositions. Then, $h_1 \cdots h_q f = 1_S$ and we can define the sequence of permutations $f_l = h_l \cdots h_1 f$ for $1 \leq l \leq q$. Since each h_i is a standard transposition, we have $inv(f_{l+1}) - inv(f_l) = 1$ or $inv(f_{l+1}) - inv(f_l) = -1$. If

$$|\{l \mid 1 \leq l \leq q-1 \text{ and } inv(f_{l+1}) - inv(f_l) = 1\}| = r,$$

then $|\{l \mid 1 \leq l \leq q-1 \text{ and } inv(f_{l+1}) - inv(f_l) = -1\}| = q - r$, so $inv(f) + r - (q-r) = 0$, which means that $q = inv(f) + 2r$. This implies the desired conclusion.

An important characteristic of permutations is their *parity*. Namely, the permutation parity is defined as the parity of the number of their inversions: a permutation $f \in PERM_n$ is *even (odd)* if $inv(f)$ is an even (odd) number.

Theorem 3.8 implies that any factorization of a permutation as a product m standard transpositions determines whether the permutation is odd or even.

Note that any transposition is an odd permutation. Indeed, if $f \in PERM_n$ is a transposition of i and j, where $i < j$ we have

$$f = (1, 2, \ldots, i-1, j, i+1, \ldots, j-1, i, j+1, \ldots, n).$$

The number j generates $j - i$ inversions, and each of the numbers $i+1, \ldots, j-1$ generates one inversion because they are followed by i. Thus, the total number of inversions is $j - i + (j - i - 1) = 2(j-i) - 1$, which is obviously an odd number.

Theorem 3.9 *A cyclic permutation f of length k is the composition of $k-1$ transpositions.*

Proof Let $(j_1, \ldots, j_k)$ be the cycle of length k of f. It is immediate that f is the product of the $k-1$ transpositions $(j_1, j_2), (j_2, j_3), \ldots, (j_{k-1}, j_1)$.

Thus, the parity of a cyclic permutation of even length is odd.

Corollary 3.10 *Let $f \in PERM_n$ be a permutation that has c_ℓ cycles of length ℓ for $\ell \geq 1$. The parity of f is the parity of the number $c_2 + c_4 + \cdots$; in other words, the parity of a permutation is given by the parity of the number of its even cycles.*

Proof By Theorem 3.9 a cyclic transposition of length ℓ is the composition of $\ell - 1$ transpositions. Thus, if f has c_ℓ cycles of length ℓ, then f is a product of $\sum_{\ell \geqslant 1} c_\ell(\ell - 1)$ transpositions. It is clear that t he parity of this sum is determined by those terms where $\ell - 1$ is impar. Thus the parity of f is given by the parity of $c_2 + 3c_4 + 5c_6 + \cdots$ and this equals the parity of $c_2 + c_4 + c_6 + \cdots$.

3.3 The Power Set of a Finite Set

Theorem 3.11 *The set of subsets of a set that contains n elements consists of* 2^n *subsets.*

Proof Let S be a set that contains n elements. By Theorem 1.55 there is a bijection $\Psi : \mathcal{P}(S) \longrightarrow (S \longrightarrow \{0, 1\})$ between the set of subsets of S and the set of indicator functions defined on S. Thus, by Theorem 1.70, the set of subsets of S has the same number of elements as the set of indicator functions defined on S, that is, 2^n.

Let S be a finite nonempty set, $S = \{s_1, \ldots, s_n\}$. We seek to count the sequences of S having length k without repetitions. Any such sequence can be regarded as an injective function $f : \{1, \ldots, k\} \longrightarrow S$. This, by counding these sequences we also determine the number of injective function between $\{1, \ldots, k\}$ and a set S with $|S| = n$.

Suppose initially that $k \geqslant 1$. For the first place in a sequence $\mathbf{s}$ of length k, we have n choices. Once an element of S has been chosen for the first place, we have $n - 1$ choices for the second place because the sequence may not contain repetitions, etc. For the kth component of $\mathbf{s}$, there are $n - 1 + k$ choices. Thus, the number of sequences of length k without repetitions is given by $n(n-1)\cdots(n-k+1)$. We denote this number by $A(n, k)$.

There exists only one sequence of length 0, namely the empty sequence, so we extend the definition of A by $A(n, 0) = 1$ for every $n \in \mathbb{N}$.

An important special case of this counting problem occurs when $k = n$. In this case, a sequence of length n without repetitions is essentially a permutation of the set S. Thus, the number of permutations of S is $n!$. We saw that when $n = 0$, $n! = 1$, which is consistent with the fact that $A(n, 0) = 1$.

Theorem 3.12 *Let S and T be two finite sets. We have*

$$|S \cup T| = |S| + |T| - |S \cap T|,$$
$$|S \oplus T| = |S| + |T| - 2 \cdot |S \cap T|.$$

Proof If $S \cap T = \emptyset$, then $S \cup T = S \oplus T$ and the equalities above are obviously true. Therefore, we may assume that $S \cap T = \{z_1, \ldots, z_p\}$, where $p \geqslant 1$. Thus, the sets S and T can be written as $S = \{x_0, \ldots, x_{m-1}, z_1, \ldots, z_p\}$ and $T = \{y_0, \ldots, y_{n-1}, z_1, \ldots, z_p\}$. The symmetric difference $S \oplus T$ can be written as $S \oplus T = \{x_0, \ldots, x_{m-1}, y_0, \ldots, y_{n-1}\}$. Since $|S| = m + p$, $T = n + p$, $|S \cup T| = m + n + p$, and $|S \oplus T| = m + p$, the equalities of the theorem follow immediately.

Let us now count the number of k-element subsets of a set that contains n elements.

Let S be a set such that $|S| = n$. Define the equivalence $\sim$ on the set $\mathbf{Seq}(S)$ by $\mathbf{s} \sim \mathbf{t}$ if there exists a bijection f such that $\mathbf{s} = \mathbf{t}f$.

It is easy to verify that $\sim$ is an equivalence, and we leave it to the reader to perform this verification. If $\mathbf{s} : \{0, \ldots, p-1\} \longrightarrow S$ and $\mathbf{t} : \{0, \ldots, q-1\} \longrightarrow S$, $\mathbf{s} \sim \mathbf{t}$, and $f : \{0, \ldots, p-1\} \longrightarrow \{0, \ldots, q-1\}$ is a bijection, then we have $p = q$, by Theorem 1.70.

If T is a subset of S such that $|T| = k$, there exists a bijection $\mathbf{t} : \{0, \ldots, k-1\} \longrightarrow T$; clearly, this is a sequence without repetitions and there exist $A(n,k)$ such sequences. If $\mathbf{u}$ is an equivalent sequence (that is, if $\mathbf{t} \sim \mathbf{u}$), then the range of this sequence is again the set T and there are $k!$ such sequences (due to the existence of the $k!$ permutations f) that correspond to the same set T. Therefore, we may conclude that $\mathcal{P}_k(S)$ contains $\frac{A(n,k)}{k!}$ elements. We denote this number by $\binom{n}{k}$ and we refer to it as the (n,k)-*binomial coefficient*. We can write $\binom{n}{k}$ using factorials:

$$\begin{aligned}\binom{n}{k} &= \frac{A(n,k)}{k!} = \frac{n(n-1)\cdots(n-k+1)}{k!} \\ &= \frac{n(n-1)\cdots(n-k+1)(n-k)\cdots 2\cdot 1}{k!(n-k)!} = \frac{n!}{k!(n-k)!}.\end{aligned}$$

Note that we have

$$\binom{n}{k} = \binom{n}{n-k},$$

an equality known as the *symmetry identity*.

We mention the following useful identities:

$$k\binom{n}{k} = n\binom{n-1}{k-1}, \tag{3.1}$$

$$\binom{n}{m} = \frac{n}{m}\binom{n-1}{m-1}. \tag{3.2}$$

Equality (3.1) can be extended as

$$k(k-1)\cdots(k-\ell)\binom{n}{k} = n(n-1)\cdots(n-\ell)\binom{n-\ell-1}{k-\ell-1} \tag{3.3}$$

for $0 \leq \ell \leq k-1$.

Consider now the n-degree polynomial in x

$$p(x) = (x+a_0)\cdots(x+a_{n-2})(x+a_{n-1}).$$

Observe that the coefficient of x^{n-k} consists of the sum of all monomials of the form $a_{i_0}\cdots a_{i_{k-1}}$, where the subscripts $i_0, \ldots, i_{k-1}$ are distinct. Thus, the coefficient of x^{n-k} contains $\binom{n}{k}$ terms corresponding to the k-element subsets of the set $\{0, \ldots, n-1\}$. Consequently, the coefficient of x^{n-k} in the power $(x+a)^n$ can be obtained from the similar coefficient in $p(x)$ by taking $a_0 = \cdots = a_{n-1} = a$; thus, the coefficient is $\binom{n}{k}a^k$. This allows us to write:

$$(x+a)^n = \sum_{k=0}^{n} \binom{n}{k} x^{n-k} a^k. \tag{3.4}$$

This equality is known as *Newton's binomial formula* and has numerous applications.

Example 3.13 If we take $x = a = 1$ in Formula (3.4) we obtain the identity

$$2^n = \sum_{k=0}^{n} \binom{n}{k}. \tag{3.5}$$

Note that this equality can be obtained directly by observing that the right member enumerates the subsets of a set having n elements by their cardinality k.

A similar interesting equality can be obtained by taking $x = 1$ and $a = -1$ in Formula (3.4). This yields

$$0 = \sum_{k=0}^{n} \binom{n}{k} (-1)^k = \binom{n}{0} + \binom{n}{2} + \binom{n}{4} + \cdots$$
$$-\binom{n}{0} - \binom{n}{2} - \binom{n}{4} - \cdots.$$

This inequality shows that each set contains an equal number of subsets having an even or odd number of elements.

Example 3.14 Consider the equality $(x+a)^n = (x+a)^{n-1}(x+a)$. The coefficient of $x^{n-k}a^k$ in the left member is $\binom{n}{k}$. In the right member $x^{n-k}a^k$ has the coefficient $\left(\binom{n-1}{k} + \binom{n-1}{k-1}\right)$, so we obtain the equality

$$\binom{n}{k} = \binom{n-1}{k} + \binom{n-1}{k-1}, \tag{3.6}$$

for $0 \leq k \leq n-1$, known as the *addition identity*.

Multinomial coefficients are generalizations of binomial coefficients that can be introduced as follows. The nth power of the sum $x_1 + \cdots + x_k$ can be written as

$$(x_1 + \cdots + x_k)^n = \sum_{(r_1,\ldots,r_k)} c(n, r_1, \ldots, r_k) x_1^{r_1} \cdots x_k^{r_k},$$

where the sum involves all $(r_1, \ldots, r_k) \in \mathbb{N}^k$ such that $\sum_{i=1}^{k} r_i = n$. By analogy with the binomial coefficients, we denote $c(n, r_1, \ldots, r_k)$ by $\binom{n}{r_1,\ldots,r_n}$. As we did with binomial coefficients in Example 3.14, starting from the equality $(x_1 + \cdots + x_k)^n = (x_1 + \cdots + x_k)^{n-1}(x_1 + \cdots + x_k)$, the coefficient of the monomial $x_1^{r_1} \cdots x_k^{r_k}$ in the right member is $\binom{n}{r_1,\ldots,r_n}$. On the left member, the same coefficient is

$$\sum_{i=1}^{k} \binom{n-1}{r_1, \ldots, r_i - 1, \ldots, r_n},$$

so we obtain the identity

$$\binom{n}{r_1, \ldots, r_n} = \sum_{i=1}^{k} \binom{n-1}{r_1, \ldots, r_i - 1, \ldots, r_n}, \tag{3.7}$$

a generalization of the identity (3.6).

3.4 The Inclusion–Exclusion Principle

Let A and B be two finite sets. It is easy to verify that

$$|A \cup B| = |A| + |B| - |A \cap B|. \tag{3.8}$$

In this section we discuss a generalization of Equality (3.8) known as the *inclusion-exclusion principle*.

Note that if U and V are two subsets of a finite set S such that $V \subseteq U$, then the function I defined by $I(x) = I_U(x) - I_V(x)$ for $x \in S$ is an indicator function, namely the indicator function of the subset $U - V$ of S.

Let a and b be two numbers that belong to the set $\{-1, 1\}$ such that the function I_{ab} defined by

$$I_{ab}(x) = aI_U(x) + bI_V(x)$$

for $x \in S$ is the indicator function of a subset W of the set S. Since $I_{ab}(x) \in \{0, 1\}$, the following cases are possible:

1. If $a = b = 1$, then we have $U \cap V = \emptyset$; otherwise (that is, if $x \in U \cap V$) we would have $aI_U(x) + bI_V(x) = 2$ and this would prevent I_{ab} from being an indicator function. Clearly, in this case, $W = U \cup V$.
2. If $a = 1$ and $b = -1$, we must have $I_V(x) \leq I_U(x)$ for every $x \in S$, which implies $V \subseteq U$. Thus, $W = U - V$.
3. The case where $a = -1$ and $b = 1$ is similar to the previous case, and we have $W = V - U$.
4. The case when $a = -1$ and $b = -1$ is possible only if $U = V = \emptyset$. In this case, $W = \emptyset$.

Note that in all these cases we have $|W| = a|U| + b|V|$. This observation is generalized by the following statement.

Theorem 3.15 *Let $U_0, \ldots, U_{n-1}$ be n subsets of a finite set S, where $n \geqslant 2$, and let $(a_0, \ldots, a_{n-1}) \in \mathbf{Seq}_n(\{-1, 1\})$ be a sequence of n numbers such that the function $I : S \longrightarrow \{0, 1\}$ defined by*

$$I(x) = a_0 I_{U_0}(x) + \cdots + a_{n-1} I_{U_{n-1}}(x)$$

for $x \in S$ is the indicator function of a subset W of S. Then,

$$|W| = a_0|U_0| + \cdots + a_{n-1}|U_{n-1}|.$$

Proof If W is a subset of S, then $\sum_{x\in S} I_W(x) = |W|$ because for each $x \in S$ its contribution to the sum $\sum_{x\in S} I_W(x)$ is equal to 1 if and only if $x \in W$. Therefore, if $I_W(x) = \sum_{i=0}^{n-1} a_i I_{U_i}(x)$ for $x \in S$, we have

$$\begin{aligned} |W| &= \sum_{x\in S} I_W(x) = \sum_{x\in S}\sum_{i=0}^{n-1} a_i I_{U_i}(x) = \sum_{i=0}^{n-1}\sum_{x\in S} a_i I_{U_i}(x) \\ &= \sum_{i=0}^{n-1} a_i \sum_{x\in S} I_{U_i}(x) = \sum_{i=0}^{n-1} a_i |U_i|. \end{aligned}$$

Corollary 3.16 (Principle of Inclusion–Exclusion) *Let $A_0, \ldots, A_{n-1}$ be n finite sets, where $n \geqslant 2$. We have*

$$\begin{aligned} \left|\bigcup_{i=0}^{n-1} A_i\right| &= \sum_{0\le i\le n-1} |A_i| - \sum_{0\le i_1<i_2\le n-1} |A_{i_1} \cap A_{i_2}| \\ &+ \sum_{0\le i_1<i_2<i_3\le n-1} |A_{i_1} \cap A_{i_2} \cap A_{i_3}| - \cdots + (-1)^{n+1}|A_0 \cap \cdots \cap A_{n-1}|. \end{aligned}$$

Proof Suppose that $A_i \subseteq S$ for $0 \le i \le n-1$, where S is a finite set. For $x \in S$, we have $x \notin A = \bigcup_{i=0}^{n-1} A_i$ if and only if $x \notin A_i$ for $0 \le i \le n-1$. This is equivalent to writing

$$1 - I_A(x) = (1 - I_{A_{i_0}}(x)) \cdots (1 - I_{A_{i_{n-1}}}(x))$$

for every $x \in S$. This equality is, in turn, equivalent to

$$\begin{aligned} I_A(x) &= \sum_{i=0}^{n-1} I_{A_i}(x) - \sum_{0\le i_1<i_2\le n-1} I_{A_{i_1}}(x) I_{A_{i_2}}(x) \\ &+ \sum_{0\le i_1<i_2<i_3\le n-1} I_{A_{i_1}}(x) I_{A_{i_2}}(x) I_{A_{i_3}}(x) - \cdots + (-1)^{n+1} I_{A_0}(x) \cdots I_{A_{n-1}}(x) \end{aligned}$$

$$= \sum_{i=0}^{n-1} I_{A_i}(x) - \sum_{0 \le i_1 < i_2 \le n-1} I_{A_{i_1} \cap A_{i_2}}(x)$$
$$+ \sum_{0 \le i_1 < i_2 < i_3 \le n-1} I_{A_{i_1} \cap A_{i_2} \cap A_{i_3}}(x) - \cdots + (-1)^{n+1} I_{A_0 \cap \cdots \cap A_{n-1}}(x).$$

By applying Theorem 3.15, we obtain the equality of the corollary.

Corollary 3.17 *Let $A_0, \ldots, A_{n-1}$ be n finite sets, where $n \geqslant 2$, and let $S = \bigcup_{i=0}^{n-1} A_i$. We have*

$$\left|\bigcap_{i=0}^{n-1} A_i\right| = |S| - \sum_{0 \le i \le n-1} |A_i| + \sum_{0 \le i_1 < i_2 \le n-1} |A_{i_1} \cap A_{i_2}|$$
$$- \sum_{0 \le i_1 < i_2 < i_3 \le n-1} |A_{i_1} \cap A_{i_2} \cap A_{i_3}| + \cdots + (-1)^n |A_0 \cap \cdots \cap A_{n-1}|.$$

Proof This follows immediately from Corollary 3.16 by observing that

$$\left|\bigcap_{i=0}^{n-1} A_i\right| = |S| - \left|\bigcup_{i=0}^{n-1} A_i\right|.$$

3.5 Locally Finite Posets and Möbius Functions

Definition 3.18 *Let $(S, \le)$ be a poset and let $x, y \in S$ be such that $x \le y$. The* closed interval *of $(S, \le)$ defined by x, y is the set*

$$[x, y] = \{t \in S \mid x \le t \le y\}.$$

In addition, we define the open interval *(x, y) as*

$$(x, y) = \{t \in S \mid x < t < y\}$$

and the semiclosed *(or* semiopen*) intervals $[x, y)$ and $(x, y]$ by*

$$[x, y) = \{t \in S \mid x \le t < y\},$$
$$(x, y] = \{t \in S \mid x < t \le y\},$$

respectively.

Note that if $x = y$, then $[x, x] = \{x\}$, while $(x, x) = \emptyset$.

Definition 3.19 *A poset* $(S, \leq)$ *is* locally finite *if every closed interval of* $(S, \leq)$ *is finite.*

Example 3.20 The poset $(\mathbb{N}, \leq)$ is locally finite. Indeed, if $[p, q]$ is a closed interval of this poset, then $[p, q]$ is a finite set that consists of $q - p + 1$ natural numbers.

Example 3.21 The poset $(\mathbb{N}, \delta)$ introduced in Example 1.27 is locally finite. Indeed, if p divides q, then $[p, q]$ is a finite set that contain all multiples of p that divide q. For example, the closed interval $[2, 12]$ contains the numbers 2, 4, 6 and 12.

Let $(S, \leq)$ be a locally finite poset and let $\mathcal{A}(S, \leq)$ be the set of all functions of the form $f : S \times S \longrightarrow \mathbb{R}$ such that $x \not\leq y$ implies $f(x, y) = 0$ for $x, y \in S$. We refer to $\mathcal{A}(S, \leq)$ as the *incidence algebra* of the poset $(S, \leq)$.

Note that if $f \in \mathcal{A}(S, \leq)$ and $x > y$ or $x \parallel y$, then $f(x, y) = 0$.

Definition 3.22 *Let* $(S, \leq)$ *be a locally finite poset and let* $f, g \in \mathcal{A}(S, \leq)$ *be two functions. Their* convolution product *is the function* $h : S \times S \longrightarrow \mathbb{R}$ *defined by*

$$h(x, y) = \begin{cases} \sum_{z \in [x,y]} f(x, z)g(z, y) & \text{if } x \leq y, \\ 0 & \text{otherwise,} \end{cases}$$

for $x, y \in S$. *The function* h *is denoted by* $f * g$.

Lemma 3.23 *The operation* $*$ *is well-defined on the set* $\mathcal{A}(S, \leq)$*; further, "*$*$*" is associative on* $\mathcal{A}(S, \leq)$ *and its unit element is the* Kronecker function k *defined by*

$$\mathsf{k}(x, y) = \begin{cases} 1 & \text{if } x = y \\ 0 & \text{otherwise,} \end{cases}$$

for $x, y \in S$.

Proof Suppose that $h = f * g$, where $f, g \in \mathcal{A}(S, \leq)$. If $x \not\leq y$, then $h(x, y) = 0$, so $h \in \mathcal{A}(S, \leq)$.

Let e, f, g be three functions of $\mathcal{A}(S, \leq)$. We claim that $(e * f) * g = e * (f * g)$. Suppose that $x \leq z$. Then, we have

$$\begin{aligned} ((e * f) * g)(x, z) &= \sum_{y \in [x,z]} (e * f)(x, y)g(y, z) \\ &= \sum_{y \in [x,z]} \left(\sum_{u \in [x,y]} e(x, u) f(u, y) \right) g(y, z) \\ &= \sum_{y \in [x,z]} \sum_{u \in [x,y]} e(x, u) f(u, y)g(y, z) \\ &= \sum_{y \in [x,z]} \sum_{u \in [x,z]} e(x, u) f(u, y)g(y, z) \end{aligned}$$

$$\text{(because if } u > y \text{ we have } f(u,y) = 0)$$
$$= \sum_{u \in [x,z]} \sum_{y \in [x,z]} e(x,u) f(u,y) g(y,z).$$

On the other hand, we can write

$$\begin{aligned}
(e * (f * g))(x,z) &= \sum_{u \in [x,z]} e(x,u)(f * g)(u,z) \\
&= \sum_{u \in [x,z]} e(x,u) \sum_{y \in [u,z]} f(u,y) g(y,z) \\
&= \sum_{u \in [x,z]} e(x,u) \sum_{y \in [x,z]} f(u,y) g(y,z), \\
&\quad \text{(because if } u > y \text{ we have } f(u,y) = 0)
\end{aligned}$$

for $x, z \in S$, which shows that $*$ is associative.

If $f \in \mathcal{A}(S, \leq)$ and $x \leq y$, then we can write

$$(f * \mathsf{k})(x,y) = \sum_{z \in [x,y]} f(x,z) \mathsf{k}(z,y) = f(x,y)$$

for $x, y \in S$. Thus, $f * \mathsf{k} = f$. A similar argument shows that $\mathsf{k} * f = f$. This allows us to conclude that k is indeed the unit with respect to the $*$ operation.

Let $\mathcal{I}(S, \leq) = \{[x,y] \mid x, y \in S \text{ and } x \leq y\} \cup \{\emptyset\}$ be the set of intervals of the poset $(S, \leq)$ to which we add the empty set. A useful point of view (see [1]) is to regard the incidence algebra of $(S, \leq)$ as consisting of formal sums of the form $\sum\{f(x,y) \cdot [x,y] \mid [x,y] \in \mathcal{I}(S, \leq) - \{\emptyset\}\}$. Define the product of two intervals as

$$[x,y][u,v] = \begin{cases} [x,v] & \text{if } y = u, \\ \emptyset & \text{otherwise.} \end{cases}$$

Further, we assume that the product of formal sums is distributive with respect to addition of these sums. Let $f, g \in \mathcal{A}(S, \leq)$ be two functions and let $\hat{f}$ and $\hat{g}$ be their corresponding formal sums,

$$\begin{aligned}
\hat{f} &= \sum\{f(x,y) \cdot [x,y] \mid [x,y] \in \mathcal{I}(S, \leq)\}, \\
\hat{g} &= \sum\{g(u,v) \cdot [u,v] \mid [u,v] \in \mathcal{I}(S, \leq)\}.
\end{aligned}$$

Then, it is immediate that

$$\hat{f}\hat{g}(x,z) = \sum_{x \leq y \leq z} f(x,y)g(y,z)[x,z],$$

so the usual product of the formal sums $\hat{f}\hat{g}$ corresponds to the convolution product of f and g.

Theorem 3.24 *Let $(S, \leq)$ be a locally finite poset. A function $f \in \mathcal{A}(S, \leq)$ has an inverse relative to the operation $*$ if and only if $f(x,x) \neq 0$ for every $x \in S$.*

Proof Suppose that there exists an inverse f' of f (that is, $f * f' = f' * f = \mathsf{k}$) which yields $(f * f')(x,x) = \mathsf{k}(x,x) = 1$ for every x. Since $(f * f')(x,x) = \sum_{z \in [x,x]} f(x,z)f'(z,x) = f(x,x)f'(x,x)$, it follows that $f(x,x) \neq 0$.

To prove the converse implication, we first show the existence of a left inverse of f; that is, a function $f' : S \times S \to \mathbb{R}$ such that $f' * f = \mathsf{k}$. For $x \leq y$, we must have $\sum_{z \in [x,y]} f'(x,z)f(z,y) = \mathsf{k}(x,y)$. This implies $f'(x,x)f(x,x) = 1$ and $\sum_{z \in [x,y]} f'(x,z)f(z,y) = 0$ if $x \neq y$. Thus, we must have

$$f'(x,x) = \frac{1}{f(x,x)}, \tag{3.9}$$

$$f'(x,y) = -\frac{1}{f(y,y)} \sum_{z \in [x,y)} f'(x,z)f(z,y), \tag{3.10}$$

when $x \leq y$ and

$$f'(x,y) = 0,$$

when $x \not\leq y$. Equalities (3.9) and (3.10) give an inductive definition of f' because the poset $(S, \leq)$ is locally finite.

To verify that f' is a left inverse of f, suppose that $x < y$. Then,

$$\begin{aligned}(f' * f)(x,y) &= \sum_{z \in [x,y]} f'(x,z)f(z,y) \\ &= \sum_{z \in [x,y)} f'(x,z)f(z,y) + f'(x,y)f(y,y) = 0.\end{aligned}$$

If $x = y$, then $(f' * f)(y,y) = 1$ and $x \not\leq y$ implies $(f' * f)(x,y) = 0$. Therefore, $f' * f = \mathsf{k}$.

The function f' is also a right inverse of f. Let $h = f * f'$. We have shown above that every function of $\mathcal{A}(S, \leq)$ has a left inverse, so let h' be the left inverse of h. Thus, we have $f * f' = h = \mathsf{k} * h = (h' * h) * h = h' * (f * f') * (f * f') = h' * f * \mathsf{k} * f' = h' * f * f' = h' * h = \mathsf{k}$, which proves that f' is also a right inverse of f. Thus, f' is the inverse of f.

If the inverse of $f \in \mathcal{A}(S, \leq)$ exists, we denote it by the common notation f^{-1}.

Corollary 3.25 *Let $(S, \leq)$ be a locally finite poset and let $\mathcal{IA}(S, \leq)$ be the set of invertible functions of $\mathcal{A}(S, \leq)$. Then $(\mathcal{IA}(S, \leq), \{\mathsf{k}, *, ^{-1}\})$ is a group.*

Proof This is a mere restatement of Theorem 3.24.

Let $(S, \leq)$ be a locally finite poset and let $\zeta : S \times S \longrightarrow \mathbb{R}$ be the *Riemann function* defined by

$$\zeta(x, y) = \begin{cases} 1 & \text{if } x \leq y, \\ 0 & \text{otherwise,} \end{cases}$$

for $x, y \in S$. Clearly, $\zeta \in \mathcal{A}(S, \leq)$, so the function ζ^{-1} exists by Corollary 3.25. This inverse, known as the *Möbius function*, is denoted by μ and its values can be computed from Equalities (3.9) and (3.10) as

$$\mu(x, x) = \frac{1}{\zeta(x, x)} = 1,$$
$$\mu(x, y) = -\sum_{z \in [x, y)} \mu(x, z)\zeta(z, y) = -\sum_{z \in [x, y)} \mu(x, z),$$

for $x < y$; for $x \not\leq y$, we have $\mu(x, y) = 0$.

Example 3.26 For the poset $(\{1, 2, 3, 4, 5, 6, 7, 8\}, \delta)$ introduced in Example 2.13 the Möbius function is given by

$$\begin{aligned}
&\mu(1, 1) = 1, \\
&\mu(1, 2) = \mu(1, 3) = \mu(1, 5) = \mu(1, 7) = -1, \\
&\mu(1, 4) = -\mu(1, 1) - \mu(1, 2) = 0, \\
&\mu(1, 6) = -\mu(1, 1) - \mu(1, 2) - \mu(1, 3) = -1 + 1 + 1 = 1, \\
&\mu(1, 8) = -\mu(1, 1) - \mu(1, 2) - \mu(1, 4) = -1 + 1 + 0 = 0,
\end{aligned}$$

and

$$\begin{aligned}
&\mu(2, 2) = 1, \mu(2, 4) = -1, \mu(2, 6) = -1, \\
&\mu(2, 8) = -\mu(2, 2) - \mu(2, 4) = -1 + 1 = 0, \\
&\mu(3, 3) = 1, \mu(3, 6) = -1, \\
&\mu(4, 4) = \mu(5, 5) = \mu(6, 6) = \mu(7, 7) = \mu(8, 8) = 1.
\end{aligned}$$

For all other pairs (p, q) with $(p, q) \notin \delta$ we have $\mu(p, q) = 0$.

The special role played by μ is discussed next.

Theorem 3.27 (Möbius Inversion Theorem) *Let $(S, \leq)$ be a locally finite poset that has the least element 0. If $f, g : S \longrightarrow \mathbb{R}$ are two real-valued functions such that $g(x) = \sum_{0 \leq z \leq x} f(z)$, then $f(x) = \sum_{0 \leq z \leq x} g(z)\mu(z, x)$ for $x \in S$.*

Proof Starting from the functions $f, g : S \longrightarrow \mathbb{R}$, define the functions $F, G \in \mathcal{A}(S, \leq)$ by

$$F(0, x) = f(x), G(0, x) = g(x)$$
$$F(u, x) = G(u, x) = 0, \text{ if } u > 0.$$

The equality $g(x) = \sum_{0 \leq z \leq x} f(z)$ can be written as

$$G(0, x) = \sum_{0 \leq z \leq x} F(0, z)\zeta(z, x),$$

where ζ is Riemann's function. We also have $G(u, x) = \sum_{u \leq z \leq x} F(u, z)\zeta(z, x)$ for $u > 0$ because in this case $G(u, x) = 0$ and $F(u, z) = 0$. Thus, $G = F * \zeta$. Since μ is the inverse of ζ in $\mathcal{IA}(S, \leq)$, it follows that $F = G * \mu$. Consequently,

$$\begin{aligned} f(x) = F(0, x) &= \sum_{0 \leq z \leq x} G(0, z)\mu(z, x) \\ &= \sum_{0 \leq z \leq x} g(z)\mu(z, x), \end{aligned}$$

which is the desired equality.

Now let $(S, \leq)$ be a poset that has the greatest element 1. By applying the Möbius inversion theorem to its dual $(S, \leq^{-1}) = (S, \geqslant)$ we obtain the following dual form of the theorem.

Theorem 3.28 (Möbius Dual Inversion Theorem) *Let* $(S, \leq)$ *be a locally finite poset that has the greatest element* 1. *If* $f, g : S \longrightarrow \mathbb{R}$ *are two real-valued functions such that*

$$g(x) = \sum_{x \leq z \leq 1} f(z),$$

then

$$f(x) = \sum_{x \leq z \leq 1} g(z)\mu(z, x)$$

for $x \in S$.

Proof This statement follows immediately from Theorem 3.27.

Example 3.29 Let M be a finite set and let $(\mathcal{P}(M), \subseteq)$ be the poset of all its subsets. The Möbius function of this poset is given by

$$\mu(A, B) = \begin{cases} (-1)^{|B|-|A|} & \text{if } A \subseteq B, \\ 0 & \text{otherwise,} \end{cases}$$

for $A, B \in \mathcal{P}(M)$.

Let $A, B \in \mathcal{P}(M)$ be such that $A \subseteq B$. We prove that $\mu(A, B) = (-1)^{|B|-|A|}$ by induction on $n = |B| - |A|$.

In the basis case $n = 0$, so $A = B$, which implies $\mu(A, B) = 1$, thus verifying the equality above. Suppose that the equality holds for sets that differ by fewer than n elements and that $|B| - |A| = n$. Then, by the definition of the Möbius function, we have

$$\mu(A, B) = -\sum_{C \in [A,B)} \mu(A, C) = -\sum_{C \in [A,B)} (-1)^{|C|-|A|}.$$

Note that there are $2^n - 1$ sets C in $[A, B)$. Namely, there are $\binom{n}{k}$ sets C such that $|C| - |A| = k$. Therefore,

$$\sum_{C \in [A,B)} (-1)^{|C|-|A|} = \sum_{k=0}^{n-1} (-1)^k \binom{n}{k}.$$

Choosing $x = -1$ in the identity $(x + 1)^n = \sum_{k=0}^{n} \binom{n}{k} x^k$ implies $0 = \sum_{k=0}^{n} \binom{n}{k} (-1)^k$, which yields the equality $\sum_{k=0}^{n-1} \binom{n}{k} (-1)^k = (-1)^{n+1}$. Thus, $\mu(A, B) = (-1)^{n+2} = (-1)^n = (-1)^{|B|-|A|}$.

It is interesting to observe that the principle of inclusion-exclusion can be obtained also from the Möbius dual inversion theorem. Let $A_0, \ldots, A_{n-1}$ be n finite sets, where $n \geqslant 2$, $S = \bigcup_{i=0}^{n-1} A_i$, and I be a subset of the set $\{0, \ldots, n-1\}$. The complement of I, $\{0, \ldots, n-1\} - I$ is denoted by $\bar{I}$.

Let B_I be the subset of S that consists of those elements that belong to every one of the sets A_i with $i \in I$ and to no other sets. Clearly, we have

$$B_I = \left(\bigcap_{i \in I} A_i \right) \cap \left(\bigcap_{i \in \bar{I}} A_i \right).$$

Note that if $I \neq I'$, then the sets B_I and $B_{I'}$ are disjoint. We claim that

$$\bigcup \{B_J \mid I \subseteq J \subseteq \{0, \ldots, n-1\}\} = \bigcap_{i \in I} A_i. \tag{3.11}$$

If $I \subseteq J$, then $B_J \subseteq \bigcap_{i \in I} A_i$. Therefore,

$$\bigcap_{i \in I} A_i \subseteq \bigcup \{B_J \mid I \subseteq J \subseteq \{0, \ldots, n-1\}\}.$$

Conversely, let $x \in \bigcap_{i \in I} A_i$ and let $J_x = \{j \in \{0, \ldots, n-1\} \mid x \in A_j\}$. It is clear that $I \subseteq J_x$ and that $x \in B_{J_x}$. Therefore, $x \in \bigcup\{B_J \mid I \subseteq J \subseteq \{0, \ldots, n-1\}\}$ and we have the reverse inclusion

$$\bigcap_{i \in I} A_i \subseteq \bigcup\{B_J \mid I \subseteq J \subseteq \{0, \ldots, n-1\}\},$$

which proves Equality (3.11). This allows us to write

$$\left|\bigcap_{i \in I} A_i\right| = \sum\{|B_J| \mid I \subseteq J \subseteq \{0, \ldots, n-1\}\}.$$

Define $f(J)$ as $|B_J|$. The last equality can now be rewritten as

$$\left|\bigcap_{i \in I} A_i\right| = \sum\{f(J) \mid I \subseteq J \subseteq \{0, \ldots, n-1\}\}.$$

By the Möbius dual inversion theorem (Theorem 3.28) applied to the poset $(\mathcal{P}(\{0, \ldots, n-1\}), \subseteq)$, we have

$$f(I) = \sum_{I \subseteq J} (-1)^{|J|-|I|} \left|\bigcap_{i \in J} A_i\right|.$$

For the special case $I = \emptyset$, we have $f(\emptyset) = \left|S - \bigcup_{0 \leq i \leq n-1} A_i\right|$ because the intersection of an empty collection of subsets of a set S equals S. Thus,

$$\left|S - \bigcup_{0 \leq i \leq n-1} A_i\right| = \sum_{J} (-1)^{|J|} \left|\bigcap_{i \in J} A_i\right|,$$

which is equivalent to Corollary 3.17.

Example 3.30 Let n be a natural number such that $n \geqslant 2$. Using the inclusion-exclusion principle we can compute the number $\phi(n)$ of positive integers that are less than n and are relatively prime with n; that is, the number of integers r such that $1 \leq r \leq n$ such that $\gcd\{n, r\} = 1$.

Suppose that $n = p_1^{a_1} p_2^{a_2} \cdots p_m^{a_m}$, where $p_1, \ldots, p_m$ are distinct prime numbers and $a_1, \ldots, a_m$ are positive integers. Let $M_i = \{r \in \mathbb{N} \mid r < n \text{ and } p_i | r\}$ for $1 \leq i \leq m$.

It is clear that $|M_i| = \frac{n}{p_i}$ for $1 \leq i \leq m$ and that

$$|M_{i_1} \cap M_{i_2} \cap \cdots \cap M_{i_k}| = \frac{n}{p_{i_1} p_{i_2} \cdots p_{i_k}}$$

for $1 \leq i_1, \ldots, i_k \leq m$.

Note that r is relatively prime with r if and only if $r \notin \bigcup_{i=1}^m M_i$.

Thus, the number that we are seeking is $n - \left|\bigcup_{i=1}^m M_i\right|$. By the inclusion-exclusion principle, we have

$$
\begin{aligned}
\phi(n) &= n - \left|\bigcup_{i=1}^m M_i\right| \\
&= n - \sum_{1 \leq i \leq m} |M_i| + \sum_{1 \leq i_1 < i_2 \leq m} |M_{i_1} \cap M_{i_2}| \\
&\quad + \cdots + (-1)^m |M_1 \cap \cdots \cap M_m| \\
&= n - \sum_{1 \leq i \leq m} \frac{n}{p_i} + \sum_{1 \leq i_1 < i_2 \leq m} \frac{n}{p_{i_1} p_{i_2}} + \\
&\quad + \cdots + (-1)^m \frac{n}{p_1 p_2 \cdots p_m} \\
&= n \prod_{i=1}^m \left(1 - \frac{1}{p_i}\right).
\end{aligned}
$$

The function ϕ is known as *Euler's function*. It is easy to see that $\phi(2) = 1, \phi(3) = 2$, $\phi(4) = 2$, etc. Furthermore, for any prime number p, we have $\phi(p) = p - 1$.

3.6 Ramsey's Theorem

Data miners should be aware of what is known today as Ramsey theory because this family of combinatorial results establishes that data sets that are sufficiently large contain spurious patterns whose existence is caused by the sheer size of the data set and do not represent "significant" structures from a data mining point of view.

We begin with a set of basic terms of Ramsey theory.

Definition 3.31 *Let $C = \{c_1, \ldots, c_k\}$ be a finite set referred to as the* set of colors. *A C-coloring of a set S is a mapping $f : S \longrightarrow C$. The set $f^{-1}(c)$ is the set of elements of S colored by c.*

A subset T of S is monochromatic *in the color c_i if $f(t) = c_i$ for every $t \in T$. A subset W of S is f-*monochromatic *if it is monochromatic in some color c_i.*

Clearly, every set of the form $f^{-1}(c)$ for a C-coloring of S is f-monochromatic. Recall that the set of subsets of size q of a set S is denoted by $\mathcal{P}_q(S)$.

Theorem 3.32 *Let S be a finite set, q a positive natural number, $f : S \longrightarrow \{c_1, c_2\}$ a coloring of the set S such that every set in $\mathcal{P}_q(S)$ is f-monochromatic, and a_1 and a_2 be two natural numbers not less than 2.*

There is a number denoted by $\mathcal{R}(a_1, a_2, q)$ such that if $|S| \geqslant \mathcal{R}(a_1, a_2, q)$, then there is $i \in \{1, 2\}$ and a subset T of S such that $|T| = a_i$ and every subset of $\mathcal{P}_q(T)$ has the color c_i.

Proof We begin by showing that $\mathcal{R}(a_1, q, q) = a_1$ for $a_1 \geqslant q$. Let S be a set of size a_1. One of the following two cases may occur:

`Case 1:` There is a subset T of S of size q that is colored by c_2. In this case, the statement holds since the T has only itself as a subset of size q.

`Case 2:` There is no subset T of S of size q that is colored by c_2. Now all subsets of size q of S have color c_1, and if T is a subset of S of size a_1, then all its subsets of size q have the color c_1 (since they are q-subsets of S).

This shows that $\mathcal{R}(a_1, q, q) = a_1$ for $a_1 \geqslant q$; similarly, $\mathcal{R}(q, a_2, q) = a_2$ for $a_2 \geqslant q$.

The argument is by induction on q.

In the basis case, $q = 1$, and we color each element individually. If S is a set of size $a_1 + a_2 - 1$, then we must have either a_1 elements colored c_1 or a_2 elements colored c_2 since otherwise, the set S would have no more than $a_1 + a_2 - 2$ elements.

For the inductive step, suppose the theorem holds for $q - 1$. Now we act by induction on $p = a_1 + a_2$. The basis case, where $a_1 = a_2 = q$, is included in the previous discussion.

Suppose that the theorem holds for $a_1 + a_2 - 1$, and let $b_1 = \mathcal{R}(a_1 - 1, a_2, q)$ and $b_2 = \mathcal{R}(a_1, a_2 - 1, q)$. Let S be a set whose size is at least $\mathcal{R}(b_1, b_2, q - 1) + 1$, and suppose that all its q-subsets are colored by c_1 or c_2. If s is a fixed element of S, then any set $U \in \mathcal{P}_q(S)$ such that $s \in U$ yields a subset $U - \{s\}$ of size $q - 1$ of the set S', where $S' = S - \{s\}$ is colored in the same color as U. Thus, we obtain a coloring of the $q - 1$ subsets of the set $S - \{s\}$ that contains at least $\mathcal{R}(b_1, b_2, q - 1)$ elements. By the inductive hypothesis, there is either a subset V of S' such that $|V| = b_1 = \mathcal{R}(a_1 - 1, a_2, q)$ and all its $q - 1$ subsets have color c_1 or there is an subset W of S' such that $|W| = b_2 = \mathcal{R}(a_1, a_2 - 1, q)$ and all its $q - 1$ subsets have color c_2.

The first case yields a coloring of S in which the q-subsets of S obtained by adding s to the $(q - 1)$-subsets of S' are colored in c_1. By the definition of $\mathcal{R}(a_1 - 1, a_2, q)$, there exists either a subset T_1 of S' that has $a_1 - 1$ elements whose q-subsets are colored c_1 or a a_2-subset T_2 of S whose q-subsets are colored c_2. The statement follows in the first situation by observing that $T_1 \cup \{s\}$ has a_1 elements. The second situation requires no further argument.

The second case is treated similarly.

Corollary 3.33 *We have the inequality*

$$\mathcal{R}(a_1, a_2, q) \leq \mathcal{R}(\mathcal{R}(a_1 - 1, a_2, q), \mathcal{R}(a_1, a_2 - 1, q)) + 1$$

for every $q \geqslant 1$ and a_1, a_2 such that $a_1, a_2 \geqslant q$.

Proof The inequality follows immediately from the proof of Theorem 3.32.

In the proof of Ramsey's theorem, we use the preliminary result contained in Theorem 3.32.

Theorem 3.34 (Ramsey's Theorem) *Let S be a finite set, q a positive natural number, $f : S \longrightarrow \{c_1, \ldots, c_k\}$ a coloring of the set S such that every set in $\mathcal{P}_q(S)$ is f-monochromatic, and $\mathbf{a} = (a_1, \ldots, a_k)$ a sequence of k positive natural numbers such that $a_i \geqslant q$ for $1 \leq i \leq k$.*

There is a number denoted by $\mathsf{Ramsey}(\mathbf{a}, q)$ such that if $|S| \geq \mathsf{Ramsey}(\mathbf{a}, q)$, then there exists a number i, $1 \leq i \leq k$, and a set T with $|T| = a_i$ such that every subset of $\mathcal{P}_q(T)$ has the color c_i.

Proof This time the proof is by induction on k, the number of colors. The basis case, $k = 2$, was discussed in Theorem 3.32. We have $\mathsf{Ramsey}((a_1, a_2), q) = \mathcal{R}(a_1, a_2, q)$.

Suppose the statement holds for $k - 1$ colors.

Let S be a set such that $|S| \geq \mathsf{Ramsey}((\mathsf{Ramsey}((a_1, \ldots, a_{k-1}), q), a_k), q)$ and let $f : S \longrightarrow \{c_1, \ldots, c_k\}$ be a coloring of S using k colors. Define the coloring $g : S \longrightarrow \{c_0, c_k\}$ by

$$g(x) = \begin{cases} c_0 & \text{if } f(x) \in \{c_1, \ldots, c_{k-1}\}, \\ c_k & \text{if } f(x) = c_k, \end{cases}$$

for $x \in S$. Using the coloring g, every q-subset of S that was colored $c_1, \ldots, c_{k-1}$ will receive the color c_0 and every q-subset of S colored c_k will remain colored by c_k. By the two-color case of Theorem 3.32, either there is a subset T such that $|T| = \mathsf{Ramsey}((a_1, \ldots, a_{k-1}), q)$ whose q-subsets are colored c_0 or a subset U such that $|U| = a_k$ whose q-subsets are colored c_k. Since f colors the q-subsets of T in any of the colors $c_1, \ldots, c_{k-1}$, the theorem follows immediately from the inductive hypothesis.

Corollary 3.35 *We have the inequality*

$$\mathsf{Ramsey}((a_1, \ldots, a_k), q) \leq \mathsf{Ramsey}((\mathsf{Ramsey}((a_1, \ldots, a_{k-1}), q), a_k), q)$$

for every $q \geqslant 1$ and a_i such that $a_i \geqslant q$ for $1 \leq i \leq k$.

Proof This result follows from the proof of Ramsey's theorem.

Note that $\mathsf{Ramsey}((\underbrace{2, \ldots, 2}_{k}), 1) = k + 1$. Indeed, if we color the elements of a set S using $|S| + 1$ colors, then there is a subset T of S that contains two elements colored with the same color. This is a well known combinatorial fact known as the *pigeonhole principle*.

3.7 Combinatorics of Partitions

Let S be a set having n elements. We are interested in the number of partitions of S that have m blocks.

We begin by counting the number of onto functions of the form $f : A \longrightarrow B$, where $|A| = n$, $|B| = m$, and $n \geqslant m$.

Lemma 3.36 *Let A and B be two sets, where $|A| = n$, $|B| = m$, and $n \geqslant m$. The number of surjective functions from A to B is given by*

$$\sum_{p=0}^{m-1} (-1)^p \binom{m}{p} (m-p)^n.$$

Proof There are m^n functions of the form $f : A \longrightarrow B$.

We begin by determining the number of functions that are not surjective. Suppose that $B = \{b_1, \ldots, b_m\}$, and let $F_j = \{f : A \longrightarrow B \mid b_j \notin f(A)\}$ for $1 \leq j \leq m$. A function is not surjective if it belongs to one of the sets F_j. Thus, we need to evaluate $|\bigcup_{j=1}^m F_j|$. An application of the inclusion-exclusion principle yields

$$\begin{aligned}\left|\bigcup_{j=1}^m F_j\right| &= \sum_{j_1=1}^m |F_{j_1}| - \sum_{j_1,j_2=1}^m |F_{j_1} \cap F_{j_2}| \\ &+ \sum_{j_1,j_2,j_3=1}^m |F_{j_1} \cap F_{j_2} \cap F_{j_3}| - \cdots - +(-1)^m |F_1 \cap F_2 \cap \cdots \cap F_m|.\end{aligned}$$

Note that the set $|F_{j-1} \cap F_{j_2} \cap \cdots \cap F_{j_p}|$ is actually the set of functions defined on A with values in the set $B - \{y_{j_1}, y_{j_2}, \ldots, y_{j_p}\}$, and there are $(m-p)^n$ such functions. Since there are $\binom{m}{p}$ choices for the set $\{j_1, j_2, \ldots, j_p\}$, it follows that there are

$$\sum_{p=1}^{m-1} (-1)^{p-1} \binom{m}{p} (m-p)^n$$

functions that are not surjective.

Thus, we can conclude that there are

$$\begin{aligned}m^n - \sum_{p=1}^{m-1} (-1)^{p-1} \binom{m}{p} (m-p)^n &= m^n + \sum_{p=1}^{m-1} (-1)^p \binom{m}{p} (m-p)^n \\ &= \sum_{p=0}^{m-1} (-1)^p \binom{m}{p} (m-p)^n\end{aligned}$$

surjective functions from A to B.

Lemma 3.37 *Let A, B be two finite sets such that $|A| = n$ and $|B| = m$ with $m \geqslant n$. There are $m!$ distinct surjective functions of the form $f : A \longrightarrow B$ that have the same kernel partition π on A.*

Proof Given a surjective function $f : A \longrightarrow B$, one can obtain a function g that has the same partition as f by defining $g(a) = h(f(a))$, where h is a permutation of the set B. Since there are $m!$ such permutations the conclusion follows.

Theorem 3.38 *The number of partitions of a set S that have m blocks, where $m \leq n$ is given by*

$$\frac{1}{m!}\sum_{p=0}^{m-1}(-1)^p\binom{m}{p}(m-p)^n.$$

Proof This statement follows from Lemmas 3.36 and 3.37.

The numbers $S(n, m)$ defined by

$$S(n,m) = \frac{1}{m!}\sum_{j=0}^{m-1}(-1)^j\binom{m}{j}(m-j)^n$$

for $m, n \in \mathbb{N}$ and $m \leq n$ are known as the *Stirling numbers!of the second kind*. The Stirling numbers of the first kind are introduced in Supplement 27.

Next we consider a notion related to set partitions, namely partitions of natural numbers.

Definition 3.39 *An* integral partition *of n is a nonincreasing sequence $\mathbf{k} = (k_1, \ldots, k_\ell)$ of positive integers such that $\sum_{i=1}^{\ell} k_i = n$.*

The set of integral partitions of n is denoted by IP_n; the set of integral partitions of n that consist of ℓ components is denoted by $IP_n(\ell)$.

Example 3.40 The sequence $\mathbf{k} = (5, 5, 3, 2, 2, 2, 1, 1)$ is an integral partition of 21.

We can regard an integral partition of n as a multiset P on the set $\{1, 2, \ldots, n\}$, where $P(k)$ is the number of entries in the sequence $\mathbf{k}$ of Definition 3.39 that equal k.

Example 3.41 The integral partition $(5, 5, 3, 2, 2, 2, 1, 1) \in IP_{21}$ defines the multiset P on the set $\{1, \ldots, 21\}$ given by

$$P(k) = \begin{cases} 2 & \text{if } k = 1 \text{ or } k = 5, \\ 3 & \text{if } k = 2, \\ 1 & \text{if } k = 3, \\ 0 & \text{in every other case.} \end{cases}$$

Fig. 3.1 Ferrers diagrams

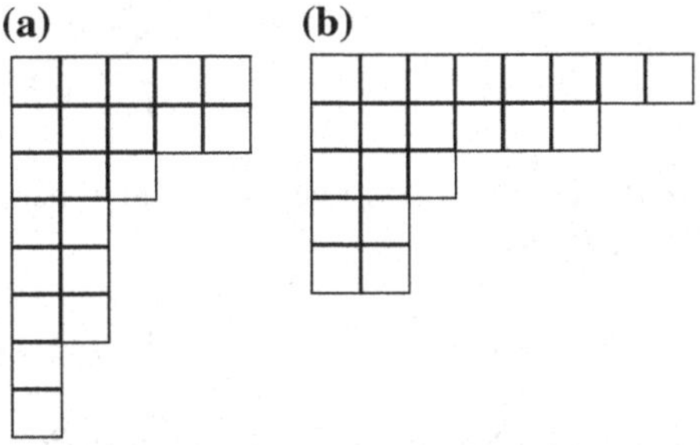

An integral partition $\mathbf{k}$ can be represented graphically by a *Ferrers diagram* that consists of a sequence of rows of squares such that each component k of $\mathbf{k}$ corresponds to a row of k cells in the diagram.

Example 3.42 The Ferrers diagram of $(5, 5, 3, 2, 2, 2, 1, 1)$ of integer 21 is shown in Fig. 3.1a.

Starting from the Ferrers diagram of $\mathbf{k} \in IP_n$, we can derive a new integral partition $\mathbf{k}' \in IP_n$ by exchanging the rows of the diagram with its columns. The new integral partition $\mathbf{k}'$ is called the *conjugate integral partition* of $\mathbf{k}$. The Ferrers diagram of the conjugate partition $\mathbf{k}'$ of $\mathbf{k}$ (where $\mathbf{k}$ is the integral partition defined in Example 3.42 is shown in Fig. 3.1b.

Theorem 3.43 *The number of integral partitions in IP_n where the largest component is ℓ equals $IP_n(\ell)$, the number of integral partitions on n with ℓ components.*

Proof This statement follows immediately by observing that the function $f : IP_n \longrightarrow IP_n$ that maps $\mathbf{k}$ into its conjugate $\mathbf{k}'$ is a bijection and the image under f of an integral partition that has ℓ components is an integral partition whose largest component is ℓ.

3.8 Combinatorics of Collections of Sets

Recall that we defined a Sperner collection of sets in Sect. 1.3 as collection of sets $\mathcal{C}$ such that $X, Y \in \mathcal{C}$ and $X \neq Y$ implies $X \not\subseteq Y$.

If $\mathcal{C}$ is a Sperner system and $\mathcal{C} \subseteq \mathcal{P}(S)$, then we say that $\mathcal{C}$ is a *Sperner system on the set S*.

The next theorem presents an inequality known as the LYM inequality, an acronym of the names of the mathematicians whose work is related to it (Lubell, Yamamoto, and Meshalkin [2–4]).

Theorem 3.44 (The LYM Inequality) *Let $\mathcal{C}$ be a Sperner system on a finite set S such that $|S| = n$. Define the function $c : \{0, 1, \ldots, n\} \longrightarrow \mathbb{N}$ by $c(k) = |\{X \in \mathcal{C} \mid |X| = k\}|$ for $0 \leq k \leq n$. We have*

$$\sum_{k=0}^{n} \frac{c(k)}{\binom{n}{k}} \leq 1.$$

Proof Let $(x_1, \ldots, x_n)$ be one of the $n!$ permutations of the set S. For $U \in \mathcal{C}$ define the set P_U as

$$P_U = \{(x_1, \ldots, x_n) \mid \{x_1, \ldots, x_m\} = U, \text{ where } m = |U|\}.$$

Since $\mathcal{C}$ is a Sperner system, we have $P_U \cap P_V = \emptyset$ for $U \neq V$ and $U, V \in \mathcal{C}$. The number of permutations in P_U is $|U|!(n - |U|)!$, so $\sum_{U \in \mathcal{C}} |U|!(n - |U|)! \leq n!$. Using the definition of the function c, we have $\sum_{k=0}^{n} c(k)|k|!(n - |k|)! \leq n!$, which yields the desired inequality.

The next statement is known as *Sperner's theorem* and was obtained in [5] using a different approach (outlined in Supplement 42).

Corollary 3.45 *Let S be a finite set such that $|S| = n$. If $\mathcal{C}$ is a Sperner system on S, then*

$$|\mathcal{C}| \leq \binom{n}{\lfloor \frac{n}{2} \rfloor}.$$

Proof The largest value of the binomial coefficient $\binom{n}{k}$ is achieved when $k = \lfloor \frac{n}{2} \rfloor$. Therefore, we have

$$\sum_{k=0}^{n} c(k)|k|!(n - |k|)! \geq \frac{\sum_{i=0}^{n} c(k)}{\binom{n}{\lfloor \frac{n}{2} \rfloor}} = \frac{|\mathcal{C}|}{\lfloor \frac{n}{2} \rfloor}.$$

By the LYM inequality we obtain $|\mathcal{C}| \leq \binom{n}{\lfloor \frac{n}{2} \rfloor}$.

The Ahlswede-Daykin inequality involves functions defined on sets and collections of sets. We use the operations "$\vee$" and "$\wedge$" between collections of sets introduced in Definition 1.18.

Let $\mathcal{E}$ be a collection of subsets of a set S and let $\phi : \mathcal{P}(S) \longrightarrow \mathbb{R}$ be a function. We define a new function (denoted by the same letter ϕ) on the set of all collections of subsets of S (that is, on $\mathcal{P}(\mathcal{P}(S))$) as

$$\phi(\mathcal{E}) = \sum \{\phi(E) \mid E \in \mathcal{E}\}.$$

This definition allows us to formulate a powerful combinatorial inequality.

Theorem 3.46 (The Ahlswede-Daykin Inequality) *Let S be a set such that $S \neq \emptyset$ and let*

$$\alpha, \beta, \gamma, \delta : \mathcal{P}(S) \longrightarrow \mathbb{N}$$

be four functions that satisfy the inequality

$$\alpha(A)\beta(B) \leq \gamma(A \cup B)\delta(A \cap B)$$

for $A, B \in \mathcal{P}(S)$. *For all collections* $\mathcal{A}, \mathcal{B}$ *of subsets of* S, *we have*

$$\alpha(\mathcal{A})\beta(\mathcal{B}) \leq \gamma(\mathcal{A} \vee \mathcal{B})\delta(\mathcal{A} \wedge \mathcal{B}).$$

Proof The argument is by induction on $n = |S|$, where $n \geqslant 1$. For the base case, $|S| = 1$, we have $\mathcal{P}(S) = \{\emptyset, S\}$. Thus, we can write

$$\alpha(\emptyset)\beta(\emptyset) \leq \gamma(\emptyset)\delta(\emptyset), \tag{3.12}$$
$$\alpha(\emptyset)\beta(S) \leq \gamma(S)\delta(\emptyset), \tag{3.13}$$
$$\alpha(S)\beta(\emptyset) \leq \gamma(S)\delta(\emptyset), \tag{3.14}$$
$$\alpha(S)\beta(S) \leq \gamma(S)\delta(S). \tag{3.15}$$

Since $\mathcal{C}, \mathcal{B} \subseteq \{\emptyset, S\}$, we need to analyze the following cases.

Case	$\mathcal{A}$	$\mathcal{B}$	$\mathcal{A} \vee \mathcal{B}$	$\mathcal{A} \wedge \mathcal{B}$
I	$\{\emptyset\}$	$\{\emptyset\}$	$\{\emptyset\}$	$\{\emptyset\}$
II	$\{\emptyset\}$	$\{S\}$	$\{S\}$	$\{\emptyset\}$
III	$\{S\}$	$\{\emptyset\}$	$\{S\}$	$\{\emptyset\}$
IV	$\{S\}$	$\{S\}$	$\{S\}$	$\{S\}$
V	$\{\emptyset\}$	$\{\emptyset, S\}$	$\{\emptyset, S\}$	$\{\emptyset\}$
VI	$\{S\}$	$\{\emptyset, S\}$	$\{S\}$	$\{\emptyset, S\}$
VII	$\{\emptyset, S\}$	$\{\emptyset\}$	$\{\emptyset, S\}$	$\{\emptyset\}$
VIII	$\{\emptyset, S\}$	$\{S\}$	$\{S\}$	$\{\emptyset, S\}$
IX	$\{\emptyset, S\}$	$\{\emptyset, S\}$	$\{\emptyset, S\}$	$\{\emptyset, S\}$

We discuss only case IX; the remaining cases are similar and are left to the reader. The inequality that we need to prove,

$$(\alpha(\emptyset) + \alpha(S))(\beta(\emptyset) + \beta(S)) \leq (\gamma(\emptyset) + \gamma(S))(\delta(\emptyset) + \delta(S))$$

follows immediately by adding Inequalities (3.12) to (3.15).

Suppose that the inequality holds for sets containing m elements, and let $S = \{s_0, \ldots, s_{m-1}, s_m\}$ be a set of size $m+1$. Define $U = \{s_0, \ldots, s_{m-1}\}$ and $V = \{s_m\}$.

The mappings $\alpha_1, \beta_1, \gamma_1, \delta_1 : \mathcal{P}(U) \longrightarrow \mathbb{N}$ are defined by

$$\alpha_1(C) = \sum\{\alpha(A) \mid A \in \mathcal{A} \text{ and } A \cap U = C\},$$

$$\beta_1(C) = \sum\{\alpha(B) \mid B \in \mathcal{B} \text{ and } B \cap U = C\},$$

$$\gamma_1(C) = \sum\{\gamma(E) \mid E \in \mathcal{A} \vee \beta \text{ and } E \cap U = C\},$$

$$\delta_1(C) = \sum\{\delta(F) \mid F \in \mathcal{A} \wedge \mathcal{B} \text{ and } F \cap U = C\}.$$

Observe that

$$\begin{aligned}\alpha_1(\mathcal{P}(U)) &= \sum\{\alpha_1(C) \mid C \in \mathcal{P}(U)\}\\ &= \sum_{C \in \mathcal{P}(U)} \sum\{\alpha(A) \mid A \in \mathcal{A} \text{ and } A \cap U = C\}\\ &= \sum\{\alpha(A) \mid A \in \mathcal{A}\}\\ &= \alpha(\mathcal{A}).\end{aligned}$$

Similarly, $\beta_1(\mathcal{P}(U)) = \beta(\mathcal{B})$, $\gamma_1(\mathcal{P}(U)) = \gamma(\mathcal{A} \vee \mathcal{B})$, and $\delta_1(\mathcal{P}(U)) = \gamma(\mathcal{A} \wedge \mathcal{B})$. Let $R \in \mathcal{P}(V)$. We have either $R = \emptyset$ or $R = \{s_m\}$.

Let $C, D \in \mathcal{P}(U)$ and let $E = C \cup D$ and $F = C \cap D$. Define the mappings $\alpha_2^C, \beta_2^D, \gamma_2^E, \delta_2^F : \mathcal{P}(V) \longrightarrow \mathbb{N}$ by

$$\alpha_2^C(R) = \begin{cases} \alpha(R \cup C) & \text{if } R \cup C \in \mathcal{A},\\ 0 & \text{otherwise,}\end{cases}$$

$$\beta_2^D(R) = \begin{cases} \beta(R \cup D) & \text{if } R \cup D \in \mathcal{B},\\ 0 & \text{otherwise,}\end{cases}$$

$$\gamma_2^E(R) = \begin{cases} \gamma(R \cup E) & \text{if } R \cup E \in \mathcal{A} \vee \mathcal{B},\\ 0 & \text{otherwise,}\end{cases}$$

$$\delta_2^F(R) = \begin{cases} \delta(R \cup F) & \text{if } R \cup F \in \mathcal{A} \wedge \mathcal{B},\\ 0 & \text{otherwise.}\end{cases}$$

We have $\alpha_1(C) = \alpha_2^C(\mathcal{P}(V))$ for every $C \subseteq U$. Indeed,

$$\begin{aligned}\alpha_2^C(\mathcal{P}(V)) &= \alpha_2^C(\emptyset) + \alpha_2^C(\{s_m\})\\ &= \begin{cases} \alpha(C) + \alpha(C \cup \{s_m\}) & \text{if } C \in \mathcal{A} \text{ and } C \cup \{s_m\} \in \mathcal{A},\\ \alpha(C) & \text{if } C \in \mathcal{A} \text{ and } C \cup \{s_m\} \notin \mathcal{A},\\ \alpha(C \cup \{s_m\}) & \text{if } C \notin \mathcal{A} \text{ and } C \cup \{s_m\} \in \mathcal{A},\\ 0 & \text{otherwise.}\end{cases}\\ &= \alpha_1(C).\end{aligned}$$

Similar arguments show that $\beta_1(D) = \beta_2^D(\mathcal{P}(V))$ for $D \in \mathcal{P}(U)$, $\gamma_1(E) = \gamma_2^E(\mathcal{P}(V))$ for $E \in \mathcal{P}(U)$, and $\delta_1(F) = \delta_2^F(\mathcal{P}(V))$ for $F \in \mathcal{P}(U)$.

We claim that $\alpha_2^C(R)\beta_2^D(Q) \leq \gamma_2^E(R \cup Q)\delta_2^F(R \cap Q)$ for all $R, Q \in \mathcal{P}(V)$. If $\alpha_2^C(R)\beta_2^D(Q) = 0$ the inequality obviously holds.

Now suppose that $\alpha_2^C(R)\beta_2^D(Q) \neq 0$, that is, $R \cup C \in \mathcal{A}$ and $Q \cup D \in \mathcal{B}$ and $\alpha_2^C(R)\beta_2^D(Q) = \alpha(R \cup C)\beta(Q \cup D)$. Note that

$$(R \cup C) \cup (Q \cup D) = (R \cup Q) \cup (C \cup D) = (R \cup Q) \cup E \in \mathcal{A} \vee \mathcal{B}$$

and

$$\begin{aligned}(R \cup C) \cap (Q \cup D) &= (R \cap Q) \cup (R \cap D) \cup (C \cap Q) \cup (C \cap D)\\ &= (R \cap Q) \cup (C \cap D) = (R \cap Q) \cup F \in \mathcal{A} \wedge \mathcal{B}\end{aligned}$$

because $R \cap D = C \cap Q = \emptyset$ (since $R, Q \in \mathcal{P}(V)$ and $C, D \in \mathcal{P}(U)$). Thus,

$$\gamma_2^E(R \cup Q)\delta_2^F(R \cap Q) = \gamma(R \cup Q \cup E)\delta((R \cap Q) \cup F).$$

By the defining property of α, β, γ, and δ, we have $\alpha(R \cup C)\beta(Q \cup D) \leq \gamma(R \cup Q \cup E)\delta((R \cap Q) \cup F)$, which yields the inequality

$$\alpha_2^C(R)\beta_2^D(Q) \leq \gamma_2^E(R \cup Q)\delta_2^F(R \cap Q)$$

for all $R, Q \in \mathcal{P}(V)$.

The inductive hypothesis (for $n = 1$) implies

$$\begin{aligned}&\alpha_1(C)\beta_1(D)\\ &\quad = \alpha_2^C(\mathcal{P}(V))\beta_2^D(\mathcal{P}(V)) \leq \gamma_2^E(\mathcal{P}(V))\delta_2^F(\mathcal{P}(V))\\ &\quad = \gamma_1(C \cup D)\delta_1(C \cap D).\end{aligned}$$

Again applying the inductive hypothesis, we can write

$$\alpha(\mathcal{A})\beta(\mathcal{B}) = \alpha_1(\mathcal{P}(U))\beta_1(\mathcal{P}(U)) \leq \gamma_1(\mathcal{P}(U))\delta_1(\mathcal{P}(U)) = \gamma(\mathcal{A} \vee \mathcal{B})\delta(\mathcal{A} \wedge \mathcal{B}).$$

Corollary 3.47 *Let $\mathcal{A}$ and $\mathcal{B}$ be two collections of subsets of S. In this case,*

$$|\mathcal{A}| \cdot |\mathcal{B}| \leq |\mathcal{A} \vee \mathcal{B}| \cdot |\mathcal{A} \wedge \mathcal{B}|.$$

Proof In the Ahlswede-Daykin inequality, choose $\alpha, \beta, \gamma, \delta : \mathcal{P}(S) \longrightarrow \mathbb{N}$ such that $\alpha(C) = \beta(C) = \gamma(C) = \delta(C) = 1$ for $C \in \mathcal{P}(S)$. The required inequality follows immediately.

Definition 3.48 *A* hereditary collection of sets *is a collection $\mathcal{I}$ such that $C \in \mathcal{I}$ and $D \subseteq C$ implies $D \in \mathcal{I}$.*

A dually hereditary collection of sets *is a collection of sets $\mathcal{F}$ such that $C \in \mathcal{F}$ and $C \subseteq D$ implies $D \in \mathcal{F}$.*

Note that if $\mathcal{I}$ is a hereditary family of subsets of a set S, then $\mathcal{P}(S) - \mathcal{I}$ is a dually hereditary family of subsets; similarly, if $\mathcal{F}$ is a dually hereditary family of subsets of S, then $\mathcal{P}(S) - \mathcal{F}$ is a hereditary family.

Theorem 3.49 *Let $\mathcal{I}$ and $\mathcal{I}'$ be two hereditary families of sets. Then,*

$$\mathcal{I} \vee \mathcal{I}' = \mathcal{I} \cap \mathcal{I}'.$$

Proof Let $C \in \mathcal{I} \vee \mathcal{I}'$. Then, $C = A \cap B$, where $A \in \mathcal{I}$ and $B \in \mathcal{I}'$. Since $C \subseteq A$ and $C \subseteq B$, the hereditary character of $\mathcal{I}$ and $\mathcal{I}'$ implies that $C \in \mathcal{I}$ and $C \in \mathcal{I}'$, so $C \in \mathcal{I} \cap \mathcal{I}'$.

Theorem 3.50 *Let $\mathcal{F}$ and $\mathcal{F}'$ be two dual hereditary families of sets. Then,*

$$\mathcal{F} \wedge \mathcal{F}' = \mathcal{F} \cap \mathcal{F}'.$$

Proof Let $C \in \mathcal{F} \wedge \mathcal{F}'$. Then, $C = A \cup B$, where $A \in \mathcal{F}$ and $B \in \mathcal{F}'$. Since $A \subseteq C$ and $B \subseteq C$, the dual hereditary character of $\mathcal{F}$ and $\mathcal{F}'$ implies that $C \in \mathcal{F}$ and $C \in \mathcal{F}'$, so $C \in \mathcal{F} \cap \mathcal{F}'$.

The inequality contained by the next corollary is known as *Kleitman's inequality*.

Corollary 3.51 *If $\mathcal{I}$ is a hereditary family and $\mathcal{F}$ is a dual hereditary family of subsets of a finite set S, then $|\mathcal{I}| \cdot |\mathcal{F}| \geq 2^{|S|} \cdot |\mathcal{I} \cap \mathcal{F}|$.*

Proof Note that $\mathcal{I}' = \mathcal{P}(S) - \mathcal{F}$ is a hereditary family. By Corollary 3.47, we have

$$\begin{aligned}
|\mathcal{I}| \cdot |\mathcal{I}'| &\leq |\mathcal{I} \vee \mathcal{I}'| \cdot |\mathcal{I} \wedge \mathcal{I}'| \\
&= |\mathcal{I} \cap \mathcal{I}'| \cdot |\mathcal{I} \wedge \mathcal{I}'| \\
&\quad \text{(by Theorem 3.49)}
\end{aligned}$$

Note that $|\mathcal{I}'| = 2^{|S|} - |\mathcal{F}|$. Thus, we can write

$$\begin{aligned}
|\mathcal{I}| \cdot \left(2^{|S|} - |\mathcal{F}|\right) &= |\mathcal{I} - (\mathcal{I} \cap \mathcal{F})| \cdot |\mathcal{I} \wedge \mathcal{I}'| \\
&= (|\mathcal{I}| - |\mathcal{I} \cap \mathcal{F}|) \cdot |\mathcal{I} \wedge \mathcal{I}'| \\
&\leq 2^{|S|} \cdot (|\mathcal{I}| - |\mathcal{I} \cap \mathcal{F}|),
\end{aligned}$$

which gives the desired inequality.

3.9 The Vapnik-Chervonenkis Dimension

The concept of the Vapnik-Chervonenkis dimension of a collection of sets was introduced in [6] and independently in [7]. Its main interest for data mining is related to one of the basic models of machine learning, the probably approximately correct learning paradigm as was shown in [8]. The subject is of great interest to probability theorists interested in empirical processes [9, 10].

Definition 3.52 *Let $\mathcal{C}$ be a collection of sets. If the trace of $\mathcal{C}$ on K, $\mathcal{C}_K$ equals $\mathcal{P}(K)$, then we say that K is* shattered *by $\mathcal{C}$.*

The Vapnik-Chervonenkis dimension *of the collection $\mathcal{C}$ (called the VC-dimension for brevity) is the largest cardinality of a set K that is shattered by $\mathcal{C}$ and is denoted by $\mathsf{VCD}(\mathcal{C})$.*

If $\mathsf{VCD}(\mathcal{C}) = d$, then there exists a set K of size d such that for each subset L of K there exists a set $C \in \mathcal{C}$ such that $L = K \cap C$.

Note that a collection $\mathcal{C}$ shatters a set K if and only if $\mathcal{C}_K$ shatters K. This allows us to assume without loss of generality that both the sets of the collection $\mathcal{C}$ and a set K shattered by $\mathcal{C}$ are subsets of a set U.

Let $\mathcal{C}$ be a collection of sets with $\mathsf{VCD}(\mathcal{C}) = d$ and let K be a set shattered by $\mathcal{C}$ with $|K| = d$. Since there exist 2^d subsets of K, there are at least 2^d subsets of $\mathcal{C}$, so $2^d \leq |\mathcal{C}|$. Consequently, $\mathsf{VCD}(\mathcal{C}) \leq \log_2 |\mathcal{C}|$. This shows that if $\mathcal{C}$ is finite, then $\mathsf{VCD}(\mathcal{C})$ is finite. As we shall see, the converse is false: there exist infinite collections $\mathcal{C}$ that have a finite VC-dimension.

If U is a finite set, then the trace of a collection $\mathcal{C} = \{C_1, \ldots, C_p\}$ of subsets of U on a subset K of U can be presented in an intuitive, tabular form. Suppose, for example, that $U = \{u_1, \ldots, u_n\}$, and let $\theta = (T_{\mathcal{C}}, u_1 u_2 \cdots u_n, \mathbf{r})$ be a table, where $\mathbf{r} = (t_1, \ldots, t_p)$. The domain of each of the attributes u_i is the set $\{0, 1\}$.

Each tuple t_k corresponds to a set C_k of $\mathcal{C}$ and is defined by

$$t_k[u_i] = \begin{cases} 1 & \text{if } u_i \in C_k, \\ 0 & \text{otherwise,} \end{cases}$$

for $1 \leq i \leq n$. Then, $\mathcal{C}$ shatters K if the content of the projection $\mathbf{r}[K]$ consists of $2^{|K|}$ distinct rows.

Example 3.53 Let $U = \{u_1, u_2, u_3, u_4\}$ and let $\mathcal{C}$ be the collection of subsets of U given by

$$\mathcal{C} = \{\{u_2, u_3\}, \{u_1, u_3, u_4\}, \{u_2, u_4\}, \{u_1, u_2\}, \{u_2, u_3, u_4\}\}.$$

The tabular representation of $\mathcal{C}$ is

$T_{\mathcal{C}}$			
u_1	u_2	u_3	u_4
0	1	1	0
1	0	1	1
0	1	0	1
1	1	0	0
0	1	1	1

The set $K = \{u_1, u_3\}$ is shattered by the collection $\mathcal{C}$ because

$$\mathbf{r}[K] = ((0, 1), (1, 1), (0, 0), (1, 0), (0, 1))$$

contains the all four necessary tuples $(0, 1)$, $(1, 1)$, $(0, 0)$, and $(1, 0)$. On the other hand, it is clear that no subset K of U that contains at least three elements can be shattered by $\mathcal{C}$ because this would require $\mathbf{r}[K]$ to contain at least eight tuples. Thus, $\mathit{VCD}(\mathcal{C}) = 2$.

Every collection of sets shatters the empty set. Also, if $\mathcal{C}$ shatters a set of size n, then it shatters a set of size p, where $p \leq n$.

For a collection of sets $\mathcal{C}$ and for $m \in \mathbb{N}$, let $\Pi_{\mathcal{C}}[m]$ be the largest number of distinct subsets of a set having m elements that can be obtained as intersections of the set with members of $\mathcal{C}$, that is,

$$\Pi_{\mathcal{C}}[m] = \max\{|\mathcal{C}_K| \mid |K| = m\}.$$

We have $\Pi_{\mathcal{C}}[m] \leq 2^m$; however, if $\mathcal{C}$ shatters a set of size m, then $\Pi_{\mathcal{C}}[m] = 2^m$.

Definition 3.54 *A* Vapnik-Chervonenkis class *(or a* VC class*) is a collection $\mathcal{C}$ of sets such that $\mathit{VCD}(\mathcal{C})$ is finite.*

Example 3.55 Let $\mathbb{R}$ be the set of real numbers and let $\mathcal{S}$ be the collection of sets $\{(-\infty, t) \mid t \in \mathbb{R}\}$. We claim that any singleton is shattered by $\mathcal{S}$. Indeed, if $S = \{x\}$ is a singleton, then $\mathcal{P}(\{x\}) = \{\emptyset, \{x\}\}$. Thus, if $t \geqslant x$, we have $(-\infty, t) \cap S = \{x\}$; also, if $t < x$, we have $(-\infty, t) \cap S = \emptyset$, so $\mathcal{S}_S = \mathcal{P}(S)$.

There is no set S with $|S| = 2$ that can be shattered by $\mathcal{S}$. Indeed, suppose that $S = \{x, y\}$, where $x < y$. Then, any member of $\mathcal{S}$ that contains y includes the entire set S, so $\mathcal{S}_S = \{\emptyset, \{x\}, \{x, y\}\} \neq \mathcal{P}(S)$. This shows that $\mathcal{S}$ is a VC class and $\mathit{VCD}(\mathcal{S}) = 1$.

Example 3.56 Consider the collection $\mathcal{I} = \{[a, b] \mid a, b \in \mathbb{R}, a \leq b\}$ of closed intervals. We claim that $\mathit{VCD}(\mathcal{I}) = 2$. To justify this claim, we need to show that there exists a set $S = \{x, y\}$ such that $\mathcal{I}_S = \mathcal{P}(S)$ and no three-element set can be shattered by $\mathcal{I}$.

For the first part of the statement, consider the intersections

Fig. 3.2 Three-point sets can be shattered by half-planes

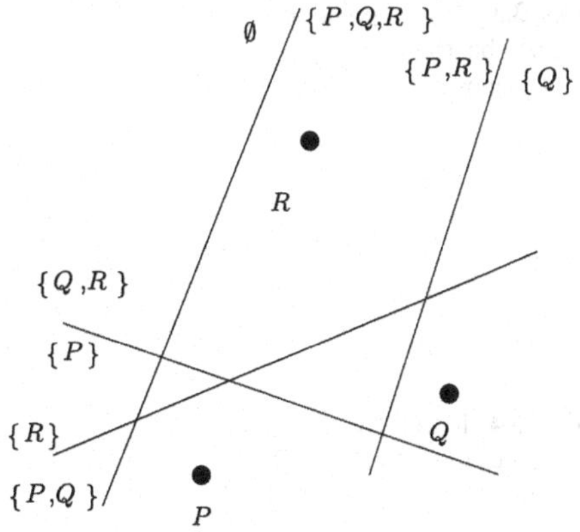

$$
\begin{aligned}
&[u, v] \cap S = \emptyset, \text{ where } v < x,\\
&[x - \epsilon, \tfrac{x+y}{2}] \cap S = \{x\},\\
&[\tfrac{x+y}{2}, y] \cap S = \{y\},\\
&[x - \epsilon, y + \epsilon] \cap S = \{x, y\},
\end{aligned}
$$

which show that $\mathcal{I}_S = \mathcal{P}(S)$.

For the second part of the statement, let $T = \{x, y, z\}$ be a set that contains three elements. Any interval that contains x and z also contains y, so it is impossible to obtain the set $\{x, z\}$ as an intersection between an interval in $\mathcal{I}$ and the set T.

Example 3.57 Let $\mathcal{H}$ be the collection of closed half-planes in $\mathbb{R}^2$, that is, the collection of sets of the form

$$\{x = (x_1, x_2) \in \mathbb{R}^2 \mid ax_1 + bx_2 - c \geqslant 0, a \neq 0 \text{ or } b \neq 0\}.$$

We claim that $\mathsf{VCD}(\mathcal{H}) = 3$.

Let P, Q, R be three points in $\mathbb{R}^2$ such that they are not located on the same line. Each line in Fig. 3.2 is marked with the sets it defines; thus, it is clear that the family of half-planes shatters the set $\{P, Q, R\}$, so $\mathsf{VCD}(\mathcal{H})$ is at least 3.

To complete the justification of the claim we need to show that no set that contains at least four points can be shattered by $\mathcal{H}$.

Let $\{P, Q, R, S\}$ be a set that contains four points such that no three points of this set are collinear. If S is located inside the triangle P, Q, R, then every half-plane that contains P, Q, R also contains S, so it is impossible to separate the subset $\{P, Q, R\}$. Thus, we may assume that no point is inside the triangle formed by the remaining three points (see Fig. 3.3). Observe that any half-plane that contains two diagonally opposite points, for example, P and R, contains either Q or S, which shows that it is impossible to separate the set $\{P, R\}$. Thus, no set that contains four points may be shattered by $\mathcal{H}$, so $\mathsf{VCD}(\mathcal{H}) = 3$.

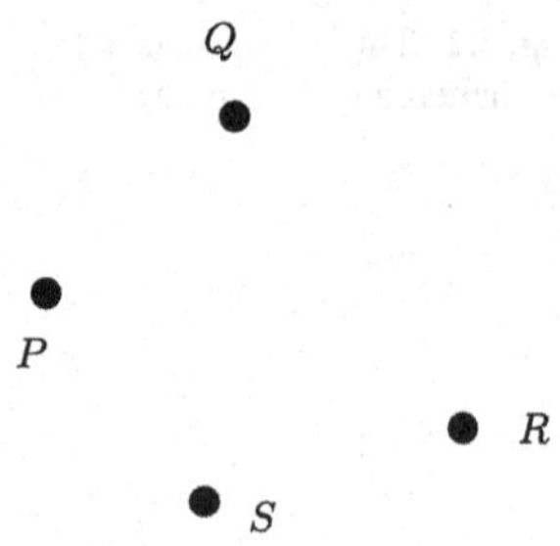

Fig. 3.3 A four-point set cannot be shattered by half-planes

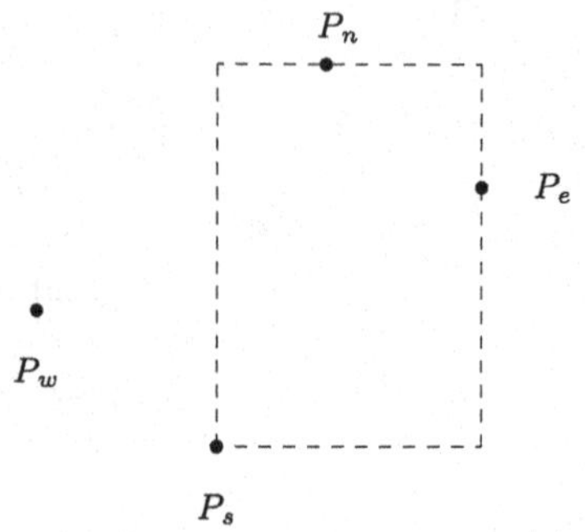

Fig. 3.4 Rectangle that separates the set $\{P_n, P_s, P_e\}$

Example 3.58 Let $\mathbb{R}^2$ be equipped with a system of coordinates and let $\mathcal{R}$ be the set of rectangles whose sides are parallel with the axes x and y. Each such rectangle has the form $[x_0, x_1] \times [y_0, y_1]$.

There is a set S with $|S| = 4$ that is shattered by $\mathcal{R}$. Indeed, let S be a set of four points in $\mathbb{R}^2$ that contains a unique "northernmost point" P_n, a unique "southernmost point" P_s, a unique "easternmost point" P_e, and a unique "westernmost point" P_w. If $L \subseteq S$ and $L \neq \emptyset$, let R_L be the smallest rectangle that contains L. For example, we show the rectangle R_L for the set $\{P_n, P_s, P_e\}$ in Fig. 3.4.

On the other hand, this collection cannot shatter a set of points that contains at least five points. Indeed, let S be a set of points such that $|S| \geqslant 5$ and, as before, let P_n be the northernmost point, etc. If the set contains more than one "northernmost" point, then we select exactly one to be P_n. Then, the rectangle that contains the set $K = \{P_n, P_e, P_s, P_w\}$ contains the entire set S, which shows the impossibility of separating the set K.

3.10 The Sauer–Shelah Theorem

If a collection of sets $\mathcal{C}$ is not a VC class (that is, if the Vapnik-Chervonenkis dimension of $\mathcal{C}$ is infinite), then $\Pi_{\mathcal{C}}[m] = 2^m$ for all $m \in \mathbb{N}$. However, we shall prove that if $\mathsf{VCD}(\mathcal{C}) = d$, then $\Pi_{\mathcal{C}}[m]$ is bounded asymptotically by a polynomial of degree d.

Definition 3.59 *Let $\mathcal{C}$ be a collection of subsets of a finite set S, let $s \in S$, and let $\mathcal{C}_{(s)}$ be the collection $\mathcal{C}_{S-\{s\}}$.*

$\mathcal{C}$ has a pair (A, B)*at* s *if* $A, B \in \mathcal{C}$, $B \subseteq A$, *and* $A - B = \{s\}$.

Note that if (A, B) is a pair of $\mathcal{C}$ at s, then $B = A - \{s\}$.

Define the subcollections $\mathcal{P}'(\mathcal{C}, s)$ and $\mathcal{P}''(\mathcal{C}, s)$ of $\mathcal{C}$ as

$$\begin{aligned}
\mathcal{P}'(\mathcal{C}, s) &= \{A \in \mathcal{C} \mid (A, B) \text{ is a pair at } s \text{ for some } B \in \mathcal{C}\} \\
&= \{A \in \mathcal{C} \mid s \in A, A - \{s\} \in \mathcal{C}\}, \\
\mathcal{P}''(\mathcal{C}, s) &= \{B \in \mathcal{C} \mid (A, B) \text{ is a pair at } s \text{ for some } A \in \mathcal{C}\} \\
&= \{B \in \mathcal{C} \mid s \notin B, B \cup \{s\} \in \mathcal{C}\}.
\end{aligned}$$

Lemma 3.60 *Let $\mathcal{C}$ be a collection of subsets of a finite set S, let $s \in S$, and let $\mathcal{C}_{(s)}$ be the collection $\mathcal{C}_{S-\{s\}}$.*

The following statements hold:

(i) *if $(A_1, B_1), (A_1, B_2), (A_2, B_1), (A_2, B_2)$ are pairs at s of $\mathcal{C}$, then $A_1 = A_2$ and $B_1 = B_2$;*
(ii) *we have $|\mathcal{P}'(\mathcal{C}, s)| = |\mathcal{P}''(\mathcal{C}, s)|$;*
(iii) *$|\mathcal{C}| - |\mathcal{C}_{(s)}| = |\mathcal{P}''(\mathcal{C}, s)|$;*
(iv) *$\mathcal{C}_{(s)} = \{C \in \mathcal{C} \mid s \notin C\} \cup \{C - \{s\} \mid C \in \mathcal{C} \text{ and } s \in C\}$;*
(v) *$|\mathcal{C}_{(s)}| = |\mathcal{C}| - |\mathcal{P}''(\mathcal{C}, s)|$.*

Proof For part (i), Definition 3.59 implies

$$B_1 \subseteq A_1, B_2 \subseteq A_1, B_1 \subseteq A_2, B_2 \subseteq A_2,$$

and $A_1 = B_1 \cup \{s\}$, $A_1 = B_2 \cup \{s\}$, $A_2 = B_1 \cup \{s\}$, and $A_2 = B_2 \cup \{s\}$. Therefore, $A_1 = B_1 \cup \{s\} = A_2$, which implies $B_1 = A_1 - \{s\} = A_2 - \{s\} = B_2$.

For part (ii), let $f : \mathcal{P}'(\mathcal{C}, s) \longrightarrow \mathcal{P}''(\mathcal{C}, s)$ be the function $f(A) = A - \{s\}$. It is easy to verify that f is a bijection and this implies $|\mathcal{P}'(\mathcal{C}, s)| = |\mathcal{P}''(\mathcal{C}, s)|$.

To prove part (iii) let $B \in \mathcal{P}''(\mathcal{C}, s)$. By the definition of $\mathcal{P}''(\mathcal{C}, s)$ we have $B \in \mathcal{C}$, $s \notin B$, and $B \cup \{s\} \in \mathcal{C}$.

Define the mapping $g : \mathcal{P}''(\mathcal{C}, s) \longrightarrow \mathcal{C} - \mathcal{C}_{(s)}$ as $g(B) = B \cup \{s\}$. Note that $B \cup \{s\} \in \mathcal{C} - \mathcal{C}_{S-\{s\}}$, so g is a well-defined function. Moreover, g is one-to one because $B_1 \cup \{s\} = B_2 \cup \{s\}$, $s \notin B_1$, and $s \notin B_2$ imply $B_1 = B_2$. Also, g is a surjection because if $D \in \mathcal{C} - \mathcal{C}_{S-\{s\}}$, then $s \in D$ and $D - \{s\} \in \mathcal{P}''(\mathcal{C}, s)$. Thus, g is a bijection, which implies $|\mathcal{C}| - |\mathcal{C}_{(s)}| = |\mathcal{C} - \mathcal{C}_{(s)}| = |\mathcal{P}''(\mathcal{C}, s)$.

To prove Part (iv) let $C \in \mathcal{C}_{(s)}$. If $s \notin C$, then C belongs to the first collection of the union; otherwise, that is if $s \in \mathcal{C}$, $C - \{s\}$ belongs to the second collection and the equality follows.

Finally, for Part (v), by applying Part (iv) we can write

$$\begin{aligned}
|\mathcal{C}_{(s)}| = &|\{C \in \mathcal{C} \mid s \notin \mathcal{C}\}| + |\{C - \{s\} \mid C \in \mathcal{C} \text{ and } s \in C\}| \\
&-|\{C \in \mathcal{C} \mid s \notin \mathcal{C}\} \cap \{C - \{s\} \mid C \in \mathcal{C} \text{ and } s \in C\}|
\end{aligned}$$

$$= |\{C \in \mathcal{C} \mid s \notin C\}| + |\{C - \{s\} \mid C \in \mathcal{C} \text{ and } s \in C\}| - |\mathcal{P}''(\mathcal{C}, s)|$$
$$= |\mathcal{C}| - |\mathcal{P}''(\mathcal{C}, s)|.$$

Lemma 3.61 *Let $\mathcal{C}$ be a collection of sets of a non-empty finite set S and let s_0 be an element of S. If $\mathcal{P}''(\mathcal{C}, s_0)$ shatters a subset T of $S - \{s_0\}$, then $\mathcal{C}$ shatters $T \cup \{s_0\}$.*

Proof Since $\mathcal{P}''(\mathcal{C}, s_0)$ shatters T, for every subset U of T there is $B \in \mathcal{P}''(\mathcal{C}, s_0)$ such that $U = T \cap B$. Let W be a subset of $T \cup \{s_0\}$. If $s_0 \notin W$, then $W \subseteq T$ and by the previous asumption, there exists $B \in \mathcal{P}''(\mathcal{C}, s_0)$ such that $W = (T \cup \{s_0\}) \cap B$. If $s_0 \in W$, then there exists $B_1 \in \mathcal{P}''(\mathcal{C}, s_0)$ such that for $W_1 = W - \{s_0\}$ we have $W_1 = T \cap B_1$. By the definition of $\mathcal{P}''(\mathcal{C}, s_0)$, $B_1 \cup \{s_0\} \in \mathcal{C}$ and

$$(T \cup \{s_0\}) \cap (B_1 \cup \{s_0\}) = (T \cap B) \cup \{s_0\} = W_1 \cup \{s_0\} = W.$$

Thus, $\mathcal{C}$ shatters $T \cup \{s_0\}$.

We saw that we have $\mathit{VCD}(\mathcal{C}) \leq \log_2 |\mathcal{C}|$. For collections of subsets of finite sets we have a stronger result.

Theorem 3.62 *Let $\mathcal{C}$ be a collection of sets of a non-empty finite set S with $\mathit{VCD}(\mathcal{C}) = d$. We have*

$$2^d \leq |\mathcal{C}| \leq (|S| + 1)^d.$$

Proof The first inequality, reproduced here for completeness, was discussed earlier.

For the second inequality the argument is by induction on $|S|$. The basis case, $|S| = 1$ is immediate.

Suppose that the inequality holds for collections of subsets with no more than n elements and let S be a set containing $n + 1$ elements. Let s_0 be an arbitrary but fixed element of S. By Part (iv) of Lemma 3.60 we have $|\mathcal{C}| = |\mathcal{C}_{s_0}| + |\mathcal{P}''(\mathcal{C}, s_0)|$.

The collection $\mathcal{C}_{s_0}$ consists of subsets of $S - \{s_0\}$. Since $\mathit{VCD}(\mathcal{C}) = d$, it is clear that $\mathit{VCD}(\mathcal{C}_{s_0}) \leq d$ and, by inductive hypothesis $|\mathcal{C}_{s_0}| \leq (|S - \{s_0\}| + 1)^d$.

We claim that $\mathit{VCD}(\mathcal{P}''(\mathcal{C}, s_0)) \leq d - 1$. Suppose that

$$\mathcal{P}''(\mathcal{C}, s_0) = \{B \in \mathcal{C} \mid s_0 \notin B, B \cup \{s_0\} \in \mathcal{C}\}$$

shatters a set T, where $T \subseteq S$ and $|T| \geqslant d$. Then, by Lemma 3.61, $\mathcal{C}$ would shatter $T \cup \{s_0\}$; since $|T \cup \{s_0\}| \geq d + 1$, this would lead to a contradiction. Therefore, we have $\mathit{VCD}(\mathcal{P}''(\mathcal{C}, s_0)) \leq d - 1$ and, by the inductive hypothesis, $|\mathcal{P}''(\mathcal{C}, s_0))| \leq (|S - \{s_0\}| + 1)^{d-1}$. These inequalities imply

$$\begin{aligned}
|\mathcal{C}| &= |\mathcal{C}_{s_0}| + |\mathcal{P}''(\mathcal{C}, s_0)| \\
&\leq (|S - \{s_0\}| + 1)^d + (|S - \{s_0\}| + 1)^{d-1} \\
&= (|S - \{s_0\}| + 1)^{d-1}(|S - \{s_0\}| + 2) \\
&\leq (|S| + 1)^d,
\end{aligned}$$

which completes the proof.

Lemma 3.63 *Let $\mathcal{C}$ be a collection of subsets of a finite set S, let $s \in S$, and let $\mathcal{C}_{(s)}$ be the collection $\mathcal{C}_{S-\{s\}}$. If $\mathsf{VCD}(\mathcal{P}''(\mathcal{C}, s)) = n - 1$ in $S - \{s\}$, then $\mathsf{VCD}(\mathcal{C}) = n$.*

Proof Since $\mathsf{VCD}(\mathcal{P}''(\mathcal{C}, s)) = n - 1$ in $S - \{s\}$, there exists a set $T \subseteq S - \{s\}$ with $|T| = n - 1$ that is shattered by $\mathcal{P}''(\mathcal{C}, s)$. Let $\mathcal{G}$ be the subcollection of $\mathcal{C}$ defined by:

$$\begin{aligned}\mathcal{G} &= \mathcal{P}'(\mathcal{C}, s) \cup \mathcal{P}''(\mathcal{C}, s)\\ &= \{A \in \mathcal{C} \mid A - \{s\} \in \mathcal{C}\} \cup \{B \in \mathcal{C} \mid B \cup \{s\} \in \mathcal{C}\}.\end{aligned}$$

We claim that $\mathcal{G}$ shatters $T \cup \{s\}$. Two cases may occur for a subset W of $T \cup \{s\}$:

(i) If $s \notin W$, then $W \subseteq T$ and, since T is shattered by by $\mathcal{P}''(\mathcal{C}, s)$ it follows that there exists $G \in \mathcal{G}$ such that $W = G \cap (T \cup \{s\})$.
(ii) If $s \in W$, let $G' \in \mathcal{P}''(\mathcal{C}, s)$ be set such that $W - \{s\} = T \cap G'$, which exists by the previous argument. Then, we have $G' \in \mathcal{P}''(\mathcal{C}, s)$, so $G' \cup \{s\} \in \mathcal{P}'(\mathcal{C}, s)$ and $(G' \cap \{s\}) \cap (T \cup \{s\}) = W$.

Thus, $\mathcal{G}$ shatters $T \cup \{s\}$ and so does $\mathcal{C}$.

For $n, k \in \mathbb{N}$ and $0 \leq k \leq n$ define the number $\binom{n}{\leq k}$ as

$$\binom{n}{\leq k} = \sum_{i=0}^{k} \binom{n}{i}.$$

Clearly, $\binom{n}{\leq 0} = 1$ and $\binom{n}{n} = 2^n$.

Theorem 3.64 *Let $\phi : \mathbb{N}^2 \longrightarrow \mathbb{N}$ be the function defined by*

$$\phi(d, m) = \begin{cases} 1 & \textit{if } m = 0 \textit{ or } d = 0\\ \phi(d, m-1) + \phi(d-1, m-1) & \textit{otherwise.}\end{cases}$$

We have

$$\phi(d, m) = \binom{m}{\leq d}$$

for $d, m \in \mathbb{N}$.

Proof The argument is by strong induction on $s = i + m$. The base case, $s = 0$, implies $m = 0$ and $d = 0$, and the equality is immediate. Suppose that the equality holds for $\phi(d', m')$, where $d' + m' < d + m$. We have

$$\begin{aligned}
\phi(d,m) &= \phi(d,m-1) + \phi(d-1,m-1) \\
&\quad \text{(by definition)} \\
&= \sum_{i=0}^{d} \binom{m-1}{i} + \sum_{i=0}^{d-1} \binom{m-1}{i} \\
&\quad \text{(by inductive hypothesis)} \\
&= \sum_{i=0}^{d} \binom{m-1}{i} + \sum_{i=0}^{d} \binom{m-1}{i-1} \\
&\quad \text{(since } \tbinom{m-1}{-1} = 0) \\
&= \sum_{i=0}^{d} \left(\binom{m-1}{i} + \binom{m-1}{i-1} \right) \\
&= \sum_{i=0}^{d} \binom{m}{i} = \binom{m}{\leq d},
\end{aligned}$$

which gives the desired conclusion.

Theorem 3.65 (Sauer–Shelah Theorem) *Let $\mathcal{C}$ be a collection of sets such that $\mathsf{VCD}(\mathcal{C}) = d$. If $m \in \mathbb{N}$ is a number such that $d \leq m$, then $\Pi_{\mathcal{C}}[m] \leq \phi(d,m)$.*

Proof The argument is by strong induction on $s = d + m$. For the base case, $s = 0$ we have $d = m = 0$ and this means that the collection $\mathcal{C}$ shatters only the empty set. Thus, $\Pi_{\mathcal{C}}[0] = |\mathcal{C}_{\emptyset}| = 1$, and this implies $\Pi_{\mathcal{C}}[0] = 1 = \phi(0,0)$.

Suppose that the statement holds for pairs (d', m') such that $d' + m' < s$, and let $\mathcal{C}$ be a collection of subsets of S such that $\mathsf{VCD}(\mathcal{C}) = d$ and K be a set with $|K| = m$.

Let k_0 be a fixed (but, otherwise, arbitrary) element of K. Since $|K - \{k_0\}| = m-1$, by inductive hypothesis, we have $|\mathcal{C}_{K-\{k_0\}}| \leq \phi(d, m-1)$.

Define

$$\mathcal{D} = \mathcal{P}''(\mathcal{C}_K, k_0) = \{B \in \mathcal{C}_K \mid k_0 \notin B \text{ and } B \cup \{k_0\} \in \mathcal{C}_K\}.$$

By Lemma 3.63, $\mathsf{VCD}(\mathcal{D}) \leq d - 1$, so $|\mathcal{D}| \leq \phi(d-1, m-1)$.

By Part (ii) of Lemma 3.60,

$$|\mathcal{C}_K| = |\mathcal{C}_{K-\{k_0\}}| + |\mathcal{D}| \leq \phi(d, m-1) + \phi(d-1, m-1) = \phi(d,m).$$

Lemma 3.66 *For $d \in \mathbb{N}$ and $d \geqslant 2$ we have*

$$2^{d-1} \leq \frac{d^d}{d!}.$$

Proof The argument is by induction on d. In the basis step, $d = 2$ both members are equal to 2.

Suppose the inequality holds for d. We have

$$\begin{aligned}
\frac{(d+1)^{d+1}}{(d+1)!} &= \frac{(d+1)^d}{d!} = \frac{d^d}{d!} \cdot \frac{(d+1)^d}{d^d} \\
&= \frac{d^d}{d!} \cdot \left(1 + \frac{1}{d}\right)^d \geqslant 2^d \cdot \left(1 + \frac{1}{d}\right)^d \geqslant 2^d \\
&\quad \text{(by inductive hypothesis)}
\end{aligned}$$

because

$$\left(1+\frac{1}{d}\right)^d \geqslant 1+d\frac{1}{d} = 2.$$

This concludes the proof of the inequality.

Lemma 3.67 *We have $\phi(d, m) \leq 2\frac{m^d}{d!}$ for every $m \geqslant d$ and $d \geqslant 1$.*

Proof The argument is by induction on d and n. If $d = 1$, then $\phi(1, m) = m+1 \leq 2m$ for $m \geqslant 1$, so the inequality holds for every $m \geqslant 1$, when $d = 1$.

If $m = d \geqslant 2$, then $\phi(d, m) = \phi(d, d) = 2^d$ and the desired inequality follows immediately from Lemma 3.66.

Suppose that the inequality holds for $m > d \geqslant 1$. We have

$$\begin{aligned}\phi(d, m+1) &= \phi(d, m) + \phi(d-1, m)\\ &\quad \text{(by the definition of } \phi)\\ &\leq 2\frac{m^d}{d!} + 2\frac{m^{d-1}}{(d-1)!}\\ &\quad \text{(by inductive hypothesis)}\\ &= 2\frac{m^{d-1}}{(d-1)!}\left(1+\frac{m}{d}\right).\end{aligned}$$

It is easy to see that the inequality

$$2\frac{m^{d-1}}{(d-1)!}\left(1+\frac{m}{d}\right) \leq 2\frac{(m+1)^d}{d!}$$

is equivalent to

$$\frac{d}{m}+1 \leq \left(1+\frac{1}{m}\right)^d$$

and, therefore, is valid. This yields immediately the inequality of the lemma.

The next theorem discusses the asymptotic behavior of the function ϕ:

Theorem 3.68 *The function ϕ satisfies the inequality:*

$$\phi(d, m) < \left(\frac{em}{d}\right)^d$$

for every $m \geqslant d$ and $d \geqslant 1$.

Proof By Lemma 3.67 $\phi(d, m) \leq 2\frac{m^d}{d!}$. Therefore, we need to show only that

$$2\left(\frac{d}{e}\right)^d < d!.$$

The argument is by induction on $d \geqslant 1$. The basis case, $d = 1$ is immediate. Suppose that $2\left(\frac{d}{e}\right)^d < d!$. We have

$$2\left(\frac{d+1}{e}\right)^{d+1} = 2\left(\frac{d}{e}\right)^d \left(\frac{d+1}{d}\right)^d \frac{d+1}{e}$$
$$= \left(1+\frac{1}{d}\right)^d \frac{1}{e} \cdot 2\left(\frac{d}{e}\right)^d (d+1) < 2\left(\frac{d}{e}\right)^d (d+1),$$

because

$$\left(1+\frac{1}{d}\right)^d < e.$$

The last inequality holds because the sequence $\left(\left(1+\frac{1}{d}\right)^d\right)_{d\in\mathbb{N}}$ is an increasing sequence whose limit is e. Since $2\left(\frac{d+1}{e}\right)^{d+1} < 2\left(\frac{d}{e}\right)^d (d+1)$, by inductive hypothesis we obtain:

$$2\left(\frac{d+1}{e}\right)^{d+1} < (d+1)!.$$

This proves the inequality of the theorem.

Corollary 3.69 *If m is sufficiently large we have $\phi(d, m) = O(m^d)$.*

Proof The statement is a direct consequence of Theorem 3.68.

Let $u : B_2^k \longrightarrow B_2$ be a Boolean function of k arguments and let $C_1, \ldots, C_k$ be k subsets of a set U. Define the set $u(C_1, \ldots, C_k)$ as the subset C of U whose indicator function is $I_C = u(I_{C_1}, \ldots, I_{C_k})$.

Example 3.70 If $u : B_2^2 \longrightarrow B_2$ is the Boolean function $u(a_1, a_2) = a_1 \vee a_2$, then $u(C_1, C_2)$ is $C_1 \cup C_2$; similarly, if $u(x_1, x_2) = x_1 \oplus x_2$, then $u(C_1, C_2)$ is the symmetric difference $C_1 \oplus C_2$ for every $C_1, C_2 \in \mathcal{P}(U)$.

Let $u : B_2^k \longrightarrow B_2$ and $\mathcal{C}_1, \ldots, \mathcal{C}_k$ are k family of subsets of U, the family of sets $u(\mathcal{C}_1, \ldots, \mathcal{C}_k)$ is

$$u(\mathcal{C}_1, \ldots, \mathcal{C}_k) = \{u(C_1, \ldots, C_k) \mid C_1 \in \mathcal{C}_1, \ldots, C_k \in \mathcal{C}_k\}.$$

Theorem 3.71 *Let $\alpha(k)$ be the least integer a such that $\frac{a}{\log(ea)} > k$.*

If $\mathcal{C}_1, \ldots, \mathcal{C}_k$ are k collections of subsets of the set U such that $d = \max\{\mathsf{VCD}(\mathcal{C}_i) \mid 1 \leq i \leq k\}$ and $u : B_2^2 \longrightarrow B_2$ is a Boolean function, then

$$\mathsf{VCD}(u(\mathcal{C}_1, \ldots, \mathcal{C}_k)) \leq \alpha(k) \cdot d.$$

Proof Let S be a subset of U that consists of m elements. The collection $(\mathcal{C}_i)_S$ is not larger than $\phi(d, m)$. For a set in the collection $W \in u(\mathcal{C}_1, \ldots, \mathcal{C}_k)_S$ we can write $W = S \cap u(C_1, \ldots, C_k)$, or, equivalently, $1_W = 1_S \cdot u(1_{C_1}, \ldots, 1_{C_k})$. By Exercise 25 of Chap. 11, there exists a Boolean function g_S such that

$$1_S \cdot u(1_{C_1}, \ldots, 1_{C_k}) = g_S(1_S \cdot 1_{C_1}, \ldots, 1_S \cdot 1_{C_k}) = g_S(1_{S \cap C_1}, \ldots, 1_{S \cap C_k}).$$

Since there are at most $\phi(d, m)$ distinct sets of the form $S \cap C_i$ for every i, $1 \leq i \leq k$, it follows that there are at most $(\phi(d, m))^k$ distinct sets W, hence $u(\mathcal{C}_1, \ldots, \mathcal{C}_k)[m] \leq (\phi(d, m))^k$.

Theorem 3.68 implies

$$u(\mathcal{C}_1, \ldots, \mathcal{C}_k)[m] \leq \left(\frac{em}{d}\right)^{kd}.$$

We observed that if $\Pi_{\mathcal{C}}[m] < 2^m$, then $\mathit{VCD}(\mathcal{C}) < m$. Therefore, to limit the Vapnik-Chervonenkis dimension of the collection $u(\mathcal{C}_1, \ldots, \mathcal{C}_k)$ it suffices to require that $\left(\frac{em}{d}\right)^{kd} < 2^m$.

Let $a = \frac{m}{d}$. The last inequality can be written as $(ea)^{kd} < 2^{ad}$; equivalently, we have $(ea)^k < 2^a$, which yields $k < \frac{a}{\log(ea)}$. If $\alpha(k)$ is the least integer a such that $k < \frac{a}{\log(ea)}$, then $m \leq \alpha(k)d$, which gives our conclusion.

Example 3.72 If $k = 2$, the least integer a such that $\frac{a}{\log(ea)} > 2$ is $k = 10$, as it can be seen by graphing this function; thus, if $\mathcal{C}_1, \mathcal{C}_2$ are two collection of concepts with $\mathit{VCD}(\mathcal{C}_1) = \mathit{VCD}(\mathcal{C}_2) = d$, the Vapnik-Chervonenkis dimension of the collections $\mathcal{C}_1 \vee \mathcal{C}_2$ or $\mathcal{C}_1 \wedge \mathcal{C}_2$ is not larger than $10d$.

Lemma 3.73 *Let S, T be two sets and let $f : S \longrightarrow T$ be a function. If $\mathcal{D}$ is a collection of subsets of T, U is a finite subset of S and $\mathcal{C} = f^{-1}(\mathcal{D})$ is the collection $\{f^{-1}(D) \mid D \in \mathcal{D}\}$, then $|\mathcal{C}_U| \leq |\mathcal{D}_{f(U)}|$.*

Proof Let $V = f(U)$ and denote $f \upharpoonright_U$ by g. For $D, D' \in \mathcal{D}$ we have

$$\begin{aligned}
&(U \cap f^{-1}(D)) \oplus (U \cap f^{-1}(D')) \\
&= U \cap (f^{-1}(D) \oplus f^{-1}(D')) = U \cap (f^{-1}(D \oplus D')) \\
&= g^{-1}(V \cap (D \oplus D')) = g^{-1}(V \cap D) \oplus g^{-1}(V \oplus D').
\end{aligned}$$

Thus, $C = U \cap f^{-1}(D)$ and $C' = U \cap f^{-1}(D')$ are two distinct members of $\mathcal{C}_U$, then $V \cap D$ and $V \cap D'$ are two distinct members of $\mathcal{D}_{f(U)}$. This implies $|\mathcal{C}_U| \leq |\mathcal{D}_{f(U)}|$.

Theorem 3.74 *Let S, T be two sets and let $f : S \longrightarrow T$ be a function. If $\mathcal{D}$ is a collection of subsets of T and $\mathcal{C} = f^{-1}(\mathcal{D})$ is the collection $\{f^{-1}(D) \mid D \in \mathcal{D}\}$, then $\mathit{VCD}(\mathcal{C}) \leq \mathit{VCD}(\mathcal{D})$. Moreover, if f is a surjection, then $\mathit{VCD}(\mathcal{C}) = \mathit{VCD}(\mathcal{D})$.*

Proof Suppose that $\mathcal{C}$ shatters an n-element subset $K = \{x_1, \ldots, x_n\}$ of S, so $|\mathcal{C}_K| = 2^n$ By Lemma 3.73 we have $|\mathcal{C}_K| \leq |\mathcal{D}_{f(U)}|$, so $|\mathcal{D}_{f(U)}| \geqslant 2^n$, which implies

$|f(U)| = n$ and $|\mathcal{D}_{f(U)}| = 2^n$, because $f(U)$ cannot have more than n elements. Thus, $\mathcal{D}$ shatters $f(U)$, so $\mathsf{VCD}(\mathcal{C}) \leq \mathsf{VCD}(\mathcal{C})$.

Suppose now that f is surjective and $H = \{t_1, \ldots, t_m\}$ is an m element set that is shattered by $\mathcal{D}$. Consider the set $L = \{u_1, \ldots, u_m\}$ such that $u_i \in f^{-1}(t_i)$ for $1 \leq i \leq m$. Let U be a subset of L. Since H is shattered by $\mathcal{D}$, there is a set $D \in \mathcal{D}$ such that $f(U) = H \cap D$, which implies $U = L \cap f^{-1}(D)$. Thus, L is shattered by $\mathcal{C}$ and this means that $\mathsf{VCD}(\mathcal{C}) = \mathsf{VCD}(\mathcal{D})$.

Definition 3.75 *The* density *of* $\mathcal{C}$ *is the number*

$$\mathsf{dens}(\mathcal{C}) = \inf\{s \in \mathbb{R}_{>0} \mid \Pi_{\mathcal{C}}[m] \leq c \cdot m^s \text{ for every } m \in \mathbb{N}\},$$

for some positive constant c.

Theorem 3.76 *Let S, T be two sets and let $f : S \longrightarrow T$ be a function. If $\mathcal{D}$ is a collection of subsets of T and $\mathcal{C} = f^{-1}(\mathcal{D})$ is the collection $\{f^{-1}(D) \mid D \in \mathcal{D}\}$, then $\mathsf{dens}(\mathcal{C}) \leq \mathsf{dens}(\mathcal{D})$. Moreover, if f is a surjection, then $\mathsf{dens}(\mathcal{C}) = \mathsf{dens}(\mathcal{D})$.*

Proof Let L be a subset of S such that $|L| = m$. Then, $|\mathcal{C}_L| \leq |\mathcal{D}_{f(L)}|$. In general, we have $|f(L)| \leq m$, so $|\mathcal{D}_{f(L)}| \leq \mathcal{D}[m] \leq cm^s$. Therefore, by Lemma 3.73, we have $|\mathcal{C}_L| \leq |\mathcal{D}_{f(L)}| \leq \mathcal{D}[m] \leq cm^s$, which implies $\mathsf{dens}(\mathcal{C}) \leq \mathsf{dens}(\mathcal{D})$.

If f is a surjection, then, for every finite subset M of T such that $|M| = m$ there is a subset L of S such that $|L| = |M|$ and $f(L) = M$. Therefore, $\mathcal{D}[m] \leq \Pi_{\mathcal{C}}[m]$ and this implies $\mathsf{dens}(\mathcal{C}) = \mathsf{dens}(\mathcal{D})$.

If $\mathcal{C}, \mathcal{D}$ are two collections of sets such that $\mathcal{C} \subseteq \mathcal{D}$, then $\mathsf{VCD}(\mathcal{C}) \leq \mathsf{VCD}(\mathcal{D})$ and $\mathsf{dens}(\mathcal{C}) \leq \mathsf{dens}(\mathcal{D})$.

Theorem 3.77 *Let $\mathcal{C}$ be a collection of subsets of a set S and let $\mathcal{C}' = \{S - C \mid C \in \mathcal{C}\}$. Then, for every $K \in \mathcal{P}(S)$ we have $|\mathcal{C}_K| = |\mathcal{C}'_K|$.*

Proof We prove the statement by showing the existence of a bijection $f : \mathcal{C}_K \longrightarrow \mathcal{C}'_K$. If $U \in \mathcal{C}_K$, then $U = K \cap C$, where $C \in \mathcal{C}$. Then $S - C \in \mathcal{C}'$ and we define $f(U) = K \cap (S - C) = K - C \in \mathcal{C}'_K$. The function f is well-defined because if $K \cap C_1 = K \cap C_2$, then $K - C_1 = K - (K \cap C_1) = K - (K \cap C_2) = K - C_2$.

It is clear that if $f(U) = f(V)$ for $U, V \in \mathcal{C}_K$, $U = K \cap C_1$, and $V = K \cap C_2$, then $K - C_1 = K - C_2$, so $K \cap C_1 = K \cap C_2$ and this means that $U = V$. Thus, f is injective. If $W \in \mathcal{C}'_K$, then $W = K \cap C'$ for some $C' \in \mathcal{C}$. Since $C' = S - C$ for some $C \in \mathcal{C}$, it follows that $W = K - C$, so $W = f(U)$, where $U = K \cap C$.

Corollary 3.78 *Let $\mathcal{C}$ be a collection of subsets of a set S and let $\mathcal{C}' = \{S - C \mid C \in \mathcal{C}\}$. We have $\mathsf{dens}(\mathcal{C}) = \mathsf{dens}(\mathcal{C}')$ and $\mathsf{VCD}(\mathcal{C}) = \mathsf{VCD}(\mathcal{C}')$.*

Proof This statement follows immediately from Theorem 3.77.

Theorem 3.79 *For every collection of sets we have $\mathsf{dens}(\mathcal{C}) \leq \mathsf{VCD}(\mathcal{C})$. Furthermore, if $\mathsf{dens}(\mathcal{C})$ is finite, then $\mathcal{C}$ is a VC-class.*

Proof If $\mathcal{C}$ is not a VC-class the inequality $\mathsf{dens}(\mathcal{C}) \le \mathit{VCD}(\mathcal{C})$ is clearly satisfied. Suppose now that $\mathcal{C}$ is a VC-class and $\mathit{VCD}(\mathcal{C}) = d$. By Sauer–Shelah Theorem (Theorem 3.65) we have $\Pi_{\mathcal{C}}[m] \le \phi(d, m)$; then, by Theorem 3.68, we obtain $\Pi_{\mathcal{C}}[m] \le \left(\frac{em}{d}\right)^d$, so $\mathsf{dens}(\mathcal{C}) \le d$.

Suppose now that $\mathsf{dens}(\mathcal{C})$ is finite. Since $\Pi_{\mathcal{C}}[m] \le cm^s \le 2^m$ for m sufficiently large, it follows that $\mathit{VCD}(\mathcal{C})$ is finite, so $\mathcal{C}$ is a VC-class.

Let $\mathcal{D}$ be a finite collection of subsets of a set S. In Supplement 6 of Chap. 1 the partition $\pi_{\mathcal{D}}$ was defined as consisting of the nonempty sets of the form $\{D_1^{a_1} \cap D_2^{a_2} \cap \cdots \cap D_r^{a_r}$, where $(a_1, a_2, \ldots, a_r) \in \{0, 1\}^r$.

Definition 3.80 *A collection $\mathcal{D} = \{D_1, \ldots, D_r\}$ of subsets of a set S is* independent *if the partition $\pi_{\mathcal{D}}$ has the maximum numbers of blocks, that is, it consists of 2^r blocks.*

If $\mathcal{D}$ is independent, then the Boolean subalgebra generated by $\mathcal{D}$ in the Boolean algebra $(\mathcal{P}(S), \{\cap, \cup, \bar{\ }, \emptyset, S\})$ contains 2^{2^r} sets, because this subalgebra has 2^r atoms. Thus, if $\mathcal{D}$ shatters a subset T with $|T| = p$, then the collection $\mathcal{D}_T$ contains 2^p sets, which implies $2^p \le 2^{2^r}$, or $p \le 2^r$.

Let $\mathcal{C}$ be a collection of subsets of a set S. The *independence number of* $\mathcal{C}$, $I(\mathcal{C})$ is:

$$\begin{aligned} I(\mathcal{C}) = \sup\{r \mid & \{C_1, \ldots, C_r\} \\ & \text{is independent for some finite } \{C_1, \ldots, C_r\} \subseteq \mathcal{C}\}. \end{aligned}$$

The next theorem is an analog of Theorem 3.74 for the independence number of a collection.

Theorem 3.81 *Let S, T be two sets and let $f : S \longrightarrow T$ be a function. If $\mathcal{D}$ is a collection of subsets of T and $\mathcal{C} = f^{-1}(\mathcal{D})$ is the collection $\{f^{-1}(D) \mid D \in \mathcal{D}\}$, then $I(\mathcal{C}) \le I(\mathcal{D})$. Moreover, if f is a surjection, then $I(\mathcal{C}) = I(\mathcal{D})$.*

Proof Let $\mathcal{E} = \{D_1, \ldots, D_p\}$ be an independent finite subcollection of $\mathcal{D}$. The partition $\pi_{\mathcal{E}}$ contains 2^r blocks. By Supplement 30 of Chap. 11, the number of atoms of the subalgebra generated by $\{f^{-1}(D_1), \ldots, f^{-1}(D_p)\}$ is not greater than 2^r. Therefore, $I(\mathcal{C}) \le I(\mathcal{D})$; from the same supplement it follows that if f is surjective, then $I(\mathcal{C}) = I(\mathcal{D})$.

Theorem 3.82 *If $\mathcal{C}$ is a collection of subsets of a set S such that $\mathit{VCD}(\mathcal{C}) \geqslant 2^n$, then $I(\mathcal{C}) \geqslant n$.*

Proof Suppose that $\mathit{VCD}(\mathcal{C}) \geqslant 2^n$, that is, there exists a subset T of S that is shattered by $\mathcal{C}$ and has at least 2^n elements. Then, the collection $\mathcal{C}_T$ contains at least 2^{2^n} sets, which means that the Boolean subalgebra of $\mathcal{P}(T)$ generated by $\mathcal{T}_C$ contains at least 2^n atoms. This implies that the subalgebra of $\mathcal{P}(S)$ generated by $\mathcal{C}$ contains at least this number of atoms, so $I(\mathcal{C}) \geqslant n$.

Exercises and Supplements

1. Prove by induction on n that for a finite set S with $|S| = n$, we have $|\mathcal{P}(S)| = 2^n$.
2. Prove that if S is a finite set such that $|S| = n$, then there are 2^{n^2} binary relations on S.
3. Prove that there are $2^{n(n-1)}$ binary reflexive relations on a finite set S that has n elements.
4. Prove that the number of antisymmetric relations on a finite set that has n elements is $2^n \cdot 3^{\frac{n(n-1)}{2}}$.
5. Let $\mathbf{x} = (x_0, \ldots, x_{n-1})$ be a sequence in $\mathbf{Seq}(\mathbb{R})$, where $n \geqslant 2$. The sequence is said to be *unimodal* if there exists $j, 0 \leq j \leq n-1$ such that $x_0 \leq x_1 \leq \cdots \leq x_j$ and $x_j \geqslant x_{j+1} \geqslant \cdots \geq x_n$.
   Prove that if $\mathbf{x} \in \mathbf{Seq}(\mathbb{R}_{>0})$ and $x_{p-1}x_{p+1} \leq x_p^2$ for $1 \leq p \leq n-2$, then $\mathbf{x}$ is a unimodal sequence.

6. Let f and g be two transpositions on a set S. Prove that there is $i \in \{1, 2, 3\}$ such that $(fg)^i = 1_S$.
7. Let $x \in \mathbb{R}$ and let $m \in \mathbb{N}$. Prove that if $x \geqslant 0$ and $m \geqslant 1$, then

$$\frac{x^{m-1}}{(m-1)!} + \frac{x^m}{m!} \leq \frac{(x+1)^m}{m!}.$$

8. Let S be subset of $\mathbb{N}$ such that $S \subseteq \{1, 2, \ldots, 2n\}$ and $|S| = n + 1$. Prove that S contains two numbers p and q such that $(p, q) \in \delta$.
   Solution: Any number $m \in S$ can be written as $m = 2^k r$, where $1 \leq r \leq 2n - 1$ and r is odd. We refer to r as the *odd part* of m. Note that there are n odd numbers between 1 and $2n-1$. Since $|S| = n+1$, and there are only n odd parts, it follows, by the pidgeonhole principle, that there are two numbers $p, q \in S$ that have the same odd part. The smaller such number divides the larger one.
9. Starting from Newton's binomial formula, prove the following identities:

$$\sum_{k=0}^{n} (n-k)\binom{n}{k} = n2^{n-1},$$

$$\sum_{k=0}^{n} \frac{\binom{n}{k}}{n-k+1} = \frac{2^{n+1} - 1}{n+1}.$$

10. Prove that

$$\binom{m+n}{k} = \sum_{i=0}^{k} \binom{m}{i}\binom{n}{k-i}$$

   for $m, n, k \in \mathbb{N}$ and $k \leq m + n$.
11. Prove that

$$\binom{m+n+p}{q_1+q_2+q_3} \geqslant \binom{m}{q_1}\binom{n}{q_2}\binom{p}{q_3}$$

for $m, n, p, q_1, q_2, q_3 \in \mathbb{N}$, where $m \geqslant q_1$, $n \geqslant q_2$, and $p \geqslant q_3$.

12. Prove that $\max\{\binom{n}{k} \mid 0 \leq k \leq n\} = \binom{n}{\lfloor \frac{n}{2} \rfloor} = \binom{n}{\lceil \frac{n}{2} \rceil}$.
13. Prove that
$$\frac{2^{2n}}{2n+1} \leq \binom{2n}{n} \leq 2^{2n}$$
for $n \geqslant 0$.
14. Prove the inequality
$$\binom{n}{i-1}\binom{n}{i+1} \leq \left(\binom{n}{i}\right)^2$$
for $1 \leq i \leq n-1$.
15. Let $T_n = \sum_{i=0}^{n}(-1)^i \binom{n-i}{i}$ for $n \geqslant 0$. Prove that $T_n = T_{n-1} - T_{n-2}$ for $n \geqslant 2$.
16. Prove that
$$\sum_{i=0}^{n} \binom{n}{i}^2 = \binom{2n}{n}.$$
17. Let $\mathcal{C}$ be a collection of subsets of a finite set S.
   (a) Prove that if $C \cap D \neq \emptyset$ for every pair (C, D) of members of $\mathcal{C}$, then $|\mathcal{C}| \leq 2^{|S|-1}$.
   (b) Prove that if $C \cup D \subset S$ for every pair (C, D) of members of $\mathcal{C}$, then $|\mathcal{C}| \leq 2^{|S|-1}$.
18. Let $\mathcal{C}$ be a collection of subsets of a finite set S such that $\mathcal{C} \subseteq \mathcal{P}_k(S)$ for some $k < |S|$. The *shadow* of $\mathcal{C}$ is the collection $\Delta\mathcal{C} = \{D \in \mathcal{P}_{k-1}(S) \mid D \subseteq C \text{ for some } C \in \mathcal{C}\}$. The *shade* of $\mathcal{C}$ is the collection $\nabla\mathcal{C} = \{D \in \mathcal{P}_{k+1}(S) \mid C \subseteq T \text{ for some } C \in \mathcal{C}\}$. Prove that:
   (a) $|\Delta\mathcal{C}| \geqslant \frac{k}{n-k+1}|\mathcal{C}|$ for $k > 0$;
   (b) $|\nabla\mathcal{C}| \geqslant \frac{n-k}{k+1}|\mathcal{C}|$ for $k < n$;
   (c) $\frac{|\Delta\mathcal{C}|}{\binom{n}{k-1}} \geqslant \frac{|\mathcal{C}|}{\binom{n}{k}}$ for $k > 0$;
   (d) $\frac{|\nabla\mathcal{C}|}{\binom{n}{k+1}} \geq \frac{|\mathcal{C}|}{\binom{n}{k}}$ for $k < n$;
   (e) if $k \leq \frac{n-1}{2}$, then $|\nabla\mathcal{C}| \geqslant |\mathcal{C}|$;
   (f) if $k \geqslant \frac{n+1}{2}$, then $|\Delta\mathcal{C}| \geqslant |\mathcal{C}|$.
19. A *derangement* is a permutation $f : \{1, \ldots, n\} \longrightarrow \{1, \ldots, n\}$ such that $f(i) \neq i$ for $1 \leq i \leq n$. Denote by $E_{\{i_1 \cdots i_k\}}$ the set of permutations f of $PERM_n$ such that $f(i_p) = i_p$ for $1 \leq p \leq k$.
   (a) Prove that $|E_{\{i_1 \cdots i_k\}}| = (n-k)!$.

(b) By applying the inclusion-exclusion principle, prove that the number of derangements is

$$D_n = n! \left(1 - \frac{1}{1!} + \frac{1}{2!} + \cdots + (-1)^n \frac{1}{n!}\right).$$

(c) Use a combinatorial argument to prove that

$$n! = \sum_{j=0}^{n} \binom{n}{j} D_{n-j}.$$

20. Let S be a set, $\mathcal{C}$ be a collection of subsets of S and T be a subset of S. Denote by $d(\mathcal{C}, T)$ the collection of sets in $\mathcal{C}$ that are disjoint from T, $\{C \in \mathcal{C} \mid C \cap T = \emptyset\}$. If c_k is the number of ways to choose a subcollection $\mathcal{D}$ of $\mathcal{C}$ such that $|\mathcal{D}| = k$ and $\bigcup \mathcal{S} = S$, then prove that $c_k = \sum_{T \subseteq S} (-1)^{|T|} |d(\mathcal{C}, T)|^k$.
Solution: Note that there are $|d(\mathcal{C}, T)|^k$ ways to pick k sets $C_1, \ldots, C_k$ of $\mathcal{C}$ that are disjoint from T. If $\bigcup C_i = S$, then at least one of these sets must intersect T, which implies $T = \emptyset$. Thus, every cover contributes only to the term $(-1)^0 |d(\mathcal{C}, \emptyset)|^k$. If $\bigcup C_i = V \subset S$, then the C_i contribute to every term corresponding to $T = S - V$, so the total contribution of $C_1, \ldots, C_k$ is $\sum_{T \subseteq S-V} (-1)^T$, and this sum equals 0 because every nonempty set has an equal number of even and odd-sized subsets. Thus, only the collection of sets that cover the entire set S contributes to c_k.
21. Let P be a subset of $\{1, \ldots, 2n\}$ such that $|P| = n + 1$. Prove that there exists a pair of numbers in $P \times P$ whose components are relatively prime numbers.
22. Prove that $\mathcal{R}(m, p, q) \leq \binom{m+p-2}{m-1}$ for $m \geqslant 2$ and $p \geqslant 2$.
23. Let $\mathbf{n} = (n_0, n_1, \ldots) \in \mathbf{Seq}_\infty(\mathbb{N})$ be a sequence of natural numbers. Prove that $\mathbf{n}$ contains a subsequence that is either strictly increasing, strictly decreasing, or constant. Extend this result to countable, totally ordered sets.
Hint: Consider the complete graph on the set $\{n_0, n_1, \ldots\}$, and color each edge (n_i, n_j) with $i < j$ with red if $n_i < n_j$, blue if $n_i > n_j$, and white if $n_i = n_j$.
24. Let $\mathbf{a} = (a_1, \ldots, a_n)$ be a sequence in $\mathbf{Seq}_n(\mathbb{N})$ such that $r \leq \min a_i$ and let $p = \max a_i$. If $\mathbf{p} = (p, \ldots, p) \in \mathbf{Seq}_n(\mathbb{N})$, prove that $\mathsf{Ramsey}(\mathbf{p}, r) \geqslant \mathsf{Ramsey}(\mathbf{a}, r)$.
25. The *left shift* and the *right shift* on $PERM_n$ are the mappings $lshift, rshift : PERM_n \longrightarrow PERM_n$ defined by

$$lshift(a_1, a_2, \ldots, a_n) = (a_2, \ldots, a_n, a_1),$$
$$rshift(a_1, a_2, \ldots, a_n) = (a_n, a_1 \ldots, a_{n-1}),$$

for every $(a_1, \ldots, a_n) \in PERM_n$, respectively.

(a) Prove that *lshift* and *rshift* are inverse to each other.

(b) Two permutations $f, g \in PERM_n$ are equivalent, $f \equiv g$, if there exists an integer k such that $lshift^{(k)}(f) = g$. Here $h^{(k)}$ denotes the kth iteration of h. Prove that "$\equiv$" is an equivalence on $PERM_n$.

26. Prove that if $0 \leq p \leq (n+1)! - 1$, then there exists a unique sequence $\mathbf{a} = (a_1, \ldots, a_n) \in \mathbf{Seq}(nn)$ such that $0 \leq a_i \leq i$ and $p = a_1 \cdot 1! + a_2 \cdot 2! + \cdots + a_n \cdot n!$.
27. Consider the polynomial $[x]_n = x(x-1)\cdots(x-n+1)$ called the *factorial power of x with exponent n*. The coefficients of this polynomial

$$[x]_n = s(n,n)x^n + s(n,n-1)x^{n-1} + \cdots + s(n,i)x^i + \cdots + s(n,0)$$

are known as *the Stirling numbers of the first kind*. Prove that

$$[x]_n = (-1)^n[n-x-1]_n;$$
$$s(n,0) = 0;$$
$$s(n,n) = 1;$$
$$s(n+1,k) = s(n,k-1) - ns(n,k).$$

Hint: To prove the last equality, observe that $[x]_{n+1} = [x]_n(x-n)$ and seek the coefficient of x_k in both sides.
28. Prove that for $p, n \in \mathbb{N}$ we have

$$[p]_n = \begin{cases} 0 & \text{if } 0 \leq p \leq n-1, \\ \frac{p!}{(p-n)!} & \text{if } p \geqslant n. \end{cases}$$

Furthermore, prove that

$$e = \sum_{p=0}^{\infty} \frac{[p]_n}{p!}$$

for any non-negative integer n.
29. Let S, U be two finite sets with $|U| = n$. Prove that the number of functions of the form $f : S \longrightarrow U$ that have the same kernel partition $\pi = \{H_1, \ldots, H_k\} \in PART(S)$ with $k \leq n$ equals the number of injective functions from a set with k elements into U, and therefore this number is $[n]_k = n(n-1)\cdots(n-k+1)$.
30. Let $\nu(\pi)$ the number of blocks of a partition $\pi \in PART(S)$, where $|S| = m$. Prove that for every $u \in \mathbb{N}$ we have $\sum\{[u]_{\nu(\pi)} \mid \pi \in PART(S)\} = u^m$.
Solution: Let U be a finite set such that $|U| = u$. There are u^m functions of the form $f : S \longrightarrow U$ and each such function has a kernel partition with k blocks with $k \leq n$. As observed before, there exist $[u]_k$ such parttions, which yields the above equality.
31. Prove that every polynomial x^n can be written as

$$x^n = \sum_{i=1}^{n} S(n,i)[x]_i,$$

where $S(n,i)$ are the Stirling numbers of the second kind.

32. Prove that the Stirling numbers of the second kind satisfy the equalities:
    (a) $S(n,2) = 2^{n-1} - 1$ and $S(n,n-1) = \binom{n}{2}$;
    (b) $S(n,i) = iS(n-1,i) + S(n-1,i-1)$.
    Solution: We discuss only the equality (b). A partition of a set $A = \{a_1, \ldots, a_n\}$ into i blocks can be obtained from a partition of $A - \{a_n\}$ into i blocks and adding a_n to one of these blocks or by placing a_n in a block by itself. The first term of the second member of the equality counts partitions obtained by the first construction, while the second term corresponds to the second construction.

33. The total number of partitions of a set having n elements is denoted by B_n and is known as the nth *Bell number*. Clearly, $B_n = \sum_{m=0}^{n} S(n,m)$. Prove that:

$$B_n = \sum_{m=0}^{n-1} \binom{n-1}{m} B_m.$$

34. Prove the inequality

$$S(n,i-1)S(n,i+1) \leq (S(n,i))^2$$

for $1 \leq i \leq n-1$.

35. Consider the equation $x_1 + \cdots + x_p = n$, where $n \geqslant 1$. Prove that:
    (a) the number of solutions in natural numbers is $\binom{n+p-1}{p-1}$;
    (b) the number of solutions in positive integers is $\binom{n-1}{p-1}$.
    Solution: Let $S = \{a, b\}$ be a two-element set and let $\mathbf{a} = (a, \ldots, a) \in \mathbf{Seq}_n(S)$ be a sequence of length n. Let $\mathbf{a}' \in \mathbf{Seq}_{n+p-1}$ be the sequence obtained from $\mathbf{a}$ by inserting $p-1$ elements b. The number of a symbols between any two consecutive bs yields a solution in natural numbers if adjacent b symbols are allowed and there are $\binom{n+p-1}{p-1}$ configurations of $\mathbf{a}'$, each corresponding to a choice of $p-1$ positions out of a total of $n+p-1$ possible places. Further, any solution of the equation can be obtained in this manner. The second part follows immediately from the first part.

36. Prove that $|IP_n(\ell)|$ equals the number of solutions of the equation $y_1 + \cdots + y_\ell = n - \ell$ such that $y_1 \geqslant y_2 \geqslant \cdots \geqslant y_\ell \geqslant 0$. Infer that $|IP_n(\ell)| = \sum_{k=1}^{\ell} |IP_{n-\ell}(k)|$.
37. Let $f : \{1, \ldots, n\} \longrightarrow \{1, \ldots, n\}$ be a function and D_f be the set $D_f = \{f(i+1) - f(i) \mid 1 \leq i \leq n-1\} \cup \{f(1) - 1, n - f(n)\}$.
    Prove that f is monotonic if and only if D_f is a non-negative solution of the

equation $x_1 + x_2 + \cdots + x_n + x_{n+1} = n$. Infer the number of monotonic transformations of the set $\{1, \ldots, n\}$.

38. Prove that the number of Ferrers diagrams that can be placed in an $m \times n$ rectangle is $\binom{m+n}{n}$.

39. The definition of binomial coefficients can be extended to pairs of the form (x, k), where $x \in \mathbb{C}$ and $k \in \mathbb{Z}$ by writing

$$\binom{z}{k} = \begin{cases} \frac{[z]_k}{k!} & \text{if } k \geqslant 0, \\ 0 & \text{if } k < 0, \end{cases}$$

where $[z]_k$ is the polynomial introduced in Exercise 27. Show that the equality $\binom{n}{k} = \binom{n}{n-k}$ fails when $n < 0$; however, the addition identity (3.6)

$$\binom{z}{k} = \binom{z-1}{k} + \binom{z-1}{k-1},$$

holds for every complex number z and every integer k.

40. A partition $\mathbf{k} \in IP_n$ is *self-conjugate* if $\mathbf{k}' = \mathbf{k}$. Using Ferrers diagrams prove that the number of self-conjugate partitions in IP_n equals the number of partitions of n into distinct odd numbers.

41. Let S be a set having n elements.

    (a) A collection $\mathcal{A}$ of subsets of S has the *intersecting property* if $A, B \in \mathcal{A}$ implies $A \cap B \neq \emptyset$. Prove that if $\mathcal{A}$ has the intersecting property, then $|\mathcal{A}| \leq 2^{n-1}$.

    (b) Prove that there are collections of subsets of S that have the intersecting property and contain 2^{n-1} subsets.
    Solution: For Part 1, note that, for any subset B, at most one of the sets B, $S - B$ may belong to $\mathcal{A}$. Therefore, $\mathcal{A}$ may not contain more than half of the members of $\mathcal{P}(S)$, so $|\mathcal{A}| \leq 2^{n-1}$. A collection of sets that answers Part (b) is the set of subsets of S that contain a fixed element a of S. Clearly, there are 2^{n-1} such sets.

42. Prove Sperner's theorem using Supplement 8 of Chap. 1.
Solution: Let $\mathcal{C}$ be a Sperner system on a finite set S such that $|S| = n$ and let $c : \{0, 1, \ldots, n\} \longrightarrow \mathbb{N}$ be defined by $c(k) = |\{X \in \mathcal{C} \mid |X| = k\}$ for $0 \leq k \leq n$. Suppose that $i_0 = \min\{i \mid 1 \leq i \leq n \mid c(i) > 0\} < \frac{n-1}{2}$. By Part (e) of Supplement 18 in Chap. 1, for each of the sets X in $\mathcal{C}$ with $|X| = i_0$, there exists a set X' in the shade of $\{X \in \mathcal{C} \mid |X| = i_0\}$ such that $|X'| = i_0 + 1$ and $X' \notin \mathcal{C}$. By replacing each X with the corresponding X', we obtain a new Sperner system with the same number of sets as $\mathcal{C}$. The process is repeated until a Sperner system $\mathcal{C}'$ that contains no sets with fewer than $\frac{n-1}{2}$ elements is obtained and $\mathcal{C}' = |\mathcal{C}|$. Thus each set in $\mathcal{C}'$ has at least $\frac{n+1}{2}$ elements. Now the process is reversed using the $\Delta\mathcal{C}'$ by replacing every set of $\mathcal{C}'$ of size $\frac{n+1}{2}$ by a set of size

$\lfloor \frac{n}{2} \rfloor$. The Sperner system $\mathcal{C}''$ has the same size as $\mathcal{C}$ and consists of sets of size $\frac{n+1}{2}$, so $|\mathcal{C}| \leq \binom{n}{\lfloor \frac{n}{2} \rfloor}$.

43. Let $\mathcal{C}$ and $\mathcal{D}$ be two collections of subsets of a set S. In Definition 1.18, we introduced the collection $\mathcal{D} - \mathcal{C}$ as $\mathcal{C} - \mathcal{D} = \{U - V \mid U \in \mathcal{C}, V \in \mathcal{D}\}$. Prove that $|\mathcal{C} - \mathcal{C}| \geqslant |\mathcal{C}|$.

    Solution: Let $\mathcal{D}$ be a collection of subsets of a set S and let $\mathcal{D}' = \{S - D \mid D \in \mathcal{D}\}$. Observe that $|\mathcal{D}| = |\mathcal{D}'|$.

    If $\mathcal{C}$ and $\mathcal{D}$ are two collections of subsets of S, we have $|\mathcal{C} \vee \mathcal{D}'| = |(\mathcal{C} \vee \mathcal{D}')'| = |\mathcal{C}' \wedge \mathcal{D}|$. By the previous observation and by Corollary 3.47, we can write

$$\begin{aligned} |\mathcal{C}| \cdot |\mathcal{D}| = |\mathcal{C}| \cdot |\mathcal{D}'| &\leq |\mathcal{C} \vee \mathcal{D}'| \cdot |\mathcal{C} \wedge \mathcal{D}'| \\ &= |\mathcal{C}' \wedge \mathcal{D}| \cdot |\mathcal{C} \wedge \mathcal{D}'| = |\mathcal{D} - \mathcal{C}| \cdot |\mathcal{C} - \mathcal{D}|. \end{aligned}$$

    If we now choose $\mathcal{D} = \mathcal{C}$, the previous inequality yields $|\mathcal{C}|^2 \leq |\mathcal{C} - \mathcal{C}|^2$, which gives the desired inequality.

44. Prove that if $\mathcal{C}$ and $\mathcal{D}$ are hereditary or dually hereditary families of subsets of a finite set S, then $|\mathcal{C}| \cdot |\mathcal{D}| \leq 2^n \cdot |\mathcal{C} \cap \mathcal{D}|$.
45. Prove that for the Möbius function of the poset $(\mathbb{N}, \delta)$ we have:

$$\mu(1, n) = \begin{cases} 1 & \text{if } n \text{ is a product of an even number of distinct primes,} \\ -1 & \text{if } n \text{ is a product of an odd number of distinct primes,} \\ 0 & \text{otherwise,} \end{cases}$$

    for $n \in \mathbb{N}$.

46. If μ is the Möbius function of the poset $(\mathbb{N}, \delta)$ prove that

$$\sum\{\mu(1, m) \mid (m, n) \in \delta\} = \begin{cases} 1 & \text{if } n = 1, \\ 0 & \text{otherwise.} \end{cases}$$

    Solution: For $n = 1$, the equality is immediate. Suppose that $n > 1$. Only numbers m that are products of distinct prime numbers contribute to the sum $\sum\{\mu(1, m) \mid (m, n) \in \delta\}$. If $n = p_1^{a_1} \cdots p_r^r$, then this sum equals $\sum_{i=0}^r \binom{r}{i}(-1)^i = 0$.

47. Compute the Möbius function of the poset $(PART(1, 2, 3, 4), \leq)$ considered in Example 2.74 of Chap. 2.
48. Prove that for a collection $\mathcal{C}$ of subsets of a set U we have $\mathit{VCD}(\mathcal{C}) = 0$ only if $|\mathcal{C}| \leq 1$.
49. Let $\mathcal{C}$ be a collection of sets such that $\mathit{VCD}(\mathcal{C}) \geq 2$. Prove that:

    (a) $\mathcal{C}$ contains two sets A, B such that $|A \oplus B| = 1$;
    (b) both $\mathcal{P}'(\mathcal{C}, s)$ and $\mathcal{P}''(\mathcal{C}, s)$ are non-empty for some $s \in S$.

50. Let $\mathcal{C}$, $\mathcal{D}$ be two collections of subsets of a set S. Prove that for every $m \in \mathbb{N}$ we have $\Pi_{\mathcal{C} \cup \mathcal{D}}[m] = \max\{\Pi_{\mathcal{C}}[m], \Pi_{\mathcal{D}}[m]\}$.
51. Let S be a nonempty set and let $\mathcal{C} = \{\{x\} \mid x \in S\}$. Prove that $\mathsf{VCD}(\mathcal{C}) = 1$.
52. Let S be a nonempty set. Prove that if $\mathcal{C}$ is a collection of subsets of S such that $|\mathcal{C}| \geqslant 2$, then $\mathsf{VCD}(\mathcal{C}) \geqslant 1$.
53. Let U be a finite set and let $\mathcal{C}$ be a collection of subsets of U such that $|\mathcal{C}| \geqslant 2$. Prove that $\mathsf{VCD}(\mathcal{C}) > \frac{\ln |\mathcal{C}|}{1+\ln |U|}$.
Solution: Observe that $\Pi_{\mathcal{C}}[|U|] = |\mathcal{C}|$. Therefore, by Sauer–Shelah Theorem (Theorem 3.65) and by Theorem 3.68, we have
$$|\mathcal{C}| \leq \left(\frac{e|U|}{d}\right)^d,$$
where d is the VC dimension of the collection $\mathcal{C}$. The last inequality implies
$$\ln |\mathcal{C}| \leq d(1 + \ln |U| - \ln d),$$
so $\ln |\mathcal{C}| \leq d(1 + \ln |U|)$, which gives the desired inequality.
54. Prove that if $\mathcal{C}$ is a chain of subsets of a set S, then $\mathsf{VCD}(\mathcal{C}) = 1$.
55. Let $\mathcal{C}$ be a collection of subsets of S. Prove that if T is a subset of S, then $\mathsf{VCD}(\mathcal{C}_T) \leq \mathsf{VCD}(\mathcal{C})$.
56. Let $\mathcal{C}$ be a collection of sets such that $C, C' \in \mathcal{C}$ and $C \neq C'$ implies $C \cap C' = \emptyset$. Prove that $\mathsf{VCD}(\mathcal{C}) = 1$.
57. Let S be a set and let $\mathcal{C}_1, \ldots, \mathcal{C}_n$ be n chains in the poset $(\mathcal{P}(S), \subseteq)$. Define the collection $\mathcal{C}$ as $\mathcal{C} = \{\bigcap_{i=1}^n C_i \mid C_i \in \mathcal{C}_i, 1 \leq i \leq n\}$. Prove that $\mathsf{VCD}(\mathcal{C}) \leq n$.
Solution: Let T be a subset of S such that $|T| = n + 1$. Clearly, T has $n + 1$ subsets that have n elements.
For each i at most one n-element subset of T is the intersection of the form $T \cap C$, where $C \in \mathcal{C}_i$. Indeed, if we would have two distinct n-element sets of the form $T \cap C'$ and $T \cap C''$, where $C', C'' \in \mathcal{C}$ this would imply the existence of $x' \in (T \cap C') - (T \cap C'')$ and of $x'' \in (T \cap C'') - (T \cap C')$, which would mean that $x' \in C' - C''$ and $x'' \in C'' - C'$, thus contradicting the $\mathcal{C}_i$ is a chain of sets. Let U_i be this n-element when it exists.
Let W be an n-element subset of T such that $W = T \cap C$ for some $C = \bigcap_{i=1}^n C_i \in \mathcal{C}$. Then, either $C_j \cap T = W$ or $C_i \cap T = T$ for $1 \leq j \leq n$ and $C_i \cap T = W$ for at least one i, $1 \leq i \leq n$. Therefore, $W = U_i$ for some i, $1 \leq i \leq n$, which shows that at most n subsets of T that contain n elements can be obtained as intersections of T with the elements of $\mathcal{C}$. Thus, T is not shattered by $\mathcal{C}$ and $\mathsf{VCD}(\mathcal{C}) \leq n$.
58. For $1 \leq i \leq n$ and $a \in \mathbb{R}$ let $C_{i,a} = \{\mathbf{x} \in \mathbb{R}^n \mid \mathbf{x} = (x_1, \ldots, x_n), x_i \leq a\}$. The chain of sets $\mathcal{C}_i$ is defined by $\{C_{i,a} \mid a \in \mathbb{R}\}$ for $1 \leq i \leq n$. Prove that $\mathcal{C} = \bigcap_{i=1}^n \mathcal{C}_i$ shatters the set $B = \{\mathbf{e}_1, \ldots, \mathbf{e}_n\}$, where $\mathbf{e}_i = (0, \ldots, 0, 1, 0, \ldots, 0)$ has 1 as its ith component for $1 \leq i \leq n$, so $\mathsf{VCD}(\mathcal{C}) = n$.

59. The statement included here is a generalization of Example 3.58. Prove that the Vapnik-Chervonenkis dimension of the collection of rectangular subsets of $\mathbb{R}^n$ given by

$$\mathcal{C} = \left\{ \prod_{i=1}^{n} [a_i, b_i] \mid a_i, b_i \in \hat{\mathbb{R}}, a_i \leq b_i, \text{ for } 1 \leq i \leq n \right\}$$

is $2n$. If $a_i = -\infty$ for all a_i, $1 \leq i \leq n$, then $VCD(\mathcal{C}) = n$.
60. Let S be a set that contains at least two elements and let $\mathcal{C}$ be a collection of subsets S. Suppose that for every two-element subset of S, $T = \{t_1, t_2\}$, there exist $U, V \in \mathcal{C}$ such that $T \subseteq U$ and $T \cap V = \emptyset$. Then $VCD(\mathcal{C}) = 1$ if and only if $\mathcal{C}$ is a chain.
Solution: Suppose that $VCD(\mathcal{C}) = 1$ but $\mathcal{C}$ is not a chain. Then, $\mathcal{C}$ contains two sets C', C'' such that neither $C' \subseteq C''$ nor $C'' \subseteq C'$. Let $c' \in C' - C''$ and $c'' \in C'' - C'$. Then, the two element set $T = \{c', c''\}$ is shattered by $\mathcal{C}$, which implies $VCD(\mathcal{C}) \geqslant 2$. The reverse implication follows from Supplement 54.
61. Prove that if $\mathcal{C}$ is a collection of subsets of a set S such that $\{\emptyset, S\} \subseteq \mathcal{C}$, then $VCD(\mathcal{C}) = 1$ if and only if $\mathcal{C}$ is a chain.
62. Let S be a finite set and let $\mathcal{C}$ be a collection of subsets of S. Prove that $|\mathcal{C}| = |\{K \in \mathcal{P}(S) \mid \mathcal{C} \text{ shatters } K\}|$.
Solution: Let x be an element of S and let $\phi_x : \mathcal{C} \longrightarrow \mathcal{P}$ be the injective mapping introduced in Supplement 26 of Chap. 1. We claim that if $\phi_x(\mathcal{C}) = \{\phi_x(C) \mid C \in \mathcal{C}\}$ shatters K, then $\mathcal{C}$ shatters K. If $x \notin K$, then $\mathcal{C}_K = \phi_x(\mathcal{C})_K$, so the statement obviously holds. If $x \in K$ and $L \subseteq K - \{x\}$, then there is $F \in \phi_x(C)$ such that $F \cap K = L \cup \{x\}$ and $T = \phi_x(C)$ for some $C \in \mathcal{C}$. Since $x \in F$, both F and $F - \{x\}$ belong to $\mathcal{C}$, so $\mathcal{C}$ shatters K.
Define $w(\mathcal{C}) = \sum\{|C| \mid C \in \mathcal{C}\}$. Let $\mathcal{C}'$ be a collection of sets obtained from $\mathcal{C}$ by applying transforms of the form ϕ_x, such that $w(\mathcal{C}')$ is minimal. For $C \in \mathcal{C}'$ and $x \in K$ we must have $C - \{x\} \in \mathcal{C}'$ because otherwise $w(\phi_x(\mathcal{C}')) < w(\mathcal{C}')$, contradicting the minimality of $\mathcal{C}'$. Thus, $\mathcal{C}'$ is hereditary, so it shatters any set it contains. Since $|\mathcal{C}'| = |\mathcal{C}|$ (by Supplement 26 of Chap. 1, and $\mathcal{C}$ shatters at least as many sets as $\mathcal{C}$ we obtain the desired equality.
63. Let $(P, \leq)$ and $(Q, \leq)$ be two locally finite posets having the Möbius function μ_P and μ_Q, respectively. In Exercise 30 of Chap. 1 we saw that $(P \times Q, \leq)$ is a locally finite poset. Prove that its Möbius function $\mu_{P \times P}$ is given by

$$\mu_{P \times P}((x, y), (x', y')) = \mu_P(x, x')\mu(y, y'),$$

where $x \leq x'$ and $y \leq y'$.

Bibliographical Comments

Supplement 20 was obtained in [11]. The inequality from Supplement 43 was obtained in [12]. Exercise 12 contains a result of T. Calders [13].

There are several well-known and comprehensive references on combinatorics that contain rich collections of ideas [1, 14, 15].

Theorem 3.71 appears in [16]. Supplements 53–61 contain results obtained in [17]. Note that in [17] the Vapnik-Chervonenkis dimension of a collection of set is defined as the smallest n such that no n-element set is shattered by $\mathcal{C}$, so values of $VCD(\mathcal{C})$ in [17] are obtained by increasing by one the value of the VCD adopted here (and in the vast majority of publications).

The notion of density of a collection of sets was introduced by Assouad in [18]. Supplement 62 originates in [19].

Exercise 30 is a result of Rota (see [20]).

References

1. R.P. Stanley, *Enumerative Combinatorics*, vol. 1 (Cambridge University Press, Cambridge, 1997)
2. D. Lubell, A short proof of Sperner theorem. J. Comb. Theory **1**, 299 (1966)
3. K. Yamamoto, Logarithmic order of free distributive lattices. J. Math. Soc. Japan **6**, 343–353 (1954)
4. L.D. Meshalkin, A generalization of Sperner's theorem on the number of subsets of a finite set. Theory Probab. Appl. **8**, 203–204 (1963). (in Russian)
5. E. Sperner, Ein Satz über Untermengen einer endlichen Menge. Math. Z. **27**, 544–548 (1928)
6. V.N. Vapnik, A.Y. Chervonenkis, On the uniform convergence of relative frequencies of events to their probabilities. Theory Probab. Appl. **16**, 264–280 (1971)
7. N. Sauer, On the density of families of sets. J. Comb. Theory (A) **13**, 145–147 (1972)
8. A. Blumer, A. Ehrenfeucht, D. Haussler, M. Warmuth, Learnability and the Vapnik-Chervonenkis dimension. J. ACM **36**, 929–965 (1989)
9. R.M. Dudley, *Uniform Central Limit Theorems* (Cambridge University Press, Cambridge, 1999)
10. D. Pollard, *Empirical Processes: Theory and Applications* (Institute of Mathematical Statistics, Hayward, CA, 1990)
11. A. Björklund, T. Husfeldt, Inclusion-exclusion algorithms for counting set partitions, *Proceedings of the 47th Annual IEEE Symposium on Foundations of Computer Science*, Berkeley, CA (IEEE Computer Society Press, Los Alamitos, CA, 2006), pp. 575–582
12. J. Marica, J. Schönheim, Differences of sets and a problem of Graham. Can. Math. Bulletin **12**, 635–637 (1969)
13. T. Calders. Axiomatization and deduction rules for the frequency of item sets. PhD thesis, Universiteit Antwerpen, 2003.
14. R.P. Stanley, *Enumerative Combinatorics*, vol. 2 (Cambridge University Press, Cambridge, 1999)
15. R.L. Graham, D.E. Knuth, O. Patashnick, *Concrete Mathematics—A Foundation for Computer Science* (Addison-Wesley, Reading, MA, 1989)
16. M. Vidyasagar, *Learning and Generalization with Applications to Neural Networks* (Springer, London, 2003)

17. R.S. Wenocur, R.M. Dudley, Some special Vapnik-Chervonenkis classes. Discrete Math. **33**, 313–318 (1981)
18. P. Assouad, Densité et dimension. Annales de l'Institut Fourier **33**, 233–282 (1983)
19. M. Ledoux, M. Talagrand, *Isoperimetry and Processes in Probability in Banach Spaces* (Springer, Berlin, 2002)
20. G.-C. Rota, The number of partitions of a set. Am. Math. Monthly **71**, 498–504 (1964)

Chapter 4
Topologies and Measures

4.1 Introduction

Topology is an area of mathematics that investigates both the local and the global structure of space. The term "topology" is derived from the Greek words $\tau\acute{o}\pi o\zeta$ (*topos*, place) and $\lambda\acute{o}\gamma o\zeta$ (*logos*, reason) and was introduced in [1]. We present in this chapter an introduction to point-set topology that is important for a subsequent discussion of various notions of dimensions of sets. Data mining makes use of topology in formulating searching algorithms that take into account the local properties of data sets.

4.2 Topologies

The term "topology" is used both to designate a mathematical discipline and to name the fundamental notion of this discipline, which is introduced next.

Definition 4.1 *A* topology *on a set S is a family $\mathcal{O}$ of subsets of S that satisfies the following conditions:*

(i) $\emptyset \in \mathcal{O}$ *and* $S \in \mathcal{O}$;
(ii) *for every collection* $\mathcal{C}$ *such that* $\mathcal{C} \subseteq \mathcal{O}$, $\bigcup \mathcal{C} \in \mathcal{O}$;
(iii) *if* $\mathcal{D}$ *is a finite collection and* $\mathcal{D} \subseteq \mathcal{O}$, *then* $\bigcap \mathcal{D} \in \mathcal{O}$.

The sets that belong to $\mathcal{O}$ are referred to as the open sets *of the topology $\mathcal{O}$. The pair $(S, \mathcal{O})$ is referred to as a* topological space.

It is easy to see that Part (iii) of Definition 4.1 is equivalent to

(iii′) if $U, U' \in \mathcal{O}$, then $U \cap U' \in \mathcal{O}$.

Actually, the first condition of Definition 4.1 is superfluous. Indeed, since we deal here with collections of subsets of S, by Part (iii) of the definition, the intersection of

D. A. Simovici and C. Djeraba, *Mathematical Tools for Data Mining*,
Advanced Information and Knowledge Processing, DOI: 10.1007/978-1-4471-6407-4_4,

the empty collection of subsets of S belongs to $\mathcal{O}$, and this intersection is S. On the other hand, by Part (ii), the union of the empty collection (which is the empty set) belongs to $\mathcal{O}$, so Part (i) is a consequence of the remaining parts of the definition.

Example 4.2 The pair $(S, \mathcal{P}(S))$ is a topological space. The topology $\mathcal{P}(S)$ is known as the *discrete topology*.

The collection $\{\emptyset, S\}$ is the *indiscrete topology*.

Example 4.3 The pair $(\emptyset, \{\emptyset\})$ is a topological space as the reader can easily verify. We refer to $(\emptyset, \{\emptyset\})$ as the *empty topological space*.

Example 4.4 Let $\mathcal{O}$ be the collection of subsets of $\mathbb{R}$ defined by $L \in \mathcal{O}$ if for every $x \in L$ there exists $\epsilon \in \mathbb{R}_{>0}$ such that $|u - x| < \epsilon$ implies $u \in L$. We claim that $\mathcal{O}$ is a topology on $\mathbb{R}$.

Indeed, it is immediate that $\emptyset$ and $\mathbb{R}$ belong to $\mathcal{O}$.

Let $\mathcal{C}$ be such that $\mathcal{C} \subseteq \mathcal{O}$ and let $x \in \bigcup \mathcal{C}$. There exists $L \in \mathcal{C}$ such that $x \in L$ and, therefore, by the definition of $\mathcal{O}$, there is $\epsilon > 0$ such that $|u - x| < \epsilon$ implies $u \in L$. Thus, $u \in \bigcup \mathcal{C}$, so $\bigcup \mathcal{C} \in \mathcal{O}$.

Suppose now that $\mathcal{D}$ is a finite subcollection of $\mathcal{O}$, $\mathcal{D} = \{D_1, \ldots, D_n\}$, and let $x \in \bigcap \mathcal{D}$. Since $x \in D_i$ for $1 \leqslant i \leqslant n$, there exists $\epsilon_1, \ldots, \epsilon_n$ such that $|u - x| < \epsilon_i$ implies $u \in D_i$ for every i, $1 \leqslant i \leqslant n$. Therefore, by defining $\epsilon = \min\{\epsilon_i \mid 1 \leqslant i \leqslant n\}$, it follows that $|x - u| \leqslant \epsilon$ implies $u \in \bigcap \mathcal{D}$, which proves that $\bigcap \mathcal{D} \in \mathcal{O}$. We conclude that $\mathcal{O}$ is a topology on $\mathbb{R}$. This topology is called the *usual topology* on $\mathbb{R}$. Unless stated otherwise, we assume that the set of real numbers is equipped with the usual topology.

Example 4.5 Example 4.4 can be extended to $\mathbb{R}^n$ by defining the $\mathcal{O}$ as consisting of subsets L of $\mathbb{R}^n$ such that for every $\mathbf{x} \in L$ there exists $\epsilon \in \mathbb{R}_{>0}$ such that $d_2(\mathbf{u}, \mathbf{x}) < \epsilon$ implies $\mathbf{u} \in L$. It is easy to verify that $(\mathbb{R}^n, \mathcal{O})$ is a topological space.

For each topology $\mathcal{O}$ on a set S, we define

Definition 4.6 *The collection of* closed sets *of a topology* $\mathcal{O}$ *is* $\mathsf{closed}(\mathcal{O}) = \{S - X \mid X \in \mathcal{O}\}$; *the* collection of neighborhoods *of an element* x *of* S *is*

$$\mathsf{neigh}_x(\mathcal{O}) = \{U \in \mathcal{P}(S) \mid \text{there is } W \in \mathcal{O} \text{ such that } x \in W \subseteq U\}.$$

Theorem 4.7 *The following statements hold for any topological space* $(S, \mathcal{O})$:

(i) $\emptyset$ *and* S *are closed sets;*
(ii) *for every collection* $\mathcal{C}$ *of closed sets,* $\bigcap \mathcal{C}$ *is a closed set;*
(iii) *for every finite collection* $\mathcal{D}$ *of closed sets,* $\bigcup \mathcal{D}$ *is a closed set.*

Proof This is an immediate consequence of Definition 4.1.

4.3 Closure and Interior Operators in Topological Spaces

Theorem 4.7 implies that for every topological space $(S, \mathcal{O})$ the collection $\mathsf{closed}(\mathcal{O})$ of closed sets is a closure system on S. For the closure operator attached to this closure system denoted by $\mathbf{K}_{S,\mathcal{O}}$, we have the supplementary property:

$$\mathbf{K}_{S,\mathcal{O}}(H \cup L) = \mathbf{K}_{S,\mathcal{O}}(H) \cup \mathbf{K}_{S,\mathcal{O}}(L) \tag{4.1}$$

for all subsets H, L of S.

Since $H, L \subseteq H \cup L$, we have $\mathbf{K}_{S,\mathcal{O}}(H) \subseteq \mathbf{K}_{S,\mathcal{O}}(H \cup L)$ and $\mathbf{K}_{S,\mathcal{O}}(L) \subseteq \mathbf{K}_{S,\mathcal{O}}(H \cup L)$ due to the monotonicity of $\mathbf{K}_{S,\mathcal{O}}$. Therefore,

$$\mathbf{K}_{S,\mathcal{O}}(H) \cup \mathbf{K}_{S,\mathcal{O}}(L) \subseteq \mathbf{K}_{S,\mathcal{O}}(H \cup L).$$

To prove the reverse inclusion, note that the set $\mathbf{K}_{S,\mathcal{O}}(H) \cup \mathbf{K}_{S,\mathcal{O}}(L)$ is a closed set by the third part of Theorem 4.7 and $H \cup L \subseteq \mathbf{K}_{S,\mathcal{O}}(H) \cup \mathbf{K}_{S,\mathcal{O}}(L)$. Therefore, the closure of $H \cup L$ is a subset of $\mathbf{K}_{S,\mathcal{O}}(H) \cup \mathbf{K}_{S,\mathcal{O}}(L)$, so $\mathbf{K}_{S,\mathcal{O}}(H \cup L) \subseteq \mathbf{K}_{S,\mathcal{O}}(H) \cup \mathbf{K}_{S,\mathcal{O}}(L)$, which implies Equality (4.1).

Also, note that $\mathbf{K}_{S,\mathcal{O}}(\emptyset) = \emptyset$ because the empty set itself is closed.

If there is no risk of confusion, we denote the closure operator $\mathbf{K}_{S,\mathcal{O}}$ simply by $\mathbf{K}$.

Note that Equality (4.1) is satisfied for every $H, L \in \mathcal{P}(S)$ if and only if the union of two $\mathbf{K}$-closed sets is $\mathbf{K}$-closed. Indeed, suppose that Equality (4.1) is satisfied, and let U and V be two $\mathbf{K}$-closed sets. Since $U = \mathbf{K}(U)$ and $V = \mathbf{K}(V)$, it follows that $U \cup V = \mathbf{K}(U) \cup \mathbf{K}(V) = \mathbf{K}(U \cup V)$, which shows that $U \cup V$ is $\mathbf{K}$-closed. Conversely, suppose that the union of two $\mathbf{K}$-closed sets is $\mathbf{K}$-closed. Then, $\mathbf{K}(U) \cup \mathbf{K}(V)$ is $\mathbf{K}$-closed and contains $U \cup V$. Therefore, $\mathbf{K}(U \cup V) \subseteq \mathbf{K}(U) \cup \mathbf{K}(V)$. The reverse equality follows from the monotonicity of $\mathbf{K}$.

Theorem 4.8 *Let S be a set and let $\mathbf{K} : \mathcal{P}(S) \longrightarrow \mathcal{P}(S)$ be a closure operator that satisfies Equality (4.1) for every $H, L \in \mathcal{P}(S)$ and $\mathbf{K}(\emptyset) = \emptyset$. The collection $\mathcal{O}_{\mathbf{K}} = \{S - U \mid U \in \mathcal{C}_{\mathbf{K}}\}$ is a topology on S.*

Proof We have $\mathbf{K}(S) = S$, so both $\emptyset$ and S are $\mathbf{K}$-closed sets, which implies $\emptyset, S \in \mathcal{O}_{\mathbf{K}}$.

Suppose that $\mathcal{C} = \{L_i \mid i \in I\} \subseteq \mathcal{O}_{\mathbf{K}}$. Since $S - L_i \in \mathcal{C}_{\mathbf{K}}$, it follows that $\bigcap\{S - L_i \mid i \in I\} = S - \bigcup_{i \in I} L_i \in \mathcal{C}_{\mathbf{K}}$. Thus, $\bigcup_{i \in I} L_i \in \mathcal{O}_{\mathbf{K}}$.

Finally, suppose that $\mathcal{D} = \{D_1, \ldots, D_n\}$ is a finite collection of subsets such that $\mathcal{D} \subseteq \mathcal{O}_{\mathbf{K}}$. Since $S - D_i \in \mathcal{C}_{\mathbf{K}}$ we have $S - \bigcup_{i=1}^{n} D_i = \bigcap_{i=1}^{n}(S - D_i) \in \mathcal{C}_{\mathbf{K}}$, hence $\bigcup_{i=1}^{n} D_i \in \mathcal{O}_{\mathbf{K}}$. This proves that $\mathcal{O}_{\mathbf{K}}$ is indeed a topology.

Theorem 4.9 *Let $(S, \mathcal{O})$ be a topological space and let U and W be two subsets of S. If U is open and $U \cap W = \emptyset$, then $U \cap \mathbf{K}(W) = \emptyset$.*

Proof $U \cap W = \emptyset$ implies $W \subseteq S - U$. Since U is open, the set $S - U$ is closed, so $\mathbf{K}(W) \subseteq \mathbf{K}(S - U) = S - U$. Therefore, $U \cap \mathbf{K}(W) = \emptyset$.

Often, we use the contrapositive of this statement: if U is an open set such that $U \cap \mathbf{K}(W) \neq \emptyset$ for some set W, then $U \cap W \neq \emptyset$.

Example 4.10 In the topological space $(\mathbb{R}, \mathcal{O})$, every open interval (a, b) with $a < b$ is an open set. Indeed, if $x \in (a, b)$ and $|x-u| < \epsilon$, where $\epsilon = \frac{1}{2}\min\{|x-a|, |x-b|\}$, then $u \in (a, b)$. A similar argument shows that the half-lines $(b, +\infty)$ and $(-\infty, a)$ are open sets for $a, b \in \mathbb{R}$. Therefore, $(-\infty, a) \cup (b, +\infty)$ is an open set which implies that its complement, the interval $[a, b]$, is closed. Also, $(-\infty, b]$ and $[a, \infty)$ are closed sets (as complements of the open sets (b, ∞) and (a, ∞), respectively).

Open sets of the topological space $(\mathbb{R}, \mathcal{O})$, where $\mathcal{O}$ is the usual topology on the set of real numbers have the following useful characterization.

Theorem 4.11 *A subset U of $\mathbb{R}$ is open in the topological space $(\mathbb{R}, \mathcal{O})$ if and only if it equals the union of a countable collection of disjoint open intervals.*

Proof Since every open interval (finite or not) is an open set, it follows that the union of a countable collection of disjoint open intervals is open.

To prove the converse, let U be an open set. Note that U can be written as a union of open intervals since for each $x \in U$ there exists $\epsilon > 0$ such that $x \in (x-\epsilon, x+\epsilon) \subseteq U$.

Define the relation θ_U on the set U by $x\theta_U y$ if there exist $a, b \in \mathbb{R}$ such that $\{x, y\} \subseteq (a, b) \subseteq U$, where (a, b) is the open interval determined by a, b. We claim that θ_U is an equivalence relation on U.

Since U is open, $x \in U$ implies the existence of a positive number ϵ such that $\{x\} \subseteq (x - \epsilon, x + \epsilon) \subseteq U$ for every $x \in U$, so θ_U is reflexive. The symmetry of θ_U is immediate. To prove its transitivity, let $x, y, z \in U$ be such that $x\theta_U z$ and $z\theta_U y$. There are $a, b, c, d \in \mathbb{R}$ such that $\{x, z\} \subseteq (a, b) \subseteq U$ and $\{z, y\} \subseteq (c, d) \subseteq U$. Since $z \in (a, b) \cap (c, d)$, it follows that $(a, b) \cup (c, d)$ is an interval (e, e') such that $\{x, y\} \subseteq (e, e') \subseteq U$, which shows that $x\theta_U y$. Thus, θ_U is an equivalence on U.

We claim that each equivalence class $[x]_{\theta_U}$ is an open interval or a set of the form $(a, +\infty)$ or a set of the form $(-\infty, b)$. Indeed, suppose that $u, v \in [x]_{\theta_U}$ (that is, $u\theta_U x$ and $v\theta_U x$) and that $t \in (u, v)$. We now prove that $t\theta_U x$.

There are two open intervals (a, b) and (c, d) such that $\{u, x\} \subseteq (a, b) \subseteq U$ and $\{x, v\} \subseteq (c, d) \subseteq U$. Again, $(a, b) \cup (c, d)$ is an open interval (e, e') and we have $(u, v) \subseteq (e, e') \subseteq U$. Thus, if $[x]_{\theta_U}$ contains two numbers u and v, it also contains the interval (u, v) determined by these numbers.

To prove that $[x]_{\theta_U}$ has the desired form, we shall prove that this set has no least element and no greatest element. Suppose that $[x]_{\theta_U}$ has a least element y. Then, there exist a and b such that $a < y < x < b$ and $(a, b) \subseteq U$. Since y is supposed to be the least element of $[x]_{\theta_U}$, if $a < z < y$, we have $z \notin [x]_{\theta_U}$. This contradicts $y\theta_U z$ and $y\theta_U x$. In a similar manner, it is possible to show that $[x]_{\theta_U}$ has no largest element.

Finally, we prove that the partition that corresponds to θ_U is countable. Select a rational number $r_x \in [x]_{\theta_U} \cap \mathbb{Q}$. Since the equivalence classes $[x]_{\theta_U}$ are pairwise disjoint, it follows that $[x]_{\theta_U} \neq [y]_{\theta_U}$ implies $r_x \neq r_y$. Thus, we have an injection $r : U/\theta_U \longrightarrow \mathbb{Q}$ given by $r([x]_{\theta_U}) = r_x$ for $x \in U$. By Theorem 1.126, the set U/θ_U is countable.

Example 4.12 Let S be an infinite set. The family of sets

$$\mathcal{O} = \{\emptyset\} \cup \{L \in \mathcal{P}(S) \mid S - U \text{ is finite}\}$$

is a topology on S. We refer to $\mathcal{O}$ as the *cofinite topology* on S.

Note that both $\emptyset$ and S belong to $\mathcal{O}$. Further, if $\mathcal{C}$ is a subcollection of $\mathcal{O}$, then $S - \bigcup \mathcal{C} = \bigcap\{(S - L) \mid L \in \mathcal{C}\}$, which is a finite set because it is a subset of every finite set $S - L$, where $L \in \mathcal{C}$.

Also, if $U, V \in \mathcal{O}$, then $S - (U \cap V) = (S - U) \cup (S - V)$, which shows that $S - (U \cap V)$ is a finite set. Thus, $U \cap V \in \mathcal{O}$.

Example 4.13 Let $(S, \leqslant)$ be a partially ordered set. A subset T of S is *upward closed* if $x \in T$ and $x \leqslant y$ implies $y \in T$. The collection of upwards closed sets $\mathcal{O}^{\uparrow}$ is a topology on S.

It is clear that both $\emptyset$ and S belong to $\mathcal{O}^{\uparrow}$. Further, if $\{L_i \mid i \in I\}$ is a family of upwards closed sets, then $\bigcup\{L_i \mid i \in I\}$ is also an upwards closed set. Indeed, suppose that $x \in \bigcup\{L_i \mid i \in I\}$ and $x \leqslant y$. There exists L_i such that $x \in L_i$ and therefore $y \in L_i$, which implies $y \in \bigcup\{L_i \mid i \in I\}$. Moreover, it is easy to see that any intersection of sets from $\mathcal{O}^{\uparrow}$ belongs to $\mathcal{O}^{\uparrow}$, not just a finite intersection (which would suffice for $\mathcal{O}^{\uparrow}$ to be a topology). This topology is known as the *Alexandrov topology* on the poset $(S, \leqslant)$.

Definition 4.14 *A topology* $\mathcal{O}$ *is* finer *than a topology* $\mathcal{O}'$ *or, equivalently,* $\mathcal{O}'$ *is a* coarser *than* $\mathcal{O}$, *if* $\mathcal{O}' \subseteq \mathcal{O}$.

Every topology on a set S is finer than the indiscrete topology on S; the discrete topology $\mathcal{P}(S)$ (which has the largest collection of open sets) is finer than any topology on S.

Theorem 4.15 *Let* $(S, \mathcal{O})$ *be a topological space and let* T *be a subset of* S. *The collection* $\mathcal{O} \upharpoonright_T$ *defined by* $\mathcal{O} \upharpoonright_T = \{L \cap T \mid L \in \mathcal{O}\}$ *is a topology on the set* T.

Proof We leave the proof of this theorem to the reader as an exercise.

Definition 4.16 *If* U *is a subset of* S, *where* $(S, \mathcal{O})$ *is a topological space, then we refer to the topological space* $(U, \mathcal{O} \upharpoonright_U)$ *as a* subspace *of the topological space* $(S, \mathcal{O})$.

To simplify notation, we denote the subspace $(U, \mathcal{O} \upharpoonright_U)$ just by U.

Theorem 4.17 *Let* $(S, \mathcal{O})$ *be a topological space and let* $(T, \mathcal{O} \upharpoonright_T)$ *be a subspace of this space. Then, a set* H *is closed in* $(T, \mathcal{O} \upharpoonright_T)$ *if and only if there exists a closed set* H_0 *in* $(S, \mathcal{O})$ *such that* $H = T \cap H_0$.

Proof Suppose that H is closed in $(T, \mathcal{O} \upharpoonright_T)$. Then, the set $T - H$ is open in this space and therefore there exists an open set L_0 in $(S, \mathcal{O})$ such that $T - H = T \cap L_0$. This is equivalent to $H = T - (T \cap L_0) = T \cap (S - L_0)$. We define H_0 as the closed set $S - L_0$.

Conversely, suppose that $H = T \cap H_0$, where H_0 is a closed set in S. Since $T - H = T \cap (S - H_0)$ and $S - H_0$ is an open set in $(S, \mathcal{O})$, it follows that $T - H$ is open in the subspace and therefore H is closed.

Corollary 4.18 *Let $(S, \mathcal{O})$ be a topological space and let $T \subseteq S$. Denote by $\mathbf{K}_S$ and $\mathbf{K}_T$ the closure operators of $(S, \mathcal{O})$ and $(T, \mathcal{O} \upharpoonright_T)$, respectively. For every subset W of T, we have $\mathbf{K}_T(W) = \mathbf{K}_S(W) \cap T$.*

Proof The set $\mathbf{K}_S(W)$ is closed in S, so $\mathbf{K}_S(W) \cap T$ is closed in T by Theorem 4.17. Since $W \subseteq \mathbf{K}_S(W) \cap T$, it follows that $\mathbf{K}_T(W) \subseteq \mathbf{K}_S(W) \cap T$.

To prove the converse inclusion, observe that we can write $\mathbf{K}_T(W) = T \cap H$, where H is a closed set in S because $\mathbf{K}_T(W)$ is a closed set in T. Since $W \subseteq H$, it follows that $\mathbf{K}_S(W) \subseteq H$, so $\mathbf{K}_S(W) \cap T \subseteq H \cap T = \mathbf{K}_T(W)$.

Corollary 4.19 *Let $(S, \mathcal{O})$ be a topological space and let $T \subseteq S$. If $U \subseteq S$, then $\mathbf{K}_T(U \cap T) \subseteq \mathbf{K}_S(U) \cap T$.*

Proof By applying Corollary 4.18 to the subset $U \cap T$ of T we have $\mathbf{K}_T(U \cap T) = \mathbf{K}_S(U \cap T) \cap T$. The needed inclusion follows from the monotonicity of $\mathbf{K}_S$.

Definition 4.20 *A set U is* dense *in a topological space $(S, \mathcal{O})$ if $\mathbf{K}(U) = S$. A topological space is* separable *if there exists a countable set U that is dense in $(S, \mathcal{O})$.*

Theorem 4.21 *If T is a subspace of a separable topological space $(S, \mathcal{O})$, then T itself is separable.*

Proof Since S is separable, there exists a countable set U such that $\mathbf{K}_S(U) = S$. On the other hand, $\mathbf{K}_T(U \cap T) = \mathbf{K}_S(U \cap T) \cap T \subseteq \mathbf{K}_S(U) \cap T = S \cap T = T$, which implies that the countable set $U \cap T$ is dense in T. Thus, T is separable.

Theorem 4.22 *If T is a separable subspace of a topological space $(S, \mathcal{O})$, then so is $\mathbf{K}_S(T)$.*

Proof Let U be a countable subset of T that is dense in T, that is, $\mathbf{K}_T(U) = T$. We need to prove that $\mathbf{K}_{\mathbf{K}_S(T)}(U) = \mathbf{K}_S(T)$ to prove that U is dense in $\mathbf{K}_S(T)$ also.

By Corollary 4.18, we have

$$\mathbf{K}_{\mathbf{K}_S(T)}(U) = \mathbf{K}_S(U) \cap \mathbf{K}_S(T) = \mathbf{K}_S(U)$$

due to the monotonicity of $\mathbf{K}_S$.

Note that $T = \mathbf{K}_T(U) = \mathbf{K}_S(U) \cap T$, so $T \subseteq \mathbf{K}_S(U)$, which implies $\mathbf{K}_S(T) \subseteq \mathbf{K}_S(U)$. Since $\mathbf{K}_S$ is monotonic, we have the reverse inclusion $\mathbf{K}_S(U) \subseteq \mathbf{K}_S(T)$, so $\mathbf{K}_S(U) = \mathbf{K}_S(T)$. This allows us to conclude that $\mathbf{K}_{\mathbf{K}_S(T)}(U) = \mathbf{K}_S(T)$, so U is dense in $\mathbf{K}_S(T)$.

Theorem 4.23 *Let $(S, \mathcal{O})$ be a topological space. The set U is dense in $(S, \mathcal{O})$ if and only if $U \cap L \neq \emptyset$ for every non-empty open set L.*

Proof Suppose that U is dense, so $\mathbf{K}(U) = S$. Since $\mathbf{K}(U \cap L) = \mathbf{K}(U) \cap \mathbf{K}(L) = S \cap \mathbf{K}(L) = \mathbf{K}(L)$, $U \cap L = \emptyset$ would imply $\mathbf{K}(L) = \mathbf{K}(\emptyset) = \emptyset$, which is a contradiction because $\emptyset \neq L \subseteq \mathbf{K}(L)$.

Conversely, suppose that U has a non-empty intersection with every non-empty open set L. Since $\mathbf{K}(U)$ is closed, $S - \mathbf{K}(U)$ is open. Observe that $U \cap (S - \mathbf{K}(U)) = \emptyset$, so the open set $S - \mathbf{K}(U)$ must be empty. Therefore, we have $\mathbf{K}(U) = S$.

Theorem 4.24 *The following statements hold for any topological space* $(S, \mathcal{O})$ *and* $x \in S$*:*

(i) *if* $U, V \in \mathsf{neigh}_x(\mathcal{O})$, *then* $U \cap V \in \mathsf{neigh}_x(\mathcal{O})$;
(ii) *if* $U \in \mathsf{neigh}_x(\mathcal{O})$ *and* $U \subseteq W \subseteq S$, *then* $W \in \mathsf{neigh}_x(\mathcal{O})$;
(iii) *a set* L *is open if and only if* L *is a neighborhood of all its points.*

Proof The first two parts follow immediately from Definition 4.7. We discuss here only the third statement.

If L is open, it is immediate that L is a neighborhood of all its points. Conversely, suppose that L is a neighborhood of all its members. Then, for each $x \in L$ there exists $W_x \in \mathcal{O}$ such that $x \in W_x \subseteq L$. Therefore,

$$L = \bigcup_{x \in L} \{x\} \subseteq \bigcup_{x \in L} W_x \subseteq L,$$

which implies $L = \bigcup_{x \in L} W_x$. This in turn implies $L \in \mathcal{O}$.

In Chap. 2, we discussed the notion of an interior system of sets on a set S and the notion of an interior operator. Since $\emptyset$ is an open set in any topological space $(S, \mathcal{O})$ and any union of open sets is an open set, it follows that the topology itself is an interior system on S. In addition, an interior system of open sets is closed to finite intersection. Definition 4.25 which follows is a restatement of the definition of the interior operator associated to an interior system contained by Theorem 1.185.

Definition 4.25 *Let* $(S, \mathcal{O})$ *be a topological space. The* interior *of a set* U, $U \subseteq S$, *is the set* $\mathbf{I}(U) = \bigcup\{L \in \mathcal{O} \mid L \subseteq U\}$.

The interior $\mathbf{I}(U)$ of a set U is the largest open set included in U, because the union of any collection of open sets is an open set. Furthermore, a set is open in a topological space if and only if it equals its interior.

Theorem 4.26 *Let* $(S, \mathcal{O})$ *be a topological space and let* U *be a subset of* S. *The closure* $\mathbf{K}(S - U)$ *of the set* $S - U$ *equals* $S - \mathbf{I}(U)$.

Proof Since $\mathbf{I}(U)$ is an open set, the set $S - \mathbf{I}(U)$ is closed. Note that $S - U \subseteq S - \mathbf{I}(U)$. Therefore, $\mathbf{K}(S - U) \subseteq S - \mathbf{I}(U)$.

Conversely, the inclusion $S - U \subseteq \mathbf{K}(S - U)$ implies $S - \mathbf{K}(S - U) \subseteq U$. Since $S - \mathbf{K}(S - U)$ is an open set included in U and $\mathbf{I}(U)$ is the largest such set, it follows that $S - \mathbf{K}(S - U) \subseteq \mathbf{I}(U)$, which implies $S - \mathbf{I}(U) \subseteq \mathbf{K}(S - U)$.

Corollary 4.27 *For every subset U of a topological space $(S, \mathcal{O})$, we have $\mathbf{I}(U) = S - \mathbf{K}(S - U)$ and $\mathbf{K}(U) = S - \mathbf{I}(S - U)$.*

Proof The first equality is immediate; the second follows from Theorem 4.26 by replacing U by $S - U$.

Theorem 4.28 *The following statements are equivalent for a topological space $(S, \mathcal{O})$:*

(i) *every countable intersection of dense open sets is a dense set;*
(ii) *every countable union of closed sets that have an empty interior has an empty interior.*

Proof (i) implies (ii): Let $H_1, \ldots, H_n, \ldots$ be a sequence of closed sets with $\mathbf{I}(H_i) = \emptyset$ for $n \geqslant 1$. Then, for the open sets L_i given by $L_i = S - H_i$, we have $\mathbf{K}(L_i) = \mathbf{K}(S - H_i) = S - \mathbf{I}(H_i) = S$, so every set L_i is dense. By (i), we have $\mathbf{K}(\bigcap_{i \geqslant 1} L_i) = S$, so

$$\mathbf{I}\left(\bigcup_{i \geqslant 1} H_i\right) = S - \mathbf{K}\left(S - \bigcup_{i \geqslant 1} H_i\right)$$
$$= S - \mathbf{K}\left(\bigcap_{i \geqslant 1} (S - H_i)\right) = S - \mathbf{K}\left(\bigcap_{i \geqslant 1} L_i\right) = \emptyset,$$

which shows that (ii) holds.

(ii) implies (i): this argument is similar to the preceding one and we omit it.

A topological space that satisfies one of the equivalent conditions of this theorem is called a *Baire space*. As we shall see in Chap. 8 (Theorem 8.55), a very important category of topological spaces, the complete topological metric spaces, are Baire spaces.

Definition 4.29 *Let $(S, \mathcal{O})$ be a topological space. The* border *of a set U, where $U \in \mathcal{P}(S)$, is the set $\partial_S K = \mathbf{K}(U) \cap \mathbf{K}(S - U)$.*

If S is clear from the context, then we omit the subscript and denote the border of U just by ∂U.

The border itself is obviously a closed set, as it is an intersection of two closed sets.

Note that, by using Corollary 4.27, the border of a set can be expressed also in term of interiors:

$$\partial U = (S - \mathbf{I}(S - U)) \cap (S - \mathbf{I}(U)) = S - (\mathbf{I}(S - U) \cup \mathbf{I}(U)). \tag{4.2}$$

Theorem 4.30 *The border of a subset U of a topological space $(S, \mathcal{O})$ consists of those elements s of S such that for every open set L that contains s we have both $L \cap U \neq \emptyset$ and $L \cap (S - U) \neq \emptyset$.*

Proof Let $x \in \partial U$ and let L be an open set such that $x \in L$. By Equality (4.2), we have both $x \notin \mathbf{I}(S - U)$ and $x \notin \mathbf{I}(U)$. Therefore, $L \not\subseteq S - U$ and $L \not\subseteq U$, which imply $L \cap U \neq \emptyset$ and $L \cap (S - U) \neq \emptyset$.

Conversely, suppose that, for every open set L that contains s, we have both $L \cap U \neq \emptyset$ and $L \cap (S - U) \neq \emptyset$. This implies $x \notin \mathbf{I}(U)$ and $s \notin \mathbf{I}(S - U)$, so $x \in \partial U$ by Equality (4.2).

Theorem 4.31 *Let $(S, \mathcal{O})$ be a topological space, $(T, \mathcal{O} \restriction_T)$ be a subspace, and W be a subset of S. The border $\partial_T(W \cap T)$ of $W \cap T$ in the subspace T is a subset of the intersection $\partial_S(W) \cap T$, where $\partial_S(W)$ is the border of W in S.*

Proof By Definition 4.29, we have

$$\begin{aligned}\partial_T(W \cap T) =&\mathbf{K}_T(W \cap T) \cap \mathbf{K}_T(T - (W \cap T))\\ =&\mathbf{K}_T(W \cap T) \cap \mathbf{K}_T(T - W)\\ \subseteq&(\mathbf{K}_S(W) \cap T) \cap \mathbf{K}_T(T - W)\\ &\text{(by Corollary 4.18).}\end{aligned}$$

Again, by Corollary 4.18, we have $\mathbf{K}_T(T - W) = \mathbf{K}_T(T \cap (S - W)) \subseteq \mathbf{K}_S(S - W) \cap T$, and this allows us to write

$$\partial_T(W \cap T) \subseteq (\mathbf{K}_S(W) \cap T) \cap \mathbf{K}_S(S - W) \cap T = \partial_S(W) \cap T,$$

which is the desired conclusion.

The next statement relates three important sets that we defined for each subset U of a topological space $(S, \mathcal{O})$.

Theorem 4.32 *Let $(S, \mathcal{O})$ be a topological space. For every subset U of S, we have $\mathbf{K}(U) = \mathbf{I}(U) \cup \partial U$.*

Proof By Equality (4.2), we have $\partial U = (S - \mathbf{I}(S - U)) \cap (S - \mathbf{I}(U))$. Therefore,

$$\begin{aligned}\partial U \cup \mathbf{I}(U) =&(S - \mathbf{I}(S - U)) \cap \mathbf{I}(U)\\ &\text{(by Corollary 4.27)}\\ =&\mathbf{K}(U) \cap \mathbf{I}(U)\\ &\text{(because } \mathbf{I}(U) \subseteq \mathbf{K}(U))\\ =&\mathbf{K}(U).\end{aligned}$$

Corollary 4.33 *Let $(S, \mathcal{O})$ be a topological space and let $(T, \mathcal{O} \restriction_T)$ be a subspace of $(S, \mathcal{O})$. For any subset U of S, we have $\partial_T(U \cap T) \subseteq \partial_S(U)$.*

Proof Let $t \in \partial_T(U \cap T)$. By Theorem 4.30, for every open set $L \in \mathcal{O} \restriction_T$ such that $t \in L$ we have both $L \cap (U \cap T) \neq \emptyset$ and $L \cap (T - (U \cap T)) \neq \emptyset$.

If L_1 is an open set of $(S, \mathcal{O})$ that contains S, then $L_1 \cap T$ is an open set of $(T, \mathcal{O} \restriction_T)$ that contains t, so for L_1 we have both $(L_1 \cap T) \cap (U \cap T) \neq \emptyset$ and $(L_1 \cap T) \cap (T - (U \cap T)) \neq \emptyset$. This immediately implies $L_1 \cap U \neq \emptyset$ and $L_1 \cap (S - U) \neq \emptyset$, that is, $t \in \partial_S(U)$.

Definition 4.34 *Let $(S, \mathcal{O})$ be a topological space. A subset U of S is* clopen *if it is both open and closed.*

Clearly, in every topological space $(S, \mathcal{O})$, both $\emptyset$ and S are clopen sets.

Theorem 4.35 *Let $(S, \mathcal{O})$ be a topological space. A set U is clopen if and only if $\partial U = \emptyset$.*

Proof Suppose that U is clopen. Then $U = \mathbf{K}(U)$; moreover, $S - U$ is also closed (because U is open) and therefore $S - U = \mathbf{K}(S - U)$. Thus, $\mathbf{K}(U) \cap \mathbf{K}(S - U) = U \cap (S - U) = \emptyset$, so $\partial U = \emptyset$.

Conversely, suppose that $\partial U = \emptyset$. Then, since $\mathbf{K}(U) \cap \mathbf{K}(S - U) = \emptyset$, it follows that $\mathbf{K}(U) \subseteq S - \mathbf{K}(S - U)$. Therefore, $\mathbf{K}(U) \subseteq S - (S - U) = U$, which implies $\mathbf{K}(U) = U$. Thus, U is closed. Furthermore, by Equality (4.2), $\partial U = \emptyset$ also implies $\mathbf{I}(S - L) \cup \mathbf{I}(L) = S$, so $S - \mathbf{I}(S - L) \subseteq \mathbf{I}(L)$. By Corollary 4.27, we have $\mathbf{K}(L) \subseteq \mathbf{I}(L)$, so $L \subseteq \mathbf{I}(L)$. Thus, $L = \mathbf{I}(L)$, so L is also an open set.

Definition 4.36 *Let $(S, \mathcal{O})$ be a topological space and let U be a subset of S. An element t of S is an* accumulation point *or a* cluster point *of the set U if, for every open set L such that $t \in L$, the set $U \cap (L - \{t\})$ is not empty.*

The set of all accumulation points of a set U is the derived set *of U and is denoted by U'.*

Lemma 4.37 *Let $(S, \mathcal{O})$ be a topological space and let U be a subset of S. We have $\partial U \subseteq U \cup U'$.*

Proof By Theorem 4.30, if $x \in \partial U$, then for every open set L such that $x \in L$, we have both $L \cap U \neq \emptyset$ and $L \cap (S - U) \neq \emptyset$.

If $U \cap (L - \{x\}) \neq \emptyset$ for every open set L, then $x \in U'$. Otherwise, there is an open set L_0 such that $x_0 \in L$ and $U \cap (L_0 - \{x\}) = \emptyset$. This can happen only if $x \in U$. Therefore, in either case, $x \in U \cup U'$, which gives the desired inclusion.

Theorem 4.38 *Let $(S, \mathcal{O})$ be a topological space and let U be a subset of S. We have $\mathbf{K}(U) = U \cup U'$ for every subset U of S.*

Proof By Theorem 4.32 and Lemma 4.37, we have $\mathbf{K}(U) = \mathbf{I}(U) \cup \partial U \subseteq \mathbf{I}(U) \cup U \cup U' = U \cup U'$ because $\mathbf{I}(U) \subseteq U$.

Let x be an accumulation point of U. If $x \in U$, then clearly $x \in \mathbf{K}(U)$. Otherwise, $x \notin U$ and we claim that in this case $x \in \mathbf{K}(U)$. Indeed, if x were not an element of $\mathbf{K}(U)$, it would belong to the open set $S - \mathbf{K}(U)$. This would imply that the set $U \cap (S - \mathbf{K}(U) - \{x\})$ is not empty, which is a contradiction. This yields the reverse inclusion, $U \cup U' \subseteq \mathbf{K}(U)$.

4.4 Bases

Let $\mathfrak{O} = \{\mathcal{O}_i \mid i \in I\}$ be a family of topologies defined on a set S that contains the discrete topology $\mathcal{P}(S)$. We claim that $\mathfrak{O}$ is a closure system on $\mathcal{P}(S)$. The first condition of Definition 1.166 is satisfied due to the definition of $\mathfrak{O}$. It is easy to verify that for every subfamily $\mathfrak{O}'$ of $\mathfrak{O}$, $\bigcap \mathfrak{O}'$ is a topology, so $\mathfrak{O}$ is indeed a closure system.

Thus, if $\mathcal{S}$ is a family of subsets of S, there exists the smallest topology that includes $\mathcal{S}$.

Theorem 4.39 *The topology* $\mathsf{TOP}(\mathcal{S})$ *generated by a family* $\mathcal{S}$ *of subsets of* S *consists of unions of finite intersections of the members of* $\mathcal{S}$.

Proof Let $\mathcal{E}$ be the collection of all unions of finite intersections of the members of $\mathcal{S}$. It is clear that $\mathcal{S} \subseteq \mathcal{E}$. We claim that $\mathcal{E}$ is a topology that contains $\mathcal{S}$.

Note that the intersection of the empty collection of sets in $\mathcal{S}$ is S, so $S \in \mathcal{E}$; also, the union of an empty collection of finite intersections is $\emptyset$, so $\emptyset \in \mathcal{E}$.

Every $U \in \mathcal{E}$ can be written as $U = \bigcup\{V_j \mid j \in J_U\}$, where the sets V_j are finite intersections of sets of $\mathcal{S}$. Therefore, it is immediate that any union of sets of this form belongs to $\mathcal{E}$.

Suppose that $\{U_i \mid i \in I\}$ is a finite collection of parts of S, where $U_i = \bigcup\{V_j \in \mathcal{S} \mid j \in J_i\}$ and that each V_j can be written as $V_j = \bigcap\{W_{jh} \in \mathcal{S} \mid h \in H_j\}$, where each set H_j is finite. One can prove by induction on $p = |I|$ that $\bigcap\{U_i \mid i \in I\} \in \mathcal{E}$. To simplify the presentation, we discuss here only the case where $|I| = 2$. So, if $U_i = \bigcup\{V_j \in \mathcal{S} \mid j \in J_i\}$ for $i = 1, 2$, we have

$$\begin{aligned} U_1 \cap U_2 &= \bigcup\{V_{j_1} \in \mathcal{S} \mid j_1 \in J_1\} \cap \bigcup\{V_{j_2} \in \mathcal{S} \mid j_2 \in J_2\} \\ &= \bigcap_{j_1, j_2} (V_{j_1} \cap V_{j_2}). \end{aligned}$$

Since each intersection $V_{j_1} \cap V_{j_2}$ is in turn a finite intersection of sets of $\mathcal{S}$, it follows that $U_1 \cap U_2 \in \mathcal{S}$.

Thus, $\mathsf{TOP}(\mathcal{S})$ is contained in $\mathcal{E}$ because $\mathsf{TOP}(\mathcal{S})$ is the coarsest topology that contains $\mathcal{S}$. This gives the desired conclusion.

Corollary 4.40 *Let* $\mathcal{B}$ *be a collection of subsets of the set* S *such that for every finite subcollection* $\mathcal{D}$ *of* $\mathcal{B}$, $x \in \bigcap \mathcal{D}$ *implies the existence of a set* $B \in \mathcal{B}$ *such that* $x \in B \subseteq \bigcap \mathcal{D}$. *Then,* $\mathsf{TOP}(\mathcal{B})$, *the topology generated by* $\mathcal{B}$, *consists of sets that are unions of subcollections of* $\mathcal{B}$.

Proof By Theorem 4.39, $\mathsf{TOP}(\mathcal{B})$ consists of unions of finite intersections of the members of $\mathcal{B}$. Therefore, unions of sets of $\mathcal{B}$ belong to $\mathsf{TOP}(\mathcal{B})$.

Conversely, let $U \in \mathcal{B}$, that is, $U = \bigcup\{V_i \mid i \in I\}$, where each V_i is a finite intersection of members of $\mathcal{B}$. For every $x \in V_i$, there exists a set $B_{x,i} \in \mathcal{B}$ such that $x \in B_{x,i} \subseteq V_i$. Therefore, $V_i = \bigcup_{x \in V_i} B_{x,i}$, and this implies that U is indeed a union of sets from $\mathcal{B}$.

Definition 4.41 *Let S be a set. A collection $\mathcal{S}$ is a* subbasis *for a topology $\mathcal{O}$ if $\mathcal{O} = \mathsf{TOP}(\mathcal{S})$.*

A collection $\mathcal{B}$ of subsets is a basis *for a topology if, for every finite subcollection $\mathcal{D}$ of $\mathcal{B}$, if $x \in \bigcap \mathcal{D}$, then there exists a set $B \in \mathcal{B}$ such that $x \in B \subseteq \bigcap \mathcal{D}$.*

Corollary 4.40 implies that, for a basis $\mathcal{B}$, we have $\bigcup \mathcal{B} = S$. Indeed, consider the intersection of the empty collection of parts of $\mathcal{B}$, which equals S. Then, for every $x \in S$, there is a set $B \in \mathcal{B}$ such that $x \in B \subseteq S$, which of course implies $\bigcup \mathcal{B} = S$.

Clearly, every set of $\mathcal{B}$ is an open set in the topological space $(S, \mathsf{TOP}(\mathcal{B}))$.

Starting from a topology, we find a basis using the following theorem.

Theorem 4.42 *Let $(S, \mathcal{O})$ be a topological space. If $\mathcal{B}$ is a collection of open subsets of S such that for every $x \in S$ and every open set $L \in \mathcal{O}$ there exists a set $B \in \mathcal{B}$ such that $x \in B \subseteq L$, then $\mathcal{B}$ is a basis for $(S, \mathcal{O})$.*

Proof This statement is an immediate consequence of Definition 4.41.

Theorem 4.43 *Let $(S, \mathcal{O})$ be a topological space. The following statements involving a family $\mathcal{B}$ of subsets of S are equivalent:*

(i) *$\mathcal{B}$ is a basis for $(S, \mathcal{O})$;*
(ii) *for every $x \in S$ and $U \in \mathsf{neigh}_x(\mathcal{O})$, there exists $B \in \mathcal{B}$ such that $x \in B \subseteq U$;*
(iii) *for every open set L, there is a subcollection $\mathcal{C}$ of $\mathcal{B}$ such that $L = \bigcup \mathcal{C}$.*

Proof (i) implies (ii): Let $\mathcal{B}$ be a basis for $(S, \mathcal{O})$ and let $U \in \mathsf{neigh}_x(\mathcal{O})$. There exists an open set L such that $x \in L \subseteq U$. Since $\mathcal{B}$ is a basis, there exists a set $B \in \mathcal{B}$ such that $x \in B \subseteq L \subseteq U$, which is what we aimed to prove.

(ii) implies (iii): Suppose that the second statement holds, and let L be an open set. Since L is a neighborhood for all its elements, for every $x \in L$ there exists $B_x \in \mathcal{B}$ such that $\{x\} \subseteq B_x \subseteq L$. Therefore, $L = \bigcup\{B_x \mid x \in L\}$.

(iii) implies (i): Part (iii) implies Part (i) immediately.

Corollary 4.44 *Let U be a subspace of a topological space $(S, \mathcal{O})$. If $\mathcal{B}$ is a basis of $(S, \mathcal{O})$, then $\mathcal{B}_U = \{U \cap B \mid B \in \mathcal{B}\}$ is a basis of the subspace U.*

Proof Let K be an open subset in the subspace U. There is an open set L in $(S, \mathcal{O})$ such that $K = U \cap L$. Since $\mathcal{B}$ is a basis for $(S, \mathcal{O})$, by the third part of Theorem 4.43, there is a subcollection $\mathcal{C}$ of $\mathcal{B}$ such that $L = \bigcup \mathcal{C}$, which implies $K = \bigcup\{U \cap C \mid C \in \mathcal{C}\}$. Thus, $\mathcal{B}_U$ is a basis for U.

Example 4.45 The collection of open intervals $\{(a, b) \mid a, b \in \mathbb{R} \text{ and } a < b\}$ is a basis for the topological space $(\mathbb{R}, \mathcal{O})$ by Theorem 4.11. Further, the collection $\mathcal{S} = \{(a, +\infty) \mid a \in \mathbb{R}\} \cup \{(-\infty, b) \mid b \in \mathbb{R}\}$ is a subbasis of this topology because every member (a, b) of the basis can be written as $(a, b) = (-\infty, b) \cap (a, +\infty)$.

Definition 4.46 *A topological space satisfies the* first axiom of countability *if for every $x \in S$ there is a countable family of open sets $\mathcal{L}_x = \{L_n \mid n \in \mathbb{N}\}$ such that $x \in \bigcap\{L_n \mid n \in \mathbb{N}\}$ and for every open set L that contains x there is a set $L_n \in \mathcal{L}_x$ such that $L_n \subseteq L$.*

A topological space satisfies the second axiom of countability *if it has a countable basis.*

It is clear that the second axiom of countability implies the first, and we will deal mostly with this second axiom. Furthermore, by Corollary 4.44, every subspace of a topological space that satisfies the second axiom of countability satisfies this axiom itself.

Theorem 4.47 *Let $(S, \mathcal{O})$ be a topological space. If $(S, \mathcal{O})$ has a countable basis, then $(S, \mathcal{O})$ is separable.*

Proof Let $\{B_n \mid n \in \mathbb{N}\}$ be a countable basis for $(S, \mathcal{O})$ and let x_n be an element of B_n for $n \in \mathbb{N}$. We claim that $S = \mathbf{K}(\{x_n \mid n \in \mathbb{N}\})$, which is equivalent to $S - \mathbf{K}(\{x_n \mid n \in \mathbb{N}\}) = \emptyset$.

Indeed, observe that $S - \mathbf{K}(\{x_n \mid n \in \mathbb{N}\})$ is a non-empty open set; therefore, there exists $m \in \mathbb{N}$ such that $B_m \subseteq S - \mathbf{K}(\{x_n \mid n \in \mathbb{N}\})$, so $x_m \in S - \mathbf{K}(\{x_n \mid n \in \mathbb{N}\}) \subseteq S - \{x_n \mid n \in \mathbb{N}\}$, which is a contradiction. Therefore, the countable set $\{x_n \mid n \in \mathbb{N}\}$ is dense in $(S, \mathcal{O})$.

The notion of an open cover of a topological space is introduced next.

Definition 4.48 *A* cover *of a topological space $(S, \mathcal{O})$ is a collection of sets $\mathcal{C}$ such that $\bigcup \mathcal{C} = S$.*

If $\mathcal{C}$ is a cover of $(S, \mathcal{O})$ and every set $C \in \mathcal{C}$ is open (closed), then we refer to $\mathcal{C}$ as an open cover *(a* closed cover, *respectively).*

A subcover of an open cover $\mathcal{C}$ *is a collection $\mathcal{D}$ such that $\mathcal{D} \subseteq \mathcal{C}$ and $\bigcup \mathcal{D} = S$.*

Theorem 4.49 *If a topological space $(S, \mathcal{O})$ satisfies the second axiom of countability, then every basis $\mathcal{B}$ for $(S, \mathcal{O})$ contains a countable collection $\mathcal{B}_0$ that is a basis for $(S, \mathcal{O})$.*

Proof Let $\mathcal{B}' = \{L_i \mid i \in \mathbb{N}\}$ be a countable basis for $(S, \mathcal{O})$ and let $\mathcal{C}_i$ be the subcollection of $\mathcal{B}$ defined by $\mathcal{C}_i = \{V \in \mathcal{B} \mid V \subseteq L_i\}$ for $i \in \mathbb{N}$. Since $\mathcal{B}$ is a basis for $(S, \mathcal{O})$, it is clear that $\mathcal{C}_i$ is an open cover for L_i; that is, $\bigcup \mathcal{C}_i = L_i$ for every $i \in \mathbb{N}$. Since each subspace L_i has a countable basis, $\mathcal{C}_i$ contains a countable subcover $\mathcal{C}'_i$ of L_i. The collection $\mathcal{B}_0 = \bigcup\{\mathcal{C}'_i \mid i \in \mathbb{N}\}$ is countable and is a basis for $(S, \mathcal{O})$ that is included in $\mathcal{B}$.

Corollary 4.50 *If a topological space $(S, \mathcal{O})$ has a countable basis, then every open cover of $(S, \mathcal{O})$ contains a countable subcover.*

Proof This fact follows directly from Theorem 4.49.

4.5 Compactness

Definition 4.51 *A topological space* $(S, \mathcal{O})$ *is* compact *if every open cover* $\mathcal{C}$ *of this space contains a finite subcover.*

Another useful concept is the notion of a family of sets with the finite intersection property.

Definition 4.52 *A collection* $\mathcal{C}$ *of subsets of a set* S *has the* finite intersection property *(f.i.p.) if* $\bigcap \mathcal{D} \neq \emptyset$ *for every finite subcollection* $\mathcal{D}$ *of* $\mathcal{C}$.

Theorem 4.53 *The following three statements concerning a topological space* $(S, \mathcal{O})$ *are equivalent:*

(i) $(S, \mathcal{O})$ *is compact;*
(ii) *if* $\mathcal{D}$ *is a family of closed subsets of* S *such that* $\bigcap \mathcal{D} = \emptyset$*, then there exists a finite subfamily* $\mathcal{D}_0$ *of* $\mathcal{D}$ *such that* $\bigcap \mathcal{D}_0 = \emptyset$*;*
(iii) *if* $\mathcal{E}$ *is a family of closed sets having the f.i.p., then* $\bigcap \mathcal{E} \neq \emptyset$.

Proof The argument is left to the reader.

Another characterization of compactness that is just a variant of Part (iii) of Theorem 4.53 that applies to an arbitrary family of sets (not necessarily closed) is given next.

Theorem 4.54 *A topological space* $(S, \mathcal{O})$ *is compact if and only if for every family of subsets* $\mathcal{C}$ *that has the f.i.p.,* $\bigcap\{\mathbf{K}(C) \mid C \in \mathcal{C}\} \neq \emptyset$.

Proof If for every family of subsets $\mathcal{C}$ that has the f.i.p. we have $\bigcap\{\mathbf{K}(C) \mid C \in \mathcal{C}\} \neq \emptyset$, then, in particular, if $\mathcal{C}$ consists of closed sets, it follows that $\bigcap\{C \mid C \in \mathcal{C}\} \neq \emptyset$, which amounts to Part (iii) of Theorem 4.53, so $(S, \mathcal{O})$ is compact.

Conversely, suppose that the space $(S, \mathcal{O})$ is compact, which means that the property of Part (iii) of Theorem 4.53 holds. Suppose that $\mathcal{C}$ is an arbitrary collection of subsets of S that has the f.i.p. Then, the collection of closed subsets $\{\mathbf{K}(C) \mid C \in \mathcal{C}\}$ also has the f.i.p. because $C \in \mathbf{K}(\mathcal{C})$ for every $C \in \mathcal{C}$. Therefore, $\bigcap\{\mathbf{K}(C) \mid C \in \mathcal{C}\} \neq \emptyset$.

Example 4.55 Let $U_1 \supseteq U_2 \supseteq \cdots$ be a descending sequence of non-empty closed subsets of a compact space $(S, \mathcal{O})$. Its intersection $\bigcap_{n \geqslant 1} U_n$ is non-empty because $(S, \mathcal{O})$ is compact and $\bigcap_{p=1}^{k} U_{i_p} = U_l \neq \emptyset$, where $l = \min\{i_1, \ldots, i_k\}$ for every $k \geqslant 1$.

This implies that the topological space $(\mathbb{R}, \mathcal{O})$ introduced in Example 4.4 is not compact because $\bigcap_{n \geqslant 1}[n, \infty) = \emptyset$.

The notion of cover refinement can be used to characterize compact topological spaces. Recall that we introduced this notion in Definition 1.12.

Theorem 4.56 *A topological space $(S, \mathcal{O})$ is compact if and only if every open cover $\mathcal{C}$ is refined by some finite open cover of the space.*

Proof Suppose that $(S, \mathcal{O})$ is compact. Then, every open cover $\mathcal{C}$ contains a finite subcover $\mathcal{C}'$. Since every $C' \in \mathcal{C}'$ is a member of $\mathcal{C}$, it follows that $\mathcal{C}$ is refined by $\mathcal{C}'$.

Conversely, suppose that every open cover $\mathcal{C}$ is refined by some finite open cover $\mathcal{D} = \{D_1, \ldots, D_p\}$. Then, for every $D_i \in \mathcal{D}$ there exists a set $C_i \in \mathcal{C}$ such that $D_i \subseteq C_i$ for $1 \leqslant i \leqslant p$. Since $\bigcup_{i=1}^n D_i = S$, it follows that $\bigcup_{i=1}^n C_i = S$, so $\{C_1, \ldots, C_n\}$ is a finite subcover of $\mathcal{C}$, which means that $(S, \mathcal{O})$ is compact.

If $(T, \mathcal{O} \restriction_T)$ is a compact topological space, then we say that T is a *compact set*.

Example 4.57 Every closed interval $[x, y]$ of $\mathbb{R}$ is a compact set. Indeed, if $\mathcal{C}$ is an open cover of $[x, y]$ we can assume without loss of generality that $\mathcal{C}$ consists of open intervals $\mathcal{C} = \{(a_i, b_i) \mid i \in I\}$.

Let

$$K = \left\{ c \middle| c \in [x, y] \text{ and } [x, c] \subseteq \bigcup_{j \in J} (a_j, b_j) \text{ for some finite } J \subseteq I \right\}.$$

Observe that $K \neq \emptyset$ because $x \in K$. Indeed, we have $[x, x] = \{x\}$ and therefore $[x, x] \subseteq (a_i, b_i)$ for some $i \in I$.

We claim that $y \leqslant w = \sup K$. It is clear that $w \leqslant y$ because y is an upper bound of $[x, y]$ and therefore an upper bound of K. Suppose that $w < y$. Note that in this case there exists an open interval (a_p, b_p) for some $p \in I$ such that $w \in (a_p, b_p)$. By Theorem 2.28, for every $\epsilon > 0$, there is $z \in K$ such that $\sup K - \epsilon < z$. Choose ϵ such that $\epsilon < w - a_p$. Since the closed interval $[x, z]$ is covered by a finite collection of open intervals $[x, z] \subseteq (a_{j_1}, b_{j_1}) \cup \cdots \cup (a_{j_r}, b_{j_r})$, it follows that the interval $[x, w]$ is covered by $(a_{j_1}, b_{j_1}) \cup \cdots \cup (a_{j_r}, b_{j_r}) \cup (a_p, b_p)$. This leads to a contradiction because the open interval (a_p, b_p) contains numbers in K that are greater than w. So we have $w = y$, which shows that $[x, y]$ can be covered by a finite family of open intervals extracted from $\mathcal{C}$.

Example 4.58 The open interval $(0, 1)$ is not compact. Indeed, it is easy to see that the collection of open sets $\left\{\left(\frac{1}{n}, 1 - \frac{1}{n}\right)\right\}$ is an open cover of $(0, 1)$. However, no finite sub-collection of this collection of sets is an open cover of $(0, 1)$.

Example 4.58 suggests the interest of the following definition.

Definition 4.59 *A subset T of a topological space is* relatively compact *if its closure $\mathbf{K}(T)$ is compact.*

Example 4.60 The set $(0, 1)$ is a relatively compact subset of $\mathbb{R}$ but not a compact one.

Theorem 4.61 *If $(S, \mathcal{O})$ is a compact topological space, any closed subset T of S is compact.*

Proof Let T be a closed subset of $(S, \mathcal{O})$. We need to show that the subspace $(T, \mathcal{O} \upharpoonright_T)$ is compact. Let $\mathcal{C}$ be an open cover of the space $(T, \mathcal{O} \upharpoonright_T)$. Then, $\mathcal{C} \cup \{S - T\}$ is a open cover of $(S, \mathcal{O})$. The compactness of $(S, \mathcal{O})$ means that there exists a finite subcover $\mathcal{D}$ of $(S, \mathcal{O})$ such that $\mathcal{D} \subseteq \mathcal{C} \cup \{S - T\}$. It follows immediately that $\mathcal{D} - \{S - T\}$ is a finite subcover of $\mathcal{C}$ for $(T, \mathcal{O} \upharpoonright_T)$.

A topological space $(S, \mathcal{O})$ is *locally compact* if for every $x \in S$ there exists an open set $L \in \mathcal{O}$ such that $x \in L$ and $\mathbf{K}(L)$ is a compact set.

Theorem 4.62 *If $(S, \mathcal{O})$ is a compact topological space, then, for every infinite subset U of S we have $U' \neq \emptyset$ (the* Bolzano-Weierstrass property*).*

Proof Let $U = \{x_i \mid i \in I\}$ be an infinite subset of S. Suppose that U has no accumulation point. For every $s \in S$, there is an open set L_s such that $s \in L_s$ and $U \cap (L_s - \{s\}) = \emptyset$. Clearly the collection $\{L_s \mid s \in S\}$ is an open cover of S, so it contains a finite subcover $\{L_{s_1}, \ldots, L_{s_p}\}$. Thus, $S = L_{s_1} \cup \cdots \cup L_{s_p}$. Note that each L_{s_i} contains at most one element of U (which happens when $s_i \in U$), which implies that U is finite. This contradiction means that $U' \neq \emptyset$.

4.6 Continuous Functions

The notion of continuous function which is central to topology is introduced next.

Definition 4.63 *Let $(S_1, \mathcal{O}_1)$ and $(S_2, \mathcal{O}_2)$ be two topological spaces. A function $f : S_1 \longrightarrow S_2$ is* continuous *if, for every open set $V \in \mathcal{O}_2$, we have $f^{-1}(V) \in \mathcal{O}_1$.*

If $f : S_1 \longrightarrow S_2$ is a continuous function between the topological spaces $(S_1, \mathcal{O}_1)$ and $(S_2, \mathcal{O}_2)$ and $\mathcal{O}'_1$ and $\mathcal{O}'_2$ are topologies on S_1 and S_2, respectively, such that $\mathcal{O}'_2 \subseteq \mathcal{O}_2$ and $\mathcal{O}_1 \subseteq \mathcal{O}'_1$, then f is also a continuous function between the topological spaces $(S_1, \mathcal{O}'_1)$ and $(S_2, \mathcal{O}'_2)$. Therefore, any function defined on the topological space $(S, \mathcal{P}(S))$ (equipped with the discrete topology) with values in an arbitrary topological space $(S', \mathcal{O}')$ is continuous; similarly, any function $f : S \longrightarrow S'$ between a topological space $(S, \mathcal{O})$ and $(S', \{\emptyset, S'\})$ (equipped with the discrete topology) is continuous.

Theorem 4.64 *Let $(S, \mathcal{O})$, $(T, \mathcal{O}')$, and $(U, \mathcal{O}'')$ be three topological spaces and let $f : S \longrightarrow T$ and $g : T \longrightarrow U$ be two continuous functions. Then, the function $gf : S \longrightarrow T$ is continuous.*

Proof This statement is an immediate consequence of Definition 4.63 and Theorem 1.61.

Several equivalent characterizations of continuous functions are given next.

Theorem 4.65 *Let $(S, \mathcal{O})$ and $(T, \mathcal{O}')$ be two topological spaces and let $f : S \longrightarrow T$ be a function. The following statements are equivalent:*

(i) *f is continuous;*
(ii) *for every closed set L, $L \subseteq T$, the set $f^{-1}(L)$ is a closed set in $(S, \mathcal{O})$;*
(iii) *$f(\mathbf{K}_1(H)) \subseteq \mathbf{K}_2(f(H))$ for every $H \subseteq S$, where $\mathbf{K}_1$ and $\mathbf{K}_2$ are the closure operators of the topological spaces $(S, \mathcal{O})$ and $(T, \mathcal{O}')$, respectively;*
(iv) *for every $x \in S$ and $V \in \mathsf{neigh}_{f(x)}(\mathcal{O}')$, there exists $U \in \mathsf{neigh}_x(\mathcal{O})$ such that $f(U) \subseteq V$.*

Proof To prove that (i) implies (ii), let f be a continuous function and let C be the open set given by $C = T - L$. By (i), $f^{-1}(C)$ is open in $(S, \mathcal{O})$ and therefore $S - f^{-1}(C)$ is closed in $(S, \mathcal{O})$. Since

$$S - f^{-1}(C) = S - f^{-1}(T - L) = f^{-1}(L)$$

(see Exercise 21), we have shown the desired implication.

To prove that (ii) implies (iii), we start from the fact that $H \subseteq f^{-1}(f(H))$. Therefore, $H \subseteq f^{-1}(\mathbf{K}_2(f(H)))$. Since $\mathbf{K}_2(f(H))$ is closed, it follows that $f^{-1}(\mathbf{K}_2(f(H)))$ is also closed. Thus, $\mathbf{K}_1(H) \subseteq f^{-1}(\mathbf{K}_2(f(H)))$.

We now show that (iii) implies (iv). Let V be a neighborhood of $f(x)$ in $(T, \mathcal{O}')$ and let W be an open set such that $f(x) \in W \subseteq V$. Define the set $U \subseteq S$ as $U = S - f^{-1}(T - W)$. Since $f(x) \in W$, $f(x) \notin T - W$, $x \notin f^{-1}(T - W)$ and therefore $x \in U$.

By (iii), we have

$$f\left(\mathbf{K}_1(f^{-1}(T - W))\right) \subseteq \mathbf{K}_2(f(f^{-}1(T - W))) \subseteq \mathbf{K}_2(T - W) = T - W,$$

because $T - W$ is a closed set. Consequently, $\mathbf{K}_1(f^{-1}(T - W)) \subseteq f^{-1}(T - W)$, so $\mathbf{K}_1(f^{-1}(T - W)) = f^{-1}(T - W)$, which implies that $f^{-1}(T - W)$ is a closed set. This means that U is an open set, and hence it is a neighborhood of x. Then, $f(U) = f\left(S - f^{-1}(T - W)\right) = f(f^{-1}(W)) \subseteq W$.

Finally, to show that (iv) implies (i), let V be an open set in $(T, \mathcal{O}')$ and $x \in f^{-1}(V)$, so $f(x) \in V$. Since V is open, it is a neighborhood of $f(x)$, so by (iv) there exists $U \in \mathsf{neigh}_x(\mathcal{O})$ such that $f(U) \subseteq V$, which implies $U \subseteq f^{-1}(V)$ and $f^{-1}(V)$ is a neighborhood of x. By Theorem 4.24, $f^{-1}(V)$ is open so f is continuous.

Definition 4.66 *Let $(S, \mathcal{O})$ and $(T, \mathcal{O}')$ be two topological spaces. A bijection $f : S \longrightarrow T$ is a* homeomorphism *if both f and its inverse f^{-1} are continuous functions.*

If a homeomorphism exists between the topological spaces $(S, \mathcal{O})$ and $(S, \mathcal{O}')$, we say that these spaces are homeomorphic.

Theorem 4.67 *A bijection $f : S \longrightarrow T$ between two topological spaces $(S, \mathcal{O})$ and $(T, \mathcal{O}')$ is a homeomorphism if and only if $U \in \mathcal{O}$ is equivalent to $f(U) \in \mathcal{O}'$.*

Proof Suppose that f is a homeomorphism between $(S, \mathcal{O})$ and $(T, \mathcal{O}')$. If $U \in \mathcal{O}$ the continuity of f^{-1} implies that $(f^{-1})^{-1}(U) = f(U) \in \mathcal{O}'$; on the other hand, if $f(U) \in \mathcal{O}'$, then, since $U = f^{-1}(f(U))$, the continuity of f yields $U \in \mathcal{O}$.

Conversely, suppose that for the bijection $f : S \longrightarrow T$, $U \in \mathcal{O}$ if and only if $f(U) \in \mathcal{O}'$. Suppose that $V \in \mathcal{O}'$; since f is a bijection, there is $W \subseteq S$ such that $V = f(W)$ and $W \in \mathcal{O}$ by hypothesis. Observe that $f^{-1}(V) = W$, so f is continuous. To prove that f^{-1} is continuous, note that we need to verify that $(f^{-1})^{-1}(Z)$ is an open set in $(S, \mathcal{O})$ for any set $Z \in \mathcal{O}'$, which is effectively the case because $(f^{-1})^{-1}(Z) = f(Z)$.

Any property of $(S, \mathcal{O})$ that can be expressed using the open sets of this topological space is preserved in topological spaces $(T, \mathcal{O}')$ that are homeomorphic to $(S, \mathcal{O})$. Therefore, such a property is said to be *topological*.

The collection of all pairs of topological spaces that are homeomorphic is an equivalence relation on the class of topological spaces as can be easily shown.

Example 4.68 We prove that all open intervals of $\mathbb{R}$, bounded or not, are homeomorphic.

Let (a, b) and (c, d) be two bounded intervals of $\mathbb{R}$ and let $f : (a, b) \longrightarrow (c, d)$ be the linear function defined by $f(x) = px + q$, where $p = \frac{d-c}{b-a}$ and $q = \frac{bc-ad}{b-a}$. It is easy to verify that f is a homeomorphism, so any two bounded intervals of $\mathbb{R}$ are homeomorphic; in particular, any bounded interval (a, b) is homeomorphic with $(0, 1)$.

Any two unbounded intervals (a, ∞) and (b, ∞) are homeomorphic; the mapping $g(x) = \frac{b}{a}x$ is a homeomorphism between these sets. Similarly, any two unbounded intervals of the form $(-\infty, a)$ and $(-\infty, b)$ are homeomorphic, and so are (a, ∞) and $(-\infty, b)$.

The function $h : (0, 1) \longrightarrow (0, \infty)$ defined by $h(x) = \tan \frac{\pi x}{2}$ is a homeomorphism, whose inverse mapping is $h^{-1}(x) = \frac{2}{\pi} \arctan x$ so $(0, 1)$ is homeomorphic with $(0, \infty)$. Finally, $(-1, 1)$ is homeomorphic to $(-\infty, \infty)$ since the mapping $h_1 : (-1, 1) \longrightarrow (-\infty, \infty)$ defined by $h(x) = \tan \frac{\pi x}{2}$ for $x \in (-1, 1)$ is a homeomorphism.

Compactness is preserved by continuous functions as we show next.

Theorem 4.69 *Let $(S, \mathcal{O})$ and $(T, \mathcal{O}')$ be two topological spaces and let $f : S \longrightarrow T$ be a continuous function. If $(S, \mathcal{O})$ is compact, then $f(S)$ is compact in $(T, \mathcal{O}')$.*

Proof Let $\mathcal{D} = \{D_i \mid i \in I\}$ be an open cover of $f(S)$. Then $f^{-1}(D_i)$ is an open set in $(S, \mathcal{O})$ because f is continuous and the collection $\mathcal{C} = \{f^{-1}(D_i) \mid i \in I\}$ is an open cover of S. Since $(S, \mathcal{O})$ is compact, there exists a finite subcover $\mathcal{C}_1 = \{f^{-1}(D_i) \mid i \in I_1\}$ of S (I_1 is a finite subset of I). Since $S = \bigcup\{f^{-1}(D_i) \mid i \in I_1\}$, we have

$$
\begin{aligned}
f(S) &= f\left(\bigcup\{f^{-1}(D_i) \mid i \in I_1\}\right) \\
&= \bigcup\{f(f^{-1}(D_i)) \mid i \in I_1\} = \bigcup\{D_i \mid i \in I_1\},
\end{aligned}
$$

which shows that $\mathcal{D}$ contains a finite subcover of $f(S)$.

Using the notion of the neighborhood of an element, it is possible to localize the notion of continuity.

Definition 4.70 *Let $(S, \mathcal{O})$ and $(T, \mathcal{O}')$ be two topological spaces. A function $f : S \longrightarrow T$ is* continuous at s, *where $s \in S$, if for every neighborhood V of $f(s)$ there exists a neighborhood U of s such that $f(U) \subseteq V$.*

Theorem 4.71 *Let $(S, \mathcal{O})$ and $(T, \mathcal{O}')$ be two topological spaces. A function $f : S \longrightarrow T$ is continuous if and only if it is continuous at every element s of S.*

Proof This statement follows immediately from Definition 4.71 and from the last part of Theorem 4.65.

4.7 Connected Topological Spaces

We now discuss a formalization of the notion of a "one-piece" topological space.

Theorem 4.72 *Let $(S, \mathcal{O})$ be a topological space. The following statements are equivalent:*

(i) *there exists a clopen subset K of S such that $K \notin \{\emptyset, S\}$;*
(ii) *there exist two non-empty open subsets L, L' of S that are complementary;*
(iii) *there exist two non-empty closed subsets H, H' of S that are complementary.*

Proof (i) implies (ii): If K is clopen and $K \notin \{\emptyset, S\}$, then both K and $\bar{K}$ are non-empty open sets.

(ii) implies (iii): Suppose that L and L' are two non-empty complementary open subsets of S. Then, L and L' are in the same time closed because the complements of each set is open.

(iii) implies (i): If H and H' are complementary closed sets, then each of them is also open because the complements of each set is closed. Thus, both sets are clopen.

Definition 4.73 *A topological space $(S, \mathcal{O})$ is* disconnected *if it satisfies any of the equivalent conditions of Theorem 4.72. Otherwise, $(S, \mathcal{O})$ is said to be* connected.

A subset T of a connected topological space is connected *if the subspace T is connected.*

Theorem 4.74 *Let T be a subset of S, where $(S, \mathcal{O})$ is a topological space. The following statements are equivalent:*

(i) *T is connected;*
(ii) *there are no open sets L_1, L_2 in $(S, \mathcal{O})$ such that $T \subseteq L_1 \cup L_2$, and $T \cap L_1$, and $T \cap L_2$ are non-empty and disjoint;*

(iii) *there are no closed sets H_1, H_2 in $(S, \mathcal{O})$ such that $T \subseteq H_1 \cup H_2$, and $T \cap H_1$ and $T \cap H_2$ are non-empty and disjoint;*
(iv) *there is no clopen set in $(S, \mathcal{O})$ that has a non-empty intersection with T.*

Proof The equivalence of the statements follows immediately from the definition of the subspace topology.

Theorem 4.75 *Let $\mathcal{C} = \{C_i \mid i \in I\}$ be a family of connected subsets of a topological space $(S, \mathcal{O})$. If $C_i \cap C_j \neq \emptyset$ for every $i, j \in I$ such that $i \neq j$, then $\bigcup \mathcal{C}$ is connected.*

Proof Suppose that $C = \bigcup \mathcal{C}$ is not connected. Then C contains two complementary open subsets L' and L''. For every $i \in I$, the sets $C_i \cap L'$ and $C_i \cap L''$ are complementary and open in C_i. Since each C_i is connected, we have either $C_i \cap L' = \emptyset$ or $C_i \cap L'' = \emptyset$ for every $i \in I$. In the first case, $C_i \subseteq L''$, while in the second, $C_i \subseteq L'$. Thus, the collection $\mathcal{C}$ can be partitioned into two subcollections, $\mathcal{C} = \mathcal{C}' \cup \mathcal{C}''$, where $\mathcal{C}' = \{C_i \in \mathcal{C} \mid C_i \subseteq L'\}$ and $\mathcal{C}'' = \{C_i \in \mathcal{C} \mid C_i \subseteq L''\}$. Clearly, two sets $C_i \in \mathcal{C}'$ and $C_j \in \mathcal{C}''$ are disjoint because the sets L' and L'' are disjoint, and this contradicts the hypothesis.

Corollary 4.76 *Let $(S, \mathcal{O})$ be a topological space and let $x \in S$. The collection $\mathcal{C}_x$ of connected subsets of S that contain x has $K_x = \bigcup \mathcal{C}_x$ as its largest element.*

Proof This follows immediately from Theorem 4.75.

We refer to K_x as *the connected component* of x.

Theorem 4.77 *Let T be a connected subset of a topological space $(S, \mathcal{O})$, and suppose that W is a subset of S such that $T \subseteq W \subseteq \mathbf{K}(T)$. Then W is connected.*

Proof Suppose that W is not connected (that is, $W = U \cup U'$, where U and U' are two nonempty, disjoint, and open sets in W). There exist two open sets L, L' in S such that $U = W \cap L$ and $U' = W \cap L'$. Since $T \subseteq W$, the sets $T \cap U$ and $T \cap U'$ are open in T, disjoint, and their union equals T. Thus, we have either $T \cap U = \emptyset$ or $T \cap U' = \emptyset$ because T is connected.

If $T \cap U = \emptyset$, then $T \cap L = (T \cap W) \cap L = T \cap (W \cap L) = T \cap U = \emptyset$, so $T \subseteq \bar{L}$. Since $\bar{L}$ is closed, $\mathbf{K}(T) \subseteq \bar{L}$, which implies $W \subseteq \bar{L}$, which implies $U = W \cap L = \emptyset$. This contradicts the assumption made earlier about U. A similar contradiction follows from $T \cap U' = \emptyset$. Thus, W is connected.

Corollary 4.78 *If T is a connected subset of a topological space $(S, \mathcal{O})$, then $\mathbf{K}(T)$ is also connected.*

Proof This statement is a special case of Theorem 4.76.

Theorem 4.79 *Let $(S, \mathcal{O})$ be a topological space. The collection of all connected components of S is a partition of S that consists of closed sets.*

Proof Corollary 4.78 implies that each connected component K_x is closed. Suppose that K_x and K_y are two connected components that are not disjoint. Then, by Theorem 4.75, $K_x \cup K_y$ is connected. Since $x \in K_x \cup K_y$, it follows that $K_x \cup K_y \subseteq K_x$ because K_x is the maximal connected set that contains x, so $K_y \subseteq K_x$. Similarly, $K_x \subseteq K_y$, so $K_x = K_y$.

Example 4.80 The topological space $(\mathbb{R}, \mathcal{O})$ is connected. Suppose that K is a clopen set in $\mathbb{R}$ distinct from $\mathbb{R}$ and $\emptyset$, and let $x \in \mathbb{R} - K$.

Suppose that the set $K \cap [x, \infty)$ is nonempty. Then, this set is closed and bounded below and therefore has a least element u. Since $K \cap [x, \infty) = K \cap (x, \infty)$ is also open, there exists $\epsilon > 0$ such that $(u - \epsilon, u + \epsilon) \subseteq K \cap [x, \infty)$, which contradicts the fact that u is the least element of $K \cap [x, \infty)$. A similar contradiction is obtained if we assume that $K \cap (-\infty, x] \neq \emptyset$, so $\mathbb{R}$ cannot contain a clopen set distinct from $\mathbb{R}$ or $\emptyset$.

Theorem 4.81 *The image of a connected topological space through a continuous function is a connected set.*

Proof Let $(S_1, \mathcal{O}_1)$ and $(S_2, \mathcal{O}_2)$ be two topological spaces and let $f : S_1 \longrightarrow S_2$ be a continuous function, where S_1 is connected. If $f(S_1)$ were not connected, we would have two nonempty open subsets L and L' of $f(S_1)$ that are complementary. Then, $f^{-1}(L)$ and $f^{-1}(L')$ would be two nonempty, open sets in S_1 which are complementary, which contradicts the fact that S_1 is connected.

A characterization of connected spaces is given next.

Theorem 4.82 *Let $(S, \mathcal{O})$ be a topological space and let $(\{0, 1\}), \mathcal{P}(\{0, 1\})$ be a two-element topological space equipped with the non-discrete topology. Then, S is connected if and only if every continuous application $f : S \longrightarrow \{0, 1\}$ is constant.*

Proof Suppose that S is connected. Both $f^{-1}(0)$ and $f^{-1}(1)$ are clopen sets in S because both $\{0\}$ and $\{1\}$ are clopen in the discrete topology. Thus, we have either $f^{-1}(0) = \emptyset$ and $f^{-1}(1) = S$, or $f^{-1}(0) = S$ and $f^{-1}(1) = \emptyset$. In the first case, f is the constant function $f(x) = 1$; in the second, it is the constant function $f(x) = 0$.

Conversely, suppose that the condition is satisfied for every continuous function $f : S \longrightarrow \{0, 1\}$ and suppose $(S, \mathcal{O})$ is not connected. Then, there exist two nonempty disjoint open subsets L and L' that are complementary. Let $f = 1_L$ be the indicator function of L, which is continuous because both L and L' are open. Thus, f is constant and this implies either $L = \emptyset$ and $L' = S$ or $L = S$ and $L' = \emptyset$, so S is connected.

Example 4.83 Theorem 4.82 allows us to prove that the connected subsets of $\mathbb{R}$ are exactly the intervals.

Suppose that T is a connected subset of S but is not an interval. Then, there are three numbers x, y, z such that $x < y < z$, $x, z \in T$ but $y \notin T$. Define the function

$f : T \longrightarrow \{0, 1\}$ by $f(u) = 0$ if $u < y$ and $f(u) = 1$ if $y < u$. Clearly, f is continuous but is not constant, and this contradicts Theorem 4.82. Thus, T must be an interval.

Suppose now that T is an open interval of $\mathbb{R}$. We saw that T is homeomorphic to $\mathbb{R}$ (see Example 4.68), so T is indeed connected. If T is an arbitrary interval, its interior $\mathbf{I}(T)$ is an open interval and, since $\mathbf{I}(T) \subseteq T \subseteq \mathbf{K}(\mathbf{I}(T))$, it follows that T is connected.

Definition 4.84 *A topological space* $(S, \mathcal{O})$ *is* totally disconnected *if, for every* $x \in S$*, the connected component of* x *is* $K_x = \{x\}$.

Example 4.85 Any topological space equipped with the discrete topology is totally disconnected.

Theorem 4.86 *Let* $(S, \mathcal{O})$ *be a topological space and let* T *be a subset of* S.

If for every pair of distinct points $x, y \in T$ *there exist two disjoint closed sets* H_x *and* H_y *such that* $T \subseteq H_x \cup H_y$, $x \in H_x$, *and* $y \in H_y$, *then* T *is totally disconnected.*

Proof Let K_x be the connected component of x, and suppose that $y \in K_x$ and $y \neq x$, that is, $K_x = K_y = K$. Then, $K \cap H_x$ and $K \cap H_y$ are nonempty disjoint closed sets and $K = (K \cap H_x) \cup (K \cap H_y)$, which contradicts the connectedness of K. Therefore, $K_x = \{x\}$ for every $x \in T$ and T is totally disconnected.

4.8 Separation Hierarchy of Topological Spaces

We introduce a hierarchy of topological spaces that is based on separation properties of these spaces.

Definition 4.87 *Let* $(S, \mathcal{O})$ *be a topological space and let* x *and* y *be two arbitrary, distinct elements of* S. *This topological space is:*

(i) a T_0 space *if there exists* $U \in \mathcal{O}$ *such that one member of the set* $\{x, y\}$ *belongs to* U *and the other to* $S - U$;
(ii) a T_1 space *if there exist* $U, V \in \mathcal{O}$ *such that* $x \in U - V$ *and* $y \in V - U$;
(iii) a T_2 space *or a* Hausdorff space *if there exist* $U, V \in \mathcal{O}$ *such that* $x \in U$ *and* $y \in V$ *and* $U \cap V = \emptyset$;
(iv) a T_3 space *if for every closed set* H *and* $x \in S - H$ *there exist* $U, V \in \mathcal{O}$ *such that* $x \in U$ *and* $H \subseteq V$ *and* $U \cap V = \emptyset$;
(v) a T_4 space *if for all disjoint closed sets* H, L *there exist* $U, V \in \mathcal{O}$ *such that* $H \subseteq U$, $L \subseteq V$, *and* $U \cap V = \emptyset$.

Theorem 4.88 *A topological space* $(S, \mathcal{O})$ *is a* T_1 *space if and only if every singleton* $\{x\}$ *is a closed set.*

Proof Suppose that $(S, \mathcal{O})$ is a T_1, space and for every $y \in S - \{x\}$ let U_y and V_y be two open sets such as $x \in U_y - V_y$ and $y \in V_y - U_y$. Then, $x \in \bigcup_{y \neq x} U_y$ and $x \notin \bigcup_{y \neq x} V_y$, so $y \in \bigcup_{y \neq x} V_y \subseteq S - \{x\}$. Thus, $S - \{x\}$ is an open set, so $\{x\}$ is closed.

Conversely, suppose that each singleton $\{u\}$ is closed. Let $x, y \in S$ be two distinct elements of S. Note that the sets $S - \{x\}$ and $S - \{y\}$ are open and $x \in (S - \{y\}) - (S - \{x\})$ and $y \in (S - \{x\}) - (S - \{y\})$, which shows that $(S, \mathcal{O})$ is a T_1-space.

Theorem 4.89 *Let $(S, \mathcal{O})$ be a T_4-separated topological space. If H is a closed set and L is an open set such that $H \subseteq L$, then there exists an open set U such that $H \subseteq U \subseteq \mathbf{K}(U) \subseteq L$.*

Proof Observe that H and $S - L$ are two disjoint closed sets under the assumptions of the theorem. Since $(S, \mathcal{O})$ is a T_4-separated topological space, there exist $U, V \in \mathcal{O}$ such that $H \subseteq U$, $S - L \subseteq V$ and $U \cap V = \emptyset$. This implies $U \subseteq S - V \subseteq L$. Since $S - V$ is closed, we have

$$H \subseteq U \subseteq \mathbf{K}(U) \subseteq \mathbf{K}(S - V) = S - V \subseteq L,$$

which proves that U satisfies the conditions of the theorem.

The next theorem is in some sense a reciprocal result of Theorem 4.61, which holds in the realm of Hausdorff spaces.

Theorem 4.90 *Each compact subset of a Hausdorff space $(S, \mathcal{O})$ is closed.*

Proof Let H be a compact subset of $(S, \mathcal{O})$ and let y be an element of the set $S - H$. It suffices to show that the set $S - H$ is open. For every $x \in H$, we have two open subsets U_x and V_x such that $x \in U_x$, $y \in V_x$ and $U_x \cap V_x = \emptyset$. The collection $\{U_x \mid x \in H\}$ is an open cover of H and the compactness of H implies the existence of a finite subcover $U_{x_1}, \ldots, U_{x_n}$ of H. Consider the open set $V = \bigcap_{i=1}^{n} V_{x_i}$, which is disjoint from each of the sets $U_{x_1}, \ldots, U_{x_n}$ and, therefore, it is disjoint from H. Thus, for every $y \in S - H$ there exists an open set V such that $y \in V \subseteq S - H$, which implies that $S - H$ is open.

Corollary 4.91 *In a Hausdorff space $(S, \mathcal{O})$, each finite subset is a closed set.*

Proof Since every finite subset of S is compact, the statement follows immediately from Theorem 4.90.

It is clear that every T_2 space is a T_1 space and each T_1 space is a T_0 space. However, this hierarchy does not hold beyond T_2. This requires the introduction of two further classes of topological spaces.

Definition 4.92 *A topological space $(S, \mathcal{O})$ is* regular *if it is both a T_1 and a T_3 space; $(S, \mathcal{O})$ is* normal *if it is both a T_1 and a T_4 space.*

Theorem 4.93 *Every regular topological space is a T_2 space and every normal topological space is a regular one.*

Proof Let $(S, \mathcal{O})$ be a topological space that is regular and let x and y be two distinct points in S. By Theorem 4.88, the singleton $\{y\}$ is a closed set. Since $(S, \mathcal{O})$ is a T_3, space, two open sets U and V exist such that $x \in U$, $\{y\} \subseteq V$, and $U \cap V = \emptyset$, so $(S, \mathcal{O})$ is a T_2 space. We leave the second part of the theorem to the reader.

4.9 Products of Topological Spaces

Theorem 4.94 *Let $\{(S_i, \mathcal{O}_i) \mid i \in I\}$ be a family of topological spaces indexed by the set I. Define on the set $S = \prod_{i \in I} S_i$ the collection of sets $\mathcal{B} = \{\bigcap_{j \in J} p_j^{-1}(L_j) \mid L_j \in \mathcal{O}_j$ and J finite$\}$. Then, $\mathcal{B}$ is a basis.*

Proof Note that every set $\bigcap_{j \in J} p_j^{-1}(L_j)$ has the form $\prod_{i \in I - J} \times \prod_{j \in J} L_j$. We need to observe only that a finite intersection of sets in $\mathcal{B}$ is again a set in $\mathcal{B}$. Therefore, $\mathcal{B}$ is a basis.

Definition 4.95 *The topology $\mathsf{TOP}(\mathcal{B})$ generated on the set S by $\mathcal{B}$ is called the* product of the topologies $\mathcal{O}_i$ *and is denoted by $\prod_{i \in I} \mathcal{O}_i$.*

The topological space $\{(S_i, \mathcal{O}_i) \mid i \in I\}$ is the product of the collection of topological spaces $\{(S_i, \mathcal{O}_i) \mid i \in I\}$.

The product of the topologies $\{\mathcal{O}_i \mid i \in I\}$ can be generated starting from the subbasis $\mathcal{S}$ that consists of sets of the form $D_{j,L} = \{t \mid t \in \prod_{i \in I} \mid t(j) \in L\}$, where $j \in I$ and L is an open set in $(S_j, \mathcal{O}_j)$. It is easy to see that any set in the basis $\mathcal{B}$ is a finite intersection of sets of the form $D_{j,L}$.

Example 4.96 Let $\mathbb{R}^n = \mathbb{R} \times \cdots \times \mathbb{R}$, where the product involves n copies of $\mathbb{R}$ and $n \geqslant 1$. In Example 4.45, we saw that the collection of open intervals $\{(a, b) \mid a, b \in \mathbb{R}$ and $a < b\}$ is a basis for the topological space $(\mathbb{R}, \mathcal{O})$. Therefore, a basis of the topological space $(\mathbb{R}^n, \mathcal{O} \times \cdots \mathcal{O})$ consists of parallelepipeds of the form $(a_1, b_1) \times \cdots \times (a_n, b_n)$, where $a_i < b_i$ for $1 \leqslant i \leqslant n$.

Theorem 4.97 *Let $\{(S_i, \mathcal{O}_i) \mid i \in I\}$ be a collection of topological spaces. Each projection $p_\ell : \prod_{i \in I} S_i \longrightarrow S_\ell$ is a continuous function for $\ell \in I$. Moreover, the product topology is the coarsest topology on S such that projections are continuous.*

Proof Let L be an open set in $(S_\ell, \mathcal{O}_\ell)$. We have

$$p_\ell^{-1}(L) = \left\{ t \in \prod_{i \in I} S_i \mid t(\ell) \in L \right\},$$

which has the form $\prod_{i \in I} K_i$, where each set K_i is open because

$$K_i = \begin{cases} S_i & \text{if } i \neq \ell, \\ L & \text{if } i = \ell, \end{cases}$$

for $i \in I$. Thus, $p_\ell^{-1}(L)$ is open and p_ℓ is continuous.

The proof of the second part of the theorem is left to the reader.

A preliminary result to a theorem that refers to the compactness of products of topological spaces is shown next.

Lemma 4.98 *Let $\mathcal{C}$ be a collection of subsets of $S = \prod_{i \in I} S_i$ such that $\mathcal{C}$ has the f.i.p. and $\mathcal{C}$ is maximal with this property.*

We have $\bigcap \mathcal{D} \in \mathcal{C}$ for every finite subcollection $\mathcal{D}$ of $\mathcal{C}$. Furthermore, if $T \cap C \neq \emptyset$ for every $C \in \mathcal{C}$, then $T \in \mathcal{C}$.

Proof Let $\mathcal{D} = \{D_1, \ldots, D_n\}$ be a finite subcollection of $\mathcal{C}$ and let $D = \bigcap \mathcal{D} \neq \emptyset$. Note that the intersection of every finite subcollection of $\mathcal{C} \cup \{D\}$ is also nonempty. The maximality of $\mathcal{C}$ implies $D \in \mathcal{C}$, which proves the first part of the lemma.

For the second part of the lemma, observe that the intersection of any finite subcollection of $\mathcal{D} \cup \{T\}$ is not empty. Therefore, as above, $T \in \mathcal{C}$.

Theorem 4.99 (Tychonoff's Theorem) *Let $\{(S_i, \mathcal{O}_i) \mid i \in I\}$ be a collection of topological spaces such that $S_i \neq \emptyset$ for every $i \in I$. Then, $(\prod_{i \in I} S_i, \prod_{i \in I} \mathcal{O}_i)$ is compact if and only if each topological space $(S_i, \mathcal{O}_i)$ is compact for $i \in I$.*

Proof If $(\prod_{i \in I} S_i, \mathcal{O})$ is compact, then, by Theorem 4.69, it is clear that each of the topological spaces $(S_i, \mathcal{O}_i)$ is compact because each projection p_i is continuous.

Conversely, suppose that each of the topological spaces $(S_i, \mathcal{O}_i)$ is compact.

Let $\mathcal{E}$ be a family of sets in $S = \prod_{i \in I} S_i$ that has the f.i.p. and let $(\mathfrak{C}, \subseteq)$ be the partially ordered set whose elements are collections of subsets of S that have the f.i.p. and contain the family $\mathcal{E}$.

Let $\{\mathcal{C}_i \mid i \in I\}$ be a chain in $(\mathfrak{C}, \subseteq)$. It is easy to verify that $\bigcup\{\mathcal{C}_i \mid i \in I\}$ has the f.i.p., so every chain in $(\mathfrak{C}, \subseteq)$ has an upper bound. Therefore, by Zorn's Lemma (see Theorem 2.81), the poset $(\mathfrak{C}, \subseteq)$ contains a maximal collection $\mathcal{C}$ that has the f.i.p.and contains $\mathcal{E}$. We aim to find an element $t \in \prod_{i \in I} S_i$ that belongs to $\bigcap\{\mathbf{K}(C) \mid C \in \mathcal{C}\}$ because, in this case, the same element belongs to $\bigcap\{\mathbf{K}(C) \mid C \in \mathcal{E}\}$ and this would imply, by Theorem 4.54, that $(S, \mathcal{O})$ is compact.

Let $\mathcal{C}_i$ be the collection of closed subsets of S_i defined by

$$\mathcal{C}_i = \{\mathbf{K}_i(p_i(C)) \mid C \in \mathcal{C}\}$$

for $i \in I$, where $\mathbf{K}_i$ is the closure of the topological space $(S_i, \mathcal{O}_i)$.

It is clear that each collection $\mathcal{C}_i$ has the f.i.p. in S_i. Indeed, since $\mathcal{C}$ has the f.i.p., if $\{C_1, \ldots, C_n\} \subseteq \mathcal{C}$ and $x \in \bigcap_{k=1}^n C_k$, then $p_i(x) \in \bigcap_{k=1}^n \mathbf{K}(p_i(C_k))$, so $\mathcal{C}_i$ has the f.i.p. Since $(S_i, \mathcal{O}_i)$ is compact, we have $\bigcap \mathcal{C}_i \neq \emptyset$, by Part (iii) of Theorem 4.53.

Let $t_i \in \bigcap \mathcal{C}_i = \bigcap\{\mathbf{K}_i(p_i(C)) \mid C \in \mathcal{C}\}$ and let $t \in S$ be defined by $t(i) = t_i$ for $i \in I$.

Let $D_{j,L} = \{u \mid u \in \prod_{i\in I} \mid u(j) \in L\}$, a set of the subbasis of the product topology that contains t, defined earlier, where L is an open set in $(S_j, \mathcal{O}_j)$. Since $g(j) \in L$, the set L has a nonempty intersection with every set $\mathbf{K}_i(p_i(C))$, where $C \in \mathcal{C}$. On the other hand, since $p_i(D_{j,L}) = S_i$ for $i \neq j$, it follows that for every $i \in I$ we have $p_i(D_{j,L}) \cap \bigcap_{C\in\mathcal{C}} \mathbf{K}_i(p_i(C)) \neq \emptyset$. Therefore, $p_i(D_{j,L})$ has a nonempty intersection with every set of the form $\mathbf{K}_i(p_i(C))$, where $C \in \mathcal{C}$. By the contrapositive of Theorem 4.9, this means that $p_i(D_{j,L}) \cup p_i(C) \neq \emptyset$ for every $i \in I$ and $C \in \mathcal{C}$. This in turn means that $D_{j,L} \cup C \neq \emptyset$ for every $C \in \mathcal{C}$. By Lemma 4.98, it follows that $D_{j,L} \in \mathcal{C}$. Since every set that belongs to the basis of the product topology is a finite intersection of sets of the form $D_{j,L}$, it follows that any member of the basis has a nonempty intersection with every set of $\mathcal{C}$. This implies that g belongs to $\bigcup\{\mathbf{K}(C) \mid C \in \mathcal{C}\}$, which implies the compactness of $(\prod_{i\in I} S_i, \prod_{i\in I} \mathcal{O}_i)$.

Example 4.100 In Example 4.57, we have shown that every closed interval $[x, y]$ of $\mathbb{R}$ where $x < y$ is compact. By Theorem 4.99, any subset of $\mathbb{R}^n$ of the form $[x_1, y_1] \times \cdots \times [x_n, y_n]$ is compact.

4.10 Fields of Sets

In this section, we introduce collections of sets that play an important role in measure and probability theory.

Definition 4.101 *Let S be a set. A* field of sets on S *is a family of subsets $\mathcal{E}$ of S that satisfies the following conditions:*

(i) $S \in \mathcal{E}$;
(ii) *if $U \in \mathcal{E}$, then $\bar{U} = S - U \in \mathcal{E}$;*
(iii) *if $U_0, \ldots, U_{n-1}$ belong to $\mathcal{E}$, then $\bigcup_{0\leqslant i\leqslant n-1} U_i$ belongs to $\mathcal{E}$.*

A σ-field of sets *on S is a family of subsets $\mathcal{E}$ of S that satisfies conditions (i) and (ii) and, in addition, satisfies the following condition:*

(iii′) if $\{U_i \mid i \in \mathbb{N}\}$ is a countable family of sets included in $\mathcal{E}$, then $\bigcup_{i\in\mathbb{N}} U_i$ belongs to $\mathcal{E}$.

Clearly, every σ-field is also a field on S.

If $\mathcal{E}$ is a σ-field of sets on S, we refer to the pair $(S, \mathcal{E})$ as a measurable space.

Example 4.102 The collection $\mathcal{E}_0 = \{\emptyset, S\}$ is a σ-field on S; moreover, for every σ-field $\mathcal{E}$ on S, we have $\mathcal{E}_0 \subseteq \mathcal{E}$.

The set $\mathcal{P}(S)$ of all subsets of a set S is a σ-field on S.

If T is a subset of S, then the collection $\{\emptyset, T, S - T, S\}$ is a σ-field on S.

Theorem 4.103 *The class of all fields (σ-fields) of sets on S is a closure system on $\mathcal{P}(S)$.*

Proof Let $\mathfrak{E} = \{\mathcal{E}_i \mid i \in I\}$ be a collection of fields of sets on S. Since $S \in \mathcal{E}_i$ for every $i \in I$, it follows that $S \in \bigcap\{\mathcal{E}_i \mid i \in I\}$.

Suppose that $A \in \bigcap \mathfrak{E}$. Since $A \in \mathcal{E}_i$ for every $i \in I$, it follows that $\bar{A} \in \mathcal{E}_i$ for every $i \in I$, which implies that $\bar{A} \in \bigcap\{\mathcal{E}_i \mid i \in I\}$.

Finally, if $\{A_i \mid 1 \leqslant i \leqslant n\} \in \bigcap\{\mathcal{E}_i \mid i \in I\}$, it is easy to see that $\bigcup_{i=1}^n A_i \in \bigcap\{\mathcal{E}_i \mid i \in I\}$.

A similar argument proves that the class of all σ-fields of sets is also a closure system on $\mathcal{P}(S)$.

Example 4.104 Let A be a subset of the set S. The σ-field generated by the collection $\{A\}$ is $\{\emptyset, A, \bar{A}, S\}$.

Definition 4.105 *Let $(S, \mathcal{O})$ be a topological space. A subset T of S is said to be a* Borel set *if it belongs to the σ-field generated by the topology $\mathcal{O}$.*

The σ-field of Borel sets of $(S, \mathcal{O})$ is denoted by $\mathcal{B}_{\mathcal{O}}$.

It is clear that all open sets are Borel sets. Also, every closed set, as a complement of an open set, is a Borel set.

Example 4.106 We identify several families of Borel subsets of the topological space $(\mathbb{R}, \mathcal{O})$.

It is clear that every open interval (a, b) and every set (a, ∞) or $(-\infty, a)$ is a Borel set for $a, b \in \mathbb{R}$ because they are open sets. The closed intervals of the form $[a, b]$ are Borel sets because they are closed sets in the topological space.

Since $[a, b) = (-\infty, b) - (-\infty, a)$, it follows that the half-open intervals of this form are also Borel sets.

For every $a \in \mathbb{R}$, we have $\{a\} \in \mathcal{B}_{\mathcal{O}}$ because $\{a\} = [a, b) - (a, b)$ for every $b \in \mathbb{R}$ such that $b > a$. Therefore, every countable subset $\{a_n \mid n \in \mathbb{N}\}$ of $\mathbb{R}$ is a Borel set.

Example 4.107 Let $\pi = \{B_i \mid i \in I\}$ be a countable partition of a set S. The σ-field generated by π is

$$\mathcal{E}_\pi = \left\{ \bigcup_{i \in J} B_i \mid J \subseteq I \right\}.$$

Clearly, every block B_i belongs to $\mathcal{E}_\pi$, so $\pi \subseteq \mathcal{E}_\pi$.

To verify that $\mathcal{E}_\pi$ is a σ-field, note first that we have $S \in \mathcal{E}_\pi$ since $S = \bigcup_{i \in I} B_i$. If $A \in \mathcal{E}_{pi}$, then $A = \bigcup_{i \in J} B_i$ for some subset J of I, so $\bar{A} = \bigcup_{i \in I - J} B_i$, which shows that $\bar{A} \in \mathcal{E}_\pi$. Let $\{A_\ell \mid \ell \in L\}$ be a family of sets included in $\mathcal{E}_\pi$. For each set A_ℓ, there exists a set J_ℓ such that $A_\ell = \bigcup\{B_i \mid i \in J_\ell\}$. Therefore,

$$\bigcup_{\ell \in L} A_\ell = \bigcup \left\{ B_i \mid i \in \bigcup_{\ell \in L} J_\ell \right\},$$

which shows that $\bigcup_{\ell \in L} A_\ell \in \mathcal{E}_\pi$. This proves that $\mathcal{E}_\pi$ is a σ-field. Moreover, any σ-field on S that includes π also includes $\mathcal{E}_\pi$, which concludes the argument.

Theorem 4.108 *Let $(S, \mathcal{E})$ be a measurable space. The following statements hold:*

(i) $\emptyset \in \mathcal{E}$;
(ii) *if* $\{A_i \mid i \in \mathbb{N}\} \subseteq \mathcal{E}$, *then* $\bigcap_{i \in \mathbb{N}} A_i \in \mathcal{E}$;
(iii) *if* $A, B \in \mathcal{E}$, *then* $A - B$ *and* $A \oplus B$ *belong to* $\mathcal{E}$.

Proof The first statement follows from the fact that $\emptyset = \bar{S}$.

Let $\{A_i \mid i \in \mathbb{N}\}$ be a family of subsets of S such that $A_i \in \mathcal{E}$ for $i \in \mathbb{N}$. Since $\overline{A_i} \in \mathcal{E}$, we have $\bigcup\{\overline{A_i} \mid i \in \mathbb{N}\} \in \mathcal{E}$. Thus,

$$\overline{\bigcup\{\overline{A_i} \mid i \in \mathbb{N}\}} = \bigcap\{A_i \mid i \in \mathbb{N}\} \in \mathcal{E},$$

which yields the second part of the theorem.

The third statement of the theorem is immediate.

Corollary 4.109 *Let $(S, \mathcal{E})$ be a measurable space and let $\{U_n \mid n \in \mathcal{E}\}$ be a sequence of members of $\mathcal{E}$. Then, both* $\liminf\{U_n \mid n \in \mathbb{N}\}$ *and* $\limsup\{U_n \mid n \in \mathbb{N}\}$ *belong to $\mathcal{E}$.*

Proof This statement follows immediately from Definition 4.101 and from Theorem 4.108.

Note that if $(S, \mathcal{E})$ is a measurable space (that is, if $\mathcal{E}$ is a σ-field on S), then condition (iii′) of Definition 101 amounts to $\mathcal{E}_\sigma \subseteq \mathcal{E}$. Moreover, by Part (ii) of Theorem 108, we also have $\mathcal{E}_\delta \subseteq \mathcal{E}$.

Example 4.110 Let S be an arbitrary set and let $\mathcal{B}$ be the family of sets that consists of sets that are either countable or complements of countable sets. We claim that $(S, \mathcal{B})$ is a measurable space.

Note that $S \in \mathcal{B}$ because S is the complement of $\emptyset$, which is countable. Next, if $A \in \mathcal{B}$ is countable, $\bar{A}$ is a complement of a countable set, so $\bar{A} \in \mathcal{B}$; otherwise, if A is not countable, then it is the complement of a countable set, which means that $\bar{A}$ is countable, so $\bar{A} \in \mathcal{B}$.

Let A and B be two sets of $\mathcal{B}$. If both are countable, then $A \cup B \in \mathcal{B}$. If $\bar{A}$ and $\bar{B}$ are countable, then $\overline{A \cup B} = \overline{A} \cap \overline{B}$, so $A \cup B \in \mathcal{B}$ because it has a countable complement. If A is countable and $\overline{B}$ is countable, then $\overline{A} \cap \overline{B}$ is countable because it is a subset of $\overline{B}$. Therefore, $A \cup B \in \mathcal{B}$ as a complement of a countable set. The case where $\overline{A}$ and B are countable is treated similarly. Thus, in any case, the union of two sets of $\mathcal{B}$ belongs to $\mathcal{B}$.

Finally, we have to prove that if $\{A_i \mid i \in \mathbb{N}\}$ is a family of sets included in $\mathcal{B}$, then the set $A = \bigcup_{i \in \mathbb{N}} A_i$ belongs to $\mathcal{B}$. Indeed, let us split the set I into I' and I'', where $i \in I'$ if the set A_i is countable and $i \in I''$ if the complement $\overline{A_i} = S - A_i$ is countable. Note that both $A' = \bigcup_{i \in I'} A_i$ and $A'' = \bigcap_{i \in I''} \overline{A_i}$ are countable sets (by Theorem 1.125 and by the fact that every subset of a countable set is countable, respectively), and that $A = A' \cup \overline{A''}$. Since both A' and $\overline{A''}$ belong to $\mathcal{B}$, it follows that $A \in \mathcal{B}$.

Definition 4.111 *Let $(S, \mathcal{D})$ and $(T, \mathcal{E})$ be two measurable spaces. A function $f : S \longrightarrow T$ is said to be* measurable *if $f^{-1}(V) \in \mathcal{D}$ for every $V \in \mathcal{E}$.*

It is easy to verify that if $(S_i, \mathcal{E}_i)$ are measurable spaces for $1 \leqslant i \leqslant 3$ and $f : S_1 \longrightarrow S_2$, $g : S_2 \longrightarrow S_3$ are measurable functions, then their composition gf is also a measurable function.

Theorem 4.112 *Let S and T be two sets and let $f : S \longrightarrow T$ be a function. If $\mathcal{E}$ is a σ-field on T, then the collection $f^{-1}(\mathcal{E})$ defined by $f^{-1}(\mathcal{E}) = \{f^{-1}(V) \mid V \in \mathcal{E}\}$ is a σ-field on S.*

Proof Since $T \in \mathcal{E}$, it is clear that $S = f^{-1}(T)$ belongs to $f^{-1}(\mathcal{E})$.

Suppose that $U \in f^{-1}(\mathcal{E})$; that is, $U = f^{-1}(W)$ for some $W \in \mathcal{E}$. Since $S - U = f^{-1}(T) - f^{-1}(W) = f^{-1}(T - W)$ (by Theorem 1.65), it follows that $S - U \in f^{-1}(\mathcal{E})$. Similarly, if $\{W_i \mid i \in \mathbb{N}\}$ is a countable family of sets included in $\mathcal{E}$, then $\{f^{-1}(W_i) \mid i \in \mathbb{N}\}$ is a countable family of sets included in $f^{-1}(\mathcal{E})$ and $\bigcup\{f^{-1}(W_i) \mid i \in \mathbb{N}\}$ belongs to $f^{-1}(\mathcal{E})$ by Theorem 1.63. Thus, $f^{-1}(\mathcal{E})$ is a σ-field on S.

Corollary 4.113 *Let $f : S \longrightarrow T$ be a function, where $(T, \mathcal{E})$ is a measurable space. Then, $f^{-1}(\mathcal{E})$ is the least σ-field of subsets of S such that f is is a measurable function between S and $(T, \mathcal{E})$.*

Proof Suppose that $\mathcal{D}$ is a σ-field on S such that f is measurable. Then, $f^{-1}(E) \in \mathcal{D}$ for every $E \in \mathcal{E}$, so $f^{-1}(\mathcal{E}) \subseteq \mathcal{D}$. The statement follows immediately since $f^{-1}(\mathcal{E})$ is a σ-field of sets.

Theorem 4.114 *Let S and T be two sets and let $f : S \longrightarrow T$ be a function. If $\mathcal{E}$ is a σ-field on S, then the collection $\mathcal{E}' = \{W \in \mathcal{P}(T) \mid f^{-1}(W) \in \mathcal{E}\}$ is a σ-field on T.*

Proof The proof is straightforward and is left to the reader as an exercise.

Theorem 4.115 *Let $(S, \mathcal{O})$ and $(T, \mathcal{O}')$ be two topological spaces and let $f : S \longrightarrow T$ be a continuous function. Then, f is measurable relative to the measurable spaces $(S, \mathcal{B}_{\mathcal{O}})$ and $(T, \mathcal{B}_{\mathcal{O}'})$, where $\mathcal{B}_{\mathcal{O}}$ and $\mathcal{B}_{\mathcal{O}'}$ are the collections of Borel sets in $(S, \mathcal{O})$ and $(T, \mathcal{O}')$, respectively.*

Proof The collection of sets $\mathcal{E}' = \{W \in \mathcal{P}(T) \mid f^{-1}(W) \in \mathcal{B}_{\mathcal{O}}\}$ is a σ-field on T. Since f is continuous, it is clear that $\mathcal{E}'$ contains every open set in $\mathcal{O}'$, so the σ-field of Borel sets $\mathcal{B}_{\mathcal{O}'}$ that is generated by $\mathcal{O}'$ is contained in $\mathcal{E}'$. Thus, for every Borel set U in T, $f^{-1}(U) \in \mathcal{B}_{\mathcal{O}}$, which allows us to conclude that f is indeed measurable.

Next, we describe the σ-field generated by a countable partition of a set.

Theorem 4.116 *Let $\pi = \{B_i \mid i \in I\}$ be a countable partition of a set S. In other words, we assume that the set of indices I of the blocks of π is countable.*

The σ-field generated by π is the collection of sets:

$$\left\{ \bigcup_{i \in J} B_i \mid J \subseteq I \right\}.$$

Proof Let $\mathcal{E}_\pi$ be the σ-field generated by π. Clearly, we have

$$\pi \subseteq \left\{ \bigcup_{i \in J} B_i \mid J \subseteq I \right\} \subseteq \mathcal{E}_\pi.$$

The collection $\{\bigcup_{i \in J} B_i \mid J \subseteq I\}$ is a σ-field. Indeed, we have $S = \bigcup\{B \mid B \in \pi\}$, so $S \in \{\bigcup_{i \in J} B_i \mid J \subseteq I\}$.

Suppose that $A = \bigcup\{B_i \mid i \in J\}$. Then $\bar{A} = \{B_i \mid i \in I - J\}$, which shows that $\bar{A} \in \{\bigcup_{i \in J} B_i \mid J \subseteq I\}$.

Suppose that $A_0, \ldots, A_n, \ldots$ belong to $\mathcal{E}$, so $A_k = \bigcup\{B_i \mid i \in J_k\}$, where $J_k \subseteq I$ for $k \in \mathbb{N}$. Then, $\bigcup_{k \geqslant 0} A_k = \bigcup\{B_i \mid i \in \bigcup_{k \geqslant 0} J_k\}$, which implies that $\bigcup_{k \geqslant 0} A_k \in \{\bigcup_{i \in J} B_i \mid J \subseteq I\}$.

This implies that $\mathcal{E}_\pi = \{\bigcup_{i \in J} B_i \mid J \subseteq I\}$.

We now give a technical result that concerns σ-fields.

Theorem 4.117 *Let $(S, \mathcal{E})$ be a measurable space and let $\{U_i \in \mathcal{E} \mid i \in \mathbb{N}\}$ be a family of sets from $\mathcal{E}$. There exists a family of sets $\{V_i \in \mathcal{E} \mid i \in \mathbb{N}\}$ that satisfies the following conditions:*

(i) *if $i, j \in \mathbb{N}$ and $i \neq j$, then $V_i \cap V_j = \emptyset$;*
(ii) *$V_i \subseteq U_i$ for $i \in \mathbb{N}$;*
(iii) *$\bigcup\{V_i \mid i \in \mathbb{N}\} = \bigcup\{U_i \mid i \in \mathbb{N}\}$.*

Proof The sets V_n are defined inductively by

$$V_0 = U_0,$$
$$V_i = U_i - \bigcup\{U_j \mid 0 \leqslant j \leqslant i - 1\}.$$

It is clear that the first two conditions of the theorem are satisfied; we prove the last part of the theorem.

For $x \in \bigcup\{U_i \mid i \in \mathbb{N}\}$, let i_x be the least i such that $x \in U_i$; clearly, $x \notin U_j$ for $j < i$, so $x \in V_i$. Thus, $\bigcup\{U_i \mid i \in \mathbb{N}\} \subseteq \bigcup\{V_i \mid i \in \mathbb{N}\}$. The reverse inclusion follows immediately from the fact that $V_i \subseteq U_i$ for every $i \in \mathbb{N}$.

4.11 Measures

Measurable spaces provide the natural framework for introducing the notion of measure.

Definition 4.118 *Let* $(S, \mathcal{E})$ *be a measurable space. A* measure *is a function* $m : \mathcal{E} \longrightarrow \hat{\mathbb{R}}_{\geqslant 0}$ *that satisfies the following conditions:*

(i) $m(\emptyset) = 0$;
(ii) for every countable collection $U_0, U_1, \ldots$ *of sets in* $\mathcal{E}$ *that are pairwise disjoint, we have*

$$m\left(\bigcup_{n \in \mathbb{N}} U_n\right) = \sum_{n \in \mathbb{N}} m(U_n).$$

The second property of the definition is the additivity of measures.
The triple $(S, \mathcal{E}, m)$ *is a* measure space.

In particular, if the collection $U_0, U_1, \ldots$ consists of two disjoint sets U and V, then

$$m(U \cup V) = m(U) + m(V). \tag{4.3}$$

Observe that if $U, V \in \mathcal{E}$ and $U \subseteq V$, then $V = U \cup (V - U)$, so by the additivity property, $m(V) = m(U) + m(V - U) \geqslant m(U)$. This shows that $U \subseteq V$ implies $m(U) \leqslant m(V)$ (the *monotonicity* of measures).

Let X and Y be two subsets of $\mathcal{E}$. Since $X \cup Y = X \cup (Y - X)$, $Y = (Y - X) \cup (Y \cap X)$, and the pairs of sets X, $(Y - X)$ and $(Y - X)$, $(Y \cap X)$ are disjoint, we can write

$$\begin{aligned} m(X \cup Y) &= m(X) + m(Y - X) \\ &= m(X) + m(Y) - m(X \cap Y). \end{aligned} \tag{4.4}$$

The resulting equality

$$m(X \cup Y) + m(X \cap Y) = m(X) + m(Y) \tag{4.5}$$

for $X, Y \in \mathcal{E}$ is known as the *modularity property* of measures.

Example 4.119 Let S be a finite set and let $\mathcal{E} = \mathcal{P}(S)$. The mapping $m : \mathcal{P}(S) \longrightarrow \mathbb{R}$ given by $m(U) = |U|$ is a measure on $\mathcal{P}(S)$, as can be verified immediately.

Example 4.120 Let S be a set and let s be a fixed element of S. Define the mapping $m_s : \mathcal{P}(S) \longrightarrow \hat{\mathbb{R}}_{\geqslant 0}$ by

$$m_s(U) = \begin{cases} 1 & \text{if } s \in U, \\ 0 & \text{otherwise.} \end{cases}$$

It is easy to verify that m_s is a measure defined on $\mathcal{P}(S)$. Indeed, we have $m_s(\emptyset) = 0$. If $U_0, U_1, \ldots$ is a countable collection of pairwise disjoint sets, then s may belong to at most one of these sets. If there is a set U_i such that $s \in U_i$, $s \in \bigcup_{n\in\mathbb{N}} U_i$, so $m_s\left(\bigcup_{n\in\mathbb{N}} U_n\right) = \sum_{n\in\mathbb{N}} m_s(U_n) = 1$. If no such set U_i exists, then $m_s\left(\bigcup_{n\in\mathbb{N}} U_n\right) = \sum_{n\in\mathbb{N}} m_s(U_n) = 0$. In either case, the second condition of Definition 4.118 is satisfied.

The behavior of measures with respect to limits of sequences of sets is discussed next.

Theorem 4.121 *Let $(S, \mathcal{E}, m)$ be a measure space. If $(U_0, U_1, \ldots)$ is an increasing or a decreasing sequence of sets from $\mathcal{E}$, then $m(\lim U_n) = \lim m(U_n)$.*

Proof Suppose that $U_0 \subset U_1 \subset \cdots$ is an increasing sequence of sets, so $m(\lim U_n) = m(\bigcup_n U_n)$. By Theorem 4.117, there exists a sequence $V_0 \subset V_1 \subset \cdots$ of disjoint sets in $\mathcal{E}$ such that $\bigcup U_n = \bigcup V_n$ and $V_0 = U_0$, and $V_n = U_n - V_{n-1}$ for $n \geqslant 1$. Then,

$$\begin{aligned} m(\lim U_n) &= m\left(\bigcup_n V_n\right) = m(V_0) + \sum_{n\geqslant 1} m(V_n) \\ &= \lim_{n\to\infty}\left(m(V_0) + \sum_{i=1}^{n} m(V_i)\right) \\ &= \lim_{n\to\infty} m\left(V_0 \cup \bigcup_{n\geqslant 1} V_i\right) = \lim_{n\to\infty} m(U_i). \end{aligned}$$

Suppose now that $U_0 \supset U_1 \supset \cdots$ is a decreasing sequence of sets, so $m(\lim U_n) = m(\bigcap_n U_n)$.

Define the sequence of sets $W_0, W_1, \ldots$ by $W_n = U_0 - U_n$ for $n \in \mathbb{N}$. Since this sequence is increasing, we have $m(\bigcup_{n\in\mathbb{N}} W_n) = \lim m(W_n)$ by the first part of the theorem. Thus, we can write

$$m\left(\bigcup_{n\in\mathbb{N}} W_n\right) = \lim m(W_n) = m(U_0) - \lim m(U_n).$$

Since

$$\begin{aligned} m\left(\bigcup_{n\in\mathbb{N}} W_n\right) &= m\left(\bigcup_{n\in\mathbb{N}} (U_0 - U_n)\right) \\ &= m\left(U_0 - \bigcap_{n\in\mathbb{N}} U_n\right) = m(U_0) - m\left(\bigcap_{n\in\mathbb{N}} U_n\right), \end{aligned}$$

it follows that $m(\lim U_n) = m\left(\bigcap_{n\in\mathbb{N}} U_n\right) = \lim m(U_n)$.

Definition 4.122 *An* outer measure *on a set* S *is a function* $\mu : \mathcal{P}(S) \longrightarrow \hat{\mathbb{R}}_{\geqslant 0}$ *that satisfies the following properties:*

(i) $\mu(\emptyset) = 0$.
(ii) μ *is countably subadditive; that is,*

$$\mu\left(\bigcup_{n\in\mathbb{N}} E_n\right) \leqslant \sum\{\mu(E_n) \mid n \in \mathbb{N}\}$$

for every countable family $\{E_n \in \mathcal{P}(S) \mid n \in \mathbb{N}\}$ *of subsets of* S.
(iii) μ *is monotonic.*

A subset T *of* S *is* μ-measurable *if* $\mu(H) = \mu(H \cap T) + \mu(H \cap \overline{T})$ *for every set* $H \in \mathcal{P}(S)$.

Lemma 4.123 *Let* S *be a set and let* μ *be an outer measure on a set* S. *A set* T *is* μ*-measurable if and only if* $\mu(H) \geqslant \mu(H \cap T) + \mu(H \cap \overline{T})$ *for every* $H \in \mathcal{P}(S)$ *such that* $\mu(H) < \infty$.

Proof The necessity of the condition is obvious. Suppose therefore that the condition is satisfied. Since μ is subadditive, we have

$$\mu(H) \leqslant \mu(H \cap T) + \mu(H \cap \overline{T}),$$

which implies $\mu(H) = \mu(H \cap T) + \mu(H \cap \overline{T})$.

Theorem 4.124 *Let* μ *be an outer measure on a set* S. *The collection of* μ*-measurable sets is a* σ*-field* $\mathcal{E}_\mu$ *on* S.

Proof It is immediate that $\emptyset \in \mathcal{E}_\mu$. Suppose that $T_0, T_1, \ldots$ is a sequence of μ-measurable sets. Then, $\mu(H) = \mu(H \cap T_0) + \mu(H \cap \overline{T}_0)$ for every subset H of S.

By substituting $H \cap T_0$ and $H \cap \overline{T}_0$ for H, we obtain

$$\mu(H \cap T_0) = \mu(H \cap T_0 \cap T_1) + \mu(H \cap T_0 \cap \overline{T}_1),$$
$$\mu(H \cap \overline{T}_0) = \mu(H \cap \overline{T}_0 \cap T_1) + \mu(H \cap \overline{T}_0 \cap \overline{T}_1),$$

which yields

$$\begin{aligned}\mu(H) =& \mu(H \cap T_0 \cap T_1) + \mu(H \cap T_0 \cap \overline{T}_1)\\ &+ \mu(H \cap \overline{T}_0 \cap T_1) + \mu(H \cap \overline{T}_0 \cap \overline{T}_1).\end{aligned} \tag{4.6}$$

Replacing H by $H \cap (T_0 \cup T_1)$, we obtain the equality

$$\mu(H \cap (T_0 \cup T_1)) = \mu(H \cap T_0 \cap T_1) + \mu(H \cap T_0 \cap \overline{T}_1) + \mu(H \cap \overline{T}_0 \cap T_1). \tag{4.7}$$

Therefore,

$$\mu(H) = \mu(H \cap (T_0 \cup T_1)) + \mu(H \cap \overline{T_0 \cup T_1}),$$

which shows that $T_0 \cup T_1$ is μ-measurable. An easy argument by induction shows that $\bigcup_{i=0}^{n} T_i$ is μ-measurable for every $n \in \mathbb{N}$.

By replacing H in Equality (4.6) by $H \cap \overline{T_0 - T_1} = H \cap (\overline{T}_0 \cup T_1)$, we have

$$\mu(H \cap (\overline{T}_0 \cup T_1)) = \mu(H \cap T_0 \cap T_1) + \mu(H \cap \overline{T}_0 \cap T_1) + \mu(H \cap \overline{T}_0 \cap \overline{T}_1),$$

which allows us to write $\mu(H) = \mu(H \cap \overline{T_0 - T_1}) + \mu(H \cap (T_0 \cap -T_1))$. Thus, $T_0 - T_1$ is μ-measurable.

If U_0 and U_1 are two disjoint μ-measurable sets, then Equality (4.7) implies

$$\mu(H \cap (U_0 \cup U_1)) = \mu(H \cap U_0) + \mu(H \cap U_1)$$

for every H. Again, an inductive argument allows us to show that if $T_0, \ldots, T_n$ are pairwise disjoint, μ-measurable sets, then

$$\mu\left(H \cap \bigcup_{i=0}^{n} U_i\right) = \sum_{i=0}^{n} \mu(H \cap U_i). \tag{4.8}$$

Define $W_n = \bigcup_{i=0}^{n} T_i$. We have seen that W_n is μ-measurable for every $n \in \mathbb{N}$. Thus, we have

$$\begin{aligned}
\mu(H) &= \mu(H \cap W_n) + \mu(H \cap \overline{W}_n) \\
&= \mu\left(H \cap \left(\bigcup_{i=0}^{n} T_i\right)\right) + \mu(H \cap \overline{W}_n) \\
&\geqslant \mu\left(H \cap \left(\bigcup_{i=0}^{n} T_i\right)\right) + \mu(H \cap \overline{W}),
\end{aligned}$$

where $W = \bigcup_{i \geqslant 0} T_i$. By Equality (4.8), we have

$$\mu(H) \geqslant \sum_{i \geqslant 0}^{n} \mu(H \cap T_i) + \mu(H \cap \overline{W}) \tag{4.9}$$

for every $n \in \mathbb{N}$. Therefore,

$$\mu(H) \geqslant \sum_{i \geqslant 0}^{\infty} \mu(H \cap T_i) + \mu(H \cap \overline{W}),$$

hence $\mu(H) \geqslant \mu(H \cap W) + \mu(H \cap \overline{W})$. By Lemma 4.123, the set W is μ-measurable. Note also that we have shown that

$$\mu(H) = \sum_{i \geqslant 0}^{n} \mu(H \cap T_i) + \mu(H \cap \overline{W}) = \mu(H \cap W) + \mu(H \cap \overline{W}). \qquad (4.10)$$

Suppose now that the sets $T_0, T_1, \ldots$ are not disjoint. Consider the sequence of pairwise disjoint sets $V_0, V_1, \ldots$ defined by

$$V_0 = T_0,$$
$$V_n = T_n - \bigcup_{i=0}^{n-1} T_i,$$

for $n \geqslant 1$. The measurability of each set V_n is immediate and, by the previous argument, $\bigcup_{n \in \mathbb{N}} V_n$ is μ-measurable. Since $\bigcup_{n \in \mathbb{N}} V_n = \bigcup_{n \in \mathbb{N}} T_n$, it follows that $\bigcup_{n \in \mathbb{N}} T_n$ is μ-measurable. We conclude that the collection of μ-measurable sets is a σ-field.

Corollary 4.125 *Let S be a set and let $\mu : \mathcal{P}(S) \longrightarrow \hat{\mathbb{R}}_{\geqslant 0}$ be an outer measure on S. The restriction $\mu \upharpoonright_{\mathcal{E}_\mu}$ to the σ-field $\mathcal{E}_\mu$ is a measure.*

Proof Let $T_0, T_1, \ldots$ be a sequence of sets in $\mathcal{E}_\mu$ that are pairwise disjoint. Choosing $H = W$ in Equality (4.10), we have $\mu(W) = \sum_{i \geqslant 0}^{n} \mu(T_i)$, which proves that $\mu \upharpoonright_{\mathcal{E}_\mu}$ is indeed a measure.

Corollary 4.126 *Let μ be an outer measure and let $U_0, U_1, \ldots$ be a sequence of μ-measurable sets. Then, both* $\liminf U_n$ *and* $\limsup U_n$ *are μ-measurable sets.*

Proof This statement follows immediately from Theorem 4.124 and from Corollary 4.109.

Theorem 4.128, which follows gives a technique for constructing outer measures known as *Munroe's Method I* or simply as *Method I* (see [2–4]).

First, we need the following definition.

Definition 4.127 *A* sequential cover *of a set S is a collection $\mathcal{C}$ of subsets of S such that $\emptyset \in \mathcal{C}$, and for every subset T of S there is a countable subcollection $\mathcal{D} = \{D_0, D_1, \ldots\}$ of $\mathcal{C}$ such that $T \subseteq \bigcup_{n=0}^{\infty} D_n$.*

The family of all countable *collections of sets from $\mathcal{C}$ that are covers of a set $W \in \mathcal{P}(S)$ is denoted by $\mathfrak{D}_{\mathcal{C},W}$. If the collection $\mathcal{C}$ is clear from the context, the subscript $\mathcal{C}$ is omitted.*

Theorem 4.128 *Let S be a set, $\mathcal{C}$ a sequential cover of the set S, and $f : \mathcal{C} \longrightarrow \hat{\mathbb{R}}_{\geqslant 0}$ a nonnegative function defined on $\mathcal{C}$ such that $f(\emptyset) = 0$.*

The function $\mu_f : \mathcal{P}(S) \longrightarrow \hat{\mathbb{R}}_{\geqslant 0}$ given by

$$\mu_f(T) = \inf \left\{ \sum_{U \in \mathcal{D}} f(U) \mid \mathcal{D} \in \mathfrak{D}_{\mathcal{C},T} \right\}$$

for $T \in \mathcal{P}(S)$ is an outer measure on S.

Proof Since $\emptyset$ is covered by the empty collection, and an empty sum has the value 0, it follows that $\mu_f(\emptyset) = 0$.

If $T, T' \in \mathcal{P}(S)$ and $T \subseteq T'$, then any cover of T' is also a cover of T; that is, $\mathfrak{D}_{\mathcal{C},T'} \subseteq \mathfrak{D}_{\mathcal{C},T}$. Therefore, $\mu_f(T) \leqslant \mu_f(T')$.

Let $\{T_n \mid n \in \mathbb{N}\}$ be a countable collection of subsets of S. If $\mu_f(T_n) = +\infty$ for one of the members of this collection, then the subadditivity of μ_f,

$$\mu_f\left(\bigcup_{n\in\mathbb{N}} U_n\right) \leqslant \sum_{n\in\mathbb{N}} \mu_f(U_n),$$

is satisfied. Therefore, we assume now that the value $\mu_f(T_n)$ is finite for each $n \in \mathbb{N}$.

The definition of $\mu_f(T_n)$ as an infimum allows us to assume the existence of a collection of sets $\mathcal{D}_n \in \mathfrak{D}_{U_n}$ such that

$$\sum_{U\in\mathcal{D}_n} f(U) \leqslant \mu_f(T_n) + \frac{\epsilon}{2^n}.$$

Consider the collection $\mathcal{D} = \bigcup_{n\in\mathbb{N}} \mathcal{D}_n$. $\mathcal{D}$ is a cover for $\bigcup_{n\in\mathbb{N}} T_n$. Therefore, by the definition of μ_f, we have

$$\begin{aligned}
\mu_f\left(\bigcup_{n\in\mathbb{N}} T_n\right) &\leqslant \sum \{f(U) \mid U \in \mathcal{D}\} \\
&\leqslant \sum_{n\in\mathbb{N}} \sum_{U\in\mathcal{D}_n} f(U) \\
&\leqslant \sum_{n\in\mathbb{N}} \mu_f(T_n) + \epsilon \sum_{n\in\mathbb{N}} \frac{1}{2^n} \\
&= \sum_{n\in\mathbb{N}} \mu_f(T_n) + 2\epsilon.
\end{aligned}$$

Since this inequality holds for every ϵ, it follows that

$$\mu_f\left(\bigcup_{n\in\mathbb{N}} T_n\right) \leqslant \sum \{\mu_f(T_n) \mid n \in \mathbb{N}\},$$

which proves that μ_f is subadditive. We conclude that μ_f is an outer measure.

Corollary 4.129 *Let S be a set, $\mathcal{C}$ a sequential cover of the set S, and $f : \mathcal{C} \longrightarrow \hat{\mathbb{R}}_{\geqslant 0}$ a function such that $f(\emptyset) = 0$. The outer measure μ_f is the unique outer measure on S that satisfies the following properties:*

(i) $\mu_f(U) \leqslant f(U)$ *for every* $U \in \mathcal{C}$, *and*
(ii) if μ' *is an outer measure such that* $\mu'(U) \leqslant f(U)$ *for every* $U \in \mathcal{C}$ *then* $\mu'(T) \leqslant \mu_f(T)$ *for every* $T \in \mathcal{P}(S)$.

Proof Since $\{\emptyset, U\}$ is a cover for U, the inequality $\mu_f(U) \leqslant f(U)$ is immediate for every $U \in \mathcal{C}$.

Let μ' be an outer measure such that $\mu'(U) \leqslant f(U)$ for every $U \in \mathcal{C}$ and let $\mathcal{D}$ be a sequential cover of a set $T \in \mathcal{P}(S)$. Then, we have

$$\mu'(T) \leqslant \mu'\left(\bigcup \mathcal{D}\right) \leqslant \{\mu'(U) \mid U \in \mathcal{D}\} \leqslant \sum\{f(U) \mid U \in \mathcal{D}\}$$

so $\mu'(T) \leqslant \mu_f(T)$. The uniqueness of μ_f follows by changing the roles of μ_f and μ'.

Corollary 4.130 *Let S be a set, $\mathcal{C}'$ and $\mathcal{C}$ two sequential covers of S such that $\mathcal{C}' \subseteq \mathcal{C}$, and $f : \mathcal{C} \longrightarrow \hat{\mathbb{R}}_{\geqslant 0}$ a function such that $f(\emptyset) = 0$. If μ'_f and μ_f are the outer measures that correspond to the collections $\mathcal{C}'$ and $\mathcal{C}$, respectively, then $\mu_f(T) \leqslant \mu'_f(T)$ for $T \in \mathcal{P}(S)$.*

Proof Observe that if $\mathfrak{D}'_T \subseteq \mathfrak{D}_T$, where $\mathfrak{D}'_T$ and $\mathfrak{D}_T$ are the families of countable collections of sets from $\mathcal{C}'$ and $\mathcal{C}$, respectively, that are covers of T, then the definitions of μ'_f and μ_f immediately imply the desired inequality.

Example 4.131 Theorem 4.128 allows us to introduce a very important outer measure on $\mathbb{R}$. Let $\mathcal{C}$ be the collection of open intervals of $\mathbb{R}$ to which the empty set is added.

Define the function $f : \mathcal{C} \longrightarrow \mathbb{R}$ by $f(a, b) = b - a$ for every open interval $(a, b) \in \mathcal{C}$ and $f(\emptyset) = 0$. For a subset T of $\mathbb{R}$, the value of the outer measure $\mu(T) = m_f(T)$ is

$$\mu(T) = \inf\left\{\sum_{n\in\mathbb{N}}(b_n - a_n) \in \mathcal{C} \mid T \subseteq \bigcup_n (a_n, b_n)\right\},$$

where the infimum is considered over all countable collections of open intervals (a_n, b_n) that cover the set T. This is the *Lebesgue outer measure* of the set T.

Let μ be the Lebesgue outer measure on $\mathbb{R}$. We have $\mu([a, b]) = b - a$. Since $[a, b] \subseteq (a-\epsilon, b+\epsilon)$ for every $\epsilon > 0$, it follows that $\mu([a, b]) < b-a+2\epsilon$ for every $\epsilon > 0$, so $\mu([a, b]) \leqslant b-a$. On the other hand, $(a, b) \subseteq [a, b]$, so $\mu([a, b]) \geqslant b-a$, which yields $\mu([a, b]) = b - a$.

This type of measure can be generalized to $\mathbb{R}^n$ by defining $\mathcal{C}$ as the collection of n-dimensional intervals of the form $I = (a_1, b_1) \times \cdots \times (a_n, b_n)$ to which we add the empty set and letting $f(I)$ be the volume $vol(I) = \prod_{i=1}^{n} |b_i - a_i|$ of I. Thus, the Lebesgue measure of a set $T \subseteq \mathbb{R}^n$ is

$$\mu(T) = \inf\left\{\sum vol(I) \mid I \in \mathcal{C}, T \subseteq \bigcup I\right\}, \tag{4.11}$$

Definition 4.132 *An outer measure μ on a set S is* regular *if for every $T \in \mathcal{P}(S)$ there exists a μ-measurable set U such that $T \subseteq U$ and $\mu(T) = \mu(U)$.*

Example 4.133 Let μ be the Lebesgue outer measure on $\mathbb{R}^n$ and let T be a subset of $\mathbb{R}^n$. For every $m \in \mathbb{N}$ there exists a countable collection of intervals $\{I_m^k \mid k \in \mathbb{N}\}$ such that

$$\mu(T) \leqslant \sum_{k \in \mathbb{N}} \mu(I_m^k) < \mu(T) + \frac{1}{m} < \mu\left(\bigcup_{k \in \mathbb{N}} I_m^k\right) + \frac{1}{m}.$$

Let $U = \bigcap_{m \in \mathbb{N}} \bigcap_{k \in \mathbb{N}} I_m^k$. Clearly, U is μ-measurable and $T \subseteq U$, so $\mu(T) \leqslant \mu(U)$. Since $U \subseteq \bigcap_{k \in \mathbb{N}} I_m^k$, we have

$$\mu(U) \leqslant \mu\left(\bigcap_{k \in \mathbb{N}} I_m^k\right) \leqslant \sum_{k \in \mathbb{N}} \mu\left(\bigcap_{k \in \mathbb{N}} I_m^k\right) \leqslant \mu(T) + \frac{1}{m},$$

so $\mu(U) \leqslant \mu(T)$. Consequently, $\mu(U) = \mu(T)$, which proves that the Lebesgue outer measure on $\mathbb{R}^n$ is regular.

Theorem 4.134 *Let S be a set and let $(S_0, S_1, \ldots)$ be a sequence of subsets of S. If μ is a regular outer measure on S, then $\mu(\liminf_n S_n) \leqslant \liminf_n \mu(S_n)$.*

Proof Since μ is regular, for each $n \in \mathbb{N}$ there exists a μ-measurable set U_n such that $S_n \subseteq U_n$ and $\mu(S_n) = \mu(U_n)$. Then, $\liminf_n \mu(S_n) \liminf_n \mu(U_n)$. Since $\liminf_n \mu(U_n)$ is measurable (by Corollary 4.126), we have

$$\mu(\liminf_n S_n) \leqslant \mu(\liminf_n U_n) \leqslant \liminf_n \mu(U_n) = \liminf_n \mu(S_n).$$

Corollary 4.135 *Let μ be an outer measure on a set S. If $\mathbf{S} = (S_0, S_1, \ldots)$ is an expanding sequence of subsets of S, then $\mu(\lim_n S_n) = \lim_n \mu(S_n)$.*

Proof Since $\mathbf{S}$ is an expanding sequence $\lim_n S_n = \bigcup_n S_n$, then $\mu(\lim_n S_n) \geqslant \mu(S_n)$ for $n \in \mathbb{N}$, so $\mu(\lim_n S_n) \geqslant \lim_n \mu(S_n)$. On the other hand, Theorem 4.134 implies $\mu(\lim S_n) \leqslant \lim \mu(S_n)$, which gives the desired equality.

For finite regular outer measures, the measurability condition can be simplified, as shown next.

Theorem 4.136 *Let μ be a regular outer measure on a set S such that $\mu(S)$ is finite. A subset T of S is measurable if and only if $\mu(S) = \mu(T) + \mu(\overline{T})$, where $\overline{T} = S - T$.*

Proof The condition is clearly necessary. To prove its sufficiency, let T be a subset of S such that $\mu(S) = \mu(T) + \mu(\overline{T})$. By Lemma 4.123, to prove that T is measurable, it suffices to show that if H is a set with $\mu(H) < \infty$, then $\mu(H) \geqslant \mu(H \cap T) + \mu(H \cap \overline{T})$.

The regularity of μ implies the existence of a μ-measurable set K such that $H \subseteq K$ and $\mu(H) = \mu(K)$. Since K is measurable, we have

$$\mu(H) = \mu(H \cap K) + \mu(H \cap \overline{K}),$$
$$\mu(\overline{H}) = \mu(\overline{H} \cap K) + \mu(\overline{H} \cap \overline{K}).$$

This implies

$$\begin{aligned}\mu(S) &= \mu(T) + \mu(\overline{T}) \\ &= \mu(T \cap K) + \mu(T \cap \overline{K}) + \mu(\overline{T} \cap K) + \mu(\overline{T} \cap \overline{K}) \\ &\geqslant \mu(K) + \mu(\overline{K}) = \mu(S).\end{aligned}$$

Thus,

$$\mu(T \cap K) + \mu(T \cap \overline{K}) + \mu(\overline{T} \cap K) + \mu(\overline{T} \cap \overline{K}) = \mu(K) + \mu(\overline{K}) = \mu(S).$$

Since $\mu(\overline{K}) \leqslant \mu(T \cap \overline{K}) + \mu(\overline{T} \cap \overline{K})$, it follows that $\mu(K \cap T) + \mu(K \cap \overline{T}) \leqslant \mu(K)$. Since $H \cap T \subseteq K \cap T$ and $H \cap \overline{T} \subseteq K \cap \overline{T}$, we have $\mu(H \cap T) + \mu(H \cap \overline{T}) \leqslant \mu(K) = \mu(H)$, which shows that T is indeed μ-measurable.

Exercises and Supplements

1. Prove that the family of subsets $\{(-n, n) \mid n \in \mathbb{N}\} \cup \{\emptyset, \mathbb{R}\}$ is a topology on $\mathbb{R}$.
2. Let S be a set and let s_0 be an element of S. Prove that the family of subsets $\mathcal{O}_{s_0} = \{L \in \mathcal{P}(S) \mid s_0 \in L\} \cup \{\emptyset\}$ is a topology on S.
3. Let $(S, \mathcal{O})$ be a topological space, L be an open set in $(S, \mathcal{O})$, and H be a closed set.
   (a) Prove that a set V is open in the subspace $(L, \mathcal{O}\restriction_L)$ if and only if V is open in $(S, \mathcal{O})$ and $V \subseteq L$.
   (b) Prove that a set W is closed in the subspace $(H, \mathcal{O}\restriction_H)$ if and only if W is closed in $(S, \mathcal{O})$ and $W \subseteq H$.
4. Let $(S, \mathcal{O})$ be a topological space where $\mathcal{O} = \{\emptyset, U, V, S\}$, where U and V are two subsets of S. Prove that either $\{U, V\}$ is a partition of S or one of the sets $\{U, V\}$ is included in the other.

5. Let $(S, \mathcal{O})$ be a topological space and let $\mathbf{I}$ be its interior operator. Prove that the poset of open sets $(\mathcal{O}, \subseteq)$ is a complete lattice, where $\sup \mathcal{L} = \bigcup \mathcal{L}$ and $\inf \mathcal{L} = \mathbf{I}\left(\bigcap \mathcal{L}\right)$ for every family of open sets $\mathcal{L}$.
6. Let $(S, \mathcal{O})$ be a topological space, let $\mathbf{K}$ be its interior operator and let $\mathcal{K}$ be its collection of closed sets. Prove that $(\mathcal{K}, \subseteq)$ is a complete lattice, where $\sup \mathcal{L} = \mathbf{K}\left(\bigcup \mathcal{L}\right)$ and $\inf \mathcal{L} = \bigcap \mathcal{L}$ for every family of closed sets.
7. Prove that if U, V are two subsets of a topological space $(S, \mathcal{O})$, then $\mathbf{K}(U \cap V) \subseteq \mathbf{K}(U) \cap \mathbf{K}(V)$. Formulate an example where this inclusion is strict.
8. Let T be a subspace of the topological space $(S, \mathcal{O})$. Let $\mathbf{K}_S$, $\mathbf{I}_S$, and ∂_S be the closure, interior and border operators associated to S and $\mathbf{K}_T$, $\mathbf{I}_T$ and ∂_T the corresponding operators associated to T. Prove that

   (a) $\mathbf{K}_T(U) = \mathbf{K}_S(U) \cap T$,
   (b) $\mathbf{I}_S(U) \subseteq \mathbf{I}_T(U)$, and
   (c) $\partial_T U \subseteq \partial_S U$

   for every subset U of T.
9. Let $(S, \mathcal{O})$ be a topological space and let $\mathbf{K}$ and $\mathbf{I}$ be its associated closure and interior operator, respectively. Define the mappings $\phi, \psi : \mathcal{P}(S) \longrightarrow \mathcal{P}(S)$ by $\phi(U) = \mathbf{I}(\mathbf{K}(U))$ and $\psi(U) = \mathbf{K}(\mathbf{I}(U))$ for $U \in \mathcal{P}(S)$.

   (a) Prove that $\phi(U)$ is an open set and $\psi(U)$ is a closed set for every set $U \in \mathcal{P}(S)$.
   (b) Prove that $\psi(H) \subseteq H$ for every closed set H and $L \subseteq \phi(L)$ for every open set L.
   (c) Prove that $\phi(\phi(U)) = \phi(U)$ and $\psi(\psi(U)) = \psi(U)$ for every $U \in \mathcal{P}(S)$.
   (d) Let $(\mathbf{J}_1, \ldots, \mathbf{J}_n)$ be a sequence such that $\mathbf{J}_i \in \{\mathbf{K}, \mathbf{I}\}$. Prove that there are at most seven distinct sets of the form $\mathbf{J}_n(\cdots(\mathbf{J}_1(U))\cdots)$ for every set $U \in \mathcal{P}(S)$, and give an example of a topological space $(S, \mathcal{O})$ and a subset U of S such that these seven sets are pairwise distinct.

10. Let $\mathcal{S}$ be the set of subsets of $\mathbb{R}$ such that, for every $U \in \mathcal{S}$, $x \in U$ implies $-x \in U$. Prove that $\{\emptyset\} \cup \mathcal{S}$ is a topology on $\mathbb{R}$.
11. Let $(S, \mathcal{O})$ be a topological space, and U and U' be two subsets of S.

    (a) Prove that $\partial(U \cup V) \subseteq \partial U \cup \partial V$.
    (b) Prove that $\partial U = \partial(S - U)$.

12. Let $(S, \mathcal{O})$ be a topological space. The subsets X and Y are said to be *separated* if $X \cap \mathbf{K}(Y) = \mathbf{K}(X) \cap Y = \emptyset$.

    (a) Prove that X and Y are separated sets in $(S, \mathcal{O})$ if and only if they are disjoint and clopen in the subspace $X \cup Y$.
    (b) Prove that two disjoint open sets or two disjoint closed sets in $(S, \mathcal{O})$ are separated.

13. Let $\mathcal{B}$ be a base for a topological space $(S, \mathcal{O})$. Prove that if $\mathcal{B}'$ is a collection of subsets of S such that $\mathcal{B} \subseteq \mathcal{B}' \subseteq \mathcal{O}$, then $\mathcal{B}'$ is a basis for $\mathcal{O}$.

14. Let $(S, \mathcal{O})$ be a topological space, U and U' two subsets of S, and $\mathcal{B}$ and $\mathcal{B}'$ two bases in the subspaces $(U, \mathcal{O}\upharpoonright_U)$ and $(U', \mathcal{O}\upharpoonright_{U'})$, respectively. Prove that $\mathcal{B} \vee \mathcal{B}'$ is a basis in the subspace $U \cup U'$.

    Solution: Let M be an open set in the subspace $U \cup V$. By the definition of the subspace topology, there exists an open set $L \in \mathcal{O}$ such that $M = L \cap (U \cup V) = (L \cap U) \cup (L \cap V)$, so L is the union of two open sets, $L \cap U$ and $L \cap U'$, in the subspaces U and U'. Since $\mathcal{B}$ is a basis in U, there is a subcollection $\mathcal{B}_1$ such that $L \cap U = \bigcup \mathcal{B}_1$. Similarly, $\mathcal{B}'$ contains a subcollection $\mathcal{B}'_1$ such that $L \cap U = \bigcup \mathcal{B}'_1$. Therefore, $M = \bigcup \mathcal{B}_1 \cup \bigcup \mathcal{B}'_1 = \bigcup \mathcal{B}_1 \vee \mathcal{B}'_1$.
15. Let S be an uncountable set and let $(S, \mathcal{O})$ be the cofinite topology on S.
    (a) Prove that every nonfinite set is dense.
    (b) Prove that there is no countable basis for this topological space. What does this say about Theorem 4.47?
16. Let $\mathcal{C}$ be the family of open intervals $\mathcal{C} = \{(a, b) \mid a, b \in \mathbb{R} \text{ and } ab > 0\}$. Prove that:
    (a) Every open set L of $(\mathbb{R}, \mathcal{O})$ contains a member of $\mathcal{C}$.
    (b) $\mathcal{C}$ is not a basis for the topology $\mathcal{O}$.
17. Let $\mathcal{C}$ be a chain of subsets of a set S such that $\bigcup \mathcal{C} = S$. Prove that $\mathcal{C}$ is the basis of a topology.
18. Prove that if $(S, \mathcal{O})$ is a topological space such that $\mathcal{O}$ is finite, then $(S, \mathcal{O})$ is compact.
19. Prove that the topological space $(\mathbb{R}, \mathcal{O})$ introduced in Example 4.4 is not compact.
20. Let $(S, \mathcal{O})$ be a compact space and let $\mathbf{H} = (H_0, H_1, \ldots)$ be a non-increasing sequence of nonempty and closed subsets of S. Prove that $\bigcap_{i \in \mathbb{N}} H_i$ is nonempty.
21. Let $(S_1, \mathcal{O}_1)$ and $(S_2, \mathcal{O}_2)$ be two topological spaces and let $f : S_1 \longrightarrow S_2$ be a continuous surjective function. Prove that if $(S_2, \mathcal{O}_2)$ is compact, then $(S_1, \mathcal{O}_1)$ is compact.
22. Let $f : \mathbb{R} \longrightarrow \mathbb{R}$ be a continuous function defined on the topological space $(\mathbb{R}, \mathcal{O})$. Prove that if $f(q) = 0$ for every $q \in \mathbb{Q}$, then $f(x) = 0$ for every $x \in \mathbb{R}$.
23. Let $f : \mathbb{R} \longrightarrow \mathbb{R}$ be a continuous function in x_0. Prove that if $f(x_0) > 0$, then there exists an open interval (a, b) such that $x_0 \in (a, b)$ and $f(x) > 0$ for every $x \in (a, b)$.
24. Let $(S, \mathcal{O}_{s_0})$ be the topological space defined in Exercise 2, where $s_0 \in S$. Prove that any continuous function $f : S \longrightarrow \mathbb{R}$ is a constant function.
25. Let $(S, \mathcal{O})$ and $(T, \mathcal{O}')$ be two topological spaces and let $\mathcal{B}'$ be a basis of $(T, \mathcal{O}')$. Prove that $f : S \longrightarrow T$ is continuous if and only if $f^{-1}(B) \in \mathcal{O}$ for every $B \in \mathcal{B}'$.

Let $(S_1, \mathcal{O}_1)$ and $(S_2, \mathcal{O}_2)$ be two topological spaces and let $f : S_1 \longrightarrow S_2$ be a function. Then f is an *open function* if $f(L)$ is open for every open set L, where $L \in \mathcal{O}_1$; the function f is a *closed function* if $f(H)$ is closed for every closed set H in S_1.

26. Let $(S_1, \mathcal{O}_1)$ and $(S_2, \mathcal{O}_2)$ be two topological spaces and let $\mathbf{K}_i$ and $\mathbf{I}_i$ be the closure and interior operators of the space S_i for $i = 1, 2$.

    (a) Prove that $f : S_1 \longrightarrow S_2$ is an open function if and only if $f(\mathbf{I}_1(U)) \subseteq \mathbf{I}_2(f(U))$ for every $U \in \mathcal{P}(S_1)$.
    (b) Prove that $f : S_1 \longrightarrow S_2$ is a closed function if and only if $\mathbf{K}_2(f(U)) \subseteq f(\mathbf{K}_1(U))$ for every $U \in \mathcal{P}(S_1)$.
    (c) Prove that a bijection $f : S_1 \longrightarrow S_2$ is open if and only if it is closed.
27. Prove that the function $f : \mathbb{R} \longrightarrow \mathbb{R}$ defined by $f(x) = x^2$ for $x \in \mathbb{R}$ is continuous but not open.
28. Prove that if $a < b$ and $c < d$, then the subspaces $[a, b]$ and $[c, d]$ are homeomorphic.
29. Let $(S, \mathcal{O})$ be a connected topological space and $f : S \longrightarrow \mathbb{R}$ be a continuous function. Prove that if $x, y \in S$, then for every $r \in [f(x), f(y)]$ there is $z \in S$ such that $f(z) = r$.
30. Let a and b be two real numbers such that $a \leqslant b$. Prove that if $f : [a, b] \longrightarrow [a, b]$ is a continuous function, then there is $c \in [a, b]$ such that $f(c) = c$.
31. Prove that a topological space $(S, \mathcal{O})$ is connected if and only if $\partial T = \emptyset$ implies $T \in \{\emptyset, S\}$ for every $T \in \mathcal{P}(S)$.

Let $(S, \mathcal{O})$ be a topological space and let x and y be two elements of S. A *continuous path* between x and y is a continuous function $f : [0, 1] \longrightarrow S$ such that $f(0) = x$ and $f(1) = y$. We refer to x as the *origin* and to y as the *destination* of f.

$(S, \mathcal{O})$ is said to be *arcwise connected* if any two points x and y are the origin and destination of a continuous path.

32. Prove that any arcwise connected topological space is connected.
33. Let $(S, \mathcal{O})$ be a T_0 topological space. Define the relation "$\leqslant$" on S by $x \leqslant y$ if $x \in \mathbf{K}(\{y\})$. Prove that $\leqslant$ is a partial order.
34. Let $(S, \mathcal{O})$ be a T_4 topological space.

    (a) Let H and H' be two closed sets and L be an open set such that $H \cap H' \subseteq L$. Prove that there exists two open sets U and U' such that $H \subseteq U$, $H' \subseteq U'$, and $L = U \cap U'$.
    (b) If $\{H_1, \ldots, H_p\}$ is a collection of closed sets such that $p \geqslant 2$ and $\bigcap_{i=1}^{p} H_i = \emptyset$, prove that there exists a family of open sets $\{U_1, \ldots, U_p\}$ such that $\bigcap_{i=1}^{p} U_i = \emptyset$ and $H_i \subseteq U_i$ for $1 \leqslant i \leqslant p$.

    Solution: Observe that the sets $H - L$ and $H' - L$ are closed and disjoint sets. Since $(S, \mathcal{O})$ is T_4, there are two disjoint open sets V and V' such that $H - L \subseteq V$ and $H' - L \subseteq V'$. Define the open sets U and U' as $U = V \cup L$ and $U' = V' \cup L$. It is clear that U and U' satisfy the requirements of the statement.

    The second part is an extension of Definition 4.87. The argument is by induction on p. The base case, $p = 2$, follows immediately from the definition of T_4 spaces.

Suppose that the statement holds for p, and let $\{H_1, \ldots, H_{p+1}\}$ be a collection of closed sets such that $\bigcap_{i=1}^{p+1} H_i = \emptyset$.

By applying the inductive hypothesis to the collection of p closed sets $\{H_1, \ldots, H_{p-1}, H_p \cap H_{p+1}\}$, we obtain the existence of the open sets $U_1, \ldots, U_{p-1}, U$ such that $H_i \subseteq U_i$ for $1 \leqslant i \leqslant p-1$, $H_p \cap H_{p+1} \subseteq U$, and $\left(\bigcap_{j=1}^{p-1} U_j\right) \cap U = \emptyset$. By the first part of this supplement, we obtain the existence of two open sets U_p and U_{p+1} such that $H_p \subseteq U_p$, $H_{p+1} \subseteq U_{p+1}$, and $U = U_p \cap U_{p+1}$. Note that $\bigcap_{j=1} U_j = \emptyset$, which concludes the argument.

35. Let $(S, \mathcal{O})$ be a T_4 topological space and let $\mathcal{L} = \{L_1, \ldots, L_p\}$ be an open cover of S.

   (a) Prove that for every k, $1 \leqslant k \leqslant p$ there exist k open sets $V_1, \ldots, V_k$ such that the collection $\{S - \mathbf{K}(V_1), \ldots, S - \mathbf{K}(V_k), L_{k+1}, \ldots, L_p\}$ is an open cover of S and for the closed sets $H_j = S - V_j$ we have $H_j \subseteq L_j$ for $1 \leqslant j \leqslant k$.

   (b) Conclude that for every open cover $\mathcal{L} = \{L_1, \ldots, L_p\}$ of S there is a closed cover $\mathcal{H} = \{H_1, \ldots, H_p\}$ of S such that $H_i \subseteq L_i$ for $1 \leqslant i \leqslant p$.

Solution: The proof of the first part is by induction on k, $1 \leqslant k \leqslant p$. For the base case, $k = 1$, observe that $S - L_1 \subseteq \bigcup_{j=2}^{p} L_j$ because $\mathcal{L}$ is a cover. Since $(S, \mathcal{O})$ is a T_4 space, there exists an open set V_1 such that $S - L_1 \subseteq V_1 \subseteq \mathbf{K}(V_1) \subseteq \bigcup_{j=2}^{p} L_j$. For $H_1 = S - V_1$, it is clear that $H_1 \subseteq L_1$ and $\{S - \mathbf{K}(V_1), L_2, \ldots, L_p\}$ is an open cover of S.

Suppose that the statement holds for k. This implies

$$S - L_{k+1} \subseteq \bigcup_{j=1}^{k} (S - \mathbf{K}(V_j)) \cup \bigcup_{j=k+2}^{p} L_j.$$

Again, by the property of T_4 spaces, there is an open set V_{k+1} such that

$$S - L_{k+1} \subseteq V_{k+1} \subseteq \mathbf{K}(V_{k+1}) \bigcup_{j=1}^{k} (S - \mathbf{K}(V_j)) \cup \bigcup_{j=k+2}^{p} L_j.$$

Thus, $\{S - \mathbf{K}(V_1), \ldots, S - \mathbf{K}(V_k), S - \mathbf{K}(V_{K+1}), L_{k+2}, \ldots, L_p\}$ is an open cover of S and $H_{k+1} = S - V_{k+1} \subseteq L_{k+1}$, which concludes the inductive step.

The second part follows immediately from the first by taking $k = p$. Indeed, since $\{S - \mathbf{K}(V_1), \ldots, S - \mathbf{K}(V_p)\}$ is a cover of S and $S - \mathbf{K}(V_i) \subseteq H_i$ for $1 \leqslant i \leqslant p$, it follows immediately that $\mathcal{H}$ is a cover of S.

36. Let $(S, \mathcal{O})$ be a T_4 topological space, $\mathcal{L} = \{L_1, \ldots, L_p\}$ be an open cover of S, and $\mathcal{H} = \{H_1, \ldots, H_p\}$ be a closed cover of S such that $H_i \subseteq L_i$ for $1 \leqslant i \leqslant p$ and $\bigcap \mathcal{H} = \emptyset$.

(a) Prove that for every k, $1 \leqslant k \leqslant p$ there exist k open sets $M_1, \ldots, M_k$ such that:
   (i) $H_j \subseteq M_j$ and $\mathbf{K}(M_j) \subseteq L_j$ for $1 \leqslant j \leqslant k$,
   (ii) the collection $\{M_1, \ldots, M_k, L_{k+1}, \ldots, L_p\}$ is an open cover of S, and
   (iii) $\bigcap_{i=1}^{k} \mathbf{K}(M_i) \cap \bigcap_{i=k+1}^{p} H_i = \emptyset$.

(b) Prove that there exists an open cover $\mathcal{M} = \{M_1, \ldots, M_p\}$ of S such that $M_i \subseteq L_i$ for $1 \leqslant i \leqslant p$ and $\bigcap \mathcal{M} = \emptyset$.

Solution: The proof of the first part is by induction on k, $1 \leqslant k \leqslant p$. For the base case, $k = 1$, observe that $H_1 \cap \bigcap_{i=2}^{p} H_i = \emptyset$ implies $H_1 \subseteq S - \bigcap_{i=2}^{p} H_i$, so $H_1 \subseteq L_1 \cap \left(S - \bigcap_{i=2}^{p} H_i\right)$. This implies the existence of an open set M_1 such that

$$H_1 \subseteq M_1 \subseteq \mathbf{K}(M_1) \subseteq L_1 \cap \left(S - \bigcap_{i=2}^{p} H_i \right),$$

which implies $\mathbf{K}(M_1) \subseteq L_1$ and $\mathbf{K}(M_1) \cap \bigcap_{i=2}^{p} H_i = \emptyset$.
Suppose that the statement holds for k. We have $H_{k+1} \subseteq L_{k+1}$ and, by the inductive hypothesis, $H_{k+1} \subseteq S - \left(\bigcap_{i=1}^{k} \mathbf{K}(M_i) \cap \bigcap_{i=k+2}^{p} H_i\right)$. Thus,

$$H_{k+1} \subseteq L_{k+1} \cap \left(S - \left(\bigcap_{i=1}^{k} \mathbf{K}(M_i) \cap \bigcap_{i=k+2}^{p} H_i \right)\right).$$

By the T_4 separation property, there exists an open set M_{k+1} such that

$$H_{k+1} \subseteq M_{k+1} \subseteq \mathbf{K}(M_{k+1}) \subseteq L_{k+1} \cap \left(S - \left(\bigcap_{i=1}^{k} \mathbf{K}(M_i) \cap \bigcap_{i=k+2}^{p} H_i \right)\right),$$

which implies $\mathbf{K}(M_{k+1}) \subseteq L_{k+1}$ and $\bigcap_{i=1}^{k+1} \mathbf{K}(M_i) \cap \bigcap_{i=k+2}^{p} H_i = \emptyset$.

The second part of the supplement follows directly from the first part.

37. Prove that if $(S, \mathcal{P}(S))$ and $(S', \mathcal{P}(S'))$ are two discrete topological spaces, then their product is a discrete topological space.
38. Let $(S, \mathcal{O})$, $(S, \mathcal{O}')$ be two topological spaces. Prove that the collection

$$\{S \times L' \mid L' \in \mathcal{O}'\} \cup \{L \times S' \mid L \in \mathcal{O}\}$$

is a subbase for the product topology $\mathcal{O} \times \mathcal{O}'$.

39. Let $(S, \mathcal{O})$, $(S, \mathcal{O}')$ be two topological spaces and let $(S \times S', \mathcal{O} \times \mathcal{O}')$ be their product.

(a) Prove that for all sets T, T' such that $T \subseteq S$ and $T' \subseteq S'$, $\mathbf{K}(T \times T') = \mathbf{K}(T) \times \mathbf{K}(T')$ and $\mathbf{I}(T \times T') = \mathbf{I}(T) \times \mathbf{I}(T')$.
(b) Prove that $\partial(T \times T') = (\partial(T) \times \mathbf{k}(T')) \cup (\mathbf{k}(T) \times \partial T')$.

40. Prove that the following classes of topological spaces are closed with respect to the product of topological spaces:

(a) the class of spaces that satisfy the first axiom of countability;
(b) the class of spaces that satisfy the second axiom of countability;
(c) the class of separable spaces.

41. Prove that, for a topological space $(S, \mathcal{O})$, the following statements are equivalent:

(a) $(S, \mathcal{O})$ is connected.
(b) If $S = L_1 \cup L_2$ and $L_1 \cap L_2 = \emptyset$, where L_1 and L_2 are open, then $L_1 = \emptyset$ or $L_2 = \emptyset$.
(c) If $S = H_1 \cup H_2$ and $H_1 \cap H_2 = \emptyset$, where H_1 and H_2 are closed, then $H_1 = \emptyset$ or $H_2 = \emptyset$.
(d) If K is a clopen set, then $K = \emptyset$ or $K = S$.

42. Prove that any subspace of a totally disconnected topological space is totally disconnected, and prove that a product of totally disconnected topological spaces is totally disconnected.
43. Let S be a set and let $\mathcal{C}$ be a collection of subsets of S. Define the collections of sets

$$\mathcal{C}' = \mathcal{C} \cup \{S - T \mid T \in \mathcal{C}\},$$
$$\mathcal{C}'' = \left\{\bigcap \mathcal{D} \mid \mathcal{D} \subseteq \mathcal{C}'\right\},$$
$$\mathcal{C}''' = \left\{\bigcup \mathcal{D} \mid \mathcal{D} \subseteq \mathcal{C}''\right\}.$$

Prove that $\mathcal{C}'''$ equals the σ-field generated by $\mathcal{C}$.
44. Let S and T be two sets and let $f : S \longrightarrow T$ be a function. Prove that if $\mathcal{E}'$ is a σ-field on T, then $\{f^{-1}(V) \mid V \in \mathcal{E}'\}$ is a σ-field on A.
45. Prove that any σ-field $\mathcal{E}$ contains the empty set; further, prove that if $\mathbf{s} = (S_0, S_1, \ldots)$ is a sequence of sets of $\mathcal{E}$, then both $\liminf \mathbf{s}$ and $\limsup \mathbf{s}$ belong to $\mathcal{E}$.
46. Let S be an infinite set and let $\mathcal{E}$ be the collection $\mathcal{E} = \{E \in \mathcal{P}(S) \mid E \text{ is finite or cofinite}\}$. Prove that $\mathcal{E}$ is a field of sets on X but not a σ-field.
47. Let S be a set and let $\mathcal{E}$ be a σ-field. Define the function $m : \mathcal{E} \longrightarrow \hat{\mathbb{R}}_{\geqslant 0}$ by

$$m(U) = \begin{cases} |U| & \text{if } U \text{ is finite,} \\ \infty & \text{otherwise,} \end{cases}$$

for $U \in \mathcal{P}(S)$. Prove that m is a measure.

48. Let $x, y, a_1, b_1, \ldots, a_n, b_n$ be n real numbers such that $x \leqslant y$ and $a_i \leqslant b_i$ for $1 \leqslant i \leqslant n$. Prove, by induction on n, that if $[x, y] \subseteq \bigcup_{i=1}^{n}(a_i, b_i)$, then $y - x \leqslant \sum_{i=1}^{n}(b_i - a_i)$.
49. Let $(S, \mathcal{E}, m)$ be a measure space. Prove that if $\mathbf{s} = (S_0, S_1, \ldots)$ is a sequence of sets such that $\sum_i m(S_i) < \infty$, then $m(\liminf \mathbf{s}) = 0$ (*the Borel-Cantelli lemma*).

    Solution: Let $T_p = \bigcup_{i=p}^{\infty} S_i$ for $p \in \mathbb{N}$. By the subadditivity of m, we have $m(T_p) \leqslant \sum_{i=p}^{\infty} m(S_i)$, and therefore $\lim_{p\to\infty} m(T_p) = 0$ because of the convergence of the series $\sum_i m(S_i)$. Since $\liminf \mathbf{s} = \bigcap_{p=0}^{\infty} \bigcup_{i=p}^{\infty} S_i = \bigcap_{p=0}^{\infty} T_p$, it follows that $m(\liminf \mathbf{s}) \leqslant m(T_p)$ for every $p \in \mathbb{N}$, so $m(\liminf \mathbf{s}) \leqslant \inf_p m(T_p) = 0$, which implies $m(\liminf \mathbf{s}) = 0$.
50. Let I be a bounded interval of $\mathbb{R}$. Prove that if K is a compact subset of $\mathbb{R}$ such that $K \subseteq I$, then $\mu(I) = \mu(K) + \mu(I - K)$, where μ is the Lebesgue outer measure.
51. Let $\{(S_i, \mathcal{E}_i, m_i) \mid i \in I\}$ be a collection of measure spaces such that $S_i S_| = \emptyset$ if $i \neq j$ for $i, j \in I$. Define the triplet $(\bigcup_{i\in I} S_i, \mathcal{E}, m)$, where

$$\mathcal{E} = \left\{ U \mid U \subseteq \bigcup_{i\in I} S_i, U \cap S_i \in \mathcal{E} \text{ for } i \in I \right\},$$

    and $m : \mathcal{E} \longrightarrow \hat{\mathbb{R}}_{\geqslant 0}$ is given by $m(U) = \sum_{i\in I} m_i(U \cap S_i)$ for $U \in \mathcal{E}$. Prove that $(\bigcup_{i\in I} S_i, \mathcal{E}, m)$ is a measure space and that $m(U)$ is finite if and only if there exists a countable subset J of I such that if $j \in J$, then μ_j is finite and $\mu_i = 0$ if $i \in I - J$.
52. The measure space $(S, \mathcal{E}, m)$ is *complete* if for every $W \in \mathcal{E}$ such that $m(W) = 0$, $U \subseteq W$ implies $U \in \mathcal{E}$. In Corollary 4.125 we saw that for every outer measure $\mu : \mathcal{P}(S) \longrightarrow \hat{\mathbb{R}}_{\geqslant 0}$ the triple $(S, \mathcal{E}_\mu, \mu)$ is a measure space. Prove that this space is complete.
53. Let $(S, \mathcal{E}, m)$ be a measure space. Define $\mathcal{E}' = \{U \cup T \mid U \in \mathcal{E}, T \subseteq W \in \mathcal{E} \text{ and } m(W) = 0\}$, and $m' : \mathcal{E}' \longrightarrow \overline{\mathbb{R}}_{\geqslant 0}$ by $m'(U \cup T) = m(U)$ for every set T such that $T \subseteq W \in \mathcal{E}$ and $m(W) = 0$.

    (a) Prove that $\mathcal{E}'$ is a σ-field that contains $\mathcal{E}$.
    (b) Prove that m' is a measure. The measure m' is known as the *completion of* m.

Bibliographical Comments

There are several excellent classic references on general topology [4–7]. A very readable introduction to topology is [8]. Fundamental references on measure theory are [2, 3].

Pioneering work in applying topology in data mining has been done in [9, 10]. Important references in measure theory are [2, 3, 11].

References

1. J.B. Listing, *Vorstudien zur Topologie* (Vanderhoek and Ruprecht, Göttingen, 1848)
2. M.E. Munroe, *Introduction to Measure and Integration* (Addison-Wesley, Reading, MA, 1959)
3. P.R. Halmos, *Measure Theory* (Van Nostrand, New York, 1959)
4. G. Edgar, *Measure, Topology, and Fractal Geometry* (Springer-Verlag, New York, 1990)
5. J.L. Kelley, *General Topology* (Van Nostrand, Princeton, NJ, 1955)
6. J. Dixmier, *General Topology* (Springer-Verlag, New York, 1984)
7. M. Eisenberg, *Topology* (Holt, Rinehart and Winston Inc, New York, 1974)
8. R. Engelking, K. Siekluchi, *Topology—A Geometric Approach* (Heldermann Verlag, Berlin, 1992)
9. B.-U. Pagel, F. Korn, C. Faloutsos, Deflating the dimensionality curse using multiple fractal dimensions. Paper presented in international conference on data, engineering, pp. 589–598, 2000
10. F. Korn, B.-U. Pagel, C. Faloutsos, On the dimensionality curse and the self similarity blessing. IEEE Trans. Knowl. Data Eng. **13**, 96–111 (2001)
11. M.E. Taylor, *Measure Theory and Integration* (American Mathematical Society, Providence, R.I., 2006)

Chapter 5
Linear Spaces

5.1 Introduction

Linear spaces are among the most important and widely used mathematical structures. Linear spaces consist of elements called *vectors* and are associated with a field (in most cases, the real field $\mathbb{R}$ or the complex field $\mathbb{C}$). The elements of this field are referred to as *scalars*. Two fundamental operations: vector addition and multiplication with scalars are defined such that certain axioms given are satisfied.

Linear spaces were introduced in their modern form by the Italian mathematician G. Peano in the second part of the 19th century; precursor ideas can be traced to more than two centuries before in connection with analytic geometry problems.

Definition 5.1 *Let L be a nonempty set and let $\mathcal{F} = (F, \{0, +, -, \cdot, \})$ be a field whose carrier is a set F. An F-*linear space *is a triple $(L, +, \cdot)$ such that $(L, \{\mathbf{0}, +, -\})$ is an Abelian group and $\cdot : F \times L \longrightarrow L$ is an operation such that the following conditions are satisfied*

(i) $a \cdot (b \cdot \boldsymbol{x}) = (a \cdot b) \cdot \boldsymbol{x}$,
(ii) $1 \cdot \boldsymbol{x} = \boldsymbol{x}$,
(iii) $a \cdot (\boldsymbol{x} + \boldsymbol{y}) = a \cdot \boldsymbol{x} + a \cdot \boldsymbol{y}$, *and*
(iv) $(a + b) \cdot \boldsymbol{x} = a \cdot \boldsymbol{x} + b \cdot \boldsymbol{x}$

for every $a, b \in F$ and $\boldsymbol{x}, \boldsymbol{y} \in L$.

If $\mathcal{F}$ is the field of real numbers (the field of complex numbers), then we will refer to any F-linear space as a *real linear space* (*complex linear space*).

The commutative binary operation of L is denoted by the same symbol "+" as the corresponding operation of the field F. The multiplication by a scalar, $\cdot$: $F \times L \longrightarrow L$ is also referred to as an *external operation* since its two arguments belong to two different sets, F and L. Again, this operation is denoted by the same symbol used for denoting the multiplication on F; if there is no risk of confusion, we shall write $a\mathbf{x}$ instead of $a \cdot \mathbf{x}$.

D. A. Simovici and C. Djeraba, *Mathematical Tools for Data Mining*,
Advanced Information and Knowledge Processing, DOI: 10.1007/978-1-4471-6407-4_5,

The elements of the set L will be denoted using bold letters $\mathbf{x}$, $\mathbf{y}$, $\mathbf{z}$, etc. The members of the field will be denoted by small letters from the beginning of the alphabet.

The additive element $\mathbf{0}$ is a special element called the *zero element*; every F-linear space must contain at least this element.

Example 5.2 The set $\mathbb{R}^n$ of n-tuples of real numbers is a real linear space under the definitions

$$\mathbf{x}+\mathbf{y} = \begin{pmatrix} x_1 + y_1 \\ \vdots \\ x_n + y_n \end{pmatrix} \text{ and } a \cdot \mathbf{x} = \begin{pmatrix} a \cdot x_1 \\ \vdots \\ a \cdot x_n \end{pmatrix}$$

of the operations $+$ and $\cdot$, where

$$\mathbf{x} = \begin{pmatrix} x_1 \\ \vdots \\ x_n \end{pmatrix} \text{ and } \mathbf{y} = \begin{pmatrix} y_1 \\ \vdots \\ y_n \end{pmatrix}.$$

In this linear space, the zero of the Abelian group is the n-tuple

$$\mathbf{0}_n = \begin{pmatrix} 0 \\ \vdots \\ 0 \end{pmatrix}.$$

Similarly, the set $\mathbb{C}^n$ of n-tuples of complex numbers is a complex linear space under the same formal definitions of vector sum and scalar multiplication as $\mathbb{R}^n$, where $a \in \mathbb{C}$ in this case.

Example 5.3 The set of infinite sequences of complex numbers $\mathbf{Seq}_\infty(\mathbb{C})$ can be organized as a linear space by defining the addition of two sequences

$$\mathbf{x} = (x_0, x_1, \ldots) \text{ and } \mathbf{y} = (y_0, y_1, \ldots)$$

as $\mathbf{x}+\mathbf{y} = (x_0+y_0, x_1+y_1, \ldots)$, and the multiplication by $c\mathbf{x}$ as $c\mathbf{x} = (cx_0, cx_1, \ldots)$ for $c \in \mathbb{C}$.

Example 5.4 The set of complex-valued functions defined on a set S is a complex linear space. The addition of functions is given by $(f+g)(s) = f(s)+g(s)$, and the multiplication of a function with a complex number is defined by $(af)(s) = af(s)$ for $s \in S$ and $a \in \mathbb{C}$.

Example 5.5 Let C be the set of real-valued continuous functions defined on $\mathbb{R}$,

$$C = \{f : \mathbb{R} \longrightarrow \mathbb{R} \mid f \text{ is continuous}\}.$$

Define $f + g$ by $(f + g)(x) = f(x) + g(x)$ and $(a \cdot f)(x) = a \cdot f(x)$ for $x \in \mathbb{R}$. The triple $(C, +, \cdot)$ is a real linear space.

Definition 5.6 *Let L be an F-linear space. A* linear combination of K *(where $K \subseteq L$) is a member of L of the form $c_1\boldsymbol{x}_1 + \cdots + c_n\boldsymbol{x}_n$, where $c_1, \ldots, c_n \in F$.*

A subset $K = \{\boldsymbol{x}_1, \ldots, \boldsymbol{x}_n\}$ of L is linearly independent *if $c_1\boldsymbol{x}_1 + \cdots + c_n\boldsymbol{x}_n = \boldsymbol{0}$ implies $c_1 = \cdots = c_n = 0$. If K is not linearly independent, we refer to K as a* linearly dependent *set.*

If $\mathbf{x} \neq \mathbf{0}$, then the set $\{\mathbf{x}\}$ is linearly independent. Of course, the set $\{\mathbf{0}\}$ is not linearly independent because $1\mathbf{0} = \mathbf{0}$. If K is a linearly independent subset of a linear space, then any subset of K is linearly independent.

Example 5.7 Let

$$\mathbf{e}_i = \begin{pmatrix} 0 \\ \vdots \\ 1 \\ \vdots \\ 0 \end{pmatrix}$$

be a vector that has a unique nonzero component equal to 1 in place i, where $1 \leqslant i \leqslant n$. The set $E = \{\mathbf{e}_1, \ldots, \mathbf{e}_n\}$ is linearly independent. Indeed, suppose that $c_1\mathbf{e}_1 + \cdots + c_n\mathbf{e}_n = \mathbf{0}$. This is equivalent to

$$\begin{pmatrix} c_1 \\ \vdots \\ c_n \end{pmatrix} = \begin{pmatrix} 0 \\ \vdots \\ 0 \end{pmatrix}$$

that is, with $c_1 = \cdots = c_n = 0$. Thus, E is linearly independent.

Theorem 5.8 *Let L be an F-linear space. A subset K of L is linearly independent if and only if for every $\boldsymbol{x} \in L$ there exists a linear combination of K, $\boldsymbol{x} = \sum_i c_i\boldsymbol{x}_i$ such that the coefficients c_i are uniquely determined.*

Proof Suppose that $\mathbf{x} = c_1\mathbf{x}_1 + \cdots + c_n\mathbf{x}_n = c_1{}'\mathbf{x}_1 + \cdots + c_n{}'\mathbf{x}_n$ and there exists i such that $c_i \neq c_i{}'$. This implies $\sum_{i=1}^n (c_i - c_i{}')\mathbf{x}_i = \mathbf{0}$, which contradicts the linear independence of K.

Definition 5.9 *A subset S of a linear space $(L, +, \cdot)$* spans *the space L (or S generates the linear space) if every $\boldsymbol{x} \in L$ is a linear combination of S.*

A basis *of the linear space $(L, +, \cdot)$ is a linearly independent subset that spans the linear space.*

In view of Theorem 5.8, a set B is a basis if every $\mathbf{x} \in L$ can be written uniquely as a linear combination of elements of B.

Definition 5.10 *A* subspace *of a F-linear space $(L, +, \cdot)$ is a nonempty subset U of L such that $\boldsymbol{x}, \boldsymbol{y} \in U$ implies $\boldsymbol{x} + \boldsymbol{y} \in U$ and $a \cdot \boldsymbol{x} \in U$ for every $a \in F$.*

If U is a subspace of an F-linear space, then U can be regarded as an F-linear space and various notions introduced for linear spaces are applicable to U.

The set $\{\mathbf{0}\}$ is a subspace of any F-linear space $(L, +, \cdot)$ included in every subspace of L.

Example 5.11 We saw in Example 5.5 that the set of real-valued continuous functions C defined on $\mathbb{R}$ is a real linear space.

The set of even real-valued continuous functions defined on $\mathbb{R}$ given by $E = \{f \in C \mid f(x) = f(-x)\}$ is a subspace of C. Indeed, note that for $f, g \in E$ we have

$$(f + g)(-x) = f(-x) + g(-x) = f(x) + g(x) = (f + g)(x)$$

and $(af)(-x) = af(-x) = af(x) = (af)(x)$, so $f, g \in E$ implies $f + g \in E$ and $af \in E$. Similarly, it is possible to show that the set of odd real-valued continuous functions defined on $\mathbb{R}$ $D = \{f \in C \mid f(-x) = -f(x)\}$ is a subspace of C.

Example 5.12 In Example 5.2 we saw that $\mathbb{R}^n$ can be regarded as an $\mathbb{R}$-linear space, while $\mathbb{C}^n$ is a $\mathbb{C}$-linear space. Even though $\mathbb{R}^n \subseteq \mathbb{C}^n$, $\mathbb{R}^n$ is not a subspace of $\mathbb{C}^n$ because for $a \in \mathbb{C}$ and $\mathbf{x} \in \mathbb{R}^n$ we do not have $a\mathbf{x} \in \mathbb{R}^n$.

If $\{K_i \mid i \in I\}$ is a nonempty collection of subspaces of a linear space, then $\bigcap\{K_i \mid i \in I\}$ is also a linear subspace. Thus, the family of subspaces of a linear space is a closure system. If U is a subset of a linear space $(L, +, \cdot)$ and $\mathbf{K}$ is the corresponding closure operator for the closure system of linear spaces, then we say that U is *spanning* the subspace $\mathbf{K}(U)$ of L. The subspace $\mathbf{K}(U)$ is said to be spanned by U. We denote this subspace by $\langle U \rangle$.

Theorem 5.13 *Let $\mathcal{L} = (L, +, \cdot)$ be an F-linear space. The following statements are equivalent:*

(i) *The finite set $K = \{\boldsymbol{x}_1, \ldots, \boldsymbol{x}_n\}$ is spanning the linear space $(L, +, \cdot)$ and K is minimal with this property.*
(ii) *K is a finite basis for $(L, +, \cdot)$.*
(iii) *The finite set K is linearly independent, and K is maximal with this property.*

Proof (i) implies (ii): We need to prove that K is linearly independent. Suppose that this is not the case. Then, there exist $c_1, \ldots, c_n \in F$ such that $c_1\mathbf{x}_1 + \cdots + c_n\mathbf{x}_n = \mathbf{0}$ and at least one of $c_1, \ldots, c_n$, say c_i, is nonzero. Then, $\mathbf{x}_i = -\frac{c_1}{c_i}\mathbf{x}_1 - \cdots - \frac{c_n}{c_i}\mathbf{x}_n$, and this implies that $K - \{\mathbf{x}_i\}$ also spans the linear space, thus contradicting the minimality of K.

(ii) implies (i): Let K be a finite basis. Suppose that K' is a proper subset of K that spans L. Then, if $\mathbf{z} \in K - K'$, z' is a linear combination of elements of K', which contradicts the fact that K is a basis.

We leave to the reader the proof of the equivalence between (ii) and (iii).

Corollary 5.14 *Every linear space that is spanned by a finite subset has a finite basis. Further, if B is a finite basis for an F-linear space $(L, +, \cdot)$, then each finite subset U of L such that $|U| = |B| + 1$ is linearly dependent.*

Proof This statement follows directly from Theorem 5.13.

Corollary 5.15 *If B and B' are two finite bases for a linear space $(L, +, \cdot)$, then* $|B| = |B'|$.

Proof If B is a finite basis, then $|B|$ is the maximum number of linearly independent elements in L. Thus, $|B'| \leqslant |B|$. Reversing the roles of B and B', we obtain $|B| \leqslant |B'|$, so $|B| = |B'|$.

Thus, the number of elements of a finite basis of L is a characteristic of L and does not depend on any particular basis.

Definition 5.16 *A linear space $(L, +, \cdot)$ is n-*dimensional *if there exists a basis of L such that $|B| = n$. The number n is the* dimension *of L and is denoted by* $\dim(L)$.

Theorem 5.17 *Let L be a finite-dimensional F-linear space and let $U = \{\boldsymbol{u}_1, \ldots, \boldsymbol{u}_k\}$ be a linearly independent subset of L. There exists an extension of U that is a basis of L.*

Proof If $\langle U \rangle = L$, then U is a basis of L. If this is not the case, let $\mathbf{w}_1 \in L - \langle U \rangle$. The set $U \cup \{\mathbf{w}_1\}$ is linearly independent and we have the strict inclusion $\langle U \rangle \subset \langle U \cup \{\mathbf{w}_1\} \rangle$. The subspace $\langle U \cup \{\mathbf{w}\} \rangle$ is $(k+1)$-dimensional. This argument can be repeated no more than $n - k$ times, where $n = \dim(L)$. Thus, $U \cup \{\mathbf{w}_1, \ldots, \mathbf{w}_{n-k}\}$ is a basis for L that extends U.

Definition 5.18 *Let L be an F-linear space and let U, V be subspaces of L. The* sum *of the subspaces U and V is the set $U + V$ defined by*

$$U + V = \{\boldsymbol{u} + \boldsymbol{v} \mid \boldsymbol{u} \in U \text{ and } \boldsymbol{v} \in V\}.$$

It is easy to verify that $U + V$ is also a subspace of L.

Theorem 5.19 *Let U, V be two subspaces of the finite-dimensional F-linear space L. We have* $\dim(U + V) + \dim(U \cap V) = \dim(U) + \dim(V)$.

Proof Suppose that $\{\mathbf{w}_1, \ldots, \mathbf{w}_k\}$ is a basis for $U \cap V$, where $k = \dim(U \cap V)$. This basis can be extended to a basis $\{\mathbf{w}_1, \ldots, \mathbf{w}_k, u_{k+1}, \ldots, u_p\}$ for U and to a basis $\{\mathbf{w}_1, \ldots, \mathbf{w}_k, v_{k+1}, \ldots, v_q\}$ for V.

Define $B = \{\mathbf{w}_1, \ldots, \mathbf{w}_k, u_{k+1}, \ldots, u_p, v_{k+1}, \ldots, v_q\}$. It is clear that $\langle B \rangle = U + V$. Suppose that there exist $c_1, \ldots, c_{p+q-k}$ such that

$$c_1 \boldsymbol{w}_1 + \cdots + c_k \boldsymbol{w}_k + c_{k+1} u_{k+1} + \cdots + c_p u_p + c_{p+1} v_{k+1} + \cdots + c_{p+q-k} v_q = \mathbf{0}.$$

The last equality implies

$$c_1 \mathbf{w}_1 + \cdots + c_k \mathbf{w}_k + c_{k+1} u_{k+1} + \cdots + c_p u_p = -c_{p+1} v_{k+1} - \cdots - c_{p+q-k} v_q.$$

Therefore, $c_1 \mathbf{w}_1 + \cdots + c_k \mathbf{w}_k + c_{k+1} u_{k+1} + \cdots + c_p u_p$ belongs to $U \cap V$, which implies $c_{k+1} = \cdots = c_p = 0$. Since

$$c_1\mathbf{w}_1 + \cdots + c_k\mathbf{w}_k + c_{p+1}v_{k+1} + \cdots + c_{p+q-k}v_q = \mathbf{0},$$

and $\{\mathbf{w}_1, \ldots, \mathbf{w}_k, v_{k+1}, \ldots, v_q\}$ is a basis for V, it follows that $c_1 = \cdots = c_k = c_{p+1} = \cdots = c_{p+q-k} = 0$.

This allows to conclude that $\dim(U+V) = p+q-k$ and this implies the equality of the theorem.

5.2 Linear Mappings

Linear mappings between linear spaces are functions that are compatible with the algebraic operations of linear spaces.

Definition 5.20 *Let L and K be two F-linear spaces. A* linear mapping *is a function $h : L \longrightarrow K$ such that $h(ax+by) = ah(\boldsymbol{x})+bh(\boldsymbol{y})$ for every $a, b \in F$ and $\boldsymbol{x}, \boldsymbol{y} \in L$.*

An affine mapping *is a function $f : L \longrightarrow K$ such that there exists a linear mapping $h : L \longrightarrow K$ and $\boldsymbol{b} \in K$ such that $f(\boldsymbol{x}) = h(\boldsymbol{x}) + \boldsymbol{b}$ for $\boldsymbol{x} \in L$.*

Linear mappings are also referred to as *homomorphisms*, as *morphisms*, or *linear operators* and it is the latter term that we usually use. The set of morphisms between two F-linear spaces L and K is denoted by $\mathsf{Hom}(L, K)$. The set of affine mappings between two F-linear spaces L and K is denoted by *Aff*(L, K).

A linear mapping $h : L \longrightarrow K$ is a morphism between the Abelian additive groups of the linear spaces; therefore, $h(\mathbf{0}_L) = \mathbf{0}_K$ and $h(-\mathbf{x}) = -h(\mathbf{x})$ for $\mathbf{x} \in L$.

Theorem 5.21 *Let L and K be two F-linear spaces having $\boldsymbol{0}_L$ and $\boldsymbol{0}_K$ as their zero elements, respectively. A morphism $h \in \mathsf{Hom}(L, K)$ is injective if and only if $h(\boldsymbol{x}) = \boldsymbol{0}_K$ implies $\boldsymbol{x} = \boldsymbol{0}_L$.*

Proof Let h be a morphism such that $h(\mathbf{x}) = \mathbf{0}_K$ implies $\mathbf{x} = \mathbf{0}_L$. If $h(\mathbf{x}) = h(\mathbf{y})$, by the linearity of h we have $h(\mathbf{x} - \mathbf{y}) = \mathbf{0}_K$, which implies $\mathbf{x} - \mathbf{y} = \mathbf{0}_L$, that is, $\mathbf{x} = \mathbf{y}$. Thus, h is injective.

Conversely, suppose that h is injective. If $\mathbf{x} \neq \mathbf{0}_L$, then $h(\mathbf{x}) \neq h(\mathbf{0}_L) = \mathbf{0}_K$. Thus, $h(\mathbf{x}) = \mathbf{0}_K$ implies $\mathbf{x} = \mathbf{0}_L$.

An *endomorphism* of an F-linear space L is a morphism $h : L \longrightarrow L$. The set of endomorphisms of L is denoted by $\mathsf{Endo}(L)$. Often, we refer to endomorphisms of L as *linear operators* on L.

The term *linear form* is reserved for linear mappings between an F-linear space and the field F itself, where F is considered as an F-linear space.

Example 5.22 For $a \in \mathbb{R}$ define the mapping $\mathsf{h}_a : \mathbb{R}^n \longrightarrow \mathbb{R}^n$ by $\mathsf{h}_a(\mathbf{x}) = a\mathbf{x}$ for $\mathbf{x} \in \mathbb{R}^n$. It is easy to verify that h_a is a linear operator on $\mathbb{R}^n$. This mapping is known as a *homotety on* $\mathbb{R}^n$.

If $a = 1$, then h_1 is given by $\mathsf{h}_1(\mathbf{x}) = \mathbf{x}$ for $\mathbf{x} \in \mathbb{R}^n$; this is the identity morphism of $\mathbb{R}^n$, which is usually denoted by 1_{R^n}.

For $a = 0$ we obtain the *zero morphism of* $\mathbb{R}^n$ given by $\mathsf{h}_0(\mathbf{x}) = \mathbf{0}$ for $\mathbf{x} \in \mathbb{R}^n$.

Example 5.23 The *translation* generated by $\mathbf{z} \in \mathbb{R}^n$ is the mapping $\mathsf{t}_{\mathbf{z}} : \mathbb{R}^n \longrightarrow \mathbb{R}^n$ defined by $\mathsf{t}_{\mathbf{z}}(\mathbf{x}) = \mathbf{x} + \mathbf{z}$ is a bijection but not a morphism unless $\mathbf{z} = \mathbf{0}$. Its inverse is $\mathsf{t}_{-\mathbf{z}}$.

Definition 5.24 *Let U and V be two subsets of $\mathbb{R}^n$. We define the subset $U + V$ of $\mathbb{R}^n$ as*

$$U + V = \{\boldsymbol{u} + \boldsymbol{v} \mid \boldsymbol{u} \in U \text{ and } \boldsymbol{v} \in V\}.$$

For $a \in \mathbb{R}$, the set aU is

$$aU = \{a\boldsymbol{u} \mid \boldsymbol{u} \in U\}.$$

If L, K are two linear spaces, then the set $\mathsf{Hom}(L, K)$ is never empty because the zero morphism $h_0 : L \longrightarrow K$ given by $h_0(\mathbf{x}) = \mathbf{0}_L$ for $\mathbf{x} \in K$ is always an element of $\mathsf{Hom}(L, K)$.

Definition 5.25 *Let L and K be two F-linear spaces. If $f, g \in \mathsf{Hom}(L, K)$, the* sum *$f + g$ is defined by $(f + g)(\boldsymbol{x}) = f(\boldsymbol{x}) + g(\boldsymbol{x})$ for $\boldsymbol{x} \in L$.*

The sum of two linear mappings is also a linear mapping because

$$\begin{aligned}(f + g)(a\mathbf{x} + b\mathbf{y}) &= f(a\mathbf{x} + b\mathbf{y}) + g(a\mathbf{x} + b\mathbf{y}) \\ &= af(\mathbf{x}) + bf(\mathbf{y}) + ag(\mathbf{x}) + bg(\mathbf{y}) \\ &= f(a\mathbf{x} + b\mathbf{y}) + g(a\mathbf{x} + b\mathbf{y}),\end{aligned}$$

for all $a, b \in F$ and $\mathbf{x}, \mathbf{y} \in L$.

Theorem 5.26 *Let M, P, Q be three F-linear spaces. The following properties of compositions of linear mappings hold:*

(i) *If $f \in \mathsf{Hom}(M, P)$ and $g \in \mathsf{Hom}(P, Q)$, then $gf \in \mathsf{Hom}(M, Q)$.*
(ii) *If $f \in \mathsf{Hom}(M, P)$ and $g_0, g_1 \in \mathsf{Hom}(P, Q)$, then*

$$f(g_0 + g_1) = fg_0 + fg_1.$$

(iii) *If $f_0, f_1 \in \mathsf{Hom}(M, P)$ and $g \in \mathsf{Hom}(P, Q)$, then*

$$(f_0 + f_1)g = f_0g + f_1g.$$

Proof We prove only the second part of the theorem and leave the proofs of the remaining parts to the reader.

Let $\mathbf{x} \in M$. Then, $f(g_0 + g_1)(\mathbf{x}) = f((g_0 + g_1)(\mathbf{x})) = f(g_0(\mathbf{x}) + g_1(\mathbf{x})) = f(g_0(\mathbf{x})) + f(g_1(\mathbf{x}))$ for $\mathbf{x} \in M$, which yields the desired equality.

We leave to the reader to verify that for any F-linear spaces M and P the algebra $(\mathsf{Hom}(M, P), \{h_0, +, -\})$ is an Abelian group that has the zero morphism h_0 as its zero-ary operations and the addition of linear mappings as its binary operation; the opposite of a linear mapping h is the mapping $-h$.

Moreover, $(\mathsf{Endo}(M), \{h_0, 1_M, +, -, \cdot\})$ is a unitary ring, where the multiplication is defined as the composition of linear mappings.

If M and P are F-linear spaces, $\mathsf{Hom}(M, P)$ is itself an linear space, where the multiplication of a morphism h by a scalar c is the morphism ch defined by $(ch)(\mathbf{x}) = c \cdot h(\mathbf{x})$. Indeed, the mapping ch is linear because

$$\begin{aligned}(ch)(a\mathbf{x} + b\mathbf{y}) &= c(ah(\mathbf{x}) + bh(\mathbf{y})) = cah(\mathbf{x}) + cbh(\mathbf{y})) \\ &= ach(\mathbf{x}) + bch(\mathbf{y})) = a(ch)(\mathbf{x}) + b(ch)(\mathbf{y})),\end{aligned}$$

for every $a, b, c \in F$ and $\mathbf{x}, \mathbf{y} \in M$.

Definition 5.27 *Let h be an endomorphism of a linear space M. The mth* iteration *of h (for $m \in \mathbb{N}$) is defined as*

(i) $h^0 = 1_M$;
(ii) $h^{m+1}(\mathbf{x}) = h(h^m(\mathbf{x}))$ *for* $m \in \mathbb{N}$.

For every $m \geqslant 1$, h^m is an endomorphism of M; this can be shown by a straightforward proof by induction on m.

Theorem 5.28 *Let L, M be two F-linear spaces and let $h : L \longrightarrow M$ be a morphism. Then, the sets*

$$\begin{aligned}\mathsf{Ker}(h) &= \{\mathbf{x} \in L \mid h(\mathbf{x}) = \mathbf{0}_M\}, \\ \mathsf{Img}(h) &= \{\mathbf{y} \in M \mid \mathbf{y} = h(\mathbf{x}) \text{ for some } \mathbf{x} \in L\}\end{aligned}$$

are subspaces of L and M, respectively.

Proof Let $\mathbf{u}, \mathbf{v} \in \mathsf{Ker}(h)$. Since $h(\mathbf{u}) = h(\mathbf{v}) = \mathbf{0}_M$ it follows that

$$h(a\mathbf{u} + b\mathbf{v}) = ah(\mathbf{u}) + bh(\mathbf{v}) = \mathbf{0}_M,$$

for $a, b \in F$, so $a\mathbf{u} + b\mathbf{v} \in \mathsf{Ker}(h)$. This shows that $\mathsf{Ker}(h)$ is indeed a subspace of L.

Let now $\mathbf{s}, \mathbf{t} \in \mathsf{Img}(h)$. There exist $\mathbf{x}, \mathbf{y} \in L$ such that $\mathbf{s} = h(\mathbf{x})$ and $\mathbf{t} = h(\mathbf{y})$. Therefore, $a\mathbf{s} + b\mathbf{t} = ah(\mathbf{x}) + bh(\mathbf{y}) = h(a\mathbf{x} + b\mathbf{y})$, hence $a\mathbf{s} + b\mathbf{t} \in \mathsf{Img}(h)$. This implies that $\mathsf{Img}(h)$ is a subspace of M.

Definition 5.29 *Let L be a F-linear space and let $M_1, \ldots, M_p$ be p linear subspaces of L. L is the* direct sum *of $M_1, \ldots, M_p$ if for every $\mathbf{x}$ of L there exists a unique sequence $(\mathbf{y}_1, \ldots, \mathbf{y}_p)$ such that $\mathbf{y}_i \in M_i$ for $1 \leqslant i \leqslant p$ and $\mathbf{x} = \mathbf{y}_1 + \cdots + \mathbf{y}_p$. This is denoted by $L = M_1 \boxplus M_2 \boxplus \cdots \boxplus M_p$.*

Observe that if $L = M_1 \boxplus M_2 \boxplus \cdots \boxplus M_p$, then the function $h_i : L \longrightarrow M_i$ given by $h_i(\mathbf{x}) = \mathbf{y}_i$ is well-defined due to the uniqueness of the sequence $(\mathbf{y}_1, \ldots, \mathbf{y}_p)$ for $1 \leqslant i \leqslant p$.

It is easy to verify that each h_i is a linear mapping. Indeed, if $a, b \in F$ and $\mathbf{u}, \mathbf{v} \in L$ can be uniquely written as $\mathbf{u} = \mathbf{y}_1 + \cdots + \mathbf{y}_p$ and $\mathbf{v} = \mathbf{z}_1 + \cdots + \mathbf{z}_p$, where $\mathbf{y}_i, \mathbf{z}_i \in M_i$ for $1 \leqslant i \leqslant p$, then $au + bv = (a\mathbf{y}_1 + b\mathbf{x}_1) + \cdots + (a\mathbf{y}_p + b\mathbf{z}_p)$. Since each M_i is a subspace, $a\mathbf{y}_i + b\mathbf{z}_i \in M_i$ for $1 \leqslant i \leqslant p$ and the uniqueness of the decomposition implies that $h_i(a\mathbf{u} + b\mathbf{v}) = ah_i(\mathbf{u}) + bh_i(\mathbf{v})$.

Each morphism h_i is idempotent, that is, $h_i(h_i(\mathbf{x})) = h_i(\mathbf{x})$ for $\mathbf{x} \in L$. Indeed, $h_i(\mathbf{x}) \in M$ and applying the uniqueness of the decomposition to $h_i(\mathbf{x})$ we obtain $h_i(h_i(\mathbf{x})) = h_i(\mathbf{x})$.

Theorem 5.30 *Let L be a linear space that is the direct sum $L = M_1 \boxplus M_2 \boxplus \cdots \boxplus M_p$. Then $M_i \cap M_j = \{0\}$ for every i, j such that $i \neq j$ and $1 \leqslant i, j \leqslant p$.*

Proof Suppose that $\mathbf{t} \neq \mathbf{0}$ belongs to $M_i \cap M_j$ for some i, j such that $i \neq j$ and $1 \leqslant i, j \leqslant p$. Let $\mathbf{x} = \mathbf{y}_1 + \cdots + \mathbf{y}_p$ be the unique decomposition of $\mathbf{x} \in L$ as a sum of vectors from the subspaces M_i. Since $\mathbf{t} \neq 0$ and $\mathbf{t} \in M_i \cap M_j$, we would have the distinct decompositions

$$\mathbf{x} = \mathbf{y}_1 + \cdots + (\mathbf{y}_i + \mathbf{t}) + \cdots + (\mathbf{y}_j - \mathbf{t}) + \cdots + \mathbf{y}_p$$
$$\mathbf{x} = \mathbf{y}_1 + \cdots + (\mathbf{y}_i - \mathbf{t}) + \cdots + (\mathbf{y}_j + \mathbf{t}) + \cdots + \mathbf{y}_p,$$

contradicting the uniqueness of the decomposition of $\mathbf{x}$.

Theorem 5.31 *Let L be a linear space. If $B_1, \ldots, B_p$ be bases in the subspaces $M_1, \ldots, M_p$, then $L = M_1 \boxplus M_2 \boxplus \cdots \boxplus M_p$ if and only if $B = \bigcup_{i=1}^{p}$ is a basis for L.*

Proof Suppose that $L = M_1 \boxplus M_2 \boxplus \cdots \boxplus M_p$. Each $\mathbf{x} \in L$ can be uniquely written as a sum $\mathbf{x} = \mathbf{y}_1 + \cdots + \mathbf{y}_p$, where $\mathbf{y}_i \in M_i$ for $1 \leqslant i \leqslant p$. It is clear that B spans L, so we need to show only that B is linearly independent.

Let $B_i = \{\mathbf{y}_1^i, \ldots, \mathbf{y}_{k_i}^i\}$ for $1 \leqslant i \leqslant p$. Suppose that

$$\sum_{i=1}^{p} \sum_{j=1}^{k_i} c_j^i y_j^i = \mathbf{0}.$$

Since $\mathbf{0}$ can be regarded as a sum of p copies of itself (where each copy is in M_i for $1 \leqslant i \leqslant p$, we have $\sum_{j=1}^{k_i} c_j^i y_j^i = \mathbf{0}$, so $c_j^i = 0$ for $1 \leqslant j \leqslant k_i$ and for $1 \leqslant i \leqslant p$. Thus, B is linearly independent and, therefore, it is a basis.

We leave to the reader the proof of the reverse implication.

If U, V are subspaces of L such that $L = U \boxplus V$, then U, V are said to be *complementary subspaces* of L.

Theorem 5.32 *Let $h : L \longrightarrow L$ be an idempotent endomorphism of the F-linear space L. Then, $L = \mathsf{Ker}(h) \boxplus \mathsf{Img}(h)$.*

Proof If $\mathbf{t} \in \mathsf{Ker}(h) \cap \mathsf{Img}(h)$, we have $h(\mathbf{t}) = \mathbf{0}$ and $t = h(\mathbf{z})$ for some $\mathbf{z} \in L$. Thus, $\mathbf{t} = h(\mathbf{z}) = h(h(\mathbf{z})) = h(\mathbf{t}) = \mathbf{0}$, which implies $\mathsf{Ker}(h) \cap \mathsf{Img}(h) = \{\mathbf{0}\}$.

Since $h(\mathbf{x}) = h(h(\mathbf{x}))$ for every $\mathbf{x} \in L$, it follows that $h(\mathbf{x} - h(\mathbf{x})) = \mathbf{0}$, so $\mathbf{y} = \mathbf{x} - h(\mathbf{x}) \in \mathsf{Ker}(h)$. This allows us to write $\mathbf{x} = \mathbf{y} + \mathbf{z}$, where $z = h(\mathbf{x}) \in \mathsf{Img}(h)$, which shows that every element $\mathbf{x}$ of L can be written as a sum of an element in $\mathsf{Ker}(h)$ and an element in $\mathsf{Img}(h)$.

Suppose now that $\mathbf{x} = \tilde{\mathbf{y}} + \tilde{\mathbf{z}}$, where $\tilde{\mathbf{y}} \in \mathsf{Ker}(h)$ and $\tilde{\mathbf{z}} \in \mathsf{Img}(h)$. Since $\mathbf{y} - \tilde{\mathbf{y}} = \tilde{\mathbf{z}} - \mathbf{z} = \mathbf{0}$, it follows that the expression of $\mathbf{t}$ as a sum of two vectors in $\mathsf{Ker}(h)$ and $\mathsf{Img}(h)$ is unique, so $L = \mathsf{Ker}(h) \boxplus \mathsf{Img}(h)$.

Theorem 5.33 *Let U and V be two subspaces of an F-linear space L. If $L = U \boxplus V$, then there exists an idempotent endomorphism h of L such that $U = \mathsf{Ker}(h)$ and $V = \mathsf{Img}(h)$.*

Proof Let $\mathbf{x}$ be a vector in L and let $\mathbf{x} = \mathbf{u} + \mathbf{v}$ be the decomposition of $\mathbf{x}$, where $\boldsymbol{u} \in U$ and $\boldsymbol{v} \in V$. Define the mapping $h : L \longrightarrow L$ as $h(\mathbf{x}) = \boldsymbol{v}$. The uniqueness of the decomposition of h implies that h is well-defined.

Note that a vector $\boldsymbol{u} \in U$ has the decomposition $\boldsymbol{u} = \boldsymbol{u} + \mathbf{0}$, so $h(\boldsymbol{u}) = \mathbf{0}$. Thus, $U = \mathsf{Ker}(h)$. Since $h(h(\mathbf{x})) = h(\mathbf{v}) = h(\mathbf{u} + \mathbf{v}) = h(\mathbf{x})$, it follows that h is idempotent.

If $\mathbf{w} \in W$, its decomposition is $\mathbf{w} = \mathbf{0} + \mathbf{w}$, so $h(\mathbf{w}) = \mathbf{w}$. Therefore, $W = \mathsf{Img}(h)$.

5.3 Matrices

Definition 5.34 *Let F be a field. A* matrix on F *is a function*

$$A : \{1, \ldots, m\} \times \{1, \ldots, n\} \longrightarrow F.$$

The pair (m, n) is the format *of the matrix A.*

If $A : \{1, \ldots, m\} \times \{1, \ldots, n\} \longrightarrow F$ is a matrix on F, we say that A is an $(m \times n)$-matrix on F. The set of all such matrices will be denoted by $F^{m \times n}$.

A matrix $A \in F^{m \times n}$ can be written as

$$\begin{pmatrix} A(1,1) & A(1,2) & \ldots & A(1,n) \\ A(2,1) & A(2,2) & \ldots & A(2,n) \\ \vdots & \vdots & \ldots & \vdots \\ A(m,1) & A(m,2) & \ldots & A(m,n) \end{pmatrix}.$$

Alternatively, a matrix $A \in F^{m \times n}$ can be regarded as consisting of m rows, where each row is a sequence of the form

$$(A(i,1), A(i,2), \ldots, A(i,n)),$$

for $1 \leqslant i \leqslant n$, or as a collection of n columns of the form

$$\begin{pmatrix} A(1,j) \\ A(2,j) \\ \vdots \\ A(m,j) \end{pmatrix},$$

where $1 \leqslant j \leqslant m$.

Example 5.35 Let $F = \{0, 1\}$. The matrix

$$\begin{pmatrix} 1 & 0 & 1 \\ 0 & 1 & 1 \end{pmatrix},$$

is a (3×2)-matrix on the set F.

The element $A(i, j)$ of the matrix A will be denoted by a_{ij} or by A_{ij} and the matrix itself will be writen as $A = (a_{ij})$.

Definition 5.36 *A* square matrix *on F is an $(n \times n)$-matrix on F for some $n \geqslant 1$.*

Let $A \in F^{n \times n}$. The *main diagonal* of the matrix A is the sequence $(a_{11}, \ldots, a_{nn})$. The set $\{a_{ij} \mid 1 \leqslant i, j \leqslant n \text{ and } i - j = k\}$ consists of elements located on the kth diagonal above the main diagonal, while $\{a_{ij} \mid j - i = k\}$ consists of elements located on the kth diagonal blow the main diagonal for $1 \leqslant k \leqslant n - 1$.

A square matrix $A \in F^{n \times n}$ is *diagonal* if $i \neq j$ implies $a_{ij} = 0$ for $1 \leqslant i, j \leqslant n$. A diagonal matrix $A \in F^{n \times n}$ having the diagonal elements $d_1, \ldots, d_n$ is be denoted as $A = \mathsf{diag}(d_1, \ldots, d_n)$.

A matrix $A \in F^{n \times n}$ is *upper triangular* (*lower triangular*) if $i > j$ implies $a_{ij} = 0$ ($j > i$ implies $a_{ij} = 0$).

Example 5.37 The matrices $A, B \in F^{4 \times 4}$ defined by

$$A = \begin{pmatrix} a_{11} & a_{12} & a_{13} & a_{14} \\ 0 & a_{22} & a_{23} & a_{24} \\ 0 & 0 & a_{33} & a_{34} \\ 0 & 0 & 0 & a_{44} \end{pmatrix} \text{ and } B = \begin{pmatrix} b_{11} & 0 & 0 & 0 \\ b_{21} & b_{22} & 0 & 0 \\ b_{31} & b_{32} & b_{33} & 0 \\ b_{41} & b_{42} & b_{43} & a_{44} \end{pmatrix}$$

are upper triangular and lower triangular, respectively.

Definition 5.38 *Let $A = (a_{ij}) \in \mathbb{R}^{m \times n}$ be a matrix on the set of real numbers. Its* transpose *is the matrix $A' \in \mathbb{R}^{n \times m}$ defined by $(A')_{ij} = a_{ji}$ for $1 \leqslant i \leqslant n$ and $1 \leqslant j \leqslant m$.*

If $A' = A$, we say that A is a symmetric matrix.
A matrix A is skew-symmetric *if $A' = -A$.*

It is easy to verify that $(A')' = A$.
A similar notion exists for complex matrices.

Definition 5.39 *Let* $A = (a_{ij}) \in \mathbb{C}^{m\times n}$. *Its* Hermitian conjugate *is the matrix* $A^{\mathsf{H}} \in \mathbb{C}^{n\times m}$ *defined by* $(A^{\mathsf{H}})_{ij} = \overline{a_{ji}}$ *for* $1 \leqslant i \leqslant n$ *and* $1 \leqslant j \leqslant m$.

If $A^{\mathsf{H}} = A$, *we say that* A *is a* Hermitian matrix.

A *is* skew-Hermitian *if* $A^{\mathsf{H}} = -A$.

Example 5.40 The transpose of the matrix

$$A = \begin{pmatrix} 1 & 0 & 2 \\ 0 & -1 & 3 \end{pmatrix} \in \mathbb{R}^{2\times 3}$$

is the matrix

$$A' = \begin{pmatrix} 1 & 0 \\ 0 & -1 \\ 2 & 3 \end{pmatrix} \in \mathbb{R}^{3\times 2}.$$

The Hermitian conjugate of the matrix

$$L = \begin{pmatrix} 1+i & 0 & 2-3i \\ 2 & 1-i & i \end{pmatrix} \in \mathbb{C}^{2\times 3}$$

is the matrix

$$L^{\mathsf{H}} = \begin{pmatrix} 1-i & 2 \\ 0 & 1+i \\ 2+3i & -i \end{pmatrix} \in \mathbb{C}^{3\times 2}.$$

A complex matrix having real entries is symmetric if and only if it is Hermitian.

Dissimilarities defined on finite sets can be represented by matrices. If $S = \{x_1, \ldots, x_n\}$ is a finite set and $d : S \times S \longrightarrow \mathbb{R}_{\geqslant 0}$ is a dissimilarity, let $M_d \in (\mathbb{R}_{\geqslant 0})^{n\times n}$ be the matrix defined by $M_{ij} = d(x_i, x_j)$ for $1 \leqslant i, j \leqslant n$. Clearly, all main diagonal elements of M_d are 0 and the matrix M is symmetric.

Example 5.41 Let S be the set $\{x_1, x_2, x_3, x_4\}$. The discrete metric on S is represented by the 4×4-matrix

$$M_d = \begin{pmatrix} 0 & 1 & 1 & 1 \\ 1 & 0 & 1 & 1 \\ 1 & 1 & 0 & 1 \\ 1 & 1 & 1 & 0 \end{pmatrix}.$$

If $x_1, x_2, x_3 \in \mathbb{R}$ are three real numbers the matrix that represents the distance $e(x_i, x_j) = |x_i - x_j|$ measured on the real line is

$$M_e = \begin{pmatrix} 0 & |x_1 - x_2| & |x_1 - x_3| \\ |x_1 - x_2| & 0 & |x_2 - x_3| \\ |x_1 - x_3| & |x_2 - x_3| & 0 \end{pmatrix}.$$

Definition 5.42 *The* $(n \times n)$-unit matrix *on the field* $\mathcal{F} = (F, \{0, +, -, \cdot\})$ *is the square matrix* $I_n \in F^{n \times n}$ *given by*

$$I_n = \begin{pmatrix} 1 & 0 & 0 & \cdots & 0 \\ 0 & 1 & 0 & \cdots & 0 \\ 0 & 0 & 1 & \cdots & 0 \\ \vdots & \vdots & \vdots & \cdots & \vdots \\ 0 & 0 & 0 & \cdots & 1 \end{pmatrix},$$

whose entries located outside its main diagonal are 0s.

The $(m \times n)$-zero matrix *is the* $(m \times n)$*-matrix* $O_{m,n} \in F^{n \times n}$ *given by*

$$O_{m,n} = \begin{pmatrix} 0 & 0 & 0 & \cdots & 0 \\ 0 & 0 & 0 & \cdots & 0 \\ 0 & 0 & 0 & \cdots & 0 \\ \vdots & \vdots & \vdots & \cdots & \vdots \\ 0 & 0 & 0 & \cdots & 0 \end{pmatrix}.$$

Definition 5.43 *Let* $A, B \in F^{m \times n}$ *be two matrices that have the same format. The* sum *of the matrices* A *and* B *is the matrix* $A + B$ *having the same format and defined by*

$$(A + B)_{ij} = a_{ij} + b_{ij}$$

for $1 \leqslant i \leqslant m$ *and* $1 \leqslant j \leqslant n$.

Example 5.44 The sum of the matrices $A, B \in \mathbb{R}^{2 \times 3}$ given by

$$A = \begin{pmatrix} 1 & -2 & 3 \\ 0 & 2 & -1 \end{pmatrix} \text{ and } B = \begin{pmatrix} -1 & 2 & 3 \\ 1 & 4 & 2 \end{pmatrix}$$

is the matrix

$$A + B = \begin{pmatrix} 0 & 0 & 6 \\ 1 & 6 & 1 \end{pmatrix}.$$

It is easy to verify that the matrix sum is an associative and commutative operation on $F^{m \times n}$; that is,

$$A + (B + C) = (A + B) + C,$$
$$A + B = B + A,$$

for all $A, B, C \in F^{m \times n}$.

The zero matrix $O_{m,n}$ acts as an additive unit on the set $F^{m \times n}$; that is,

$$A + O_{m,n} = O_{m,n} + A$$

for every $A \in S^{m\times n}$.

The additive inverse, or the opposite of a matrix $A \in F^{m\times n}$, is the matrix $-A$ given by $(-A)_{ij} = -A_{ij}$ for $1 \leqslant i \leqslant m$ and $1 \leqslant j \leqslant n$.

Example 5.45 The opposite of $A \in \mathbb{R}^{2\times 3}$, given by

$$A = \begin{pmatrix} 1 & -2 & 3 \\ 0 & 2 & -1 \end{pmatrix}$$

is the matrix

$$-A = \begin{pmatrix} -1 & 2 & -3 \\ 0 & -2 & 1 \end{pmatrix}.$$

It is immediate that $A + (-A) = O_{2,3}$.

The set of matrices $F^{m\times n}$ is an F-linear space. Furthermore, it is easy to see that the sets of symmetric matrices and skew-symmetric matrices are subspaces of the linear space of square matrices $\mathbb{R}^{n\times n}$ and the sets of Hermitian and skew-Hermitian matrices are subspaces of $\mathbb{C}^{n\times n}$.

Definition 5.46 *Let $A \in F^{m\times n}$ and $B \in F^{n\times p}$ be two matrices. The* product *of the matrices A, B is the matrix $C \in F^{m\times p}$ defined by $c_{ik} = \sum_{j=1}^{n} a_{ij} b_{jk}$, where $1 \leqslant i \leqslant m$ and $1 \leqslant k \leqslant p$. The product of the matrices A, B will be denoted by AB.*

The matrix product is a partial operation because in order to multiply two matrices A and B, they must have the formats $m \times n$ and $n \times p$, respectively. In other words, the number of columns of the first matrix must equal the number of rows of the second matrix.

Theorem 5.47 *Matrix multiplication is associative.*

Proof Let $A \in F^{m\times n}$, $B \in F^{n\times p}$, and $C \in F^{p\times r}$ be three matrices. We prove that $(AB)C = A(BC)$.

By applying the definition of the matrix product, we have

$$\begin{aligned}((AB)C)_{i\ell} &= \sum_{k=1}^{p} (AB)_{ik} C_{k\ell} = \sum_{k=1}^{p} \left(\sum_{j=1}^{n} A_{ij} B_{jk} \right) C_{k\ell} \\ &= \sum_{j=1}^{n} A_{ij} \sum_{k=1}^{p} B_{jk} C_{k\ell} = \sum_{j=1}^{n} A_{ij} (BC)_{j\ell} = (A(BC))_{i\ell}\end{aligned}$$

for $1 \leqslant i \leqslant m$ and $1 \leqslant \ell \leqslant r$, which shows that matrix multiplication is indeed associative.

Theorem 5.48 *If $A \in F^{m\times n}$, then $I_m A = A I_n = A$.*

Proof The statement follows immediately from the definition of a matrix product.

Note that if $A \in F^{n\times n}$, then $I_n A = AI_n = A$, so I_n is a unit relative to matrix multiplication considered as an operation on the set of square matrices $F^{n\times n}$.

The product of matrices is *not* commutative. Indeed, consider the matrices $A, B \in \mathbb{R}^{2\times 2}$ defined by

$$A = \begin{pmatrix} 0 & 1 \\ 2 & 3 \end{pmatrix} \text{ and } B = \begin{pmatrix} -1 & 1 \\ 1 & 0 \end{pmatrix}.$$

We have

$$AB = \begin{pmatrix} 1 & 0 \\ 1 & 2 \end{pmatrix} \text{ and } BA = \begin{pmatrix} 2 & 2 \\ 0 & 1 \end{pmatrix},$$

so $AB \neq BA$.

For $A \in \mathbb{C}^{n\times n}$ the power A^n, where $n \in \mathbb{N}$ is defined inductively by $A^0 = I_n$ and $A^{n+1} = A^n\, A$. It is immediate that $A^1 = A$. This allows us to define the matrix $f(A)$, where f is a polynomial with complex coefficients, $f(x) = a^n x^n + a_{n-1}x^{n-1} + \cdots + a_0$, as

$$f(A) = a^n A^n + a_{n-1}A^{n-1} + \cdots + a_0 I_n.$$

The product of two lower (upper) triangular matrices lower (upper) triangular matrix. Therefore, any power of a lower (upper) triangular matrix is a triangular matrix.

Theorem 5.49 *If $T \in \mathbb{C}^{m\times m}$ is an upper (a lower) triangular matrix and f is a polynomial, then $f(T)$ is an upper (a lower) triangular matrix. Furthermore, if the diagonal elements of T are $t_{11}, t_{22}, \ldots, t_{mm}$, then the diagonal elements of $f(T)$ are $f(t_{11}), f(t_{22}), \ldots, f(t_{mm})$, respectively.*

Proof Since every power T^k of T is an upper (a lower) triangular matrix, and the sum of upper (lower) triangular matrices is upper (lower) triangular, if follows that $f(T)$ is an upper triangular (a lower triangular) matrix.

An easy argument by induction on k (left to the reader) shows that if the diagonal elements of T are $t_{11}, t_{22}, \ldots, t_{mm}$, then the diagonal elements of T^k are $t_{11}^k, t_{22}^k, \ldots, t_{mm}^k$. The second part of the theorem follows immediately.

Definition 5.50 *Let $A = (a_{ij}) \in \mathbb{C}^{n\times n}$ be a square matrix. The* trace *of A is the number $trace(A)$ given by*

$$trace(A) = a_{11} + a_{22} + \cdots + a_{nn}.$$

Theorem 5.51 *Let A and B be two square matrices in $\mathbb{C}^{n\times n}$. We have:*

(i) $trace(aA) = a\, trace(A)$;
(ii) $trace(A + B) = trace(A) + trace(B)$, *and*
(iii) $trace(AB) = trace(BA)$.

Proof The first two parts are direct consequences of the definition of the trace. For the last part we can write:

$$trace(AB) = \sum_{i=1}^{n} (AB)_{ii} = \sum_{i=1}^{n} \sum_{j=1}^{n} a_{ij} b_{ji}.$$

Exchanging the subscripts i and j and, then the order of the summations, we have

$$\sum_{i=1}^{n} \sum_{j=1}^{n} a_{ij} b_{ji} = \sum_{j=1}^{n} \sum_{i=1}^{n} a_{ji} b_{ij} = \sum_{i=1}^{n} \sum_{j=1}^{n} b_{ij} a_{ji} = \sum_{i=1}^{n} (BA)_{ii},$$

which proves the desired equality.

Let A, B, C be three matrices in $\mathbb{C}^{n \times n}$. We have

$$trace(ABC) = trace((AB)C) = trace(C(AB)) = trace(CAB),$$

and

$$trace(ABC) = trace(A(BC)) = trace((BC)A) = trace(BCA).$$

However, it is important to notice that the third part of Theorem 5.51 *cannot* be extended to arbitary permutations of a product of matrices. Consider, for example the matrices

$$A = \begin{pmatrix} 1 & 0 \\ 1 & 1 \end{pmatrix},\ B = \begin{pmatrix} 1 & 1 \\ 1 & 0 \end{pmatrix},\ \text{and } C = \begin{pmatrix} 1 & 1 \\ 0 & 1 \end{pmatrix}.$$

We have

$$ABC = \begin{pmatrix} 1 & 2 \\ 2 & 3 \end{pmatrix} \text{ and } ACB = \begin{pmatrix} 2 & 1 \\ 3 & 1 \end{pmatrix},$$

so $trace(ABC) = 4$ and $trace(ACB) = 3$.

Definition 5.52 *A matrix $A \in \mathbb{C}^{m \times n}$ is* non-negative *if all its entries a_{ij} are real numbers and $a_{ij} \geqslant 0$ for $1 \leqslant i \leqslant m$ and $1 \leqslant j \leqslant n$. This is denoted by $A \geqslant O_{m,n}$.*

A is positive *if all its entries are real numbers, and $a_{ij} > 0$ for $1 \leqslant i \leqslant m$ and $1 \leqslant j \leqslant n$. This is denoted by $A > 0_{m,n}$.*

If $B, C \in \mathbb{R}^{m \times n}$ we write $B \geqslant C$ ($B > C$) if $B - C \geq O_{m,n}$ ($B - C > O_{m,n}$, respectively).

The sets of non-negative (non-positive, positive, negative) $m \times n$-matrices is denoted by $\mathbb{R}^{m \times n}_{\geqslant 0}$ $\left(\mathbb{R}^{m \times n}_{\leqslant 0}, \mathbb{R}^{m \times n}_{>0}, \mathbb{R}^{m \times n}_{<0}\right.$, respectively$\left.\right)$.

Example 5.53 The diagonal matrix I_n is non-negative but not positive.

Definition 5.54 *A matrix* $A \in S^{n\times n}$ *is* nilpotent *if there is* $m \in \mathbb{N}$ *such that* $A^m = O_{n,n}$. *The* nilpotency *of* A *is the number* $\mathsf{nilp}(A) = \min\{m \in \mathbb{N} \mid A^m = O_{n,n}\}$.

In other words, if $A \in S^{n\times n}$ is a nilpotent matrix, we have $\mathsf{nilp}(A) = m$ if and only if $A^m = O_{n,n}$ but $A^{m-1} \neq O_{n,n}$.

Example 5.55 Let a and b be two positive numbers in $\mathbb{R}$. The matrix $A \in \mathbb{R}^{3\times 3}$ given by

$$A = \begin{pmatrix} 0 & a & 0 \\ 0 & 0 & b \\ 0 & 0 & 0 \end{pmatrix}$$

is nilpotent because

$$A^2 = \begin{pmatrix} 0 & 0 & ab \\ 0 & 0 & 0 \\ 0 & 0 & 0 \end{pmatrix} \text{ and } A^3 = \begin{pmatrix} 0 & 0 & 0 \\ 0 & 0 & 0 \\ 0 & 0 & 0 \end{pmatrix}$$

Thus, $\mathsf{nilp}(A) = 3$.

Definition 5.56 *A matrix* $A \in S^{n\times n}$ *is* idempotent *if* $A^2 = A$.

Example 5.57 The matrix

$$A = \begin{pmatrix} 0.5 & 1 \\ 0.25 & 0.5 \end{pmatrix}$$

is idempotent, as the reader can easily verify.

Let $A \in F^{m\times n}$ be a matrix and suppose that $m = m_1 + \cdots + m_p$ and $n = n_1 + \cdots + n_q$, where F is the real or the complex field. A *partitioning of* A is a collection of matrices $A_{hk} \in F^{m_h\times n_k}$ such that A_{hk} is the contiguous submatrix

$$A\begin{bmatrix} m_1 + \cdots + m_{h-1} + 1, \ldots, m_1 + \cdots + m_{h-1} + m_h \\ n_1 + \cdots + n_{k-1} + 1, \ldots, n_1 + \cdots + n_k \end{bmatrix},$$

for $1 \leqslant h \leqslant p$ and $1 \leqslant k \leqslant q$.

If $\{A_{hk} \mid 1 \leqslant h \leqslant p \text{ and } 1 \leqslant k \leqslant q\}$ is a partitioning of A, A is written as

$$A = \begin{pmatrix} A_{11} & A_{12} & \cdots & A_{1q} \\ A_{21} & A_{22} & \cdots & A_{2q} \\ \vdots & \vdots & \cdots & \vdots \\ A_{p1} & A_{p2} & \cdots & A_{pq} \end{pmatrix}.$$

The matrices A_{hk} are referred to as the *blocks* of the partitioning. All blocks located in a column must have the number of columns; all blocks located in a row must have the same number of rows.

Example 5.58 The matrix $A \in F^{5\times 6}$ given by

$$A = \begin{pmatrix} a_{11} & a_{12} & a_{13} & a_{14} & a_{15} & a_{16} \\ a_{21} & a_{22} & a_{23} & a_{24} & a_{25} & a_{26} \\ a_{31} & a_{32} & a_{33} & a_{34} & a_{35} & a_{36} \\ a_{41} & a_{42} & a_{43} & a_{44} & a_{45} & a_{46} \\ a_{51} & a_{52} & a_{53} & a_{54} & a_{55} & a_{56} \end{pmatrix}$$

can be partitioned as

$$\left(\begin{array}{ccc|c|cc} a_{11} & a_{12} & a_{13} & a_{14} & a_{15} & a_{16} \\ a_{21} & a_{22} & a_{23} & a_{24} & a_{25} & a_{26} \\ a_{31} & a_{32} & a_{33} & a_{34} & a_{35} & a_{36} \\ \hline a_{41} & a_{42} & a_{43} & a_{44} & a_{45} & a_{46} \\ a_{51} & a_{52} & a_{53} & a_{54} & a_{55} & a_{56} \end{array}\right)$$

Thus, if we introduce the matrices

$$A_{11} = \begin{pmatrix} a_{11} & a_{12} & a_{13} \\ a_{21} & a_{22} & a_{23} \\ a_{31} & a_{32} & a_{33} \end{pmatrix}, \; A_{12} = \begin{pmatrix} a_{14} \\ a_{24} \\ a_{34} \end{pmatrix}, \; A_{13} = \begin{pmatrix} a_{15} & a_{16} \\ a_{25} & a_{26} \\ a_{35} & a_{36} \end{pmatrix},$$

$$A_{21} = \begin{pmatrix} a_{41} & a_{42} & a_{43} \\ a_{51} & a_{52} & a_{53} \end{pmatrix}, \; A_{22} = \begin{pmatrix} a_{45} \\ a_{55} \end{pmatrix}, \; A_{23} = \begin{pmatrix} a_{45} & a_{46} \\ a_{55} & a_{56} \end{pmatrix},$$

the matrix A can be written as

$$A = \begin{pmatrix} A_{11} & A_{12} & A_{13} \\ A_{21} & A_{22} & A_{23} \end{pmatrix}.$$

Definition 5.59 *A matrix is $A \in \mathbb{C}^{n\times n}$ is* normal *if $A^{H}A = AA^{H}$ and is* unitary *if $A^{H}A = AA^{H} = I_n$. Every unitary matrix is normal.*

Theorem 5.60 *A matrix $A \in \mathbb{C}^{n\times n}$ is normal and upper triangular (or lower triangular) if and only if A is a diagonal matrix.*

Proof Suppose that A is both normal and upper triangular. The normality of A implies $(A^{\mathsf{H}}A)_{pp} = (AA^{\mathsf{H}})_{pp}$ for $1 \leqslant p \leqslant n$. We show, by induction on p that all non-diagonal elements of A are 0.

For the base step $p = 1$ we have $\bar{a}_{11}a_{11} = \sum_{j=1}^{n} a_{1j}\bar{a}_{j1} = a_{11}\bar{a}_{11} + \sum_{j=2}^{n} a_{1j}\bar{a}_{j1}$. Since $\bar{a}_{11}a_{11} = a_{11}\bar{a}_{11} = |a_{11}|^2$, it follows that $\sum_{j=2}^{n} a_{1j}\bar{a}_{j1} = \sum_{j=2}^{n} |a_{1j}|^2 = 0$, so $a_{1j} = 0$ for $2 \leqslant j \leqslant n$, which implies that all non-diagonal elements of the first line of A are 0.

For the inductive step suppose that all non-diagonal elements of the first $p-1$ rows are 0. Then

$$(A^{\mathsf{H}}A)_{pp} = \sum_{i=1}^{n} \bar{a}_{ip}a_{ip} = \sum_{i=p}^{n} \bar{a}_{ip}a_{ip} = \bar{a}_{pp}a_{pp},$$

by the inductive hypothesis and the fact that A is upper diagonal. Therefore,

$$\bar{a}_{pp}a_{pp} = \sum_{j=p}^{n} a_{pj}\bar{a}_{pj} = a_{pp}\bar{a}_{pp} + \sum_{j=p+1}^{n} a_{pj}\bar{a}_{pj},$$

which implies $\sum_{j=p+1}^{n} a_{pj}\bar{a}_{pj} = \sum_{j=p+1}^{n} |a_{pj}|^2 = 0$. This, in turn, yields $a_{p\ p+1} = \cdots = a_{pn} = 0$.

The argument is similar for the lower diagonal case.

Clearly, any diagonal matrix is normal and both upper triangular and lower triangular.

Let $A \in \mathbb{C}^{m\times n}$ be a matrix. The *matrix of the absolute values of* A is the matrix $\mathsf{abs}(A) \in \mathbb{R}^{m\times n}$ defined by

$$(\mathsf{abs}(A))_{ij} = |a_{ij}|$$

for $1 \leqslant i \leqslant m$ and $1 \leqslant j \leqslant n$. In particular, if $\mathbf{x} \in \mathbb{C}^n$, we have $(\mathsf{abs}(\mathbf{x}))_j = |x_j|$.

Theorem 5.61 *Let $A \in \mathbb{C}^{m\times n}$ and $B \in \mathbb{C}^{n\times p}$ be two matrices. We have $\mathsf{abs}(AB) \leqslant \mathsf{abs}(A)\mathsf{abs}(B)$.*

Proof Since $(AB)_{ik} = \sum_{j=1}^{n} a_{ij}b_{jk}$, it follows that

$$|(AB)_{ik}| = \left|\sum_{j=1}^{n} a_{ij}b_{jk}\right| \leqslant \sum_{j=1}^{n} |a_{ij}b_{jk}| = \sum_{j=1}^{n} |a_{ij}|\ |b_{jk}|,$$

for $1 \leqslant i \leqslant m$ and $1 \leqslant k \leqslant p$. This amounts to $\mathsf{abs}(AB) \leqslant \mathsf{abs}(A)\mathsf{abs}(B)$.

Theorem 5.62 *For $A \in \mathbb{C}^{n\times n}$ we have $\mathsf{abs}(A^k) \leqslant (\mathsf{abs}(A))^k$ for every $k \in \mathbb{N}$.*

Proof The proof is by induction on k. The base case, $k = 0$, is immediate. Suppose that the inequality holds for k. We have

$$\begin{aligned} \mathsf{abs}(A^{k+1}) &= \mathsf{abs}(A^k\ A) \\ &\leqslant \mathsf{abs}(A^k)\mathsf{abs}(A) \\ &\quad \text{(by Theorem 5.61)} \\ &\leqslant (\mathsf{abs}(A))^k\mathsf{abs}(A) \\ &\quad \text{(by the inductive hypothesis)} \\ &= (\mathsf{abs}(A))^{k+1}, \end{aligned}$$

which completes the induction case.

Partitioning matrices is useful because matrix operations can be performed on block submatrices in a manner similar to scalar operations as we show next.

Theorem 5.63 *Let $A \in F^{m\times n}$ and $B \in F^{n\times p}$ be two matrices. Suppose that the matrices A, B are partitioned as*

$$A = \begin{pmatrix} A_{11} & \cdots & A_{1k} \\ \vdots & \ldots & \vdots \\ A_{h1} & \cdots & A_{hk} \end{pmatrix} \text{ and } B = \begin{pmatrix} B_{11} & \cdots & B_{1\ell} \\ \vdots & \ldots & \vdots \\ B_{k1} & \cdots & B_{k\ell} \end{pmatrix},$$

where $A_{rs} \in F^{m_r \times n_s}$, $B_{st} \in F^{n_s \times p_t}$ for $1 \leqslant r \leqslant h$, $1 \leqslant s \leqslant k$ and $1 \leqslant t \leqslant \ell$. Then, the product $C = AB$ can be partitioned as

$$C = \begin{pmatrix} C_{11} & \ldots & C_{1\ell} \\ \vdots & \ldots & \vdots \\ C_{h1} & \cdots & C_{hl} \end{pmatrix},$$

where $C_{uv} = \sum_{t=1}^{k} A_{ut} B_{tv}$, $1 \leqslant u \leqslant h$, and $1 \leqslant v \leqslant \ell$.

Proof Note that $m_1 + \cdots + m_h = m$ and $p_1 + \cdots + p_\ell = p$. For a pair (i, j) such that $1 \leqslant i \leqslant m$ and $1 \leqslant j \leqslant n$ let u be the least number such that $i \leqslant m_1 + \cdots + m_u$ and let v be the least number such that $j \leqslant p_1 + \cdots + p_v$. The definition of u and v implies $m_1 + \cdots + m_{u-1} + 1 \leqslant i \leqslant m_1 + \cdots + m_u$ and $p_1 + \cdots + p_{v-1} + 1 \leqslant j \leqslant p_1 + \cdots + p_v$. This implies that the c_{ij} element of the product is located in the submatrix $C_{uv} = \sum_{t=1}^{k} A_{ut} B_{tv}$ of C. By the definition of the matrix product we have

$$\begin{aligned} c_{ij} &= \sum_{g=1}^{n} a_{ig} b_{gj} \\ &= \sum_{g=1}^{n_1} a_{ig} b_{gj} + \sum_{g=n_1+1}^{n_1+n_2} a_{ig} b_{gj} + \cdots + \sum_{g=n_1+\cdots+n_{k-1}+1}^{n_1+\cdots+n_s} a_{ig} b_{gj} \end{aligned}$$

Observe that the vectors $(a_{i1}, \ldots, a_{in_1})$ and $(b_{1j}, \ldots, b_{n_1 j})'$ represent the line number $i - (m_1 + \cdots + m_{u-1} + 1)$ and the column number $j - (p_1 + \cdots + p_{v-1} + 1)$ of the matrix A_{u1} and B_{1v}, etc. Similarly,

$$(a_{i,n_1+\cdots+n_{k-1}+1}, \ldots, a_{i,n_1+\cdots+n_s})$$

and

$$(b_{n_1+\cdots+n_{k-1}+1,j}, \ldots, b_{n_1+\cdots+n_s,j})'$$

represent the line number $i - (m_1 + \cdots + m_{u-1} + 1)$ and the column number $j - (p_1 + \cdots + p_{v-1} + 1)$ of the matrix A_{uk} and B_{kv}, which shows that c_{ij} is computed correctly as an element of the block C_{uv}.

Next, we explore the relationship between linear mappings and matrices. Let $h \in \mathsf{Hom}(\mathbb{C}^m, \mathbb{C}^n)$ be a linear transformation between the linear spaces $\mathbb{C}^m$ and $\mathbb{C}^n$, let $R = \{\boldsymbol{r}_1, \ldots, \boldsymbol{r}_m\}$ be a basis in $\mathbb{C}^m$, and let $S = \{\boldsymbol{s}_1, \ldots, \boldsymbol{s}_n\}$ be a basis in $\mathbb{C}^n$.

Since $h(\boldsymbol{r}_j) \in \mathbb{C}^n$ we can write:

$$h(\boldsymbol{r}_j) = a_{1j}\boldsymbol{s}_1 + a_{2j}\boldsymbol{s}_2 + \cdots + a_{nj}\boldsymbol{s}_n.$$

Definition 5.64 *The* matrix $A_h \in \mathbb{C}^{n\times m}$ associated to the linear mapping $h : \mathbb{C}^m \longrightarrow \mathbb{C}^n$ *is the matrix that has*

$$h(\boldsymbol{r}_j) = \begin{pmatrix} a_{1j} \\ a_{2j} \\ \vdots \\ a_{nj} \end{pmatrix}$$

as its jth *column for* $1 \leqslant j \leqslant m$.

Let $\mathbf{v} \in \mathbb{C}^m$ be a vector such that $\mathbf{v} = v_1\boldsymbol{r}_1 + \cdots + v_m\boldsymbol{r}_m$. Then, the image of $\mathbf{v}$ under h is

$$\begin{aligned} h(\mathbf{v}) &= h\left(\sum_{j=1}^m v_j\boldsymbol{r}_j\right) = \sum_{j=1}^m v_j h(\boldsymbol{r}_j) \\ &= \sum_{j=1}^m v_j \begin{pmatrix} a_{1j} \\ a_{2j} \\ \vdots \\ a_{nj} \end{pmatrix} = \begin{pmatrix} \sum_{j=1}^m a_{1j}v_j \\ \sum_{j=1}^m a_{2j}v_j \\ \vdots \\ \sum_{j=1}^m a_{nj}v_j \end{pmatrix}, \end{aligned}$$

which is easily seen to equal $A_h\mathbf{v}$.

As we saw above, the matrix A_h attached to $h : \mathbb{C}^m \longrightarrow \mathbb{C}^n$ depends on the bases chosen for the linear spaces $\mathbb{C}^m$ and $\mathbb{C}^n$.

Let A_h^R and A_h^S be the matrices associated to h that correspond to the bases R and S, respectively. We have $A_h^R = (h(\boldsymbol{r}_1) \cdots h(\boldsymbol{r}_n))$ and $A_h^S = (h(\boldsymbol{s}_1) \cdots h(\boldsymbol{s}_n))$. The Equalities (5.3) can be written succinctly as

$$A_h^R = PA_h^S, \tag{5.1}$$

where P is the matrix

$$P = \begin{pmatrix} p_{11} & \cdots & p_{1n} \\ \vdots & \cdots & \vdots \\ p_{n1} & \cdots & p_{nn} \end{pmatrix},$$

whose entries have been introduced in Equalities (5.2).

Let $h : \mathbb{C}^n \longrightarrow \mathbb{C}^n$ be an endomorphism of $\mathbb{C}^n$ and let $R = \{\boldsymbol{r}_1, \ldots, \boldsymbol{r}_n\}$ and $S = \{\boldsymbol{s}_1, \ldots, \boldsymbol{s}_n\}$ be two bases of $\mathbb{C}^n$. The vectors $\boldsymbol{s}_i$ can be expressed as linear combinations of the vectors $\boldsymbol{r}_1, \ldots, \boldsymbol{r}_n$:

$$\boldsymbol{s}_i = p_{i1}\boldsymbol{r}_1 + \cdots + p_{in}\boldsymbol{r}_n, \tag{5.2}$$

for $1 \leqslant i \leqslant n$, which implies

$$h(\boldsymbol{s}_i) = p_{i1}h(\boldsymbol{r}_1) + \cdots + p_{in}h(\boldsymbol{r}_n). \tag{5.3}$$

for $1 \leqslant i \leqslant n$.

Matrix multiplication corresponds to the composition of linear mappings, as we show next.

Theorem 5.65 *Let $h \in \mathsf{Hom}(\mathbb{C}^m, \mathbb{C}^n)$ and $g \in \mathsf{Hom}(\mathbb{C}^n, \mathbb{C}^p)$. Then,*

$$A_{gh} = A_g A_h.$$

Proof If $\boldsymbol{p}_1, \ldots, \boldsymbol{p}_m$ is a basis for $\mathbb{C}^m$, then $A_{gh}(\boldsymbol{p}_i) = gh(\boldsymbol{p}_i) = g(h(\boldsymbol{p}_i)) = g(A_h\boldsymbol{p}_i) = A_g(A_h(\boldsymbol{p}_i))$ for every i, where $1 \leqslant i \leqslant n$. This proves that $A_{gh} = A_g A_h$.

Thus, if h is an idempotent endomorphism of a linear space the matrix A_h is idempotent.

Starting from a matrix $A \in \mathbb{C}^{n \times m}$ we can define a *linear operator associated to* A, $h_A : \mathbb{C}^m \longrightarrow \mathbb{C}^n$ as $h_A(\mathbf{x}) = A\mathbf{x}$ for $\mathbf{x} \in \mathbb{C}^m$. If $\boldsymbol{p}_1, \ldots, \boldsymbol{p}_m$ is a basis for $\mathbb{C}^m$, then $h_A(\boldsymbol{p}_i)$ is the ith column of the matrix A.

It is immediate that $A_{h_A} = A$ and $h_{A_h} = h$.

Attributes of a matrix A are transferred to the linear operator h_A. For example, if A is Hermitian we say that h_A is Hermitian.

The association of matrices in $\mathbb{C}^{n \times m}$ with linear operators, described in Definition 5.64, suggests the association of certain subspaces of the linear spaces $\mathbb{C}^n$ and $\mathbb{C}^m$ to A.

Definition 5.66 *Let $A \in \mathbb{C}^{n \times m}$ be a matrix. The* range of A *is the subspace $\mathsf{Img}(h_A)$ of $\mathbb{C}^n$. The* null space of A *is the subspace $\mathsf{Ker}(h_A)$.*

The range of A and the null space of A are denoted by $\mathsf{Ran}(A)$ and $\mathsf{NullSp}(A)$, respectively.

Clearly, $C_{A,n} = \mathsf{Ran}(A)$. The null space of $A \in \mathbb{C}^{m \times n}$ consists of those $\mathbf{x} \in \mathbb{C}^n$ such that $A\mathbf{x} = \mathbf{0}$.

Let $\{\boldsymbol{p}_1, \ldots, \boldsymbol{p}_m\}$ be a basis of $\mathbb{C}^m$. Since $\mathsf{Ran}(A) = \mathsf{Img}(h_A)$ it follows that this subspace is generated by the set $\{h_A(\boldsymbol{p}_1), \ldots, h_A(\boldsymbol{p}_m)\}$, that is, by the columns of the matrix A. For this reason the subspace $\mathsf{Ran}(A)$ is also known as the *column subspace* of A.

Several important facts concerning idempotent endomorphisms that were previously presented can now be formulated in terms of matrices. For example, Theorem 5.32 applied to $\mathbb{C}^n$ states that if A is an idempotent matrix, then $\mathbb{C}^n = \mathsf{NullSp}(A) \boxplus \mathsf{Ran}(A)$. Conversely, by Theorem 5.33 if U and W are two subspaces of $\mathbb{C}^n$ such that $\mathbb{C}^n = U \boxplus W$, then there exists an idempotent matrix $A \in \mathbb{C}^{n\times n}$ such that $U = \mathsf{NullSp}(A)$ and $W = \mathsf{Ran}(A)$.

Let $A \in \mathbb{C}^{n\times n}$ be a square matrix. Suppose that there exist two matrices U and V such that $AU = I_n$ and $VA = I_n$. This implies

$$V = VI_n = V(AU) = (VA)U = I_nU = U.$$

Thus, if $AU = VA = I_n$, the two matrices involved, U and V, must be equal.

Definition 5.67 *A matrix $A \in \mathbb{C}^{n\times n}$ is invertible if there exists a matrix $B \in \mathbb{C}^{n\times n}$ such that $AB = BA = I_n$.*

Suppose that C is another matrix such that $AC = CA = I_n$. By the associativity of the matrix product we have $C = CI_n = C(AB) = (CA)B = I_nB = B$. Therefore, if A is invertible there is exactly one matrix B such that $AB = BA = I_n$. We denote the matrix B by A^{-1} and we refer to it as the *inverse of the matrix* A.

Note that $A \in \mathbb{C}^{n\times n}$ is a unitary matrix if and only if $A^{-1} = A^{\mathsf{H}}$.

Theorem 5.68 *If $A, B \in \mathbb{C}^{n\times n}$ are two invertible matrices, then the product AB is invertible and $(AB)^{-1} = B^{-1}A^{-1}$.*

Proof Applying the definition of the inverse of a matrix we obtain

$$(AB)(B^{-1}A^{-1}) = A(BB^{-1})A^{-1} = AI_nA^{-1} = AA^{-1} = I_n,$$

which implies $(AB)^{-1} = B^{-1}A^{-1}$.

Theorem 5.69 *If $A \in \mathbb{C}^{n\times n}$ is invertible, then A^{H} is invertible and $(A^{H})^{-1} = (A^{-1})^{H}$.*

Proof Since $AA^{-1} = I_n$, we have $(A^{-1})^{\mathsf{H}}A^{\mathsf{H}} = I_n$, which shows that $(A^{-1})^{\mathsf{H}}$ is the inverse of A^{H} and $(A^{\mathsf{H}})^{-1} = (A^{-1})^{\mathsf{H}}$.

Example 5.70 Let

$$A = \begin{pmatrix} a_{11} & a_{12} \\ a_{21} & a_{22} \end{pmatrix}$$

be a matrix in $\mathbb{R}^{2\times 2}$. We seek to determine conditions under which A is invertible. Suppose that

$$X = \begin{pmatrix} x_{11} & x_{12} \\ x_{21} & x_{22} \end{pmatrix}$$

is a matrix in $\mathbb{R}^{2\times 2}$ such that $AX = I_2$. This matrix equality amounts to four scalar equalities:

$$a_{11}x_{11} + a_{12}x_{21} = 1,\ a_{11}x_{12} + a_{12}x_{22} = 0,$$
$$a_{21}x_{11} + a_{22}x_{21} = 1,\ a_{21}x_{12} + a_{22}x_{22} = 0,$$

which, under certain conditions, can be solved with respect to $x_{11}, x_{12}, x_{21}, x_{22}$.

By multiplying the first equality by a_{22} and the third by $-a_{12}$ and adding the resulting equalities we obtain $(a_{11}a_{22} - a_{12}a_{21})x_{11} = -a_{22}$. Thus, if $a_{11}a_{22} - a_{12}a_{21} \neq 0$, we have $x_{11} = -\frac{a_{22}}{a_{11}a_{22}-a_{12}a_{21}}$. The same condition, $a_{11}a_{22} - a_{12}a_{21} \neq 0$, suffices to allow us to obtain the value of the remaining components of X, as the reader can easily verify. Thus, A is an invertible matrix if and only if $a_{11}a_{22} - a_{12}a_{21} \neq 0$.

Definition 5.71 *A* stochastic matrix *is a matrix* $A \in \mathbb{R}^{n\times n}$ *such that* $a_{ij} \geqslant 0$ *for* $1 \leqslant i, j \leqslant n$ *and* $\sum_{j=1}^{n} a_{ij} = 1$ *for every* i, $1 \leqslant i \leqslant n$.

A doubly stochastic matrix *is a matrix* $A \in \mathbb{R}^{n\times n}$ *such that both* A *and* A' *are stochastic.*

The rows of a stochastic matrix can be regarded as discrete probability distributions.

Example 5.72 The matrix $A \in \mathbb{R}^{3\times 3}$ defined by

$$A = \begin{pmatrix} \frac{1}{2} & 0 & \frac{1}{2} \\ \frac{1}{3} & \frac{1}{2} & \frac{1}{6} \\ 0 & \frac{2}{3} & \frac{1}{3} \end{pmatrix}$$

is a stochastic matrix.

Example 5.73 Let

$$\phi : \begin{pmatrix} 1 & \cdots & k & \cdots & n \\ a_1 & \cdots & a_k & \cdots & a_n \end{pmatrix},$$

be a permutation of the set $\{1, \ldots, n\}$, where $a_k = \phi(k)$ for $1 \leqslant k \leqslant n$.

The matrix of this permutation is the square matrix $P_\phi = (p_{ij}) \in \{0, 1\}^{n\times n}$, where

$$p_{ij} = \begin{cases} 1 & \text{if } j = \phi(i), \\ 0 & \text{otherwise,} \end{cases} \tag{5.4}$$

for $1 \leqslant i, j \leqslant n$.

Note that the matrix of the permutation $1_{1,\ldots,n}$ is the matrix I_n.

Also, if ϕ, ψ are two permutations of the set $\{1, \ldots, n\}$, then $P_{\psi\phi} = P_\phi P_\psi$. Indeed, since $(P_\phi P_\psi)_{ij} = \sum_{k=1}^{n} (P_\phi)_{ik}(P_\psi)_{kj}$, observe that only the term $(P_\phi)_{ik}(P_\psi)_{kj}$ in which $k = \phi(i)$ and $j = \psi(k)$ is different from 0. Thus, $(P_\phi P_\psi)_{ij} \neq 0$ if and only if $j = \psi(\phi(i))$, which means that $P_{\psi\phi} = P_\phi P_\psi$.

Thus, if ϕ and ϕ^{-1} are two inverse permutations in $PERM_n$, we have $P_\phi P_{\phi^{-1}} = I_n$, so P_ϕ is invertible and $P_\phi^{-1} = P_{\phi^{-1}}$.

For instance, if $\phi \in PERM_4$ is

$$\phi : \begin{pmatrix} 1 & 2 & 3 & 4 \\ 3 & 1 & 4 & 2 \end{pmatrix},$$

then

$$P_\phi = \begin{pmatrix} 0 & 0 & 1 & 0 \\ 1 & 0 & 0 & 0 \\ 0 & 0 & 0 & 1 \\ 0 & 1 & 0 & 0 \end{pmatrix}$$

Its inverse is

$$P_\phi^{-1} = P_{\phi^{-1}} = \begin{pmatrix} 0 & 1 & 0 & 0 \\ 0 & 0 & 0 & 1 \\ 1 & 0 & 0 & 0 \\ 0 & 0 & 1 & 0 \end{pmatrix}.$$

It is easy to verify that the inverse of a permutation matrix P_ϕ coincides with its transpose $(P_\phi)'$.

Observe that if $A \in \mathbb{R}^{n\times n}$ having the rows $\boldsymbol{r}_1, \ldots, \boldsymbol{r}_n$ and P_ϕ is a permutation matrix, then $P_\phi A$ is the matrix whose rows are $\boldsymbol{r}_{\phi(1)}, \boldsymbol{r}_{\phi(2)}, \ldots, \boldsymbol{r}_{\phi(n)}$. Similarly, if the columns of A are $\mathbf{c}_1, \ldots, \mathbf{c}_n$, the columns of the matrix AP_ϕ are $\mathbf{c}_{\phi(1)}, \ldots, \mathbf{c}_{\phi(n)}$. In other words, $P_\phi A$ is obtained from A be permuting its rows according to the permutation ϕ and AP_ϕ is obtained from A by permuting the columns according to the same permutation.

Since every column and row of a permutation matrix contains exactly one 1, it follows that each such matrix is also a doubly-stochastic matrix.

Theorem 5.74 *Let $A \in \mathbb{R}^{n\times n}$ be a lower (upper) triangular matrix such that $a_{ii} \neq 0$ for $1 \leqslant i \leqslant n$. The matrix A is invertible and its inverse is a lower (upper) triangular matrix having diagonal elements equal to the reciprocal of the diagonal elements of A.*

Proof Let A be a lower triangular matrix

$$A = \begin{pmatrix} a_{11} & 0 & 0 & \cdots & 0 \\ a_{21} & a_{22} & 0 & \cdots & 0 \\ \vdots & \vdots & \vdots & \cdots & \vdots \\ a_{n1} & a_{n2} & a_{n3} & \cdots & a_{nn} \end{pmatrix},$$

where $a_{ii} \neq 0$ for $1 \leqslant i \leqslant n$. The proof is by induction on $n \geqslant 1$.

The base case, $n = 1$ is immediate, since the inverse of the matrix (a_{11}) is $\left(\frac{1}{a_{11}}\right)$.

Suppose that the statement holds for matrices in $\mathbb{R}^{(n-1)\times(n-1)}$. Then A can be written as

$$A = \left(\begin{array}{c|c} B & \mathbf{0}_{n-1} \\ \hline a_{n1}\ a_{n2}\ \cdots\ a_{n\,n-1} & a_{nn} \end{array} \right),$$

where $B \in \mathbb{R}^{(n-1)\times(n-1)}$ is a lower triangular matrix. By the inductive hypothesis, this matrix is invertible, its inverse B^{-1} is also lower triangular and the diagonal elements of B^{-1} are the reciprocal elements of the corresponding diagonal elements of B. The matrix

$$\left(\begin{array}{c|c} B^{-1} & \mathbf{0}_{n-1} \\ \hline \boldsymbol{v} & \frac{1}{a_{nn}} \end{array} \right)$$

is the inverse of A, where $\boldsymbol{v} = -\frac{1}{a_{nn}}\mathbf{a}'B^{-1}$, and $\mathbf{a}' = (a_{n1}, a_{n2}, \ldots, a_{n\,n-1})$, as the reader can easily verify.

A similar argument can be used for upper triangular matrices.

Theorem 5.75 *Let $A \in \mathbb{R}^{n\times n}$ be an invertible matrix. Then, its transpose A' is invertible and $(A')^{-1} = (A^{-1})'$.*

Proof Observe that $A'(A^{-1})' = (A^{-1}A)' = I_n' = I_n$. Therefore, $(A')^{-1} = (A^{-1})'$.

If $A \in \mathbb{R}^{n\times n}$ is invertible we have $AA^{-1} = I_n$, so $trace(A)trace(A^{-1}) = trace(I_n) = n$. This implies

$$trace(A^{-1}) = \frac{n}{trace(A)}. \tag{5.5}$$

Theorem 5.76 *Let $\{\boldsymbol{r}_1, \ldots, \boldsymbol{r}_n\}$ be a basis in $\mathbb{C}^n$. A matrix $A \in \mathbb{C}^{n\times n}$ is invertible if and only if the set of vectors $\{A\boldsymbol{r}_1, \ldots, A\boldsymbol{r}_n\}$ is a basis in $\mathbb{C}^n$.*

Proof Suppose that A is an invertible matrix. Note that $A\mathbf{x}_i = A\mathbf{x}_j$ implies $\mathbf{x}_i = \mathbf{x}_j$, so $\{A\boldsymbol{r}_1, \ldots, A\boldsymbol{r}_n\}$ consists on n distinct vectors. We claim that the set $\{A\boldsymbol{r}_1, \ldots, A\boldsymbol{r}_n\}$ is linearly independent. Indeed, suppose that $c_1A\boldsymbol{r}_1 + \cdots + c_nA\boldsymbol{r}_n = \mathbf{0}_n$ such that not all coefficients c_i equal 0. Then, by multiplying by A^{-1} to the left we obtain $c_1\boldsymbol{r}_1 + \cdots + c_n\boldsymbol{r}_n = \mathbf{0}_n$, which contradicts the fact that $\{\boldsymbol{r}_1, \ldots, \boldsymbol{r}_n\}$ is a basis. Thus, $\{A\boldsymbol{r}_1, \ldots, A\boldsymbol{r}_n\}$ is a linearly independent that consists of n vectors, which means that this set is a basis in $\mathbb{C}^n$.

Conversely, suppose that for any basis $\{\boldsymbol{r}_1, \ldots, \boldsymbol{r}_n\}$ of $\mathbb{C}^n$, $\{A\boldsymbol{r}_1, \ldots, A\boldsymbol{r}_n\}$ is a basis in $\mathbb{C}^n$. Each of the vectors $\boldsymbol{r}_i$ can be uniquely expressed as a linear combination of $A\boldsymbol{r}_1, \ldots, A\boldsymbol{r}_n$. In particular, for the *standard basis* $\{\mathbf{e}_1, \ldots, \mathbf{e}_n\}$, each of the vectors $\mathbf{e}_i$ can be uniquely expressed as a linear combination of the vectors $A\mathbf{e}_1 = \mathbf{a}_1, \ldots, A\mathbf{e}_n = \mathbf{a}_n$, where $\mathbf{a}_1, \ldots, \mathbf{a}_n$ are the columns of the matrix A. In other words, we have the equalities

$$\mathbf{e}_i = b_{i1}\mathbf{a}_1 + \cdots + b_{in}\mathbf{a}_n$$

for $1 \leqslant i \leqslant n$. In a succinct form, these equalities can be written as $I_n = BA$, where B is the matrix of the coefficient b_{ij}, which shows that A is an invertible matrix.

Theorem 5.77 *Let $\boldsymbol{u}_1, \ldots, \boldsymbol{u}_n$ and $\boldsymbol{w}_1, \ldots, \boldsymbol{w}_n$ be two bases of $\mathbb{C}^n$. There exists an invertible matrix $P \in \mathbb{C}^{n \times n}$ such that*

$$(\boldsymbol{u}_1 \ \cdots \ \boldsymbol{u}_n) = (\boldsymbol{w}_1 \ \cdots \ \boldsymbol{w}_n)P.$$

Proof Since $\mathbf{w}_1, \ldots, \mathbf{w}_n$ is a basis of $\mathbb{C}^n$ each vector $\boldsymbol{u}_i$ is a unique linear combination of the vectors $\mathbf{w}_1, \ldots, \mathbf{w}_n$, that is

$$\mathbf{u}_i = p_{1i}\mathbf{w}_1 + \cdots + p_{ni}\mathbf{w}_n = (\mathbf{w}_1 \ \cdots \ \mathbf{w}_n)\begin{pmatrix} p_{1i} \\ \vdots \\ p_{ni} \end{pmatrix},$$

for $1 \leqslant i \leqslant n$, so the equality of the theorem holds for the matrix $P = (p_{ij})$. We have to show that P is an invertible matrix.

Assume that $P\mathbf{t} = \mathbf{0}_n$. The equality of the theorem implies

$$(\mathbf{u}_1 \ \cdots \ \mathbf{u}_n)\begin{pmatrix} t_1 \\ \vdots \\ t_n \end{pmatrix} = (\mathbf{w}_1 \ \cdots \ \mathbf{w}_n)P\mathbf{t} = \mathbf{0}_n.$$

which implies $t_1\boldsymbol{u}_1 + \cdots + t_n\boldsymbol{u}_n = \mathbf{0}_n$. Since $\boldsymbol{u}_1, \ldots, \boldsymbol{u}_n$ is a basis we obtain $t_1 = \cdots = t_n = 0$, so $\mathbf{t} = \mathbf{0}_n$, which implies that P is an invertible matrix.

Corollary 5.78 *Let $z \in \mathbb{C}^n$ and assume that $z \in \mathbb{C}^n$ can be expressed relatively to the bases $\boldsymbol{u}_1, \ldots, \boldsymbol{u}_n$ and $\boldsymbol{w}_1, \ldots, \boldsymbol{w}_n$ as $z = \sum_{i=1}^n x_i\boldsymbol{u}_i$ and as $z = \sum_{i=1}^n y_i\boldsymbol{w}_i$, respectively. If $(\boldsymbol{u}_1 \ \cdots \ \boldsymbol{u}_n) = (\boldsymbol{w}_1 \ \cdots \ \boldsymbol{w}_n)P$, then*

$$P\begin{pmatrix} y_1 \\ \vdots \\ y_n \end{pmatrix} = \begin{pmatrix} x_1 \\ \vdots \\ x_n \end{pmatrix}.$$

Proof We have

$$z = (\boldsymbol{u}_1 \ \cdots \ \boldsymbol{u}_n)\begin{pmatrix} x_1 \\ \vdots \\ x_n \end{pmatrix} = (\mathbf{w}_1 \ \cdots \ \mathbf{w}_n)P\begin{pmatrix} y_1 \\ \vdots \\ y_n \end{pmatrix}.$$

The linear independence of $\mathbf{w}_1, \ldots, \mathbf{w}_n$ implies the desired equality.

5.4 Rank

The subspace $\mathsf{Ran}(A)$ of a matrix A is generated by the columns of this matrix. An analogous space is $\mathsf{rows}(A)$, the linear space spanned by the rows of A.

Definition 5.79 *Let $A \in \mathbb{C}^{m\times n}$ be a matrix. Its* column rank *is the number*

$$\textit{c-rank}(A) = \dim(\mathsf{Ran}(A)).$$

The row rank *of A is the number r-rank(A) equal to the dimension of the row space of A.*

Theorem 5.80 *Let $A \in \mathbb{C}^{m\times n}$ be a matrix. We have r-rank$(A) =$ c-rank(A).*

Proof Let $\boldsymbol{r}_1, \ldots, \boldsymbol{r}_m$ be the rows of A and let $\mathbf{c}_1, \ldots, \mathbf{c}_n$ be the columns of the same, so

$$A = \begin{pmatrix} \boldsymbol{r}_1 \\ \vdots \\ \boldsymbol{r}_m \end{pmatrix} = (\mathbf{c}_1 \ \cdots \ \mathbf{c}_n).$$

Suppose that *r-rank*$(A) = r$. There exists a basis $\boldsymbol{b}_1, \ldots, \boldsymbol{b}_r$ of the subspace $\mathsf{rows}(A)$ such that every row $\boldsymbol{r}_i$ of A can be written as a linear combination: $\boldsymbol{r}_i = \boldsymbol{u}_i B$, where $\boldsymbol{u}_i = (u_{i1} \ \cdots u_{ir})$ for $1 \leqslant i \leqslant m$ and

$$B = \begin{pmatrix} \boldsymbol{b}_1 \\ \vdots \\ \boldsymbol{b}_r \end{pmatrix} \in \mathbb{C}^{r\times n}.$$

Since $a_{ij} = \boldsymbol{r}_i \mathbf{e}_j$, we can write $a_{ij} = \boldsymbol{r}_i \mathbf{e}_j = \boldsymbol{u}_i B \mathbf{e}_j$.

Let $U \in \mathbb{C}^{m\times r}$ be the matrix whose rows are $\boldsymbol{u}_1, \ldots, \boldsymbol{u}_m$ and let $\boldsymbol{d}_1, \ldots, \boldsymbol{d}_r$ be the columns of this matrix. The jth column of A can be written as

$$\begin{pmatrix} a_{1j} \\ \vdots \\ a_{mj} \end{pmatrix} = \begin{pmatrix} \boldsymbol{u}_1 B\mathbf{e}_j \\ \vdots \\ \boldsymbol{u}_m B\mathbf{e}_j \end{pmatrix} = UB\mathbf{e}_j = (\boldsymbol{d}_1 \cdots \boldsymbol{d}_r) B\mathbf{e}_j$$
$$= (\boldsymbol{d}_1 \cdots \boldsymbol{d}_r) \begin{pmatrix} b_{1j} \\ \vdots \\ b_{rj} \end{pmatrix} = b_{1j}\boldsymbol{d}_1 + \cdots + b_{rj}\boldsymbol{d}_r.$$

This shows that the column space of A is generated by the set $\{\boldsymbol{d}_1, \ldots, \boldsymbol{d}_r\}$, so *c-rank*$(A) \leqslant r$. The same argument applied to A' implies that the $r =$ *c-rank*$(A') \leqslant$ *r-rank*$(A') =$ *c-rank*(A), so *c-rank*$(A) = r$, which concludes the argument.

Definition 5.81 *The* rank *of a matrix A is the number denoted by $rank(A)$ given by $rank(A) = \dim(\mathsf{Ran}(A)) = \dim(\mathsf{Img}(h_A))$.*

In other words, the rank of A is the maximal size of a set of linearly independent columns of A. By Theorem 5.80, the rank of A equals the maximal size of a set of linearly independent rows of A.

Consider the matrices $T_n^{i\leftrightarrow j}$, $T_n^{i\overset{+}{\leftarrow}j}$ and $T_{n,i}^{(a)}$ in $\mathbb{R}^{n\times n}$ defined by:

$$\begin{aligned} T_n^{i\leftrightarrow j} &= (\mathbf{e}_1 \ \cdots \ \mathbf{e}_{i-1} \ \mathbf{e}_j \ \mathbf{e}_{i+1} \ \cdots \ \mathbf{e}_{j-1} \ \mathbf{e}_i \ \mathbf{e}_{j+1} \ \cdots \ \mathbf{e}_n) \\ T_n^{i\overset{+}{\leftarrow}j} &= (\mathbf{e}_1 \ \cdots \ \mathbf{e}_{i-1} \ \mathbf{e}_i + \mathbf{e}_j \ \mathbf{e}_i \ \cdots \ \mathbf{e}_n), \end{aligned}$$

and

$$T_{n,i}^{(a)} = \mathsf{diag}(1, 1, \ldots, 1, a, 1, \ldots, 1),$$

where a occupies the ith diagonal position.

By multiplying a matrix $A \in \mathbb{C}^{m\times n}$ at the right with any of these matrices, certain transformations on the set of columns of A take place. Namely, the matrix $AT_n^{i\leftrightarrow j}$ is obtained from A by permuting the ith and the jth column. The matrix $AT_n^{i\overset{+}{\leftarrow}j}$ results by adding the jth column to the ith column. Finally, $AT_{n,i}^{(a)}$ is obtained from A by multiplying the ith column by a, where $a \neq 0$.

Similar effects are obtained on the rows of A by multiplying A at the left with $T_m^{i\leftrightarrow j}$, $(T_m^{i\overset{+}{\leftarrow}j})'$ and $T_{m,i}^{(a)}$.

Example 5.82 Let

$$A = \begin{pmatrix} a_{11} & a_{12} & a_{13} & a_{14} \\ a_{21} & a_{22} & a_{23} & a_{24} \\ a_{31} & a_{32} & a_{33} & a_{34} \end{pmatrix}.$$

We have

$$T_3^{2\leftrightarrow 3}A = \begin{pmatrix} 1 & 0 & 0 \\ 0 & 0 & 1 \\ 0 & 1 & 0 \end{pmatrix} \begin{pmatrix} a_{11} & a_{12} & a_{13} & a_{14} \\ a_{21} & a_{22} & a_{23} & a_{24} \\ a_{31} & a_{32} & a_{33} & a_{34} \end{pmatrix} = \begin{pmatrix} a_{11} & a_{12} & a_{13} & a_{14} \\ a_{31} & a_{32} & a_{33} & a_{34} \\ a_{21} & a_{22} & a_{23} & a_{24} \end{pmatrix},$$

and

$$AT_4^{2\leftrightarrow 3} = \begin{pmatrix} a_{11} & a_{12} & a_{13} & a_{14} \\ a_{21} & a_{22} & a_{23} & a_{24} \\ a_{31} & a_{32} & a_{33} & a_{34} \end{pmatrix} \begin{pmatrix} 1 & 0 & 0 & 0 \\ 0 & 0 & 1 & 0 \\ 0 & 1 & 0 & 0 \\ 0 & 0 & 0 & 1 \end{pmatrix} = \begin{pmatrix} a_{11} & a_{13} & a_{12} & a_{14} \\ a_{21} & a_{23} & a_{22} & a_{24} \\ a_{31} & a_{33} & a_{32} & a_{34} \end{pmatrix}.$$

For $A \in \mathbb{C}^{m\times n}$ the column space of the matrices $AT_n^{i\leftrightarrow j}$, $AT_n^{i\overset{+}{\leftarrow}j}$ and $AT_{n,i}^{(a)}$ is the same as the column space of A and the row space of $T_m^{i\leftrightarrow j}A$, $(T_m^{i\overset{+}{\leftarrow}j})'A$ and $T_{m,i}^{(a)}A$

is the same as the row space of A. Therefore, any of the matrices

$$AT_n^{i\leftrightarrow j}, AT_n^{i\overset{+}{\leftarrow}j}, AT_{n,i}^{(a)}, T_{m,i}^{i\leftrightarrow j}A, (T_m^{i\overset{+}{\leftarrow}j})'A, T_{m,i}^{(a)}A$$

has the same rank as A.

Theorem 5.83 *If $A \in \mathbb{C}^{m\times n}$ we have*

$$\dim(\mathit{NullSp}(A)) + \mathit{rank}(A) = n. \tag{5.6}$$

Proof Suppose that $\dim(\mathsf{NullSp}(A)) = q$, $\{\mathbf{v}_1, \ldots, \mathbf{v}_q\}$ is a basis of $\mathsf{NullSp}(A)$, and that $B = \{\mathbf{v}_1, \ldots, \mathbf{v}_q, \mathbf{v}_{q+1}, \ldots, \mathbf{v}_n\}$ is its extension to a basis B of $\mathbb{C}^m$.

If $\mathbf{y} \in \mathsf{Ran}(A)$, then $\mathbf{y} = A\mathbf{x}$ for some $\mathbf{x} \in \mathbb{C}^n$. Since $\mathbf{x}$ is a linear combination of the basis $\{\mathbf{v}_1, \ldots, \mathbf{v}_n\}$ we can write

$$\mathbf{x} = a_1\mathbf{v}_1 + \cdots + a_q\mathbf{v}_q + a_{q+1}\mathbf{v}_{q+1} + \cdots + a_n\mathbf{v}_n,$$

so

$$\begin{aligned}\mathbf{y} &= a_1A\mathbf{v}_1 + \cdots + a_qA\mathbf{v}_q + a_{q+1}A\mathbf{v}_{q+1} + \cdots + a_nA\mathbf{v}_n\\ &= a_{q+1}A\mathbf{v}_{q+1} + \cdots + a_nA\mathbf{v}_n.\end{aligned}$$

Therefore, $\{A\mathbf{v}_{q+1}, \ldots, A\mathbf{v}_n\}$ spans $\mathsf{Ran}(A)$. We claim that this set of vectors is linearly independent. Suppose this is not the case. Then, there exist $n - q$ numbers $d_{q+1}, \ldots, d_n$ such that

$$d_{q+1}A\mathbf{v}_{q+1} + \cdots + d_nA\mathbf{v}_n = \mathbf{0},$$

so $A(d_{q+1}\mathbf{v}_{q+1} + \cdots + d_n\mathbf{v}_n) = \mathbf{0}$, which means that $\mathbf{w} = d_{q+1}\mathbf{v}_{q+1} + \cdots + d_n\mathbf{v}_n \in \mathsf{NullSp}(A)$. Therefore, $\mathbf{w}$ is a linear combination of $\{\mathbf{v}_1, \ldots, \mathbf{v}_q\}$, which implies the existence of $d_1, \ldots, d_q$ such that

$$d_{q+1}\mathbf{v}_{q+1} + \cdots + d_n\mathbf{v}_n = d_1\mathbf{v}_1 + \cdots + d_q\mathbf{v}_q.$$

This contradicts the fact that B is a basis, so the set $\{A\mathbf{v}_{q+1}, \ldots, A\mathbf{v}_n\}$ is linearly independent and, therefore, is a basis of $\mathsf{Ran}(A)$. This implies that $\mathit{rank}(A) = n - q$, which is the desired equality.

Example 5.84 For the matrix

$$A = \begin{pmatrix}1 & 0 & 2\\ 1 & -1 & 1\\ 2 & 1 & 5\\ 1 & 2 & 4\end{pmatrix}$$

we have $rank(A) = 2$. Indeed, if $\mathbf{c}_1, \mathbf{c}_2, \mathbf{c}_3$ are its columns, then it is easy to see that $\{\mathbf{c}_1, \mathbf{c}_2\}$ is a linearly independent set, and $\mathbf{c}_3 = 2\mathbf{c}_1 + \mathbf{c}_2$. Thus, the maximal size of a set of linearly independent columns of A is 2.

Example 5.85 Let $A \in \mathbb{C}^{n\times m}$ and $B \in \mathbb{C}^{p\times q}$. For the matrix $C \in \mathbb{C}^{(n+p)\times(m+q)}$ defined by

$$C = \begin{pmatrix} A & O_{n,q} \\ O_{p,m} & B \end{pmatrix}$$

we have $rank(C) = rank(A) + rank(B)$.

Suppose that $rank(C) = \ell$ and let $\mathbf{c}_1, \ldots, \mathbf{c}_\ell$ be a maximal set of linearly independent columns of C. Without loss of generality we may assume that the first k columns are among the first m columns of A and the remaining $\ell - k$ columns are among the last q columns of C. The first k columns of C correspond to k linearly independent columns of A, while the last $\ell - k$ columns correspond to $\ell - k$ linearly independent columns of B. Thus, $rank(C) = k \leqslant rank(A) + rank(B)$.

Conversely, suppose that $rank(A) = s$ and $rank(B) = t$ and let $\mathbf{a}_{i_1}, \ldots, \mathbf{a}_{i_s}$ be a maximal set of linearly independent columns of A and let $\mathbf{b}_{j_1}, \ldots, \mathbf{b}_{j_t}$ be a maximal set of linearly independent columns of B. Then, it is easy to see that the vectors

$$\begin{pmatrix} \mathbf{a}_{i_1} \\ \mathbf{0}_n \end{pmatrix}, \ldots, \begin{pmatrix} \mathbf{a}_{i_s} \\ \mathbf{0}_n \end{pmatrix}, \ldots, \begin{pmatrix} \mathbf{0}_n \\ \mathbf{b}_{j_1} \end{pmatrix}, \ldots, \begin{pmatrix} \mathbf{0}_n \\ \mathbf{b}_{j_t} \end{pmatrix}$$

constitute a linearly independent set of columns of C, so $rank(A) + rank(B) \leqslant rank(C)$. Thus, $rank(C) = rank(A) + rank(B)$.

Example 5.86 Let $\mathbf{x}$ and $\mathbf{y}$ be two vectors in $\mathbb{C}^n - \{\mathbf{0}\}$. The matrix $\mathbf{x}\mathbf{y}^{\mathsf{H}}$ has rank 1. Indeed, if $\mathbf{y}^{\mathsf{H}} = (y_1, y_2, \ldots, y_n)$, then $\mathbf{x}\mathbf{y}^{\mathsf{H}} = (y_1\mathbf{x}\ y_2\mathbf{x}\ \cdots\ y_n\mathbf{x})$, which implies that the maximum number of linearly independent columns of $\mathbf{x}\mathbf{y}^{\mathsf{H}}$ is 1.

The above discussion also shows that if $A \in \mathbb{C}^{n\times m}$, then $rank(A) \leqslant \min\{m, n\}$.

Theorem 5.87 *Let $A \in \mathbb{C}^{m\times n}$ be a matrix. We have $rank(A) = rank(\overline{A})$.*

Proof Suppose that $A = (\mathbf{a}_1, \ldots, \mathbf{a}_n)$ and that the set $\{\mathbf{a}_{i_1}, \ldots, \mathbf{a}_{i_p}\}$ is a set of linearly independent columns of A. Then, the set $\{\overline{\mathbf{a}_{i_1}}, \ldots, \overline{\mathbf{a}_{i_p}}\}$ is a set of linearly independent columns of $\overline{A}$. This implies $rank(\overline{A}) = rank(A)$.

Corollary 5.88 *We have $rank(A) = rank(A^{\mathsf{H}})$ for every matrix $A \in \mathbb{C}^{m\times n}$.*

Proof Since $A^{\mathsf{H}} = \overline{A'}$, the statement follows immediately.

Definition 5.89 *A matrix $A \in \mathbb{C}^{n\times m}$ is a* full-rank *matrix if $rank(A) = \min\{m, n\}$.*

If $A \in \mathbb{C}^{m\times n}$ is a full-rank matrix and $m \geqslant n$, then the n columns of the matrix are linearly independent; similarly, if $n \geqslant m$, the m rows of the matrix are linearly independent.

A matrix that is not a full-rank is said to be *degenerate*. A degenerate square matrix is said to be *singular*. A *non-singular matrix* $A \in \mathbb{C}^{n\times n}$ is a matrix that is not singular and, therefore has $rank(A) = n$.

Theorem 5.90 *A matrix $A \in \mathbb{C}^{n\times n}$ is non-singular if and only if it is invertible.*

Proof Suppose that A is non-singular, that is, $rank(A) = n$. In other words the set of columns $\{\mathbf{c}_1, \ldots, \mathbf{c}_n\}$ of A is linearly independent, and therefore, is a basis of $\mathbb{C}^n$. Then, each of the vectors $\mathbf{e}_i$ can be expressed as a unique combination of the columns of A, that is

$$\mathbf{e}_i = b_{1i}\mathbf{c}_1 + b_{2i}\mathbf{c}_2 + \cdots + b_{ni}\mathbf{c}_n,$$

for $1 \leqslant i \leqslant n$. These equalities can be written as

$$(\mathbf{c}_1 \;\cdots\; \mathbf{c}_n)\begin{pmatrix} b_{11} & \cdots & b_{1n} \\ b_{21} & \cdots & b_{2n} \\ \vdots & \cdots & \vdots \\ b_{n1} & \cdots & b_{nn} \end{pmatrix} = I_n.$$

Consequently, the matrix A is invertible and

$$A^{-1} = \begin{pmatrix} b_{11} & \cdots & b_{1n} \\ b_{21} & \cdots & b_{2n} \\ \vdots & \cdots & \vdots \\ b_{n1} & \cdots & b_{nn} \end{pmatrix}.$$

Suppose now that A is invertible and that

$$d_1\mathbf{c}_1 + \cdots + d_n\mathbf{c}_n = \mathbf{0}.$$

This is equivalent to

$$A\begin{pmatrix} d_1 \\ \vdots \\ d_n \end{pmatrix} = \mathbf{0}.$$

Multiplying both sides by A^{-1} implies

$$\begin{pmatrix} d_1 \\ \vdots \\ d_n \end{pmatrix} = \mathbf{0},$$

so $d_1 = \cdots = d_n = 0$, which means that the set of columns of A is linearly independent, so $rank(A) = n$.

Corollary 5.91 *A matrix $A \in \mathbb{C}^{n\times n}$ is non-singular if and only if $A\mathbf{x} = \mathbf{0}$ implies $\mathbf{x} = \mathbf{0}$ for $\mathbf{x} \in \mathbb{C}^n$.*

Proof If A is non-singular then, by Theorem 5.90, A is invertible. Therefore, $A\mathbf{x} = \mathbf{0}$ implies $A^{-1}(Ax) = A^{-1}\mathbf{0}$, so $\mathbf{x} = \mathbf{0}$.

Conversely, suppose that $A\mathbf{x} = \mathbf{0}$ implies $\mathbf{x} = \mathbf{0}$. If $A = (\mathbf{c}_1 \cdots \mathbf{c}_n)$ and $\mathbf{x} = (x_1, \ldots, x_n)'$, the previous implication means that $x_1\mathbf{c}_1 + \cdots + x_n\mathbf{c}_n = \mathbf{0}$ implies $x_1 = \cdots = x_n = 0$, so $\{\mathbf{c}_1, \ldots, \mathbf{c}_n\}$ is linearly independent. Therefore, $rank(A) = n$, so A is non-singular.

Let $A \in \mathbb{C}^{n\times m}$ be a matrix. It is easy to see that the square matrix $B = A^{\mathsf{H}}A \in \mathbb{C}^{m\times m}$ is Hermitian.

Theorem 5.92 *Let $A \in \mathbb{C}^{n\times m}$ be a matrix and let $B = A^{H}A$. The matrices A and B have the same rank.*

Proof We prove that $\mathsf{NullSp}(A) = \mathsf{NullSp}(B)$. If $Au = \mathbf{0}$, then $Bu = A^{\mathsf{H}}(Au) = \mathbf{0}$, so $\mathsf{NullSp}(A) \subseteq \mathsf{NullSp}(B)$. If $\mathbf{v} \in \mathsf{NullSp}(B)$, then $A^{\mathsf{H}}Av = 0$, which implies that $\mathbf{v}^{\mathsf{H}}A^{\mathsf{H}}A\mathbf{v} = 0$. This, in turn can be written as $(A\mathbf{v})^{\mathsf{H}}(A\mathbf{v}) = 0$, so, by a previous observation, we have $A\mathbf{v} = \mathbf{0}$, which means that $\mathbf{v} \in \mathsf{NullSp}(A)$. We conclude that $\mathsf{NullSp}(A) = \mathsf{NullSp}(A^{\mathsf{H}}A)$. The equalities

$$\dim(\mathsf{NullSp}(A)) + rank(A) = m,$$
$$\dim(\mathsf{NullSp}(A)) + rank(A^{\mathsf{H}}A) = m,$$

imply that $rank(A^{\mathsf{H}}A) = m$.

Corollary 5.93 *Let $A \in \mathbb{C}^{n\times m}$ be a matrix of full-rank. If $m \geqslant n$, then the matrix $A^{H}A$ is non-singular; if $n \leqslant m$, then AA^{H} is non-singular.*

Proof Suppose that $m \geqslant n$. Then, $rank(A^{\mathsf{H}}A) = rank(A) = m$ because A is a full-rank matrix. Thus, $A^{\mathsf{H}}A \in \mathbb{C}^{m\times m}$ is non-singular. The argument for the second part of the corollary is similar.

Example 5.94 Let $A = (\mathbf{a}_1 \cdots \mathbf{a}_m) \in \mathbb{C}^{n\times m}$. Since $AA^{\mathsf{H}} = \mathbf{a}_1\mathbf{a}_1^{\mathsf{H}} + \cdots + \mathbf{a}_m\mathbf{a}_m^{\mathsf{H}}$ it follows that the rank of the matrix $\mathbf{a}_1\mathbf{a}_1^{\mathsf{H}} + \cdots + \mathbf{a}_m\mathbf{a}_m^{\mathsf{H}}$ equals the rank of the matrix A and, therefore, it cannot exceed m.

Theorem 5.95 (Sylvester's Rank Theorem) *Let $A \in \mathbb{C}^{m\times n}$ and $B \in \mathbb{C}^{n\times p}$ be two matrices. We have*

$$rank(AB) = rank(B) - \dim(\mathsf{NullSp}(A) \cap \mathsf{Ran}(B)).$$

Proof Both $\mathsf{NullSp}(A)$ and $\mathsf{Ran}(B)$ are subspaces of $\mathbb{C}^n$, so $\mathsf{NullSp}(A) \cap \mathsf{Ran}(B)$ is a subspace of $\mathbb{C}^n$. If $\mathbf{u}_1, \ldots, \mathbf{u}_k$ is a basis of the subspace $\mathsf{NullSp}(A) \cap \mathsf{Ran}(B)$, then exists a basis $\mathbf{u}_1, \ldots, \mathbf{u}_k, \mathbf{u}_{k+1}, \ldots, \mathbf{u}_l$ of the subspace $\mathsf{Ran}(B)$.

The set $\{A\mathbf{u}_{k+1}, \ldots, A\mathbf{u}_l\}$ is linearly independent. Indeed, suppose that there exists a linear combination

$$a_1 A\mathbf{u}_{k+1} + \cdots + a_{l-k} A\mathbf{u}_l = \mathbf{0}.$$

Then, $A(a_1\mathbf{u}_{k+1} + \cdots + a_{l-k}\mathbf{u}_l) = \mathbf{0}$, so $a_1\mathbf{u}_{k+1} + \cdots + a_{l-k}\mathbf{u}_l \in \mathsf{NullSp}(A)$. Since $\mathbf{u}_{k+1}, \ldots, \mathbf{u}_l \in \mathsf{Ran}(B)$, it follows that $a_1\mathbf{u}_{k+1}+\cdots+a_{l-k}\mathbf{u}_l \in \mathsf{NullSp}(A) \cap \mathsf{Ran}(B)$. Since $\mathbf{u}_1, \ldots, \mathbf{u}_k$ is a basis of the subspace $\mathsf{NullSp}(A) \cap \mathsf{Ran}(B)$, we have

$$a_1\mathbf{u}_{k+1} + \cdots + a_{l-k}\mathbf{u}_l = d_1\mathbf{u}_1 + \cdots + d_k\mathbf{u}_k$$

for some $d_1, \ldots, d_k \in \mathbb{C}$, which implies

$$a_1\mathbf{u}_{k+1} + \cdots + a_{l-k}\mathbf{u}_l - d_1\mathbf{u}_1 - \cdots - d_k\mathbf{u}_k = \mathbf{0}.$$

Since $\mathbf{u}_1, \ldots, \mathbf{u}_k, u_{k+1}, \ldots, \mathbf{u}_l$ is a basis of $\mathsf{Ran}(B)$, it follows that $a_1 = \cdots = a_{l-k} = d_1 = \cdots = d_k = 0$, so $A\mathbf{u}_{k+1}, \ldots, A\mathbf{u}_l$ is indeed linear independent.

Next, we show that $A\mathbf{u}_{k+1}, \ldots, A\mathbf{u}_l$ spans the subspace $\mathsf{Ran}(AB)$. Since $\mathbf{u}_j \in \mathsf{Ran}(B)$ it is clear that $A\mathbf{u}_j \in \mathsf{Ran}(AB)$ for $k + 1 \leqslant j \leqslant l$. If $\mathbf{w} \in \mathsf{Ran}(AB)$, then $\mathbf{w} = AB\mathbf{x}$ for some $\mathbf{x} \in \mathbb{C}^p$. Since $B\mathbf{x} \in \mathsf{Ran}(B)$ we can write $B\mathbf{x} = b_1\mathbf{u}_1 + \cdots + b_k\mathbf{u}_k + b_{k+1}\mathbf{u}_{k+1} + \cdots + b_l\mathbf{u}_l$, which implies

$$\mathbf{w} = AB\mathbf{x} = b_{k+1}A\mathbf{u}_{k+1} + \cdots + b_lA\mathbf{u}_l,$$

because $A\mathbf{u}_1 = \cdots = A\mathbf{u}_k = \mathbf{0}$, as $\mathbf{u}_1, \ldots, u_k$ belong to $\mathsf{NullSp}(A)$. Thus, $A\mathbf{u}_{k+1}, \ldots, A\mathbf{u}_l$ spans the subspace $\mathsf{Ran}(AB)$, which allows us to conclude that this linearly independent set is a basis for this subspace that contains $l - k$ elements. This allows us to conclude that $rank(AB) = \dim(\mathsf{Ran}(AB)) = rank(B) - \dim(\mathsf{NullSp}(A) \cap \mathsf{Ran}(B))$.

Corollary 5.96 *Let $A \in \mathbb{C}^{m\times n}$. If $R \in \mathbb{C}^{m\times m}$ and $Q \in \mathbb{C}^{n\times n}$ are invertible matrices then*

$$rank(A) = rank(RA) = rank(AQ) = rank(RAQ).$$

Proof Note that $rank(R) = m$ and $rank(Q) = n$. Thus, $\mathsf{NullSp}(R) = \{\mathbf{0}_m\}$ and $\mathsf{NullSp}(Q) = \{\mathbf{0}_n\}$. By Sylvester's Rank Theorem we have

$$\begin{aligned} rank(RA) &= rank(A) - \dim(\mathsf{NullSp}(R) \cap \mathsf{Ran}(A)) \\ &= rank(A) - \dim(\{\mathbf{0}\}) = rank(A). \end{aligned}$$

On the other hand, we have

$$\begin{aligned} rank(AQ) &= rank(Q) - \dim(\mathsf{NullSp}(A) \cap \mathsf{Ran}(Q)) \\ &= n - \dim(\mathsf{NullSp}(A)) = rank(A), \end{aligned}$$

because $\mathsf{Ran}(Q) = \mathbb{C}^n$.

The last equality of the theorem follows from the first two.

Corollary 5.97 *Let $A \in \mathbb{C}^{m\times n}$ and $B \in \mathbb{C}^{n\times p}$ be two matrices. We have*

$$\dim(\mathsf{NullSp}(AB)) = \dim(\mathsf{NullSp}(B)) + \dim(\mathsf{NullSp}(A) \cap \mathsf{Ran}(B)).$$

Proof By Equality (5.6) we have:

$$\begin{aligned} \dim(\mathsf{NullSp}(AB)) + \mathit{rank}(AB) &= p, \\ \dim(\mathsf{NullSp}(B)) + \mathit{rank}(B) &= p. \end{aligned}$$

An application of Sylvester's Rank Theorem implies

$$\dim(\mathsf{NullSp}(AB)) = \dim(\mathsf{NullSp}(B)) + \dim(\mathsf{NullSp}(A) \cap \mathsf{Ran}(B)).$$

Corollary 5.98 *Let $A \in \mathbb{C}^{m\times n}$ and $B \in \mathbb{C}^{n\times p}$ be two matrices. We have*

$$rank(A) + rank(B) - n \leqslant rank(AB) \leqslant \min\{rank(A), rank(B)\},$$

and

$$\begin{aligned} \max\{\dim(\mathsf{NullSp}(A)), \dim(\mathsf{NullSp}(B))\} &\leqslant \dim(\mathsf{NullSp}(AB)) \\ &\leqslant \dim(\mathsf{NullSp}(A)) + \dim(\mathsf{NullSp}(B)). \end{aligned}$$

Proof Since $\dim(\mathsf{NullSp}(A) \cap \mathit{rank}(B)) \leqslant \dim(\mathsf{NullSp}(A)) = n - \mathit{rank}(A)$ it follows that $\mathit{rank}(AB) \geqslant \mathit{rank}(B) - (n - \mathit{rank}(A)) = \mathit{rank}(A) + \mathit{rank}(B) - n$.

For the second inequality, observe that Sylvester's Rank Theorem implies immediately $\mathit{rank}(AB) \leqslant \mathit{rank}(B)$. Also, $\mathit{rank}(AB) = \mathit{rank}((AB)') = \mathit{rank}(B'A') \leqslant \mathit{rank}(A') = \mathit{rank}(A)$, so $rank(AB) \leqslant \min\{rank(A), rank(B)\}$.

The second part of the Corollary follows from the first part.

Corollary 5.99 *If $A \in \mathbb{C}^{m\times n}$ is a full-rank matrix with $m \geqslant n$, then $\mathit{rank}(AB) = \mathit{rank}(B)$ for any $B \in \mathbb{C}^{n\times p}$.*

Proof Since $m \geqslant n$, we have $\mathit{rank}(A) = n$; therefore, the n columns of A are linearly independent so $\mathsf{NullSp}(A) = \{\mathbf{0}\}$. By Sylvester's Rank Theorem we have $\mathit{rank}(AB) = \mathit{rank}(B)$.

Theorem 5.100 (The Full-Rank Factorization Theorem) *Let $A \in \mathbb{C}^{m\times n}$ be a matrix with $\mathit{rank}(A) = r > 0$. There exists $B \in \mathbb{C}^{m\times r}$ and $C \in \mathbb{C}^{r\times n}$ such that $A = BC$.*

Furthermore, if $A = DE$, where $D \in \mathbb{C}^{m\times r}$, $E \in \mathbb{C}^{r\times n}$, then both D and E are full-rank matrices, that is, we have $\mathit{rank}(D) = \mathit{rank}(E) = r$.

Proof Let $\{\boldsymbol{b}_1, \ldots, \boldsymbol{b}_r\} \subseteq \mathbb{C}^m$ be a basis for the $\mathsf{Ran}(A)$. Define $B = (\boldsymbol{b}_1 \ \cdots \ \boldsymbol{b}_r) \in \mathbb{C}^{m\times r}$. The columns of A, $\mathbf{a}_1, \ldots, \mathbf{a}_n$ can be written as $\mathbf{a}_i = c_{1i}\boldsymbol{b}_1 + \cdots c_{ri}\boldsymbol{b}_r$ for $1 \leqslant i \leqslant n$, which amounts to

$$A = (\mathbf{a}_1 \ \cdots \ \mathbf{a}_n) = (\boldsymbol{b}_1 \ \cdots \ \boldsymbol{b}_r) \begin{pmatrix} c_{11} \cdots c_{1r} \\ \vdots \ \ \dots \ \ \vdots \\ c_{r1} \ \cdots \ c_r \end{pmatrix}.$$

Thus, $A = BC$, where

$$C = \begin{pmatrix} c_{11} \cdots c_{1r} \\ \vdots \ \ \dots \ \ \vdots \\ c_{r1} \ \cdots \ c_r \end{pmatrix}.$$

Suppose now that $A = DE$, where $D \in \mathbb{C}^{m \times r}$, $E \in \mathbb{C}^{r \times n}$. It is clear that we have both $rank(D) \leqslant r$ and $rank(E) \leqslant r$. On another hand, by Corollary 5.98, $r = rank(A) = rank(DE) \leqslant \min\{rank(D), rank(E)\}$ implies $r \leqslant rank(D)$ and $r \leqslant rank(E)$, so $rank(D) = rank(E) = r$.

Corollary 5.101 *Let $A \in \mathbb{C}^{m \times n}$ be a matrix such that $rank(A) = r > 0$, and let $A = BC$ be a full-rank factorization of A.*

If the columns of B constitute a basis of the column space of A then C is uniquely determined. Furthermore, if the rows of C constitute a basis of the row space of A and, then B is uniquely determined.

Proof This statement is an immediate consequence of the full-rank factorization theorem.

Corollary 5.102 *If $A \in \mathbb{C}^{m \times n}$ is a matrix with $rank(A) = r > 0$, then A can be written as*

$$A = \boldsymbol{b}_1\boldsymbol{c}_1' + \cdots + \boldsymbol{b}_r\boldsymbol{c}_r',$$

where $\{\boldsymbol{b}_1, \ldots, \boldsymbol{b}_r\} \subseteq \mathbb{C}^m$ and $\{\mathbf{c}_1, \ldots, \boldsymbol{c}_r\} \subseteq \mathbb{C}^n$ are linearly independent sets.

Proof The corollary follows from Theorem 5.100 by adopting the set of columns of B as $\{\boldsymbol{b}_1, \ldots, \boldsymbol{b}_r\}$ and the transposed rows of C as $\{\mathbf{c}_1, \ldots, \mathbf{c}_r\}$.

Theorem 5.103 *Let $A \in \mathbb{C}^{m \times n}$ be a full-rank matrix. If $m \geqslant n$, then there exists a matrix $D \in \mathbb{C}^{n \times m}$ such that $DA = I_n$. If $n \geqslant m$, then there exists a matrix $E \in \mathbb{C}^{n \times m}$ such that $AE = I_m$.*

Proof Suppose that $A = (\mathbf{a}_1 \cdots \mathbf{a}_n) \in \mathbb{C}^{m \times n}$ is a full-rank matrix and $m \geqslant n$. Then, the n columns of A are linearly independent and we can extend the set of columns to a basis of $\mathbb{C}^m$, $\{\mathbf{a}_1, \ldots, \mathbf{a}_n, \boldsymbol{d}_1, \ldots, \boldsymbol{d}_{m-n}\}$. The matrix $T = (\mathbf{a}_1 \ \cdots \ \mathbf{a}_n \ \boldsymbol{d}_1 \ \cdots \ \boldsymbol{d}_{m-n})$ is invertible, so there exists

$$T^{-1} = \begin{pmatrix} \mathbf{t}_1 \\ \vdots \\ \mathbf{t}_n \\ \mathbf{t}_{n+1} \vdots \\ \mathbf{t}_m \end{pmatrix}$$

such that $T^{-1}T = I_m$. If we define

$$D = \begin{pmatrix} \mathbf{t}_1 \\ \vdots \\ \mathbf{t}_n \end{pmatrix}$$

it is immediate that $DA = I_n$.

The argument for the second part is similar.

Definition 5.104 *Let $A \in \mathbb{C}^{m\times n}$. A* left inverse *of A is a matrix $D \in \mathbb{C}^{n\times m}$ such that $DA = I_n$. A* right inverse *of A is a matrix $E \in \mathbb{C}^{n\times m}$ such that $AE = I_m$.*

Theorem 5.103 can now be restated as follows. Let $A \in \mathbb{C}^{m\times n}$ be a full-rank matrix. If $m \geqslant n$, then A has a left inverse; if $n \geqslant m$, then A has a right inverse.

Corollary 5.105 *Let $A \in \mathbb{C}^{n\times n}$ be a square matrix. The following statements are equivalent.*

(i) *A has a left inverse;*
(ii) *A has a right inverse;*
(iii) *A has an inverse.*

Proof It is clear that (iii) implies both (i) and (ii). Suppose now that A has a left inverse, so $DA = I_n$. Then, the columns of A, $\mathbf{c}_1, \ldots, \mathbf{c}_n$ are linearly independent, for if $a_1\mathbf{c}_1 + \cdots a_n\mathbf{c}_n = \mathbf{0}$, we have $a_1D\mathbf{c}_1 + \cdots + a_nD\mathbf{c}_n = a_1\mathbf{e}_1 + \cdots + a_n\mathbf{c}_n = \mathbf{0}$, which implies $a_1 = \cdots = a_n = 0$. Thus, $rank(A) = n$, so A has an inverse.

In a similar manner (using the rows of A) we can show that (ii) implies (iii).

Theorem 5.106 *Let $A \in \mathbb{C}^{m\times n}$ be a matrix with $rank(A) = r > 0$. There exists a non-singular matrix $G \in \mathbb{C}^{m\times m}$ and a non-singular matrix $H \in \mathbb{C}^{n\times n}$ such that*

$$A = G\begin{pmatrix} I_r & O_{r,n-r} \\ O_{m-r,r} & O_{m-r,n-r} \end{pmatrix} H.$$

Proof By the Full-Rank Factorization Theorem (Theorem 5.100) there are two full-rank matrices $B \in \mathbb{C}^{m\times r}$ and $C \in \mathbb{C}^{r\times n}$ such that $A = BC$. Let $\{\boldsymbol{b}_1, \ldots, \boldsymbol{b}_r\}$ be the columns of B and let $\mathbf{c}'_1, \ldots, \mathbf{c}'_r$ be the rows of C. It is clear that both sets of vectors are linearly independent and, therefore, for the first set there exist $\boldsymbol{b}_{r+1}, \ldots, \boldsymbol{b}_m$ such that $\{\boldsymbol{b}_1, \ldots, \boldsymbol{b}_m\}$ is a basis of $\mathbb{C}^m$; for the second set we have the vectors $\mathbf{c}'_{r+1}, \ldots, \mathbf{c}_n{}'$ such that $\{\mathbf{c}'_1, \ldots, \mathbf{c}'_n\}$ is a basis for $\mathbb{R}^n$. Define $G = (\boldsymbol{b}_1, \cdots, \boldsymbol{b}_m)$ and

$$H = \begin{pmatrix} \mathbf{c}'_1 \\ \vdots \\ \mathbf{c}'_n \end{pmatrix}.$$

Clearly, both G and H are non-singular and

$$A = G \begin{pmatrix} I_r & O_{r,n-r} \\ O_{m-r,r} & O_{m-r,n-r} \end{pmatrix} H.$$

Lemma 5.107 *If $A \in \mathbb{C}^{m\times n}$ is a matrix and $\mathbf{x} \in \mathbb{C}^m$, $\mathbf{y} \in \mathbb{C}^n$ are two vectors such that $\mathbf{x}^{\mathsf{H}} A\mathbf{y} \neq 0$, then $rank(A\mathbf{y}\mathbf{x}^{\mathsf{H}} A) = 1$.*

Proof By the associative property of matrix product we have $A\mathbf{y}\mathbf{x}^{\mathsf{H}}A = A(\mathbf{y}\mathbf{x}^{\mathsf{H}})A$, so $rank(A\mathbf{y}\mathbf{x}^{\mathsf{H}}A) \leqslant \min\{rank(\mathbf{y}\mathbf{x}^{\mathsf{H}}, rank(A)\} = 1$, by Corollary 5.98.

We claim that $A\mathbf{y}\mathbf{x}^{\mathsf{H}}A \neq O_{m,n}$. Suppose that $A\mathbf{y}\mathbf{x}^{\mathsf{H}}A = O_{m,n}$. This implies $\mathbf{x}^{\mathsf{H}}A\mathbf{y}\mathbf{x}^{\mathsf{H}}A\mathbf{y} = 0$. If $z = \mathbf{x}^{\mathsf{H}}A\mathbf{y}$, the previous equality amounts to $z^2 = 0$, which yields $z = \mathbf{x}^{\mathsf{H}}A\mathbf{y} = 0$. This contradicts the hypothesis of the lemma, so $A\mathbf{y}\mathbf{x}^{\mathsf{H}}A \neq O_{m,n}$, which implies $rank(A\mathbf{y}\mathbf{x}^{\mathsf{H}}A) \geqslant 1$. This allows us to conclude that $rank(A\mathbf{y}\mathbf{x}^{\mathsf{H}}A) = 1$.

The rank-1 matrix $A\mathbf{y}\mathbf{x}^{\mathsf{H}}A$ discussed in Lemma 5.107 plays a central role in the next statement.

Theorem 5.108 (Wedderburn's Theorem) *Let $A \in \mathbb{C}^{m\times n}$ be a matrix. If $\mathbf{x} \in \mathbb{C}^m$ and $\mathbf{y} \in \mathbb{C}^n$ are two vectors such that $\mathbf{x}^{\mathsf{H}}A\mathbf{y} \neq 0$ and B is the matrix*

$$B = A - \frac{1}{\mathbf{x}^{\mathsf{H}}A\mathbf{y}} A\mathbf{y}\mathbf{x}^{\mathsf{H}}A,$$

then $rank(B) = rank(A) - 1$.

Proof Observe that if $\mathbf{z} \in \mathsf{NullSp}(A)$, then $A\mathbf{z} = \mathbf{0}$. Therefore, we have

$$B\mathbf{z} = -\frac{1}{\mathbf{x}^{\mathsf{H}}A\mathbf{y}} A\mathbf{y}\mathbf{x}^{\mathsf{H}}A\mathbf{z} = \mathbf{0},$$

so $\mathsf{NullSp}(A) \subseteq \mathsf{NullSp}(B)$. Conversely, if $\mathbf{z} \in \mathsf{NullSp}(B)$, we have

$$A\mathbf{z} - \frac{1}{\mathbf{x}^{\mathsf{H}}A\mathbf{y}} A\mathbf{y}\mathbf{x}^{\mathsf{H}}A\mathbf{z} = \mathbf{0},$$

which can be written as

$$A\mathbf{z} = \frac{1}{\mathbf{x}^{\mathsf{H}}A\mathbf{y}} A\mathbf{y}(\mathbf{x}^{\mathsf{H}}A\mathbf{z}) = \frac{\mathbf{x}^{\mathsf{H}}A\mathbf{z}}{\mathbf{x}^{\mathsf{H}}A\mathbf{y}} A\mathbf{y}.$$

Thus, we obtain $A(\mathbf{z} - k\mathbf{y}) = \mathbf{0}$, where $k = \frac{\mathbf{x}^{\mathsf{H}}A\mathbf{z}}{\mathbf{x}^{\mathsf{H}}A\mathbf{y}}$. Since $A\mathbf{y} \neq \mathbf{0}$, this shows that a basis of $\mathsf{NullSp}(B)$ can be obtained by adding $\mathbf{y}$ to a basis of $\mathsf{NullSp}(A)$. Therefore, $\dim(\mathsf{NullSp}(B)) = \dim(\mathsf{NullSp}(A)) + 1$, so $rank(B) = rank(A) - 1$.

Theorem 5.109 *A square matrix $A \in \mathbb{C}^{n\times n}$ generates an increasing sequence of null spaces*

$$\{\mathbf{0}\} = \mathsf{NullSp}(A^0) \subseteq \mathsf{NullSp}(A^1) \subseteq \cdots \subseteq \mathsf{NullSp}(A^k) \subseteq \cdots$$

and a decreasing sequence of subspaces

$$\mathbb{C}^n = \mathit{Ran}(A^0) \supseteq \mathit{Ran}(A^1) \supseteq \cdots \supseteq \mathit{Ran}(A^k) \supseteq \cdots$$

Furthermore, there exists a number ℓ such that

$$\mathit{NullSp}(A^0) \subset \mathit{NullSp}(A^1) \subset \cdots \subset \mathit{NullSp}(A^\ell) = \mathit{NullSp}(A^{\ell+1}) = \cdots$$

and

$$\mathit{Ran}(A^0) \supset \mathit{Ran}(A^1) \supset \cdots \supset \mathit{Ran}(A^\ell) = \mathit{Ran}(A^{\ell+1}) = \cdots$$

Proof The proof of the existence of the increasing sequence of null subspaces and the decreasing sequence of ranges is immediate. Since $\mathsf{NullSp}(A^k) \subseteq \mathbb{C}^n$ for every k there exists a least number p such that $\mathsf{Ran}(A^p) = \mathsf{Ran}(A^{p+1})$. Therefore, $\mathsf{Ran}(A^{p+i}) = A^i\mathsf{Ran}(A^p) = A^i\mathsf{Ran}(A^{p+1}) = \mathsf{Ran}(A^{p+i+1})$ for every $i \in \mathbb{N}$. Thus, once two consecutive subspaces $\mathsf{Ran}(A^\ell)$ and $\mathsf{Ran}(A^{\ell+1})$ are equal the sequence of range subspaces stops growing.

By Equality (5.6), we have $\dim(\mathsf{Ran}(A^k)) + \dim(\mathsf{NullSp}(A^k)) = n$, so the sequence of null spaces stabilizes at the same number ℓ.

Definition 5.110 *The* index *of a square matrix $A \in \mathbb{C}^{n\times n}$ is the number ℓ defined in Theorem 5.109.*

We denote the index of a matrix $A \in \mathbb{C}^{n\times n}$ by $\mathsf{index}(A)$.

Observe that if $A \in \mathbb{C}^{n\times n}$ is a non-singular matrix, then $\mathsf{index}(A) = 0$ because in this case $\mathbb{C}^n = \mathsf{Ran}(A^0) = \mathsf{Ran}(A)$.

Theorem 5.111 *Let $A \in \mathbb{C}^{n\times n}$ be a square matrix. The following statements are equivalent:*

(i) $\mathit{Ran}(A^k) \cap \mathit{NullSp}(A^k) = \{\mathbf{0}\}$;
(ii) $\mathbb{C}^n = \mathit{Ran}(A^k) \boxplus \mathit{NullSp}(A^k)$;
(iii) $k \geqslant \mathit{index}(A)$.

Proof We prove this theorem by showing that (i) and (ii) are equivalent, (i) implies (iii), and (iii) implies (ii).

Suppose that the first statement holds. The set

$$T = \{\mathbf{t} \in V \mid \mathbf{t} = \boldsymbol{u} + \mathbf{v}, \boldsymbol{u} \in \mathsf{Ran}(A^k), \mathbf{v} \in \mathsf{NullSp}(A^k)\}$$

is a subspace of $\mathbb{C}^n$ and $\dim(T) = \dim(\mathsf{Ran}(A^k)) + \dim(\mathsf{NullSp}(A^k)) = n$. Therefore, $T = \mathbb{C}^n$, so $\mathbb{C}^n = \mathsf{Ran}(A^k) \boxplus \mathsf{NullSp}(A^k)$. The second statement clearly implies the first.

Suppose now that $\mathbb{C}^n = \mathsf{Ran}(A^k) \boxplus \mathsf{NullSp}(A^k)$. Then,

$$\mathsf{Ran}(A^k) = A^k\mathbb{C}^n = A\mathsf{Ran}(A^k) = \mathsf{Ran}(A^{k+1})$$

so $k \geqslant \mathsf{index}(A)$.

Conversely, if $k \geqslant \mathsf{index}(A)$ and $\mathbf{x} \in \mathsf{Ran}(A^k) \cap \mathsf{NullSp}(A^k)$, then $\mathbf{x} = A^k\mathbf{y}$ and $A^k\mathbf{x} = \mathbf{0}$, so $A^{2k}\mathbf{y} = \mathbf{0}$. Thus, $\mathbf{y} \in \mathsf{NullSp}(A^{2k}) = \mathsf{NullSp}(A^k)$, which means that $\mathbf{x} = A^k\mathbf{y} = \mathbf{0}$. Thus, the first statement holds.

5.5 Multilinear Forms

The notion of linear mapping can be extended as follows.

Definition 5.112 *Let $\mathcal{F}$ be a field and let $\{M_1, \ldots, M_n\}$ be a family of n F-linear spaces.*

*An F-*multilinear *mapping is a mapping $f : M_1 \times \cdots \times M_n \longrightarrow M$, where M is an F-linear space that is linear in each of its arguments. In other words, f satisfies the conditions*

$$f(\boldsymbol{x}_1, \ldots, \boldsymbol{x}_{i-1}, \sum_{j=1}^{k} a_j \boldsymbol{x}_i^j, \boldsymbol{x}_{i+1}, \ldots, \boldsymbol{x}_n)$$

$$= \sum_{j=1}^{k} a_j f(\boldsymbol{x}_1, \ldots, \boldsymbol{x}_{i-1}, \boldsymbol{x}_i^j, \boldsymbol{x}_{i+1}, \ldots, \boldsymbol{x}_n),$$

for every $\boldsymbol{x}_i, \boldsymbol{x}_i^j \in M_i$ and $a_1, \ldots, a_k \in F$.

*If M is the field F itself, then we refer to f as an n-*linear form. *For the special case $n = 2$ we use the terms* bilinear mapping *or* bilinear form.

We introduce next a class of multilinear forms that plays a central role in this chapter.

Definition 5.113 *Let $\mathcal{F} = (F, \{0, 1, +, -, \cdot\})$ be a field and let M be an F-linear space. An F-multilinear form $f : M^n \longrightarrow F$ is* skew-symmetric *if $\boldsymbol{x}_i = \boldsymbol{x}_j$ for $1 \leqslant i \neq j \leqslant n$ implies $f(\boldsymbol{x}_1, \ldots, \boldsymbol{x}_i, \ldots, \boldsymbol{x}_j, \ldots, \boldsymbol{x}_n) = 0$.*

The next statement shows that when two arguments of f are interchanged, then the value of f is multiplied by -1.

Theorem 5.114 *Let L be an F-linear space and let $f : L^n \longrightarrow F$ be a skew-symmetric F-multilinear form. We have*

$$f(\boldsymbol{x}_1, \ldots, \boldsymbol{x}_i, \ldots, \boldsymbol{x}_j, \ldots, \boldsymbol{x}_n) = -f(\boldsymbol{x}_1, \ldots, \boldsymbol{x}_j, \ldots, \boldsymbol{x}_i, \ldots, \boldsymbol{x}_n),$$

for $\boldsymbol{x}_1, \ldots, \boldsymbol{x}_n \in L$.

Proof Since f is a multilinear form we have:

$$\begin{aligned}
&f(\mathbf{x}_1, \ldots, \mathbf{x}_i + \mathbf{x}_j, \ldots, \mathbf{x}_i + \mathbf{x}_j, \ldots, \mathbf{x}_n)\\
&= f(\mathbf{x}_1, \ldots, \mathbf{x}_i, \ldots, \mathbf{x}_i, \ldots, \mathbf{x}_n) + f(\mathbf{x}_1, \ldots, \mathbf{x}_i, \ldots, \mathbf{x}_j, \ldots, \mathbf{x}_n)\\
&\quad + f(\mathbf{x}_1, \ldots, \mathbf{x}_j, \ldots, \mathbf{x}_i, \ldots, \mathbf{x}_n) + f(\mathbf{x}_1, \ldots, \mathbf{x}_j, \ldots, \mathbf{x}_j, \ldots, \mathbf{x}_n)\\
&= f(\mathbf{x}_1, \ldots, \mathbf{x}_i, \ldots, \mathbf{x}_j, \ldots, \mathbf{x}_n) + f(\mathbf{x}_1, \ldots, \mathbf{x}_j, \ldots, \mathbf{x}_j, \ldots, \mathbf{x}_n).
\end{aligned}$$

By the defining property of skew-symmetry we have the equalities

$$\begin{aligned}
f(\mathbf{x}_1, \ldots, \mathbf{x}_i + \mathbf{x}_j, \ldots, \mathbf{x}_i + \mathbf{x}_j, \ldots, \mathbf{x}_n) &= 0,\\
f(\mathbf{x}_1, \ldots, \mathbf{x}_i, \ldots, \mathbf{x}_i, \ldots, \mathbf{x}_n) &= 0,\\
f(\mathbf{x}_1, \ldots, \mathbf{x}_j, \ldots, \mathbf{x}_j, \ldots, \mathbf{x}_n) &= 0,
\end{aligned}$$

which yield

$$f(\mathbf{x}_1, \ldots, \mathbf{x}_i, \ldots, \mathbf{x}_j, \ldots, \mathbf{x}_n) = -f(\mathbf{x}_1, \ldots, \mathbf{x}_j, \ldots, \mathbf{x}_i, \ldots, \mathbf{x}_n),$$

for $\mathbf{x}_1, \ldots, \mathbf{x}_n \in L$.

Corollary 5.115 *Let $\mathcal{F}$ be a field, L be an F-linear space and let $f : L^n \longrightarrow F$ be a skew-symmetric F-multilinear form.*

If $\boldsymbol{x}_i = \boldsymbol{x}_j$ for $i \neq j$, then $f(\boldsymbol{x}_1, \ldots, \boldsymbol{x}_i, \ldots, \boldsymbol{x}_j, \ldots, \boldsymbol{x}_n) = 0$.

Proof This follows immediately from Theorem 5.114.

Theorem 5.114 has the following useful extension.

Theorem 5.116 *Let L be an F-linear space and let $f : L^n \longrightarrow F$ be a skew-symmetric F-multilinear form.*

If $\phi \in PERM_n$ is a permutation given by

$$\phi : \begin{pmatrix} 1 & \cdots & i & \cdots & n \\ j_1 & \cdots & j_i & \cdots & j_n \end{pmatrix},$$

then $f(\boldsymbol{x}_{j_1}, \ldots, \boldsymbol{x}_{j_n}) = (-1)^{inv(\phi)} f(\boldsymbol{x}_1, \ldots, \boldsymbol{x}_n)$ for $\boldsymbol{x}_1, \ldots, \boldsymbol{x}_n \in M$.

Proof The argument is by induction on $p = inv(\phi)$. The basis case, $p = 0$ is immediate because in this case, ϕ is the identity mapping.

Suppose that the argument holds for permutations that have no more than p inversions and let ϕ be a permutation that has $p + 1$ inversions. Then, as we saw in the proof of Theorem 3.8, there exists a standard transposition ψ such that for the permutation ϕ' defined as $\phi' = \psi\phi$ we have $inv(\phi') = inv(\phi) - 1$. Suppose that ϕ' is the permutation

$$\phi' : \begin{pmatrix} 1 & 2 & \cdots & \ell & \ell+1 & \cdots & n \\ j_1 & j_2 & \cdots & j_\ell & j_{\ell+1} & \cdots & j_n \end{pmatrix}$$

and ψ is the standard transposition that exchanges j_ℓ and $j_{\ell+1}$, so

$$\phi : \begin{pmatrix} 1 & 2 & \cdots & \ell & \ell+1 & \cdots & n \\ j_1 & j_2 & \cdots & j_{\ell+1} & j_\ell & \cdots & j_n \end{pmatrix}$$

By the inductive hypothesis,

$$f(\mathbf{x}_{j_1}, \ldots, \mathbf{x}_{j_\ell}, \mathbf{x}_{j_{\ell+1}}, \ldots, \mathbf{x}_{j_n}) = (-1)^{inv(\phi')} f(\mathbf{x}_1, \ldots, \mathbf{x}_n)$$

and

$$\begin{aligned} & f(\mathbf{x}_{j_1}, \ldots, \mathbf{x}_{j_{\ell+1}}, \mathbf{x}_{j_\ell}, \ldots, \mathbf{x}_{j_n}) \\ & = -f(\mathbf{x}_{j_1}, \ldots, \mathbf{x}_{j_\ell}, \mathbf{x}_{j_{\ell+1}}, \ldots, \mathbf{x}_{j_n}) \\ & = -(-1)^{inv(\phi')} f(\mathbf{x}_1, \ldots, \mathbf{x}_n) = (-1)^{inv(\phi)} f(\mathbf{x}_1, \ldots, \mathbf{x}_n), \end{aligned}$$

which concludes the argument.

Theorem 5.117 *Let $\mathcal{F}$ be a field, L be an F-linear space, $f : L^n \longrightarrow F$ be a skew-symmetric F-multilinear form, and let $a \in F$.*

If $i \neq j$ and $\boldsymbol{x}_1, \ldots, \boldsymbol{x}_n \in M^n$, then

$$f(\boldsymbol{x}_1, \ldots, \boldsymbol{x}_n) = f(\boldsymbol{x}_1, \ldots, \boldsymbol{x}_i + a\boldsymbol{x}_j, \ldots, \boldsymbol{x}_n).$$

Proof Suppose that $i < j$. Then, by the linearity of f we have

$$\begin{aligned} & f(\mathbf{x}_1, \ldots, \mathbf{x}_i + a\mathbf{x}_j, \ldots, \mathbf{x}_n) \\ & = f(\mathbf{x}_1, \ldots, \mathbf{x}_i, \ldots, \mathbf{x}_n) + af(\mathbf{x}_1, \ldots, \mathbf{x}_j, \ldots, \mathbf{x}_j, \ldots, \mathbf{x}_n) \\ & = f(\mathbf{x}_1, \ldots, \mathbf{x}_i, \ldots, \mathbf{x}_n), \end{aligned}$$

by Corollary 5.115.

Theorem 5.118 *Let L be an F-linear space and let $f : L^n \longrightarrow \mathbb{R}$ be a skew-symmetric linear form on L. If $\{\boldsymbol{x}_1, \ldots, \boldsymbol{x}_n\}$ is a linearly dependent subset of L, then $f(\boldsymbol{x}_1, \ldots, \boldsymbol{x}_n) = 0$.*

Proof Suppose that $\{\mathbf{x}_1, \ldots, \mathbf{x}_n\}$ is linearly dependent set, that is, one of the vectors can be expressed as a linear combination of the remaining vectors, say $\mathbf{x}_n = a_1\mathbf{x}_1 + \cdots + a_{n-1}\mathbf{x}_{n-1}$. Then,

$$\begin{aligned} f(\mathbf{x}_1, \ldots, \mathbf{x}_{n-1}, \mathbf{x}_n) &= f(\mathbf{x}_1, \ldots, \mathbf{x}_{n-1}, a_1\mathbf{x}_1 + \cdots + a_{n-1}\mathbf{x}_{n-1}) \\ &= \sum_{i=1}^{n-1} a_i f(\mathbf{x}_1, \ldots, \mathbf{x}_i, \ldots, \mathbf{x}_{n-1}, \mathbf{x}_i) = 0, \end{aligned}$$

by Corollary 5.115.

Theorem 5.119 *Let L be an n-dimensional linear space and let $\{\boldsymbol{u}_1, \ldots, \boldsymbol{u}_n\}$ be a basis of L. There exists a unique, skew-symmetric multilinear form $d_n : L^n \longrightarrow \mathbb{R}$ such that $d_n(\boldsymbol{u}_1, \ldots, \boldsymbol{u}_n) = 1$.*

Proof Let $\mathbf{x}_1, \ldots, \mathbf{x}_n$ be n vectors such that

$$\mathbf{x}_i = a_{i1}\boldsymbol{u}_1 + a_{i2}\boldsymbol{u}_2 + \cdots + a_{in}\boldsymbol{u}_n$$

for $1 \leqslant i \leqslant n$. If d_n is a skew symmetric multilinear form, $d_n : L^n \longrightarrow \mathbb{R}$, then

$$\begin{aligned} & d_n(\mathbf{x}_1, \mathbf{x}_2, \ldots, \mathbf{x}_n) \\ & = d_n\left(\sum_{j_1=1}^{n} a_{1j_1}\boldsymbol{u}_{j_1}, \sum_{j_2=1}^{n} a_{2j_2}\boldsymbol{u}_{j_2}, \ldots, \sum_{j_n=1}^{n} a_{nj_n}\boldsymbol{u}_{j_n}\right) \\ & = \sum_{j_1=1}^{n}\sum_{j_2=1}^{n}\cdots\sum_{j_n=1}^{n} a_{1j_1}a_{2j_2}\cdots a_{nj_n} d_n(\boldsymbol{u}_{j_1}, \boldsymbol{u}_{j_2}, \ldots, \boldsymbol{u}_{j_n}) \end{aligned}$$

We need to retain only the terms of this sum in which the arguments of $d_n(\mathbf{x}_{j_1}, \mathbf{x}_{j_2}, \ldots, \mathbf{x}_{j_n})$ are pairwise distinct (because term where $j_p = j_q$ for $p \neq q$ is zero, by Corollary 5.115). In other words, only the terms in which the list $(j_1, \ldots, j_n)$ is a permutation of $(1, \ldots, n)$ have a non-zero contribution to the sum. By Theorem 5.116, we can write

$$\begin{aligned} & d_n(\mathbf{x}_1, \mathbf{x}_2, \ldots, \mathbf{x}_n) \\ & = d_n(\boldsymbol{u}_1, \boldsymbol{u}_2, \ldots, \boldsymbol{u}_n) \textstyle\sum_{j_1,\ldots,j_n} (-1)^{inv(j_1,\ldots,j_n)} a_{1j_1}a_{2j_2}\cdots a_{nj_n}. \end{aligned}$$

where the sum extends to all $n!$ permutations $(j_1, \ldots, j_n)$ of $(1, \ldots, n)$. Since $d_n(\boldsymbol{u}_1, \ldots, \boldsymbol{u}_n) = 1$, it follows that

$$d_n(\mathbf{x}_1, \mathbf{x}_2, \ldots, \mathbf{x}_n) = \sum_{j_1,\ldots,j_n} (-1)^{inv(j_1,\ldots,j_n)} a_{1j_1}a_{2j_2}\cdots a_{nj_n}.$$

Note that $d_n(\mathbf{x}_1, \mathbf{x}_2, \ldots, \mathbf{x}_n)$ is expressed using the elements of the matrix A, where

$$A = \begin{pmatrix} a_{11} & a_{12} & \cdots & a_{1n} \\ a_{21} & a_{22} & \cdots & a_{2n} \\ \vdots & \vdots & \cdots & \vdots \\ a_{n1} & a_{n2} & \cdots & a_{nn} \end{pmatrix}.$$

5.6 Linear Systems

Consider the following set of linear equalities

$$\begin{aligned} a_{11}x_1 + \ldots + a_{1n}x_n &= b_1, \\ a_{21}x_1 + \ldots + a_{2n}x_n &= b_2, \\ \vdots \quad & \quad \vdots \\ a_{m1}x_1 + \ldots + a_{mn}x_n &= b_m, \end{aligned}$$

where a_{ij} and b_i belong to a field F. This set constitutes a *system of linear equations* and solving it means finding $x_1, \ldots, x_n$ that satisfy all equalities.

The system can be written succinctly in a matrix form as $A\mathbf{x} = \boldsymbol{b}$, where

$$A = \begin{pmatrix} a_{11} & \cdots & a_{1n} \\ a_{21} & \cdots & a_{2n} \\ \vdots & \cdots & \vdots \\ a_{m1} & \cdots & a_{mn} \end{pmatrix}, \boldsymbol{b} = \begin{pmatrix} b_1 \\ b_2 \\ \vdots \\ b_m \end{pmatrix}, \text{ and } \mathbf{x} = \begin{pmatrix} x_1 \\ x_2 \\ \vdots \\ x_n \end{pmatrix}.$$

In terms of linear transformations, solving this linear system amounts to determining those vectors $\mathbf{x}$ such that $h_A(\mathbf{x}) = \boldsymbol{b}$.

If the set of solutions of a system $A\mathbf{x} = \boldsymbol{b}$ is not empty we say that the system is *consistent*. Note that $A\mathbf{x} = \boldsymbol{b}$ is consistent if and only if $\boldsymbol{b} \in \mathsf{Ran}(A)$.

Let $A\mathbf{x} = \boldsymbol{b}$ be a linear system in matrix form, where $A \in \mathbb{C}^{m \times n}$. The matrix $(A \;\; \boldsymbol{b}) \in \mathbb{C}^{m \times (n+1)}$ is the *augmented matrix* of the system $A\boldsymbol{x} = \boldsymbol{b}$.

Theorem 5.120 *Let $A \in \mathbb{C}^{m \times n}$ be a matrix and let $\boldsymbol{b} \in \mathbb{C}^{n \times 1}$. The linear system $A\boldsymbol{x} = \boldsymbol{b}$ is consistent if and only if $rank(A\ \boldsymbol{b}) = rank(A)$.*

Proof If $A\mathbf{x} = \boldsymbol{b}$ is consistent and $\mathbf{x}$ is a solution of this system, then $\boldsymbol{b} = x_1\mathbf{c}_1 + \cdots + x_n\mathbf{c}_n$, where $\mathbf{c}_1, \ldots, \mathbf{c}_n$ are the columns of A. This implies $rank(A\ \ \boldsymbol{b}) = rank(A)$.

Conversely, if $rank(A\ \boldsymbol{b}) = rank(A)$, the vector $\boldsymbol{b}$ is a linear combination of the columns of A, which means that $A\mathbf{x} = \boldsymbol{b}$ is a consistent system.

Definition 5.121 *An* homogeneous linear system *is a linear system of the form $A\boldsymbol{x} = \boldsymbol{0}_m$, where $A \in \mathbb{C}^{m \times n}$, $\boldsymbol{x} \in \mathbb{C}^{n,1}$ and $\boldsymbol{0} \in \mathbb{C}^{m \times 1}$.*

Clearly, any homogeneous system $A\mathbf{x} = \mathbf{0}_m$ is consistent and has the solution $\mathbf{x} = \mathbf{0}_n$. This solution is referred to as the *trivial solution*. The set of solutions of such a system is $\mathsf{NullSp}(A)$, the null space of the matrix A.

Let $\boldsymbol{u}$ and $\mathbf{v}$ be two solutions of the system $A\mathbf{x} = \boldsymbol{b}$. Then $A(\boldsymbol{u} - \mathbf{v}) = \mathbf{0}_m$, so $\mathbf{z} = \boldsymbol{u} - \mathbf{v}$ is a solution of the homogeneous system $A\mathbf{x} = \mathbf{0}_m$, so $\mathbf{z} \in \mathsf{NullSp}(A)$. Thus, the set of solutions of $A\mathbf{x} = \boldsymbol{b}$ can be obtained as a translation of the null space of A by any particular solution of $A\mathbf{x} = \boldsymbol{b}$. In other words the set of solution of $A\mathbf{x} = \boldsymbol{b}$ is $\{\boldsymbol{x} + \boldsymbol{z} \mid \boldsymbol{z} \in \mathsf{NullSp}(A)\}$.

Thus, for $A \in \mathbb{C}^{m \times n}$, the system $A\mathbf{x} = \boldsymbol{b}$ has a unique solution if and only if $\mathsf{NullSp}(A) = \{\mathbf{0}_n\}$, that is, according to Equality (5.6), if $rank(A) = n$.

Theorem 5.122 *Let $A \in \mathbb{C}^{n \times n}$. Then, A is invertible (which is to say that $rank(A) = n$) if and only if the system $A\mathbf{x} = \boldsymbol{b}$ has a unique solution for every $\boldsymbol{b} \in \mathbb{C}^n$.*

Proof If A is invertible, then $\mathbf{x} = A^{-1}\boldsymbol{b}$, so the system $A\mathbf{x} = \boldsymbol{b}$ has a unique solution.

Conversely, if the system $A\mathbf{x} = \boldsymbol{b}$ has a unique solution for every $\boldsymbol{b} \in \mathbb{C}^n$, let $\mathbf{c}_1, \ldots, \mathbf{c}_n$ be the solution of the systems $A\mathbf{x} = \mathbf{e}_1, \ldots, A\mathbf{x} = \mathbf{e}_n$, respectively. Then, we have

$$A(\mathbf{c}_1 \quad \cdots \quad \mathbf{c}_n) = I_n,$$

which shows that A is invertible and $A^{-1} = (\mathbf{c}_1 | \cdots | \mathbf{c}_n)$.

Corollary 5.123 *An homogeneous linear system $A\mathbf{x} = \boldsymbol{0}$, where $A \in \mathbb{C}^{n \times n}$ has a non-trivial solution if and only if A is a singular matrix.*

Proof This statement follows from Theorem 5.112.

Thus, by calculating the inverse of A we can solve any linear system of the form $A\mathbf{x} = \boldsymbol{b}$.

Definition 5.124 *A matrix $A \in \mathbb{C}^{n \times n}$ is* diagonally dominant *if $|a_{ii}| > \sum\{|a_{ik}| \mid 1 \leqslant k \leqslant n \text{ and } k \neq i\}$.*

Theorem 5.125 *A diagonally dominant matrix is non-singular.*

Proof Suppose that $A \in \mathbb{C}^{n \times n}$ is a diagonally dominant matrix that is singular. By Corollary 5.123, the homogeneous system $A\mathbf{x} = \mathbf{0}$ has a non-trivial solution $\mathbf{x} \neq \mathbf{0}$. Let x_k be a component of $\mathbf{x}$ that has the largest absolute value. Since $\mathbf{x} \neq \mathbf{0}$, we have $|x_k| > 0$. We can write

$$a_{kk}x_k = -\sum\{a_{kj}x_j \mid 1 \leqslant j \leqslant n \text{ and } j \neq k\},$$

which implies

$$\begin{aligned}|a_{kk}|\,|x_k| &= \left|\sum\{a_{kj}x_j \mid 1 \leqslant j \leqslant n \text{ and } j \neq k\}\right| \\ &\leqslant \sum\{|a_{kj}|\,|x_j| \mid 1 \leqslant j \leqslant n \text{ and } j \neq k\}\Big| \\ &\leqslant |x_k| \sum\{|a_{kj}| \mid 1 \leqslant j \leqslant n \text{ and } j \neq k\}.\end{aligned}$$

Thus, we obtain

$$|a_{kk}| \leqslant \sum\{|a_{kj}| \mid 1 \leqslant j \leqslant n \text{ and } j \neq k\},$$

which contradicts the fact that A is diagonally dominant.

5.7 Determinants

Determinants are a class of numerical multilinear functions defined on the set of square matrices. They play an important role in theoretical considerations of linear algebra and are useful for symbolic computations. As we shall see, determinants can be used to solve certain small and well-behaved linear system; however, they are of limited use for large or numerically difficult linear systems.[1]

Definition 5.126 *Let $A = (a_{ij}) \in \mathbb{C}^{n\times n}$ be a square matrix. The* determinant of A *is the number $\sum_{j_1,\ldots,j_n} (-1)^{inv(j_1,\ldots,j_n)} a_{1j_1} a_{2j_2} \cdots a_{nj_n}$.*

The determinant of A is denoted either by $\det(A)$ *or by:*

$$\begin{vmatrix} a_{11} & a_{12} & \cdots & a_{1n} \\ a_{21} & a_{22} & \cdots & a_{2n} \\ \vdots & \vdots & \cdots & \vdots \\ a_{n1} & a_{n2} & \cdots & a_{nn} \end{vmatrix}.$$

Example 5.127 We have $\det(I_n) = 1$ since $(I_n)_{ij} = 1$ if $j = i$ and $(I_n)_{ij} = 0$ otherwise. Thus, there exists only one non-zero term in the sam

$$\det(I_n) = \sum_{j_1,\ldots,j_n} (-1)^{inv(j_1,\ldots,j_n)} (I_n)_{1j_1} (I_n)_{2j_2} \cdots (I_n)_{nj_n},$$

which is obtained when $j_i = i$ for $1 \leqslant i \leqslant n$, and this unique term is 1.

Example 5.128 Let $A \in \mathbb{R}^{3\times 3}$ be the matrix

$$A = \begin{pmatrix} a_{11} & a_{12} & a_{13} \\ a_{21} & a_{22} & a_{23} \\ a_{31} & a_{32} & a_{33} \end{pmatrix}.$$

The number $\det(A)$ is the sum of six terms corresponding to the six permutations of the set $\{1, 2, 3\}$, as shown below.

Thus, we have

$$\begin{aligned} \det(A) &= a_{11}a_{22}a_{33} + a_{13}a_{21}a_{32} + a_{12}a_{23}a_{31} \\ &= -a_{12}a_{21}a_{33} - a_{13}a_{22}a_{31} - a_{11}a_{23}a_{32}. \end{aligned}$$

[1] Historically, determinants appeared long before matrices related to solving linear systems. In modern times, determinants were introduced by Leibniz at the end of the 17th century and Cramer formula appeared in 1750. The term "determinant" was introduced by Gauss in 1801. The term "matrix", the Latin word for womb, was introduced in 1848 by James Joseph Sylvester (1814–1897), a British mathematician whose name is linked to many fundamental results in linear algebra. The term was suggested by the role of matrices as generators of determinants.

Permutation ϕ	$inv(\phi)$	Term
(1, 2, 3)	0	$a_{11}a_{22}a_{33}$
(3, 1, 2)	2	$a_{13}a_{21}a_{32}$
(2, 3, 1)	2	$a_{12}a_{23}a_{31}$
(2, 1, 3)	1	$-a_{12}a_{21}a_{33}$
(3, 2, 1)	3	$-a_{13}a_{22}a_{31}$
(1, 3, 2)	1	$-a_{11}a_{23}a_{32}$

The number of terms of a determinant of order n is $n!$; this number grows very fast with n. For instance, for $n = 10$, we have $10! = 3,682,800$ terms. Thus, direct computations of determinants are very expensive.

Theorem 5.129 *Let $A \in \mathbb{C}^{n\times n}$ be a matrix. We have* $\det(A') = \det(A)$.

Proof The definition of A' allows us to write

$$\det(A') = \sum_{j_1,\ldots,j_n} (-1)^{inv(j_1,\ldots,j_n)} a_{j_1 1} a_{j_2 2} \cdots a_{j_n n},$$

where the sum extends to all permutations of $(1,\ldots,n)$. Due to the commutativity of numeric multiplication we can rearrange the term $a_{j_1 1}a_{j_2 2}\cdots a_{j_n n}$ as $a_{1k_1}a_{2k_2}\cdots a_{nk_n}$, where

$$\phi : \begin{pmatrix} 1 & 2 & \cdots & n \\ j_1 & j_2 & \cdots & j_n \end{pmatrix} \text{ and } \psi : \begin{pmatrix} 1 & 2 & \cdots & n \\ k_1 & k_2 & \cdots & k_n \end{pmatrix}$$

are inverse permutations. Since both ϕ and ψ have the same parity, it follows that

$$(-1)^{inv(j_1,\ldots,j_n)} a_{j_1 1} a_{j_2 2} \cdots a_{j_n n} = (-1)^{inv(k_1,\ldots,k_n)} a_{1k_1} a_{2k_2} \cdots a_{nj_n},$$

which implies $\det(A') = \det(A)$.

Corollary 5.130 *If $A \in \mathbb{C}^{n\times n}$, then* $\det(A^{\mathsf{H}}) = \overline{\det(A)}$. *Furthermore, if A is a Hermitian matrix,* $\det(A)$ *is a real number.*

Proof Let $\bar{A}$ be the matrix obtained from A by replacing each a_{ij} by its conjugate. Since conjugation of complex numbers permutes with both the sum and product of complex numbers it follows that $\det(\bar{A}) = \overline{\det(A)}$. Thus, $\det(A^{\mathsf{H}}) = \det(\bar{A})' = \det(\bar{A}) = \overline{\det(A)}$.

The second part of the corollary follows from the equality $\det(A) = \overline{\det(A)}$.

Corollary 5.131 *If $A \in \mathbb{C}^{n\times n}$ is a unitary matrix, then* $|\det(A)| = 1$.

Proof Since A is unitary we have $A^{\mathsf{H}}A = AA^{\mathsf{H}} = I_n$. By Theorem 5.133, $\det(AA^{\mathsf{H}}) = \det(A)\det(A^{\mathsf{H}}) = \det(A)\overline{\det(A)} = |\det(A)|^2 = 1$. Thus, $|\det(A)| = 1$.

Theorem 5.132 *The following properties of* $\det(A)$ *hold for any* $A \in \mathbb{C}^{n\times n}$*:*

(i) $\det(A)$ *is a linear function of the rows of* A *(of the columns of* A*);*
(ii) *if two rows (columns) are permuted, then* $\det(A)$ *is changing signs;*
(iii) *if* A *has two equal rows (columns), then* $\det(A) = 0$*;*
(iv) *if a row of of a matrix, multiplied by a constant, is added to another row, then* $\det(A)$ *remains unchanged; the same holds if instead of rows we consider columns;*
(v) *if a row (column) equals* $\mathbf{0}_n$*, then* $\det(A) = 0$.

Proof We begin with the above statements that involve columns of A.

To prove Part (i) let

$$A = \begin{pmatrix} \mathbf{a}_1 \\ \vdots \\ \mathbf{a}_{k-1} \\ \beta \boldsymbol{b}_k + \gamma \mathbf{c}_k \\ \mathbf{a}_{k+1} \\ \vdots \\ \mathbf{a}_n \end{pmatrix}, B = \begin{pmatrix} \mathbf{a}_1 \\ \vdots \\ \mathbf{a}_{k-1} \\ \boldsymbol{b}_k \\ \mathbf{a}_{k+1} \\ \vdots \\ \mathbf{a}_n \end{pmatrix} \text{ and } C = \begin{pmatrix} \mathbf{a}_1 \\ \vdots \\ \mathbf{a}_{k-1} \\ \mathbf{c}_k \\ \mathbf{a}_{k+1} \\ \vdots \\ \mathbf{a}_n \end{pmatrix},$$

where $\boldsymbol{b}_j, \mathbf{c}_j$ are row vectors and $\beta, \gamma \in \mathbb{R}$. By the definition of $\det(A)$ we have

$$\begin{aligned} \det(A) &= \sum_{j_1,\ldots,j_n} (-1)^{inv(j_1,\ldots,j_n)} a_{1j_1} \cdots (\beta b_{kj_k} + \gamma c_{kj_k}) \cdots a_{nj_n} \\ &= \beta \sum_{j_1,\ldots,j_n} (-1)^{inv(j_1,\ldots,j_n)} a_{1j_1} \cdots b_{kj_k} \cdots a_{nj_n} \\ &\quad + \gamma \sum_{j_1,\ldots,j_n} (-1)^{inv(j_1,\ldots,j_n)} a_{1j_1} \cdots c_{kj_k} \cdots a_{nj_n} \\ &= \beta \det(B) + \gamma \det(C), \end{aligned}$$

which proves that $\det(\cdot)$ is linear.

Let now

$$A = \begin{pmatrix} \mathbf{a}_1 \\ \vdots \\ \mathbf{a}_p \\ \vdots \\ \mathbf{a}_q \\ \vdots \\ \mathbf{a}_n \end{pmatrix} \text{ and let } \tilde{A} = \begin{pmatrix} \mathbf{a}_1 \\ \vdots \\ \mathbf{a}_q \\ \vdots \\ \mathbf{a}_p \\ \vdots \\ \mathbf{a}_n \end{pmatrix}$$

be the matrix obtained by swaping the pth and the qth row of A.

By the definition of determinants,

$$\det(\tilde{A}) = \sum_{j_1,\ldots,j_n} (-1)^{inv(j_1,\ldots,j_p,\ldots,j_q,\ldots,j_n)} a_{1j_1} \cdots a_{qj_q} \cdots a_{pj_p} \cdots a_{nj_n}.$$

Note that the permutation $(j_1, \ldots, j_p, \ldots, j_q, \ldots, j_n)$ is obtained by the composition of $(j_1, \ldots, j_q, \ldots, j_p, \ldots, j_n)$ with the transposition that swaps j_p with j_q. Therefore, $\det(A) = -\det(\tilde{(A)})$, which proves Part (ii).

If two rows of A are equal, then by swapping these rows we get $\det(A) = -\det(A)$, so $\det(A) = 0$, which proves Part (iii).

Part (iv) follows from the first three parts; the last part is a direct consequence of the definition of $\det(A)$.

The corresponding statements concerning rows of A follow from Theorem 5.129 because the rows of A are the transposed columns of A'.

Theorem 5.133 *Let $A, B \in \mathbb{C}^{n\times n}$ be two matrices. We have* $\det(AB) = \det(A)\det(B)$.

Proof Let $\mathbf{a}_1, \ldots, \mathbf{a}_n$ and $\boldsymbol{b}_1, \ldots, \boldsymbol{b}_n$ be the rows of the matrices A and B respectively, where $\mathbf{a}_i = (a_{i1}, \ldots, a_{in})$ for $1 \leqslant i \leqslant n$. Then, the rows $\mathbf{c}_1, \ldots, \mathbf{c}_n$ of the matrix $C = AB$ are given by $\mathbf{c}_i = a_{i1}\boldsymbol{b}_1 + \cdots + a_{in}\boldsymbol{b}_n$, as it can be easily seen.

If $d_n : (\mathbb{C}^n)^n \longrightarrow \mathbb{C}$ is the skew-symmetric multilinear that defines the determinant whose existence and uniqueness was shown in Theorem 5.119, then we have

$$\begin{aligned}\det(AB) &= d_n(\mathbf{c}_1, \ldots, \mathbf{c}_i, \ldots, \mathbf{c}_n)\\ &= d_n\left(\sum_{j_1=1}^{n} a_{1j_1}\boldsymbol{b}_{j_1}, \ldots, \sum_{j=1}^{n} a_{ij_i}\boldsymbol{b}_{j_i}, \ldots, \sum_{j_n=1}^{n} a_{nj_n}\boldsymbol{b}_{j_n}\right)\\ &= \sum_{j_1=1}^{n}\cdots\sum_{j_1=1}^{n}\cdots\sum_{j_1=1}^{n} a_{1j_1}\cdots a_{ij_i}\cdots a_{nj_n} d_n(\boldsymbol{b}_{j_1}, \ldots, \boldsymbol{b}_{j_i}, \ldots, \boldsymbol{b}_{j_n}),\end{aligned}$$

due to the linearity of d_n. Observe now that only the sequences $(j_1, \ldots, j_n)$ that represent permutations of the set $\{1, \ldots, n\}$ contribute to the sum because d_n is skew-symmetric. Furthermore, if $(j_1, \ldots, j_n)$ represents a permutation ϕ, then $d_n(\boldsymbol{b}_{j_1}, \ldots, \boldsymbol{b}_{j_i}, \ldots, \boldsymbol{b}_{j_n}) = (-1)^{inv(\phi)} d_n(\boldsymbol{b}_1, \ldots, \boldsymbol{b}_n)$. Thus, we can write

$$\begin{aligned}&\det(AB)\\ &= \textstyle\sum_{j_1=1}^{n}\cdots\sum_{j_1=1}^{n}\cdots\sum_{j_1=1}^{n} a_{1j_1}\cdots a_{ij_i}\cdots a_{nj_n} d_n(\boldsymbol{b}_{j_1}, \ldots, \boldsymbol{b}_{j_i}, \ldots, \boldsymbol{b}_{j_n})\\ &= \left(\textstyle\sum_{j_1=1}^{n}\cdots\sum_{j_1=1}^{n}\cdots\sum_{j_1=1}^{n}(-1)^{inv(j_1,\ldots,j_n)} a_{1j_1}\cdots a_{ij_i}\cdots a_{nj_n}\right) d_n(\boldsymbol{b}_1, \ldots, \boldsymbol{b}_n)\\ &= \det(A)\det(B).\end{aligned}$$

Corollary 5.134 *Let $A \in \mathbb{R}^{n\times n}$. We have*

$$\det(AT_n^{i\leftrightarrow j}) = -\det(A), \det(AT_n^{i\overset{+}{\leftarrow} j}) = \det(A), \det(AT_{n,i}^{(a)}) = a\det(A).$$
$$\det(T_n^{i\leftrightarrow j}A) = -\det(A), \det((T_n^{i\overset{+}{\leftarrow} j})'A) = \det(A), \det(T_{n,i}^{(a)}A) = a\det(A).$$

Proof Note that $\det(T_n^{i\leftrightarrow j} = -1$, $\det(T_n^{i\overset{+}{\leftarrow} j}) = 1$ and $\det(T_{n,i}^{(a)} = a$. The statement follows immediately from Theorem 5.133.

Lemma 5.135 *Let* $B \in \mathbb{R}^{(n+1)\times(n+1)}$ *be*

$$B = \begin{pmatrix} 1 & 0 & 0 & \cdots & 0 \\ 0 & a_{11} & a_{12} & \cdots & a_{1n} \\ \vdots & \vdots & \vdots & \cdots & \vdots \\ 0 & a_{n1} & a_{n2} & \cdots & a_{nn} \end{pmatrix}.$$

We have $\det(B) = \det(A)$, *where*

$$A = \begin{pmatrix} a_{11} & a_{12} & \cdots & a_{1n} \\ \vdots & \vdots & \cdots & \vdots \\ a_{n1} & a_{n2} & \cdots & a_{nn} \end{pmatrix}$$

Proof If $B = (b_{ij})$, then

$$b_{1j} = \begin{cases} 1 & \text{if } j = 1 \\ 0 & \text{otherwise,} \end{cases} \quad \text{and } b_{i1} = \begin{cases} 1 & \text{if } i = 1 \\ 0 & \text{otherwise.} \end{cases}$$

Also, if $i > 1$ and $j > 1$, then $b_{ij} = a_{i-1,j-1}$ for $2 \leqslant i, j \leqslant n+1$. By the definition of the determinant, each term of the sum that defines $\det(B)$ must include an element of the first row. However, only the first element of this row is non-zero, so

$$\begin{aligned} \det(B) &= \sum_{(j_1, j_2, \ldots, j_{n+1})} (-1)^{inv(j_1, j_2, \ldots, j_{n+1})} b_{1j_1} b_{2j_2} \cdots b_{n+1, j_{n+1}}, \\ &= \sum_{(j_2, \ldots, j_{n+1})} (-1)^{inv(1, j_2, \ldots, j_{n+1})} a_{1, j_2-1} \cdots a_{n-1, j_n-1} a_{n, j_{n+1}-1}, \end{aligned}$$

where $(j_2, \ldots, j_{n+1})$ is a permutation of the set $\{2, \ldots, n+1\}$. Since

$$inv(1, j_2, \ldots, j_{n+1}) = inv(j_2, \ldots, j_{n+1}),$$

it follows that

$$\det(B) = \sum_{(j_2, \ldots, j_{n+1})} (-1)^{inv(j_2, \ldots, j_{n+1})} a_{1, j_2-1} a_{2, j_3-1} \cdots a_{n, j_{n+1}-1}.$$

Observe now that if $(j_2, \ldots, j_{n+1})$ is a permutation of the set $\{2, \ldots, n+1\}$, then $(k_1, \ldots, k_n)$, where $k_i = j_{i+1} - 1$ for $1 \leqslant i \leqslant n$ is a permutation of $(1, \ldots, n)$ that has the same number of inversions as $(j_2, \ldots, j_{n+1})$. Therefore,

$$\det(B) = \sum_{(k_1,\ldots,k_n)} (-1)^{inv(k_1,\ldots,k_n)} a_{1k_1} a_{2k_2} \ldots a_{nk_n} = \det(A).$$

Lemma 5.136 *Let $A \in \mathbb{R}^{n\times n}$ be a matrix partitioned as:*

$$A = \left(\begin{array}{ccc|ccc} a_{11} & \cdots & a_{1q} & a_{1,q+1} & \cdots & a_{1n} \\ \vdots & \cdots & \vdots & \vdots & \cdots & \vdots \\ a_{p1} & \cdots & a_{pq} & a_{p,q+1} & \cdots & a_{pn} \\ \hline a_{p+1,1} & \cdots & a_{p+1,q} & a_{p+1,q+1} & \cdots & a_{p+1,n} \\ \vdots & \cdots & \vdots & \vdots & \cdots & \vdots \\ a_{n1} & \cdots & a_{nq} & a_{n,q+1} & \cdots & a_{nn} \end{array}\right)$$

and let $B \in \mathbb{R}^{(n+1)\times(n+1)}$ be defined by

$$B = \left(\begin{array}{ccc|c|ccc} a_{11} & \cdots & a_{1q} & 0 & a_{1,q+1} & \cdots & a_{1n} \\ \vdots & \cdots & \vdots & \vdots & \vdots & \cdots & \vdots \\ a_{p1} & \cdots & a_{pq} & 0 & a_{p,q+1} & \cdots & a_{pn} \\ \hline 0 & \cdots & 0 & 1 & 0 & \cdots & 0 \\ \hline a_{p+1,1} & \cdots & a_{p+1,q} & 0 & a_{p+1,q+1} & \cdots & a_{p+1,n} \\ \vdots & \cdots & \vdots & \vdots & \vdots & \cdots & \vdots \\ a_{n1} & \cdots & a_{nq} & 0 & a_{n,q+1} & \cdots & a_{nn} \end{array}\right).$$

Then, $\det(B) = (-1)^{p+q} \det(A)$.

Proof By permuting the $(p+1)$st row of B with each of the p rows preceding it in the matrix B and, then, by permuting the $(q+1)$st column with each of the q columns preceding it we obtain the matrix C given by

$$C = \begin{pmatrix} 1 & 0 & 0 & 0 & 0 & 0 & 0 \\ 0 & a_{11} & \cdots & a_{1q} & a_{1,q+1} & \cdots & a_{1n} \\ \vdots & \vdots & \cdots & \vdots & \vdots & \cdots & \vdots \\ 0 & a_{p1} & \cdots & a_{pq} & a_{p,q+1} & \cdots & a_{pn} \\ 0 & a_{p+1,1} & \cdots & a_{p+1,q} & a_{p+1,q+1} & \cdots & a_{p+1,n} \\ \vdots & \vdots & \cdots & \vdots & \vdots & \cdots & \vdots \\ 0 & a_{n1} & \cdots & a_{nq} & a_{n,q+1} & \cdots & a_{nn} \end{pmatrix}.$$

By the third part of Theorem 5.132, each of these row or column permutations multiplies $\det(B)$ by -1, so $\det(C) = (-1)^{p+q}\det(B)$ By Lemma 5.135 we have $\det(C) = \det(A)$, so $\det(B) = (-1)^{p+q}\det(A)$.

Definition 5.137 *Let $A \in \mathbb{C}^{m\times n}$. A* minor *of* order k *of A is a determinant of the form*

$$det\left(A\begin{bmatrix} i_1 & \cdots & i_k \\ j_1 & \cdots & j_k \end{bmatrix}\right).$$

A principal minor of order k *of A is a determinant of the form:*

$$\det\left(A\begin{bmatrix} i_1 & \cdots & i_k \\ i_1 & \cdots & i_k \end{bmatrix}\right)$$

The leading principal minor *of order k is the determinant*

$$\det\left(A\begin{bmatrix} 1 & \cdots & k \\ 1 & \cdots & k \end{bmatrix}\right).$$

For $A \in \mathbb{C}^{n\times n}$, $\det(A)$ is the unique principal minor of order n, and that the principal minors of order 1 of A are just the diagonal entries of A: $a_{11}, \ldots, a_{nn}$.

Theorem 5.138 *Let $A \in \mathbb{C}^{n\times n}$. Define the matrix $A_{ij} \in \mathbb{C}^{(n-1)\times(n-1)}$ as*

$$A_{ij} = A\begin{bmatrix} 1 & \cdots & i-1 & i+1 & \cdots & n \\ 1 & \cdots & j-1 & j+1 & \cdots & h \end{bmatrix},$$

that is, the matrix obtained from A by removing the ith *row and the* jth *column. Then, we have*

$$\sum_{j=1}^{n}(-1)^{i+j}a_{ij}\det(A_{\ell j}) = \begin{cases} \det(A) & \text{if } i = \ell, \\ 0 & \text{otherwise}, \end{cases}$$

for every i, ℓ, $1 \leqslant i, \ell \leqslant n$.

Proof Let $\mathbf{x}_i$ be the ith row of A, which can be expressed as

$$\mathbf{x}_i = \sum_{j=1}^{n} a_{ij}\mathbf{e}_j,$$

where $\mathbf{e}_1, \ldots, \mathbf{e}_n$ is a basis of $\mathbb{R}^n$ such that $d_n(\mathbf{e}_1, \ldots, \mathbf{e}_n) = 1$.

By the linearity of d_n we have

$$\begin{aligned} d_n(A) &= d_n(\mathbf{x}_1, \ldots, \mathbf{x}_n) \\ &= d_n(\mathbf{x}_1, \ldots, \mathbf{x}_{i-1}, \sum_{j=1}^{n} a_{ij}\mathbf{e}_j, \mathbf{x}_{i+1}, \ldots, \mathbf{x}_n) \\ &= \sum_{j=1}^{n} a_{ij} d_n(\mathbf{x}_1, \ldots, \mathbf{x}_{i-1}, \mathbf{e}_j, \mathbf{x}_{i+1}, \ldots, \mathbf{x}_n) \end{aligned}$$

The determinant $d_n(\mathbf{x}_1, \ldots, \mathbf{x}_{i-1}, \mathbf{e}_j, \mathbf{x}_{i+1}, \ldots, \mathbf{x}_n)$ corresponds to a matrix $D^{(i,j)}$ obtained from A by replacing the ith row by the sequence

$$(0, \ldots, 0, 1, 0, \ldots, 0),$$

whose unique non-zero component is on the jth position. Next, by multiplying the ith row by $-a_{kj}$ and adding the result to the kth row for $1 \leqslant k \leqslant i-1$ and $i+1 \leqslant k \leqslant n$ yields a matrix $E^{(i,j)}$ that coincides with the matrix A with the following exceptions:

(i) the elements of row i are 0 with the exception of the jth element of this row that equals 1, and
(ii) the elements of column j are 0 with the exception of the element mentioned above.

Clearly, $\det(D^{(i,j)}) = \det(E^{(i,j)})$. By applying Lemma 5.136 we obtain $\det(E^{(i,j)}) = (-1)^{i+j} \det(A_{ij})$, so

$$d_n(A) = \sum_{j=1}^{n} a_{ij} d_n(E^{(i,j)}) = \sum_{j=1}^{n} (-1)^{i+j} a_{ij} \det(A_{ij}),$$

which is the first case of the desired formula.

Suppose now that $i \neq \ell$. The same determinant could be computed by using an expansion on the ℓth row:

$$d_n(A) = \sum_{j=1}^{n} (-1)^{i+j} a_{\ell j} \det(A_{\ell j}).$$

Then, $\sum_{j=1}^{n} (-1)^{i+j} a_{ij} \det(A_{\ell j})$ is the determinant of a matrix obtained from A by replacing the ℓth row by the ith row and such a determinant is 0 because the new matrix has two identical rows. This proves the second case of the equality of the theorem.

The equality of the theorem is known as the *Laplace expansion of* $\det(A)$ *by row* i.

Since the determinant of a matrix A equals the determinant of A', $\det(A)$ can be expanded by the jth row as

$$\det(A) = \sum_{i=1}^{n} (-1)^{i+j} a_{ij} \det(A_{ij})$$

for every $1 \leqslant j \leqslant n$. Thus, we have

$$\sum_{i=1}^{n} (-1)^{i+j} a_{ij} \det(A_{\ell j}) = \begin{cases} \det(A) & \text{if } i = \ell, \\ 0 & \text{if } i \neq \ell. \end{cases}$$

This formula is *Laplace expansion of* $\det(A)$ *by column* j.

The number *cof*$(a_{ij}) = (-1)^{i+j} \det(A_{ij})$ is the *cofactor* of a_{ij} in either kind of Laplace expansion. Thus, the both types of Laplace expansions can be succinctly expressed by the equalities

$$\det(A) = \sum_{j=1}^{n} a_{ij} \mathit{cof}(a_{ij}) = \sum_{i=1}^{n} a_{ij} \mathit{cof}(a_{ij}), \tag{5.7}$$

for all $i, j \in \{1, \ldots, n\}$.

Cofactors of the form *cof*(a_{ii}) are known as *principal cofactors* of A.

Example 5.139 Let $\mathbf{a} = (a_1, \ldots, a_n)$ be a sequence of n real numbers. The *Vandermonde determinant* $V_{\mathbf{a}}$ is defined by

$$V_{\mathbf{a}} = \begin{vmatrix} 1 & a_1 & a_1^2 & \cdots & a_1^{n-1} \\ 1 & a_2 & a_2^2 & \cdots & a_2^{n-1} \\ \vdots & \vdots & & \cdots & \vdots \\ 1 & a_n & a_n^2 & \cdots & a_n^{n-1} \end{vmatrix}$$

By subtracting the first line from the remaining lines we have

$$V_{\mathbf{a}} = \begin{vmatrix} 1 & a_1 & a_1^2 & \cdots & a_1^{n-1} \\ 0 & a_2 - a_1 & a_2^2 - a_1^2 & \cdots & a_2^{n-1} - a_1^{n-1} \\ \vdots & \vdots & & \cdots & \vdots \\ 0 & a_n - a_1 & a_n^2 - a_1^2 & \cdots & a_n^{n-1} - a_1^{n-1} \end{vmatrix} = \begin{vmatrix} a_2 - a_1 & a_2^2 - a_1^2 & \cdots & a_2^{n-1} - a_1^{n-1} \\ \vdots & \vdots & \cdots & \vdots \\ a_n - a_1 & a_n^2 - a_1^2 & \cdots & a_n^{n-1} - a_1^{n-1} \end{vmatrix}.$$

Factoring now $a_{i+1} - a_1$ from the ith line of the new determinant for $1 \leqslant i \leqslant n$ yields

$$V_{\mathbf{a}} = (a_2 - a_1) \cdots (a_n - a_1) \begin{vmatrix} 1 & a_2 + a_1 & \cdots & \sum_{i=0}^{n-2} a_2^{n-2-i} a_1^i \\ \vdots & \vdots & \cdots & \vdots \\ 1 & a_n + a_1 & \cdots & \sum_{i=0}^{n-2} a_n^{n-2-i} a_1^i \end{vmatrix}$$

Consider two successive columns of this determinant:

$$\mathbf{c}_k = \begin{pmatrix} \sum_{i=0}^{k-1} a_2^{k-1-i} a_1^i \\ \vdots \\ \sum_{i=0}^{k-1} a_n^{k-1-i} a_1^i \end{pmatrix} \text{ and } \mathbf{c}_{k+1} = \begin{pmatrix} \sum_{i=0}^{k} a_2^{k-i} a_1^i \\ \vdots \\ \sum_{i=0}^{k} a_n^{k-i} a_1^i \end{pmatrix}$$

Observe that

$$\mathbf{c}_{k+1} = \begin{pmatrix} a_2^k \\ \vdots \\ a_n^k \end{pmatrix} + a_1 \mathbf{c}_k,$$

it follows that my subtracting from each column $\mathbf{c}_{k+1}$ be the previous column multiplied by a_1 (from right to left) we obtain

$$\begin{aligned} V_{\mathbf{a}} &= (a_2 - a_1) \cdots (a_n - a_1) \begin{vmatrix} 1 & a_2 & \cdots & a_2^{n-2} \\ \vdots & \vdots & \ldots & \vdots \\ 1 & a_n & \cdots & a_n^{n-2} \end{vmatrix} \\ &= (a_2 - a_1) \cdots (a_n - a_1) V_{(a_2, \ldots, a_n)}. \end{aligned}$$

By applying repeatedly this formula, it follows that $V_{\mathbf{a}} = \prod_{p>q} (a_p - a_q)$, where $1 \leqslant p, q \leqslant n$.

Theorem 5.133 can be extended to products of rectangular matrices.

Theorem 5.140 *Let $A \in \mathbb{C}^{m \times n}$ and $B \in \mathbb{C}^{n \times m}$ be two matrices, where $m \leqslant n$. We have*

$$\begin{aligned} \det(AB) = \sum \Bigg\{ & \det\left(A \begin{bmatrix} 1 & \cdots & m \\ k_1 & \cdots & k_m \end{bmatrix} \right) \det\left(B \begin{bmatrix} k_1 & \cdots & k_m \\ 1 & \cdots & m \end{bmatrix} \right) \\ & \Bigg| 1 \leqslant k_1 < k_2 < \cdots < k_m \leqslant n \Bigg\}. \end{aligned}$$

This equality is known as the Cauchy-Binet formula.

Proof Let $\mathbf{a}_1, \ldots, \mathbf{a}_n$ be the rows of the matrix A and let $C = AB$. The first column of the matrix AB equals $\sum_{k_1=1}^{n} \mathbf{a}_{k_1} b_{k_1 1}$ Since $\det(C)$ is linear we can write

$$\det(C) = \sum_{k_1=1}^{n} b_{k_1 1} \begin{vmatrix} a_{k_1 1} & c_{12} & \cdots & c_{1n} \\ a_{k_1 2} & c_{22} & \cdots & c_{2n} \\ \vdots & \vdots & \ldots & \vdots \\ a_{k_1 m} & c_{m2} & \cdots & c_{mn} \end{vmatrix}$$

Similarly, the second row of C equals $\sum_{k_2=1}^{n} \mathbf{a}_{k_2} b_{k_2 1}$. A further decomposition yields the sum

$$\det(C) = \sum_{k_1=1}^{n} \sum_{k_2=1}^{n} b_{k_1 1} b_{k_2 2} \begin{vmatrix} a_{k_1 1} & a_{k_2 1} & \cdots & c_{1n} \\ a_{k_1 2} & a_{k_2 2} & \cdots & c_{2n} \\ \vdots & \vdots & \cdots & \vdots \\ a_{k_1 m} & a_{k_2 2} & \cdots & c_{mn} \end{vmatrix},$$

and so on. Eventually, we can write

$$\det(C) = \sum_{k_1=1}^{n} \cdots \sum_{k_m=1}^{n} b_{k_1 1} \cdots b_{k_m m} \begin{vmatrix} a_{1k_1} & \cdots & a_{1k_m} \\ \vdots & \vdots & \vdots \\ a_{mk_1} & \cdots & a_{mk_m} \end{vmatrix}.$$

due to the multilinearity of the determinants. Only terms involving distinct numbers $k_1, \ldots, k_m$ can be retained in this sum because any such term with $k_p = k_q$ equals 0. Suppose that $\{k_1, \ldots, k_m\} = \{h_1, \ldots, h_m\}$, where $h_1 < \cdots < h_m$ and ϕ is the bijection defined by $k_i = \phi(h_i)$ for $1 \leqslant i \leqslant m$. Then,

$$\begin{vmatrix} a_{1k_1} & \cdots & a_{1k_m} \\ \vdots & \vdots & \vdots \\ a_{mk_1} & \cdots & a_{mk_m} \end{vmatrix} = (-1)^{inv(k_1,\ldots,k_m)} \begin{vmatrix} a_{1h_1} & \cdots & a_{1h_m} \\ \vdots & \vdots & \vdots \\ a_{mh_1} & \cdots & a_{mh_m} \end{vmatrix},$$

which allows us to write

$$\det(C) = \sum_{h_1 < \cdots < h_m} \begin{vmatrix} a_{1h_1} & \cdots & a_{1h_m} \\ \vdots & \vdots & \vdots \\ a_{mh_1} & \cdots & a_{mh_m} \end{vmatrix} \sum_{\phi} (-1)^{inv(\phi)} b_{\phi(h_1)1} \cdots b_{\phi(h_m)m},$$

where ϕ is a permutation of the set $\{h_1, \ldots, h_m\}$. The last equality is equivalent to the Cauchy-Binet formula.

Example 5.141 Let $A \in \mathbb{C}^{2\times n}$ and $B \in \mathbb{C}^{n\times 2}$ be the matrices

$$A = \begin{pmatrix} a_1 & \cdots & a_n \\ b_1 & \cdots & b_n \end{pmatrix} \text{ and } B = \begin{pmatrix} c_1 & d_1 \\ \vdots & \vdots \\ c_n & d_n \end{pmatrix}.$$

Note that

$$AB = \begin{pmatrix} \sum_{i=1}^{n} a_i c_i & \sum_{i=1}^{n} a_i d_i \\ \sum_{i=1}^{n} b_i c_i & \sum_{i=1}^{n} b_i d_i \end{pmatrix}.$$

By applying Binet-Cauchy formula we obtain:

$$\left(\sum_{i=1}^n a_i c_i\right)\left(\sum_{i=1}^n b_i d_i\right) - \left(\sum_{i=1}^n a_i d_i\right)\left(\sum_{i=1}^n b_i c_i\right)$$
$$= \sum_{1 \leqslant i < j \leqslant n} (a_i b_j - a_j b_i)(c_i d_j - d_i c_j).$$

This equality is known as *Lagrange's Identity*.

Let $C \in \mathbb{C}^{n \times n}$ be a square matrix and let $\boldsymbol{b}$ be a vector in $\mathbb{C}^n$. Denote by $(C \stackrel{i}{\leftarrow} \boldsymbol{b})$ the matrix obtained from C by replacing the ith column by $\boldsymbol{b}$.

Example 5.142 Let $A \in \mathbb{C}^{n \times n}$ be a matrix. Then, $(A \stackrel{q}{\leftarrow} \mathbf{e}_p)$ is the (p, q)-minor of A and $(-1)^{p+q} \det(A \stackrel{q}{\leftarrow} \mathbf{e}_p)$ is the cofactor of a_{pq}.

Theorem 5.143 *Let $\{g_{ij} : \mathbb{R} \longrightarrow \mathbb{R} \mid 1 \leqslant i, j \leqslant n\}$ be a collection of n^2 differentiable functions and let $G(x)$ the matrix defined by*

$$G(x) = \begin{pmatrix} g_{11}(x) & \cdots & g_{1n}(x) \\ \vdots & \cdots & \vdots \\ g_{n1}(x) & \cdots & g_{nn}(x) \end{pmatrix}$$

for $x \in \mathbb{R}$. The derivative of the function $\det(G(x))$ *is given by*

$$(\det(G(x))' = \sum_{i=1}^n \det(G(x) \stackrel{i}{\leftarrow} \boldsymbol{g}_i(x)'),$$

where

$$\boldsymbol{g}_i(x)' = \begin{pmatrix} g'_{1i} \\ \vdots \\ g'_{ni} \end{pmatrix}$$

is the column of the derivatives of the functions positioned in column i of the matrix $G(x)$, for $1 \leqslant i \leqslant n$.

Proof By the definition of determinants we have:

$$\det(G(x)) = \sum_{j_1,\ldots,j_n} (-1)^{inv(j_1,\ldots,j_n)} g_{1j_1}(x) g_{2j_2}(x) \cdots g_{nj_n}(x)$$

Therefore, we can write

$$\det(G(x))' = \sum_{j_1,\ldots,j_n} (-1)^{inv(j_1,\ldots,j_n)} g'_{1j_1}(x) g_{2j_2}(x) \cdots g_{nj_n}(x)$$
$$+ \sum_{j_1,\ldots,j_n} (-1)^{inv(j_1,\ldots,j_n)} g_{1j_1}(x) g_{2j_2}(x)' \cdots g_{nj_n}(x)$$

$$+ \sum_{j_1,\ldots,j_n} (-1)^{inv(j_1,\ldots,j_n)} g_{1j_1}(x) g_{2j_2}(x) \cdots g_{nj_n}(x)'$$
$$= \sum_{i=1}^{n} \det(G(x) \overset{i}{\leftarrow} \mathbf{g}_i(x)'),$$

which concludes the argument.

Example 5.144 Let $A = (a_{ij}) \in \mathbb{R}^{3\times 3}$ and let $G(x) = \det(A - xI_3)$. We have

$$\begin{aligned}
&(\det(G(x))' \\
&= \det(G(x) \overset{1}{\leftarrow} (-\mathbf{e}_1)) + \det(G(x) \overset{2}{\leftarrow} (-\mathbf{e}_2) + \det(G(x) \overset{3}{\leftarrow} (-\mathbf{e}_3)) \\
&= \begin{vmatrix} -1 & a_{12} & a_{13} \\ 0 & a_{22} - x & a_{23} \\ 0 & a_{32} & a_{33} - x \end{vmatrix} + \begin{vmatrix} a_{11} - x & 0 & a_{13} \\ a_{21} & -1 & a_{23} \\ a_{31} & 0 & a_{33} - x \end{vmatrix} + \begin{vmatrix} a_{11} - x & a_{12} & 0 \\ a_{21} & a_{22} - x & 0 \\ a_{31} & a_{32} & -1 \end{vmatrix} \\
&= -\begin{vmatrix} a_{22} - x & a_{23} \\ a_{32} & a_{33} - x \end{vmatrix} - \begin{vmatrix} a_{11} - x & a_{13} \\ a_{31} & a_{33} - x \end{vmatrix} - \begin{vmatrix} a_{11} - x & a_{12} \\ a_{21} & a_{22} - x \end{vmatrix}.
\end{aligned}$$

The same technique is applied to compute the second derivative

$$\begin{aligned}
(\det(G(x))'' = &- \begin{vmatrix} -1 & a_{23} \\ 0 & a_{33} - x \end{vmatrix} - \begin{vmatrix} a_{22} - x & 0 \\ a_{32} & -1 \end{vmatrix} \\
&- \begin{vmatrix} -1 & a_{13} \\ 0 & a_{33} - x \end{vmatrix} - \begin{vmatrix} a_{11} - x & 0 \\ a_{31} & -1 \end{vmatrix} \\
&- \begin{vmatrix} -1 & a_{12} \\ 0 & a_{22} - x \end{vmatrix} - \begin{vmatrix} a_{11} - x & 0 \\ a_{21} & -1 \end{vmatrix} \\
= &\; 2(a_{11} + a_{22} + a_{33} - 3x).
\end{aligned}$$

Note that $G(0) = \det(A)$, $-G'(0)$ equals the sum of order 2 principal minors of A, while $G''(0)$ is twice the sum of order 1 principal minors of A.

This observation can be generalized to square matrices of any size: if $A \in \mathbb{R}^{n\times n}$, then for the kth derivative of the function $G(x) = \det(A - xI_n)$ we have

$$G^{(k)}(0) = (-1)^k k! S_{n-k}(A),$$

where $S_p(A)$ is the sum of all order-p principal minors of A (see Exercise 52).

Example 5.145 Let $Q_n(a, b)$ be the determinant of order n:

$$Q_n(a, b) = \begin{vmatrix} a & 1 & 1 & 1 & \cdots & 1 \\ 1 & b & 1 & 0 & \cdots & 0 \\ 1 & 0 & b & 0 & \cdots & 0 \\ \vdots & \vdots & \vdots & \vdots & \cdots & \vdots \\ 1 & 0 & 0 & 0 & \cdots & b \end{vmatrix}$$

We have $Q_2(a,b) = ab-1$. To compute the value of $Q_n(a,b)$ note that, by expanding this determinant by its last column we have:

$$Q_n(a,b) = (-1)^{n+1} \begin{vmatrix} 1 & b & 0 & 0 & \cdots & 0 \\ 1 & 0 & b & 0 & \cdots & 0 \\ \vdots & \vdots & \vdots & \vdots & \cdots & \vdots \\ 1 & 0 & 0 & 0 & \cdots & b \\ 1 & 0 & 0 & 0 & \cdots & 0 \end{vmatrix} + bQ_{n-1}(a,b) = -b^{n-2} + bQ_{n-1}(a,b).$$

It is easy to verify that $Q_n(a,b) = b^{n-1}a - (n-1)b^{n-2}$ for $n \geqslant 2$.

Theorem 5.146 *Let $A, B \in \mathbb{C}^{n\times n}$. Then $\det(A+B)$ is equal to the sum of the determinants of the 2^n matrices obtained by replacing each subset of the columns of A by the corresponding subset of columns of B.*

Proof Let $A = (\mathbf{a}_1 \ \cdots \ \mathbf{a}_n)$ and $B = (\boldsymbol{b}_1 \ \cdots \ \boldsymbol{b}_n)$. Since $\det(A+B) = \det(\mathbf{a}_1 + \boldsymbol{b}_1 \ \cdots \ \mathbf{a}_n + \boldsymbol{b}_n)$, by the linearity of determinants, we can write

$$\begin{aligned}
&\det(\mathbf{a}_1 + \boldsymbol{b}_1 \ \cdots \ \mathbf{a}_n + \boldsymbol{b}_n) \\
&= \det(\mathbf{a}_1 \ \mathbf{a}_2 + \boldsymbol{b}_2 \ \cdots, \mathbf{a}_n + \boldsymbol{b}_n) + \det(\boldsymbol{b}_1 \ \mathbf{a}_2 + \boldsymbol{b}_2 \ \cdots, \mathbf{a}_n + \boldsymbol{b}_n) \\
&= \det(\mathbf{a}_1 \ \mathbf{a}_2 \ \cdots, \mathbf{a}_n + \boldsymbol{b}_n) + \det(\mathbf{a}_1 \ \boldsymbol{b}_2 \ \ldots, \mathbf{a}_n + \boldsymbol{b}_n) \\
&\quad + \det(\boldsymbol{b}_1 \ \mathbf{a}_2 \ \cdots, \mathbf{a}_n + \boldsymbol{b}_n) + \det(\boldsymbol{b}_1 \ \boldsymbol{b}_2 \ \ldots, \mathbf{a}_n + \boldsymbol{b}_n) \\
&= \cdots \\
&= \det(\mathbf{a}_1 \ \mathbf{a}_2 \ \cdots \ \mathbf{a}_n) + \det(\mathbf{a}_1 \ \boldsymbol{b}_2 \ \cdots \ \mathbf{a}_n) + \cdots + \det(\boldsymbol{b}_1 \ \boldsymbol{b}_2 \ \cdots \ \boldsymbol{b}_n).
\end{aligned}$$

In principle, determinants can be used for solving linear systems of equation and we discuss a formula that allows us to do just that. Let $A \in \mathbb{C}^{n\times n}$, $\boldsymbol{b} \in \mathbb{C}^n$ and consider the linear system $A\mathbf{x} = \boldsymbol{b}$. The columns of the matrix A are denoted by $\mathbf{a}_1, \ldots, \mathbf{a}_n$, that is, $A = (\mathbf{a}_1 \ \cdots \ \mathbf{a}_n)$. Note that

$$\begin{aligned}
A(I_n \overset{i}{\leftarrow} \mathbf{x}) &= A(\mathbf{e}_1 \ \mathbf{e}_2 \ \cdots \ \mathbf{e}_{i-1} \ \mathbf{x} \ \mathbf{e}_{i+1} \ \cdots \ \mathbf{e}_n) \\
&= (\mathbf{a}_1 \ \mathbf{a}_2 \ \cdots \ \mathbf{a}_{i-1} \ \boldsymbol{b} \ \mathbf{a}_{i+1} \ \cdots \ \mathbf{a}_n) \\
&= (A \overset{i}{\leftarrow} \boldsymbol{b}).
\end{aligned} \tag{5.8}$$

By expanding the determinant

$$\det(I_n \overset{i}{\leftarrow} \mathbf{x}) = \begin{vmatrix} 1 & 0 & \cdots & x_1 & \cdots & 0 \\ 0 & 1 & \cdots & x_2 & \cdots & 0 \\ \vdots & \vdots & \cdots & \vdots & \cdots & \vdots \\ 0 & 0 & \cdots & x_i & \cdots & 0 \\ \vdots & \vdots & \cdots & \vdots & \cdots & \vdots \\ 0 & 0 & \cdots & x_n & \cdots & 1 \end{vmatrix}$$

by its ith row we obtain $\det(I_n \overset{i}{\leftarrow} \mathbf{x}) = x_i$. Thus, computing the determinants on both sides of Equality (5.8) we have

$$\det(A)x_i = \det(A \overset{i}{\leftarrow} \boldsymbol{b}),$$

for $1 \leqslant i \leqslant n$. This method for computing the components of the solution of the system $A\mathbf{x} = \boldsymbol{b}$ is known as *Cramer's formula*.

Definition 5.147 *Let $A \in \mathbb{R}^{n\times n}$ be a square matrix. The* adjoint matrix *of A is the matrix*

$$adj(A) = \begin{pmatrix} cof(a_{11}) & \cdots & cof(a_{n1}) \\ \vdots & \cdots & \vdots \\ cof(a_{1n}) & \cdots & cof(a_{nn}) \end{pmatrix},$$

that is, the transposed matrix of the matrix whose entries are the cofactors of the elements of A.

Example 5.148 For the matrix

$$A = \begin{pmatrix} 1 & 0 & 2 \\ 0 & -1 & 1 \\ 1 & 2 & 1 \end{pmatrix}$$

the matrix of cofactors is

$$C = \begin{pmatrix} -3 & 1 & 1 \\ 4 & -1 & -2 \\ 2 & -1 & -1 \end{pmatrix},$$

so the adjoint matrix is

$$adj(A) = \begin{pmatrix} -3 & 4 & 2 \\ 1 & -1 & -1 \\ 1 & -2 & -1 \end{pmatrix}.$$

Theorem 5.149 *If $A \in \mathbb{R}^{n\times n}$, then $A\ adj(A) = \det(A)I_n$.*

Proof Equalities (5.7) allow us to write:

$$(A\ adj(A))_{ij} = \sum_{k=1}^{n} a_{ik}(adj(A))_{kj} = \sum_{k=1}^{n} a_{ik}cof(a_{jk}) = \begin{cases} \det(A) & \text{if } i = j \\ 0 & \text{otherwise.} \end{cases}$$

Therefore, $A\ adj(A) = \det(A)I_n$.

This allows us to give an explicit formula for computing the inverse of a non-singular matrix A as

$$A^{-1} = \frac{1}{\det(A)} adj(A). \tag{5.9}$$

By Theorem 5.133, if A is a non-singular matrix, we have $\det(A)\det(A^{-1}) = 1$.

Corollary 5.150 *If* $A \in \mathbb{R}^{n\times n}$ *then* $\det(A) \neq 0$ *if and only rank*$(A) = n$.

Proof By Theorem 5.112 $rank(A) = n$ if and only if A is invertible; A is invertible if and only if $\det(A) \neq 0$.

Example 5.151 For the matrix $A \in \mathbb{R}^{3\times 3}$ introduced in Example 5.148 we have $det(A) = -1$ and

$$A^{-1} = \frac{1}{\det(A)} adj(A) = \begin{pmatrix} 3 & -4 & -2 \\ -1 & 1 & 1 \\ -1 & 2 & 1 \end{pmatrix}.$$

5.8 Partitioned Matrices and Determinants

Lemma 5.152 *Let* $A \in \mathbb{C}^{n\times n}$, $B \in \mathbb{C}^{m\times m}$ *be two square matrices and let* $D \in \mathbb{C}^{(m+n)\times(m+n)}$ *be the matrix*

$$D = \begin{pmatrix} A & O_{m,n} \\ C & B \end{pmatrix},$$

where $C \in \mathbb{C}^{m\times n}$. *Then, we have* $\det(D) = \det(A)\det(B)$.

If the matrix $F \in \mathbb{C}^{(m+n)\times(m+n)}$ *is*

$$F = \begin{pmatrix} A & E \\ O_{m,n} & B \end{pmatrix},$$

where $E \in \mathbb{C}^{n\times m}$, *then* $\det(F) = \det(A)\det(B)$.

Proof Suppose that $D = (d_{ij})$. The definition of $\det(D)$ implies that

$$\det(D) = \sum_{j_1,\ldots,j_n,j_{n+1},\ldots,j_{n+m}} (-1)^{inv(j_1,\ldots,j_{n+m})} d_{1j_1}\cdots d_{n+m\ j_{n+m}}.$$

Each term of this sum involves factors chosen from each row (specified by the first subscript of d_{ij}). Note that any term

$$d_{1j_1}\cdots d_{nj_n} d_{n+1\ j_{n+1}}\cdots d_{n+m\ j_{n+m}}$$

in which any of the first n subscripts $j_1,\ldots,j_n$ is at least equal to $n+1$ equals 0. Therefore, non-zero terms are those in which $(j_1,\ldots,j_n)$ is a permutation of $(1,\ldots,n)$. In such terms $(j_{n+1},\ldots,j_{n+m})$ is a permutation of $(n+1,\ldots,n+m)$. By the definition of the matrix D the product $d_{1j_1}\cdots d_{nj_n}$ is actually $a_{1j_1}\cdots a_{nj_n}$; the product $d_{n+1\,j_{n+1}} \cdots d_{n+m\,j_{n+m}}$ equals $b_{1k_1} \cdots b_{mk_m}$, where $k_1 = j_{n+1} - n, \ldots, k_m = j_{n+m} - n$ and

$$inv(j_1, \ldots, j_n, j_{n+1}, \ldots, j_{n+m}) = inv(j_1, \ldots, j_n) + inv(k_1, \ldots, k_m).$$

This allows us to write

$$\det(D) = \left(\sum_{j_1 \cdots j_n} a_{1j_1} \cdots a_{nj_n} \right) \cdot \left(\sum_{k_1 \cdots k_m} b_{1k_1} \cdots b_{mk_m} \right) = \det(A)\det(B).$$

The proof of the second part of the lemma is similar.

Theorem 5.153 *Let A be an block upper (or lower) triangular partitioned matrix given by*

$$A = \begin{pmatrix} A_{11} & A_{12} & \cdots & A_{1m} \\ O & A_{22} & \cdots & A_{2m} \\ \vdots & \vdots & \cdots & \vdots \\ O & O & \cdots & A_{mm} \end{pmatrix},$$

where $A_{ii} \in \mathbb{R}^{p_i \times p_i}$ for $1 \leqslant i \leqslant m$. Then,

$$\det(A) = \det(A_{11})\det(A_{22}) \cdots \det(A_{mm}).$$

If A is a block lower triangular matrix

$$A = \begin{pmatrix} A_{11} & O & \cdots & O \\ A_{21} & A_{22} & \cdots & O \\ \vdots & \vdots & \cdots & \vdots \\ A_{m1} & A_{m2} & \cdots & A_{mm} \end{pmatrix},$$

the same equality holds.

Proof The argument is by induction on m, where $m \geqslant 2$. The base case, $m = 2$ was shown in Lemma 5.152. Suppose that the statement holds for partitioned upper diagonal matrices having $m - 1$ diagonal blocks. Note that if

$$A = \begin{pmatrix} A_{11} & A_{12} & \cdots & A_{1m} \\ O & A_{22} & \cdots & A_{2m} \\ \vdots & \vdots & \cdots & \vdots \\ O & O & \cdots & A_{mm} \end{pmatrix},$$

we can also regard A as a partitioned matrix

$$A = \begin{pmatrix} B & C \\ O & A_{mm} \end{pmatrix},$$

where B is a partitioned upper diagonal matrices having $m - 1$ diagonal blocks. Then, by the base case and by the inductive hypothesis we have

$$\det(A) = \det(B)\det(A_{mm}) = \det(A_1)\cdots\det(A_{m-1})\det(A_{mm}).$$

Theorem 5.154 *Let $A \in \mathbb{C}^{m\times m}$, $D \in \mathbb{C}^{n\times n}$ be two square matrices and let $E \in \mathbb{C}^{(m+n)\times(m+n)}$ be the matrix*

$$E = \begin{pmatrix} A & B \\ C & D \end{pmatrix},$$

where $B \in \mathbb{C}^{m\times n}$ and $C \in \mathbb{C}^{n\times m}$. If the matrix A is invertible, then $\det(E) = \det(A)\det(D - CA^{-1}B)$.

Proof If A is invertible, then

$$\begin{pmatrix} I_m & O \\ -CA^{-1} & I_n \end{pmatrix}\begin{pmatrix} A & B \\ C & D \end{pmatrix} = \begin{pmatrix} A & B \\ 0 & D - CA^{-1}B \end{pmatrix},$$

which implies

$$\det(E) = \det\begin{pmatrix} A & B \\ 0 & D - CA^{-1}B \end{pmatrix}.$$

Theorem 5.153 implies the desired equality.

Theorem 5.155 *Let $A, B, C, D \in \mathbb{C}^{m\times m}$ be four square matrices such that $AC = CA$ and let $E \in \mathbb{C}^{2m\times 2m}$ be the matrix*

$$E = \begin{pmatrix} A & B \\ C & D \end{pmatrix}.$$

We have $\det(E) = det(AD - CB)$.

Proof Suppose initially that A is invertible, so $\det(E) = \det(A^{-1})\det(D - CA^{-1}B)$ by Theorem 5.154. Then,

$$\begin{aligned}\det(E) &= \det(A)\det(D - CA^{-1}B) = \det(AD - ACA^{-1}B) \\ &= \det(AD - CAA^{-1}B) = \det(AD - CB).\end{aligned}$$

If A is not invertible, that is, if A is singular consider the continuous function $f : \mathbb{R} \longrightarrow \mathbb{R}$ defined by $f(x) = \det(A + xI)$ for $x \in \mathbb{R}$. There exists $\delta > 0$ such that if $x \in (0, \delta)$, then $f(x) \neq 0$, which means that $\det(A + xI) \neq 0$, which implies that $A + xI$ is an invertible matrix. Note that if $AC = CA$, then $(A + xI)C = C(A + xI)$, so the first part of the argument can be applied to the matrix

$$E_x = \begin{pmatrix} A + xI & B \\ C & D \end{pmatrix}.$$

This implies $\det(E_x) = \det((A + xI)D - CB)$ if $x \in (0, \delta)$. If x tends towards 0, by the continuity of $f(x)$ it follows that

$$\det(E) = \lim_{x \to 0} \det((A + xI)D - CB) = \det(AD - BC),$$

which concludes our argument.

The argument presented in the second part of the proof of Theorem 5.155 is known as a *continuity argument.*

Theorem 5.156 *Let $A \in \mathbb{C}^{n \times n}$ be a square matrix. The rank of A equals the largest size of a non-zero minor of A.*

Proof Let $r = rank(A)$ and let s be the largest size of a non-zero minor of A that is the determinant of the submatrix $S = A \begin{bmatrix} i_1 & \cdots & i_s \\ j_1 & \cdots & j_s \end{bmatrix}$. By permuting the rows and the columns of A we obtain a matrix B of the same rank as A such that $B \begin{bmatrix} 1 & \cdots & s \\ 1 & \cdots & s \end{bmatrix} = S$. Since any permutation of rows or columns preserves the non-nullity of a determinant we have

$$\det \left(B \begin{bmatrix} 1 & \cdots & s \\ 1 & \cdots & s \end{bmatrix} \right) \neq 0.$$

Thus, the rows of S are linearly independent and, therefore, the first s rows of B are linearly independent, so $s \leqslant r$.

Since $rank(A) = r$, A has r linearly independent rows. By permuting the rows of A, these rows can be brought in the first r position in a matrix B which has the same rank r as A. Then, r linearly independent columns of B are brought on the first r position to result into a matrix C. Let $P = C \begin{bmatrix} 1, \ldots, r \\ 1, \ldots, r \end{bmatrix}$. Since P is of rank r it follows that $\det(P) \neq 0$, so $r \leqslant s$.

5.9 The Kronecker and Hadamard products

Definition 5.157 *Let $A \in \mathbb{C}^{m \times n}$ and $B \in \mathbb{C}^{p \times q}$ be two matrices. The* Kronecker product *of these matrices is the matrix $A \otimes B \in \mathbb{C}^{mp \times nq}$ defined by*

$$A \otimes B = \begin{pmatrix} a_{11}B & a_{12}B & \cdots & a_{1n}B \\ a_{21}B & a_{22}B & \cdots & a_{2n}B \\ \vdots & \vdots & \ddots & \vdots \\ a_{m1}B & a_{m2}B & \cdots & a_{mn}B \end{pmatrix}.$$

Example 5.158 Consider the matrices

$$A = \begin{pmatrix} a_{11} & a_{12} \\ a_{21} & a_{22} \end{pmatrix} \text{ and } B = \begin{pmatrix} b_{11} & b_{12} & b_{13} \\ b_{21} & b_{22} & b_{23} \\ b_{31} & b_{32} & b_{33} \end{pmatrix}$$

Their Kronecker product is

$$A \otimes B = \left(\begin{array}{ccc|ccc} a_{11}b_{11} & a_{11}b_{12} & a_{11}b_{13} & a_{12}b_{11} & a_{12}b_{12} & a_{12}b_{13} \\ a_{11}b_{21} & a_{11}b_{22} & a_{11}b_{23} & a_{12}b_{21} & a_{12}b_{22} & a_{12}b_{23} \\ a_{11}b_{31} & a_{11}b_{32} & a_{11}b_{33} & a_{12}b_{31} & a_{12}b_{32} & a_{12}b_{33} \\ \hline a_{21}b_{11} & a_{21}b_{12} & a_{21}b_{13} & a_{22}b_{11} & a_{22}b_{12} & a_{22}b_{13} \\ a_{21}b_{21} & a_{21}b_{22} & a_{21}b_{23} & a_{22}b_{21} & a_{22}b_{22} & a_{22}b_{23} \\ a_{21}b_{31} & a_{21}b_{32} & a_{21}b_{33} & a_{22}b_{31} & a_{22}b_{32} & a_{22}b_{33} \end{array}\right).$$

The next theorem contains a few elementary properties of Kronecker's product.

Theorem 5.159 *For any matrices A, B, C, D we have:*

(i) $(A \otimes B)' = A' \otimes B'$,
(ii) $(A \otimes B) \otimes C = A \otimes (B \otimes C)$,
(iii) $(A \otimes B)(C \otimes D) = (AC \otimes BD)$,
(iv) $A \otimes B + A \otimes C = A \otimes (B + C)$,
(v) $A \otimes D + B \otimes D = (A + B) \otimes D$,
(vi) $(A \otimes B)' = A' \otimes B'$,
(vii) $(A \otimes B)^H = A^H \otimes B^H$,

when the usual matrix sum and multiplication are well-defined in each of the above equalities.

Proof The proof is straightforward and is left to the reader.

Example 5.160 Let $\mathbf{x} \in \mathbb{C}^n$ and $\mathbf{y} \in \mathbb{C}^m$. We have

$$\mathbf{x} \otimes \mathbf{y} = \begin{pmatrix} x_1\mathbf{y} \\ \vdots \\ x_n\mathbf{y} \end{pmatrix} = \begin{pmatrix} y_1\mathbf{x} \\ \vdots \\ y_m\mathbf{x} \end{pmatrix} \in \mathbb{C}^{mn}.$$

Theorem 5.161 *If $A \in \mathbb{C}^{n\times n}$ and $B \in \mathbb{C}^{m\times m}$ are two invertible matrices, then $A \otimes B$ is invertible and $(A \otimes B)^{-1} = A^{-1} \otimes B^{-1}$.*

Proof Since

$$(A \otimes B)(A^{-1} \otimes B^{-1}) = (AA^{-1} \otimes BB^{-1}) = I_n \otimes I_m,$$

the theorem follows by noting that $I_n \otimes I_m = I_{nm}$.

Theorem 5.162 *The Kronecker product $A \otimes B$ of two normal (unitary) matrices A, B is a normal (a unitary) matrix.*

Proof By Theorem 5.159 we can write

$$\begin{aligned}(A \otimes B)'(A \otimes B) &= (A' \otimes B')(A \otimes B)\\ &= (A'A \otimes B'B)\\ &= (AA' \otimes BB')\\ &\quad \text{(because both } A \text{ and } B \text{ are normal)}\\ &= (A \otimes B)(A \otimes B)',\end{aligned}$$

which implies that $A \otimes B$ is normal.

Definition 5.163 *Let $A \in \mathbb{C}^{m\times m}$ and $B \in \mathbb{C}^{n\times n}$ be two square matrices. Their* Kronecker sum *is the matrix $A \oplus B \in \mathbb{C}^{mn\times mn}$ defined by*

$$A \oplus B = (A \otimes I_n) + (I_m \otimes B).$$

The Kronecker difference *is the matrix $A \ominus B \in \mathbb{C}^{mn\times mn}$ defined by*

$$A \ominus B = (A \otimes I_n) - (I_m \otimes B).$$

Definition 5.164 *Let $A, B \in \mathbb{C}^{m\times n}$. The* Hadamard product *of A and B is the matrix $A \odot B \in \mathbb{C}^{m\times n}$ defined by*

$$A \odot B = \begin{pmatrix} a_{11}b_{11} & a_{12}b_{12} & \cdots & a_{1n}b_{1n}\\ a_{21}b_{21} & a_{22}b_{22} & \cdots & a_{2n}b_{2n}\\ \vdots & \vdots & \ddots & \vdots\\ a_{m1}b_{m1} & a_{m2}b_{m2} & \cdots & a_{mn}b_{mn}\end{pmatrix}.$$

The Hadamard quotient *$A \oslash B$ is defined only if $b_{ij} \neq 0$ for $1 \leqslant i \leqslant m$ and $1 \leqslant j \leqslant n$. In this case*

$$A \oslash B = \begin{pmatrix} \frac{a_{11}}{b_{11}} & \frac{a_{12}}{b_{12}} & \cdots & \frac{a_{1n}}{b_{1n}}\\ \frac{a_{21}}{b_{21}} & \frac{a_{22}}{b_{22}} & \cdots & \frac{a_{2n}}{b_{2n}}\\ \vdots & \vdots & \ddots & \vdots\\ \frac{a_{m1}}{b_{m1}} & \frac{a_{m2}}{b_{m2}} & \cdots & \frac{a_{mn}}{b_{mn}}\end{pmatrix}.$$

Theorem 5.165 *If $A, B, C \in \mathbb{C}^{m\times n}$ and $c \in \mathbb{C}$ we have*

(i) $A \odot B = B \odot A$;
(ii) $A \odot J_{m,n} = J_{m,n} \odot A = A$;
(iii) $A \odot (B + C) = A \odot B + A \odot C$;
(iv) $A \odot (cB) = c(A \odot B)$.

Proof The proof is straightforward and is left to the reader.

Note that the Hadamard product of two matrices $A, B \in \mathbb{C}^{m \times n}$ is a submatrix of the Kronecker product $A \otimes B$.

Example 5.166 Let $A, B \in \mathbb{C}^{2 \times 3}$ be the matrices

$$A = \begin{pmatrix} a_{11} & a_{12} & a_{13} \\ a_{21} & a_{22} & a_{23} \end{pmatrix} \text{ and } B = \begin{pmatrix} b_{11} & b_{12} & b_{13} \\ b_{21} & b_{22} & b_{23} \end{pmatrix}.$$

The Kronecker product of these matrices is $A \otimes B \in \mathbb{C}^{4 \times 9}$ given by:

$$A \otimes B = \begin{pmatrix} a_{11}b_{11} & a_{11}b_{12} & a_{11}b_{13} & a_{12}b_{11} & a_{12}b_{12} & a_{12}b_{13} & a_{13}b_{11} & a_{13}b_{12} & a_{13}b_{13} \\ a_{11}b_{21} & a_{11}b_{22} & a_{11}b_{23} & a_{12}b_{21} & a_{12}b_{22} & a_{12}b_{23} & a_{13}b_{21} & a_{13}b_{22} & a_{13}b_{23} \\ a_{21}b_{11} & a_{21}b_{12} & a_{21}b_{13} & a_{22}b_{11} & a_{22}b_{12} & a_{22}b_{13} & a_{23}b_{11} & a_{23}b_{12} & a_{23}b_{13} \\ a_{21}b_{21} & a_{21}b_{22} & a_{21}b_{23} & a_{22}b_{21} & a_{22}b_{22} & a_{22}b_{23} & a_{23}b_{21} & a_{23}b_{22} & a_{23}b_{23} \end{pmatrix}.$$

The Hadamard product of the same matrices is

$$A \odot B = \begin{pmatrix} a_{11}b_{11} & a_{12}b_{12} & a_{13}b_{13} \\ a_{21}b_{21} & a_{22}b_{22} & a_{23}b_{23} \end{pmatrix},$$

and we can regard the Hadamard product as a submatrix of the Kronecker product $A \otimes B$,

$$A \odot B = (A \otimes B) \begin{bmatrix} 1, 5, 9 \\ 4, 4, 4 \end{bmatrix}.$$

5.10 Topological Linear Spaces

We are examining now the interaction between the algebraic structure of linear spaces and topologies that can be defined on linear spaces that are compatible in a certain sense with the algebraic structure. Compatibility, in this case, is defined as the continuity of addition and scalar multiplication.

Definition 5.167 *Let F be the real field $\mathbb{R}$ or complex field $\mathbb{C}$. An F-*topological linear space *is a topological space $(V, \mathcal{O})$ such that*

(i) *V is an F-linear space;*
(ii) *the vector addition is a continuous function between V^2 and V;*
(iii) *the scalar multiplication is a continuous function between $F \times V$ and V.*

Unless stated otherwise, we assume that the field F is either the real or the complex field.

Theorem 5.168 *Let $(V, \mathcal{O})$ be an F-*topological linear space *and let $z \in V$. The translation mapping $\mathsf{t}_z : V \longrightarrow V$ is a homeomorphism.*

Proof It is immediate that $\mathsf{t_z}$ is a bijection whose inverse is $\mathsf{t_{-z}}$. The continuity of both $\mathsf{t_z}$ and $\mathsf{t_{-z}}$ follows from the continuity of the vector addition of V.

Example 5.169 If $a \neq 0$, then each homotety h_a of a topological linear space $(V, \mathcal{O})$ is a homeomorphism. Indeed, the inverse of h_a is $\mathsf{h}_{a^{-1}}$. The continuity of both h_a and $\mathsf{h}_{a^{-1}}$ follows from the continuity of the scalar multiplication of V.

Theorem 5.170 *Let $(V, \mathcal{O})$ be a topological linear space. If W is a neighborhood of $\mathbf{0}$, then $\mathsf{t_x}(W)$ is a neighborhood of $\mathbf{x}$. Moreover, every neighborhood of x can be obtained by a translation of a neighborhood of $\mathbf{0}$.*

Proof Since W is a neighborhood of the origin, there exists an open subset L of V such that $\mathbf{0} \in L \subseteq W$. This implies $\mathbf{x} = \mathsf{t_x}(\mathbf{0}) \in \mathsf{t_x}(L) \subseteq \mathsf{t_x}(W)$. Since every translation is a homeomorphism of $(V, \mathcal{O})$ it follows that $\mathsf{t_x}(L)$ is an open set and this, in turn, implies that $\mathsf{t_x}(W)$ is a neighborhood of $\mathbf{x}$.

Conversely, let U be a neighborhood of $\mathbf{x}$ and let K be an open set such that $\mathbf{x} \in K \subseteq U$. Then, we have $\mathbf{0} = \mathsf{t_{-x}}(\mathbf{x}) \in \mathsf{t_{-x}}(K) \subseteq \mathsf{t_{-x}}(U)$. Since $\mathsf{t_{-x}}(K)$ is an open set, it follows that $\mathsf{t_{-x}}(K)$ is a neighborhood of $\mathbf{0}$ and the desired conclusion follows from the fact that $U = \mathsf{t_x}(\mathsf{t_{-x}}(U))$.

Theorem 5.170 shows that in a topological linear space the neighborhoods of any point are obtained by translating the neighborhoods of $\mathbf{0}$.

Corollary 5.171 *If $\mathcal{F}_x$ is a fundamental system of neighborhoods of $\mathbf{x}$ in the topological linear space $(V, \mathcal{O})$, then $\mathcal{F}_x$ can be obtained by a translation of a fundamental system of neighborhoods $\mathcal{F}_0$ of $\mathbf{0}$.*

Proof This statement follows immediately from Theorem 5.170.

The next theorem shows that a linear function between two topological linear spaces is continuous if and only if it is continuous in the zero element of the first space.

Theorem 5.172 *Let $(V_1, \mathcal{O}_1)$ and $(V_2, \mathcal{O}_2)$ be two topological F-linear spaces having $\mathbf{0}_1$ and $\mathbf{0}_2$ as zero elements, respectively. A linear operator $f \in \mathsf{Hom}(V_1, V_2)$ is continuous in $\mathbf{x} \in V_1$ if and only if it is continuous in $\mathbf{0}_1 \in V_1$.*

Proof Let f be a function that is continuous in a point $\mathbf{x} \in V_1$. If $U \in neigh_{\mathbf{0}_2}(\mathcal{O}_2)$, then $f(\mathbf{x}) + U$ is a neighborhood of $f(\mathbf{x})$. Since f is continuous, there exists a neighborhood W of $\mathbf{x}$ such that $f(W) \subseteq f(\mathbf{x}) + U$.

Observe that the set $-\mathbf{x} + W$ is a neighborhood of $\mathbf{0}_1$. Moreover, any neighborhood of $\mathbf{0}_1$ has this form. If $\mathbf{t} \in -\mathbf{x} + W$, then $\mathbf{t} + \mathbf{x} \in W$ and, therefore, $f(\mathbf{t}) + f(\mathbf{x}) = f(\mathbf{t} + \mathbf{x}) \in f(\mathbf{x}) + U$. This shows that $f(\mathbf{t}) \in U$, which proves that f is continuous in $\mathbf{0}_1$.

Conversely, suppose that f is continuous in $\mathbf{0}_1$. Let $\mathbf{x} \in V_1$ and let $Z \in neigh_{f(\mathbf{x})}(\mathcal{O}_2)$. The set $-f(\mathbf{x}) + Z$ is a neighborhood of $\mathbf{0}_2$ in V_2. The continuity of f in $\mathbf{0}_1$ implies the existence of a neighborhood T of $\mathbf{0}_1$ such that $f(T) \subseteq -f(\mathbf{x}) + Z$. Note that $\mathbf{x} + T$ is a neighborhood of $\mathbf{x}$ in V_1 and every neighborhood of $\mathbf{x}$ in V_1 has this form. Since $f(\mathbf{x} + T) \subseteq Z$, it follows that f is continuous in $\mathbf{x}$.

Corollary 5.173 *Let $(V_1, \mathcal{O}_1)$ and $(V_2, \mathcal{O}_2)$ be two topological F-linear spaces. A linear operator $f \in \mathsf{Hom}(V_1, V_2)$ is either continuous on V_1 or is discontinuous in every point of V_1.*

Proof This statement is a direct consequence of Theorem 5.172.

Theorem 5.174 *Let C and D be two subsets of $\mathbb{R}^n$ such that C is compact and D is closed. Then the set $C + D = \{\boldsymbol{x} + \boldsymbol{y} \mid \boldsymbol{x} \in C, \boldsymbol{y} \in D\}$ is closed.*

Proof Let $\mathbf{x} \in \mathbf{K}(C + D)$. There exists a sequence $(\mathbf{x}_0, \mathbf{x}_1, \ldots)$ such that $\mathbf{x}_i \in C + D$ and $\lim_{n\to\infty} \mathbf{x}_n = \mathbf{x}$. The definition of $C + D$ means that there is a sequence $(\boldsymbol{u}_0, \boldsymbol{u}_1, \ldots) \in \mathbf{Seq}_\infty(C)$ and a sequence $(\boldsymbol{v}_0, \boldsymbol{v}_1, \ldots) \in \mathbf{Seq}_\infty(D)$ such that $\mathbf{x}_i = \boldsymbol{u}_i + \boldsymbol{v}_i$ for $i \in \mathbb{N}$.

Since C is compact, the sequence $(\mathbf{v}_0, \mathbf{v}_1, \ldots)$ contains a convergent subsequence $(\boldsymbol{u}_{i_0}, \boldsymbol{u}_{i_1}, \ldots)$. Let $\boldsymbol{u} = \lim_{m\to\infty} \boldsymbol{u}_{i_m}$. Clearly, $\lim_{m\to\infty} \mathbf{x}_{i_m} = \mathbf{x}$. Since D is a closed set, $\lim_{m\to\infty} \mathbf{v}_{i_m} = \mathbf{x} - \boldsymbol{u} \in D$. Therefore, $\mathbf{x} = \boldsymbol{u} + \mathbf{v} \in C + D$, so $\mathbf{K}(C+D) = C+D$, which means that $C + D$ is closed.

Exercises and Supplements

1. Let L be an F-linear space. Prove that $0\mathbf{x} = \mathbf{0}$ and $a\mathbf{0} = \mathbf{0}$ for every $a \in F$ and $\mathbf{x} \in L$.
2. Prove that a subset K of a linear space is linearly dependent if and only of there is $\mathbf{x} \in K$ that can be expressed as a linear combination of $K - \{\mathbf{x}\}$.
3. Let L, M be two F-linear spaces, $h : L \longrightarrow M$ be a linear mapping, and let $X = \{\mathbf{x}_1, \ldots, \mathbf{x}_m\}$ be a subset of L. Prove that if $\{h(\mathbf{x}_1), \ldots, h(\mathbf{x}_m)\}$ is a linearly independent set in M, then X is linearly independent in L.
4. Let U, V be two subspaces of the finite-dimensional F-linear space L. Prove that there exists a vector $\mathbf{w} \neq \mathbf{0}$ in both U and V only if $\dim(U)+\dim(V) > \dim(L)$.
5. Let L be a finite-dimensional F-linear space and let U, V be two subspaces of L. Show that $U + V = U \boxplus V$ if and only if $\dim(U + V) = \dim(U) + \dim(V)$.
6. Let U, V, W be subspaces of a finite-dimensional F-linear space L. Prove that
$$\dim(U \cap V \cap W) \geqslant \dim(U) + \dim(V) - 2\dim(L).$$
7. Let $\mathbf{w}$ be a vector in $\mathbb{R}^n$. Prove that the set
$$P_{\mathbf{w}} = \{\mathbf{x} \in \mathbb{R}^n \mid \mathbf{w}'\mathbf{x} = 0\}$$
is a subspace of $\mathbb{R}^n$.
8. Let POL be the real linear space of all polynomials in the variable x and let POL_n be the set of all polynomials of degree at most n. Prove that
   (a) any sequence of polynomials of degrees $0, 1, 2, \cdots$ is a basis for POL;

(b) POL_n is a subspace of dimension $n+1$ of POL.

9. Prove that there exists a unique linear function $L : \text{POL} \longrightarrow \mathbb{R}$ such that $L(1) = 1$ and $L([x]_k) = 1$ for $k \geqslant 1$, where the polynomial $[x]_k = x(x-1)\cdots(x-k+1)$ was introduced in Exercise 27 of Chap. 3.
10. Prove that

$$B_{n+1} = \frac{1}{e}\left(1^n + \frac{2^n}{1!} + \frac{3^n}{2!} + \cdots\right)$$

(Dobinski's Formula), where B_n is the nth Bell number introduced in Exercise 33 of Chap. 3.

Solution: As observed in Exercise 28 of Chap. 3, we have $e = \sum_{k=0}^{\infty} \frac{[k]_n}{k!}$. By applying L (defined in Exercise 9) to both members of the equality of Exercise 30 of Chap. 3 we have

$$\sum\{L([u]_k) \mid k = |\pi|, \pi \in PART(A)\} = L(u^m),$$

which implies $B_n = L(u^m)$.

If $p(u)$ is a polynomial, taking into account that L is linear and that the polynomials of the form $[u]_n$ constitute a basis for POL, we can write $p(u) = \sum_n a_n[u]_n$. Therefore,

$$\begin{aligned} L(p(u)) &= \sum_n a_n L([u]_n) = \sum_n a_n \frac{1}{e} \sum_k \frac{[k]_n}{k!} \\ &= \frac{1}{e} \sum_n \sum_k a_n \frac{[k]_n}{k!} = \frac{1}{e} \sum_k \sum_n a_n \frac{[k]_n}{k!} \\ &= \frac{1}{e} \sum_k \frac{p(k)}{k!} \end{aligned}$$

for any polynomial p. Choosing $p(u) = u^n$ yields Dobinski's formula.

An *affine subspace* of a F-linear space $(L, +, \cdot)$ is a nonempty subset U of L such that there exists $\boldsymbol{u} \in L$ such that the set $U - \{\boldsymbol{u}\} = \{\mathbf{x} - \boldsymbol{u} \mid \mathbf{x} \in U\}$ is a linear subspace.

11. Let $\mathbf{x}_0, \mathbf{a} \in \mathbb{R}^n$. The *line that passes through* $\mathbf{x}_0$ *and has the direction* $\mathbf{a}$ is the set

$$L_{\mathbf{x}_0,\mathbf{a}} = \{\mathbf{x} \in \mathbb{R}^n \mid \mathbf{x} = \mathbf{x}_0 + t\mathbf{a} \text{ for some } t \in \mathbb{R}\}.$$

Prove that $L_{\mathbf{x}_0,\mathbf{a}}$ is an affine subspace of $\mathbb{R}^n$.

12. Let $\mathbf{x}_0$ and $\mathbf{w}$ be two vectors in $\mathbb{R}^n$. Prove that the set

$$H_{\mathbf{x}_0,\mathbf{w}} = \{\mathbf{x} \in \mathbb{R}^n \mid \mathbf{w}'(\mathbf{x} - \mathbf{x}_0) = 0\}$$

is an affine subset of $\mathbb{R}^n$.

13. Let $\mathbf{e}_1, \ldots, \mathbf{e}_m$ be the vectors of the standard basis of $\mathbb{C}^m$. Prove that $\sum_{k=1}^{m} \mathbf{e}_k \mathbf{e}_k' = I_m$.
14. Let A, B, C be three matrices such that $A = BC$. Prove that
    (a) $\mathsf{Ran}(A) \subseteq \mathsf{Ran}(B)$; also, if C is a square invertible matrix, show that $\mathsf{Ran}(A) = \mathsf{Ran}(B)$;
    (b) $\mathsf{NullSp}(C) \subseteq \mathsf{NullSp}(A)$; also, if B is a square invertible matrix, $\mathsf{NullSp}(C) = \mathsf{NullSp}(A)$.
15. Let $A \in \mathbb{R}^{n\times n}$. Prove that:
    (a) the matrices
    $$B = \frac{1}{2}(A + A') \text{ and } C = \frac{1}{2}(A - A')$$
    are symmetric and skew-symmetric, respectively;
    (b) any square real matrix $A \in \mathbb{R}^{n\times n}$ can be uniquelly written as the sum of a symmetric and a skew-symmetric matrix.

    Conclude that $\mathbb{R}^{n\times n}$ is the direct sum of the subspace of symmetric matrices and the subspace of skew-symmetric matrices.
16. Prove that there exist 2^{n^2} matrices in $\{0, 1\}^{n\times n}$.
17. Let $A, B \in \mathbb{R}^{n\times n}$ such that there exists a non-singular matrix $X \in \mathbb{C}^{n\times n}$ such that $AX = XB$. Prove that there exists a non-singular matrix $Y \in \mathbb{R}^{n\times n}$ such that $AY = YB$.
18. Let $A \in C^{n\times n}$. Prove that:
    (a) the matrices
    $$B = \frac{1}{2}(A + A') \text{ and } C = \frac{1}{2}(A - A')$$
    are Hermitian and skew-Hermitian, respectively;
    (b) any square complex matrix $A \in C^{n\times n}$ can be uniquelly written as the sum of a Hermitian and a skew-Hermitian matrix.

    Conclude that $C^{n\times n}$ is the direct sum of the subspace of Hermitian matrices and the subspace of skew-Hermitian matrices.
19. Let K be a finite set that spans the F-linear space $(L, +, \cdot)$ and let H be a subset of L that is linearly independent. There exists a basis B such that $H \subseteq B \subseteq K$.
20. Let $C, D \in C^{n\times m}$ and let $\{\mathbf{t}_1, \ldots, \mathbf{t}_m\}$ be a basis in C^m. Prove that if $C\mathbf{t}_i = D\mathbf{t}_i$ for $1 \leqslant i \leqslant m$, then $C = D$.
21. Let $A \in C^{n\times n}$ be the matrix

$$A = \begin{pmatrix} \mathbf{0}'_{n-1} & & -a_0 \\ 1\,0 & \cdots 0 & -a_1 \\ 0\,1 & \cdots 0 & -a_2 \\ \vdots\,\vdots & \cdots \vdots & \vdots \\ 0\,0 & \cdots 1 & -a_{n-1} \end{pmatrix}$$

Prove that

(a) $A\mathbf{e}_k = \mathbf{e}_{k+1}$ for $1 \leqslant k \leqslant n-1$ and $A\mathbf{e}_n = \mathbf{a}$, where $\mathbf{a}$ is the last column of A.
(b) if $p(t) = t^n + a_{n-1}t^{n-1} + \cdots + a_1 t + a_0$ we have $p(A)\mathbf{e}_i = \mathbf{0}_n$ for $1 \leqslant i \leqslant n$, and therefore, $p(A) = O_{n,n}$;
(c) there is no polynomial q of degree less than n such that $q(A) = O_{n,n}$.

The matrix A is referred as the *companion matrix* of the polynomial p.

22. Let S be a finite set $S = \{x_1, \ldots, x_n\}$ and let $*$ be a binary operation on S. For t in S, define the matrices $\mathbf{L}_t, \mathbf{M}_t \in S^{n\times n}$ as $(\mathbf{L}_t)_{ij} = u$ if $(x_i * t) * x_j = u$ and $(\mathbf{R}_t)_{ij} = v$ if $x_i * (t * x_j) = u$.

Prove that "$*$" is an associative operation on S if and only if for every $t \in S$ we have $\mathbf{L}_t = \mathbf{R}_t$.

23. Let $\mathbf{A} = (a_{ij})$ be an $(m \times n)$-matrix of real numbers. Prove that

$$\max_j \min_i a_{ij} \leqslant \min_i \max_j a_{ij}$$

(the *minimax inequality*).

Solution: Note that $a_{ij_0} \leqslant \max_j a_{ij}$ for every i and j_0, so $\min_i a_{ij_0} \leqslant \min_i \max_j a_{ij}$, again for every j_0. Thus, $\max_j \min_i a_{ij} \leqslant \min_i \max_j a_{ij}$.

24. Let $A \in C^{n\times n}$ be a matrix and let $\mathbf{x}, \mathbf{y} \in C^n$ be two vectors such that $A\mathbf{x} = a\mathbf{x}$ and $A'\mathbf{y} = a\mathbf{y}$ for some $a \in C$, and $\mathbf{x}'\mathbf{y} = 1$. Let L be the rank-1 matrix $L = \mathbf{x}\mathbf{y}'$. Prove that:

(a) $L\mathbf{x} = \mathbf{x}$ and $\mathbf{y}'L = \mathbf{y}'$;
(b) L is an idemponent matrix;
(c) $A^m L = LA^m = a^m L$ for $m \geqslant 1$;
(d) $L(A - aL) = O_{n,n}$;
(e) $(A - aL)^m = A^m - a^m L$ for $m \geqslant 1$.

25. Let $A \in \mathbb{R}^{n\times n}$ be a matrix such that $A > O_{n,n}$ and let $\mathbf{x} \in \mathbb{R}^n$ be a non-negative vector such that $\mathbf{x} \neq \mathbf{0}_n$. Prove that $A\mathbf{x} > 0$.

Solution: It is clear that $A\mathbf{x} \geqslant \mathbf{0}_n$. Suppose that there exists i, $1 \leqslant i \leqslant n$ such that $(A\mathbf{x})_i = \sum_{j=1}^n a_{ij}x_j = 0$. Since A is positive and $\mathbf{x}$ is non-negative, it follows that $a_{ij}x_j = 0$ for $1 \leqslant j \leqslant n$; this is possible only if $x_j = 0$ for $1 \leqslant j \leqslant n$, that is, if $\mathbf{x} = \mathbf{0}_n$. This contradiction shows that $A\mathbf{x} > 0$.

26. Prove that the product of two doubly-stochastic matrices is a doubly-stochastic matrix.

27. Let $A \in \mathbb{R}^{n \times n}$ be a matrix such that for every doubly-stochastic matrix $B \in \mathbb{R}^{n \times n}$ we have $AB = BA$. Prove that there exist $a, b \in \mathbb{R}$ such that $A = aI_n + bJ_n$.
28. Let $S = \{x_1, \ldots, x_n\}$ be a finite set. If $\pi = \{B_1, \ldots, B_m\}$ and $\sigma = \{C_1, \ldots, C_p\}$ are two partitions on S, prove that for the matrix $Q = M_\pi' M_\sigma \in \mathbb{N}^{m \times p}$ we have $q_{hk} = |B_h \cap C_k|$ for $1 \leqslant h \leqslant m$ and $1 \leqslant k \leqslant p$.
29. Let $\phi \in PERM_n$ be

$$\phi : \begin{pmatrix} 1 & \cdots & i & \cdots & n \\ a_1 & \cdots & a_i & \cdots & a_n \end{pmatrix},$$

and let $v_p(\phi) = |\{(i_k, i_l) \mid i_l = p, k < l, i_k > i_l\}$ be the number of inversions of ϕ that have p as their second component, for $1 \leqslant p \leqslant n$. Prove that

(a) $v_p \leqslant n - p$ for $1 \leqslant p \leqslant n$;
(b) for every sequence of numbers $(v_1, \ldots, v_n) \in \mathbb{N}^n$ such that $v_p \leqslant n - p$ for $1 \leqslant p \leqslant n$ there exists a unique permutation ϕ that has $(v_1, \ldots, v_n) \in \mathbb{N}^n$ as its sequence of inversions.

30. Let p be the polynomial $p(x_1, \ldots, x_n) = \prod_{i<j}(x_i - x_j)$. For a permutation $\phi \in PERM_n$,

$$\phi : \begin{pmatrix} 1 & \cdots & i & \cdots & n \\ a_1 & \cdots & a_i & \cdots & a_n \end{pmatrix},$$

define the number p_ϕ as $p_\phi = p(a_1, \ldots, a_n)$. Prove that

(a) $(-1)^{inv}(\phi) = \frac{p_\phi}{p_{\iota_n}}$ for any permutation ϕ;
(b) $(-1)^{inv(\psi\phi)} = (-1)^{inv(\psi)}(-1)^{inv(\phi)}$.

31. exer:nov1413a Let $A \in \mathbb{C}^{n \times n}$ be a matrix. Prove that for every permutation matrix P_ϕ, A is a symmetric matrix if and only if $P_\phi^{-1} A P_\phi$ is a symmetric matrix.
32. Let $A \in \mathbb{R}^{n \times n}$ be a symmetric matrix. Prove that if $trace(A) = 0$ and the sum of principal minors of order 2 equals 0, then $A = O_{n,n}$.

Solution: A principal minor of order 2 has the form $m_{ij} = a_{ii}a_{jj} - a_{ij}^2$ for $1 \leqslant i < j \leqslant n$ (because A is a symmetric matrix) and there are $\binom{n}{2}$ such minors. Therefore, the sum of these minors is $M = \sum_{i<j} a_{ii}a_{jj} - \sum_{i<j} a_{ij}^2 = 0$. Since $trace(A) = 0$ we have $trace(A)^2 = \sum_{i=1}^n a_{ii}^2 + 2\sum_{i<j} a_{ii}a_{jj} = 0$. These equalities imply

$$0 = \sum_{i=1}^{n} a_{ii}^2 + 2\sum_{i<j} a_{ii}a_{jj} = \sum_{i=1}^{n} a_{ii}^2 + 2\sum_{i<j} a_{ij}^2$$

which yields $a_{11} = \cdots = a_{nn} = a_{12} = \cdots = a_{n-1\,n} = 0$. Thus, $A = O_{n,n}$.

33. Let D_n be the determinant having n rows, defined by

$$D_n = \begin{vmatrix} 1 & 0 & 0 & \cdots & 0 & 0 \\ a & 1 & 0 & \cdots & 0 & 0 \\ \vdots & \vdots & \vdots & \vdots & \vdots & \vdots \\ 0 & 0 & 0 & \cdots & 1 & 0 \\ 0 & 0 & 0 & \cdots & a & 1 \end{vmatrix},$$

where $n \geqslant 2$. Prove that $D_n = 1$ for every $n \geqslant 2$.

34. Using elementary properties of determinants prove that

$$\begin{vmatrix} a^2 + x^2 & ab + xy & ac + xz \\ ab + xy & b^2 + y^2 & bc + yz \\ ac + xz & bc + yz & c^2 + z^2 \end{vmatrix} = 0.$$

35. Let $T_n(a)$ be the tri-diagonal determinant

$$T_n(a) = \begin{vmatrix} a & 1 & 0 & 0 & \cdots & 0 & 0 \\ 1 & a & 1 & 0 & \cdots & 0 & 0 \\ 0 & 1 & a & 1 & \cdots & 0 & 0 \\ \vdots & \vdots & \vdots & \vdots & \vdots & 0 & 0 \\ 0 & 0 & 0 & 0 & \cdots & 1 & a \end{vmatrix}$$

having n rows, where $n \geqslant 2$.

Prove that $T_2(a) = a^2 - 1$ and that $T_{n+2}(a) = aT_{n+1}(a) - T_n(a)$ for $n \geqslant 2$.

36. Let $A \in \mathbb{R}^{n \times n}$ and $c \in \mathbb{R}$. Prove that $\det(cA) = c^n \det(A)$.

A matrix $A \in C^{n \times n}$ is *unimodular* if $|\det(A)| = 1$.

37 Let $A \in \mathbb{R}^{n \times n}$ be a matrix such that $a_{ij} \in \mathbb{Z}$. Prove that A is nonsingular and the matrix A^{-1} has integer entries if and only if A is unimodular.

Solution: Since $A^{-1} = \frac{1}{\det(A)} adj(A)$, it is clear that for a unimodular matrix whose elements are integers, the entries of A^{-1} are integers.

Conversely, if A^{-1} exists and its entries are integers, $A^{-1}A = I_n$ implies $\det(A)\det(A^{-1}) = 1$, which, in turn implies $\det(A) \in \{-1, 1\}$.

38. Let $A \in \mathbb{Z}^{m \times n}$ be a matrix having integer entries. Prove that there exist unimodular matrices $C \in \mathbb{R}^{m \times m}$, $D \in \mathbb{R}^{n \times n}$ such that $CAD = \text{diag}(z_1, \ldots, z_r, 0, \ldots, 0)$, where $r \leqslant \min\{m, n\}$ and $z_1, \ldots, z_r$ are positive integers such that $z_i | z_{i+1}$ for $1 \leqslant i \leqslant r - 1$. This decomposition of A is known as the *Smith normal form of A*.

Solution: Let $a = \min_{i,j}\{|a_{ij}| \mid a_{ij} \neq 0, 1 \leqslant i \leqslant m, 1 \leqslant j \leqslant n\}$. Using row and column transpositions (which can be achieved by multiplications at left and at right by matrices of the form $T_n^{i \leftrightarrow j}$ place a in the top leftmost position of the matrix $T_1 A T_2$, where T_1 and T_2 are products of matrices of the form $T_n^{i \leftrightarrow j}$.

Suppose initially that a divides all entries of the matrix A. Then, by multiplications at left and right with matrices of the form $T_n^{i \overset{+}{\leftarrow} j}$ and $T_{n,i}^{(a)}$ we can place 0s everywhere in the first row and the first column except in the position $(1, 1)$ yielding a matrix of the form

$$\begin{pmatrix} 1 & \mathbf{0}'_{n-1} \\ \mathbf{0}_{n-1} & \tilde{A} \end{pmatrix}.$$

Applying the same argument to $\tilde{A}$, we reach the desired conclusion.

Suppose that a_{1j} is not divisible by a_{11}. Then, we have $a_{1j} = a_{11}q + r$ with $0 < |r| < |a_{11}|$. After subtracting the first column multiplied by q from the jth column, we obtain r in position $(1, j)$. Then, we apply the argument used in the previous case. Since $|r| < |a_{11}|$ the process must end. The case, when a_{j1} is not divisible by a_{11} can be treated similarly. Thus, we have shown that there exist invertible matrices $T_1, \ldots, T_p, S_1, \ldots, S_q$ such that $T_p \cdots T_1 A S_1 \cdots S_q = \mathsf{diag}(z_1, \ldots, z_r, 0, \ldots, 0)$. By choosing $C = (T_p \cdots T_1)^{-1}$ and $D = (S_1 \cdots S_q)^{-1}$ we reach the desired conclusion.

39. Prove that if a matrix $A \in C^{n \times n}$ is skew-Hermitian and n is an odd number, then $\Re(\det(A)) = 0$.
40. Prove that $trace(J_{n,n} A J_{n,n} B) = trace(J_{n,n} A) trace(J_{n,n} B)$, where $A, B \in C^{n \times n}$.
41. Prove that $\det(aI_n + bJ_{n,n}) = a^n + na^{n-1}b$ for $n \geqslant 1$.
42. Let $A \in C^{n \times n}$ and let $x, y \in C$. Prove that

$$(xI_n - A)^{-1} - (yI_n - A)^{-1} = (y - x)(xI_n - A)^{-1}(yI_n - A)^{-1}.$$

43. Let $A \in \mathbb{R}^{n \times n}$ be a skew-symmetric matrix and let $B = A + bJ_{n,n}$, where $b \in \mathbb{R}$. Prove that if n is even, then $\det(A) = det(B)$.
44. Let $U, V \in C^{n \times m}$ be two matrices. Prove that $\det(I_n + UV^{\mathsf{H}}) = \det(I_m + V^{\mathsf{H}}U)$ (Sylvester's Identity).

    Solution: Starting from the matrix equalities

$$\begin{pmatrix} I_n & -U \\ V^{\mathsf{H}} & I_m \end{pmatrix} = \begin{pmatrix} I_n & O_{n,m} \\ V^{\mathsf{H}} & I_m \end{pmatrix} \cdot \begin{pmatrix} I_n & -U \\ O_{m,n} & I_m + V^{\mathsf{H}}U \end{pmatrix}$$
$$= \begin{pmatrix} I_n & -U \\ O_{m,n} & I_m \end{pmatrix} \cdot \begin{pmatrix} I_n + UV^{\mathsf{H}} & O_{n,m} \\ V^{\mathsf{H}} & I_m \end{pmatrix},$$

which are immediate, and taking the determinants of both sides the Sylvester's Identity follows immediately taking into account that

$$\det\begin{pmatrix} I_n & U \\ O_{m,n} & I_m + V^{\mathsf{H}}U \end{pmatrix} = \det(I_m + V^{\mathsf{H}}U)$$

and

$$\det\begin{pmatrix} I_n + UV^{\mathsf{H}} & O_{n,m} \\ V^{\mathsf{H}} & I_m \end{pmatrix} = \det(I_n + UV^{\mathsf{H}}).$$

We used the fact that the determinant of a block upper triangular matrix equals the product of the determinants of matrices situated on the diagonal.

45. Let $\mathbf{x}, \mathbf{y} \in C^n$. Prove that:
    (a) $\det(I_n + \mathbf{x}\mathbf{y}^{\mathsf{H}}) = 1 + \mathbf{y}^{\mathsf{H}}\mathbf{x}$ and $\det(I_n - \mathbf{x}\mathbf{y}^{\mathsf{H}}) = 1 - \mathbf{y}^{\mathsf{H}}\mathbf{x}$;
    (b) if $A \in C^{n\times n}$ is an invertible matrix, then $\det(A + \mathbf{x}\mathbf{y}^{\mathsf{H}}) = \det(A)(1 + \mathbf{y}^{\mathsf{H}}A^{-1}\mathbf{x})$.

    Solution: The first part is a direct consequence of Supplement 44 by taking $m = 1$.
    For the second part we have $A + \mathbf{x}\mathbf{y}^{\mathsf{H}} = A(I_n + A^{-1}\mathbf{x}\mathbf{y}^{\mathsf{H}})$, which yields the desired inequality.

46. Let $A \in \mathbb{R}^{m\times n}$, where $m \leqslant n$. Prove that

$$\det(AA') = \sum\left\{\det\left(A\begin{bmatrix} 1 & \cdots & m \\ k_1 & \cdots & k_m \end{bmatrix}\right)^2 \middle| 1 \leqslant k_1 < k_2 < \cdots < k_m \leqslant n\right\}.$$

Solution: This equality follows from Cauchy-Binet formula.

47. Let $v_1, \ldots, v_n$ be n complex numbers such that $\sum_{k=1}^{n} |v_k|^2 = 2$. Prove that

$$\begin{vmatrix} 1 - v_1\bar{v}_1 & v_1\bar{v}_2 & v_1\bar{v}_3 & \cdots & v_1\bar{v}_n \\ v_2\bar{v}_1 & 1 - v_2\bar{v}_2 & v_2\bar{v}_3 & \cdots & v_2\bar{v}_n \\ \vdots & \vdots & \vdots & \cdots & \vdots \\ v_n\bar{v}_1 & v_n\bar{v}_2 & v_n\bar{v}_3 & \cdots & 1 - v_n\bar{v}_n \end{vmatrix} = -1.$$

Solution: Note that the matrix whose determinant is to be computed equals $I_n - \mathbf{v}\mathbf{v}^{\mathsf{H}}$. Thus, by Supplement 45, we have $\det(I_n - \mathbf{v}\mathbf{v}^{\mathsf{H}}) = 1 + \mathbf{v}^{\mathsf{H}}\mathbf{v} = -1$.

48. Let $D = \mathsf{diag}(a_1, \ldots, a_n)$ and let $A = D + \mathbf{1}_n\mathbf{1}_n'$. Prove that

$$\det(A) = \prod_{i=1}^{n} a_i \left(1 + \sum_{i=1}^{n} \frac{1}{a_i}\right).$$

Solution: By Part (b) of Supplement 45, we have

$$\det(A) = \det(D + \mathbf{1}_n\mathbf{1}_n^{\mathsf{H}}) = \det(D)(1 + \mathbf{1}_n^{\mathsf{H}}D^{-1}\mathbf{1}_n),$$

which amounts to the formula we need to prove.

49. Let $Z \in C^{n\times n}$, $W \in C^{m\times m}$ and let $U, V \in C^{n\times m}$ be four matrices such that each of the matrices W, Z, $Z + UWV'$, and $W^{-1} + UWV'$ have an inverse.

Prove that

$$(Z + UWV')^{-1} = Z^{-1} - Z^{-1}U(W^{-1} + V'Z^{-1}U)^{-1}V'Z^{-1}$$

(the *Woodbury–Sherman–Morrison identity*).
Solution: Consider the system of matrix equations:

$$ZX + UY = I_n, \; V'X - W^{-1}Y = O_{m,n},$$

where $X \in \mathbb{R}^{n\times n}$ and $Y \in \mathbb{R}^{m\times n}$

The second equation implies $V'X = W^{-1}Y$, so $UWV'X = UY$. Substituting UY in the first equation yields $ZX + UWV'X = I$, so $(Z + UWV')X = I$, which implies

$$X = (Z + UWV')^{-1}. \quad (5.10)$$

On the other hand, we have $X = Z^{-1}(I - UY)$ from the first equation. Substituting X in the second equation yields $V'Z^{-1}(I - UY) = W^{-1}Y$, which is equivalent to

$$\begin{aligned} V'Z^{-1} &= +W^{-1}Y + V'Z^{-1}UY \\ &= (W^{-1} + V'Z^{-1}U)Y. \end{aligned}$$

Thus, we have $Y = (W^{-1} + V'Z^{-1}U)^{-1}V'Z^{-1}$. Substituting the values of X and Y in the first equality implies

$$ZX + U(W^{-1} + V'Z^{-1}U)^{-1}V'Z^{-1} = I.$$

Therefore, $ZX = I - U(W^{-1} + V'Z^{-1}U)^{-1}V'Z^{-1}$, which implies

$$X = Z^{-1} - Z^{-1}U(W^{-1} + V'Z^{-1}U)^{-1}V'Z^{-1}. \quad (5.11)$$

The Woodbury–Sherman–Morrison identity follows immediately from Equalities (5.10) and (5.11).

50. Using the notations introduced in the previous Supplement prove the Woodbury–Sherman–Morrison identity for determinants:

$$\det(Z + UWV') = \det(Z)\det(W)\det(W^{-1} + V'Z^{-1}U).$$

Solution: We can write:

$$\begin{aligned}
&\det(I_n - U(W^{-1} + V'Z^{-1}U)^{-1}V'Z^{-1})\\
&= \det(I_n - V'Z^{-1}U(W^{-1} + V'Z^{-1}U)^{-1})\\
&\quad \text{(by Sylvester's identity)}\\
&= \det((W^{-1} + V'Z^{-1}U)(W^{-1} + V'Z^{-1}U)^{-1}\\
&\quad - V'Z^{-1}U(W^{-1} + V'Z^{-1}U)^{-1})\\
&= \det((W^{-1} + V'Z^{-1}U - V'Z^{-1}U)\det(W^{-1} + V'Z^{-1}U)^{-1}\\
&= \det(W^{-1})\det(W^{-1} + V'Z^{-1}U)^{-1}.
\end{aligned}$$

This allows us to write $\det(Z)\det(X) = \det(I_n - U(W^{-1} + V'Z^{-1}U)^{-1}V'Z^{-1})$. Since $X = (Z + UWV')^{-1}$, it follows that

$$\frac{det(Z)}{\det(Z + UWV')} = \frac{1}{\det(W^{-1} + V'Z^{-1}U)\det(W)},$$

which is the desired equality.

51. Prove that if $U, V \in C^{n\times m}$ are two matrices such that $I_n + UV'$ and $I_m + V'U$ are invertible matrices, then $\det(I_n + UV') = \det(I_m + V'U)$.
52. Let $A \in \mathbb{R}^{n\times n}$. Prove that for the kth derivative of the function $G(x) = \det(A - xI_n)$ we have

$$G^{(k)}(0) = (-1)^k k! S_{n-k}(A),$$

where $S_p(A)$ is the sum of all order-p principal minors of A.
53. Let $\boldsymbol{x} = (x_1, \ldots, x_n)$ and $\boldsymbol{y} = (y_1, \ldots, y_n)$ be two sequences of real numbers such that $x_i + y_j \neq 0$ for $1 \leqslant i, j \leqslant n$ and $n \geqslant 2$. The *Cauchy matrix* of these sequences is the matrix $C_{\mathbf{x},\mathbf{y}}$ given by

$$(C_{\mathbf{x},\mathbf{y}})_{ij} = \frac{1}{x_i + y_j}$$

for $1 \leqslant i, j \leqslant n$. Prove that

$$\det(C_{\mathbf{x},\mathbf{y}}) = \frac{\prod_{1\leqslant j<i\leqslant n}(x_i - x_j)\prod_{1\leqslant j<i\leqslant n}(y_i - y_j)}{\prod_{1\leqslant i,j\leqslant n}(x_i + y_j)}.$$

54. The *n-Hilbert matrix* H_n is a special Cauchy matrix, where $x_i = y_i = i - \frac{1}{2}$ for $1 \leqslant i \leqslant n$. Let $h_n = 1!2!\cdots(n-1)!$ for $n \in \mathbb{N}$ and $n \geqslant 1$. Prove that:

    (a) the determinant of the Hilbert matrix H_n is $\det(H_n) = \frac{h_n^4}{h_{2n}}$;
    (b) the number $\frac{1}{\det(H_n)}$ is an integer.

55. Let $A \in \mathbb{R}^{n\times n}$ be defined by $a_{ij} = i + j$ for $1 \leqslant i, j \leqslant n$. Prove that if $n \geqslant 3$, $\det(A) = 0$.
56. Let $A \in C^{n\times n}$. Prove that $rank(A) = n$ if and only if $rank(A^{sH}) = n$.

57. Let $A \in C^{m\times n}$ be a matrix, $\mathbf{x} \in C^n$ and $\mathbf{y} \in C^m$. Prove that if $rank(A - aA\mathbf{x}\mathbf{y}^{\mathsf{H}}A) = rank(A) - 1$, then $\mathbf{y}^{\mathsf{H}}A\mathbf{x} \neq 0$ and $a = \frac{1}{\mathbf{y}^{\mathsf{H}}A\mathbf{x}}$.
58. Let $A \in C^{m\times n}$, $B \in C^{n\times p}$, and $C \in C^{p\times q}$ be three matrices. Prove the following inequality

$$rank(AB) + rank(BC) \leqslant rank(B) + rank(ABC),$$

known as Frobenius' Inequality.
59. Let $A \in C^{m\times n}$ be a matrix, $\boldsymbol{u} \in C^m$ and $\boldsymbol{v} \in C^n$ be two vectors, $a \in C - \{0\}$, and let $B = A - \frac{1}{a}\boldsymbol{u}\mathbf{v}^{\mathsf{H}}$. We have $rank(B) < rank(A)$, if and only if there are vectors $\mathbf{x} \in C^m$ and $\mathbf{y} \in C^n$ such that $\boldsymbol{u} = A\mathbf{y}$, $\mathbf{v} = A^{\mathsf{H}}\mathbf{x}$, and $a = \mathbf{x}^{\mathsf{H}}A\mathbf{y}$, in which case $rank(B) = rank(A) - 1$.
60. Let d_n be the determinant defined by

$$d_n = \begin{vmatrix} \cos\alpha & \cos 2\alpha & \ldots & \cos n\alpha \\ \cos(n+1)\alpha & \cos(n+2)\alpha & \ldots & \cos 2n\alpha \\ \vdots & \vdots & \ldots & \vdots \\ \cos(n^2-n+1)\alpha & \cos(n^2-n+2)\alpha & \ldots & \cos n^2\alpha \end{vmatrix}$$

Prove that for $n \geqslant 3$ we have $d_n = 0$.
Hint: add the third column to the first column.

Let $A \in C^{m\times n}$ be a partitioned matrix given by

$$A = \begin{pmatrix} A_{11} & A_{12} \\ A_{21} & A_{22} \end{pmatrix},$$

where A_{11} is an invertible square matrix, $A_{11} \in C^{p\times p}$. Note that $A_{21} \in C^{(m-p)\times p}$, $A_{12} \in C^{p\times(n-p}$, and $A_{22} \in C^{(m-p)\times(n-p)}$. Therefore, the matrix $B = A_{22} - A_{21}A_{11}^{-1}A_{12} \in C^{(m-p)\times(n-p)}$ is well-defined. We refer to B as the *Schur's complement of* A_{11} relative to A and is denoted by A/A_{11}.
61. Let $A \in C^{n\times n}$ be a square partitioned matrix given by

$$A = \begin{pmatrix} A_{11} & A_{12} \\ A_{21} & A_{22} \end{pmatrix},$$

where A_{11} is an invertible matrix. Prove that
(a) we have $\det(A/A_{11}) = \frac{\det(A)}{\det(A_{11})}$;
(b) we have the equalities:

$$A = \begin{pmatrix} I & O \\ A_{21}A_{11}^{-1} & I \end{pmatrix} \begin{pmatrix} A_{11} & O \\ O & A_{22} - A_{21}A_{11}^{-1}A_{12} \end{pmatrix} \begin{pmatrix} I & A_{11}^{-1}A_{12} \\ O & I \end{pmatrix}.$$

and

$$\det(A) = \det(A_{11}) \det(A_{22} - A_{21} A_{11}^{-1} A_{12}).$$

62. Let

$$A = \begin{pmatrix} A_{11} & \cdots & A_{1m} \\ \vdots & \vdots & \vdots \\ A_{m1} & \cdots & A_{mm} \end{pmatrix}$$

be a partitioned matrix such that each matrix

$$B_p = \begin{pmatrix} A_{11} & \cdots & A_{1p} \\ \vdots & \vdots & \vdots \\ A_{p1} & \cdots & A_{pp} \end{pmatrix}$$

is invertible for $1 \leqslant p \leqslant m - 1$.

Prove that A can be uniquely written as a product $A = LDU$ such that the following conditions are satisfied:

(a) L is a lower block triangular matrix whose diagonal blocks are unit matrices;
(b) D is a block diagonal matrix, $D = \mathsf{diag}(D_1, \ldots, D_m)$ whose diagonal blocks are invertible matrices such that $D_1 = A_{11} = B_1$, and $D_j = B_j / B_{j-1}$ for $1 \leqslant p \leqslant m$;
(c) U is an upper block diagonal matrix whose diagonal blocks are unit matrices.

63. A matrix $A \in C^{n \times n}$ is *strongly non-singular* if all its principal matrices are non-singular. Prove that if $A \in C^{n \times n}$ is strongly non-singular, then there exists a unique factorization $A = LDU$ such that the following conditions are satisfied:

(a) L is a lower block triangular matrix whose diagonal elements are equal to 1;
(b) D is a diagonal matrix, $D = \mathsf{diag}(d_1, \ldots, d_n)$ such that $d_1 = a_{11}$ and $d_k = \frac{A_k}{A_{k-1}}$, where $A_k = A \begin{bmatrix} 1 \cdots k \\ 1 \cdots k \end{bmatrix}$;
(c) U is an upper diagonal matrix whose diagonal elements are equal to 1.
Hint: This follows immediately from Supplement 62.

64. Prove that if $H \in \mathbb{R}^{n \times n}$ is a Hadamard matrix, then $|\det(H)| = n^{\frac{n}{2}}$.
65. Prove that if $H \in \mathbb{R}^{n \times n}$ is a Hadamard matrix and $n > 2$, then n is a multiple of 4.
66. Let $a_1, \ldots, a_n$ be n numbers and let $b_{ij} = |a_i - a_j|$ for $1 \leqslant i, j \leqslant n$. Compute the determinant of the matrix $B = (b_{ij})$.
67 Let $S = \{(x_i, y_i) \in C^2 \mid 1 \leqslant i \leqslant n\}$ be a set of n pairs of complex numbers such that $i \neq j$ implies $x_i \neq x_j$ and and let $p(x) = c_0 + c_1 x + \cdots + c_{n-1} x^{n-1}$ be a polynomial such that $p(x_i) = y_i$ for $1 \leqslant i \leqslant n$. Prove that

(a) the coefficients c_i of p are given by the formula

$$c_i = \frac{V_{\mathbf{x}} \stackrel{i}{\leftarrow} \mathbf{y}}{V_{\mathbf{x}}},$$

where $V_{\mathbf{x}} \stackrel{i}{\leftarrow} \mathbf{y}$ is the determinant obtained from $V_{\mathbf{x}}$ by replacing the ith column by $\mathbf{y}$;

(b) the polynomial $p(x)$ can be written as

$$p(x) = \sum_{j=1}^{n} y_j p_j(x),$$

where $p_j(x) = \prod\{\frac{x-x_k}{x_j-x_k} \mid k \in \{1, \ldots, n\} - \{j\}\}$. The polynomial p is known as the *Lagrange interpolation polynomial* for S.

68. Let $A \in C^{n\times n}$ be a matrix such that $A\mathbf{1}_n = A'\mathbf{1}_n = \mathbf{0}_n$. Prove that all cofactors $cof(a_{ij})$ of A are equal.
69. Prove that if $H \in \mathbb{R}^{n\times n}$ is a Hadamard matrix, then

$$\begin{pmatrix} H & H \\ H & -H \end{pmatrix} \in \mathbb{R}^{2n\times 2n}$$

is also a Hadamard matrix.
70. Let

$$M = \left(\begin{array}{c|c} A & B \\ \hline C & D \end{array}\right)$$

be a partitioned matrix, where $A \in \mathbb{R}^{m\times m}$, $B \in \mathbb{R}^{m\times p}$, $C \in \mathbb{R}^{p\times m}$, and $D \in \mathbb{R}^{p\times p}$. Prove that if $U \in \mathbb{R}^{m\times m}$ and $V \in \mathbb{R}^{p\times p}$ are orthogonal matrices, then

$$\begin{aligned}\det(M) &= \det\begin{pmatrix} UA & UB \\ C & D \end{pmatrix} = \det\begin{pmatrix} A & BV \\ C & DV \end{pmatrix} \\ &= \det\begin{pmatrix} AU & B \\ CU & D \end{pmatrix} = \det\begin{pmatrix} A & B \\ VC & VD \end{pmatrix}.\end{aligned}$$

71. Let $A \in \mathbb{R}^{n\times n}$ be a skew-symmetric matrix. If n is an odd number, prove that $\det(A) = 0$.
72. A *generalized inverse* (or a g-inverse) of a matrix $A \in C^{m\times n}$ is a matrix $B \in C^{n\times m}$ such that $ABA = A$. Prove that if B is a g-inverse and $A\mathbf{x} = \boldsymbol{b}$ has a solution, then $\mathbf{x} = G\boldsymbol{b}$ is one of these solutions.
73. Let $A \in C^{m\times n}$ be a matrix such that $A = \begin{pmatrix} A_{11} & A_{12} \\ A_{21} & A_{22} \end{pmatrix}$, where $A_{11} \in C^{r\times r}$ and $rank(A_{11}) = rank(A) = r$. Prove that:

(a) there exists a matrix $S \in C^{r \times n-r}$ such that $A_{12} = A_{11}S$ and $A_{22} = A_{21}S$;
(b) the matrix $B = \begin{pmatrix} A_{11}^{-1} & O_{r,m-r} \\ O_{n-r,r} & 0_{n-r,m-r} \end{pmatrix}$ is a g-inverse of A.

74. Let $A \in \mathbb{R}^{n \times n}$ be a matrix such that $a_{ii} > 0$ for $1 \leqslant i \leqslant n$, such that $a_{ii} > \sum\{|a_{ik}| \mid 1 \leqslant k \leqslant n \text{ and } k \neq i\}$ and $1 \leqslant i \leqslant n$. Prove that $\det(A) > 0$.
75. Prove that for $A \in C^{m \times n}$, there exists at most one matrix $M \in C^{n \times m}$ such that the matrices MA and AM are Hermitian, $AMA = A$, and $MAM = M$.

The matrix M introduced above is a special g-inverse referred to as the *Moore-Penrose pseudoinverse of* A and is denoted by $A^{\dagger}$.

76. Let $A \in C^{n \times n}$. Prove that if A is invertible, then $A^{\dagger}$ exists and equals A^{-1}.
77. Give an example of a matrix that is not invertible but has a Moore-Penrose pseudoinverse.
Hint: Consider the matrix $O_{m,n}$.
78. Prove that if $A^{\dagger}$ exists then $(A^{\dagger})^{\dagger} = A$.
79. Let $GL(n, C)$ be the set of invertible matrices in $C^{n \times n}$. Prove that

(a) the algebra $(GL(n, C), \{I_n, \cdot, ^{-1}\}$, where $\cdot$ is the usual matrix multiplication is a group (this is known as the linear group);
(b) the mapping $\phi : GL(2, C) \longrightarrow GL(3, C)$ given by

$$\phi\begin{pmatrix} a_{11} & a_{12} \\ a_{21} & a_{22} \end{pmatrix} = \begin{pmatrix} a_{11}^2 & 2a_{11}a_{12} & a_{12}^2 \\ a_{11}a_{21} & a_{11}a_{22} + a_{12}a_{22} & a_{12}a_{22} \\ a_{21}^2 & 2a_{21}a_{22} & a_{22}^2 \end{pmatrix}$$

is a group mprphism.

80. Let $A \in C^{m \times n}$ be a matrix with $rank(A) = r > 0$ and let $A = BC$ be a full-rank factorization of A, where $B \in C^{m \times r}$ and $C \in C^{r \times n}$ are full-rank matrices. Prove that

(a) the matrices $B^{\mathsf{H}}B \in C^{r \times r}$ and $CC^{\mathsf{H}} \in C^{r \times r}$ are non-singular;
(b) the matrix $B^{\mathsf{H}}AC^{\mathsf{H}}$ is non-singular;
(c) the Moore-Penrose pseudoinverse of A is given by

$$A^{\dagger} = C^{\mathsf{H}}(CC^{\mathsf{H}})^{-1}(B^{\mathsf{H}}B)^{-1}B^{\mathsf{H}}.$$

A *data matrix* is a matrix $D \in \mathbb{R}^{m \times n}$. The columns of D, $\boldsymbol{v}_1, \ldots, \boldsymbol{v}_n$ are the *features* of the data; its rows $\boldsymbol{u}'_1, \ldots, \boldsymbol{u}'_m$ are the *observations*.

The *mean* of D is the vector $\tilde{D} = \frac{1}{m}D'\mathbf{1}_m \in \mathbb{R}^n$. D is *centered* if $\tilde{D} = \mathbf{0}_n$.

81. Let $D \in \mathbb{R}^{m \times n}$ be a data matrix and let $H_m = I_m - \frac{1}{m}\mathbf{1}_m\mathbf{1}'_m$. Prove that $H_m D$ is a centered data matrix.
Solution: We have

$$\widetilde{(H_m D)} = \frac{1}{m} D' H_m' \mathbf{1}_m = \frac{1}{m} D' \left(I_m - \frac{1}{m} \mathbf{1}_m \mathbf{1}_m' \right)' \mathbf{1}_m$$
$$= \frac{1}{m} D' \left(I_m - \frac{1}{m} \mathbf{1}_m \mathbf{1}_m' \right) \mathbf{1}_m = \frac{1}{m} D' \left(\mathbf{1}_m - \frac{1}{m} \mathbf{1}_m \mathbf{1}_m' \mathbf{1}_m \right) = \mathbf{0}_n.$$

82. Prove that the centering matrix $H_m = I_m - \frac{1}{m}\mathbf{1}_m\mathbf{1}_m'$ is both symmetric and idempotent; further, prove that $H_m\mathbf{1}_m = \mathbf{0}_m$.

Bibliographical Comments

MacLane and Birkhoff [1] is a fundamental reference for algebra. Artin's book [2] contains a vast amount of material from many areas of algebra presented in a lucid manner. Anther basic referrence is [3]. Concise and important sources are [4, 5] and [6].

Exercise 9 is a result of Rota (see [7]). The Russian literature has produced several readable books [8, 9] which were translated into English.

References

1. S. MacLane, G. Birkhoff, *Algebra*, 3rd edn. (Chelsea Publishing Company, New York, 1993)
2. M. Artin, Algebra, 1st edn. (Prentice-Hall, NJ, 1991) (A second edition was published in 2010)
3. F. R. Gantmacher, The Theory of Matrices (2 vols.). (American Mathematical Society, Providence, RI, 1977)
4. L. Mirsky, *An Introduction to Linear Algebra* (Dover, NY, 1990). Reprint edition
5. M. Marcus, H. Minc, *A Survey of Matrix Theory and Matrix Inequalities* (Dover, NY, 2010). Reprint edition
6. M. Fiedler, *Special Matrices and Their Applications in Numerical Mathematics*, 2nd edn. (Dover Publications, Mineola, NY, 2008)
7. G.-C. Rota, The number of partitions of a set. Am. Math. Mon. **71**, 498–504 (1964)
8. I.R. Shafarevich, A.O. Remizov, *Linear Algebra and Geometry* (Springer, Heidelberg, 2013)
9. A.I. Kostrikin, Y.I. Manin, *Linear Algebra and Geometry (Algebra, Logic and Applications)* (Gordon and Breach Science Publishers, New York, 1989)

Chapter 6
Norms and Inner Products

6.1 Introduction

The notion of norm is introduced for evaluating the magnitude of vectors and, in turn, allows the definition of certain metrics on linear spaces equipped with norms.

After presenting some useful inequalities on linear spaces, we introduce norms and the topologies they induce on linear spaces. Then, we discuss inner products, angles between vectors, and the orthogonality of vectors.

We study unitary, orthogonal, positive definite and positive semidefinite matrices that describe important classes of linear transformations. The notion of orthogonality leads to the study of projection on subspaces and the Gram-Schmidt orthogonalization algorithm.

6.2 Inequalities on Linear Spaces

We begin with a technical result.

Lemma 6.1 *Let $p, q \in \mathbb{R} - \{0, 1\}$ be two numbers such that $\frac{1}{p} + \frac{1}{q} = 1$ and $p > 1$. Then, for every $a, b \in \mathbb{R}_{\geqslant 0}$, we have*

$$ab \leqslant \frac{a^p}{p} + \frac{b^q}{q},$$

where the equality holds if and only if $a = b^{-\frac{1}{1-p}}$.

Proof Let $f : [0, \infty) \longrightarrow \mathbb{R}$ be the function defined by $f(x) = x^p - px + p - 1$. Note that $f(1) = 0$ and that $f'(x) = p(x^p - 1)$. This implies that f has a minimum in $x = 1$ and, therefore, $x^p - px + p - 1 \geqslant 0$ for $x \in [0, \infty)$. Substituting $ab^{-\frac{1}{p-1}}$ for x yields the desired inequality.

D. A. Simovici and C. Djeraba, *Mathematical Tools for Data Mining*,
Advanced Information and Knowledge Processing, DOI: 10.1007/978-1-4471-6407-4_6,

Theorem 6.2 (The Hölder Inequality) *Let $a_1, \ldots, a_n$ and $b_1, \ldots, b_n$ be $2n$ non-negative numbers, and let p and q be two numbers such that $\frac{1}{p} + \frac{1}{q} = 1$ and $p > 1$. We have*

$$\sum_{i=1}^{n} a_i b_i \leqslant \left(\sum_{i=1}^{n} a_i^p\right)^{\frac{1}{p}} \cdot \left(\sum_{i=1}^{n} b_i^q\right)^{\frac{1}{q}}.$$

Proof Define the numbers

$$x_i = \frac{a_i}{\left(\sum_{i=1}^{n} a_i^p\right)^{\frac{1}{p}}} \text{ and } y_i = \frac{b_i}{\left(\sum_{i=1}^{n} b_i^q\right)^{\frac{1}{q}}}$$

for $1 \leqslant i \leqslant n$. Lemma 6.1 applied to x_i, y_i yields

$$\frac{a_i b_i}{\left(\sum_{i=1}^{n} a_i^p\right)^{\frac{1}{p}} \left(\sum_{i=1}^{n} b_i^q\right)^{\frac{1}{q}}} \leqslant \frac{1}{p} \frac{a_i^p}{\sum_{i=1}^{n} a_i^p} + \frac{1}{q} \frac{b_i^p}{\sum_{i=1}^{n} b_i^p}.$$

Adding these inequalities, we obtain

$$\sum_{i=1}^{n} a_i b_i \leqslant \left(\sum_{i=1}^{n} a_i^p\right)^{\frac{1}{p}} \left(\sum_{i=1}^{n} b_i^q\right)^{\frac{1}{q}}.$$

Theorem 6.3 *Let $a_1, \ldots, a_n$ and $b_1, \ldots, b_n$ be $2n$ real numbers and let p and q be two numbers such that $\frac{1}{p} + \frac{1}{q} = 1$ and $p > 1$. We have*

$$\left|\sum_{i=1}^{n} a_i b_i\right| \leqslant \left(\sum_{i=1}^{n} |a_i|^p\right)^{\frac{1}{p}} \cdot \left(\sum_{i=1}^{n} |b_i|^q\right)^{\frac{1}{q}}.$$

Proof By Theorem 6.2, we have

$$\sum_{i=1}^{n} |a_i||b_i| \leqslant \left(\sum_{i=1}^{n} |a_i|^p\right)^{\frac{1}{p}} \cdot \left(\sum_{i=1}^{n} |b_i|^q\right)^{\frac{1}{q}}.$$

The needed equality follows from the fact that

$$\left|\sum_{i=1}^{n} a_i b_i\right| \leqslant \sum_{i=1}^{n} |a_i||b_i|.$$

Corollary 6.4 (The Cauchy Inequality) *Let $a_1, \ldots, a_n$ and $b_1, \ldots, b_n$ be $2n$ real numbers. We have*

$$\left|\sum_{i=1}^{n} a_i b_i\right| \leqslant \sqrt{\sum_{i=1}^{n} |a_i|^2} \cdot \sqrt{\sum_{i=1}^{n} |b_i|^2}.$$

Proof The inequality follows immediately from Theorem 6.3 by taking $p = q = 2$.

Theorem 6.5 (Minkowski's Inequality) *Let $a_1, \ldots, a_n$ and $b_1, \ldots, b_n$ be $2n$ nonnegative numbers. If $p \geqslant 1$, we have*

$$\left(\sum_{i=1}^{n}(a_i + b_i)^p\right)^{\frac{1}{p}} \leqslant \left(\sum_{i=1}^{n} a_i^p\right)^{\frac{1}{p}} + \left(\sum_{i=1}^{n} b_i^p\right)^{\frac{1}{p}}.$$

If $p < 1$, the inequality sign is reversed.

Proof For $p = 1$, the inequality is immediate. Therefore, we can assume that $p > 1$. Note that

$$\sum_{i=1}^{n}(a_i + b_i)^p = \sum_{i=1}^{n} a_i(a_i + b_i)^{p-1} + \sum_{i=1}^{n} b_i(a_i + b_i)^{p-1}.$$

By Hölder's inequality for p, q such that $p > 1$ and $\frac{1}{p} + \frac{1}{q} = 1$, we have

$$\begin{aligned}\sum_{i=1}^{n} a_i(a_i + b_i)^{p-1} &\leqslant \left(\sum_{i=1}^{n} a_i^p\right)^{\frac{1}{p}} \left(\sum_{i=1}^{n}(a_i + b_i)^{(p-1)q}\right)^{\frac{1}{q}} \\ &= \left(\sum_{i=1}^{n} a_i^p\right)^{\frac{1}{p}} \left(\sum_{i=1}^{n}(a_i + b_i)^p\right)^{\frac{1}{q}}.\end{aligned}$$

Similarly, we can write

$$\sum_{i=1}^{n} b_i(a_i + b_i)^{p-1} \leqslant \left(\sum_{i=1}^{n} b_i^p\right)^{\frac{1}{p}} \left(\sum_{i=1}^{n}(a_i + b_i)^p\right)^{\frac{1}{q}}.$$

Adding the last two inequalities yields

$$\sum_{i=1}^{n}(a_i + b_i)^p \leqslant \left(\left(\sum_{i=1}^{n} a_i^p\right)^{\frac{1}{p}} + \left(\sum_{i=1}^{n} b_i^p\right)^{\frac{1}{p}}\right) \left(\sum_{i=1}^{n}(a_i + b_i)^p\right)^{\frac{1}{q}},$$

which is equivalent to the desired inequality

$$\left(\sum_{i=1}^{n}(a_i+b_i)^p\right)^{\frac{1}{p}} \leqslant \left(\sum_{i=1}^{n}a_i^p\right)^{\frac{1}{p}} + \left(\sum_{i=1}^{n}b_i^p\right)^{\frac{1}{p}}.$$

6.3 Norms on Linear Spaces

Definition 6.6 *Let L be a linear space (real or complex). A* seminorm *on L is a mapping $\nu : L \longrightarrow \mathbb{R}$ that satisfies the following conditions:*

(i) *$\nu(\boldsymbol{x}+\boldsymbol{y}) \leqslant \nu(\boldsymbol{x})+\nu(\boldsymbol{y})$ (subadditivity), and*
(ii) *$\nu(a\boldsymbol{x}) = |a|\nu(\boldsymbol{x})$ (positive homogeneity),*

for $\boldsymbol{x}, \boldsymbol{y} \in L$ and every scalar a.

By taking $a = 0$ in the second condition of the definition we have $\nu(\mathbf{0}) = 0$ for every seminorm on a real or complex space.

A seminorm can be defined on every linear space L. Indeed, if B is a basis of L, $B = \{\mathbf{v}_i \mid i \in I\}$, J is a finite subset of I, and $\mathbf{x} = \sum_{i \in I} x_i \mathbf{v}_i$, define $\nu_J(\mathbf{x})$ as

$$\nu_J(\mathbf{x}) = \begin{cases} 0 & \text{if } \mathbf{x} = \mathbf{0}, \\ \sum_{j \in J} |a_j| & \text{otherwise} \end{cases}$$

for $\mathbf{x} \in L$. We leave to the reader the verification of the fact that ν_J is indeed a seminorm.

Theorem 6.7 *If L is a real or complex linear space and $\nu : L \longrightarrow \mathbb{R}$ is a seminorm on L, then $\nu(\boldsymbol{x}-\boldsymbol{y}) \geqslant |\nu(\boldsymbol{x}) - \nu(\boldsymbol{y})|$ for $\boldsymbol{x}, \boldsymbol{y} \in L$.*

Proof We have $\nu(\mathbf{x}) \leqslant \nu(\mathbf{x}-\mathbf{y}) + \nu(\mathbf{y})$, so $\nu(\mathbf{x}) - \nu(\mathbf{y}) \leqslant \nu(\mathbf{x}-\mathbf{y})$. Since $\nu(\mathbf{x}-\mathbf{y}) = |-1|\nu(\mathbf{y}-\mathbf{x}) \geqslant \nu(\mathbf{y}) - \nu(\mathbf{x})$, we have $-(\nu(\mathbf{x}) - \nu(\mathbf{y})) \leqslant \nu(\mathbf{x}) - \nu(\mathbf{y})$.

Corollary 6.8 *If $\nu : L \longrightarrow \mathbb{R}$ is a seminorm on the linear space L, then $\nu(\boldsymbol{x}) \geqslant 0$ for $\boldsymbol{x} \in L$.*

Proof By choosing $\mathbf{y} = \mathbf{0}$ in the inequality of Theorem 6.7 we have $\nu(\mathbf{x}) \geqslant |\nu(\mathbf{x})| \geqslant 0$.

Definition 6.9 *Let L be a real or complex linear space. A* norm *on L is a seminorm $\nu : L \longrightarrow \mathbb{R}$ such that $\nu(\boldsymbol{x}) = 0$ implies $\boldsymbol{x} = \boldsymbol{0}$ for $\boldsymbol{x} \in L$.*

The pair (L, ν) is referred to as a normed linear space.

Example 6.10 The set of real-valued continuous functions defined on the interval $[-1, 1]$ is a real linear space. The addition of two such functions f, g, is defined by

$(f+g)(x) = f(x)+g(x)$ for $x \in [-1, 1]$; the multiplication of f by a scalar $a \in \mathbb{R}$ is $(af)(x) = af(x)$ for $x \in [-1, 1]$.

Define $\nu(f) = \sup\{|f(x)| \mid x \in [-1, 1]\}$. Since $|f(x)| \leqslant \nu(f)$ and $|g(x)| \leqslant \nu(g)$ for $x \in [-1, 1]$, it follows that $|(f+g)(x)| \leqslant |f(x)| + |g(x)| \leqslant \nu(f) + \nu(g)$. Thus, $\nu(f+g) \leqslant \nu(f)+\nu(g)$. We leave to the reader the verification of the remaining properties of Definition 6.6.

We denote $\nu(f)$ by $\| f \|$.

Corollary 6.11 *For $p \geqslant 1$, the function $\nu_p : \mathbb{C}^n \longrightarrow \mathbb{R}_{\geqslant 0}$ defined by*

$$\nu_p(\mathbf{x}) = \left(\sum_{i=1}^{n} |x_i|^p\right)^{\frac{1}{p}},$$

is a norm on the linear space $(\mathbb{C}^n, +, \cdot)$.

Proof Let $\mathbf{x}, \mathbf{y} \in \mathbb{C}^n$. Minkowski's inequality applied to the nonnegative numbers $a_i = |x_i|$ and $b_i = |y_i|$ amounts to

$$\left(\sum_{i=1}^{n} (|x_i| + |y_i|)^p\right)^{\frac{1}{p}} \leqslant \left(\sum_{i=1}^{n} |x_i|^p\right)^{\frac{1}{p}} + \left(\sum_{i=1}^{n} |y_i|^p\right)^{\frac{1}{p}}.$$

Since $|x_i + y_i| \leqslant |x_i| + |y_i|$ for every i, we have

$$\left(\sum_{i=1}^{n} (|x_i + y_i|)^p\right)^{\frac{1}{p}} \leqslant \left(\sum_{i=1}^{n} |x_i|^p\right)^{\frac{1}{p}} + \left(\sum_{i=1}^{n} |y_i|^p\right)^{\frac{1}{p}},$$

that is, $\nu_p(\mathbf{x}+\mathbf{y}) \leqslant \nu_p(\mathbf{x}) + \nu_p(\mathbf{y})$. Thus, ν_p is a norm on $\mathbb{C}^n$.

We refer to ν_p as a *Minkowski norm* on $\mathbb{C}^n$.

The normed linear space $(\mathbb{C}^n, \nu_p)$ is denoted by ℓ^n_p.

Example 6.12 Consider the mappings $\nu_1, \nu_\infty : \mathbb{C}^n \longrightarrow \mathbb{R}$ given by

$$\nu_1(\mathbf{x}) = |x_1| + |x_2| + \cdots + |x_n| \text{ and } \nu_\infty(\mathbf{x}) = \max\{|x_1|, |x_2|, \ldots, |x_n|\},$$

for every $\mathbf{x} \in \mathbb{C}^n$. Both ν_1 and ν_∞ are norms on $\mathbb{C}^n$; the corresponding linear spaces are denoted by ℓ^n_1 and ℓ^n_∞.

To verify that ν_∞ is a norm we start from the inequality $|x_i + y_i| \leqslant |x_i| + |y_i| \leqslant \nu_\infty(\mathbf{x}) + \nu_\infty(\mathbf{y})$ for $1 \leqslant i \leqslant n$. This in turn implies

$$\nu_\infty(\mathbf{x}+\mathbf{y}) = \max\{|x_i + y_i| \mid 1 \leqslant i \leqslant n\} \leqslant \nu_\infty(\mathbf{x}) + \nu_\infty(\mathbf{y}),$$

which gives the desired inequality.

This norm can be regarded as a limit case of the norms ν_p. Indeed, let $\mathbf{x} \in \mathbb{C}^n$ and let $M = \max\{|x_i| \mid 1 \leqslant i \leqslant n\} = |x_{l_1}| = \cdots = |x_{l_k}|$ for some $l_1, \ldots, l_k$, where $1 \leqslant l_1, \ldots, l_k \leqslant n$. Here $x_{l_1}, \ldots, x_{l_k}$ are the components of $\mathbf{x}$ that have the maximal absolute value and $k \geqslant 1$. We can write

$$\lim_{p\to\infty} \nu_p(\mathbf{x}) = \lim_{p\to\infty} M\left(\sum_{i=1}^{n}\left(\frac{|x_i|}{M}\right)^p\right)^{\frac{1}{p}} = \lim_{p\to\infty} M(k)^{\frac{1}{p}} = M,$$

which justifies the notation ν_∞.

We will frequently use the alternative notation $\| \mathbf{x} \|_p$ for $\nu_p(\mathbf{x})$. We refer to the norm ν_2 as the *Euclidean norm*.

Example 6.13 Let $\mathbf{x} = \begin{pmatrix} x_1 \\ x_2 \end{pmatrix} \in \mathbb{C}^2$ be a unit vector in the sense of the Euclidean norm. We have $|x_1|^2 + |x_2|^2 = 1$. Since x_1 and x_2 are complex numbers we can write $x_1 = r_1 e^{i\alpha_1}$ and $x_2 = r_2 e^{i\alpha_2}$, where $r_1^2 + r_2^2 = 1$. Thus, there exists $\theta \in (0, \pi/2)$ such that $r_1 = \cos\theta$ and $r_2 = \sin\theta$, which allows us to write

$$\mathbf{x} = \begin{pmatrix} e^{i\alpha_1}\cos\theta \\ e^{i\alpha_2}\sin\theta \end{pmatrix}.$$

Theorem 6.14 *Each norm $\nu : L \longrightarrow \mathbb{R}_{\geqslant 0}$ on a real or complex linear space $(L, +, \cdot)$ generates a metric on the set L defined by $d_\nu(\boldsymbol{x}, \boldsymbol{y}) = \| \boldsymbol{x} - \boldsymbol{y} \|$ for $\boldsymbol{x}, \boldsymbol{y} \in L$.*

Proof Note that if $d_\nu(\mathbf{x}, \mathbf{y}) = \| \mathbf{x} - \mathbf{y} \| = 0$, it follows that $\mathbf{x} - \mathbf{y} = \mathbf{0}$, so $\mathbf{x} = \mathbf{y}$.

The symmetry of d_ν is obvious and so we need to verify only the triangular axiom. Let $\mathbf{x}, \mathbf{y}, \mathbf{z} \in L$. We have

$$\nu(\mathbf{x} - \mathbf{z}) = \nu(\mathbf{x} - \mathbf{y} + \mathbf{y} - \mathbf{z}) \leqslant \nu(\mathbf{x} - \mathbf{y}) + \nu(\mathbf{y} - \mathbf{z})$$

or, equivalently, $d_\nu(\mathbf{x}, \mathbf{z}) \leqslant d_\nu(\mathbf{x}, \mathbf{y}) + d_\nu(\mathbf{y}, \mathbf{z})$, for every $\mathbf{x}, \mathbf{y}, \mathbf{z} \in L$, which concludes the argument.

We refer to d_ν as the *metric induced by the norm ν on the linear space* $(L, +, \cdot)$.

For $p \geqslant 1$, then d_p denotes the metric d_{ν_p} induced by the norm ν_p on the linear space $(\mathbb{C}^n, +, \cdot)$ known as the *Minkowski metric* on $\mathbb{R}^n$.

The metrics d_1, d_2 and d_∞ defined on $\mathbb{R}^n$ are given by

$$d_1(\mathbf{x}, \mathbf{y}) = \sum_{i=1}^{n} |x_i - y_i|, \tag{6.1}$$

$$d_2(\mathbf{x}, \mathbf{y}) = \sqrt{\sum_{i=1}^{n} |x_i - y_i|^2}, \tag{6.2}$$

$$d_\infty(\mathbf{x}, \mathbf{y}) = \max\{|x_i - y_i| \mid 1 \leqslant i \leqslant n\}, \tag{6.3}$$

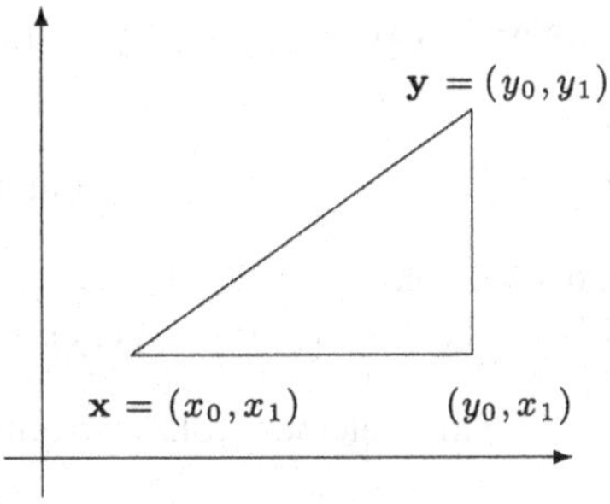

Fig. 6.1 The distances $d_1(\mathbf{x}, \mathbf{y})$ and $d_2(\mathbf{x}, \mathbf{y})$

for $\mathbf{x}, \mathbf{y} \in \mathbb{R}^n$.

These metrics are visualized in Fig. 6.1 for the special case of $\mathbb{R}^2$. If $\mathbf{x} = (x_0, x_1)$ and $\mathbf{y} = (y_0, y_1)$, then $d_1(\mathbf{x}, \mathbf{y})$ is the sum of the lengths of the two legs of the triangle, $d_2(\mathbf{x}, \mathbf{y})$ is the length of the hypotenuse of the right triangle and $d_\infty(\mathbf{x}, \mathbf{y})$ is the largest of the lengths of the legs.

Theorem 6.16 to follow allows us to compare the norms ν_p (and the metrics of the form d_p) that were introduced on $\mathbb{R}^n$. We begin with a preliminary result.

Lemma 6.15 *Let $a_1, \ldots, a_n$ be n positive numbers. If p and q are two positive numbers such that $p \leqslant q$, then $\left(a_1^p + \cdots + a_n^p\right)^{\frac{1}{p}} \geqslant \left(a_1^q + \cdots + a_n^q\right)^{\frac{1}{q}}$.*

Proof Let $f : \mathbb{R}^{>0} \longrightarrow \mathbb{R}$ be the function defined by $f(r) = \left(a_1^r + \cdots + a_n^r\right)^{\frac{1}{r}}$. Since

$$\ln f(r) = \frac{\ln\left(a_1^r + \cdots + a_n^r\right)}{r},$$

it follows that

$$\frac{f'(r)}{f(r)} = -\frac{1}{r^2}\left(a_1^r + \cdots + a_n^r\right) + \frac{1}{r} \cdot \frac{a_1^r \ln a_1 + \cdots + a_n^r \ln a_r}{a_1^r + \cdots + a_n^r}.$$

To prove that $f'(r) < 0$, it suffices to show that

$$\frac{a_1^r \ln a_1 + \cdots + a_n^r \ln a_r}{a_1^r + \cdots + a_n^r} \leqslant \frac{\ln\left(a_1^r + \cdots + a_n^r\right)}{r}.$$

This last inequality is easily seen to be equivalent to

$$\sum_{i=1}^{n} \frac{a_i^r}{a_1^r + \cdots + a_n^r} \ln \frac{a_i^r}{a_1^r + \cdots + a_n^r} \leqslant 0,$$

which holds because

$$\frac{a_i^r}{a_1^r + \cdots + a_n^r} \leqslant 1$$

for $1 \leqslant i \leqslant n$.

Theorem 6.16 *Let p and q be two positive numbers such that $p \leqslant q$. We have $\| \boldsymbol{u} \|_p \geqslant \| \boldsymbol{u} \|_q$ for $\boldsymbol{u} \in \mathbb{C}^n$.*

Proof This statement follows immediately from Lemma 6.15.

Corollary 6.17 *Let p, q be two positive numbers such that $p \leqslant q$. For every $\boldsymbol{x}, \boldsymbol{y} \in \mathbb{C}^n$, we have $d_p(\boldsymbol{x}, \boldsymbol{y}) \geqslant d_q(\boldsymbol{x}, \boldsymbol{y})$.*

Proof This statement follows immediately from Theorem 6.16.

Theorem 6.18 *Let $p \geqslant 1$. We have $\| \boldsymbol{x} \|_\infty \leqslant \| \boldsymbol{x} \|_p \leqslant n \ \| \boldsymbol{x} \|_\infty$ for $\boldsymbol{x} \in \mathbb{C}^n$.*

Proof The first inequality is an immediate consequence of Theorem 6.16. The second inequality follows by observing that

$$\| \mathbf{x} \|_p = \left(\sum_{i=1}^n |x_i|^p \right)^{\frac{1}{p}} \leqslant n \max_{1 \leqslant i \leqslant n} |x_i| = n \ \| \mathbf{x} \|_\infty .$$

Corollary 6.19 *Let p and q be two numbers such that $p, q \geqslant 1$. For $\boldsymbol{x} \in \mathbb{C}^n$ we have:*

$$\frac{1}{n} \| \boldsymbol{x} \|_q \leqslant \| \boldsymbol{x} \|_p \leqslant n \ \| \boldsymbol{x} \|_q .$$

Proof Since $\| \mathbf{x} \|_\infty \leqslant \| \mathbf{x} \|_p$ and $\| x \|_q \leqslant n \ \| \mathbf{x} \|_\infty$, it follows that $\| x \|_q \leqslant n \ \| \mathbf{x} \|_p$. Exchanging the roles of p and q, we have $\| x \|_p \leqslant n \ \| \mathbf{x} \|_q$, so

$$\frac{1}{n} \| \mathbf{x} \|_q \leqslant \| \mathbf{x} \|_p \leqslant n \ \| \mathbf{x} \|_q$$

for every $\mathbf{x} \in \mathbb{C}^n$.

For $p = 1$ and $q = 2$ and $\mathbf{x} \in \mathbb{R}^n$ we have the inequalities

$$\frac{1}{n} \sqrt{\sum_{i=1}^n x_i^2} \leqslant \sum_{i=1}^n |x_i| \leqslant n \sqrt{\sum_{i=1}^n x_i^2}. \tag{6.4}$$

Corollary 6.20 *For every $\boldsymbol{x}, \boldsymbol{y} \in \mathbb{C}^n$ and $p \geqslant 1$, we have $d_\infty(\boldsymbol{x}, \boldsymbol{y}) \leqslant d_p(\boldsymbol{x}, \boldsymbol{y}) \leqslant n d_\infty(\boldsymbol{x}, \boldsymbol{y})$. Further, for $p, q > 1$, there exist $c, d \in \mathbb{R}_{>0}$ such that*

$$c \ d_q(\boldsymbol{x}, \boldsymbol{y}) \leqslant d_p(\boldsymbol{x}, \boldsymbol{y}) \leqslant c \ d_q(\boldsymbol{x}, \boldsymbol{y})$$

for $\boldsymbol{x}, \boldsymbol{y} \in \mathbb{C}^n$.

Proof This follows from Theorem 6.18 and from Corollary 6.20.

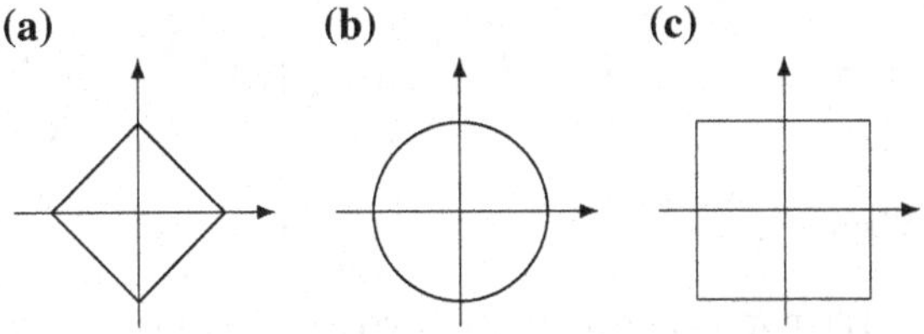

Fig. 6.2 Spheres $B_{d_p}(\mathbf{0}, 1)$ for $p = 1, 2, \infty$

Corollary 6.17 implies that if $p \leqslant q$, then the closed sphere $B_{d_p}(\mathbf{x}, r)$ is included in the closed sphere $B_{d_q}(\mathbf{x}, r)$. For example, we have

$$B_{d_1}(\mathbf{0}, 1) \subseteq B_{d_2}(\mathbf{0}, 1) \subseteq B_{d_\infty}(\mathbf{0}, 1).$$

In Fig. 6.2 (a–c) we represent the closed spheres $B_{d_1}(\mathbf{0}, 1)$, $B_{d_2}(\mathbf{0}, 1)$, and $B_{d_\infty}(\mathbf{0}, 1)$.

An useful consequence of Theorem 6.2 is the following statement:

Theorem 6.21 *Let $x_1, \ldots, x_m$ and $y_1, \ldots, y_m$ be $2m$ nonnegative numbers such that $\sum_{i=1}^m x_i = \sum_{i=1}^m y_i = 1$ and let p and q be two positive numbers such that $\frac{1}{p} + \frac{1}{q} = 1$. We have*

$$\sum_{j=1}^m x_j^{\frac{1}{p}} y_j^{\frac{1}{q}} \leqslant 1.$$

Proof The Hölder inequality applied to $x_1^{\frac{1}{p}}, \ldots, x_m^{\frac{1}{p}}$ and $y_1^{\frac{1}{q}}, \ldots, y_m^{\frac{1}{q}}$ yields the needed inequality

$$\sum_{j=1}^m x_j^{\frac{1}{p}} y_j^{\frac{1}{q}} \leqslant \sum_{j=1}^m x_j \sum_{j=1}^m y_j = 1$$

The linear space $\mathbf{Seq}_\infty(\mathbb{C})$ discussed in Example 5.3 can be equipped with norms similar to the $\{\nu_p \mid p > 1\}$ family.

For $\mathbf{x} = (x_0, x_1, \ldots)$ is a sequence of complex numbers define

$$\nu_p(\mathbf{x}) = \left(\sum_{i=0}^{\infty} |x_i|^p \right)^{\frac{1}{p}}.$$

Theorem 6.22 *The set of sequences $\mathbf{x} \in \mathbf{Seq}_\infty(\mathbb{C})$ such that $\nu_p(\mathbf{x})$ is finite is a normed linear space.*

Proof In Example 5.3 we saw that $\mathbf{Seq}_\infty(\mathbb{C})$ can be organized as a linear space. Let $\mathbf{x}, \mathbf{y} \in \mathbf{Seq}_\infty(\mathbb{C})$ be two sequences such that $\nu_p(\mathbf{x})$ and $\nu_p(\mathbf{y})$ are finite. By Minkowski's inequality, if $p \geqslant 1$ we have

$$\left(\sum_{i=1}^{n}|x_i + y_i|^p\right)^{\frac{1}{p}} \leqslant \left(\sum_{i=1}^{n}(|x_i| + |y_i|)^p\right)^{\frac{1}{p}} \leqslant \left(\sum_{i=1}^{n}|x_i|^p\right)^{\frac{1}{p}} + \left(\sum_{i=1}^{n}|y_i|^p\right)^{\frac{1}{p}}.$$

When n tends to ∞ we have $\nu_p(\mathbf{x}+\mathbf{y}) \leqslant \nu_p(\mathbf{x}) + \nu_p(y)$, so ν_p is a norm.

The normed linear space $(\mathbf{Seq}_\infty(\mathbb{C}), \nu_p)$ is denoted by ℓ_p.

6.4 Inner Products

Inner product spaces are linear spaces equipped with an additional operation that associates to a pair of vectors a scalar called their *inner product*.

Definition 6.23 *Let L be a $\mathbb{C}$-linear space. An* inner product *on L is a function $f : L \times L \longrightarrow \mathbb{C}$ that has the following properties:*

(i) *$f(ax + by, z) = af(x, z) + bf(y, z)$ (linearity in the first argument);*
(ii) *$f(x, y) = \overline{f(y, x)}$ for $y, x \in L$ (conjugate symmetry);*
(iii) *if $x \neq 0$, then $f(x, x)$ is a positive real number (positivity),*
(iv) *$f(x, x) = 0$ if and only if $x = 0$ (definiteness),*

for every $x, y, z \in L$ and $a, b \in \mathbb{C}$.
The pair (L, f) is called an inner product space.

For the second argument of a scalar product we have the property of *conjugate linearity*, that is,

$$f(\mathbf{z}, a\mathbf{x} + b\mathbf{y}) = \bar{a}f(\mathbf{z}, \mathbf{x}) + \bar{b}f(\mathbf{z}, \mathbf{y})$$

for every $\mathbf{x}, \mathbf{y}, \mathbf{z} \in L$ and $a, b \in \mathbb{C}$. Indeed, by the conjugate symmetry property we can write

$$\begin{aligned} f(\mathbf{z}, a\mathbf{x} + b\mathbf{y}) &= \overline{f(a\mathbf{x} + b\mathbf{y}, \mathbf{z})} = \overline{af(\mathbf{x}, \mathbf{z}) + bf(\mathbf{y}, \mathbf{z})} \\ &= \bar{a}\overline{f(\mathbf{x}, \mathbf{z})} + \bar{b}\overline{f(\mathbf{y}, \mathbf{z})} = \bar{a}f(\mathbf{z}, \mathbf{x}) + \bar{b}f(\mathbf{z}, \mathbf{y}). \end{aligned}$$

Observe that conjugate symmetry property on inner products implies that for $\mathbf{x} \in L$, $f(\mathbf{x}, \mathbf{x})$ is a real number because $f(\mathbf{x}, \mathbf{x}) = \overline{f(\mathbf{x}, \mathbf{x})}$.

When L is a real linear space the definition of the inner product becomes simpler because the conjugate of a real number a is a itself. Thus, for real linear spaces, the conjugate symmetry is replaced by the plain symmetry property, $f(\mathbf{x}, \mathbf{y}) = f(\mathbf{y}, \mathbf{x})$, for $\mathbf{x}, \mathbf{y} \in L$ and f is linear in both arguments.

Example 6.24 Let $\mathbb{C}^n$ be the linear space of n-tuples of complex numbers. If $a_1, \ldots, a_n$ are n real, positive numbers, then the function $f : \mathbb{C}^n \times \mathbb{C}^n \longrightarrow \mathbb{C}$ defined by $f(\mathbf{x}, \mathbf{y}) = a_1x_1\bar{y}_1 + a_2x_2\bar{y}_2 + \cdots + a_nx_n\bar{y}_n$ is an inner product on $\mathbb{C}^n$, as the reader can easily verify.

If $a_1 = \cdots = a_n = 1$, we have the *Euclidean inner product*:

$$f(\mathbf{x}, \mathbf{y}) = x_1\bar{y}_1 + \cdots + x_n\bar{y}_n = \mathbf{y}^{\mathsf{H}}\mathbf{x}.$$

For the linear space $\mathbb{R}^n$, the *Euclidean inner product* is

$$f(\mathbf{x}, \mathbf{y}) = x_1 y_1 + \cdots + x_n y_n = \mathbf{y}'\mathbf{x} = \mathbf{x}'\mathbf{y},$$

where $\mathbf{x}, \mathbf{y} \in \mathbb{R}^n$.

To simplify notations we denote an inner product $f(\mathbf{x}, \mathbf{y})$ by $(\mathbf{x}, \mathbf{y})$ when there is no risk of confusion.

A fundamental property of the inner product defined on $\mathbb{C}^n$ in Example 6.24 is the equality

$$(A\mathbf{x}, \mathbf{y}) = (\mathbf{x}, A^{\mathsf{H}}\mathbf{y}), \tag{6.5}$$

which holds for every $A \in \mathbb{C}^{n\times n}$ and $\mathbf{x}, \mathbf{y} \in \mathbb{C}^n$. Indeed, we have

$$\begin{aligned}(A\mathbf{x}, \mathbf{y}) &= \sum_{i=1}^{n}(A\mathbf{x})_i\bar{y}_i = \sum_{i=1}^{n}\sum_{j=1}^{n} a_{ij}x_j\bar{y}_i = \sum_{j=1}^{n} x_j \sum_{i=1}^{n} a_{ij}\bar{y}_i \\ &= \sum_{j=1}^{n} x_j \sum_{i=1}^{n} \overline{\bar{a}_{ij} y_i} = (\mathbf{x}, A^{\mathsf{H}}\mathbf{y}).\end{aligned}$$

More generally we have the following definition.

Definition 6.25 *A matrix $B \in \mathbb{C}^{n\times n}$ is the* adjoint *of a matrix $A \in \mathbb{C}^{n\times n}$ relative to the inner product $(\cdot, \cdot)$ if $(A\mathbf{x}, \mathbf{y}) = (\mathbf{x}, B\mathbf{y})$ for every $\mathbf{x}, \mathbf{y} \in \mathbb{C}^n$.*

A matrix is *self-adjoint* if it equals its own adjoint, that is if $(A\mathbf{x}, \mathbf{y}) = (\mathbf{x}, A\mathbf{y})$ for every $\mathbf{x}, \mathbf{y} \in \mathbb{C}^n$. Thus, a Hermitian matrix is self-adjoint relative to the inner product $(\mathbf{x}, \mathbf{y}) = \mathbf{x}^{\mathsf{H}}\mathbf{y}$ for $\mathbf{x}, \mathbf{y} \in \mathbb{C}^n$. If we use the Euclidean inner product we omit the reference to this product and refer to the adjoint of A relative to this product simply as the adjoint of A.

Example 6.26 An inner product on $\mathbb{C}^{n\times n}$, the linear space of matrices of format $n \times n$, can be defined as $(X, Y) = \mathit{trace}(XY^{\mathsf{H}})$ for $X, Y \in \mathbb{C}^{n\times n}$.

A linear form on $\mathbb{R}^n$ can be expressed using the Euclidean inner product. Let f be a linear form defined on $\mathbb{R}^n$ and let $\mathbf{x} \in \mathbb{R}^n$. If $\mathbf{u}_1, \ldots, \mathbf{u}_n$ is a basis in $\mathbb{R}^n$ and $\mathbf{x} = x_1\mathbf{u}_1 + \cdots + x_n\mathbf{u}_n$, then $f(\mathbf{x}) = x_1 f(\mathbf{u}_1) + \cdots + x_n f(\mathbf{u}_n)$ and $f(\mathbf{x})$ can be written as $f(\mathbf{x}) = (\mathbf{x}, \mathbf{a})$, where $a_i = f(\mathbf{u}_i)$ for $1 \leqslant i \leqslant n$. The vector $\mathbf{a}$ is uniquely determined for a linear form. Indeed, suppose that there exists $\mathbf{b} \in \mathbb{R}^n$ such that $f(\mathbf{x}) = (\mathbf{x}, \mathbf{b})$ for $\mathbf{x} \in \mathbb{R}^n$. Since $(\mathbf{x}, \mathbf{a}) = (\mathbf{x}, \mathbf{b})$ for every $\mathbf{x} \in \mathbb{R}^n$, it follows that $(\mathbf{x}, \mathbf{a} - \mathbf{b}) = 0$. Choosing $\mathbf{x} = \mathbf{a} - \mathbf{b}$, it follows that $\| \mathbf{a} - \mathbf{b} \|_2 = 0$, so $\mathbf{a} = \mathbf{b}$.

Theorem 6.27 *Any inner product on a linear space L generates a norm on that space defined by* $\| x \| = \sqrt{(x, x)}$ *for* $x \in L$.

Proof We need to verify that the norm satisfies the conditions of Definition 6.6. Applying the properties of the inner product we have

$$\begin{aligned} \| \mathbf{x}+\mathbf{y} \|^2 &= (\mathbf{x}+\mathbf{y}, \mathbf{x}+\mathbf{y}) = (\mathbf{x}, \mathbf{x}) + 2(\mathbf{x}, \mathbf{y}) + (\mathbf{y}, \mathbf{y}) \\ &= \| \mathbf{x} \|^2 + 2(\mathbf{x}, \mathbf{y}) + \| \mathbf{y} \|^2 \leqslant \| \mathbf{x} \|^2 + 2 \| \mathbf{x} \| \| \mathbf{y} \| + \| \mathbf{y} \|^2 \\ &= (\| \mathbf{x} \| + \| \mathbf{y} \|)^2. \end{aligned}$$

Because $\| \mathbf{x} \| \geqslant 0$ it follows that $\| \mathbf{x}+\mathbf{y} \| \leqslant \| \mathbf{x} \| + \| \mathbf{y} \|$, which is the subadditivity property.

If $a \in \mathbb{C}$, then $\| a\mathbf{x} \| = \sqrt{(a\mathbf{x}, a\mathbf{x})} = \sqrt{a\bar{a}(\mathbf{x}, \mathbf{x})} = \sqrt{|a|^2(\mathbf{x}, \mathbf{x})} = |a|\sqrt{(\mathbf{x}, \mathbf{x})} = |a| \| \mathbf{x} \|$.

Finally, from the definiteness property of the inner product it follows that $\| \mathbf{x} \| = 0$ if and only if $\mathbf{x} = \mathbf{0}$, which allows us to conclude that $\| \cdot \|$ is indeed a norm.

Observe that if the inner product $(\mathbf{x}, \mathbf{y})$ of the vectors $\mathbf{x} \in \mathbb{C}^n$ and $y \in \mathbb{C}^n$ is defined as in Example 6.24 with $a_1 = \cdots = a_n = 1$, then

$$\| \mathbf{x} \|^2 = (\mathbf{x}, \mathbf{x}) = x_1\bar{x}_1 + x_2\bar{x}_2 + \cdots + x_n\bar{x}_n = \sum_{i=1}^{2} |x_i|^2,$$

which shows that the norm induced by the inner product is precisely $\| \mathbf{x} \|_2$.

Not every norm can be induced by an inner product. A characterization of this type of norms in linear spaces is presented next.

This equality shown in the next theorem is known as the *parallelogram equality*.

Theorem 6.28 *Let L be a real linear space. A norm* $\| \cdot \|$ *is induced by an inner product if and only if*

$$\| x+y \|^2 + \| x-y \|^2 = 2(\| x \|^2 + \| y \|^2),$$

for every $x, y \in L$.

Proof Suppose that the norm is induced by an inner product. In this case we can write for every $\mathbf{x}$ and $\mathbf{y}$:

$$\begin{aligned} (\mathbf{x}+\mathbf{y}, \mathbf{x}+\mathbf{y}) &= (\mathbf{x}, \mathbf{x}) + 2(\mathbf{x}, \mathbf{y}) + (\mathbf{y}, \mathbf{y}), \\ (\mathbf{x}-\mathbf{y}, \mathbf{x}-\mathbf{y}) &= (\mathbf{x}, \mathbf{x}) - 2(\mathbf{x}, \mathbf{y}) + (\mathbf{y}, \mathbf{y}). \end{aligned}$$

Thus,

$$(\mathbf{x}+\mathbf{y}, \mathbf{x}+\mathbf{y}) + (\mathbf{x}-\mathbf{y}, \mathbf{x}-\mathbf{y}) = 2(\mathbf{x}, \mathbf{x}) + 2(\mathbf{y}, \mathbf{y}),$$

which can be written in terms of the norm generated as the inner product as

$$\| \mathbf{x}+\mathbf{y} \|^2 + \| \mathbf{x}-\mathbf{y} \|^2 = 2(\| \mathbf{x} \|^2 + \| \mathbf{y} \|^2).$$

Conversely, suppose that the condition of the theorem is satisfied by the norm $\| \cdot \|$. Consider the function $f : V \times V \longrightarrow \mathbb{R}$ defined by

$$f(\mathbf{x}, \mathbf{y}) = \frac{1}{4}\left(\| \mathbf{x}+\mathbf{y} \|^2 - \| \mathbf{x}-\mathbf{y} \|^2\right), \tag{6.6}$$

for $\mathbf{x}, \mathbf{y} \in L$. The symmetry of f is immediate, that is, $f(\mathbf{x}, \mathbf{y}) = f(\mathbf{y}, \mathbf{x})$ for $\mathbf{x}, \mathbf{y} \in L$.

The definition of f implies

$$f(\mathbf{0}, \mathbf{y}) = \frac{1}{4}\left(\| \mathbf{y} \|^2 - \| -\mathbf{y} \|^2\right) = 0. \tag{6.7}$$

We prove that f is a bilinear form that satisfies the conditions of Definition 6.23.

Starting from the parallelogram equality we can write

$$\| \mathbf{u}+\mathbf{v}+\mathbf{y} \|^2 + \| \mathbf{u}+\mathbf{v}-\mathbf{y} \|^2 = 2(\| \mathbf{u}+\mathbf{v} \|^2 + \| \mathbf{y} \|^2),$$
$$\| \mathbf{u}-\mathbf{v}+\mathbf{y} \|^2 + \| \mathbf{u}-\mathbf{v}-\mathbf{y} \|^2 = 2(\| \mathbf{u}-\mathbf{v} \|^2 + \| \mathbf{y} \|^2).$$

Subtracting these equality yields:

$$\| \mathbf{u}+\mathbf{v}+\mathbf{y} \|^2 + \| \mathbf{u}+\mathbf{v}-\mathbf{y} \|^2 - \| \mathbf{u}-\mathbf{v}+\mathbf{y} \|^2 - \| \mathbf{u}-\mathbf{v}-\mathbf{y} \|^2$$
$$= 2(\| \mathbf{u}+\mathbf{v} \|^2 - \| \mathbf{u}-\mathbf{v} \|^2).$$

This equality can be written as

$$f(\mathbf{u}+\mathbf{y}, \mathbf{v}) + f(\mathbf{u}-\mathbf{y}, \mathbf{v}) = 2f(\mathbf{u}, \mathbf{v}).$$

Choosing $\mathbf{y} = \mathbf{u}$ implies

$$f(2\mathbf{u}, \mathbf{v}) = 2f(\mathbf{u}, \mathbf{v}), \tag{6.8}$$

due to Equality (6.7).

Let $\mathbf{t} = \mathbf{u}+\mathbf{y}$ and $\mathbf{s} = \mathbf{u}-\mathbf{y}$. Since $\mathbf{u} = \frac{1}{2}(\mathbf{t}+\mathbf{s})$ and $\mathbf{y} = \frac{1}{2}(\mathbf{t}-\mathbf{s})$ we have

$$f(\mathbf{t}, \mathbf{v}) + f(\mathbf{s}, \mathbf{v}) = 2f\left(\frac{1}{2}(\mathbf{t}+\mathbf{s}), \mathbf{v}\right) = f(\mathbf{t}+\mathbf{s}, \mathbf{v}),$$

by Equality (6.8).

Next, we show that $f(a\mathbf{x}, \mathbf{y}) = af(\mathbf{x}, \mathbf{y})$ for $a \in \mathbb{R}$ and $\mathbf{x}, \mathbf{y} \in L$. Consider the function $\phi : \mathbb{R} \longrightarrow \mathbb{R}$ defined by $\phi(a) = f(a\mathbf{x}+\mathbf{y})$.

The basic properties of norms imply that

$$\Big| \| a\mathbf{x}+\mathbf{y} \| - \| b\mathbf{x}+\mathbf{y} \| \Big| \leqslant \| (a-b)\mathbf{x} \|$$

for every $a, b \in \mathbb{R}$ and $\mathbf{x}, \mathbf{y} \in L$. Therefore, the function $\phi, \psi : \mathbb{R} \longrightarrow \mathbb{R}$ given by $\phi(a) = \| a\mathbf{x} + \mathbf{y} \|$ and $\psi(a) = \| a\mathbf{x} - \mathbf{y} \|$ for $a \in \mathbb{R}$ are continuous. The continuity of these functions implies that the function f defined by Equality (6.6) is continuous relative to a.

Define the set:

$$S = \{a \in \mathbb{R} \mid f(a\mathbf{x}, \mathbf{y}) = af(\mathbf{x}, \mathbf{y})\}.$$

Clearly, we have $1 \in S$. Further, if $a, b \in S$, then $a + b \in S$ and $a - b \in S$, which implies $\mathbb{Z} \subseteq S$.

If $b \neq 0$ and $b \in S$, then, by substituting $\mathbf{x}$ by $\frac{1}{b}\mathbf{x}$ in the equality $f(b\mathbf{x}, \mathbf{y}) = bf(\mathbf{x}, \mathbf{y})$ we have $f(\mathbf{x}, \mathbf{y}) = bf(\frac{1}{b}\mathbf{x}, \mathbf{y})$, so $\frac{1}{b}f(\mathbf{x}, \mathbf{y}) = f(\frac{1}{b}\mathbf{x}, \mathbf{y})$. Thus, if $a, b \in S$ and $b \neq 0$, we have $f(\frac{a}{b}\mathbf{x}, \mathbf{y}) = \frac{a}{b}f(\mathbf{x}, \mathbf{y})$, so $\mathbb{Q} \subseteq S$. Consequently, $S = \mathbb{R}$. This allows us to conclude that f is linear in its first argument. The symmetry of f implies the linearity in its second argument, so f is bilinear.

Observe that $f(\mathbf{x}, \mathbf{x}) = \| \mathbf{x} \|^2$. The definition of norms implies that $f(\mathbf{x}, \mathbf{x}) = 0$ if and only if $\mathbf{x} = \mathbf{0}$ and if $\mathbf{x} \neq \mathbf{0}$, then $f(\mathbf{x}, \mathbf{x}) > 0$. Thus, f is indeed an inner product and $\| \mathbf{x} \| = \sqrt{f(\mathbf{x}, \mathbf{x})}$.

Theorem 6.29 *Let $\mathbf{x}, \mathbf{y} \in \mathbb{R}^n$ be two vectors such that $x_1 \geqslant x_1 \geqslant \cdots \geqslant x_n$, $y_1 \geqslant y_2 \geqslant \cdots \geqslant y_n$. For every permutation matrix P we have $\mathbf{x}'\mathbf{y} \geqslant \mathbf{x}'(P\mathbf{y})$.*

If $x_1 \geqslant x_1 \geqslant \cdots \geqslant x_n$ and $y_1 \leqslant y_2 \leqslant \cdots \leqslant y_n$, then for every permutation matrix P we have $\mathbf{x}'\mathbf{y} \leqslant \mathbf{x}'(P\mathbf{y})$.

Proof Let ϕ be the permutation that corresponds to the permutation matrix P and suppose that $\phi = \psi_p \ldots \psi_1$, where $p = inv(\phi)$ and $\psi_1, \ldots, \psi_p$ are standard transpositions that correspond to all standard inversions of ϕ.

Let ψ be a standard transposition of $\{1, \ldots, n\}$,

$$\psi : \begin{pmatrix} 1 & \cdots & i & i+1 & \cdots & n \\ 1 & \cdots & i+1 & i & \cdots & n \end{pmatrix}.$$

We have

$$\mathbf{x}'(P\mathbf{y}) = x_1 y_1 + \cdots + x_{i-1} y_{i-1} + x_i y_{i+1} + x_{i+1} y_i + \cdots + x_n y_n,$$

so the inequality $\mathbf{x}'\mathbf{y} \geqslant \mathbf{x}'(P\mathbf{y})$ is equivalent to

$$x_i y_i + x_{i+1} y_{i+1} \geqslant x_i y_{i+1} + x_{i+1} y_i.$$

This, in turn is equivalent to $(x_{i+1} - x_i)(y_{i+1} - y_i) \geqslant 0$, which obviously holds in view of the hypothesis.

As we observed previously, $P_\phi = P_{\psi_1} \cdots P_{\psi_p}$, so

$$\mathbf{x}'\mathbf{y} \geqslant \mathbf{x}'(P_{\psi_p}\mathbf{y}) \geqslant \mathbf{x}'(P_{\psi_{p-1}} P_{\psi_p}\mathbf{y}) \geqslant \cdots \geqslant \mathbf{x}'(P_{\psi_1} \cdots P_{\psi_p}\mathbf{y}) = \mathbf{x}(P\mathbf{y}),$$

which concludes the proof of the first part of the theorem.

To prove the second part of the theorem apply the first part to the vectors $\mathbf{x}$ and $-\mathbf{y}$.

Corollary 6.30 *Let* $\boldsymbol{x}, \boldsymbol{y} \in \mathbb{R}^n$ *be two vectors such that* $x_1 \geqslant x_1 \geqslant \cdots \geqslant x_n$, $y_1 \geqslant y_2 \geqslant \cdots \geqslant y_n$. *For every permutation matrix* P *we have*

$$\| \boldsymbol{x} - \boldsymbol{y} \|_F \leqslant \| \boldsymbol{x} - P\boldsymbol{y} \|_F .$$

If $x_1 \geqslant x_1 \geqslant \cdots \geqslant x_n$ *and* $y_1 \leqslant y_2 \leqslant \cdots \leqslant y_n$, *then for every permutation matrix* P *we have*

$$\| \boldsymbol{x} - \boldsymbol{y} \|_F \geqslant \| \boldsymbol{x} - P\boldsymbol{y} \|_F .$$

Proof Note that

$$\begin{aligned} \| \mathbf{x} - \mathbf{y} \|_F^2 &= \| \mathbf{x} \|_F^2 + \| \mathbf{y} \|_F^2 - 2\mathbf{x}'\mathbf{y}, \\ \| \mathbf{x} - P\mathbf{y} \|_F^2 &= \| \mathbf{x} \|_F^2 + \| P\mathbf{y} \|_F^2 - 2\mathbf{x}'(P\mathbf{y}) \\ &= \| \mathbf{x} \|_F^2 + \| \mathbf{y} \|_F^2 - 2\mathbf{x}'(P\mathbf{y}) \end{aligned}$$

because $\| P\mathbf{y} \|_F^2 = \| \mathbf{y} \|_F^2$. Then, by Theorem 6.29, $\| \mathbf{x} - \mathbf{y} \|_F \leqslant \| \mathbf{x} - P\mathbf{y} \|_F$.

The argument for the second part of the corollary is similar.

6.5 Orthogonality

The Cauchy-Schwarz Inequality implies that $|(\mathbf{x}, \mathbf{y})| \leqslant \| \mathbf{x} \|_2 \| \mathbf{y} \|_2$. Equivalently, this means that

$$-1 \leqslant \frac{(\mathbf{x}, \mathbf{y})}{\| \mathbf{x} \|_2 \| \mathbf{y} \|_2} \leqslant 1.$$

This double inequality allows us to introduce the notion of angle between two vectors $\mathbf{x}, \mathbf{y}$ of a real linear space L.

Definition 6.31 *The* angle *between the vectors* $\boldsymbol{x}$ *and* $\boldsymbol{y}$ *is the number* $\alpha \in [0, \pi]$ *defined by*

$$\cos \alpha = \frac{(\boldsymbol{x}, \boldsymbol{y})}{\| \boldsymbol{x} \|_2 \| \boldsymbol{y} \|_2}.$$

This angle will be denoted by $\angle(\boldsymbol{x}, \boldsymbol{y})$.

Example 6.32 Let $\mathbf{u} = (u_1, u_2) \in \mathbb{R}^2$ be a unit vector. Since $u_1^2 + u_2^2 = 1$, there exists $\alpha \in [0, 2\pi]$ such that $u_1 = \cos \alpha$ and $u_2 = \sin \alpha$. Thus, for any two unit vectors in $\mathbb{R}^2$, $\mathbf{u} = (\cos \alpha, \sin \alpha)$ and $\mathbf{v} = (\cos \beta, \sin \beta)$ we have $(\mathbf{u}, \mathbf{v}) = \cos \alpha \cos \beta + \sin \alpha \sin \beta = \cos(\alpha - \beta)$, where $\alpha, \beta \in [0, 2\pi]$. Consequently, $\angle(\mathbf{u}, \mathbf{v})$ is the angle in the interval $[0, \pi]$ that has the same cosine as $\alpha - \beta$.

Theorem 6.33 (The Cosine Theorem) *Let x and y be two vectors in $\mathbb{R}^n$ equipped with the Euclidean inner product. We have:*

$$\| x - y \|^2 = \| x \|^2 + \| y \|^2 - 2 \| x \| \| y \| \cos\alpha,$$

where $\alpha = \angle(x, y)$.

Proof Since the norm is induced by the inner product we have

$$\begin{aligned} \| \mathbf{x} - \mathbf{y} \|^2 &= (\mathbf{x} - \mathbf{y}, \mathbf{x} - \mathbf{y}) \\ &= (\mathbf{x}, \mathbf{x}) - 2(\mathbf{x}, \mathbf{y}) + (\mathbf{y}, \mathbf{y}) \\ &= \| \mathbf{x} \|^2 - 2 \| \mathbf{x} \| \| \mathbf{y} \| \cos\alpha + \| \mathbf{y} \|^2, \end{aligned}$$

which is the desired equality.

The notion of angle between two vectors allows the introduction of the notion of *orthogonality*.

Definition 6.34 *Let L be an inner product space. Two vectors x and y of L are* orthogonal *if $(x, y) = 0$.*

A pair of orthogonal vectors $(\mathbf{x}, \mathbf{y})$ is denoted by $\mathbf{x} \perp \mathbf{y}$. If $T \subseteq V$, then the set $T^\perp$ is defined by

$$T^\perp = \{\mathbf{v} \in L \mid \mathbf{v} \perp \mathbf{t} \text{ for every } \mathbf{t} \in T\}$$

Note that $T \subseteq U$ implies $U^\perp \subseteq T^\perp$.

If S, T are two subspaces of an inner product space, then S and T are *orthogonal* if $\mathbf{s} \perp \mathbf{t}$ for every $\mathbf{s} \in S$ and every $\mathbf{t} \in T$. This is denoted as $S \perp T$.

Theorem 6.35 *Let L be an inner product space and let T be a subset of L. The set $T^\perp$ is a subspace of L. Furthermore, $\langle T \rangle^\perp = T^\perp$.*

Proof Let $\mathbf{x}$ and $\mathbf{y}$ be two members of T. We have $(\mathbf{x}, \mathbf{t}) = (\mathbf{y}, \mathbf{t}) = 0$ for every $\mathbf{t} \in T$. Therefore, for every $a, b \in F$, by the linearity of the inner product we have $(a\mathbf{x} + b\mathbf{y}, \mathbf{t}) = a(\mathbf{x}, \mathbf{t}) + b(\mathbf{y}, \mathbf{t}) = 0$, for $\mathbf{t} \in T$, so $a\mathbf{x} + b\mathbf{t} \in T^\perp$. Thus, $T^\perp$ is a subspace of L.

By a previous observation, since $T \subseteq \langle T \rangle$, we have $\langle T \rangle^\perp \subseteq T^\perp$. To prove the converse inclusion, let $\mathbf{z} \in T^\perp$.

If $\mathbf{y} \in \langle T \rangle$, $\mathbf{y}$ is a linear combination of vectors of T, $\mathbf{y} = a_1\mathbf{t}_1 + \cdots + a_m\mathbf{t}_m$, so $(\mathbf{y}, \mathbf{z}) = a_1(\mathbf{t}_1, \mathbf{z}) + \cdots + a_m(\mathbf{t}_m, \mathbf{z}) = 0$. Therefore, $\mathbf{z} \perp \mathbf{y}$, which implies $\mathbf{z} \in \langle T \rangle^\perp$. This allows us to conclude that $\langle T \rangle^\perp = T^\perp$.

We refer to $T^\perp$ as the *orthogonal complement* of T.

Note that $T \cap T^\perp \subseteq \{\mathbf{0}\}$. If T is a subspace, then this inclusion becomes an equality, that is, $T \cap T^\perp = \{\mathbf{0}\}$.

Theorem 6.36 *Let T be a subspace of the finite-dimensional linear space L. We have* $\dim(T) + \dim(T^\perp) = \dim(L)$.

Proof This statement follows directly from Theorem 5.19.

If $\mathbf{x}$ and $\mathbf{y}$ are orthogonal, by Theorem 6.33 we have

$$\parallel \mathbf{x} - \mathbf{y} \parallel^2 = \parallel \mathbf{x} \parallel^2 + \parallel \mathbf{y} \parallel^2,$$

which is the well-known *Pythagora's Theorem.*

Theorem 6.37 *Let T be a subspace of $\mathbb{C}^n$. We have $(T^\perp)^\perp = T$.*

Proof Observe that $T \subseteq (T^\perp)^\perp$. Indeed, if $\mathbf{t} \in T$, then $(\mathbf{t}, \mathbf{z}) = 0$ for every $\mathbf{z} \in T^\perp$, so $\mathbf{t} \in (T^\perp)^\perp$.

To prove the reverse inclusion, let $\mathbf{x} \in (T^\perp)^\perp$. Theorem 6.39 implies that we can write $\mathbf{x} = \mathbf{u} + \mathbf{v}$, where $\mathbf{u} \in T$ and $\mathbf{v} \in T^\perp$, so $\mathbf{x} - \mathbf{u} = \mathbf{v} \in T^\perp$.

Since $T \subseteq (T^\perp)^\perp$, we have $\mathbf{u} \in (T^\perp)^\perp$, so $\mathbf{x} - \mathbf{u} \in (T^\perp)^\perp$. Consequently, $\mathbf{x} - \mathbf{u} \in T^\perp \cap (T^\perp)^\perp = \{\mathbf{0}\}$, so $\mathbf{x} = \mathbf{u} \in T$. Thus, $(T^\perp)^\perp \subseteq T$, which concludes the argument.

Corollary 6.38 *Let Z be a subset of $\mathbb{C}^n$. We have $(Z^\perp)^\perp = \langle Z \rangle$.*

Proof Let Z be a subset of $\mathbb{C}^n$. Since $Z \subseteq \langle Z \rangle$ it follows that $\langle Z \rangle^\perp \subseteq Z^\perp$. Let now $\mathbf{y} \in Z^\perp$ and let $\mathbf{z} = a_1\mathbf{z}_1 + \cdots + a_p\mathbf{z}_p \in \langle Z \rangle$, where $\mathbf{z}_1, \ldots, \mathbf{z}_p \in Z$. Since

$$(\mathbf{y}, \mathbf{z}) = a_1(\mathbf{y}, \mathbf{z}_1) + \cdots + a_p(\mathbf{y}, \mathbf{z}_p) = 0,$$

it follows that $\mathbf{y} \in \langle Z \rangle^\perp$. Thus, we have $Z^\perp = \langle Z \rangle^\perp$.

This allows us to write $(Z^\perp)^\perp = (\langle Z \rangle^\perp)^\perp$. Since $\langle Z \rangle$ is a subspace of $\mathbb{C}^n$, by Theorem 6.37, we have $(\langle Z \rangle^\perp)^\perp = \langle Z \rangle$, so $(Z^\perp)^\perp = \langle Z \rangle$.

Theorem 6.39 *Let U be a subspace of $\mathbb{C}^n$. Then, $\mathbb{C}^n = U \boxplus U^\perp$.*

Proof If $U = \{\mathbf{0}\}$, then $U^\perp = \mathbb{C}^n$ and the statement is immediate. Therefore, we can assume that $U \neq \{\mathbf{0}\}$.

In Theorem 6.35 we saw that $U^\perp$ is a subspace of $\mathbb{C}^n$. Thus, we need to show that $\mathbb{C}^n$ is the direct sum of the subspaces U and $U^\perp$. We need to verify only that every $\mathbf{x} \in \mathbb{C}^n$ can be uniquely written as a sum $\mathbf{x} = \mathbf{u} + \mathbf{v}$, where $\mathbf{u} \in U$ and $\mathbf{v} \in U^\perp$.

Let $\mathbf{u}_1, \ldots, \mathbf{u}_m$ be an orthonormal basis of U, that is, a basis such that

$$(\mathbf{u}_i, \mathbf{u}_j) = \begin{cases} 1 & \text{if } i = j, \\ 0 & \text{otherwise,} \end{cases}$$

for $1 \leqslant i, j \geqslant m$. Define $\mathbf{u} = (\mathbf{x}, \mathbf{u}_1)\mathbf{u}_1 + \cdots + (\mathbf{x}, \mathbf{u}_m)\mathbf{u}_m$ and $\mathbf{v} = \mathbf{x} - \mathbf{u}$.

The vector $\mathbf{v}$ is orthogonal to every vector $\mathbf{u}_i$ because

$$(\mathbf{v}, \mathbf{u}_i) = (\mathbf{x} - \mathbf{u}, \mathbf{u}_i) = (\mathbf{x}, \mathbf{u}_i) - (\mathbf{u}, \mathbf{u}_i) = 0.$$

Therefore $\mathbf{v} \in U^{\perp}$ and $\mathbf{x}$ has the necessary decomposition. To prove that the decomposition is unique suppose that $\mathbf{x} = \mathbf{s} + \mathbf{t}$, where $\mathbf{s} \in U$ and $\mathbf{t} \in U_{\perp}$. Since $\mathbf{s} + \mathbf{t} = \mathbf{u} + \mathbf{v}$ we have $\mathbf{s} - \mathbf{u} = \mathbf{v} - \mathbf{t} \in U \cap U^{\perp} = \{\mathbf{0}\}$, which implies $\mathbf{s} = \mathbf{u}$ and $\mathbf{t} = \mathbf{v}$.

Let $W = \{\mathbf{w}_1, \ldots, \mathbf{w}_n\}$ be a basis in the real n-dimensional inner product space L. If $\mathbf{x} = x_1\mathbf{w}_1 + \cdots + x_n\mathbf{w}_n$ and $\mathbf{y} = y_1\mathbf{w}_1 + \cdots + y_n\mathbf{w}_n$, then

$$(\mathbf{x}, \mathbf{y}) = \sum_{i=1}^{n}\sum_{j=1}^{n} x_i y_j (\mathbf{w}_i, \mathbf{w}_j),$$

due to the bilinearity of the inner product.

Let $A = (a_{ij}) \in \mathbb{R}^{n \times n}$ be the matrix defined by $a_{ij} = (\mathbf{w}_i, \mathbf{w}_j)$ for $1 \leqslant i, j \leqslant n$. The symmetry of the inner product implies that the matrix A itself is symmetric. Now, the inner product can be expressed as

$$(\mathbf{x}, \mathbf{y}) = (x_1, \ldots, x_n)A \begin{pmatrix} y_1 \\ \vdots \\ y_n \end{pmatrix}.$$

We refer to A as the *matrix associated* with W.

Definition 6.40 *An* orthogonal set of vectors *in an inner product space* $(V, (\cdot, \cdot))$ *is a subset* W *of* L *such that for every* $\boldsymbol{u}, \boldsymbol{v} \in W$ *we have* $\boldsymbol{u} \perp \boldsymbol{v}$.

If, in addition, $\| \boldsymbol{u} \| = 1$ *for every* $\boldsymbol{u} \in W$, *then we say that* W *is* orthonormal.

Theorem 6.41 *If* W *is a set of non-zero orthogonal vectors, then* W *is linearly independent.*

Proof Let $a_1\mathbf{w}_1 + \cdots + a_n\mathbf{w}_n = \mathbf{0}$ for a linear combination of elements of W. This implies $a_i \| \mathbf{w}_i \|^2 = 0$, so $a_i = 0$ because $\| \mathbf{w}_i \|^2 \neq 0$, and this holds for every i, where $1 \leqslant i \leqslant n$. Thus, W is linearly independent.

Corollary 6.42 *Let* L *be an* n*-dimensional linear space. If* W *is an orthogonal (orthonormal) set and* $|W| = n$, *then* W *is an orthogonal (orthonormal) basis of* L.

Proof This statement is an immediate consequence of Theorem 6.41.

Theorem 6.43 *Let* S *be a subspace of* $\mathbb{C}^n$ *such that* $\dim(S) = k$. *There exists a matrix* $A \in \mathbb{C}^{n \times k}$ *having orthonormal columns such that* $S = \mathsf{Ran}(A)$.

Proof Let $\mathbf{v}_1, \ldots, \mathbf{v}_k$ be an orthonormal basis of S. Define the matrix A as $A = (\mathbf{v}_1, \ldots, \mathbf{v}_k)$. We have $\mathbf{x} \in S$, if and only if $\mathbf{x} = a_1\mathbf{v}_1 + \cdots + a_k\mathbf{v}_k$, which is equivalent to $\mathbf{x} = A\mathbf{a}$. This amounts to $\mathbf{x} \in \mathsf{Ran}(A)$, so $S = \mathsf{Ran}(A)$.

For an orthonormal basis in an n-dimensional space, the associated matrix is the diagonal matrix I_n. In this case, we have

$$(\mathbf{x}, \mathbf{y}) = x_1 y_1 + x_2 y_2 + \cdots + x_n y_n$$

for $\mathbf{x}, \mathbf{y} \in L$.

Observe that if $W = \{\mathbf{w}_1, \ldots, \mathbf{w}_n\}$ is an orthonormal set and $\mathbf{x} \in \langle W \rangle$, which means that $\mathbf{x} = a_1\mathbf{w}_1 + \cdots + a_n\mathbf{w}_n$, then $a_i = (\mathbf{x}, \mathbf{w}_i)$ for $1 \leqslant i \leqslant n$.

Definition 6.44 *Let $W = \{w_1, \ldots, w_n\}$ be an orthonormal set and let $x \in \langle W \rangle$. The equality*

$$x = (x, w_1)w_1 + \cdots + (x, w_n)w_n \tag{6.9}$$

is the Fourier expansion of x *with respect to the orthonormal set W.*

Furthermore, we have *Parseval's equality*:

$$\| \mathbf{x} \|^2 = (\mathbf{x}, \mathbf{x}) = \sum_{i=1}^{n} (\mathbf{x}, \mathbf{w}_i)^2. \tag{6.10}$$

Thus, if $1 \leqslant q \leqslant n$ we have

$$\sum_{i=1}^{q} (\mathbf{x}, \mathbf{w}_i)^2 \leqslant \| \mathbf{x} \|^2 . \tag{6.11}$$

It is easy to see that a square matrix $C \in \mathbb{C}^{n \times n}$ is unitary if and only if its set of columns is an orthonormal set in $\mathbb{C}^n$.

Example 6.45 Let

$$C = \begin{pmatrix} x_1 + ix_2 & y_1 + iy_2 \\ u_1 + iu_2 & v_1 + iv_2 \end{pmatrix} \in \mathbb{C}^2$$

be a unitary matrix, where $x_i, y_i, u_i, v_i \in \mathbb{R}$ for $i \in \{1, 2\}$. We have

$$C^{\mathsf{H}}C = \begin{pmatrix} x_1 - ix_2 & u_1 - iu_2 \\ y_1 - iy_2 & v_1 - iv_2 \end{pmatrix} \begin{pmatrix} x_1 + ix_2 & y_1 + iy_2 \\ u_1 + iu_2 & v_1 + iv_2 \end{pmatrix} = \begin{pmatrix} 1 & 0 \\ 0 & 1 \end{pmatrix}.$$

This equality implies the equalities

$$\begin{aligned} x_1^2 + x_2^2 + u_1^2 + u_2^2 &= 1, \\ y_1^2 + y_2^2 + v_1^2 + v_2^2 &= 1, \\ x_1 y_1 + x_2 y_2 + u_1 v_1 + u_2 v_2 &= 0, \\ x_1 y_2 - x_2 y_1 + u_1 v_2 - u_2 v_1 &= 0. \end{aligned}$$

It is easy to verify that the matrices

$$X = \begin{pmatrix} 0 & 1 \\ 1 & 0 \end{pmatrix}, Y = \begin{pmatrix} 0 & -i \\ i & 0 \end{pmatrix}, Z = \begin{pmatrix} 1 & 0 \\ 0 & -1 \end{pmatrix},$$

known as the *Pauli matrices* are both Hermitian and unitary.

Definition 6.46 *Let $w \in \mathbb{R}^n - \{0\}$ and let $a \in \mathbb{R}$. The* hyperplane *determined by* **w** *and a is the set $H_{w,a} = \{x \in \mathbb{R}^n \mid w'x = a\}$.*

If $\mathbf{x}_0 \in H_{\mathbf{w},a}$, then $\mathbf{w}'\mathbf{x}_0 = a$, so $H_{\mathbf{w},a}$ is also described by the equality

$$H_{\mathbf{w},a} = \{\mathbf{x} \in \mathbb{R}^n \mid \mathbf{w}'(\mathbf{x} - \mathbf{x}_0) = 0\}.$$

Any hyperplane $H_{\mathbf{w},a}$ partitions $\mathbb{R}^n$ into three sets:

$$\begin{aligned} H^{>}_{\mathbf{w},a} &= \{\mathbf{x} \in \mathbb{R}^n \mid \mathbf{w}'\mathbf{x} > a\}, \\ H^{0}_{\mathbf{w},a} &= H_{\mathbf{w},a}, \\ H^{<}_{\mathbf{w},a} &= \{\mathbf{x} \in \mathbb{R}^n \mid \mathbf{w}'\mathbf{x} < a\}. \end{aligned}$$

The sets $H^{>}_{\mathbf{w},a}$ and $H^{<}_{\mathbf{w},a}$ are the *positive* and *negative open* half-spaces determined by $H_{\mathbf{w},a}$, respectively. The sets

$$\begin{aligned} H^{\geqslant}_{\mathbf{w},a} &= \{\mathbf{x} \in \mathbb{R}^n \mid \mathbf{w}'\mathbf{x} \geqslant a\}, \\ H^{\leqslant}_{\mathbf{w},a} &= \{\mathbf{x} \in \mathbb{R}^n \mid \mathbf{w}'\mathbf{x} \leqslant a\}. \end{aligned}$$

are the *positive* and *negative closed* half-spaces determined by $H_{\mathbf{w},a}$, respectively.

If $\mathbf{x}_1, \mathbf{x}_2 \in H_{\mathbf{w},a}$, then $\mathbf{w} \perp \mathbf{x}_1 - \mathbf{x}_2$. This justifies referring to **w** as the *normal to the hyperplane* $H_{\mathbf{w},a}$. Observe that a hyperplane is fully determined by a vector $\mathbf{x}_0 \in H_{\mathbf{w},a}$ and by **w**.

Let $\mathbf{x}_0 \in \mathbb{R}^n$ and let $H_{\mathbf{w},a}$ be a hyperplane. We seek $\mathbf{x} \in H_{\mathbf{w},a}$ such that $\| \mathbf{x} - \mathbf{x}_0 \|_2$ is minimal. Finding **x** amounts to minimizing the function $f(\mathbf{x}) = \| \mathbf{x} - \mathbf{x}_0 \|_2^2 = \sum_{i=1}^{n} (x_i - x_{0i})^2$ subjected to the constraint $\mathbf{w}_1\mathbf{x}_1 + \cdots + w_n\mathbf{x}_n - a = 0$. Using the Lagrangean $\Lambda(\mathbf{x}) = f(\mathbf{x}) + \lambda(\mathbf{w}'\mathbf{x} - a)$ and the multiplier λ we impose the conditions

$$\frac{\partial \Lambda}{\partial x_i} = 0 \text{ for } 1 \leqslant i \leqslant n$$

which amount to

$$\frac{\partial f}{\partial x_i} + \lambda w_i = 0$$

for $1 \leqslant i \leqslant n$. These equalities yield $2(x_i - x_{0i}) + \lambda \mathbf{w}_i = 0$, so we have $x_i = x_{0i} - \frac{1}{2}\lambda \mathbf{w}_i$. Consequently, we have $\mathbf{x} = \mathbf{x}_0 - \frac{1}{2}\lambda \mathbf{w}$. Since $\mathbf{x} \in H_{\mathbf{w},a}$ this implies

$$\mathbf{w}'\mathbf{x} = \mathbf{w}'\mathbf{x}_0 - \frac{1}{2}\lambda \mathbf{w}'\mathbf{w} = a.$$

Thus,

$$\lambda = 2\frac{\mathbf{w}'\mathbf{x}_0 - a}{\mathbf{w}'\mathbf{w}} = 2\frac{\mathbf{w}'\mathbf{x}_0 - a}{\| \mathbf{w} \|_2^2}.$$

We conclude that the closest point in $H_{\mathbf{w},a}$ to $\mathbf{x}_0$ is

$$\mathbf{x} = \mathbf{x}_0 - \frac{\mathbf{w}'\mathbf{x}_0 - a}{\| \mathbf{w} \|_2^2}\mathbf{w}.$$

The smallest distance between $\mathbf{x}_0$ and a point in the hyperplane $H_{\mathbf{w},a}$ is given by

$$\| \mathbf{x}_0 - \mathbf{x} \| = \frac{\mathbf{w}'\mathbf{x}_0 - a}{\| \mathbf{w} \|_2}.$$

If we define the distance $d(H_{\mathbf{w},a}, \mathbf{x}_0)$ between $\mathbf{x}_0$ and $H_{\mathbf{w},a}$ as this smallest distance we have

$$d(H_{\mathbf{w},a}, \mathbf{x}_0) = \frac{\mathbf{w}'\mathbf{x}_0 - a}{\| \mathbf{w} \|_2}. \tag{6.12}$$

6.6 Unitary and Orthogonal Matrices

Lemma 6.47 *Let $A \in \mathbb{C}^{n \times n}$. If $\mathbf{x}^{\mathsf{H}} A \mathbf{x} = 0$ for every $\mathbf{x} \in \mathbb{C}^n$, then $A = O_{n,n}$.*

Proof If $\mathbf{x} = \mathbf{e}_k$, then $\mathbf{x}^{\mathsf{H}} A \mathbf{x} = a_{kk}$ for every k, $1 \leqslant k \leqslant n$, so all diagonal entries of A equal 0. Choose now $\mathbf{x} = \mathbf{e}_k + \mathbf{e}_j$. Then,

$$\begin{aligned}(\mathbf{e}_k + \mathbf{e}_j)^{\mathsf{H}} A (\mathbf{e}_k + \mathbf{e}_j) &= \mathbf{e}_k^{\mathsf{H}} A \mathbf{e}_k + \mathbf{e}_k^{\mathsf{H}} A \mathbf{e}_j + \mathbf{e}_j^{\mathsf{H}} A \mathbf{e}_k + \mathbf{e}_j^{\mathsf{H}} A \mathbf{e}_j \\ &= \mathbf{e}_k^{\mathsf{H}} A \mathbf{e}_j + \mathbf{e}_j^{\mathsf{H}} A \mathbf{e}_k = a_{kj} + a_{jk} = 0.\end{aligned}$$

Similarly, if we choose $\mathbf{x} = \mathbf{e}_k + i\mathbf{e}_j$ we obtain:

$$\begin{aligned}(\mathbf{e}_k + i\mathbf{e}_j)^{\mathsf{H}} A (\mathbf{e}_k + i\mathbf{e}_j) &= (\mathbf{e}_k^{\mathsf{H}} - i\mathbf{e}_j^{\mathsf{H}}) A (\mathbf{e}_k + i\mathbf{e}_j) \\ &= \mathbf{e}_k^{\mathsf{H}} A \mathbf{e}_k - i\mathbf{e}_j^{\mathsf{H}} A \mathbf{e}_k + i\mathbf{e}_k^{\mathsf{H}} A \mathbf{e}_j + \mathbf{e}_j^{\mathsf{H}} A \mathbf{e}_j \\ &= -i a_{jk} + i a_{kj} = 0.\end{aligned}$$

The equalities $a_{kj} + a_{jk} = 0$ and $-a_{jk} + a_{kj} = 0$ imply $a_{kj} = a_{jk} = 0$. Thus, all off-diagonal elements of A are also 0, hence $A = O_{n,n}$.

Theorem 6.48 *A matrix $U \in \mathbb{C}^{n\times n}$ is unitary if and only if $\| U\boldsymbol{x} \|_2 = \| \boldsymbol{x} \|_2$ for every $\boldsymbol{x} \in \mathbb{C}^n$.*

Proof If U is unitary we have $\| U\mathbf{x} \|_2^2 = (U\mathbf{x})^{\mathsf{H}} U\mathbf{x} = \mathbf{x}^{\mathsf{H}} U^{\mathsf{H}} U\mathbf{x} = \| \mathbf{x} \|_2^2$ because $U^{\mathsf{H}} U = I_n$. Thus, $\| U\mathbf{x} \|_2 = \| \mathbf{x} \|_2$.

Conversely, let U be a matrix such that $\| U\mathbf{x} \|_2 = \| \mathbf{x} \|_2$ for every $\mathbf{x} \in \mathbb{C}^n$. This implies $\mathbf{x}^{\mathsf{H}} U^{\mathsf{H}} U\mathbf{x} = \mathbf{x}^{\mathsf{H}}\mathbf{x}$, hence $\mathbf{x}^{\mathsf{H}}(U^{\mathsf{H}} U - I_n)\mathbf{x} = 0$ for $\mathbf{x} \in \mathbb{C}^n$. By Lemma 6.47 this implies $U^{\mathsf{H}} U = I_n$, so U is a unitary matrix.

Corollary 6.49 *The following statements that concern a matrix $U \in \mathbb{C}^{n\times n}$ are equivalent:*

(i) *U is unitary;*
(ii) *$\| U\boldsymbol{x} - U\boldsymbol{y} \|_2 = \| \boldsymbol{x} - \boldsymbol{y} \|_2$ for $\boldsymbol{x}, \boldsymbol{y} \in \mathbb{C}^n$;*
(iii) *$(U\boldsymbol{x}, U\boldsymbol{y}) = (\boldsymbol{x}, \boldsymbol{y})$ for $\boldsymbol{x}, \boldsymbol{y} \in \mathbb{C}^n$.*

Proof This statement is a direct consequence of Theorem 6.48.

The counterpart of unitary matrices in the set of real matrices are introduced next.

Definition 6.50 *A matrix $A \in \mathbb{R}^{n\times n}$ is* orthogonal *if it is unitary.*

In other words, $A \in \mathbb{R}^{n\times n}$ is orthogonal if and only if $A'A = AA' = I_n$. Clearly, A is orthogonal if and only if A' is orthogonal.

Theorem 6.51 *If $A \in \mathbb{R}^{n\times n}$ is an orthogonal matrix, then* $\det(A) \in \{-1, 1\}$.

Proof By Corollary 5.131, $|\det(A)| = 1$. Since $\det(A)$ is a real number, it follows that $\det(A) \in \{-1, 1\}$.

Corollary 6.52 *Let A be a matrix in $\mathbb{R}^{n\times n}$. The following statements are equivalent:*

(i) *A is orthogonal;*
(ii) *A is invertible and $A^{-1} = A'$;*
(iii) *A' is invertible and $(A')^{-1} = A$;*
(iv) *A' is orthogonal.*

Proof The equivalence between these statements is an immediate consequence of definitions.

Corollary 6.52 implies that the columns of a square matrix form an orthonormal set of vectors if and only if the set of rows of the matrix is an orthonormal set.

Theorem 6.48 specialized to orthogonal matrices shows that a matrix A is orthogonal if and only if it preserves the length of vectors.

Theorem 6.53 *Let S be an r-dimensional subspace of $\mathbb{R}^n$ and let $\{\boldsymbol{u}_1, \ldots, \boldsymbol{u}_r\}$, $\{\boldsymbol{v}_1, \ldots, \boldsymbol{v}_r\}$ be two orthonormal bases of the space S. The orthogonal matrices $B = (\boldsymbol{u}_1 \; \cdots \; \boldsymbol{u}_r) \in \mathbb{R}^{n\times r}$ and $C = (\boldsymbol{v}_1 \; \cdots \; \boldsymbol{v}_r) \in \mathbb{R}^{n\times r}$ of the any two such bases are related by the equality $B = CT$, where $T = C'B \in \mathbb{C}^{r\times r}$ is an orthogonal matrix.*

Proof Since the columns of B form a basis for S, each vector $\mathbf{v}_i$ can be written as

$$\mathbf{v}_i = \mathbf{v}_1 t_{1i} + \cdots + \mathbf{v}_r t_{ri}$$

for $1 \leqslant i \leqslant r$. Thus, $B = CT$. Since B and C are orthogonal, we have

$$B^{\mathsf{H}} B = T^{\mathsf{H}} C^{\mathsf{H}} C T = T^{\mathsf{H}} T = I_r,$$

so T is an orthogonal matrix and because it is a square matrix, it is also a unitary matrix. Furthermore, we have $C^{\mathsf{H}} B = C^{\mathsf{H}} C T = T$, which concludes the argument.

Definition 6.54 *A* rotation matrix *is an orthogonal matrix $R \in \mathbb{R}^{n\times n}$ such that* $\det(R) = 1$. *A* reflexion matrix *is an orthogonal matrix $R \in \mathbb{R}^{n\times n}$ such that* $\det(R) = -1$.

Example 6.55 In the two dimensional case, $n = 2$, a rotation is a matrix $R \in \mathbb{R}^{2\times 2}$,

$$R = \begin{pmatrix} r_{11} & r_{12} \\ r_{21} & r_{22} \end{pmatrix}$$

such that

$$\begin{aligned} &r_{11}^2 + r_{21}^2 = 1, r_{12}^2 + r_{22}^2 = 1, \\ &r_{11}r_{12} + r_{21}r_{22} = 0, r_{11}r_{22} - r_{12}r_{21} = 1. \end{aligned}$$

These equalities implies

$$r_{22}(r_{11}r_{12} + r_{21}r_{22}) - r_{12}(r_{11}r_{22} - r_{12}r_{21}) = -r_{12},$$

or

$$r_{21}(r_{22}^2 + r_{12}^2) = -r_{12},$$

so $r_{21} = -r_{12}$.

If $r_{21} = -r_{21} = 0$, the above equalities imply that either $r_{11} = r_{22} = 1$ or $r_{11} = r_{22} = -1$. Otherwise, the equality $r_{11}r_{12} + r_{21}r_{22} = 0$ implies $r_{11} = r_{22}$.

Since $r_{11}^2 \leqslant 1$ it follows that there exits θ such that $r_{11} = \cos\theta$. This shows that R has the form

$$R = \begin{pmatrix} \cos\theta & \sin\theta \\ -\sin\theta & \cos\theta \end{pmatrix}.$$

The vector $\mathbf{y} = R\mathbf{x}$, where

$$\mathbf{y} = \begin{pmatrix} x_1 \cos\theta + x_2 \sin\theta \\ -x_1 \sin\theta + x_2 \cos\theta \end{pmatrix},$$

is obtained from $\mathbf{x}$ by a clockwise rotation through an angle θ. It is easy to see that $\det(R) = 1$, so the term "rotation matrix" is clearly justified for R. To mark the dependency of R on θ we will use the notation

$$R(\theta) = \begin{pmatrix} \cos\theta & \sin\theta \\ -\sin\theta & \cos\theta \end{pmatrix}.$$

A extension of this example is the *Givens matrix* $G(p, q, \theta) \in \mathbb{R}^{n\times n}$ defined as

$$\begin{matrix} \begin{pmatrix} 1 & \cdots & \cdots & \cdots & \cdots & \cdots & 0 \\ \vdots & \cdots & \vdots & \cdots & \vdots & \cdots & \vdots \\ 0 & \cdots & \cos\theta & \cdots & \sin\theta & \cdots & 0 \\ \vdots & \cdots & \vdots & \cdots & \vdots & \cdots & \vdots \\ 0 & \cdots & -\sin\theta & \cdots & \cos\theta & \cdots & 0 \\ \vdots & \cdots & \vdots & \cdots & \vdots & \cdots & \vdots \\ 0 & \cdots & \cdots & \cdots & \cdots & \cdots & 1 \end{pmatrix} & \begin{matrix} \\ \\ p \\ \\ q \\ \\ \\ \end{matrix} \\ \begin{matrix} \quad p & \quad q \end{matrix} & \end{matrix}.$$

We can write

$$G(p, q, \theta) = (\mathbf{e}_1 \ \cdots \ \cos\theta\,\mathbf{e}_p - \sin\theta\,\mathbf{e}_q \ \cdots \ \sin\theta\,\mathbf{e}_p + \cos\theta\ \mathbf{e}_q \ \cdots \ \mathbf{e}_n).$$

It is easy to verify that $G(p, q, \theta)$ is a rotation matrix since it is orthogonal and $\det(G(p, q, \theta)) = 1$.

Since

$$G(p, q, \theta)\begin{pmatrix} v_1 \\ \vdots \\ v_p \\ \vdots \\ v_q \\ \vdots \\ v_n \end{pmatrix} = \begin{pmatrix} v_1 \\ \vdots \\ \cos\theta\, v_p + \sin\theta\, v_q \\ \vdots \\ -\sin\theta\, v_p + \cos\theta\, v_q \\ \vdots \\ v_n \end{pmatrix},$$

the the multiplication of a vector $\mathbf{v}$ be a Givens matrix amounts to a clockwise rotation by θ in the plane of the coordinates (v_p, v_q).

If $v_p \neq 0$, then the rotation described by the Givens matrix can be used to zero the qth component of the resulting vector by taking θ such that $\tan\theta = \frac{v_q}{v_p}$.

It is easy to see that $R(\theta)^{-1} = R(-\theta)$ and that $R(\theta_2)R(\theta_1) = R(\theta_1 + \theta_2)$.

Example 6.56 Let $\mathbf{v} \in \mathbb{C}^n - \{\mathbf{0}_n\}$ be a unit vector. The *Householder matrix* $H_{\mathbf{v}} \in \mathbb{C}^{n\times n}$ is defined by $H_{\mathbf{v}} = I_n - 2\mathbf{v}\mathbf{v}^{\mathsf{H}}$.

The matrix $H_{\mathbf{v}}$ is clearly Hermitian. Moreover, we have

$$HH^{\mathsf{H}} = HH = \left(I_n - 2\mathbf{v}\mathbf{v}^{\mathsf{H}}\right)^2 = I_n - 4\mathbf{v}\mathbf{v}^{\mathsf{H}} + 4\mathbf{v}(\mathbf{v}^{\mathsf{H}}\mathbf{v})\mathbf{v}^{\mathsf{H}} = I_n,$$

so $H_{\mathbf{v}}$ is unitary and involutive. Since $\det(H_{\mathbf{v}}) = -1$, $H_{\mathbf{v}}$ is a reflexion. For a unit vector $\mathbf{v} \in \mathbb{R}^n$, $H_{\mathbf{v}}$ is an orthogonal and involutive matrix.

The vector $H_{\mathbf{v}}\mathbf{w}$ is a reflexion of the vector $\mathbf{w}$ relative to the hyperplane $H_{\mathbf{v},0}$ defined by $\mathbf{v}^{\mathsf{H}}\mathbf{x} = 0$, because the vector

$$\mathbf{w} - H_{\mathbf{v}}\mathbf{w} = (I_n - H_{\mathbf{v}})\mathbf{w} = 2\mathbf{v}(\mathbf{v}^{\mathsf{H}}\mathbf{w})$$

is orthogonal to the hyperplane $\mathbf{v}'\mathbf{x} = 0$. Furthermore, the vector $H_{\mathbf{v}}\mathbf{w}$ has the same norm as $\mathbf{w}$.

Theorem 6.57 *Let $A \in \mathbb{C}^{n\times n}$ and $B \in \mathbb{C}^{k\times k}$ be two matrices. If there exists $U \in \mathbb{C}^{n\times k}$ having an orthonormal set of columns such that $AU = UB$, then there exists $V \in \mathbb{C}^{n\times(n-k)}$ such that $(U\ V) \in \mathbb{C}^{n\times n}$ is a unitary matrix and*

$$(U\ V)^{\mathsf{H}}A(UV) = \begin{pmatrix} B & U^{\mathsf{H}}AV \\ O_{n-k,k} & V^{\mathsf{H}}AV \end{pmatrix}.$$

Proof Since U has an orthonormal set of columns, there exists $V \in \mathbb{C}^{n\times(n-k)}$ such that $(U\ V)$ is a unitary matrix. We have

$$U^{\mathsf{H}}AU = U^{\mathsf{H}}UB = B \text{ and } V^{\mathsf{H}}AU = V^{\mathsf{H}}UB = O_{n-k,k}B = O_{n-k,k}.$$

The equality of the theorem follows immediately.

6.7 The Topology of Normed Linear Spaces

A normed space can be equipped with the topology of a metric space, using the metric defined by the norm. Since this topology is induced by a metric, any normed space is a Hausdorff space. Further, if $\mathbf{v} \in L$, then the collection of subsets $\{C_d(\mathbf{v}, r) \mid r > 0\}$ is a fundamental system of neighborhoods for $\mathbf{v}$.

By specializing the definition of local continuity of functions between metric spaces, a function $f : L \longrightarrow M$ between two normed spaces (L, ν) and (M, ν') is continuous in $\mathbf{x}_0 \in L$ if for every $\epsilon > 0$ there exists $\delta > 0$ such that $\nu(\mathbf{x} - \mathbf{x}_0) < \delta$ implies $\nu'(f(\mathbf{x}) - f(\mathbf{x}_0)) < \epsilon$.

A sequence $(\mathbf{x}_0, \mathbf{x}_1, \ldots)$ of elements of L converges to $\mathbf{x}$ if for every $\epsilon > 0$ there exists $n_\epsilon \in \mathbb{N}$ such that $n \geqslant n_\epsilon$ implies $\nu(\mathbf{x}_n - \mathbf{x}) < \epsilon$.

Theorem 6.58 *In a normed linear space (L, ν), the norm, the multiplication by scalars and the vector addition are continuous functions.*

Proof Let $(\mathbf{x}_0, \mathbf{x}_1, \ldots)$ be a sequence in L such that $\lim_{n\to\infty} \mathbf{x}_n = \mathbf{x}$. By Theorem 6.7 we have $\nu(\mathbf{x}_n - \mathbf{x}) \geqslant |\nu(\mathbf{x}_n) - \nu(\mathbf{x})|$, which implies $\lim_{n\to\infty} \nu(\mathbf{x}_n) = \nu(\mathbf{x})$. Thus, the norm is continuous.

Suppose now that $\lim_{n\to\infty} a_n = a$ and $\lim_{n\to\infty} \mathbf{x}_n = \mathbf{x}$, where (a_n) is a sequence of scalars. Since the sequence $(\mathbf{x}_n)$ is bounded, we have

$$\begin{aligned}\nu(a\mathbf{x} - a_n\mathbf{x}_n) &\leqslant \nu(a\mathbf{x} - a_n\mathbf{x}) + \nu(a_n\mathbf{x} - a_n\mathbf{x}_n)\\ &\leqslant |a - a_n|\nu(\mathbf{x}) + a_n\nu(\mathbf{x} - \mathbf{x}_n),\end{aligned}$$

which implies that $\lim_{n\to\infty} a_n\mathbf{x}_n = a\mathbf{x}$. This shows that the multiplication by scalars is a continuous function.

To prove that the vector addition is continuous, let $(\mathbf{x}_n)$ and $(\mathbf{y}_n)$ be two sequences in L such that $\lim_{n\to\infty} \mathbf{x}_n = \mathbf{x}$ and $\lim_{n\to\infty} \mathbf{y}_n = \mathbf{y}$. Note that

$$\nu\left((\mathbf{x} + \mathbf{y}) - (\mathbf{x}_n + \mathbf{y}_n)\right) \leqslant \nu(\mathbf{x} - \mathbf{x}_n) + \nu(\mathbf{y} - \mathbf{y}_n),$$

which implies that $\lim_{n\to\infty}(\mathbf{x}_n + \mathbf{y}_n) = \mathbf{x} + \mathbf{y}$. Thus, the vector addition is continuous.

Definition 6.59 *Two norms ν and ν' on a linear space L are* equivalent *if they generate the same topology.*

Theorem 6.60 *Let L be a linear space and let $\nu : L \longrightarrow \mathbb{R}_{\geqslant 0}$ and $\nu' : L \longrightarrow \mathbb{R}_{\geqslant 0}$ be two norms on L that generate the topologies $\mathcal{O}$ and $\mathcal{O}'$ on L, respectively.*

The topology $\mathcal{O}'$ is finer than the topology $\mathcal{O}$ (that is, $\mathcal{O} \subseteq \mathcal{O}'$) if and only if there exists $c \in \mathbb{R}_{>0}$ such that $\nu(\mathbf{v}) \leqslant c\nu'(\mathbf{v})$ for every $\mathbf{v} \in L$.

Proof Suppose that $\mathcal{O} \subseteq \mathcal{O}'$. Then, any open sphere $C_\nu(\mathbf{0}, r_0) = \{\mathbf{x} \in L \mid \nu(\mathbf{x}) < r_0\}$ (in $\mathcal{O}$) must be an open set in $\mathcal{O}'$. Therefore, there exists an open sphere $C_{\nu'}(\mathbf{0}, r_1)$ such that $C_{\nu'}(\mathbf{0}, r_1) \subseteq C_\nu(\mathbf{0}, r_0)$. This means that for $r_0 \in \mathbb{R}_{\geqslant 0}$ and $\mathbf{v} \in L$ there exists $r_1 \in \mathbb{R}_{\geqslant 0}$ such that $\nu'(\mathbf{v}) < r_1$ implies $\nu(\mathbf{v}) < r_0$ for every $\mathbf{u} \in L$. In particular, for $r_0 = 1$, there is $k > 0$ such that $\nu'(\mathbf{v}) < k$ implies $\nu(\mathbf{v}) < 1$, which is equivalent to $c\nu'(\mathbf{v}) < 1$ implies $\nu(\mathbf{v}) < 1$, for every $\mathbf{v} \in L$ and $c = \frac{1}{k}$.

For $\mathbf{w} = \frac{1}{c+\epsilon}\frac{\mathbf{v}}{\nu'(\mathbf{v})}$, where $\epsilon > 0$ it follows that

$$c\nu'(\mathbf{w}) = c\nu'\left(\frac{1}{c+\epsilon}\frac{\mathbf{v}}{\nu'(\mathbf{v})}\right) = \frac{c}{c+\epsilon} < 1,$$

so

$$\nu(\mathbf{w}) = \nu\left(\frac{1}{c+\epsilon}\frac{\mathbf{v}}{\nu'(\mathbf{v})}\right) = \frac{1}{c+\epsilon}\frac{\nu(\mathbf{v})}{\nu'(\mathbf{v})} < 1.$$

Since this inequality holds for every $\epsilon > 0$ it follows that $\nu(\mathbf{v}) \leqslant c\nu'(\mathbf{v})$.

Conversely, suppose that there exists $c \in \mathbb{R}_{>0}$ such that $\nu(\mathbf{v}) \leqslant c\nu'(\mathbf{v})$ for every $\mathbf{v} \in L$. Since

$$\left\{\mathbf{v} \mid \nu'(\mathbf{v}) \leqslant \frac{r}{c}\right\} \subseteq \{\mathbf{v} \mid \nu(\mathbf{v}) \leqslant r\},$$

for $\mathbf{v} \in L$ and $r > 0$ it follows that $\mathcal{O} \subseteq \mathcal{O}'$.

Corollary 6.61 *Let ν and ν' be two norms on a linear space L. Then, ν and ν' are equivalent if and only if there exist $a, b \in \mathbb{R}_{>0}$ such that $a\nu(\boldsymbol{v}) \leqslant \nu'(\boldsymbol{v}) \leqslant b\nu(\boldsymbol{v})$ for $\boldsymbol{v} \in V$.*

Proof This statement follows directly from Theorem 6.60.

Example 6.62 By Corollary 6.19 any two norms ν_p and ν_q, on $\mathbb{R}^n$ (with $p, q \geqslant 1$) are equivalent.

Continuous linear operators between normed spaces have a simple characterization.

Theorem 6.63 *Let (L, ν) and (L', ν') be two normed F-linear spaces where F is either $\mathbb{R}$ or $\mathbb{C}$. A linear operator $f : L \longrightarrow L'$ is continuous if and only if there exists $M \in \mathbb{R}_{>0}$ such that $\nu'(f(\boldsymbol{x})) \leqslant M\nu(\boldsymbol{x})$ for every $\boldsymbol{x} \in L$.*

Proof Suppose that $f : L \longrightarrow L'$ satisfies the condition of the theorem. Then,

$$f\left(C_\nu\left(\mathbf{0}, \frac{r}{M}\right)\right) \subseteq C_{\nu'}(\mathbf{0}, r),$$

for every $r > 0$, which means that f is continuous in $\mathbf{0}$ and, therefore, it is continuous everywhere (by Theorem 5.172).

Conversely, suppose that f is continuous. Then, there exists $\delta > 0$ such that $f(C_\nu(\mathbf{0}, \delta)) \subseteq C_{\nu'}(f(x), 1)$, which is equivalent to $\nu(\mathbf{x}) < \delta$ implies $\nu'(f(\mathbf{x})) < 1$. Let $\epsilon > 0$ and let $\mathbf{z} \in L$ be defined by

$$\mathbf{z} = \frac{\delta}{\nu(\mathbf{x}) + \epsilon}\mathbf{x}.$$

We have $\nu(\mathbf{z}) = \frac{\delta\nu(\mathbf{x})}{\nu(\mathbf{x}) + \epsilon} < \delta$. This implies $\nu'(f(\mathbf{z})) < 1$, which is equivalent to

$$\frac{\delta}{\nu(\mathbf{x}) + \epsilon}\nu'(f(\mathbf{x})) < 1$$

because of the linearity of f. This means that

$$\nu'(f(\mathbf{x})) < \frac{\nu(\mathbf{x}) + \epsilon}{\delta}$$

for every $\epsilon > 0$, so $\nu'(f(\mathbf{x})) \leqslant \frac{1}{\delta}\nu(\mathbf{x})$.

Lemma 6.64 *Let* (L, ν) *and* (L', ν') *be two normed* F*-linear spaces where* F *is either* $\mathbb{R}$ *or* $\mathbb{C}$. *A linear function* $f : L \longrightarrow L'$ *is not injective if and only if there exists* $\boldsymbol{u} \in L - \{\boldsymbol{0}\}$ *such that* $f(\boldsymbol{u}) = \boldsymbol{0}$.

Proof It is clear that the condition of the lemma is sufficient for failing injectivity. Conversely, suppose that f is not injective. There exist $\mathbf{t}, \mathbf{v} \in L$ such that $\mathbf{t} \neq \mathbf{v}$ and $f(\mathbf{t}) = f(\mathbf{v})$. The linearity of f implies $f(\mathbf{t} - \mathbf{v}) = 0$. By defining $\mathbf{u} = \mathbf{t} - \mathbf{v} \neq \mathbf{0}$, we have the desired element $\mathbf{u}$.

Theorem 6.65 *Let* (L, ν) *and* (L', ν') *be two normed* F*-linear spaces where* F *is either* $\mathbb{R}$ *or* $\mathbb{C}$. *A linear function* $f : L \longrightarrow L'$ *is injective if and only if there exists* $m \in \mathbb{R}_{>0}$ *such that* $\nu'(f(\boldsymbol{x})) \geqslant m\nu(\boldsymbol{x})$ *for every* $\boldsymbol{x} \in V_1$.

Proof Suppose that f is not injective. By Lemma 6.64, there exists $\mathbf{u} \in L - \{\mathbf{0}\}$ such that $f(\mathbf{u}) = \mathbf{0}$, so $\nu'(f(\mathbf{u})) < m\nu(\mathbf{u})$ for any $m > 0$. Thus, the condition of the theorem is sufficient for injectivity.

Suppose that f is injective, so the inverse function $f^{-1} : L' \longrightarrow L$ is a linear function. By Theorem 6.63, there exists $M > 0$ such that

$$\nu(f^{-1}(\mathbf{y})) \leqslant M\nu'(\mathbf{y})$$

for every $\mathbf{y} \in L'$. Choosing $\mathbf{y} = f(\mathbf{x})$ yields $\nu(\mathbf{x}) \leqslant M\nu'(f(\mathbf{x})$, so $\nu'(f(\mathbf{x})) \geqslant m\nu(\mathbf{x})$ for $m = \frac{1}{M}$, which concludes the argument.

Corollary 6.66 *Every linear function* $f : \mathbb{C}^m \longrightarrow \mathbb{C}^n$ *is continuous.*

Proof Suppose that both $\mathbb{C}^m$ and $\mathbb{C}^n$ are equipped with the norm ν_1. If $\mathbf{x} \in \mathbb{C}^m$ we can write $\mathbf{x} = x_1\mathbf{e}_1 + \cdots + x_m\mathbf{x}_m$ and the linearity of f implies

$$\begin{aligned}\nu_1(f(\mathbf{x})) &= \nu_1\left(f\left(\sum_{i=1}^m x_i\mathbf{e}_i\right)\right) = \nu_1\left(\sum_{i=1}^m x_i f(\mathbf{e}_i)\right)\\ &\leqslant \sum_{i=1}^m |x_i|\nu_1(f(\mathbf{e}_i)) \leqslant M\sum_{i=1}^m |x_i| = M\nu_1(\mathbf{x}),\end{aligned}$$

where $M = \sum_{i=1}^m \nu_1(f(\mathbf{e}_i))$. By Theorem 6.63, the continuity of f follows.

Next, we introduce a norm on the linear space $\mathsf{Hom}(\mathbb{C}^m, \mathbb{C}^n)$ of linear functions from $\mathbb{C}^m$ to $\mathbb{C}^n$. Recall that if $f : \mathbb{C}^m \longrightarrow \mathbb{C}^n$ is a linear function and ν, ν' are norms on $\mathbb{C}^m$ and $\mathbb{C}^n$ respectively, then there exists a non-negative constant m such that $\nu'(f(\mathbf{x})) \leqslant M\nu(\mathbf{x})$ for every $\mathbf{x} \in \mathbb{C}^m$. Define the *norm of* f, $\mu(f)$, as

$$\mu(f) = \inf\{M \in \mathbb{R}_{\geqslant 0} \mid \nu'(f(\mathbf{x})) \leqslant M\nu(\mathbf{x}) \text{ for every } \mathbf{x} \in \mathbb{C}^m\}. \qquad (6.13)$$

Theorem 6.67 *The mapping* μ *defined by Equality* (6.13) *is a norm on the linear space of linear functions* $\mathsf{Hom}(\mathbb{C}^m, \mathbb{C}^n)$.

Proof Let f, g be two functions in $\mathsf{Hom}(\mathbb{C}^m, \mathbb{C}^n)$. There exist M_f and M_g in $\mathbb{R}_{\geqslant 0}$ such that $\nu'(f(\mathbf{x})) \leqslant M_f \nu(\mathbf{x})$ and $\nu'(g(\mathbf{x})) \leqslant M_g \nu(\mathbf{x})$ for every $\mathbf{x} \in V$. Thus,

$$\nu'((f+g)(\mathbf{x})) = \nu'(f(\mathbf{x}) + g(\mathbf{x})) \leqslant \nu'(f(\mathbf{x})) + \nu'(g(\mathbf{x})) \leqslant (M_f + M_g)\nu(\mathbf{x}),$$

so

$$M_f + M_g \in \{M \in \mathbb{R}_{\geqslant 0} \mid \nu'((f+g)(\mathbf{x})) \leqslant M\nu(\mathbf{x}) \text{ for every } \mathbf{x} \in V\}.$$

Therefore,

$$\mu(f+g) \leqslant \mu(f) + \mu(g).$$

We leave to the reader the verification of the remaining norm properties of μ.

Since the norm μ defined by Equality (6.13) depends on the norms ν and ν' we denote it by $N(\nu, \nu')$.

Theorem 6.68 *Let $f : \mathbb{C}^m \longrightarrow \mathbb{C}^n$ and $g : \mathbb{C}^n \longrightarrow \mathbb{C}^p$ and let $\mu = N(\nu, \nu')$, $\mu' = N(\nu', \nu'')$ and $\mu'' = N(\mu, \mu'')$, where ν, ν', ν'' are norms on $\mathbb{C}^m$, $\mathbb{C}^n$, and $\mathbb{C}^p$, respectively. We have $\mu''(gf) \leqslant \mu(f)\mu'(g)$.*

Proof Let $\mathbf{x} \in \mathbb{C}^m$. We have $\nu'(f(\mathbf{x})) \leqslant (\mu(f)+\epsilon')\nu(\mathbf{x})$ for every $\epsilon' > 0$. Similarly, for $\mathbf{y} \in \mathbb{C}^n$, $\nu''(g(\mathbf{y})) \leqslant (\mu'(g)+\epsilon'')\nu'(\mathbf{y})$ for every $\epsilon'' > 0$. These inequalities imply

$$\nu''(g(f(\mathbf{x})) \leqslant (\mu'(g) + \epsilon'')\nu'(f(\mathbf{x})) \leqslant (\mu'(g) + \epsilon'')\mu(f) + \epsilon')\nu(\mathbf{x}).$$

Thus, we have $\mu''(gf) \leqslant (\mu'(g) + \epsilon'')\mu(f) + \epsilon')$, for every ϵ' and ϵ''. This allows us to conclude that $\mu''(fg) \leqslant \mu(f)\mu'(g)$.

Equivalent definitions of the norm $\mu = N(\nu, \nu')$ are given next.

Theorem 6.69 *Let $f : \mathbb{C}^m \longrightarrow \mathbb{C}^n$ and let ν and ν' be two norms defined on $\mathbb{C}^m$ and $\mathbb{C}^n$, respectively. If $\mu = N(\nu, \nu')$, we have*

(i) $\mu(f) = \inf\{M \in \mathbb{R}_{\geqslant 0} \mid \nu'(f(\boldsymbol{x})) \leqslant M\nu(\boldsymbol{x}) \textit{ for every } \boldsymbol{x} \in \mathbb{C}^m\}$;
(ii) $\mu(f) = \sup\{\nu'(f(\boldsymbol{x})) \mid \nu(\boldsymbol{x}) \leqslant 1\}$;
(iii) $\mu(f) = \max\{\nu'(f(\boldsymbol{x})) \mid \nu(\boldsymbol{x}) \leqslant 1\}$;
(iv) $\mu(f) = \max\{\nu'(f(\boldsymbol{x})) \mid \nu(\boldsymbol{x}) = 1\}$;
(v) $\mu(f) = \sup\left\{\frac{\nu'(f(\boldsymbol{x}))}{\nu(\boldsymbol{x})} \mid \boldsymbol{x} \in \mathbb{C}^m - \{\boldsymbol{0}_m\}\right\}$.

Proof The first equality is the definition of $\mu(f)$.

Let ϵ be a positive number. By the definition of the infimum, there exists M such that $\nu'(f(\mathbf{x})) \leqslant M\nu(\mathbf{x})$ for every $\mathbf{x} \in \mathbb{C}^m$ and $M \leqslant \mu(f) + \epsilon$. Thus, for any $\mathbf{x}$ such that $\nu(\mathbf{x}) \leqslant 1$, we have $\nu'(f(\mathbf{x})) \leqslant M \leqslant \mu(f) + \epsilon$. Since this inequality holds for every ϵ, it follows that $\nu'(f(\mathbf{x})) \leqslant \mu(f)$ for every $\mathbf{x} \in \mathbb{C}^m$ with $\nu(\mathbf{x}) \leqslant 1$.

Furthermore, if ϵ' is a positive number, we claim that there exists $\mathbf{x}_0 \in \mathbb{C}^m$ such that $\nu(\mathbf{x}_0) \leqslant 1$ and $\mu(f) - \epsilon' \leqslant \nu'(f(\mathbf{x}_0)) \leqslant \mu(f)$. Suppose that this is not the case. Then, for every $\mathbf{z} \in \mathbb{C}^n$ with $\nu(\mathbf{z}) \leqslant 1$ we have $\nu'(f(\mathbf{z})) \leqslant \mu(f) - \epsilon'$. If $\mathbf{x} \in \mathbb{C}^n$, then $\nu\left(\frac{1}{\nu(\mathbf{x})}\mathbf{x}\right) = 1$, so $\nu'(f(\mathbf{x})) \leqslant (\mu(f) - \epsilon')\nu(\mathbf{x})$, which contradicts the definition of $\mu(f)$. This allows us to conclude that $\mu(f) = \sup\{\nu'(f(\mathbf{x})) \mid \nu(\mathbf{x}) \leqslant 1\}$, which proves the second equality.

Observe that the third equality (where we replaced sup by max) holds because the closed sphere $B(\mathbf{0}, 1)$ is a compact set in $\mathbb{R}^n$. Thus, we have

$$\mu(A) = \max\{\nu'(f(\mathbf{x})) \mid \nu(\mathbf{x}) \leqslant 1\}. \tag{6.14}$$

For the fourth equality, since $\{\mathbf{x} \mid \nu(\mathbf{x}) = 1\} \subseteq \{\mathbf{x} \mid \nu(\mathbf{x}) \leqslant 1\}$, it follows that

$$\max\{\nu'(f(\mathbf{x})) \mid \nu(\mathbf{x}) = 1\} \leqslant \max\{\nu'(f(\mathbf{x})) \mid \nu(\mathbf{x}) \leqslant 1\} = \mu(f).$$

By the third equality there exists a vector $\mathbf{z} \in \mathbb{R}^n - \{\mathbf{0}\}$ such that $\nu(\mathbf{z}) \leqslant 1$ and $\nu'(f(\mathbf{z})) = \mu(f)$. Thus, we have

$$\mu(f) = \nu(\mathbf{z})\nu\left(f\left(\frac{\mathbf{z}}{\nu(\mathbf{z})}\right)\right) \leqslant \nu\left(f\left(\frac{\mathbf{z}}{\nu(\mathbf{z})}\right)\right).$$

Since $\nu\left(\frac{\mathbf{z}}{\nu(\mathbf{z})}\right) = 1$, it follows that $\mu(A) \leqslant \max\{\nu'(f(\mathbf{x})) \mid \nu(\mathbf{x}) = 1\}$. This yields the desired conclusion.

Finally, to prove the last equality observe that for every $\mathbf{x} \in \mathbb{C}^m - \{\mathbf{0}_m\}$, $\frac{1}{\nu(\mathbf{x})}\mathbf{x}$ is a unit vector. Thus, $\nu'(f(\frac{1}{\nu(\mathbf{x})}\mathbf{x}) \leqslant \mu(f)$, by the fourth equality. On the other hand, by the third equality, there exists $\mathbf{x}_0$ such that $\nu(\mathbf{x}_0) = 1$ and $\nu'(f(\mathbf{x}_0)) = \mu(f)$. This concludes the argument.

Definition 6.70 *A normed linear space is* complete *if it is complete as a metric space, that is, if every Cauchy sequence is convergent. A* Banach space *is a complete normed space.*

By Theorem 8.52, if T is a closed subspace of a a Banach space S, then T is complete; the reverse implication is immediate, so a subspace of a Banach space is closed if and only if it is complete.

Let $(\mathbf{x}_0, \mathbf{x}_1, \ldots)$ be a sequence in a normed linear space (L, ν). A *series* in (L, ν) is a sequence $(\mathbf{s}_0, \mathbf{s}_1, \ldots)$ such that $\mathbf{s}_n = \sum_{i=0}^{n} \mathbf{x}_i$ for $n \in \mathbb{N}$. We refer to the elements $\mathbf{x}_i$ as the terms of the series and to $\mathbf{s}_n$ as the nth *partial sum* of the series. The series will be often be denoted by $\sum_{i=0}^{\infty} \mathbf{x}_i$.

If $\lim_{n\to\infty} \mathbf{s}_i = \mathbf{s}$ we say that $\mathbf{s}$ is the *sum* of the series $\sum_{i=0}^{\infty} \mathbf{x}_i$.

A series $\sum_{i=0}^{\infty} \mathbf{x}_i$ in a linear normed space (V, ν) is *absolutely convergent* if the numerical series $\sum_{i=0}^{\infty} \nu(\mathbf{x}_i)$ is convergent.

Theorem 6.71 *In a Banach space (L, ν) every absolutely convergent series is convergent.*

Proof Let $\sum_{i=0}^{\infty} \mathbf{x}_i$ be an absolutely convergent series in (V, ν). We show that the sequence of its partial sums $(\mathbf{s}_0, \mathbf{s}_1, \ldots)$ is a Cauchy sequence. Let $\epsilon > 0$ and let n_0 be a number such that $\sum_{n=n_0}^{\infty} \nu(\mathbf{x}_i) < \epsilon$. Then, if $m > n \geqslant n_0$ we can write

$$\nu(\mathbf{s}_m - \mathbf{s}_n) = \nu\left(\sum_{i=n+1}^{m} \mathbf{x}_i\right) \leqslant \sum_{i=n+1}^{m} \nu(\mathbf{x}_i) < \epsilon,$$

which proves that the sequence of partial sum is a Cauchy sequence, which implies its convergence.

Theorem 6.72 *Let (L, ν) be a linear normed space. If every absolutely convergent series is convergent, then (L, ν) is a Banach space.*

Proof Let $(\mathbf{x}_0, \mathbf{x}_1, \ldots)$ be a Cauchy sequence in L.For every $k \in \mathbb{N}$, there exists p_k such that if $m, n \geqslant p_k$, we have $\nu(\mathbf{x}_m - \mathbf{x}_n) < 2^{-k}$.

Define $\mathbf{y}_n = \mathbf{x}_n - \mathbf{x}_{n-1}$ for $n \geqslant 1$ and $\mathbf{y}_0 = \mathbf{x}_0$. Then $\mathbf{x}_n$ is the partial sum of the sequence $(\mathbf{y}_0, \mathbf{y}_1, \ldots)$, $\mathbf{x}_n = \mathbf{y}_0 + \mathbf{y}_1 + \cdots + \mathbf{y}_n$. If $n \geqslant p_k$, $\nu(\mathbf{y}_{n+1}) = \nu(\mathbf{x}_{n+1} - \mathbf{x}_n) < 2^{-k}$, which implies that the series $\sum_{i=0}^{\infty} \mathbf{y}_n$ is absolutely convergent, so the sequence $(\mathbf{x}_0, \mathbf{x}_1, \ldots)$ is convergent. This shows that (V, ν) is a Banach space.

6.8 Norms for Matrices

In Sect. 5.3 we saw that the set $\mathbb{C}^{m \times n}$ is a linear space. The introduction of norms for matrices can be done by treating matrices as vectors, or by regarding matrices as representations of linear operators.

The mapping vectm introduced next allows us to treat matrices as vectors.

Definition 6.73 *The* $(m \times n)$-vectorization mapping *is the mapping* $\mathsf{vectm} : \mathbb{C}^{m \times n}$ $\mathbb{C}\mathbb{R}^{mn}$ *defined by*

$$\mathsf{vectm}(A) = \begin{pmatrix} a_{11} \\ \vdots \\ a_{m1} \\ \vdots \\ a_{1n} \\ \vdots \\ a_{mn} \end{pmatrix},$$

obtained by reading A column-wise.

Using vector norms on $\mathbb{C}^{mn}$ we can define vectorial norms of matrices.

Definition 6.74 *Let ν be a vector norm on the space $\mathbb{C}^{mn}$. The* vectorial matrix norm $\mu^{(m,n)}$ on $\mathbb{C}^{m \times n}$ *is the mapping* $\mu^{(m,n)} : \mathbb{C}^{m \times n} \longrightarrow \mathbb{R}_{\geqslant 0}$ *defined by* $\mu^{(m,n)}(A) = \nu(\mathsf{vectm}(A))$, *for* $A \in \mathbb{C}^{m \times n}$.

Vectorial norms of matrices are defined without regard for matrix products. The link between linear transformations of finite-dimensional linear spaces and matrices suggest the introduction of an additional condition. Since every matrix $A \in \mathbb{C}^{m \times n}$ corresponds to a linear transformation $h_A : \mathbb{C}^m \longrightarrow \mathbb{C}^n$, if ν and ν' are norms on $\mathbb{C}^m$ and $\mathbb{C}^n$, respectively, it is normal to define a norm on $\mathbb{C}^{m \times n}$ as $\mu(A) = \mu(h_A)$, where $\mu = N(\nu, \nu')$ is a norm on space of linear transformations between $\mathbb{C}^m$ and $\mathbb{C}^n$.

Suppose that ν, ν' and ν'' are vector norms defined on $\mathbb{C}^m$, $\mathbb{C}^n$ and $\mathbb{C}^p$, respectively. In Theorem 6.68 we saw that $\mu''(gf) \leqslant \mu(f)\mu'(g)$, where $\mu = N(\nu, \nu')$, $\mu' = N(\nu', \nu'')$ and $\mu'' = N(\mu, \mu'')$, so $\mu''(AB) \leqslant \mu(A)\mu'(B)$. This leads us the following definition.

Definition 6.75 *A* consistent family of matrix norms *is a family of functions* $\mu^{(m,n)} : \mathbb{C}^{m \times n} \longrightarrow \mathbb{R}_{\geqslant 0}$, *where* $m, n \in \mathbb{P}$ *that satisfies the following conditions:*

(i) $\mu^{(m,n)}$ *is a norm on* $\mathbb{C}^{(m,n)}$ *for* $m, n \in \mathbb{P}$, *and*
(ii) $\mu^{(m,p)}(AB) \leqslant \mu^{(m,n)}(A)\mu^{(n,p)}(B)$ *for every matrix* $A \in \mathbb{C}^{m \times n}$ *and* $B \in \mathbb{C}^{n \times p}$ *(the* submultiplicative property*).*

If the format of the matrix A is clear from context or is irrelevant, then we shall write $\mu(A)$ instead of $\mu^{(m,n)}(A)$.

Example 6.76 Let $P \in \mathbb{C}^{n \times n}$ be an idempotent matrix. If μ is a matrix norm, then either $\mu(P) = 0$ or $\mu(P) \geqslant 1$.

Indeed, since P is idempotent we have $\mu(P) = \mu(P^2)$. By the submultiplicative property, $\mu(P^2) \leqslant (\mu(P))^2$, so $\mu(P) \leqslant (\mu(P))^2$. Consequently, if $\mu(P) \neq 0$, then $\mu(P) \geqslant 1$.

Some vectorial matrix norms turn out to be actual matrix norms; others fail to be matrix norms. This point is illustrated by the next two examples.

Example 6.77 Consider the vectorial matrix norm μ_1 induced by the vector norm ν_1. We have $\mu_1(A) = \sum_{i=1}^{n} \sum_{j=1}^{m} |a_{ij}|$ for $A \in \mathbb{C}^{m \times n}$. Actually, this is a matrix norm.

Indeed, for $A \in \mathbb{C}^{m \times p}$ and $B \in \mathbb{C}^{p \times n}$ we have:

$$\begin{aligned}
\mu_1(AB) &= \sum_{i=1}^{m} \sum_{j=1}^{n} \left| \sum_{k=1}^{p} a_{ik} b_{kj} \right| \leqslant \sum_{i=1}^{m} \sum_{j=1}^{n} \sum_{k=1}^{p} |a_{ik} b_{kj}| \\
&\leqslant \sum_{i=1}^{m} \sum_{j=1}^{n} \sum_{k'=1}^{p} \sum_{k''=1}^{p} |a_{ik'}||b_{k''j}| \\
&\quad \text{(because we added extra non-negative terms to the sums)} \\
&= \left(\sum_{i=1}^{m} \sum_{k'=1}^{p} |a_{ik'}| \right) \cdot \left(\sum_{j=1}^{n} \sum_{k''=1}^{p} |b_{k''j}| \right) = \mu_1(A)\mu_1(B).
\end{aligned}$$

We denote this vectorial matrix norm by the same notation as the corresponding vector norm, that is, by $\| A \|_1$.

The vectorial matrix norm μ_2 induced by the vector norm ν_2 is also a matrix norm. Indeed, using the same notations we have:

$$(\mu_2(AB))^2 = \sum_{i=1}^{m}\sum_{j=1}^{n}\left|\sum_{k=1}^{p} a_{ik}b_{kj}\right|^2 \leqslant \sum_{i=1}^{m}\sum_{j=1}^{n}\left(\sum_{k=1}^{p}|a_{ik}|^2\right)\left(\sum_{l=1}^{p}|b_{lj}|^2\right)$$

$$\text{(by Cauchy-Schwarz inequality)}$$

$$\leqslant (\mu_2(A))^2(\mu_2(B))^2.$$

The vectorial norm of $A \in \mathbb{C}^{m\times n}$ $\mu_2(A) = \left(\sum_{i=1}^{n}\sum_{j=1}^{m}|a_{ij}|^2\right)^{\frac{1}{2}}$, denoted also by $\| A \|_F$, is known as the *Frobenius norm*.

It is easy to see that for real matrices we have

$$\| A \|_F^2 = trace(AA') = trace(A'A) \tag{6.15}$$

and for complex matrices the corresponding equality is

$$\| A \|_F^2 = trace(AA^{\mathsf{H}}) = trace(A^{\mathsf{H}}A). \tag{6.16}$$

Note that $\| A^{\mathsf{H}} \|_F^2 = \| A \|_F^2$ for every A.

Example 6.78 The vectorial norm μ_∞ induced by the vector norm ν_∞ is denoted by $\| A \|_\infty$ and is given by $\| A \|_\infty = \max_{i,j}|a_{ij}|$ for $A \in \mathbb{C}^{n\times n}$. This is *not* a matrix norm. Indeed, let a, b be two positive numbers and consider the matrices

$$A = \begin{pmatrix} a & a \\ a & a \end{pmatrix} \text{ and } B = \begin{pmatrix} b & b \\ b & b \end{pmatrix}.$$

We have $\| A \|_\infty = a$ and $\| B \|_\infty = b$. However, since

$$AB = \begin{pmatrix} 2ab & 2ab \\ 2ab & 2ab \end{pmatrix},$$

we have $\| AB \|_\infty = 2ab$ and the submultiplicative property of matrix norms is violated.

By regarding matrices as transformations between linear spaces, we can define norms for matrices that turn out to be matrix norms.

Definition 6.79 *Let ν_m be a norm on $\mathbb{C}^m$ and ν_n be a norm on $\mathbb{C}^n$ and let $A \in \mathbb{C}^{n\times m}$ be a matrix. The* operator norm *of A is the number $\mu^{(n,m)}(A) = \mu^{(n,m)}(h_A)$, where $\mu^{(n,m)} = N(\nu_m, \nu_n)$.*

Theorem 6.80 *Let $\{\nu_n \mid n \geqslant 1\}$ be a family of vector norms, where ν_n is a vector norm on $\mathbb{C}^n$. The family of norms $\{\mu^{(n,m)} \mid n, m \geqslant 1\}$ is consistent.*

Proof It is easy to see that the family of norms $\{\mu^{(n,m)} \mid n, m \geqslant 1\}$ satisfies the first three conditions of Definition 6.75. For the fourth condition of Definition 6.75 and $A \in \mathbb{C}^{n\times m}$ and $B \in \mathbb{C}^{m\times p}$, we can write:

$$\begin{aligned}
\mu^{(n,p)}(AB) &= \sup\{\nu_n((AB)\mathbf{x}) \mid \nu_p(\mathbf{x}) \leqslant 1\} \\
&= \sup\{\nu_n(A(B\mathbf{x})) \mid \nu_p(\mathbf{x}) \leqslant 1\} \\
&= \sup\left\{\nu_n\left(A\frac{B\mathbf{x}}{\nu_m(B\mathbf{x})}\right)\nu_m(B\mathbf{x}) \middle| \nu_p(\mathbf{x}) \leqslant 1\right\} \\
&\leqslant \mu^{(n,m)}(A)\sup\{\nu_m(B\mathbf{x}) \middle| \nu_p(\mathbf{x}) \leqslant 1\} \\
&\qquad \left(\text{because } \nu_m\left(\frac{B\mathbf{x}}{\nu(B\mathbf{x})}\right) = 1\right) \\
&= \mu^{(n,m)}(A)\mu^{(m,p)}(B).
\end{aligned}$$

Theorem 6.69 implies the following equivalent definitions of $\mu^{(n,m)}(A)$.

Theorem 6.81 *Let ν_n be a norm on $\mathbb{C}^n$ for $n \geqslant 1$. The following equalities hold for $\mu^{(n,m)}(A)$, where $A \in \mathbb{C}^{(n,m)}$.*

$$\begin{aligned}
\mu^{(n,m)}(A) &= \inf\{M \in \mathbb{R}_{\geqslant 0} \mid \nu_n(A\boldsymbol{x}) \leqslant M\nu_m(\boldsymbol{x}) \text{ for every } \boldsymbol{x} \in \mathbb{C}^m\} \\
&= \sup\{\nu_n(A\boldsymbol{x}) \mid \nu_m(\boldsymbol{x}) \leqslant 1\} = \max\{\nu_n(A\boldsymbol{x}) \mid \nu_m(\boldsymbol{x}) \leqslant 1\} \\
&= \max\{\nu'(f(\boldsymbol{x})) \mid \nu(\boldsymbol{x}) = 1\} = \sup\left\{\frac{\nu'(f(\boldsymbol{x}))}{\nu(\boldsymbol{x})} \mid \boldsymbol{x} \in \mathbb{C}^m - \{\boldsymbol{0}_m\}\right\}.
\end{aligned}$$

Proof The theorem is simply a reformulation of Theorem 6.69.

Corollary 6.82 *Let μ be the matrix norm on $\mathbb{C}^{n\times n}$ induced by the vector norm ν. We have $\nu(A\boldsymbol{u}) \leqslant \mu(A)\nu(\boldsymbol{u})$ for every $\boldsymbol{u} \in \mathbb{C}^n$.*

Proof The inequality is obviously satisfied when $\mathbf{u} = \mathbf{0}_n$. Therefore, we may assume that $\mathbf{u} \neq \mathbf{0}_n$ and let $\mathbf{x} = \frac{1}{\nu(\mathbf{u})}\mathbf{u}$. Clearly, $\nu(\mathbf{x}) = 1$ and Equality (6.14) implies that

$$\nu\left(A\frac{1}{\nu(\mathbf{u})}\mathbf{u}\right) \leqslant \mu(A)$$

for every $\mathbf{u} \in \mathbb{C}^n - \{\mathbf{0}_n\}$. This implies immediately the desired inequality.

If μ is a matrix norm induced by a vector norm on $\mathbb{R}^n$, then $\mu(I_n) = \sup\{\nu(I_n\mathbf{x}) \mid \nu(\mathbf{x}) \leqslant 1\} = 1$.

The operator matrix norm induced by the vector norm $\|\cdot\|_p$ is denoted by $|||\cdot|||_p$.

Example 6.83 To compute $|||A|||_1 = \sup\{\| A\mathbf{x} \|_1 \mid \| \mathbf{x} \|_1 \leqslant 1\}$, where $A \in \mathbb{R}^{n\times n}$, suppose that the columns of A are the vectors $\mathbf{a}_1, \ldots, \mathbf{a}_n$, that is

$$\mathbf{a}_j = \begin{pmatrix} a_{1j} \\ a_{2j} \\ \vdots \\ a_{nj} \end{pmatrix}.$$

Let $\mathbf{x} \in \mathbb{R}^n$ be a vector whose components are $x_1, \ldots, x_n$. Then, $\mathbf{Ax} = x_1\mathbf{a}_1 + \cdots + x_n\mathbf{a}_n$, so

$$\begin{aligned} \| A\mathbf{x} \|_1 = \| x_1\mathbf{a}_1 + \cdots + x_n\mathbf{a}_n \|_1 &\leqslant \sum_{j=1}^{n} |x_j| \, \| \mathbf{a}_j \|_1 \\ &\leqslant \max_j \| \mathbf{a}_j \|_1 \sum_{j=1}^{n} |x_j| = \max_j \| \mathbf{a}_j \|_1 \cdot \| \mathbf{x} \|_1 . \end{aligned}$$

Thus, $|||A|||_1 \leqslant \max_j \| \mathbf{a}_j \|_1$.

Let $\mathbf{e}_j$ be the vector whose components are 0 with the exception of its jth component that is equal to 1. Clearly, we have $\| \mathbf{e}_j \|_1 = 1$ and $\mathbf{a}_j = A\mathbf{e}_j$. This, in turn implies $\| \mathbf{a}_j \|_1 = \| A\mathbf{e}_j \|_1 \leqslant |||A|||_1$ for $1 \leqslant j \leqslant n$. Therefore, $\max_j \| \mathbf{a}_j \|_1 \leqslant |||A|||_1$, so

$$|||A|||_1 = \max_j \| \mathbf{a}_j \|_1 = \max_j \sum_{i=1}^{n} |a_{ij}|.$$

In other words, $|||A|||_1$ equals the maximum column sum of the absolute values.

Example 6.84 Let $A \in \mathbb{R}^{n \times n}$. We have

$$\begin{aligned} \| A\mathbf{x} \|_\infty = \max_{1 \leqslant i \leqslant n} \left| \sum_{j=1}^{n} a_{ij} x_j \right| &\leqslant \max_{1 \leqslant i \leqslant n} \sum_{j=1}^{n} |a_{ij} x_j| \\ &\leqslant \max_{1 \leqslant i \leqslant n} \| \mathbf{x} \|_\infty \sum_{j=1}^{n} |a_{ij}|. \end{aligned}$$

Consequently, if $\| \mathbf{x} \|_\infty \leqslant 1$ we have $\| A\mathbf{x} \|_\infty \leqslant \max_{1 \leqslant i \leqslant n} \sum_{j=1}^{n} |a_{ij}|$. Thus, $|||A|||_\infty \leqslant \max_{1 \leqslant i \leqslant n} \sum_{j=1}^{n} |a_{ij}|$.

The converse inequality is immediate if $A = O_{n,n}$. Therefore, assume that $A \neq O_{n \times n}$, and let $(a_{p1}, \ldots, a_{pn})$ be any row of A that has at least one element distinct from 0. Define the vector $\mathbf{z} \in \mathbb{R}^n$ by

$$z_j = \begin{cases} \frac{|a_{pj}|}{a_{pj}} & \text{if } a_{pj} \neq 0, \\ 1 & \text{otherwise,} \end{cases}$$

for $1 \leqslant j \leqslant n$. It is clear that $z_j \in \{-1, 1\}$ for every j, $1 \leqslant j \leqslant n$ and, therefore, $\| \mathbf{z} \|_\infty = 1$. Moreover, we have $|a_{pj}| = a_{pj} z_j$ for $1 \leqslant j \leqslant n$. Therefore, we can write:

$$\begin{aligned}\sum_{j=1}^{n} |a_{pj}| = \sum_{j=1}^{n} a_{pj} z_j \leqslant \left| \sum_{j=1}^{n} a_{pj} z_j \right| \leqslant \max_{1 \leqslant i \leqslant n} \left| \sum_{j=1}^{n} a_{ij} z_j \right| \\ = \| A\mathbf{z} \|_\infty \leqslant \max\{\| A\mathbf{x} \|_\infty \mid \| \mathbf{x} \|_\infty \leqslant 1\} = |||A|||_\infty.\end{aligned}$$

Since this holds for every row of A, it follows that $\max_{1 \leqslant i \leqslant n} \sum_{j=1}^{n} |a_{ij}| \leqslant |||A|||_\infty$, which proves that $|||A|||_\infty = \max_{1 \leqslant i \leqslant n} \sum_{j=1}^{n} |a_{ij}|$. In other words, $|||A|||_\infty$ equals the maximum row sum of the absolute values.

Example 6.85 Let $D = \mathsf{diag}(d_1, \ldots, d_n) \in \mathbb{C}^{n \times n}$ be a diagonal matrix. If $\mathbf{x} \in \mathbb{C}^n$ we have

$$D\mathbf{x} = \begin{pmatrix} d_1 x_1 \\ \vdots \\ d_n x_n \end{pmatrix},$$

so

$$\begin{aligned}|||D|||_2 &= \max\{\| D\mathbf{x} \|_2 \mid \| \mathbf{x} \|_2 = 1\} \\ &= \max\{\sqrt{(d_1 x_1)^2 + \cdots + (d_n x_n)^2} \mid x_1^2 + \cdots + x_n^2 = 1\} \\ &= \max\{|d_i| \mid 1 \leqslant 1 \leqslant n\}.\end{aligned}$$

Norms that are invariant with respect to multiplication by unitary matrices are known as *unitarily invariant norms*.

Theorem 6.86 *Let $U \in \mathbb{C}^{n \times n}$ be a unitary matrix. The following statements hold:*

(i) $\| Ux \|_2 = \| x \|_2$ *for every* $\mathbf{x} \in \mathbb{C}^n$;
(ii) $|||UA|||_2 = |||A|||_2$ *for every* $A \in \mathbb{C}^{n \times p}$;
(iii) $\| UA \|_F = \| A \|_F$ *for every* $A \in \mathbb{C}^{n \times p}$.

Proof For the first part of the theorem note that $\| U\mathbf{x} \|_2^2 = (U\mathbf{x})^{\mathsf{H}} U\mathbf{x} = \mathbf{x}^{\mathsf{H}} U^{\mathsf{H}} U\mathbf{x} = \mathbf{x}^{\mathsf{H}}\mathbf{x} = \| \mathbf{x} \|_2^2$, because $U^{\mathsf{H}} A = I_n$.

The proof of the second part is shown next:

$$\begin{aligned}|||UA|||_2 &= \max\{\| (UA)\mathbf{x} \|_2 \mid \| \mathbf{x} \|_2 = 1\} = \max\{\| U(A\mathbf{x}) \|_2 \mid \| \mathbf{x} \|_2 = 1\} \\ &= \max\{\| A\mathbf{x} \|_2 \mid \| \mathbf{x} \|_2 = 1\} = |||A|||_2.\end{aligned}$$

For the Frobenius norm note that

$$\| UA \|_F = \sqrt{trace((UA)^{\mathsf{H}} UA)} = \sqrt{trace(A^{\mathsf{H}} U^{\mathsf{H}} UA)} = \sqrt{trace(A^{\mathsf{H}} A)} = \| A \|_F,$$

by Equality (6.15).

Corollary 6.87 *If $U \in \mathbb{C}^{n\times n}$ is a unitary matrix, then $|||U|||_2 = 1$.*

Proof Since $|||U|||_2 = \sup\{\| U\mathbf{x} \|_2 \mid \| \mathbf{x} \|_2 \leqslant 1\}$, by Part (ii) of Theorem 6.86,

$$|||U|||_2 = \sup\{\| \mathbf{x} \|_2 \mid \| \mathbf{x} \|_2 \leqslant 1\} = 1.$$

Corollary 6.88 *Let $A, U \in \mathbb{C}^{n\times n}$. If U is an unitary matrix, then*

$$\| U^{\mathsf{H}}AU \|_F = \| A \|_F .$$

Proof Since U is a unitary matrix, so is U^{H}. By Part (iii) of Theorem 6.86,

$$\| U^{\mathsf{H}}AU \|_F = \| AU \|_F = \| U^{\mathsf{H}}A^{\mathsf{H}} \|_F^2 = \| A^{\mathsf{H}} \|_F^2 = \| A \|_F^2,$$

which proves the corollary.

Example 6.89 Let $S = \{\mathbf{x} \in \mathbb{R}^n \mid \| \mathbf{x} \|_2 = 1\}$ be the surface of the sphere in $\mathbb{R}^n$. The image of S under the linear transformation h_U that corresponds to the unitary matrix U is S itself. Indeed, by Theorem 6.86, $\| h_U(\mathbf{x}) \|_2 = \| \mathbf{x} \|_2 = 1$, so $h_U(\mathbf{x}) \in S$ for every $\mathbf{x} \in S$. Also, note that h_U restricted to S is a bijection because $h_{U^{\mathsf{H}}}(h_U(\mathbf{x})) = \mathbf{x}$ for every $\mathbf{x} \in \mathbb{R}^n$.

Theorem 6.90 *Let $A \in \mathbb{R}^{n\times n}$. We have $|||A|||_2 \leqslant \| A \|_F$.*

Proof Let $\mathbf{x} \in \mathbb{R}^n$. We have

$$A\mathbf{x} = \begin{pmatrix} \mathbf{r}_1\mathbf{x} \\ \vdots \\ \mathbf{r}_n\mathbf{x} \end{pmatrix},$$

where $\mathbf{r}_1, \ldots, \mathbf{r}_n$ are the rows of the matrix A. Thus,

$$\frac{\| A\mathbf{x} \|_2}{\| \mathbf{x} \|_2} = \frac{\sqrt{\sum_{i=1}^n (\mathbf{r}_i\mathbf{x})^2}}{\| \mathbf{x} \|_2}.$$

By Cauchy-Schwarz inequality we have $(\mathbf{r}_i\mathbf{x})^2 \leqslant \| \mathbf{r}_i \|_2^2 \| \mathbf{x} \|_2^2$, so

$$\frac{\| A\mathbf{x} \|_2}{\| \mathbf{x} \|_2} \leqslant \sqrt{\sum_{i=1}^n \| \mathbf{r}_i \|_2^2} = \| A \|_F .$$

This implies $|||A|||_2 \leqslant \| A \|_F$.

6.9 Projection on Subspaces

If U, W are two complementary subspaces of $\mathbb{C}^n$, then there exist idempotent endomorphisms g and h of $\mathbb{C}^n$ such that $W = \mathsf{Ker}(g)$, $U = \mathsf{Img}(g)$ and $U = \mathsf{Ker}(h)$ and $W = \mathsf{Img}(h)$. The vector $g(\mathbf{x})$ is the *oblique projection of* $\mathbf{x}$ *on* U *along the subspace* W and $h(\mathbf{x})$ is the *oblique projection* of $\mathbf{x}$ on W along the subspace U.

If g and h are represented by the matrices B_U and B_W, respectively, it follows that these matrices are idempotent, $g(\mathbf{x}) = B_U\mathbf{x} \in U$, and $h(\mathbf{x}) = B_W\mathbf{x} \in W$ for $\mathbf{x} \in \mathbb{C}^n$. Also, $B_U B_W = B_W B_U = O_{n,n}$.

Let U and W be two complementary subspaces of $\mathbb{C}^n$, $\{\mathbf{u}_1, \ldots, \mathbf{u}_p\}$ be a basis for U, and let $\{\mathbf{w}_1, \ldots, \mathbf{w}_q\}$ be a basis for W, where $p + q = n$. Clearly, $\{\mathbf{u}_1, \ldots, \mathbf{u}_p, \mathbf{w}_1, \ldots, \mathbf{w}_q\}$ is a basis for $\mathbb{C}^n$ and every $\mathbf{x} \in \mathbb{C}^n$ can be written as

$$\mathbf{x} = x_1\mathbf{u}_1 + \cdots + x_p\mathbf{u}_p + x_{p+1}\mathbf{w}_1 + \cdots + x_{p+q}\mathbf{w}_q.$$

Let $B \in \mathbb{C}^{n\times n}$ be the matrix $B = (\mathbf{u}_1 \ \cdots \ \mathbf{u}_p \ \mathbf{w}_1 \ \cdots \ \mathbf{w}_q)$, which is clearly invertible. Note that

$$B_U\mathbf{u}_i = \mathbf{u}_i,\ B_U\mathbf{w}_j = \mathbf{0}_n,\ B_W\mathbf{u}_i = \mathbf{0}_n,\ B_W\mathbf{w}_j = \mathbf{w}_j,$$

for $1 \leqslant i \leqslant p$ and $1 \leqslant j \leqslant q$. Therefore, we have

$$B_U B = (\mathbf{u}_1 \ \cdots \ \mathbf{u}_p \ \mathbf{0}_n \ \cdots \ \mathbf{0}_n) = B\begin{pmatrix} I_p & O_{p,n-p} \\ O_{n-p,p} & O_{n-p,n-p} \end{pmatrix},$$

so

$$B_U = B\begin{pmatrix} I_p & O_{p,n-p} \\ O_{n-p,p} & O_{n-p,n-p} \end{pmatrix} B^{-1}.$$

Similarly, we can show that

$$B_W = B\begin{pmatrix} O_{n-q,n-q} & O_{n-q,q} \\ O_{q,n-q} & I_q \end{pmatrix} B^{-1}.$$

Note that $B_U + B_W = BI_nB^{-1} = I_n$. Thus, the oblique projection on U along W is given by

$$g(\mathbf{x}) = B_U\mathbf{x} = B\begin{pmatrix} I_p & O_{p,n-p} \\ O_{n-p,p} & O_{n-p,n-p} \end{pmatrix} B^{-1}\mathbf{x}.$$

The similar oblique projection on W along U is

$$h(\mathbf{x}) = B_U\mathbf{x} = B\begin{pmatrix} O_{n-q,n-q} & O_{n-q,q} \\ O_{q,n-q} & I_q \end{pmatrix} B^{-1}\mathbf{x},$$

for $\mathbf{x} \in \mathbb{C}^n$. Observe that $g(\mathbf{x}) + h(\mathbf{x}) = \mathbf{x}$, so the projection on W along U is $h(\mathbf{x}) = \mathbf{x} - g(\mathbf{x}) = (I_n - B_U)\mathbf{x}$.

A special important type of projections involves pairs of orthogonal subspaces. Let U be a subspace of $\mathbb{C}^n$ with $\dim U = p$ and let $B_U = \{\mathbf{u}_1, \ldots, \mathbf{u}_p\}$ be an orthonormal basis of U. Taking into account that $(U^\perp)^\perp = U$, by Theorem 5.33 there exists an idempotent endomorphism g of $\mathbb{C}^n$ such that $U = \mathsf{Img}(g)$ and $U^\perp = \mathsf{Ker}(g)$. The proof of Theorem 6.39 shows that this endomorphism is defined by $g(\mathbf{x}) = (\mathbf{x}, \mathbf{u}_1)\mathbf{u}_1 + \cdots + (\mathbf{x}, \mathbf{u}_m)\mathbf{u}_m$.

Definition 6.91 *Let $\{\boldsymbol{u}_1, \ldots, \boldsymbol{u}_m\}$ be an orthonormal basis of an m-dimensional subspace of U of $\mathbb{C}^n$. The* orthogonal projection of the vector $\boldsymbol{x} \in \mathbb{C}^n$ on the subspace U *is the vector* $\mathsf{proj}_U(\boldsymbol{x}) = (\boldsymbol{x}, \boldsymbol{u}_1)\boldsymbol{u}_1 + \cdots + (\boldsymbol{x}, \boldsymbol{u}_m)\boldsymbol{u}_m$.

If $\{\mathbf{z}_1, \ldots, \mathbf{z}_m\}$ is an orthogonal basis of the space U, then an orthonormal basis of the same subspace is $\{\frac{1}{\|\mathbf{z}_1\|}\mathbf{z}_1, \ldots, \frac{1}{\|\mathbf{z}_m\|}\mathbf{z}_m\}$. Thus, if $\mathbf{u}_i = \frac{1}{\|\mathbf{z}_i\|}\mathbf{z}_i$ for $1 \leqslant i \leqslant m$, we have $(\mathbf{x}, \mathbf{u}_i)\mathbf{u}_i = \frac{(\mathbf{x}, \mathbf{z}_i)}{\|\mathbf{z}_i\|^2}\mathbf{z}_i$ for $1 \leqslant i \leqslant m$, and the orthogonal projection $\mathsf{proj}_U(\mathbf{x})$ can be written as

$$\mathsf{proj}_U(\mathbf{x}) = \frac{(\mathbf{x}, \mathbf{z}_1)}{\| \mathbf{z}_1 \|^2}\mathbf{z}_1 + \cdots + \frac{(\mathbf{x}, \mathbf{z}_m)}{\| \mathbf{z}_m \|^2}\mathbf{z}_m. \tag{6.17}$$

In particular, the projection of $\mathbf{x}$ on the 1-dimensional subspace generated by $\mathbf{z}$ is denoted by $\mathsf{proj}_{\mathbf{z}}(\mathbf{x})$ and is given by

$$\mathsf{proj}_{\mathbf{z}}(\mathbf{x}) = \frac{(\mathbf{x}, \mathbf{z})}{\| \mathbf{z} \|^2}\mathbf{z} = \frac{\mathbf{x}^{\mathsf{H}}\mathbf{z}}{\mathbf{z}^{\mathsf{H}}\mathbf{z}}\mathbf{z}. \tag{6.18}$$

Theorem 6.92 *Let U be an m-dimensional subspace of $\mathbb{R}^n$ and let $\boldsymbol{x} \in \mathbb{R}^n$. The vector $\boldsymbol{y} = \boldsymbol{x} - \mathsf{proj}_U(\boldsymbol{x})$ belongs to the subspace $U^\perp$.*

Proof Let $B_U = \{\mathbf{u}_1, \ldots, \mathbf{u}_m\}$ be an orthonormal basis of U. Note that

$$\begin{aligned}(\mathbf{y}, \mathbf{u}_j) &= (\mathbf{x}, \mathbf{u}_j) - \left(\sum_{i=1}^{m}(\mathbf{x}, \mathbf{u}_i)\mathbf{u}_i, \mathbf{u}_j\right) \\ &= (\mathbf{x}, \mathbf{u}_j) - \sum_{i=1}^{m}(\mathbf{x}, \mathbf{u}_i)(\mathbf{u}_i, \mathbf{u}_j) = 0,\end{aligned}$$

due to the orthogonality of the basis B_U. Therefore, $\mathbf{y}$ is orthogonal on every linear combination of B_U, that is on the subspace U.

Theorem 6.93 *Let U be an m-dimensional subspace of $\mathbb{C}^n$ having the orthonormal basis $\{\boldsymbol{u}_1, \ldots, \boldsymbol{u}_m\}$. The orthogonal projection proj_U is given by $\mathsf{proj}_U(\boldsymbol{x}) = B_U B_U^{\mathsf{H}}\boldsymbol{x}$ for $\boldsymbol{x} \in \mathbb{C}^n$, where $B_U \in \mathbb{R}^{n\times m}$ is the matrix $B_U = (\boldsymbol{u}_1 \ \cdots \ \boldsymbol{u}_m) \in \mathbb{C}^{n\times m}$.*

Proof We can write

$$\mathsf{proj}_U(\mathbf{x}) = \sum_{i=1}^{m} \mathbf{u}_i(\mathbf{u}_i^{\mathsf{H}}\mathbf{x}) = (\mathbf{u}_1 \cdots \mathbf{u}_m) \begin{pmatrix} \mathbf{u}_1^{\mathsf{H}} \\ \vdots \\ \mathbf{u}_m^{\mathsf{H}} \end{pmatrix} \mathbf{x} = B_U B_U^{\mathsf{H}} \mathbf{x}.$$

Since the basis $\{\mathbf{u}_1, \ldots, \mathbf{u}_m\}$ is orthonormal, we have $B_U^{\mathsf{H}} B_U = I_m$. Observe that the matrix $B_U B_U^{\mathsf{H}} \in \mathbb{C}^{n \times n}$ is symmetric and idempotent because

$$(B_U B_U^{\mathsf{H}})(B_U B_U^{\mathsf{H}}) = B_U (B_U^{\mathsf{H}} B_U) B_U^{\mathsf{H}} = B_U B_U^{\mathsf{H}}.$$

Corollary 6.94 *Let U be an m-dimensional subspace of $\mathbb{C}^n$ having the orthonormal basis $\{\boldsymbol{u}_1, \ldots, \boldsymbol{u}_m\}$. We have*

$$\mathsf{proj}_U(\boldsymbol{x}) = \sum_{i=1}^{m} \mathsf{proj}_{\boldsymbol{u}_i}(\boldsymbol{x}).$$

Proof This statement follows directly from Theorem 6.93.

For an m-dimensional subspace U of $\mathbb{C}^n$ we denote by $P_U = B_U B_U^{\mathsf{H}} \in \mathbb{C}^{n \times n}$, where B_U is a matrix of an orthonormal basis of U as defined before. P_U is the *projection matrix* of the subspace U.

Corollary 6.95 *For every non-zero subspace U, the matrix P_U is a Hermitian matrix, and therefore, a self-adjoint matrix.*

Proof Since $P_U = B_U B_U^{\mathsf{H}}$ where B_U is a matrix of an orthonormal basis of the subspace S, it is immediate that $P_U^{\mathsf{H}} = P_U$.

The self-adjointness of P_U means that $(\mathbf{x}, P_U \mathbf{y}) = (P_U \mathbf{x}, \mathbf{y})$ for every $\mathbf{x}, \mathbf{y} \in \mathbb{C}^n$.

Corollary 6.96 *Let U be an m-dimensional subspace of $\mathbb{C}^n$ having the orthonormal basis $\{\boldsymbol{u}_1, \ldots, \boldsymbol{u}_m\}$.*

If $B_U = (\boldsymbol{u}_1 \cdots \boldsymbol{u}_m) \in \mathbb{C}^{n \times m}$, then for every $\boldsymbol{x} \in \mathbb{C}$ we have the decomposition $\boldsymbol{x} = P_U \boldsymbol{x} + P_{U^\perp} \boldsymbol{x}$, where $P_U = B_U B_U^{\mathsf{H}}$ and $P_{U^\perp} = I_n - P_U = I_n - B_U B_{U^\perp}$.

Proof This statement follows immediately from Theorem 6.93.

It is possible to give a direct argument for the independence of the projection matrix P_U relative to the choice of orthonormal basis in U.

Theorem 6.97 *Let U be an m-dimensional subspace of $\mathbb{C}^n$ having the orthonormal bases $\{\boldsymbol{u}_1, \ldots, \boldsymbol{u}_m\}$ and $\{\boldsymbol{v}_1, \ldots, \boldsymbol{v}_m\}$ and let $B_U = (\boldsymbol{u}_1 \cdots \boldsymbol{u}_m) \in \mathbb{C}^{n \times m}$ and $\tilde{B}_U = (\boldsymbol{v}_1 \cdots \boldsymbol{v}_m) \in \mathbb{C}^{n \times m}$. The matrix $B_U^{\mathsf{H}} \tilde{B}_U \in \mathbb{C}^{m \times m}$ is unitary and $\tilde{B}_U \tilde{B}_U^{\mathsf{H}} = B_U B_U^{\mathsf{H}}$.*

Proof Since the both sets of columns of B_U and $\tilde{B}_U$ are bases for U, there exists a unique square matrix $Q \in \mathbb{C}^{m \times m}$ such that $B_U = \tilde{B}_U Q$. The orthonormality of B_U and $\tilde{B}_U$ implies $B_U^{\mathsf{H}} B_U = \tilde{B}_U^{\mathsf{H}} \tilde{B}_U = I_m$. Thus, we can write $I_m = B_U^{\mathsf{H}} B_U = Q^{\mathsf{H}} \tilde{B}_U^{\mathsf{H}} \tilde{B}_U Q = Q^{\mathsf{H}} Q$, which shows that Q is unitary. Furthermore, $B_U^{\mathsf{H}} \tilde{B}_U = Q^{\mathsf{H}} \tilde{B}_U^{\mathsf{H}} \tilde{B}_U = Q^{\mathsf{H}}$ is unitary and $B_U B_U^{\mathsf{H}} = \tilde{B}_U Q Q^{\mathsf{H}} \tilde{B}_U^{\mathsf{H}} = \tilde{B}_U \tilde{B}_U^{\mathsf{H}}$.

In Example 6.76 we have shown that if P is an idempotent matrix, then for every matrix norm μ we have $\mu(P) = 0$ or $\mu(P) \geqslant 1$. For orthogonal projection matrices of the form P_U, where U is a non-zero subspace we have $|||P_U|||_2 = 1$. Indeed, we can write $\mathbf{x} = (\mathbf{x} - \mathsf{proj}_U(\mathbf{x})) + \mathsf{proj}_U(\mathbf{x})$, so

$$\| \mathbf{x} \|_2^2 = \| \mathbf{x} - \mathsf{proj}_U(\mathbf{x}) \|_2^2 + \| \mathsf{proj}_U(\mathbf{x}) \|_2^2 \geqslant \| \mathsf{proj}_U(\mathbf{x}) \|_2^2 .$$

Thus, $\| \mathbf{x} \|_2 \geqslant \| P_U(\mathbf{x}) \|$ for any $\mathbf{x} \in \mathbb{C}^n$, which implies

$$|||P_U|||_2 = \sup \left\{ \frac{\| P_U\mathbf{x} \|_2}{\| \mathbf{x} \|_2} \middle| \mathbf{x} \in \mathbb{C}^n - \{\mathbf{0}\} \right\} \leqslant 1.$$

This implies $|||P_U|||_2 = 1$.

The next theorem shows that the best approximation of a vector $\mathbf{x}$ is a subspace U (in the sense of Euclidean distance) is the orthogonal projection on $\mathbf{x}$ on U.

Theorem 6.98 *Let U be an m-dimensional subspace of $\mathbb{C}^n$ and let $\boldsymbol{x} \in \mathbb{C}^n$. The minimal value of $d_2(\boldsymbol{x}, \boldsymbol{u})$, the Euclidean distance between $\boldsymbol{x}$ and an element $\boldsymbol{u}$ of the subspace U is achieved when $\boldsymbol{u} = \mathsf{proj}_U(\boldsymbol{x})$.*

Proof We saw that $\mathbf{x}$ can be uniquely written as $\mathbf{x} = \mathbf{y} + \mathsf{proj}_U(\mathbf{x})$, where $\mathbf{y} \in U^{\perp}$. Let now $\mathbf{u}$ be an arbitrary member of U. We have

$$d_2(\mathbf{x}, \mathbf{u})^2 = \| \mathbf{x} - \mathbf{u} \|_2^2 = \| (\mathbf{x} - \mathsf{proj}_U(\mathbf{x})) + (\mathsf{proj}_U(\mathbf{x}) - \mathbf{u}) \|_2^2 .$$

Since $\mathbf{x} - \mathsf{proj}_U(\mathbf{x}) \in U^{\perp}$ and $\mathsf{proj}_U(\mathbf{x}) - \mathbf{u} \in U$, it follows that these vectors are orthogonal. Thus, we can write

$$d_2(\mathbf{x}, \mathbf{u})^2 = \| (\mathbf{x} - \mathsf{proj}_U(\mathbf{x})) \|_2^2 + \| (\mathsf{proj}_U(\mathbf{x}) - \mathbf{u}) \|_2^2,$$

which implies that $d_2(\mathbf{x}, \mathbf{u}) \geqslant d_2(\mathbf{x} - \mathsf{proj}_U(\mathbf{x}))$.

The orthogonal projections associated with subspaces allow us to define a metric on the collection of subspaces of $\mathbb{C}^n$. Indeed, if S and T are two subspaces of $\mathbb{C}^n$ we define $d_F(S, T) = \| P_S - P_T \|_F$. When using the vector norm $\| \cdot \|_2$ and the metric induced by this norm on $\mathbb{C}^n$ we denote the corresponding metric on subspaces by d_2.

Example 6.99 Let $\mathbf{u}, \mathbf{w}$ be two distinct unit vectors in the linear space L. The orthogonal projection matrices of $\langle \mathbf{u} \rangle$ and $\langle \mathbf{w} \rangle$ are $\mathbf{u}\mathbf{u}'$ and $\mathbf{w}\mathbf{w}'$, respectively. Thus,

$$d_F(\langle \mathbf{u} \rangle, \langle \mathbf{w} \rangle) = \| \mathbf{u}\mathbf{u}' - \mathbf{v}\mathbf{v}' \|_F .$$

Suppose now that $L = \mathbb{R}^2$. Since $\mathbf{u}$ and $\mathbf{w}$ are unit vectors in $\mathbb{R}^2$ there exist $\alpha, \beta \in [0, 2\pi]$ such that

$$\mathbf{u} = \begin{pmatrix} \cos\alpha \\ \sin\alpha \end{pmatrix} \text{ and } \mathbf{w} = \begin{pmatrix} \cos\beta \\ \sin\beta \end{pmatrix}.$$

Thus, we can write

$$\mathbf{u}\mathbf{u}' - \mathbf{v}\mathbf{v}' = \begin{pmatrix} \cos^2\alpha - \cos^2\beta & \cos\alpha\sin\alpha - \cos\beta\sin\beta \\ \cos\alpha\sin\alpha - \cos\beta\sin\beta & \sin^2\alpha - \sin^2\beta \end{pmatrix}.$$

and $d_F(\langle\mathbf{u}\rangle, \langle\mathbf{w}\rangle) = \sqrt{2}|\sin(\alpha - \beta)|$.

We could use any matrix norm in the definition of the distance between subspaces. For example, we could replace the Frobenius norm by $|||\cdot|||_1$ or by $|||\cdot|||_2$.

Let S be a subspace of $\mathbb{C}^n$ and let $\mathbf{x} \in \mathbb{C}^n$. The *distance between* $\mathbf{x}$ *and* S *defined by the norm* $\|\cdot\|$ is

$$d(\mathbf{x}, S) = \|\, \mathbf{x} - \mathsf{proj}_S(\mathbf{x}) \,\| = \|\, \mathbf{x} - P_S\mathbf{x} \,\| = \|\, (I - P_S)\mathbf{x} \,\| .$$

Theorem 6.100 *Let S and T be two non-zero subspaces of $\mathbb{C}^n$ and let*

$$\begin{aligned} \delta_S &= \max\{d_2(\boldsymbol{x}, T) \mid \boldsymbol{x} \in S, \|\, \boldsymbol{x} \,\|_2 = 1\}, \\ \delta_T &= \max\{d_2(\boldsymbol{x}, S) \mid \boldsymbol{x} \in T, \|\, \boldsymbol{x} \,\|_2 = 1\}. \end{aligned}$$

We have $d_2(S, T) = \max\{\delta_S, \delta_T\}$.

Proof If $\mathbf{x} \in S$ and $\|\, \mathbf{x} \,\|_2 = 1$ we have

$$\begin{aligned} d_2(\mathbf{x}, T) &= \|\, \mathbf{x} - P_T\mathbf{x} \,\|_2 = \|\, P_S\mathbf{x} - P_T\mathbf{x} \,\|_2 \\ &= \|\, (P_S - P_T)\mathbf{x} \,\|_2 \leqslant |||P_S - P_T|||_2. \end{aligned}$$

Therefore, $\delta_S \leqslant |||P_S - P_T|||_2$. Similarly, $\delta_T \leqslant |||P_S - P_T|||_2$, so $\max\{\delta_S, \delta_T\} \leqslant d_2(S, T)$.

Note that

$$\begin{aligned} \delta_S &= \max\{\|\, (I - P_T)\mathbf{x} \,\|_2 \mid \mathbf{x} \in S, \|\, \mathbf{x} \,\|_2 = 1\}, \\ \delta_T &= \max\{\|\, (I - P_S)\mathbf{x} \,\|_2 \mid \mathbf{x} \in T, \|\, \mathbf{x} \,\|_2 = 1\}, \end{aligned}$$

so, taking into account that $P_S\mathbf{x} \in S$ and $P_T\mathbf{x} \in T$ for every $\mathbf{x} \in \mathbb{C}^n$ we have

$$\|\, (I - P_S)P_T\mathbf{x} \,\|_2 \leqslant \delta_S \,\|\, P_T\mathbf{x} \,\|_2, \|\, (I - P_T)P_S\mathbf{x} \,\|_2 \leqslant \delta_T \,\|\, P_S\mathbf{x} \,\|_2 .$$

We have

$$\begin{aligned}\| P_T(I - P_S)\mathbf{x} \|_2^2 &= (P_T(I - P_S)\mathbf{x}, P_T(I - P_S)\mathbf{x})\\ &= ((P_T)^2(I - P_S)\mathbf{x}, (I - P_S)\mathbf{x})\\ &= (P_T(I - P_S)\mathbf{x}, (I - P_S)\mathbf{x})\\ &= (P_T(I - P_S)\mathbf{x}, (I - P_S)^2\mathbf{x})\\ &= ((I - P_S)P_T(I - P_S)\mathbf{x}, (I - P_S)\mathbf{x})\end{aligned}$$

because both P_S and $I - P_S$ are idempotent and self-adjoint. Therefore,

$$\begin{aligned}\| P_T(I - P_S)\mathbf{x} \|_2^2 &\leqslant \| (I - P_S)P_T(I - P_S)\mathbf{x} \|_2 \| (I - P_S)\mathbf{x} \|_2\\ &\leqslant \delta_T \| P_T(I - P_S)\mathbf{x} \|_2 \| (I - P_S)\mathbf{x} \|_2 .\end{aligned}$$

This allows us to infer that

$$\| P_T(I - P_S)\mathbf{x} \|_2 \leqslant \delta_T \| (I - P_S)\mathbf{x} \|_2 .$$

We discuss now four fundamental subspaces associated to a matrix $A \in \mathbb{C}^{m\times n}$. The range and the null space of A, $\mathsf{Ran}(A) \subseteq \mathbb{C}^m$ and $\mathsf{NullSp}(A) \subseteq \mathbb{C}^n$ have been already discussed. We add now two new subspaces: $\mathsf{Ran}(A^{\mathsf{H}}) \subseteq \mathbb{C}^n$ and $\mathsf{NullSp}(A^{\mathsf{H}}) \subseteq \mathbb{C}^m$.

Theorem 6.101 *For every matrix $A \in \mathbb{C}^{m\times n}$ we have $(\mathsf{Ran}(A))^{\perp} = \mathsf{NullSp}(A^{\mathsf{H}})$.*

Proof The statement follows from the equivalence of the following statements:

(i) $\mathbf{x} \in (\mathsf{Ran}(A))^{\perp}$;
(ii) $(\mathbf{x}, A\mathbf{y}) = 0$ for all $y \in \mathbb{C}^n$;
(iii) $\mathbf{x}^{\mathsf{H}}A\mathbf{y} = 0$ for all $y \in \mathbb{R}^n$;
(iv) $\mathbf{y}^{\mathsf{H}}A^{\mathsf{H}}\mathbf{x} = 0$ for all $y \in \mathbb{R}^n$;
(v) $A^{\mathsf{H}}\mathbf{x} = \mathbf{0}$;
(vi) $\mathbf{x} \in \mathsf{NullSp}(A^{\mathsf{H}})$.

Corollary 6.102 *For every matrix $A \in \mathbb{C}^{m\times n}$ we have $(\mathsf{Ran}(A^{\mathsf{H}}))^{\perp} = \mathsf{NullSp}(A)$.*

Proof This statement follows from Theorem 6.101 by replacing A by A^{H}.

Corollary 6.103 *For every matrix $A \in \mathbb{C}^{m\times n}$ we have*

$$\begin{aligned}\mathbb{C}^m &= \mathsf{Ran}(A) \boxplus \mathsf{NullSp}(A^{\mathsf{H}})\\ \mathbb{C}^n &= \mathsf{NullSp}(A) \boxplus \mathsf{Ran}(A^{\mathsf{H}}).\end{aligned}$$

Proof By Theorem 6.39 we have $\mathbb{C}^m = \mathsf{Ran}(A) \boxplus \mathsf{Ran}(A)^{\perp}$ and $\mathbb{C}^n = \mathsf{NullSp}(A) \boxplus \mathsf{NullSp}(A)^{\perp}$. Taking into account Theorem 6.101 and Corollary 6.103 we obtain the desired equalities.

6.10 Positive Definite and Positive Semidefinite Matrices

Definition 6.104 *A matrix* $A \in \mathbb{C}^{n\times n}$ *is* positive definite *if* $\mathbf{x}^{\mathsf{H}}A\mathbf{x}$ *is a real positive number for every* $\mathbf{x} \in \mathbb{C}^n - \{\mathbf{0}\}$.

Theorem 6.105 *If* $A \in \mathbb{C}^{n\times n}$ *is positive definite, then* A *is Hermitian.*

Proof Let $A \in \mathbb{C}^{n\times n}$ be a matrix. Since $\mathbf{x}^{\mathsf{H}}A\mathbf{x}$ is a real number it follows that it equals its conjugate, so $\mathbf{x}^{\mathsf{H}}A\mathbf{x} = \mathbf{x}^{\mathsf{H}}A^{\mathsf{H}}\mathbf{x}$ for every $\mathbf{x} \in \mathbb{C}^n$. There exists a unique pair of Hermitian matrices H_1 and H_2 such that $A = H_1 + iH_2$, which implies $A^{\mathsf{H}} = H_1^{\mathsf{H}} - iH_2^{\mathsf{H}}$. Thus, we have

$$\mathbf{x}^{\mathsf{H}}(H_1 + iH_2)\mathbf{x} = \mathbf{x}^{\mathsf{H}}(H_1^{\mathsf{H}} - iH_2^{\mathsf{H}})\mathbf{x} = \mathbf{x}^{\mathsf{H}}(H_1 - iH_2)\mathbf{x},$$

because H_1 and H_2 are Hermitian. This implies $\mathbf{x}^{\mathsf{H}}H_2\mathbf{x} = 0$ for every $\mathbf{x} \in \mathbb{C}^n$, which, in turn, implies $H_2 = O_{n,n}$. Consequently, $A = H_1$, so A is indeed Hermitian.

Definition 6.106 *A matrix* $A \in \mathbb{C}^{n\times n}$ *is* positive semidefinite *if* $\mathbf{x}^{\mathsf{H}}A\mathbf{x}$ *is a non-negative real number for every* $\mathbf{x} \in \mathbb{C}^n - \{\mathbf{0}\}$.

Positive definiteness (positive semidefiniteness) is denoted by $A \succ 0$ ($A \succeq 0$, respectively).

The definition of positive definite (semidefinite) matrix can be specialized for real matrices as follows.

Definition 6.107 *A symmetric matrix* $A \in \mathbb{R}^{n\times n}$ *is* positive definite *if* $\mathbf{x}'A\mathbf{x} > 0$ *for every* $\mathbf{x} \in \mathbb{R}^n - \{\mathbf{0}\}$.

If A *satisfies the weaker inequality* $\mathbf{x}'A\mathbf{x} \geqslant 0$ *for every* $\mathbf{x} \in \mathbb{R}^n - \{\mathbf{0}\}$, *then we say that* A *is* positive semidefinite.

$A \succ 0$ *denotes that* A *is positive definite and* $A \succeq 0$ *means that* A *is positive semidefinite.*

Note that in the case of real-valued matrices we need to require explicitely the symmetry of the matrix because, unlike the complex case, the inequality $\mathbf{x}'A\mathbf{x} > 0$ for $\mathbf{x} \in \mathbb{R}^n - \{\mathbf{0}_n\}$ does *not* imply the symmetry of A. For example, consider the matrix

$$A = \begin{pmatrix} a & b \\ -b & a \end{pmatrix},$$

where $a, b \in \mathbb{R}$ and $a > 0$. We have

$$\mathbf{x}'A\mathbf{x} = (x_1\ x_2)\begin{pmatrix} a & b \\ -b & a \end{pmatrix}\begin{pmatrix} x_1 \\ x_2 \end{pmatrix} = a(x_1^2 + x_2^2) > 0$$

if $\mathbf{x} \neq \mathbf{0}_2$.

Example 6.108 The symmetric real matrix

$$A = \begin{pmatrix} a & b \\ b & c \end{pmatrix}$$

is positive definite if and only if $a > 0$ and $b^2 - ac < 0$. Indeed, we have $\mathbf{x}'A\mathbf{x} > 0$ for every $\mathbf{x} \in \mathbb{R}^2 - \{\mathbf{0}\}$ if and only if $ax_1^2 + 2bx_1x_2 + cx_2^2 > 0$, where $\mathbf{x}' = (x_1 \ \ x_2)$; elementary algebra considerations lead to $a > 0$ and $b^2 - ac < 0$.

A positive definite matrix is non-singular. Indeed, if $A\mathbf{x} = \mathbf{0}$, where $A \in \mathbb{R}^{n\times n}$ is positive definite, then $\mathbf{x}^{\mathsf{H}}A\mathbf{x} = 0$, so $\mathbf{x} = \mathbf{0}$. By Corollary 5.91, A is non-singular.

Example 6.109 If $A \in \mathbb{C}^{m\times n}$, then the matrices $A^{\mathsf{H}}A \in \mathbb{C}^{n\times n}$ and $AA^{\mathsf{H}} \in \mathbb{C}^{m\times m}$ are positive semidefinite. For $\mathbf{x} \in \mathbb{C}^n$ we have

$$\mathbf{x}^{\mathsf{H}}(A^{\mathsf{H}}A)\mathbf{x} = (\mathbf{x}^{\mathsf{H}}A^{\mathsf{H}})(A\mathbf{x}) = (A\mathbf{x})^{\mathsf{H}}(A\mathbf{x}) = \parallel A\mathbf{x} \parallel_2^2 \geqslant 0.$$

The argument for AA^{H} is similar.

If $rank(A) = n$, then the matrix $A^{\mathsf{H}}A$ is positive definite because $\mathbf{x}^{\mathsf{H}}(A^{\mathsf{H}}A)\mathbf{x} = 0$ implies $A\mathbf{x} = \mathbf{0}$, which, in turn, implies $\mathbf{x} = \mathbf{0}$.

Theorem 6.110 *If $A \in \mathbb{C}^{n\times n}$ is a positive definite matrix, then any principal submatrix $B = A\begin{bmatrix} i_1 & \cdots & i_k \\ i_1 & \cdots & i_k \end{bmatrix}$ is a positive definite matrix.*

Proof Let $\mathbf{x} \in \mathbb{C}^n - \{\mathbf{0}\}$ be a vector such that all components located on positions other than $i_1, \ldots, i_k$ equal 0 and let $\mathbf{y} = \mathbf{x}\begin{bmatrix} i_1 & \cdots & i_k \\ & 1 & \end{bmatrix} \in \mathbb{C}^k$ be the vector obtained from $\mathbf{x}$ by retaining only the components located on positions $i_1, \ldots, i_k$. Since $\mathbf{y}^{\mathsf{H}}B\mathbf{y} = \mathbf{x}^{\mathsf{H}}A\mathbf{x} > 0$ it follows that $B \succ 0$.

Corollary 6.111 *If $A \in \mathbb{C}^{n\times n}$ is a positive definite matrix, then any diagonal element a_{ii} is a real positive number for $1 \leqslant i \leqslant n$.*

Proof This statement follows immediately from Theorem 6.110 by observing that every diagonal element of A is an 1×1 principal submatrix of A.

Theorem 6.112 *If $A, B \in \mathbb{C}^{n\times n}$ are two positive semidefinite matrices and a, b are two non-negative numbers, then $aA + bB \succeq 0$.*

Proof The statement holds because $\mathbf{x}^{\mathsf{H}}(aA + bB)\mathbf{x} = a\mathbf{x}^{\mathsf{H}}A\mathbf{x} + b\mathbf{x}^{\mathsf{H}}B\mathbf{x} \geqslant 0$, due to the fact that A and B are positive semidefinite.

Theorem 6.113 *Let $A \in \mathbb{C}^{n\times n}$ be a positive definite matrix and let $S \in \mathbb{C}^{n\times m}$. The matrix $S^{\mathsf{H}}AS$ is positive semidefinite and has the same rank as S. Moreover, if $rank(S) = m$, then $S^{\mathsf{H}}AS$ is positive definite.*

Proof Since A is positive definite, it is Hermitian and $(S^{\mathsf{H}}AS)^{\mathsf{H}} = S^{\mathsf{H}}AS$ implies that $S^{\mathsf{H}}AS$ is a Hermitian matrix.

Let $\mathbf{x} \in \mathbb{C}^m$. We have $\mathbf{x}^{\mathsf{H}}S^{\mathsf{H}}AS\mathbf{x} = (S\mathbf{x})^{\mathsf{H}}A(S\mathbf{x}) \geqslant 0$ because A is positive definite. Thus, the matrix $S^{\mathsf{H}}AS$ is positive semidefinite.

If $S\mathbf{x} = \mathbf{0}$, then $S^{\mathsf{H}}AS\mathbf{x} = \mathbf{0}$; conversely, if $S^{\mathsf{H}}AS\mathbf{x} = \mathbf{0}$, then $\mathbf{x}^{\mathsf{H}}S^{\mathsf{H}}AS\mathbf{x} = 0$, so $S\mathbf{x} = \mathbf{0}$. This allows us to conclude that $\mathsf{NullSp}(S) = \mathsf{NullSp}(S^{\mathsf{H}}AS)$. Therefore, by Equality (5.6), we have $rank(S) = rank(S^{\mathsf{H}}AS)$.

Suppose now that $rank(S) = m$ and that $\mathbf{x}^{\mathsf{H}}S^{\mathsf{H}}AS\mathbf{x} = 0$. Since A is positive definite we have $S\mathbf{x} = \mathbf{0}$ and this implies $\mathbf{x} = \mathbf{0}$, because of the assumption made relative to $rank(S)$. Consequently, $S^{\mathsf{H}}AS$ is positive definite.

Corollary 6.114 *Let $A \in \mathbb{C}^{n\times n}$ be a positive definite matrix and let $S \in \mathbb{C}^{n\times n}$. If S is non-singular, then so is $S^{H}AS$.*

Proof This is an immediate consequence of Theorem 6.113.

Theorem 6.115 *A Hermitian matrix $B \in \mathbb{C}^{n\times n}$ is positive definite if and only if the mapping $f : \mathbb{C}^n \times \mathbb{C}^n \longrightarrow \mathbb{C}$ given by $f(x, y) = x^{H}By$ for $x, y \in \mathbb{C}^n$ defines an inner product on $\mathbb{C}^n$.*

Proof Suppose that B defines an inner product on $\mathbb{C}^n$. Then, by Property (iii) of Definition 6.22 we have $f(\mathbf{x}, \mathbf{x}) > 0$ for $\mathbf{x} \neq \mathbf{0}$, which amounts to the positive definiteness of B.

Conversely, if B is positive definite, then f satisfies the condition from Definition 6.22. We show here only that f has the conjugate symmetry property.

We can write $\overline{f(\mathbf{y}, \mathbf{x})} = \overline{\mathbf{y}^{\mathsf{H}}B\mathbf{x}} = \mathbf{y}'\overline{B}\overline{\mathbf{x}}$, for $\mathbf{x}, \mathbf{y} \in \mathbb{C}^n$. Since B is Hermitian, $\overline{B} = \overline{B^{\mathsf{H}}} = B'$, so $\overline{f(\mathbf{y}, \mathbf{x})} = \mathbf{y}'B'\overline{\mathbf{x}}$. Observe that $\mathbf{y}'B'\overline{\mathbf{x}}$ is a number (that is, an 1×1 matrix), so $(\mathbf{y}'B'\overline{\mathbf{x}})' = \mathbf{x}^{\mathsf{H}}B\mathbf{y} = f(\mathbf{x}, \mathbf{y})$.

Corollary 6.116 *A symmetric matrix $B \in \mathbb{R}^{n\times n}$ is positive definite if and only if the mapping $f : \mathbb{R}^n \times \mathbb{R}^n \longrightarrow \mathbb{R}$ given by $f(x, y) = x'By$ for $x, y \in \mathbb{R}^n$ defines an inner product on $\mathbb{R}^n$.*

Proof This follows immediately from Theorem 6.115.

Definition 6.117 *Let $L = (v_1, \ldots, v_m)$ be a sequence of vectors in $\mathbb{R}^n$. The* Gram matrix *of L is the matrix $G_L = (g_{ij}) \in \mathbb{R}^{m\times m}$ defined by $g_{ij} = v_i'v_j$ for $1 \leqslant i, j \leqslant m$.*

Note that if $A_L = (\mathbf{v}_1 \ \cdots \ \mathbf{v}_m) \in \mathbb{R}^{n\times m}$, then $G_L = A_L'A_L$. Also, note that G_L is a symmetric matrix.

Theorem 6.118 *Let $L = (v_1, \ldots, v_m)$ be a sequence of m vectors in $\mathbb{R}^n$, where $m \leqslant n$. If L is linearly independent, then the Gram matrix G_L is positive definite.*

Proof Suppose that L is linearly independent. Let $\mathbf{x} \in \mathbb{R}^m$. We have $\mathbf{x}'G_L\mathbf{x} = \mathbf{x}'A_L'A_L\mathbf{x} = (A_L\mathbf{x})'A_L\mathbf{x} = \parallel A_L\mathbf{x} \parallel_2^2$. Therefore, if $\mathbf{x}'G_L\mathbf{x} = 0$, we have $A_L\mathbf{x} = \mathbf{0}$, which is equivalent to $x_1\mathbf{v}_1 + \cdots + x_n\mathbf{v}_n = 0$. Since $\{\mathbf{v}_1, \ldots, \mathbf{v}_m\}$ is linearly independent it follows that $x_1 = \cdots = x_m = 0$, so $\mathbf{x} = \mathbf{0}$. Thus, A is indeed, positive definite.

The Gram matrix of an arbitrary sequence of vectors is positive semidefinite, as the reader can easily verify.

Definition 6.119 *Let* $L = (v_1, \ldots, v_m)$ *be a sequence of m vectors in* $\mathbb{R}^n$*, where* $m \leqslant n$*. The* Gramian *of L is the number* $\det(G_L)$.

Theorem 6.120 *If* $L = (v_1, \ldots, v_m)$ *is a sequence of m vectors in* $\mathbb{R}^n$*. Then, L is linearly independent if and only if* $\det(G_L) \neq 0$.

Proof Suppose that $\det(G_L) \neq 0$ and that L is not linearly independent. In other words, the numbers $a_1, \ldots, a_m$ exist such that at least one of them is not 0 and $a_1\mathbf{x}_1 + \cdots + a_m\mathbf{x}_m = \mathbf{0}$. This implies the equalities

$$a_1(\mathbf{x}_1, \mathbf{x}_j) + \cdots + a_m(\mathbf{x}_m, \mathbf{x}_j) = \mathbf{0},$$

for $1 \leqslant j \leqslant m$, so the system $G_L\mathbf{a} = \mathbf{0}$ has a non-trivial solution in $a_1, \ldots, a_m$. This implies $\det(G_L) = 0$, which contradicts the initial assumption.

Conversely, suppose that L is linearly independent and $\det(G_L) = 0$. Then, the linear system

$$a_1(\mathbf{x}_1, \mathbf{x}_j) + \cdots + a_m(\mathbf{x}_m, \mathbf{x}_j) = \mathbf{0},$$

for $1 \leqslant j \leqslant m$, has a non-trivial solution in $a_1, \ldots, a_m$. If $\mathbf{w} = a_1\mathbf{x}_1 + \cdots a_m\mathbf{x}_m$, this amounts to $(\mathbf{w}, \mathbf{x}_i) = 0$ for $1 \leqslant i \leqslant n$. This implies $(\mathbf{w}, \mathbf{w}) = \parallel \mathbf{w} \parallel_2^2 = 0$, so $\mathbf{w} = 0$, which contradicts the linear independence of L.

Theorem 6.121 (Cholesky's Decomposition Theorem) *Let* $A \in \mathbb{C}^{n \times n}$ *be a Hermitian positive definite matrix. There exists a unique upper triangular matrix R with real, positive diagonal elements such that* $A = R^H R$.

Proof The argument is by induction on $n \geqslant 1$. The base step, $n = 1$, is immediate.

Suppose that a decomposition exists for all Hermitian positive matrices of order n, and let $A \in \mathbb{C}^{(n+1)\times(n+1)}$ be a symmetric and positive definite matrix. We can write

$$A = \begin{pmatrix} a_{11} & \mathbf{a}^{\mathsf{H}} \\ \mathbf{a} & B \end{pmatrix},$$

where $B \in \mathbb{C}^{n \times n}$. By Theorem 6.110, $a_{11} > 0$ and $B \in \mathbb{C}^{n \times n}$ is a Hermitian positive definite matrix. It is easy to verify the identity:

$$A = \begin{pmatrix} \sqrt{a_{11}} & \mathbf{0} \\ \frac{1}{\sqrt{a_{11}}}\mathbf{a} & I_n \end{pmatrix} \begin{pmatrix} 1 & \mathbf{0}' \\ \mathbf{0} & B - \frac{1}{a_{11}}\mathbf{a}\mathbf{a}^{\mathsf{H}} \end{pmatrix} \begin{pmatrix} \sqrt{a_{11}} & \frac{1}{\sqrt{a_{11}}}\mathbf{a}^{\mathsf{H}} \\ 0 & I_n \end{pmatrix}. \tag{6.19}$$

Let $R_1 \in \mathbb{C}^{n \times n}$ be the upper triangular non-singular matrix

$$R_1 = \begin{pmatrix} \sqrt{a_{11}} & \frac{1}{\sqrt{a_{11}}}\mathbf{a}^{\mathsf{H}} \\ 0 & I_n \end{pmatrix}.$$

This allows us to write

$$A = R_1^{\mathsf{H}} \begin{pmatrix} 1 & \mathbf{0}' \\ \mathbf{0} & A_1 \end{pmatrix} R_1,$$

where $A_1 = B - \frac{1}{a_{11}} \mathbf{a}\mathbf{a}^{\mathsf{H}}$. Since

$$\begin{pmatrix} 1 & \mathbf{0}' \\ \mathbf{0} & A_1 \end{pmatrix} = (R_1^{-1})^{\mathsf{H}} A R_1^{-1},$$

by Theorem 6.113, the matrix

$$\begin{pmatrix} 1 & \mathbf{0}' \\ \mathbf{0} & A_1 \end{pmatrix}$$

is positive definite, which allows us to conclude that the matrix $A_1 = B - \frac{1}{a_{11}} \mathbf{a}\mathbf{a}^{\mathsf{H}} \in \mathbb{C}^{n \times n}$ is a Hermitian positive definite matrix.

By the inductive hypothesis, A_1 can be factored as

$$A_1 = P^{\mathsf{H}} P,$$

where P is an upper triangular matrix. This allows us to write

$$\begin{pmatrix} 1 & \mathbf{0}' \\ \mathbf{0} & A_1 \end{pmatrix} = \begin{pmatrix} 1 & \mathbf{0}' \\ \mathbf{0} & P^{\mathsf{H}} \end{pmatrix} \begin{pmatrix} 1 & \mathbf{0}' \\ \mathbf{0} & P \end{pmatrix}$$

Thus,

$$A = R_1^{\mathsf{H}} \begin{pmatrix} 1 & \mathbf{0}' \\ \mathbf{0} & P^{\mathsf{H}} \end{pmatrix} \begin{pmatrix} 1 & \mathbf{0}' \\ \mathbf{0} & P \end{pmatrix} R_1$$

If R is defined as

$$R = \begin{pmatrix} 1 & \mathbf{0}' \\ \mathbf{0} & P \end{pmatrix} R_1 = \begin{pmatrix} 1 & \mathbf{0}' \\ \mathbf{0} & P \end{pmatrix} \begin{pmatrix} \sqrt{a_{11}} & \frac{1}{\sqrt{a_{11}}} \mathbf{a}^{\mathsf{H}} \\ 0 & I_n \end{pmatrix} = \begin{pmatrix} \sqrt{a_{11}} & \frac{1}{\sqrt{a_{11}}} \mathbf{a}^{\mathsf{H}} \\ 0 & P \end{pmatrix},$$

then $A = R^{\mathsf{H}} R$ and R is clearly an upper triangular matrix.

We refer to the matrix R as the *Cholesky factor* of A.

Corollary 6.122 *If $A \in \mathbb{C}^{n \times n}$ is a Hermitian positive definite matrix, then* $\det(A) > 0$.

Proof By Corollary 5.130, $\det(A)$ is a real number. By Theorem 6.121, $A = R^{\mathsf{H}} R$, where R is an upper triangular matrix with real, positive diagonal elements, so $\det(A) = \det(R^{\mathsf{H}}) \det(R) = (\det(R))^2$. Since $\det(R)$ is the product of its diagonal elements, $\det(R)$ is a real, positive number, which implies $\det(A) > 0$.

Example 6.123 Let A be the symmetric matrix

$$A = \begin{pmatrix} 3 & 0 & 2 \\ 0 & 2 & 1 \\ 2 & 1 & 2 \end{pmatrix}.$$

We leave to the reader to verify that this matrix is indeed positive definite starting from Definition 6.104.

By Equality (6.19), the matrix A can be written as

$$A = \begin{pmatrix} \sqrt{3} & 0 & 0 \\ 0 & 1 & 0 \\ \frac{2}{\sqrt{3}} & 0 & 1 \end{pmatrix} \begin{pmatrix} 1 & 0 & 0 \\ 0 & 2 & 1 \\ 0 & 1 & \frac{2}{3} \end{pmatrix} \begin{pmatrix} \sqrt{3} & 0 & \frac{2}{\sqrt{3}} \\ 0 & 1 & 0 \\ 0 & 0 & 1 \end{pmatrix},$$

because

$$A_1 = \begin{pmatrix} 2 & 1 \\ 1 & 2 \end{pmatrix} - \frac{1}{3}\begin{pmatrix} 0 \\ 2 \end{pmatrix}(0\ 2) = \begin{pmatrix} 2 & 1 \\ 1 & \frac{2}{3} \end{pmatrix}.$$

Applying the same equality to A_1 we have

$$A_1 = \begin{pmatrix} \sqrt{2} & 0 \\ \frac{1}{\sqrt{2}} & 1 \end{pmatrix} \begin{pmatrix} 1 & 0 \\ 0 & \frac{1}{6} \end{pmatrix} \begin{pmatrix} \sqrt{2} & \frac{1}{\sqrt{2}} \\ 0 & 1 \end{pmatrix}.$$

Since the matrix $(\frac{1}{6})$ can be factored directly we have

$$\begin{aligned} A_1 &= \begin{pmatrix} \sqrt{2} & 0 \\ \frac{1}{\sqrt{2}} & 1 \end{pmatrix} \begin{pmatrix} 1 & 0 \\ 0 & \frac{1}{\sqrt{6}} \end{pmatrix} \begin{pmatrix} 1 & 0 \\ 0 & \frac{1}{\sqrt{6}} \end{pmatrix} \begin{pmatrix} \sqrt{2} & \frac{1}{\sqrt{2}} \\ 0 & 1 \end{pmatrix} \\ &= \begin{pmatrix} \sqrt{2} & 0 \\ \frac{1}{\sqrt{2}} & \frac{1}{\sqrt{6}} \end{pmatrix} \begin{pmatrix} \sqrt{2} & \frac{1}{\sqrt{2}} \\ 0 & \frac{1}{\sqrt{6}} \end{pmatrix}. \end{aligned}$$

In turn, this implies

$$\begin{pmatrix} 1 & 0 & 0 \\ 0 & 2 & 1 \\ 0 & 1 & \frac{2}{3} \end{pmatrix} = \begin{pmatrix} 1 & 0 & 0 \\ 0 & \sqrt{2} & 0 \\ 0 & \frac{1}{\sqrt{2}} & \frac{1}{\sqrt{6}} \end{pmatrix} \begin{pmatrix} 1 & 0 & 0 \\ 0 & \sqrt{2} & \frac{1}{\sqrt{2}} \\ 0 & 0 & \frac{1}{\sqrt{6}} \end{pmatrix},$$

which produces the Cholesky final decomposition of A:

$$A = \begin{pmatrix} \sqrt{3} & 0 & 0 \\ 0 & 1 & 0 \\ \frac{2}{\sqrt{3}} & 0 & 1 \end{pmatrix} \begin{pmatrix} 1 & 0 & 0 \\ 0 & \sqrt{2} & 0 \\ 0 & \frac{1}{\sqrt{2}} & \frac{1}{\sqrt{6}} \end{pmatrix} \begin{pmatrix} 1 & 0 & 0 \\ 0 & \sqrt{2} & \frac{1}{\sqrt{2}} \\ 0 & 0 & \frac{1}{\sqrt{6}} \end{pmatrix} \begin{pmatrix} \sqrt{3} & 0 & \frac{2}{\sqrt{3}} \\ 0 & 1 & 0 \\ 0 & 0 & 1 \end{pmatrix}$$

$$= \begin{pmatrix} \sqrt{3} & 0 & 0 \\ 0 & \sqrt{2} & 0 \\ \frac{2}{\sqrt{3}} & \frac{1}{\sqrt{2}} & \frac{1}{\sqrt{6}} \end{pmatrix} \begin{pmatrix} \sqrt{3} & 0 & \frac{2}{\sqrt{3}} \\ 0 & \sqrt{2} & \frac{1}{\sqrt{2}} \\ 0 & 0 & \frac{1}{\sqrt{6}} \end{pmatrix}.$$

Cholesky's Decomposition Theorem can be extended to positive semi-definite matrices.

Theorem 6.124 (Cholesky's Decomposition Theorem for Positive Semidefinite Matrices) *Let $A \in \mathbb{C}^{n\times n}$ be a Hermitian positive semidefinite matrix. There exists an upper triangular matrix R with real, non-negative diagonal elements such that $A = R^{\mathsf{H}}R$.*

Proof The argument is similar to the one used for Theorem 6.121 and is omitted.

Observe that for positive semidefinite matrices, the diagonal elements of R are non-negative numbers and the uniqueness of R does not longer hold.

Example 6.125 Let $A = \begin{pmatrix} 1 & -1 \\ -1 & 1 \end{pmatrix}$. Since $\mathbf{x}'A\mathbf{x} = (x_1 - x_2)$ it is clear that A is a positive semidefinite but not a positive definite matrix. Let R be a matrix of the form

$$R = \begin{pmatrix} r_1 & r \\ 0 & r_2 \end{pmatrix}$$

such that $A = R'R$. It is easy to see that the last equality is equivalent to $r_1^2 = r_2^2 = 1$ and $rr_1 = -1$. Thus, we have for distinct Cholesky factors: matrices

$$\begin{pmatrix} 1 & -1 \\ 0 & 1 \end{pmatrix}, \begin{pmatrix} 1 & -1 \\ 0 & -1 \end{pmatrix}, \begin{pmatrix} -1 & 1 \\ 0 & 1 \end{pmatrix}, \begin{pmatrix} -1 & 1 \\ 0 & -1 \end{pmatrix}.$$

Theorem 6.126 *A Hermitian matrix $A \in \mathbb{C}^{n\times n}$ is positive definite if and only if all its leading principal minors are positive.*

Proof By Theorem 6.110, if A is positive definite, then every principal submatrix is positive definite, so by Corollary 6.122, each principal minor of A is positive.

Conversely, suppose that $A \in \mathbb{C}^{n\times n}$ is an Hermitian matrix having positive leading principal minors. We prove by induction on n that A is positive definite.

The base case, $n = 1$ is immediate. Suppose that the statement holds for matrices in $\mathbb{C}^{(n-1)\times(n-1)}$. Note that A can be written as

$$A = \begin{pmatrix} B & \mathbf{b} \\ \mathbf{b}^{\mathsf{H}} & a \end{pmatrix},$$

where $B \in \mathbb{C}^{(n-1)\times(n-1)}$ is a Hermitian matrix. Since the leading minors of B are the first $n-1$ leading minors of A it follows, by the inductive hypothesis, that B is positive definite. Thus, there exists a Cholesky decomposition $B = R^{\mathsf{H}}R$, where R is an upper triangular matrix with real, positive diagonal elements. Since R is invertible, let $\mathbf{w} = (R^{\mathsf{H}})^{-1}b$.

The matrix B is invertible. By Theorem 5.154, we have $\det(A) = \det(B)(a - \mathbf{b}^{\mathsf{H}}B^{-1}\mathbf{b}) > 0$. Since $\det(B) > 0$ it follows that $a \geqslant \mathbf{b}^{\mathsf{H}}B^{-1}\mathbf{b}$. We observed that if B is positive definite, then so is B^{-1}. Therefore, $a \geqslant 0$ is and we can write $a = c^2$ for some positive c. This allows us to write

$$A = \begin{pmatrix} R^{\mathsf{H}} & \mathbf{0} \\ \mathbf{w}^{sH} & c \end{pmatrix} \begin{pmatrix} R & \mathbf{w} \\ \mathbf{0}^{\mathsf{H}} & c \end{pmatrix} = C^{\mathsf{H}}C,$$

where C is the upper triangular matrix with positive e

$$C = \begin{pmatrix} R & \mathbf{w} \\ \mathbf{0}^{\mathsf{H}} & c \end{pmatrix}$$

This implies immediately the positive definiteness of A.

Let $A, B \in \mathbb{C}^{n\times n}$. We write $A \succ B$ if $A - B \succ 0$, that is, if $A - B$ is a positive definite matrix. Similarly, we write $A \succeq B$ if $A - B \succeq O$, that is, if $A - B$ is positive semidefinite.

Theorem 6.127 *Let $A_0, A_1, \ldots, A_m$ be $m+1$ matrices in $\mathbb{C}^{n\times n}$ such that A_0 is positive definite and all matrices are Hermitian. There exists $a > 0$ such that for any $t \in [-a, a]$ the matrix $B_m(t) = A_0 + A_1 t + \cdots + A_m t^m$ is positive definite.*

Proof Since all matrices $A_0, \ldots, A_m$ are Hermitian, note that $\mathbf{x}^{\mathsf{H}}A_i\mathbf{x}$ are real numbers for $0 \leqslant i \leqslant m$. Therefore, $p_m(t) = \mathbf{x}^{\mathsf{H}}B_m(t)\mathbf{x}$ is a polynomial in t with real coefficients and $p_m(0) = \mathbf{x}^{\mathsf{H}}A_0\mathbf{x}$ is a positive number if $\mathbf{x} \neq \mathbf{0}$. Since p_m is a continuous function there exists an interval $[-a, a]$ such that $t \in [-a, a]$ implies $p_m(t) > 0$ if $\mathbf{x} \neq \mathbf{0}$. This shows that $B_m(t)$ is positive definite.

6.11 The Gram-Schmidt Orthogonalization Algorithm

The Gram-Schmidt algorithm starts with a basis $\{\mathbf{u}_1, \ldots, \mathbf{u}_m\}$ of an m-dimensional space U of $\mathbb{C}^n$ and generates an orthonormal basis $\{\mathbf{q}_1, \ldots, \mathbf{q}_m\}$ of the same subspace. Clearly, we have $m \leqslant n$.

The algorithm starts with the sequence of vectors $(\mathbf{u}_1, \ldots, \mathbf{u}_m)$ and constructs the sequence of orthonormal vectors $(\mathbf{q}_1, \ldots, \mathbf{q}_m)$ such that

$$\langle \mathbf{u}_1, \ldots, \mathbf{u}_k \rangle = \langle \mathbf{q}_1, \ldots, \mathbf{q}_k \rangle$$

for $1 \leqslant k \leqslant m$ as follows:

$$\begin{aligned}
\mathbf{z}_1 &= \mathbf{u}_1, & \mathbf{q}_1 &= \frac{1}{\|\mathbf{z}_1\|}\mathbf{z}_1,\\
\mathbf{z}_2 &= \mathbf{u}_2 - \mathsf{proj}_{\mathbf{z}_1}(\mathbf{u}_2), & \mathbf{q}_2 &= \frac{1}{\|\mathbf{z}_2\|}\mathbf{z}_2,\\
\mathbf{z}_3 &= \mathbf{u}_3 - \mathsf{proj}_{\mathbf{z}_1}(\mathbf{u}_3) - \mathsf{proj}_{\mathbf{z}_2}(\mathbf{u}_3), & \mathbf{q}_3 &= \frac{1}{\|\mathbf{z}_3\|}\mathbf{z}_3,\\
&\vdots & &\vdots\\
\mathbf{z}_m &= \mathbf{u}_m - \mathsf{proj}_{\mathbf{z}_1}(\mathbf{u}_m) - \cdots - \mathsf{proj}_{\mathbf{z}_{m-1}}(\mathbf{u}_m), & \mathbf{q}_m &= \frac{1}{\|\mathbf{z}_m\|}\mathbf{z}_m.
\end{aligned}$$

The algorithm can be written in pseudo-code as shown in Algorithm 6.11.1.

Lemma 6.128 *The sequence $(z_1, \ldots, z_m)$ constructed by the Gram-Schmidt algorithm consists of pairwise orthogonal vectors; furthermore, the sequence $(q_1, \ldots, q_m)$ consists of pairwise orthonormal vectors.*

Proof Note that

$$\mathbf{z}_k = \mathbf{u}_k - \sum_{j=1}^{k-1} \mathsf{proj}_{\langle \mathbf{z}_1,\ldots,\mathbf{z}_{k-1}\rangle}(\mathbf{u}_k)$$

This implies that $\mathbf{z}_k$ is orthogonal on the subspace $\langle \mathbf{z}_1, \ldots, \mathbf{z}_{k-1}\rangle$, that is, on all its predecessors in the sequence. The second part of the lemma follows immediately.

Lemma 6.129 *We have $\langle q_1, \ldots, q_k\rangle = \langle u_1, \ldots, u_k\rangle$ for $1 \leqslant k \leqslant m$.*

Proof The proof is by induction on k. The base step, $k = 1$, is immediate.

Suppose that the equality holds for k. Since $\mathbf{q}_j$ and $\mathbf{z}_j$ determine the same one-dimensional subspace, we have $\mathsf{proj}_{\mathbf{z}_j}(\mathbf{u}_k) = \mathsf{proj}_{\mathbf{q}_j}(\mathbf{u}_k)$. Thus,

$$\begin{aligned}
\mathbf{q}_{k+1} &= \frac{1}{\| \mathbf{z}_{k+1} \|}\left(u_{k+1} - \sum_{j=1}^{k} \mathsf{proj}_{\mathbf{q}_j}(\mathbf{u}_{k+1})\right)\\
&= \frac{1}{\| \mathbf{z}_{k+1} \|}\left(u_{k+1} - \mathsf{proj}_{\langle \mathbf{q}_1,\ldots,\mathbf{q}_k\rangle}(\mathbf{u}_{k+1})\right).
\end{aligned}$$

By the inductive hypothesis,

$$\mathbf{q}_{k+1} = \frac{1}{\| \mathbf{z}_{k+1} \|}\left(u_{k+1} - \mathsf{proj}_{\langle \mathbf{u}_1,\ldots,\mathbf{u}_k\rangle}(\mathbf{u}_{k+1})\right),$$

which implies $\mathbf{q}_{k+1} \in \langle \mathbf{u}_1, \ldots, \mathbf{u}_{k+1}\rangle$. Thus, $\langle \mathbf{q}_1, \ldots, \mathbf{q}_{k+1}\rangle \subseteq \langle \mathbf{u}_1, \ldots, \mathbf{u}_{k+1}\rangle$.

For the reverse inclusion, note that

$$\begin{aligned}\mathbf{u}_{k+1} &= \mathbf{z}_{k+1} + \sum_{j=1}^{k} \mathsf{proj}_{\mathbf{z}_j}(\mathbf{u}_{k+1}) \\ &= \mathbf{z}_{k+1} + \mathsf{proj}_{\langle \mathbf{z}_1,\ldots,\mathbf{z}_k\rangle}(\mathbf{u}_{k+1}) \\ &= \parallel \mathbf{z}_{k+1} \parallel \mathbf{q}_{k+1} + \mathsf{proj}_{\langle \mathbf{q}_1,\ldots,\mathbf{q}_k\rangle}(\mathbf{u}_{k+1}),\end{aligned}$$

which implies that $\mathbf{u}_{k+1} \in \langle \mathbf{q}_1, \ldots, \mathbf{q}_{k+1}\rangle$. Therefore,

$$\langle \mathbf{u}_1, \ldots, \mathbf{u}_{k+1}\rangle \subseteq \langle \mathbf{q}_1, \ldots, \mathbf{q}_{k+1}\rangle,$$

which concludes the argument.

Algorithm 6.11.1: Gram-Schmidt Orthogonalization Algorithm

Data: A basis $\{\mathbf{u}_1, \ldots, \mathbf{u}_m\}$ for a subspace U of $\mathbb{C}^n$
Result: An orthonormal basis $\{\mathbf{q}_1, \ldots, \mathbf{q}_m\}$ for U

$\mathbf{z}_1 = \mathbf{u}_1$; $\mathbf{q}_1 = \frac{1}{\|\mathbf{z}_1\|}\mathbf{z}_1$;
for $k = 2$ *to* m **do**
  $\mathbf{z}_k = \mathbf{u}_k - \sum_{j=1}^{k-1} \mathsf{proj}_{\mathbf{z}_j}(\mathbf{u}_k)$;
  $\mathbf{q}_k = \frac{1}{\|\mathbf{z}_k\|}\mathbf{z}_k$;
end
return $Q = (\mathbf{q}_1 \cdots \mathbf{q}_m)$;

Theorem 6.130 *Let $(\boldsymbol{q}_1, \ldots, \boldsymbol{q}_m)$ be the sequence of vectors constructed by the Gram-Schmidt algorithm starting from the basis $\{\boldsymbol{u}_1, \ldots, \boldsymbol{u}_m\}$ of an m-dimensional subspace U of $\mathbb{C}^n$. The set $\{\boldsymbol{q}_1, \ldots, \boldsymbol{q}_m\}$ is an orthogonal basis of U and $\langle \boldsymbol{q}_1, \ldots, \boldsymbol{q}_k\rangle = \langle \boldsymbol{u}_1, \ldots \boldsymbol{u}_k\rangle$ for $1 \leqslant k \leqslant m$.*

Proof This statement follows immediately from Lemmas 6.128 and 6.129.

Example 6.131 Let $A \in \mathbb{R}^{3\times 2}$ be the matrix

$$A = \begin{pmatrix} 1 & 1 \\ 0 & 0 \\ 1 & 3 \end{pmatrix}.$$

It is easy to see that $rank(A) = 2$. We have $\{\mathbf{u}_1, \mathbf{u}_2\} \subseteq \mathbb{R}^3$ and we construct an orthogonal basis for the subspace generated by these columns. The matrix W is intialized to $O_{3,2}$.

By Algorithm 6.11.1, we begin by defining $\mathbf{z}_1 = \mathbf{u}_1$ and

$$\mathbf{q}_1 = \frac{1}{\parallel \mathbf{z}_1 \parallel}\mathbf{z}_1 = \begin{pmatrix} \frac{\sqrt{2}}{2} \\ 0 \\ \frac{\sqrt{2}}{2} \end{pmatrix}$$

because $\| \mathbf{z}_1 \| = \sqrt{2}$. Next, we have

$$\begin{aligned}\mathbf{z}_2 &= \mathbf{u}_2 - \mathsf{proj}_{\mathbf{z}_1}(\mathbf{u}_2) = \mathbf{u}_2 - 2\mathbf{z}_1\\ &= \begin{pmatrix}-1\\0\\1\end{pmatrix},\end{aligned}$$

which implies

$$\mathbf{q}_2 = \begin{pmatrix}-\frac{\sqrt{2}}{2}\\0\\\frac{\sqrt{2}}{2}\end{pmatrix}.$$

Thus, the orthonormal basis we are seeking consists of the vectors

$$\begin{pmatrix}\frac{\sqrt{2}}{2}\\0\\\frac{\sqrt{2}}{2}\end{pmatrix} \text{ and } \begin{pmatrix}-\frac{\sqrt{2}}{2}\\0\\\frac{\sqrt{2}}{2}\end{pmatrix}.$$

Using the Gram-Schmidt algorithm we can factor a matrix as a product of a matrix having orthogonal columns and an upper triangular matrix. This useful matrix decomposition (described in the next theorem) is known as the *QR-decomposition.*

Theorem 6.132 (Reduced QR Decomposition) *Let $U \in \mathbb{C}^{n\times m}$ be a full-rank matrix, where $m \leqslant n$. There exists a matrix $Q \in \mathbb{C}^{n\times m}$ having a set of orthonormal columns and an upper triangular matrix $R \in \mathbb{C}^{m\times m}$ such that $U = QR$. Furthermore, the diagonal entries of R are non-zero.*

Proof Let $U = (\mathbf{u}_1 \ \cdots \ \mathbf{u}_m)$. We use the same notations as above. Since $\mathbf{u}_k \in \langle \mathbf{q}_1, \ldots, \mathbf{q}_k\rangle$ for $1 \leqslant k \leqslant m$, it follows that we can write the equalities

$$\begin{aligned}\mathbf{u}_1 &= \mathbf{q}_1 r_{11},\\ \mathbf{u}_2 &= \mathbf{q}_1 r_{12} + \mathbf{q}_2 r_{22},\\ \mathbf{u}_3 &= \mathbf{q}_1 r_{13} + \mathbf{q}_2 r_{23} + \mathbf{q}_3 r_{33},\\ &\vdots\\ \mathbf{u}_m &= \mathbf{q}_1 r_{1m} + \mathbf{q}_2 r_{2m} + \mathbf{q}_3 r_{3m} + \cdots + \mathbf{q}_m r_{mm}.\end{aligned}$$

In matrix form these equalities are

$$(\mathbf{u}_1, \ldots, \mathbf{u}_m) = (\mathbf{q}_1, \ldots, \mathbf{q}_m) \begin{pmatrix} r_{11} & r_{12} & \cdots & r_{1m} \\ 0 & r_{22} & \cdots & r_{2m} \\ 0 & 0 & \cdots & r_{3m} \\ \vdots & \vdots & \vdots & \vdots \\ 0 & 0 & \cdots & r_{mm} \end{pmatrix}.$$

Thus, we have $Q = (\mathbf{q}_1 \ \cdots \ \mathbf{q}_m)$ and

$$R = \begin{pmatrix} r_{11} & r_{12} & \cdots & r_{1m} \\ 0 & r_{22} & \cdots & r_{2m} \\ 0 & 0 & \cdots & r_{3m} \\ \vdots & \vdots & \vdots & \vdots \\ 0 & 0 & \cdots & r_{mm} \end{pmatrix}.$$

Note that $r_{kk} \neq 0$ for $1 \leqslant k \leqslant m$. Indeed, if we were to have $r_{kk} = 0$ this would imply $\mathbf{u}_k = \mathbf{q}_1 r_{1k} + \mathbf{q}_2 r_{2k} + \mathbf{q}_3 r_{3k} + \cdots + \mathbf{q}_{k-1} r_{k-1\,k-1}$, which would contradict the equality $\langle \mathbf{q}_1, \ldots, \mathbf{q}_k \rangle = \langle \mathbf{u}_1, \ldots, \mathbf{u}_k \rangle$.

Theorem 6.133 **(Full QR Decomposition)** *Let $U \in \mathbb{C}^{n\times m}$ be a full-rank matrix, where $m \leqslant n$. There exists a unitary matrix $Q \in \mathbb{C}^{m\times m}$ and an upper triangular matrix $R \in \mathbb{C}^{n\times m}$ such that $U = QR$.*

Proof Let $U = Q_1 R_1$ be the reduced QR decomposition of U, where $Q_1 = (\mathbf{q}_1 \ \cdots \ \mathbf{q}_m)$ is a matrix having an orthonormal set of columns. This set can be extended to an orthonormal basis $\{\mathbf{q}_1, \ldots, \mathbf{q}_m, \mathbf{q}_{m+1}, \ldots, \mathbf{q}_n\}$. of $\mathbb{C}^n$. If $Q = (\mathbf{q}_1; \cdots \ \mathbf{q}_m \ \mathbf{q}_{m+1} \ \cdots \ \mathbf{q}_n)$ and R is the matrix obtained from R by adding $n - m$ rows equal to $\mathbf{0}$, then $U = QR$ is the desired full decomposition of U.

Corollary 6.134 *Let $U \in \mathbb{C}^{n\times n}$ be a non-singular square matrix. There exists a unitary matrix $Q \in \mathbb{C}^{n\times n}$ and an upper triangular matrix $R \in \mathbb{C}^{n\times n}$ such that $U = QR$.*

Proof The corollary is a direct consequence of Theorem 6.133.

Theorem 6.135 *If $L = (\boldsymbol{v}_1, \ldots, \boldsymbol{v}_m)$ is a sequence of m vectors in $\mathbb{R}^n$. We have*

$$\det(G_L) \leqslant \prod_{j=1}^{m} \| \boldsymbol{v}_j \|_2^2 .$$

The equality takes place only if the vectors of L are pairwise orthogonal.

Proof Suppose that L is linearly independent and construct the orthonormal set $\{\mathbf{y}_1, \ldots, \mathbf{y}_m\}$, where $\mathbf{y}_j = b_{j1}\mathbf{v}_1 + \cdots + b_{jj}\mathbf{v}_j$ for $1 \leqslant j \leqslant m$, using the Gram-Schmidt algorithm. Since $b_{jj} \neq 0$ it follows that we can write

$$\mathbf{v}_j = c_{j1}\mathbf{y}_1 + \cdots + c_{jj}\mathbf{y}_j$$

for $1 \leqslant j \leqslant m$ so that $(\mathbf{v}_j, \mathbf{y}_p) = 0$ if $j < p$ and $(\mathbf{v}_j, \mathbf{y}_p) = c_{jp}$ if $p \leqslant j$. Thus, we have

$$(\mathbf{v}_1, \ldots, \mathbf{v}_m) = (\mathbf{y}_1, \ldots, \mathbf{y}_m) \begin{pmatrix} (\mathbf{v}_1, \mathbf{y}_1) & (\mathbf{v}_2, \mathbf{y}_1) & \cdots & (\mathbf{v}_m, \mathbf{y}_1) \\ 0 & (\mathbf{v}_2, \mathbf{y}_2) & \cdots & (\mathbf{v}_m, \mathbf{y}_2) \\ 0 & 0 & \cdots & (\mathbf{v}_m, \mathbf{y}_3) \\ \vdots & \vdots & \vdots & \vdots \\ 0 & 0 & \cdots & (\mathbf{v}_m, \mathbf{y}_m) \end{pmatrix}.$$

This implies

$$\begin{pmatrix} (\mathbf{v}_1, \mathbf{v}_1) & \cdots & (\mathbf{v}_1, \mathbf{v}_m) \\ \vdots & \cdots & \vdots \\ (\mathbf{v}_m, \mathbf{v}_1) & \cdots & (\mathbf{v}_m, \mathbf{v}_m) \end{pmatrix} = \begin{pmatrix} \mathbf{v}_1' \\ \vdots \\ \mathbf{v}_m' \end{pmatrix} (\mathbf{v}_1, \ldots, \mathbf{v}_m)$$

$$= \begin{pmatrix} (\mathbf{v}_1, \mathbf{y}_1) & 0 & 0 \\ (\mathbf{v}_2, \mathbf{y}_1) & (\mathbf{v}_2, \mathbf{y}_2) & 0 \\ \vdots & \vdots & \vdots \\ (\mathbf{v}_m, \mathbf{y}_1) & (\mathbf{v}_m, \mathbf{y}_2) & (\mathbf{v}_m, \mathbf{y}_m) \end{pmatrix} \begin{pmatrix} (\mathbf{v}_1, \mathbf{y}_1) & (\mathbf{v}_2, \mathbf{y}_1) & \cdots & (\mathbf{v}_m, \mathbf{y}_1) \\ 0 & (\mathbf{v}_2, \mathbf{y}_2) & \cdots & (\mathbf{v}_m, \mathbf{y}_2) \\ \vdots & \vdots & \vdots & \vdots \\ 0 & 0 & \cdots & (\mathbf{v}_m, \mathbf{y}_m) \end{pmatrix}.$$

Therefore, we have

$$\det(G_L) = \prod_{i=1}^{m} (\mathbf{v}_i, \mathbf{y}_i)^2 \leqslant \prod_{i=1}^{m} (\mathbf{v}_i, \mathbf{v}_i)^2,$$

because $(\mathbf{v}_i, \mathbf{y}_i)^2 \leqslant (\mathbf{v}_i, \mathbf{v}_i)^2 (\mathbf{y}_i, \mathbf{y}_i)^2$ and $(\mathbf{y}_i, \mathbf{y}_i) = 1$ for $1 \leqslant i \leqslant m$.

To have $\det(G_L) = \prod_{i=1}^{m} (\mathbf{v}_i, \mathbf{v}_i)^2$ we must have $\mathbf{v}_i = k_i \mathbf{y}_i$, that is, the vectors $\mathbf{v}_i$ must be pairwise orthogonal.

Definition 6.136 *A* Hadamard matrix *is a matrix $H \in \mathbb{R}^{nn}$ such that $h_{ij} \in \{1, -1\}$ for $1 \leqslant i, j \leqslant n$ and $HH' = nI_n$.*

Example 6.137 The matrices

$$A = \begin{pmatrix} 1 & 1 \\ 1 & -1 \end{pmatrix} \text{ and } B = \begin{pmatrix} 1 & 1 & 1 & 1 \\ 1 & 1 & -1 & -1 \\ 1 & -1 & -1 & 1 \\ 1 & -1 & 1 & -1 \end{pmatrix}$$

are Hadamard matrices in $\mathbb{R}^{2\times 2}$ and $\mathbb{R}^{4\times 4}$, respectively.

Corollary 6.138 *Let $A \in \mathbb{R}^{n\times n}$ be a matrix such that $|a_{ij}| \leqslant 1$ for $1 \leqslant i, j \leqslant n$. Then, $|\det(A)| \leqslant n^{\frac{n}{2}}$ and the equality holds only if A is a Hadamard matrix.*

Proof Let $\mathbf{a}_i = (a_{i1}, \ldots, a_{in})$ be the ith row of A. We have $\| \mathbf{a}_i \|_2 \leqslant \sqrt{n}$, so $|(\mathbf{a}_i, \mathbf{a}_j)| \leqslant n$ by Cauchy-Schwartz inequality, for $1 \leqslant i, j \leqslant n$.

Note that $G_L = A'A$, where L is the set of rows of A. Consequently, $\det(A)^2 = \det(G_L) \leqslant \prod_{j=1}^{n} \parallel \mathbf{v}_j \parallel_2^2$, so

$$|\det(A)| \leqslant \prod_{j=1}^{n} \parallel \mathbf{v}_j \parallel_2 \leqslant n^{\frac{n}{2}}.$$

To have the equality we must have $\parallel \mathbf{v}_j \parallel = \sqrt{n}$. This is possible only if $v_{jk} \in \{-1, 1\}$. This fact together with orthogonality of the vectors $\mathbf{v}_1, \ldots, \mathbf{v}_n$ implies that A is a Hadamard matrix.

We saw that orthonormal sets of vectors are linearly independent. This allows us to extend orthonormal set of vectors to orthonormal bases.

Theorem 6.139 *Let L be a finite-dimensional linear space. If U is an orthonormal set of vectors, then there exists a basis T of L that consists of orthonormal vectors such that $U \subseteq T$.*

Proof Let $U = \{\mathbf{u}_1, \ldots, \mathbf{u}_m\}$ be an orthonormal set of vectors in L. There is an extension of U, $Z = \{\mathbf{u}_1, \ldots, \mathbf{u}_m, \mathbf{u}_{m+1}, \ldots, \mathbf{u}_n\}$ to a basis of L, where $n = \dim(V)$, by Theorem 5.17. Now, apply the Gram-Schmidt algorithm to the set U to produce an orthonormal basis $W = \{\mathbf{w}_1, \ldots, \mathbf{w}_n\}$ for the entire space L. It is easy to see that $\mathbf{w}_i = \mathbf{u}_i$ for $1 \leqslant i \leqslant m$, so $U \subseteq W$ and W is the orthonormal basis of L that extends the set U.

Corollary 6.140 *If A is an $(m \times n)$-matrix with $m \geqslant n$ having orthonormal set of columns, then there exists an $(m \times (m-n))$-matrix B such that $(A \;\; B)$ is an orthogonal (unitary) square matrix.*

Proof This follows directly from Theorem 6.139.

Corollary 6.141 *Let U be a subspace of an n-dimensional linear space L such that $\dim(U) = m$, where $m < n$. Then $\dim(U^{\perp}) = n - m$.*

Proof Let $\mathbf{u}_1, \ldots, \mathbf{u}_m$ be an orthonormal basis of U, and let

$$\mathbf{u}_1, \ldots, \mathbf{u}_m, \mathbf{u}_{m+1}, \ldots, \mathbf{u}_n$$

be its completion to an orthonormal basis for L, which exists by Theorem 6.139. Then, $\mathbf{u}_{m+1}, \ldots, \mathbf{u}_n$ is a basis of the orthogonal complement $U^{\perp}$, so $\dim(U^{\perp}) = n - m$.

Theorem 6.142 *A subspace U of $\mathbb{R}^n$ is m-dimensional if and only if is the set of solution of an homogeneous linear system $A\mathbf{x} = \mathbf{0}$, where $A \in \mathbb{R}^{(n-m)\times n}$ is a full-rank matrix.*

Proof Suppose that U is an m-dimensional subspace of $\mathbb{R}^n$. If $\mathbf{v}_1, \ldots, \mathbf{v}_{n-m}$ is a basis of the orthogonal complement of U, then $\mathbf{v}_i'\mathbf{x} = 0$ for every $\mathbf{x} \in U$ and $1 \leqslant i \leqslant n-m$. These conditions are equivalent to the equality

$$(\mathbf{v}'_1\ \mathbf{v}'_2\ \cdots\ \mathbf{v}'_{n-m})\mathbf{x} = \mathbf{0},$$

which shows that U is the set of solution of an homogeneous linear $A\mathbf{x} = \mathbf{0}$, where $A = (\mathbf{v}'_1\ \mathbf{v}'_2\ \cdots\ \mathbf{v}'_{n-m})$.

Conversely, if $A \in \mathbb{R}^{(n-m)\times n}$ is a full-rank matrix, then the set of solutions of the homogeneous system $A\mathbf{x} = \mathbf{b}$ is the null subspace of A and, therefore is an m-dimensional subspace.

Exercises and Supplements

1. Let ν be a norm on $\mathbb{C}^n$. Prove that there exists a number $k \in \mathbb{R}$ such that for any vector $\mathbf{x} \in \mathbb{C}^n$ we have $\nu(\mathbf{x}) \leqslant k \sum_{i=1}^n |x_i|$.
2. Prove that $\nu(\mathbf{x}+\mathbf{y})^2 + \nu(\mathbf{x}-\mathbf{y})^2 \leqslant 4(\nu(\mathbf{x})^2 + \nu(\mathbf{y})^2)$ for every vector norm ν on $\mathbb{R}^n$ and $\mathbf{x}, \mathbf{y} \in \mathbb{R}^n$.
3. Let $\mathbf{a} \in \mathbb{R}^n$ be a vector such that $\mathbf{a} \geqslant \mathbf{0}_n$. Prove that
$$\left(\sum_{i=1}^n a_i\right)^2 \leqslant n \sum_{i=1}^n a_i^2.$$
4. Let (S, d) be a dissimilarity space, where d is a definite dissimilarity. Define the set
$$P(x, y, z) = \{p \in \mathbb{R}_{\geqslant 0} \mid d(x, y)^p \leqslant d(x, z)^p + d(z, y)^p\}$$
for $x, y, z \in S$. Prove that
   (a) $P(x, y, z) \neq \emptyset$ for $x, y, z \in S$;
   (b) if $p \in P(x, y, z)$ and $q \leqslant p$, then $q \in P(x, y, z)$;
   (c) if $\sup P(x, y, z) \geqslant 1$ for all $x, y, z \in S$, then d is a metric on S;
   (d) if $\sup P(x, y, z) = \infty$ for all $x, y, z \in S$, then d is an ultrametric on S.
5. Let $u, v \in \mathbb{C}$. Prove that:
   (a) $|\bar{u}v - u\bar{v}|^2 = 2|u|^2|v|^2 - \bar{u}^2v^2 - u^2\bar{v}^2$;
   (b) if $\mathbf{x} \in \mathbb{C}$, $\parallel \mathbf{x} \parallel = 1$, and $s = \sum_{i=1}^n x_i^2$, we have $\sum_{i=1}^n \sum_{j=1}^n |\bar{x}_i x_j - x_j \bar{x}_i|^2 \leqslant 2 - 2|s|^2$.
6. Let $\{\mathbf{q}_1, \ldots, \mathbf{q}_n\} \subseteq \mathbb{C}^n$ be an orthonormal set of n vectors. If $\{I, J\}$ is a partition of $\{1, \ldots, n\}$, $I = \{i_1, \ldots, i_p\}$ and $J = \{j_1, \ldots, j_q\}$, prove that the subspaces $S = \langle \mathbf{q}_{i_1}, \ldots, \mathbf{q}_{i_p} \rangle$ and $T = \langle \mathbf{q}_{j_1}, \ldots, \mathbf{q}_{j_q} \rangle$ are complementary.
7. Let $Q = (\mathbf{q}_1\ \cdots\ \mathbf{q}_k) \in \mathbb{C}^{n\times k}$ be a matrix having a set of orthonormal columns. Prove that $I_n - QQ^{\mathsf{H}} = \prod_{j=1}^k (I_n - \mathbf{q}_j \mathbf{q}_j^{\mathsf{H}})$.

Solution: The equality to be shown amounts to $I_n - \sum_{j=1}^{k} \mathbf{q}_j \mathbf{q}_j^{\mathsf{H}} = \prod_{j=1}^{k}(I_n - \mathbf{q}_j \mathbf{q}_j^{\mathsf{H}})$, and the argument is by induction on k. The base case, $k = 1$ is immediate. Suppose that the equality holds for k and let $\{\mathbf{q}_1, \ldots, \mathbf{q}_k, \mathbf{q}_{k+1}\}$ be a set of othonormal vectors. We have

$$\begin{aligned}\prod_{j=1}^{k+1}(I_n - \mathbf{q}_j \mathbf{q}_j^{\mathsf{H}}) &= \prod_{j=1}^{k}(I_n - \mathbf{q}_j \mathbf{q}_j^{\mathsf{H}})(I_n - \mathbf{q}_{k+1} \mathbf{q}_{k+1}^{\mathsf{H}}) \\ &= \left(I_n - \sum_{j=1}^{k} \mathbf{q}_j \mathbf{q}_j^{\mathsf{H}}\right)(I_n - \mathbf{q}_{k+1} \mathbf{q}_{k+1}^{\mathsf{H}}) \\ &\quad \text{(by the inductive hypothesis)} \\ &= I_n - \sum_{j=1}^{k} \mathbf{q}_j \mathbf{q}_j^{\mathsf{H}} - \mathbf{q}_{k+1} \mathbf{q}_{k+1}^{\mathsf{H}} + \left(\sum_{j=1}^{k} \mathbf{q}_j \mathbf{q}_j^{\mathsf{H}}\right) \mathbf{q}_{k+1} \mathbf{q}_{k+1}^{\mathsf{H}}.\end{aligned}$$

Since $\mathbf{q}_j^{\mathsf{H}} \mathbf{q}_{k+1} = 0$ for $1 \leqslant j \leqslant k$, the inductive step is concluded.

8. Let $X = \{x_1, \ldots, x_m\}$ be a set and let S be a subset of X. The characteristic vector of S is $\mathbf{c}_S \in \{0, 1\}^m$ whose components $c_1, \ldots, c_m$ are defined by $c_i = 1$ if $x_i \in S$ and $c_i = 0$, otherwise. Prove that

   (a) $\| \mathbf{c}_S \|_2 = |S|$;
   (b) if $S, T \subseteq X$, then $\mathbf{c}_S' \mathbf{c}_T = |S \cap T|$;
   (c) if $S \subseteq X$, then

$$\sum\{(c_i - c_j)^2 \mid 1 \leqslant i < j \leqslant m\} = |S| \cdot |X - S|.$$

   Solution: We solve only the third part. Without loss of generality assume that $S = \{x_1, \ldots, x_p\}$. Then, $c_i = 1$ for $1 \leqslant i \leqslant p$ and $c_i = 0$ for $p+1 \leqslant i \leqslant m$. The contribution of the terms of the form $(c_i - c_j)^2$, where $1 \leqslant i \leqslant p$ equals $m - p$ and there are p such terms. Thus, $\sum\{(c_i - c_j)^2 \mid 1 \leqslant i < j \leqslant m\} = p(m - p) = |S| \cdot |X - S|$.

9. Let $\mathbf{x}, \mathbf{y} \in \mathbb{C}^n$. Prove that $trace(\mathbf{x}\mathbf{y}^{\mathsf{H}}) = \overline{\mathbf{x}^{\mathsf{H}}\mathbf{y}}$.
10. Let $\mathbf{x}, \mathbf{y} \in \mathbb{C}^n$ and let M, P be the matrices $M = \mathbf{x}\mathbf{x}^{\mathsf{H}}$ and $P = \mathbf{y}\mathbf{y}^{\mathsf{H}}$. Prove that for the inner product on $\mathbb{C}^{n \times n}$, defined in Example 6.26, we have $(M, P) = |\mathbf{x}^{\mathsf{H}}\mathbf{y}|^2$.
11. Let $\mathbf{x} \in \mathbb{R}^n$. Prove that for every $\epsilon > 0$ there exists $\mathbf{y} \in \mathbb{R}^n$ such that the components of the vector $\mathbf{x} + \mathbf{y}$ are distinct and $\| \mathbf{y} \|_2 < \epsilon$.

    Solution: Partition the set $\{1, \ldots, n\}$ into the blocks $B_1, \ldots, B_k$ such that all components of $\mathbf{x}$ that have an index in B_j have a common value c_j. Suppose that $|B_j| = p_j$. Then, $\sum_{j=1}^{k} p_j = n$ and the numbers $\{c_1, c_2, \ldots, c_k\}$ are pairwise distinct. Let $d = \min_{i,j} |c_i - c_j|$. The vector $\mathbf{y}$ can be defined as follows. If $B_j = \{i_1, \ldots, i_{p_j}\}$, then

$$y_{i_1} = \eta \cdot 2^{-1}, y_{i_2} = \eta \cdot 2^{-2}, \ldots, y_{i_{p_j}} = \eta \cdot 2^{-p},$$

where $\eta > 0$, which makes the numbers $c_j + y_{i_1}, c_j + y_{i_2}, \ldots, c_j + y_{i_{p_j}}$ pairwise distinct. It suffices to take $\eta < d$ to ensure that the components of $\mathbf{x} + \mathbf{y}$ are pairwise distinct. Also, note that $\parallel \mathbf{y} \parallel_2^2 \leqslant \sum_{j=1}^{k} p_j \frac{\eta^2}{4} = \frac{n\eta^2}{4}$. It suffices to choose η such that $\eta < \min\{d, \frac{2\epsilon}{n}\}$ to ensure that $\parallel \mathbf{y} \parallel_2 < \epsilon$.

12. Prove that the norms ν_1 and ν_∞ on $\mathbb{R}^n$ are not generated by an inner product.
13. Prove that if $0 < p < 1$, ν_p is not a norm on $\mathbb{R}^n$.
14. The number $\zeta(\mathbf{x})$, defined as the number of non-zero components of the vector $\mathbf{x} \in \mathbb{R}^n$, is refered to as the *zero-norm* of $\mathbf{x}$ (even though it is not a norm in the sense of Definition 6.9) and is used in the study of linear models in machine learning (see [1]). Prove that $\zeta(\mathbf{x}) = \lim_{p \to 0} (\nu_p(\mathbf{x}))^p$ and that $\lim_{p \to 0} n^{-\frac{1}{p}} \nu_p(\mathbf{x})$ equals the geometric mean of the absolute values of the components of $\mathbf{x}$.
15. Let ν be a norm on $\mathbb{R}^n$ that satisfies the parallelogram equality. Prove that the function $p : \mathbb{R}^n \times \mathbb{R}^n \longrightarrow \mathbb{R}$ given by

$$p(\mathbf{x}, \mathbf{y}) = \frac{1}{4} \left(\nu(\mathbf{x} + \mathbf{y})^2 - \nu(\mathbf{x} - \mathbf{y})^2 \right)$$

is an inner product on $\mathbb{R}^n$.

16. Let $\{\mathbf{w}_1, \ldots, \mathbf{w}_k\} \subseteq \mathbb{C}^n$ be a set of unit vectors such that $\mathbf{w}_i \perp \mathbf{w}_j$ for $i \neq j$ and $1 \leqslant i, j \leqslant k$. If $W_k = (\mathbf{w}_1 \ \cdots \ \mathbf{w}_k) \in \mathbb{C}^{n \times k}$, prove that

$$I_n - W_k W_k^{\mathsf{H}} = (I_n - \mathbf{w}_k \mathbf{w}_k^{\mathsf{H}}) \cdots (I_n - \mathbf{w}_1 \mathbf{w}_1^{\mathsf{H}}).$$

17. Let $\mu^{(m,n)} : \mathbb{C}^{m \times n} \longrightarrow \mathbb{R}_{\geqslant 0}$ be a vectorial matrix norm. Prove that for every $A \in \mathbb{C}^{m \times n}$ there exists a constant $k \in \mathbb{R}$ such that $\mu^{(m,n)}(A) \leqslant k \sum_{i=1}^{m} \sum_{j=1}^{n} |a_{ij}|$.
18. We use here the notations introduced in Theorem 6.57. Prove that if $A \in \mathbb{C}^{n \times n}$ is a Hermitian matrix such that $AU = UB$ for some $U \in \mathbb{C}^{n \times k}$ having an orthonormal set of columns, then we can write:

$$(U \ V)^{\mathsf{H}} A (U V) = \begin{pmatrix} B & O_{k,n-k} \\ O_{n-k,k} & V^{\mathsf{H}} A V \end{pmatrix}$$

for some matrix $V \in \mathbb{C}^{n \times (n-k)}$ such that $(U \ V) \in \mathbb{C}^{n \times n}$ is a unitary matrix.

19. Prove that a matrix $A \in \mathbb{C}^{n \times n}$ is normal if and only if $\parallel A\mathbf{x} \parallel_2 = \parallel A^{\mathsf{H}}\mathbf{x} \parallel_2$ for every $\mathbf{x} \in \mathbb{C}^n$.
20. Prove that for every matrix $A \in \mathbb{C}^{n \times n}$ we have $|||A|||_2 = |||A^{\mathsf{H}}|||_2$.
21. Let $A \in \mathbb{R}^{m \times n}$. Prove that there exists i, $1 \leqslant i \leqslant n$ such that $\parallel A\mathbf{e}_i \parallel_2^2 \geqslant \frac{1}{n} \parallel A \parallel_F^2$.
22. Let $U \in \mathbb{C}^{n \times n}$ be a matrix whose set of columns is orthonormal and let $V \in \mathbb{C}^{n \times n}$ be a matrix whose set of rows is orthonormal. Prove that $|||UAV|||_F = |||A|||_F$.

23. Let $M \in \mathbb{R}^{n\times n}$ be a positive definite matrix. Prove that $\mathbf{x}'M\mathbf{x}+\mathbf{y}'M\mathbf{y} \geqslant \mathbf{x}'M\mathbf{y}+\mathbf{y}'M\mathbf{x}$ for $\mathbf{x}, \mathbf{y} \in \mathbb{R}^n$.
24. Let $H \in \mathbb{C}^{n\times n}$ be a non-singular matrix. Prove that the function $f : \mathbb{C}^{n\times n} \longrightarrow \mathbb{R}_{\geqslant 0}$ defined by $f(X) = \parallel HXH^{-1} \parallel_2$ for $X \in \mathbb{C}^{n\times n}$ is a matrix norm.
25. Let $A \in \mathbb{C}^{m\times n}$ and $B \in \mathbb{C}^{p\times q}$ be two matrices. Prove that $\parallel A \otimes B \parallel_F^2 = trace(A'A \otimes B'B)$.
26. Let $\mathbf{x}_0, \mathbf{x}, \mathbf{y} \in \mathbb{R}^n$. Prove that if $t \in [0, 1]$ and $\mathbf{u} = t\mathbf{x}+(1-t)\mathbf{y}$, then $\parallel \mathbf{x}_0 - \mathbf{u} \parallel_2 \leqslant \max\{\parallel \mathbf{x}_0 - \mathbf{x} \parallel_2, \parallel \mathbf{x}_0 - \mathbf{y} \parallel_2\}$.
27. Let $\mathbf{u}_1, \ldots, \mathbf{u}_m$ be m unit vectors in $\mathbb{R}^2$, such that $\parallel \mathbf{u}_i - \mathbf{u}_j \parallel = 1$. Prove that $m \leqslant 6$.
28. Prove that if $A \in \mathbb{C}^{n\times n}$ is an invertible matrix, then $\mu(A) \geqslant \frac{1}{\mu(A^{-1})}$ for any matrix norm μ.
29. Let $\parallel \cdot \parallel$ be an unitarily invariant norm. Prove that $\parallel A - I_n \parallel \leqslant \parallel A - U \parallel \leqslant \parallel A + I_n \parallel$ for every Hermitian matrix $A \in \mathbb{C}^{n\times n}$ and every unitary matrix U.
30. Let $A \in \mathbb{C}^{n\times n}$ be an invertible matrix and let $\parallel \cdot \parallel$ be a norm on $\mathbb{C}^n$. Prove that

$$|||A^{-1}||| = \frac{1}{\min\{\parallel A\mathbf{x} \parallel \mid \parallel \mathbf{x} \parallel = 1\}},$$

where $|||\cdot|||$ is the matrix norm generated by $\parallel \cdot \parallel$.

31. Let $Y \in \mathbb{C}^{n\times p}$ be a matrix that has an orthonormal set of columns, that is, $Y^{\mathsf{H}}Y = I_p$. Prove that:

(a) $\parallel Y \parallel_F = p$;
(b) for every matrix $R \in \mathbb{C}^{p\times q}$ we have $\parallel YR \parallel_F = \parallel R \parallel_F$.

32. Let $D \in \mathbb{R}^{n\times n}$ be a diagonal matrix such that $d_{ii} \geqslant 0$ for $1 \leqslant i \leqslant n$. Prove that if X is an orthogonal matrix, then $trace(XD) \leqslant trace\,(D)$.

33. Let

$$X = \begin{pmatrix} 0 & 1 \\ 1 & 0 \end{pmatrix}, Y = \begin{pmatrix} 0 & -i \\ i & 0 \end{pmatrix}, Z = \begin{pmatrix} 1 & 0 \\ 0 & -1 \end{pmatrix},$$

be the Pauli matrices defined in Example 6.45 Prove that

(a) the Pauli matrices are pairwise orthogonal;
(b) we have the equalities $XY = iZ$, $YZ = iX$, and $ZX = iY$;
(c) if W is a Hermitian matrix in $\mathbb{C}^{2\times 2}$ such that $traceW = 0$, then W is a linear combination of the Pauli matrices.

34. Prove that if $\mathbf{x}, \mathbf{y} \in \mathbb{C}^n$ are such that $\parallel \mathbf{x} \parallel_2 = \parallel \mathbf{y} \parallel_2$, then $\mathbf{x}+\mathbf{y} \perp \mathbf{x}-\mathbf{y}$.
35. If S is a subspace of $\mathbb{C}^n$, prove that $(S^\perp)^\perp = S$.
36. Prove that every permutation matrix P_ϕ is orthogonal.
37. Let $A \in \mathbb{R}^{n\times n}$ be a symmetric matrix. Prove that

$$(\mathbf{x}, A\mathbf{x}) - (\mathbf{y}, A\mathbf{y}) = (A(\mathbf{x}-\mathbf{y}), \mathbf{x}+\mathbf{y})$$

for every $\mathbf{x}, \mathbf{y} \in \mathbb{R}^n$.

38. Let $\mathbf{x}, \mathbf{y} \in \mathbb{R}^n$ be two unit vectors. Prove that

$$|\sin \angle(\mathbf{x}, \mathbf{y})| = \frac{\| \mathbf{x}+\mathbf{y} \|_2 \| \mathbf{x}-\mathbf{y} \|_2}{2}.$$

39. Let $\mathbf{u}$ and $\mathbf{v}$ be two unit vectors in $\mathbb{R}^n$. Prove that

    (a) if $\alpha = \angle(\mathbf{u}, \mathbf{v})$, then $\| \mathbf{u} - \mathbf{v}\cos\alpha \|_2 = \sin\alpha$;
    (b) $\mathbf{v}\cos\alpha$ is the closest vector in $\langle \mathbf{v} \rangle$ to $\mathbf{u}$.

40. Let $\{\mathbf{v}_1, \ldots, \mathbf{v}_p\} \subseteq \mathbb{R}^n$ be a collection of p unit vectors such that $\angle(\mathbf{v}_i, \mathbf{v}_j) = \theta$, where $0 < \theta \leqslant \frac{\pi}{2}$ for every pair $(\mathbf{v}_i, \mathbf{v}_j)$ such that $1 \leqslant i, j \leqslant p$ and $i \neq j$. Prove that $p \leqslant \frac{n(n+1)}{2}$.
41. Let $\mathbb{C}^{n \times n}$ be the linear space of complex matrices. Prove that:

    (a) the set of Hermitian matrices $\mathcal{H}$ and the set of skew-Hermitian matrices $\mathcal{K}$ in $\mathbb{C}^{n \times n}$ are subspaces of $\mathbb{C}^{n \times n}$;
    (b) if $\mathbb{C}^{n \times n}$ is equipped with the inner product defined in Example 6.26, then $\mathcal{K} = \mathcal{H}^{\perp}$.

42. Give an example of a matrix that has positive elements but is not positive definite.
43. Let $M \in \mathbb{R}^{n \times n}$ be a positive definite matrix. Prove that $\mathbf{x}'M\mathbf{x}+\mathbf{y}'M\mathbf{y} \geqslant \mathbf{x}'M\mathbf{y}+\mathbf{y}'M\mathbf{x}$ for $\mathbf{x}, \mathbf{y} \in \mathbb{R}^n$.
44. Prove that if $A \in \mathbb{R}^{n \times n}$ is a positive definite matrix, then A is invertible and A^{-1} is also positive definite.
45. Let $A \in \mathbb{C}^{n \times n}$ be a positive definite Hermitian matrix. If $A = B + iC$, where $B, C \in \mathbb{R}^{n \times n}$, prove that the real matrix

$$D = \begin{pmatrix} B & -C \\ C & B \end{pmatrix}$$

is positive definite.

46. Let L be a real linear space and let $\| \cdot \|$ be a norm generated by an inner product defined on L. L is said to be *symmetric relative to the norm* if $\| a\mathbf{x} - \mathbf{y} \| = \| \mathbf{x} - a\mathbf{y} \|$ for $a \in \mathbb{R}$ and $\mathbf{x}, \mathbf{y} \in V$ such that $\| \mathbf{x} \| = \| \mathbf{y} \| = 1$.

    (a) Prove that if a norm on a linear vector space L is induced by an inner product, then L is symmetric relative to that norm.
    (b) Prove that L satisfies the *Ptolemy inequality* $\| \mathbf{x} - \mathbf{y} \| \| \mathbf{z} \| \leqslant \| \mathbf{y} - \mathbf{z} \| \| \mathbf{x} \| + \| \mathbf{z} - \mathbf{x} \| \| \mathbf{y} \|$, for $\mathbf{x}, \mathbf{y}, \mathbf{z} \in V$ if and only if L is symmetric.

47. Let $H_{\mathbf{u}}$ be the Householder matrix corresponding to the unit vector $\mathbf{u} \in \mathbb{R}^n$. If $\mathbf{x} \in \mathbb{R}^n$ is written as $\mathbf{x} = \mathbf{y} + \mathbf{z}$, where $\mathbf{y} = a\mathbf{u}$ and $\mathbf{z} \perp \mathbf{u}$, then $H_{\mathbf{u}}\mathbf{x}$ is obtained by a reflection of $\mathbf{x}$ relative to the hyperplane that is perpendicular on $\mathbf{u}$, that is, $H_{\mathbf{u}}\mathbf{x} = -\mathbf{u} + \mathbf{v}$.

48. Let $Y \in \mathbb{R}^{n \times k}$ be a matrix such that $Y'Y = I_k$, where $k \leqslant n$. Prove that the matrix $I_n - YY'$ is positive semidefinite.
49. Prove that if $A, B \in \mathbb{R}^{2\times 2}$ are two rotation matrices, then $AB = BA$.
50. Let $\mathbf{u} \in \mathbb{R}^n$ be a unit vector. A *rotation with axis* $\mathbf{u}$ is an orthogonal matrix A such that $A\mathbf{u} = \mathbf{u}$. Prove that if $\mathbf{v} \perp \mathbf{u}$, then $A\mathbf{v} \perp \mathbf{u}$ and $A'\mathbf{v} \perp \mathbf{u}$.
51. Let $\mathbf{u}$, $\mathbf{v}$ and $\mathbf{w}$ be three unit vectors in $\mathbb{R}^2 - \{\mathbf{0}\}$. Prove that $\angle(\mathbf{u}, \mathbf{v}) \leqslant \angle(\mathbf{u}, \mathbf{w}) + \angle(\mathbf{w}, \mathbf{v})$.
52. Let $A \in \mathbb{R}^{m\times n}$ be a matrix such that $rank(A) = n$. Prove that the R-factor of the QR-decomposition of $A = QR$ has positive diagonal elements, it equals the Cholesky factor of $A'A$, and therefore is uniquely determined.
53. Let $A \in \mathbb{C}^{n\times m}$ be a full-rank matrix such that $m \geqslant n$. Prove that A can be factored as $A = LQ$, where $L \in \mathbb{C}^{n\times n}$ and $Q \in \mathbb{C}^{n\times m}$, such that the columns of Q constitute an orthonormal basis for $\mathsf{Ran}(A^{\mathsf{H}})$, and $L = (\ell_{ij})$ is an lower triangular invertible matrix such that its diagonal elements are real non-negative numbers, that is, $\ell_{ii} \geqslant 0$ for $1 \leqslant i \leqslant n$.

Let

$$D = \begin{pmatrix} \mathbf{u}'_1 \\ \vdots \\ \mathbf{u}'_m \end{pmatrix} = (\mathbf{v}_1 \ \cdots \ \mathbf{v}_n) \in \mathbb{R}^{m\times n}$$

be a data matrix and let $\mathbf{z} \in \mathbb{R}^n$. The *inertia of D relative to* $\mathbf{z}$ is the number $I_{\mathbf{z}}(D) = \sum_{j=1}^{m} \parallel \mathbf{u}_j - \mathbf{z} \parallel_2^2$.

54. Let

$$D = \begin{pmatrix} \mathbf{u}'_1 \\ \vdots \\ \mathbf{u}'_m \end{pmatrix}$$

be a data matrix. Prove that

$$I_{\mathbf{z}}(D) - I_{\tilde{D}}(D) = m \parallel \tilde{D} - \mathbf{z} \parallel_2^2,$$

for every $\mathbf{z} \in \mathbb{R}^n$. Conclude that the minimal value of the inertia $I_{\mathbf{z}}(D)$ is achieved for $\mathbf{z} = \tilde{D}$.

The *standard deviation of a vector* $\mathbf{v} \in \mathbb{R}^m$ is the number

$$s_{\mathbf{v}} = \sqrt{\frac{1}{m-1}\sum_{i=1}^{m}(v_i - \tilde{v})^2},$$

where $\tilde{v} = \frac{1}{m}\sum_{i=1}^{m} v_i$ is the mean of the components of $\mathbf{v}$. The *variance* is $var(\mathbf{v}) = s_{\mathbf{v}}^2$.

The *standard deviation of a data matrix* $D \in \mathbb{R}^{m\times n}$, where $D = (\mathbf{v}_1 \ \cdots \ \mathbf{v}_n)$ is the row $\mathbf{s} = (s_{\mathbf{v}_1}, \ldots, s_{\mathbf{v}_n}) \in \mathbb{R}^n$.

Let $\mathbf{u}$ and $\mathbf{w}$ be two vectors in $\mathbb{R}^m$, where $m > 1$, having the means $\tilde{u}$ and $\tilde{w}$, and the standard deviations s_u and s_v, respectively. The *covariance coefficient* of $\mathbf{u}$ and $\mathbf{w}$ is the number

$$cov(\mathbf{u}, \mathbf{w}) = \frac{1}{m-1} \sum_{i=1}^{m-1} (u_i - \tilde{u})(w_i - \tilde{w}).$$

The *correlation coefficient* of $\mathbf{u}$ and $\mathbf{w}$ is the number

$$\rho(\mathbf{u}, \mathbf{w}) = \frac{cov(\mathbf{u}, \mathbf{w})}{s_u s_w}.$$

The *covariance matrix* is of a data matrix $D \in \mathbb{R}^{m \times n}$ is

$$cov(D) = \frac{1}{m-1} \hat{D}' \hat{D} \in \mathbb{R}^{n \times n}.$$

55. Prove that for $\mathbf{v} \in \mathbb{R}^m$ we have

$$var(\mathbf{v}) = \frac{1}{m-1} \left(\| \mathbf{v} \|^2 - m\tilde{v}^2 \right).$$

56. Let $D \in \mathbb{R}^{m \times n}$ be a data matrix, where

$$D = \begin{pmatrix} \mathbf{u}_1' \\ \vdots \\ \mathbf{u}_m' \end{pmatrix} = (\mathbf{v}_1 \ \cdots \ \mathbf{v}_n).$$

Prove that the mean square distance between column vectors $\mathbf{v}_1, \ldots, \mathbf{v}_n$ is equal to twice the sum of row variances, $\sum_{i=1}^m var(\mathbf{u}_i)$.

Solution: The mean square distance between the columns of D is

$$\begin{aligned}
\frac{2}{n(n-1)} \sum_{i<j} \| \mathbf{v}_i - \mathbf{v}_j \|^2 &= \frac{2}{n(n-1)} \left(\sum_{j=1}^n \| \mathbf{v}_j \|^2 - 2 \sum_{i<j} \mathbf{v}_i' \mathbf{v}_j \right) \\
&= \frac{2}{n(n-1)} \left((n-1) \| D \|_F^2 + \| D \|_F^2 - \mathbf{1}_n' D D' \mathbf{1}_n \right) \\
&= \frac{2}{n(n-1)} \left(n \| D \|_F^2 - \mathbf{1}_n' D D' \mathbf{1}_n \right).
\end{aligned}$$

Since each vector $\mathbf{u}_k$ belongs to $\mathbb{R}^n$, the the sum of row variances is

$$\sum_{k=1}^m var(\mathbf{u}_k) = \sum_{k=1}^m \frac{1}{n-1} \left(\| \mathbf{u}_k \|^2 - n\tilde{u}_k^2 \right) = \frac{1}{n-1} \| D \|_F^2 - \frac{n}{n-1} \sum_{k=1}^n \tilde{u}_k^2.$$

Taking into account that

$$s\tilde{u}_k = \frac{1}{n}\mathbf{1}'_n D'\mathbf{e}_k = \frac{1}{n}e'_k D\mathbf{1}_n,$$

we have $\tilde{u}_k^2 = \frac{1}{n^2}\mathbf{1}'_n D'\mathbf{e}_k e'_k D\mathbf{1}_n$, which implies

$$\sum_{k=1}^{n} \tilde{u}_k^2 = \frac{1}{n^2}\mathbf{1}'_n D'\left(\sum_{k=1}^{n} \mathbf{e}_k\mathbf{e}'_k\right) D\mathbf{1}_n = \frac{1}{n^2}\mathbf{1}'_n D' D\mathbf{1}_n,$$

because $\sum_{k=1}^{n} \mathbf{e}_k\mathbf{e}'_k = I_m$. The desired equality follows immediately.

57. Prove that $-1 \leq \rho(\mathbf{u}, \mathbf{w}) \leq 1$ for any vectors $\mathbf{u}, \mathbf{w} \in \mathbb{R}^m$.
58. Prove that the covariance matrix of a data matrix $D \in \mathbb{R}^{m\times n}$ can be written as $cov(D) = \frac{1}{m-1}D'H_m D$; if D is centered, then $cov(D) = \frac{1}{m-1}D'D$.
59. Let $D \in \mathbb{R}^{m\times n}$ be a centered data matrix and let $R \in \mathbb{R}^{n\times n}$ be an orthogonal matrix. Prove that $cov(DR) = R'cov(X)R$.

Bibliographical Comments

The notion of vector inner product, which is fundamental for linear algebra, functional analysis and other mathematical disciplines was introduced by Hermann Günther Grassmann (1809–1877) is his fundamental work "Die Lineale Ausdehnungslehre, ein neuer Zweig der Mathematik".

Almost every advanced linear algebra reference deals with inner products and norms and their applications at the level that we need. We recommend especially [2] and the two volumes [3] and [4].

References

1. J. Weston, A. Elisseeff, B. Schölkopf, M. Tipping, Use of zero-norm with linear models and kernel methods. J. Mach. Learn. **3**, 1439–1461 (2003)
2. C.D. Meyer, *Matrix Analysis and Applied Linear Algebra* (Society for Industrial and Applied Mathematics, Philadelphia, SIAM, 2000)
3. R.A. Horn, C.R. Johnson, *Matrix Analysis* (Cambridge University Press, Cambridge, 1985)
4. R.A. Horn, C.R. Johnson, *Topics in Matrix Analysis* (Cambridge University Press, Cambridge, 2008)

Chapter 7
Spectral Properties of Matrices

7.1 Introduction

The existence of directions that are preserved by linear transformations (which are referred to as eigenvectors) has been discovered by L. Euler in his study of movements of rigid bodies. This work was continued by Lagrange, Cauchy, Fourier, and Hermite. The study of eigenvectors and eigenvalues acquired increasing significance through its applications in heat propagation and stability theory. Later, Hilbert initiated the study of eigenvalue in functional analysis (in the theory of integral operators). He introduced the terms of eigenvalue and eigenvector. The term *eigenvalue* is a German-English hybrid formed from the German word *eigen* which means "own" and the English word "value". It is interesting that Cauchy referred to the same concept as *characteristic value* and the term *characteristic polynomial* of a matrix (which we introduce in Definition 7.1) was derived from this naming.

We present the notions of geometric and algebraic multiplicities of eigenvalues, examine properties of spectra of special matrices, discuss variational characterizations of spectra and the relationships between matrix norms and eigenvalues. We conclude this chapter with a section dedicated to singular values of matrices.

7.2 Eigenvalues and Eigenvectors

Let $A \in \mathbb{C}^{n \times n}$ be a square matrix. An *eigenpair* of A is a pair $(\lambda, \mathbf{x}) \in \mathbb{C} \times (\mathbb{C}^n - \{\mathbf{0}\})$ such that $A\mathbf{x} = \lambda\mathbf{x}$. We refer to λ is an *eigenvalue* and to $\mathbf{x}$ is an *eigenvector*. The set of eigenvalues of A is the *spectrum* of A and will be denoted by $\mathsf{spec}(A)$.

If $(\lambda, \mathbf{x})$ is an eigenpair of A, the linear system $A\mathbf{x} = \lambda\mathbf{x}$ has a non-trivial solution in $\mathbf{x}$. An equivalent homogeneous system is $(\lambda I_n - A)\mathbf{x} = \mathbf{0}$ and this system has a non-trivial solution only if $\det(\lambda I_n - A) = 0$.

D. A. Simovici and C. Djeraba, *Mathematical Tools for Data Mining*,
Advanced Information and Knowledge Processing, DOI: 10.1007/978-1-4471-6407-4_7,

Definition 7.1 *The* characteristic polynomial *of the matrix A is the polynomial p_A defined by $p_A(\lambda) = \det(\lambda I_n - A)$ for $\lambda \in \mathbb{C}$.*

Thus, the eigenvalues of A are the roots of the characteristic polynomial of A.

Lemma 7.2 *Let $A = (\boldsymbol{a}_1 \ \cdots \ \boldsymbol{a}_n) \in \mathbb{C}^n$ and let B be the matrix obtained from A by replacing the column $\boldsymbol{a}_j$ by $\boldsymbol{e}_j$. Then, we have*

$$\det(B) = \det\left(A\begin{bmatrix}1 & \cdots & j-1 & j+1 & \cdots & n\\ 1 & \cdots & j-1 & j+1 & \cdots & n\end{bmatrix}\right).$$

Proof The result follows immediately by expanding B on the jth column.

The result obtained in Lemma 7.2 can be easily extended as follows. If B is the matrix obtained from A by replacing the columns $\mathbf{a}_{j_1}, \ldots, \mathbf{a}_{j_k}$ by $\mathbf{e}_{j_1}, \ldots, \mathbf{e}_{j_k}$ and $\{i_1, \ldots, i_p\} = \{1, \ldots, n\} - \{j_1, \ldots, j_k\}$, then

$$\det(B) = \det\left(A\begin{bmatrix}i_1 & \cdots & i_p\\ i_1 & \cdots & i_p\end{bmatrix}\right). \tag{7.1}$$

In other words, $\det(B)$ equals a principal p-minor of A.

Theorem 7.3 *Let $A \in \mathbb{C}^{n\times n}$ be a matrix. Its characteristic polynomial p_A can be written as*

$$p_A(\lambda) = \sum_{k=0}^{n}(-1)^k a_k \lambda^{n-k},$$

where a_k is the sum of the principal minors of order k of A.

Proof By Theorem 5.146 the determinant

$$p_A(\lambda) = \det(\lambda I_n - A) = (-1)^n \det(A - \lambda I_n)$$

can be written as a sum of 2^n determinants of matrices obtained by replacing each subset of the columns of A by the corresponding subset of columns of $-\lambda I_n$. If the subset of columns of $-\lambda I_n$ involved are $-\lambda\mathbf{e}_{j_1}, \ldots, -\lambda\mathbf{e}_{j_k}$ the result of the substitution is $(-1)^k\lambda^k \det\left(A\begin{bmatrix}i_1 \cdots i_p\\ i_1 \cdots i_p\end{bmatrix}\right)$, where $\{i_1, \ldots, i_p\} = \{1, \ldots, n\} - \{j_1, \ldots, j_k\}$. The total contribution of sets of k columns of $-\lambda I_n$ is $(-1)^k\lambda^k a_{n-k}$. Therefore,

$$p_A(\lambda) = (-1)^n \sum_{k=0}^{n}(-1)^k\lambda^k a_{n-k}.$$

Replacing k by $n-k$ as the summation index yields

$$p_A(\lambda) = (-1)^n \sum_{k=0}^{n}(-1)^{n-k}\lambda^{n-k} a_k = \sum_{k=0}^{n}(-1)^k a_k \lambda^{n-k}.$$

Definition 7.4 *Two matrices $A, B \in \mathbb{C}^{n\times n}$ are* similar *if there exists an invertible matrix $P \in \mathbb{C}^{n\times n}$ such that $B = PAP^{-1}$. This is denoted by $A \sim B$.*

If there exists a unitary matrix U such that $B = UAU^{-1}$, then A is unitarily similar *to B. This is denoted by $A \sim_u B$.*

The matrices A, B are congruent *if $B = SAS^{\mathsf{H}}$ for some non-singular matrix S. This is denoted by $A \approx B$. If $A, B \in \mathbb{R}^{n\times n}$, we say that they are* t-congruent *if $B = SAS'$ for some invertible matrix S; this is denoted by $A \approx_t B$.*

For real matrices the notions of t-congruence and congruence are identical.

It is easy to verify that $\sim$, $\sim_u$ and $\approx$ are equivalence relations on $\mathbb{C}^{n\times n}$ and $\approx_t$ is an equivalence on $\mathbb{R}^{n\times n}$.

Similar matrices have the same characteristic polynomial. Indeed, suppose that $B = PAP^{-1}$. We have

$$\begin{aligned} p_B(\lambda) &= \det(\lambda I_n - B) = \det(\lambda I_n - PAP^{-1}) \\ &= \det(\lambda P I_n P^{-1} - PAP^{-1}) = \det(P(\lambda I_n - A)P^{-1}) \\ &= \det(P)\det(\lambda I_n - A)\det(P^{-1}) = \det(\lambda I_n - A) = p_A(\lambda), \end{aligned}$$

because $\det(P)\det(P^{-1}) = 1$. Thus, similar matrices have the same eigenvalues.

Example 7.5 Let A be the matrix

$$A = \begin{pmatrix} \cos\theta & -\sin\theta \\ \sin\theta & \cos\theta \end{pmatrix}.$$

We have

$$p_A = \det(\lambda I_2 - A) = (\lambda - \cos\theta)^2 + \sin^2\theta = \lambda^2 - 2\lambda\cos\theta + 1.$$

The roots of this polynomial are $\lambda_1 = \cos\theta + i\sin\theta$ and $\lambda_2 = \cos\theta - i\sin\theta$, so they are complex numbers.

We regard A as a complex matrix with real entries. If we were to consider A as a real matrix, we would not be able to find real eigenvalues for A unless θ were equal to 0.

Definition 7.6 *The* algebraic multiplicity *of the eigenvalue λ of a matrix $A \in \mathbb{C}^{n\times n}$ is the multiplicity of λ as a root of the characteristic polynomial p_A of A.*

The algebraic multiplicity of λ is denoted by $\mathsf{algm}(A, \lambda)$. If $\mathsf{algm}(A, \lambda) = 1$ we say that λ is a simple eigenvalue.

Example 7.7 Let $A \in \mathbb{R}^{2\times 2}$ be the matrix

$$A = \begin{pmatrix} 0 & -1 \\ 1 & 2 \end{pmatrix}$$

The characteristic polynomial of A is

$$p_A(\lambda) = \begin{vmatrix} \lambda & 1 \\ -1 & \lambda - 2 \end{vmatrix} = \lambda^2 - 2\lambda + 1.$$

Therefore, A has the eigenvalue 1 with $\mathsf{algm}(A, 1) = 2$.

Example 7.8 Let $P(a) \in \mathbb{C}^{n \times n}$ be the matrix $P(a) = (a-1)I_n + J_n$. To find the eigenvalues of $P(a)$ we need to solve the equation

$$\begin{vmatrix} \lambda - a & -1 & \cdots & -1 \\ -1 & \lambda - a & \cdots & -1 \\ \vdots & \vdots & \cdots & \vdots \\ -1 & -1 & \cdots & \lambda - a \end{vmatrix} = 0.$$

By adding the first $n-1$ columns to the last and factoring out $\lambda - (a + n - 1)$, we obtain the equivalent equation

$$(\lambda - (a + n - 1)) \begin{vmatrix} \lambda - a & -1 & \cdots & 1 \\ -1 & \lambda - a & \cdots & 1 \\ \vdots & \vdots & \cdots & \vdots \\ -1 & -1 & \cdots & 1 \end{vmatrix} = 0.$$

Adding the last column to the first $n-1$ columns and expanding the determinant yields the equation $(\lambda - (a + n - 1))(\lambda - a + 1)^{n-1} = 0$, which shows that $P(a)$ has the eigenvalue $a + n - 1$ with $\mathsf{algm}(P(a), a + n - 1) = 1$ and the eigenvalue $a - 1$ with $\mathsf{algm}(P(a), a - 1) = n - 1$.

In the special case when $a = 1$ we have $P(1) = J_{n,n}$. Thus, $J_{n,n}$ has the eigenvalue $\lambda_1 = n$ with algebraic multiplicity 1 and the eigenvalue 0 with algebraic multiplicity $n - 1$.

Theorem 7.9 *The eigenvalues of Hermitian complex matrices are real numbers.*

Proof Let $A \in \mathbb{C}^{n \times n}$ be a Hermitian matrix and let λ be an eigenvalue of A. We have $A\mathbf{x} = \lambda\mathbf{x}$ for some $\mathbf{x} \in \mathbb{C}^n - \{\mathbf{0}_n\}$, so $\mathbf{x}^{\mathsf{H}} A^{\mathsf{H}} = \overline{\lambda}\mathbf{x}^{\mathsf{H}}$. Since $A^{\mathsf{H}} = A$, we have

$$\lambda \mathbf{x}^{\mathsf{H}}\mathbf{x} = \mathbf{x}^{\mathsf{H}} A \mathbf{x} = \mathbf{x}^{\mathsf{H}} A^{\mathsf{H}} \mathbf{x} = \overline{\lambda} \mathbf{x}^{\mathsf{H}}\mathbf{x}.$$

Since $\mathbf{x} \neq \mathbf{0}$ implies $\mathbf{x}^{\mathsf{H}}\mathbf{x} \neq 0$, it follows that $\overline{\lambda} = \lambda$. Thus, λ is a real number.

Corollary 7.10 *The eigenvalues of symmetric real matrices are real numbers.*

Proof This is a direct consequence of Theorem 7.9.

Theorem 7.11 *The eigenvectors of a complex Hermitian matrix corresponding to distinct eigenvalues are orthogonal to each other.*

Proof Let $(\lambda, \boldsymbol{u})$ and $(\mu, \boldsymbol{v})$ be two eigenpairs of the Hermitian matrix $A \in \mathbb{C}^{n\times n}$, where $\lambda \neq \mu$. Since A is Hermitian, $\lambda, \mu \in \mathbb{R}$. Since $A\boldsymbol{u} = \lambda\boldsymbol{u}$ we have $\boldsymbol{v}^{\mathsf{H}}A\boldsymbol{u} = \lambda\boldsymbol{v}^{\mathsf{H}}\boldsymbol{u}$. The last equality can be written as $(A\boldsymbol{v})^{\mathsf{H}}\boldsymbol{u} = \lambda\boldsymbol{v}^{\mathsf{H}}\boldsymbol{u}$, or as $\mu\boldsymbol{v}^{\mathsf{H}}\boldsymbol{u} = \lambda\boldsymbol{v}^{\mathsf{H}}\boldsymbol{u}$. Since $\mu \neq \lambda$, $\boldsymbol{v}^{\mathsf{H}}\boldsymbol{u} = 0$, so $\boldsymbol{u}$ and $\boldsymbol{v}$ are orthogonal.

The statement clearly holds if we replace complex Hermitian matrices by real symmetric matrices.

Corollary 7.12 *The eigenvectors of a Hermitian matrix corresponding to distinct eigenvalues form a linearly independent set.*

Proof This statement follows from Theorems 6.41 and 7.11.

The next statement is a result of Issai Schur (1875–1941), a mathematician born in Russia, who studied and worked in Germany.

Theorem 7.13 (Schur's Triangularization Theorem) *Let $A \in \mathbb{C}^{n\times n}$ be a square matrix. There exists an upper-triangular matrix $T \in \mathbb{C}^{n\times n}$ such that $A \sim_u T$.*

The diagonal elements of T are the eigenvalues of A; moreover, each eigenvalue λ of A occurs in the sequence of diagonal elements of T a number of $\mathsf{algm}(A, \lambda)$ times. The columns of U are unit eigenvectors of A.

Proof The argument is by induction on n. The base case, $n = 1$, is immediate.

Suppose that the statement holds for matrices in $\mathbb{C}^{(n-1)\times(n-1)}$ and let $A \in \mathbb{C}^{n\times n}$. If $(\lambda, \mathbf{x})$ is an eigenpair of A with $\| \mathbf{x} \|_2 = 1$, let H_v be a Householder matrix that transforms $\mathbf{x}$ into $\mathbf{e}_1$. Since we also have $H_v\mathbf{e}_1 = \mathbf{x}$, $\mathbf{x}$ is the first column of H_v and we can write $H_v = (\mathbf{x}\ K)$, where $K \in \mathbb{C}^{n\times(n-1)}$. Consequently,

$$H_vAH_v = H_vA(\mathbf{x}\ K) = H_v(\lambda\mathbf{x}\ H_vAK) = (\lambda\mathbf{e}_1\ H_vAK).$$

Since H_v is Hermitian and $H_v = (\mathbf{x}\ K)$, it follows that

$$H_v^{\mathsf{H}} = \begin{pmatrix} \mathbf{x}^{\mathsf{H}} \\ K^{\mathsf{H}} \end{pmatrix} = H_v.$$

Therefore,

$$H_vAH_v = \begin{pmatrix} \lambda & \mathbf{x}^{\mathsf{H}}AK \\ \mathbf{0}_{n-1} & K^{\mathsf{H}}AK \end{pmatrix}.$$

Since $K^{\mathsf{H}}AK \in \mathbb{C}^{(n-1)\times(n-1)}$, by the inductive hypothesis, there exists a unitary matrix W and an upper triangular matrix S such that $W^{\mathsf{H}}(K^{\mathsf{H}}AK)W = S$. Note that the matrix

$$U = H_v \begin{pmatrix} 1 & \mathbf{0}'_{n-1} \\ \mathbf{0}_{n-1} & W \end{pmatrix}$$

is unitary and

$$U^{\mathsf{H}}AU^{\mathsf{H}} = \begin{pmatrix} \lambda & x^{\mathsf{H}}AKW \\ \mathbf{0}_{n-1} & W^{\mathsf{H}}K^{\mathsf{H}}AKW \end{pmatrix} = \begin{pmatrix} \lambda & \mathbf{x}^{\mathsf{H}}AKW \\ \mathbf{0}_{n-1} & S \end{pmatrix}.$$

The last matrix is clearly upper triangular.

Since $A \sim_u T$, A and T have the same characteristic polynomials and, therefore, the same eigenvalues, with identical multiplicities. Note that the factorization of A can be written as $A = UDU^{\mathsf{H}}$ because $U^{-1} = U^{\mathsf{H}}$. Since $AU = UD$, each column $\boldsymbol{u}_i$ of U is an eigenvector of A that corresponds to the eigenvalue λ_i for $1 \leqslant i \leqslant n$.

Corollary 7.14 *If $A \in \mathbb{R}^{n\times n}$ is a matrix such that* $\mathsf{spec}(A) = \{0\}$*, then A is nilpotent.*

Proof By Schur's Triangularization Theorem, A is unitarily similar to a strictly upper triangular matrix, $A = UTU^{\mathsf{H}}$, so $A^n = UT^nU^{\mathsf{H}}$. Since $\mathsf{spec}(T) = \{0\}$, we have $T^n = O$, so $A^n = O$.

Corollary 7.15 *Let $A \in \mathbb{C}^{n\times n}$ and let f be a polynomial. If* $\mathsf{spec}(A) = \{\lambda_1, \ldots, \lambda_n\}$ *(including multiplicities), then* $\mathsf{spec}(f(A)) = \{f(\lambda_1), \ldots, f(\lambda_n)\}$.

Proof By Schur's Triangularization Theorem there exists a unitary matrix $U \in \mathbb{C}^{n\times n}$ and an upper-triangular matrix $T \in \mathbb{C}^{n\times n}$ such that $A = UTU^{\mathsf{H}}$ and the diagonal elements of T are the eigenvalues of A, $\lambda_1, \ldots, \lambda_n$. Therefore $f(A) = Uf(T)U^{\mathsf{H}}$, and by Theorem 5.49, the diagonal elements of $f(T)$ are $f(\lambda_1), \ldots, f(\lambda_m)$. Since $f(A) \sim_u f(T)$, we obtain the desired conclusion because two similar matrices have the same eigenvalues with the same algebraic multiplicities.

Definition 7.16 *A matrix $A \in \mathbb{C}^{n\times n}$ is* diagonalizable *(unitarily diagonalizable) if there exists a diagonal matrix* $D = \mathsf{diag}(d_1, \ldots, d_n)$ *such that $A \sim D$ ($A \sim_u D$).*

Theorem 7.17 *A matrix $A \in \mathbb{C}^{n\times n}$ is diagonalizable if and only if there exists a linearly independent set $\{\boldsymbol{v}_1, \ldots, \boldsymbol{v}_n\}$ of n eigenvectors of A.*

Proof Let $A \in \mathbb{C}^{n\times n}$ such that there exists a set $\{\boldsymbol{v}_1, \ldots, \boldsymbol{v}_n\}$ of n eigenvectors of A that is linearly independent and let P be the matrix $(\boldsymbol{v}_1 \ \boldsymbol{v}_2 \ \cdots \ \boldsymbol{v}_n)$ that is clearly invertible. We have:

$$\begin{aligned} P^{-1}AP &= P^{-1}(A\boldsymbol{v}_1 \ A\boldsymbol{v}_2 \ \cdots \ A\boldsymbol{v}_n) = P^{-1}(\lambda_1\boldsymbol{v}_1 \ \lambda_2\boldsymbol{v}_2 \ \cdots \ \lambda_n\boldsymbol{v}_n) \\ &= P^{-1}P\begin{pmatrix} \lambda_1 & 0 & \cdots & 0 \\ 0 & \lambda_2 & \cdots & 0 \\ \vdots & \vdots & \cdots & \vdots \\ 0 & 0 & \cdots & \lambda_n \end{pmatrix} = \begin{pmatrix} \lambda_1 & 0 & \cdots & 0 \\ 0 & \lambda_2 & \cdots & 0 \\ \vdots & \vdots & \cdots & \vdots \\ 0 & 0 & \cdots & \lambda_n \end{pmatrix}. \end{aligned}$$

Therefore, we have $A = PDP^{-1}$, where

$$D = \begin{pmatrix} \lambda_1 & 0 & \cdots & 0 \\ 0 & \lambda_2 & \cdots & 0 \\ \vdots & \vdots & \cdots & \vdots \\ 0 & 0 & \cdots & \lambda_n \end{pmatrix},$$

so $A \sim D$.

Conversely, suppose that A is diagonalizable, so $AP = PD$, where D is a diagonal matrix and P is an invertible matrix, and let $\boldsymbol{v}_1, \ldots, \boldsymbol{v}_n$ be the columns of the matrix P. We have $A\boldsymbol{v}_i = d_{ii}\boldsymbol{v}_i$ for $1 \leqslant i \leqslant n$, so each $\boldsymbol{v}_i$ is an eigenvector of A. Since P is invertible, its columns are linear independent.

Corollary 7.18 *If $A \in \mathbb{C}^{n\times n}$ is diagonalizable then the columns of any matrix P such that $D = P^{-1}AP$ is a diagonal matrix are eigenvectors of A. Furthermore, the diagonal entries of D are the eigenvalues that correspond to the columns of P.*

Proof This statement follows from the proof of Theorem 7.17.

Corollary 7.19 *A matrix $A \in \mathbb{C}^{n\times n}$ is unitarily diagonalizable if and only if there exists a set $\{\boldsymbol{v}_1, \ldots, \boldsymbol{v}_n\}$ of n orthonormal eigenvectors of A.*

Proof This statement follows from the proof of Theorem 7.17 by observing that if $\{\boldsymbol{v}_1, \ldots, \boldsymbol{v}_n\}$ is a set n orthonormal eigenvectors of A, then $P = (\boldsymbol{v}_1, \ldots, \boldsymbol{v}_n)$ is a unitary matrix that gives a unitary diagonalization of A. Conversely, if P is an unitary matrix such that $A = PDP^{-1}$ its set of columns consists of orthogonal unitary eigenvectors of A.

Corollary 7.20 *Let $A \in \mathbb{R}^{n\times n}$ be a symmetric matrix. There exists a orthonormal matrix U and a diagonal matrix T such that $A = UTU^{-1}$. The diagonal elements of T are the eigenvalues of A; moreover, each eigenvalue λ of A occurs in the sequence of diagonal elements of T a number of $\mathsf{algm}(A, \lambda)$ times.*

Proof As the previous corollary, this follows from the proof of Theorem 7.17.

Theorem 7.21 *Let $A \in \mathbb{C}^{n\times n}$ be a block diagonal matrix,*

$$A = \begin{pmatrix} A_{11} & O & \cdots & O \\ O & A_{22} & \cdots & O \\ \vdots & \vdots & \cdots & \vdots \\ O & O & \cdots & A_{mm} \end{pmatrix}.$$

A is diagonalizable if and only if every matrix A_{ii} is diagonalizable for $1 \leqslant i \leqslant m$.

Proof Suppose that A is a block diagonal matrix which is diagonalizable. Furthermore, suppose that $A_{ii} \in \mathbb{C}^{n_i \times n_i}$ for $1 \leqslant i \leqslant n$ and $\sum_{i=1}^{m} n_i = n$. There exists an invertible matrix $P \in \mathbb{C}^{n\times n}$ such that $P^{-1}AP$ is a diagonal matrix $D = \mathsf{diag}(\lambda_1, \ldots, \lambda_n)$. Let $\mathbf{p}_1, \ldots, \mathbf{p}_n$ be the columns of P, which are eigenvectors of A. Each vector $\mathbf{p}_i$ is divided into m blocks $\mathbf{p}_i^j$ with $1 \leqslant j \leqslant m$, where $\mathbf{p}_i^j \in \mathbb{C}^{n_j}$. Thus, P can be written as

$$P = \begin{pmatrix} \mathbf{p}_1^1 & \mathbf{p}_2^1 & \cdots & \mathbf{p}_n^1 \\ \mathbf{p}_1^2 & \mathbf{p}_2^2 & \cdots & \mathbf{p}_n^2 \\ \vdots & \vdots & \cdots & \vdots \\ \mathbf{p}_1^m & \mathbf{p}_2^m & \cdots & \mathbf{p}_n^m \end{pmatrix}.$$

The equality $A\mathbf{p}_i = \lambda_i \mathbf{p}_i$ can be expressed as

$$\begin{pmatrix} A_{11} & O & \cdots & O \\ O & A_{22} & \cdots & O \\ \vdots & \vdots & \cdots & \vdots \\ O & O & \cdots & A_{mm} \end{pmatrix} \begin{pmatrix} \mathbf{p}_i^1 \\ \mathbf{p}_i^2 \\ \vdots \\ \mathbf{p}_i^m \end{pmatrix} = \lambda_i \begin{pmatrix} \mathbf{p}_i^1 \\ \mathbf{p}_i^2 \\ \vdots \\ \mathbf{p}_i^m \end{pmatrix},$$

which shows that $A_{jj}\mathbf{p}_i^j = \lambda_i \mathbf{p}_i^j$ for $1 \leqslant j \leqslant m$. Let $M^j = (\mathbf{p}_1^j \ \mathbf{p}_2^j \ \cdots \ \mathbf{p}_n^j) \in \mathbb{C}^{n_j \times n}$. We claim that $rank(M^j) = n_j$. Indeed if $rank(M^j)$ were less than n_j, we would have fewer that n independent rows M^j for $1 \leqslant j \leqslant m$. This, however, would imply that the rank of P is less then n, which contradicts the invertibility of P. Since there are n_j linearly independent eigenvectors of A_{jj}, it follows that each block A_{jj} is diagonalizable.

Conversely, suppose that each A_{jj} is diagonalizable, that is, there exists a invertible matrix Q_j such that $Q_j^{-1} A_{jj} Q_j$ is a diagonal matrix. Then, it is immediate to verify that the block diagonal matrix

$$Q = \begin{pmatrix} Q_1 & O & \cdots & O \\ O & Q_2 & \cdots & O \\ \vdots & \vdots & \cdots & \vdots \\ O & O & \cdots & Q_m \end{pmatrix}$$

is invertible and $Q^{-1}AQ$ is a diagonal matrix.

Theorem 7.22 *Let $A \in \mathbb{C}^{n \times n}$ be a matrix and let $(\lambda_1, \boldsymbol{v}_1), \ldots, (\lambda_k, \boldsymbol{v}_k)$ be k eigenpairs of A, where $\lambda_1, \ldots, \lambda_k$ are pairwise distinct. Then, $\{\boldsymbol{v}_1, \ldots, \boldsymbol{v}_k\}$ is a linearly independent set.*

Proof Suppose that $\{\boldsymbol{v}_1, \ldots, \boldsymbol{v}_k\}$ is not linearly independent. Then, there exists a linear combination of this set that equals $\mathbf{0}$, that is

$$c_1 \boldsymbol{v}_{i_1} + \cdots + c_r \boldsymbol{v}_{i_r} = \mathbf{0}_n, \tag{7.2}$$

at least one coefficient is not 0, and $r > 1$. Choose this linear combination to involve the minimal number of terms. We have

$$A(c_1\boldsymbol{v}_{i_1} + c_2\boldsymbol{v}_{i_2} + \cdots + c_r\boldsymbol{v}_{i_r}) = c_1\lambda_{i_1}\boldsymbol{v}_1 + c_2\lambda_{i_2}\boldsymbol{v}_{i_2} + \cdots + c_r\lambda_{i_r}\boldsymbol{v}_{i_r} = \mathbf{0}_n.$$

By multiplying Equality (7.2) by λ_{i_1} we have

$$c_1\lambda_{i_1}\boldsymbol{v}_{i_1} + c_2\lambda_{i_1}\boldsymbol{v}_{i_2} + \cdots + c_r\lambda_{i_1}\boldsymbol{v}_{i_r} = \mathbf{0}_n.$$

It follows that $c_2(\lambda_{i_2} - \lambda_{i_1})\boldsymbol{v}_{i_2} + \cdots + c_r(\lambda_{i_r} - \lambda_{i_1})\boldsymbol{v}_{i_r} = \mathbf{0}_n$. Since there exists a coefficient $\lambda_{i_p} - \lambda_{i_1}$ that is non-zero, this contradicts the minimality of r.

Corollary 7.23 *If $A \in \mathbb{C}^{n\times n}$ has n distinct eigenvalues, then A is diagonalizable.*

Proof If $\mathsf{spec}(A) = \{\lambda_1, \ldots, \lambda_n\}$ consists of n complex numbers and $\boldsymbol{v}_1, \ldots, \boldsymbol{v}_n$ are corresponding eigenvectors, then, by Theorem 7.22, $\{\boldsymbol{v}_1, \ldots, \boldsymbol{v}_n\}$ is linearly independent. The statement follows immediately from Theorem 7.17.

7.3 Geometric and Algebraic Multiplicities of Eigenvalues

For a matrix $A \in \mathbb{C}^{n\times n}$ let $S_{A,\lambda}$ be the subspace $\mathsf{NullSp}(\lambda I_n - A)$. We refer to $S_{A,\lambda}$ as *invariant subspace of A and λ* or as the *eigenspace* of λ.

Definition 7.24 *Let $A \in \mathbb{C}^{n\times n}$ and let $\lambda \in \mathsf{spec}(A)$. The* geometric multiplicity *of λ is the dimension $\mathsf{geomm}(A, \lambda)$ of $S_{A,\lambda}$.*

Example 7.25 The geometric multiplicity of 1 as an eigenvalue of the matrix

$$A = \begin{pmatrix} 0 & -1 \\ 1 & 2 \end{pmatrix},$$

considered in Example 7.7, is 1. Indeed if $\mathbf{x}$ is an eigenvector that corresponds to this value we have $-x_2 = x_1$ and $x_1 + 2x_2 = x_2$, which means that any such eigenvector has the form $a\mathbf{1}_2$. Thus, $\dim(S_{A,1}) = 1$.

The definition of the geometric multiplicity of $\lambda \in \mathsf{spec}(A)$ implies

$$\mathsf{geomm}(A, \lambda) = \dim(\mathsf{NullSp}(A - \lambda I_n)) = n - rank(A - \lambda I_n). \tag{7.3}$$

Theorem 7.26 *Let $A \in \mathbb{R}^{n\times n}$. We have $0 \in \mathsf{spec}(A)$ if and only if A is a singular matrix. Moreover, in this case, $\mathsf{geomm}(A, 0) = n - rank(A) = \dim(\mathsf{NullSp}(A))$. If $\mathsf{algm}(A, 0) = 1$, then $rank(A) = n - 1$.*

Proof The statement is an immediate consequence of Equality (7.3).

Theorem 7.27 *Let $A \in \mathbb{C}^{n\times n}$ be a square matrix and let $\lambda \in \mathsf{spec}(A)$. We have $\mathsf{geomm}(A, \lambda) \leqslant \mathsf{algm}(A, \lambda)$.*

Proof By the definition of $\mathsf{geomm}(A, \lambda)$ we have

$$\mathsf{geomm}(A, \lambda) = \dim(\mathsf{NullSp}(\lambda I_n - A)) = n - rank(\lambda I_n - A).$$

Let $\boldsymbol{u}_1, \ldots, \boldsymbol{u}_m$ be an orthonormal basis of $S_{A,\lambda}$, where $m = \mathsf{geomm}(A, \lambda)$ and let $U = (\boldsymbol{u}_1 \cdots \boldsymbol{u}_m)$. We have $(\lambda I_n - A)U = O_{n,n}$, so $AU = \lambda U = U(\lambda I_m)$. Thus, by Theorem 6.57 there exists a matrix V such that we have

$$A \sim \begin{pmatrix} \lambda I_m & U^{\mathsf{H}} A V \\ O & V^{\mathsf{H}} A V \end{pmatrix},$$

where $U \in \mathbb{C}^{n\times m}$ and $V \in \mathbb{C}^{n\times(n-m)}$. Therefore, A has the same characteristic polynomial as

$$B = \begin{pmatrix} \lambda I_m & U^{\mathsf{H}}AV \\ O & V^{\mathsf{H}}AV \end{pmatrix},$$

which implies $\mathsf{algm}(A, \lambda) = \mathsf{algm}(B, \lambda)$. Since the algebraic multiplicity of λ in B is at least equal to m it follows that $\mathsf{algm}(A, \lambda) \geqslant m = \mathsf{geomm}(A, \lambda)$.

Definition 7.28 *An eigenvalue λ of a matrix A is* simple *if $\mathsf{algm}(A, \lambda) = 1$. If $\mathsf{geomm}(A, \lambda) = \mathsf{algm}(A, \lambda)$, then we refer to λ as a* semisimple eigenvalue.

The matrix A is defective *if there exists at least one eigenvalue that is not semi-simple. Otherwise, A is said to be* non-defective.

A is a non-derogatory *matrix if $\mathsf{geomm}(A, \lambda) = 1$ for every eigenvalue λ.*

Note that if λ is a simple eigenvalue of A, then $\mathsf{geomm}(A, \lambda) = \mathsf{algm}(A, \lambda) = 1$, so λ is semi-simple.

Theorem 7.29 *Each eigenvalue of a symmetric matrix $A \in \mathbb{R}^{n\times n}$ is semi-simple.*

Proof We saw that each symmetric matrix has real eigenvalues and is orthonormally diagonalizable (by Corollary 7.20). Starting from the real Schur factorization $A = UTU^{-1}$, where U is an orthonormal matrix and $T = \mathsf{diag}(t_{11}, \ldots, t_{nn})$ is a diagonal matrix we can write $AU = UT$. If we denote the columns of U by $\boldsymbol{u}_1, \ldots, \boldsymbol{u}_n$, then we can write

$$(A\boldsymbol{u}_1, \ldots, A\boldsymbol{u}_n) = (t_{11}\boldsymbol{u}_1, \ldots, t_{nn}\boldsymbol{u}_n),$$

so $A\boldsymbol{u}_i = t_{ii}\boldsymbol{u}_i$ for $1 \leqslant i \leqslant n$. Thus, the diagonal elements of T are the eigenvalues of A and the columns of U are corresponding eigenvectors. Since these eigenvectors are pairwise orthogonal, the dimension of the invariant subspace that corresponds to an eigenvalue equals the algebraic multiplicity of the eigenvalue, so each eigenvalue is semi-simple.

Example 7.30 Let $A \in \mathbb{R}^{2\times 2}$ be the matrix

$$A = \begin{pmatrix} a & b \\ 0 & a \end{pmatrix},$$

where $a, b \in \mathbb{R}$ are such that $ab \neq 0$. The characteristic polynomial of A is $p_A(\lambda) = (a - \lambda)^2$, so $\mathsf{spec}(A) = \{a\}$ and $\mathsf{algm}(A, a) = 2$.

Let $\mathbf{x}$ be a characteristic vector of A that corresponds to a. We have $ax_1 + bx_2 = ax_1$ and $ax_2 = x_2$, which implies $x_2 = 0$. Thus, the invariant subspace is

$$\left\{ \begin{pmatrix} x_1 \\ x_2 \end{pmatrix} \in \mathbb{R}^2 \middle| x_2 = 0 \right\},$$

which is one-dimensional, so $\mathsf{geomm}(A, a) = 1$. Thus, a is not semi-simple.

7.4 Spectra of Special Matrices

Theorem 7.31 *Let A be an block upper triangular partitioned matrix given by*

$$A = \begin{pmatrix} A_{11} & A_{12} & \cdots & A_{1m} \\ O & A_{22} & \cdots & A_{2m} \\ \vdots & \vdots & \cdots & \vdots \\ O & O & \cdots & A_{mm} \end{pmatrix},$$

where $A_{ii} \in \mathbb{R}^{p_i \times p_i}$ for $1 \leqslant i \leqslant m$. Then, $\mathsf{spec}(A) = \bigcup_{i=1}^m \mathsf{spec}(A_{ii})$.

If A is a block lower triangular matrix

$$A = \begin{pmatrix} A_{11} & O & \cdots & O \\ A_{21} & A_{22} & \cdots & O \\ \vdots & \vdots & \cdots & \vdots \\ A_{m1} & A_{m2} & \cdots & A_{mm} \end{pmatrix},$$

the same equality holds.

Proof Let A be a block upper triangular matrix. Its characteristic equation is $\det(\lambda I_n - A) = 0$. Observe that the matrix $\lambda I_n - A$ is also an block upper triangular matrix:

$$\lambda I_n - A = \begin{pmatrix} \lambda I_{p_1} - A_{11} & O & \cdots & O \\ -A_{21} & \lambda I_{p_2} - A_{22} & \cdots & O \\ \vdots & \vdots & \cdots & \vdots \\ -A_{m1} & -A_{m2} & \cdots & \lambda I_{p_m} - A_{mm} \end{pmatrix}.$$

By Theorem 5.153 the characteristic polynomial of A can be written as

$$p_A(\lambda) = \prod_{i=1}^m \det(\lambda I_{p_i} - A_{ii}) = \prod_{i=1}^m p_{A_{ii}}(\lambda).$$

Therefore, $\mathsf{spec}(A) = \bigcup_{i=1}^m \mathsf{spec}(A_{ii})$.

The argument for block lower triangular matrices is similar.

Corollary 7.32 *Let $A \in \mathbb{R}^{n \times n}$ be a block diagonal matrix given by*

$$A = \begin{pmatrix} A_{11} & O & \cdots & O \\ O & A_{22} & \cdots & O \\ \vdots & \vdots & \cdots & \vdots \\ O & O & \cdots & A_{mm} \end{pmatrix},$$

where $A_{ii} \in \mathbb{R}^{n_i \times n_i}$ for $1 \leqslant i \leqslant m$. We have $\mathsf{spec}(A) = \bigcup_{i=1}^{m} \mathsf{spec}(A_{ii})$ *and* $\mathsf{algm}(A, \lambda) = \sum_{i=1}^{m} \mathsf{algm}(A_i, \lambda)$. *Moreover,* $\boldsymbol{v} \neq \boldsymbol{0}_n$ *is an eigenvector of* A *if and only if we can write*

$$\boldsymbol{v} = \begin{pmatrix} \boldsymbol{v}_1 \\ \vdots \\ \boldsymbol{v}_m \end{pmatrix},$$

where each vector $\boldsymbol{v}_i$ *is either an eigenvector of* A_i *or* $\boldsymbol{0}_{n_i}$ *for* $1 \leqslant i \leqslant m$ *and there exists* i *such that* $\boldsymbol{v}_i \neq \boldsymbol{0}_{n_i}$.

Proof This statement follows immediately from Theorem 7.31.

Theorem 7.33 *Let* $A = (a_{ij}) \in \mathbb{C}^{n \times n}$ *be an upper (lower) triangular matrix. Then,* $\mathsf{spec}(A) = \{a_{ii} \mid 1 \leqslant i \leqslant n\}$.

Proof It is easy to see that the the characteristic polynomial of A is $p_A(\lambda) = (\lambda - a_{11}) \cdots (\lambda - a_{nn})$, which implies immediately the theorem.

Corollary 7.34 *If* $A \in \mathbb{C}^{n \times n}$ *is an upper triangular matrix and* λ *is an eigenvalue such that the diagonal entries of* A *that equal* λ *occur in* $a_{i_1 i_1}, \ldots, a_{i_p i_p}$, *then* $S_{A,\lambda}$ *is a* p*-dimensional subspace of* $\mathbb{C}^n$ *generated by* $\boldsymbol{e}_{i_1}, \ldots \boldsymbol{e}_{i_p}$.

Proof This statement is immediate.

Corollary 7.35 *We have*

$$\mathsf{spec}(\mathsf{diag}(d_1, \ldots, d_n)) = \{d_1, \ldots, d_n\}.$$

Proof This statement is a direct consequence of Theorem 7.33.

Note that if $\lambda \in \mathsf{spec}(A)$ we have $A\mathbf{x} = \lambda\mathbf{x}$, $A^2\mathbf{x} = \lambda^2\mathbf{x}$ and, in general, $A^k\mathbf{x} = \lambda^k\mathbf{x}$ for $k \geqslant 1$. Thus, $\lambda \in \mathsf{spec}(A)$ implies $\lambda^k \in \mathsf{spec}(A^k)$ for $k \geqslant 1$.

Theorem 7.36 *If* $A \in \mathbb{C}^{n \times n}$ *is a nilpotent matrix, then* $\mathsf{spec}(A) = \{0\}$.

Proof Let $A \in \mathbb{C}^{n \times n}$ be a nilpotent matrix such that $\mathsf{nilp}(A) = k$. By a previous observation if $\lambda \in \mathsf{spec}(A)$, then $\lambda^k \in \mathsf{spec}(A^k) = \mathsf{spec}(O_{n,n}) = \{0\}$. Thus, $\lambda = 0$.

Theorem 7.37 *If* $A \in \mathbb{C}^{n \times n}$ *is an idempotent matrix, then* $\mathsf{spec}(A) \subseteq \{0, 1\}$.

Proof Let $A \in \mathbb{C}^{n \times n}$ be an idempotent matrix, λ be an eigenvalue of A, and let $\mathbf{x}$ be an eigenvector of λ. We have $P^2\mathbf{x} = P\mathbf{x} = \lambda\mathbf{x}$; on another hand, $P^2\mathbf{x} = P(P\mathbf{x}) = P(\lambda\mathbf{x}) = \lambda P(\mathbf{x}) = \lambda^2\mathbf{x}$, so $\lambda^2 = \lambda$, which means that $\lambda \in \{0, 1\}$.

Theorem 7.38 *At least one eigenvalue of a stochastic matrix is equal to* 1 *and all eigenvalues lie on or inside the unit circle.*

Proof Let $A \in \mathbb{R}^{n \times n}$ be a stochastic matrix. Then, $1 \in \mathsf{spec}(A)$ and $\mathbf{1}$ is an eigenvector that corresponds to the eigenvalue 1 as the reader can easily verify.

If λ is an eigenvalue of A and $A\mathbf{x} = \lambda\mathbf{x}$, then $\lambda x_i = \sum_{i=1}^{n} a_{ij} x_j$ for $1 \leqslant n \leqslant n$, which implies

$$|\lambda||x_i| \leqslant \sum_{i=1}^{n} a_{ij}|x_j|.$$

Since $\mathbf{x} \neq \mathbf{0}$, let x_p be a component of $\mathbf{x}$ such that $|x_p| = \max\{|x_i| \mid 1 \leqslant i \leqslant n\}$. Choosing $i = p$ we have

$$|\lambda| \leqslant \sum_{i=1}^{n} a_{ij} \frac{|x_j|}{|x_p} \leqslant \sum_{i=1}^{n} a_{ij} = 1,$$

which shows that all eigenvalues of A lie on or inside the unit circle.

Theorem 7.39 *All eigenvalues of a unitary matrix are located on the unit circle.*

Proof Let $A \in \mathbb{R}^{n \times n}$ be an unitary matrix and let λ be an eigenvalue of A. By Theorem 6.86, if $\mathbf{x}$ is an eigenvector that corresponds to λ we have

$$\| \mathbf{x} \| = \| A\mathbf{x} \| = \| \lambda\mathbf{x} \| = |\lambda| \, \| \mathbf{x} \|,$$

which implies $|\lambda| = 1$.

Next, we show that unitary diagonalizability is a characteristic property of normal matrices.

Theorem 7.40 (Spectral Theorem for Normal Matrices) *A matrix $A \in \mathbb{C}^{n \times n}$ is normal if and only if there exists a unitary matrix U and a diagonal matrix D such that*

$$A = UDU^{\mathsf{H}}, \tag{7.4}$$

the columns of U are unit eigenvectors and the diagonal elements of D are the eigenvalues of A that correspond to these eigenvectors.

Proof Suppose that A is a normal matrix. By Schur's Triangularization Theorem there exists a unitary matrix $U \in \mathbb{C}^{n \times n}$ and an upper-triangular matrix $T \in \mathbb{C}^{n \times n}$ such that $A = UTU^{-1}$. Thus, $T = U^{-1}AU = U^{\mathsf{H}}AU$ and $T^{\mathsf{H}} = U^{\mathsf{H}}A^{\mathsf{H}}U$. Therefore,

$$\begin{aligned} T^{\mathsf{H}}T &= U^{\mathsf{H}}A^{\mathsf{H}}UU^{\mathsf{H}}AU = U^{\mathsf{H}}A^{\mathsf{H}}AU \\ &\quad \text{(because } U \text{ is unitary)} \\ &= U^{\mathsf{H}}AA^{\mathsf{H}}U \\ &\quad \text{(because } A \text{ is normal)} \\ &= U^{\mathsf{H}}AUU^{\mathsf{H}}A^{\mathsf{H}}U = TT^{\mathsf{H}}, \end{aligned}$$

so T is a normal matrix. By Theorem 5.60, T is a diagonal matrix, so D's role is played by T.

We leave the proof of the converse implication to the reader.

Let $U = (\boldsymbol{u}_1 \cdots \boldsymbol{u}_n)$. Since

$$
\begin{aligned}
A = UDU^{\mathsf{H}} &= (\boldsymbol{u}_1 \ \cdots \ \boldsymbol{u}_n) \begin{pmatrix} \lambda_1 & 0 & \cdots & 0 \\ 0 & \lambda_2 & \cdots & 0 \\ \vdots & \vdots & \cdots & \vdots \\ 0 & 0 & \cdots & \lambda_n \end{pmatrix} \begin{pmatrix} \boldsymbol{u}_1^{\mathsf{H}} \\ \vdots \\ \boldsymbol{u}_n^{\mathsf{H}} \end{pmatrix} \\
&= \lambda_1 \boldsymbol{u}_1 \boldsymbol{u}_1^{\mathsf{H}} + \cdots + \lambda_n \boldsymbol{u}_n \boldsymbol{u}_n^{\mathsf{H}}, \qquad (7.5)
\end{aligned}
$$

it follows that $A\boldsymbol{u}_i = \lambda_i \boldsymbol{u}_i$ for $1 \leqslant i \leqslant n$, which proves the statement.

The equivalent Equalities (7.4) or (7.5) are referred to as *spectral decompositions* of the normal matrix A.

Theorem 7.41 (Spectral Theorem for Hermitian Matrices) *If the matrix $A \in \mathbb{C}^{n\times n}$ is Hermitian or skew-Hermitian, A can be written as $A = UDU^{\mathsf{H}}$, where U is a unitary matrix and D is a diagonal matrix having the eigenvalues of A as its diagonal elements.*

Proof This statement follows from Theorem 7.40 because any Hermitian or skew-Hermitian matrix is normal.

Corollary 7.42 *The rank of a Hermitian matrix is equal to the number of non-zero eigenvalues.*

Proof The statement of the corollary obviously holds for any diagonal matrix. If A is a Hermitian matrix, by Theorem 7.41, we have $\mathit{rank}(A) = \mathit{rank}(D)$, where D is a diagonal matrix having the eigenvalues of A as its diagonal elements. This implies the statement of the corollary.

Let $A \in \mathbb{C}^{n\times n}$ be a Hermitian matrix of rank p. By Theorem 7.41 A can be written as $A = UDU^{\mathsf{H}}$, where U is a unitary matrix, $D = \mathsf{diag}(\lambda_1, \ldots, \lambda_p, 0, \ldots, 0)$ having as non-zero diagonal elements $\lambda_1, \ldots, \lambda_p$ the eigenvalues of A and $\lambda_1 \geqslant \cdots \geqslant \lambda_p > 0$. Thus, if $W \in \mathbb{C}^{n\times p}$ is a matrix that consists of the first p columns of U we can write

$$
\begin{aligned}
A &= W \begin{pmatrix} \lambda_1 & 0 & \cdots & 0 \\ 0 & \lambda_2 & \cdots & 0 \\ \vdots & \vdots & \vdots & \vdots \\ 0 & 0 & \cdots & \lambda_p \end{pmatrix} W' \\
&= (\boldsymbol{u}_1 \ \cdots \ \boldsymbol{u}_p) \begin{pmatrix} \lambda_1 & 0 & \cdots & 0 \\ 0 & \lambda_2 & \cdots & 0 \\ \vdots & \vdots & \vdots & \vdots \\ 0 & 0 & \cdots & \lambda_p \end{pmatrix} \begin{pmatrix} \boldsymbol{u}_1^{\mathsf{H}} \\ \vdots \\ \boldsymbol{u}_p^{\mathsf{H}} \end{pmatrix}
\end{aligned}
$$

$$= \lambda_1 \boldsymbol{u}_1 \boldsymbol{u}_1^{\mathsf{H}} + \cdots + \lambda_p \boldsymbol{u}_p \boldsymbol{u}_p^{\mathsf{H}}.$$

If A is not Hermitian, $rank(A)$ may differ from the number of non-zero-eigenvalues. For example, the matrix

$$A = \begin{pmatrix} 0 & 1 \\ 0 & 0 \end{pmatrix}$$

has no non-zero eigenvalues. However, its rank is 1.

The spectral decomposition (7.5) of Hermitian matrices,

$$A = \lambda_1 \boldsymbol{u}_1 \boldsymbol{u}_1^{\mathsf{H}} + \cdots + \lambda_n \boldsymbol{u}_n \boldsymbol{u}_n^{\mathsf{H}}$$

allows us to extend functions of the form $f : \mathbb{R} \longrightarrow \mathbb{R}$ to Hermitian matrices. Since the eigenvalues of a Hermitian matrix are real numbers, it makes sense to define $f(A)$ as

$$f(A) = f(\lambda_1) \boldsymbol{u}_1 \boldsymbol{u}_1^{\mathsf{H}} + \cdots + f(\lambda_n) \boldsymbol{u}_n \boldsymbol{u}_n^{\mathsf{H}}.$$

In particular, if A is positive semi-definite, we have $\lambda_i \geqslant 0$ for $1 \leqslant i \leqslant n$ and we can define

$$\sqrt{A} = \sqrt{\lambda_1} \boldsymbol{u}_1 \boldsymbol{u}_1^{\mathsf{H}} + \cdots + \sqrt{\lambda_n} \boldsymbol{u}_n \boldsymbol{u}_n^{\mathsf{H}}.$$

Definition 7.43 *Let $A \in \mathbb{C}^{n \times n}$ be a Hermitian matrix. The triple $\mathfrak{I}(A) = (n_+(A), n_-(A), n_0(A))$, where $n_+(A)$ is the number of positive eigenvalues, $n_-(A)$ is the number of negative eigenvalues, and $n_0(A)$ is the number of zero eigenvalues is the* inertia *of the matrix A.*

The number $sig(A) = n_+(A) - n_-(A)$ is the signature *of A.*

Example 7.44 If $A = \mathsf{diag}(4, -1, 0, 0, 1)$, then $\mathfrak{I}(A) = (2, 1, 2)$ and $sig(A) = 1$.

Let $A \in \mathbb{C}^{n \times n}$ be a Hermitian matrix. By Theorem 7.41 A can be written as $A = U^{\mathsf{H}} D U$, where U is a unitary matrix and $D = \mathsf{diag}(\lambda_1, \ldots, \lambda_n)$ is a diagonal matrix having the eigenvalues of A (which are real numbers) as its diagonal elements. Without loss of generality we may assume that the positive eigenvalues of A are $\lambda_1, \ldots, \lambda_{n_+}$, followed by the negative values $\lambda_{n_+ + 1}, \ldots, \lambda_{n_+ + n_-}$, and the zero eigenvalues $\lambda_{n_+ + n_- + 1}, \ldots, \lambda_n$.

Let θ_j be the numbers defined by

$$\theta_j = \begin{cases} \sqrt{\lambda_j} & \text{if } 1 \leqslant j \leqslant n_+, \\ \sqrt{-\lambda_j} & \text{if } n_+ + 1 \leqslant j \leqslant n_+ + n_-, \\ 1 & \text{if } n_+ + n_- + 1 \leqslant j \leqslant n \end{cases}$$

for $1 \leqslant j \leqslant n$. If $T = \text{diag}(\theta_1, \ldots, \theta_n)$, then we can write $D = T^{\mathsf{H}}GT$, where G is a diagonal matrix, $G = (g_1, \ldots, g_n)$ defined by

$$g_j = \begin{cases} 1 & \text{if } \lambda_j > 0, \\ -1 & \text{if } \lambda_j < 0, \\ 0 & \text{if } \lambda_j = 0, \end{cases}$$

for $1 \leqslant j \leqslant n$. This allows us to write $A = U^{\mathsf{H}}DU = U^{\mathsf{H}}T^{\mathsf{H}}GTU = (TU)^{\mathsf{H}}G(TU)$. The matrix TU is nonsingular, so $A \approx G$. The matrix G defined above is the *inertia matrix* of A and these definitions show that any Hermitian matrix is congruent to its inertia matrix.

For a Hermitian matrix $A \in \mathbb{C}^{n\times n}$ let $\mathsf{S}_+(A)$ be the subspace of $\mathbb{C}^n$ generated by $n_+(A)$ orthonormal eigenvectors that correspond to the positive eigenvalues of A. Clearly, we have $\dim(\mathsf{S}_+(A)) = n_+(A)$. This notation is used in the proof of the next theorem.

Theorem 7.45 (Sylvester's Inertia Theorem) *Let A, B be two Hermitian matrices, $A, B \in \mathbb{C}^{n\times n}$. We $A \approx B$ if and only if $\mathfrak{I}(A) = \mathfrak{I}(B)$.*

Proof If $\mathfrak{I}(A) = \mathfrak{I}(B)$, then we have $A = S^{\mathsf{H}}GS$ and $B = T^{\mathsf{H}}GT$, where both S and T are nonsingular matrices. Since $A \approx G$ and $B \approx G$, we have $A \approx B$.

Conversely, suppose that $A \approx B$, that is, $A = S^{\mathsf{H}}BS$, where S is a nonsingular matrix. We have $\mathit{rank}(A) = \mathit{rank}(B)$, so $n_0(A) = n_0(B)$. To prove that $\mathfrak{I}(A) = \mathfrak{I}(B)$ it suffices to show that $n_+(A) = n_+(B)$.

Let $m = n_+(A)$ and let $\mathbf{v}_1, \ldots, \mathbf{v}_m$ be m orthonormal eigenvectors of A that correspond to the m positive eigenvalues of this matrix, and let $\mathsf{S}_+(A)$ be the subspace generated by these vectors. If $\mathbf{v} \in \mathsf{S}_+(A) - \{\mathbf{0}\}$, then we have $\mathbf{v} = a_1\mathbf{v}_1 + \cdots + a_m\mathbf{v}_m$, so

$$\mathbf{v}^{\mathsf{H}}A\mathbf{v} = \left(\sum_{j=1}^{m} a_j\mathbf{v}_j\right)^{\mathsf{H}} A \left(\sum_{j=1}^{m} a_j\mathbf{v}_j\right) = \sum_{j=1}^{m} |a_j|^2 > 0.$$

Therefore, $\mathbf{x}^{\mathsf{H}}S^{\mathsf{H}}BS\mathbf{x} > 0$, so if $\mathbf{y} = S\mathbf{x}$, then $\mathbf{y}^{\mathsf{H}}B\mathbf{y} > 0$, which means that $\mathbf{y} \in \mathsf{S}_+(B)$. This shows that $\mathsf{S}_+(A)$ is isomorphic to a subspace of $\mathsf{S}_+(B)$, so $n_+(A) \leqslant n_+(B)$. The reverse inequality can be shown in the same manner, so $n_+(A) = n_+(B)$.

We can add an interesting detail to the full-rank decomposition of a matrix.

Corollary 7.46 *If $A \in \mathbb{C}^{m\times n}$ and $A = CR$ is the full-rank decomposition of A with $\mathit{rank}(A) = k$, $C \in \mathbb{C}^{m\times k}$, and $R \in \mathbb{C}^{k\times n}$, then C may be chosen to have orthogonal columns and R to have orthogonal rows.*

Proof Since the matrix $A^{\mathsf{H}}A \in \mathbb{C}^{n\times n}$ is Hermitian, by Theorem 7.41, there exists an unitary matrix $U \in \mathbb{C}^{n\times k}$ such that $A^{\mathsf{H}}A = U^{\mathsf{H}}DU$, where $D \in \mathbb{C}^{k\times k}$ is a non-negative diagonal matrix. Let $C = AU^{\mathsf{H}} \in C^{n\times k}$ and $R = U$. Clearly, $CR = A$, and

R has orthogonal rows because U is unitary. Let $\mathbf{c}_p, \mathbf{c}_q$ be two columns of C, where $1 \leqslant p, q \leqslant k$ and $p \neq q$. Since $\mathbf{c}_p = A\boldsymbol{u}_p$ and $\mathbf{c}_q = A\boldsymbol{u}_q$, where $\boldsymbol{u}_p, \boldsymbol{u}_q$ are the corresponding columns of U, we have

$$\mathbf{c}_p^{\mathsf{H}}\mathbf{c}_q = \boldsymbol{u}_p^{\mathsf{H}} A^{\mathsf{H}} A \boldsymbol{u}_q = \boldsymbol{u}_p^{\mathsf{H}} U^{\mathsf{H}} D U \boldsymbol{u}_q = \mathbf{e}_p^{\mathsf{H}} D \mathbf{e}_q = 0,$$

because $p \neq q$.

7.5 Variational Characterizations of Spectra

Let $A \in \mathbb{C}^{n \times n}$ be a Hermitian matrix. By the Spectral Theorem for Hermitian Matrices (Theorem 7.41) A can be factored as $A = UDU^{\mathsf{H}}$, where U is a unitary matrix and $D = \mathsf{diag}(\lambda_1, \ldots, \lambda_n)$. We assume that $\lambda_1 \geqslant \cdots \geqslant \lambda_n$. The columns of U constitute a family of orthonormal vectors $\{\boldsymbol{u}_1, \ldots, \boldsymbol{u}_n\}$ and $(\lambda_k, \boldsymbol{u}_k)$ are the eigenpairs of A.

Theorem 7.47 *Let A be a Hermitian matrix, $\lambda_1 \geqslant \cdots \geqslant \lambda_n$ be its eigenvalues having the orthonormal eigenvectors $\boldsymbol{u}_1, \ldots, \boldsymbol{u}_n$, respectively.*

Define the subspace $M = \langle \boldsymbol{u}_p, \ldots, \boldsymbol{u}_q \rangle$, where $1 \leqslant p \leqslant q \leqslant n$. If $\boldsymbol{x} \in M$ and $\| \boldsymbol{x} \|_2 = 1$, we have $\lambda_q \leqslant \boldsymbol{x}^{\mathsf{H}} A \boldsymbol{x} \leqslant \lambda_p$.

Proof If $\mathbf{x}$ is a unit vector in M, then $\mathbf{x} = a_p \boldsymbol{u}_p + \cdots + a_q \boldsymbol{u}_q$, so $\mathbf{x}^{\mathsf{H}} \boldsymbol{u}_i = \overline{a_i}$ for $p \leqslant i \leqslant q$. Since $\| \mathbf{x} \|_2 = 1$, we have $|a_p|^2 + \cdots + |a_q|^2 = 1$. This allows us to write:

$$\begin{aligned}
\mathbf{x}^{\mathsf{H}} A \mathbf{x} &= \mathbf{x}^{\mathsf{H}} (a_p A \boldsymbol{u}_p + \cdots + a_q A \boldsymbol{u}_q) \\
&= \mathbf{x}^{\mathsf{H}} (a_p \lambda_p \boldsymbol{u}_p + \cdots + a_q \lambda_q \boldsymbol{u}_q) \\
&= \mathbf{x}^{\mathsf{H}} (|a_p|^2 \lambda_p + \cdots + |a_q|^2 \lambda_q).
\end{aligned}$$

Since $|a_p|^2 + \cdots + |a_q|^2 = 1$, the desired inequalities follow immediately.

Corollary 7.48 *Let A be a Hermitian matrix, $\lambda_1 \geqslant \cdots \geqslant \lambda_n$ be its eigenvalues having the orthonormal eigenvectors $\boldsymbol{u}_1, \ldots, \boldsymbol{u}_n$, respectively. The following statements hold for a unit vector $\boldsymbol{x}$:*

(i) if $\boldsymbol{x} \in \langle \boldsymbol{u}_1, \ldots, \boldsymbol{u}_i \rangle$, then $\boldsymbol{x}^{\mathsf{H}} A \boldsymbol{x} \geqslant \lambda_i$;
(ii) if $\boldsymbol{x} \in \langle \boldsymbol{u}_1, \ldots, \boldsymbol{u}_{i-1} \rangle^{\perp}$, then $\boldsymbol{x}^{\mathsf{H}} A \boldsymbol{x} \leqslant \lambda_i$.

Proof The first statement follows directly from Theorem 7.47.

For the second statement observe that $\mathbf{x} \in \langle \boldsymbol{u}_1, \ldots, \boldsymbol{u}_{i-1} \rangle^{\perp}$ is equivalent to $\mathbf{x} \in \langle \boldsymbol{u}_i, \ldots, \boldsymbol{u}_n \rangle$; again, the second inequality follows from Theorem 7.47.

Theorem 7.49 (Rayleigh-Ritz Theorem) *Let A be a Hermitian matrix and let $(\lambda_1, \boldsymbol{u}_1), \ldots, (\lambda_n, \boldsymbol{u}_n)$ be the eigenpairs of A, where $\lambda_1 \geqslant \cdots \geqslant \lambda_n$. If $\boldsymbol{x}$ is a unit vector, we have $\lambda_n \leqslant \boldsymbol{x}^{\mathsf{H}} A \boldsymbol{x} \leqslant \lambda_1$.*

Proof This statement follows from Theorem 7.47 by observing that the subspace generated by $\boldsymbol{u}_1, \ldots, \boldsymbol{u}_n$ is the entire space $\mathbb{C}^n$.

Next we discuss an important generalization of Rayleigh-Ritz Theorem.

Let $\mathcal{S}_p^n$ be the collection of p-dimensional subspaces of $\mathbb{C}^n$. Note that $\mathcal{S}_0^n = \{\{\mathbf{0}_n\}\}$ and $\mathcal{S}_n^n = \{\mathbb{C}^n\}$.

Theorem 7.50 (Courant-Fisher Theorem) *Let $A \in \mathbb{C}^{n \times n}$ be a Hermitian matrix having the eigenvalues $\lambda_1 \geqslant \cdots \geqslant \lambda_n$. We have*

$$\begin{aligned}\lambda_k &= \max_{U \in \mathcal{S}_k^n} \min\{\boldsymbol{x}^H A \boldsymbol{x} \mid \boldsymbol{x} \in U \textit{and} \parallel \boldsymbol{x} \parallel_2 = 1\} \\ &= \min_{U \in \mathcal{S}_{n-k+1}^n} \max\{\boldsymbol{x}^H A \boldsymbol{x} \mid \boldsymbol{x} \in U \textit{and} \parallel \boldsymbol{x} \parallel_2 = 1\}.\end{aligned}$$

Proof Let $A = U^{\mathsf{H}} \mathsf{diag}(\lambda_1, \ldots, \lambda_n) U$ be the factorization of A provided by Theorem 7.41), where $U = (\boldsymbol{u}_1 \ \cdots \ \boldsymbol{u}_n)$.

If $U \in \mathcal{S}_k^n$ and $W = \langle \boldsymbol{u}_k, \ldots, \boldsymbol{u}_n \rangle \in \mathcal{S}_{n-k+1}^n$, then there is a non-zero vector $\mathbf{x} \in U \cap W$ because $\dim(U) + \dim(W) = n + 1$; we can assume that $\parallel \mathbf{x} \parallel_2 = 1$. Therefore, by Theorem 7.47 we have $\lambda_k \geqslant \mathbf{x}^{\mathsf{H}} A \mathbf{x}$, and, therefore, for any $U \in \mathcal{S}_k^n$, $\lambda_k \geqslant \min\{\mathbf{x}^{\mathsf{H}} A \mathbf{x} \mid \mathbf{x} \in U \text{and} \parallel \mathbf{x} \parallel_2 = 1\}$. This implies $\lambda_k \geqslant \max_{U \in \mathcal{S}_k^n} \min\{\mathbf{x}^{\mathsf{H}} A \mathbf{x} \mid \mathbf{x} \in U \text{and} \parallel \mathbf{x} \parallel_2 = 1\}$.

The same Theorem 7.47 implies that for a unit vector $\mathbf{x} \in \langle \boldsymbol{u}_1, \ldots, \boldsymbol{u}_k \rangle \in \mathcal{S}_k^n$ we have $\mathbf{x}^{\mathsf{H}} A \mathbf{x} \geqslant \lambda_k$ and $\boldsymbol{u}_k^{\mathsf{H}} A \boldsymbol{u}_k = \lambda_k$. Therefore, for $W = \langle \boldsymbol{u}_1, \ldots, \boldsymbol{u}_k \rangle \in \mathcal{S}_k^n$ we have $\min\{\mathbf{x}^{\mathsf{H}} A \mathbf{x} \mid \mathbf{x} \in W, \parallel \mathbf{x} \parallel_2 = 1\} \geqslant \lambda_k$, so $\max_{W \in \mathcal{S}_k^n} \min\{\mathbf{x}^{\mathsf{H}} A \mathbf{x} \mid \mathbf{x} \in W, \parallel \mathbf{x} \parallel_2 = 1\} \geqslant \lambda_k$. The inequalities proved above yield

$$\lambda_k = \max_{U \in \mathcal{S}_k^n} \min\{\mathbf{x}^{\mathsf{H}} A \mathbf{x} \mid \mathbf{x} \in U \text{and} \parallel \mathbf{x} \parallel_2 = 1\}.$$

For the second equality, let $U \in \mathcal{S}_{n-k+1}^n$. If $W = \langle \boldsymbol{u}_1, \ldots, \boldsymbol{u}_k \rangle$, there is a non-zero unit vector $\mathbf{x} \in U \cap W$ because $\dim(U) + \dim(W) \geqslant n + 1$. By Theorem 7.47 we have $\mathbf{x}^{\mathsf{H}} A \mathbf{x} \leqslant \lambda_k$. Therefore, for any $U \in \mathcal{S}_{n-k+1}^n$, $\lambda_k \geqslant \max\{\mathbf{x}^{\mathsf{H}} A \mathbf{x} \mid \mathbf{x} \in U \text{and} \parallel \mathbf{x} \parallel_2 = 1\}$. This implies $\lambda_k \geqslant \min_{U \in \mathcal{S}_{n-k+1}^n} \max\{\mathbf{x}^{\mathsf{H}} A \mathbf{x} \mid \mathbf{x} \in U \text{and} \parallel \mathbf{x} \parallel_2 = 1\}$.

Theorem 7.47 implies that for a unit vector $\mathbf{x} \in \langle \boldsymbol{u}_k, \ldots, \boldsymbol{u}_n \rangle \in \mathcal{S}_{n-k+1}^n$ we have $\lambda_k \leqslant \mathbf{x}^{\mathsf{H}} A \mathbf{x}$ and $\lambda_k = \boldsymbol{u}_k^{\mathsf{H}} A \boldsymbol{u}_k$. Thus, $\lambda_k \leqslant \max\{\mathbf{x}^{\mathsf{H}} A \mathbf{x} \mid \mathbf{x} \in U \text{and} \parallel \mathbf{x} \parallel_2 = 1\}$. Consequently, $\lambda_k \leqslant \min_{U \in \mathcal{S}_{n-k+1}^n} \max\{\mathbf{x}^{\mathsf{H}} A \mathbf{x} \mid \mathbf{x} \in U \text{ and } \parallel \mathbf{x} \parallel_2 = 1\}$, which completes the proof of the second equality of the theorem.

An equivalent formulation of Courant-Fisher Theorem is given next.

Theorem 7.51 *Let $A \in \mathbb{C}^{n \times n}$ be a Hermitian matrix having the eigenvalues $\lambda_1 \geqslant \cdots \geqslant \lambda_n$. We have*

$$\begin{aligned}\lambda_k &= \max_{\boldsymbol{w}_1, \ldots, \boldsymbol{w}_{n-k}} \min\{\boldsymbol{x}^H A \boldsymbol{x} \mid \boldsymbol{x} \perp \boldsymbol{w}_1, \ldots, \boldsymbol{x} \perp \boldsymbol{w}_{n-k} \textit{and} \parallel \boldsymbol{x} \parallel_2 = 1\} \\ &= \min_{\boldsymbol{w}_1, \ldots, \boldsymbol{w}_{k-1}} \max\{\boldsymbol{x}^H A \boldsymbol{x} \mid \boldsymbol{x} \perp \boldsymbol{w}_1, \ldots, \boldsymbol{x} \perp \boldsymbol{w}_{k-1} \textit{and} \parallel \boldsymbol{x} \parallel_2 = 1\}.\end{aligned}$$

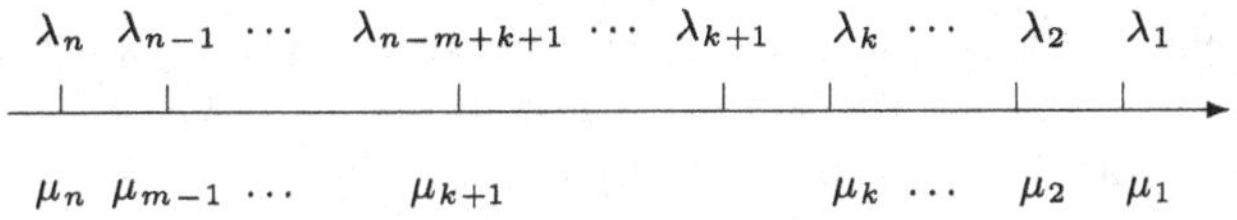

Fig. 7.1 Tight interlacing

Proof The equalities of the Theorem follow from the Courant-Fisher theorem taking into account that if $U \in \mathcal{S}^n_k$, then $U^\perp = \langle \boldsymbol{w}_1, \ldots, \boldsymbol{w}_{n-k}\rangle$ for some vectors $\boldsymbol{w}_1, \ldots, \boldsymbol{w}n - k$, and if $U \in \mathcal{S}^n_{n-k+1}$, then $U = \langle \boldsymbol{w}_1, \ldots \boldsymbol{w}_{k-1}\rangle$ for some vectors $\boldsymbol{w}_1, \ldots, \boldsymbol{w}k - 1$ in $\mathbb{C}^n$.

Definition 7.52 *Consider two non-increasing sequences of real numbers* $(\lambda_1, \ldots, \lambda_n)$ *and* $(\mu_1, \ldots, \mu_m)$ *with* $m < n$. *We say that the second sequence* interlace *the first if* $\lambda_i \geqslant \mu_i \geqslant \lambda_{n-m+i}$ *for* $1 \leqslant i \leqslant m$ (see Fig. 7.1).

The interlacing is tight *if there exists* $k \in \mathbb{N}$, $0 \leqslant k \leqslant m$ *such that* $\lambda_i = \mu_i$ *for* $0 \leqslant i \leqslant k$ *and* $\lambda_{n-m+1} = \mu_i$ *for* $k + 1 \leqslant i \leqslant m$.

The next statement is known as the Interlacing Theorem. The variant included here was obtained in [1].

Theorem 7.53 (Interlacing Theorem) *Let* $A \in \mathbb{C}^{n\times n}$ *be a Hermitian matrix,* $S \in \mathbb{C}^{n\times m}$ *be a matrix such that* $S^{\mathsf{H}}S = I_m$, *and let* $B = S^{\mathsf{H}}AS \in \mathbb{C}^{m\times m}$, *where* $m \leqslant n$.

Assume that A has the eigenvalues $\lambda_1 \geqslant \cdots \geqslant \lambda_n$ *with the orthonormal eigenvectors* $\boldsymbol{u}_1, \ldots, \boldsymbol{u}_n$, *respectively, and B has the eigenvalues* $\mu_1 \geqslant \cdots \geqslant \mu_m$ *with the respective eigenvectors* $\boldsymbol{v}_1, \ldots, \boldsymbol{v}_m$. *The following statements hold:*

(i) $\lambda_i \geqslant \mu_i \geqslant \lambda_{n-m+i}$;
(ii) *if* $\mu_i = \lambda_i$ *or* $\mu_i = \lambda_{n-m+i}$ *for some* i, $1 \leqslant i \leqslant m$, *then B has an eigenvector* $\boldsymbol{v}$ *such that* $(\mu_i, \boldsymbol{v})$ *is an eigenpair of B and* $(\mu_i, S\boldsymbol{v})$ *is an eigenpair of A;*
(iii) *if for some integer* ℓ, $\mu_i = \lambda_i$ *for* $1 \leqslant i \leqslant \ell$ *(or* $\mu_i = \lambda_{n-m+i}$ *for* $\ell \leqslant i \leqslant m$*), then* $(\mu_i, S\boldsymbol{v}_i)$ *are eigenpairs of A for* $1 \leqslant i \leqslant \ell$ *(respectively,* $\ell \leqslant i \leqslant m$*);*
(iv) *if the interlacing is tight* $SB = AS$.

Proof Note that $B^{\mathsf{H}} = S^{\mathsf{H}}A^{\mathsf{H}}S = S^{\mathsf{H}}AS = B$, so B is also Hermitian.

Since $\dim(\langle \boldsymbol{v}_1, \ldots, \boldsymbol{v}_i\rangle) = i$ and $\dim(\langle S^{\mathsf{H}}\boldsymbol{u}_1, \ldots, S^{\mathsf{H}}\boldsymbol{u}_{i-1}\rangle^\perp) = n - i + 1$, there exist non-zero vectors in the intersection of these subspaces. Let $\mathbf{t}$ be a unit vector in $\langle \boldsymbol{v}_1, \ldots, \boldsymbol{v}_i\rangle \cap \langle S^{\mathsf{H}}\boldsymbol{u}_1, \ldots, S^{\mathsf{H}}\boldsymbol{u}_{i-1}\rangle^\perp$.

Since $\mathbf{t} \in \langle S^{\mathsf{H}}\boldsymbol{u}_1, \ldots, S^{\mathsf{H}}\boldsymbol{u}_{i-1}\rangle^\perp$ we have $\mathbf{t}^{\mathsf{H}}S^{\mathsf{H}}\boldsymbol{u}_\ell = (S\mathbf{t})^{\mathsf{H}}\boldsymbol{u}_\ell = 0$ for $1 \leqslant \ell \leqslant i - 1$. Thus, $S\mathbf{t} \in \langle \boldsymbol{u}_1, \ldots, \boldsymbol{u}_{i-1}\rangle^\perp$, so, by Theorem 7.47, it follows that $\lambda_i \geqslant (S\mathbf{t})^{\mathsf{H}}A(S\mathbf{t})$. On other hand,

$$(S\mathbf{t})^{\mathsf{H}}A(S\mathbf{t}) = \mathbf{t}^{\mathsf{H}}(S^{\mathsf{H}}AS)\mathbf{t} = \mathbf{t}^{\mathsf{H}}B\mathbf{t}$$

and $\mathbf{t} \in \langle \boldsymbol{v}_1, \ldots, \boldsymbol{v}_i\rangle$, which yield $\mathbf{t}^{\mathsf{H}}B\mathbf{t} \geqslant \mu_i$, again, by Theorem 7.47. Thus, we conclude that

$$\lambda_i \geqslant (S\mathbf{t})^{\mathsf{H}}A(S\mathbf{t}) = \mathbf{t}^{\mathsf{H}}B\mathbf{t} \geqslant \mu_i. \tag{7.6}$$

Note that the matrices $-A$ and $-B$ have the eigenvalues $-\lambda_n \geqslant \cdots \geqslant -\lambda_1$ and $-\mu_m \geqslant \cdots \geqslant -\mu_1$. The ith eigenvalue in the list $-\lambda_n \geqslant \cdots \geqslant -\lambda_1$ is $-\lambda_{n-m+i}$, so, by applying the previous argument to the matrices $-A$ and $-B$ we obtain: $\mu_i \geqslant \lambda_{n-m+i}$, which concludes the proof of (i).

If $\lambda_i = \mu_i$, since $B\boldsymbol{v}_i = \mu_i \boldsymbol{v}_i$, it follows that $S^{\mathsf{H}} A S \boldsymbol{v}_i = \mu_i \boldsymbol{v}_i$, so $A(S\boldsymbol{v}_i) = \mu_i (S\boldsymbol{v}_i)$ because S is a unitary matrix. Thus, $(\mu_i, S\boldsymbol{v}_i)$ is an eigenpair of A, which proves Part (ii).

Part (iii) follows directly from Part (ii) and its proof.

Finally, if the interlacing is tight, $S\boldsymbol{v}_1, \ldots, S\boldsymbol{v}_m$ is an orthonormal set of eigenvectors of A corresponding to the eigenvalues $\mu_1, \ldots, \mu_m$, so $SB\boldsymbol{v}_i = \mu_i S\boldsymbol{v}_i = AS\boldsymbol{v}_i$ for $1 \leqslant i \leqslant m$. Since $SB, AS \in \mathbb{C}^{n\times m}$ and the vectors $\boldsymbol{v}_1, \ldots, \boldsymbol{v}_m$ form a basis in $\mathbb{C}^m$, we have $SB = AS$.

Example 7.54 Let $R = \{r_1, \ldots, r_m\}$ be a subset of $\{1, \ldots, n\}$. Define the matrix $S_R \in \mathbb{C}^{n\times m}$ by $S_R = (\mathbf{e}_{r_1} \cdots \mathbf{e}_{r_m})$.

For example, if $n = 4$ and $R = \{2, 4\}$ we have the matrix

$$S_R = \begin{pmatrix} 0 & 0 \\ 1 & 0 \\ 0 & 0 \\ 0 & 1 \end{pmatrix}.$$

It is immediate that $S_R^{\mathsf{H}} S_R = I_m$ and that $S_R^{\mathsf{H}} A S_R$ is the principal submatrix $A\begin{bmatrix} R \\ R \end{bmatrix}$, defined by the intersection of rows $r_1, \ldots, r_m$ with the columns $r_1, \ldots, r_m$.

Corollary 7.55 *Let $A \in \mathbb{C}^{n\times n}$ be a Hermitian matrix and let $B = A\begin{bmatrix} R \\ R \end{bmatrix} \in \mathbb{C}^{m\times m}$ be a principal submatrix of A. If A has the eigenvalues $\lambda_1 \geqslant \cdots \geqslant \lambda_n$ and B has the eigenvalues $\mu_1 \geqslant \cdots \geqslant \mu_m$, then $\lambda_i \geqslant \mu_i \geqslant \lambda_{n-m+i}$ for $1 \leqslant i \leqslant m$.*

Proof This statement follows immediately from Theorem 7.53, by taking $S = S_R$.

Theorem 7.56 *Let $A, B \in \mathbb{C}^{n\times}$ be two Hermitian matrices and let $E = B - A$. Suppose that the eigenvalues of A, B, E these are $\alpha_1 \geqslant \cdots \geqslant \alpha_n$, $\beta_1 \geqslant \cdots \geqslant \beta_n$, and $\epsilon_1 \geqslant \cdots \geqslant \epsilon_n$, respectively. Then, we have $\epsilon_n \leqslant \beta_i - \alpha_i \leqslant \epsilon_1$.*

Proof Note that E is also Hermitian, so all matrices involved have real eigenvalues. By Courant-Fisher Theorem,

$$\beta_k = \min_{W} \max_{\mathbf{x}} \{\mathbf{x}^{\mathsf{H}} B \mathbf{x} \mid \| \mathbf{x} \|_2 = 1 \text{ and } \boldsymbol{w}_i^{\mathsf{H}} \mathbf{x} = 0 \text{ for } 1 \leqslant i \leqslant k-1\},$$

where $W = \{\boldsymbol{w}_1, \ldots, \boldsymbol{w}_{k-1}\}$. Thus,

$$\beta_k \leqslant \max_{\mathbf{x}} \mathbf{x}^{\mathsf{H}} B \mathbf{x} = \max_{\mathbf{x}} (\mathbf{x}^{\mathsf{H}} A \mathbf{x} + \mathbf{x}^{\mathsf{H}} E \mathbf{x}). \tag{7.7}$$

Let U be a unitary matrix such that $U^{\mathsf{H}}AU = \mathsf{diag}(\alpha_1, \ldots, \alpha_n)$. Choose $\boldsymbol{w}_i = U\mathbf{e}_i$ for $1 \leqslant i \leqslant k-1$. We have $\boldsymbol{w}_i^{\mathsf{H}}\mathbf{x} = \mathbf{e}_i^{\mathsf{H}}U^{\mathsf{H}}\mathbf{x} = 0$ for $1 \leqslant i \leqslant k-1$.

Define $\mathbf{y} = U^{\mathsf{H}}\mathbf{x}$. Since U is an unitary matrix, $\| \mathbf{y} \|_2 = \| \mathbf{x} \|_2 = 1$. Observe that $\mathbf{e}_i^{\mathsf{H}}\mathbf{y} = y_i = 0$ for $1 \leqslant i \leqslant k$. Therefore, $\sum_{i=k}^{n} y_i^2 = 1$. This, in turn implies $\mathbf{x}^{\mathsf{H}}A\mathbf{x} = \mathbf{y}^{\mathsf{H}}U^{\mathsf{H}}AU\mathbf{y} = \sum_{i=k}^{n} \alpha_i y_i^2 \leqslant \alpha_k$.

From the Inequality (7.7) it follows that

$$\beta_k \leqslant \alpha_k + \max_{\mathbf{x}} \mathbf{x}^{\mathsf{H}}E\mathbf{x} \leqslant \alpha_k + \epsilon_n.$$

Since $A = B - E$, by inverting the roles of A and B we have $\alpha_k \leqslant \beta_k - \epsilon_1$, or $\epsilon_1 \leqslant \beta_k - \alpha_k$, which completes the argument.

Lemma 7.57 *Let $T \in \mathbb{C}^{n\times n}$ be a upper triangular matrix and let $\lambda \in \mathsf{spec}(T)$ be an eigenvalue such that the diagonal entries that equal λ occur in $t_{i_1 i_1}, \ldots, t_{i_p i_p}$. Then, the invariant subspace $S_{T,\lambda}$ is the p-dimensional subspace generated by $\boldsymbol{e}_{i_1}, \ldots, \boldsymbol{e}_{i_p}$.*

Proof The argument is straightforward and is omitted.

Lemma 7.58 *Let $T \in \mathbb{C}^{n\times n}$ be an upper triangular matrix and let $p_T(\lambda) = \lambda^n + a_1\lambda^{n-1} + \cdots + a_{n-1}\lambda + a_n$ be its characteristic polynomial. Then,*

$$p_T(T) = T^n + a_1 T^{n-1} + \cdots + a_{n-1}T + a_n I_n = O_{n,n}.$$

Proof We have

$$p_T(T) = (T - \lambda_1 I_n) \cdots (T - \lambda_n I_n).$$

Observe that for any matrix $A \in \mathbb{C}^{n\times n}$, $\lambda_j, \lambda_k \in \mathsf{spec}(A)$, and every eigenvector $\boldsymbol{v}$ of A in S_{A,λ_k} we have

$$(\lambda_j I_n - A)\boldsymbol{v} = (\lambda_j - \lambda_k)\boldsymbol{v}.$$

Therefore, for $\boldsymbol{v} \in S_{T,\lambda_k}$ we have:

$$p_T(T)\boldsymbol{v} = (\lambda_1 I_n - T) \cdots (\lambda_n I_n - T)\boldsymbol{v} = \mathbf{0},$$

because $(\lambda_k I - T)\boldsymbol{v} = \mathbf{0}$.

By Lemma 7.57, $p_T(T)\mathbf{e}_i = \mathbf{0}$ for $1 \leqslant i \leqslant n$, so $p_T(T) = O_{n,n}$.

Theorem 7.59 (Cayley-Hamilton Theorem) *If $A \in \mathbb{C}^{n\times n}$ is a matrix, then $p_A(A) = O_{n,n}$.*

Proof By Schur's Triangularization Theorem there exists a unitary matrix $U \in \mathbb{C}^{n\times n}$ and an upper-triangular matrix $T \in \mathbb{C}^{n\times n}$ such that $A = UTU^{\mathsf{H}}$ and the diagonal elements of T are the eigenvalues of A. Taking into account that U is unitary we can write:

$$\begin{aligned}
p_A(A) &= (\lambda_1 I_n - A)(\lambda_2 I_n - A)\cdots(\lambda_n I_n - A) \\
&= (\lambda_1 UU^{\mathsf{H}} - UTU^{\mathsf{H}})\cdots(\lambda_n UU^{\mathsf{H}} - UTU^{\mathsf{H}}) \\
&= U(\lambda_1 I_n - T)U^{\mathsf{H}}U(\lambda_2 I_n - T)U^{\mathsf{H}}\cdots U(\lambda_n I_n - T)U^{\mathsf{H}} \\
&= U(\lambda_1 I_n - T)(\lambda_2 I_n - T)\cdots(\lambda_n I_n - T)U^{\mathsf{H}} \\
&= Up_T(T)U^{\mathsf{H}} = O_{n,n},
\end{aligned}$$

by Lemma 7.58.

Theorem 7.60 (Ky Fan's Theorem) *Let $A \in \mathbb{C}^{n\times n}$ be a Hermitian matrix such that $\mathsf{spec}(A) = \{\lambda_1, \ldots, \lambda_n\}$, where $\lambda_1 \geqslant \lambda_2 \geqslant \cdots \geqslant \lambda_n$. Also, let $V \in \mathbb{C}^{n\times n}$ be a matrix, $V = (\boldsymbol{v}_1, \ldots, \boldsymbol{v}_n)$ whose set of columns constitutes an orthonormal set of eigenvectors of A.*

For every $q \in \mathbb{N}$ such that $1 \leqslant q \leqslant n$, the sums $\sum_{i=1}^{q} \lambda_i$ and $\sum_{i=1}^{q} \lambda_{n+1-i}$ are the maximum and minimum of $\sum_{j=1}^{q} \boldsymbol{x}_j^{\mathsf{H}} A\boldsymbol{x}_j$, where $\{\boldsymbol{x}_1, \ldots, \boldsymbol{x}_q\}$ is an orthonormal set of vectors in $\mathbb{C}^n$, respectively. The maximum (minimum) is achieved when $\boldsymbol{x}_1, \ldots, \boldsymbol{x}_q$ are the first (last) columns of V.

Proof Let $\{\mathbf{x}_1, \ldots, \mathbf{x}_n\}$ be an orthonormal set of eigenvectors of A and let $\mathbf{x}_i = \sum_{k=1}^{n} b_{ki}\boldsymbol{v}_k$ be the expression of $\mathbf{x}_i$ using the columns of V as a basis for $1 \leqslant i \leqslant n$. Since each $\mathbf{x}_i$ is a unit vector we have

$$\parallel \mathbf{x}_i \parallel^2 = \mathbf{x}_i^{\mathsf{H}}\mathbf{x}_i = \sum_{k=1}^{n} |b_{ki}|^2 = 1$$

for $1 \leqslant i \leqslant n$. Also, note that

$$\mathbf{x}_i^{\mathsf{H}}\boldsymbol{v}_r = \left(\sum_{k=1}^{n} \overline{b_{ki}}\boldsymbol{v}_k^{\mathsf{H}}\right)\boldsymbol{v}_r = \overline{b_{ri}},$$

due to the orthonormality of the set of columns of V. We have

$$\begin{aligned}
\mathbf{x}_i^{\mathsf{H}}A\mathbf{x}_i &= \mathbf{x}_i^{\mathsf{H}}A\sum_{k=1}^{n} b_{ki}\boldsymbol{v}_k = \sum_{k=1}^{n} b_{ki}\mathbf{x}_i^{\mathsf{H}}A\boldsymbol{v}_k \\
&= \sum_{k=1}^{n} b_{ki}\mathbf{x}_i^{\mathsf{H}}\lambda_k\boldsymbol{v}_k = \sum_{k=1}^{n} \lambda_k b_{ki}\overline{b_{ki}} = \sum_{k=1}^{n} |b_{ki}|^2\lambda_k \\
&= \lambda_q\sum_{k=1}^{n} |b_{ki}|^2 + \sum_{k=1}^{q}(\lambda_k - \lambda_q)|b_{ki}|^2 + \sum_{k=q+1}^{n}(\lambda_k - \lambda_q)|b_{ki}|^2
\end{aligned}$$

$$\leqslant \lambda_q + \sum_{k=1}^{q} (\lambda_k - \lambda_q)|b_{ki}|^2.$$

The last inequality implies

$$\sum_{i=1}^{q} \mathbf{x}_i^{\mathsf{H}} A \mathbf{x}_i \leqslant q\lambda_q + \sum_{i=1}^{q}\sum_{k=1}^{q} (\lambda_k - \lambda_q)|b_{ki}|^2.$$

Therefore,

$$\sum_{i=1}^{q} \lambda_i - \sum_{i=1}^{q} \mathbf{x}_i^{\mathsf{H}} A \mathbf{x}_i \geqslant \sum_{i=1}^{q} (\lambda_i - \lambda_q)\left(1 - \sum_{k=1}^{q} |b_{ki}|^2\right). \tag{7.8}$$

By Inequality (6.10), we have $\sum_{k=1}^{q} |b_{ik}|^2 \leqslant \| \mathbf{x}_i \|^2 = 1$, so

$$\sum_{i=1}^{q} (\lambda_i - \lambda_q)\left(1 - \sum_{k=1}^{q} |b_{ki}|^2\right) \geqslant 0.$$

The left member of Inequality (7.8) becomes 0 when $\mathbf{x}_i = \boldsymbol{v}_i$, so $\sum_{i=1}^{q} \mathbf{x}_i^{\mathsf{H}} A \mathbf{x}_i \leqslant \sum_{i=1}^{q} \lambda_i$. The maximum of $\sum_{i=1}^{q} \mathbf{x}_i^{\mathsf{H}} A \mathbf{x}_i$ is obtained when $\mathbf{x}_i = \boldsymbol{v}_i$ for $1 \leqslant i \leqslant q$, that is, when X consists of the first q columns of V.

The argument for the minimum is similar.

Theorem 7.61 *Let $A \in \mathbb{C}^{n \times n}$ be a Hermitian matrix. If A is positive semidefinite, then all its eigenvalues are non-negative; if A is positive definite then its eigenvalues are positive.*

Proof Since A is Hermitian all its eigenvalues are real numbers. Suppose that A is positive semidefinite, that is, $\mathbf{x}^{\mathsf{H}} A \mathbf{x} \geqslant 0$ for $\mathbf{x} \in \mathbb{C}^n$. If $\lambda \in \mathsf{spec}(A)$, then $A\boldsymbol{v} = \lambda \boldsymbol{v}$ for some eigenvector $\boldsymbol{v} \neq \mathbf{0}$. The positive semi-definiteness of A implies $\boldsymbol{v}^{\mathsf{H}} A \boldsymbol{v} = \lambda \boldsymbol{v}^{\mathsf{H}} \boldsymbol{v} = \lambda \| \boldsymbol{v} \|_2^2 \geqslant 0$, which implies $\lambda \geqslant 0$. It is easy to see that if A is positive definite, then $\lambda > 0$.

Theorem 7.62 *Let $A \in \mathbb{C}^{n \times n}$ be a Hermitian matrix. If A is positive semidefinite, then all its principal minors are non-negative real numbers. If A is positive definite then all its principal minors are positive real numbers.*

Proof Since A is positive semidefinite, every sub-matrix $A\begin{bmatrix} i_1 & \cdots & i_k \\ i_1 & \cdots & i_k \end{bmatrix}$ is a Hermitian positive semidefinite matrix by Theorem 6.110, so every principal minor is a non-negative real number. The second part of the theorem is proven similarly.

Corollary 7.63 *Let $A \in \mathbb{C}^{n \times n}$ be a Hermitian matrix. The following statements are equivalent.*

(i) *A is positive semidefinite;*
(ii) *all eigenvalues of A are non-negative numbers;*
(iii) *there exists a Hermitian matrix* $C \in \mathbb{C}^{n\times n}$ *such that* $C^2 = A$*;*
(iv) *A is the Gram matrix of a sequence of vectors, that is,* $A = B^{\mathsf{H}}B$ *for some* $B \in \mathbb{C}^{n\times n}$.

Proof (i) implies (ii): This was shown in Theorem 7.61.

(ii) implies (iii): Suppose that A is a matrix such that all its eigenvalues are the non-negative numbers $\lambda_1, \ldots, \lambda_n$. By Theorem 7.41, A can be written as $A = U^{\mathsf{H}}DU$, where U is a unitary matrix and

$$D = \begin{pmatrix} \lambda_1 & 0 & \cdots & 0 \\ 0 & \lambda_2 & \cdots & 0 \\ \vdots & \vdots & \cdots & \vdots \\ 0 & 0 & \cdots & \lambda_n \end{pmatrix}.$$

Define the matrix $\sqrt{D}$ as

$$\sqrt{D} = \begin{pmatrix} \sqrt{\lambda_1} & 0 & \cdots & 0 \\ 0 & \sqrt{\lambda_2} & \cdots & 0 \\ \vdots & \vdots & \cdots & \vdots \\ 0 & 0 & \cdots & \sqrt{\lambda_n} \end{pmatrix}.$$

Clearly, we gave $(\sqrt{D})^2 = D$. Now we can write $A = U\sqrt{D}U^{\mathsf{H}}U\sqrt{D}U^{\mathsf{H}}$, which allows us to define the desired matrix C as $C = U\sqrt{D}U^{\mathsf{H}}$.

(iii) implies (iv): Since C is itself a Hermitian matrix, this implication is obvious.

(iv) implies (i): Suppose that $A = B^{\mathsf{H}}B$ for some matrix $B \in \mathbb{C}^{n\times k}$. Then, for $\mathbf{x} \in \mathbb{C}^n$ we have $\mathbf{x}^{\mathsf{H}}A\mathbf{x} = \mathbf{x}^{\mathsf{H}}B^{\mathsf{H}}B\mathbf{x} = (B\mathbf{x})^{\mathsf{H}}(B\mathbf{x}) = \| B\mathbf{x} \|_2^2 \geqslant 0$, so A is positive semidefinite.

7.6 Matrix Norms and Spectral Radii

Definition 7.64 *Let* $A \in \mathbb{C}^{n\times n}$*. The* spectral radius *of A is the number* $\rho(A) = \max\{|\lambda| \mid \lambda \in \mathsf{spec}(A)\}$.

If $(\lambda, \mathbf{x})$ is an eigenpair of A, then $|\lambda|\|\mathbf{x}\| = \|A\mathbf{x}\| \leqslant \|A\|\|\mathbf{x}\|$, so $|\lambda| \leqslant \|A\|$, which implies $\rho(A) \leqslant \|A\|$ for any matrix norm $\|\cdot\|$. Moreover, we can prove the following statement.

Theorem 7.65 *Let* $A \in \mathbb{C}^{n\times n}$*. The spectral radius* $\rho(A)$ *is the infimum of the set that consists of numbers of the form* $\|A\|$*, where* $\|\cdot\|$ *ranges over all matrix norms defined on* $\mathbb{C}^{n\times n}$.

Proof Since we have shown that $\rho(A)$ is a lower bound of the set of numbers mentioned in the statement, we need to prove only that for every $\epsilon > 0$ there exists a matrix norm $|||\cdot|||$ such that $|||A||| \leqslant \rho(A) + \epsilon$.

By Schur's Triangularization Theorem there exists a unitary matrix U and an upper triangular matrix T such that $A = UTU^{-1}$ such that the diagonal elements of T are $\lambda_1, \ldots, \lambda_n$.

For $\alpha \in \mathbb{R}_{>0}$ let $S_\alpha = \mathsf{diag}(\alpha, \alpha^2, \ldots, \alpha^n)$. We have

$$S_\alpha T S_\alpha^{-1} = \begin{pmatrix} \lambda_1 & \alpha^{-1}t_{12} & \alpha^{-2}t_{12} & \cdots & \alpha^{-(n-1)}t_{1n} \\ 0 & \lambda_1 & \alpha^{-1}t_{23} & \cdots & \alpha^{-(n-2)}t_{2n} \\ \vdots & \vdots & \vdots & \cdots & \vdots \\ \vdots & \vdots & \vdots & \cdots & \lambda_n \end{pmatrix}.$$

If α is sufficiently large, $|||S_\alpha T S_\alpha^{-1}|||_1 \leqslant \rho(A) + \epsilon$ because, in this case, the sum of the absolute values of the supradiagonal elements can be arbitrarily small. Let $M = (US_\alpha)^{-1}$. For the matrix norm $\mu_M(\cdot)$ (see Exercise 7) we have $\mu_M(A) = |||US_\alpha A S_\alpha^{-1} U^{-1}|||_1$. If α is sufficiently large we have $\mu_M(A) \leqslant \rho(A) + \epsilon$.

Let $A, B \in \mathbb{C}^{n\times n}$. We leave to the reader to verify that if $\mathsf{abs}(A) \leqslant \mathsf{abs}(B)$, then $\| A \|_2 \leqslant \| B \|_2$; also, $\| A \|_2 = \| \mathsf{abs}(A) \|_2$.

Theorem 7.66 *Let $A, B \in \mathbb{C}^{n\times n}$. If $\mathsf{abs}(A) \leqslant B$, then $\rho(A) \leqslant \rho(\mathsf{abs}(A)) \leqslant \rho(B)$.*

Proof By Theorem 5.62 we have $\mathsf{abs}(A^k) \leqslant (\mathsf{abs}(A))^k \leqslant B^k$ for every $k \in \mathbb{N}$. Therefore, $\| \mathsf{abs}(A^k) \| \leqslant (\mathsf{abs}(A))^k \leqslant B^k$, so $\| A^k \|_2 \leqslant \| \mathsf{abs}(A)^k \|_2 \leqslant \| B^k \|_2$, which implies $\| A^k \|_2^{\frac{1}{k}} \leqslant \| \mathsf{abs}(A)^k \|_2^{\frac{1}{k}} \leqslant \| B^k \|_2^{\frac{1}{k}}$.

By letting k tend to ∞ we obtain the double inequality of the theorem.

Corollary 7.67 *If $A, B \in \mathbb{C}^{n\times n}$ are two matrices such that $O_{n,n} \leqslant A \leqslant B$, then $\rho(A) \leqslant \rho(B)$.*

Proof The corollary follows immediately from Theorem 7.66 by observing that under the hypothesis, $A = \mathsf{abs}(A)$.

Theorem 7.68 *Let $A \in \mathbb{C}^{n\times n}$. We have $\lim_{k\to\infty} = O_{n,n}$ if and only if $\rho(A) < 1$.*

Proof Suppose that $\lim_{k\to\infty} = O_{n,n}$. Let $(\lambda, \mathbf{x})$ be an eigenpair of A, so $\mathbf{x} \neq \mathbf{0}_n$ and $A\mathbf{x} = \lambda\mathbf{x}$. This implies $A^k\mathbf{x} = \lambda^k\mathbf{x}$, so $\lim_{k\to\infty} \lambda^k\mathbf{x} = \mathbf{0}_n$. Thus, $\lim_{k\to\infty} \lambda^k\mathbf{x} = \mathbf{0}_n$, which implies $\lim_{k\to\infty} \lambda^k = 0$. Thus, $|\lambda| < 1$ for every $\lambda \in \mathsf{spec}(A)$, so $\rho(A) < 1$.

Conversely, suppose that $\rho(A) < 1$. By Theorem 7.65, there exists a matrix norm $|||\cdot|||$ such that $|||A||| < 1$. Thus, $\lim_{k\to\infty} A^k = O_{n,n}$.

7.7 Singular Values of Matrices

Definition 7.69 *Let $A \in \mathbb{C}^{m\times n}$ be a matrix. A* singular triplet *of A is a triplet $(\sigma, \boldsymbol{u}, \boldsymbol{v})$ such that $\sigma \in \mathbb{R}_{>0}$, $\boldsymbol{u} \in \mathbb{C}^n$, $\boldsymbol{v} \in \mathbb{C}^m$, $A\boldsymbol{u} = \sigma\boldsymbol{v}$ and $A^{\mathsf{H}}\boldsymbol{v} = \sigma\boldsymbol{u}$. The number σ is a* singular value *of A, $\boldsymbol{u}$ is a* left singular vector *and $\boldsymbol{v}$ is a* right singular vector.

For a singular triplet $(\sigma, \boldsymbol{u}, \boldsymbol{v})$ of A we have $A^{\mathsf{H}}A\boldsymbol{u} = \sigma A^{\mathsf{H}}\boldsymbol{v} = \sigma^2\boldsymbol{u}$ and $AA^{\mathsf{H}}\boldsymbol{v} = \sigma A\boldsymbol{u} = \sigma^2\boldsymbol{v}$. Therefore, σ^2 is both an eigenvalue of AA^{H} and an eigenvalue of $A^{\mathsf{H}}A$.

Example 7.70 Let A be the real matrix

$$A = \begin{pmatrix} \cos\alpha & \sin\alpha \\ \cos\beta & \sin\beta \end{pmatrix}.$$

We have $\det(A) = \sin(\beta-\alpha)$, so the eigenvalues of $A'A$ are the roots of the equation $\lambda^2 - 2\lambda + \sin^2(\beta-\alpha) = 0$, that is, $\lambda_1 = 1+\cos(\beta-\alpha)$ and $\lambda_2 = 1-\cos(\beta-\alpha)$. Therefore, the singular values of A are $\sigma_1 = \sqrt{2}\left|\cos\frac{\beta-\alpha}{2}\right|$ and $\sigma_2 = \sqrt{2}\left|\sin\frac{\beta-\alpha}{2}\right|$.

It is easy to see that a unit left singular vector that corresponds to the eigenvalue $1+\cos(\beta-\alpha)$ is

$$\boldsymbol{u} = \begin{pmatrix} \cos\frac{\alpha+\beta}{2} \\ \sin\frac{\alpha+\beta}{2} \end{pmatrix},$$

which corresponds to the average direction of the rows of A.

We noted that the eigenvalues of a positive semi-definite matrix are non-negative numbers. Since both AA^{H} and $A^{\mathsf{H}}A$ are positive semi-definite matrices for $A \in \mathbb{C}^{m\times n}$ (see Example 6.109), the spectra of these matrices consist of non-negative numbers $\lambda_1, \ldots, \lambda_n$. Furthermore, AA^{H} and $A^{\mathsf{H}}A$ have the same rank r and therefore, the same number r of non-zero eigenvalues $\lambda_1, \ldots, \lambda_r$. Accordingly, the singular values of A have the form $\sqrt{\lambda_1} \geqslant \cdots \geqslant \sqrt{\lambda_r}$. We will use the notation $\sigma_i = \sqrt{\lambda_i}$ for $1 \leqslant i \leqslant r$ and will assume that $\sigma_1 \geqslant \cdots \geqslant \sigma_r > 0$.

Theorem 7.71 *Let $A \in \mathbb{C}^{n\times n}$ be a matrix having the singular values $\sigma_1 \geqslant \cdots \geqslant \sigma_n$. If λ is an eigenvalue value of A, then $\sigma_n \leqslant |\lambda| \leqslant \sigma_1$.*

Proof Let $\boldsymbol{u}$ be an unit eigenvector for the eigenvalue λ. Since $A\boldsymbol{u} = \lambda\boldsymbol{u}$ it follows that $(A^{\mathsf{H}}A\boldsymbol{u}, \boldsymbol{u}) = (A\boldsymbol{u}, A\boldsymbol{u}) = \overline{\lambda}\lambda(\boldsymbol{u}, \boldsymbol{u}) = \overline{\lambda}\lambda = |\lambda|^2$. The matrix $A^{\mathsf{H}}A$ is Hermitian and its largest and smallest eigenvalues are σ_1^2 and σ_n^2, respectively. Thus, $\sigma_n \leqslant |\lambda| \leqslant \sigma_1$.

Theorem 7.72 (SVD Theorem) *If $A \in \mathbb{C}^{m\times n}$ is a matrix and $\text{rank}(A) = r$, then A can be factored as $A = UDV^{\mathsf{H}}$, where $U \in \mathbb{C}^{m\times m}$ and $V \in \mathbb{C}^{n\times n}$ are unitary matrices, and $D = \mathsf{diag}(\sigma_1, \ldots, \sigma_r, 0, \ldots, 0) \in \mathbb{R}^{m\times n}$, where $\sigma_1 \geqslant \ldots \geqslant \sigma_r$ are real positive numbers.*

Proof We saw that the square matrix $A^{\mathsf{H}}A \in \mathbb{C}^{n\times n}$ has the same rank r as the matrix A and is positive semidefinite. Therefore, there are r positive eigenvalues of this matrix, denoted by $\sigma_1^2, \ldots, \sigma_r^2$, where $\sigma_1 \geqslant \sigma_2 \geqslant \cdots \geqslant \sigma_r > 0$ and let $\boldsymbol{v}_1, \ldots, \boldsymbol{v}_r$ be the corresponding pairwise orthogonal unit eigenvectors in $\mathbb{C}^n$.

We have $A^{\mathsf{H}}A\boldsymbol{v}_i = \sigma_i^2\boldsymbol{v}_i$ for $1 \leqslant i \leqslant r$. Define $V = (\boldsymbol{v}_1 \ \cdots \ \boldsymbol{v}_r \ \boldsymbol{v}_{r+1} \ \cdots \ \boldsymbol{v}_n)$ by completing the set $\{\boldsymbol{v}_1, \ldots, \boldsymbol{v}_r\}$ to an orthogonal basis

$$\{\boldsymbol{v}_1, \ldots, \boldsymbol{v}_r, \boldsymbol{v}_{r+1}, \ldots, \boldsymbol{v}_n\}$$

for $\mathbb{C}^n$. If $V_1 = (\boldsymbol{v}_1 \ \cdots \ \boldsymbol{v}_r)$ and $V_2 = (\boldsymbol{v}_{r+1} \ \cdots \ \boldsymbol{v}_n)$, we can write $V = (V_1 \ V_2)$.

The equalities involving the eigenvectors can now be written as $A^{\mathsf{H}}AV_1 = V_1E^2$, where $E = \mathsf{diag}(\sigma_1, \ldots, \sigma_r)$.

Define $U_1 = AV_1E^{-1} \in \mathbb{C}^{m\times r}$. We have $U_1^{\mathsf{H}} = S^{-1}V_1^{\mathsf{H}}A^{\mathsf{H}}$, so

$$U_1^{\mathsf{H}}U_1 = S^{-1}V_1^{\mathsf{H}}A^{\mathsf{H}}AV_1E^{-1} = E^{-1}V_1^{\mathsf{H}}V_1E^2E^{-1} = I_r,$$

which shows that the columns of U_1 are pairwise orthogonal unit vectors. Consequently, $U_1^{\mathsf{H}}AV_1E^{-1} = I_r$, so $U_1^{\mathsf{H}}AV_1 = E$.

If $U_1 = (\boldsymbol{u}_1 \ \cdots, \boldsymbol{u}_r)$, let $U_2 = (\boldsymbol{u}_{r+1}, \ldots, \boldsymbol{u}_m)$ be the matrix whose columns constitute the extension of the set $\{\boldsymbol{u}_1 \ \cdots, \boldsymbol{u}_r\}$ to an orthogonal basis of $\mathbb{C}^m$. Define $U \in \mathbb{C}^{m\times m}$ as $U = (U_1 \ U_2)$. Note that

$$\begin{aligned} U^{\mathsf{H}}AV &= \begin{pmatrix} U_1^{\mathsf{H}} \\ U_2^{\mathsf{H}} \end{pmatrix} A(V_1 \ V_2) = \begin{pmatrix} U_1^{\mathsf{H}}AV_1 & U_1^{\mathsf{H}}AV_2 \\ U_2^{\mathsf{H}}AV_1 & U_2^{\mathsf{H}}AV_2 \end{pmatrix} \\ &= \begin{pmatrix} U_1^{\mathsf{H}}AV_1 & U_1^{\mathsf{H}}AV_2 \\ U_2^{\mathsf{H}}AV_1 & U_2^{\mathsf{H}}AV_2 \end{pmatrix} = \begin{pmatrix} U_1^{\mathsf{H}}AV_1 & O \\ O & O \end{pmatrix} = \begin{pmatrix} E & O \\ O & O \end{pmatrix}, \end{aligned}$$

which is the desired decomposition.

Corollary 7.73 *Let $A \in \mathbb{C}^{m\times n}$ be a matrix such that rank$(A) = r$. If $\sigma_1 \geqslant \ldots \geqslant \sigma_r$ are non-zero singular values, then*

$$A = \sigma_1\boldsymbol{u}_1\boldsymbol{v}_1^H + \cdots + \sigma_r\boldsymbol{u}_r\boldsymbol{v}_r^H, \tag{7.9}$$

where $(\sigma_i, \boldsymbol{u}_i, \boldsymbol{v}_i)$ are singular triplets of A for $1 \leqslant i \leqslant r$.

Proof This follows directly from Theorem 7.72.

The value of a unitarily invariant norm of a matrix depends only on its singular values.

Corollary 7.74 *Let $A \in \mathbb{C}^{m\times n}$ be a matrix and let $A = UDV^H$ be the singular value decomposition of A. If $\| \cdot \|$ is a unitarily invariant norm, then*

$$\| A \| = \| D \| = \| \mathit{diag}(\sigma_1, \ldots, \sigma_r, 0, \ldots, 0) \| .$$

Proof This statement is a direct consequence of Theorem 7.72 because the matrices $U \in \mathbb{C}^{m\times m}$ and $V \in \mathbb{C}^{n\times n}$ are unitary.

As we saw in Theorem 6.86, $|||\cdot|||_2$ and $\|\cdot\|_F$ are unitarily invariant. Therefore, the Frobenius norm can be written as

$$\| A \|_F = \sqrt{\sum_{i=1}^{r} \sigma_r^2}.$$

and $|||A|||_2 = \sigma_1$.

Theorem 7.75 *Let A and B be two matrices in $\mathbb{C}^{m\times n}$. If $A \sim_u B$, then they have the same singular values.*

Proof Suppose that $A \sim_u B$, that is, $A = W_1^{\mathsf{H}} B W_2$ for some unitary matrices W_1 and W_2. If A has the SVD $A = U^{\mathsf{H}}\mathsf{diag}(\sigma_1, \ldots, \sigma_r, 0, \ldots, 0)V$, then

$$B = W_1 A W_2^{\mathsf{H}} = (W_1 U^{\mathsf{H}})\mathsf{diag}(\sigma_1, \ldots, \sigma_r, 0, \ldots, 0)(V W_2^{\mathsf{H}}).$$

Since $W_1 U^{\mathsf{H}}$ and $V W_2^{\mathsf{H}}$ are both unitary matrices, it follows that the singular values of B are the same as the singular values of A.

Let $\boldsymbol{v} \in \mathbb{C}^n$ be an eigenvector of the matrix $A^{\mathsf{H}}A$ that corresponds to a non-zero, positive eigenvalue σ^2, that is, $A^{\mathsf{H}}A\boldsymbol{v} = \sigma^2\boldsymbol{v}$.

Define $\boldsymbol{u} = \frac{1}{\sigma}A\boldsymbol{v}$. We have $A\boldsymbol{v} = \sigma\boldsymbol{u}$. Also,

$$A^{\mathsf{H}}\boldsymbol{u} = A^{\mathsf{H}}\left(\frac{1}{\sigma}A\boldsymbol{v}\right) = \sigma\boldsymbol{v}.$$

This implies $AA^{\mathsf{H}}\boldsymbol{u} = \sigma^2\boldsymbol{u}$, so $\boldsymbol{u}$ is an eigenvector of AA^{H} that corresponds to the same eigenvalue σ^2.

Conversely, if $\boldsymbol{u} \in \mathbb{C}^m$ is an eigenvector of the matrix AA^{H} that corresponds to a non-zero, positive eigenvalue σ^2, we have $AA^{\mathsf{H}}\boldsymbol{u} = \sigma^2\boldsymbol{u}$. Thus, if $\boldsymbol{v} = \frac{1}{\sigma}A\boldsymbol{u}$ we have $A\boldsymbol{v} = \sigma\boldsymbol{u}$ and $\boldsymbol{v}$ is an eigenvector of $A^{\mathsf{H}}A$ for the eigenvalue σ^2.

The Courant-Fisher Theorem (Theorem 7.50) allows the formulation of a similar result for singular values.

Theorem 7.76 *Let $A \in \mathbb{C}^{m\times n}$ be a matrix such that $\sigma_1 \geqslant \sigma_2 \geqslant \cdots \geqslant \sigma_r$ is the non-increasing sequence of singular values of A. For $1 \leqslant k \leqslant r$ we have*

$$\sigma_k = \min_{\dim(S)=n-k+1} \max\{\| A\boldsymbol{x} \|_2 \mid \boldsymbol{x} \in S \text{ and } \| \boldsymbol{x} \|_2 = 1\}$$

$$\sigma_k = \max_{\dim(T)=k} \min\{\| A\boldsymbol{x} \|_2 \mid \boldsymbol{x} \in T \text{ and } \| \boldsymbol{x} \|_2 = 1\},$$

where S and T range over subspaces of $\mathbb{C}^n$.

Proof We give the argument only for the second equality of the theorem; the first can be shown in a similar manner.

We saw that σ_k equals the square root of kth largest absolute value of the eigenvalue $|\lambda_k|$ of the matrix $A^{\mathsf{H}}A$. By Courant-Fisher Theorem, we have

$$\begin{aligned}\lambda_k &= \max_{dim(T)=k} \min_{\mathbf{x}}\{\mathbf{x}^{\mathsf{H}}A^{\mathsf{H}}A\mathbf{x} \mid \mathbf{x} \in T \text{and} \parallel \mathbf{x} \parallel_2 = 1\}\\ &= \max_{dim(T)=k} \min_{\mathbf{x}}\{\parallel A\mathbf{x} \parallel_2^2 \mid \mathbf{x} \in T \text{and} \parallel \mathbf{x} \parallel_2 = 1\},\end{aligned}$$

which implies the second equality of the theorem.

The equalities established in Theorem 7.76 can be rewritten as

$$\begin{aligned}\sigma_k &= \min_{\boldsymbol{w}_1,\ldots,\boldsymbol{w}_{k-1}} \max\{\parallel A\mathbf{x} \parallel_2 \mid \mathbf{x} \perp \boldsymbol{w}_1, \ldots, \mathbf{x} \perp \boldsymbol{w}_{k-1} \text{ and } \parallel \mathbf{x} \parallel_2 = 1\}\\ &= \max_{\boldsymbol{w}_1,\ldots,\boldsymbol{w}_{n-k}} \min\{\parallel A\mathbf{x} \parallel_2 \mid \mathbf{x} \perp \boldsymbol{w}_1, \ldots, \mathbf{x} \perp \boldsymbol{w}_{n-k} \text{ and } \parallel \mathbf{x} \parallel_2 = 1\}.\end{aligned}$$

Corollary 7.77 *The smallest singular value of a matrix $A \in \mathbb{C}^{m\times n}$ equals*

$$\min\{\parallel A\boldsymbol{x} \parallel_2 \mid \boldsymbol{x} \in \mathbb{C}^n \textit{and} \parallel \boldsymbol{x} \parallel_2 = 1\}.$$

The largest singular value of a matrix $A \in \mathbb{C}^{m\times n}$ equals

$$\max\{\parallel A\boldsymbol{x} \parallel_2 \mid \boldsymbol{x} \in \mathbb{C}^n \textit{and} \parallel \boldsymbol{x} \parallel_2 = 1\}.$$

Proof The corollary is a direct consequence of Theorem 7.76.

The SVD allows us to find the best approximation of of a matrix by a matrices of limited rank. The central result of this section is Theorem 7.79.

Lemma 7.78 *Let $A = \sigma_1\boldsymbol{u}_1\boldsymbol{v}_1^H + \cdots + \sigma_r\boldsymbol{u}_r\boldsymbol{v}_r^H$ be the SVD of a matrix $A \in \mathbb{R}^{m\times n}$, where $\sigma_1 \geqslant \cdots \geqslant \sigma_r > 0$. For every k, $1 \leqslant k \leqslant r$ the matrix $B(k) = \sum_{i=1}^{k}\sigma_i\boldsymbol{u}_i\boldsymbol{v}_i^H$ has rank k.*

Proof The null space of the matrix $B(k)$ consists of those vectors $\mathbf{x}$ such that $\sum_{i=1}^{k}\sigma_i\boldsymbol{u}_i\boldsymbol{v}_i^{\mathsf{H}}\mathbf{x} = \mathbf{0}$. The linear independence of the vectors $\boldsymbol{u}_i$ and the fact that $\sigma_i > 0$ for $1 \leqslant i \leqslant r$ implies the equalities $\boldsymbol{v}_i^{\mathsf{H}}\mathbf{x} = \mathbf{0}$ for $1 \leqslant i \leqslant r$. Thus,

$$\mathsf{NullSp}(B(k)) = \mathsf{NullSp}\left((\boldsymbol{v}_1 \ \cdots \ \boldsymbol{v}_k)\right).$$

Since $\boldsymbol{v}_1, \ldots, \boldsymbol{v}_k$ are linearly independent it follows that $\dim(\mathsf{NullSp}(B(k)) = n-k$, which implies $rank(B(k)) = k$ for $1 \leqslant k \leqslant r$.

Theorem 7.79 (Eckhart-Young Theorem) *Let $A \in \mathbb{C}^{m\times n}$ be a matrix whose sequence of non-zero singular values is $(\sigma_1, \ldots, \sigma_r)$. Assume that $\sigma_1 \geqslant \cdots \geqslant \sigma_r > 0$ and that A can be written as*

$$A = \sigma_1 \boldsymbol{u}_1 \boldsymbol{v}_1^H + \cdots + \sigma_r \boldsymbol{u}_r \boldsymbol{v}_r^H.$$

Let $B(k) \in \mathbb{C}^{m \times n}$ be the matrix defined by

$$B(k) = \sum_{i=1}^{k} \sigma_i \boldsymbol{u}_i \boldsymbol{v}_i^H.$$

If $r_k = \inf\{\|A - X\|_2 \mid X \in \mathbb{C}^{m \times n}$ and $rank(X) \leqslant k\}$, then

$$\|A - B(k)\|_2 = r_k = \sigma_{k+1},$$

for $1 \leqslant k \leqslant r$, where $\sigma_{r+1} = 0$ and $B(k)$ is the best approximation of A among the matrices of rank no larger than k in the sense of the norm $\|\cdot\|_2$.

Proof Observe that

$$A - B(k) = \sum_{i=k+1}^{r} \sigma_i \boldsymbol{u}_i \boldsymbol{v}_i^H,$$

and the largest singular value of the matrix $\sum_{i=k+1}^{r} \sigma_i \boldsymbol{u}_i \boldsymbol{v}_i^H$ is σ_{k+1}. Since σ_{k+1} is the largest singular value of $A - B(k)$ we have $\|A - B(k)\|_2 = \sigma_{k+1}$ for $1 \leqslant k \leqslant r$.

We prove now that for every matrix $X \in \mathbb{C}^{m \times n}$ such that $rank(X) \leqslant k$, we have $\|A - X\|_2 \geqslant \sigma_{k+1}$. Since $\dim(\mathsf{NullSp}(X)) = n - rank(X)$, it follows that $\dim(\mathsf{NullSp}(X)) \geqslant n - k$. If T is the subspace of $\mathbb{R}^n$ spanned by $\boldsymbol{v}_1, \ldots, \boldsymbol{v}_{k+1}$, we have $\dim(T) = k + 1$. Since $\dim(\mathsf{NullSp}(X)) + \dim(T) > n$, the intersection of these subspaces contains a non-zero vector and, without loss of generality, we can assume that this vector is a unit vector $\mathbf{x}$.

We have $\mathbf{x} = a_1 \boldsymbol{v}_1 + \cdots a_k \boldsymbol{v}_k + a_{k+1} \boldsymbol{v}_{k+1}$ because $\mathbf{x} \in T$. The orthogonality of $\boldsymbol{v}_1, \ldots, \boldsymbol{v}_k, \boldsymbol{v}_{k+1}$ implies $\| \mathbf{x} \|_2^2 = \sum_{i=1}^{k+1} |a_i|^2 = 1$.

Since $\mathbf{x} \in \mathsf{NullSp}(X)$, we have $X\mathbf{x} = \mathbf{0}$, so

$$(A - X)\mathbf{x} = A\mathbf{x} = \sum_{i=1}^{k+1} a_i A \boldsymbol{v}_i = \sum_{i=1}^{k+1} a_i \sigma_i \boldsymbol{u}_i.$$

Thus, we have

$$\|(A - X)\mathbf{x}\|_2^2 = \sum_{i=1}^{k+1} |\sigma_i a_i|^2 \geqslant \sigma_{k+1}^2 \sum_{i=1}^{k+1} |a_i|^2 = \sigma_{k+1}^2,$$

because $\boldsymbol{u}_1, \ldots, \boldsymbol{u}_k$ are also orthonormal. This implies $\|A - X\|_2 \geqslant \sigma_{k+1} = \|A - B(k)\|_2$.

It is interesting to observe that the matrix $B(k)$ provides an optimal approximation of A not only with respect to $\|\cdot\|_2$ but also relative to the Frobenius norm.

Theorem 7.80 *Using the notations introduced in Theorem 7.79, $B(k)$ is the best approximation of A among matrices of rank no larger than k in the sense of the Frobenius norm.*

Proof Note that $\parallel A - B(k) \parallel_F^2 = \parallel A \parallel_F^2 - \sum_{i=1}^k \sigma_i^2$. Let X be a matrix of rank k, which can be written as $X = \sum_{i=1}^k \mathbf{x}_i \mathbf{y}_i^{\mathsf{H}}$. Without loss of generality we may assume that the vectors $\mathbf{x}_1, \ldots, \mathbf{x}_k$ are orthonormal. If this is not the case, we can use the Gram-Schmidt algorithm to express then as linear combinations of orthonormal vectors, replace these expressions in $\sum_{i=1}^k \mathbf{x}_i \mathbf{y}_i^{\mathsf{H}}$ and rearrange the terms. Now, the Frobenius norm of $A - X$ can be written as

$$\begin{aligned}\parallel A - X \parallel_F^2 &= trace\left(\left(A - \sum_{i=1}^k \mathbf{x}_i \mathbf{y}^{\mathsf{H}}\right)^{\mathsf{H}} \left(A - \sum_{i=1}^k \mathbf{x}_i \mathbf{y}^{\mathsf{H}}\right)\right) \\ &= trace\left(A^{\mathsf{H}}A + \sum_{i=1}^k (\mathbf{y}_i - A^{\mathsf{H}}\mathbf{x}_i)(\mathbf{y}_i - A^{\mathsf{H}}\mathbf{x}_i)^{\mathsf{H}} - \sum_{i=1}^k A^{\mathsf{H}}\mathbf{x}_i\mathbf{x}_i^{\mathsf{H}}A\right).\end{aligned}$$

Taking into account that $\sum_{i=1}^k (\mathbf{y}_i - A^{\mathsf{H}}\mathbf{x}_i)(\mathbf{y}_i - A^{\mathsf{H}}\mathbf{x}_i)^{\mathsf{H}}$ is a real non-negative number and that $\sum_{i=1}^k A^{\mathsf{H}}\mathbf{x}_i\mathbf{x}_i^{\mathsf{H}}A = \parallel A\mathbf{x}_i \parallel_F^2$ we have

$$\parallel A - X \parallel_F^2 \geq trace\left(A^{\mathsf{H}}A - \sum_{i=1}^k A^{\mathsf{H}}\mathbf{x}_i\mathbf{x}_i^{\mathsf{H}}A\right) = \parallel A \parallel_F^2 - trace\left(\sum_{i=1}^k A^{\mathsf{H}}\mathbf{x}_i\mathbf{x}_i^{\mathsf{H}}A\right).$$

Let $A = U\mathsf{diag}(\sigma_1, \ldots, \sigma_n)V^{\mathsf{H}}$ be the singular value decomposition of A. If $V = (V_1 \; V_2)$, where V_1 has k columns $\boldsymbol{v}_1, \ldots, \boldsymbol{v}_k$, $D_1 = \mathsf{diag}(\sigma_1, \ldots, \sigma_k)$ and $D_2 = \mathsf{diag}(\sigma_{k+1}, \ldots, \sigma_n)$, then we can write

$$\begin{aligned}A^{\mathsf{H}}A = VD^{\mathsf{H}}U^{\mathsf{H}}UDV^{\mathsf{H}} &= (V_1 \; V_2)\begin{pmatrix} D_1^2 & O \\ O & D_2^2 \end{pmatrix}\begin{pmatrix} V_1^{\mathsf{H}} \\ V_2^{\mathsf{H}} \end{pmatrix} \\ &= V_1D_1^2V_1^{\mathsf{H}} + V_2D_2^2V_2^{\mathsf{H}}.\end{aligned}$$

and $A^{\mathsf{H}}A = VD^2V^{\mathsf{H}}$. These equalities allow us to write:

$$\begin{aligned}\parallel A\mathbf{x}_i \parallel_F^2 &= trace(\mathbf{x}_i^{\mathsf{H}}A^{\mathsf{H}}A\mathbf{x}_i) \\ &= trace\left(\mathbf{x}_i^{\mathsf{H}}V_1D_1^2V_1^{\mathsf{H}}\mathbf{x}_i + \mathbf{x}_i^{\mathsf{H}}V_2D_2^2V_2^{\mathsf{H}}\mathbf{x}_i\right) \\ &= \parallel D_1V_1^{\mathsf{H}}\mathbf{x}_i \parallel_F^2 + \parallel D_2V_2^{\mathsf{H}}\mathbf{x}_i \parallel_F^2 \\ &= \sigma_k^2 + \left(\parallel D_1V_1^{\mathsf{H}}\mathbf{x}_i \parallel_F^2 - \sigma_k^2 \parallel V_1^{\mathsf{H}}\mathbf{x}_i \parallel_F^2\right) \\ &\quad - \left(\sigma_k^2 \parallel V_2^{\mathsf{H}}\mathbf{x}_i \parallel_F^2 - \parallel D_2V_2^{\mathsf{H}}\mathbf{x}_i \parallel_F^2)\right) - \sigma_k^2(1 - \parallel V^{\mathsf{H}}\mathbf{x}_i \parallel).\end{aligned}$$

Since $\| V^{\mathsf{H}}\mathbf{x}_i \|_F^1 = 1$ (because $\mathbf{x}_i$ is an unit vector and V is an unitary matrix) and $\sigma_k^2 \| V_2^{\mathsf{H}}\mathbf{x}_i \|_F^2 - \| D_2 V_2^{\mathsf{H}}\mathbf{x}_i \|_F^2 \geqslant 0$, it follows that

$$\| A\mathbf{x}_i \|_F^2 \leqslant \sigma_k^2 + \left(\| D_1 V_1^{\mathsf{H}}\mathbf{x}_i \|_F^2 - \sigma_k^2 \| V_1^{\mathsf{H}}\mathbf{x}_i \|_F^2 \right).$$

Consequently,

$$\begin{aligned}
\sum_{i=1}^{k} \| A\mathbf{x}_i \|_F^2 &\leqslant k\sigma_k^2 + \sum_{i=1}^{k} \left(\| D_1 V_1^{\mathsf{H}}\mathbf{x}_i \|_F^2 - \sigma_k^2 \| V_1^{\mathsf{H}}\mathbf{x}_i \|_F^2 \right) \\
&= k\sigma_k^2 + \sum_{i=1}^{k} \sum_{j=1}^{k} (\sigma_j^2 - \sigma_k^2) |\mathbf{v}_j^{\mathsf{H}}\mathbf{x}_i|^2 \\
&= \sum_{j=1}^{k} \left(\sigma_k^2 + (\sigma_j^2 - \sigma_k^2) \sum_{i=1}^{k} |\mathbf{v}_j \mathbf{x}_i|^2 \right) \\
&\leqslant \sum_{j=1}^{k} (\sigma_k^2 + (\sigma_j^2 - \sigma_k^2)) = \sum_{j=1}^{k} \sigma_j^2,
\end{aligned}$$

which concludes the argument.

Definition 7.81 *Let $A \in \mathbb{C}^{m\times n}$. The* numerical rank *of A is the function $\mathsf{nr}_A : [0, \infty) \longrightarrow \mathbb{N}$ given by*

$$\mathsf{nr}_A(d) = \min\{rank(B) \mid \|\!|A - B|\!\|_2 \leqslant d\}$$

for $d \geqslant 0$.

Theorem 7.82 *Let $A \in \mathbb{C}^{m\times n}$ be a matrix having the sequence of non-zero singular values $\sigma_1 \geqslant \sigma_2 \geqslant \cdots \geqslant \sigma_r$. Then, $\mathsf{nr}_A(d) = k < r$ if and only if $\sigma_k > d \geqslant \sigma_{k+1}$.*

Proof Let d be a number such that $\sigma_k > d \geqslant \sigma_{k+1}$. Equivalently, by Eckhart-Young Theorem, we have

$$\|\!|A - B(k-1)|\!\|_2 > d \geqslant \|\!|A - B(k)|\!\|_2,$$

Since $\|\!|A - B(k-1)|\!\|_2 = \min\{\|\!|A - X|\!\|_2 \mid rank(X) = k - 1\} > d$, it follows that $\min\{rank(B) \mid \|\!|A - B|\!\|_2 \leqslant d\} = k$, so $\mathsf{nr}_A(d) = k$.

Conversely, suppose that $\mathsf{nr}_A(d) = k$. This means that the minimal rank of a matrix B such that $\|\!|A - B|\!\|_2 \leqslant d$ is k. Therefore, $\|\!|A - B(k-1)|\!\|_2 > d$. On another hand, $d \geqslant \|\!|A - B(k)|\!\|_2$ because there exists a matrix C of rank k such that $d \geqslant \|\!|A - C|\!\|_2$, so $d \geqslant \|\!|A - B(k)|\!\|_2 = \sigma_{k+1}$. Thus, $\sigma_k > d \geqslant \sigma_{k+1}$.

Exercises and Supplements

1. Prove that if $(a, \mathbf{x})$ is an eigenpair of a matrix $A \in \mathbb{C}^{n\times n}$ if and only if $(a - b, \mathbf{x})$ is an eigenpair of the matrix $A - bI_n$.
2. Let $A \in \mathbb{R}^{n\times n}$ be a matrix and let $(a, \mathbf{x})$ be an eigenpair of A and $(a, \mathbf{y})$ be an eigenpair of A' such that $\mathbf{x}'\mathbf{y} = 1$. If $L = \mathbf{x}\mathbf{y}'$, prove that

   (a) every non-zero eigenvalue of $A - aL$ is also an eigenvalue of A and every eigenpair $(\lambda, \mathbf{t})$ of $A - aL$ is an eigenpair of A;
   (b) if $a \neq 0$ is an eigenvalue of A with $\mathsf{geomm}(A, a) = 1$, then a is not an eigenvalue of $A - aL$.

   Solution: Let λ a non-zero eigenvalue of $A - aL$. We have $(A - aL)\mathbf{t} = \lambda\mathbf{t}$ for some $\mathbf{t} \neq \mathbf{0}_n$. By Part (d) of Exercise 24 of Chap. 5, $L(A - aL) = O_{n,n}$, so $\lambda L\mathbf{t} = \mathbf{0}_n$, so $L\mathbf{t} = \mathbf{0}$, which implies $A\mathbf{t} = \lambda\mathbf{t}$.

   For the second part suppose that $a \neq 0$ were an eigenvalue of $A - aL$ and let $(a, \boldsymbol{w})$ be an eigenpair of this matrix. By the first part, $A\boldsymbol{w} = a\boldsymbol{w}$. Since $\mathsf{geomm}(A, a) = 1$, there exists $b \in \mathbb{C} - \{0\}$ such that $\boldsymbol{w} = b\mathbf{x}$. This allows us to write

$$\begin{aligned} a\boldsymbol{w} &= (A - aL)\boldsymbol{w} = (A - aL)b\mathbf{x} = ab\mathbf{x} - abL\mathbf{x} = ab\mathbf{x} - ab(\mathbf{x}\mathbf{y}')\mathbf{x} \\ &= ab\mathbf{x} - ab\mathbf{x}(\mathbf{y}'\mathbf{x}) = \mathbf{0}_n, \end{aligned}$$

   because $\mathbf{y}'\mathbf{x} = 1$. Since $a \neq 0$ and $\mathbf{x} \neq \mathbf{0}_n$, this is impossible.

3. Let $A \in \mathbb{R}^{n\times n}$ be a matrix, $(\lambda, \mathbf{x})$ be an eigenpair of A and $(\lambda, \mathbf{y})$ be an eigenpair of A' such that:

   (i) $\mathbf{x}'\mathbf{y} = 1$;
   (ii) λ be an eigenvalue of A with $|\lambda| = \rho(A)$ and λ is unique with this property.

   If $L = \mathbf{x}\mathbf{y}'$, and $\theta \in \mathsf{spec}(A)$ is such that $|\theta| < \rho(A)$ and $|\theta|$ is maximal with this property, prove that:

   (a) $\rho(A - \lambda L) \leqslant |\theta| < \rho(A)$;
   (b) $\left(\frac{1}{\lambda}A\right)^m = L + \left(\frac{1}{\lambda}A - L\right)^m$ and $\lim_{m\to\infty}\left(\frac{1}{\lambda}A\right)^m = L$.

   Solution: By Part (a) of Supplement 2, every non-zero eigenvalue of $A - \lambda L$ is an eigenvalue of A. Therefore, either $\rho(A - \lambda L) = 0$ or $\rho(A - \lambda L) = |\lambda'|$ for some $\lambda' \in \mathsf{spec}(A)$. Therefore, in either case, $\rho(A - \lambda L) \leqslant |\theta| < \rho(A)$.

Since $(A - \lambda L)^m = A^m - \lambda^m L$, we have $\left(\frac{1}{\lambda}A\right)^m = L + \left(\frac{1}{\lambda}A - L\right)^m$. Note that $\rho\left(\frac{1}{\lambda}A - L\right) = \frac{\rho(A-\lambda L)}{\rho(A)} \leqslant \frac{|\theta|}{\rho(A)} < 1$. Therefore, by Theorem 7.68, $\lim_{m\to\infty}\left(\frac{1}{\lambda}A\right)^m = L$.

4. Let $A \in \mathbb{R}^{n\times n}$, where n is an odd number. Prove that A has at least one real eigenvalue.
5. Prove that the eigenvalues of an upper triangular (or lower triangular) matrix are its diagonal entries.

Let $\mathsf{s}_k : \mathbb{C}^n \longrightarrow \mathbb{C}$ be the kth symmetric function of n arguments defined by

$$\mathsf{s}^n_k(z_1, \ldots, z_n) = \sum_{i_1,\ldots,i_k} \left\{ \prod_{j=1}^{k} \mathbf{z}_{i_j} \mid 1 \leqslant i_1 < \cdots < i_k \leqslant n \right\},$$

for $z_1, \ldots, z_n \in \mathbb{C}$. For example, we have

$$\mathsf{s}^3_1(z_1, z_2, z_3) = z_1 + z_2 + z_3,$$
$$\mathsf{s}^3_2(z_1, z_2, z_3) = z_1z_2 + z_1z_3 + z_2z_3,$$
$$\mathsf{s}^3_3(z_1, z_2, z_3) = z_1z_2z_3.$$

6. Prove that

$$(t - z_1)\cdots(t - z_n) = t^n - \mathsf{s}^n_1(z_1, \ldots, z_n)t^{n-1} + \mathsf{s}^n_2(z_1, \ldots, z_n)t^{n-2} - \cdots + (-1)^n\, \mathsf{s}^n_n(z_1, \ldots, z_n).$$

Solution: The equality follows by observing that the coefficient of t^{n-k} in $(t - z_1)\cdots(t - z_n)$ equals $(-1)^k\, \mathsf{s}^n_k(z_1, \ldots, z_n)$.
7. Let $M \in \mathbb{C}^{n\times n}$ be an invertible matrix and let $|||\cdot|||$ be a matrix norm. Prove that the mapping $\mu_M : \mathbb{C}^{n\times n} \longrightarrow \mathbb{R}_{\geqslant 0}$ given by $\mu_M(A) = |||M^{-1}AM|||$ is a matrix norm.
8. Let $A \in \mathbb{C}^{n\times n}$ be a matrix. Prove that:

   (a) if $\lambda \in \mathsf{spec}(A)$, then $1 + \lambda \in \mathsf{spec}(I_n + A)$;
   (b) $\mathsf{algm}(I_n + A, 1 + \lambda) = \mathsf{algm}(A, \lambda)$;
   (c) $\rho(I_n + A) \leqslant 1 + \rho(A)$.

   Solution: Suppose that $\lambda \in \mathsf{spec}(A)$ and $\mathsf{algm}(A, \lambda) = k$. Then λ is root of multiplicity k of $p_A(\lambda) = \det(\lambda I_n - A)$. Since $p_{I_n+A}(\lambda) = \det(\lambda I_n - I_n - A)$, it follows that $p_{I_n+A}(1 + \lambda) = \det(\lambda I_n - A) = p_A(\lambda)$. Thus $1 + \lambda \in \mathsf{spec}(I_n + A)$ and $\mathsf{algm}(I_n + A, 1 + \lambda) = \mathsf{algm}(A, \lambda)$.

   We have $\rho(I_n + A) = \max\{|1 + \lambda| \mid \lambda \in \mathsf{spec}(A)\} \leqslant 1 + \max\{|\lambda| \mid \lambda \in \mathsf{spec}(A) = 1 + \rho(A)$.

9. Let $A \in \mathbb{R}^{n\times n}$ be a symmetric matrix having the eigenvalues $\lambda_1 \geqslant \cdots \geqslant \lambda_n$. Prove that

$$\left\| A - \frac{\lambda_1 + \lambda_n}{2} I_n \right\|_2 = \frac{\lambda_1 - \lambda_n}{2}.$$

10. Let $A \in \mathbb{R}^{n \times n}$ be a matrix and let $c \in \mathbb{R}$. Prove that for $\mathbf{x}, \mathbf{y} \in \mathbb{R}^n - \{\mathbf{0}_n\}$ we have $\mathsf{ral}_A(\mathbf{x}) - \mathsf{ral}_A(\mathbf{y}) = \mathsf{ral}_B(\mathbf{x}) - \mathsf{ral}_B(\mathbf{y})$, where $B = A + cI_n$.
11. Let $A \in \mathbb{R}^{n \times n}$ be a symmetric matrix having the eigenvalues $\lambda_1 \geqslant \cdots \geqslant \lambda_n$ and let $\mathbf{x}$ and $\mathbf{y}$ be two vectors in $\mathbb{R}^n - \{\mathbf{0}_n\}$. Prove that

$$|\mathsf{ral}_A(\mathbf{x}) - \mathsf{ral}_A(\mathbf{y})| \leqslant (\lambda_1 - \lambda_n) \sin \angle(\mathbf{x}, \mathbf{y}).$$

Solution: Assume that $\parallel \mathbf{x} \parallel = \parallel \mathbf{y} \parallel = 1$ and let $B = A - \frac{\lambda_1 + \lambda_n}{2} I_n$. By Exercise 10, we have

$$|\mathsf{ral}_A(\mathbf{x}) - \mathsf{ral}_A(\mathbf{y})| = |\mathsf{ral}_B(\mathbf{x}) - \mathsf{ral}_B(\mathbf{y})| == |\mathbf{x}'B\mathbf{x} - \mathbf{y}'B\mathbf{y}| = |B(\mathbf{x} - \mathbf{y})'(\mathbf{x} + \mathbf{y})|.$$

By Cauchy-Schwarz Inequality we have

$$|B(\mathbf{x} - \mathbf{y})'(\mathbf{x} + \mathbf{y})| \leqslant 2 \parallel B \parallel \frac{\parallel \mathbf{x} - \mathbf{y} \parallel \parallel \mathbf{x} + \mathbf{y} \parallel}{2} = (\lambda_1 - \lambda_n) \sin \angle(\mathbf{x}, \mathbf{y}).$$

12. Prove that if A is a unitary matrix and $1 \notin \mathsf{spec}(A)$, then there exists a skew-Hermitian S such that $A = (I_n - S)(I_n + S)^{-1}$.
13. Let $f : \mathbb{R}^{n \times n} \longrightarrow \mathbb{R}$ be a function such that $f(AB) = f(BA)$ for $A, B \in \mathbb{R}^{n \times n}$. Prove that if $A \sim B$, then $f(A) = f(B)$.
14. Let $a, b \in \mathbb{C} - \{0\}$ and let $B_r(\lambda, a) \in \mathbb{C}^{r \times r}$ be the matrix defined by

$$B_r(\lambda, a) = \begin{pmatrix} \lambda & a & 0 & \cdots & 0 \\ 0 & \lambda & a & \cdots & 0 \\ \vdots & \vdots & \ddots & \ddots & \vdots \\ 0 & 0 & 0 & \cdots & a \\ 0 & 0 & 0 & \cdots & \lambda \end{pmatrix} \in \mathbb{C}^{r \times r}.$$

Prove that

(a) $B_n(\lambda, a) \sim B_n(\lambda, b)$;
(b) $B_r(\lambda)$ is given by

$$(B_r(\lambda)^k = \begin{pmatrix} \lambda^k & \binom{k}{1}\lambda^{k-1} & \binom{k}{2}\lambda^{k-2} & \cdots & \binom{k}{r-1}\lambda^{k-r+1} \\ 0 & \lambda^k & \binom{k}{1}\lambda^{k-1} & \cdots & \binom{k}{r-2}\lambda^{k-r+2} \\ \vdots & \vdots & \vdots & \cdots & \vdots \\ 0 & 0 & 0 & \cdots & \lambda^k \end{pmatrix};$$

(c) if $|\lambda| < 1$, then $\lim_{k \to \infty} B_r(\lambda)^k = O$.

15. Let $A \in \mathbb{R}^{n\times n}$ be a symmetric real matrix. Prove that

$$\nabla \mathsf{ral}_A(\mathbf{x}) = \frac{2}{\mathbf{x}'\mathbf{x}}(A\mathbf{x} - \mathsf{ral}_A(\mathbf{x})\mathbf{x}).$$

Also, show that the eigenvectors of A are the stationary points of the function $\mathsf{ral}_A(\mathbf{x})$.

16. Let $A, B \in \mathbb{C}^{n\times n}$ be two Hermitian matrices. Prove that AB is a Hermitian matrix if and only if $AB = BA$.
17. Let $A \in \mathbb{R}^{3\times 3}$ be a symmetric matrix. Prove that if $trace(A) \neq 0$, the sum of principal minors of order 2 equals 0, and $\det(A) = 0$, then $rank(A) = 1$.

    Solution: The characteristic polynomial of A is $p_A(\lambda) = \lambda^3 - trace(A)\lambda^2 = 0$. Thus, $\mathsf{spec}(A) = \{trace(A), 0\}$, where $\mathsf{algm}(A, 0) = 2$, so $rank(A) = 1$.
18. Let $A \in \mathbb{R}^{3\times 3}$ be a symmetric matrix. Prove that if the sum of principal minors of order 2 does not equal 0 but $\det(A) = 0$, then $rank(A) = 2$.
19. Let $A \in \mathbb{C}^{n\times n}$ be a Hermitian matrix, $\boldsymbol{u} \in \mathbb{C}^n$ be a vector and let a be a complex number. Define the Hermitian matrix B as

$$B = \begin{pmatrix} A & \boldsymbol{u} \\ \boldsymbol{u}^{\mathsf{H}} & a \end{pmatrix}.$$

Let $\alpha_1 \leqslant \cdots \leqslant \alpha_n$ be the eigenvalues of A and let $\beta_1 \leqslant \cdots \leqslant \beta_n \leqslant \beta_{n+1}$ be the eigenvalues of B. Prove that

$$\beta_1 \leqslant \alpha_1 \leqslant \beta_2 \leqslant \cdots \leqslant \beta_n \leqslant \alpha_n \leqslant \beta_{n+1}.$$

Solution: Since $B \in \mathbb{C}^{(n+1)\times(n+1)}$, by Courant-Fisher Theorem we have

$$\begin{aligned}\beta_{k+1} &= \min_W \max_{\mathbf{x}} \{\mathbf{x}^{\mathsf{H}} B\mathbf{x} \mid \parallel \mathbf{x} \parallel_2 = 1 \text{and} \mathbf{x} \in \langle W\rangle^{\perp}\} \\ &= \max_Z \min_{\mathbf{x}} \{\mathbf{x}^{\mathsf{H}} B\mathbf{x} \mid \parallel \mathbf{x} \parallel_2 = 1 \text{and} \mathbf{x} \in \langle Z\rangle^{\perp}\},\end{aligned}$$

where W ranges of sets of k non-zero arbitrary vectors, and let Z be a subset of $\mathbb{C}^n$ that consists of $n - k$ non-zero arbitrary vectors in $\mathbb{C}^{n+1}$.

Let U be a set of k non-zero vectors in $\mathbb{C}^n$ and let Y be a set of $n - k - 1$ vectors in $\mathbb{C}^n$. Define the subsets W_U and Z_Y of $\mathbb{C}^{n+1}$ as

$$W_U = \left\{ \begin{pmatrix} \boldsymbol{u} \\ 0 \end{pmatrix} \middle| \boldsymbol{u} \in U \right\}$$

and

$$Z_Y = \left\{ \begin{pmatrix} \mathbf{y} \\ 0 \end{pmatrix} \middle| \mathbf{y} \in Y \right\} \cup \{\mathbf{e}_{n+1}\}.$$

By restricting the sets W and Z to sets of the form W_U and Z_Y we obtain the double inequality

$$\max_{Z_Y} \min_{\mathbf{x}} \{\mathbf{x}^{\mathsf{H}} B\mathbf{x} \mid \parallel \mathbf{x} \parallel_2 = 1 \text{ and } \mathbf{x} \in \langle Z_Y \rangle^{\perp}\}$$
$$\leqslant \beta_{k+1} \leqslant \min_{W_U} \max_{\mathbf{x}} \{\mathbf{x}^{\mathsf{H}} B\mathbf{x} \mid \parallel \mathbf{x} \parallel_2 = 1 \text{ and } \mathbf{x} \in \langle W_U \rangle^{\perp}\}.$$

Note that, if $\mathbf{x} \in \langle Z_Y \rangle^{\perp}$, then we have $\mathbf{x} \perp \mathbf{e}_{n+1}$, so $x_{n+1} = 0$. Therefore,

$$\mathbf{x}^{\mathsf{H}} B\mathbf{x} = (\mathbf{y}^{\mathsf{H}} 0) \begin{pmatrix} A & \boldsymbol{u} \\ \boldsymbol{u}^{\mathsf{H}} & a \end{pmatrix} \begin{pmatrix} \mathbf{y} \\ 0 \end{pmatrix} = \mathbf{y}^{\mathsf{H}} A\mathbf{y}.$$

Consequently,

$$\max_{Z_Y} \min_{\mathbf{x}} \{\mathbf{x}^{\mathsf{H}} B\mathbf{x} \mid \parallel \mathbf{x} \parallel_2 = 1 \text{and} \mathbf{x} \in \langle Z_Y \rangle^{\perp}\}$$
$$= \max_{Y} \min_{\mathbf{y}} \{\mathbf{y}^{\mathsf{H}} A\mathbf{y} \mid \parallel \mathbf{y} \parallel_2 = 1 \text{and} \mathbf{y} \in \langle Y \rangle^{\perp}\} = \alpha_k.$$

This allows us to conclude that $\alpha_k \leqslant \beta_{k+1}$ for $1 \leqslant k \leqslant n$.

On another hand, if $\mathbf{x} \in \langle W_U \rangle^{\perp}$ and

$$\mathbf{x} = \begin{pmatrix} \boldsymbol{u} \\ 0 \end{pmatrix}$$

then $\mathbf{x}^{\mathsf{H}} B\mathbf{x} = \boldsymbol{u}^{\mathsf{H}} A\boldsymbol{u}$ and $\parallel \mathbf{x} \parallel_2 = \parallel \boldsymbol{u} \parallel_2$. Now we can write

$$\min_{W_U} \max_{\mathbf{x}} \{\mathbf{x}^{\mathsf{H}} B\mathbf{x} \mid \parallel \mathbf{x} \parallel_2 = 1 \text{ and } \mathbf{x} \in \langle W_U \rangle^{\perp}\}$$
$$= \min_{U} \max_{\boldsymbol{u}} \{\boldsymbol{u}^{\mathsf{H}} A\boldsymbol{u} \mid \parallel \boldsymbol{u} \parallel_2 = 1 \text{ and } \boldsymbol{u} \in \langle U \rangle^{\perp}\} = \alpha_{k+1},$$

so $\beta_{k+1} \leqslant \alpha_{k+1}$ for $1 \leqslant k \leqslant n-1$.

20. Let $A, B \in \mathbb{C}^{n \times n}$ be two matrices such that $AB = BA$. Prove that A and B have a common eigenvector.

Solution: Let $\lambda \in \mathsf{spec}(A)$ and let $\{\mathbf{x}_1, \ldots, \mathbf{x}_k\}$ be a basis for $\mathsf{NullSp}(A - \lambda I_n)$. Observe that the matrices $A - \lambda I_n$ and B commute because

$$(A - \lambda I_n)B = AB - \lambda B \text{ and } B(A - \lambda I_n) = BA - \lambda B.$$

Therefore, we have

$$(A - \lambda I_n)B\mathbf{x}_i = B(A - \lambda I_n)\mathbf{x}_i = \mathbf{0},$$

so $(A - \lambda I_n)BX = O_{n,n}$, where $X = (\mathbf{x}_1, \ldots, \mathbf{x}_k)$. Consequently, $ABX = \lambda BX$. Let $\mathbf{y}_1, \ldots, \mathbf{y}_m$ be the columns of the matrix BX. The last equality implies that $A\mathbf{y}_i = \lambda\mathbf{y}_i$, so $\mathbf{y}_i \in \mathsf{NullSp}(A - \lambda I_n)$. Since X is a basis of $\mathsf{NullSp}(A - \lambda I_n)$ it follows that each $\mathbf{y}_i$ is a linear combination of the columns of X so there exists a matrix P such that $(\mathbf{y}_1 \cdots \mathbf{y}_m) = (\mathbf{x}_1 \cdots \mathbf{x}_k)P$, which is equivalent to $BX = XP$. Let $\boldsymbol{w}$ be an eigenvector of P. We have $P\boldsymbol{w} = \mu\boldsymbol{w}$. Consequently, $BX\boldsymbol{w} = XP\boldsymbol{w} = \mu X\boldsymbol{w}$, which proves that $X\boldsymbol{w}$ is an eigenvector of B. Also, $A(X\boldsymbol{w}) = A(BX\boldsymbol{w}) = (\lambda BX)\boldsymbol{w} = \lambda\mu X\boldsymbol{w}$, so $X\boldsymbol{w}$ is also an eigenvector of A.

21. Let $A \in \mathbb{C}^{m\times n}$ and $B \in \mathbb{C}^{n\times m}$ be two matrices. Prove that the set of non-zero eigenvalues of the matrices $AB \in \mathbb{C}^{m\times m}$ and $BA \in \mathbb{C}^{n\times n}$ are the same and $\mathsf{algm}(AB, \lambda) = \mathsf{algm}(BA, \lambda)$ for each such eigenvalue.

Solution: Consider the following straightforward equalities:

$$\begin{pmatrix} I_m & -A \\ O_{n,m} & \lambda I_n \end{pmatrix}\begin{pmatrix} \lambda I_m & A \\ B & I_n \end{pmatrix} = \begin{pmatrix} \lambda I_m - AB & O_{m,n} \\ -\lambda B & \lambda I_n \end{pmatrix}$$
$$\begin{pmatrix} -I_m & O_{m,n} \\ -B & \lambda I_n \end{pmatrix}\begin{pmatrix} \lambda I_m & A \\ B & I_n \end{pmatrix} = \begin{pmatrix} -\lambda I_m & -A \\ O_{n,m} & \lambda I_n - BA \end{pmatrix}.$$

Observe that

$$\det\left(\begin{pmatrix} I_m & -A \\ O_{n,m} & \lambda I_n \end{pmatrix}\begin{pmatrix} \lambda I_m & A \\ B & I_n \end{pmatrix}\right) = \det\left(\begin{pmatrix} -I_m & O_{m,n} \\ -B & \lambda I_n \end{pmatrix}\begin{pmatrix} \lambda I_m & A \\ B & I_n \end{pmatrix}\right),$$

and therefore,

$$\det\begin{pmatrix} \lambda I_m - AB & O_{m,n} \\ -\lambda B & \lambda I_n \end{pmatrix} = \det\begin{pmatrix} -\lambda I_m & -A \\ O_{n,m} & \lambda I_n - BA \end{pmatrix}.$$

The last equality amounts to $\lambda^n p_{AB}(\lambda) = \lambda^m p_{BA}(\lambda)$. Thus, for $\lambda \neq 0$ we have $p_{AB}(\lambda) = p_{BA}(\lambda)$, which gives the desired conclusion.

22. Let $\mathbf{a} \in \mathbb{C}^n - \{\mathbf{0}_n\}$. Prove that the matrix $\mathbf{a}\mathbf{a}^{\mathsf{H}} \in \mathbb{C}^{n\times n}$ has one eigenvalue distinct from 0, and this eigenvalue is equal to $\parallel \mathbf{a} \parallel^2$.

23. Let $A \in \mathbb{R}^{n\times n}$ be a symmetric matrix such that $a_{ij} \in \{0, 1\}$ for $1 \leqslant i, j \leqslant n$. If $d = \frac{|\{a_{ij} \mid a_{ij}=1\}|}{n^2}$, prove that $n\sqrt{d} \geqslant \lambda_1 \geqslant nd$, where λ_1 is the largest eigenvalue of A.

Solution: By Rayleigh-Ritz Theorem (Theorem 7.49) we have $\lambda_1 \mathbf{1}'_n\mathbf{1}_n \geqslant \mathbf{1}'_n A\mathbf{1}_n$. Since $\mathbf{1}'_n\mathbf{1}_n = n$ and $\mathbf{1}'_n A\mathbf{1}_n = n^2 d$ it follows that $\lambda_1 \geqslant nd$. On another hand we have $\sum_{i=1}^n \lambda_i^2 \leqslant \parallel A \parallel_F^2 = n^2 d$, so $\lambda_1 \leqslant n\sqrt{d}$.

24. Let $U = (\boldsymbol{u}_1, \ldots, \boldsymbol{u}_r) \in \mathbb{C}^{n\times r}$ be a matrix having an orthonormal set of columns. For $A \in \mathbb{C}^{n\times n}$ define the matrix $A_U \in \mathbb{C}^{r\times r}$ by $A_U = U^{\mathsf{H}}AU$.

(a) Prove that if A is a Hermitian matrix, then A_U is also a Hermitian matrix.

(b) If A is a Hermitian matrix having the eigenvalues $\lambda_1 \leqslant \cdots \leqslant \lambda_n$ and A_U has the eigenvalues $\mu_1 \leqslant \cdots \leqslant \mu_r$, prove that

$$\lambda_k \leqslant \mu_k \leqslant \lambda_{k+n-r}.$$

(c) Prove that

$$\sum_{i=1}^{r} \lambda_i = \min\{trace(A_U) \mid U^{\mathsf{H}}U = I_r\}$$

$$\sum_{i=n-r+1}^{n} \lambda_i = \max\{trace(A_U) \mid U^{\mathsf{H}}U = I_r\}.$$

Solution: Observe that $(A_U)^{\mathsf{H}} = U^{\mathsf{H}}A^{\mathsf{H}}U = U^{\mathsf{H}}AU = A_U$, so A_U is indeed Hermitian and its eigenvalues $\mu_1, \ldots, \mu_r$ are real numbers.

Extend the set of columns $\boldsymbol{u}_1, \ldots, \boldsymbol{u}_r$ of U to an orthonormal basis

$$\{\boldsymbol{u}_1, \ldots, \boldsymbol{u}_r, \boldsymbol{u}_{r+1}, \ldots, \boldsymbol{u}_n\}$$

and let $W \in \mathbb{C}^{n\times n}$ be the matrix whose columns are $\boldsymbol{u}_1, \ldots, \boldsymbol{u}_n$. Since W is a unitary matrix, $\mathsf{spec}(W^{\mathsf{H}}AW) = \mathsf{spec}(A)$ and A_U is a principal submatrix of $\mathsf{spec}(W^{\mathsf{H}}AW)$. The second part follows from Theorem 7.53.

The first equality of the third part follows from the fact that the second part implies

$$\sum_{i=1}^{r} \lambda_i \leqslant \sum_{i=1}^{r} \mu_i = trace(A_U),$$

where $A_U \in \mathbb{C}^{r\times r}$. If the columns $\boldsymbol{u}_1, \ldots, \boldsymbol{u}_r$ of U are chosen as orthonormal eigenvectors the above inequality becomes an equality. In this case we have $U^{\mathsf{H}}U = I_r$ and the first equality of the third part follows. The argument for the second equality is similar.

25. Let $A, B \in \mathbb{C}^{n\times n}$ be two Hermitian matrices, where $\mathsf{spec}(A) = \{\xi_1, \ldots, \xi_n\}$, $\mathsf{spec}(B) = \{\zeta_1, \ldots, \zeta_n\}$, and $\mathsf{spec}(A+B) = \{\lambda_1, \ldots, \lambda_n\}$. Also, suppose that $\xi_1 \leqslant \cdots \leqslant \xi_n$, $\zeta_1 \leqslant \cdots \leqslant \zeta_n$, and $\lambda_1 \leqslant \cdots \leqslant \lambda_n$. Prove that for $r \leqslant n$ we have

$$\sum_{i=1}^{r} \lambda_i \leqslant \sum_{i=1}^{r} \xi_i + \sum_{i=1}^{r} \zeta_i$$

and

$$\sum_{i=n-r+1}^{n} \lambda_i \geqslant \sum_{i=n-r+1}^{n} \xi_i + \sum_{i=n-r+1}^{n} \zeta_i.$$

Solution: Supplement 24 implies that

$$\begin{aligned}\sum_{i=1}^{r} \lambda_i &= \min\{trace((A+B)_U) \mid U^{\mathsf{H}}U = I_r\}\\ &\geqslant \min\{trace(A_U) \mid U^{\mathsf{H}}U = I_r\} + \min\{trace(B_U) \mid U^{\mathsf{H}}U = I_r\}\\ &= \sum_{i=1}^{r} \xi_i + \sum_{i=1}^{r} \zeta_i .\end{aligned}$$

For the second part we can write

$$\begin{aligned}\sum_{i=n-r+1}^{n} \lambda_i &= \max\{trace((A+B)_U) \mid U^{\mathsf{H}}U = I_r\}\\ &\leqslant \max\{trace(A_U) \mid U^{\mathsf{H}}U = I_r\} + \max\{trace(B_U) \mid U^{\mathsf{H}}U = I_r\}\\ &= \sum_{i=n-r+1}^{n} \xi_i + \sum_{i=n-r+1}^{n} \zeta_i .\end{aligned}$$

26. Let $A \in \mathbb{R}^{2\times 2}$ be the matrix

$$A = \begin{pmatrix} a & b \\ c & d \end{pmatrix}.$$

Prove that A is diagonalizable if and only if $(a-d)^2 + 4bc \neq 0$.

27. Let $A \in \mathbb{C}^{n\times n}$ be a matrix. Prove that the following statements are equivalent:

    (a) A is a rank 1 matrix;
    (b) A has exactly one non-zero eigenvalue λ with $\mathsf{algm}(A, \lambda) = 1$;
    (c) There exist $\mathbf{x}, \mathbf{y} \in \mathbb{C}^n - \{\mathbf{0}\}$ such that $A = \mathbf{x}\mathbf{y}^{\mathsf{H}}$ and $\mathbf{x}^{\mathsf{H}}\mathbf{y}$ is an eigenvalue of A.

28. Prove that the characteristic polynomial of the companion matrix of a polynomial p is p itself.

29. Let $A \in \mathbb{C}^{n\times n}$, $B \in \mathbb{C}^{k\times k}$, and $X \in \mathbb{C}^{n\times k}$ be three matrices such that $AX = XB$. Prove that

    (a) $\mathsf{Ran}(X)$ is an invariant subspace of A;
    (b) if $\boldsymbol{v}$ is an eigenvector of B, then $X\boldsymbol{v}$ is an eigenvector of A;
    (c) if $rank(X) = k$, then $\mathsf{spec}(B) \subseteq \mathsf{spec}(A)$.

The next supplements present a result known as *Weyl's Theorem* (Supplement 30) and several of its important consequences. For a Hermitian matrix $A \in \mathbb{C}^{n\times n}$ we denote its eigenvalues arranged in increasing order as $\lambda_1(A) \leqslant \cdots \leqslant \lambda_n(A)$.

30. Let A and B be two Hermitian matrices in $\mathbb{C}^{n\times n}$. Prove that

$$\lambda_1(B) \leqslant \lambda_k(A+B) - \lambda_k(A) \leqslant \lambda_n(B)$$

for $1 \leqslant k \leqslant n$.

Solution: By the Rayleigh-Ritz Theorem we have

$$\lambda_1(B) \leqslant \mathbf{x}^{\mathsf{H}} B\mathbf{x} \leqslant \lambda_n(B),$$

for $\mathbf{x} \neq \mathbf{0}$ and $\| \mathbf{x} \|_2 = 1$. Then, by the Courant-Fisher Theorem,

$$\lambda_k(A + B) = \min_W \max_{\mathbf{x}} \{\mathbf{x}^{\mathsf{H}}(A + B)\mathbf{x} \mid \| \mathbf{x} \|_2 = 1 \text{ and } \mathbf{x} \in \langle W \rangle^{\perp}\},$$

where the minimum is taken over sets W that contain $k - 1$ vectors. Since $\mathbf{x}^{\mathsf{H}}(A + B)\mathbf{x} = \mathbf{x}^{\mathsf{H}} A\mathbf{x} + \mathbf{x}^{\mathsf{H}} B\mathbf{x}$, it follows that

$$\begin{aligned}\lambda_k(A + B) &\geqslant \min_W \max_{\mathbf{x}} \{\mathbf{x}^{\mathsf{H}} A\mathbf{x} + \lambda_n(B) \mid \| \mathbf{x} \|_2 = 1 \text{ and } \mathbf{x} \in \langle W \rangle^{\perp}\} \\ &= \lambda_k(A) + \lambda_n(B).\end{aligned}$$

Similarly, we have

$$\begin{aligned}\lambda_k(A + B) &\leqslant \min_W \max_{\mathbf{x}} \{\mathbf{x}^{\mathsf{H}} A\mathbf{x} + \lambda_1(B) \mid \| \mathbf{x} \|_2 = 1 \text{ and } \mathbf{x} \in \langle W \rangle^{\perp}\} \\ &= \lambda_k(A) + \lambda_1(B).\end{aligned}$$

31. Let A and E be Hermitian matrices in $\mathbb{C}^{n \times n}$. Prove that $|\lambda_p(A+E) - \lambda_p(A)| \leqslant \rho(E) = |||E|||_2$ for $1 \leqslant p \leqslant n$.

Solution: By Weyl's inequalities (Supplement 30) we have

$$\lambda_1(E) \leqslant \lambda_p(A + E) - \lambda_p(A) \leqslant \lambda_n(E)$$

By Definition 7.64, this implies $|\lambda_p(A+E) - \lambda_p(A)| \leqslant \rho(E) = |||E|||_2$ because A is Hermitian.

32. Let $A \in \mathbb{C}^{n \times n}$ be a Hermitian matrix and let $\boldsymbol{w} \in \mathbb{C}^n$. Then, $\lambda_k(A + \boldsymbol{w}\boldsymbol{w}^{\mathsf{H}}) \leqslant \lambda_{k+1}(A) \leqslant \lambda_{k+2}(A + \boldsymbol{w}\boldsymbol{w}^{\mathsf{H}})$ and $\lambda_k(A) \leqslant \lambda_{k+1}(A + \boldsymbol{w}\boldsymbol{w}^{\mathsf{H}}) \leqslant \lambda_{k+2}(A)$ for $1 \leqslant k \leqslant n - 2$.

Solution: Let W ranging over the subsets of $\mathbb{C}^n$ that consist of $n-k-2$ vectors. By Courant-Fisher Theorem (Theorem 7.50),

$$\begin{aligned}&\lambda_{k+2}(A + \boldsymbol{w}\boldsymbol{w}^{\mathsf{H}}) \\ &= \min_W \max_{\mathbf{x}} \{\mathbf{x}^{\mathsf{H}}(A + \boldsymbol{w}\boldsymbol{w}^{\mathsf{H}})\mathbf{x} \mid \| \mathbf{x} \|_2 = 1 \text{ and } \mathbf{x} \in \langle W \rangle^{\perp}\} \\ &\geqslant \min_W \max_{\mathbf{x}} \{\mathbf{x}^{\mathsf{H}}(A + \boldsymbol{w}\boldsymbol{w}^{\mathsf{H}})\mathbf{x} \mid \| \mathbf{x} \|_2 = 1, \mathbf{x} \in \langle W \rangle^{\perp} \text{ and } \mathbf{x} \perp \boldsymbol{w}\} \\ &= \min_W \max_{\mathbf{x}} \{\mathbf{x}^{\mathsf{H}} A\mathbf{x} \mid \| \mathbf{x} \|_2 = 1, \mathbf{x} \in \langle W \rangle^{\perp} \text{ and } \mathbf{x} \perp \boldsymbol{w}\}\end{aligned}$$

$$\text{(because } \mathbf{x}^{\mathsf{H}} \boldsymbol{w} = \boldsymbol{w}^{\mathsf{H}} \mathbf{x} = 0)$$
$$\geqslant \min_{W_1} \max_{\mathbf{x}} \{\mathbf{x}^{\mathsf{H}} A \mathbf{x} \mid \| \mathbf{x} \|_2 = 1 \text{ and } \mathbf{x} \in \langle W_1 \rangle^{\perp}\} = \lambda_{k+1}(A),$$

where W_1 ranges over the sets that contain $n - k - 1$ vectors.

For $2 \leqslant k \leqslant n - 2$, the same Courant-Fisher Theorem yields

$$\lambda_k(A + \boldsymbol{w}\boldsymbol{w}^{\mathsf{H}}) = \max_{Z} \min_{\mathbf{x}} \{\mathbf{x}^{\mathsf{H}}(A + \boldsymbol{w}\boldsymbol{w}^{\mathsf{H}})\mathbf{x} \mid \| \mathbf{x} \|_2 = 1 \text{and} \mathbf{x} \in \langle Z \rangle^{\perp}\},$$

where Z is a set that contains $k - 1$ vectors. This implies

$$\lambda_k(A + \boldsymbol{w}\boldsymbol{w}^{\mathsf{H}})$$
$$\leqslant \max_{Z} \min_{\mathbf{x}} \{\mathbf{x}^{\mathsf{H}}(A + \boldsymbol{w}\boldsymbol{w}^{\mathsf{H}})\mathbf{x} \mid \| \mathbf{x} \|_2 = 1 \text{ and } \mathbf{x} \in \langle Z \rangle^{\perp} \text{and} \mathbf{x} \perp \boldsymbol{w}\}$$
$$= \max_{Z} \min_{\mathbf{x}} \{\mathbf{x}^{\mathsf{H}} A \mathbf{x} \mid \| \mathbf{x} \|_2 = 1 \text{ and } \mathbf{x} \in \langle Z \rangle^{\perp} \text{and} \mathbf{x} \perp \boldsymbol{w}\}$$
$$\leqslant \max_{Z_1} \min_{\mathbf{x}} \{\mathbf{x}^{\mathsf{H}} A \mathbf{x} \mid \| \mathbf{x} \|_2 = 1 \text{ and } \mathbf{x} \in \langle Z_1 \rangle^{\perp}\} = \lambda(k + 1, A),$$

where Z_1 ranges over sets that contain k vectors.

33. Let A and B be two Hermitian matrices in $\mathbb{C}^{n \times n}$. If $rank(B) \leqslant r$, prove that $\lambda_k(A + B) \leqslant \lambda_{k+r}(A) \leqslant \lambda_{k+2r}(A + B)$ for $1 \leqslant k \leqslant n - 2r$ and $\lambda_k(A) \leqslant \lambda_{k+r}(A + B) \leqslant \lambda_{k+2r}(A)$.

Solution: If B is a Hermitian matrix of rank no larger than r, then $B = U\mathsf{diag}(\beta_1, \ldots, \beta_r, 0, \ldots, 0)U^{\mathsf{H}}$, where $U = (\boldsymbol{u}_1 \ \cdots \ \boldsymbol{u}_n)$ is a unitary matrix. This amounts to $B = \beta_1 \boldsymbol{u}_1 \boldsymbol{u}_1' + \cdots + \beta_r \boldsymbol{u}_r \boldsymbol{u}_r'$. Conversely, every Hermitian matrix of rank no larger than r can be written in this form.

Let W range over the subsets of $\mathbb{R}^n$ that consist of $n - k - 2r$ vectors. We have

$$\lambda_{k+2r}(A + B)$$
$$= \min_{W} \max_{\mathbf{x}} \{\mathbf{x}^{\mathsf{H}}(A + B)\mathbf{x} \mid \| \mathbf{x} \|_2 = 1 \text{ and } \mathbf{x} \in \langle W \rangle^{\perp}\}$$
$$\geqslant \min_{W} \max_{\mathbf{x}} \{\mathbf{x}^{\mathsf{H}}(A + B)\mathbf{x} \mid \| \mathbf{x} \|_2 = 1, \mathbf{x} \in \langle W \cup \{\boldsymbol{u}_1, \ldots, \boldsymbol{u}_r\} \rangle^{\perp}\}$$
$$= \min_{W} \max_{\mathbf{x}} \{\mathbf{x}^{\mathsf{H}} A \mathbf{x} \mid \| \mathbf{x} \|_2 = 1, \mathbf{x} \in \langle W \cup \{\boldsymbol{u}_1, \ldots, \boldsymbol{u}_r\} \rangle^{\perp}\}$$
$$\text{(because } \mathbf{x}^{\mathsf{H}} \boldsymbol{u}_i = \boldsymbol{u}_i^{\mathsf{H}} \mathbf{x} = 0)$$
$$\geqslant \min_{W_1} \max_{\mathbf{x}} \{\mathbf{x}^{\mathsf{H}} A \mathbf{x} \mid \| \mathbf{x} \|_2 = 1 \text{ and } \mathbf{x} \in \langle W_1 \rangle^{\perp}\} = \lambda_{k+r}(A),$$

where W_1 ranges over the subsets of $\mathbb{R}^n$ that contain $n - k - r$ vectors.

The proof of the remaining inequalities follows the same pattern as above and generalize the results and proofs of Supplement 32.

34. Let A be a Hermitian matrix in $\mathbb{C}^{n \times n}$ such that $A = UDU^{\mathsf{H}}$, where $U = (\boldsymbol{u}_1 \ \cdots \ \boldsymbol{u}_n)$ is an unitary matrix and $D = (\lambda_1(A), \ldots, \lambda_n(A))$. If $A_j =$

$\sum_{i=j+1}^{n} \lambda_i(A) \boldsymbol{u}_i \boldsymbol{u}_i'$ for $0 \leqslant j \leqslant n-1$, prove that the largest eigenvalue of the matrix $A - A_{n-k}$ is $\lambda_{n-k}(A)$.

Solution: Since $A - A_{n-k} = \sum_{i=1}^{n-k} \lambda_i(A) \boldsymbol{u}_i \boldsymbol{u}_i'$ the statement follows immediately.

35. Let A and B be two Hermitian matrices in $\mathbb{C}^{n \times n}$. Using the same notations as in Supplement 30 prove that for any i, j such that $1 \leqslant i, j \leqslant n$ and $i + j \geqslant n + 1$ we have
$$\lambda_{j+k-n}(A + B) \leqslant \lambda_j(A) + \lambda_k(B).$$
Also, if $i + j \leqslant n + 1$, then
$$\lambda_j(A) + \lambda_k(B) \leqslant \lambda_{j+k-1}(A + B).$$

The *field of values of a matrix* $A \in \mathbb{C}^{n \times n}$ is the set of numbers $F(A) = \{\mathbf{x} A \mathbf{x}^{\mathsf{H}} \mid \mathbf{x} \in \mathbb{C}^n \text{and} \parallel \mathbf{x} \parallel_2 = 1\}$.

36. Prove that $\mathsf{spec}(A) \subseteq F(A)$ for any $A \in \mathbb{C}^{n \times n}$.
37. If $U \in \mathbb{C}^{n \times n}$ is a unitary matrix and $A \in \mathbb{C}^{n \times n}$, prove that $F(UAU^{\mathsf{H}}) = F(A)$.
38. Prove that $A \sim B$ implies $f(A) \sim f(B)$ for every $A, B \in \mathbb{C}^{n \times n}$ and every polynomial f.
39. Let $A \in \mathbb{C}^{n \times n}$ be a matrix such that $\mathsf{spec}(A) = \{\lambda_1, \ldots, \lambda_n\}$. Prove that $\sum_{i=1}^{n} |\lambda_i|^2 \leqslant \sum_{i=1}^{n} \sum_{j=1}^{n} |a_{ij}|^2$; furthermore, prove that A is normal if and only if $\sum_{i=1}^{n} |\lambda_i|^2 = \sum_{i=1}^{n} \sum_{j=1}^{n} |a_{ij}|^2$.
40. Let $A \in \mathbb{C}^{n \times n}$ such that $A \geqslant O_{n,n}$. Prove that if $\mathbf{1}_n$ is an eigenvector of A, then $\rho(A) = |||A|||_\infty$ and if $\mathbf{1}_n$ is an eigenvector of A', then $\rho(A) = |||A|||_1$.

    Solution: If $\mathbf{1}_n$ is an eigenvector of A, then $A\mathbf{1}_n = \lambda \mathbf{1}_n$, so $\sum_{j=1}^{n} a_{ij} = \lambda$ for every i, $1 \leqslant i \leqslant n$. This means that all rows of A have the same sum λ and, therefore, $\lambda = |||A|||_\infty$, as we saw in Example 6.84. This implies $\rho(A) = |||A|||_\infty$. The argument for the second part is similar.
41. Prove that the matrix $A \in \mathbb{C}^{n \times n}$ is normal if and only if there exists a polynomial p such that $p(A) = A^{\mathsf{H}}$.
42. Prove that the matrix $A \in \mathbb{C}^{n \times n}$ is normal if and only if there exist $B, C \in \mathbb{C}^{n \times n}$ such that $A = B + iC$ and $BC = CB$.
43. Let $A \in \mathbb{C}^{n \times n}$ be a matrix and let $\mathsf{spec}(A) = \{\lambda_1, \ldots, \lambda_n\}$. Prove that

    (a) $\sum_{p=1}^{n} |\lambda_p|^2 \leqslant \parallel A \parallel_F^2$;

    (b) the equality $\sum_{p=1}^{n} |\lambda_p|^2 = \parallel A \parallel_F^2$ holds if and only if A is normal.

    Solution: By Schur's Triangularization Theorem there exists a unitary matrix $U \in \mathbb{C}^{n \times n}$ and an upper-triangular matrix $T \in \mathbb{C}^{n \times n}$ such that $A = U^{\mathsf{H}} T U$ and the diagonal elements of T are the eigenvalues of A. Thus,
$$\parallel A \parallel_F^2 = \parallel T \parallel_F^2 = \sum_{p=1}^{n} |\lambda_p|^2 + \sum_{i<j} |t_{ij}|^2,$$

which implies the desired inequality.

The equality of the second part follows from the Spectral Theorem for Normal Matrices. The converse implication can be obtained noting that by the first part, the equality of the second part implies $t_{ij} = 0$ for $i < j$, which means that T is actually a diagonal matrix.

44. Let A and B be two normal matrices in $\mathbb{C}^{n\times n}$. Prove that if AB is a normal matrix, then so is BA.

Solution: By Supplement 21 the matrices AB and BA have the same non-zero eigenvalues. Since A and B are normal we have $A^{\mathsf{H}}A = AA^{\mathsf{H}}$ and $B^{\mathsf{H}}B = BB^{\mathsf{H}}$. Thus, we can write

$$\begin{aligned}
\| AB \|_F^2 &= trace((AB)^{\mathsf{H}}AB) = trace(B^{\mathsf{H}}A^{\mathsf{H}}AB) = trace(B^{\mathsf{H}}AA^{\mathsf{H}}B) \\
&\quad \text{(because A is a normal matrix)} \\
&= trace((B^{\mathsf{H}}A)(A^{\mathsf{H}}B)) = trace((A^{\mathsf{H}}B)(B^{\mathsf{H}}A)) \\
&\quad \text{(by the third part of Theorem 5.51)} \\
&= trace(A^{\mathsf{H}}(BB^{\mathsf{H}})A)) = trace(A^{\mathsf{H}}(B^{\mathsf{H}}B)A)) \\
&\quad \text{(because B is a normal matrix)} \\
&= trace((BA)^{\mathsf{H}}BA) = \| BA \|_F^2 .
\end{aligned}$$

Since AB is a normal matrix, if $\mathsf{spec}(AB) = \{\lambda_1, \ldots, \lambda_p\}$, we have $\sum_{p=1}^n |\lambda_p|^2 = \| AB \|_F^2 = \| BA \|_F^2$.

Taking into account the equalities shown above, it follows that BA is a normal matrix by Supplement 43.

45. Let $A \in \mathbb{C}^{n\times n}$ be a matrix with $\mathsf{spec}(A) = \{\lambda_1, \ldots, \lambda_n\}$. Prove that A is normal if and only if its singular values are $|\lambda_1|, \ldots, |\lambda_n|$.
46. Let A be a non-negative matrix in $\mathbb{C}^{n\times n}$ and let $\boldsymbol{u} = A\mathbf{1}_n$ and $\boldsymbol{v} = A'\mathbf{1}_n$. Prove that

$$\max\{\min u_i, \min v_j\} \leqslant \rho(A) \leqslant \min\{\max u_i, \max v_j\}.$$

Solution: Note that $|||A|||_\infty = \max u_i$ and $|||A|||_1 = \max v_j$. By Theorem 7.65, we have $\rho(A) \leqslant \min\{\max u_i, \max v_j\}$.

Let $a = \min u_i$. If $a > 0$ define the non-negative matrix $B \in \mathbb{R}^{n\times n}$ as $b_{ij} = \frac{a a_{ij}}{u_i}$. We have $A \geqslant B \geqslant O_{n,n}$. By Corollary 7.67 we have $\rho(A) \geqslant \rho(B) = a$; the same equality, $\rho(A) \geqslant a$ holds trivially when $a = 0$. In a similar manner, one could prove that $\min v_j \leqslant \rho(A)$, so $\max\{\min u_i, \min v_j\} \leqslant \rho(A)$.
47. Let A be a non-negative matrix in $\mathbb{C}^{n\times n}$ and let $\mathbf{x} \in \mathbb{R}^n$ be a vector such that $\mathbf{x} > \mathbf{0}_n$. Prove that

(a) $\min\left\{\frac{(A\mathbf{x})_j}{x_j} \mid 1 \leqslant j \leqslant n\right\} \leqslant \rho(A) \leqslant \max\left\{\frac{(A\mathbf{x})_j}{x_j} \mid 1 \leqslant j \leqslant n\right\}$;

(b) $\min\left\{x_j \sum_{i=1}^n \frac{a_{ij}}{x_j} \mid 1 \leqslant j \leqslant n\right\} \leqslant \rho(A) \leqslant \max\left\{x_j \sum_{i=1}^n \frac{a_{ij}}{x_j} \mid 1 \leqslant j \leqslant n\right\}$.

Solution: Define the diagonal matrix $S = \mathsf{diag}(x_1, \ldots, x_n)$. Its inverse is $S^{-1} = \mathsf{diag}(\frac{1}{x_1}, \ldots, \frac{1}{x_n})$. The matrix $B = S^{-1}AS$ is non-negative and we have $b_{ij} = \frac{x_i a_{ij}}{x_j}$. This implies

$$u = B\mathbf{1}_n = \begin{pmatrix} x_1 \sum_{j=1}^n \frac{a_{1j}}{x_j} \\ \vdots \\ x_n \sum_{j=1}^n \frac{a_{nj}}{x_j} \end{pmatrix} \text{ and } v = B'\mathbf{1}_n = \begin{pmatrix} \frac{\sum_{i=1}^n x_i a_{1i}}{x_1} \\ \vdots \\ \frac{\sum_{i=1}^n x_i a_{ni}}{x_n} \end{pmatrix} = \begin{pmatrix} \frac{(A\mathbf{x})_1}{x_1} \\ \vdots \\ \frac{(A\mathbf{x})_n}{x_n} \end{pmatrix}.$$

Thus, by applying the inequalities of Supplement 46 we obtain the desired inequalities.

48. Let $A \in \mathbb{R}^{n\times n}$ and $\mathbf{x} \in \mathbb{R}^n$ such that $A \geqslant O_{n,n}$ and $\mathbf{x} > 0$. Prove that if $a, b \in \mathbb{R}_{\geqslant 0}$ are such that $a\mathbf{x} \leqslant A\mathbf{x} \leqslant b\mathbf{x}$, then $a \leqslant \rho(A) \leqslant b$.

Solution: Since $a\mathbf{x} \leqslant A\mathbf{x}$ we have $a \leqslant \min_{1\leqslant i\leqslant n} \frac{(A\mathbf{x})_i}{x_i}$, so $a \leqslant \rho(A)$ by Supplement 47. Similarly, we hace $\max_{1\leqslant i\leqslant n} \frac{(A\mathbf{x})_i}{x_i} \leqslant b$, so $\rho(A) \leqslant b$.

49. Let $A \in \mathbb{C}^{n\times n}$ be a matrix such that $A \geqslant O_{n,n}$. Prove that if there exists $k \in \mathbb{N}$ such that $A^k > O_{n,n}$, then $\rho(A) > 0$.
50. Let $A \in \mathbb{C}^{n\times n}$ be a matrix such that $A \geqslant O_{n,n}$. If $A \neq O_{n,n}$ and there exists an eigenvector $\mathbf{x}$ of A such that $\mathbf{x} > \mathbf{0}_n$, prove that $\rho(A) > 0$.
51. Prove that $A \in \mathbb{C}^{n\times n}$ is positive semidefinite if and only if there is a set $U = \{v_1, \ldots, v_n\} \subseteq \mathbb{C}^n$ such that $A = \sum_{i=1}^n v_i v_i^{\mathsf{H}}$. Furthermore, prove that A is positive definite if and only if there exists a linearly independent set U as above.
52. Prove that A is positive definite if and only if A^{-1} is positive definite.
53. Prove that if $A \in \mathbb{C}^{n\times n}$ is a positive semidefinite matrix, then A^k is positive semidefinite for every $k \geqslant 1$.
54. Let $A \in \mathbb{C}^{n\times n}$ be a Hermitian matrix and let $p_A(\lambda) = \lambda^n + c_1\lambda^{n-1} + \cdots + c_m\lambda^{n-m}$ be its characteristic polynomial, where $c_m \neq 0$. Then, A is positive semidefinite if and only if $c_i \neq 0$ for $0 \leqslant k \leqslant m$ (where $c_0 = 1$) and $c_j c_{j+1} < 0$ for $0 \leqslant j \leqslant m-1$.
55. Let $A \in \mathbb{C}^{n\times n}$ be a positive semidefinite matrix. Prove that for every $k \geqslant 1$ there exists a positive semidefinite matrix B having the same rank as A such that

(a) $B^k = A$;
(b) $AB = BA$;
(c) B can be expressed as a polynomial in A.

Solution: Since A is Hermitian, its eigenvalues are real nonnegative numbers and, by the Spectral Theorem for Hermitian matrices, there exists a unitary matrix $U \in \mathbb{C}^{n\times n}$ such that $A = U^{\mathsf{H}}\mathsf{diag}(\lambda_1, \ldots, \lambda_n)U$. Let $B = U^{\mathsf{H}}\mathsf{diag}(\lambda_1^{\frac{1}{k}}, \ldots, \lambda_n^{\frac{1}{k}})U$, where $\lambda_i^{\frac{1}{k}}$ is a non-negative root of order k of λ_i. Thus, $B^k = A$, B is clearly positive semidefinite, $\mathit{rank}(B) = \mathit{rank}(A)$, and $AB = BA$.

Let

$$p(x) = \sum_{j=1}^{n} \lambda_j^{\frac{1}{k}} \prod_{k=1,k\neq j}^{n} \frac{x-\lambda_k}{\lambda_j - \lambda_k}$$

be a Lagrange interpolation polynomial such that $p(\lambda_j) = \lambda_j^{\frac{1}{k}}$ (see Exercise 67). Then,

$$p(\mathsf{diag}(\lambda_1, \ldots, \lambda_n)) = \mathsf{diag}(\lambda_1^{\frac{1}{k}}, \ldots, \lambda_n^{\frac{1}{k}}),$$

so

$$\begin{aligned} p(A) &= p(U^{\mathsf{H}}\mathsf{diag}(\lambda_1, \ldots, \lambda_n)U) = U^{\mathsf{H}} p(\mathsf{diag}(\lambda_1, \ldots, \lambda_n))U \\ &= U^{\mathsf{H}}\mathsf{diag}(\lambda_1^{\frac{1}{k}}, \ldots, \lambda_n^{\frac{1}{k}})U = B. \end{aligned}$$

56. Let $A \in \mathbb{R}^{n\times n}$ be a symmetric matrix. Prove that there exists $b \in \mathbb{R}$ such that $A + b(\mathbf{1}\mathbf{1}' - I_n)$ is positive semi-definite, where $\mathbf{1} \in \mathbb{R}^n$.

Solution: We need to find b such that for every $\mathbf{x} \in \mathbb{R}^n$ we will have

$$\mathbf{x}'(A + b(\mathbf{1}\mathbf{1}' - I_n))\mathbf{x} \geqslant 0.$$

We have $\mathbf{x}'(A+b(\mathbf{1}\mathbf{1}'-I_n))\mathbf{x} = \mathbf{x}'A\mathbf{x}+b\mathbf{x}'\mathbf{1}\mathbf{1}'\mathbf{x}-b\mathbf{x}'\mathbf{x} \geqslant 0$, which amounts to

$$\mathbf{x}'A\mathbf{x} + b\left(\left(\sum_{i=1}^{n} x_i\right)^2 - \| \mathbf{x} \|_2^2\right) \geqslant 0.$$

Since A is symmetric, by Rayleigh-Ritz Theorem, we have $\mathbf{x}'A\mathbf{x} \geqslant \lambda_1 \| \mathbf{x} \|_2^2$, where λ_1 is the least eigenvalue of A. Therefore, it suffices to take $b \leqslant \lambda_1$ to satisfy the equality for every $\mathbf{x}$.

57. If $A \in \mathbb{R}^{n\times n}$ is a positive definite matrix prove that there exist $c, d > 0$ such that $c \| \mathbf{x} \|_2^2 \leqslant \mathbf{x}'A\mathbf{x} \leqslant d \| \mathbf{x} \|_2^2$, for every $\mathbf{x} \in \mathbb{R}^n$.
58. Let $A = \mathsf{diag}(A_1, \ldots, A_p)$ and $B = (B_1, \ldots, B_q)$ be two block-diagonal matrices. Prove that $sep_F(A, B) = \min\{sep_F(A_i, B_j) \mid 1 \leqslant i \leqslant p \text{and} 1 \leqslant j \leqslant q\}$.
59. Let $A \in \mathbb{C}^{n\times n}$ be a Hermitian matrix. Prove that if for any $\lambda \in \mathsf{spec}(A)$ we have $\lambda > -a$, then the matrix $A + aI$ is positive-semidefinite.
60. Let $A \in \mathbb{C}^{m\times m}$ and $B \in \mathbb{C}^{n\times n}$ be two matrices that have the eigenvalues $\lambda_1, \ldots, \lambda_m$ and $\mu_1, \ldots, \mu_n$, respectively. Prove that:

    (a) if A and B are positive definite, then so is $A \otimes B$;

(b) if $m = n$ and A, B are symmetric positive definite, the Hadamard product $A \odot B$ is positive definite.

Solution: For the second part recall that the Hadamard product $A \odot B$ of two square matrices of the same format is a principal submatrix of $A \otimes B$. Then, apply Theorem 7.53.

61. Let $A \in \mathbb{C}^{n\times n}$ be a Hermitian matrix. Prove that if A is positive semidefinite, then all its eigenvalues are non-negative; if A is positive definite then its eigenvalues are positive.

Solution: Since A is Hermitian, all its eigenvalues are real numbers. Suppose that A is positive semidefinite, that is, $\mathbf{x}^{\mathsf{H}} A\mathbf{x} \geqslant 0$ for $\mathbf{x} \in \mathbb{C}^n$. If $\lambda \in \mathsf{spec}(A)$, then $A\boldsymbol{v} = \lambda\boldsymbol{v}$ for some eigenvector $\boldsymbol{v} \neq \mathbf{0}$. The positive semi-definiteness of A implies $\boldsymbol{v}^{\mathsf{H}} A\boldsymbol{v} = \lambda\boldsymbol{v}^{\mathsf{H}}\boldsymbol{v} = \lambda \parallel \boldsymbol{v} \parallel_2^2 \geqslant 0$, which implies $\lambda \geqslant 0$. It is easy to see that if A is positive definite, then $\lambda > 0$.

62. Let $A \in \mathbb{C}^{n\times n}$ be a Hermitian matrix. Prove that if A is positive semidefinite, then all its principal minors are non-negative real numbers; if A is positive definite then all its principal minors are positive real numbers.

Solution: Since A is positive semidefinite, every sub-matrix $A\begin{bmatrix} i_1 & \cdots & i_k \\ i_1 & \cdots & i_k \end{bmatrix}$ is a Hermitian positive semidefinite matrix by Theorem 6.110, so every principal minor is a non-negative real number. The second part is proven similarly.

63. Let $A \in \mathbb{C}^{n\times n}$ be a Hermitian matrix. Prove that he following statements are equivalent:

(a) A is positive semidefinite;
(b) all eigenvalues of A are non-negative numbers;
(c) there exists a Hermitian matrix $C \in \mathbb{C}^{n\times n}$ such that $C^2 = A$;
(d) A is the Gram matrix of a sequence of vectors, that is, $A = B^{\mathsf{H}} B$ for some $B \in \mathbb{C}^{n\times n}$.

Solution: (a) implies (b): This is stated in Exercise 61.

(b) implies (c): Suppose that A is a matrix such that all its eigenvalues are the non-negative numbers $\lambda_1, \ldots, \lambda_n$. By Theorem 7.41, A can be written as $A = U^{\mathsf{H}} DU$, where U is a unitary matrix and

$$D = \begin{pmatrix} \lambda_1 & 0 & \cdots & 0 \\ 0 & \lambda_2 & \cdots & 0 \\ \vdots & \vdots & \cdots & \vdots \\ 0 & 0 & \cdots & \lambda_n \end{pmatrix}.$$

Define the matrix $\sqrt{D}$ as

$$\sqrt{D} = \begin{pmatrix} \sqrt{\lambda_1} & 0 & \cdots & 0 \\ 0 & \sqrt{\lambda_2} & \cdots & 0 \\ \vdots & \vdots & \cdots & \vdots \\ 0 & 0 & \cdots & \sqrt{\lambda_n} \end{pmatrix}.$$

Clearly, we gave $(\sqrt{D})^2 = D$. Now we can write $A = U\sqrt{D}U^{\mathsf{H}}U\sqrt{D}U^{\mathsf{H}}$, which allows us to define the desired matrix C as $C = U\sqrt{D}U^{\mathsf{H}}$.

(c) implies (d): Since C is itself a Hermitian matrix, this implication is obvious.

(d) implies (a): Suppose that $A = B^{\mathsf{H}}B$ for some matrix $B \in \mathbb{C}^{n\times k}$. Then, for $\mathbf{x} \in \mathbb{C}^n$ we have $\mathbf{x}^{\mathsf{H}}A\mathbf{x} = \mathbf{x}^{\mathsf{H}}B^{\mathsf{H}}B\mathbf{x} = (B\mathbf{x})^{\mathsf{H}}(B\mathbf{x}) = \parallel B\mathbf{x} \parallel_2^2 \geqslant 0$, so A is positive semidefinite.

64. Let $A \in \mathbb{R}^{n\times n}$ be a real matrix that is symmetric and positive semidefinite such that $A\mathbf{1}_n = \mathbf{0}_n$. Prove that $2\max_{1\leqslant i\leqslant n}\sqrt{a_{ii}} \leqslant \sum_{j=1}^n \sqrt{a_{jj}}$.

 Solution: By Supplement 63(d), A is the Gram matrix of a sequence of vectors $B = (\mathbf{b}_1, \ldots, \mathbf{b}_n)$, so $A = B'B$. Since $A\mathbf{1}_n = \mathbf{0}_n$, it follows that $(B\mathbf{1}_n)'(B\mathbf{1}_n) = 0$, so $B\mathbf{1}_n = \mathbf{0}_n$. Thus, $\sum_{i=1}^n \mathbf{b}_i = \mathbf{0}_n$. Then, we have $\parallel \mathbf{b}_i \parallel_2 = \parallel -\sum_{j\neq i} \mathbf{b}_j \parallel_2 \leqslant \sum_{j\neq i} \parallel \mathbf{b}_j \parallel_2$, which implies $2\max_{1\leqslant i\leqslant n} \parallel \mathbf{b}_i \parallel_2 \leqslant \sum_{j=1}^n \parallel \mathbf{b}_j \parallel_2$. This, is equivalent to the inequality to be shown.

65. Let $A \in \mathbb{R}^{n\times n}$ be a matrix and let $(\lambda, \mathbf{x})$ be an eigenpair of A. Prove that

 (a) $2\Im(\lambda) = \mathbf{x}^{\mathsf{H}}(A - A')\mathbf{x}$;
 (b) if $\alpha = \frac{1}{2}\max\{|a_{ij} - a_{ji}| \mid 1 \leqslant i, j \leqslant n\}$, then $2|\Im(\lambda)| \leqslant \alpha \sum\left\{|\bar{x}_i x_j - x_i\bar{x}_j| \middle| 1 \leqslant i, j \leqslant n, i \neq j\right\}$;
 (c) $|\Im(\lambda)| \leqslant \alpha\sqrt{\frac{n(n-1)}{2}}$.

 The inequality of Part (c) is known as *Bendixon Inequality*.

 Hint: apply the results of Exercise 5 of Chap. 6.

66. Let $A \in \mathbb{C}^{m\times m}$ and $B \in \mathbb{C}^{n\times n}$ be two matrices that have the eigenvalues $\lambda_1, \ldots, \lambda_m$ and $\mu_1, \ldots, \mu_n$, respectively. Prove that the Kronecker product $A\otimes B$ has the eigenvalues $\lambda_1\mu_1, \ldots, \lambda_1\mu_n, \ldots, \lambda_m\mu_1, \ldots, \lambda_m\mu_n$.

 Solution: Suppose that $A\boldsymbol{v}_i = \lambda_i\boldsymbol{v}_i$ and $B\boldsymbol{u}_j = \mu_j\boldsymbol{u}_j$. Then, $(A \otimes B)(\boldsymbol{v}_i \otimes \boldsymbol{u}_j) = (A\boldsymbol{v}_i) \otimes (B\boldsymbol{u}_j) = \lambda_i\mu_j(\boldsymbol{v}_i \times \boldsymbol{u}_j)$.

67. Let $A \in \mathbb{C}^{n\times n}$ and $B \in \mathbb{C}^{m\times m}$. Prove that $trace(A \otimes B) = trace(A)trace(B) = trace(B \otimes A)$ and $\det(A \otimes B) = (\det(A))^m(\det(B))^n = \det(B \otimes A)$.

68. Let $A \in \mathbb{C}^{n\times n}$ and $B \in \mathbb{C}^{m\times m}$ be two matrices. If $\mathsf{spec}(A) = \{\lambda_1, \ldots, \lambda_n\}$ and $\mathsf{spec}(B) = \{\mu_1, \ldots, \mu_m\}$, prove that $\mathsf{spec}(A \oplus B) = \{\lambda_i + \mu_j \mid 1 \leqslant i \leqslant n, 1 \leqslant j \leqslant m\}$ and $\mathsf{spec}(A \ominus B) = \{\lambda_i - \mu_j \mid 1 \leqslant i \leqslant n, 1 \leqslant j \leqslant m\}$.

Solution: Let $\mathbf{x}$ and $\mathbf{y}$ be two eigenvectors of A and B that correspond to the eigenvalues λ and μ, respectively. Since $A \oplus B = (A \otimes I_m) + (I_n \otimes B)$ we have

$$\begin{aligned}(A \oplus B)(\mathbf{x} \otimes \mathbf{y}) &= (A \otimes I_m)(\mathbf{x} \otimes \mathbf{y}) + (I_n \otimes B)(\mathbf{x} \otimes \mathbf{y}) \\ &= (A\mathbf{x} \otimes \mathbf{y}) + (\mathbf{x} \otimes B\mathbf{y}) \\ &= \lambda(\mathbf{x} \otimes \mathbf{y}) + \mu(\mathbf{x} \otimes \mathbf{y}) \\ &= (\lambda + \mu)(\mathbf{x} \otimes \mathbf{y}).\end{aligned}$$

By replacing B by $-B$ we obtain the spectrum of $A \ominus B$.

69. Let $A \in \mathbb{C}^{n\times n}$ be matrix and let $r_i = \sum\{a_{ij} \mid 1 \leqslant j \leqslant n \text{and} j \neq i\}$, for $1 \leqslant i \leqslant n$. Prove that $\mathsf{spec}(A) \subseteq \bigcup_{i=1}^{n}\{z \in \mathbb{C} \mid |z - a_{ii}| \leqslant r_i\}$.

A disk of the form $D_i(A) = \{z \in \mathbb{C} \mid |z - a_{ii}| \leqslant r_i\}$ is called a *Gershgorin disk*.

Solution: Let $\lambda \in \mathsf{spec}(A)$ and let suppose that $A\mathbf{x} = \lambda\mathbf{x}$, where $\mathbf{x} \neq \mathbf{0}$. Let p be such that $|x_p| = \max\{|x_i| \mid 1 \leqslant i \leqslant n\}$. Then, $\sum_{j=1}^{n} a_{pj}x_j = \lambda x_p$, which is the same as $\sum_{j=1, j\neq p}^{n} a_{pj}x_j = (\lambda - a_{pp})x_p$. This, in turn, implies

$$\begin{aligned}|x_p||\lambda - a_{pp}| &= \left|\sum_{j=1,j\neq p}^{n} a_{pj}x_j\right| \leqslant \sum_{j=1,j\neq p}^{n} |a_{pj}||x_j| \\ &\leqslant |x_p| \sum_{j=1,j\neq p}^{n} |a_{pj}| = |x_p|r_p.\end{aligned}$$

Therefore, $|\lambda - a_{pp}| \leqslant r_p$ for some p.

70. If $A, B \in \mathbb{R}^{m\times m}$ is a symmetric matrices and $A \sim B$, prove that $\mathfrak{I}(A) = \mathfrak{I}(B)$.
71. If A is a symmetric block diagonal matrix, $A = \mathsf{diag}(A_1, \ldots, A_k)$, then $\mathfrak{I}(A) = \sum_{i=1}^{k} \mathfrak{I}(A_i)$.
72. Let $A = \begin{pmatrix} B & \mathbf{c} \\ \mathbf{c}' & b \end{pmatrix} \in \mathbb{R}^{m\times m}$ be a symmetric matrix, where $B \in \mathbb{R}^{(m-1)\times(m-1)}$ such that there exists $\boldsymbol{u} \in \mathbb{R}^{m-1}$ for which $B\boldsymbol{u} = \mathbf{0}_{m-1}$ and $\mathbf{c}'\boldsymbol{u} \neq 0$. Prove that $\mathfrak{I}(A) = \mathfrak{I}(B) + (1, 1, -1)$.

Solution: It is clear that $\boldsymbol{u} \neq \mathbf{0}_{m-1}$ and we may assume that $u_1 \neq 0$. We can write

$$\boldsymbol{u} = \begin{pmatrix} u_1 \\ \boldsymbol{v} \end{pmatrix}, B = \begin{pmatrix} b_{11} & \mathbf{d}' \\ \mathbf{d} & D \end{pmatrix}, \text{and } \mathbf{c} = \begin{pmatrix} c_1 \\ \mathbf{e} \end{pmatrix},$$

where $\boldsymbol{v} \in \mathbb{R}^{m-2}$, $D \in \mathbb{R}^{(m-2)\times(m-2)}$, and $\mathbf{d}, \mathbf{e} \in \mathbb{R}^{m-2}$. Define $k = \mathbf{c}'\boldsymbol{u} = c_1u_1 + \mathbf{e}'\boldsymbol{v} \neq 0$ and

$$P = \begin{pmatrix} u_1 & \boldsymbol{v}' & 0 \\ \mathbf{0}_{m-2} & I_{m-2} & \mathbf{0}_{m-2} \\ 0 & \mathbf{0}'_{m-2} & 1 \end{pmatrix}.$$

The equality $B\boldsymbol{u} = \mathbf{0}_{m-1}$ can be written as $b_{11}u_1 + \mathbf{d}'\boldsymbol{v} = 0$ and $\mathbf{d}u_1 + D\boldsymbol{v} = \mathbf{0}_{m-2}$. With these notations, A can be written as

$$A = \begin{pmatrix} b_{11} & \mathbf{d}' & c_1 \\ \mathbf{d} & D & \mathbf{e} \\ \mathbf{c}_1 & \mathbf{e}' & b \end{pmatrix}$$

and we have

$$PAP' = \begin{pmatrix} 0 & \mathbf{0}_{m-2} & k \\ \mathbf{0}'_{m-2} & D & \mathbf{e} \\ k & \mathbf{e}' & b \end{pmatrix}$$

For

$$Q = \begin{pmatrix} 1 & \mathbf{0}' & 0 \\ \frac{1}{k}\mathbf{e} & I_{m-2} & \mathbf{0}_{m-2} \\ -\frac{2}{k}b & \mathbf{0}'_{m-2} & 1 \end{pmatrix}$$

we have

$$(QP)A(QP)' = \begin{pmatrix} 0 & \mathbf{0}'_{m-2} & k \\ \mathbf{0} & D & \mathbf{0}_{m-2} \\ k & \mathbf{0}'_{m-2} & 0 \end{pmatrix}.$$

Let R be the permutation matrix

$$R = \begin{pmatrix} 0 & \mathbf{0}'_{m-2} & 1 \\ 1 & \mathbf{0}'_{m-2} & 0 \\ \mathbf{0}_{m-2} & I_{m-2} & \mathbf{0}_{m-2} \end{pmatrix}.$$

Observe that

$$R(QP)A(QP)'R' = \begin{pmatrix} 0 & k & \mathbf{0}'_{m-2} \\ k & 0 & \mathbf{0}'_{m-2} \\ \mathbf{0}_{m-2} & \mathbf{0}_{m-2} & D \end{pmatrix},$$

which implies that $\mathfrak{I}(A) = \mathfrak{I}(D) + (1, 1, 0)$ because the eigenvalues of the matrix $\begin{pmatrix} 0 & k \\ k & 0 \end{pmatrix}$ are k and $-k$.

On the other hand, if

$$S = \begin{pmatrix} u_1 & \boldsymbol{v} \\ \mathbf{0}_{m-2} & I_{m-2} \end{pmatrix},$$

we have $SDS' = \begin{pmatrix} 0 & \mathbf{0}'_{m-2} \\ \mathbf{0}_{m-2} & D \end{pmatrix}$, which implies $\mathfrak{I}(B) = \mathfrak{I}(D) + (0, 0, 1)$. This yields $\mathfrak{I}(A) = \mathfrak{I}(B) + (1, 1, -1)$.

73. Let $A \in \mathbb{C}^{n \times n}$ and let $U \in \mathbb{C}^{n \times n}$ an unitary matrix and T an upper-triangular matrix $T \in \mathbb{C}^{n \times n}$ such that $A = U^{\mathsf{H}} T U$ whose existence follows by Schur's Triangularition Theorem. Let $B = \frac{1}{2}(A + A^{\mathsf{H}})$ and $C = \frac{1}{2}(A - A^{\mathsf{H}})$ be the matrices introduced in Exercise 18 of Chap. 5.

    Prove that

$$\sum_{i=1}^{n} \Re(\lambda_i)^2 \leqslant \| B \|_F^2 \text{ and} \sum_{i=1}^{n} \Im(\lambda_i)^2 \leqslant \| C \|_F^2 .$$

Bibliographical Comments

The proof of Theorem 7.56 is given in [2]. Hoffman-Wielandt theorem was shown in [3]. Ky Fan's Theorem appeared in [4].

Supplement 23 is a result of Juhasz which appears in [5]; Supplement 11 originated in [6], and Supplement 72 is a result of Fiedler [7].

As before, the two volumes [8] and [9] are highly recommended for a deep understanding of spectral aspects of linear algebra.

References

1. W.H. Haemers, Interlacing values and graphs. Linear Algebra and Applications, **226**, 593–616 (1995)
2. J.H. Wilkerson, *The Algebraic Eigenvalue Problem* (Clarendon Press-Oxford, London, 1965)
3. A.J. Hoffman, H.W. Wielandt, The variation of the spectrum of a normal matrix. Duke Math J **20**, 37–39 (1953)
4. Ky Fan, On a theorem of Weil concerning eigenvalues of linear transformations-I. Proc. Natl. Acad. Sci. **35**, 652–655 (1949)
5. F. Juhász, On the spectrum of a random graph, in *Colloquia Mathematica Societatis János Bolyai*, vol. 25 (Szeged, 1978), pp. 313–316
6. A.V. Knyazev, M.E. Argentati, On proximity of rayleigh quotients for different vectors and ritz values generated by different trial subspaces. Linear Algebra Appl. **415**, 82–95 (2006)
7. M. Fiedler, Eigenvectors of acyclic matrices. Czech. Math. J. **25**, 607–618 (1975)
8. R.A. Horn, C.R. Johnson, *Matrix Analysis* (Cambridge University Press, Cambridge, 1985)
9. R.A. Horn, C.R. Johnson, *Topics in Matrix Analysis* (Cambridge University Press, Cambridge, 2008)

Chapter 8
Metric Spaces Topologies and Measures

8.1 Introduction

The study of topological properties of metric spaces allows us to present an introduction to the dimension theory of these spaces, a topic that is relevant for data mining due to its role in understanding the complexity of searching in data sets that have a natural metric structure.

8.2 Metric Space Topologies

Metrics spaces are naturally equipped with topologies using a mechanism that we describe next.

Theorem 8.1 *Let (S, d) be a metric space. The collection $\mathcal{O}_d$ defined by*

$$\mathcal{O}_d = \{L \in \mathcal{P}(S) \mid \text{ for each } x \in L \text{ there exists } \epsilon > 0 \text{ such that } C_d(x, \epsilon) \subseteq L\}$$

is a topology on the set S.

Proof We have $\emptyset \in \mathcal{O}_d$ because there is no x in $\emptyset$, so the condition of the definition of $\mathcal{O}_d$ is vacuously satisfied. The set S belongs to $\mathcal{O}_d$ because $C_d(x, \epsilon) \subseteq S$ for every $x \in S$ and every positive number ϵ.

If $\{U_i \mid i \in I\} \subseteq \mathcal{O}_d$ and $x \in \bigcup\{U_i \mid i \in I\}$, then $x \in U_j$ for some $j \in I$. Then, there exists $\epsilon > 0$ such that $C(x, \epsilon) \subseteq U_j$ and therefore $C(x, \epsilon) \subseteq \bigcup\{U_i \mid i \in I\}$. Thus, $\bigcup\{U_i \mid i \in I\} \in \mathcal{O}_d$.

Finally, let $U, V \in \mathcal{O}_d$ and let $x \in U \cap V$. Since $U \in \mathcal{O}_d$, there exists $\epsilon > 0$ such that $C(x, \epsilon) \subseteq U$. Similarly, there exists ϵ' such that $C(x, \epsilon') \subseteq V$. If $\epsilon_1 = \min\{\epsilon, \epsilon'\}$, then $C(x, \epsilon_1) \subseteq C(x, \epsilon) \cap C(x, \epsilon') \subseteq U \cap V$, so $U \cap V \in \mathcal{O}_d$. This concludes the argument.

D. A. Simovici and C. Djeraba, *Mathematical Tools for Data Mining*,
Advanced Information and Knowledge Processing, DOI: 10.1007/978-1-4471-6407-4_8,

Theorem 8.1 justifies the following definition.

Definition 8.2 *Let d be a metric on a set S. The* topology induced by d *is the family of sets $\mathcal{O}_d$.*

We refer to the pair $(S, \mathcal{O}_d)$ as a *topological metric space.*

Example 8.3 The usual topology of the set of real numbers $\mathbb{R}$ introduced in Example 4.4 is actually induced by the metric $d : \mathbb{R} \times \mathbb{R} \longrightarrow \mathbb{R}_{\geqslant 0}$ given by $d(x, y) = |x - y|$ for $x, y \in \mathbb{R}$. Recall that, by Theorem 4.4, every open set of this space is the union of a countable set of disjoint open intervals.

The next statement explains the terms "open sphere" and "closed sphere", which we have used previously.

Theorem 8.4 *Let $(S, \mathcal{O}_d)$ be a topological metric space. If $t \in S$ and $r > 0$, then any open sphere $C(t, r)$ is an open set and any closed sphere $B(t, r)$ is a closed set in the topological space $(S, \mathcal{O}_d)$.*

Proof Let $x \in C(t, r)$, so $d(t, x) < r$. Choose ϵ such that $\epsilon < r - d(t, x)$. We claim that $C(x, \epsilon) \subseteq C(t, r)$. Indeed, let $z \in C(x, \epsilon)$. We have $d(x, z) < \epsilon < r - d(t, x)$. Therefore, $d(z, t) \leqslant d(z, x) + d(x, t) < r$, so $z \in C(t, r)$, which implies $C(x, \epsilon) \subseteq C(t, r)$. We conclude that $C(t, r)$ is an open set.

To show that the closed sphere $B(t, r)$ is a closed set, we will prove that its complement $S - B(t, r) = \{u \in S \mid d(u, t) > r\}$ is an open set. Let $v \in S - B(t, r)$. Now choose ϵ such that $\epsilon < d(v, t) - r$. It is easy to see that $C(v, \epsilon) \subseteq S - B(t, r)$, which proves that $S - B(t, r)$ is an open set.

Corollary 8.5 *The collection of all open spheres in a topological metric space $(S, \mathcal{O}_d)$ is a basis.*

Proof This statement follows immediately from Theorem 8.4.

The definition of open sets in a topological metric space implies that a subset L of a topological metric space $(S, \mathcal{O}_d)$ is closed if and only if for every $x \in S$ such that $x \notin L$ there is $\epsilon > 0$ such that $C(x, \epsilon)$ is disjoint from L. Thus, if $C(x, \epsilon) \cap L \neq \emptyset$ for every $\epsilon > 0$ and L is a closed set, then $x \in L$.

The closure and the interior operators $\mathbf{K}_{\mathcal{O}_d}$ and $\mathbf{I}_{\mathcal{O}_d}$ in a topological metric space $(S, \mathcal{O}_d)$ are described next.

Theorem 8.6 *In a topological metric space $(S, \mathcal{O}_d)$, we have*

$$\mathbf{K}_{\mathcal{O}_d}(U) = \{x \in S \mid C(x, \epsilon) \cap U \neq \emptyset \text{ for every } \epsilon > 0\}$$

and

$$\mathbf{I}_{\mathcal{O}_d}(U) = \{x \in S \mid C(x, \epsilon) \subseteq U \text{ for some } \epsilon > 0\}$$

for every $U \in \mathcal{P}(S)$.

Proof Let $\mathbf{K} = \mathbf{K}_{\mathcal{O}_d}$. If $C(x, \epsilon) \cap U \neq \emptyset$ for every $\epsilon > 0$, then clearly $C(x, \epsilon) \cap \mathbf{K}(U) \neq \emptyset$ for every $\epsilon > 0$ and therefore $x \in \mathbf{K}(U)$ by a previous observation.

Now let $x \in \mathbf{K}(U)$ and let $\epsilon > 0$. Suppose that $C(x, \epsilon) \cap U = \emptyset$. Then, $U \subseteq S - C(x, \epsilon)$ and $S - C(x, \epsilon)$ is a closed set. Therefore, $\mathbf{K}(U) \subseteq S - C(x, \epsilon)$. This is a contradiction because $x \in \mathbf{K}(U)$ and $x \notin S - C(x, \epsilon)$.

The second part of the theorem follows from the first part and from Corollary 4.27.

If the metric topology $\mathcal{O}_d$ is clear from the context, then we will denote the closure operator $\mathbf{K}_{\mathcal{O}_d}$ simply by $\mathbf{K}$.

Corollary 8.7 *The subset U of the topological metric space $(S, \mathcal{O}_d)$ is closed if and only if $C(x, \epsilon) \cap U \neq \emptyset$ for every $\epsilon > 0$ implies $x \in U$.*

The border ∂U is given by

$$\partial L = \{x \in S \mid \textit{for every } \epsilon > 0, C(x, \epsilon) \cap L \neq \emptyset, \textit{ and } C(x, \epsilon) \cap (S - L) \neq \emptyset\}.$$

Proof This corollary follows immediately from Theorem 8.6.

Theorem 8.8 *Let T be a subset of a topological metric space $(S, \mathcal{O}_d)$. We have $diam(T) = diam(\mathbf{K}(T))$.*

Proof Since $T \subseteq \mathbf{K}(T)$, it follows immediately that $diam(T) \leqslant diam(\mathbf{K}(T))$, so we have to prove only the reverse inequality.

Let $u, v \in \mathbf{K}(T)$. For every positive number ϵ, we have $C(u, \epsilon) \cap T \neq \emptyset$ and $C(v, \epsilon) \cap T \neq \emptyset$. Thus, there exists $x, y \in T$ such that $d(u, x) < \epsilon$ and $d(v, y) < \epsilon$. Thus, $d(u, v) \leqslant d(u, x) + d(x, y) + d(y, v) \leqslant 2\epsilon + diam(T)$ for every ϵ, which implies $d(u, v) \leqslant diam(T)$ for every $u, v \in \mathbf{K}(T)$. This yields $diam(\mathbf{K}(T)) \leqslant diam(T)$.

A metric topology can be defined, as we shall see, by more than one metric.

Definition 8.9 *Two metrics d and d' defined on a set S are* topologically equivalent *if the topologies $\mathcal{O}_d$ and $\mathcal{O}_{d'}$ are equal.*

Example 8.10 Let d and d' be two metrics defined on a set S. If there exist two numbers $a, b \in \mathbb{R}_{>0}$ such that $a\, d(x, y) \leqslant d'(x, y) \leqslant b\, d(x, y)$, for $x, y \in S$, then $\mathcal{O}_d = \mathcal{O}'_d$.

Let $C_d(x, r)$ be an open sphere centered in x, defined by d. The previous inequalities imply

$$C_d\left(\frac{r}{b}\right) \subseteq C_{d'}(x, r) \subseteq C_d\left(x, \frac{r}{a}\right).$$

Let $L \in \mathcal{O}_d$. By Definition 8.2, for each $x \in L$ there exists $\epsilon > 0$ such that $C_d(x, \epsilon) \subseteq L$. Then, $C'_d(x, a\epsilon) \subseteq C_d(x, \epsilon) \subseteq L$, which implies $L \in \mathcal{O}_{d'}$. We leave it to the reader to prove the reverse inclusion $\mathcal{O}_{d'} \subseteq \mathcal{O}_d$.

By Corollary 6.17, any two Minkowski metrics d_p and d_q on $\mathbb{R}^n$ are topologically equivalent.

8.3 Continuous Functions in Metric Spaces

Continuous functions between topological spaces were introduced in Definition 4.63. Next, we give a characterization of continuous functions between topological metric spaces.

Theorem 8.11 *Let $(S, \mathcal{O}_d)$ and $(T, \mathcal{O}_e)$ be two topological metric spaces. The following statements concerning a function $f : S \longrightarrow T$ are equivalent:*

(i) *f is a continuous function;*
(ii) *for $x \in S$ and $\epsilon > 0$, there exists $\delta > 0$ such that $f(C_d(x, \delta)) \subseteq C_e(f(x), \epsilon)$.*

Proof (i) implies (ii): Suppose that f is a continuous function. Since $C_e(f(x), \epsilon)$ is an open set in $(T, \mathcal{O}_e)$, the set $f^{-1}(C_e(f(x), \epsilon)$ is an open set in $(S, \mathcal{O}_d)$. Clearly, $x \in f^{-1}(C_e(f(x), \epsilon))$, so by the definition of the metric topology there exists $\delta > 0$ such that $C_d(x, \delta) \subseteq f^{-1}(C_e(f(x), \epsilon)$, which yields $f(C_d(x, \delta)) \subseteq C_e(f(x), \epsilon)$.

(ii) implies (i): Let V be an open set of $(T, \mathcal{O}_e)$. If $f^{-1}(V)$ is empty, then it is clearly open. Therefore, we may assume that $f^{-1}(V)$ is not empty. Let $x \in f^{1-}(V)$. Since $f(x) \in V$ and V is open, there exists $\epsilon > 0$ such that $C_e(f(x), \epsilon) \subseteq V$. By Part (ii) of the theorem, there exists $\delta > 0$ such that $f(C_d(x, \delta)) \subseteq C_e(f(x), \epsilon)$, which implies $x \in C_d(x, \delta) \subseteq f^{-1}(V)$. This means that $f^{-1}(V)$ is open, so f is continuous.

In general, for a continuous function f from $(S, \mathcal{O}_d)$ and $(T, \mathcal{O}_e)$, the number δ depends both on x and on ϵ. If δ is dependent only on ϵ, then we say that f is *uniformly continuous*. Thus, f is uniformly continuous if for every $\epsilon > 0$ there exists δ such that $d(u, v) < \delta$ implies $e(f(u), f(v)) < \epsilon$.

Example 8.12 The function $f : \mathbb{R} \longrightarrow \mathbb{R}$ defined by $f(u) = u^2$ for $u \in \mathbb{R}$ is continuous but not uniformly continuous. Indeed, suppose that f would be uniformly continuous and let ϵ be a positive number such that $|f(u)-f(v)| = |(u-v)(u+v)| < \epsilon$. This implies $|u - v| < \frac{\epsilon}{|u+v|}$ and, if u and v are sufficiently large, δ must be arbitrarily close to 0 to ensure that $|u - v| < \delta$ implies $|f(u) - f(v)| < \epsilon$.

Theorem 8.13 *Let $(S, \mathcal{O}_d)$ and $(T, \mathcal{O}_e)$ be two topological metric spaces, let $f : S \longrightarrow T$ be a function, and let $\mathbf{u} = (u_0, u_1, \ldots)$ and $\mathbf{v} = (v_0, v_1, \ldots)$ in $\mathbf{Seq}_\infty(S)$. The following statements are equivalent:*

(i) *f is uniformly continuous;*
(ii) *if $\lim_{n\to\infty} d(u_n, v_n) = 0$, then $\lim_{n\to\infty} e(f(u_n), f(v_n)) = 0$;*
(iii) *if $\lim_{n\to\infty} d(u_n, v_n) = 0$, we have $\lim_{k\to\infty} e(f(u_{n_k}), f(v_{n_k})) = 0$, where $(u_{n_0}, u_{n_1}, \ldots)$ and $(v_{n_0}, v_{n_1}, \ldots)$ are two arbitrary subsequences of $\mathbf{u}$ and $\mathbf{v}$, respectively.*

Proof (i) implies (ii): For $\epsilon > 0$, there exists δ such that $d(u, v) < \delta$ implies $e(f(u), f(v)) < \epsilon$. Therefore, if $\mathbf{u}$ and $\mathbf{v}$ are sequences as above, there exists n_δ such that $n > n_\delta$ implies $d(u_n, v_n) < \delta$, so $e(f(u_n), f(v_n)) < \epsilon$. Thus, $\lim_{n\to\infty} e(f(u_n), f(v_n)) = 0$.

(ii) implies (iii): This implication is obvious.

(iii) implies (i): Suppose that f satisfies (iii) but is not uniformly continuous. Then, there exists $\epsilon > 0$ such that for every $\delta > 0$ there exist $u, v \in X$ such that $d(u, v) < \delta$ and $e(f(u), f(v)) > \epsilon$. Let u_n, v_n be such that $d(u_n, v_n) < \frac{1}{n}$ for $n \geqslant 1$. Then, $\lim_{n\to\infty} d(u_n, v_n) = 0$ but $e(f(u_n), f(v_n))$ does not converge to 0.

Example 8.14 The function $f : \mathbb{R} \longrightarrow \mathbb{R}$ given by $f(x) = x \sin x$ is continuous but not uniformly continuous. Indeed, let $u_n = n\pi$ and $v_n = n\pi + \frac{1}{n}$. Note that $\lim_{n\to\infty} |u_n - v_n| = 0$, $f(u_n) = 0$, and $f(v_n) = (n\pi + \frac{1}{n})\sin(n\pi + \frac{1}{n}) = (n\pi + \frac{1}{n})(-1)^n \sin\frac{1}{n}$. Therefore,

$$\lim_{n\to\infty} |f(u_n) - f(v_n)| = \lim_{n\to\infty} \left(n\pi + \frac{1}{n}\right) \sin\frac{1}{n} = \pi \lim_{n\to\infty} \frac{n}{\sin\frac{1}{n}} = \pi,$$

so f is not uniformly continuous.

A local continuity property is introduced next.

Definition 8.15 *Let $(S, \mathcal{O}_d)$ and $(T, \mathcal{O}_e)$ be two topological metric spaces and let $x \in S$.*

A function $f : S \longrightarrow T$ is continuous in x *if for every $\epsilon > 0$ there exists $\delta > 0$ such that $f(C(x, \delta)) \subseteq C(f(x), \epsilon)$.*

It is clear that f is continuous if it is continuous in every $x \in S$.

The definition can be restated by saying that f is continuous in x if for every $\epsilon > 0$ there is $\delta > 0$ such that $d(x, y) < \delta$ implies $e(f(x), f(y)) < \epsilon$.

Definition 8.16 *Let (S, d) be a metric space and let U be a subset of S. The* distance from an element x to U *is the number*

$$d(x, U) = \inf\{d(x, u) \mid u \in U\}.$$

Note that if $x \in U$, then $d(x, U) = 0$.

Theorem 8.17 *Let (S, d) be a metric space and let U be a subset of S. The function $f : S \longrightarrow \mathbb{R}$ given by $f(x) = d(x, U)$ for $x \in S$ is continuous.*

Proof Since $d(x, z) \leq d(x, y) + d(y, z)$, we have $d(x, U) \leqslant d(x, y) + d(y, U)$. By exchanging x and y we also have $d(y, U) \leqslant d(x, y) + d(x, U)$ and, together, these inequalities yield $|d(x, U) - d(y, U)| \leqslant d(x, y)$. Therefore, if $d(x, y) < \epsilon$, it follows that $|d(x, U) - d(y, U)| \leqslant \epsilon$, which implies the continuity of $d(x, U)$.

Theorem 8.18 *Let (S, d) be a metric space. The following statements hold:*

(i) *$d(u, V) = 0$ if and only if $u \in \mathbf{K}(V)$, and*
(ii) *$d(u, V) = d(u, \mathbf{K}(V))$*

for every $u, u' \in S$ and $V \subseteq S$.

Proof Suppose that $d(u, V) = 0$. Again, by the definition of $d(u, V)$, for every $\epsilon > 0$ there exists $v \in V$ such that $d(u, v) < \epsilon$, which means that $C(u, \epsilon) \cap V \neq \emptyset$. By Theorem 8.6, we have $u \in \mathbf{K}(V)$. The converse implication is immediate, so (i) holds.

To prove (ii), observe that $V \subseteq \mathbf{K}(V)$ implies that $d(u, \mathbf{K}(V)) \leqslant d(u, V)$, so we need to show only the reverse inequality.

Let w be an arbitrary element of $\mathbf{K}(V)$. By Theorem 8.6, for every $\epsilon > 0$, $C(w, \epsilon) \cap V \neq \emptyset$. Let $v \in C(w, \epsilon) \cap V$. We have

$$d(u, v) \leqslant d(u, w) + d(w, v) \leqslant d(u, w) + \epsilon,$$

so $d(u, V) \leqslant d(u, w) + \epsilon$. Since this inequality holds for every ϵ, $d(u, V) \leqslant d(u, w)$ for every $w \in \mathbf{K}(V)$, so $d(u, V) \leqslant d(u, \mathbf{K}(V))$. This allows us to conclude that $d(u, V) = d(u, \mathbf{K}(V))$.

Theorem 8.18 can be restated using the function $d_U : S \longrightarrow \mathbb{R}_{\geqslant 0}$ defined by $d_U(x) = d(x, V)$ for $u \in S$. Thus, for every subset U of S and $x, y \in S$, we have $|d_U(x) - d_U(y)| \leqslant d(x, y)$, $d_U(x) = 0$ if and only if $x \in \mathbf{K}(U)$, and $d_U = d_{\mathbf{K}(V)}$. The function d_U is continuous.

8.4 Separation Properties of Metric Spaces

A dissimilarity $d : S \times S \longrightarrow \hat{\mathbb{R}}_{\geqslant 0}$ can be extended to the set of subsets of S by defining $d(U, V)$ as

$$d(U, V) = \inf\{d(u, v) \mid u \in U \text{and } v \in V\}$$

for $U, V \in \mathcal{P}(S)$. The resulting extension is also a dissimilarity. However, even if d is a metric, then its extension is not, in general, a metric on $\mathcal{P}(S)$ because it does not satisfy the triangular inequality. Instead, we prove that if d is a metric, then for every U, V, W we have

$$d(U, W) \leqslant d(U, V) + \mathit{diam}(V) + d(V, W).$$

Indeed, by the definition of $d(U, V)$ and $d(V, W)$, for every $\epsilon > 0$, there exist $u \in U$, $v, v' \in V$, and $w \in W$ such that

$$\begin{aligned} d(U, V) &\leqslant d(u, v) \leqslant d(U, V) + \tfrac{\epsilon}{2}, \\ d(V, W) &\leqslant d(v', w) \leqslant d(V, W) + \tfrac{\epsilon}{2}. \end{aligned}$$

By the triangular axiom, we have $d(u, w) \leqslant d(u, v) + d(v, v') + d(v', w)$. Hence, $d(u, w) \leqslant d(U, V) + \mathit{diam}(V) + d(V, W) + \epsilon$, which implies $d(U, W) \leqslant d(U, V) + \mathit{diam}(V) + d(V, W) + \epsilon$ for every $\epsilon > 0$. This yields the needed inequality.

Definition 8.19 *Let (S, d) be a metric space. The sets $U, V \in \mathcal{P}(S)$ are* separate *if $d(U, V) > 0$.*

The notions of an open sphere and a closed sphere in a metric space (S, d) are extended by defining the sets $C(T, r)$ and $B(T, r)$ as

$$C(T, r) = \{u \in S \mid d(u, T) < r\},$$
$$B(T, r) = \{u \in S \mid d(u, T) \leqslant r\},$$

for $T \in \mathcal{P}(S)$ and $r \geqslant 0$, respectively.

The next statement is a generalization of Theorem 8.4.

Theorem 8.20 *Let $(S, \mathcal{O}_d)$ be a topological metric space. For every set T, $T \subseteq S$, and every $r > 0$, $C(T, r)$ is an open set and $B(T, r)$ is a closed set in $(S, \mathcal{O}_d)$.*

Proof Let $u \in C(T, r)$. We have $d(u, T) < r$, or, equivalently, $\inf\{d(u, t) \mid t \in T\} < r$. We claim that if ϵ is a positive number such that $\epsilon < \frac{r}{2}$, then $C(u, \epsilon) \subseteq C(T, r)$.

Let $z \in C(u, \epsilon)$. For every $v \in T$, we have $d(z, v) \leqslant d(z, u) + d(u, v) < \epsilon + d(u, v)$. From the definition of $d(u, T)$ as an infimum, it follows that there exists $v' \in T$ such that $d(u, v') < d(u, V) + \frac{\epsilon}{2}$, so $d(z, v') < d(u, T) + \epsilon < r + \epsilon$. Since this inequality holds for every $\epsilon > 0$, it follows that $d(z, v') < r$, so $d(z, T) < r$, which proves that $C(u, \epsilon) \subseteq C(T, r)$. Thus, $C(T, r)$ is an open set.

Suppose now that $s \in \mathbf{K}(B(T, r))$. By Part (ii) of Theorem 8.18, we have $d(s, B(T, r)) = 0$, so $\inf\{d(s, w) \mid w \in B(T, r)\} = 0$. Therefore, for every $\epsilon > 0$, there is $w \in B(T, r)$ such that $d(s, w) < \epsilon$. Since $d(w, T) \leqslant r$, it follows from the first part of Theorem 8.18 that $|d(s, T) - d(w, T)| \leqslant d(s, w) < \epsilon$ for every $\epsilon > 0$. This implies $d(s, T) = d(w, T)$, so $s \in B(T, r)$. This allows us to conclude that $B(T, r)$ is indeed a closed set.

Theorem 8.21 (Lebesgue's Lemma) *Let $(S, \mathcal{O}_d)$ be a topological metric space that is compact and let $\mathcal{C}$ be an open cover of this space. There exists $r \in \mathbb{R}_{>0}$ such that for every subset U with $diam(U) < r$ there is a set $L \in \mathcal{C}$ such that $U \subseteq L$.*

Proof Suppose that the statement is not true. Then, for every $k \in \mathbb{P}$, there exists a subset U_k of S such that $diam(U_k) < \frac{1}{k}$ and U_k is not included in any of the sets L of $\mathcal{C}$. Since $(S, \mathcal{O}_d)$ is compact, there exists a finite subcover $\{L_1, \ldots, L_p\}$ of $\mathcal{C}$.

Let x_{ik} be an element in $U_k - L_i$. For every two points x_{ik}, x_{jk}, we have $d(x_{ik}, x_{jk}) \leqslant \frac{1}{k}$ because both belong to the same set U_k. By Theorem 8.61, the compactness of S implies that any sequence $\mathbf{x}_i = (x_{i1}, x_{i2}, \ldots)$ contains a convergent subsequence. Denote by x_i the limit of this subsequence, where $1 \leqslant i \leqslant p$. The inequality $d(x_{ik}, x_{jk}) \leqslant \frac{1}{k}$ for $k \geqslant 1$ implies that $d(x_i, x_j) = 0$ so $x_i = x_j$ for $1 \leqslant i, j \leqslant p$. Let x be their common value. Then x does not belong to any of the sets L_i, which contradicts the fact that $\{L_1, \ldots, L_p\}$ is an open cover.

Theorem 8.22 *Every topological metric space $(S, \mathcal{O}_d)$ is a Hausdorff space.*

Proof Let x and y be two distinct elements of S, so $d(x, y) > 0$. Choose $\epsilon = \frac{d(x,y)}{3}$. It is clear that for the open spheres $C(x, \epsilon)$ and $C(y, \epsilon)$, we have $x \in C(x, \epsilon)$, $y \in C(y, \epsilon)$, and $C(x, \epsilon) \cap C(y, \epsilon) = \emptyset$, so $(S, \mathcal{O}_d)$ is indeed a Hausdorff space.

Corollary 8.23 *Every compact subset of a topological metric space is closed.*

Proof This follows directly from Theorems 8.22 and 4.90.

Corollary 8.24 *If S is a finite set and d is a metric on S, then the topology $\mathcal{O}_d$ is the discrete topology.*

Proof Let $S = \{x_1, \ldots, x_n\}$ be a finite set. We saw that every singleton $\{x_i\}$ is a closed set. Therefore, every subset of S is closed as a finite union of closed sets.

Theorem 8.25 *Every topological metric space $(S, \mathcal{O}_d)$ is a T_4 space.*

Proof We need to prove that for all disjoint closed sets H_1 and H_2 of S there exist two open disjoint sets V_1 and V_2 such that $H_1 \subseteq V_1$ and $H_2 \subseteq V_2$.

Let $x \in H_1$. Since $H_1 \cap H_2 = \emptyset$, it follows that $x \notin H_2 = \mathbf{K}(H_2)$, so $d(x, H_2) > 0$ by Part (ii) of Theorem 8.18. By Theorem 8.20, the set $C\left(H_1, \frac{d(x,L)}{3}\right)$ is an open set and so is

$$Q_H = \bigcup \left\{ C\left(H_1, \frac{d(x, L)}{3}\right) \mid x \in H_1 \right\}.$$

The open set Q_{H_2} is defined in a similar manner as

$$Q_{H_2} = \bigcup \left\{ C\left(H_2, \frac{d(y, H_1)}{3}\right) \mid y \in H_2 \right\}.$$

The sets Q_{H_1} and Q_{H_2} are disjoint because $t \in Q_{H_1} \cap Q_{H_2}$ implies that there is $x_1 \in H_1$ and $x_2 \in H_2$ such that $d(t, x_1) < \frac{d(x_1, H_2)}{3}$ and $d(t, x_2) < \frac{d(x_2, H_1)}{3}$. This, in turn, would imply

$$d(x_1, x_2) < \frac{d(x_1, H_2) + d(x_2, H_1)}{3} \leqslant \frac{2}{3} d(x_1, x_2),$$

which is a contradiction. Therefore, $(S, \mathcal{O}_d)$ is a T_4 topological space.

Corollary 8.26 *Every metric space is normal.*

Proof By Theorem 8.22, a metric space is a T_2 space and therefore a T_1 space. The statement then follows directly from Theorem 8.25.

Corollary 8.27 *Let H be a closed set and L be an open set in a topological metric space $(S, \mathcal{O}_d)$ such that $H \subseteq L$. Then, there is an open set V such that $H \subseteq V \subseteq \mathbf{K}(V) \subseteq L$.*

Proof The closed sets H and $S - L$ are disjoint. Therefore, since $(S, \mathcal{O})$ is normal, there exist two disjoint open sets V and W such that $H \subseteq V$ and $S - L \subseteq W$. Since $S - W$ is closed and $V \subseteq S - W$, it follows that $\mathbf{K}(V) \subseteq S - W \subseteq L$. Thus, we obtain $H \subseteq V \subseteq \mathbf{K}(V) \subseteq L$.

A stronger form of Theorem 8.25, where the disjointness of the open sets is replaced by the disjointness of their closures, is given next.

Theorem 8.28 *Let $(S, \mathcal{O}_d)$ be a metric space. For all disjoint closed sets H_1 and H_2 of S, there exist two open sets V_1 and V_2 such that $H_1 \subseteq V_1$, $H_2 \subseteq V_2$, and $\mathbf{K}(V_1) \cap \mathbf{K}(V_2) = \emptyset$.*

Proof By Theorem 8.25, we obtain the existence of the disjoint open sets Q_{H_1} and Q_{H_2} such that $H_1 \subseteq Q_{H_1}$ and $H_2 \subseteq Q_{H_2}$. We claim that the closures of these sets are disjoint.

Suppose that $s \in \mathbf{K}(Q_{H_1}) \cap \mathbf{K}(Q_{H_2})$. Then, we have $C\left(s, \frac{\epsilon}{12}\right) \cap Q_{H_1} \neq \emptyset$ and $C\left(s, \frac{\epsilon}{12}\right) \cap Q_{H_2} \neq \emptyset$. Thus, there exist $t \in Q_{H_1}$ and $t' \in Q_{H_2}$ such that $d(t, s) < \frac{\epsilon}{12}$ and $d(t', s) < \frac{\epsilon}{12}$.

As in the proof of the previous theorem, there is $x_1 \in H_1$ and $y_1 \in H_2$ such that $d(t, x_1) < \frac{d(x_1, H_2)}{3}$ and $d(t', y_1) < \frac{d(y_1, H_1)}{3}$. Choose t and t' above for $\epsilon = d(x_1, y_1)$. This leads to a contradiction because

$$d(x_1, y_1) \leqslant d(x_1, t) + d(t, s) + d(s, t') + d(t', y_1) \leqslant \frac{5}{6} d(x_1, y_1).$$

Corollary 8.29 *Let $(S, \mathcal{O}_d)$ be a metric space. If $x \in L$, where L is an open subset of S, then there exists two open sets V_1 and V_2 in S such that $x \in V_1$, $S - L \subseteq V_2$, and $\mathbf{K}(V_1) \cap \mathbf{K}(V_2) = \emptyset$.*

Proof The statement follows by applying Theorem 8.28 to the disjoint closed sets $H_1 = \{x\}$ and $H_2 = S - L$.

Recall that the Bolzano-Weierstrass property of topological spaces was introduced in Theorem 4.62. Namely, a topological space $(S, \mathcal{O})$ has the Bolzano-Weierstrass property if every infinite subset T of S has at least one accumulation point. For metric spaces, this property is equivalent to compactness, as we show next.

Theorem 8.30 *Let $(S, \mathcal{O}_d)$ be a topological metric space. The following three statements are equivalent:*

(i) *$(S, \mathcal{O}_d)$ is compact.*
(ii) *$(S, \mathcal{O}_d)$ has the Bolzano-Weierstrass property.*
(iii) *Every countable open cover of $(S, \mathcal{O}_d)$ contains a finite subcover.*

Proof (i) implies (ii): by Theorem 4.62.

(ii) implies (iii): Let $\{L_n \mid n \in \mathbb{N}\}$ be a countable open cover of S. Without loss of generality, we may assume that none of the sets L_n is included in $\bigcup_{p=1}^{n-1} L_p$; indeed, if this is not the case, we can discard L_n and still have a countable open cover.

Let $x_n \in L_n - \bigcup_{p=1}^{n-1} L_p$ and let $U = \{x_n \mid n \in \mathbb{N}\}$. Since $(S, \mathcal{O}_d)$ has the Bolzano-Weierstrass property, we have $U' \neq \emptyset$, so there exists an accumulation point z of U. In every open set L that contains z, there exists $x_n \in U$ such that $x_n \neq z$.

Since $\{L_n \mid n \in \mathbb{N}\}$ is an open cover, there exists L_m such that $z \in L_m$. Suppose that the set L_m contains only a finite number of elements $x_{n_1}, \ldots, x_{n_k}$, and let $d = \min\{d(z, x_{n_i}) \mid 1 \leqslant i \leqslant k\}$. Then, $L_m \cap C\left(z, \frac{d}{2}\right)$ is an open set that contains no elements of U with the possible exception of z, which contradicts the fact that z is an accumulation point. Thus, L_m contains an infinite subset of U, which implies that there exists $x_q \in L_m$ for some $q > m$. This contradicts the definition of the elements x_n of U. We conclude that there exists a number r_0 such that $L_r - \bigcup_{i=0}^{r-1} L_i = \emptyset$ for $r \geqslant r_0$, so $S = L_0 \cup \cdots \cup L_{r_0-1}$, which proves that $L_0, \ldots, L_{r_0-1}$ is a finite subcover.

(iii) implies (i). Let ϵ be a positive number. Suppose that there is an infinite sequence $\mathbf{x} = (x_0, \ldots, x_n, \ldots)$ such that $d(x_i, x_j) > \epsilon$ for every $i, j \in \mathbb{N}$ such that $i \neq j$. Consider the open spheres $C(x_i, \epsilon)$ and the set

$$C = S - \mathbf{K}\left(\bigcup_{i\in\mathbb{N}} C\left(x_i, \frac{\epsilon}{2}\right)\right).$$

We will show that $\{C\} \cup \{C(x_i, \epsilon) \mid i \in \mathbb{N}\}$ is a countable open cover of S.

Suppose that $x \in S - C$; that is $x \in \mathbf{K}\left(\bigcup_{i\in\mathbb{N}} C\left(x_i, \frac{\epsilon}{2}\right)\right)$. By Theorem 4.38, we have either that $x \in \bigcup_{i\in\mathbb{N}} C\left(x_i, \frac{\epsilon}{2}\right)$ or x is an accumulation point of that set.

In the first case, $x \in \bigcup_{i\in\mathbb{N}} C(x_i, \epsilon)$ because $C\left(x_i, \frac{\epsilon}{2}\right) \subseteq C(x_i, \epsilon)$. If x is an accumulation point of $\bigcup_{i\in\mathbb{N}} C\left(x_i, \frac{\epsilon}{2}\right)$, given any open set L such that $x \in L$, then L must intersect at least one of the spheres $C\left(x_i, \frac{\epsilon}{2}\right)$. Suppose that $C\left(x, \frac{\epsilon}{2}\right) \cap C\left(x_i, \frac{\epsilon}{2}\right) \neq \emptyset$, and let t be a point that belongs to this intersection. Then, $d(x, x_i) < d(x, t) + d(t, x_i) < \frac{\epsilon}{2} + \frac{\epsilon}{2} = \epsilon$, so $x \in C(x_i, \epsilon)$.

Therefore, $\{C\} \cup \{C(x_i, \epsilon) \mid i \in \mathbb{N}\}$ is a countable open cover of S. Since every countable open cover of $(S, \mathcal{O}_d)$ contains a finite subcover, it follows that this open cover contains a finite subcover. Observe that there exists an open sphere $C(x_i, \epsilon)$ that contains infinitely many x_n because none of these elements belongs to C. Consequently, for any two of these points, the distance is less than ϵ, which contradicts the assumption we made initially about the sequence $\mathbf{x}$.

Choose $\epsilon = \frac{1}{k}$ for some $k \in \mathbb{N}$ such that $k \geq 1$. Since there is no infinite sequence of points such that every two distinct points are at a distance greater than $\frac{1}{k}$, it is possible to find a finite sequence of points $\mathbf{x} = (x_0, \ldots, x_{n-1})$ such that $i \neq j$ implies $d(x_i, x_j) > \frac{1}{k}$ for $0 \leqslant i, j \leqslant n-1$ and for every other point $x \in S$ there exists x_i such that $d(x_i, x) \leqslant \frac{1}{k}$.

Define the set $L_{k,m,i}$ as the open sphere $C\left(x_i, \frac{1}{m}\right)$, where x_i is one of the points that belongs to the sequence above determined by k and $m \in \mathbb{N}$ and $m \geqslant 1$. The collection $\{L_{k,m,i} \mid m \geqslant 1, 0 \leqslant i \leqslant n-1\}$ is clearly countable. We will prove that each open set of $(S, \mathcal{O}_d)$ is a union of sets of the form $L_{k,m,i}$; in other words, we will show that this family of sets is a basis for $(S, \mathcal{O}_d)$.

Let L be an open set and let $z \in L$. Since L is open, there exists $\epsilon > 0$ such that $z \in C(z, \epsilon) \subseteq L$. Choose k and m such that $\frac{1}{k} < \frac{1}{m} < \frac{\epsilon}{2}$. By the definition of the sequence $\mathbf{x}$, there is x_i such that $d(z, x_i) < \frac{1}{k}$. We claim that

$$L_{k,m,i} = C\left(x_i, \frac{1}{m}\right) \subseteq L.$$

Let $y \in L_{k,m,i}$. Since $d(z, y) \leqslant d(z, x_i) + d(x_i, y) < \frac{1}{k} + \frac{1}{m} < \epsilon$, it follows that $L_{k,m,i} \subseteq C(z, \epsilon) \subseteq L$. Since $d(y, z) < \frac{1}{k} < \frac{1}{m}$, we have $z \in L_{k,m,i}$. This shows that L is a union of sets of the form $L_{k,m,i}$, so this family of sets is a countable open cover of S. It follows that there exists a finite open cover of $(S, \mathcal{O}_d)$ because every countable open cover of $(S, \mathcal{O}_d)$ contains a finite subcover.

Theorem 8.31 *Let d and d' be two metrics on a set S such that there exist $c_0, c_1 \in \mathbb{R}_{>0}$ for which $c_0 d(x, y) \leqslant d'(x, y) \leqslant c_1 d(x, y)$ for every $x, y \in S$. Then, the topologies $\mathcal{O}_d$ and $\mathcal{O}_{d'}$ coincide.*

Proof Suppose that $L \in \mathcal{O}_d$, and let $x \in L$. There exists $\epsilon > 0$ such that $C_d(x, \epsilon) \subseteq L$. Note that $C_{d'}(x, c_1\epsilon) \subseteq C_d(x, \epsilon)$. Thus, $C_{d'}(x, \epsilon') \subseteq L$, where $\epsilon' = c_1\epsilon$, which shows that $L \in \mathcal{O}_{d'}$. In a similar manner, one can prove that $\mathcal{O}_{d'} \subseteq \mathcal{O}_d$, so the two topologies are equal.

If d and d' are two metrics on a set S such that $\mathcal{O}_d = \mathcal{O}_{d'}$, we say that d and d' are topologically equivalent. Corollary 6.20 implies that all metrics d_p on $\mathbb{R}^n$ with $p \geqslant 1$ are topologically equivalent.

In Sect. 4.2, we saw that if a topological space has a countable basis, then the space is separable (Theorem 4.47) and each open cover of the basis contains a countable subcover (Corollary 4.50). For metric spaces, these properties are equivalent, as we show next.

Theorem 8.32 *Let $(S, \mathcal{O}_d)$ be a topological metric space. The following statements are equivalent:*

(i) *$(S, \mathcal{O}_d)$ has a countable basis;*
(ii) *$(S, \mathcal{O}_d)$ is a separable;*
(iii) *every open cover of $(S, \mathcal{O}_d)$ contains a countable subcover.*

Proof By Theorem 4.47 and Corollary 4.50, the first statement implies (ii) and (iii). Therefore, it suffices to prove that (iii) implies (ii) and (ii) implies (i).

To show that (iii) implies (ii), suppose that every open cover of $(S, \mathcal{O}_d)$ contains a countable subcover. The collection of open spheres $\{C\left(x, \frac{1}{n}\right) \mid x \in S, n \in \mathbb{N}_{>0}\}$ is an open cover of S and therefore there exists a countable set $T_n \subseteq S$ such that $\mathcal{C}_n = \{C\left(x, \frac{1}{n}\right) \mid x \in T_n, n \in \mathbb{N}_{>0}\}$ is an open cover of S. Let $C = \bigcup_{n \geqslant 1} T_n$. By Theorem 1.125, C is a countable set.

We claim that C is dense in $(S, \mathcal{O}_d)$. Indeed, let $s \in S$ and choose n such that $n > \frac{1}{\epsilon}$. Since $\mathcal{C}_n$ is an open cover of S, there is $x \in T_n$ such that $s \in C\left(x, \frac{1}{n}\right) \subseteq C(x, \epsilon)$. Since $T_n \subseteq C$, it follows that C is dense in $(S, \mathcal{O}_d)$. Thus, $(S, \mathcal{O}_d)$ is separable.

To prove that (ii) implies (i), let $(S, \mathcal{O}_d)$ be a separable space. There exists a countable set U that is dense in $(S, \mathcal{O}_d)$. Consider the countable collection

$$\mathcal{C} = \left\{C\left(u, \frac{1}{n}\right) \mid u \in U, n \geqslant 1\right\}.$$

If L is an open set in $(S, \mathcal{O}_d)$ and $x \in L$, then there exists $\epsilon > 0$ such that $C(x, \epsilon) \subseteq L$. Let n be such that $n > \frac{2}{\epsilon}$. Since U is dense in $(S, \mathcal{O}_d)$, we know that $x \in \mathbf{K}(U)$, so there exists $y \in S(x, \epsilon) \cap U$ and $x \in C\left(y, \frac{1}{n}\right) \subseteq C\left(x, \frac{2}{n}\right) \subseteq C(x, \epsilon) \subseteq L$. Thus, $\mathcal{C}$ is a countable basis.

Theorem 8.33 *Let $(S, \mathcal{O}_d)$ be a topological metric space. Every closed set of this space is a countable intersection of open sets, and every open set is a countable union of closed sets.*

Proof Let H be a closed set and let U_n be the open set

$$U_n = \bigcup_{n \geqslant 1} \left\{C\left(x, \frac{1}{n}\right) \mid x \in F\right\}.$$

It is clear that $H \subseteq \bigcap_{n \geqslant 1} U_n$. Now let $u \in \bigcap_{n \geqslant 1} U_n$ and let ϵ be an arbitrary positive number. For every $n \geqslant 1$, there is an element $x_n \in H$ such that $d(u, x_n) < \frac{1}{n}$. Thus, if $\frac{1}{n} < \epsilon$, we have $x_n \in H \cap C(u, \epsilon)$, so $C(u, \epsilon) \cap H \neq \emptyset$. By Corollary 8.7, it follows that $u \in H$, which proves the reverse inclusion $\bigcap_{n \geqslant 1} U_n \subseteq H$. This shows that every closed set is a countable union of open sets.

If L is an open set, then its complement is closed and, by the first part of the theorem, it is a countable intersection of open sets. Thus, L itself is a countable union of closed sets.

Definition 8.34 *Let $(S, \mathcal{O}_d)$ be a topological metric space. A G_δ-set is a countable intersection of open sets. An F_δ-set is a countable union of open sets.*

Now, Theorem 8.33 can be restated by saying that every closed set of a topological metric space is a G_δ-set and every open set is an F_δ-set.

Theorem 8.35 *Let U be a G_δ-set in the topological metric space $(S, \mathcal{O}_d)$. If T is a G_δ-set in the subspace U, then T is a G_δ-set in S.*

Proof Since T is a G_δ-set in the subspace U, we can write $T = \bigcap_{n \in \mathbb{N}} L_n$, where each L_n is an open set in the subspace U. By the definition of the subspace topology, for each L_n there exists an open set in S such that $L_n = L'_n \cap U$, so

$$T = \bigcap_{n \in \mathbb{N}} L_n = \bigcap_{n \in \mathbb{N}} (L'_n \cap U) = U \cap \bigcap_{n \in \mathbb{N}} L'_n.$$

Since U is a countable intersection of open sets of S, the last equality shows that T is a countable intersection of open sets of S and hence a G_δ-set in S.

8.5 Sequences in Metric Spaces

Definition 8.36 *Let $(S, \mathcal{O}_d)$ be a topological metric space and let $\mathbf{x} = (x_0, \ldots, x_n, \ldots)$ be a sequence in $\mathbf{Seq}_\infty(S)$.*

The sequence $\mathbf{x}$ converges *to an element x of S if for every $\epsilon > 0$ there exists $n_\epsilon \in \mathbb{N}$ such that $n \geqslant n_\epsilon$ implies $x_n \in C(x, \epsilon)$.*

A sequence $\mathbf{x}$ is convergent *if it converges to an element x of S.*

Theorem 8.37 *Let $(S, \mathcal{O}_d)$ be a topological metric space and let $\mathbf{x} = (x_0, \ldots, x_n, \ldots)$ be a sequence in $\mathbf{Seq}_\infty(S)$. If $\mathbf{x}$ is convergent, then there exists a unique x such that $\mathbf{x}$ converges to x.*

Proof Suppose that there are two distinct elements x and y of the set S that satisfy the condition of Definition 8.36. We have $d(x, y) > 0$. Define $\epsilon = \frac{d(x,y)}{3}$. By definition, there exists n_ϵ such that $n \geqslant n_\epsilon$ implies $d(x, x_n) < \epsilon$ and $d(x_n, y) < \epsilon$. By applying the triangular inequality, we obtain

$$d(x, y) \leqslant d(x, x_n) + d(x_n, y) < 2\epsilon = \frac{2}{3} d(x, y),$$

which is a contradiction.

If the sequence $\mathbf{x} = (x_0, \ldots, x_n, \ldots)$ converges to x, this is denoted by $\lim_{n\to\infty} x_n = x$.

An alternative characterization of continuity of functions can be formulated using convergent sequences.

Theorem 8.38 *Let $(S, \mathcal{O}_d)$ and $(T, \mathcal{O}_e)$ be two topological metric spaces and let $f : S \longrightarrow T$. The function f is continuous in x if and only if for every sequence $\mathbf{x} = (x_0, \ldots, x_n, \ldots)$ such that $\lim_{n\to\infty} x_n = x$ we have $\lim_{n\to\infty} f(x_n) = f(x)$.*

Proof Suppose that f is continuous in x, and let $\mathbf{x} = (x_0, \ldots, x_n, \ldots)$ be a sequence such that $\lim_{n\to\infty} x_n = x$. Let $\epsilon > 0$. By Definition 8.36, there exists $\delta > 0$ such that $f(C(x, \delta)) \subseteq C(f(x), \epsilon)$. Since $\lim_{n\to\infty} x_n = x$, there exists n_δ such that $n \geq n_\delta$ implies $x_n \in C(x, \delta)$. Then, $f(x_n) \in f(C(x, \delta)) \subseteq C(f(x), \epsilon)$. This shows that $\lim_{n\to\infty} f(x_n) = f(x)$.

Conversely, suppose that for every sequence $\mathbf{x} = (x_0, \ldots, x_n, \ldots)$ such that $\lim_{n\to\infty} x_n = x$, we have $\lim_{n\to\infty} f(x_n) = f(x)$. If f were not continuous in x, we would have an $\epsilon > 0$ such that for all $\delta > 0$ we would have $y \in C(x, \delta)$ but $f(y) \notin C(f(x), \epsilon)$. Choosing $\delta = \frac{1}{n}$, let $y_n \in S$ such that $y_n \in C\left(x, \frac{1}{n}\right)$ and $f(y_n) \notin C(f(x), \epsilon)$. This yields a contradiction because we should have $\lim_{n\to\infty} f(y_n) = f(x)$.

8.5.1 Sequences of Real Numbers

Theorem 8.39 *Let $\mathbf{x} = (x_0, \ldots, x_n, \ldots)$ be a sequence in $(\mathbb{R}, \mathcal{O})$, where $\mathcal{O}$ is the usual topology on $\mathbb{R}$.*

If $\mathbf{x}$ is an increasing (decreasing) sequence and there exists a number $b \in \mathbb{R}$ such that $x_n \leqslant b$ ($x_n \geqslant b$, respectively), then the sequence $\mathbf{x}$ is convergent.

Proof Since the set $\{x_n \mid n \in \mathbb{N}\}$ is bounded above, its supremum s exists by the Completeness Axiom for $\mathbb{R}$ given in Sect. 2. We claim that $\lim_{n\to\infty} x_n = s$. Indeed, by Theorem 2.28, for every $\epsilon > 0$ there exists $n_\epsilon \in \mathbb{N}$ such that $s - \epsilon < x_{n_\epsilon} \leqslant s$. Therefore, by the monotonicity of the sequence and its boundedness, we have $s - \epsilon < x_n \leqslant s$ for $n \geqslant n_\epsilon$, so $x_n \in C(x, \epsilon)$, which proves that $\mathbf{x}$ converges to s.

We leave it to the reader to show that any decreasing sequence in $(\mathbb{R}, \mathcal{O})$ that is bounded below is convergent.

If $\mathbf{x}$ is an increasing sequence and there is no upper bound for $\mathbf{x}$, this means that for every $b \in \mathbb{R}$ there exists a number n_b such that $n \geqslant n_b$ implies $x_n > b$. If this is the case, we say that $\mathbf{x}$ is a sequence *divergent to* $+\infty$ and we write $\lim_{n\to\infty} x_n = +\infty$. Similarly, if $\mathbf{x}$ is a decreasing sequence and there is no lower bound for it, this means that for every $b \in \mathbb{R}$ there exists a number n_b such that $n \geqslant n_b$ implies $x_n < b$. In this case, we say that $\mathbf{x}$ is a sequence *divergent to* $-\infty$ and we write $\lim_{n\to\infty} x_n = -\infty$.

Theorem 8.39 and the notion of a divergent sequence allow us to say that $\lim_{n\to\infty} x_n$ exists for every increasing or decreasing sequence; this limit may be a real number or $\pm\infty$ depending on the boundedness of the sequence.

Theorem 8.40 *Let $[a_0, b_0] \supseteq [a_1, b_1] \supset \cdots \supset [a_n, b_n] \supset \cdots$ be a sequence of nested closed intervals of real numbers. There exists a closed interval $[a, b]$ such that $a = \lim_{n\to\infty} a_n$, $b = \lim_{n\to\infty} b_n$, and*

$$[a, b] = \bigcap_{n\in\mathbb{N}} [a_n, b_n].$$

Proof The sequence $a_0, a_1, \ldots, a_n, \ldots$ is clearly increasing and bounded because we have $a_n \leqslant b_m$ for every $n, m \in \mathbb{N}$. Therefore, it converges to a number $a \in \mathbb{R}$ and $a \leqslant b_m$ for every $m \in \mathbb{N}$. Similarly, $b_0, b_1, \ldots, b_n, \ldots$ is a decreasing sequence that is bounded below, so it converges to a number b such that $a_n \leqslant b$ for $n \in \mathbb{N}$. Consequently, $[a, b] \subseteq \bigcap_{n\in\mathbb{N}} [a_n, b_n]$.

Conversely, let c be a number in $\bigcap_{n\in\mathbb{N}} [a_n, b_n]$. Since $c \geqslant a_n$ for $n \in \mathbb{N}$, it follows that $c \geqslant \sup\{a_n \mid n \in \mathbb{N}\}$, so $c \geqslant a$. A similar argument shows that $c \leqslant b$, so $c \in [a, b]$, which implies the reverse inclusion $\bigcap_{n\in\mathbb{N}} [a_n, b_n] \subseteq [a, b]$.

In Example 4.57, we saw that every closed interval $[a, b]$ of $\mathbb{R}$ is a compact set. This allows us to prove the next statement.

Theorem 8.41 **(Bolzano-Weierstrass Theorem)** *A bounded sequence of real numbers has a convergent subsequence.*

Proof Let $\mathbf{x} = (x_0, \ldots, x_n, \ldots)$ be a bounded sequence of real numbers. The boundedness of $\mathbf{x}$ implies the existence of a closed interval $D_0 = [a_0, b_0]$ such that $\{x_n \mid n \in \mathbb{N}\} \subseteq [a_0, b_0]$.

Let $c = \frac{a_0+b_0}{2}$ be the midpoint of D_0. At least one of the sets $\mathbf{x}^{-1}([a_0, c_0])$, $\mathbf{x}^{-1}([c_0, b_0])$ is infinite. Let $[a_1, b_1]$ be one of $[a_0, c_0]$ or $[c_0, b_0]$, for which $\mathbf{x}^{-1}([a_0, c_0]), \mathbf{x}^{-1}([c_0, b_0])$ is infinite.

Suppose that we have constructed the interval $D_n = [a_n, b_n]$ having $c_n = \frac{a_n+b_n}{2}$ as its midpoint such that $\mathbf{x}^{-1}(D_n)$ is infinite. Then, $D_{n+1} = [a_{n+1}, b_{n+1}]$ is obtained from D_n as one of the intervals $[a_n, c_n]$ or $[c_n, b_n]$ that contains x_n for infinitely many n.

Thus, we obtain a descending sequence of closed intervals $[a_0, b_0] \supset [a_1, b_1] \supset \cdots$ such that each interval contains an infinite set of members of the sequence $\mathbf{x}$. By Theorem 8.40, we have $[a, b] = \bigcup_{n \in \mathbb{N}} [a_n, b_n]$, where $a = \lim_{n\to\infty} a_n$ and $b = \lim_{n\to\infty} b_n$. Note that $b_n - a_n = \frac{b_0-a_0}{2^n}$, so $a = \lim_{n\to\infty} a_n = \lim_{n\to\infty} b_n = b$.

The interval D_0 contains at least one member of $\mathbf{x}$, say x_{n_0}. Since D_1 contains infinitely many members of $\mathbf{x}$, there exists a member x_{n_1} of $\mathbf{x}$ such that $n_1 > n_0$. Continuing in this manner, we obtain a subsequence $x_{n_0}, x_{n_1}, \ldots, x_{n_p}, \ldots$. Since $a_p \leqslant x_{n_p} \leqslant b_p$, it follows that the sequence $(x_{n_0}, x_{n_1}, \ldots, x_{n_p}, \ldots)$ converges to a.

Let $\mathbf{x} = (x_0, x_1, \ldots)$ be a sequence of real numbers. Consider the sequence of sets $S_n = \{x_n, x_{n+1}, \ldots\}$ for $n \in \mathbb{N}$. It is clear that $S_0 \supseteq S_1 \supseteq \cdots \subseteq S_n \supseteq \cdots$. Therefore, we have the increasing sequence of numbers $\inf S_0 \leqslant \inf S_1 \leqslant \cdots \leqslant \inf S_n \leqslant \cdots$; we define $\liminf \mathbf{x}$ as $\lim_{n\to\infty} \inf S_n$. On the other hand, we have the decreasing sequence $\sup S_0 \geqslant \sup S_1 \geqslant \cdots \geqslant \sup S_n \geqslant \cdots$ of numbers; we define $\limsup \mathbf{x}$ as $\lim_{n\to\infty} \sup S_n$.

Example 8.42 Let $\mathbf{x}$ be the sequence defined by $x_n = (-1)^n$ for $n \in \mathbb{N}$. It is clear that $\sup S_n = 1$ and $\inf S_n = -1$. Therefore, $\limsup \mathbf{x} = 1$ and $\liminf \mathbf{x} = -1$.

Theorem 8.43 *For every sequence* $\mathbf{x}$ *of real numbers, we have* $\liminf \boldsymbol{x} \leqslant \limsup \boldsymbol{x}$.

Proof Let $S_n = \{x_n, x_{n+1}, \ldots\}$, $y_n = \inf S_n$, and $z_n = \sup S_n$ for $n \in \mathbb{N}$. If $p \geqslant n$, we have $y_n \leqslant y_p \leqslant z_p \leqslant z_n$, so $y_n \leqslant z_p$ for every n, p such that $p \geqslant n$. Since $z_1 \geqslant z_2 \geqslant \cdots \geqslant z_p$, it follows that $y_n \leqslant z_p$ for every $p \in \mathbb{N}$. Therefore, $\limsup \mathbf{x} = \lim_{p\to\infty} z_p \geqslant y_n$ for every $n \in \mathbb{N}$, which in turn implies $\liminf \mathbf{x} = \lim_{n\to\infty} y_n \leqslant \limsup \mathbf{x}$.

Corollary 8.44 *Let* $\boldsymbol{x} = (x_0, x_1, \ldots, x_n, \ldots)$ *be a sequence of real numbers. We have* $\liminf \boldsymbol{x} = \limsup \boldsymbol{x} = \ell$ *if and only if* $\lim_{n\to\infty} x_n = \ell$.

Proof Suppose that $\liminf \mathbf{x} = \limsup \mathbf{x} = \ell$ and that it is not the case that $\lim_{n\to\infty} x_n = \ell$. This means that there exists $\epsilon > 0$ such that, for every $m \in \mathbb{N}$, $n \geqslant m$ implies $|x_n - \ell| \geqslant \epsilon$, which is equivalent to $x_n \geqslant \ell + \epsilon$ or $x_n \leqslant \ell - \epsilon$. Thus, at least one of the following cases occurs:

(i) there are infinitely many n such that $x_n \geqslant \ell+\epsilon$, which implies that $\limsup x_n \geqslant \ell+\epsilon$, or
(ii) there are infinitely many n such that $x_n \leqslant \ell-\epsilon$, which implies that $\liminf x_n \geqslant \ell-\epsilon$.

Either case contradicts the hypothesis, so $\lim_{n\to\infty} x_n = \ell$.

Conversely, suppose that $\lim_{n\to\infty} x_n = \ell$. There exists n_ϵ such that $n \geqslant n_\epsilon$ implies $\ell-\epsilon < x_n < \ell+\epsilon$. Thus, $\sup\{x_n \mid n \geqslant n_\epsilon\} \leqslant \ell+\epsilon$, so $\limsup \mathbf{x} \leqslant \ell+\epsilon$. Similarly, $y-\epsilon \leqslant \liminf \mathbf{x}$ and the inequality

$$\ell-\epsilon \leqslant \liminf \mathbf{x} \leqslant \limsup \mathbf{x} \leqslant \ell+\epsilon,$$

which holds for every $\epsilon > 0$, implies $\liminf \mathbf{x} = \limsup \mathbf{x} = \ell$.

8.5.1.1 Sequences and Open and Closed Sets

Theorem 8.45 *Let $(S, \mathcal{O}_d)$ be a topological metric space. A subset U of S is open if and only if for every $x \in U$ and every sequence $(x_0,\ldots,x_n,\ldots)$ such that $\lim_{n\to\infty} x_n = x$ there is m such that $n \geqslant m$ implies $x_n \in U$.*

Proof Suppose U is an open set. Since $x \in U$, there exists $\epsilon > 0$ such that $C(x,\epsilon) \subseteq U$. Let $(x_0,\ldots,x_n,\ldots)$ be such that $\lim_{n\to\infty} = x$. By Definition 8.36, there exists n_ϵ such that $n \geqslant n_\epsilon$ implies $x_n \in C(x,\epsilon) \subseteq U$.

Conversely, suppose that the condition is satisfied and that U is not open. Then, there exists $x \in U$ such that for every $n \geqslant 1$ we have $C\left(x,\frac{1}{n}\right) - U \neq \emptyset$. Choose $x_{n-1} \in C\left(x,\frac{1}{n}\right)$ for $n \geqslant 1$. It is clear that the sequence $(x_0,\ldots,x_n,\ldots)$ converges to x. However, none of the members of this sequence belong to U. This contradicts our supposition, so U must be an open set.

Theorem 8.46 *Let $(S, \mathcal{O}_d)$ be a topological metric space. A subset W of S is closed if and only if for every sequence $\mathbf{x} = (x_0,\ldots,x_n,\ldots) \in \mathbf{Seq}_\infty(W)$ such that $\lim_{n\to\infty} x_n = x$ we have $x \in W$.*

Proof If W is a closed set and $\mathbf{x} = (x_0,\ldots,x_n,\ldots)$ is a sequence whose members belong to W, then none of these members belong to $S - W$. Since $S - W$ is an open set, by Theorem 8.46, it follows that $x \notin S - W$; that is, $x \in W$.

Conversely, suppose that for every sequence $(x_0,\ldots,x_n,\ldots)$ such that $\lim_{n\to\infty} = x$ and $x_n \in W$ for $n \in \mathbb{N}$ we have $x \in W$. Let $v \in S - W$, and suppose that for every $n \geqslant 1$ the open sphere $C\left(v,\frac{1}{n}\right)$ is not included in $S - W$. This means that for each $n \geqslant 1$ there is $z_{n-1} \in C\left(v,\frac{1}{n}\right) \cap W$. We have $\lim_{n\to\infty} z_n = v$; this implies $v \in W$. This contradiction means that there is $n \geqslant 1$ such that $C\left(v,\frac{1}{n}\right) \subseteq V$, so V is an open set. Consequently, $W = S - V$ is a closed set.

8.6 Completeness of Metric Spaces

Let $\mathbf{x} = (x_0, \ldots, x_n, \ldots)$ be a sequence in the topological metric space $(S, \mathcal{O}_d)$ such that $\lim_{n\to\infty} x_n = x$. If $m, n > n_{\frac{\epsilon}{2}}$, we have $d(x_m, x_n) \leqslant d(x_m, x) + d(x, x_n) < \frac{\epsilon}{2} + \frac{\epsilon}{2} = \epsilon$. In other words, if $\mathbf{x}$ is a sequence that converges to x, then given a positive number ϵ we have members of the sequence closer than ϵ if we go far enough in the sequence. This suggests the following definition:

Definition 8.47 *A sequence $\mathbf{x} = (x_0, \ldots, x_n, \ldots)$ in the topological metric space $(S, \mathcal{O}_d)$ is a* Cauchy sequence *if for every $\epsilon > 0$ there exists $n_\epsilon \in \mathbb{N}$ such that $m, n \geqslant n_\epsilon$ implies $\rho(x_m, x_n) < \epsilon$.*

Theorem 8.48 *Every convergent sequence in a topological metric space $(S, \mathcal{O}_d)$ is a Cauchy sequence.*

Proof Let $\mathbf{x} = (x_0, x_1, \ldots)$ be a convergent sequence and let $x = \lim_{n\to\infty} \mathbf{x}$. There exists $n'_{\frac{\epsilon}{2}}$ such that if $n > n'_{\frac{\epsilon}{2}}$, then $d(x_n, x) < \frac{\epsilon}{2}$. Thus, if $m, n \geqslant n_\epsilon = n'_{\frac{\epsilon}{2}}$, it follows that

$$d(x_m, x_n) \leqslant d(x_m, x) + d(x, x_n) < \frac{\epsilon}{2} + \frac{\epsilon}{2} = \epsilon,$$

which means that $\mathbf{x}$ is a Cauchy sequence.

Example 8.49 The converse of Theorem 8.48 is not true, in general, as we show next.

Let $((0, 1), d)$ be the metric space equipped with the metric $d(x, y) = |x - y|$ for $x, y \in (0, 1)$. The sequence defined by $x_n = \frac{1}{n+1}$ for $n \in \mathbb{N}$ is a Cauchy sequence. Indeed, it suffices to take $m, n \geqslant \frac{1}{\epsilon} - 1$ to obtain $|x_n - x_m| < \epsilon$; however, the sequence x_n is not convergent to an element of $(0, 1)$.

Definition 8.50 *A topological metric space is* complete *if every Cauchy sequence is convergent.*

Example 8.51 The topological metric space $(\mathbb{R}, \mathcal{O}_d)$, where $d(x, y) = |x - y|$ for $x, y \in\in \mathbb{R}$, is complete.

Let $\mathbf{x} = (x_0, x_1, \ldots)$ be a Cauchy sequence in $\mathbb{R}$. For every $\epsilon > 0$, there exists $n_\epsilon \in \mathbb{N}$ such that $m, n \geqslant n_\epsilon$ implies $|x_m - x_n| < \epsilon$. Choose $m_0 \in \mathbb{N}$ such that $m_0 \geqslant n_\epsilon$. Thus, if $n \geqslant n_\epsilon$, then $x_{m_0} - \epsilon < x_n < x_{m_0} + \epsilon$, which means that $\mathbf{x}$ is a bounded sequence. By Theorem 8.41, the sequence $\mathbf{x}$ contains a bounded subsequence $(x_{i_0}, x_{i_1}, \ldots)$ that is convergent. Let $\ell = \lim_{k\to\infty} x_{i_k}$. It is not difficult to see that $\lim_{x_n} x_n = \ell$, which shows that $(\mathbb{R}, \mathcal{O}_d)$ is complete.

Theorem 8.52 *Let $(S, \mathcal{O}_d)$ be a complete topological metric space. If T is a closed subset of S, then the subspace T is complete.*

Proof Let T be a closed subset of S and let $\mathbf{x} = (x_0, x_1, \ldots)$ be a Cauchy sequence in this subspace. The sequence $\mathbf{x}$ is a Cauchy sequence in the complete space S, so there exists $x = \lim_{n\to\infty} x_n$. Since T is closed, we have $x \in T$, so T is complete.

Conversely, suppose that T is complete. Let $x \in \mathbf{K}(T)$. There exists a sequence $\mathbf{x} = (x_0, x_1, \ldots) \in \mathbf{Seq}_\infty(T)$ such that $\lim_{n\to\infty} x_n = x$. Then, $\mathbf{x}$ is a Cauchy sequence in T, so there is a limit t of this sequence in T. The uniqueness of the limit implies $x = t \in T$, so T is a closed set.

Theorem 8.53 *There is no clopen set in the topological space* $(\mathbb{R}, \mathcal{O})$ *except the empty set and the set* $\mathbb{R}$.

Proof Suppose that L is a clopen subset of $\mathbb{R}$ that is distinct from $\emptyset$ and $\mathbb{R}$. Then, there exist $x \in L$ and $y \notin L$. Starting from x and y, we define inductively the terms of two sequences $\mathbf{x} = (x_0, \ldots, x_n, \ldots)$ and $\mathbf{y} = (y_0, \ldots, y_n, \ldots)$ as follows. Let $x_0 = x$ and $y_0 = y$. Suppose that x_n and y_n are defined. Then,

$$x_{n+1} = \begin{cases} \frac{x_n+y_n}{2} & \text{if } \frac{x_n+y_n}{2} \in L, \\ x_n & \text{otherwise,} \end{cases}$$

and

$$y_{n+1} = \begin{cases} \frac{x_n+y_n}{2} & \text{if } \frac{x_n+y_n}{2} \notin L, \\ y_n & \text{otherwise.} \end{cases}$$

It is clear that $\{x_n \mid n \in \mathbb{N}\} \subseteq L$ and $\{y_n \mid n \in \mathbb{N}\} \subseteq \mathbb{R} - L$. Moreover, we have

$$|y_{n+1} - x_{n+1}| = \frac{|y_n - x_n|}{2} = \cdots = \frac{|y - x|}{2^{n+1}}.$$

Note that

$$|x_{n+1} - x_n| \leqslant |y_n - x_n| \leqslant \frac{|y - x|}{2^n}.$$

This implies that $\mathbf{x}$ is a Cauchy sequence and therefore there is $x = \lim_{n\to\infty} x_n$; moreover, the sequence $\mathbf{y}$ also converges to x, so x belongs to ∂L, which is a contradiction.

Theorem 8.54 *In a complete topological metric space* $(S, \mathcal{O}_d)$, *every descending sequence of closed sets* $V_0 \supset \cdots \supset V_n \supset V_{n+1} \supset \cdots$ *such that* $\lim_{n\to\infty} diam(V_n) = 0$ *has a nonempty intersection, that is,* $\bigcap_{n\in\mathbb{N}} V_n \neq \emptyset$.

Proof Consider a sequence $x_0, x_1, \ldots, x_n, \ldots$ such that $x_n \in V_n$. This is a Cauchy sequence. Indeed, let $\epsilon > 0$. Since $\lim_{n\to\infty} diam(V_n) = 0$, there exists n_ϵ such that if $m, n > n_\epsilon$ we have $x_m, x_n \in V_{\min\{m,n\}}$. Since $\min\{m, n\} \geqslant n_\epsilon$, it follows that $d(x_m, x_n) \leqslant diam(V_{\min m,n}) < \epsilon$. Since the space $(S, \mathcal{O}_d)$ is complete, it follows that there exists $x \in S$ such that $\lim_{n\to\infty} x_n = x$. Note that all members of the sequence above belong to V_m, with the possible exception of the first m members. Therefore, by Theorem 8.46, $x \in V_m$, so $x \in \bigcap_{n\in\mathbb{N}} V_n$, so $\bigcap_{n\in\mathbb{N}} V_n \neq \emptyset$.

Recall that the definition of Baire spaces was introduced on page 154.

Theorem 8.55 *Every complete topological metric space is a Baire space.*

Proof We prove that if $(S, \mathcal{O}_d)$ is complete, then it satisfies the first condition of Theorem 4.28.

Let $L_1, \ldots, L_n, \ldots$ be a sequence of open subsets of S that are dense in S and let L be an open, nonempty subset of S. We construct inductively a sequence of closed sets $H_1, \ldots, H_n, \ldots$ that satisfy the following conditions:

(i) $H_1 \subseteq L_0 \cap L$,
(ii) $H_n \subseteq L_n \cap H_{n-1}$ for $n \geqslant 2$,
(iii) $\mathbf{I}(H_n) \neq \emptyset$, and
(iv) $diam(H_n) \leqslant \frac{1}{n}$

for $n \geqslant 2$.

Since L_1 is dense in S, by Theorem 4.23, $L_1 \cap L \neq \emptyset$, so there is a closed sphere of diameter less than 1 enclosed in $L_1 \cap L$. Define H_1 as this closed sphere.

Suppose that H_{n-1} was constructed. Since $\mathbf{I}(H_{n-1}) \neq \emptyset$, the open set $L_n \cap \mathbf{I}(H_{n-1})$ is not empty because L_n is dense in S. Thus, there is a closed sphere H_n included in $L_n \cap \mathbf{I}(H_{n-1})$, and therefore included in $L_n \cap H_{n-1}$, such that $diam(H_n) < \frac{1}{n}$. Clearly, we have $\mathbf{I}(H_n) \neq \emptyset$. By applying Theorem 8.54 to the descending sequence of closed sets $H_1, \ldots, H_n, \ldots$, the completeness of the space implies that $\bigcap_{n \geqslant 1} H_n \neq \emptyset$. If $s \in \bigcap_{n \geqslant 1} H_n$, then it is clear that $x \in \bigcap_{n \geqslant 1} L_n$ and $x \in L$, which means that the set $\bigcap_{n \geqslant 1} L_n$ has a nonempty intersection with every open set L. This implies that $\bigcap_{n \geqslant 1} L_n$ is dense in S.

The notion of precompactness that we are about to introduce is weaker than the notion of compactness formulated for general topological spaces.

Definition 8.56 *Let (S, d) be a metric space. A finite subset $\{x_1, \ldots, x_n\}$ is an* r-net *on (S, d) if $S = \bigcup_{i=1}^{n} C(x_i, r)$.*

Observe that, for every positive number r the family of open spheres $\{C(x, r) \mid x \in S\}$ is an open cover of the space S.

Definition 8.57 *A topological metric space $(S, \mathcal{O}_d)$ is* precompact *if, for every positive number r, the open cover $\{C(x, r) \mid x \in S\}$ contains an r-net $\{C(x_1, r), \ldots, C(x_n, r)\}$.*

Clearly, compactness implies precompactness.

Using the notion of an r-net, it is possible to give the following characterization to precompactness.

Theorem 8.58 *$(S, \mathcal{O}_d)$ is precompact if and only if for every positive number r there exists an r-net N_r on $(S, \mathcal{O}_d)$.*

Proof This statement is an immediate consequence of the definition of precompactness.

Next, we show that precompactness is inherited by subsets.

Theorem 8.59 *If $(S, \mathcal{O}_d)$ is a precompact topological metric space and $T \subseteq S$, then the subspace $(T, \mathcal{O}_d \upharpoonright_T)$ is also precompact.*

Proof Since $(S, \mathcal{O}_d)$ is precompact, for every $r > 0$ there exists a finite open cover $\mathcal{C}_{r/2} = \{C\left(s_i, \frac{r}{2}\right) \mid s_i \in S, 1 \leqslant i \leqslant n\}$. Let $\mathcal{C}' = \{C\left(s_{i_j}, \frac{r}{2}\right) \mid 1 \leqslant j \leqslant m\}$ be a minimal subcollection of $\mathcal{C}_{r/2}$ that consists of those open spheres that cover T; that is,

$$T \subseteq \bigcup \left\{C\left(s_{i_j}, \frac{r}{2}\right) \mid 1 \leqslant j \leqslant m\right\}.$$

The minimality of $\mathcal{C}'$ implies that each set $C\left(s_{i_j}, \frac{r}{2}\right)$ contains an element y_j of T. By Exercise 68 of Chap. 14, we have $C\left(s_{i_j}, \frac{r}{2}\right) \subseteq C(y_j, r)$ and this implies that the set $\{y_1, \ldots, y_m\}$ is an r-net for the set T.

If the subspace $(T, \mathcal{O}_d \upharpoonright_T)$ of $(S, \mathcal{O}_d)$ is precompact, we say that the set T is *precompact.*

The next corollary shows that there is no need to require the centers of the spheres involved in the definition of the precompactness of a subspace to be located in the subspace.

Corollary 8.60 *Let $(S, \mathcal{O}_d)$ be a topological metric space (not necessarily precompact) and let T be a subset of S. The subspace $(T, \mathcal{O}_d \upharpoonright_T)$ is precompact if and only if for every positive number r there exists a finite subcover $\{C(x_1, r), \ldots, C(x_n, r) \mid x_i \in S$ for $1 \leqslant i \leqslant n\}$.*

Proof The argument has been made in the proof of Theorem 8.59.

The next theorem adds two further equivalent characterizations of compact metric spaces to the ones given in Theorem 8.30.

Theorem 8.61 *Let $(S, \mathcal{O}_d)$ be a topological metric space. The following statements are equivalent.*

(i) *$(S, \mathcal{O}_d)$ is compact;*
(ii) *every sequence $\mathbf{x} \in \mathbf{Seq}_\infty(S)$ contains a convergent subsequence;*
(iii) *$(S, \mathcal{O}_d)$ is precompact and complete.*

Proof (i) implies (ii): Let $(S, \mathcal{O}_d)$ be a compact topological metric space and let $\mathbf{x}$ be a sequence in $\mathbf{Seq}_\infty(S)$. By Theorem 8.30, $(S, \mathcal{O}_d)$ has the Bolzano-Weierstrass property, so the set $\{x_n \mid n \in \mathbb{N}\}$ has an accumulation point t. For every $k \geqslant 1$, the set $\{x_n \mid n \in \mathbb{N}\} \cap C\left(t, \frac{1}{k}\right)$ contains an element x_{n_k} distinct from t. Since $d(t, x_{n_k}) < \frac{1}{k}$ for $k \geqslant 1$, it follows that the subsequence $(x_{n_1}, x_{n_2}, \ldots)$ converges to t.

(ii) implies (iii): Suppose that every sequence $\mathbf{x} \in \mathbf{Seq}_\infty(S)$ contains a convergent subsequence and that $(S, \mathcal{O}_d)$ is not precompact. Then, there exists a positive number r such that S cannot be covered by any collection of open spheres of radius r.

Let x_0 be an arbitrary element of S. Note that $C(x_0, r) - S \neq \emptyset$ because otherwise the $C(x_0, r)$ would constitute an open cover for S. Let x_1 be an arbitrary element in $C(x_0, r) - S$. Observe that $d(x_0, x_1) \geqslant r$. The set $(C(x_0, r) \cup C(x_1, r)) - S$ is not empty. Thus, for any $x_2 \in (C(x_0, r) \cup C(x_1, r)) - S$, we have $d(x_0, x_2) \geqslant r$ and $d(x_0, x_1) \geqslant r$, etc. We obtain in this manner a sequence $x_0, x_1, \ldots, x_n, \ldots$ such that $d(x_i, x_j) \geqslant r$ when $i \neq j$. Clearly, this sequence cannot contain a convergent sequence, and this contradiction shows that the space must be precompact.

To prove that $(S, \mathcal{O}_d)$ is complete, consider a Cauchy sequence $\mathbf{x} = (x_0, x_1, \ldots, x_n, \ldots)$. By hypothesis, this sequence contains a convergent subsequence $(x_{n_0}, x_{n_1}, \ldots)$. Suppose that $\lim_{k \to \infty} x_{n_k} = l$. Since $\mathbf{x}$ is a Cauchy sequence, there is $n'_{\frac{\epsilon}{2}}$ such that $n, n_k \geqslant n'_{\frac{\epsilon}{2}}$ implies $d(x_n, x_{n_k}) < \frac{\epsilon}{2}$. The convergence of the subsequence $(x_{n_0}, x_{n_1}, \ldots)$ means that there exists $n''_{\frac{\epsilon}{2}}$ such that $n_k \geqslant n''_{\frac{\epsilon}{2}}$ implies $d(x_{n_k}, l) < \frac{\epsilon}{2}$. Choosing $n_k \geqslant n''_{\frac{\epsilon}{2}}$, if $n \geqslant n'_{\frac{\epsilon}{2}} = n_\epsilon$, we obtain

$$d(x_n, l) \leqslant d(x_n, x_{n_k}) + d(x_{n_k}, l) < \frac{\epsilon}{2} + \frac{\epsilon}{2} = \epsilon,$$

which proves that $\mathbf{x}$ is convergent. Consequently, $(S, \mathcal{O}_d)$ is both precompact and complete.

(iii) implies (i): Suppose that $(S, \mathcal{O}_d)$ is both precompact and complete but not compact, which means that there exists an open cover $\mathcal{C}$ of S that does not contain any finite subcover.

Since $(S, \mathcal{O}_d)$ is precompact, there exists a $\frac{1}{2}$-net, $\{x_1^1, \ldots, x_{n_1}^1\}$. For each of the closed spheres $B(x_i^1, \frac{1}{2})$, $1 \leqslant i \leqslant n_1$, the trace collection $\mathcal{C}_{B(x_i^1, \frac{1}{2})}$ is an open cover. There is a closed sphere $B(x_j^1, \frac{1}{2})$ such that the open cover $\mathcal{C}_{B(x_j^1, \frac{1}{2})}$ does not contain any finite subcover of $B(x_j^1, \frac{1}{2})$ since $(S, \mathcal{O}_d)$ was assumed not to be compact. Let $z_1 = x_j^1$.

By Theorem 8.59, the closed sphere $B(z_1, \frac{1}{2})$ is precompact. Thus, there exists a $\frac{1}{2^2}$-net $\{x_1^2, \ldots, x_{n_2}^2\}$ of $B(z_1, \frac{1}{2})$. There exists a closed sphere $B(x_k^2, \frac{1}{2^2})$ such that the open cover $\mathcal{C}_{B(x_k^2, \frac{1}{2^2})}$ does not contain any finite subcover of $B(x_k^2, \frac{1}{2^2})$. Let $z_2 = x_k^2$; note that $d(z_1, z_2) \leqslant \frac{1}{2}$.

Thus, we construct a sequence $\mathbf{z} = (z_1, z_2, \ldots)$ such that $d(z_{n+1}, z_n) \leqslant \frac{1}{2^n}$ for $n \geqslant 1$.

Observe that

$$\begin{aligned} d(z_{n+p}, z_n) &\leqslant d(z_{n+p}, z_{n+p-1}) + d(z_{n+p-1}, z_{n+p-2}) + \cdots + d(z_{n+1}, z_n) \\ &\leqslant \frac{1}{2^{n+p-1}} \frac{1}{2^{n+p-2}} + \cdots + \frac{1}{2^n} \\ &= \frac{1}{2^{n-1}} \left(1 - \frac{1}{2^p}\right). \end{aligned}$$

Thus, the sequence $\mathbf{z}$ is a Cauchy sequence and there exists $z = \lim_{n \to \infty} z_n$, because $(S, \mathcal{O}_d)$ is complete.

Since $\mathcal{C}$ is an open cover, there exists a set $L \in \mathcal{C}$ such that $z \in L$. Let r be a positive number such that $C(z, r) \subseteq L$. Let n_0 be such that $d(z_n, z) < \frac{r}{2}$ and $\frac{1}{2^n} \leqslant \frac{r}{2}$. If $x \in B(z_n, \frac{1}{2^n})$, then $d(x, z) \leqslant d(x, z_n) + d(z_n, z) < \frac{1}{2^n} + \frac{r}{2} \leqslant r$, so $B(z_n, \frac{1}{2^n}) \subseteq C(z, r) \subseteq L$. This is a contradiction because the spheres $B(z_n, \frac{1}{2^n})$ were defined such that $\mathcal{C}_{B(z_n, \frac{1}{2^n})}$ did not contain any finite subcover. Thus, $(S, \mathcal{O}_d)$ is compact.

Theorem 8.62 *A subset T of $(\mathbb{R}^n, \mathcal{O})$ is compact if and only if it is closed and bounded.*

Proof Let T be a compact set. By Corollary 8.23. T is closed. Let r be a positive number and let $\{C(t, r) \mid t \in T\}$ be a cover of T. Since T is compact, there exists a finite collection $\{C(t_i, r) \mid 1 \leqslant i \leqslant p\}$ such that $T \subseteq \bigcup\{C(t_i, r) \mid 1 \leqslant i \leqslant p\}$. Therefore, if $x, y \in T$, we have $d(x, y) \leqslant 2 + \max\{d(t_i, t_j) \mid 1 \leqslant i, j \leqslant p\}$, which implies that T is also bounded.

Conversely, suppose that T is closed and bounded. The boundedness of T implies the existence of a parallelepiped $[x_1, y_1] \times \cdots \times [x_n, y_n]$ that includes T, and we saw in Example 4.100 that this parallelepiped is compact. Since T is closed, it is immediate that T is compact by Theorem 4.61.

Corollary 8.63 *Let $(S, \mathcal{O})$ be a compact topological space and let $f : S \longrightarrow \mathbb{R}$ be a continuous function, where $\mathbb{R}$ is equipped with the usual topology. Then, f is bounded and there exist $u_0, u_1 \in S$ such that $f(u_0) = \inf_{x \in S} f(x)$ and $f(u_1) = \sup_{x \in S} f(x)$.*

Proof Since S is compact and f is continuous, the set $f(S)$ is a compact subset of $\mathbb{R}$ and, by Theorem 8.62, is bounded and closed.

Both $\inf_{x \in S} f(x)$ and $\sup_{x \in S} f(x)$ are cluster points of $f(S)$; therefore, both belong to $f(S)$, which implies the existence of u_0 and u_1.

Theorem 8.64 (Heine's Theorem) *Let $(S, \mathcal{O}_d)$ be a compact topological metric space and let $(T, \mathcal{O}_e)$ be a metric space. Every continuous function $f : S \longrightarrow T$ is uniformly continuous on S.*

Proof Let $\mathbf{u} = (u_0, u_1, \ldots)$ and $\mathbf{v} = (v_0, v_1, \ldots)$ be two sequences in $\mathbf{Seq}_\infty(S)$ such that $\lim_{n\to\infty} d(u_n, v_n) = 0$. By Theorem 8.61, the sequence $\mathbf{u}$ contains a convergent subsequence $(u_{p_0}, u_{p_1}, \ldots)$. If $x = \lim_{n\to\infty} u_{p_n}$, then $\lim_{n\to\infty} v_{p_n} = x$. The continuity of f implies that $\lim_{n\to\infty} e(f(u_{p_n}), f(v_{p_n})) = e(f(x), f(x)) = 0$, so f is uniformly continuous by Theorem 8.13.

8.7 Contractions and Fixed Points

Definition 8.65 *Let (S, d) and (T, d') be two metric spaces. A function $f : S \longrightarrow T$ is a* similarity *if there exists a number $r > 0$ for which $d'(f(x), f(y)) = rd(x, y)$ for every $x, y \in S$. If the two metric spaces coincide, we refer to f as a* self-similarity *of (S, d).*

The number r is called the ratio *of the similarity f and is denoted by ratio(f).*
An isometry *is a similarity of ratio 1. If an isometry exists between the metric spaces (S, d) and (T, d'), then we say that these spaces are* isometric.
If there exists $r > 0$ such that $d'(f(x), f(y)) \leqslant rd(x, y)$ for all $x, y \in S$, then we say that f is a Lipschitz function. *Furthermore, if this inequality is satisfied for a number $r < 1$, then f is a* contraction.

Example 8.66 Let $(\mathbb{R}, d)$ be the metric space defined by $d(x, y) = |x - y|$. Any linear mapping (that is, any mapping of the form $f(x) = ax + b$ for $x \in \mathbb{R}$) is a similarity having ratio a.

Theorem 8.67 *Let $(S, \mathcal{O}_d)$ and $(T, \mathcal{O}_{d'})$ be two metric spaces. Every Lipschitz function $f : S \longrightarrow T$ is uniformly continuous.*

Proof Suppose that $d'(f(x), f(y)) \leqslant rd(x, y)$ for $x, y \in S$ and let ϵ be a positive number. Define $\delta = \frac{\epsilon}{r}$. If $z \in f(C(x, \delta))$, there exists $y \in C(x, \delta)$ such that $z = f(y)$. This implies $d'(f(x), z) = d(f(x), f(y)) < rd(x, y) < r\delta = \epsilon$, so $z \in C(f(x), \epsilon)$. Thus, $f(C(x, \delta)) \subseteq C(f(x), \epsilon)$, which means that f is uniformly continuous.

Theorem 8.67 implies that every similarity is uniformly continuous.
Let $f : S \longrightarrow S$ be a function. We define inductively the functions $f^{(n)} : S \longrightarrow S$ for $n \in \mathbb{N}$ by

$$f^{(0)}(x) = x$$

and

$$f^{(n+1)}(x) = f(f^{(n)}(x))$$

for $x \in S$. The function $f^{(n)}$ is the nth *iteration of the function f*.

Example 8.68 Let $f : \mathbb{R} \longrightarrow \mathbb{R}$ be the function defined by $f(x) = ax + b$ for $x \in \mathbb{R}$, where $a, b \in \mathbb{R}$ and $a \neq 1$. It is easy to verify that $f^{(n)}(x) = a^n x + \frac{a^n - 1}{a-1} \cdot b$ for $x \in \mathbb{R}$.

Definition 8.69 *Let $f : S \longrightarrow S$ be a function. A* fixed point *of f is a member x of the set S that satisfies the equality $f(x) = x$.*

Example 8.70 The function f defined in Example 8.68 has the fixed point $x_0 = \frac{b}{1-a}$.

Theorem 8.71 (Banach Fixed Point Theorem) *Let $(S, \mathcal{O}_d)$ be a complete topological metric space and let $f : S \longrightarrow S$ be a contraction on S. There exists a unique fixed point $u \in S$ for f, and for any $x \in S$ we have $\lim_{n\to\infty} f^{(n)}(x) = u$.*

Proof Since f is a contraction, there exists a positive number r, $r < 1$, such that $d(f(x), f(y)) \leqslant rd(x, y)$ for $x, y \in S$. Note that each such function has at most one fixed point. Indeed, suppose that we have both $u = f(u)$ and $v = f(v)$ and $u \neq v$, so $d(u, v) > 0$. Then, $d(f(u), f(v)) = d(u, v) \leqslant rd(u, v)$, which is absurd because $r < 1$.

The sequence $\mathbf{s} = (x, f(x), \ldots, f^{(n)}(x), \ldots)$ is a Cauchy sequence. Indeed, observe that

$$d(f^{(n)}(x), f^{(n+1)}(x)) \leqslant r d(f^{(n-1)}(x), f^{(n)}(x)) \leqslant \cdots \leqslant r^n d(x, f(x)).$$

For $n \leqslant p$, this implies

$$\begin{aligned} d(f^{(n)}(x), f^{(p)}(x)) &\leqslant d(f^{(n)}(x), f^{(n+1)}(x)) + d(f^{(n+1)}(x), f^{(n+2)}(x)) + \\ &\quad \cdots + d(f^{(p-1)}(x), f^{(p)}(x)) \\ &\leqslant r^n d(x, f(x) + \cdots + r^{p-1} d(x, f(x)) \\ &\leqslant \frac{r^n}{1-r} d(x, f(x)), \end{aligned}$$

which shows that the sequence $\mathbf{s}$ is indeed a Cauchy sequence. By the completeness of $(S, \mathcal{O}_d)$, there exists $u \in S$ such that $u = \lim_{n\to\infty} f^{(n)}(x)$. The continuity of f implies

$$u = \lim_{n\to\infty} f^{(n+1)}(x) = \lim_{n\to\infty} f(f^{(n)}(x)) = f(u),$$

so u is a fixed point of f. Since $d(f^{(n)}(x), f^{(p)}(x)) \leqslant \frac{r^n}{1-r} d(x, f(x))$, we have

$$\lim_{p\to\infty} d(f^{(n)}(x), f^{(p)}(x)) = d(f^{(n)}(x), u) \leqslant \frac{r^n}{1-r} d(x, f(x))$$

for $n \in \mathbb{N}$.

8.7.1 The Hausdorff Metric Hyperspace of Compact Subsets

Lemma 8.72 Let (S, d) be a metric space and let U and V be two subsets of S. If $r \in \mathbb{R}_{\geqslant 0}$ is such that $U \subseteq C(V, r)$ and $V \subseteq C(U, r)$, then we have $|d(x, U) - d(x, V)| \leqslant r$ for every $x \in S$.

Proof Since $U \subseteq C(V, r)$, for every $u \in U$ there is $v \in V$ such that $d(u, v) < r$. Therefore, by the triangular inequality, it follows that for every $u \in U$ there is $v \in V$ such that $d(x, u) < d(x, v) + r$, so $d(x, U) < d(x, v) + r$. Consequently, $d(x, U) \leqslant d(x, V) + r$. In a similar manner, we can show that $V \subseteq C(U, r)$ implies $d(x, V) \leqslant d(x, U) + r$. Thus, $|d(x, U) - d(x, V)| \leqslant r$ for every $x \in S$.

Let $(S, \mathcal{O}_d)$ be a topological metric space. Denote by $\mathcal{K}(S, \mathcal{O}_d)$ the collection of all nonempty, compact subsets of $(S, \mathcal{O}_d)$, and define the mapping $\delta : \mathcal{K}(S, \mathcal{O}_d)^2 \longrightarrow \mathbb{R}_{\geqslant 0}$ by

$$\delta(U, V) = \inf\{r \in \mathbb{R}_{\geqslant 0} \mid U \subseteq C(V, r) \text{ and } V \subseteq C(U, r)\}$$

for $U, V \in \mathcal{K}(S, \mathcal{O}_d)$.

Lemma 8.73 Let U and V be two compact subsets of a topological metric space $(S, \mathcal{O}_d)$. We have

$$\sup_{x \in S} |d(x, U) - d(x, V)| = \max\left\{\sup_{x \in V} d(x, U), \sup_{x \in U} d(x, V)\right\}.$$

Proof Let $x \in S$. There is $v_0 \in V$ such that $d(x, v_0) = d(x, V)$ because V is a compact set. Then, the compactness of U implies that there is $u_0 \in U$ such that $d(u_0, v_0) = d(v_0, U)$. We have

$$\begin{aligned} d(x, U) - d(x, V) &= d(x, U) - d(x, v_0) \\ &\leqslant d(x, u_0) - d(x, v_0) \\ &\leqslant d(u_0, v_0) \leqslant \sup_{x \in V} d(U, x). \end{aligned}$$

Similarly, $d(x, U) - d(x, V) \leqslant \sup_{x \in U} d(x, V)$, which implies

$$\sup_{x \in S} |d(x, U) - d(x, V)| \leqslant \max\left\{\sup_{x \in V} d(x, U), \sup_{x \in U} d(x, V)\right\}.$$

On the other hand, since $U \subseteq S$, we have

$$\sup_{x \in S} |d(x, U) - d(x, V)| \geqslant \sup_{x \in U} |d(x, U) - d(x, V)| = \sup_{x \in U} d(x, V)$$

and, similarly, $\sup_{x \in S} |d(x, U) - d(x, V)| \geqslant \sup_{x \in V} d(x, U)$, and these inequalities prove that

$$\sup_{x \in S} |d(x, U) - d(x, V)| \geqslant \max\left\{\sup_{x \in V} d(x, U), \sup_{x \in U} d(x, V)\right\},$$

which concludes the argument.

An equivalent useful definition of δ is given in the next theorem.

Theorem 8.74 *Let (S, d) be a metric space and let U and V be two compact subsets of S. We have the equality*

$$\delta(U, V) = \sup_{x \in S} |d(x, U) - d(x, V)|.$$

Proof Observe that we have both $U \subseteq C(V, \sup_{x \in U} d(x, V))$ and $V \subseteq C(U, \sup_{x \in V} d(x, U))$. Therefore, we have

$$\delta(U, V) \leqslant \max\{\sup_{x \in V} d(x, U), \sup_{x \in U} d(x, V)\}.$$

Combining this observation with Lemma 8.73 yields the desired equality.

Theorem 8.75 *Let $(S, \mathcal{O}_d)$ be a complete topological metric space. The mapping $\delta : \mathcal{K}(S, \mathcal{O}_d)^2 \longrightarrow \mathbb{R}_{\geqslant 0}$ is a metric on $\mathcal{K}(S, \mathcal{O}_d)$.*

Proof It is clear that $\delta(U, U) \geqslant 0$ and that $\delta(U, V) = \delta(V, U)$ for every $U, V \in \mathcal{K}(S, \mathcal{O}_d)$. Suppose that $\delta(U, V) = 0$; that is, $d(x, U) = d(x, V)$ for every $x \in S$. If $x \in U$, then $d(x, U) = 0$, so $d(x, V) = 0$. Since V is closed, by Part (ii) of Theorem 8.18, we have $x \in V$, so $U \subseteq V$. The reverse inclusion can be shown in a similar manner.

To prove the triangular inequality, let $U, V, W \in \mathcal{K}(S, \mathcal{O}_d)$. Since

$$|d(x, U) - d(x, V)| \leqslant |d(x, U) - d(x, V)| + |d(x, V) - d(x, W)|,$$

for every $x \in S$, we have

$$\begin{aligned} \sup_{x\in S} |d(x, U) - d(x, V)| &\leqslant \sup_{x\in S} (|d(x, U) - d(x, V)| + |d(x, V) - d(x, W)|) \\ &\leqslant \sup_{x\in S} |d(x, U) - d(x, V)| + \sup_{x\in S} |d(x, V) - d(x, W)|, \end{aligned}$$

which implies the triangular inequality

$$\delta(U, V) \leqslant \delta(U, W) + \delta(W, V).$$

The metric δ is known as the *Hausdorff metric*, and the metric space $(\mathcal{K}(S, \mathcal{O}_d), \delta)$ is known as the *Hausdorff metric hyperspace* of $(S, \mathcal{O}_d)$.

Theorem 8.76 *If $(S, \mathcal{O}_d)$ is a complete topological metric space, then so is the Hausdorff metric hyperspace $(\mathcal{K}(S, \mathcal{O}_d), \delta)$.*

Proof Let $\mathbf{U} = (U_0, U_1, \ldots)$ be a Cauchy sequence in $(\mathcal{K}(S, \mathcal{O}_d), \delta)$ and let $U = \mathbf{K}(\bigcup_{n\in\mathbb{N}} U_n)$. It is clear that U consists of those elements x of S such that $x = \lim_{n\to\infty} x_n$ for some sequence $\mathbf{x} = (x_0, x_1, \ldots)$, where $x_n \in U_n$ for $n \in \mathbb{N}$.

The set U is precompact. Indeed, let $\epsilon > 0$ and let n_0 be such that $\delta(U_n, U_{n_0}) \leqslant \epsilon$ for $n \geqslant n_0$. Let N be an ϵ-net for the compact set $H = \bigcup_{n\leqslant n_0} U_n$. Clearly, $H \subseteq C(N, \epsilon)$. Since $\delta(U_n, U_{n_0}) \leqslant \epsilon$, it follows that $U \subseteq C(H, \epsilon)$, so $U \subseteq C(N, 2\epsilon)$. This shows that U is precompact. Since U is closed in the complete space $(S, \mathcal{O}_d)$, it follows that U is compact.

Let ϵ be a positive number. Since $\mathbf{U}$ is a Cauchy sequence, there exists $n_{\frac{\epsilon}{2}}$ such that $m, n \geqslant n_{\frac{\epsilon}{2}}$ implies $\delta(U_m, U_n) < \frac{\epsilon}{2}$; that is, $\sup_{s\in S} |d(s, U_m) - d(s, U_n)| < \frac{\epsilon}{2}$. In particular, if $x_m \in U_m$, then $d(x_m, U_n) = \inf_{y\in U_m} d(x, y) < \frac{\epsilon}{2}$, so there exists $y \in U_n$ such that $d(x_m, y) < \frac{\epsilon}{2}$.

For $x \in U$, there exists a sequence $\mathbf{x} = (x_0, x_1, \ldots)$ such that $x_n \in U_n$ for $n \in \mathbb{N}$ and $\lim_{n\to\infty} x_n = x$. Therefore, there exists a number $n'_{\frac{\epsilon}{2}}$ such that $p \geqslant n'_{\frac{\epsilon}{2}}$ implies $d(x, x_p) < \frac{\epsilon}{2}$. This implies $d(x, y) \leqslant d(x, x_p) + d(x_p, y) \leqslant \epsilon$ if $n \geqslant \max\{n_{\frac{\epsilon}{2}}, n'_{\frac{\epsilon}{2}}\}$, and therefore $U \subseteq C(U_n, \epsilon)$.

Let $y \in U_n$. Since $\mathbf{U}$ is a Cauchy sequence, there exists a subsequence $\mathbf{U}' = (U_{k_0}, U_{k_1}, \ldots)$ of $\mathbf{U}$ such that $k_0 = q$ and $\delta(U_{k_j}, U_n) < 2^j \epsilon$ for all $n \geqslant k_j$.

Define the sequence $\mathbf{z} = (z_0, z_1, \ldots)$ by choosing z_k arbitrarily for $k < q$, $z_q = y$, and $z_k \in U_k$ for $k_j < k < k_{j+1}$ such that $d(z_k, z_{k_j}) < 2^{-j}\epsilon$. The sequence $\mathbf{z}$ is a Cauchy sequence in S, so there exists $z = \lim_{k\to\infty} z_k$ and $z \in U$. Since $d(y, z) = \lim_{k\to\infty} d(y, z_k) < \epsilon$, it follows that $y \in C(U, \epsilon)$. Therefore, $\delta(U, U_n) < \epsilon$, which proves that $\lim_{n\to\infty} U_n = U$. We conclude that $(\mathcal{K}(S, \mathcal{O}_d), \delta)$ is complete.

8.8 Measures in Metric Spaces

In this section, we discuss the interaction between metrics and measures defined on metric spaces.

Definition 8.77 *Let (S, d) be a metric space. A* Carathéodory outer measure *on (S, d) is an outer measure on S, $\mu : \mathcal{P}(S) \longrightarrow \hat{\mathbb{R}}_{\geqslant 0}$ such that, for every two sets U and V of the topological space $(S, \mathcal{O}_d)$ such that $d(U, V) > 0$, we have $\mu(U \cup V) = \mu(U) + \mu(V)$.*

Example 8.78 The Lebesgue outer measure introduced in Example 4.131 is a Carathéodory outer measure.

Indeed, let U and V be two disjoint subsets of $\mathbb{R}^n$ such that $r = d(U, V) > 0$ and let $\mathcal{D}$ be the family of n-dimensional intervals that covers $U \cup V$. Without loss of generality we may assume that the diameter of each of these intervals is less than r.

If $D = \bigcup \mathcal{D}$, we have $D = D_U \cup D_V \cup D'$, where D_U is the union of those intervals that cover U, D_V is the similar set for V, and D' is the union of those intervals that are disjoint from U or V, that is, $D' \cap (U \cup V) = \emptyset$. Since $vol(D_U) + vol(D_V) \leqslant vol(D)$, we have $\mu(U) + \mu(V) \leqslant \mu(U \cup V)$, so μ is a Carathéodory outer measure.

Theorem 8.79 *Let (S, d) be a metric space. The outer measure μ on S is a Carathéodory outer measure if and only if every closed set of $(S, \mathcal{O}_d)$ is μ-measurable.*

Proof Suppose that every closed set is μ-measurable, and let U and V be two subsets of S such that $d(U, V) > 0$. Consider the closed set $\mathbf{K}(C(U, r))$, where $r = \frac{d(u,v)}{2}$. Since this is a μ-measurable set, we have

$$\mu(U \cup V) = \mu((U \cup V) \cap C(U, r)) + \mu((U \cup V) \cap \mathbf{K}(C(U, r))) = \mu(U) + \mu(V),$$

so μ is a Carathéodory outer measure.

Conversely, suppose that μ is a Carathéodory outer measure; that is, $d(U, V) > 0$ implies $\mu(U \cup V) = \mu(U) + \mu(V)$.

Let U be an open set, L be a subset of U, and $L_1, L_2, \ldots$ be a sequence of sets defined by

$$L_n = \left\{ t \in L \mid d(t, \mathbf{K}(U)) \geqslant \frac{1}{n} \right\}$$

for $n \geqslant 1$. Note that $L_1, L_2, \ldots$ is an increasing sequence of sets, so the sequence $\mu(L_1), \mu(L_2), \ldots$ is increasing. Therefore, $\lim_{n\to\infty} \mu(L_i)$ exists and $\lim_{n\to\infty} \mu(L_i) \leqslant \mu(L)$. We claim that $\lim_{n\to\infty} \mu(L_i) = \mu(L)$.

Since every set L_n is a subset of L, it follows that $\bigcup_{n\geqslant 1} L_n \subseteq L$. Let $t \in L \subseteq U$. Since U is an open set, there exists $\epsilon > 0$ such that $C(t, \epsilon) \subseteq U$, so $d(t, \mathbf{K}(U)) \geqslant \frac{1}{n}$ if $n > \frac{1}{\epsilon}$. Thus, for sufficiently large values of n, we have $t \in L_n$, so $L \subseteq \bigcup_{n\geqslant 1} L_n$. This shows that $L = \bigcup_{n\geqslant 1} L_n$.

Consider the sequence of sets $M_n = L_{n+1} - L_n$ for $n \geqslant 1$. Clearly, we can write

$$L = L_{2n} \cup \bigcup_{k=2n}^{\infty} M_k = L_{2n} \cup \bigcup_{p=n}^{\infty} M_{2p} \cup \bigcup_{p=n}^{\infty} M_{2p+1},$$

so

$$\mu(L) \leqslant \mu(L_{2n}) + \sum_{p=n}^{\infty} \mu(M_{2p}) + \sum_{p=n}^{\infty} \mu(M_{2p+1}).$$

If both series $\sum_{p=1}^{\infty} \mu(M_{2p})$ and $\sum_{p=1}^{\infty} \mu(M_{2p+1})$ are convergent, then

$$\lim_{n\to\infty} \sum_{p=n}^{\infty} \mu(M_{2p}) = 0 \text{ and } \lim_{n\to\infty} \sum_{p=n}^{\infty} \mu(M_{2p+1}) = 0,$$

and so $\mu(L) \leqslant \lim_{n\to\infty} \mu(L_{2n})$.

If the series $\sum_{p=n}^{\infty} \mu(M_{2p})$ is divergent, let $t \in M_{2p} \subseteq L_{2p+1}$. If $z \in \mathbf{K}(U)$, then $d(t, z) \geqslant \frac{1}{2p+1}$ by the definition of L_{2p+1}. Let $y \in M_{2p+2} \subseteq L_{2p+3}$. We have

$$\frac{1}{2p+2} > d(y, z) > \frac{1}{2p+3},$$

so

$$d(t, y) \geqslant t(t, z) - d(y, z) \geqslant \frac{1}{2p+1} - \frac{1}{2p+2},$$

which means that $d(M_{2p}, M_{2p+2}) > 0$ for $p \geqslant 1$. Since μ is a Carathéodory outer measure, we have

$$\sum_{p=1}^{n} \mu(M_{2p}) = \mu\left(\bigcup_{p=1}^{n} M_{2p}\right) \leqslant \mu(L_{2n}).$$

This implies $\lim_{n\to\infty} \mu(L_n) = \lim_{n\to\infty} \mu(L_{2n}) = \infty$, so we have in all cases $\lim_{n\to\infty} \mu(A_n) = \mu(L)$.

Let F be a closed set in $(S, \mathcal{O}_d)$ and let V be an arbitrary set. The set $V \cup \mathbf{K}(F)$ is contained in the set $\mathbf{K}(F) = F$, so, by the previous argument, there exists a sequence of sets L_n such that $d(L_n, F) \geqslant \frac{1}{n}$ for each n and $\lim_{n\to\infty} \mu(L_n) = \mu(V \cap \mathbf{K}(F))$. Consequently, $\mu(V) \geqslant \mu((V \cap F) \cup L_n) = \mu(V \cup F) + \mu(L_n)$. Taking the limit, we obtain $\mu(V) \geqslant \mu(V \cap F) + \mu(V \cap \mathbf{K}(F))$, which proves that F is μ-measurable.

Corollary 8.80 *Let (S, d) be a metric space. Every Borel subset of S is μ-measurable, where μ is a Carathéodory outer measure on S.*

Proof Since every closed set is μ-measurable relative to a Carathéodory outer measure, it follows that every Borel set is μ-measurable with respect to such a measure.

Thus, we can conclude that every Borel subset of S is Lebesgue measurable.

Let (S, d) be a metric space and let $\mathcal{C}$ be a countable collection of subsets of S. Define

$$\mathcal{C}_r = \{C \in \mathcal{C} \mid diam(C) < r\},$$

and assume that for every $x \in S$ and $r > 0$ there exists $C \in \mathcal{C}_r$ such that $x \in C$. Thus, the collection $\mathcal{C}_r$ is a sequential cover for S, and for every function $f : \mathcal{C} \to \hat{\mathbb{R}}_{\geqslant 0}$ we can construct an outer measure $\mu_{f,r}$ using Method I (the method described in Theorem 4.128). This construction yields an outer measure that is not necessarily a Carathéodory outer measure.

By Corollary 4.130, when r decreases, $\mu_{f,r}$ increases. This allows us to define

$$\hat{\mu}_f = \lim_{r\to 0} \mu_{f,r}.$$

We shall prove that the measure $\hat{\mu}_f$ is a Carathéodory outer measure.

Since each measure $\mu_{f,r}$ is an outer measure, it follows immediately that $\hat{\mu}_f$ is an outer measure.

Theorem 8.81 *Let (S, d) be a metric space, $\mathcal{C}$ be a countable collection of subsets of S, and $f : \mathcal{C} \to \hat{\mathbb{R}}_{\geqslant 0}$. The measure $\hat{\mu}_f$ is a Carathéodory outer measure.*

Proof Let U and V be two subsets of S such that $d(U, V) > 0$. We need to show only that $\hat{\mu}_f(U \cup V) \geqslant \hat{\mu}_f(U) + \hat{\mu}_f(V)$.

Choose r such that $0 < r < d(U, V)$, and let $\mathcal{D}$ be an open cover of $U \cup V$ that consists of sets of $\mathcal{C}_r$. Each set of $\mathcal{D}$ can intersect at most one of the set U and V. This observation allows us to write $\mathcal{D}$ as a disjoint union of two collections, $\mathcal{D} = \mathcal{D}_U \cup \mathcal{D}_V$, where $\mathcal{D}_U$ is an open cover for U and $\mathcal{D}_V$ is an open cover for V. Then,

$$\begin{aligned}\sum\{f(D) \mid D \in \mathcal{D}\} &= \sum\{f(D) \mid D \in \mathcal{D}_U\} + \sum\{f(D) \mid D \in \mathcal{D}_V\} \\ &\geqslant \mu_{f,r}(U) + \mu_{f,r}(V).\end{aligned}$$

This implies $\mu_{f,r}(U \cup V) \geqslant \mu_{f,r}(U) + \mu_{f,r}(V)$, which yields $\hat{\mu}_f(U \cup V) \geqslant \hat{\mu}_f(U) + \hat{\mu}_f(V)$ by taking the limit for $r \to 0$.

The construction of the Carathéodory outer measure $\hat{\mu}_f$ described earlier is knows as *Munroe's Method II* or simply *Method II* (see [1, 2]).

8.9 Embeddings of Metric Spaces

Searching in multimedia databases and visualization of the objects of such databases is facilitated by representing objects in a k-dimensional space, as observed in [3]. In general, the starting point is the matrix of distances between objects, and the aim of the representation is to preserve as much as possible the distances between objects.

Definition 8.82 *Let* (S, d) *and* (S', d') *be two metric spaces. An* embedding *of* (S, d) *in* (S', d') *is a function* $f : S \longrightarrow S'$. *The embedding* f *is an* isometry *if* $d'(f(x), f(y)) = cd(x, y)$ *for some positive constant number* c. *If* f *is an isometry, we refer to it as an* isometric embedding.

If an isometric embedding $f : S \longrightarrow S'$ *exists, then we say that* (S, d) *is* isometrically embedded in (S, d).

Note that an isometry is an injective function for if $f(x) = f(y)$, then $d'(f(x), f(y)) = 0$, which implies $d(x, y) = 0$. This, in turn, implies $x = y$.

Example 8.83 Let S be a set that consists of four objects, $S = \{o_1, o_2, o_3, o_4\}$, that are equidistant in the metric space (S, d); in other words, we assume that $d(o_i, o_j) = k$ for every pair of distinct objects (o_i, o_j).

The subset $U = \{o_1, o_2\}$ of S can be isometrically embedded in $\mathbb{R}^1$; the isometry $h : U \longrightarrow \mathbb{R}^1$ is defined by $h(o_1) = (0)$ and $h(o_2) = (k)$.

For the subset $\{o_1, o_2, o_3\}$ define the embedding $f : \{o_1, o_2, o_3\} \longrightarrow \mathbb{R}^2$ by

$$f(o_1) = (0, 0), f(o_2) = (k, 0), f(o_3) = (c_1, c_2),$$

subject to the conditions

$$c_1^2 + c_2^2 = k^2, (c_1 - k)^2 + c_2^2 = k^2.$$

These equalities yield $c_1 = \frac{k}{2}$ and $c_2^2 = \frac{3k^2}{4}$. Choosing the positive solution of the last equality yields $f(o_3) = (\frac{k}{2}, \frac{k\sqrt{3}}{2})$.

To obtain an isometric embedding g of S in $\mathbb{R}^3$, we seek the mapping $g : S \longrightarrow \mathbb{R}^3$ as

$$g(o_1) = (0, 0, 0), g(o_2) = (k, 0, 0), g(o_3) = \left(\frac{k}{2}, \frac{k\sqrt{3}}{2}, 0\right), g(o_4) = (e_1, e_2, e_3),$$

where

$$e_1^2 + e_2^2 + e_3^2 = k^2,$$
$$(e_1 - k)^2 + e_2^2 + e_3^2 = k^2,$$
$$\left(e_1 - \frac{k}{2}\right)^2 + \left(e_2 - \frac{k\sqrt{3}}{2}\right)^2 + e_3^2 = k^2.$$

The first two equalities imply $e_1 = \frac{k}{2}$; this, in turn, yields

$$e_2^2 + e_3^2 = \frac{3k^2}{4},$$
$$\left(e_2 - \frac{k\sqrt{3}}{2}\right)^2 + e_3^2 = k^2.$$

Subtracting these equalities, one gets $e_2 = \frac{k\sqrt{3}}{6}$. Finally, we have $e_3^2 = \frac{2k^2}{3}$. Choosing the positive solution, we obtain the embedding

$$g(o_1) = (0, 0, 0), g(o_2) = (k, 0, 0), g(o_3) = \left(\frac{k}{2}, \frac{k\sqrt{3}}{2}, 0\right), g(o_4) = \left(\frac{k}{2}, \frac{k\sqrt{3}}{6}, \frac{k\sqrt{6}}{3}\right).$$

Example 8.84 Let (S, d) be a finite metric space such that $|S| = n$. We show that there exists an isometric embedding of (S, d) into $(\mathbb{R}^{n-1}, d_\infty)$, where d_∞ was defined by Equality (6.3).

Indeed, suppose that $S = \{x_1, \ldots, x_n\}$, and define $f : S \longrightarrow \mathbb{R}^{n-1}$ as

$$f(x_i) = (d(x_1, x_i), \ldots, d(x_{n-1}, x_i))$$

for $1 \leqslant i \leqslant n$. We prove that $d(x_i, x_j) = d_\infty(f(x_i), f(x_j))$ for $1 \leqslant i, j \leqslant n$, which will imply that f is an isometry with $c = 1$.

By the definition of d_∞, we have

$$d_\infty(f(x_i), f(x_j)) = \max_{1 \leqslant k \leqslant n-1} |d(x_k, x_i) - d(x_k, x_j)|.$$

Note that for every k we have $|d(x_k, x_i) - d(x_k, x_j)| \leqslant d(x_i, x_j)$ (see Exercise 62). Moreover, for $k = i$, we have $|d(x_i, x_i) - d(x_i, x_j)| = d(x_i, x_j)$, so $\max_{1 \leqslant k \leqslant n-1} |d(x_k, x_i) - d(x_k, x_j)| = d(x_i, x_j)$. The isometry whose existence was established in this example is known as the *Fréchet isometry* and was obtained in [4].

Example 8.85 We now prove the existence of an isometry between the metric spaces $(\mathbb{R}^2, d_\infty)$ and $(\mathbb{R}^2, d_1)$.

Consider the function $f : \mathbb{R}^2 \longrightarrow \mathbb{R}^2$ defined by

$$f(u, v) = \left(\frac{u - v}{2}, \frac{u + v}{2}\right)$$

for $(u, v) \in \mathbb{R}^2$.

Since $\max\{a, b\} = \frac{1}{2}(|a - b| + |a + b|)$ for every $a, b \in \mathbb{R}$, it is easy to see that

$$\max\{|u - u'|, |v - v'|\} = \frac{1}{2}\left|u - u' - (v - v')\right| + \left|u - u' + (v - v')\right|,$$

which is equivalent to

$$d_\infty((u, v), (u', v')) = d_1\left(\left(\frac{u - v}{2}, \frac{u + v}{2}\right), \left(\frac{u' - v'}{2}, \frac{u' + v'}{2}\right)\right).$$

The last equality shows that f is an isometry between $(\mathbb{R}^2, d_\infty)$ and $(\mathbb{R}^2, d_1)$.

Exercises and Supplements

1. Prove that any subset U of a topological metric space $(S, \mathcal{O}_d)$ that has a finite diameter is included in a closed sphere $B(x, diam(U))$ for some $x \in U$.
2. Let d_u be the metric defined in Exercise 69 of Chap. 1, where $S = \mathbb{R}^2$ and d is the usual Euclidean metric on $\mathbb{R}^2$.

   (a) Prove that if $x \neq u$, the set $\{x\}$ is open.
   (b) Prove that the topological metric space $(\mathbb{R}^2, \mathcal{O}_{d_u})$ is not separable.

3. Let $(S, \mathcal{O}_d)$ be a topological metric space. Prove that, for every $s \in S$ and every positive number r, we have $\mathbf{K}(C(x, r)) \subseteq B(x, r)$.
4. Let $(S, \mathcal{O}_d)$ be a topological metric space and let U and V be two subsets of S such that $\delta(U, V) \leqslant r$. Prove that if $\mathcal{D} = \{D_i \mid i \in I\}$ is a cover for V, then the collection $\mathcal{D}' = \{C(D_i, r) \mid i \in I\}$ is a cover for U.
5. Prove that if N_r is an r-net for a subset T of a topological metric space $(S, \mathcal{O}_d)$, then $T \subseteq C(N_r, r)$.
   Solution: Suppose $N_r = \{y_i \mid 1 \leqslant i \leqslant n\}$, so $T \subseteq \bigcup_{i=1}^n C(y_i, r)$. Thus, for each $t \in T$ there is $y_j \in N_r$ such that $d(y_j, t) < r$. This implies that $d(t, N_r) = \inf\{d(t, y) \mid y \in N_r\} < r$, so $t \in C(N_r, r)$.
6. Prove that if N_r is an r-net for each of the sets of a collection of subsets of a metric space (S, d), then N_r is an r-net for $\bigcup \mathcal{C}$.
7. Let U and V be two subsets of a metric space (S, d) such that $\delta(U, V) \leqslant c$. Prove that every r-net for V is an $(r + c)$-net for U.
8. Let $(S, \mathcal{O}_d)$ be a topological metric space and let $f : S \longrightarrow \mathbb{R}$ be the function defined by $f(x) = d(x_0, x)$ for $x \in S$. Prove that f is a continuous function between the topological spaces $(S, \mathcal{O}_d)$ and $(\mathbb{R}, \mathcal{O})$.
9. Define the function $f : \mathbb{R} \longrightarrow (0, 1)$ by

$$f(x) = \frac{1}{1 + e^{-x}}$$

for $x \in \mathbb{R}$. Prove that f is a homeomorphism between the topological spaces $(\mathbb{R}, \mathcal{O}_d)$ and $((0, 1), \mathcal{O}_d \upharpoonright_{(0,1)})$. Conclude that completeness is not a topological property.

10. Let X and Y be two separated sets in the topological metric space $(S, \mathcal{O}_d)$ (recall that the notion of separated sets was introduced in Exercise 12 of Chap. 1). Prove that there are two disjoint open sets L_1, L_2 in $(S, \mathcal{O}_d)$ such that $X \subseteq L_1$, $Y \subseteq L_2$.

    Solution: By Theorem 8.18, the functions d_X and d_Y are continuous, $\mathbf{K}(X) = d_X^{-1}(0)$, and $\mathbf{K}(Y) = d_Y^{-1}(0)$. Since X and Y are separated, we have $X \cap d_Y^{-1}(0) = Y \cap d_X^{-1}(0) = \emptyset$. The disjoint sets $L_1 = \{s \in S \mid d_X(s) - d_Y(s) < 0\}$ and $L_2 = \{s \in S \mid d_X(s) - d_Y(s) > 0\}$ are open due to the continuity of d_X and d_Y, and $X \subseteq L_1$ and $Y \subseteq L_2$.

11. This is a variant of the T_4 separation property of topological metric spaces formulated for arbitrary sets instead of closed sets. Let $(S, \mathcal{O}_d)$ be a topological metric space and let U_1 and U_2 be two subsets of S such that $U_1 \cap \mathbf{K}(U_2) = \emptyset$ and $U_2 \cap \mathbf{K}(U_1) = \emptyset$. There exists two open, disjoint subsets V_1 and V_2 of S such that $U_i \subseteq V_i$ for $i = 1, 2$.

    Solution: Define the disjoint open sets

$$V_1 = \{x \in S \mid d(x, U_1) < d(x, U_2)\},$$
$$V_2 = \{x \in S \mid d(x, U_2) < d(x, U_1)\}.$$

    We have $U_1 \subseteq V_1$ because, for $x \in U_1$, $d(x, U_1) = 0$ and $d(x, U_2) > 0$ since $x \notin \mathbf{K}(U_2)$. Similarly, $U_2 \subseteq V_2$.

12. Prove that, for every sequence $\mathbf{x}$ of real numbers, we have $\liminf \mathbf{x} = -\limsup(-\mathbf{x})$.
13. Find $\liminf \mathbf{x}$ and $\limsup \mathbf{x}$ for $\mathbf{x} = (x_0, \ldots, x_n, \ldots)$, where

    (a) $x_n = (-1)^n \cdot n$,
    (b) $x_n = \frac{(-1)^n}{n}$.

14. Let $(S, \mathcal{O}_d)$ be a topological metric space. Prove that $\mathbf{x} = (x_0, x_1, \ldots)$ is a Cauchy sequence if and only if for every $\epsilon > 0$ there exists $n \in \mathbb{N}$ such that $d(x_n, x_{n+m}) < \epsilon$ for every $m \in \mathbb{N}$.
15. Let $(S, \mathcal{O}_d)$ be a topological metric space. Prove that if every bounded subset of S is compact, then $(S, \mathcal{O}_d)$ is complete.
16. Prove that if $(S_1, d_1), \ldots, (S_n, d_n)$ are complete metric spaces, then their product is a complete metric space.
17. Prove that a topological metric space $(S, \mathcal{O}_d)$ is complete if and only if for every nonincreasing sequence of nonempty closed sets $\mathbf{S} = (S_0, S_1, \ldots)$ such that $\lim_{n\to\infty} diam(S_n) = 0$, we have $\bigcap_{i\in\mathbb{N}} S_i \neq \emptyset$.
18. Let (S, d) and (T, d') be two metric spaces. Prove that every similarity $f : S \longrightarrow T$ is a homeomorphism between the metric topological spaces $(S, \mathcal{O}_d)$ and $(T, \mathcal{O}_{d'})$.

19. Let $(S, \mathcal{O}_d)$ be a complete topological metric space and let $f : B(x_0, r) \longrightarrow S$ be a contraction such that $d(f(x), f(y)) \leqslant kd(x, y)$ for $x, y \in B(x_0, r)$ and $k \in (0, 1)$.
    (a) Prove that if $d(f(x_0), x_0)$ is sufficiently small, then the sequence $\mathbf{x} = (x_0, x_1, \ldots)$, where $x_{i+1} = f(x_i)$ for $i \in \mathbb{N}$, consists of points located in $B(x_0, r)$.
    (b) Prove that $y = \lim_{n\to\infty} x_n$ exists and $f(y) = y$.
20. Let $f : \mathbb{R}^n \longrightarrow \mathbb{R}^n$ be a self-similarity of the topological metric $(\mathbb{R}^n, d_2)$ having similarity ratio r. Prove that if H is a Lebesgue-measurable set, then $f(H)$ is also Lebesgue-measurable and $\mu(f(H)) = r^n\mu(H)$, where μ is the Lebesgue outer measure.
    Solution: Suppose initially that $r > 0$. Since f is a homeomorphism, the image on an n-dimensional interval $I = \prod_{i=1}^n(a_i, b_i)$ is the n-dimensional interval $f(I) = \prod_{i=1}^n(f(a_i), f(b_i))$. The definition of f implies $vol(f(I)) = r^n vol(I)$. Since
$$\mu(f(H)) = \inf\left\{\sum vol(f(I)) \mid I \in \mathcal{C}, H \subseteq \bigcup I\right\},$$
    it follows that $\mu(f(H)) \leqslant r^n\mu(H)$. Note that the inverse of f is a self similarity with ratio $\frac{1}{r}$, which implies $\mu(f(H)) \geqslant r^n\mu(H)$. Thus, $\mu(f(H)) = r^n\mu(H)$.
21. If U is a subset of $\mathbb{R}^n$ and μ is the Lebesgue outer measure, then $\mu(U) = \inf\{\mu(L) \mid U \subseteq L, L \text{ is open }\}$.
    Solution: The monotonicity of μ implies $\mu(U) \leqslant \inf\{\mu(L) \mid U \subseteq L, L \text{ is open }\}$. If $\mu(U) = \infty$, the reverse inequality is obvious. Suppose therefore that $\mu(U) < \infty$. By Equality (4.11) of Chap. 4, we have
$$\mu(U) = \inf\left\{\sum vol(I_j) \mid j \in J, U \subseteq \bigcup_{j\in J} I_j\right\},$$
    so there exists a collection of n-dimensional open intervals $\{I_j \mid j \in \mathbb{N}\}$ such that $\sum_{j\in\mathbb{N}} \mu(I_j) < \mu(U) + \epsilon$. Thus, $\mu(U) = \inf\{\mu(L) \mid U \subseteq L, L \text{ is open}\}$.
22. A subset U of $\mathbb{R}^n$ is Lebesgue measurable if and only if for every $\epsilon > 0$ there exist an open set L and a closed set H such that $H \subseteq U \subseteq L$ and $\mu(L - H) < \epsilon$.
23. Let $\mathcal{D}$ be a gauge on a set S and let $\mathcal{O}_{\mathcal{D}}$ be the collection of subsets T of S such that for each $t \in T$, there is some finite subset $\mathcal{D}_0$ of $\mathcal{D}$ and some positive number r such that $\bigcap\{C_d(x, r) \mid d \in \mathcal{D}_0\} \subseteq T$. Prove that $\mathcal{O}_{\mathcal{D}}$ is a topology on S.
24. Let $\mathcal{D} = \{d_1, d_2\}$ be a gauge on a set S. In Supplement 71 of Chap. 1 we saw that $d = \max\{d_1, d_2\}$ and $e = d_1 + d_2$ are semimetrics on S. Prove that the topologies determined by either d or e coincide with $\mathcal{O}_{\mathcal{D}}$.
    A *uniformity* on a set S is a collection $\mathcal{U}$ of binary relations on X such that $\iota_X \in \mathcal{U}$, $\xi \in \mathcal{U}$ and $\xi \subseteq \zeta$ implies $\zeta \in \mathcal{U}$, $\xi_1, \xi_2 \in \mathcal{U}$ implies $\xi_1 \cap \xi_2 \in \mathcal{U}$, if

$\xi \in \mathcal{U}$, there is ρ such that $\rho^2 \subseteq \xi$, and $\xi \in \mathcal{U}$ implies $\xi^{-1} \in \mathcal{U}$. A *uniform space* is a pair $(S, \mathcal{U})$, where $\mathcal{U}$ is a uniformity on S. The members of $\mathcal{U}$ are *entourages* of $(S, \mathcal{U})$.

25. Let (S, d) be a metric space and let $\eta_r = \{(x, y) \in S \times S \mid d(x, y) < r\}$, where $r > 0$. Prove that the collection

$$\mathcal{U}_d = \{\zeta \in \mathcal{P}(S \times S) \mid \eta_r \subseteq \zeta\}$$

is a uniformity on S.

26. Let S be a set and let $s_0 \in S$. Define the function $d : S \times S \longrightarrow \{0, 1\}$ as

$$d(x, y) = \begin{cases} 0 & \text{if } \{x, y\} = \{s_0\} \text{ or } s_0 \notin \{x, y\} \\ 1 & \text{if exactly one of } x, y \text{ equals } s_0. \end{cases}$$

Prove that the uniformity $\mathcal{U}_d$ is

$$\mathcal{U}_d = \{\zeta \in \mathcal{P}(S \times S) \mid (S - \{s_0\}) \times (S - \{s_0\}) \subseteq \zeta \text{ and } (s_0, s_0) \in \zeta\}.$$

27. Let $(S, \mathcal{U})$ be a uniform space, ξ be an entourage of $\mathcal{U}$ and x be an element in S. Define the x-section of ξ as the set $\xi[x] = \{y \in S \mid (x, y) \in \xi\}$. Let $\mathcal{O} = \{L \in \mathcal{P}(S) \mid \text{ for every } x \in L, \xi[x] \subseteq L \text{ for some } \xi \in U\}$. Prove that $\mathcal{O}$ is a topology on S.

Bibliographical Comments

A number of topology texts emphasize the study of topological metric spaces. We mention [5] and [6]. The proof of Theorem 8.30 originates in [7]. Supplement 2 is a result of K. Falconer [8].

References

1. M.E. Munroe, Introduction to Measure and Integration (Addison-Wesley, Reading, 1959).
2. G. Edgar, *Measure, Topology, and Fractal Geometry* (Springer, New York, 1990)
3. C. Faloutsos, K.I. Lin, Fastmap: A fast algorithm for indexing, data-mining and visualization of traditional and multimedia datasets, in *Proceedings of SIGMOD, San Jose, California*, ed. by Michael J. Carey, Donovan A. Schneider (ACM Press, NY, 1995), pp. 163–174
4. M. Fréchet, Les dimensions d'un ensemble abstrait. Mathematische Annalen **68**, 145–178 (1910)
5. R. Engelking, K. Siekluchi, *Topology–A Geometric Approach* (Heldermann Verlag, Berlin, 1992)
6. M. Eisenberg, *Topology* (Holt, Rinehart and Winston Inc, New York , 1974)
7. D.W. Khan, *Topology–An Introduction to Point-Set and Algebraic Areas*, 2nd edn. (Dover, New York, 1995)
8. K. Falconer, Fractal Geometry (Wiley, New York, second edition, 2003).

Chapter 9
Convex Sets and Convex Functions

9.1 Introduction

Convex sets and functions have been studied since the nineteenth century; the twentieth century literature on convexity began with Bonnesen and Fenchel's book [1], subsequently reprinted as [2].

Convexity has extensive application in optimization and, therefore, it is important for machine learning and data mining.

9.2 Convex Sets

Let $\mathbf{x}, \mathbf{y} \in \mathbb{R}^n$. The *closed segment* determined by $\mathbf{x}$ and $\mathbf{y}$ is the set

$$[\mathbf{x}, \mathbf{y}] = \{(1-a)\mathbf{x} + a\mathbf{y} \mid 0 \leqslant a \leqslant 1\}.$$

The *half-closed segment* determined by $\mathbf{x}$ and $\mathbf{y}$ is the set

$$[\mathbf{x}, \mathbf{y}) = \{(1-a)\mathbf{x} + a\mathbf{y} \mid 0 \leqslant a < 1\}.$$

Definition 9.1 *A subset C of $\mathbb{R}^n$ is* convex *if, for all $\boldsymbol{x}, \boldsymbol{y} \in C$ we have $[\boldsymbol{x}, \boldsymbol{y}] \subseteq C$.*

Example 9.2 The convex subsets of $\mathbb{R}$ are the intervals of $\mathbb{R}$.

A related concept is given next.

Definition 9.3 *A subset D of $\mathbb{R}^n$ is* affine *if, for all $\boldsymbol{x}, \boldsymbol{y} \in C$ and all $a \in \mathbb{R}$, we have $(1-a)\boldsymbol{x} + a\boldsymbol{y} \in D$.*

In other words, D is an affine set if every point on the line determined by $\mathbf{x}$ and $\mathbf{y}$ belongs to C.

Note that D is a subspace of $\mathbb{R}^n$ if $\mathbf{0} \in D$ and D is an affine set.

D. A. Simovici and C. Djeraba, *Mathematical Tools for Data Mining*, Advanced Information and Knowledge Processing, DOI: 10.1007/978-1-4471-6407-4_9,

Theorem 9.4 *Let D be a non-empty affine set in $\mathbb{R}^n$. There exists translation t_u and a unique subspace L of $\mathbb{R}^n$ such that $D = \mathsf{t}_u(L)$.*

Proof Let $L = \{\mathbf{x} - \mathbf{y} \mid \mathbf{x}, \mathbf{y} \in D\}$ and let $\mathbf{x}_0 \in D$. We have $\mathbf{0} = \mathbf{x}_0 - \mathbf{x}_0 \in L$ and it is immediate that L is an affine set. Therefore, L is a subspace.

Suppose that $D = \mathsf{t}_\mathbf{u}(L) = \mathsf{t}_\mathbf{v}(K)$, where both L and K are subspaces of $\mathbb{R}^n$. Since $\mathbf{0} \in K$, it follows that there exists $\mathbf{w} \in L$ such that $\mathbf{u} + \mathbf{w} = \mathbf{v}$. Similarly, since $\mathbf{0} \in L$, it follows that there exists $\mathbf{t} \in K$ such that $\mathbf{u} = \mathbf{v} + \mathbf{t}$. Consequently, since $\mathbf{w} + \mathbf{t} = \mathbf{0}$, both $\mathbf{w}$ and $\mathbf{t}$ belong to both subspaces L and K.

If $\mathbf{s} \in L$, it follows that $\mathbf{u}+\mathbf{s} = \mathbf{v}+\mathbf{z}$ for some $\mathbf{z} \in K$. Therefore, $\mathbf{s} = (\mathbf{v}-\mathbf{u})+\mathbf{z} \in K$ because $\mathbf{w} = \mathbf{v}-\mathbf{u} \in K$. This implies $L \subseteq K$. The reverse inclusion can be shown similarly.

Affine sets arise in conjunction with solving linear systems, as we show next.

Theorem 9.5 *Let $A \in \mathbb{R}^{m\times n}$ and let $\boldsymbol{b} \in \mathbb{R}^m$. The set $S = \{\boldsymbol{x} \in \mathbb{R}^n \mid A\boldsymbol{x} = \boldsymbol{b}\}$ is an affine subset of $\mathbb{R}^n$. Conversely, every affine subset of $\mathbb{R}^n$ is the set of solutions of a system of the form $A\boldsymbol{x} = \boldsymbol{b}$.*

Proof It is immediate that the set of solutions of a linear system is affine. Conversely, let S be an affine subset of $\mathbb{R}^n$ and let L be the linear subspace such that $S = \mathbf{u} + L$. Let $\{\mathbf{a}_1, \ldots, \mathbf{a}_m\}$ be a basis of $L^\perp$. We have

$$L = \{\mathbf{x} \in \mathbb{R}^n \mid \mathbf{a}_i'\mathbf{x} = 0 \text{ for } 1 \leqslant i \leqslant m\} = \{\mathbf{x} \in \mathbb{R}^n \mid A\mathbf{x} = \mathbf{0}\},$$

where A is a matrix whose rows are $\mathbf{a}_1', \ldots, \mathbf{a}_m'$. By defining $\mathbf{b} = A\mathbf{u}$ we have

$$S = \{\mathbf{u} + \mathbf{x} \mid A\mathbf{x} = \mathbf{0}\} = \{\mathbf{y} \in \mathbb{R}^n \mid A\mathbf{y} = \mathbf{b}\}.$$

Definition 9.6 *Let U be a subset of $\mathbb{R}^n$. A* convex combination *of U is a vector of the form $a_1\boldsymbol{x}_1 + \cdots + a_k\boldsymbol{x}_k$, where $\boldsymbol{x}_1, \ldots, \boldsymbol{x}_k \in U$, $a_i \geqslant 0$ for $1 \leqslant i \leqslant k$, and $a_1 + \cdots + a_k = 1$.*

If the conditions $a_i \geqslant 0$ are dropped, we have an affine combination of U. In other words, $\boldsymbol{x}$ is an affine combination of U if there exist $a_1, \ldots, a_k \in \mathbb{R}$ such that $\boldsymbol{x} = a_1\boldsymbol{x}_1 + \cdots + a_k\boldsymbol{x}_k$, for $\boldsymbol{x}_1, \ldots, \boldsymbol{x}_k \in U$, and $\sum_{i=1}^k a_i = 1$.

Definition 9.7 *Let U be a subset of $\mathbb{R}^n$. A subset $\{\boldsymbol{x}_1, \ldots, \boldsymbol{x}_n\}$ is* affinely dependent *if $\boldsymbol{0} = a_1\boldsymbol{x}_1 + \cdots + a_n\boldsymbol{x}_n$, at least one of the numbers $a_1, \ldots, a_n$ is nonzero, and $\sum_{i=1}^n a_i = 0$. If no such affine combination exists, then $\boldsymbol{x}_1, \ldots, \boldsymbol{x}_n$ are* affinely independent.

Theorem 9.8 *The set $U = \{\boldsymbol{x}_1, \ldots, \boldsymbol{x}_n\}$ is affinely independent if and only if the set $V = \{\boldsymbol{x}_1 - \boldsymbol{x}_n, \ldots, \boldsymbol{x}_{n-1} - \boldsymbol{x}_n\}$ is linearly independent.*

Proof Suppose that U is affinely independent but V is linearly dependent; that is, $\mathbf{0} = b_1(\mathbf{x}_1 - \mathbf{x}_n) + \cdots + b_{n-1}(\mathbf{x}_{n-1} - \mathbf{x}_n)$ such that not all numbers b_i are 0. This implies $b_1\mathbf{x}_1 + \cdots + b_{n-1}\mathbf{x}_{n-1} - \left(\sum_{i=1}^{n-1} b_i\right)\mathbf{x}_n = \mathbf{0}$, which contradicts the affine independence of U.

Conversely, suppose that V is linearly independent but U is not affinely independent. In this case, $\mathbf{0} = a_1\mathbf{x}_1 + \cdots + a_n\mathbf{x}_n$ such that at least one of the numbers $a_1, \ldots, a_n$ is nonzero and $\sum_{i=1}^{n} a_i = 0$. This implies $a_n = -\sum_{i=1}^{n-1} a_i$, so $\mathbf{0} = a_1(\mathbf{x}_1 - \mathbf{x}_n) + \cdots + a_{n-1}(\mathbf{x}_{n-1} - \mathbf{x}_n)$. Observe that at least one of the numbers $a_1, \ldots, a_{n-1}$ must be distinct from 0 because otherwise we would have $a_1 = \cdots = a_{n-1} = a_n = 0$. This contradicts the linear independence of V, so U is affinely independent.

Example 9.9 Let $\mathbf{x}_1, \mathbf{x}_2$ be vectors in $\mathbb{R}^2$. The line that passes through $\mathbf{x}_1$ and $\mathbf{x}_2$ consists of all $\mathbf{x}$ such that $\mathbf{x}-\mathbf{x}_1$ and $\mathbf{x}-\mathbf{x}_2$ are collinear; that is, $a(\mathbf{x}-\mathbf{x}_1)+b(\mathbf{x}-\mathbf{x}_2) = 0$ for some $a, b \in \mathbb{R}$ such that $a + b \neq 0$. Thus, we have $\mathbf{x} = a_1\mathbf{x}_1 + a_2\mathbf{x}_2$, where $a_1 = \frac{a}{a+b}$, $a_2 = \frac{b}{a+b}$ and $a_1 + a_2 = 1$, so $\mathbf{x}$ is an affine combination of $\mathbf{x}_1$ and $\mathbf{x}_2$. It is easy to see that the segment of line contained between $\mathbf{x}_1$ and $\mathbf{x}_2$ is given by a convex combination of $\mathbf{x}_1$ and $\mathbf{x}_2$; that is, by an affine combination $a_1\mathbf{x}_1 + a_2\mathbf{x}_2$ such that $a_1, a_2 \geqslant 0$.

Theorem 9.10 *The intersection of any collection of convex (affine) sets in $\mathbb{R}^n$ is a convex (affine) set.*

Proof Let $\mathcal{C} = \{C_i \mid i \in I\}$ be a collection of convex sets and let $C = \bigcap \mathcal{C}$. Suppose that $\mathbf{x}_1, \ldots, \mathbf{x}_k \in C$, $a_i \geqslant 0$ for $1 \leqslant i \leqslant k$, and $a_1 + \cdots + a_k = 1$. Since $\mathbf{x}_1, \ldots, \mathbf{x}_k \in C_i$, it follows that $a_1\mathbf{x}_1 + \cdots + a_k\mathbf{x}_k \in C_i$ for every $i \in I$. Thus, $a_1\mathbf{x}_1 + \cdots + a_k\mathbf{x}_k \in C$, which proves the convexity of C. The argument for the affine sets is similar.

Corollary 9.11 *The families of convex sets and affine sets of $\mathbb{R}^n$ are closure systems.*

Proof This statement follows immediately from Theorem 9.10 by observing that $\mathbb{R}^n$ itself is convex (affine).

Corollary 9.11 allows us to define the *convex hull* (or the *convex closure*) of a subset U of $\mathbb{R}^n$ as the closure $\mathbf{K}_{conv}(U)$ of U relative to the closure system of the convex subsets of $\mathbb{R}^n$. Similarly, the affine hull of U is the closure $\mathbf{K}_{aff}(U)$.

Theorem 9.12 *Every affine subset S of $\mathbb{R}^n$ is the intersection of a finite collections of hyperplanes.*

Proof By Theorem 9.5, S can be written as $S = \{\mathbf{x} \in \mathbb{R}^n \mid A\mathbf{x} = \mathbf{b}\}$, where $A \in \mathbb{R}^{m\times n}$ and $\mathbf{b} \in \mathbb{R}^m$. Therefore, $\mathbf{x} \in S$ if and only if $\mathbf{a}_i'\mathbf{x} = b_i$, where $\mathbf{a}_i$ is the ith row of A. Thus, $S = \bigcap_{i=1}^{m} H_{\mathbf{a}_i, b_i}$.

Definition 9.13 *A* polytope *in $\mathbb{R}^n$ is the convex hull of a finite set of points in $\mathbb{R}^n$.*

It is easy to see that every polytope is a closed and bounded subset of $\mathbb{R}^n$. A polytope P that is the convex hull of $k + 1$ affinely independent points is called a *k-simplex* or a *simplex of dimension k*.

Example 9.14 A two-dimensional simplex is defined starting from three points $\mathbf{x}_1, \mathbf{x}_2, \mathbf{x}_3$ in $\mathbb{R}^2$ such that none of these points is an affine combination of the other two (no point is collinear with the others two). Thus, the two-dimensional simplex generated by $\mathbf{x}_1, \mathbf{x}_2, \mathbf{x}_3$ is the full triangle determined by $\mathbf{x}_1, \mathbf{x}_2, \mathbf{x}_3$.

In general, an n-dimensional simplex is the convex hull of a set of $n + 1$ points $\mathbf{x}_1, \ldots, \mathbf{x}_{n+1}$ in $\mathbb{R}^n$ such that no point is an affine combination of the remaining n points.

Let S be the n-dimensional simplex generated by the points $\mathbf{x}_1, \ldots, \mathbf{x}_{n+1}$ in $\mathbb{R}^n$ and let $\mathbf{x} \in S$. If $\mathbf{x} \in S$, then $\mathbf{x}$ is a convex combination of $\mathbf{x}_1, \ldots, \mathbf{x}_n, \mathbf{x}_{n+1}$. In other words, there exist $a_1, \ldots, a_n, a_{n+1}$ such that $a_1, \ldots, a_n, a_{n+1} \in (0, 1)$, $\sum_{i=1}^{n+1} a_i = 1$, and $\mathbf{x} = a_1\mathbf{x}_1 + \cdots + a_n\mathbf{x}_n + a_{n+1}\mathbf{x}_{n+1}$.

The numbers $a_1, \ldots, a_n, a_{n+1}$ are the *baricentric coordinates* of $\mathbf{x}$ relative to the simplex S and are uniquely determined by $\mathbf{x}$. Indeed, if we have

$$\mathbf{x} = a_1\mathbf{x}_1 + \cdots + a_n\mathbf{x}_n + a_{n+1}\mathbf{x}_{n+1} = b_1\mathbf{x}_1 + \cdots + b_n\mathbf{x}_n + b_{n+1}\mathbf{x}_{n+1},$$

and $a_i \neq b_i$ for some i, this implies

$$(a_1 - b_1)\mathbf{x}_1 + \cdots + (a_n - b_n)\mathbf{x}_n + (a_{n+1} - b_{n+1})\mathbf{x}_{n+1} = \mathbf{0},$$

which contradicts the affine independence of $\mathbf{x}_1, \ldots, \mathbf{x}_{n+1}$.

In Chap. 6 we saw that a hyperplane $H_{\mathbf{w},a}$ partitions $\mathbb{R}^n$ into the sets $H^{>}_{\mathbf{w},a}$, $H^{0}_{\mathbf{w},a} = H_{\mathbf{w},a}$, and $H^{<}_{\mathbf{w},a}$.

Definition 9.15 *Let $P \subseteq \mathbb{R}^n$ be a polytope. A hyperplane $H_{\mathbf{w},a}$* supports *P, or $H_{\mathbf{w},a}$ is a* supporting hyperplane *of P, if $P \cap H_{\mathbf{w},a} \neq \emptyset$ and we have either $P \subseteq H^{>}_{\mathbf{w},a}$ or $P \subseteq H^{0}_{\mathbf{w},a}$.*

If $H_{\mathbf{w},a}$ is a supporting hyperplane of P, then $P \cap H_{\mathbf{w},a}$ is a face *of the polytope P.*

A face of a polytope is necessarily convex. The dimension of a face of a polytope of dimension d ranges from 0 to $d - 1$. Faces of dimension 0 are called *vertices* and faces of dimension 1 are called *edges*. The empty set is defined to be a face of dimension -1 and the entire polytope is also a face of itself. Faces distinct from $\emptyset$ and P are *proper faces*.

The next statement plays a central role in the study of convexity. We reproduce the proof given in [3].

Theorem 9.16 (Carathéodory's Theorem) *If U is a subset of $\mathbb{R}^n$, then for every $\mathbf{x} \in \mathbf{K}_{conv}(U)$ we have $\mathbf{x} = \sum_{i=1}^{n+1} a_i\mathbf{x}_i$, where $\mathbf{x}_i \in U$, $a_i \geqslant 0$ for $1 \leqslant i \leqslant n+1$, and $\sum_{i=1}^{n+1} a_i = 1$.*

Proof Consider $\mathbf{x} \in \mathbf{K}_{conv}(U)$. We can write $\mathbf{x} = \sum_{i=1}^{p+1} a_i\mathbf{x}_i$, where $\mathbf{x}_i \in U$, $a_i \geqslant 0$ for $1 \leqslant i \leqslant p+1$, and $\sum_{i=1}^{p+1} a_i = 1$. Let p be the smallest number which allows this kind of expression for $\mathbf{x}$. We prove the theorem by showing that $p \leqslant n$.

Suppose that $p \geqslant n + 1$. Then, the set $\{\mathbf{x}_1, \ldots, \mathbf{x}_{p+1}\}$ is affinely dependent, so there exist $b_1, \ldots, b_{p+1}$ not all zero such that $\mathbf{0} = \sum_{i=1}^{p+1} b_i \mathbf{x}_i$ and $\sum_{i=1}^{p+1} b_i = 0$. Without loss of generality, we can assume $b_{p+1} > 0$ and $\frac{a_{p+1}}{b_{p+1}} \leqslant \frac{a_i}{b_i}$ for all i such that $1 \leqslant i \leqslant p$ and $b_i > 0$. Define

$$c_i = b_i \left(\frac{a_i}{b_i} - \frac{a_{p+1}}{b_{p+1}} \right)$$

for $1 \leqslant i \leqslant p$. We have

$$\sum_{i=1}^{p} c_i = \sum_{i=1}^{p} a_i - \frac{a_{p+1}}{b_{p+1}} \sum_{i=1}^{p} b_i = 1.$$

Furthermore, $c_i \geqslant 0$ for $1 \leqslant i \leqslant p$. Indeed, if $b_i \leqslant 0$, then $c_i \geqslant a_i \geqslant 0$; if $b_i > 0$, then $c_i \geqslant 0$ because $\frac{a_{p+1}}{b_{p+1}} \leqslant \frac{a_i}{b_i}$ for all i such that $1 \leqslant i \leqslant p$ and $b_i > 0$. Thus, we have

$$\sum_{i=1}^{p} c_i \mathbf{x}_i = \sum_{i=1}^{p} \left(a_i - \frac{a_p}{b_p} b_i \right) \mathbf{x}_i = \sum_{i=1}^{p} a_i \mathbf{x}_i = \mathbf{x},$$

which contradicts the choice of p.

A finite set of points P in $\mathbb{R}^2$ is a *convex polygon* if no member p of P lies in the convex hull of $P - \{p\}$.

Theorem 9.17 *A finite set of points P in $\mathbb{R}^2$ is a convex polygon if and only if no member p of P lies in a two-dimensional simplex formed by three other members of P.*

Proof The argument is straightforward and is left to the reader as an exercise.

Theorem 9.18 (Radon's Theorem) *Let $P = \{\mathbf{x}_i \in \mathbb{R}^n \mid 1 \leqslant i \leqslant n+2\}$ be a set of $n+2$ points in $\mathbb{R}^n$. Then, there are two disjoint subsets R and Q of P such that $\mathbf{K}_{conv}(R) \cap \mathbf{K}_{conv}(Q) \neq \emptyset$.*

Proof Since $n+2$ points in $\mathbb{R}^n$ are affinely dependent, there exist $a_1, \ldots, a_{n+2}$ not all equal to 0 such that

$$\sum_{i=1}^{n+2} a_i \mathbf{x}_i = \mathbf{0} \tag{9.1}$$

and $\sum_{i=1}^{n+2} a_i = 0$. Without loss of generality, we can assume that the first k numbers are positive and the last $n+2-k$ are not. Let $a = \sum_{i=1}^{k} a_i > 0$ and let $b_j = \frac{a_j}{a}$ for $1 \leqslant j \leqslant k$. Similarly, let $c_l = -\frac{a_l}{a}$ for $k+1 \leqslant l \leqslant n+2$. Equality (9.1) can now be written as

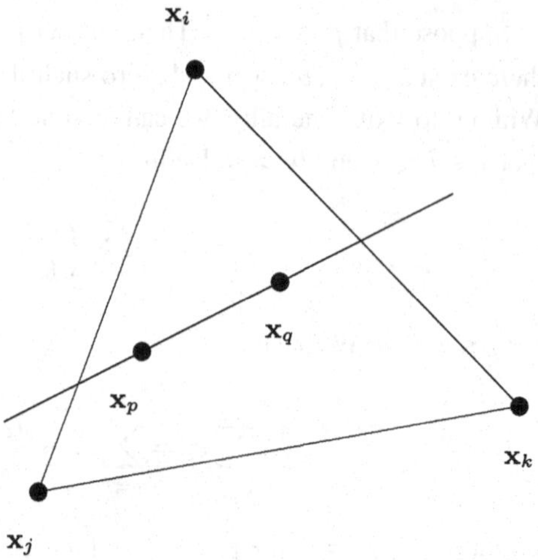

Fig. 9.1 A five-point configuration in $\mathbb{R}^2$

$$\sum_{j=1}^{k} b_j \mathbf{x}_j = \sum_{l=k+1}^{n+2} c_l \mathbf{x}_l.$$

Since the numbers b_j and c_l are nonnegative and $\sum_{j=1}^{k} b_j = \sum_{l=k+1}^{n+2} c_l = 1$, it follows that $\mathbf{K}_{conv}(\{\mathbf{x}_1, \ldots, \mathbf{x}_k\}) \cap \mathbf{K}_{conv}(\{\mathbf{x}_{k+1}, \ldots, \mathbf{x}_{n+2}\}) \neq \emptyset$.

Theorem 9.19 (Klein's Theorem) *If $P \subseteq \mathbb{R}^2$ is a set of five points such that no three of them are collinear, then P contains four points that form a convex quadrilateral.*

Proof Let $P = \{\mathbf{x}_i \mid 1 \leqslant i \leqslant 5\}$. If these five points form a convex polygon, then any four of them form a convex quadrilateral. If exactly one point is in the interior of a convex quadrilateral formed by the remaining four points, then the desired conclusion is reached.

Suppose that none of the previous cases occur. Then, two of the points, say $\mathbf{x}_p, \mathbf{x}_q$, are located inside the triangle formed by the remaining points $\mathbf{x}_i, \mathbf{x}_j, \mathbf{x}_k$. Note that the line $\mathbf{x}_p\mathbf{x}_q$ intersects two sides of the triangle $\mathbf{x}_i\mathbf{x}_j\mathbf{x}_k$, say $\mathbf{x}_i\mathbf{x}_j$ and $\mathbf{x}_i\mathbf{x}_k$ (see Fig. 9.1). Then $\mathbf{x}_p\mathbf{x}_q\mathbf{x}_k\mathbf{x}_j$ is a convex quadrilateral.

A beautiful application of Theorem 3.34 is known as the Erdös-Szekeres theorem. We need the following preliminary observation.

Lemma 9.20 *Let P be a set of points in $\mathbb{R}^2$. If every four-point subset of P is a convex polygon, then the set P itself is a convex polygon.*

Proof This is a direct consequence of Theorem 9.17. □

Theorem 9.21 (The Erdös-Szekeres Theorem) *For every number $n \in \mathbb{N}$, $n \geqslant 3$, there exists a number $E(n)$ such that any set P of points in the plane such that $|P| = E(n)$ and no three points of P are collinear contains an n-point convex polygon.*

Proof A four-point subset of P may or may not be a convex polygon. Thus, the four-point subsets may be colored with two colors: c_1 for convex polygons and c_2 for the other four-point sets.

Choose $E(n) = \mathsf{Ramsey}((n, 5), 4)$, which involves coloring all sets in $\mathcal{P}_4(P)$ with the colors c_1 and c_2. Note that by Klein's theorem, (Theorem 9.19), no five point set can be colored in c_2 (which would mean that none of its four-point sets is convex). Therefore, there exists an n-element set K that can be colored by c_1, and, by Lemma 9.20, the set K is convex. □

9.3 Convex Functions

Definition 9.22 *Let S be a non-empty convex subset of $\mathbb{R}^n$. A function $f : S \longrightarrow \mathbb{R}$ is convex if $f(t\mathbf{x}+(1-t)\mathbf{y}) \leqslant tf(\mathbf{x})+(1-t)f(\mathbf{y})$ for every $\mathbf{x}, \mathbf{y} \in S$ and $t \in [0, 1]$.*

If $f(t\mathbf{x} + (1-t)\mathbf{y}) < tf(\mathbf{x}) + (1-t)f(\mathbf{y})$ for every $\mathbf{x}, \mathbf{y} \in S$ and $t \in (0, 1)$ then f is said to be strictly convex.

The function $g : S \longrightarrow \mathbb{R}$ is concave *if $-g$ is convex.*

Example 9.23 Any norm ν on $\mathbb{R}^n$ is convex. Indeed, we have

$$\nu(t\mathbf{x} + (1-t)\mathbf{y}) \leqslant \nu(t\mathbf{x}) + \nu((1-t)\mathbf{y}) = t\nu(\mathbf{x}) + (1-t)\nu(\mathbf{y})$$

for $\mathbf{x}, \mathbf{y} \in \mathbb{R}^n$ and $t \in (0, 1)$.

Example 9.24 Let $f : (0, \infty) \longrightarrow \mathbb{R}$ be defined by $f(x) = x^2$. The definition domain of f is clearly convex and

$$\begin{aligned} & tx^2 + (1-t)y^2 - (tx + (1-t)y)^2 \\ & = t(1-t)x^2 + t(1-t)y^2 - 2t(1-t)xy \\ & = t(1-t)(x-y)^2 \geqslant 0, \end{aligned}$$

which implies that f is indeed convex.

Let $f : S \longrightarrow \mathbb{R}$ be a convex function, where S is a convex subset of $\mathbb{R}^n$. As a notational convenience, define the function $\hat{f} : \mathbb{R}^n \longrightarrow \hat{\mathbb{R}}$ as

$$\hat{f}(\mathbf{x}) = \begin{cases} f(\mathbf{x}) & \text{if } \mathbf{x} \in S, \\ +\infty & \text{otherwise.} \end{cases}$$

Then, f is convex if and only if $\hat{f}$ is convex, that is, it satisfies the inequality $\hat{f}(t\mathbf{x}+(1-t)\mathbf{y}) \leqslant t\hat{f}(\mathbf{x})+(1-t)\hat{f}(\mathbf{y})$ for every $\mathbf{x}, \mathbf{y} \in \mathbb{R}^n$. We extended the usual definition of real-number operations on $\mathbb{R}$ by $t\infty = \infty t = \infty$ for $t > 0$. If there is no risk of confusion we denote $\hat{f}$ simply by f.

Definition 9.25 *Let $f : S \longrightarrow \mathbb{R}$ be a function defined on a convex set S, where $S \subseteq \mathbb{R}^n$. Its* epigraph *is the set*

$$\mathsf{epi}(f) = \{(\boldsymbol{x}, y) \in S \times \mathbb{R} \mid f(\boldsymbol{x}) \leqslant y\} \subseteq \mathbb{R}^{n+1}.$$

The hypograph *of f is the set*

$$\mathsf{hyp}(f) = \{(\boldsymbol{x}, y) \in S \times \mathbb{R} \mid y \leqslant f(\boldsymbol{x})\} \subseteq \mathbb{R}^{n+1}.$$

The intersection

$$\mathsf{epi}(f) \cap \mathsf{hyp}(f) = \{(\mathbf{x}, y) \in S \times \mathbb{R} \mid y = f(\mathbf{x})\} \subseteq \mathbb{R}^{n+1}.$$

is the *graph* of the function f.

Theorem 9.26 *Let $f : S \longrightarrow \mathbb{R}$ be a function defined on the convex subset S of $\mathbb{R}^n$. Then, f is convex on S if and only if its epigraph is a convex subset of $\mathbb{R}^{n+1}$; f is concave if and only if its hypograph is a convex subset of $\mathbb{R}^{n+1}$.*

Proof Let f be a convex function on S. We have $f(t\mathbf{x}+(1-t)\mathbf{y}) \leqslant tf(\mathbf{x})+(1-t)f(\mathbf{y})$ for every $\mathbf{x}, \mathbf{y} \in S$ and $t \in [0, 1]$.

If $(\mathbf{x}_1, y_1), (\mathbf{x}_2, y_2) \in \mathsf{epi}(f)$ we have $f(\mathbf{x}_1) \leqslant y_1$ and $f(\mathbf{x}_2) \leqslant y_2$. Therefore,

$$\begin{aligned} f(t\mathbf{x}_1 + (1-t)\mathbf{x}_2) &\leqslant tf(\mathbf{x}_1) + (1-t)f(\mathbf{x}_2) \\ &\leqslant ty_1 + (1-t)y_2, \end{aligned}$$

so $(t\mathbf{x}_1 + (1-t)\mathbf{x}_2, ty_1 + (1-t)y_2) = t(\mathbf{x}_1, y_1) + (1-t)(\mathbf{x}_2, y_2) \in \mathsf{epi}(f)$ for $t \in [0, 1]$. This shows that $\mathsf{epi}(f)$ is convex.

Conversely, suppose that $\mathsf{epi}(f)$ is convex, that is, if $(\mathbf{x}_1, y_1) \in \mathsf{epi}(f)$ and $(\mathbf{x}_2, y_2) \in \mathsf{epi}(f)$, then

$$t(\mathbf{x}_1, y_1) + (1-t)(\mathbf{x}_2, y_2) = (t\mathbf{x}_1 + (1-t)\mathbf{x}_2, ty_1 + (1-t)y_2) \in \mathsf{epi}(f)$$

for $t \in [0, 1]$. By the definition of the epigraph, this is equivalent to $f(\mathbf{x}_1) \leqslant y_1$, $f(\mathbf{x}_2) \leqslant y_2$ implies $f(t\mathbf{x}_1 + (1-t)\mathbf{x}_2) \leqslant ty_1 + (1-t)y_2$. Choosing $y_1 = f(\mathbf{x}_1)$ and $y_2 = f(\mathbf{x}_2)$ yields $f(t\mathbf{x}_1 + (1-t)\mathbf{x}_2) \leqslant tf(\mathbf{x}_1) + (1-t)f(\mathbf{x}_2)$, which means that f is convex.

The second part of the theorem follows by applying the first part to the function $-f$.

Definition 9.27 *A convex function is* closed *if $\mathsf{epi}(f)$ is a closed set.*

Theorem 9.28 *Let C be a convex subset of $\mathbb{R}^n$, b be a number in $\mathbb{R}$, and let $\mathcal{F} = \{f_i \mid f_i : C \longrightarrow \mathbb{R}, i \in I\}$ be a family of convex functions such that $f_i(\boldsymbol{x}) \leqslant b$ for every $i \in I$ and $\boldsymbol{x} \in C$. Then, the function $f : C \longrightarrow \mathbb{R}$ defined by*

$$f(x) = \sup\{f_i(x) \mid i \in I\}$$

for $\mathbf{x} \in C$ *is a convex function.*

Proof Since the family of function $\mathcal{F}$ is upper bounded, the definition of f is correct. Let $\mathbf{x}, \mathbf{y} \in C$. We have $t\mathbf{x} + (1-t)\mathbf{y} \in C$ because C is convex.

For every $i \in I$ we have $f_i(t\mathbf{x}+(1-t)\mathbf{y}) \leqslant tf_i(\mathbf{x}) + (1-t)f_i(\mathbf{y})$. The definition of f implies $f_i(\mathbf{x}) \leqslant f(\mathbf{x})$ and $f_i(\mathbf{y}) \leqslant f(\mathbf{y})$, so $tf_i(\mathbf{x}) + (1-t)f_i(\mathbf{y}) \leqslant tf(\mathbf{x}) + (1-t)f(\mathbf{y})$ for $i \in I$ and $t \in [0, 1]$.

The definition of f implies $f(t\mathbf{x}+(1-t)\mathbf{y}) \leqslant tf(\mathbf{x}) + (1-t)f(\mathbf{y})$ for $\mathbf{x}, \mathbf{y} \in C$ and $t \in [0, 1]$, so f is convex on C.

Definition 9.29 *Let* $f : S \longrightarrow \mathbb{R}$ *be a convex function. Its* conjugate *is the function* $f^* : \mathbb{R}^n \longrightarrow \mathbb{R}$ *given by* $f^*(y) = \sup\{y'x - f(x) \mid x \in \mathbb{R}^n\}$.

Note that for each $\mathbf{y} \in \mathbb{R}^n$ the function $g_{\mathbf{y}} = \mathbf{y}'\mathbf{x} - f(\mathbf{x})$ is a convex function. Therefore, by Theorem 9.28, f^* is a convex function.

9.3.1 Convexity of One-Argument Functions

The next theorem allows us to reduce convexity of functions of n arguments to convexity of one-argument functions.

Theorem 9.30 *Let* $f : \mathbb{R}^n \longrightarrow \hat{\mathbb{R}}$ *be a function. The function* f *is convex if and only if the function* $\phi_{x,h} : \mathbb{R} \longrightarrow \hat{\mathbb{R}}$ *given by* $\phi_{x,h}(t) = f(x + th)$ *is a convex function for every* $\boldsymbol{x}$ *and* $\boldsymbol{h}$ *in* $\mathbb{R}^n$.

Proof Suppose that f is convex. We have

$$\begin{aligned}
\phi_{\mathbf{x},\mathbf{h}}(ta + (1-t)b) &= f(\mathbf{x} + (ta + (1-t)b)\mathbf{h}) \\
&= f(t(\mathbf{x} + a\mathbf{h}) + (1-t)(\mathbf{x} + b\mathbf{h})) \\
&\leqslant tf(\mathbf{x} + a\mathbf{h}) + (1-t)f(\mathbf{x} + b\mathbf{h}) \\
&= t\phi_{\mathbf{x},\mathbf{h}}(a) + (1-t)\phi_{\mathbf{x},\mathbf{h}}(b),
\end{aligned}$$

which shows that $\phi_{\mathbf{x},\mathbf{h}}$ is indeed convex. The converse implication follows in a similar manner.

Since each set of the form $L_{\mathbf{x},\mathbf{h}} = \{\mathbf{x} + t\mathbf{h} \mid t \in \mathbb{R}\}$ is a line in $\mathbb{R}^n$ if $\mathbf{h} \neq \mathbf{0}$ and $\phi_{\mathbf{x},\mathbf{h}}$ is the restriction of f to $L_{\mathbf{x},\mathbf{h}}$, it follows that $f : \mathbb{R}^n \longrightarrow \hat{\mathbb{R}}$ is convex if and only if its restriction to any line $L_{\mathbf{x},\mathbf{h}}$ is an one-argument convex function or is ∞.

For $n = 1$, $f : \mathbb{R} \longrightarrow \mathbb{R}$ is *convex* if its graph on an interval is located below the chord determined by the endpoints of the interval.

Lemma 9.31 *Let $f : I \longrightarrow \mathbb{R}$ be a function, where I is an open interval. The following statements are equivalent for $a < b < c$, where $a, b, c \in I$:*

(i) $(c-a)f(b) \leqslant (b-a)f(c) + (c-b)f(a)$;
(ii) $\frac{f(b)-f(a)}{b-a} \leqslant \frac{f(c)-f(a)}{c-a}$;
(iii) $\frac{f(c)-f(a)}{c-a} \leqslant \frac{f(c)-f(b)}{c-b}$.

Proof (i) is equivalent to (ii): Suppose that (i) holds. Then we have

$$(c-a)f(b) - (c-a)f(a) \leqslant (b-a)f(c) + (c-b)f(a) - (c-a)f(a),$$

which is equivalent to

$$(c-a)(f(b) - f(a)) \leqslant (b-a)(f(c) - f(a)). \tag{9.2}$$

By dividing both sides by $(b-a)(c-a) > 0$ we obtain Inequality (ii).

Conversely, note that (ii) implies Inequality (9.2). By adding $(c-a)f(a)$ to both sides of this inequality we obtain (i).

In a similar manner it is possible to prove the equivalence between (i) and (iii).

Theorem 9.32 *Let I be an open interval and let $f : I \longrightarrow \mathbb{R}$ is a function. Each of the conditions of Lemma 9.31 for $a < b < c$ in I is equivalent to the convexity of f.*

Proof Let $f : I \longrightarrow \mathbb{R}$ be a convex function and let $a, b, c \in I$ be such that $a < b < c$. Define $t = \frac{c-b}{c-a}$. Clearly $0 < t < 1$ and by the convexity property,

$$\begin{aligned} f(b) &= f(at + (1-t)c) \leqslant tf(a) + (1-t)f(c) \\ &= \frac{c-b}{c-a} f(a) + \frac{b-a}{c-a} f(c), \end{aligned}$$

which yields the first inequality of Lemma 9.31.

Conversely, suppose that the first inequality of Lemma 9.31 is satisfied. Choose $a = x, c = y$ and $b = tx + (1-t)y$ for $t \in (0, 1)$. We have $(c-a)f(b) = (y-x)$ $f(tx+(1-t)y)$ and $(b-a)f(c)+(c-b)f(a) = (1-t)(y-x)f(y)+t(y-x)f(x)$ Taking into account that $y > x$, we obtain the inequality $f(tx + (1-t)y) \leqslant tf(x) + (1-t)f(y)$, which means that f is convex.

Theorem 9.33 *Let I be an open interval and let $f : \mathbb{R} \longrightarrow \mathbb{R}$ is a convex function. The function $g(x, h)$ defined for $x \in I$ and $h \in \mathbb{R} - \{0\}$ as*

$$g(x, h) = \frac{f(x+h) - f(x)}{h}$$

is increasing with respect to each of its arguments.

Proof We need to examine three cases: $0 < h_1 < h_2$, $h_1 < h_2 < 0$, and $h_1 < 0 < h_2$.

In the first case choose $a = x, b = x + h_1$ and $c = x + h_2$ in the second inequality of Lemma 9.31, where all three numbers $x, x + h_1$ and $x + h_2$ belong to I. We obtain $\frac{f(x+h_1)-f(x)}{h_1} \leqslant \frac{f(x+h_2)-f(x)}{h_2}$, which shows that $g(x, h_1) \leqslant g(x, h_2)$.

If $h_1 < h_2 < 0$, choose $a = x + h_1, b = x + h_2$ and $c = x$ in the last inequality of Lemma 9.31. This yields: $\frac{f(x)-f(x+h_1)}{-h_1} \leqslant \frac{f(x)-f(x+h_2)}{-h_2}$, that is $g(x, h_1) \leqslant g(x, h_2)$.

In the last case, $h_1 < 0 < h_2$, begin by noting that the last two inequalities of Lemma 9.31 imply

$$\frac{f(b) - f(a)}{b - a} \leqslant \frac{f(c) - f(b)}{c - b}.$$

By taking $a = x + h_1, b = x$, and $c = x + h_2$ in this inequality we obtain

$$\frac{f(x) - f(x + h_1)}{-h_1} \leqslant \frac{f(x + h_2) - f(x)}{h_2},$$

which is equivalent to $g(x, h_1) \leqslant g(x, h_2)$. This g is increasing with respect to its second argument.

To prove the monotonicity in the first argument let x_1, x_2 be in I such that $x_1 < x_2$ and let h be a number such that both $x_1 + h$ and $x_2 + h$ belong to I. Since g is monotonic in its second argument we have

$$g(x_1, h) = \frac{f(x_1 + h) - f(x_1)}{h} \leqslant \frac{f(x_2 + h) - f(x_1)}{h + (x_2 - x_1)}$$

and

$$\begin{aligned}\frac{f(x_2 + h) - f(x_1)}{h + (x_2 - x_1)} &= \frac{f(x_1) - f(x_2 + h)}{-h - (x_2 - x_1)} \\ &= \frac{f((x_2 + h) - h - (x_2 - x_1)) - f(x_2 + h)}{-h - (x_2 - x_1)} \\ &\leqslant \frac{f((x_2 + h) - h) - f(x_2 + h)}{-h} = \frac{f(x_2 + h) - f(x_2)}{h},\end{aligned}$$

which proves the monotonicity in its first argument.

Theorem 9.34 *Let $f : (a, b) \longrightarrow \mathbb{R}$ be a function such that its second derivative f'' exists on (a, b). Then, f is convex if and only if $f''(t) \geqslant 0$ for $t \in (a, b)$.*

Proof Let $x, y \in (a, b)$ such that $x < y$ and let $(h_0, h_1, \ldots, h_n, \ldots)$ be a decreasing sequence of positive numbers such that $h_0 = y - x$ and $\lim_{n\to\infty} h_n = 0$. Then, $\lim_{n\to\infty}(x + h_n) = x$. By Theorem 9.33, the the sequence $\frac{f(x+h_n)-f(x)}{h_n}$ is decreasing, so

$$f'(x) = \lim_{n\to\infty} \frac{f(x + h_n) - f(x)}{h_n} \leqslant \frac{f(x + h_0) - f(x)}{h_0} = \frac{f(y) - f(x)}{y - x}.$$

Table 9.1 Examples of convex or concave functions

Function	Second derivative	Convexity property
x^r for $r > 0$	$r(r-1)x^{r-2}$	Concave for $r < 1$ Convex for $r \geqslant 1$
$\ln x$	$-\frac{1}{x^2}$	Concave
$x \ln x$	$\frac{1}{x}$	Convex
e^x	e^x	Convex

Similarly, by considering an increasing sequence $(k_0, k_1, \ldots, k_n, \ldots)$ with $k_0 = x-y$ such that $\lim_{n\to\infty} = 0$, and the corresponding sequence $\lim_{n\to\infty}(y + k_n) = y$, it is possible to show that $\frac{f(y)-f(x)}{y-x} \leqslant f'(y)$. Therefore, $f'(x) \leqslant f'(y)$ for $x, y \in (a, b)$, which implies $f''(t) \geqslant 0$ for $t \in (a, b)$.

Conversely, suppose that $f''(t) > 0$ for $x \in (a, b)$ and let $x, y \in (a, b)$ such that $x < y$ and $z = ax + (1-a)y$. Since f' is nondecreasing, $f(t) \leqslant f(z)$ for $t \in [x, z]$ By the fundamental theorem of integral calculus, $f(z) - f(x) = \int_x^z f'(t)dt \leqslant f'(z)(z-x)$. Similarly, $f(y) - f(z) = \int_z^y f'(t)dt \geqslant f'(z)(y-z)$, which yields the inequalities

$$\begin{aligned}
f(z) &= af(z) + (1-a)f(z) \\
&\leqslant a(f(x) + f'(z)(z-x)) + (1-a)(f(y) - f'(z)(y-z)) \\
&= af(x) + (1-a)f(y) + f'(z)(a(z-x) - (1-a)(y-z)).
\end{aligned}$$

Since $z = ax + (1-a)y$ we have $a(z-x) - (1-a)(y-z) = 0$, which implies $f(z) \leqslant af(x) + (1-a)f(y)$. Thus, f is convex.

The functions listed in the Table 9.1, defined on the set $\mathbb{R} \geqslant 0$, provide examples of convex (or concave) functions.

9.3.2 Jensen's Inequality

Theorem 9.35 (Jensen's Theorem) *Let f be a function that is convex on an interval I. If $t_1, \ldots, t_n \in [0, 1]$ are n numbers such that $\sum_{i=1}^n t_i = 1$, then*

$$f\left(\sum_{i=1}^n t_i x_i\right) \leqslant \sum_{i=1}^n t_i f(x_i)$$

for every $x_1, \ldots, x_n \in I$.

Proof The argument is by induction on n, where $n \geqslant 2$. The basis step, $n = 2$, follows immediately from Definition 9.22.

Suppose that the statement holds for n, and let $u_1, \ldots, u_n, u_{n+1}$ be $n+1$ numbers such that $\sum_{i=1}^{n+1} u_i = 1$. We have

$$\begin{aligned}&f(u_1x_1+\cdots+u_{n-1}x_{n-1}+u_nx_n+u_{n+1}x_{n+1})\\&=f\left(u_1x_1+\cdots+u_{n-1}x_{n-1}+(u_n+u_{n+1})\frac{u_nx_n+u_{n+1}x_{n+1}}{u_n+u_{n+1}}\right).\end{aligned}$$

By the inductive hypothesis, we can write

$$\begin{aligned}&f(u_1x_1+\cdots+u_{n-1}x_{n-1}+u_nx_n+u_{n+1}x_{n+1})\\&\leqslant u_1f(x_1)+\cdots+u_{n-1}f(x_{n-1})+(u_n+u_{n+1})f\left(\frac{u_nx_n+u_{n+1}x_{n+1}}{u_n+u_{n+1}}\right).\end{aligned}$$

Next, by the convexity of f, we have

$$f\left(\frac{u_nx_n+u_{n+1}x_{n+1}}{u_n+u_{n+1}}\right)\leqslant\frac{u_n}{u_n+u_{n+1}}f(x_n)+\frac{u_{n+1}}{u_n+u_{n+1}}f(x_{n+1}).$$

Combining this inequality with the previous inequality gives the desired conclusion.

Of course, if f is a concave function and $t_1,\ldots,t_n\in[0,1]$ are n numbers such that $\sum_{i=1}^n t_i=1$, then

$$f\left(\sum_{i=1}^n t_ix_i\right)\geqslant\sum_{i=1}^n t_if(x_i). \tag{9.3}$$

Example 9.36 We saw that the function $f(x)=\ln x$ is concave. Therefore, if $t_1,\ldots,t_n\in[0,1]$ are n numbers such that $\sum_{i=1}^n t_i=1$, then

$$\ln\left(\sum_{i=1}^n t_ix_i\right)\geqslant\sum_{i=1}^n t_i\ln x_i.$$

This inequality can be written as

$$\ln\left(\sum_{i=1}^n t_ix_i\right)\geqslant\ln\prod_{i=1}^n x_i^{t_i},$$

or equivalently

$$\sum_{i=1}^n t_ix_i\geqslant\prod_{i=1}^n x_i^{t_i},$$

for $x_1,\ldots,x_n\in(0,\infty)$.

In the special case where $t_1=\cdots=t_n=\frac{1}{n}$, we have the inequality that relates the arithmetic to the geometric average on n positive numbers:

$$\frac{x_1 + \cdots + x_n}{n} \geqslant \left(\prod_{i=1}^{n} x_i \right)^{\frac{1}{n}}. \tag{9.4}$$

Let $\mathbf{w} = (w_1, \ldots, w_n) \in \mathbb{R}^n$ be such that $\sum_{i=1}^{n} w_i = 1$. For $r \neq 0$, the **w**-*weighted mean of order r* of a sequence of n positive numbers $\mathbf{x} = (x_1, \ldots, x_n) \in \mathbb{R}^n_{>0}$ is the number

$$\mu^r_{\mathbf{w}}(\mathbf{x}) = \left(\sum_{i=1}^{n} w_i x_i^r \right)^{\frac{1}{r}}.$$

Of course, $\mu^r_{\mathbf{w}}(\mathbf{x})$ is not defined for $r = 0$; we will give as special definition

$$\mu^0_{\mathbf{w}}(\mathbf{x}) = \lim_{r \to 0} \mu^r_{\mathbf{w}}(\mathbf{x}).$$

We have

$$\begin{aligned}\lim_{r \to 0} \ln \mu^r_{\mathbf{w}}(\mathbf{x}) &= \lim_{r \to 0} \frac{\ln \sum_{i=1}^{n} w_i x_i^r}{r} = \lim_{r \to 0} \frac{\sum_{i=1}^{n} w_i x_i^r \ln x_i}{\sum_{i=1}^{n} w_i x_i^r} \\ &= \sum_{i=1}^{n} w_i \ln x_i = \ln \prod_{i=1}^{n} x_i^{w_i}.\end{aligned}$$

Thus, if we define $\mu^0_{\mathbf{w}}(\mathbf{x}) = \prod_{i=1}^{n} x_i^{w_i}$, the weighted mean of order r becomes a function continuous everywhere with respect to r.

For $w_1 = \cdots = w_n = \frac{1}{n}$, we have

$$\begin{aligned}\mu^{-1}_{\mathbf{w}}(\mathbf{x}) &= \frac{n x_1 \cdots x_n}{x_2 \cdots x_n + \cdots + x_1 \cdots x_{n-1}} \\ &\quad \text{(the harmonic average of } \mathbf{x}\text{)}, \\ \mu^0_{\mathbf{w}}(\mathbf{x}) &= (x_1 \cdots x_n)^{\frac{1}{n}} \\ &\quad \text{(the geometric average of } \mathbf{x}\text{)}, \\ \mu^1_{\mathbf{w}}(\mathbf{x}) &= \frac{x_1 + \cdots + x_n}{n} \\ &\quad \text{(the arithmetic average of } \mathbf{x}\text{)}.\end{aligned}$$

Theorem 9.37 *If $p < r$, we have $\mu^p_{\mathbf{w}}(\mathbf{x}) \leqslant \mu^r_{\mathbf{w}}(\mathbf{x})$.*

Proof There are three cases depending on the position of 0 relative to p and r.

In the first case, suppose that $r > p > 0$. The function $f(x) = x^{\frac{r}{p}}$ is convex, so by Jensen's inequality applied to $x_1^p, \ldots, x_n^p$, we have

$$\left(\sum_{i=1}^{n} w_i x_i^p\right)^{\frac{r}{p}} \leqslant \sum_{i=1}^{n} w_i x_i^r,$$

which implies

$$\left(\sum_{i=1}^{n} w_i x_i^p\right)^{\frac{1}{p}} \leqslant \left(\sum_{i=1}^{n} w_i x_i^r\right)^{\frac{1}{r}},$$

which is the inequality of the theorem.

If $r > 0 > p$, the function $f(x) = x^{\frac{r}{p}}$ is again convex because $f''(x) = \frac{r}{p}\left(\frac{r}{p} - 1\right) x^{\frac{r}{p}-2} \geq 0$. Thus, the same argument works as in the previous case.

Finally, suppose that $0 > r > p$. Since $0 < \frac{r}{p} < 1$, the function $f(x) = x^{\frac{r}{p}}$ is concave. Thus, by Jensen's inequality,

$$\left(\sum_{i=1}^{n} w_i x_i^p\right)^{\frac{r}{p}} \geq \sum_{i=1}^{n} w_i x_i^r.$$

Since $\frac{1}{r} < 0$, we obtain again

$$\left(\sum_{i=1}^{n} w_i x_i^p\right)^{\frac{1}{p}} \leqslant \left(\sum_{i=1}^{n} w_i x_i^r\right)^{\frac{1}{r}}.$$

Exercises and Supplements

1. Prove that if C is a convex subset of $\mathbb{R}^n$, then $\mathbf{t_u}(C)$ is also convex.
2. A *cone* in $\mathbb{R}^n$ is a set $C \subseteq \mathbb{R}^n$ such that $\mathbf{x} \in C$ and $a \in \mathbb{R}_{\geqslant 0}$ imply $a\mathbf{x} \in C$. Prove that a cone C is convex if and only if $a\mathbf{x} + b\mathbf{y} \in C$ for $a, b \in \mathbb{R}_{\geqslant 0}$ and $\mathbf{x}, \mathbf{y} \in C$.
3. Prove that D is an affine set in $\mathbb{R}^n$ if and only if $\mathbf{u} + D = \{\mathbf{u} + \mathbf{x} \mid \mathbf{x} \in D\}$ is an affine set for every $\mathbf{u} \in \mathbb{R}^n$.
4. Let S be a convex set in $\mathbb{R}^n$ such that $|S| \geqslant n$ and let $\mathbf{x} \in S$. If $r \in \mathbb{N}$ such that $n + 1 \leqslant r \leqslant |S|$ prove that there exist $\binom{|S|-n}{r-n}$ set of points Y, $Y \subseteq S$, such that $\mathbf{x} \in \mathbf{K}_{conv}(Y)$.

   Hint: Use induction on $k = |S| - n$ and Carathéodory's Theorem.

   Let $S, T \subseteq \mathbb{R}^n$ be two subsets in $\mathbb{R}^n$. The Minkowski sum of S and T is the set

$$S + T = \{\mathbf{x} + \mathbf{y} \mid \mathbf{x} \in S, \mathbf{y} \in T\}.$$

5. Prove that for every set $S \subseteq \mathbb{R}^n$ we have $S + \{\mathbf{0}\} = S$ and $S + \emptyset = \emptyset$.

6. Let S, T be two subsets of $\mathbb{R}^n$. Prove that $\mathbf{K}_{conv}(U)(S+T) = \mathbf{K}_{conv}(U)(S) + \mathbf{K}_{conv}(U)(T)$.
7. Prove that if S is a convex set, then $aS + bS$ is a convex set; furthermore, prove that $aS + bS = (a+b)S$ for every $a, b \geqslant 0$.
8. Let T be a subset of $\mathbb{R}^n$. The *core* of T is the set $\mathsf{core}(T) = \{\mathbf{x} \in T \mid a\mathbf{x} + (1-a)\mathbf{y} \in S \text{ for } a \in (0,1), \mathbf{y} \in T\}$. Prove that the core of any set T is a convex subset of T.
9. Prove that the intersection of two convex polygons having a total of n edges is either empty or is a convex polygon with at most n edges.
10. Let $\mathcal{C} = \{C_1, \ldots, C_m\}$ be a collection of m convex sets in $\mathbb{R}^n$, where $m \geq n+1$ such that if every subcollection $\mathcal{C}'$ of $\mathcal{C}$ that contains $n+1$ sets has a non-empty intersection. Prove that $\bigcap \mathcal{C} \neq \emptyset$.

    Hint: Proceed by induction on $k = m - (n+1)$. Apply Radon's theorem in the inductive step of the proof.
11. Prove that the border of a polytope is the union of its proper faces.
12. Prove that each polytope has a finite number of faces.
13. Let $P \subseteq \mathbb{R}^n$ be a polytope and let $\{F_i \mid 1 \leqslant m \leqslant m\}$ be the set of its faces. Prove that if $H_{\mathbf{w}_i, a_i}$ is a hyperplane that supports P such that $F_i = P \cap H^{\geqslant}_{\mathbf{w}_i, a_i}$ for $1 \leqslant i \leqslant m$, then $P = \bigcap_{i=1}^{m} H^{\geqslant}_{\mathbf{w}_i, a_i}$.
14. Let $f : S \longrightarrow \mathbb{R}$ and $g : T \longrightarrow \mathbb{R}$ be two convex functions, where $S, T \subseteq \mathbb{R}^n$, and let $a, b \in \mathbb{R}_{\geqslant 0}$. Prove that $af + bg$ is a convex function.
15. Let $F : S \longrightarrow \mathbb{R}^m$ be a function, where $\emptyset \subset S \subseteq \mathbb{R}^n$. Prove that if each component f_i of F is a convex function on S and $g : \mathbb{R}^m \longrightarrow \mathbb{R}$ is a monotonic function, then the function gF defined by $gF(\mathbf{x}) = g(F(\mathbf{x}))$ for $\mathbf{x} \in S$ is convex.
16. Let $f : S \longrightarrow \mathbb{R}$ be a convex function, where $S \subseteq \mathbb{R}^n$. Define the function $g : \mathbb{R}_{>0} \times S \longrightarrow \mathbb{R}$ by $g(t, \mathbf{x}) = tf\left(\frac{x}{t}\right)$. Prove that g is a convex function.
17. Let $f : \mathbb{R} \geqslant 0 \longrightarrow \mathbb{R}$ be a convex function. Prove that if $f(0) = 0$ and f is monotonic and convex, then f is subadditive.

    Solution: By applying the convexity of f to the interval $[0, x+y]$ with $a = \frac{x}{x+y}$, we have

    $$f(a \cdot 0 + (1-a)(x+y)) \leqslant af(0) + (1-a)f(x+y),$$

    we have $f(y) \leqslant \frac{y}{x+y} f(x+y)$. Similarly, we can show that $f(x) \leqslant \frac{x}{x+y} f(x+y)$. By adding the last two inequalities, we obtain the subadditivity of f.
18. Let S be a convex subset of $\mathbb{R}^n$ and let I be an open interval of $\mathbb{R}$. If $f : S \longrightarrow \mathbb{R}$ and $g : I \longrightarrow \mathbb{R}$ are convex functions such that $g(I) \subseteq S$ and g is non-decreasing, prove that gf is a convex function on S.
19. Let $S \subseteq \mathbb{R}^n$ be a convex set, $S \neq \emptyset$. Define the *support function of* S, $h_S : \mathbb{R}^n \longrightarrow \mathbb{R}$ by $h_S(\mathbf{x}) = \sup\{\mathbf{s}'\mathbf{x} \mid \mathbf{s} \in S\}$. Prove that $\mathrm{Dom}(h_S)$ is a cone in $\mathbb{R}^n$, h_S is a convex function and $h_S(a\mathbf{x}) = ah_S(\mathbf{x})$ for $a \geqslant 0$.

20. Let $f : S \to \mathbb{R}$ be a function, where S is a convex and open subset of $\mathbb{R}^n$ and $f(\mathbf{x}) > 0$ for $\mathbf{x} \in S$. Prove that if $\log f$ is convex, then so is f.
21. The *a-level set of a convex function* $f : S \longrightarrow \mathbb{R}$, where $S \subseteq \mathbb{R}^n$, is the set $L_{a,f} = \{\mathbf{x} \in S \mid f(\mathbf{x}) \leqslant a\}$.

    (a) Prove that if S is a convex set and f is a convex function, then every level set $L_{a,f}$ is a convex set.
    (b) Give an example of a non-convex function whose level sets are convex.

22. Let C be a convex subset of $\mathbb{R}^n$. Prove that if $\mathbf{x} \in \mathbf{I}(C)$ and $y \in \mathbf{K}(C)$, then $[\mathbf{x}, \mathbf{y}) \subseteq \mathbf{I}(C)$.
23. If $C \subseteq \mathbb{R}^n$ is convex, prove that both $\mathbf{I}(C)$ and $\mathbf{K}_{conv}(C)$ are convex sets. Further, prove that if C is an open set, then $\mathbf{K}_{conv}(C)$ is open.
24. Prove that for any subset U of $\mathbb{R}^n$ we have $\mathbf{K}_{conv}(\mathbf{K}(U)) \subseteq \mathbf{K}(\mathbf{K}_{conv}(U))$. If U is bounded, then $\mathbf{K}_{conv}(\mathbf{K}(U)) = \mathbf{K}(\mathbf{K}_{conv}(U))$.
25. Let $f : S \longrightarrow \mathbb{R}$, where $S \subseteq \mathbb{R}^{n \times n}$ is the set of symmetric real matrices and $f(A)$ is the largest eigenvalue of A. Prove that f is a convex function.
26. Let $\mathcal{M}_1 = \{YY' \mid Y \in \mathbb{R}^{n \times k} \text{ and } Y'Y = I_k\}$ and

$$\mathcal{M}_2 = \{W \in \mathbb{R}^{n \times n} \mid W = W', trace(W) = k \text{ and } W \text{ and } I_n - W \text{ are positive semidefinite}\}.$$

Prove that

(a) we have $\mathcal{M}_2 = \mathbf{K}_{conv}(\mathcal{M}_1)$;
(b) $\mathcal{M}_1$ is the set of extreme points of the polytope $\mathcal{M}_2$.

Solution: Every convex combination of elements of $\mathcal{M}_1$ lies in $\mathcal{M}_2$. Indeed, let $Z = a_1 Y_1 Y_1' + \cdots + a_p Y_p Y_p'$ be a convex combination of $\mathcal{M}_1$. It is immediate that Z is a symmetric matrix. Furthermore, by Theorem 5.51 we have

$$trace(Z) = \sum_{h=1}^{p} a_h trace(Y_h Y_h') = \sum_{h=1}^{p} a_h trace(Y_h' Y_h) = \sum_{h=1}^{p} a_h trace(I_k) = k.$$

because $\sum_{h=1}^{p} a_h = 1$. The positive semi-definiteness of YY' follows from Example 6.109, while the positive semi-definiteness of $I_n - YY'$ follows from Supplement 48. Thus, $\mathbf{K}_{conv}(\mathcal{M}_1) \subseteq \mathcal{M}_2$.

Conversely, let $W \in \mathcal{M}_2$. By the spectral theorem for Hermitian matrices (Theorem 7.41) applied to real symmetric matrices, W can be written as $W = U'DU$, where U is an unitary matrix. Clearly, all eigenvalues of W belong to the interval $[0, 1]$.

If the columns of the matrix U' are $\mathbf{u}_1, \ldots, \mathbf{u}_n$ and the eigenvalues of W are $\lambda_1, \ldots, \lambda_n$, then

$$W = (\mathbf{u}_1 \ \cdots \ \mathbf{u}_n) \begin{pmatrix} \lambda_1 & 0 & \cdots & 0 \\ 0 & \lambda_2 & \cdots & 0 \\ \vdots & \vdots & \cdots & \vdots \end{pmatrix} \begin{pmatrix} \mathbf{u}_1' \\ \vdots \\ \mathbf{u}_n' \end{pmatrix},$$

which allows us to write $W = \lambda_1 \mathbf{u}_1 \mathbf{u}_1' + \ldots + \lambda_r \mathbf{u}_r \mathbf{u}_r'$, where W has rank r, $\lambda_1, \ldots, \lambda_r$ are the non-zero eigenvalues of W, and $\sum_{i=1}^{r} \lambda_i = trace(W) = k$. Note that the rank of W is at least k because its eigenvalues reside in the interval $[0, 1]$ and their sum is k.

If the rank of W equals k, then $W = \mathbf{u}_1 \mathbf{u}_1' + \ldots + \mathbf{u}_k \mathbf{u}_k'$ because all eigenvalues equal 1. This allows us to write $W = ZZ'$, where $Z \in \mathbb{R}^{n \times k}$ is the matrix $Z = (\mathbf{u}_1 \ \cdots \ \mathbf{u}_k)$. Since $Z'Z = I_k$ it follows that in this case $W \in \mathcal{M}_1$. In other words, $\mathcal{M}_1$ is exactly the subset of $\mathcal{M}_2$ that consists of rank k matrices.

If $rank(W) = r > k$ we have $W = \lambda_1 \mathbf{u}_1 \mathbf{u}_1' + \ldots + \lambda_r \mathbf{u}_r \mathbf{u}_r'$. Starting from the r matrices $\mathbf{u}_i \mathbf{u}_i'$ we can form $\binom{r}{k}$ matrices of rank k of the form $\sum_{i \in I} \mathbf{u}_i \mathbf{u}_i'$ by considering all subsets I of $\{1, \ldots, r\}$ that contain k elements. We have $W = \sum_{j=1}^{r} \lambda_j \mathbf{u}_j \mathbf{u}_j' = \sum_{I, |I|=k} \alpha_I \sum_{i \in I} \mathbf{u}_i \mathbf{u}_i'$. If we match the coefficients of $\mathbf{u}_i \mathbf{u}_i'$ we have $\lambda_i = \sum_{I, i \in I, |I|=k} \alpha_I$. If we add these equalities we obtain

$$k = \sum_{i=1}^{r} \sum_{I, i \in I, |I|=k} \alpha_I.$$

We choose α_I to depend on the cardinality of I and take into account that each α_I occurs k times in the previous sum. This implies $\sum_{I, i \in I, |I|=k} \alpha_I = 1$, so each W is a convex combination of matrices of rank k, so $\mathbf{K}_{\text{conv}}(\mathcal{M}_1) = \mathcal{M}_2$. No matrix of rank greater than k can be an extreme point. Since every convex and compact set has extreme elements, only matrices of rank k can play this role. Since the definition of $\mathcal{M}_2$ makes no distinction between the k-rank matrices, it follows that the set of extreme elements coincides with $\mathcal{M}_1$.

27. Prove that for a normal matrix $A \in \mathbb{C}^{n \times n}$, $F(A)$ equals the convex hull of $\mathsf{spec}(A)$. Infer that A is Hermitian if and only if $F(A)$ is an interval of $\mathbb{R}$.

Solution: Since A is normal, by the Spectral Theorem for Normal Matrices (Theorem 7.40) there exists a unitary matrix U and a diagonal matrix D such that $A = U^{\mathsf{H}} D U$ and the diagonal elements of D are the eigenvalues of A. Then, by Exercise 37, $F(A) = F(D)$. Therefore, $z \in F(A)$ if $z = \mathbf{x} D \mathbf{x}^{\mathsf{H}}$ for some $\mathbf{x} \in \mathbb{C}^n$ such that $\| \mathbf{x} \|_2 = 1$, so $z = \sum_{k=1}^{n} |x_k|^2 \lambda_k$, where $\mathsf{spec}(A) = \mathsf{spec}(D) = \{\lambda_1, \ldots, \lambda_n\}$, which proves that $F(A)$ is included in the convex closure of $\mathsf{spec}(A)$. The reverse inclusion is immediate.

28. Let $C_1, \ldots, C_k$ be k convex subsets of $\mathbb{R}^n$, where $k \geqslant n+2$. Prove that if any $n+1$ of these sets have a common point, then all the sets have a common point. This fact is known as Helly's Theorem.

Solution: For $i \in \{1, \ldots, k\}$ there exists $\mathbf{x}_i \in C_1 \cap \cdots \cap C_{i-1} \cap C_{i+1} \cap \cdots \cap C_k$. This results in a set $\{\mathbf{x}_1, \ldots, \mathbf{x}_k\}$ of more than $n+2$ vectors that are affinely dependent. By Radon's Theorem we obtain that after a suitable renumbering we have

$$\mathbf{x} \in \mathbf{K}_{conv}(\{\mathbf{x}_1, \ldots, \mathbf{x}_j\}) \cap \mathbf{K}_{conv}(\{\mathbf{x}_{j+1}, \ldots, \mathbf{x}_k\})$$

for some j, where $1 \leqslant j \leqslant k-1$. Since each of the points $\mathbf{x}_1, \ldots, \mathbf{x}_j$ belong to $C_{j+1} \cap \cdots \cap C_k$ we have

$$\mathbf{x} \in \mathbf{K}_{conv}(\{\mathbf{x}_1, \ldots, \mathbf{x}_j\}) \subseteq C_{j+1} \cap \cdots \cap C_k.$$

Similarly, $x \in \mathbf{K}_{conv}(\{\mathbf{x}_{j+1}, \ldots, \mathbf{x}_k\}) \subseteq C_1 \cap \cdots \cap C_j$.

29. Let $\mathcal{C}$ be a finite collection of convex subsets in $\mathbb{R}^n$ and let C be a convex subset of $\mathbb{R}^n$. Prove that if any $n+1$ subsets of $\mathcal{C}$ are intersected by some translation of C, then all sets of $\mathcal{C}$ are intersected by some translation of C.

Let S be an interval of $\mathbb{R}$ and let $\mathbf{x} = (x_1, \ldots, x_n)$ and $\mathbf{y} = (y_1, \ldots, y_n)$ be two sequences of numbers from S such that $x_1 \geqslant x_2 \geqslant \cdots \geqslant x_n$, $y_1 \geqslant y_2 \geqslant \cdots \geqslant y_n$. If $\sum_{i=1}^k x_i \geqslant \sum_{i=1}^k y_i$ for $1 \leqslant k < n$ and $\sum_{i=1}^n x_i = \sum_{i=1}^n y_i$ we say that $\mathbf{x}$ *majorizes* $\mathbf{y}$ and we write $\mathbf{x} \succeq \mathbf{y}$.

30. Let S be an interval of $\mathbb{R}$, $x_1, \ldots, x_n, y_1, \ldots, y_n \in S$ and let $f : S \longrightarrow \mathbb{R}$ be a convex function. Prove the if $(x_1, \ldots, x_n) \succeq (y_1, \ldots, y_n)$, then

$$\sum_{i=1}^n f(x_i) \geqslant \sum_{i=1}^n f(y_i).$$

Solution: The proof is by induction on $n \geqslant 2$. For the base step, $n = 2$, we have $x_1 \geqslant x_2$, $y_1 \geqslant y_2$, $x_1 \geqslant y_1$, and $x_1 + x_2 \geqslant y_1 + y_2$, which imply $x_1 \geqslant y_1 \geqslant y_2 \geqslant x_2$. Therefore, there exists $p \in [0, 1]$ such that $y_1 = px_1 + (1-p)x_2$ and $y_2 = (1-p)x_2 + px_2$, so $f(y_1) \leqslant pf(x_1) + (1-p)f(x_2)$ and $f(y_1) \leqslant (1-p)f(x_1) + pf(x_2)$ because f is convex. By adding these inequalities we obtain the inequality for $n = 2$.

Suppose that the inequality holds for sequences of length n and let

$$(x_1, \ldots, x_{n+1}), (y_1, \ldots, y_{n+1})$$

be two sequences that satisfy the previous conditions. Then, $x_1 + \cdots + x_n \geqslant y_1 + \cdots + y_n$, so there exists a non-negative number z such that

$$x_1 + \cdots + x_n = y_1 + \cdots + (y_n + z).$$

Since $x_1 + \cdots + x_n + x_{n+1} = y_1 + \cdots + y_n + y_{n+1}$, it follows that $z + x_{n+1} = y_{n+1}$.

By the inductive hypothesis, $f(x_1) + \cdots + f(x_n) \geqslant f(y_1) + \cdots + f(y_n + z)$, so

$$f(x_1) + \cdots + f(x_n) + f(x_{n+1}) \geqslant f(y_1) + \cdots + f(y_n + z) + f(x_{n+1}).$$

Since $y_n + z \geqslant y_n \geqslant y_{n+1} \geqslant x_{n+1}$ and $y_n + z + x_{n+1} = y_n + y_{n+1}$, by using again the base case, we obtain $f(y_n + z) + f(x_{n+1}) \geqslant f(y_n) + f(y_{n+1})$, which yields the desired conclusion.

31. A reciprocal result of the inequality introduced in Supplement 30 holds as well. Namely, prove that if $n \geqslant 2$, and the sequences $x_1 \geqslant \cdots \geqslant x_n$ and $y_1 \geqslant \cdots \geqslant y_n$ from an interval S of $\mathbb{R}$ satisfy the inequality $\sum_{i=1}^n f(x_i) \geqslant \sum_{i=1}^n f(y_i)$ for every convex function, then $\sum_{i=1}^k x_i \geqslant \sum_{i=1}^k y_i$ for $1 \leqslant k \leqslant n-1$ and $\sum_{i=1}^n x_i \geqslant \sum_{i=1}^n y_i$.

Solution: The choice $f(x) = x$ for $x \in S$ yields the inequality $x_1 + \cdots + x_n \geqslant y_1 + \cdots + y_n$; the choice $f(x) = -x$ yields $-x_1 - \cdots - x_n \geqslant -y_1 - \cdots - y_n$, so $x_1 + \cdots + x_n = y_1 + \cdots + y_n$. Let now $f_k : S \longrightarrow \mathbb{R}$ be the convex function given by

$$f_k(x) = \begin{cases} 0 & \text{if } x < x_k \\ x - x_k & \text{if } x \geqslant x_k \end{cases}$$

Using f_k we obtain the inequality

$$\begin{aligned} x_1 + \cdots + x_k - kx_k &= h(x_1) + \cdots + h(x_n) \\ &\geqslant h(y_1) + \cdots + h(y_n) \geqslant h(y_1) + \cdots + h(y_k) \\ &\geqslant y_1 + \cdots + y_k - kx_k, \end{aligned}$$

which implies $x_1 + \cdots + x_k \geqslant y_1 + \cdots + y_k$.

32. Let $f : \mathbb{R} \longrightarrow \mathbb{R}$ be a convex, increasing function.

(a) Prove that for any two sequences $(x_1, \ldots, x_n)$ and $(y_1, \ldots, y_n)$ of real numbers such that $y_1 \geqslant \cdots \geqslant y_n$ and $x_1 + \cdots + x_k \geqslant y_1 + \cdots + y_k$ for $1 \leqslant k \leqslant n$ we have

$$f(x_1) + \cdots + f(x_n) \geqslant f(y_1) + \cdots + f(y_n).$$

(b) Prove that for any two sequences of positive real numbers $(a_1, \ldots, a_n)$ and $(b_1, \ldots, b_n)$ such that $a_1 \geqslant \cdots \geqslant a_n$ and $b_1 \geqslant \cdots \geqslant b_n$,

$$a_1 \cdots a_k \geqslant b_1 \cdots b_k$$

for $1 \leqslant k \leqslant n$, we have

$$f(a_1) + \cdots + f(a_n) \geqslant f(b_1) + \cdots + f(b_n).$$

Solution: This statement can be obtained from the inequality shown in Supplement 30 as follows. Let $c = x_1 + \cdots + x_n - (y_1 + \cdots + y_n)$. Clearly, we have $c \geqslant 0$. Let y_{n+1} be a number such that $y_n \geqslant y_{n+1}$ and consider the sequences $(x_1, \ldots, x_n, y_{n+1} - c)$ and $(y_1, \ldots, y_n, y_{n+1})$.

The inequality of Supplement 30 is applicable to these sequences and it yields

$$f(x_1) + \cdots + f(x_n) + f(y_{n+1} - c) \geqslant f(y_1) + \cdots + f(y_n) + f(y_{n+1}),$$

Since f is an increasing function we have

$$f(x_1) + \cdots + f(x_n) + f(y_{n+1}) \geqslant f(y_1) + \cdots + f(y_n) + f(y_{n+1}),$$

which amounts to the equality to be proven.

To prove the second part apply the first part to the numbers $x_i = \log a_i$ and $b_i = \log b_i$.

Bibliographical Comments

The books [4–8] contain a vast amount of results in convexity theory. References that focus on geometric aspects are [3, 9].

The inequality of Supplement 30 is known as the HLPK inequality, an acronym of the authors Hardy, Littlewook, Polya [10] and Karamata [11] (who discovered independently this inequality).

References

1. T. Bonnesen, W. Fenchel, *Theorie der Konvexen Körper* (Ergebnisse der Mathematik und ihrer Grenzgebiete, Berlin, 1934)
2. T. Bonnesen, W. Fenchel. Theorie der Konvexen Körper. American Mathematical Society, reprinted german edition, 1997
3. B. Grünbaum, *Convex Polytopes* (Wiley Interscience, London, 1967)
4. R.T. Rockafellar, *Convex Analysis* (Princeton University Press, Princeton, 1970)

5. S. Boyd, L. Vandenberghe, *Convex Optimization* (Cambridge University Press, Cambridge, 2004)
6. D.P. Bertsekas, A. Nedić, A.E. Ozdaglar, *Convex Analysis and Optimization* (Athena Scientific, Cambridge, 2003)
7. A. Borvinok, *A Course in Convexity* (American Mathematical Society, Providence, 2002)
8. S.R. Lay, *Convex Sets and Their Applications* (Wiley, New York, 1982)
9. R. Schneider, *Convex Bodies: The Brun-Minkowski Theory* (Cambridge University Press, Cambridge, 1993)
10. G.H. Hardy, J.E. Littlewood, G. Polya, Some simple inequalities satisfied by convex functions. Messenger Math. **58**, 145–152 (1929)
11. J. Karamata, Sur une inégalité relative aux fonctions convexes. Publications Mathématiques de l'Université de Belgrade **1**, 145–148 (1932)

Chapter 10
Graphs and Matrices

10.1 Introduction

Graphs model relations between elements of sets. The term "graph" is suggested by the fact that these mathematical structures can be graphically represented. We discuss two types of graphs: directed graphs, which are suitable for representing arbitrary binary relations, and undirected graphs that are useful for representing symmetric binary relations. Special attention is paid to trees, a type of graph that plays a prominent role in many data mining tasks.

Graph mining has become an important research direction in data mining, as illustrated by the recent collections of articles [1, 2].

10.2 Graphs and Directed Graphs

Definition 10.1 *An* undirected graph, *or simply a* graph, *is a pair* $\mathcal{G} = (V, E)$, *where* V *is a set of* vertices *or* nodes *and* E *is a collection of two-element subsets of* V. *If* $\{x, y\} \in E$, *we say that* $e = \{x, y\}$ *is an* edge *of* $\mathcal{G}$ *that joins* x *to* y. *The vertices* x *and* y *are the* endpoints *of the edge* e.

A graph $\mathcal{G} = (V, E)$ *is* finite *if both* V *and* E *are finite. The number of vertices* $|V|$ *is referred to as the* order of the graph.

If u is an endpoint of an edge e, we say that e is *incident* to u. To simplify the notation, we denote an edge $e = \{x, y\}$ by (x, y). If $e = (x, y)$ is an edge, we say that x and y are *adjacent* vertices. If e and e' are two distinct edges in a graph $\mathcal{G}$, then $|e \cap e'| \leqslant 1$.

Graphs can be drawn by representing each vertex by a point in a plane and each edge (x, y) by an arc joining x and y.

Example 10.2 In Fig. 10.1 we show the graph $\mathcal{G} = (\{v_i \mid 1 \leqslant i \leqslant 8\}, E)$, where $E = \{(v_1, v_2), (v_1, v_3), (v_2, v_3), (v_4, v_5), (v_5, v_6), (v_6, v_7), (v_7, v_8), (v_5, v_8)\}$.

D. A. Simovici and C. Djeraba, *Mathematical Tools for Data Mining*,
Advanced Information and Knowledge Processing, DOI: 10.1007/978-1-4471-6407-4_10,

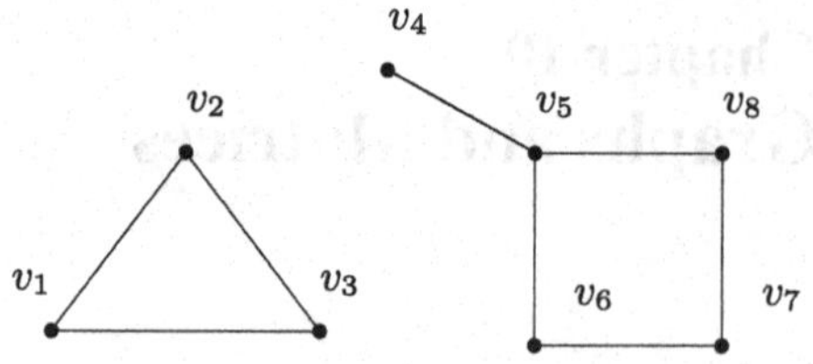

Fig. 10.1 Graph $\mathcal{G} = (\{v_i \mid 1 \leqslant i \leqslant 8\}, E)$

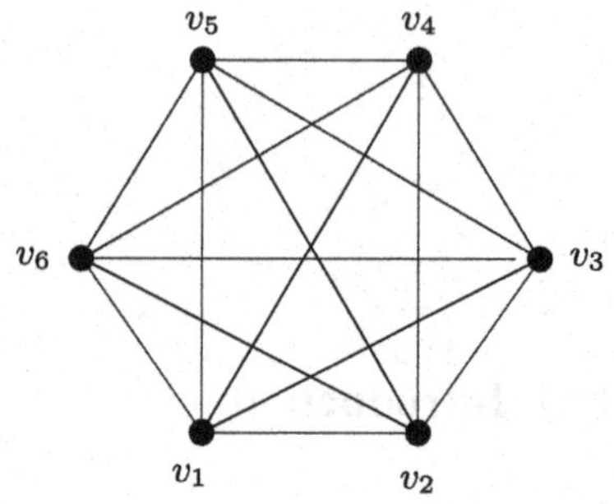

Fig. 10.2 The complete graph $\mathcal{K}_6$

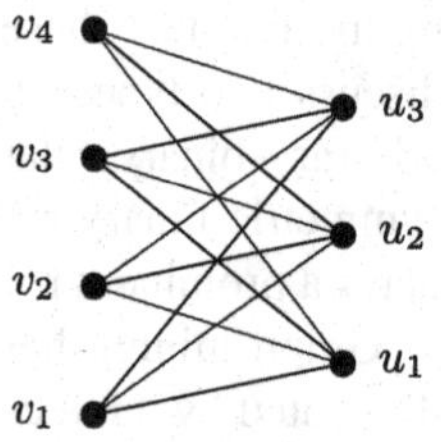

Fig. 10.3 The graph $\mathcal{K}_{4,3}$

Definition 10.3 *A graph* $\mathcal{G} = (V, E)$ *is* complete, *if for every* $u, v \in E$ *such that* $u \neq v$, *we have* $(u, v) \in E$.

The complete graph $\mathcal{G} = (\{1, \ldots, m\}, E)$ will be denoted by $\mathcal{K}_m$.

Example 10.4 The graph shown in Fig. 10.2 is complete.

Example 10.5 Let $\mathcal{K}_{p,q}$ be the graph

$$(\{v_1, \ldots, v_p\} \cup \{u_1, \ldots, u_q\}, \{v_1, \ldots, v_p\} \times \{u_1, \ldots, v_q\}).$$

Note that the set of edges consists of all pairs (u_i, v_j) for $1 \leqslant i \leqslant p$ and $1 \leqslant j \leqslant q$. The graph $\mathcal{K}_{4,3}$ is shown in Fig. 10.3.

Definition 10.6 *A* subgraph *of a graph* $\mathcal{G} = (V, E)$ *is a graph* $\mathcal{G}' = (V', E')$, *where* $V' \subseteq V$ *and* $E' \subseteq E$.

The subgraph of $\mathcal{G}$ induced by a set of vertices U *is the subgraph* $\mathcal{G}_U = (U, E_U)$, *where* $E_U = \{e \in E \mid e = (u, u') \text{ and } u, u' \in U\}$.

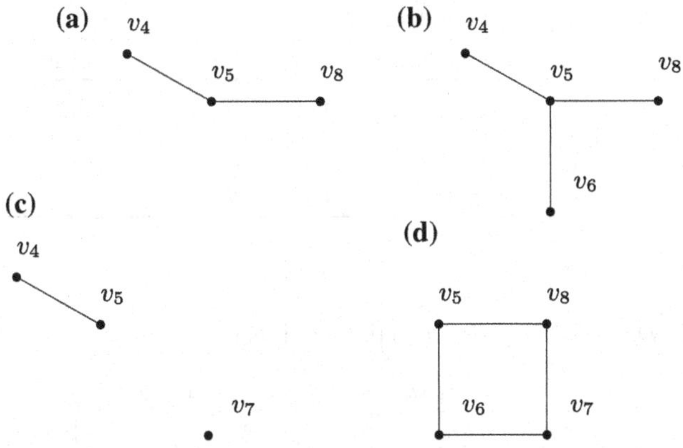

Fig. 10.4 Subgraphs of the graph $\mathcal{G}$

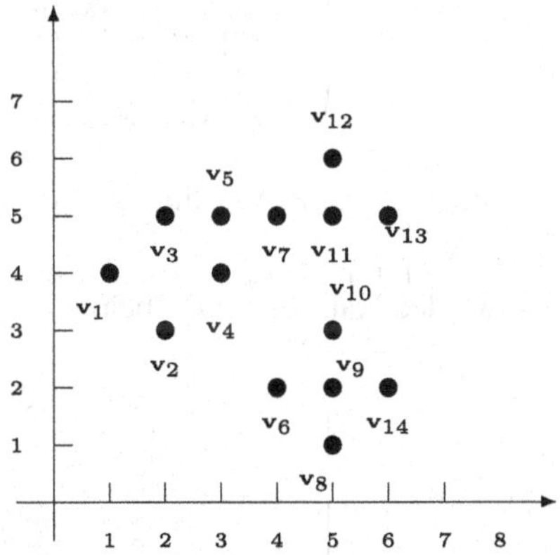

Fig. 10.5 Set of poiFnts

Example 10.7 Consider the graph $\mathcal{G}$ shown in Fig. 10.1. Figure 10.4a–d show the subgraphs $\mathcal{G}$ induced by $\{v_4, v_5, v_8\}$,$\{v_4, v_5, v_6, v_8\}$, $\{v_4, v_5, v_7\}$, and $\{v_5, v_6, v_7, v_9\}$, respectively.

Definition 10.8 *Let V be a finite set where $V = \{v_1, \ldots, v_m\}$, s be a similarity on V, and let α be a positive number. The* α-similarity graph *of (V, s) is the graph $\mathcal{G}_\alpha = (V, E_\alpha)$, where*

$$E_\alpha = \{(u, v) \in V \times V \mid u \neq v \text{ and } s(u, v) \geqslant \alpha\}.$$

Example 10.9 If $\alpha = 0$, the similarity graph of (V, s) is the complete graph on V; if $\alpha = 1$, $\mathcal{G}_1$ consists of m isolated vertices.

Example 10.10 Consider the set of points in $\mathbb{R}^2$ shown in Fig. 10.5 whose coordinates are shown in Table 10.1.

Table 10.1 Coordinates of points to be clustered

Point	Coordinates	Point	Coordinates	Point	Coordinates	Point	Coordinates
$\mathbf{v}_1$	(1, 4)	$\mathbf{v}_5$	(3, 5)	$\mathbf{v}_9$	(5, 2)	$\mathbf{v}_{13}$	(6, 5)
$\mathbf{v}_2$	(2, 3)	$\mathbf{v}_6$	(4, 2)	$\mathbf{v}_{10}$	(5, 3)	$\mathbf{v}_{14}$	(6, 2)
$\mathbf{v}_3$	(2, 5)	$\mathbf{v}_7$	(4, 5)	$\mathbf{v}_{11}$	(5, 5)		
$\mathbf{v}_4$	(3, 4)	$\mathbf{v}_8$	(5, 1)	$\mathbf{v}_{12}$	(5, 6)		

The matrix D of distances between these points is

$$\begin{pmatrix}
0 & 1.41 & 1.41 & 2.00 & 2.23 & 3.60 & 3.16 & 5.00 & 4.47 & 4.12 & 4.12 & 4.47 & 5.09 & 5.38 \\
1.41 & 0 & 2.00 & 1.41 & 2.23 & 2.23 & 2.82 & 3.60 & 3.16 & 3.00 & 3.60 & 4.24 & 4.47 & 4.12 \\
1.41 & 2.00 & 0 & 1.41 & 1.00 & 3.60 & 2.00 & 5.00 & 4.24 & 3.60 & 3.00 & 3.16 & 4.00 & 5.00 \\
2.00 & 1.41 & 1.41 & 0 & 1.00 & 2.23 & 1.41 & 3.60 & 2.82 & 2.23 & 2.23 & 2.82 & 3.16 & 3.60 \\
2.23 & 2.23 & 1.00 & 1.00 & 0 & 3.16 & 1.00 & 4.47 & 3.60 & 2.82 & 2.00 & 2.23 & 3.00 & 4.24 \\
3.60 & 2.23 & 3.60 & 2.23 & 3.16 & 0 & 3.00 & 1.41 & 1.00 & 1.41 & 3.16 & 4.12 & 3.60 & 2.00 \\
3.16 & 2.82 & 2.00 & 1.41 & 1.00 & 3.00 & 0 & 4.12 & 3.16 & 2.23 & 1.00 & 1.41 & 2.00 & 3.60 \\
5.00 & 3.60 & 5.00 & 3.60 & 4.47 & 1.41 & 4.12 & 0 & 1.00 & 2.00 & 4.00 & 5.00 & 4.12 & 1.41 \\
4.47 & 3.16 & 4.24 & 2.82 & 3.60 & 1.00 & 3.16 & 1.00 & 0 & 1.00 & 3.00 & 4.00 & 3.16 & 1.00 \\
4.12 & 3.00 & 3.60 & 2.23 & 2.82 & 1.41 & 2.23 & 2.00 & 1.00 & 0 & 2.00 & 3.00 & 2.23 & 1.41 \\
4.12 & 3.60 & 3.00 & 2.23 & 2.00 & 3.16 & 1.00 & 4.00 & 3.00 & 2.00 & 0 & 1.00 & 1.00 & 3.16 \\
4.47 & 4.24 & 3.16 & 2.82 & 2.23 & 4.12 & 1.41 & 5.00 & 4.00 & 3.00 & 1.00 & 0 & 1.41 & 4.12 \\
5.09 & 4.47 & 4.00 & 3.16 & 3.00 & 3.60 & 2.00 & 4.12 & 3.16 & 2.23 & 1.00 & 1.41 & 0 & 3.00 \\
5.38 & 4.12 & 5.00 & 3.60 & 4.24 & 2.00 & 3.60 & 1.41 & 1.00 & 1.41 & 3.16 & 4.12 & 3.00 & 0
\end{pmatrix}.$$

A similarity matrix S of the set of points can be obtained by defining $s(v_i, v_j) = e^{-\frac{d^2(\mathbf{v}_i,\mathbf{v}_j)}{2}}$ for $1 \leqslant i, j \leqslant 14$. The entries of the similarity matrix (truncated to the last two decimals) are given below:

$$\begin{pmatrix}
1.00 & 0.36 & 0.36 & 0.13 & 0.08 & 0.00 & 0.00 & 0.00 & 0.00 & 0.00 & 0.00 & 0.00 & 0.00 & 0.00 \\
0.36 & 1.00 & 0.13 & 0.36 & 0.08 & 0.08 & 0.01 & 0.00 & 0.00 & 0.01 & 0.00 & 0.00 & 0.00 & 0.00 \\
0.36 & 0.13 & 1.00 & 0.36 & 0.60 & 0.00 & 0.13 & 0.00 & 0.00 & 0.00 & 0.01 & 0.00 & 0.00 & 0.00 \\
0.13 & 0.36 & 0.36 & 1.00 & 0.60 & 0.08 & 0.36 & 0.00 & 0.01 & 0.08 & 0.08 & 0.01 & 0.00 & 0.00 \\
0.08 & 0.08 & 0.60 & 0.60 & 1.00 & 0.00 & 0.60 & 0.00 & 0.00 & 0.01 & 0.13 & 0.08 & 0.01 & 0.00 \\
0.00 & 0.08 & 0.00 & 0.08 & 0.00 & 1.00 & 0.01 & 0.36 & 0.60 & 0.36 & 0.00 & 0.00 & 0.00 & 0.13 \\
0.00 & 0.01 & 0.13 & 0.36 & 0.60 & 0.01 & 1.00 & 0.00 & 0.00 & 0.08 & 0.60 & 0.36 & 0.13 & 0.00 \\
0.00 & 0.00 & 0.00 & 0.00 & 0.00 & 0.36 & 0.00 & 1.00 & 0.60 & 0.13 & 0.00 & 0.00 & 0.00 & 0.36 \\
0.00 & 0.00 & 0.00 & 0.01 & 0.00 & 0.60 & 0.00 & 0.60 & 1.00 & 0.60 & 0.01 & 0.00 & 0.00 & 0.60 \\
0.00 & 0.01 & 0.00 & 0.08 & 0.01 & 0.36 & 0.08 & 0.13 & 0.60 & 1.00 & 0.13 & 0.01 & 0.08 & 0.36 \\
0.00 & 0.00 & 0.01 & 0.08 & 0.13 & 0.00 & 0.60 & 0.00 & 0.01 & 0.13 & 1.00 & 0.60 & 0.60 & 0.00 \\
0.00 & 0.00 & 0.00 & 0.01 & 0.08 & 0.00 & 0.36 & 0.00 & 0.00 & 0.01 & 0.60 & 1.00 & 0.36 & 0.00 \\
0.00 & 0.00 & 0.00 & 0.00 & 0.01 & 0.00 & 0.13 & 0.00 & 0.00 & 0.08 & 0.60 & 0.36 & 1.00 & 0.01 \\
0.00 & 0.00 & 0.00 & 0.00 & 0.00 & 0.13 & 0.00 & 0.36 & 0.60 & 0.36 & 0.00 & 0.00 & 0.01 & 1.00
\end{pmatrix}.$$

The similarity graphs $\mathcal{G}_{0.2}$ and $\mathcal{G}_{0.4}$ are shown in Fig. 10.6a, b, respectively.

Definition 10.11 *Let* $\mathcal{G}_1 = (V_1, E_1)$ *and* $\mathcal{G}_2 = (V_2, E_2)$ *be two graphs having disjoint sets of vertices. Their* union *is the graph* $\mathcal{G}_1 \cup \mathcal{G}_2 = (V_1 \cup V_2, E_1 \cup E_2)$.

The join *of these graphs is the graph* $\mathcal{G}_1 + \mathcal{G}_2 = (V_1 \cup V_2, E_1 \cup E_2 \cup (V_1 \times V_2)$.

The complement *of* $\mathcal{G}_1$ *is the graph*

$$\overline{\mathcal{G}_1} = (V_1, \{(u, v) \in V_1 \times V_1, u \neq v \text{ and } (u, v) \notin E\}).$$

The product *of the graphs* $\mathcal{G}_1$ *and* $\mathcal{G}_2$ *is the graph* $\mathcal{G}_1 \times \mathcal{G}_2 = \{V_1 \times V_2, E\}$, *where* E *consists of pairs* $((u_1, u_2), (v_1, v_2))$ *such that either* $u_1 = v_1$ *and* $(u_2, v_2) \in E_2$ *or* $(u_1, v_1) \in E_1$ *and* $u_2 = v_2$.

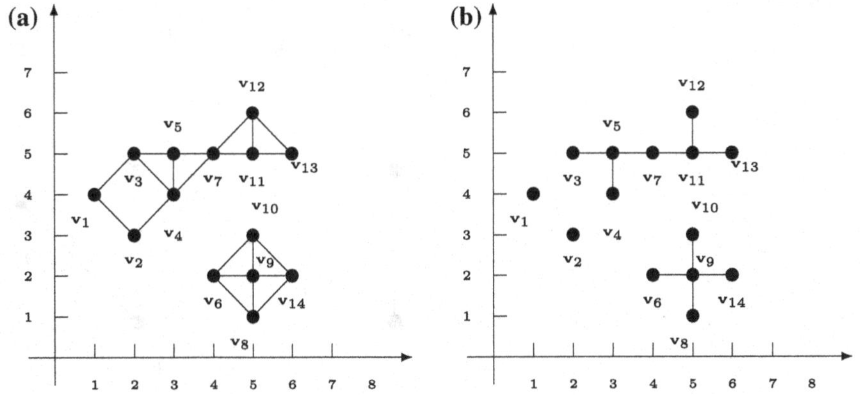

Fig. 10.6 Similarity graphs $\mathcal{G}_{0.2}$ and $\mathcal{G}_{0.4}$

Fig. 10.7 The graph $\mathcal{K}_3+\mathcal{K}_2$

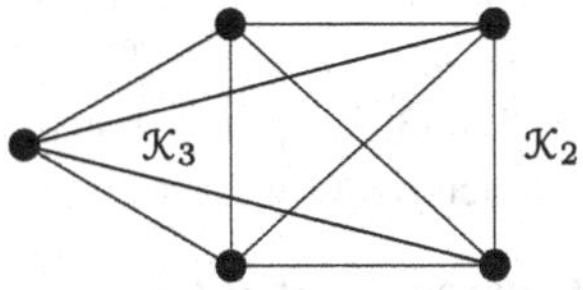

Example 10.12 The complement of a complete graph $\mathcal{K}_4$ is the graph $\overline{\mathcal{K}_4} = (\{v_1, v_2, v_3, v_4\}, \emptyset)$, that is, a graph with no edges. Thus, $\mathcal{K}_{4,3}$ can be expressed as $\mathcal{K}_{4,3} = \overline{\mathcal{K}_4} \cup \overline{\mathcal{K}_3}$. The join of the graph $\mathcal{K}_3$ and $\mathcal{K}_2$ is shown in Fig. 10.7.

If $\mathcal{G}$ is a connected graph, the graph that consist of the union of n disjoint copies of $\mathcal{G}$ is denoted by $n\mathcal{G}$.

Definition 10.13 *Let $\mathcal{G} = (V, E)$ be a graph. The* degree of a vertex v *is the number of edges* $\mathsf{d}_{\mathcal{G}}(v)$ *that are incident with* v.

When the graph $\mathcal{G}$ is clear from the context, we omit the subscript and simply write $\mathsf{d}(v)$.

The degree matrix *of $\mathcal{G} = (V, E)$, where $V = \{v_1, \ldots, v_m\}$ is the diagonal matrix* $D_{\mathcal{G}} = diag(\mathsf{d}_{\mathcal{G}}(v_1), \ldots, \mathsf{d}_{\mathcal{G}}(v_m)) \in \mathbb{N}^{m\times m}$.

If $\mathsf{d}(v) = 0$, then v is an *isolated vertex*. For a graph $\mathcal{G} = (V, E)$, we have

$$\sum\{\mathsf{d}(v) \mid v \in V\} = 2|E| \tag{10.1}$$

because, when adding the degrees of the vertices of the graph we count the number of edges twice. Since the sum of the degrees is an even number, it follows that a finite graph has an even number of vertices that have odd degrees. Also, for every vertex v, we have $\mathsf{d}(v) \leqslant |V| - 1$.

Definition 10.14 *A sequence $(d_1, \ldots, d_m) \in$ $\mathbf{Seq}_n(\mathbb{N})$ is* graphic *if there is a graph $\mathcal{G} = (\{v_1, \ldots, v_m\}, E)$ such that $\mathsf{d}(v_i) = d_i$ for $1 \leqslant i \leqslant m$.*

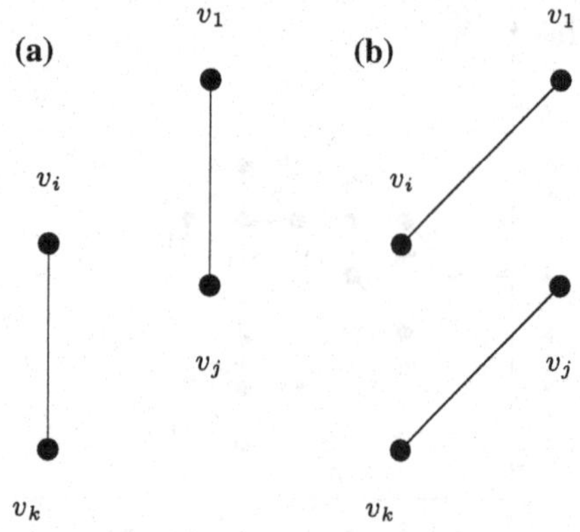

Fig. 10.8 Construction of graph $\hat{\mathcal{G}}$

Clearly, not every sequence of natural numbers is graphic since we must have $d_i \leqslant m - 1$ and $\sum_{i=1}^{m} d_i$ must be an even number. For example, the sequence $(5, 5, 4, 3, 3, 2, 1)$ is not graphic since the sum of its components is not even. A characterization of graphic sequences obtained in [3, 4] is given next.

Theorem 10.15 (The Havel-Hakimi Theorem) *Let* $\mathbf{d} = (d_1, \ldots, d_m)$ *be a sequence of natural numbers such that* $d_1 \geqslant d_2 \geqslant \cdots \geqslant d_m$, $m \geqslant 2$, $d_1 \geqslant 1$, *and* $d_i \leqslant m - 1$ *for* $1 \leqslant i \leqslant m$. *The sequence* $\mathbf{d}$ *is graphic if and only if the sequence*

$$\mathbf{e} = (d_2 - 1, d_3 - 1, \ldots, d_{d_1+1} - 1, d_{d_1+2}, \ldots, d_m)$$

is graphic.

Proof Suppose that $\mathbf{d} = (d_1, \ldots, d_m)$ is a graphic sequence, and let $\mathcal{G} = (\{v_1, \ldots, v_m\}, E)$ be a graph having $\mathbf{d}$ as the sequence of degrees of its vertices.

If there exists a vertex v_1 of degree d_1 that is adjacent with vertices having degrees $d_2, d_3, \ldots, d_{d_1+1}$, then the graph $\mathcal{G}'$ obtained from $\mathcal{G}$ by removing the vertex v_1 and the edges having v_1 as an endpoint has $\mathbf{e}$ as its degree sequence, so $\mathbf{e}$ is graphic.

If no such vertex exists, then there are vertices v_i, v_j such that $i < j$ (and thus $d_i \geqslant d_j$), such that (v_1, v_j) is an edge but (v_1, v_i) is not. Since $d_i \geqslant d_j$ there exists a vertex v_k such that v_k is adjacent to v_i but not to v_j.

Let $\hat{\mathcal{G}}$ be the graph obtained from $\mathcal{G}$ by removing the edges (v_1, v_j) and (v_i, v_k) shown in Fig. 10.8a and adding the edges (v_1, v_i) and (v_j, v_k) shown in Fig. 10.8b. Observe that the degree sequence of $\hat{\mathcal{G}}$ remains the same but the sum of the degrees of the vertices adjacent to v_1 increases. This process may be repeated only a finite number of times before ending with a graph that belongs to the first case.

Conversely, suppose that $\mathbf{e}$ is a graphic sequence, and let $\mathcal{G}_1$ be a graph that has $\mathbf{e}$ as the sequence of vertex degrees. Let $\mathcal{G}_2$ be a graph obtained from $\mathcal{G}_1$ by adding a new vertex v adjacent to vertices of degrees $d_2 - 1, d_3 - 1, \ldots, d_{d_1+1} - 1$. Clearly, the new vertex has degree d_1 and the degree sequence of the new graph is precisely $\mathbf{d}$.

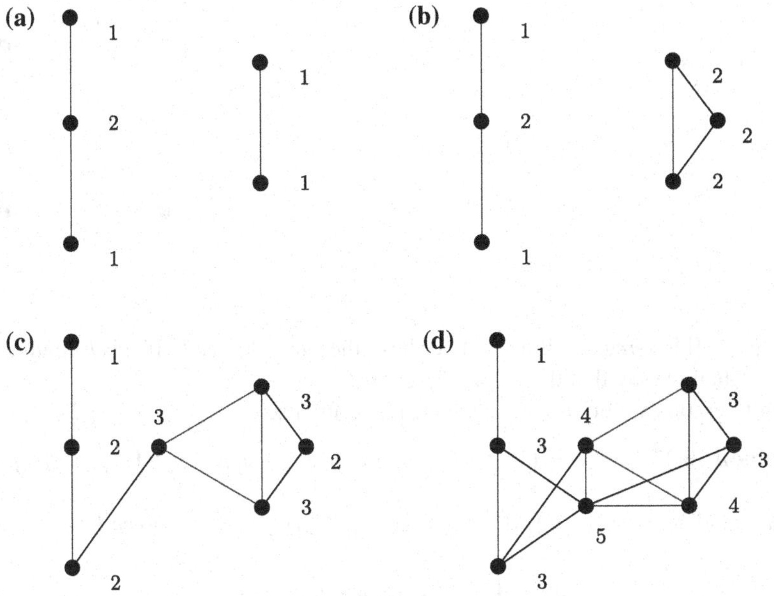

Fig. 10.9 Construction of a graph with a prescribed degree sequence

Table 10.2 Degree sequences

d_1	Sequence
	(5, 4, 4, 3, 3, 3, 3, 1)
5	(3, 3, 3, 2, 2, 2, 1)
3	(2, 2, 2, 2, 1, 1)
2	(2, 1, 1, 1, 1)

Example 10.16 Let us determine if the sequence (5, 4, 4, 3, 3, 3, 3, 1) is a graphic sequence. Note that the sum of its components is an even number. The sequence derived from it by applying the transformation of Theorem 10.15 is (3, 3, 2, 2, 2, 3, 1). Rearranging the sequence in non-increasing order, we have the same question for the sequence (3, 3, 3, 2, 2, 2, 1). A new transformation yields the sequence (2, 2, 1, 2, 2, 1). Placing the components of this sequence in increasing order yields (2, 2, 2, 2, 1, 1). A new transformation produces the shorter sequence (1, 1, 2, 1, 1). The new resulting sequence (2, 1, 1, 1, 1) can be easily seen to be the degree sequence of the graph shown in Fig. 10.9a. We show the degree of each vertex.

The process is summarized in Table 10.2.

Starting from the graph having degree sequence (2, 1, 1, 1, 1), we add a new vertex and two edges linking this vertex to two vertices of degree 1 to obtain the graph of Fig. 10.9b, which has the degree sequence (2, 2, 2, 2, 1, 1).

In the next step, a new vertex is added that is linked by three edges to vertices of degrees 2, 2, and 1. The resulting graph shown in Fig. 10.9c has the degree sequence (3, 3, 3, 2, 2, 2, 1). Finally, a vertex of degree 5 is added to produce the graph shown in Fig. 10.9d, which has the desired degree sequence.

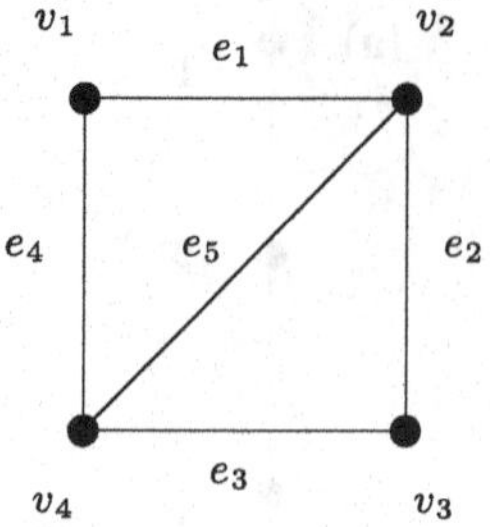

Fig. 10.10 Graph whose incidence and adjacency matrix are given in Example 10.18

A graph $\mathcal{G}$ is *k-regular* if all vertices have the same degree k. If $\mathcal{G}$ is k-regular for some k, then we say that the graph is *regular*.

Finite graphs are often represented using matrices.

Definition 10.17 *Let $\mathcal{G} = (V, E)$ be a finite graph, where $V = \{v_1, \ldots, v_m\}$ and $E = \{e_1, \ldots, e_n\}$.*

The incidence matrix *of $\mathcal{G}$ is the matrix $U_{\mathcal{G}} = (u_{ip}) \in \mathbb{R}^{m\times n}$ given by*

$$u_{ip} = \begin{cases} 1 & if\ v_i\ is\ incident\ to\ e_p, \\ 0 & otherwise, \end{cases}$$

for $1 \leqslant i \leqslant m$ and $1 \leqslant p \leqslant n$.

The adjacency matrix *of $\mathcal{G}$ is the matrix $A_{\mathcal{G}} = (a_{ij}) \in \mathbb{R}^{m\times m}$ given by*

$$a_{ij} = \begin{cases} 1 & if\ v_i\ is\ adjacent\ to\ v_j, \\ 0 & otherwise, \end{cases}$$

for $1 \leqslant i, j \leqslant m$.

Example 10.18 Let $\mathcal{G}$ be the graph shown in Fig. 10.10. Its incidence and adjacency matrices are

$$U_{\mathcal{G}} = \begin{pmatrix} 1\,0\,0\,1\,0 \\ 1\,1\,0\,0\,0 \\ 0\,1\,1\,0\,0 \\ 0\,0\,1\,1\,1 \end{pmatrix} \in \{0,1\}^{4\times 5} \text{ and } A_{\mathcal{G}} = \begin{pmatrix} 1\,0\,0\,1 \\ 1\,0\,1\,1 \\ 0\,1\,0\,1 \\ 1\,1\,1\,0 \end{pmatrix} \in \{0,1\}^{4\times 4}.$$

Definition 10.19 *A* walk *in a graph $\mathcal{G} = (V, E)$ is a sequence of vertices $\mathbf{w} = (v_0, \ldots, v_p)$ such that (v_i, v_{i+1}) is an edge of $\mathcal{G}$ for $0 \leqslant i \leqslant p-1$. The* length *of the walk $\mathbf{w}$, $\ell(\mathbf{w})$, is the number of edges in $\mathbf{w}$ counting repetitions.*

The vertices v_0 and v_p are the endpoints *of $\mathbf{w}$, and we say that the walk $\mathbf{w}$* connects *the vertices v_0 and v_p.*

A path *is a walk having distinct vertices.*

A cycle *is a walk $\mathbf{w} = (v_0, \ldots, v_p)$ such that $p \geqslant 3$ and $v_0 = v_p$.*

A cycle is simple *if all vertices are distinct with the exception of the first and the last. A cycle of length* 3 *is called a* triangle.

A graph with no cycles is said to be acyclic.

For every vertex v of a graph $\mathcal{G} = (V, E)$, there is a unique walk (v) of length 0 that joins v to itself.

Theorem 10.20 *Let $\mathcal{G} = (V, E)$ be a graph, where $|V| = m$. The numbers of walks of length k that join the vertex v_i to the vertex v_j equals $(A^k_{\mathcal{G}})_{ij}$ for $k \in \mathbb{N}$.*

Proof The argument is by induction on k. For the base step, $k = 0$, is immediate because there is a unique walk (v_i) of length 0 that joins v_i to itself and $A^0_{\mathcal{G}} = I_n$.

Suppose that the statement holds for k. Every walk $\mathbf{w}$ of length $k+1$ that joins v_i to v_j can be written as $\mathbf{w} = (v_i, \ldots, v, v_j)$, where $(v_i, \ldots, v)$ is a walk of length k and $(v, v_j) \in E$. By the inductive hypothesis, if $v = v_p$, there are $((A_{\mathcal{G}})^k)_{ip}$ walks of length k that join v_i to v_p. If $(v_p, v_j) \in E$ (which amounts to $(A_{\mathcal{G}})_{pj} = 1$), the number of walks of length $k+1$ from v_i to v_j equals $\sum_{p=1}^{m} (A_{\mathcal{G}})^k)_{ip}(A_{\mathcal{G}})_{pj}$, which equals $((A_{\mathcal{G}})^{k+1})_{ij}$.

Definition 10.21 *Let $\mathcal{G} = (V, E)$ be a graph and let $x, y \in V$ be two vertices. The* distance $d(x, y)$ between x and y *is the length of the shortest path that has x and y as its endpoints. If no such path exists, then we define $d(x, y) = \infty$.*

The diameter *of $\mathcal{G}$ is $d_{\mathcal{G}} = \max\{d(x, y) \mid x, y \in V\}$.*

Example 10.22 The distances between the vertices of the graph shown in Fig. 10.1 are given in the following table.

d	v_1	v_2	v_3	v_4	v_5	v_6	v_7	v_8
v_1	0	1	1	∞	∞	∞	∞	∞
v_2	1	0	1	∞	∞	∞	∞	∞
v_3	1	1	0	∞	∞	∞	∞	∞
v_4	∞	∞	∞	0	1	2	3	2
v_5	∞	∞	∞	1	0	1	2	1
v_6	∞	∞	∞	2	1	0	1	2
v_7	∞	∞	∞	3	2	1	0	1
v_8	∞	∞	∞	2	1	2	1	0

Theorem 10.23 *Let $\mathcal{G} = (V, E)$ be a graph and let x, y be two distinct vertices in V. Every walk that connects x to y contains a path that joins x to y.*

Proof Let $\mathbf{w}$ be a walk in $\mathcal{G}$ that joins x to y. The proof is by induction on $k = \ell(\mathbf{w})$.

If $\ell(\mathbf{w}) = 1$ or $\ell(\mathbf{w}) = 2$, then $\mathbf{w}$ is a path and the conclusion follows.

Suppose that the statement holds for walks having length less than k and let $\mathbf{w} = (v_0, v_1, \ldots, v_k)$ be a path of length k such that $x = v_0$ and $v_k = y$. If the vertices $v_0, \ldots, v_k$ are distinct, then $\mathbf{w}$ is itself a path. Otherwise, let h be the number

$$h = \min\{i \mid 1 \leqslant i \leqslant k, v_i = v_j \text{ for some } j, 0 \leqslant i < j\}.$$

Let $\mathbf{q} = (v_0, \ldots, v_h, v_{j+1}, \ldots, v_k)$, where $v_j = v_h$ and $h < j$. We have $\ell(\mathbf{q}) < k$ and, therefore, by inductive hypothesis, $\mathbf{q}$ contains a path that joints x to y. This implies that the walk $\mathbf{w}$ contains such a path.

10.2.1 Directed Graphs

Directed graphs differ from graphs in that every edge in such a graph has an orientation. The formal definition that follows captures this aspect of directed graphs by defining an edge as an ordered pair of vertices rather than a two-element set.

Definition 10.24 *A* directed graph *(or, for short, a* digraph*) is a pair* $\mathcal{G} = (V, E)$, *where* V *is a set of* vertices *or* nodes *and* $E \subseteq V \times V$ *is the set of* edges. *A digraph* $\mathcal{G} = (V, E)$ *is* finite *if* V *is a finite set.*

If $e = (u, v) \in E$, we refer to u as *the source of the edge* e and to v as *the destination of* e. The source and the destination of an edge e are denoted by $source(e)$ and $dest(e)$, respectively. Thus, we have the mappings $source : E \longrightarrow V$ and $dest : E \longrightarrow V$, which allow us to define for every subset U of the set of vertices the sets

$$out(U) = \{source^{-1}(U) - dest^{-1}(U)\},$$
$$in(U) = \{dest^{-1}(U) - source^{-1}(U)\}.$$

In other words, $out(U)$ is the set of edges that originate in U without ending in this set and $in(U)$ is the set of edges that end in U without originating in U.

Note that a digraph may contain both edges (u, v) and (v, u), and *loops* of the form (u, u).

Definition 10.25 *Let* $\mathcal{G} = (V, E)$ *be a digraph. The* out-degree of a vertex v *is the number* $\mathsf{d}_o(v) = |\{e \in E \mid v = source(e)\}|$, *and the* in-degree of a vertex v *is the number* $\mathsf{d}_i(v) = |\{e \in E \mid v = dest(e)\}|$.

Clearly, we have $\sum_{v \in V} \mathsf{d}_o(v) = \sum_{v \in V} \mathsf{d}_i(v) = |E|$.

The notion of a walk for digraphs is similar to the notion of a walk for graphs.

Definition 10.26 *Let* $\mathcal{G} = (V, E)$ *be a digraph. A* walk *in* $\mathcal{G}$ *is a sequence of vertices* $\mathbf{w} = (v_0, \ldots, v_k)$ *such that* $(v_i, v_{i+1}) \in E$ *for* $0 \leqslant i \leqslant k - 1$.

The number k *is the* length *of* w. *We refer to* $\mathbf{w}$ *as a* walk that joins v_0 to v_{k-1}. *If* $v_0 = v_k$, *we say that the walk is* closed.

If all vertices of the sequence $w = (v_0, \ldots, v_{k-1}, v_k)$ *are distinct, with the possible exception of* v_0 *and* v_k, *then* w *is a* path.

If $v_0 = v_k$, *then the path* w *is a* cycle. *A directed graph with no cycles is said to be* acyclic.

An undirected walk *in a digraph* $\mathcal{G} = (V, E)$ *is a sequence of vertices* $\mathbf{u} = (v_0, \ldots, v_k)$ *such that either* $(v_i, v_{i+1}) \in E$ *or* $(v_{i+1}, v_i) \in E$ *for every* i, $0 \leqslant i \leqslant k-1$.

Note that a walk $\mathbf{w}$ may have length 0; in this case, $\mathbf{w}$ is the null sequence of edges and the sequence of vertices of $\mathbf{w}$ consists of a single vertex.

Definition 10.27 *Let* $\mathcal{G} = (V, E)$ *be an acyclic digraph and let* $u, v \in V$. *The vertex* u *is an* ancestor *of* v, *and* v *is a* descendant *of* u *if there is a path* $\mathbf{p}$ *from* u *to* v.

Every vertex is both an ancestor and a descendant of itself due to the existence of walks of length 0. If u is an ancestor of v and $u \neq v$, then we say that u is a *proper ancestor* of v. Similarly, if v is a descendant of u and $u \neq v$, we refer to v as a *proper descendant* of u.

Definition 10.28 *A digraph* $\mathcal{G} = (V, E)$ *is* linear *if* $\mathsf{d}_i(v) = \mathsf{d}_o(v) = 1$ *for every vertex* $v \in V$.

It is not difficult to see that a linear digraph is a collection of directed cycles such that every vertex occurs in exactly one directed cycles.

Theorem 10.29 *In a finite acyclic digraph* $\mathcal{G}$, *there is a vertex whose in-degree is* 0.

Proof Let $G = (V, E)$ be a finite acyclic digraph and let $\mathbf{p}$ be a path of maximum length in $\mathcal{G}$ from x to y. If there is a vertex z on the path $\mathbf{p}$ such that $(z, x) \in E$, then the $\mathcal{G}$ has a cycle, contradicting the fact that $\mathcal{G}$ is acyclic. If x is the destination of an edge whose source is not on $\mathbf{p}$, then there exists a longer path in $\mathcal{G}$ ending in y, contradicting the maximality of $\mathbf{p}$. Thus, there is no vertex w such that $(w, x) \in E$, so $\mathsf{d}_i(x) = 0$.

Definition 10.30 *Let* $\mathcal{G} = (V, E)$ *be a directed acyclic graph, where* $|V| = m$. *A* topological sorting *of* V *is a bijection* $t : \{1, \ldots, m\} \longrightarrow V$ *such that if* $(t(i), t(j))$ *is an edge in* $\mathcal{G}$, *then* $i < j$.

An algorithm for obtaining a topological sort on the set A can be formulated using Theorem 10.29.

Algorithm 10.2.1: Topological Sort Algorithm

Data: An acyclic directed graph $\mathcal{G} = (V, E)$, where $V = \{v_1, \ldots, v_m\}$
Result: A topological sort $t : \{1, \ldots, m\} \longrightarrow V$

$i = 1$;
while *there exist vertices in* $\mathcal{G}$ **do**
  let v be a vertex of $\mathcal{G}$ with $\mathsf{d}_i(v) = 0$ and let $t(i) = v$;
  replace $\mathcal{G}$ by the subgraph of $\mathcal{G}$ generated by $V - \{v\}$;
  $i++$;
  return t;
end

Theorem 10.31 *Let V_0 be the set of vertices of a finite forest $\mathcal{G} = (V, E)$ having in-degree 0. For every vertex $v \in V - V_0$, there exists a unique vertex $v_0 \in V_0$ and a unique path that joins v_0 to v.*

Proof Since $v \in V - V_0$, we have $\mathsf{d}_i(v) = 1$ and there exists at least one edge whose destination is v. Let $\mathbf{p} = (v_0, v_1, \ldots, v_{k-1})$ be a maximal path whose destination is v (so $v_{k-1} = v$). We have $\mathsf{d}_i(v_0) = 0$. Indeed, if this is not the case, then there is a vertex v' such that an edge (v', v) exists in E and v' is distinct from every vertex of $\mathbf{p}$ because otherwise $\mathcal{G}$ would not be acyclic. This implies the existence of a path $\mathbf{p}' = (v', v_0, v_1, \ldots, v_{k-1})$, which contradicts the maximality of $\mathbf{p}$.

The path that joins a vertex of in-degree 0 to v is unique. Indeed, suppose that $\mathbf{q}$ is another path in $\mathcal{G}$ that joins a vertex of in-degree 0 to v. Let u be the first vertex having a positive in-degree that is common to both paths. The predecessors of u on $\mathbf{p}$ and $\mathbf{q}$ must be distinct, so $\mathsf{d}_i(u) > 1$. This contradicts the fact that $\mathcal{G}$ is a forest. The uniqueness of the path implies also the uniqueness of the source.

The adjacency matrix and the incidence matrix for directed graphs are introduced next.

Definition 10.32 *Let $\mathcal{G} = (V, E)$ be a directed graph, with $|V| = m$. The* adjacency matrix *of $\mathcal{G}$ is the matrix $A_{\mathcal{G}} \in \{0, 1\}^{m \times m}$ defined by*

$$(A_{\mathcal{G}})_{ij} = \begin{cases} 1 & if\ there\ exists\ an\ edge(v_i, v_j) \in E, \\ 0 & otherwise. \end{cases}$$

In general, the adjacency matrix of a directed graph is not symmetric.

Definition 10.33 *Let $\mathcal{G} = (V, E)$ be a directed graph, with $|V| = m$ and $|E| = n$. The* incidence matrix *of $\mathcal{G}$ is a matrix $U_{\mathcal{G}} \in \{-1, 0, 1\}^{m \times n}$ such that*

$$(U_{\mathcal{G}})_{ip} = \begin{cases} 1 & if\ source(e_p) = v_i, \\ -1 & if\ dest(e_p) = v_i, \\ 0 & otherwise \end{cases}$$

for $1 \leqslant i \leqslant m$ and $1 \leqslant p \leqslant n$.

Example 10.34 Let $\mathcal{G} = (V, E)$ be the directed graph shown in Fig. 10.11.

The incidence matrix $U_{\mathcal{G}}$ and the adjacency matrix $A_{\mathcal{G}}$ are given by

$$U_{\mathcal{G}} = \begin{pmatrix} 1 & 1 & 0 & 0 & 1 \\ -1 & 0 & 1 & 0 & 0 \\ 0 & 0 & -1 & -1 & -1 \\ 0 & -1 & 0 & 1 & 0 \\ 0 & 0 & 0 & 0 & 0 \end{pmatrix} \text{ and } A_{\mathcal{G}} = \begin{pmatrix} 0 & 1 & 1 & 1 \\ 0 & 0 & 1 & 0 \\ 0 & 0 & 0 & 0 \\ 0 & 0 & 1 & 0 \end{pmatrix}.$$

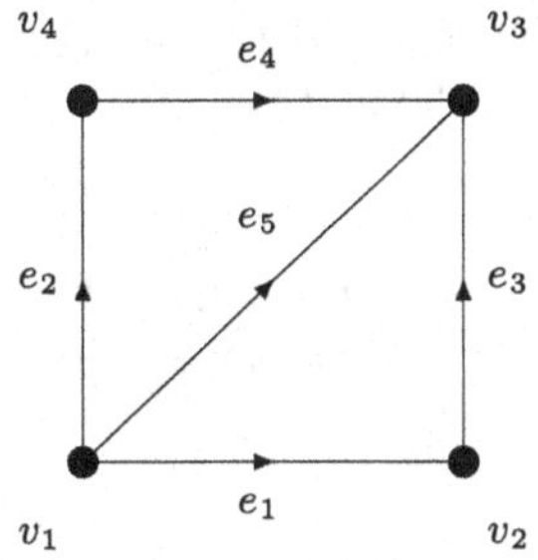

Fig. 10.11 Directed graph $\mathcal{G} = (\{v_i \mid 1 \leqslant i \leqslant 4\}, \{e_j \mid 1 \leqslant j \leqslant 5\})$

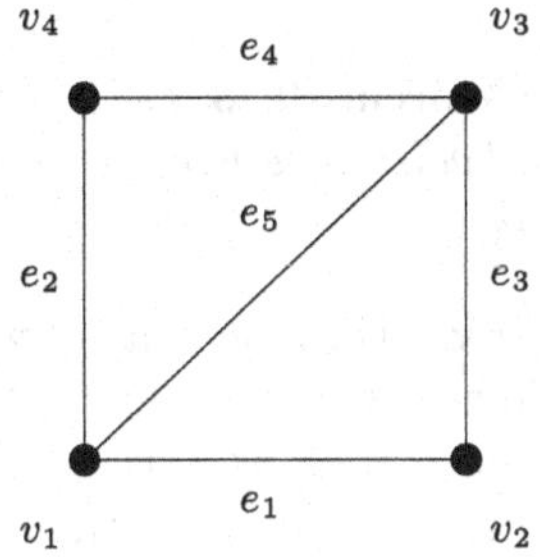

Fig. 10.12 Graph $\mathcal{G}_0 = (\{v_i \mid 1 \leqslant i \leqslant 4\}, \{e_j \mid 1 \leqslant j \leqslant 5\})$

Theorem 10.35 *The determinant of any square submatrix of the incidence matrix $U_{\mathcal{G}}$ of a directed graph $\mathcal{G}$ equals* 0, 1, *or* -1.

Proof We show that the result holds for any matrix of the form $A\begin{bmatrix} i_1, \ldots, i_k \\ j_1, \ldots, j_k \end{bmatrix}$ by induction on $k \geqslant 1$.

The basis case, $k = 1$ is immediate because an entry of $U_{\mathcal{G}}$ is 0, 1 or -1.

Suppose the statement holds for k and let C be a $(k+1) \times (k+1)$ submatrix of $U_{\mathcal{G}}$. If each column of C has an 1 or -1, then $\det(C) = 0$, which follows from the fact that the sum of the rows of C is $\mathbf{0}'_k$. The same happens if C has a zero column. So, suppose that C has a column with one non-zero entry (which must be 1 or -1). By expanding C along this column, the result follows by the inductive hypothesis.

A directed graph an be obtained from an undirected graph by assigning an orientation to the edges of the undirected graph.

Definition 10.36 *Let $\mathcal{G} = (V, E)$ be a graph. An* orientation *on $\mathcal{G}$ is a function $r : V \times V \longrightarrow \{-1, 0, 1\}$ such that $r(u, v) \neq 0$ if and only if (u, v) is an edge of V, and $r(u, v) + r(v, u) = 0$ for $u, v \in V \times V$. If $r(u, v) = 1$ (and, therefore, $r(v, u) = -1$) we say that v is the* positive end *of the edge (u, v) under this orientation and u is the* negative end *of the edge.*

The directed graph determined by the orientation r is $\mathcal{G}^r = (V, E^r)$, where E^r consists of all ordered pairs $(u, v) \in E \times E$ such that $r(u, v) = 1$.

Example 10.37 The directed graph $\mathcal{G}$ shown in Fig. 10.11 can be obtained from the graph $\mathcal{G}_0$ of Fig. 10.12 be applying the orientation defined by

$$\begin{aligned}&r(v_1, v_2) = 1, \quad r(v_2, v_1) = -1, r(v_1, v_3) = 1, r(v_3, v_1) = -1,\\&r(v_1, v_4) = 1, \quad r(v_4, v_1) = -1, r(v_2, v_3) = 1, r(v_3, v_2) = -1,\\&r(v_3, v_4) = -1, r(v_4, v_3) = 1.\end{aligned}$$

10.2.2 Graph Connectivity

Theorem 10.38 *Let $\mathcal{G} = (V, E)$ be a graph. The relation $\gamma_\mathcal{G}$ on V that consists of all pairs of vertices (x, y) such that there is a walk that joins x to y is an equivalence on V.*

Proof The reflexivity of $\gamma_\mathcal{G}$ follows from the fact that there is a walk of length 0 that joins any vertex x to itself.

If a walk $\mathbf{p} = (v_0, \ldots, v_n)$ joins x to y (which means that $x = v_0$ and $y = v_n$), then the walk $\mathbf{q} = (v_n, \ldots, v_0)$ joins y to x. Therefore, $\gamma_\mathcal{G}$ is symmetric.

Finally, suppose that $(x, y) \in \gamma_\mathcal{G}$ and $(y, z) \in \gamma_\mathcal{G}$. There is a walk $\mathbf{p} = (v_0, f \ldots, v_n)$ with $x = v_0$ and $y = v_n$ and a walk $\mathbf{q} = (v'_0, \ldots, v'_m)$ such that $y = v'_0$ and $z = v'_m$. The walk $\mathbf{r} = (v_0, \ldots, v_n = v'_0, \ldots, v'_m)$ joins x to z, so $(x, z) \in \gamma_\mathcal{G}$. Thus, $\gamma_\mathcal{G}$ is transitive, so it is an equivalence.

Definition 10.39 *Let $\mathcal{G} = (V, E)$ be a graph. The* connected components *of $\mathcal{G}$ are the equivalence classes of the relation $\gamma_\mathcal{G}$.*

A graph is connected *if it has only one connected component.*

A graph is linear *if $\mathsf{d}(v) = 1$ for every vertex v.*

It is easy to see that in a linear graph each connected component is an edge or a cycle.

Example 10.40 The sequences (v_4, v_5, v_8, v_7) and (v_4, v_5, v_6, v_7) are both walks of length 3 in the graph shown in Fig. 10.1.

The connected components of this graph are

$$\{v_1, v_2, v_3\} \text{ and } \{v_4, v_5, v_6, v_7, v_8\}.$$

Definition 10.41 A spanning subgraph *of a graph $\mathcal{G} = (V, E)$ is a subgraph of the form $\mathcal{G}' = (V, E')$; that is, a subgraph that has the same set of vertices as $\mathcal{G}$.*

Example 10.42 The graph shown in Fig. 10.13 is a spanning subgraph of the graph defined in Example 10.2.

Theorem 10.43 *If $\mathcal{G} = (V, E)$ is a connected graph, then $|E| \geqslant |V| - 1$.*

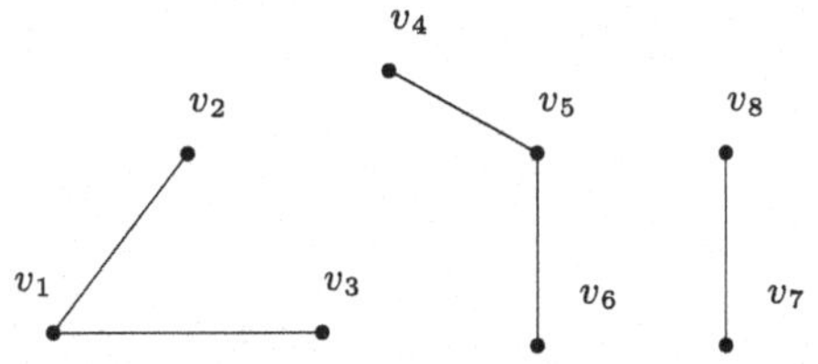

Fig. 10.13 Spanning subgraph of the graph defined in Example 10.2

Proof We prove the statement by induction on $|E|$. If $|E| = 0$, then $|V| \leqslant 1$ because $\mathcal{G}$ is connected and the inequality is clearly satisfied.

Suppose that the inequality holds for graphs having fewer than n edges, and let $\mathcal{G} = (V, E)$ be a connected graph with $|E| = n$. Let $e = (x, y)$ be an arbitrary edge and let $\mathcal{G}' = (V, E - \{e\})$ be the graph obtained by removing the edge e. The graph $\mathcal{G}'$ may have one or two connected components, so we need to consider the following cases:

1. If $\mathcal{G}'$ is connected, then, by the inductive hypothesis, we have $|E'| \geqslant |V| - 1$, which implies $|E| = |E'| + 1 \geqslant |V| - 1$.
2. If $\mathcal{G}'$ contains two connected components V_0 and V_1, let E_0 and E_1 be the set of edges whose endpoints belong to V_0 and V_1, respectively. By the inductive hypothesis, $|E_0| \geqslant |V_0| - 1$ and $|E_1| \geqslant |V_1| - 1$. Therefore,

$$|E| = |E_0| + |E_1| + 1 \geqslant |V_0| + |V_1| - 1 = |V| - 1.$$

This concludes the argument.

Corollary 10.44 *Let $\mathcal{G} = (V, E)$ be a graph that has k connected components. We have $|E| \leqslant |V| - k$.*

Proof Let V_i and E_i be the set of edges of the ith connected component of $\mathcal{G}$, where $1 \leqslant i \leqslant k$. It is clear that $V = \bigcup_{i=1}^{k} V_i$ and $E = \bigcup_{i=1}^{k} E_i$. Since the sets $V_1, \ldots, V_k$ form a partition of V and the sets $E_1, \ldots, E_k$ form a partition of E, we have $|E| = \sum_{i=1}^{k} |E|_i \leqslant \sum_{i=1}^{k} |V|_i - k = |V| - k$, which is the desired inequality.

A subset U of the set of vertices of a graph $\mathcal{G} = (V, E)$ is *complete* if the subgraph induced by it is complete.

A set of vertices W is a *clique in* $\mathcal{G}$ if it is maximally complete. In other words, W is a clique if the graph induced by W is complete and there is no set of vertices Z such that $W \subset Z$ and Z is complete.

Example 10.45 The sets $\{v_1, v_2, v_3, v_4\}$ and $\{v_3, v_5, v_6\}$ are cliques of the graph shown in Fig. 10.14.

Definition 10.46 *Let $\mathcal{G}_i = (V_i, E_i)$ be two graphs, where $i \in \{1, 2\}$. The graphs $\mathcal{G}_1$ and $\mathcal{G}_2$ are* isomorphic *if there exists a bijection $f : V_1 \longrightarrow V_2$ such that $(f(u), f(v)) \in E'$ if and only if $(u, v) \in E$. The mapping f in this case is called a* graph isomorphism. *If $\mathcal{G}_1 = \mathcal{G}_2$, then we say that f is a graph* automorphism *of $\mathcal{G}_1$.*

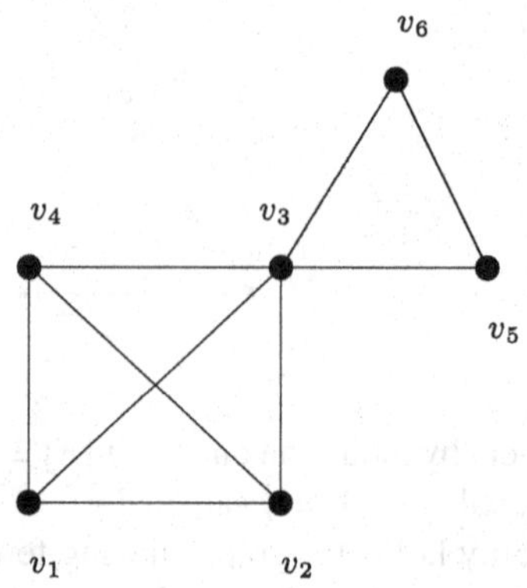

Fig. 10.14 Graph and its two cliques

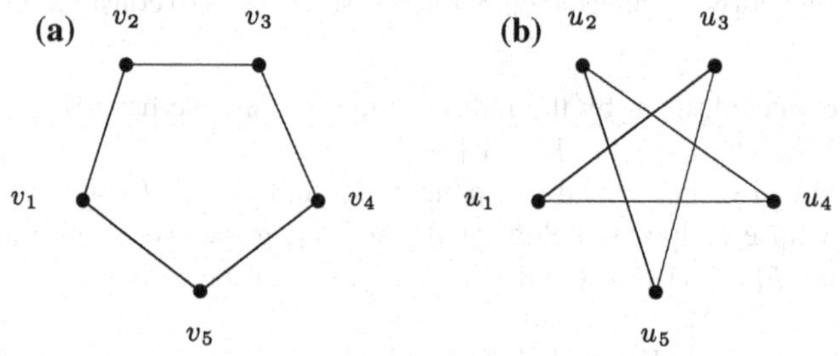

Fig. 10.15 Two isomorphic graphs

Two isomorphic graphs can be represented by drawings that differ only with respect to the labels of the vertices.

Example 10.47 $\mathcal{G}_1 = (\{v_1, v_2, v_3, v_4, v_5\}, E_1)$ and $\mathcal{G}_2 = (\{u_1, u_2, u_3, u_4, u_5\}, E_2)$ shown in Fig. 10.15a, b, respectively, are isomorphic graphs. Indeed, the function $f : \{v_1, v_2, v_3, v_4, v_5\} \longrightarrow \{u_1, u_2, u_3, u_4, u_5\}$ defined by $f(v_1) = u_1$, $f(v_2) = u_3$, $f(v_3) = u_5$, $f(v_4) = u_2$, and $f(v_5) = u_4$ can be easily seen to be a graph isomorphism.

On the other hand, both graphs shown in Fig. 10.16 have six vertices but cannot be isomorphic. Indeed, the first graph consists of one connected component, while the second has two connected components ($\{u_1, u_3, u_5\}$ and $\{u_2, u_4, u_6\}$).

If two graphs are isomorphic, they have the same degree sequences. The inverse is not true; indeed, the graphs shown in Fig. 10.16 have the same degree sequence $\mathbf{d} = (2, 2, 2, 2, 2, 2)$ but are not isomorphic.

In general, an *invariant* of graphs is a set of numbers that is the same for two isomorphic graphs. Thus, the degree sequence is an invariant of graphs.

For digraphs we have two types of connectivity.

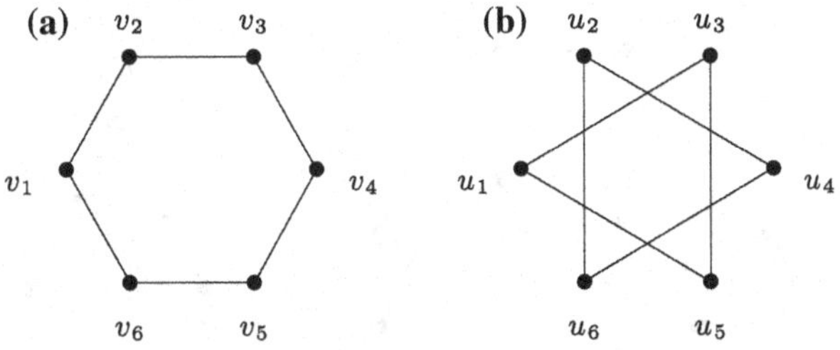

Fig. 10.16 Two graphs that are not isomorphic

Definition 10.48 *A digraph is* weakly connected *if there exists an undirected walk between any pair of vertices and is* strongly connected *it there exists a walk between every pair of vertices.*

The connectivity relation $\gamma_{\mathcal{G}}$ of a digraph $\mathcal{G} = (V, E)$ consists of all pairs $(u, v) \in V \times V$ such that there exists a path from u to v and a path from v to u. Thus, for a strongly connected digraph $\gamma_{\mathcal{G}}$ has a single equivalence class.

Theorem 10.49 *A digraph $\mathcal{G} = (V, E)$ is strongly connected if and only if there exists no partition $\{S, T\}$ of V such that all edges between S and T have their source in S and their destination in T.*

Proof Let $\mathcal{G} = (V, E)$ be a strongly connected digraph. Suppose that $\{S, T\}$ is a partition of V such that all edges between S and T have their source in S and their destination in T and let s, t be two vertices such that $s \in S$ and $t \in T$. Since $\mathcal{G}$ is strongly connected there is a walk $\mathbf{w} = (x_0, x_1, \ldots, x_k)$ that joins t to s, that is, $t = x_0$ and $x_k = s$. Let x_p be the first vertex on this path that belongs to S. If is clear that $x_{p-1} \in T$ and the edge (x_{p-1}, x_p) has its source in T and its destination in S, which contradicts the definition of the partition $\{S, T\}$.

Conversely, suppose that $\mathcal{G}$ is not strongly connected. There exists a pair of vertices (x, y) such that there is no path from x to y. Let $S \subset V$ consist of those vertices that can be reached from x and let $T = V - S$. Clearly, both S and T are non-empty subsets of V because $x \in S$, and $y \in T$. Thus, $\{S, T\}$ is a partition of V.

There is no edge in $\mathcal{G}$ that has its source in S and its destination in T. Indeed, suppose that $(u, v) \in E$ were such an edge with $u \in S$ and $v \in T$. By the definition of S there is a path $\mathbf{w}$ from x to u. Then $\mathbf{w}(u, v)$ would be a path that would reach v from x. This contradicts the definition of T. Therefore, if $\mathcal{G}$ is strongly connected, a partition $\{S, T\}$ of V such that all edges between S and T have their source in S and their destination in T cannot exist.

Definition 10.50 *Let $\mathcal{G} = (V, E)$ be a directed graph and let $V_1, \ldots, V_k$ the set of equivalence classes relative to the equivalence $\gamma_{\mathcal{G}}$.*

The condensed graph *of $\mathcal{G}$ is the digraph $\mathsf{C}(\mathcal{G}) = (\{V_1, \ldots, V_k\}, K)$, having the strong components as its vertices such that $(V_i, V_j) \in K$ if and only if there exists $v_i \in V_i$ and $v_j \in V_j$ such that $(v_i, v_j) \in E$.*

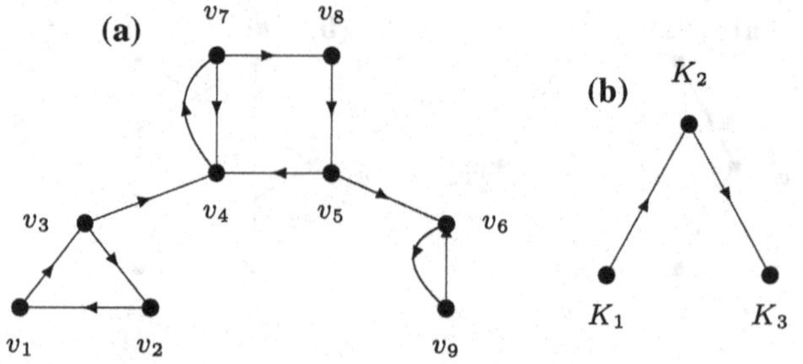

Fig. 10.17 Directed graph $\mathcal{G}$ and its condensed graph $\mathsf{C}(\mathcal{G})$

Theorem 10.51 *The condensed graph* $\mathsf{C}(\mathcal{G})$ *is acyclic.*

Proof Suppose that $(v_{i_1}, \ldots, v_{i_\ell}, v_{i_1})$ is a cycle in the graph $\mathsf{C}(\mathcal{G})$ that consists of ℓ distinct vertices. By the definition of $\mathsf{C}(\mathcal{G})$ we have a sequence of vertices $(v_{i_1}, \ldots, v_{i_\ell, v_{i_1}}$ in $\mathcal{G}$ such that the $(v_{i_p}, v_{i_{p+1}} \in E$ for $1 \leqslant p \leqslant \ell - 1$ and $(v_{i_{ell}}, v_{i_1}) \in E$. Therefore, $(v_{i_1}, \ldots, v_{i_\ell, v_{i_1}}$ is a cycle and for any two vertices u, v of this cycle there is a path from u to v and a path from v to u. In other words, for any pair of vertices (u, v) of this cycle we have $(u, v) \in \gamma_{\mathcal{G}}$, so $V_{i_1} = \cdots = V_{i_\ell}$, which contradicts our initial assumption.

Example 10.52 The strong components of the digraph $\mathcal{G}$ shown in Fig. 10.17a are $K_1 = \{v_1, v_2, v_3\}$, $K_2 = \{v_4, v_5, v_7, v_8\}$, and $K_3 = \{v_6, v_9\}$. The condensed graph $\mathsf{C}(\mathcal{G})$ shown in Fig. 10.17b has $\{K_1, K_2, K_3\}$ as its set of vertices and $\{(K_1, K_2), (K_2, K_3)\}$ as its set of edges.

10.2.3 Variable Adjacency Matrices

Let $\mathcal{G} = (V, E)$ be a digraph, where $V = \{v_1, \ldots, v_m\}$. For each edge $(v_i, v_j) \in E$ we introduce the variable e_{ij}. The *variable adjacency matrix* of the graph $\mathcal{G}$ is the matrix $A_{\mathcal{G}}(E)$ defined by

$$(A_{\mathcal{G}}(E))_{ij} = \begin{cases} e_{ij} & \text{if } (v_i, v_j) \text{ is an edge in } E, \\ 0 & \text{otherwise.} \end{cases}$$

The definition of the adjacency matrix for undirected graph is identical.

Example 10.53 The variable adjacency matrix of the graph shown in Fig. 10.18 is

$$A_{\mathcal{G}}(E) = \begin{pmatrix} 0 & e_{12} & 0 & e_{14} \\ e_{21} & 0 & e_{23} & e_{24} \\ e_{31} & e_{32} & 0 & e_{34} \\ e_{41} & 0 & 0 & 0 \end{pmatrix}.$$

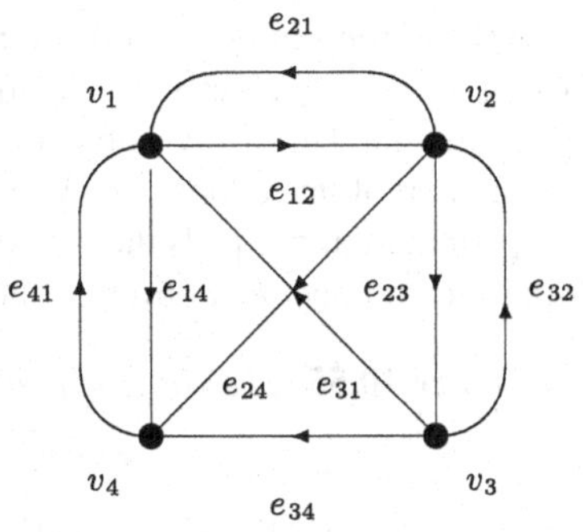

Fig. 10.18 Directed graph

By taking $e_{ij} = 1$ we get the standard adjacency matrix of $\mathcal{G}$,

$$A_{\mathcal{G}} = \begin{pmatrix} 0\,1\,0\,1 \\ 1\,0\,1\,1 \\ 1\,1\,0\,1 \\ 1\,0\,0\,0 \end{pmatrix}.$$

The determinant of $A_{\mathcal{G}}(E)$ is

$$\det(A_{\mathcal{G}}(E)) = -e_{12}e_{23}e_{34}e_{41} + e_{14}e_{41}e_{23}e_{32}.$$

A term in the expansion of the determinant $\det(A_{\mathcal{G}}(E))$ corresponds to a permutation

$$\pi = \begin{pmatrix} 1 & 2 & \cdots & m \\ j_1 & j_2 & \cdots & j_m \end{pmatrix}$$

and has the form

$$(-1)^{inv(\pi)} e_{1\,j_1} \cdots e_{m\,j_m}.$$

Therefore, such a term is nonzero if and only if the edges of a digraph corresponding to the entries in this term constitute a linear spanning subgraph that consists of a collection of cycles such that each vertex of $\mathcal{G}$ is contained in exactly one of these cycles.

Theorem 10.54 (Harary's Theorem) *Let s be the number of linear spanning subgraphs of a digraph $\mathcal{G}$. Then,* $\det(A_{\mathcal{G}}(E))$ *is given by*

$$\det(A_{\mathcal{G}}(E)) = \sum_{h=1}^{s} (-1)^{e_h} p_h,$$

where p_h is the product of the variables corresponding to the edges of the h*th linear spanning subgraph $\mathcal{H}_h$ of $\mathcal{G}$ and e_h is the number of even cycles of $\mathcal{H}_h$.*

Proof The contribution of a spanning subgraph $\mathcal{H}_h$ consists of two factors: the product p_h of the variables corresponding to the edges of $\mathcal{H}_h$, and the factor $(-1)^{e_h}$. Since the number of inversions of a cyclic permutation of odd length is even, and the number of inversions of a cyclic permutation of even length is odd, the parity of the permutation π equals the sum of the parities of the even cycles and, therefore, it equals the number e_h of even cyclic permutations.

Corollary 10.55 *The value of the determinant* $\det(A_{\mathcal{G}})$ *is*

$$\det(A_{\mathcal{G}}) = \sum_{h=1}^{s} (-1)^{e_h},$$

where e_h is the number of even cycles of the spanning graph $\mathcal{H}_h$.

Proof This follows from Theorem 10.54 by taking $e_{ij} = 1$ for every edge $(v_i, v_j) \in E$.

Example 10.56 The spanning linear subgraphs of the digraph introduced in Example 10.53 are defined by the sets of edges $(v_1, v_2), (v_2, v_3), (v_3, v_4)(v_4, v_1)$ and (v_1, v_4), $(v_4 v_1), (v_2 v_3), (v_3 v_2)$. The first linear spanning subgraph is a cycle of length 4; the second consists of two cycles of length 2.

If a digraph $\mathcal{G}$ has k strong connected components $V_1, \ldots, V_k$ and its vertices are numbered such that vertices that belong to a strong connected component are numbered consecutively, then its adjacency matrix (and its variable adjacency matrix) is a block diagonal matrix, $A_{\mathcal{C}} = \mathsf{diag}(A_1, \ldots, A_k)$, where A_i is the adjacency matrix of the subgraph of $\mathcal{G}$ induced by V_i for $1 \leqslant i \leqslant k$.

To transfer the results for digraphs to graphs, note that the adjacency matrix of a graph $\mathcal{G} = (V, E)$, which is symmetric, can be interpreted as the adjacency matrix of a digraph $\mathcal{G}' = (V, E')$, where E' contains both edges (v_i, v_j) and (v_j, v_i) whenever E contains the edge (v_i, v_j).

The digraph $\mathcal{G}'$ is said to be obtained from $\mathcal{G}$ by *symmetrization*.

Example 10.57 Let $A \in \{0, 1\}^4$ be the matrix

$$A = \begin{pmatrix} 0 & 1 & 1 & 1 \\ 1 & 0 & 1 & 1 \\ 1 & 1 & 0 & 1 \\ 1 & 1 & 1 & 0 \end{pmatrix}.$$

We have $A = A_{\mathcal{K}_4}$, where $\mathcal{K}_4$ is the graph shown in Fig. 10.19a. The digraph that has the same incidence matrix is shown in Fig. 10.19b.

The variable adjacency matrix of a graph has the same definition as the corresponding adjacency matrix of a digraph with the provision that variables of the form e_{ij} and e_{ji} are identified. For example, the variable adjacency matrix of $\mathcal{K}_4$ is

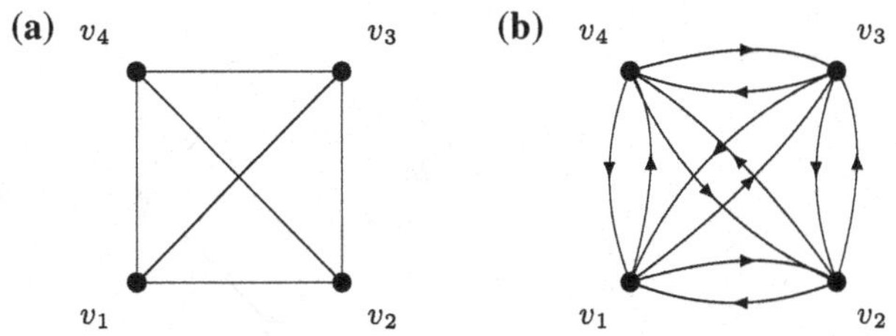

Fig. 10.19 The graph $\mathcal{K}_4$ and its symmetrized digraph

$$A_{\mathcal{G}}(E) = \begin{pmatrix} 0 & e_{12} & e_{13} & e_{14} \\ e_{12} & 0 & e_{23} & e_{24} \\ e_{13} & e_{23} & 0 & e_{34} \\ e_{14} & e_{24} & e_{34} & 0 \end{pmatrix}$$

and its determinant is

$$\begin{aligned} \det(A_{\mathcal{G}}(E)) &= e_{12}^2 e_{34}^2 + e_{13}^2 e_{24}^2 + e_{14}^2 e_{23}^2 \\ &\quad -2e12 e_{13} e_{24} e_{34} - 2e_{12} e_{14} e_{23} e_{34} - 2e_{13} e_{14} e_{23} e_{24}. \end{aligned}$$

For both digraphs and graphs the determinant of the variable adjacency matrix is the sum of the variable determinants of its linear subgraphs,

$$\det(A_{\mathcal{G}}(E)) = \sum_{h=1}^{n} \det(A_{\mathcal{H}_h}(E)).$$

Thus, to obtain a formula for the determinant of a graph it suffices to develop such a formula for the determinant of a linear graph.

Recall that the components of a linear graph $\mathcal{H}$ are either edges or cycles. Denote by $L_{\mathcal{H}}$ the set of of components of $\mathcal{H}$ that consist of two vertices and the edge joining them and let $M_{\mathcal{H}}$ remaining components of $\mathcal{H}$ which are cycles. The same symbols, $L_{\mathcal{H}}$ and $M_{\mathcal{H}}$ are used to denote the sets of variables that correspond to the edges in the two sets. Denote by $e_{\mathcal{H}}$ the number of components of $\mathcal{H}$ that consist of an even number of vertices and by $c_{\mathcal{H}}$ the number of components of $\mathcal{H}$ that contain more than two points (and, therefore, consist of a cycle).

Theorem 10.58 *Let $\mathcal{H}$ be a linear graph. We have*

$$\det(A_{\mathcal{H}}(E_1 \cup E) = (-1)^{e_{\mathcal{H}}} 2^{c_{\mathcal{H}}} \prod_{e \in L_{\mathcal{H}}} e^2 \prod_{e \in M_{\mathcal{H}}} e.$$

Proof The equality of the theorem follows from the similar equality for digraphs obtained in Theorem 10.54.

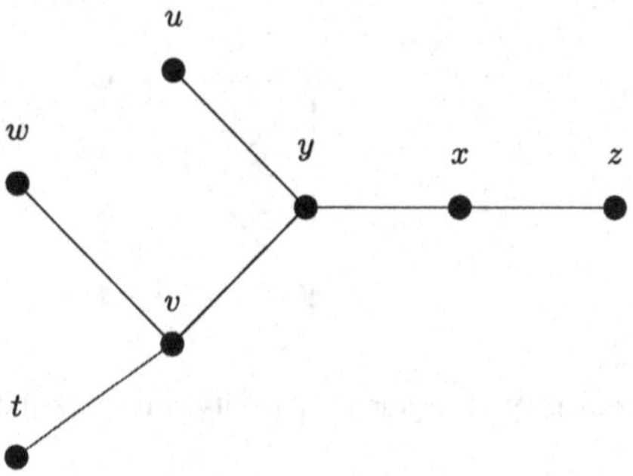

Fig. 10.20 Tree having $V = \{t, u, v, w, x, y, z\}$ as its set of vertices

Consider the digraph $\mathcal{H}'$ obtained by symmetrization from $\mathcal{H}$. The factor $(-1)^{e_{\mathcal{H}}}$ is the same as the factor given in Theorem 10.54.

Note that the symmetrized digraph $\mathcal{H}'$ may contain several linear subgraphs because every cycle in $\mathcal{H}$ yields two directed cycles in $\mathcal{H}'$. Thus, the number of linear subgraphs of $\mathcal{H}'$ is $2^{c_{\mathcal{H}}}$. For each of the linear subgraphs of $\mathcal{H}'$ the product of variables is obtained by multiplying all variables that correspond to edges that are part of a cycle and the squares of the variables that correspond to the edges.

10.3 Trees

Trees are graphs of special interest to data mining due to the presence of tree-structured data in areas such as Web and text mining and computational biology.

Definition 10.59 A tree *is a graph* $\mathcal{G} = (V, E)$ *that is both connected and acyclic.* A forest *is a graph* $\mathcal{G} = (V, E)$ *whose connected components are trees.*

Example 10.60 The graph of Fig. 10.20 is a tree having $V = \{t, u, v, w, x, y, z\}$ as its set of vertices and $E = \{(t, v), (v, w), (y, u), (y, v), (x, y), (x, z)\}$ as its set of edges.

Next, we give several equivalent characterizations of trees.

Theorem 10.61 *Let* $\mathcal{G} = (V, E)$ *be a graph. The following statements are equivalent:*

(i) $\mathcal{G}$ *is a tree.*
(ii) *Any two vertices* $x, y \in V$ *are connected by a unique path.*
(iii) $\mathcal{G}$ *is minimally connected; in other words, if an edge e is removed from E, the resulting graph* $\mathcal{G}' = (E, V - \{e\})$ *is not connected.*
(iv) $\mathcal{G}$ *is connected, and* $|E| = |V| - 1$.
(v) $\mathcal{G}$ *is an acyclic graph, and* $|E| = |V| - 1$.

Proof (i) implies (ii): Let $\mathcal{G}$ be a graph that is connected and acyclic and let u and v be two vertices of the graph. If $\mathbf{p}$ and $\mathbf{q}$ are two distinct paths that connect u to v, then $\mathbf{pq}$ is a cycle in $\mathcal{G}$, which contradicts its acyclicity. Therefore, (ii) follows.

(ii) implies (iii): Suppose that any two vertices of $\mathcal{G}$ are connected by a unique path. If $e = (u, v)$ is an edge in $\mathcal{G}$, then (u, v) is a path in $\mathcal{G}$ and therefore it must be the unique path connecting u and v. If e is removed, then it is impossible to reach v from u, and this contradicts the connectivity of $\mathcal{G}$.

(iii) implies (iv): The argument is by induction on $|V|$. If $|V| = 1$, there is no edge, so the equality is satisfied. Suppose that the statement holds for graphs with fewer than n vertices and that $|V| = n$. Choose an edge $(u, v) \in E$. If the edge (u, v) is removed, then the graph separates into two connected components $\mathcal{G}_0 = (V_0, E_0)$ and $\mathcal{G}_1 = (V_1, E_1)$ with fewer vertices because $\mathcal{G}$ is minimally connected. By the inductive hypothesis, $|E_0| = |V_0| - 1$ and $|E_1| = |V_1| - 1$. Therefore, $|E| = |E_0| + |E_1| + 1 = |V_0| - 1 + |V_1| - 1 + 1 = |V| - 1$.

(iv) implies (v): Let $\mathcal{G}$ be a connected graph such that $|E| = |V| - 1$. Suppose that $\mathcal{G}$ has a cycle $\mathbf{c} = (v_1, \ldots, v_p, v_1)$. Let $\mathcal{G}_0 = (V_0, E_0)$ be the subgraph of $\mathcal{G}$ that consists of the cycle. Clearly, $\mathcal{G}_0$ contains p vertices and an equal number of edges.

Let $U = V - \{v_1, \ldots, v_p\}$. Since $\mathcal{G}$ is connected, there is a vertex $u_1 \in U$ and a vertex v of the cycle $\mathbf{c}$ such that an edge (u_1, v) exists in the graph $\mathcal{G}$. Let $\mathcal{G}_1 = (V_1, E_1)$, where $V_1 = V_0 \cup \{u_1\}$ and $E_1 = E_0 \cup \{(u_1, v)\}$. It is clear that $|V_1| = |E_1|$. If $V - V_1 \neq \emptyset$, there exists a vertex $v_2 \in V - V_2$ and an edge (v_2, w), where $w \in V_1$. This yields the graph $\mathcal{G}_2 = (V_1 \cup \{v_2\}, E_1 \cup \{(v_2, w)\})$, which, again has an equal number of vertices and edges. The process may continue until we exhaust all vertices. Thus, we have a sequence $\mathcal{G}_0, \mathcal{G}_1, \ldots, \mathcal{G}_m$ of subgraphs of $\mathcal{G}$, where $\mathcal{G}_m = (V_m, E_m)$, $|E_m| = |V_m|$ and $V_m = V$. Since $\mathcal{G}_m$ is a subgraph of $\mathcal{G}$, we have $|V| = |V_m| = |E_m| \leqslant |E|$, which contradicts the fact that $|E| = |V| - 1$. Therefore, $\mathcal{G}$ is acyclic.

(v) implies (i): Let $\mathcal{G} = (V, E)$ be an acyclic graph such that $|E| = |V| - 1$. Suppose that $\mathcal{G}$ has k connected components, $V_1, \ldots, V_k$, and let E_i be the set of edges that connect vertices that belong to the set V_i. Note that the graphs $\mathcal{G}_i = (V_i, E_i)$ are both connected and acyclic so they are trees. We have $|E| = \sum_{i=1}^n |E_i|$ and $|V| = \sum_{i=1}^n |V_i|$. Therefore, $|E| = \sum_{i=1}^n |E_i| = \sum_{i=1}^n |V_i| - k = |V| - k$. Since $|E| = |V| - 1$ it follows that $k = 1$, so $\mathcal{G} = (V, E)$ is connected. This implies that $\mathcal{G}$ is a tree.

Corollary 10.62 *The graph $\mathcal{G} = (V, E)$ is a tree if and only if it is maximally acyclic; in other words, if an edge e is added to E, the resulting graph $\mathcal{G}' = (E, V \cup \{e\})$ contains a cycle.*

Proof Let $\mathcal{G}$ be a tree. If we add an edge $e = (u, v)$ to the E, then, since u and v are already connected by a path, we create a cycle. Thus, $\mathcal{G}$ is maximally acyclic.

Conversely, suppose that $\mathcal{G}$ is maximally acyclic. For every pair of vertices u and v in $\mathcal{G}$, two cases may occur:

1. there is an edge (u, v) in $\mathcal{G}$ or
2. there is no edge (u, v) in $\mathcal{G}$.

In the second case, adding the edge (u, v) creates a cycle, which means that there is a path in $\mathcal{G}$ that connects u to v. Therefore, in either case, there is a path connecting u to v, so $\mathcal{G}$ is a connected graph and therefore a tree.

Corollary 10.63 *If $\mathcal{G} = (V, E)$ is a connected graph, then $\mathcal{G}$ contains a subgraph that is a tree that has V as its set of vertices.*

Proof Define the graph $\mathcal{T} = (V, E')$ as a minimally connected subgraph having the set V as its set of vertices. It is immediate that $\mathcal{T}$ is a tree.

We shall refer to a tree $\mathcal{T}$ whose existence was shown in Corollary 10.63 as a *spanning tree* of $\mathcal{G}$.

Corollary 10.64 *If $\mathcal{G}$ is an acyclic graph, then $\mathcal{G}$ contains $|V| - |E|$ connected components.*

Proof This statement follows immediately from the proof of Theorem 10.61.

Definition 10.65 *A* rooted tree *is a pair $(\mathcal{T}, v_0)$, where $\mathcal{T} = (V, E)$ is a tree and v_0 is a vertex of $\mathcal{T}$ called the* root *of $\mathcal{R}$.*

If $(\mathcal{T}, v_0)$ is a rooted tree and v is an arbitrary vertex of $\mathcal{T}$ there is a unique path that joins v_0 to v. The *height* of v is the length of this path, denoted by $\mathsf{height}(v)$.

The number $\max\{\mathsf{height}(v) \mid v \in V\}$ is the *height* of the rooted tree $(\mathcal{T}, v_0)$; this number is denoted by $\mathsf{height}(\mathcal{T}, v_0)$.

Rooted trees are generally drawn with the root at the top of the picture; if (u, v) is an edge and $\mathsf{height}(u) = \mathsf{height}(v) + 1$, then u is drawn above v.

Example 10.66 Let $(\mathcal{T}, v_0)$ be the rooted tree shown in Fig. 10.21. The heights of the vertices are shown in the following table:

v	v_1	v_2	v_3	v_4	v_5	v_6	v_7	v_8
$\mathsf{height}(v)$	1	2	3	1	2	2	3	3

A rooted tree $(\mathcal{T}, v_0)$ may be regarded as a directed graph. Note that if (u, v) is an edge in a rooted tree, the heights of u and v differ by 1. If $\mathsf{height}(u) = \mathsf{height}(v) + 1$, then we say that u is an *immediate descendant of* v and that v is an immediate ascendant of u. The unoriented edge $\{u, v\}$ can now be replaced by the oriented edge (u, v).

In a rooted tree, vertices can be partitioned into sets of nodes named *levels*. Each level L_i consists of those nodes whose height in the tree equals i. In a rooted tree $(\mathcal{T}, v_0)$ of height h there exist $h + 1$ levels, $L_0, \cdots, L_h$.

Example 10.67 The directed graph that corresponds to the rooted tree from Fig. 10.21 is shown in Fig. 10.22.

The levels of this rooted tree are $L_0 = \{v_0\}$, $L_1 = \{v_1, v_4\}$, $L_2 = \{v_2, v_5, v_6\}$, and $L_3 = \{v_3, v_7, v_8\}$.

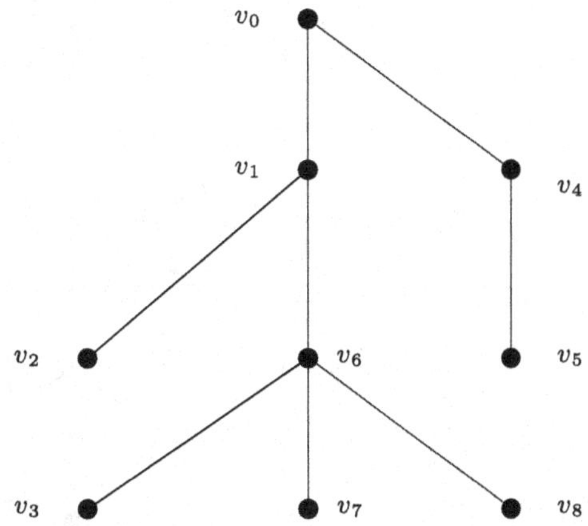

Fig. 10.21 Rooted tree

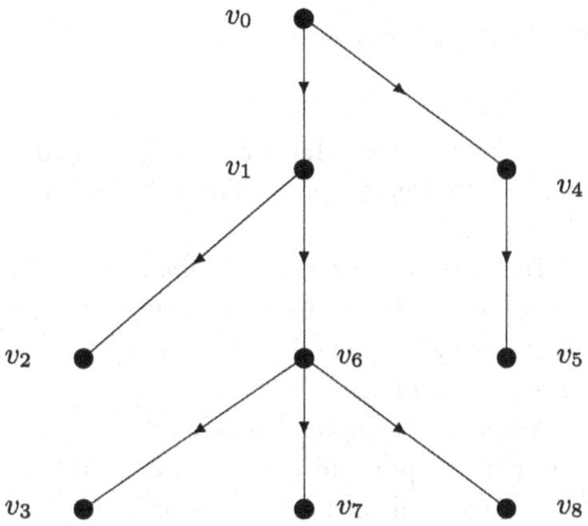

Fig. 10.22 Directed graph of a rooted tree

An *ordered rooted tree* is a triple $(\mathcal{T}, v_0, r)$, where $\mathcal{T} = (V, E)$ and v_0 have the same meaning as above, and $r : V \longrightarrow \mathbf{Seq}(V)$ is a function defined on the set of vertices of $\mathcal{T}$ such that $r(v)$ is a sequence, without repetition, of the descendants of the node v. If v is a leaf, then $r(v) = \lambda$.

Example 10.68 In Fig. 10.23, we present an ordered rooted tree that is created starting from the rooted tree from Fig. 10.22.

In general, we omit the explicit specification of the sequences of descendants for an ordered rooted tree and assume that each such sequence $r(v)$ consists of the direct descendants of v read from the graph from left to right.

Definition 10.69 *A* binary tree *is a rooted tree* $(\mathcal{T}, v_0)$ *such that each node has at most two descendants.*

The rooted tree is a subgraph of $(\mathcal{T}, v_0)$ that consists of all descendants of the left son of v_0 is the *left subtree* of the binary tree. Similarly, the set of descendants of the right son of v_0 forms the *right subtree* of $\mathcal{T}$.

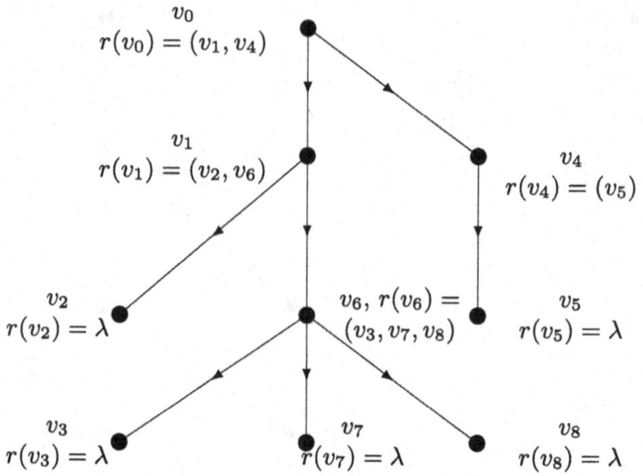

Fig. 10.23 Ordered rooted tree

In a binary tree, a level L_i may contain up to 2^i nodes; when this happens, we say that the level is *complete*. Thus, in a binary tree of height h, we may have at most 2^{h+1} nodes.

The notion of an *ordered binary tree* corresponds to binary trees in which we specify the order of the descendants of each node. If v, x, y are three nodes of an ordered binary tree and $r(v) = (x, y)$, then we say that x is the *left son of* v and y is the *right son of* v.

An *almost complete binary tree* is a binary tree such that all levels, with the possible exception of the last, are complete. The last level of an almost complete binary tree is always filled from left to right.

Note that the ratio of the number of nodes of a right subtree and the number of nodes of the left subtree of an almost complete binary tree is at most 2. Thus, the size of these subtrees is not larger than $\frac{2}{3}$ the size of the almost complete binary tree.

Example 10.70 The binary tree shown in Fig. 10.24 is an almost complete binary tree.

Given a graph $\mathcal{G} = (V, E)$ and two unconnected vertices $u, v \in V$ let $\mathcal{G} + (u, v)$ be the graph $(V, E \cup \{(u, v)\})$. If (r, s) is an edge in $\mathcal{G}$, we denote by $\mathcal{G} - (r, s)$ the graph $(V, E - \{(r, s)\})$.

Definition 10.71 *A* weighted graph *is a triple* $\mathcal{G} = (V, E, w)$*, where* (V, E) *is a graph and* $w : E \longrightarrow \mathbb{R}$ *is a* weight function. *If* $e \in E$*, we refer to* $w(e)$ *as the* weight of the edge *e*.

The weighted degree of a vertex $v \in V$ *is*

$$\mathsf{d}(v) = \sum \{w(v, t) \mid (v, t) \in E\}.$$

The weight of a set of edges *F is the number* $w(F)$ *defined by*

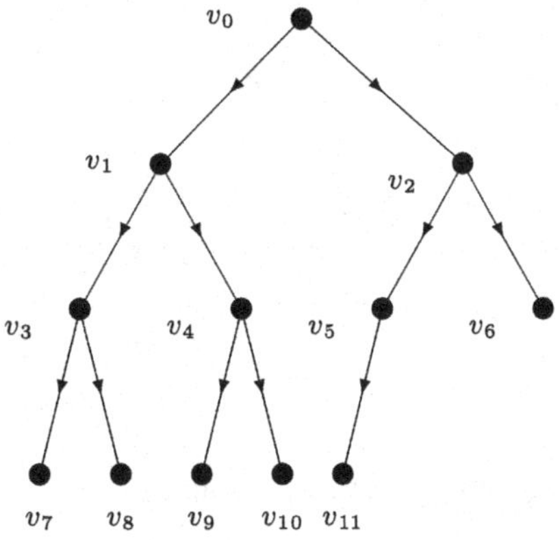

Fig. 10.24 An almost complete binary tree

$$w(F) = \sum\{w(e)|e \in F\}.$$

The adjacency matrix and the degree matrix of a weighted graph are direct generalizations of the corresponding notions for graphs.

The *adjacency matrix of a weighted graph* $\mathcal{G} = (V, E, w)$, where $V = \{v_1, \ldots, v_m\}$ is defined by

$$(A_{\mathcal{G}})_{ij} = \begin{cases} w(v_i, v_j) & \text{if } (v_i, v_j) \in E, \\ 0 & \text{otherwise.} \end{cases}$$

for $1 \leqslant i, j \leqslant m$. The *degree matrix of a weighted graph* $\mathcal{G} = (V, E, w)$ is

$$D_{\mathcal{G}} = \mathsf{diag}(\mathsf{d}(v_1), \ldots, \mathsf{d}(v_m)).$$

Clearly, if $w(e) = 1$ for every edge $e \in E$, the adjacency and the degree matrices of a weighted graph $\mathcal{G} = (V, E, w)$ coincide with the corresponding matrices of the graph (V, E).

Definition 10.72 *A* minimal spanning tree *for a weighted graph* $\mathcal{G} = (V, E, w)$ *is a spanning graph for* (V, E) *that is a tree* $\mathcal{T} = (V, F)$ *such that* $w(F)$ *is minimal.*

We present an effective construction of a minimal spanning tree for a weighted graph $\mathcal{G} = (V, E, w)$, where (V, E) is a finite, connected graph known as *Kruskal's algorithm*. Suppose that $e_1, e_2, \ldots, e_n$ is the list of all edges of $\mathcal{G}$ listed in increasing order of their weights; that is, $w(e_1) \leqslant w(e_2) \leqslant \cdots \leqslant w(e_n)$. We use the following algorithm.

We claim that $\mathcal{T} = (V, F)$ is a minimal spanning tree. Indeed, let $\mathcal{T}' = (V, F')$ be a minimal spanning tree such that $|F \cap F'|$ is maximal. Suppose that $F' \neq F$, and let $e = (x, y)$ be the first edge of F in the list of edges that does not belong to F'.

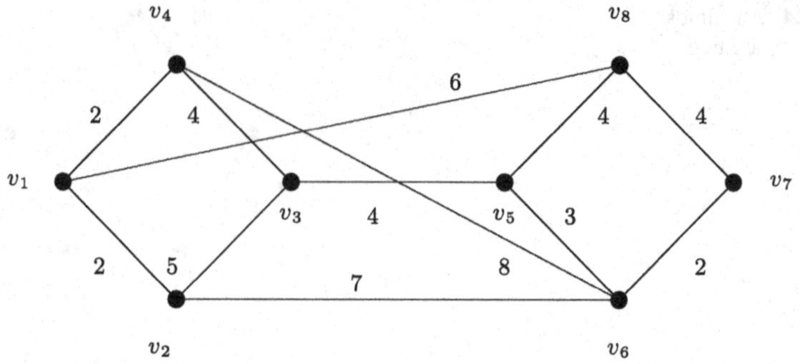

Fig. 10.25 Weighted graph (V, E, w)

The tree $\mathcal{T}'$ contains a unique path $\mathbf{p}$ that joins x to y. Note that this path cannot be included in $\mathcal{T}$ since otherwise $\mathcal{T}$ would contain a cycle formed by $\mathbf{p}$ and (x, y). Therefore, there exists an edge e' on the path $\mathbf{p}$ that does not belong to $\mathcal{T}$.

Algorithm 10.3.1: Kruskal's Algorithm

Data: a weighted graph $\mathcal{G} = (V, E, w)$
Result: a set of edges F that defines a minimal spanning tree

```
initialize F = e_1;
repeat
    select the first edge e in the list of edges such that e ∉ F and the subgraph (V, F ∪ {e}) is
    acyclic;
    F := F ∪ {e};
until no edges exist such thate ∉ F and the subgraph (V, F ∪ {e}) is acyclic;
return T = (V, F);
```

Note that the weight of edge e cannot be larger than the weight of e' because e was chosen for $\mathcal{T}$ by the algorithm and e' is not an edge of $\mathcal{T}$, which shows that e precedes e' in the previous list of edges. The set $F_1 = F' - \{e'\} \cup \{e\}$ defines a spanning tree $\mathcal{T}_1$, and since $w(F_1) = w(F') - w(e') + w(e) \leqslant w(F')$, it follows that the tree $\mathcal{T}' = (V, F')$ is a minimal spanning tree of $\mathcal{G}$. Since $|F_1 \cap F| > |F' \cap F|$, this leads to a contradiction. Thus, $F' = F$ and $\mathcal{T}$ is indeed a minimal spanning tree.

Example 10.73 Consider the weighted graph given in Fig. 10.25, whose edges are marked by the weights. The list of edges in nondecreasing order of their weights is

$$(v_1, v_2), (v_1, v_4), (v_6, v_7), (v_5, v_6),$$
$$(v_3, v_4), (v_3, v_5), (v_5, v_8), (v_7, v_8),$$
$$(v_2, v_3), (v_1, v_8), (v_2, v_6), (v_4, v_6).$$

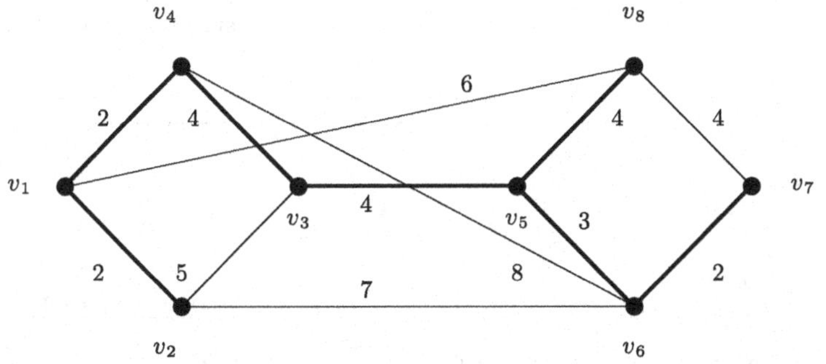

Fig. 10.26 Minimal spanning tree for the weighted graph from Fig. 10.25

The minimal spanning tree for this weighted graph is shown in thick lines in Fig. 10.26. The sequence of edges added to the set of edges of the minimal spanning tree is

$$(v_1, v_2), (v_1, v_4), (v_6, v_7), (v_5, v_6), (v_3, v_4), (v_3, v_5), (v_5, v_8).$$

An alternative algorithm known as *Prim's algorithm* is given next. In this modality of constructing the minimal spanning tree of a finite, connected graph $\mathcal{G} = (V, E)$, we construct a sequence of pairs of sets of vertices and edges that begins with a pair $(V_1, E_1) = (\{v\}, \emptyset)$, where v is an arbitrary vertex.

Suppose that we constructed the pairs $(V_1, E_1), \ldots, (V_k, E_k)$. Define the set of edges $H_k = \{(v, w) \in E | v \in V_k, w \notin V_k\}$. If (v_k, t_k) is an edge in H_k of minimal weight, then $V_{k+1} = V_k \cup \{v_k\}$ and $E_{k+1} = E_k \cup \{(v_k, t_k)\}$. The algorithm halts when $H_k = \emptyset$.

Consider the increasing sequences $V_1 \subseteq V_2 \subseteq \cdots$ and $E_1 \subseteq E_2 \subseteq \cdots$. An easy induction argument on k shows that the subgraphs (V_k, E_k) are acyclic. The sequence halts with the pair (V_n, E_n), where $H_n = \emptyset$, so $V_n = V$. Thus, (V_n, E_n) is indeed a spanning tree.

To prove that $(V_n, E_n) = (V, E_n)$ is a minimal spanning tree, we will show that for every subgraph (V_k, E_k), E_k is a subset of the set of edges of a minimal spanning tree $\mathcal{T} = (V, E)$.

The argument is by induction on k. The basis case, $k = 1$, is immediate since $E_1 = \emptyset$.

Suppose that E_k is a subset of the set of edges of a minimal spanning tree $\mathcal{T} = (V, E)$ and $E_{k+1} = E_k \cup \{(v_k, t_k)\}$.

Since $\mathcal{T}$ is a connected graph, there is a path in this graph that connects v_k to t_k. Let (r, s) be the first edge in this path that has one endpoint in V_k. By the definition of (v_k, t_k), we have $w(v_k, t_k) \leqslant w(r, s)$. Thus, if we replace (r, s) by (v_k, t_k) in $\mathcal{T}$, we obtain a minimal spanning tree whose set of edges includes E_{k+1}.

Example 10.74 We apply Prim's algorithm to the weighted graph introduced in Example 10.73 starting with the vertex v_3.

The sequences $V_1 \subseteq V_2 \subseteq \cdots$ and $E_1 \subseteq E_2 \subseteq \cdots$ are given in the following table:

k	E_k	V_k
1	$\{v_3\}$	$\emptyset$
2	$\{v_3, v_5\}$	$\{(v_3, v_5)\}$
3	$\{v_3, v_5, v_6\}$	$\{(v_3, v_5), (v_5, v_6)\}$
4	$\{v_3, v_5, v_6, v_7\}$	$\{(v_3, v_5), (v_5, v_6), (v_6, v_7)\}$
5	$\{v_3, v_5, v_6, v_7, v_4\}$	$\{(v_3, v_5), (v_5, v_6), (v_6, v_7), (v_3, v_4)\}$
6	$\{v_3, v_5, v_6, v_7, v_4, v_1\}$	$\{(v_3, v_5), (v_5, v_6), (v_6, v_7), (v_3, v_4), (v_4, v_1)\}$
7	$\{v_3, v_5, v_6, v_7, v_4, v_1, v_2\}$	$\{(v_3, v_5), (v_5, v_6), (v_6, v_7), (v_3, v_4), (v_4, v_1), (v_1, v_2)\}$
8	$\{v_3, v_5, v_6, v_7, v_4, v_1, v_2\}$	$\{(v_3, v_5), (v_5, v_6), (v_6, v_7), (v_3, v_4), (v_4, v_1), (v_1, v_2), (v_5, v_8)\}$

Definition 10.75 *Let $\mathcal{G} = (V, E, w)$ be a weighted graph. A* cut *of $\mathcal{G}$ is a two-block partition $\pi = \{S, T\}$ of V.*

The cut set *of the cut $\pi = \{S, T\}$ is the set of edges*

$$CS(\pi) = \{(u, v) \in E \mid u \in S \text{ and } v \in T\}.$$

The size of the cut π *is the number $\sum\{w(u, v) | (u, v) \in CS(\pi)|$. If $w(u, v) = 1$ for every edge in E, then the size of a cut is just $|CS(\pi)|$.*

If $s, t \in V$ are two vertices such that $s \in S$ and $t \in T$, then we refer to the cut $\pi = \{S, T\}$ as an (s, t)-cut.

For a weighted graph $\mathcal{G} = (V, E, w)$ the *separation* of π is

$$sep(\pi) = \min\{w(v_1, v_2) \mid v_1 \in V_1 \text{ and } v_2 \in V_2\}.$$

The *set of links of* π is the set of edges

$$LK(\pi) = \{(x, y) \in CS(\pi) \mid w(x, y) = sep(\pi)\}.$$

Example 10.76 Let $\pi = \{S, T\}$ be a partition of the set of vertices of $\mathcal{K}_m$. Then, $CS(\pi) = S \times T$, so the size of this cut is $|S| \cdot |T|$.

Theorem 10.77 *For every partition $\pi = \{V_1, V_2\}$ of a weighted graph (V, E, w) and minimal spanning tree $\mathcal{T}$, there exists an edge that belongs to $\mathcal{T}$ and to $LK(\pi)$.*

Proof Suppose that $\mathcal{T}$ is a minimal spanning graph that contains no edge of $LK(\pi)$. If an edge $(v_1, v_2) \in LK(\{V_1, V_2\})$ is added to $\mathcal{T}$, the resulting graph $\mathcal{G}'$ contains a unique cycle. The part of this cycle contained in $\mathcal{T}$ must contain at least one other edge $(s, t) \in CS(\{V_1, V_2\})$ because $v_1 \in V_1$ and $v_2 \in V_2$. The edge (s, t) does not belong to $LK(\{V_1, V_2\})$ by the supposition we made concerning $\mathcal{T}$. Consequently, $w(s, t) > w(v_1, v_2)$, which means that the spanning tree $\mathcal{T}_1$ obtained from $\mathcal{T}$ by removing (s, t) and adding (v_1, v_2) will have a smaller weight than $\mathcal{T}$. This would contradict the minimality of $\mathcal{T}$.

Theorem 10.78 *If (x, y) is an edge of a tree $\mathcal{T} = (V, E)$, then there exists a partition $\pi = \{V_1, V_2\}$ of V such that $CS(\{V_1, V_2\}) = \{(x, y)\}$.*

Proof Since $\mathcal{T}$ is a minimally connected graph, removing an edge (x, y) results in a graph that contains two disjoint connected components V_1 and V_2 such that $x \in V_1$ and $y \in V_2$. Then, it is clear that $\{(x, y)\} = CS(\{V_1, V_2\})$.

Theorem 10.79 *Let $\mathcal{G} = (V, E, w)$ be a weighted graph and let $\mathcal{T}$ be a minimal spanning link of $\mathcal{G}$. All minimal spanning tree edges are links of some partition of $\mathcal{G}$.*

Proof Let (x, y) be an edge in $\mathcal{T}$. If (V_1, V_2) is the partition of V that corresponds to the edge (x, y) that exists by Theorem 10.78, then, by Theorem 10.77, $\mathcal{T}$ must contain an edge from $CS(\{V_1, V_2\})$. Since $\mathcal{T}$ contains only one such edge, it follows that this edge must belong to $LK(\{V_1, V_2\})$.

Corollary 10.80 *Let $\mathcal{G} = (V, E, w)$ be a weighted graph, where. If $w : E \longrightarrow \mathbb{R}$ is an injective mapping (that is, if all weights of the edges are distinct), then the minimal spanning tree is unique. Furthermore, this minimal spanning tree has the form $\mathcal{T} = (V, L(\mathcal{G}))$, where $L(\mathcal{G})$ is the set of all links of $\mathcal{G}$.*

Proof Let $\mathcal{T} = (V, E')$ be a minimal spanning tree of $\mathcal{G}$. Since w is injective, for any partition π of V, the set $LK(\pi)$ consists of a unique edge that belongs to each minimal spanning tree. Thus, $L(\mathcal{G}) \subseteq E'$. The reverse inclusion follows immediately from Theorem 10.79, so $\mathcal{G}$ has a unique spanning tree $\mathcal{T} = (V, L(\mathcal{G}))$.

Theorem 10.81 *Let $\mathcal{G} = (V, E, w)$ be a weighted graph. If U is a nonempty subset of V such that $sep(\{U_1, U_2\}) < sep(U, V - U)$ for every two-block partition $\pi = \{U_1, U_2\} \in PART(U)$, then, for every minimal spanning tree $\mathcal{T}$ of $\mathcal{G}$, the subgraph $\mathcal{T}_U$ is a subtree of $\mathcal{T}$.*

Proof Let $\pi = \{U_1, U_2\}$ be a two block partition of U. To prove the statement, it suffices to show that every minimal spanning tree $\mathcal{T}$ of $(\mathcal{G}, w)$ contains an edge in $CS(\pi)$. This, in turn, will imply that the subgraph $\mathcal{T}_U$ of $\mathcal{T}$ determined by U has only one connected component, which means that $\mathcal{T}_U$ is a subtree of $\mathcal{T}$.

To prove that $\mathcal{T}$ contains an edge from $CS(\pi)$, it will suffice to show that $sep(U_1, U_2) < sep(U_1, V - U)$ because this implies $LK(U_1, V - U_1) \subseteq CS(U_1, U_2)$. Indeed, if this is the case, then the shortest link between a vertex in U_1 and one outside of U_1 must be an edge that joins a vertex from U_1 to a vertex in U_2.

Observe that

$$sep(U, V - U) = sep(U_1 \cup U_2, V - U) = \min\{sep(U_1, V - U), sep(U_2, V - U)\},$$

and therefore $sep(U_1, V - U) \geqslant sep(U, V - U)$.

By the hypothesis of the theorem, $sep(U, V - U) > sep(U_1, U_2)$, and therefore

$$sep(U_1, U_2) < sep(U, V - U) \leqslant sep(U_1, V - U),$$

which leads to the desired conclusion.

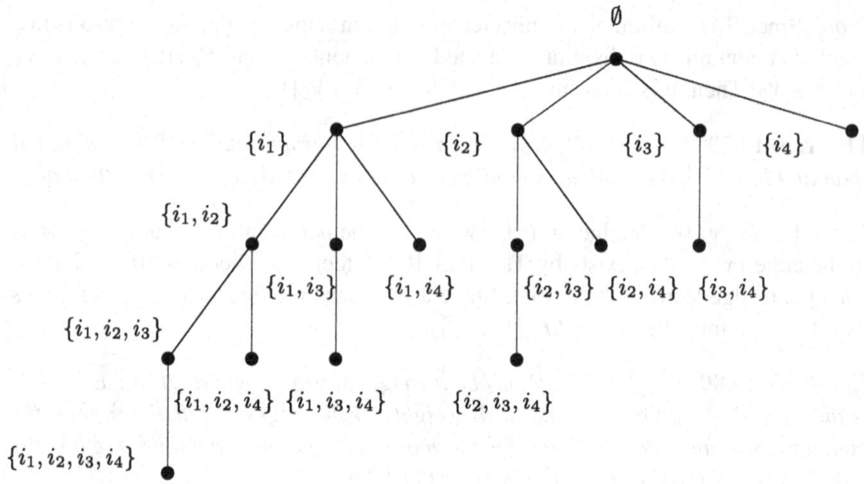

Fig. 10.27 Rymon tree for $\mathcal{P}(\{i_1, i_2, i_3, i_4\})$

Search enumeration trees were introduced by Rymon in [5] in order to provide a unified search-based framework for several problems in artificial intelligence; they are also useful for data mining algorithms.

Let S be a set and let $d : S \longrightarrow \mathbb{N}$ be an injective function. The number $d(x)$ is the *index* of $x \in S$. If $P \subseteq S$, the *view of* P is the subset

$$view(d, P) = \left\{ s \in S \mid d(s) > \max_{p \in P} d(p) \right\}.$$

Definition 10.82 *Let $\mathcal{C}$ be a hereditary collection of subsets of a set S. The graph* $\mathcal{G} = (\mathcal{C}, E)$ is a Rymon tree for $\mathcal{C}$ and the indexing function d *if*

(i) *the root of $\mathcal{G}$ is the empty set, and*
(ii) *the children of a node P are the sets of the form $P \cup \{s\}$, where $s \in view(d, P)$.*

If $S = \{s_1, \ldots, s_n\}$ and $d(s_i) = i$ for $1 \leqslant i \leqslant n$, we will omit the indexing function from the definition of the Rymon tree for $\mathcal{P}(S)$.

Example 10.83 Let $S = \{i_1, i_2, i_3, i_4\}$ and let $\mathcal{C}$ be $\mathcal{P}(S)$, which is clearly a hereditary collection of sets. Define the injective mapping d by $d(i_k) = k$ for $1 \leqslant k \leqslant 4$. The Rymon tree for $\mathcal{C}$ and d is shown in Fig. 10.27.

A key property of a Rymon tree is stated next.

Theorem 10.84 *Let $\mathcal{G}$ be a Rymon tree for a hereditary collection $\mathcal{C}$ of subsets of a set S and an indexing function d. Every set P of $\mathcal{C}$ occurs exactly once in the tree.*

Proof The argument is by induction on $p = |P|$. If $p = 0$, then P is the root of the tree and the theorem obviously holds.

Suppose that the theorem holds for sets having fewer than p elements, and let $P \in \mathcal{C}$ be such that $|P| = p$. Since $\mathcal{C}$ is hereditary, every set of the form $P - \{x\}$ with $x \in P$ belongs to $\mathcal{C}$ and, by the inductive hypothesis, occurs exactly once in the tree.

Let z be the element of P that has the largest value of the index function d. Then $view(P - \{z\})$ contains z and P is a child of the vertex $P - \{z\}$. Since the parent of P is unique, it follows that P occurs exactly once in the tree.

If a set U is located at the left of a set V in the tree $\mathcal{G}_I$, we shall write $U \sqsubset V$. Thus, we have

$$\begin{aligned}\emptyset &\sqsubset \{i_1\} \sqsubset \{i_1, i_2\} \sqsubset \{i_1, i_2, i_3, i_4\}\\ &\sqsubset \{i_1, i_2, i_4\} \sqsubset \{i_1, i_3\} \sqsubset \{i_1, i_3, i_4\}\\ &\sqsubset \{i_1, i_4\} \sqsubset \{i_2\} \sqsubset \{i_2, i_3\}\\ &\sqsubset \{i_2, i_3, i_4\} \sqsubset \{i_2, i_4\} \sqsubset \{i_3\}\\ &\sqsubset \{i_3, i_4\} \sqsubset \{i_4\}.\end{aligned}$$

Note that in the Rymon tree of a collection of the form $\mathcal{P}(S)$, the collection of sets of $\mathcal{S}_r$ that consists of sets located at distance r from the root denotes all $\binom{n}{r}$ subsets of size r of S.

Definition 10.85 *A* numbering *of a graph* $\mathcal{G} = (V, E)$ *is a bijection* $\nu : V \longrightarrow 1, \ldots, |V|$. *The pair* $(\mathcal{G}, \nu)$ *is referred to as a* numbered graph.

Theorem 10.86 *Let* $\nu : V \longrightarrow \{1, \ldots, n\}$ *be a bijection on the set* V, *where* $|V| = n$. *There are* n^{n-2} *numbered trees* $(\mathcal{T}, \nu)$ *having* V *as the set of vertices.*

Proof The best-known argument for this theorem is based on a bijection between the set of numbered trees having n vertices and the set of sequences of length $n - 2$ defined on the set $\{1, \ldots, n\}$ and has been formulated in [6].

Let $(\mathcal{T}, \nu)$ be a numbered tree having n vertices. Define a sequence of trees $(\mathcal{T}_1, \ldots, \mathcal{T}_{n-1})$ and a *Prüfer sequence* $(\ell_1, \ldots, \ell_{n-2}) \in \mathbf{Seq}_n(\mathbb{N})$ as follows. The initial tree $\mathcal{T}_1$ equals $\mathcal{T}$. The tree $\mathcal{T}_i$ will have $n - i + 1$ vertices for $1 \leqslant i \leqslant n - 1$.

The $\mathcal{T}_{i+1}$ is obtained from $\mathcal{T}_i$ by seeking the leaf x of $\mathcal{T}_i$ such that $\nu(x)$ is *minimal* and deleting the unique edge of the form (x, y). The number $\nu(y)$ is added to the Prüfer sequence. Note that the label ℓ of a vertex u will occur exactly $\mathsf{d}(u) - 1$ times in the Prüfer sequence, once for every vertex adjacent to u that is removed in the process of building the sequence of trees.

Let $L(\mathcal{T}, \nu)$ be the Prüfer sequence of $(\mathcal{T}, \nu)$. If NT_n is the set of numbered trees on n vertices, then the mapping $L : NT_n \longrightarrow \mathbf{Seq}_{n-2}(\{1, \ldots, n\})$ is a bijection.

The edges that are removed in the process of constructing the Prüfer sequences

Fig. 10.28 Enumerated tree

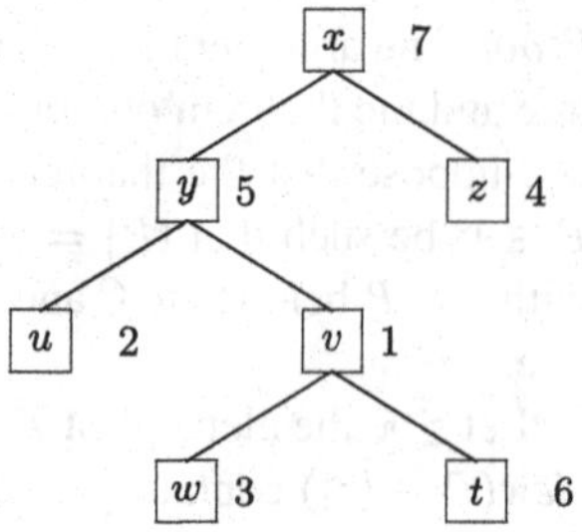

can be listed in a table:

Starting tree	Leaf	Vertex	Resulting tree
$\mathcal{T}_1$	x_1	y_1	$\mathcal{T}_2$
$\vdots$	$\vdots$	$\vdots$	$\vdots$
$\mathcal{T}_{n-2}$	x_{n-2}	y_{n-2}	$\mathcal{T}_{n-1}$
$\mathcal{T}_{n-1}$	x_{n-1}	y_{n-1}	–

Note that the edges of $\mathcal{T}_i$ are (x_j, y_j) for $i \leqslant j \leqslant n-1$.

The next to the last tree in the sequence, $\mathcal{T}_{n-2}$, has two edges and therefore three vertices. The last tree in the sequence $\mathcal{T}_{n-1}$ consists of a unique edge (x_{n-1}, y_{n-1}). Since a tree with at least two vertices has at least two leaves, the node whose label is n will never be the leaf with the minimal label. Therefore, $\nu(y_{n-1}) = n$ and n is always the last number of $L(\mathcal{T}, \nu)$.

Also, observe that the leaves of the tree $\mathcal{T}_i$ are those vertices that do not belong to $\{x_1, \ldots, x_{i-1}, y_i, \ldots, y_{n-1}\}$, which means that x_i is the vertex that has the minimal label and is not in the set above. In particular, x_1 is the vertex that has the least label and is not in $L(\mathcal{T}, \nu)$. This shows that we can uniquely determine the vertices x_i from $L(\mathcal{T}, \nu)$ and $x_1, \ldots, x_{i-1}$.

Example 10.87 Consider the tree $\mathcal{T}$ shown in Fig. 10.28.

The labels of the vertices are placed at the right of each rectangle that represents a vertex. The table that contains the sequence of edges is

Starting tree	Leaf	Vertex y	$\nu(y)$	Resulting tree
$\mathcal{T}_1$	u	y	5	$\mathcal{T}_2$
$\mathcal{T}_2$	w	v	1	$\mathcal{T}_3$
$\mathcal{T}_3$	z	x	7	$\mathcal{T}_4$
$\mathcal{T}_4$	t	v	1	$\mathcal{T}_5$
$\mathcal{T}_5$	v	y	5	$\mathcal{T}_6$
$\mathcal{T}_6$	y	x	7	–

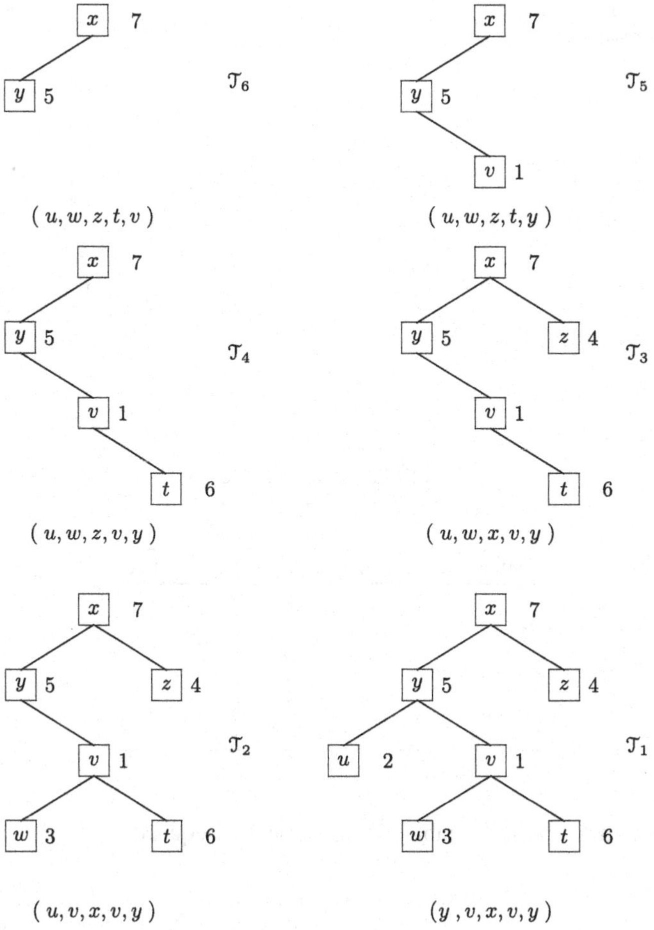

Fig. 10.29 Numbered trees with sequences $(x_1, \ldots, x_{i-1}, y_i, \ldots, y_{n-1})$

This means that $L(\mathcal{T}, \mu) = (5, 1, 7, 1, 5) = \nu^{-1}(y, v, x, v, y)$. The vertex with the smallest label that does occur in $L(\mathcal{T}, \mu) = (5, 1, 7, 1, 5)$ is u because $\ell(u) = 2$. This means that the first edge is (u, y) since $\ell(y) = 5$. The succession of trees is shown in Fig. 10.29. Under each tree, we show the sequence $(x_1, \ldots, x_{i-1}, y_i, \ldots, y_{n-1})$, which allows us to select the current leaf x_i.

Example 10.88 Suppose again that we have a tree having the set of nodes $V = \{x, y, z, u, v, w, t\}$ with the numbering given by
We reconstruct the tree that has (2, 3, 5, 2, 3) as its Prüfer sequence. The first leaf of this tree will be the vertex with the least value of ν that is not present in the sequence $\nu^{-1}(2, 3, 5, 2, 3) = (u, w, y, u, w)$; that is, v. This means that we start with the following table.

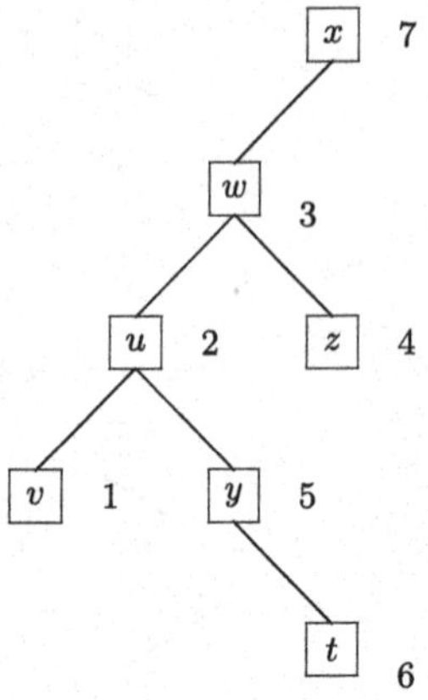

Fig. 10.30 Tree reconstructed from its Prüfer sequence

Vertex	x	y	z	u	v	w	t
ν(vertex)	7	5	4	2	1	3	6

Starting tree	Leaf	Vertex y	$\nu(y)$	Resulting tree
$\mathcal{T}_1$	x	u	2	$\mathcal{T}_2$
$\mathcal{T}_2$		w	3	$\mathcal{T}_3$
$\mathcal{T}_3$		y	5	$\mathcal{T}_4$
$\mathcal{T}_4$		u	2	$\mathcal{T}_5$
$\mathcal{T}_5$		w	3	$\mathcal{T}_6$
$\mathcal{T}_6$		x	7	–

For each step in filling in this table, we construct the sequence

$$(x_1, \ldots, x_{i-1}, y_i, \ldots, y_{n-1})$$

and choose x_i as the vertex having minimal numbering that is not in the sequence. The final table is

Starting tree	Leaf	Vertex y	$\nu(y)$	Resulting tree
$\mathcal{T}_1$	x	u	2	$\mathcal{T}_2$
$\mathcal{T}_2$	z	w	3	$\mathcal{T}_3$
$\mathcal{T}_3$	t	y	5	$\mathcal{T}_4$
$\mathcal{T}_4$	y	u	2	$\mathcal{T}_5$
$\mathcal{T}_5$	u	w	3	$\mathcal{T}_6$
$\mathcal{T}_6$	w	x	7	–

and gives the tree shown in Fig. 10.30.

10.4 Bipartite Graphs

Definition 10.89 *A* bipartite graph *is a graph* $\mathcal{G} = (V, E)$ *such that there is a two-block partition* $\pi = \{V_1, V_2\}$ *for which* $E \subseteq V_1 \times V_2$. *If* $E = V_1 \times V_2$, *then we say that* $\mathcal{G}$ *is* bipartite complete.

We refer to π *as the* bipartition *of* $\mathcal{G}$.

A bipartite complete graph, where $|V_1| = p$ and $|V_2| = q$ is denoted by $\mathcal{K}_{p,q}$. The bipartite graph $\mathcal{K}_{4,3}$ is shown in Fig. 10.3.

Theorem 10.90 *Let* $\mathcal{G} = (V, E)$ *be a bipartite graph having the bipartition* $\pi = \{V_1, V_2\}$. *Then* $\sum\{\mathsf{d}_{\mathcal{G}}(v) \mid v \in V_1\} = \sum\{\mathsf{d}_{\mathcal{G}}(v) \mid v \in V_2\}$.

Proof Let $p = |V_1|$ and $q = |V_2|$. If $p = q = 1$, the conclusion is immediate since E consists just of one edge. Thus, we can assume that $p, q > 1$. The argument is by induction on $|E|$ and the base case, $|E| = 0$ is immediate. Suppose that the conclusion holds for graphs having m edges and let (v_1, v_2) be an edge that is joining a vertex $v_1 \in V_1$ with a vertex $v_2 \in V_2$. Let $\mathcal{G}'$ be the graph obtained by removing the edge (v_1, v_2) from E. By inductive hypothesis we have $\sum\{\mathsf{d}_{\mathcal{G}'}(v) \mid v \in V_1\} = \sum\{\mathsf{d}_{\mathcal{G}'}(v) \mid v \in V_2\}$. Adding back the edge (v_1, v_2) to $\mathcal{G}'$ yields the graph $\mathcal{G}$ and we have

$$\begin{aligned} \sum\{\mathsf{d}_{\mathcal{G}}(v) \mid v \in V_1\} &= \sum\{\mathsf{d}_{\mathcal{G}'}(v) \mid v \in V_1\} + 1 \\ &= \sum\{\mathsf{d}_{\mathcal{G}'}(v) \mid v \in V_2\} + 1 = \sum\{\mathsf{d}_{\mathcal{G}}(v) \mid v \in V_2\}. \end{aligned}$$

Theorem 10.91 *A graph is bipartite if and only if it contains no cycle of odd length.*

Proof Let $\mathcal{G} = (V, E)$ be a bipartite graph having the bipartition $\{V_1, V_2\}$. Suppose that $\mathcal{G}$ contains a cycle of odd length $(v_0, v_1, \ldots, v_{2\ell})$. Without loss of generality, assume that $v_0 \in V_1$. Because $\mathcal{G}$ is bipartite it follows that $v_1 \in V_2, v_2 \in V_1, \ldots, v_{2\ell} \in V_1$. The existence of the edge $(v_{2\ell}, v_0)$ implies $v_0 \in V_2$, which contradicts the fact that $V_1 \cap V_2 = \emptyset$.

Conversely, suppose that $\mathcal{G} = (V, E)$ is a graph that contains no cycles having odd length. Assume initially that $\mathcal{G}$ is connected.

Let $u \in V$ be a fixed vertex in V. Partition V into two sets V_1 and V_2 such that V_1 is the set of vertices v such that $d(u, v)$ is odd and V_2 is the set of vertices v such that $d(u, v)$ is even. Clearly, $u \in V_2$ and $V_1 \cap V_2 = \emptyset$.

Suppose that $s, t \in V_1$ and there is an edge $(s, t) \in E$. Then, there exists a cycle $(u, \ldots, s, t, \ldots, u)$ which has odd length, which is impossible.

If $s, t \in V_2$ and there is an edge $(s, t) \in E$, then, again $(u, \ldots, s, t, \ldots, u)$ is a cycle of odd length. Thus, no edge may exists that joins vertices in the same block of the bipartition, so $\mathcal{G}$ is bipartite.

If $\mathcal{G}$ is not connected the above argument can be applied to each of its connected component.

Definition 10.92 *A* matching *in a graph* $\mathcal{G} = (V, E)$ *is a set of edges* M *such that no two edges in* M *have a vertex in common.*

M *is a* matching of a set U *of vertices if every vertex* $u \in U$ *is an endpoint of an edge in* M.

The matching number of $\mathcal{G}$ *is the size of the largest matching of* $\mathcal{G}$. *This number is denoted by* $\mathsf{match}(\mathcal{G})$.

The *set of neighbors of a set of vertices* W of a graph $\mathcal{G} = (V, E)$ is the set

$$N_{\mathcal{G}}(W) = \{v \in V - W \mid (v, w) \in E \text{ for some } w \in W\}.$$

Theorem 10.93 (Hall's Matching Theorem) *Let* $\mathcal{G} = (V, E)$ *be a bipartite graph having the bipartition* (V_0, V_1). *There exists a matching of* V_0 *in* $\mathcal{G}$ *if and only if* $|N_{\mathcal{G}}(U)| \geqslant |U|$ *for every subset* U *of* V_0.

Proof The condition of the theorem is obviously necessary. Therefore, we need to show only that the condition is sufficient. The argument is by induction on $n = |V_0|$.

The base case, $n = 1$, is immediate. Suppose that the condition is sufficient when the size of V_0 is smaller than n and let $\mathcal{G}$ be a bipartite graph such that $|V_0| = n$.

If $|N_{\mathcal{G}}(U)| > |U|$ for every set $U \subset V_0$, let (v_0, v_1) be an edge in $\mathcal{G}$, where $v_0 \in V_0$ and $v_1 \in V_1$, and let $\mathcal{G}' = (V, E - \{(v_0, v_1)\})$. We have $N_{\mathcal{G}'}(U) \geqslant N_{\mathcal{G}}(U) - 1 \geqslant |U|$, so $\mathcal{G}'$ contains a matching of $V_0 - \{v_0\}$. Adding (v_0, v_1) to this matching we obtain a matching of V_0 in $\mathcal{G}$.

If there exists a subset U of V_0 such that $N_{\mathcal{G}}(U) = |U|$, then, by the inductive hypothesis, the bipartite graph $\mathcal{H} = (U \cup N_{\mathcal{G}}(U), \{(u, v) \in E \mid u \in U \text{ and } v \in N_{\mathcal{G}}(U)\}$ contains a matching of U.

Let $\mathcal{H}'$ be the graph having the set of vertices $(V_0 - U) \cup (V_1 - N_{\mathcal{G}}(U)$ and the set of edges consisting of those edges of $\mathcal{G}$ whose endpoints are located in the sets $(V_0 - U)$ and $(V_1 - N_{\mathcal{G}}(U)$. The bipartite graph $\mathcal{H}'$ also has a matching for if S were a subset of $V_0 - U$ with $|N_{\mathcal{H}'}(S)| < |S|$, this would imply $|N_{\mathcal{G}}(S \cup U)| < |S \cup U|$, which contradicts the initial assumption. Thus, by the inductive hypothesis, $\mathcal{H}'$ has a matching of $V_0 - S$, which together with the matching for U yields a matching of V_0.

A beautiful application of Hall's Matching Theorem involves doubly stochastic matrices.

We observed in Example 5.73 that permutation matrices are doubly-stochastic matrices. Therefore, any convex combination of permutation matrices is a doubly-stochastic matrix. An important converse statement is discussed next.

Theorem 10.94 (Birkhoff-von Neumann Theorem) *If* $A \in \mathbb{R}^{n \times n}$ *is a doubly-stochastic matrix, then* A *is a convex combination of permutation matrices.*

Proof Let $A \in \mathbb{R}^{n \times n}$ be a doubly-stochastic matrix. Define a bipartite graph $\mathcal{G} = (V, E)$, where V contains a node r_i for each row of A and a node c_j for each column of A. The bipartition of the graph is $\{R, C\}$, where $R = \{r_1, \ldots, r_n\}$ and $C = \{c_1, \ldots, c_n\}\}$.

An edge (r_i, c_j) exists if E if $a_{ij} > 0$. In this case, the edge weight $w(r_i, c_j)$ is a_{ij}. Recall that we denote the set of neighbors of a set of vertices T in $\mathcal{G}$ by

$$N_{\mathcal{G}}(T) = \{v \in V - T \mid (v, t) \in E \text{ for some } t \in W\}.$$

For every vertex $v \in V$ we have

$$\sum \{w(v, t) \mid t \in N_{\mathcal{G}}(\{v\})\} = 1. \tag{10.2}$$

Indeed, if $v = r_i$, then the set of neighbors of v consists of vertices that correspond to columns of A such that $a_{ij} > 0$. Since the sum of all components of A in row i is 1, the equality follows immediately. A similar argument applied to the jth column works when $v = c_j$.

Let T be a set of vertices. We have

$$\sum \{w(t, u) \mid t \in T \text{ and } u \in N_{\mathcal{G}}(T)\} = \sum_{t \in T} \sum \{w(t, u) \mid u \in N_{\mathcal{G}}(t)\} = |T|,$$

by Equality (10.2). If $U = N_{\mathcal{G}}(T)$, then $T \subseteq N_{\mathcal{G}}(U)$, so

$$\begin{aligned} |N_{\mathcal{G}}(T)| = |U| &= \sum \{w(u, s) \mid u \in U \text{ and } s \in N_{\mathcal{G}}(U)\} \\ &\geqslant \sum \{w(u, s) \mid u \in U \text{ and } s \in T\} = |T|. \end{aligned}$$

By Hall's Matching Theorem (Theorem 10.93), there exists a matching M of R in $\mathcal{G}$ for $\{r_1, \ldots, r_n\}$. Define the matrix P by

$$p_{ij} = \begin{cases} 1 & \text{if } (r_i, c_j) \in M, \\ 0 & \text{otherwise.} \end{cases}$$

We claim that P is a permutation matrix.

For every row i of A there exists an edge (r_i, c_j) in M, so $p_{ij} = 1$. There exits only one 1 in the jth column of A for, if $p_{i_1 j} = p_{i_2 j} = 1$ for $i_1 \neq i_2$, it would follow that we have both $(i_1, j) \in M$ and $(i_2, j) \in M$ contradicting the fact that M is a matching.

Let $a = \min\{a_{ij} \mid p_{ij} \neq 0\}$. Clearly, $a > 0$ and $a = a_{pq}$ for some p and q. Let $C = A - aP$. If $C = O_{n,n}$, then A is a permutation matrix. Otherwise, note that

(i) $\sum_{j=1}^{n} c_{ij} = 1 - a$ and $\sum_{i=1}^{n} c_{ij} = 1 - a$;
(ii) $0 \leqslant c_{ij} \leqslant 1 - a$ for $1 \leqslant i \leqslant n$ and $1 \leqslant j \leqslant n$;
(iii) $c_{pq} = 0$.

Therefore, the matrix $D = \frac{1}{1-a}C$ is doubly-stochastic and we have $A = aP + (1 - a)D$, where D has at least one more zero element than A.

The equality $A = aP + (1-a)D$ shows that A is a convex combination of a permutation matrix and a doubly stochastic matrix with strictly more zero components than A. The statement follows by repeatedly applying this procedure.

Birkhoff-von Neumann Theorem can be applied to obtain a spectral bounding of the Frobenius distance between two matrices.

Theorem 10.95 (Hoffman-Wielandt Theorem) *Let $A, B \in \mathbb{C}^{n\times n}$ be two normal matrices having the eigenvalues $\alpha_1, \ldots, \alpha_n$ and $\beta_1, \ldots, \beta_n$, respectively. Then, there exist permutations ϕ and ψ in $PERM_n$ such that*

$$\sum_{i=1}^{n} |\alpha_i - \beta_{\psi(i)}|^2 \geqslant \| A - B \|_F^2 \geqslant \sum_{i=1}^{n} |\alpha_i - \beta_{\phi(i)}|^2.$$

Proof Since A and B are normal matrices they can be diagonalized as $A = U^{\mathsf{H}} D_A U$ and $B = W^{\mathsf{H}} D_B W$, where U and W are unitary matrices and C, D are diagonal matrices, $C = \mathsf{diag}(\alpha_1, \ldots, \alpha_n)$ and $D = \mathsf{diag}(\beta_1, \ldots, \beta_n)$. Then, we can write

$$\| A - B \|_F^2 = \| U^{\mathsf{H}} C U - W^{\mathsf{H}} D W \|_F^2 = trace(E^{\mathsf{H}} E),$$

where $E = U^{\mathsf{H}} C U - W^{\mathsf{H}} D W$. Note that

$$\begin{aligned} E^{\mathsf{H}} E &= (U^{\mathsf{H}} C^{\mathsf{H}} U - W^{\mathsf{H}} D^{\mathsf{H}} W)(U^{\mathsf{H}} C U - W^{\mathsf{H}} D W) \\ &= U^{\mathsf{H}} C^{\mathsf{H}} C U + W^{\mathsf{H}} D^{\mathsf{H}} D W - W^{\mathsf{H}} D^{\mathsf{H}} W U^{\mathsf{H}} C U - U^{\mathsf{H}} C^{\mathsf{H}} U W^{\mathsf{H}} D W \\ &= U^{\mathsf{H}} C^{\mathsf{H}} C U + W^{\mathsf{H}} D^{\mathsf{H}} D W - U^{\mathsf{H}} C^{\mathsf{H}} U W^{\mathsf{H}} D W - (U^{\mathsf{H}} C^{\mathsf{H}} U W^{\mathsf{H}} D W)^{\mathsf{H}} \\ &= U^{\mathsf{H}} C^{\mathsf{H}} C U + W^{\mathsf{H}} D^{\mathsf{H}} D W - 2\Re(U^{\mathsf{H}} C^{\mathsf{H}} U W^{\mathsf{H}} D W). \end{aligned}$$

Observe that

$$\begin{aligned} trace(\Re(U^{\mathsf{H}} C^{\mathsf{H}} U W^{\mathsf{H}} D W)) &= \Re(trace(U^{\mathsf{H}} C^{\mathsf{H}} U W^{\mathsf{H}} D W)) \\ &= \Re(trace(C^{\mathsf{H}} U W^{\mathsf{H}} D W U^{\mathsf{H}})). \end{aligned}$$

Thus, if Z is the unitary matrix $Z = W U^{\mathsf{H}}$, we have

$$trace(\Re(U^{\mathsf{H}} C^{\mathsf{H}} U W^{\mathsf{H}} D W)) = \Re(trace(C^{\mathsf{H}} Z^{\mathsf{H}} D Z)).$$

Since $\| C \|_F^2 = \sum_{i=1}^n \alpha_i^2$ and $\| D \|_F^2 = \sum_{i=1}^n \beta_i^2$, we have:

$$\begin{aligned} trace(E^{\mathsf{H}} E) &= \| C \|_F^2 + \| D \|_F^2 - 2\Re(trace(C^{\mathsf{H}} Z^{\mathsf{H}} D Z)) \\ &= \sum_{i=1}^{n} \alpha_i^2 + \sum_{i=1}^{n} \beta_i^2 - 2\Re\left(\sum_{i=1}^{n}\sum_{j=1}^{n} \bar{a}_i |z_{ij}|^2 \beta_j\right). \end{aligned}$$

The matrix S that has the elements $|z_{ij}|^2$ is doubly-stochastic because Z is a unitary matrix. This allows us to write:

$$\begin{aligned} \| A - B \|_F^2 &= trace(E^{\mathsf{H}} E) \\ &\geqslant \sum_{i=1}^{n} \alpha_i^2 + \sum_{i=1}^{n} \beta_i^2 - \max_S \Re \left(\sum_{i=1}^{n} \sum_{j=1}^{n} \bar{a}_i s_{ij} \beta_j \right), \end{aligned}$$

and

$$\begin{aligned} \| A - B \|_F^2 &= trace(E^{\mathsf{H}} E) \\ &\leqslant \sum_{i=1}^{n} \alpha_i^2 + \sum_{i=1}^{n} \beta_i^2 - \min_S \Re \left(\sum_{i=1}^{n} \sum_{j=1}^{n} \bar{a}_i s_{ij} \beta_j \right), \end{aligned}$$

where the maximum and the minimum are taken over the set of all doubly-stochastic matrices.

Birkhoff-von Neumann Theorem states that the polyhedron of doubly-stochastic matrices has the permutation matrices as its vertices. Therefore, the extremes of the linear function

$$f(S) = \Re \left(\sum_{i=1}^{n} \sum_{j=1}^{n} \bar{\alpha}_i s_{ij} \beta_j \right)$$

are achieved when S is a permutation matrix. Let P_ϕ the permutation matrix that gives the maximum of f and let P_ψ be the permutation matrix that gives the minimum.

If $S = P_\phi$, then $\sum_{j=1}^{n} s_{ij} \beta_j = \beta_{\phi(i)}$, so

$$\begin{aligned} \| A - B \|_F^2 &\geqslant \sum_{i=1}^{n} \alpha_i^2 + \sum_{i=1}^{n} \beta_i^2 - \Re \left(\sum_{i=1}^{n} \sum_{j=1}^{n} \bar{\alpha}_i \beta_{\phi(j)} \right) \\ &= \sum_{i=1}^{n} |\alpha_i - \beta_{\phi(i)}|^2. \end{aligned}$$

In the last equality we used the elementary equality $|a - b|^2 = |a|^2 + |b|^2 - 2\Re(\bar{a}b)$ for $a, b \in \mathbb{C}$.

Similarly, if $S = P_\psi$, we obtain the other inequality.

Corollary 10.96 *Let $A, B \in \mathbb{C}^{n \times n}$ be two Hermitian matrices having the eigenvalues $\alpha_1, \ldots, \alpha_n$ and $\beta_1, \ldots, \beta_n$, respectively, where $\alpha_1 \geqslant \cdots \geqslant \alpha_n$ and $\beta_1 \geqslant \cdots \geqslant \beta_n$. Then, $\sum_{i=1}^{n} |\alpha_i - \beta_i|^2 \leqslant \| A - B \|_F^2$.*

If $\alpha_1 \geqslant \cdots \geqslant \alpha_n$ and $\beta_1 \leqslant \cdots \leqslant \beta_n$, then $\sum_{i=1}^{n} |\alpha_i - \beta_i|^2 \geqslant \| A - B \|_F^2$.

Proof Since A and B are Hermitian, their eigenvalues are real numbers. By Hoffman-Wielandt Theorem, there exist two permutations $\phi, \psi \in PERM_n$ such that

$$\sum_{i=1}^{n} |\alpha_i - \beta_{\psi(i)}|^2 \geqslant \| A - B \|_F^2 \geqslant \sum_{i=1}^{n} |\alpha_i - \beta_{\phi(i)}|^2.$$

We have $\sum_{i=1}^{n} |\alpha_i - \beta_{\phi(i)}|^2 = \| \mathbf{a} - P_\phi \mathbf{b} \|_F^2$ and $\sum_{i=1}^{n} |\alpha_i - \beta_{\psi(i)}|^2 = \| \mathbf{a} - P_\psi \mathbf{b} \|_F^2$, where

$$\mathbf{a} = \begin{pmatrix} \alpha_1 \\ \vdots \\ \alpha_n \end{pmatrix} \text{ and } \mathbf{b} = \begin{pmatrix} \beta_1 \\ \vdots \\ \beta_n \end{pmatrix},$$

so

$$\| \mathbf{a} - P_\psi \mathbf{b} \|_F^2 \geqslant \| A - B \|_F^2 \geqslant \| \mathbf{a} - P_\phi \mathbf{b} \|_F^2 .$$

By Corollary 6.30, since the components of $\mathbf{a}$ and $\mathbf{b}$ are placed in decreasing order, we have $\| \mathbf{a} - P_\phi \mathbf{b} \|_F \geqslant \| \mathbf{a} - \mathbf{b} \|_F$, so

$$\| \mathbf{a} - \mathbf{b} \|_F^2 = \sum_{i=1}^{n} |\alpha_i - \beta_i|^2 \leqslant \| \mathbf{a} - P_\phi \mathbf{b} \|_F \leqslant \| A - B \|_F^2,$$

which proves the first inequality of the corollary.

For the second part, by Corollary 6.30 we have

$$\| A - B \|_F \leqslant \| \mathbf{a} - P_\psi \mathbf{b} \|_F \leqslant \| \mathbf{a} - \mathbf{b} \|_F,$$

Bipartite graphs allow an intuitive description of the supremum and infimum of two partitions.

Definition 10.97 *Let $\pi, \sigma \in PART(S)$, where $\pi = \{B_i \mid i \in I\}$ and $\sigma = \{C_j \mid j \in J\}$. The* graph of the pair (π, σ) *is the bipartite graph*

$$\mathcal{G}_{\pi,\sigma} = (\{B_i \mid i \in I\} \cup \{C_j \mid j \in J\}, E),$$

where E consists of those two-element sets $\{B_i, C_j\}$ such that $B_i \cap C_j \neq \emptyset$.

Example 10.98 Let $S = \{a_i \mid 1 \leqslant i \leqslant 12\}$ and let $\pi = \{B_i \mid 1 \leqslant i \leqslant 5\}$ and $\sigma = \{C_j \mid 1 \leqslant j \leqslant 4\}$, where

$$\begin{aligned} B_1 &= \{a_1, a_2\}, & C_1 &= \{a_2, a_4\}, \\ B_2 &= \{a_3, a_4, a_5\}, & C_2 &= \{a_1, a_3, a_5, a_6, a_7\}, \\ B_3 &= \{a_6, a_7\}, & C_3 &= \{a_8, a_{11}\}, \\ B_4 &= \{a_8, a_9, a_{10}\}, & C_4 &= \{a_9, a_{10}, a_{12}\}, \\ B_5 &= \{a_{11}, a_{12}\}. \end{aligned}$$

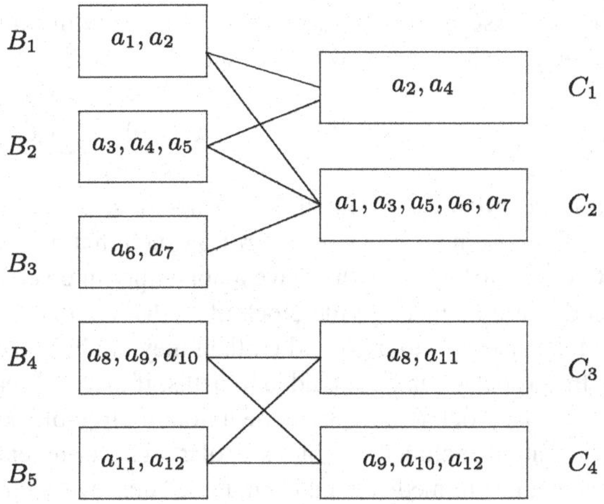

Fig. 10.31 The graph $G_{\pi,\sigma}$

The graph $G_{\pi,\sigma}$ is shown in Fig. 10.31.

Note that the blocks of $\pi \wedge \sigma$ correspond to the edges of the graph $\mathcal{G}_{\pi,\sigma}$.

Example 10.99 If π and σ are the partitions introduced in Example 10.98, then the partition $\pi \wedge \sigma$ consists of nine blocks that correspond to the edges of the graph:

$$\begin{array}{lll} B_1 \cap C_1 = \{a_2\}, & B_1 \cap C_2 = \{a_1\}, & B_2 \cap C_1 = \{a_4\}, \\ B_2 \cap C_2 = \{a_3, a_5\}, & B_3 \cap C_2 = \{a_6, a_7\}, & B_4 \cap C_3 = \{a_8\}, \\ B_4 \cap C_4 = \{a_9, a_{10}\}, & B_5 \cap C_3 = \{a_{11}\}, & B_5 \cap C_4 = \{a_{12}\}. \end{array}$$

Theorem 10.100 *Let π and $\sigma \in PART(S)$, where $\pi = \{B_i \mid i \in I\}$ and $\sigma = \{C_j \mid j \in J\}$. The* $\sup\{\pi, \sigma\}$ *exists in the poset $(PART(S), \leqslant)$, and the blocks of the partition* $\sup\{\pi, \sigma\}$ *are the unions of the blocks that belong to connected components of the graph $G_{\pi,\sigma}$.*

Proof The connected components of the graph $\mathcal{G}_{\pi,\sigma}$ form a partition of the set of vertices of the graph. Let τ be the partition of S whose blocks are the unions of the blocks that belong to connected components of the graph $G_{\pi,\sigma}$.

Let D be a block of τ and let $\{B_{i_1}, \ldots, B_{i_p}\}$ and $\{C_{j_1}, \ldots, C_{j_q}\}$ be the sets of blocks of π and σ, respectively, that are included in τ. We claim that

$$\bigcup\{B_{i_k} \mid 1 \leqslant k \leqslant p\} = \bigcup\{C_{j_h} \mid 1 \leqslant h \leqslant q\}.$$

Indeed, let $x \in \bigcup\{B_{i_k} \mid 1 \leqslant k \leqslant p\}$. There exists a block B_{i_ℓ} such that $x \in B_{i_\ell}$. Also, there is a block C of σ such that $x \in C$. Since $x \in B_{i_\ell} \cap C$, it follows that there exists an edge (B_{i_ℓ}, C) in $\mathcal{G}_{\pi,\sigma}$, so C belongs to the same connected component as B_{i_ℓ}; that is, $C = C_{j_g}$ for some g, $1 \leqslant g \leqslant q$. Therefore,

$$\bigcup\{B_{i_k} \mid 1 \leqslant k \leqslant p\} \subseteq \bigcup\{C_{j_h} \mid 1 \leqslant h \leqslant q\}.$$

The reverse inclusion can be shown in a similar manner. This proves the needed equality, which can now be written

$$D = \bigcup\{B_{i_k} \mid 1 \leqslant k \leqslant p\} = \bigcup\{C_{j_h} \mid 1 \leqslant h \leqslant q\}.$$

It is clear that we have both $\pi \leqslant \tau$ and $\sigma \leqslant \tau$.

Suppose now that τ' is a partition such that $\pi \leqslant \tau'$ and $\pi \leqslant \tau'$. Let $B \in \pi$ and $C \in \sigma$ be two blocks that have a nonempty intersection. If $x \in B \cap C$, then both B and C are included in the block of τ' that contains x. In other words, if in $\mathcal{G}_{\pi,\sigma}$ an edge exists that joins B and C, then they are both included in the same block of τ'. This property can be extended to paths: if there is a path in $\mathcal{G}_{\pi,\sigma}$ that joins a block B of π to a block C of σ, then the union of all π-blocks and of all the σ-blocks along this path is included in a block E of τ'. The argument, by induction on the length of the path, is immediate and is omitted. Thus, every block of τ, which is a union of all π-blocks that belong to a connected component (and of all σ-blocks that belong to the same connected component), is included in a block of τ'. Therefore, $\tau \leqslant \tau'$, and this proves that $\tau = \sup\{\pi, \sigma\}$.

Let $\pi, \sigma \in PART(S)$ and let $\tau = \sup\{\pi, \sigma\}$. We have $(x, y) \in \rho_\tau$ if and only if $\{x, y\}$ is enclosed in the same connected component of the graph $\mathcal{G}_{\pi,\sigma}$; that is, if and only if there exists an alternating sequence of blocks of π and σ – $B_{i_1}, C_{j_1}, B_{i_2}, C_{j_2}, \ldots, B_{i_r}, C_{j_s}$ – such that $x \in B_{i_1}$ and $y \in C_{j_s}$. This is equivalent to the existence of a sequence of elements $z_0, z_1, \ldots, z_m$ of S such that $z_0 = x$, $z_m = y$, and $(z_i, z_{i+1}) \in \rho_\pi$ or $(z_i, z_{i+1}) \in \rho_\sigma$ for every i, $0 \leqslant i \leqslant m-1$.

The partition $\sup\{\pi, \sigma\}$ will be denoted by $\pi \vee \sigma$.

Example 10.101 The graph of the partitions π, σ introduced in Example 10.98 has two connected components that correspond to the blocks,

$$D_1 = \{a_1, a_2, a_3, a_4, a_5, a_6, a_7\} = B_1 \cup B_2 \cup B_3 = C_1 \cup C_2,$$
$$D_2 = \{a_8, a_9, a_{10}, a_{11}, a_{12}\} = B_4 \cup B_5 = C_3 \cup C_4,$$

of the partition $\pi \vee \sigma$.

Theorem 10.102 *Let S be a set and let $\rho, \rho' \in EQ(S)$. We have $\pi_\rho \wedge \pi_{\rho'} = \pi_{\rho \cap \rho'}$.*

Proof Indeed, note that $\rho \cap \rho'$ is an equivalence on S, and the equivalence classes of this equivalence (that is, the blocks of the partition $\pi_{\rho\cap\rho'}$) are the nonempty intersections of the blocks of ρ and ρ'. The definition of the infimum of two partitions shows that the set of blocks of $\pi_\rho \wedge \pi_{\rho'}$ is exactly the same, which gives the equality of the theorem.

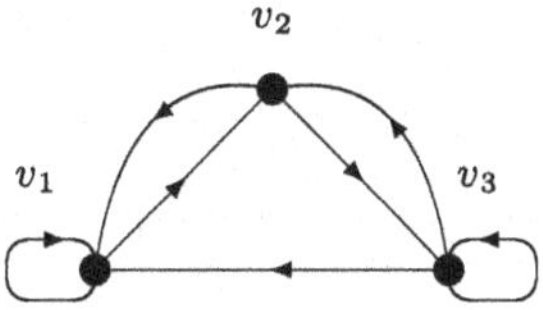

Fig. 10.32 Digraph of the matrix A

10.5 Digraphs of Matrices

The adjacency matrix is the usual matrix representation of a digraph. In this section we are concerned with the reverse process that associates a graph to a matrix.

Definition 10.103 *Let $A \in \mathbb{C}^{n\times n}$ be a square matrix. The* directed graph *of A is the digraph $\mathcal{G}_A = (\{1, \ldots, n\}, E)$, where $(i, j) \in E$ if and only if $a_{ij} \neq 0$.*

Example 10.104 The directed graph of the matrix

$$A = \begin{pmatrix} 1 & 2 & 0 \\ -1 & 0 & 1 \\ -2 & 1 & 1 \end{pmatrix}$$

is shown in Fig. 10.32.

Definition 10.105 *A matrix $A \in \mathbb{C}^{n\times n}$ is* irreducible *if its digraph $\mathcal{G}_A$ is strongly connected.*

Let $M(A)$ be the adjacency matrix of the digraph $\mathcal{G}_A$. It is clear that A is irreducible if and only if $M(A)$ is irreducible. Furthermore, the irreducibility of A is equivalent to the irreducibility of A^{H} (and of A').

Theorem 10.49 states that the digraph $\mathcal{G}_A$ is strongly connected if and only if there does not exist a partition $\{S, T\}$ of $\{1, \ldots, n\}$ such that all edges between S and T have their source in S and their destination in T. Assuming that we number the vertices of S as $1, \ldots, m$ and the vertices of T as $\{m+1, \ldots, n\}$, A is irreducible if and only if A *does not* have the form

$$A = \begin{pmatrix} U & V \\ O_{n-m,m} & W \end{pmatrix},$$

where $U \in \mathbb{C}^{m\times m}$, $V \in \mathbb{C}^{m\times(n-m)}$ and $W \in \mathbb{C}^{(n-m)\times(n-m)}$.

If the vertices of the digraph $\mathcal{G}_A$ are renumbered by replacing each number j by $\phi(j)$, where ϕ is a permutation of the set $\{1, \ldots, n\}$, the adjacency matrix of the resulting graph $\mathcal{G}'$ is $A_{\mathcal{G}'} = P_\phi A \Phi_\phi^{-1} = P_\phi A \Phi_\phi'$. Thus, a matrix $A \in \mathbb{C}^{n\times n}$ is irreducible if and only if there is there is no permutation ϕ such that

$$P_\phi A \Phi'_\phi = \begin{pmatrix} U & V \\ O_{n-m,m} & W \end{pmatrix},$$

where $U \in \mathbb{C}^{m\times m}$, $V \in \mathbb{C}^{m\times(n-m)}$ and $W \in \mathbb{C}^{(n-m)\times(n-m)}$.

Theorem 10.106 *Let A be matrix, $A \in \mathbb{C}^{n\times n}$. There exists a permutation matrix P_ϕ and $p \in \mathbb{N}$, $p \geqslant 1$ such that $P_\phi A P'_\phi$ is a upper triangular block matrix,*

$$P_\phi A P'_\phi = \begin{pmatrix} A_{11} & A_{12} & \cdots & A_{1p} \\ O & A_{22} & \cdots & A_{2p} \\ \vdots & \vdots & \vdots & \vdots \\ O & O & \cdots & A_{pp} \end{pmatrix},$$

where the diagonal blocks $A_{11}, \ldots, A_{pp}$ are irreducible.

Proof Let $K_1, \ldots, K_p$ be the strong connected components of the digraph $\mathcal{G}_A$ and let $\mathsf{C}(\mathcal{G})$ the condensed graph graph of $\mathcal{G}$. We may assume that $K_1, \ldots, K_p$ are listed in topological order since $\mathsf{C}(\mathcal{G})$ is an acyclic graph. In other words, if $\mathsf{C}(\mathcal{G})$ contains an edge (K_i, K_j), we have $i < j$. Assume initially that the vertices of K_i are $v_{\ell_i}, v_{\ell_i+1}, \ldots, v_{\ell_i+|K_i|-1}$, where $\ell_i = \sum_{j=1}^{i-1} |K_j|$ for $1 \leqslant i \leqslant p$. Under this assumption the adjacency matrix $A_\mathcal{G}$ has the form

$$A_\mathcal{G} = \begin{pmatrix} A_{11} & A_{12} & \cdots & A_{1p} \\ O & A_{22} & \cdots & A_{2p} \\ \vdots & \vdots & \vdots & \vdots \\ O & O & O & A_{pp} \end{pmatrix},$$

where A_{ii} is the adjacency matrix of the strongly connected subgraph generated by K_i; clearly, each A_{ii} is an irreducible matrix.

If the vertices of K_i are not numbered according to the previous assumptions, a permutation ϕ can be applied; thus $P_\phi A_\mathcal{G} P'_\phi$ has the necessary form.

Corollary 10.107 *Let A be symmetric matrix, $A \in \mathbb{C}^{n\times n}$. There exists a permutation matrix P_ϕ and $p \in \mathbb{N}$, $p \geqslant 1$ such that $P_\phi A P'_\phi$ is a upper diagonal block matrix,*

$$P_\phi A P'_\phi = \begin{pmatrix} A_{11} & 0 & \cdots & 0 \\ O & A_{22} & \cdots & 0 \\ \vdots & \vdots & \vdots & \vdots \\ O & O & \cdots & A_{pp} \end{pmatrix},$$

where the diagonal blocks $A_{11}, \ldots, A_{pp}$ are irreducible.

Proof This follows directly from Theorem 10.106.

Definition 10.108 *The degree of reducibility* $\mathsf{red}(A)$ *of a matrix* $A \in \mathbb{C}^{n\times n}$ *is the* $p-1$, *where* p *is the number of strongly connected components of the digraph* $\mathcal{G}_A$.

A is an irreducible matrix if and only if $\mathsf{red}(A) = 0$. Furthermore, if A is a symmetric matrix, then $\mathsf{red}(A) = p-1$ if and only if there exists a permutation matrix P_ϕ such that $P_\phi A P'_\phi$ is a block diagonal matrix that consits of p irreducible blocks $A_{11}, \ldots, A_{pp}$.

It is easy to see that a matrix A is irreducible if and only if its transpose is irreducible.

Theorem 10.109 *Let* $A \in \mathbb{R}^{n\times n}$ *be a non-negative matrix. For* $m \geqslant 1$ *we have* $(A^m)_{ij} > 0$ *if and only if there exists a path of length* m *in* $\mathcal{G}_A$ *from* i *to* j.

Proof The argument is by induction on $m \geqslant 1$. The base case, $m = 1$, is immediate. Suppose that the theorem holds for numbers less than m. Then, $(A^m)_{ij} = \sum_{k=1}^{n}(A^{m-1})_{ik}A_{kj}$. $(A^m)_{ij} > 0$ if and only if there is a positive term $(A^{m-1})_{ik}A_{kj}$ in the right-hand sum because all terms are non-negative. By the inductive hypothesis this is the case if and only if there exists a path of length $m-1$ joining i to k and an edge joining k to j, that is, a path of length m joining i to j.

Theorem 10.110 *Let* $A \in \mathbb{R}^{n\times n}$ *be an irreducible matrix such that* $A \geqslant O$. *If* $k_i > 0$ *for* $1 \leqslant i \leqslant n-1$, *then* $\sum_{i=0}^{n-1} k_i A^i > O_{n,n}$.

Proof Since A is an irreducible matrix, the digraph $\mathcal{G}_A$ is strongly connected. Thus, there exists a path of length no larger than $n-1$ that joins any two distinct vertices i and j of the digraph $\mathcal{G}_A$. By Theorem 10.109, there exists $m \leqslant n-1$ such that $(A^m)_{ij} > 0$. Since

$$\left(\sum_{i=0}^{n-1} k_i A^i\right)_{ij} = \sum_{i=0}^{n-1} k_i (A^m)_{ij},$$

and all numbers that occur in this equality are non-negative, it follows that for $i \neq j$ we have $\left(\sum_{i=0}^{n-1} k_i A^i\right)_{ij} > 0$. If $i = j$, the same inequality follows from the fact that $k_0 I_n > O_{n,n}$.

Corollary 10.111 *Let* $A \in \mathbb{R}^{n\times n}$ *be a matrix such that* $A \geqslant O_{n,n}$. *Then* A *is irreducible matrix if and only if* $(I_n + A)^{n-1} > O_{n,n}$.

Proof Suppose that A is irreducible. By choosing $k_i = \binom{n-1}{i}$ for $0 \leqslant i \leqslant n-1$ in Theorem 10.110, the desired inequality follows immediately.

Conversely, suppose that $(I_n + A)^{n-1} > O_{n,n}$ and let i, j be two vertices of the digraph $\mathcal{G}_A$. Since

$$\left((I_n + A)^{n-1}\right)_{ij}$$
$$= (I_n)_{ij} + \binom{n-1}{1}(A)_{ij} + \cdots + \binom{n-1}{k}(A^k)_{ij} + \cdots + (A^{n-1})_{ij} > 0,$$

taking into account that for $i \neq j$ we have $(I_n)_{ij} = 0$, we have $\binom{n-1}{1}(A)_{ij} + \cdots + \binom{n-1}{k}(A^k)_{ij} + \cdots + (A^{n-1})_{ij} > 0$. At least one term of this sum, say $\binom{n-1}{k}(A^k)_{ij}$, must be positive. This means that the vertices i and j in $\mathcal{G}_A$ are joined by a walk of length k. This shows that $\mathcal{G}_A$ is strongly connected, so A is irreducible.

Definition 10.112 *A matrix $A \in \mathbb{R}^{n\times n}$ is* primitive *if $A \geqslant O_{n,n}$, and there exists $m \geqslant 1$ such that $A^m > O_{n,n}$.*

Theorem 10.113 *Let $A \in \mathbb{R}^{n\times n}$ be an irreducible matrix with $A \geqslant O_{n,n}$ and let g_i be the greatest common divisor of the lengths of the cycles in $\mathcal{G}_A$ that begin and end at vertex i, $1 \leqslant i \leqslant n$. If A is primitive, then $g_i = 1$ for $1 \leqslant i \leqslant n$.*

Proof Since A is irreducible every node i is located on a cycle. By the definition of primitiveness there exists m such that $A^m > 0$, so $A^k > 0$ for $k \geqslant m$. This means that there exist cycles of length $m+1, m+2, \ldots$ starting and ending in i, so $g_i = 1$ for every vertex i.

10.6 Spectra of Non-negative Matrices

Non-negative and irreducible matrices have important interactions with graph theory. We focus initially on spectra of a special class of such matrices, namely positive matrices. The main results presented in this section were obtained at the beginning of the 20th century by the German mathematicians Oskar Perron (1880–1975) and Georg Frobenius (1849–1917).

Theorem 10.114 *Let λ be an eigenvalue of A such that $|\lambda| = \rho(A)$ and let $(\lambda, \mathbf{x})$ be an eigenpair. Then, $(\rho(A), \mathsf{abs}(\mathbf{x}))$ is an eigenpair of A.*

Proof We have

$$\begin{aligned}\rho(A)\,\mathsf{abs}(\mathbf{x}) &= |\lambda|\,\mathsf{abs}(\mathbf{x}) = \mathsf{abs}(\lambda\mathbf{x}) \\ &= \mathsf{abs}(A\mathbf{x}) \leqslant \mathsf{abs}(A)\,\mathsf{abs}(\mathbf{x}) = A\,\mathsf{abs}(\mathbf{x}),\end{aligned}$$

which implies that $\rho(A)$ is a positive eigenvalue of A and $\mathsf{abs}(\mathbf{x}) > 0$ is a positive eigenvector that corresponds to this eigenvalue.

Lemma 10.115 *Let $z_1, \ldots, z_n$ be n non-zero complex numbers such that*

$$\left|\sum_{k=1}^{n} z_k\right| = \sum_{k=1}^{n} |z_k|.$$

There exists α such that $z_k = |z_k|e^{i\alpha}$ for $1 \leqslant k \leqslant n$.

Proof Let $z_k = |z_k|e^{i\alpha_k}$ for $1 \leqslant k \leqslant n$. We have

$$\left|\sum_{k=1}^{n} z_k\right|^2 = \left(\sum_{k=1}^{n} z_k\right)\left(\overline{\sum_{k=1}^{n} z_k}\right) = \left(\sum_{p=1}^{n} |z_p|e^{i\alpha_p}\right)\left(\sum_{q=1}^{n} |z_q|e^{i\alpha_q}\right)$$
$$= \sum_{p=1}^{n}\sum_{q=1}^{n} |z_p||z_q|e^{i(\alpha_p-\alpha_q)} = \sum_{p=1}^{n} |z_p|^2 + 2\sum_{1\leqslant p,q\leqslant n} |z_p||z_q|\cos(\alpha_p - \alpha_q).$$

On the other hand,

$$\sum_{k=1}^{n} |z_k|^2 = \sum_{p=1}^{n} |z_p|^2 + 2\sum_{1\leqslant p,q\leqslant n} |z_p||z_q|.$$

The equality of the lemma is possible only if $\alpha_p = \alpha_q$ for $1 \leqslant p < q \leqslant n$, that is if there exists α such that $\alpha_1 = \cdots = \alpha_n = \alpha$. □

Lemma 10.116 *Let $A \in \mathbb{R}^{n\times n}$ be a matrix such that $A > O_{n,n}$. If $(\lambda, \boldsymbol{x})$ is an eigenpair of A such that $|\lambda| = \rho(A)$, then there exists $\theta \in \mathbb{R}$ such that $\mathsf{abs}(\boldsymbol{x}) = e^{-i\theta}\boldsymbol{x}$.*

Proof By Theorem 10.114 we have $A\,\mathsf{abs}(\mathbf{x}) = \rho(A)\,\mathsf{abs}(\mathbf{x})$. Thus, we have

$$\rho(A)\,|x_k| = |\lambda|\,|x_k| = |\lambda x_k| = \left|\sum_{p=1}^{n} a_{kp}x_p\right|$$
$$\leqslant \sum_{p=1}^{n} |a_{kp}|\,|x_p| = \sum_{p=1}^{n} a_{kp}\,|x_p| = \rho(A)|x_k|.$$

Consequently, the above inequality becomes an equality and we obtain

$$\left|\sum_{p=1}^{n} a_{kp}x_p\right| = \sum_{p=1}^{n} a_{kp}\,|x_p|.$$

By Lemma 10.115 this is possible, only if $a_{kp}x_p = |a_{kp}x_p|e^{i\alpha}$ for $1 \leqslant p \leqslant n$. Thus, $a_{kp}x_pe^{-i\alpha} = |a_{kp}x_p| > 0$ for $1 \leqslant p \leqslant n$. Since $a_{kp} > 0$, it follows that $|x_p| = x_pe^{-i\alpha} > 0$ for $1 \leqslant p \leqslant n$, which implies $\mathsf{abs}(\mathbf{x}) = \mathbf{x}e^{-i\alpha} > \mathbf{0}_n$.

Theorem 10.117 *Let $A \in \mathbb{R}^{n\times n}$ be a matrix such that $A > O_{n,n}$. If $\lambda \in \mathsf{spec}(A)$ and $\lambda \neq \rho(A)$, then $|\lambda| < \rho(A)$.*

Proof The definition of $\rho(A)$ implies $|\lambda| \leqslant \rho(A)$.

Suppose that $|\lambda| = \rho(A)$ and $A\mathbf{x} = \lambda\mathbf{x}$ for $\mathbf{x} \neq \mathbf{0}_n$. By Lemma 10.116 there exists $\theta \in \mathbb{R}$ such that $\mathsf{abs}(\mathbf{x}) = e^{-i\theta}\mathbf{x}$, so $A\ \mathsf{abs}(\mathbf{x}) = \lambda\ \mathsf{abs}(\mathbf{x})$, which implies $\lambda = \rho(A)$; this contradicts the hypothesis of the theorem.

Theorem 10.118 *Let $A \in \mathbb{R}^{n\times n}$ be a positive matrix. The geometric multiplicity of $\rho(A)$ equals* 1.

Proof Let $(\rho(A), \mathbf{x})$ and $(\rho(A), \mathbf{y})$ be two eigenpairs of A. By Lemma 10.116, there exist θ_1 and θ_2 such that $\mathsf{abs}(\mathbf{x}) = e^{-i\theta_1}\mathbf{x} > 0$ and $\mathsf{abs}(\mathbf{y}) = e^{-i\theta_2}\mathbf{y} > 0$.

Define $\mathbf{z} = \mathsf{abs}(\mathbf{y}) - b\ \mathsf{abs}(\mathbf{x})$, where $b = \min\left\{\frac{|y_i|}{|x_i|} \mid 1 \leqslant i \leqslant n\right\}$ and suppose that $\mathbf{z} \neq \mathbf{0}_n$. Then $\mathbf{z} \geqslant \mathbf{0}_n$ and there exists z_i such that $z_i = 0$, so $\mathbf{z}$ is not positive. Since $A\mathbf{z} = A(\mathsf{abs}(\mathbf{y}) - b\ \mathsf{abs}(x)) = \rho(A)\mathsf{abs}(\mathbf{y}) - b\rho(A)\mathsf{abs}(\mathbf{x}) = \rho(A)\mathbf{z}$, if $\mathbf{z} \neq \mathbf{0}_n$, it follows that $\mathbf{z} = \frac{1}{\rho(A)}A\mathbf{z} > \mathbf{0}_n$. By Supplement 25 of Chap. 5, we have $\mathbf{z} > \mathbf{0}_n$, which contradicts the fact that that $\mathbf{z}$ is not positive. Thus, $\mathbf{z} = \mathbf{0}_n$, so $\mathsf{abs}(\mathbf{y}) = b\ \mathsf{abs}(\mathbf{x})$ and $\mathbf{y} = be^{i(\theta_2-\theta_1)}\mathbf{x}$.

Definition 10.119 *Let $A \in \mathbb{R}^{n\times n}$ be a positive matrix. The* Perron vector of *A is the positive eigenvector $\mathbf{x}$ that corresponds to $\rho(A)$ such that $\parallel \mathbf{x} \parallel_1 = 1$.*

Theorem 10.118 ensures the existence of the Perron vector for any positive matrix A.

The next statement gives a stronger result (compared to Theorem 10.118):

Theorem 10.120 *Let $A \in \mathbb{R}^{n\times n}$ be a matrix such that $A > O_{n,n}$. The spectral radius $\rho(A)$ is an eigenvalue of A having algebraic multiplicity* 1.

Proof By Schur's Triangularization Theorem (Theorem 7.13) there exists a unitary matrix U and an upper-triangular matrix T such that $A = UTU^{-1}$, where diagonal elements of T are the eigenvalues of A. We saw that each eigenvalue λ of A occurs in the sequence of diagonal elements of T a number of $\mathsf{algm}(A, \lambda)$ times. Let $p = \mathsf{algm}(A, \rho(A))$, and note that for every other eigenvalue λ of A we have $|\lambda| < \rho(A)$. The matrix $\frac{1}{\rho(A)}T$ is an upper triangular matrix and the sequence of its diagonal elements is

$$\left(1, \ldots, 1, \frac{\lambda_{p+1}}{\rho(A)}, \ldots, \frac{\lambda_n}{\rho(A)}\right).$$

Therefore, $\left(\frac{1}{\rho(A)}T\right)^m$ is an upper triangular matrix whose sequence of diagonal elements is

$$\left(1, \ldots, 1, \left(\frac{\lambda_{p+1}}{\rho(A)}\right)^m, \ldots, \left(\frac{\lambda_n}{\rho(A)}\right)^m\right).$$

Let L be the matrix introduced in Supplement 3 of Chap. 7 for which $rank(L) = 1$. We saw that $\lim_{m\to\infty}\left(\frac{1}{\rho(A)}A\right)^m = L$. Since

$$\left(\frac{1}{\rho(A)}A\right)^m = U\left(\frac{1}{\rho(A)}T\right)^m U^{-1},$$

it follows that $\lim_{m\to\infty}\left(\frac{1}{\rho(A)}A\right)^m$ is an upper triangular matrix having the diagonal sequence $(1,\ldots,1,0,\ldots,0)$. Since 1 occurs p times in this sequence, $rank\left(\lim_{m\to\infty}\left(\frac{1}{\rho(A)}A\right)^m\right) = p$, and we must have $p = 1$ because $rank(L) = 1$.

For no eigenvalue of A other than $\rho(A)$ there exists an eigenvector whose components are positive. Indeed, if λ' is an eigenvalue distinct from $\rho(A)$ and $\mathbf{w}$ is an eigenvector of λ', then $\mathbf{w}$ cannot be a positive vector for this would contradict the orthogonality of $\mathbf{w}$ and Perron vector $\mathbf{u}$.

Corollary 10.121 (Perron Theorem) *Let $A \in \mathbb{R}^{n\times n}$ be a symmetric matrix with positive elements. The following statements hold:*

(i) *$\rho(A)$ is a positive number and is an eigenvalue of A;*
(ii) *there exists an eigenpair $(\rho(A), \boldsymbol{x})$ with $\boldsymbol{x} > \boldsymbol{0}_n$;*
(iii) $\mathsf{algm}(A, \rho(A)) = 1$;
(iv) *if θ is any other eigenvalue of A, then $|\theta| < \rho(A)$.*

Proof The statements of the corollary follow from the preceding theorems.

Positive matrices are clearly irreducible, non-negative matrices. Therefore, it is natural to extend Perron's Theorem to this larger class of matrices. The next statement concerns non-negative matrices.

Theorem 10.122 *Let $A \in \mathbb{R}^{n\times n}$ be a non-negative matrix. We have $\rho(A) \in \mathsf{spec}(A)$ and there exists $\boldsymbol{x} \in \mathbb{R}^n - \{\boldsymbol{0}_n\}$ such that $\boldsymbol{x} \geqslant \boldsymbol{0}_n$ and $(\rho(A), \boldsymbol{x})$ is an eigenpair of A.*

Proof Let $A(\epsilon) = A + \epsilon J_n$ for $\epsilon > 0$ and let $\mathbf{x}(\epsilon)$ be the Perron vector of the positive matrix $A(\epsilon)$. The collection of vectors $\{\mathbf{x}(\epsilon) \mid \epsilon > 0\}$ is contained in the closed sphere $B_1(\mathbf{0}_n, 1)$ (which is a compact set), so, by Theorem 8.30, there a monotonic decreasing sequence $\epsilon_1, \ldots, \epsilon_p, \ldots$ such that $\lim_{p\to\infty}\epsilon_p = 0$ and $\lim_{p\to\infty}\mathbf{x}(\epsilon_p) = \mathbf{x}$. Since $A(\epsilon_1) \geqslant \cdots \geqslant A(\epsilon_p) \geqslant \cdots \geqslant A$, by Theorem 7.66, we have $\rho(A(\epsilon_1)) \geqslant \cdots \geqslant \rho(A(\epsilon_p)) \geqslant \cdots \geqslant \rho(A)$. This implies $\lim_{k\to\infty}\rho(A(\epsilon_k)) \geqslant \rho(A)$.

Since $\mathbf{x}(\epsilon_p) > \mathbf{0}_n$, it follows that $\mathbf{x} \geqslant \mathbf{0}_n$. Moreover, since $\sum_{i=1}^n x_i = \lim_{p\to\infty}\sum_{i=1}^n x_i(\epsilon_p) = 1$, we have $\mathbf{x} \neq \mathbf{0}_n$.

The continuity of the matrix product implies

$$\begin{aligned}A\mathbf{x} &= \lim_{k\to\infty} A(\epsilon_k)\mathbf{x}(\epsilon_k) = \lim_{k\to\infty}\rho(A(\epsilon_k))\mathbf{x}(\epsilon_k)\\ &= \lim_{k\to\infty}\rho(A(\epsilon_k))\lim_{k\to\infty}\mathbf{x}(\epsilon_k) = \lim_{k\to\infty}\rho(A(\epsilon_k))\mathbf{x}.\end{aligned}$$

Consequently, $\lim_{k\to\infty}\rho(A(\epsilon_k)) \in \mathsf{spec}(A)$, so $\lim_{k\to\infty}\rho(A(\epsilon_k)) \leqslant \rho(A)$. Thus, $\lim_{k\to\infty}\rho(A(\epsilon_k)) = \rho(A)$. Thus, $(\rho(A), \mathbf{x})$ is an eigenpair of A.

Theorem 10.123 *Let $A \in \mathbb{R}^{n\times n}$ be an primitive matrix. The spectral radius $\rho(A)$ is a simple eigenvalue of A.*

Proof Suppose that $A^m > O_{n,n}$. If $\lambda_1, \ldots, \lambda_n$ are the eigenvalues of A, then $\lambda_1^m, \ldots, \lambda_n^m$ are the eigenvalues of A^m. Theorem 10.122 implies that $\rho(A)$ is an eigenvalue of A, so $\rho(A)^m$ is an eigenvalue of A^m. By Perron's Theorem, $\rho(A)^m$ is a simple eigenvalue of A^m, so $\rho(A)$ must be a simple eigenvalue of $\rho(A)$.

Theorem 10.124 (Perron-Frobenius Theorem) *Let $A \in \mathbb{C}^{n\times n}$ be a matrix that is non-negative and irreducible. Then, $\rho(A) > 0$ and $\rho(A)$ is an eigenvalue of A. Furthermore, there is a positive $\boldsymbol{x} \in \mathbb{R}^n$ such that $(\rho(A), \boldsymbol{x})$ is an eigenpair and* $\mathsf{algm}(A, \rho(A)) = 1$.

Proof Supplement 46 of Chap. 7 implies $\rho(A) > 0$ and $(\rho(A), \mathbf{x})$ is an eigenpair of A by Theorem 10.122, where $\mathbf{x} \neq \mathbf{0}_n$ and $\mathbf{x} \geqslant \mathbf{0}_n$.

Note that

$$(I_n + A)^{n-1}\mathbf{x} = (1 + \rho(A))^n\mathbf{x}.$$

By Corollary 10.111, $(I_n + A)^{n-1} > O_{n,n}$, so $(I_n + A)^{n-1}\mathbf{x} > 0$. Since $\mathbf{x} = (1 + \rho(A))^{-(n-1)}(I_n + A)^{n-1}\mathbf{x}$, it follows that $\mathbf{x} > 0$.

Finally, since $\rho(A)$ is an eigenvalue of A, $1 + \rho(A)$ is an eigenvalue of $I + A$ having the same multiplicity. Since $I + A \geqslant O_{n,n}$ and $(I + A)^{n-1} > 0$, $1 + \rho(A)$ is a simple eigenvalue of $I + A$.

The eigenvector $\mathbf{x}$ of the eigenpair $(\rho(A), \mathbf{x})$ having $\sum_{x_i} = 1$ is the *Perron vector of A*.

Perron-Frobenius Theorem is useful for the study of adjacency matrices of graphs which are clearly non-negative matrices.

10.7 Fiedler's Classes of Matrices

This section is dedicated to certain classes of matrices introduced by Fiedler in [7]. Let $\mathcal{Z}_n$ be a subset of $\mathbb{R}^{n\times n}$ that consists of those matrices whose off-diagonal elements are not larger than 0 and let $\mathcal{Z} = \bigcup_{n\geqslant 1} \mathcal{Z}_n$.

In [7] Fiedler has obtained eighteen equivalent characterizations of a subclass of $\mathcal{Z}_n$, known as the class of **K-matrices**. Further equivalent characterization can be found in [8].

We reproduce Fiedler's result in Theorems 10.125–10.128. The numbering of statements is consistent across these theorems.

Theorem 10.125 *Let $A \in \mathcal{Z}_n$ be a matrix. The following statements are equivalent:*

(i) *there exists $\boldsymbol{x} \in \mathbb{R}^n$ such that $\boldsymbol{x} \geqslant \boldsymbol{0}_n$ and $A\boldsymbol{x} > \boldsymbol{0}_n$;*
(ii) *there exists $\boldsymbol{x} \in \mathbb{R}^n$ such that $\boldsymbol{x} > \boldsymbol{0}_n$ and $A\boldsymbol{x} > \boldsymbol{0}_n$;*

(iii) *there exists a diagonal matrix* $D = \mathsf{diag}(d_1, \ldots, d_n) \in \mathbb{R}^{n \times n}$ *with positive diagonal entries such that if* $W = AD$*, then* $w_{ii} > \sum\{|w_{ik}| \mid 1 \leqslant k \leqslant n, k \neq i\}$;
(iv) *if* $B \in \mathbb{Z}_n$ *and* $B \geqslant A$*, then* B *is invertible;*
(v) *every real eigenvalue of any principal submatrix of* A *is positive;*
(vi) *all principal minors of* A *are positive;*
(vii) *for each* k*,* $1 \leqslant k \leqslant n$*, the sum of all principal minors of order* k *of* A *is positive;*
(viii) *every real eigenvalue of* A *is positive;*
(ix) *there exists a matrix* $C \geqslant O$ *and a number* $k > \rho(C)$ *such that* $A = kI_n - C$;
(x) *the matrix* A *can be written as* $A = D - E$ *such that* $D^{-1} \geqslant O$*,* $E \geqslant O$*, and* $\rho(D^{-1}E) < 1$;
(xi) A *is invertible and* $A^{-1} \geqslant O$*.*

Proof `(i) implies (ii)`: Suppose that $\mathbf{x} \geqslant \mathbf{0}_n$ and $A\mathbf{x} > \mathbf{0}_n$. Since $A\mathbf{x} > \mathbf{0}_n$ there exists $\epsilon > 0$ such that $A\mathbf{x} + \epsilon A\mathbf{1} > 0$, so $A(\mathbf{x} + \epsilon \mathbf{1}_n) > 0$. Thus, the vector $\mathbf{y} = \mathbf{x} + \epsilon \mathbf{1}$ satisfies (ii).

`(ii) implies (iii)`: Let $\mathbf{x} \in \mathbb{R}^n$ be a vector that satisfies (ii) and let $D = \mathsf{diag}(x_1, \ldots, x_n)$. We have $w_{ij} = a_{ij}x_j$. Since $A\mathbf{x} > \mathbf{0}_n$ we have

$$a_{ii}x_i > \sum\{-a_{ij}x_j \mid 1 \leqslant j \leqslant n, j \neq i\},$$

so $w_{ii} > \sum\{|w_{ij}| \mid 1 \leqslant j \leqslant n, j \neq i\}$ because $A \in \mathbb{Z}_n$.

`(iii) implies (iv)`: Let D be a diagonal matrix $D = \mathsf{diag}(d_1, \ldots, d_n) \in \mathbb{R}^{n \times n}$ with positive diagonal entries such that if $W = AD$, then $w_{ii} > \sum\{|w_{ik}| \mid 1 \leqslant k \leqslant n, k \neq i\}$.

Suppose that (iv) does not hold, that is, there exists $B \in \mathbb{Z}_n$ such that $B \geqslant A$ and B is singular. This supposition entails the existence of $\mathbf{y} \in \mathbb{R}^n$, $\mathbf{y} \neq \mathbf{0}_n$ and $B\mathbf{y} = \mathbf{0}_n$.

Define $\ell = \arg\max_{1 \leqslant j \leqslant n} \frac{|y_j|}{d_j}$. We have $\frac{|y_j|}{d_j} \leqslant \frac{|y_\ell|}{d_\ell}$ for $1 \leqslant j \leqslant n$.

Since $B\mathbf{y} = \mathbf{0}_n$, we have $b_{\ell\ell}y_\ell = \sum_{j \neq \ell}(-b_{\ell j})y_j$. By the definition of W we also have

$$a_{\ell\ell}d_\ell > \sum_{k=1, k \neq \ell}^{n} |a_{\ell k}|d_k.$$

Since $B \geqslant A$, we have $b_{\ell\ell} \geqslant a_{\ell\ell}$ and $b_{\ell j} \geqslant a_{\ell j}$ for $j \neq \ell$. Since the off-diagonal elements of both A and B are not larger than 0, this means that $|b_{\ell j}| \leqslant |a_{\ell j}|$ for $j \neq \ell$. Therefore,

$$|b_{\ell\ell}y_\ell| \leqslant \sum_{j \neq \ell} |b_{\ell j}y_j| \leqslant \sum_{j \neq \ell} |a_{\ell j}|d_j \frac{|y_j|}{d_j}$$

$$\leqslant \left(\sum_{j \neq \ell} |a_{\ell j}| d_j \right) \frac{|y_\ell|}{d_\ell} < a_{\ell\ell} \frac{|y_\ell|}{d_\ell} \leqslant b_{\ell\ell}\, |y_\ell| \leqslant |b_{\ell\ell} y_\ell|,$$

which is a contradiction. Thus, B is non-singular.

`(iv) implies (v)` : Let $A \in \mathcal{Z}_n$ and let $\{i_1, \ldots, i_m\}$ be a subset of $\{1, \ldots, n\}$ and let λ be an eigenvalue of $A \begin{bmatrix} i_1, \ldots, i_m \\ i_1, \ldots, i_m \end{bmatrix}$. Suppose that λ is negative. Then, for the matrix $B \in \mathcal{Z}_n$ defined by

$$b_{ij} = \begin{cases} a_{ii} - \lambda & \text{if } j = i, \\ a_{ij} & \text{if } i, j \in \{i_1, \ldots, i_m\} \text{ and } i \neq j \\ 0 & \text{otherwise,} \end{cases}$$

we have $B \geqslant A$. By (iv), B is non-singular. On the other hand,

$$B \begin{bmatrix} i_1, \ldots, i_m \\ i_1, \ldots, i_m \end{bmatrix} = A \begin{bmatrix} i_1, \ldots, i_m \\ i_1, \ldots, i_m \end{bmatrix} - \lambda I_m$$

so $\det \left(B \begin{bmatrix} i_1, \ldots, i_m \\ i_1, \ldots, i_m \end{bmatrix} \right) = 0$ because λ is an eigenvalue of $A \begin{bmatrix} i_1, \ldots, i_m \\ i_1, \ldots, i_m \end{bmatrix}$. Thus, $\det(B) = 0$, which leads to a contradiction.

`(v) implies (vi)` : Let

$$\det \left(A \begin{bmatrix} i_1 & \cdots & i_m \\ i_1 & \cdots & i_m \end{bmatrix} \right)$$

be a principal minor of A whose eigenvalues are positive. Since $A \begin{bmatrix} i_1 & \cdots & i_m \\ i_1 & \cdots & i_m \end{bmatrix}$ is a real matrix, complex eigenvalues occur in conjugate pairs, so the product of the complex eigenvalues is positive, which implies that the product of all eigenvalues is positive. Since this product equals $\det \left(A \begin{bmatrix} i_1 & \cdots & i_m \\ i_1 & \cdots & i_m \end{bmatrix} \right)$, statement (vi) follows.

`(vi) implies (vii)` : This implication is immediate.

`(vii) implies (viii)` : Suppose that for each k, $1 \leqslant k \leqslant n$, the sum of all principal minors of order k of A is positive. In Theorem 7.3 we saw that the characteristic equation of A has the form

$$\sum_{k=0}^{n} (-1)^k a_k \lambda^{n-k} = 0$$

where a_k is the sum of the principal minors of order k of A. Thus, the coefficients of this polynomial have alternating signs, so no real eigenvalue can be less than or equal to 0.

`(viii) implies (ix)` : Let $A \in \mathcal{Z}_n$ be a matrix whose real eigenvalues are positive. If $m = \max_{1 \leqslant i \leqslant n} a_{ii}$, the matrix $C = mI_n - A$ is non-negative. By Theorem 10.122, $\rho(C) \in \mathsf{spec}(C)$. Since $A = mI_n - C$, $m - \rho(C)$ is an eigenvalue of A, so $m > \rho(C)$ because all real eigenvalues of A are positive.

`(ix) implies (x)` : Suppose that there exists a matrix $C \geqslant O$ and a number $k > \rho(C)$ such that $A = kI_n - C$. Define $D = kI_n$ and $E = C$, so $A = D - E$. Furthermore, we have

$$\rho(D^{-1}E) = \frac{1}{k}\rho(C) < 1.$$

`(x) implies (xi)` : Suppose that the matrix $A \in \mathcal{Z}_n$ can be written as $A = D - E$ such that D is an invertible matrix, $D^{-1} \geqslant O$, $E \geqslant O$, and $\rho(D^{-1}E) < 1$. Then, $A = D(I_n - D^{-1}E)$. Since $\rho(D^{-1}E) < 1$ it follows that $A^{-1} = \sum_{k=1}^{\infty}(D^{-1}E)^k D^{-1}$. Since $D^{-1}E \geqslant O$, we have $A^{-1} \geqslant O$.

`(xi) implies (i)` : Let $A \in \mathcal{Z}_n$ be a non-singular matrix such that $A^{-1} > O$. Define $\mathbf{x} = A^{-1}\mathbf{1}_n$. Since $A^{-1} > O$, it follows that $\mathbf{x} \geqslant \mathbf{0}_n$ and $A\mathbf{x} = \mathbf{1}_n > \mathbf{0}_n$.

Theorem 10.126 *Let $A \in \mathcal{Z}_n$ be a matrix. The following statements are equivalent:*

(vi) *all principal minors of A are positive;*

(xii) *we have*

$$\det\left(A\begin{bmatrix}1 & \cdots & k\\ 1 & \cdots & k\end{bmatrix}\right) > 0$$

for every k, $1 \leqslant k \leqslant n$;

(xiii) *the matrix A can be factored as $A = LU$, where L is a lower triangular matrix and U is an upper triangular matrix such that both L and U have positive diagonal entries;*

(xiv) *the matrix A can be factored as $A = LU$, where L is a lower triangular matrix* in $\mathcal{Z}_n$ *and U is an upper triangular matrix* in $\mathcal{Z}_n$ *such that both L and U have positive diagonal entries;*

(xi) *the matrix A is non-singular and $A^{-1} \geqslant O$.*

Proof `(vi) implies (xii)` is immediate.

`(xii) implies (xiii)` : Suppose that

$$\det\left(A\begin{bmatrix}1\cdots k\\ 1\cdots k\end{bmatrix}\right) > 0$$

for $1 \leqslant k \leqslant n$. This means that the matrix A is strongly non-singular (see Supplement 63 of Chap. 5), so A can be factored as a product $A = LDU$, where L is a lower triangular matrix having 1s on its diagonal, D is a diagonal matrix, and U is an upper diagonal matrix whose diagonal elements are equal to 1. Since the diagonal elements of D are positive, by the same supplement and Part (xii), it follows that (DU) is an upper diagonal matrix having positive elements. Thus, $A = L(DU)$ is the desired factorization.

`(xiii) implies (xiv)`: Suppose that $A \in \mathcal{Z}_n$ can be factored as $A = LU$, where L is a lower triangular matrix and U is an upper triangular matrix such that both L and U have positive diagonal entries. To prove that L and U belong to $\mathcal{Z}_n$ it suffices to show that if $i \neq j$, then $l_{ij} \leqslant 0$ and $u_{ij} \leqslant 0$. We have

$$a_{ik} = \sum_{j=1}^{n} l_{ij} u_{jk} = \sum_{j=1}^{\min i,k} l_{ij} u_{jk}.$$

Thus, $a_{12} = l_{11}u_{12}$ and $a_{21} = l_{21}u_{11}$. Since $a_{12} < 0$ and $a_{21} < 0$ it follows that $u_{12} < 0$ and $l_{21} < 0$. Thus, we established that $l_{ik} \leqslant 0$ and $u_{ik} \leqslant 0$ if $i + k = 3$. We proceed by induction on $i + k \geqslant 3$. The base case $i + k = 3$ was just proven. Suppose that the property holds for $i + k < p$.

If $i > k$ we have $a_{ik} = \sum_{j=1}^{k} l_{ij} u_{jk} < 0$, so

$$l_{ik} u_{kk} = a_{ij} - \sum_{j=1}^{k-1} l_{ij} u_{jk}.$$

Note that if $j \leqslant k - 1$, then $i + j \leqslant i + k - 1$ and $j + k \leqslant 2k - 1 \leqslant i + k - 1$, so $l_{ij} \leqslant 0$ and $u_{jk} \leqslant 0$ by the inductive hypothesis. Therefore, $l_{ik} \leqslant 0$.

If $i < k$ we have $a_{ik} = \sum_{j=1}^{i} l_{ij} u_{jk}$, so

$$l_{ii} u_{ik} = a_{ik} - \sum_{j=1}^{i-1} l_{ij} u_{jk}.$$

By a similar argument we obtain $u_{ik} \leqslant 0$, so $L, U \in \mathcal{Z}_n$.

`(xiv) implies (xi)`: Suppose that $A = LU$, where $L, U \in \mathcal{Z}_n$. By Theorem 5.74, the matrix $L^{-1} \geqslant O$ and $U^{-1} \geqslant 0$. Therefore, $A^{-1} = U^{-1}L^{-1} \geqslant O$.

`(xi) implies (vi)`: This is immediate by Theorem 10.125.

Theorem 10.127 *Let $A \in \mathcal{Z}_n$ be a matrix. The following statements are equivalent:*

(xi) *A is invertible and $A^{-1} \geqslant O$;*
(xv) *the matrix A can be written as $A = D - E$ such that $D^{-1} \geqslant O$, $E \geqslant O$ and $\rho(D^{-1}E) < 1$;*
(ix) *there exists a matrix $C \geqslant O$ and a number $k > \rho(C)$ such that $A = kI_n - C$.*

Proof `(xi) implies (xv)`: Let $A \in \mathcal{Z}_n$ be an invertible matrix and $A^{-1} \geqslant O$. We can write $A = A - O$ and, therefore, we can take $D = A$ and $E = O$ and $D^{-1}E \geqslant O$. This establishes that A can be written as a difference of matrices $A = D - E$ such that $D^{-1} \geqslant O$, $E \geqslant O$ and $D^{-1}E \geqslant O$. By Theorem 10.122, $\rho(D^{-1}E)$ is an eigenvalue of $D^{-1}E$ and there exists a non-negative eigenvector $\mathbf{v} > \mathbf{0}$ that corresponds to this eigenvalue. This implies $D^{-1}E\mathbf{v} = \rho(D^{-1}E)\mathbf{v}$, so $E\mathbf{v} = \rho(D^{-1}E)D\mathbf{v}$. If $\rho(D^{-1}E)$ were at least equal to 1, this would imply

$$A\mathbf{v} = (D - E)\mathbf{v} = D\mathbf{v} - E\mathbf{v} = \left(\frac{1}{\rho(D^{-1}E)} - 1\right) E\mathbf{v} \leqslant \mathbf{0}.$$

Since $A\mathbf{v} \leqslant \mathbf{0}$, by multiplying this inequality by $A^{-1} \geqslant O$ we obtain $\mathbf{v} \leqslant \mathbf{0}$. Thus, $\rho(D^{-1}E) < 1$.

`(xv) implies (ix)`: Suppose that $A \in \mathcal{Z}_n$ can be written as $A = D - E$ such that $D^{-1} \geqslant O$ and $Q \geqslant O$, and for each such pair (D, E) we have $\rho(D^{-1}E) < 1$.

We claim that $k = \max\{a_{ii} \mid 1 \leqslant i \leqslant n\} > 0$. Indeed, suppose that this is not the case. Since $A \in \mathcal{Z}_n$ this would imply $A \leqslant O$, so $D \leqslant E$. In turn, this implies $I < D^{-1}E$ and this would contradict the fact that $\rho(D^{-1}E) < 1$. Therefore, $k > 0$.

Let $C = kI_n - A$. Since $A = kI_n - C$, $(kI_n)^{-1} = k^{-1}I_n \geqslant O$ and $C \geqslant O$ it follows that $\rho(k^{-1}C) < 1$. Therefore, $k > \rho(C)$ and we obtain (ix).

`(ix) implies (xi)`: This is immediate by Theorem 10.125.

Theorem 10.128 *Let $A \in \mathcal{Z}_n$ be a matrix. The following statements are equivalent:*

(ix) *there exists a matrix $C \geqslant O$ and a number $k > \rho(C)$ such that $A = kI_n - C$;*

(xvi) *there exists a diagonal matrix $G = \mathsf{diag}(g_{11}, \ldots, g_{nn})$ with $g_{ii} > 0$ for $1 \leqslant i \leqslant n$ such that if $B = GAG^{-1}$, then the symmetric matrix $B + B'$ is positive definite;*

(xvii) *there exists a diagonal matrix $H = \mathsf{diag}(h_{11}, \ldots, h_{nn})$ with $h_{ii} > 0$ for $1 \leqslant i \leqslant n$ such that if $C = AH$, then $C + C'$ is positive definite;*

(xviii) *for any $\lambda \in \mathsf{spec}(A)$, $\Re(\lambda) > 0$.*

Proof `(ix) implies (xvi)`: Let $A \in \mathcal{Z}_n$ be a matrix that can be written as $A = kI_n - C$, where $C \geqslant O$ and $k > \rho(C)$.

Suppose initially that C is irreducible. By Theorem 10.114 its spectral radius $\rho(C)$ is an eigenvalue and there exists a positive eigenvector $\mathbf{u}$ that corresponds to this eigenvalue, that is $C\mathbf{u} = \rho(C)\mathbf{u}$. Since C' is also irreducible and non-negative, it follows that there exists a positive eigenvector $\mathbf{v}$ such that $C'\mathbf{v} = \rho(C)\mathbf{v}$.

Let $\mathbf{w} \in \mathbb{R}^n$ be the vector defined by $w_i = \sqrt{u_i v_i}$ for $1 \leqslant i \leqslant n$ and let G be the diagonal matrix $G = \mathsf{diag}(g_1, \ldots, g_n)$, where

$$g_i = \sqrt{\frac{v_i}{u_i}}$$

for $1 \leqslant i \leqslant n$. We have

$$G\mathbf{w} = \mathbf{v} \text{ and } G^{-1}\mathbf{w} = \mathbf{u},$$

so

$$\begin{aligned}(GCG^{-1})w &= GC\mathbf{u} = \rho(C)G\mathbf{u} = \rho(C)\mathbf{w},\\ (GCG^{-1})'\mathbf{w} &= G^{-1}C'G\mathbf{w} = G^{-1}C'\mathbf{v} = \rho(C)G^{-1}\mathbf{v} = \rho(C)\mathbf{w}.\end{aligned}$$

We need to show that if $B = GAG^{-1}$, then the matrix $B + B'$ is positive definite. Clearly, $B \in \mathcal{Z}_n$ and we have

$$B\mathbf{w} = GAG^{-1}\mathbf{w} = G(kI_n - C)G^{-1}\mathbf{w} = k\mathbf{w} - GCG^{-1}\mathbf{w} = (k - \rho(C))\mathbf{w}$$
$$B'\mathbf{w} = (GAG^{-1})'\mathbf{w} = (G(kI_n - C)G^{-1})'\mathbf{w} = (k - \rho(C))\mathbf{w}.$$

Therefore,

$$\frac{1}{2}(B + B')\mathbf{w} = (k - \rho(C))\mathbf{w}$$

and $k - \rho(C) > 0$. Using the equivalence of statements (ii) and (vi) shown in Theorem 10.125, it follows that all minors of $\frac{1}{2}(B + B')$ are positive. Therefore, $\frac{1}{2}(B + B')$ is positive definite, so $B + B'$ is positive definite.

Suppose now that C is not irreducible. By Theorem 10.106 there exists a permutation matrix P such that

$$PCP' = \begin{pmatrix} C_{11} & C_{12} & \cdots & C_{1\ell} \\ O & C_{22} & \cdots & C_{2\ell} \\ \vdots & \vdots & \cdots & \vdots \\ O & O & \cdots & C_{\ell\ell} \end{pmatrix},$$

and the diagonal blocks $C_{11}, \ldots, C_{\ell\ell}$ are irreducible.

Let $M = PAP'$. We have the block upper triangular matrix given by

$$\begin{aligned} M &= PAP' = P(kI_n - C)P' = kI_n - PCP' \\ &= \begin{pmatrix} kI_1 - C_{11} & -C_{12} & \cdots & -C_{1\ell} \\ O & kI_2 - C_{22} & \cdots & -C_{2\ell} \\ \vdots & \vdots & \cdots & \vdots \\ O & O & \cdots & kI_\ell - C_{\ell\ell} \end{pmatrix}. \end{aligned}$$

By the first part of the proof there exist diagonal matrices L_i having positive diagonal entries such that, for $B_i = L_i M_{ii} L_i^{-1} = L_i(kI_i - C_{ii})L_i^{-1}$, the matrices $\frac{1}{2}(B_i + B_i')$ are positive definite for $1 \leqslant i \leqslant \ell$.

Let $L(a)$ be the block diagonal matrix

$$L(a) = \mathsf{diag}\left(L_1, \frac{1}{a}L_2, \frac{1}{a^2}L_3, \ldots, \frac{1}{a^{\ell-1}}L_\ell\right).$$

Since

$$L(a)^{-1} = \mathsf{diag}\left(L_1^{-1}, aL_2, a^2L_3, \ldots, a^{\ell-1}L_\ell\right),$$

we have

$$L(a)ML(a)^{-1} = \begin{pmatrix} B_1 & B_{12} & B_{13} & \cdots & B_{1\ell} \\ O & B_2 & B_{21} & \cdots & B_{2\ell} \\ \vdots & \vdots & \vdots & \vdots & \vdots \\ O & O & O & \cdots & B_\ell \end{pmatrix},$$

where $B_{ij} = a^{j-i} L_i(a) C_{ij} L_j^{-1}(a)$ for $i < j$.

If E is the block diagonal matrix

$$E = \begin{pmatrix} B_1 & O & \cdots & O \\ O & B_2 & \cdots & O \\ \vdots & \vdots & \vdots & \vdots \\ O & O & \cdots & B_\ell \end{pmatrix} = L(0)ML(0)^{-1},$$

then it is clear that $\frac{1}{2}(E + E')$ is positive definite. By Theorem 6.127 there exists a number $a > 0$ such that $\frac{1}{2}(L(a)(M + M')L(a)^{-1})$ is positive definite. Let D be the diagonal matrix $P'L(a)P$. Then, since

$$B = DAD^{-1} = (P'L(a)P)A(P^{-1}L(a)^{-1}) = P'L(a)A(L(a))^{-1}P$$

the matrix $\frac{1}{2}(B + B')$ is positive definite and therefore, $B + B'$ is positive definite.

`(xvi) implies (xvii)`: Suppose that there exists a diagonal matrix $G = \mathsf{diag}(g_{11}, \ldots, g_{nn})$ with $g_{ii} > 0$ for $1 \leqslant i \leqslant n$ such that if $B = GAG^{-1}$, then the symmetric matrix $B + B'$ is positive definite.

Let H be the diagonal matrix given by

$$H = \mathsf{diag}\left(\frac{1}{g_{11}^2}, \ldots, \frac{1}{g_{nn}^2}\right) = (G^{-1})^2.$$

Clearly, all diagonal elements of H are positive. For $C = AH$ we have

$$\begin{aligned} C + C' &= AH + H'A' = AH + HA' = A(G^{-1})^2 + (G^{-1})^2 A' \\ &= G^{-1}(GAG^{-1} + G^{-1}A'G)G^{-1} = G^{-1}(B + B')G^{-1}. \end{aligned}$$

It is immediate that the matrix $C + C'$ is positive definite.

`(xvii) implies (xviii)`: Suppose that there exists a diagonal matrix $H = \mathsf{diag}(h_{11}, \ldots, h_{nn})$ with $h_{ii} > 0$ for $1 \leqslant i \leqslant n$ such that for $C = AH$ the matrix $C + C' = AH + HA'$ is positive definite. Then, $\mathbf{x}^{\mathsf{H}}(AH + HA')\mathbf{x} > 0$, so $\mathbf{x}^{\mathsf{H}}AH\mathbf{x} + \mathbf{x}^{\mathsf{H}}HA'\mathbf{x} > 0$ for $\mathbf{x} \in \mathbb{C}^n - \{\mathbf{0}\}$. Therefore, $\Re(\mathbf{x}^{\mathsf{H}}AH\mathbf{x}) > 0$.

Let λ be an eigenvalue of A and let $\mathbf{y}$ be an eigenvector associated with λ. For $\mathbf{x} = H^{-1}\mathbf{y}$ we have $H\mathbf{x} = \mathbf{y}$, and $\mathbf{y}^{\mathsf{H}} = \mathbf{x}^{\mathsf{H}}H$, because H is a diagonal matrix with real positive elements on its diagonal, so

$$\begin{aligned}\Re(\mathbf{x}^{\mathsf{H}}AH\mathbf{x}) &= \Re(\mathbf{x}^{\mathsf{H}}A\mathbf{y}) = \Re(\mathbf{x}^{\mathsf{H}}\lambda\mathbf{y}) \\ &= \Re(\mathbf{y}^{\mathsf{H}}H^{-1}\lambda\mathbf{y}) = \Re(\lambda)(\mathbf{y}^{\mathsf{H}}H^{-1}\mathbf{y}) > 0,\end{aligned}$$

which implies $\Re(\lambda) > 0$.

`(xviii) implies (ix)`: This implication follows from Theorem 10.125.

Theorems 10.125–10.128 allow us to introduce the class of **K**-*matrices*.

Definition 10.129 *A matrix $A \in \mathbb{Z}_n$ is a* **K**-matrix *if it satisfies any of the eighteen equivalent conditions contained by Theorems* 10.125–10.126.

The following results of Fiedler [7] (Theorems 10.130 and 10.131) will help with the characterization of **K**-matrices as members of another important class of matrices.

Theorem 10.130 *The following properties of a matrix $A \in \mathbb{Z}_n$ are equivalent:*

(i) $A + \epsilon I_n \in \mathbf{K}$ *for every* $\epsilon > 0$;
(ii) *every real eigenvalue of any principal minor of A is non-negative.*
(iii) *all principal minors of A are non-negative;*
(iv) *the sum of all principal minors of order k is non-negative for* $1 \leqslant k \leqslant n$;
(v) *every real eigenvalue of A is non-negative;*
(vi) *A can be written as* $A = kI - C$, *where* $C \geqslant O$ *and* $k \geqslant \rho(C)$.

Proof `(i) implies (ii)`: Let A be a matrix from $\mathbb{Z}_n$ such that $A + \epsilon I_n \in \mathbf{K}$ for every $\epsilon > 0$ and let λ be an eigenvalue of a principal submatrix $A\begin{bmatrix} i_1 \cdots i_k \\ i_1 \cdots i_k \end{bmatrix}$ of A. Suppose that $\lambda < 0$ and let $B = A + \lambda I$. By Part (v) of Theorem 10.125 all real eigenvalues of the matrix

$$(A - \lambda I_n)\begin{bmatrix} i_1 \cdots i_k \\ i_1 \cdots i_k \end{bmatrix}$$

are positive. However, $A - \lambda I_n$ has the eigenvalue 0, which is a contradiction.

`(ii) implies (iii)`: The proof is virtually identical to the argument used for the implication (vi) by (v) in the proof of Theorem 10.125.

`(iii) implies (iv)`: This implication is immediate.

`(iv) implies (v)`: Suppose that for each k, $1 \leqslant k \leqslant n$, the sum of all principal minors of order k of A is non-negative.

The characteristic equation of A has the form

$$\lambda^n - S_1(A)\lambda^{n-1} + \cdots + (-1)^{n-1}S_{n-1}(A)\lambda + (-1)^n S_n(A) = 0,$$

where $S_i(A)$ is the sum of all principal minors of order i of A. If λ were a real negative root than all numbers of the left-hand side would have the same sign which would make the equality with 0 impossible. Thus, every real eigenvalue is non-negative.

`(v) implies (vi)`: Let $A \in \mathbb{Z}_n$ be a matrix whose real eigenvalues are non-negative. If $m = \max_{1 \leqslant i \leqslant n} a_{ii}$, the matrix $C = mI_n - A$ is non-negative. As we saw before, $\rho(C) \in \mathsf{spec}(C)$. Since $A = mI_n - C$, $m - \rho(C)$ is an eigenvalue of A, so $m \geqslant \rho(C)$.

`(vi) implies (i)`: Suppose that A can be written as $A = kI - C$, where $C \geqslant O$ and $k \geqslant \rho(C)$. For the matrix $B = A + \epsilon I = (k+\epsilon)I - C$ we have $k + \epsilon > \rho(C)$, so by Part (ix) of Theorem 10.125, $A + \epsilon I \in \mathbf{K}$.

Theorem 10.131 *Let A be a matrix in $\mathcal{Z}_n$. We have $A + \epsilon I_n \in \mathbf{K}$ for any $\epsilon > 0$ if and only if for every $\lambda \in \mathsf{spec}(A)$ we have $\Re(\lambda) \geqslant 0$.*

Proof This follows immediately from Theorem 10.128.

The *class* $\mathbf{K}_0$ consists of all matrices of $\mathcal{Z}_n$ that satisfy one of the equivalent conditions in Theorems 10.130 or 10.131.

Theorem 10.132 *The class of $\mathbf{K}$-matrices is a subclass of $\mathbf{K}_0$ which consists of all invertible matrices of $\mathbf{K}_0$.*

Proof Let $A \in \mathbf{K}$. Then, all principal minors of A are positive, so they are nonnegative and $\mathbf{K} \subseteq \mathbf{K}_0$.

Suppose that $A \in \mathbf{K}_0$ and A is invertible. Then, $0 \notin \mathsf{spec}(A)$, so every eigenvalue of A is positive. Therefore, by Part (viii) of Theorem 10.125, $A \in \mathbf{K}$.

10.8 Flows in Digraphs

Definition 10.133 *A* network *is a 4-tuple $\mathcal{N} = (\mathcal{G}, cap, s, t)$, where*

- *$\mathcal{G} = (V, E)$ is a finite digraph,*
- *$cap : V \times V \longrightarrow \mathbb{R}_{\geqslant 0}$ is a function called the* capacity function *such that $(u, v) \notin E$ implies $cap(u, v) = 0$, and*
- *s and t are two distinct vertices of $\mathcal{G}$, referred to as the* source *and the* sink, *respectively.*

The number $cap(e)$ is the capacity of the edge *e. If $\mathbf{p} = (v_0, \ldots, v_n)$ is a path in the graph $\mathcal{G}$ the capacity of this path is the number $cap(\mathbf{p}) = \min\{cap(v_i, v_{i+1}) \mid 0 \leqslant i \geqslant n-1\}$.*

Example 10.134 The network $\mathcal{N} = (\mathcal{G}, cap, s, t)$ is shown in Fig. 10.33. If (u, v) is an edge in $\mathcal{G}$, the number $cap(u, v)$ is written near the edge.

The capacity of the path $\mathbf{p} = (v_1, v_2, v_5, v_6)$ is 3 because the smallest capacity of an edge on this path is $cap(v_2, v_5) = 3$.

Definition 10.135 *A* flow *in the network $\mathcal{N} = (\mathcal{G}, cap, s, t)$ is a function $f : V \times V \longrightarrow \mathbb{R}$ that satisfies the following conditions:*

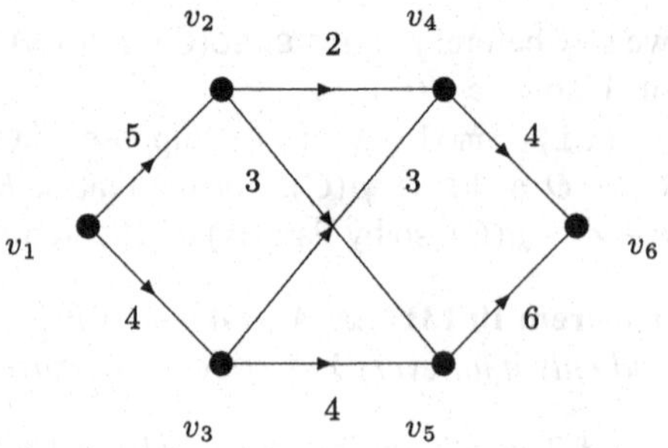

Fig. 10.33 6-vertex network

(i) *For every edge* $(u, v) \in E$ *we have* $0 \leqslant f(u, v) \leqslant cap(u, v)$.

(ii) *The function* f *is skew-symmetric, that is,* $f(u, v) = -f(v, u)$ *for every pair* $(u, v) \in V \times V$.

(iii) *The equality*

$$\sum\{f(v, x) \mid v \in V\} = 0,$$

known as Kirchhoff's law, *holds for every vertex* $x \in V - \{s, t\}$.

The value of a flow f *in a network* $\mathcal{N} = (\mathcal{G}, cap, s, t)$ *is the number*

$$val(f) = \sum\{f(s, v) \mid v \in V\},$$

that is, the net flow that exits the source.

The set of flows of a network $\mathcal{N}$ *is denoted by* $FL(\mathcal{N})$.

A flow h *in* $\mathcal{N}$ *is* maximal *if* $val(f) \leqslant val(h)$ *for every flow* $f \in FL(\mathcal{N})$.

If $f(u, v) = c(u, v)$ *for an edge of* $\mathcal{G}$, *then we say that the edge* (u, v) *is* saturated.

Let $\mathcal{N} = (\mathcal{G}, cap, s, t)$ be a network, where $\mathcal{G} = (V, E)$. If $f(u, v) = 0$ for every pair $(u, v) \in V \times V$, then f is a flow in the network. We will refer to this flow as the *zero flow* in $\mathcal{N}$.

Theorem 10.136 *Let* $\mathcal{N} = (\mathcal{G}, cap, s, t)$ *be a network,* f *and* g *be two flows in* $FL(\mathcal{N})$, *and* a *and* b *be two real numbers. Define the function* $af + bg$ *by*

$$(af + bg)(u, v) = af(u, v) + bg(u, v)$$

for every $(u, v) \in V \times V$. *If* $0 \leqslant (af + bg)(u, v) \leqslant c(u, v)$ *for* $(u, v) \in V \times V$, *then* $af + bg$ *is a flow in the network* $\mathcal{N}$.

Proof We leave this easy argument to the reader.

Note that if $v \in V - \{s, t\}$, Kirchhoff's law can be written equivalently as

$$\sum\{f(v, x) \mid v \in V, (v, x) \in E\} + \sum\{f(v, x) \mid v \in V, (x, v) \in E\} = 0,$$

which amounts to

$$\sum\{f(v, x) \mid v \in V, (v, x) \in E\} = \sum\{f(x, v) \mid v \in V, (x, v) \in E\}$$

due to the skew symmetry of f.

Theorem 10.137 *Let $\mathcal{N} = (\mathcal{G}, cap, s, t)$ be a network and let f be a flow in $\mathcal{N}$. If U is a set of vertices such that $s \in U$ and $t \notin U$, then*

$$\sum\{f(u, v) \mid (u, v) \in out(U)\} - \sum\{f(u, v) \mid (u, v) \in in(U)\} = val(f).$$

Proof If f is a flow in $\mathcal{N}$ and x is a vertex in $\mathcal{G}$, then

$$\sum\{f(x, v) \mid v \in V\} - \sum\{f(v, x) \mid v \in V\} = \begin{cases} val(f) & \text{if } x = s, \\ 0 & \text{if } x \in U - \{s\}. \end{cases}$$

Therefore,

$$\sum_{x \in U} \left(\sum\{f(x, v) \mid v \in V\} - \sum\{f(v, x) \mid v \in V\} \right) = val(f).$$

If an edge that occurs in the inner sums has both its endpoints in U, then its contribution in the first inner sum cancels with its contribution in the second inner sum. Thus, the previous equality can be written as

$$\sum\{f(u, v) \mid (u, v) \in out(U)\} - \sum\{f(u, v) \mid (u, v) \in in(U)\} = val(f).$$

Corollary 10.138 *Let $\mathcal{N} = (\mathcal{G}, cap, s, t)$ be a network and let f be a flow in $\mathcal{N}$. For every vertex x, we have*

$$\sum\{f(v, x) \mid v \in V\} - \sum\{f(x, v) \mid v \in V\} = val(f).$$

Proof Choose $U = V - \{x\}$. For the set U, we have

$$\begin{aligned} out(U) &= (V \times \{x\}) \cap E, \\ in(U) &= (\{x\} \times V) \cap E, \end{aligned}$$

so it follows that

$$\sum\{f(v,x) \mid v \in v\} - \sum\{f(x,v) \mid v \in V\} = val(f).$$

Definition 10.139 *A* cut *in a network* $\mathcal{N} = (\mathcal{G}, cap, s, t)$ *is an* (s,t)*-cut.*
The capacity of a cut (C, C') *is the number*

$$cap(C, C') = \sum\{cap(u,w) \mid u \in C \text{ and } w \in C'\}.$$

A cut with minimal capacity is a minimal cut.
If f *is a flow in* $\mathcal{N}$ *and* (C, C') *is a cut, the* value of the flow f across the cut (C, C') *is the number*

$$f(C, C') = \sum\{f(u,w) \mid u \in C \text{ and } w \in C'\}.$$

The set of cuts of a network $\mathcal{N}$ *is denoted by* $CUTS(\mathcal{N})$.

Thus, Theorem 10.137 can be rephrased as stating that the flow across any cut equals the value of the flow. An essential observation is that since the value of a flow across a cut cannot exceed the capacity of the cut, it follows that the value of any flow is less than the capacity of any cut. As we shall see, the maximum value of a flow equals the minimal capacity of a cut.

Definition 10.140 *Let* $\mathcal{N} = (\mathcal{G}, cap, s, t)$ *be a network and let* f *be a flow in* $\mathcal{N}$. *The* residual network of $\mathcal{N}$ relative to f *is the network* $RES(\mathcal{N}, f) = (\mathcal{N}, cap', s, t)$, *where* $cap'(u,v) = cap(u,v) - f(u,v)$.

Theorem 10.141 *Let* f *be a flow on the network* $\mathcal{N} = (\mathcal{G}, cap, s, t)$ *and let* g *be a maximal flow of* $\mathcal{N}$. *The value of a maximal flow on* $RES(\mathcal{N}, f)$ *is* $val(g) - val(f)$.

Proof Let f' be a flow in the residual network $RES(\mathcal{N}, f)$. It is easy to see that $f + f' \in FL(\mathcal{N})$, so $val(f') \leqslant val(g) - val(f)$. On the other hand, $h = g - f$ is a flow on $RES(\mathcal{N}, f)$ and $val(h) = val(g) - val(f)$, so h is a maximal flow having the value $val(g) - val(f)$.

Theorem 10.142 *The following statements that concern a flow* f *in a network* $\mathcal{N} = (\mathcal{G}, cap, s, t)$ *are equivalent:*

(i) f *is a maximal flow.*
(ii) *There is no path that joins* s *to* t *in the residual network* $RES(\mathcal{N}, f)$ *with a positive capacity.*
(iii) $val(f) = cap(C, C')$ *for some cut* (C, C') *in* $\mathcal{N}$.

Proof (i) implies (ii): Let f be a maximal flow in $\mathcal{N}$, and suppose that there is a path $\mathbf{p}$ in the residual network $RES(\mathcal{N}, f)$ with a positive capacity. Then, the flow g defined by

$$g(u,v) = \begin{cases} f(u,v) + cap(\mathbf{p}) & \text{if } (u,v) \text{ is an edge on } \mathbf{p}, \\ f(u,v) & \text{otherwise}, \end{cases}$$

is a flow in $\mathcal{N}$ and $val(g) = val(f) + cap(\mathbf{p})$, which contradicts the maximality of the flow f.

(ii) implies (iii): Suppose that there is no path is the residual network $RES(\mathcal{N}, f)$ that joins the source with the sink and has a positive capacity.

Let C be the set of vertices that can be reached from s via a path with positive capacity (to which we add s) in the residual network $RES(\mathcal{N}, f)$ and let $C' = V - C$. Then the pair (C, C') is a cut in $\mathcal{N}$. Observe that if $x \in C$ and $y \in C'$, then the residual capacity of the edge (x, y) is 0 by the definition of C, which means that $f(x, y) = cap(x, y)$. Thus, $val(f) = f(C, C')$.

(iii) implies (i): Since any flow value is less than the capacity of any cut, it is immediate that f is a maximal flow.

Any path that joins the source to the target of a network $\mathcal{N}$ and has a positive capacity in that network is called an *augmenting path for the network.*

Theorem 10.142 suggests the following algorithm for constructing a maximal flow in a network.

Algorithm 10.8.1: The Ford-Fulkerson Algorithm

Data: a network $\mathcal{N} = (\mathcal{G}, cap, s, t)$
Result: a maximal flow in $\mathcal{N}$

```
initialize flow f to the zero flow in N;
while there exists an augmenting path p do
    augment flow f along p
end
return f;
```

Example 10.143 To find a maximal flow, we begin with the zero flow f_0 shown in Example 10.34(a). There are several cuts in this graph having a minimal capacity equal to 9. One such cut is $\{\{v_1\}, \{v_2, v_3, v_4, v_5, v_6\}\}$; the edges that join the two sets are (v_1, v_2) and (v_1, v_3), which have the capacities 4 and 5, respectively.

The first augmenting path is (v_1, v_2, v_4, v_6), having capacity 2. The flow f_1 along this path has the value 2, it saturates the edge (v_2, v_4), and is shown in Fig. 10.34b. The next augmenting path is (v_1, v_2, v_5, v_6), which has a capacity of 3 and is shown in Fig. 10.34c. Now the edges (v_1, v_2) and (v_2, v_5) become saturated. The next flow in also shown in Fig. 10.34c. In Fig. 10.34d, we show the augmenting path (v_1, v_3, v_5, v_6) having capacity 3. This saturates the edge (v_5, v_6). Finally, the last augmenting path of capacity 1 (shown in Fig. 10.34e) is (v_1, v_3, v_4, v_6), and the corresponding flow saturates the edge (v_1, v_3). The value of the flow is 9.

Corollary 10.144 *Let $\mathcal{N} = (\mathcal{G}, cap, s, t)$ be a network such that $cap(e)$ is an integer for every edge of the graph $\mathcal{G}$. Then, a maximal flow ranges over the set of natural numbers.*

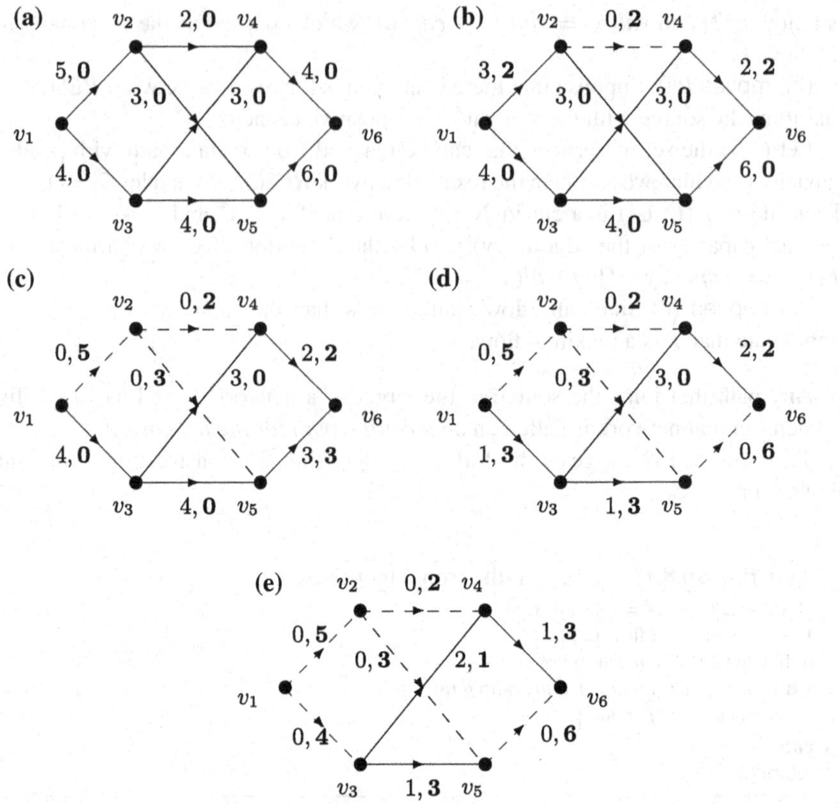

Fig. 10.34 Construction of a flow in a network

Proof The argument for the corollary is on the number n of augmenting paths used for the construction of a flow.

The basis step, $n = 0$, is immediate since the zero flow takes as values natural numbers.

Suppose that the statement holds for the flow constructed after applying $n - 1$ augmentations. Since the value of any path is a natural number and the residual capacities are integers, the values obtained after using the nth augmenting path are again integers.

Flows in networks that range over the set of integers are referred to as *integral flows*.

Network flows can be used to prove several important graph-theoretical results. We begin with a technical result.

Lemma 10.145 *Let $\mathcal{N} = (\mathcal{G}, cap, x, y)$ be a network such that $cap(u, v) = 1$ for every edge $(u, v) \in E$. If f is an integral flow in $FL(\mathcal{N})$ and $val(f) = m$, then there are m pairwise edge-disjoint simple paths from x to y.*

Proof If f is a flow in $\mathcal{N}$ that ranges over integers, then $f(u, v) \in \{0, 1\}$ for every edge $(u, v) \in E$. Also, note that the capacity of any path that joins x to y equals 1.

Let E_f be the set of edges saturated by the flow f,

$$E_f = \{(u, v) \in E \mid f(u, v) = 1\}.$$

Note that no two distinct simple paths can share an edge because this would violate Kirchhoff's law. Since each path contributes a unit of flow to f, it follows that there exist m pairwise edge-disjoint paths in $\mathcal{N}$.

Theorem 10.146 (Menger's Theorem for Directed Graphs) *Let $\mathcal{G} = (V, E)$ be a directed graph and let x and y be two vertices. The maximum number of paths that join x to y whose sets of edges are pairwise disjoint equals the minimum number of edges whose removal eliminates all paths from x to y.*

Proof Define a network $\mathcal{N} = (\mathcal{G}, cap, x, y)$ such that $cap(u, v) = 1$ for every edge $(u, v) \in E$, and let f be an maximal integral flow in $\mathcal{N}$ whose value is m. By Lemma 10.145, this number equals the number of pairwise edge-disjoint simple paths from x to y and is also equal to the minimum capacity of a cut in $\mathcal{N}$. Since this latter capacity equals the minimal number of edges whose removal eliminates all paths from x to y, we obtain the desired equality.

Menger's theorem allows us to prove the following statement involving bipartite graphs.

Definition 10.147 *A* vertex cover *of $\mathcal{G}$ is a set of vertices U such that, for every edge $e \in E$, one of its endpoints is in U.*

An edge cover *is a set of edges F such that every vertex in V is an endpoint of an edge in F.*

Theorem 10.148 (König's Theorem for Bipartite Graphs) *Let $\mathcal{G} = (V, E)$ be a bipartite graph. A maximum size of a matching of $\mathcal{G}$ equals the minimum size of an vertex cover of $\mathcal{G}$.*

Proof Suppose that $\{V_1, V_2\}$ is the partition of the vertices of the graph such that $E \subseteq V_1 \times V_2$. Define the digraph $\mathcal{G}' = (V \cup \{s, t\}, \{(x, y) \mid (x, y) \in E\} \cup \{(s, x) \mid x \in V\} \cup \{(x, t) \mid x \in V\})$ and the network $\mathcal{N} = (\mathcal{G}', cap, s, t)$. We assume that s, t are new vertices. The capacity of every edge equals 1.

A matching M of the bipartite graph $\mathcal{G}$ yields a number of $|M|$ pairwise disjoint paths in $\mathcal{N}$.

An vertex cover U in $\mathcal{G}$ produces a cut in the network $\mathcal{N}$ using the following mechanism. Define the sets of vertices U_1, U_2 as $U_i = U \cap V_i$ for $i \in \{1, 2\}$. Then, $(\{s\} \cup U_1, U_2 \cup \{t\})$ is a cut in $\mathcal{N}$ and its capacity equals $|U|$. By removing the edges having an endpoint in U, we eliminate all paths that join s to t. Thus, the current statement follows immediately from Menger's theorem.

Fig. 10.35 4-vertex graph

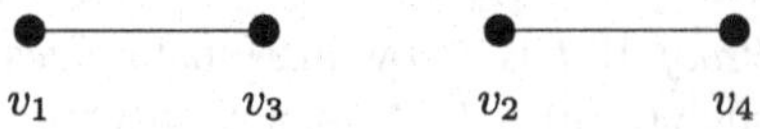

10.9 The Ordinary Spectrum of a Graph

Let $\mathcal{G} = (V, E)$ be a graph and $A_{\mathcal{G}}$ its adjacency matrix. Since $A_{\mathcal{G}}$ is a symmetric matrix, each of its eigenvalues λ are real and $\mathsf{geomm}(A_{\mathcal{G}}, \lambda) = \mathsf{algm}(A_{\mathcal{G}}, \lambda)$ by Theorem 7.29. Furthermore, since $trace(A_{\mathcal{G}}) = 0$, the sum of the eigenvalues of $A_{\mathcal{G}}$ is 0 and $\sum\{\lambda^2 \mid \lambda \in \mathsf{spec}(A_{\mathcal{G}})\} = 2n$.

Definition 10.149 *Let $G = (V, E)$ be a graph. Its* ordinary spectrum *is* $\mathsf{spec}(A_{\mathcal{G}})$.

Example 10.150 Let $G = (\{v_1, v_2, v_3, v_4\}, \{(v_1, v_3), (v_2, v_4)\})$ be the graph shown in Fig. 10.35. The adjacency matrix is

$$A_{\mathcal{G}} = \begin{pmatrix} 0 & 0 & 1 & 0 \\ 0 & 0 & 0 & 1 \\ 1 & 0 & 0 & 0 \\ 0 & 1 & 0 & 0 \end{pmatrix}.$$

Its characteristic polynomial is $p_{A_{\mathcal{G}}}$ is

$$p_{A_{\mathcal{G}}}(\lambda) = (\lambda - 1)^2(\lambda + 1)^2 = \lambda^4 - 2\lambda^2 + 1.$$

Thus, the ordinary spectrum of this graph is the set $\{1, -1\}$ and each eigenvalue has the multiplicity (algebraic and geometric) equal to 2.

Note that if $\mathcal{G} = (V, E)$ is a graph with $|V| = n$, then the components of the vector $A_{\mathcal{G}}\mathbf{1}_n$ are equal to the degrees of their respective vertices.

Example 10.151 Let $\mathcal{G} = (V, E)$ be a graph. We have $A_{\mathcal{G}}\mathbf{1} = k\mathbf{1}$ if and only if $\mathcal{G}$ is a k-regular graph. Thus, a k-regular graph has the ordinary eigenvalue k. Moreover, if $\mathcal{G}$ is k-regular graph then for every ordinary eigenvalue λ we have $|\lambda| \leqslant k$. Indeed, if $|a| > k$, then the matrix $A_{\mathcal{G}} - aI_n$ is diagonally dominant and therefore, by Theorem 5.125, non-singular, which prevents a from being an eigenvalue of $A_{\mathcal{G}}$. Thus, k is the largest eigenvalue of $A_{\mathcal{G}}$.

Example 10.152 We saw that the adjacency matrix of a bipartite graph $\mathcal{G} = (\{v_1, \ldots, v_m\}, E)$ having the bipartition $\{V_1, V_2\}$, where $V_1 = \{v_1, \ldots, v_p\}$ and $V_2 = \{v_{p+1}, \ldots, v_n\}$ has the form

$$A_{\mathcal{G}} = \begin{pmatrix} O_{p,p} & B \\ B' & O_{q,q} \end{pmatrix},$$

where $B \in \{0, 1\}^{p \times q}$.

Let λ be an eigenvalue of $A_{\mathcal{G}}$ and let $\mathbf{x}$ be the corresponding eigenvector. We can write

$$\mathbf{x} = \begin{pmatrix} \mathbf{y} \\ \mathbf{z} \end{pmatrix},$$

where y consists of the first p components of x and $\mathbf{z}$ consists of the last $n - p$ components of the same vector. These definitions imply that $B\mathbf{z} = \lambda\mathbf{y}$ and $B'\mathbf{y} = \lambda\mathbf{z}$. Let $\mathbf{t} \in \mathbb{R}^n$ be defined by

$$\mathbf{t} = \begin{pmatrix} \mathbf{y} \\ -\mathbf{z} \end{pmatrix}.$$

Since $B(-\mathbf{z}) = \lambda\mathbf{y}$ and $B'\mathbf{y} = -\lambda(-\mathbf{z})$, it follows that $\mathbf{t}$ is an eigenvector that corresponds to the eigenvalue $-\lambda$. Thus, the eigenvalues of the adjacency matrix of a bipartite graph come in pairs placed symmetrically with respect to the origin.

In fact, the property discussed in Example 10.152 is a characteristic property of bipartite graphs as we show next.

Theorem 10.153 *Let $\mathcal{G}$ be a graph. If $\mathsf{algm}(A, \lambda) = \mathsf{algm}(A, -\lambda)$ for every $\lambda \in \mathsf{spec}(A)$, then $\mathcal{G}$ is a bipartite graph.*

Proof The property of eigenvalues mentioned above implies $p_{A_{\mathcal{G}}}(-\lambda) = p_{A_{\mathcal{G}}}(\lambda)$, which means that the characteristic polynomial of $A_{\mathcal{G}}$ is a sum of even powers of λ. Therefore, $\mathcal{G}$ has no odd-length cycles, so it is a bipartite graph.

Example 10.154 The adjacency matrix of the complete graph $\mathcal{K}_n$ is $A_{\mathcal{K}} = J_{n,n} - I_n$. Note that $J_{n,n} = \mathbf{1}_n\mathbf{1}_n'$ and $\mathbf{1}_n'\mathbf{1}_n = n$. The characteristic polynomial of $A_{\mathcal{K}_n}$ is

$$\begin{aligned} p(\lambda) &= \det(\lambda I_n - J_{n,n} + I_n) = \det((1+\lambda)I_n - J_{n,n}) \\ &= \det((1+\lambda)I_n - \mathbf{1}_n\mathbf{1}_n') = (-1)^n(1+\lambda)^n \det\left(I_n - \frac{1}{1+\lambda}\mathbf{1}_n\mathbf{1}_n'\right). \end{aligned}$$

By Supplement 45 Part (a) we have

$$\begin{aligned} p_A(\lambda) &= (-1)^n(1+\lambda)^n \det\left(I_n - \frac{1}{1+\lambda}\mathbf{1}_n\mathbf{1}_n'\right) \\ &= (-1)^n(1+\lambda)^n \left(1 - \frac{1}{1+\lambda}\mathbf{1}_n'\mathbf{1}_n\right) \\ &= (-1)^n(1+\lambda)^n \left(1 - \frac{n}{1+\lambda}\right) \\ &= (-1)^n(1+\lambda)^{n-1}(1+\lambda-n). \end{aligned}$$

Thus, the eigenvalues of $\mathcal{K}_n$ are -1 and $n-1$ with $\mathsf{algm}(A_{\mathcal{K}_n}, -1) = n - 1$ and $\mathsf{algm}(A_{\mathcal{K}_n}, n-1) = 1$.

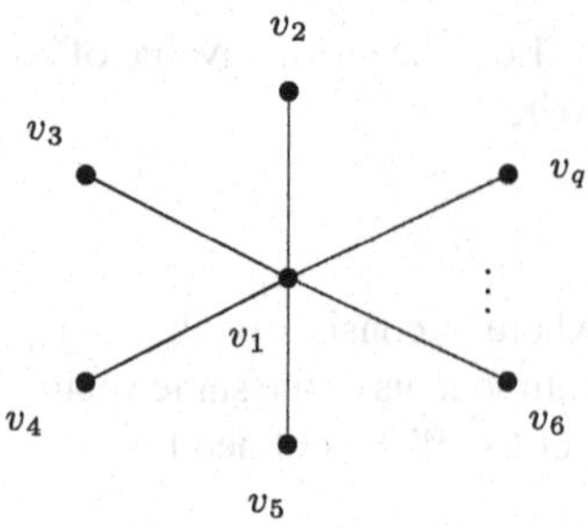

Fig. 10.36 Star graph

Example 10.155 The adjacency matrix of the complete bipartite graph $\mathcal{K}_{p,q}$ has the form

$$A_{\mathcal{K}_{p,q}} = \begin{pmatrix} O_{p,p} & J_{p,q} \\ J_{q,p} & O_{q,q} \end{pmatrix},$$

and its characteristic polynomial is

$$p(\lambda) = \det(\lambda I_{p+q} - A_{\mathcal{K}_{p,q}}) = \begin{vmatrix} \lambda I_p & -J_{p,q} \\ -J_{q,p} & \lambda I_q \end{vmatrix}.$$

By Exercise 61 of Chap. 5 we have

$$\begin{aligned} p(\lambda) &= \det(\lambda I_p) \det\left(\lambda I_q - \frac{1}{\lambda} J_{q,p} J_{p,q}\right) = \lambda^{p+q} \det\left(I_q - \frac{p}{\lambda^2} J_q\right) \\ &= (-1)^q p^q \lambda^{p-q} \det\left(J_q - \frac{\lambda^2}{p} I_q\right), \end{aligned}$$

taking into account that $J_{q,p} J_{p,q} = p J_q$. By Exercise 41 of Chap. 5,

$$p(\lambda) = (-1)^{p+q-1} \lambda^{p+q-2} (pq - \lambda^2),$$

which means that the ordinary spectrum of $\mathcal{K}_{p,q}$ consists of 0, $\sqrt{pq}$ and $-\sqrt{pq}$, where $\mathsf{algm}(A_{\mathcal{K}_{p,q}}, 0) = p + q - 2$ and $\mathsf{algm}(A_{\mathcal{K}_{p,q}}, \sqrt{pq}) = \mathsf{algm}(A_{\mathcal{K}_{p,q}}, -\sqrt{pq}) = 1$.

A special type of complete bipartite graph occurs when $p = 1$; that graph (which has $q + 1$ vertices), shown in Fig. 10.36, is known as a *star graph* and has the spectrum $0, \sqrt{q}, -\sqrt{q}$, with $\mathsf{algm}(A_{\mathcal{K}_{1,q}}, 0) = q - 1$ and $\mathsf{algm}(A_{\mathcal{K}_{1,q}}, \sqrt{q}) = \mathsf{algm}(A_{\mathcal{K}_{1,q}}, -\sqrt{q}) = 1$.

The principal minors of the adjacency matrix of a graph play an important role in the study of graphs. Recall that $p_{A_\mathcal{G}}(\lambda) = \sum_{k=0}^{n} (-1)^k a_k \lambda^{n-k}$, where a_k is the sum of the principal minors of order k of A.

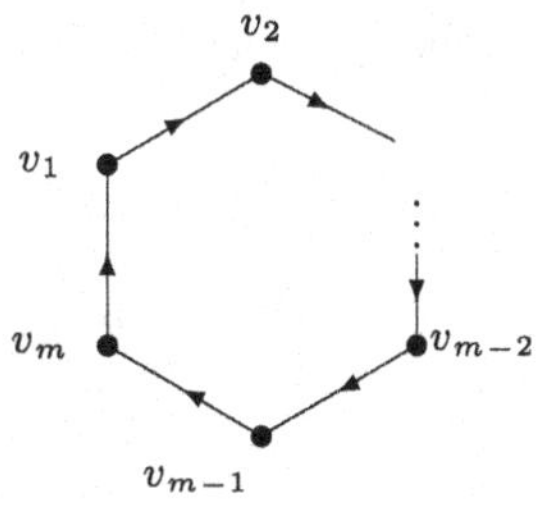

Fig. 10.37 Directed simple cycle $\mathcal{D}_m$

Since all principal minors of order 1 of $A_{\mathcal{G}}$ are 0 it follows that the sum of the eigenvalues of A is 0. Consider now a principal minor or order 2,

$$\det\left(A\begin{bmatrix} i_1 \ i_2 \\ i_1 \ i_2 \end{bmatrix}\right) = \begin{vmatrix} a_{i_1 i_1} & a_{i_1 i_2} \\ a_{i_2 i_1} & a_{i_2 i_2} \end{vmatrix}.$$

Since $a_{i_1 i_1} = a_{i_2 i_2} = 0$, this minor equals -1 if and only if there is an edge (v_{i_1}, v_{i_2}) in $\mathcal{G}$. Thus, the sum of the principal minors of order 2 is $-|E|$, and so $a_2 = -|E|$.

A principal minor of order 3 has the form

$$\det\left(A\begin{bmatrix} i_1 \ i_2 \ i_3 \\ i_1 \ i_2 \ i_3 \end{bmatrix}\right) = \begin{vmatrix} 0 & a_{i_1 i_2} & a_{i_1 i_3} \\ a_{i_2 i_1} & 0 & a_{i_2 i_3} \\ a_{i_3 i_1} & a_{i_3 i_2} & 0 \end{vmatrix} = 2a_{i_1 i_2} a_{i_2 i_3} a_{i_3 i_1}.$$

The contribution of such a minor to a_3 is 1 only when $a_{i_1 i_2} = a_{i_2 i_3} = a_{i_3 i_1} = 1$, that is, when the vertices v_{i_1}, v_{i_2} and v_{i_3} form a triangle. Thus, the number of triangles in $\mathcal{G}$ is $\frac{a_3}{2}$.

For directed graphs the definition of the ordinary spectrum is exactly the same as for undirected graphs. Note however, that unlike undirected graphs whose adjacency matrices are symmetric and, therefore, have real eigenvalues, in the case of directed graphs their spectra may consist of complex numbers.

Example 10.156 Let $\mathcal{D}_m$ be a directed simple cycle that has m vertices:

$$\mathcal{D}_m = (\{v_1, \ldots, v_m\}, \{(v_i, v_{i+1}) \mid 1 \leqslant i \leqslant m-1\} \cup \{(v_m, v_1)\})$$

shown in Fig. 10.37. Its adjacency matrix is

$$A_{\mathcal{D}_m} = \begin{pmatrix} 0 \, 1 \, 0 \, 0 \cdots 0 \\ 0 \, 0 \, 1 \, 0 \cdots 0 \\ 0 \, 0 \, 0 \, 1 \cdots 0 \\ \vdots \ \vdots \ \vdots \ \vdots \ \cdots \ \vdots \\ 1 \, 0 \, 0 \, 0 \cdots 0 \end{pmatrix}.$$

The characteristic polynomial of $A_{\mathcal{D}_m}$ is

$$p_{A_{\mathcal{D}_m}} = \begin{vmatrix} \lambda & -1 & 0 & 0 & \cdots & 0 \\ 0 & \lambda & -1 & 0 & \cdots & 0 \\ 0 & 0 & \lambda & -1 & \cdots & 0 \\ \vdots & \vdots & \vdots & \vdots & \cdots & \vdots \\ -1 & 0 & 0 & 0 & \cdots & \lambda \end{vmatrix} = \lambda^m - 1.$$

Thus, the ordinary spectrum of $\mathcal{D}_m$ consists of the m-ary complex roots $z_0 = 1, z_1, \ldots, z_{m-1}$ of 1, where

$$z_k = \cos\frac{2k\pi}{m} + i\sin\frac{2k\pi}{m} = e^{\frac{2ki\pi}{m}},$$

for $0 \leqslant k \leqslant m-1$.

Observe that

$$A_{\mathcal{D}_m}\begin{pmatrix} 1 \\ z_k \\ \vdots \\ z_k^{m-1} \end{pmatrix} = \begin{pmatrix} z_k \\ \vdots \\ z_k^{m-1} \\ 1 \end{pmatrix} = z_k \begin{pmatrix} 1 \\ \vdots \\ z_k^{m-2} \\ z_k^{m-1} \end{pmatrix},$$

which shows that each ordinary eigenvalue z_k of $\mathcal{D}_m$ has

$$\begin{pmatrix} 1 \\ z_k \\ \vdots \\ z_k^{m-1} \end{pmatrix}$$

as an eigenvector.

Exercises and Supplements

1. How many graphs exist that have a finite set of vertices V such that $|V| = n$?
2. Let $\mathcal{G} = (V, E)$ be a k-regular graph having m vertices and $E \neq \emptyset$. Prove that at least one of the numbers k and m must be even.
3. Let $\mathcal{G} = (V, E)$ be a graph such that $|E| = n$. Prove that $|V| \geqslant \frac{1+\sqrt{1+8n}}{2}$.
4. Let $\mathcal{G} = (V, E)$ be a finite graph. How many spanning subgraphs exist for $\mathcal{G}$?
5. Prove that at least one of the graphs $\mathcal{G}$ or $\overline{\mathcal{G}}$ is connected.
6. Let $\mathcal{G} = (V, E)$ be a graph. Prove that for any vertex $v \in V$ we have $\mathsf{d}_{\mathcal{G}}(v) + \mathsf{d}_{\overline{\mathcal{G}}}(v) = |V| - 1$.

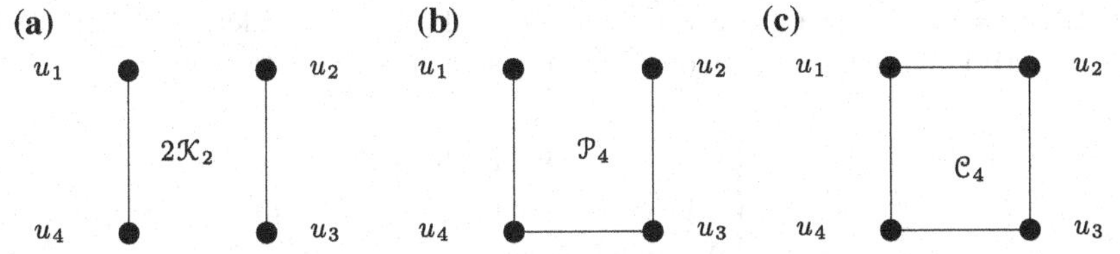

Fig. 10.38 Four-vertex graphs that are not threshold

7. Let $\mathcal{G} = (V, E)$ be a graph with $|V| = m$. Prove that the if $\lambda \in \mathsf{spec}(A_{\mathcal{G}})$ with $\mathsf{algm}(A_{\mathcal{G}}, \lambda) = p$, $p > 1$, then $-\lambda - 1 \in \mathsf{spec}(A_{\overline{\mathcal{G}}})$, and $p - 1 \leqslant \mathsf{algm}(A_{\mathcal{G}}, -\lambda - 1) \leqslant p + 1$.
8. Let $\mathcal{G} = (V, E)$ be a graph, where $|V| = m$ and let $q(\mathcal{G}) = \max_{v \in V} \mathsf{d}(v)$. Prove that the matrix $S_{\mathcal{G}}$ defined as $S_{\mathcal{G}} = I_m - \frac{1}{q(\mathcal{G})} L_{\mathcal{G}}$ is a symmetric and doubly stochastic matrix.

A matrix $A \in \{0, 1\}^{n \times m}$ is a *threshold matrix* if there exists a vector $\mathbf{a} \in \mathbb{R}^m$ and $b \in \mathbb{R}$ such that $A\mathbf{x} \leqslant \mathbf{1}_n$ if and only if $\mathbf{a}'\mathbf{x} \leqslant b$ for every $\mathbf{x} \in \{0, 1\}^m$.

The *intersection graph* of a matrix $A \in \{0, 1\}^{n \times m}$ is $\mathcal{T}_A = (\{1, \ldots, m\}, E)$, where an edge (i, j) exists in E if $\mathbf{a}_i'\mathbf{a}_j > 0$ for $1 \leqslant i < j \leqslant m$.

Recall that the characteristic vector of a subset S of a set V is denoted by $\mathbf{c}_S$ (see Supplement 8 of Chap. 6).

.

A graph $\mathcal{G} = (V, E)$, where $V = \{v_1, \ldots, v_m\}$ is *threshold* if there exists $\mathbf{a} \in \mathbb{R}^m$ and $b \in \mathbb{R}$ such that $\mathbf{a}'\mathbf{c}_S \leqslant b$ if and only if S is independent.

9. Let $A \in \{0, 1\}^{n \times m}$ be a matrix and let $\mathcal{T}_A$ be its intersection graph. Prove that:
   (a) the set S, $S \subseteq \{1, \ldots, m\}$ is independent if and only if $A\mathbf{w} \leqslant \mathbf{1}_n$;
   (b) A is a threshold matrix if and only if $\mathcal{T}_A$ is a threshold graph.
10. Prove that the graphs shown in Fig. 10.38a–c are not threshold. Furthermore, prove that a graph is not threshold if it contains a subgraph isomorphic to any of the graphs shown in Fig. 10.38.
11. Let $\mathcal{G} = (V, E)$ be a graph having the connected components $C_1, \ldots, C_k$. Prove that $\mathcal{G}$ is threshold if and only if there exits a component C_j that is a threshold graph and each component C_i such that $i \neq j$ consists of a single vertex.

    Solution: If (x, y) is an edge in a component C_p and (u, v) is an edge in a component C_q, where $p \neq q$, then the subgraph generated by $\{x, y, u, v\}$ is isomorphic to $2\mathcal{K}_2$ and $\mathcal{G}$ cannot be threshold.
12. Prove that a threshold graph can be obtained from the empty graph by applying a sequence of one of the following operations: add an isolated node, or take the complement.

Let $\mathcal{G} = (V, E)$ be a graph. The *vertex space* of $\mathcal{G}$ is the $\mathbb{C}$-linear space $\mathsf{V}_{\mathcal{G}}$ of all functions of the form $f : V \longrightarrow \mathbb{C}$; the *edge space* of $\mathcal{G}$ is the $\mathbb{C}$-linear space $\mathsf{E}_{\mathcal{G}}$ of the functions $g : E \longrightarrow \mathbb{C}$. Note that $\dim(\mathsf{V}_{\mathcal{G}}) = |V|$ and $\dim(\mathsf{E}_{\mathcal{G}}) = |E|$.

13. Let $\mathcal{G} = (V, E)$ be a graph with $|V| = m$ and $|E| = n$ and let $r : V \times V \longrightarrow \{-1, 0, 1\}$ be an orientation on $\mathcal{G}$. Define the matrix $D \in \mathbb{C}^{m \times n}$ by

$$d_{\ell j} = \begin{cases} 1 & \text{if } v_\ell \text{ is the positive end of } e_j, \\ -1 & \text{if } v_\ell \text{ is the negative end of } e_j, \\ 0 & \text{otherwise.} \end{cases}$$

Prove that $rank(D) = |E| - c$, where c is the number of connected components of $\mathcal{G}$.

14. In Definition 10.19 it is required that the length n of a cycle $\mathbf{p} = (v_0, \ldots, v_n)$ in a graph $\mathcal{G}$ be at least 3. Why is this necessary?
15. Let S be a finite set. Define the graph $\mathcal{G}_S = (\mathcal{P}(S), E)$ such that an edge (K, L) exists if $K \subset L$ and $|L| = |K| + 1$. Prove that there exist $(|M| - |K|)!$ paths that join the vertex K to M.
16. Let $\mathcal{G} = (V, E)$ be a finite graph. Define the triangular number of $\mathcal{G}$ as $t = \max\{|\Gamma(x) \cap \Gamma(y)| \mid (x, y) \in E\}$.

(a) Prove that for every two vertices $x, y \in V$ we have $\mathsf{d}(x) + \mathsf{d}(y) \leqslant |V| + t$.
(b) Show that

$$\sum\{\mathsf{d}(x) + \mathsf{d}(y) \mid (x, y) \in E\} = \sum_{x \in V} \mathsf{d}^2(x).$$

Conclude that

$$\sum_{x \in V} \mathsf{d}^2(x) \leqslant (|V| + t)|E|.$$

17. Let $m(\mathcal{G})$ be the minimum degree of a vertex of the graph $\mathcal{G}$. Prove that $\mathcal{G}$ contains a path of length $m(\mathcal{G})$ and a cycle of length at least $m(\mathcal{G}) + 1$.

Hint: Consider a path of maximal length in $\mathcal{G}$.

18. Let $\mathcal{C}$ be a collection of sets. The graph $\mathcal{G}_{\mathcal{C}}$ of $\mathcal{C}$ has $\mathcal{C}$ as the set of vertices. The set of edges consists of those pairs $C, D \in \mathcal{C}$ such that $C \neq D$ and $C \cap D \neq \emptyset$.

(a) Prove that for every graph $\mathcal{G}$ there exists a collection of sets $\mathcal{C}$ such that $\mathcal{G} = \mathcal{G}_{\mathcal{C}}$.
(b) Let $\mathcal{C} = (V, E)$ and let

$$c(\mathcal{G}) = \min\{|S| \mid \mathcal{G} = \mathcal{G}_{\mathcal{C}} \text{ for some } \mathcal{C} \subseteq \mathcal{P}(S)\}.$$

Prove that if $\mathcal{G}$ is connected and $|\mathcal{C}| \geqslant 3$, then $c(\mathcal{C}) \leqslant |E|$.

19. Let $\mathcal{G}_1$ and $\mathcal{G}_2$ be two graphs whose adjacency matrices are

$$A_{\mathcal{G}_1} = \begin{pmatrix} 0\,0\,0\,0\,0 \\ 0\,0\,1\,0\,1 \\ 0\,1\,0\,1\,0 \\ 0\,0\,1\,0\,1 \\ 0\,1\,0\,1\,0 \end{pmatrix} \text{ and } A_{\mathcal{G}_2} = \begin{pmatrix} 0\,1\,1\,1\,1 \\ 1\,0\,0\,0\,0 \\ 1\,0\,0\,0\,0 \\ 1\,0\,0\,0\,0 \\ 1\,0\,0\,0\,0 \end{pmatrix}.$$

Prove that the graphs $\mathcal{G}_1$ and $\mathcal{G}_2$ are not isomorphic. However, their ordinary spectra are identical and consist of $\{-2, 0, 2\}$, where $\mathsf{algm}(A_{\mathcal{G}_1}, 0) = \mathsf{algm}(A_{\mathcal{G}_2}, 0) = 3$.

20. Let $\mathcal{G} = (V, E)$ be a bipartite graph. Prove that $trace(A_{\mathcal{G}}^{2k+1}) = 0$ for every $k \in \mathbb{N}$.
21. Let $\mathcal{G} = (V, E)$ be a graph such that $(u, v) \in (V \times V) - E$ implies $\mathsf{d}(u) + \mathsf{d}(v) \geqslant |V|$, where V contains at least three vertices. Prove that $\mathcal{G}$ is connected.

    Solution: Suppose that u and v belong to two distinct connected components C and C' of the graph $\mathcal{G}$, where $|C| = p$ and $|C'| = q$, so $p+q \leqslant |V|$. If $x \in C$ and $y \in C'$, then there is no edge (x, y), $\mathsf{d}(x) \leqslant p-1$ and $\mathsf{d}(y) \leqslant q-1$. Thus, $\mathsf{d}(x) + \mathsf{d}(y) \leqslant p+q-2 \leqslant |V| - 2$, which contradicts the hypothesis. Thus, u, v must belong to the same connected component, so $\mathcal{G}$ is connected.
22. Prove that $A_{\mathcal{K}_n} = J_n - I_n$ and that

$$(A_{\mathcal{K}_n})^k = \frac{(n-1)^k - (-1)^k}{n} J_n + (-1)^k I_n.$$

Let $\mathcal{G} = (V, E)$ be a graph and let C be a set referred to as the *set of colors*. A *C-coloring* of $\mathcal{G}$ is a function $f : V \longrightarrow C$ such that $(u, v) \in E$ implies $f(u) \neq f(v)$. The *chromatic number* of $\mathcal{G}$ is the least number $|C|$ such that the graph has a C-coloring. The chromatic number of $\mathcal{G}$ is denoted by $\chi(\mathcal{G})$; if $\chi(\mathcal{G}) = n$, then we say that $\mathcal{G}$ is *n-chromatic*.

23. Prove that the graph $\mathcal{G}$ has $\chi(\mathcal{G}) = 1$ if and only if it is totally disconnected. Further, prove that $\chi(\mathcal{G}) = 2$ if and only if it has no odd cycles.
24. Let $\mathcal{G} = (\mathbb{N}, E)$ be a complete graph having $\mathbb{N}$ as its set of vertices and $E = (m, n) \in \mathcal{P}_2(\mathbb{N}) \mid m \neq n$.

    (a) If $f : E \longrightarrow \{c_1, c_2\}$ is a two-color coloring of the edges of $\mathcal{G}$, prove that there exists an infinite complete subgraph of $\mathcal{G}$ that is monochromatic.
    (b) Extend this statement to an r-color coloring, where $r \geqslant 3$.

    Solution: Define the sequence of infinite subsets $T_0, T_1, \ldots$ of $\mathbb{N}$ as follows. The initial set is $T_0 = \mathbb{N}$. Suppose T_i is defined. Choose $n_i \in T_i$, and let $U_{ij} = \{n \in T - \{n_i\} \mid f(n_i, n) = c_j\}$ for $j = 1, 2$. At least one of U_{i1}, U_{i2} is infinite, and T_{i+1} is chosen as one of U_{i1}, U_{i2} that is infinite.

If $i \leqslant \min\{j,k\}$, then $n_j \in T_j \subset T_{i+1}$ and $n_k \in T_k \subset T_{i+1}$, which implies $f(n_i, n_j) = f(n_i, n_k)$ because of the definition of the sets T_i. Let $U = \{n_0, n_1, \ldots\}$ and let $g : U \longrightarrow \{c_1, c_2\}$ be given by $g(n_i) = f(n_i, n_j)$ for $i < j$. It is clear that g is well-defined. Since U is an infinite set, at least one of the subsets $g^{-1}(c_1), g^{-1}(c_2)$ is infinite. Let W be one of these subsets that is infinite. Then, for $n_l, n_k \in W$, where $l < k$, we have $g(n_l) = g(n_k) = c$ and therefore $f(n_l, n_k) = c$. Thus, the subgraph induced by U is monochromatic.

25. Let $(P, \leqslant)$ be a finite partially ordered set. The *comparability graph* of $(P, \leqslant)$ is the graph (P, E), where $(u, v) \in E$ if (u, v) are comparable. Prove that the chromatic number of (P, E) equals the minimum number of antichains in a partition of P into antichains.

Let $\mathcal{G} = (V, E)$ be a graph. A *Hamiltonian path* that joins the vertex u to the vertex v is a path in $\mathcal{G}$ that joins u to v such that every vertex of V occurs in the path.

A *Hamiltonian cycle* in $\mathcal{G}$ is a simple cycle that contains all vertices of $\mathcal{G}$. Next, we present several sufficient conditions for the existence of a Hamiltonian path due to O. Ore and G. A. Dirac.

26. Let $\mathcal{G} = (V, E)$ be a graph such that $(u, v) \in (V \times V) - E$ implies $\mathsf{d}(u) + \mathsf{d}(v) \geqslant |V|$, where V contains at least three vertices. Prove that $\mathcal{G}$ contains a Hamiltonian cycle.

Solution: The graph $\mathcal{G}$ is connected. Let $\mathbf{p} = (v_1, \ldots, v_m)$ be the longest path in $\mathcal{G}$.
Suppose that $m = |V|$, which implies that $\mathbf{p}$ is a Hamiltonian path that joins v_1 to v_m. If v_m and v_1 are joined by an edge, then $(v_1, \ldots, v_m, v_1)$ is a Hamiltonian cycle.
Suppose that no edge exists between v_m and v_1, so $\mathsf{d}(v_1) + \mathsf{d}(v_m) \geqslant |V|$. The vertex v_1 is joined to $\mathsf{d}(v_1)$ vertices on the path $(v_2, \ldots, v_m)$ and there are $\mathsf{d}(v_1)$ nodes that precede these nodes on the path $\mathbf{p}$. If v_m were not joined to any of these nodes, then the set of nodes on $\mathbf{p}$ joined to v_m would come from a set of $m - 1 - \mathsf{d}(v_1)$ nodes, so we would have $\mathsf{d}(v_m) \leqslant m - 1 - \mathsf{d}(v_1)$, which would contradict the assumption that $\mathsf{d}(v_1) + \mathsf{d}(v_m) \geqslant |V|$. Thus, there exists a node v_i on $\mathbf{p}$ such that $(v_1, v_i), (v_{i-1}, v_n) \in E$. Therefore $(v_1, v_i, v_{i+1}, \ldots, v_n, v_{i-1}, v_{i-2}, \ldots, v_2, v_1$ is a Hamiltonian cycle.
Suppose that $m < |V|$. If there is an edge (w, v_1), where w is not a node of $\mathbf{p}$, then $(w, v_1, \ldots, v_m)$ is longer than $\mathbf{p}$. Thus, v_1 is joined only to nodes in $\mathbf{p}$ and so is v_m. An argument similar to the one previously used shows that there exists a simple cycle $\mathbf{q} = (y_1, \ldots, y_m, y_1)$ of length m. Since $m < |V|$ and $\mathcal{G}$ is connected, there is a node t not on $\mathbf{q}$ that is joined to some vertex y_k in $\mathbf{q}$. Then $(t, y_k, y_{k+1}, \ldots, y_m, y_1, y_2, \ldots, y_{k-1})$ is a path of length $m + 1$, which contradicts the maximality of the length of $\mathbf{p}$. Since this case cannot occur, it follows that $\mathcal{G}$ has a Hamiltonian cycle.

27. Let $\mathcal{G} = (V, E)$ be a graph such that $|V| \geqslant 3$ in which $\mathsf{d}(v) \geqslant |V|/2$.

(a) Prove that $\mathcal{G}$ is connected.
(b) Prove that $\mathcal{G}$ has a Hamiltonian cycle.

28. Let $\mathcal{G} = (V, E)$ be a graph such that $|E| \geqslant \frac{(|V|-1)(|V|-2)}{2} + 2$. Prove that $\mathcal{G}$ has a Hamiltonian cycle.
29. Prove that every subgraph of a bipartite graph is bipartite.
30. Prove that a tree is a bipartite graph.
31. Prove that the adjacency matrix of a tree that has n vertices contains $(n-1)^2+1$ zeros.
32. Prove that a tree that has at least two vertices has at least two leaves.
33. Prove that if $\mathcal{T}$ is a tree and x is a leaf, then the graph obtained by removing x and the edge that has x as an endpoint from $\mathcal{T}$ is again a tree. Also, show that if z is a new vertex, then the graph obtained from $\mathcal{T}$ by joining z with any node of $\mathcal{T}$ is also a tree.
34. Prove that Dilworth's Theorem (Theorem 2.57) is equivalent to König Theorem (Theorem 10.148), that is, each each of these statement can be proven starting with the other.

Solution: We start with König Theorem. Let $(S, \leqslant)$ is a finite nonempty poset, where $S = \{s_1, \ldots, s_n\}$ such that $width(S, \leqslant) = m$. Consider a bipartite graph

$$\mathcal{G} = (\{s_1, \ldots, s_n, s'_1, \ldots, s'_n\}, E)$$

having the bipartition $\{S, S'\}$, where $S' = \{s'_1, \ldots, s'_n\}$. The set of edges of $\mathcal{G}$ is $E = \{(s_i, s'_j) \mid s_i < s_j\}$.

By König Theorem there exists a maximal matching M of $\mathcal{G}$ and a minimal vertex cover U of $\mathcal{G}$ of the same cardinality. Let $T = \{s \in S \mid \{s, s'\} \cap U = \emptyset\}$. Note that no two elements of T are comparable in the poset $(S, \leqslant)$ for otherwise, an edge would exist between two elements of T and one of these vertices would belong to U. Thus, T is an antichain in $(S, \leqslant)$.
Define the reflexive and symmetric relation κ on S as consisting of those pairs (s_i, s_j) such that $i = j$, or $(s_i, s'_j) \in M$, or $(s_j, s'_i) \in M$. Note that if $(s_i, s_j) \in \kappa$ and $(s_j, s_k) \in \kappa$, then $(s_i, s_k) \in M$. Since M and U have the same cardinality, it follows that T has the same cardinality as the partition S/κ.
To prove König Theorem starting from Dilworth's Theorem we start with a bipartite graph $\mathcal{G} = (V, E)$ having the bipartition $\{Y, Z\}$. Define a partial order on V by taking $y \leqslant z$ when $y = z$, or when $y \in Y$ and $z \in Z$. By Dilworth's Theorem, there exists an antichain and a partition of the poset $(V, \leqslant)$ having the same size. The nontrivial chains in $(V, \leqslant)$ yield a matching in the graph; the complement of the set of vertices corresponding to the antichain yield an vertex cover with the same cardinality as the matching.

35. Let u be a vertex in a tree $\mathcal{T} = (V, E)$, where $|V| \geqslant 2$. Prove that $\sum\{d(u, v) \mid v \in V\} \leqslant \binom{|V|}{2}$.

Solution: The proof is by induction on $n = |V|$. The base case. $n = 2$ is immediate. Suppose that the statement holds for trees with fewer than n vertices and let x be a leaf of $\mathcal{T}$ such that $x \neq u$. Note that no shortest path from u to a vertex v, where $v \neq x$ exists that passes through x. Then, by inductive hypothesis, $\sum\{d(u, v) \mid v \in V\} = \sum\{d(u, v) \mid v \in V, v \neq x\} + d(u, x) \leqslant \binom{n-1}{2} + (n-1) = \binom{n}{2}$.

36. Prove that the inequality from Supplement 35 holds in arbitrary graphs using spanning trees.
37. Let $f : S \longrightarrow S$ be a function. Define the directed graph $\mathcal{G}_f = (S, E)$, where $E = \{(x, y) \in S \times S \mid y = f(x)\}$. Prove that each connected component of $\mathcal{G}_f$ consists of a cycle and a number of trees linked to the cycle.
38. Let $\mathcal{G}$ be a graph. Prove that for every square submatrix M of $U_{\mathcal{G}}$ we have $\det(M) \in \{-1, 0, 1\}$.
39. Let $\mathcal{G} = (V, E)$ be a connected graph with $|V| = n$. Prove that $rank(U_{\mathcal{G}}) = n-1$.
40. Let $\mathcal{G} = (V, E)$ be a tree such that $|V| = n$. Prove that any square submatrix $M \in \mathbb{R}^{(n-1)\times(n-1)}$ of the incidence matrix $U_{\mathcal{G}}$ is non-singular.
41. Let $\mathcal{G} = (V, E)$ be a graph with $|V| = n$, $|E| = n - 1$, where $n \geqslant 2$ and let $\mathcal{G}^r$ be an oriented graph obtained from $\mathcal{G}$ having the incidence matrix $U_{\mathcal{G}_r}$. If $B = U_{\mathcal{G}_r}\begin{bmatrix} 2 \cdots n \\ 1 \cdots n-1 \end{bmatrix}$, prove that $\det(U_{\mathcal{G}^r}) \in \{-1, 0, 1\}$ and $\det(U_{\mathcal{G}^r}) \in \{-1, 1\}$ if and only if $\mathcal{G}$ is a tree.
42. Let $\mathcal{G} = (V, E)$ be a digraph. Prove that there exists a path of length m from vertex v_i to vertex v_j, where $v_i, v_j \in V$ if and only if $(A_{\mathcal{G}}^m)_{ij} > 0$.
43. Let $\mathcal{G} = (V, E)$ be a digraph. Prove that $\mathcal{G}$ is strongly connected if and only if there exists a closed walk that contains each vertex.
44. Let $\mathcal{N} = (\mathcal{G}, cap, s, t)$ be a network. Prove that if (C_1, C_1') and (C_2, C_2') are minimal cuts, then $(C_1 \cup C_2, C_1' \cap C_2')$ is a minimal cut¡.
45. Let $\mathcal{G} = (V, E)$ be a graph such that $V = \{v_1, \ldots, v_m\}$ and $\mathcal{G}$ contains no subgraph isomorphic to $\mathcal{K}_{p,p}$. Prove that $\sum_{i=1}^{m} \binom{\mathsf{d}(v_i)}{p} \leqslant \binom{m}{p}(p-1)$.

Solution: Consider the pairs of the form $(\{v_1, \ldots, v_p\}, v)$ such that $(v_i, v) \in E$ for $1 \leqslant i \leqslant p$. Note that for a set $\{v_1, \ldots, v_p\}$ there exist at most $p - 1$ choices for v because, otherwise, $\mathcal{G}$ would contain a subgraph isomorphic to $\mathcal{K}_{p,p}$. Thus, there are no more than $\binom{m}{p}(p-1)$ pairs of the specified form.

Note that for a vertex v there exists a pair $(\{v_1, \ldots, v_p\}, v)$ only if $\mathsf{d}(v) \geqslant p$ and, in this case, there are $\binom{\mathsf{d}(v)}{p}$ such pairs. Consequently, $\sum_{i=1}^{m} \binom{\mathsf{d}(v_i)}{p} \leqslant \binom{m}{p}(p-1)$.

46. Let $\mathcal{G} = (V, E)$ be a digraph. If $U \subseteq V$, let $C(U) = \{(x, y) \in E \mid x \in U, y \notin U\}$ and let $f : \mathcal{P}(V) \longrightarrow \mathbb{R}_{\geqslant 0}$ be the function given by $f(U) = |C(U)|$ for $U \in \mathcal{P}(V)$. Prove that f satisfies the inequality

$$f(U_1 \cup U_2) + f(U_1 \cap U_2) \leqslant f(U_1) + f(U_2)$$

for every $U_1, U_2 \in \mathcal{P}(V)$.

47. Let $A, B \in \mathbb{R}^{n\times n}$ be two symmetric matrices, where

$$\mathsf{spec}(A) = \{\alpha_1, \ldots, \alpha_n\} \text{ and } \mathsf{spec}(B) = \{\beta_1, \ldots, \beta_n\}.$$

Prove that:

(a) if $\alpha_1 \geqslant \cdots \geqslant \alpha_n$ and $\beta_1 \geqslant \cdots \geqslant \beta_n$, then $trace(AB) \leqslant \sum_{i=1}^{n} \alpha_i \beta_i$;
(b) if $\alpha_1 \geqslant \cdots \geqslant \alpha_n$ and $\beta_1 \leqslant \cdots \leqslant \beta_n$, then $trace(AB) \geqslant \sum_{i=1}^{n} \alpha_i \beta_i$.

Solution: Observe that $\| A-B \|_F^2 = trace((A-B)'(A-B)) = trace((A-B)^2)$ due to the symmetry of the matrices A and B. This implies

$$\| A - B \|_F^2 = trace(A^2 - 2AB + B^2) = \| A \|_F^2 + \| A \|_F^2 - 2\, trace(AB),$$

so $2\, trace(AB) = \| A \|_F^2 + \| B \|_F^2 - \| A - B \|_F^2$.

By Corollary 10.96 of Hoffman-Wieland Theorem, we have $\| A - B \|_F^2 \geqslant \sum_{i=1}^{n} (\alpha_i - \beta_i)^2$, so

$$\begin{aligned} 2\, trace(AB) &\leqslant \| A \|_F^2 + \| B \|_F^2 - \sum_{i=1}^{n} (\alpha_i - \beta_i)^2 \\ &= \sum_{i=1}^{n} \alpha_i^2 + \sum_{i=1}^{n} \beta_i^2 - \sum_{i=1}^{n} (\alpha_i - \beta_i)^2 = 2 \sum_{i=1}^{n} \alpha_i \beta_i, \end{aligned}$$

which yields the first inequality.
If $\alpha_1 \geqslant \cdots \geqslant \alpha_n$ and $\beta_1 \leqslant \cdots \leqslant \beta_n$, then by the second part of Corollary 10.96, we have $\| A-B \|_F^2 \leqslant \sum_{i=1}^{n} (\alpha_i - \beta_i)^2$, and this implies $trace(AB) \leqslant \sum_{i=1}^{n} (\alpha_i - \beta_i)^2$.

48. Let $\mathcal{G} = (V, E)$ be a graph, where $|V| = m$ and $|E| = n$, and let $\pi = \{B_1, \ldots, B_p\}$ be a partition of V. The *community matrix of* π is the matrix $S_\pi \in \mathbb{R}^{m\times p}$ defined by $(S_\pi)_{ij} = 1$ if $v_i \in B_j$. Prove that $S'S = \mathsf{diag}(|B_1|, \ldots, |B_p|)$.

49. Let $\mathcal{G} = (V, E)$ be a graph, where $|V| = m$ and $|E| = n$, and let $\pi = \{B_1, \ldots, B_p\}$ be a partition of V such that the vertices that belong to a block are numbered consecutively, that is, $B_\ell = \{v_{m_{ell-1}+1}, \ldots, v_{m_\ell}\}$, where $|B_\ell| = m_\ell$ for $1 \leqslant \ell \leqslant p$.
Define the matrix

$$A_{\mathcal{G}^\pi} = \mathsf{diag}\left(\frac{1}{|B_1|}, \ldots, \frac{1}{|B_p|}\right) S_\pi' A_{\mathcal{G}} S_\pi.$$

A partition π is *equitable* if $A_{\mathcal{G}}^{\pi} S_\pi = S_\pi A_{\mathcal{G}}$.
Prove that π is equitable if for blocks B_r, B_s the number of neighbors that a vertex in B_r has in the block B_s is the same for every vertex in B_r.

Let $\mathcal{G}$ be a graph. The *eigenvector centrality* is a measure of the influence of a node in the graph $\mathcal{G}$. Its definition (given below) formalizes the idea that the centrality of a vertex is determined by the measures of centrality of its neighbors in $\mathcal{G}$. Google ranks web pages based on their centrality in the web graph.

50. Let $\mathcal{G} = (V, E, w)$ be a weighted graph, where $V = \{v_1, \ldots, v_n\}$. The *centrality* of a vertex v_i is a non-negative number c_i such that

$$c_i = k \sum \{w_{ij} c_j \mid v_j \in N_{\mathcal{G}}(\{v_i\})\}$$

for $1 \leqslant i \leqslant n$. A vector whose components are the numbers c_i is a *centrality vector*. Prove that:

(a) $\mathbf{c} = k A_{\mathcal{G}} \mathbf{c}$;
(b) $\mathbf{c}$ is an eigenvector that corresponds to the spectral radius of $A_{\mathcal{G}}$.

51. Let $\mathcal{G} = (V, E)$ be a graph, where $|V| = m$ and $|E| = n$. If λ_1 is the largest eigenvalue of $A_{\mathcal{G}}$, prove that

$$\lambda_1 \leqslant \sqrt{\frac{2n(m-1)}{m}}.$$

Solution: If $\lambda_1 \geqslant \lambda_2 \geqslant \cdots \geqslant \lambda_m$, we have $\lambda_1 = -\sum_{i=2}^{m} \lambda_i$, so $\lambda_1 \leqslant \sum_{i=2}^{m} |\lambda_i|$. By Inequality (6.4) of Chap. 6, we have

$$2n - \lambda_1^2 = \sum_{i=2}^{m} \lambda_i^2 \geqslant \frac{1}{m-1} \left(\sum_{i=2}^{m} |\lambda_i| \right)^2 \geqslant \frac{\lambda_1^2}{m-1}.$$

Thus, $2n \geqslant \lambda_1^2 \frac{m}{m-1}$, which gives the desired inequality.

52. Let $\mathcal{G} = (V, E)$ be a k-regular graph, U, W be two sets of vertices and let $E(U, V) = (U \times V) \cap E$ be the set of edges whose endpoints belong to U and V. Prove that

$$|E(U, V)| \leqslant \lambda \sqrt{|U| \cdot |V|} + \frac{k}{n} |U| \cdot |V|,$$

where $n = |V|$ and λ is the largest absolute value of an eigenvalue of $A_{\mathcal{G}}$.

Solution: Let $\lambda_n \leqslant \cdots \leqslant \lambda_1 = k$ be the ordinary spectrum of $\mathcal{G}$ (see Example 10.151) and let $\mathbf{v}_n, \ldots, \mathbf{v}_1$ be the orthonormal base of $\mathbb{R}^n$ that consists of the corresponding eigenvectors. We saw that $\mathbf{v}_1 = \frac{1}{\sqrt{n}} \mathbf{1}_n$. Let $\mathbf{c}_U$ and $\mathbf{c}_W$ be the characteristic vectors of U and V, respectively (see Supplement 8 of Chap. 6) and let $\mathbf{c}_U = \sum_{i=1}^{n} a_i \mathbf{v}_i$, $\mathbf{c}_W = \sum_{i=j}^{n} b_j \mathbf{v}_j$ be the representations of $\mathbf{c}_U$ and $\mathbf{c}_W$ in this basis. We have

$$
\begin{aligned}
|E(U, W)| = \mathbf{c}_U A_{\mathcal{G}} \mathbf{c}_W &= \left(\sum_{i=1}^n a_i \mathbf{v}_i\right) A_{\mathcal{G}} \left(\sum_{j=1}^n b_j \mathbf{v}_j\right) \\
&= \left(\sum_{i=1}^n a_i \mathbf{v}_i\right) \left(\sum_{j=1}^n b_j A_{\mathcal{G}} \mathbf{v}_j\right) \\
&= \left(\sum_{i=1}^n a_i \mathbf{v}_i\right) \left(\sum_{j=1}^n b_j \lambda_j \mathbf{v}_j\right) = \sum_{i=1}^n \lambda_i a_i b_i,
\end{aligned}
$$

due to the orthonormality of the basis. Since $\mathbf{c}'_U \frac{1}{\sqrt{n}} \mathbf{1}_n = |U|$ and $\mathbf{c}'_W \frac{1}{\sqrt{n}} \mathbf{1}_n = |V|$ we have

$$
|E(U, W)| = k\frac{|U| \cdot |W|}{n} + \sum_{i=2}^n \lambda_i a_i b_i = k\frac{|U| \cdot |W|}{n} + \sum_{i=2}^n \lambda_i a_i b_i.
$$

Thus, we can write

$$
\begin{aligned}
&\left| |E(U, W)| - k\frac{|U| \cdot |W|}{n} \right| \\
&\leqslant \sum_{i=2}^n |\lambda_i a_i b_i| \leqslant \lambda \sum_{i=2}^n |a_i b_i| \leqslant \lambda \parallel \mathbf{a} \parallel_2 \cdot \parallel \mathbf{b} \parallel_2 \\
&= \lambda \parallel \mathbf{c}_U \parallel_2 \cdot \parallel \mathbf{c}_W \parallel_2 = \lambda |U| \cdot |W|,
\end{aligned}
$$

which gives the desired inequality.

Bibliographical Comments

Among the many references for graph theory we mention [9–11] and [12]. We adopted the treatment of maximal flows (including Theorem 10.142) from [13], a classic reference for networks.

Supplement 28 is a result of Ore [14]; the same author (see [15]) has shown the statement in Exercise 28.

The result contained in Exercise 27 is from Dirac [16]. The notions of intersection graph and threshold graph were introduced in [17].

Exercise 41 originates in [18]. Exercise 20 contains a result of Cvetcović [19].

A great source for application of spectral theory of matrices to graphs is [20]. Our presentation of spectra of non-negative matrices follows [21].

Interesting applications of graph theory in the study of the world-wide web and search engines can be found in [22, 23].

The results that concern variable adjacency matrix of graphs and digraphs were obtained in [24]. The equivalence of Dilworth's and König Theorems was shown in [25].

References

1. D.J. Cook, L.B. Holder, *Mining Graph Data* (Wiley-Interscience, Hoboken, 2007)
2. C. Aggarwal, H. Wang (eds.), *Managing and Mining Graph Data* (Springer, New York, 2010)
3. V. Havel, A remark on the existence of finite graphs. Čas. pro Pěst. Mat. **80**, 477–480 (1955)
4. S.L. Hakimi, On the realizability of a set of integers as degrees of the vertices of a graph. SIAM J. Appl. Math. **10**, 496–506 (1962)
5. R. Rymon, Search through systematic set enumeration, in *Proceedings of the 3rd International Conference on Principles of Knowledge Representation and Reasoning*, Cambridge, MA, ed. by B. Nebel, C. Rich, W.R. Swartout (Morgan Kaufmann, San Mateo, CA, 1992), pp. 539–550
6. H. Prüfer, Neuer beweis eines satzes über permutationen. Archiv für Mathematik und Physik **27**, 142–144 (1918)
7. M. Fiedler, *Special Matrices and Their Applications in Numerical Mathematics*, 2nd edn. (Dover Publications, New York, 2008)
8. A. Berman, R.J. Plemmons, *Nonnegative Matrices in the Mathematical Sciences* (Academic Press, New York, 1979)
9. F. Harary, *Graph Theory* (Addison-Wesley, Reading, 1971)
10. C. Berge, *The Theory of Graphs* (Dover, New York, 2001)
11. B. Bollobás, *Graph Theory—An Introductory Course* (Springer-Verlag, New York, 1979)
12. R. Diestel, *Graph Theory* (Springer-Verlag, Berlin, 2005)
13. R.E. Tarjan, *Data Structures and Network Algorithms* (SIAM, Philadelphia, 1983)
14. O. Ore, Arc coverings of graphs. Ann. Mat. Pura Appl. **55**, 315-322 (1961)
15. O. Ore, Hamilton connected graphs. J. Math. Pures Appl. **42**, 21–27 (1963)
16. G.A. Dirac, Some theorems on abstract graphs. Proc. London Math. Soc. **2**, 69–81 (1952)
17. V. Chvátal, P.L. Hammer, Aggregation of inequality in integer programming. Ann. Discrete Math. **1**, 145–162 (1977)
18. J. Matoušek, J. Neštřil, *Invitation to Discrete Mathematics*, 2nd edn. (Oxford University Press, Oxford, 2011)
19. D. Cvetković, Graphs and their spectra. Publ. Elek. Fak. Univ. Beograd: Ser. Mat. Fiz. **354**, 1–50 (1971)
20. A.E. Brouwer, W.H. Haemers, *Spectra of Graphs* (Springer, New York, 2011)
21. R.A. Horn, C.R. Johnson, *Matrix Analysis* (Cambridge University Press, Cambridge, 1985)
22. A.N. Langville, C.D. Meyer, *Google's Page Rank and Beyond: The Science of Search Engine Rankings* (Princeton University Press, Princeton, 2006)
23. A. Bonato, *A Course on the Web Graph* (American Mathematical Society, Providence, 2008)
24. F. Harary, The determinant of the adjacency matrix of a graph. SIAM Rev. **4**, 202–210 (1962)
25. L. Lovasz, M.D. Plummer, *Matching Theory* (American Mathematical Society, Providence, 2009)

Chapter 11
Lattices and Boolean Algebras

11.1 Introduction

Lattices can be defined either as special partially ordered sets or as algebras. In this chapter, we present both definitions and show their equivalence. We study several special classes of lattices: modular and distributive lattices and complete lattices. The last part of the chapter is dedicated to Boolean algebras and Boolean functions.

11.2 Lattices as Partially Ordered Sets and Algebras

We begin with a simple algebraic structure.

Definition 11.1 *A* semilattice *is a semigroup $\mathcal{S} = (S, \{*\})$ such that $s * s = s$ and $s * t = t * s$ for all $s, t \in S$.*

In other words, $\mathcal{S} = (S, \{*\})$ is a semilattice if "$*$" is a commutative and idempotent operation.

Example 11.2 Let $*$ be the binary operation on the set $\mathbb{N}_1$ of positive natural numbers defined by $n * p = \gcd(n, p)$. In Example 1.142, we saw that $*$ is an associative operation. Since $\gcd(n, p) = \gcd(p, n)$ and $\gcd(n, n) = n$ for every $n \in \mathbb{N}$, it follows that $(\mathbb{N}_1, \{*\})$ is indeed a semilattice.

It is easy to see that $(\mathbb{N}_1, \{\text{lcm}\})$ is also a semilattice.

Theorem 11.3 *Let $\mathcal{S} = (S, \{*\})$ be a semilattice. The relation $x \leqslant y$ defined by $x = x * y$ for $x, y \in S$ is a partial order on S. Further, $\inf\{u, v\}$ in the partially ordered set $(S, \leqslant)$ exists for all $u, v \in S$ and $u * v = \inf\{u, y\}$.*

Proof The idempotency of $*$, $x = x * x$ implies $x \leqslant x$ for every $x \in S$; that is, the reflexivity of $\leqslant$.

Suppose that $x \leqslant y$ and $y \leqslant x$; that is, $x = x * y$ and $y = y * x$. The commutativity of $*$ implies that $x = y$, so $\leqslant$ is antisymmetric.

D. A. Simovici and C. Djeraba, *Mathematical Tools for Data Mining*, Advanced Information and Knowledge Processing, DOI: 10.1007/978-1-4471-6407-4_11,

Now let x, y, z be three elements of S such that $x \leqslant y$ and $y \leqslant z$, that is, $x = x * y$ and $y = y * z$. We have $x * z = (x * y) * z = x * (y * z) = x * y$, which proves that $x \leqslant z$. Thus, $\leqslant$ is transitive, so it is a partial order on S.

Let u and v be two arbitrary elements of S. Note that $u * v \leqslant u$ and $v * v \leqslant u$ because $(u * v) * u = u * (u * v) = (u * u) * v = u * v$ and $(u * v) * v = u * (v * v) = u * v$. Thus, $u * v$ is a lower bound of the set $\{u, v\}$. Suppose now that w is an arbitrary lower bound of $\{u, v\}$, that is, $w = w * u$ and $w = w * v$. We have $w * (u * v) = (w * u) * v = w * v = w$, which proves that $w \leqslant u * v$. This allows us to conclude that $u * v = \inf\{u, v\}$.

We also need the following converse result.

Theorem 11.4 *Let $(S, \leqslant)$ be a partially ordered set such that* $\inf\{u, v\}$ *exists for all $u, v \in S$. If $*$ is the operation defined by $u * v =$* $\inf\{u, v\}$ *for $u, v \in S$, then $(S, \{*\})$ is a semilattice.*

Proof It is immediate that $*$ is an idempotent and commutative operation. We prove here only its associativity.

Let $t, u, v \in S$ and let p, q be defined by $p = \inf\{t, \inf\{u, v\}\}$ and $q = \inf\{\inf\{t, u\}, v\}$. By the definition of infimum, we have $p \leqslant t$ and $p \leqslant \inf\{u, v\}$, so $p \leqslant u$ and $p \leqslant v$. Since $p \leqslant t$ and $p \leq u$, we have $p \leqslant \inf\{t, u\}$. This inequality, together with $p \leqslant v$, implies $p \leqslant \inf\{\inf\{t, u\}, v\}$, so $p \leqslant q$.

By the same definition of infimum, we have $q \leqslant \inf\{t, u\}$ and $q \leqslant v$. The first inequality implies $q \leqslant t$ and $q \leqslant u$. Thus, $q \leqslant \inf\{u, v\}$; together with $q \leqslant t$, these inequalities imply $q \leqslant \inf\{t, \inf\{u, v\}\} = p$. We conclude that $p = q$, which shows that $*$ is indeed an associative operation.

The next statement is closely related to the previous theorem.

Theorem 11.5 *Let $(S, \leqslant)$ be a partially ordered set such that* $\sup\{u, v\}$ *exists for all $u, v \in S$. If $\star$ is the operation defined by $u \star v =$* $\sup\{u, v\}$ *for $u, v \in S$, then $(S, \{\star\})$ is a semilattice.*

Proof This statement follows from Theorem 11.4 by duality.

Example 11.6 The partially ordered sets $(P, \leqslant)$ and $(Q, \leqslant)$, whose Hasse diagrams are given in Fig. 11.1, are semilattices because $\sup\{u, v\}$ exists for any pair of elements in each of these sets. The operation $\star$ is described by the the following table.

$(P, \star)$	a	b	c	$(Q, \star)$	x	y	z
a	a	c	c	x	x	y	z
b	c	b	c	y	y	y	z
c	c	c	c	z	z	z	z

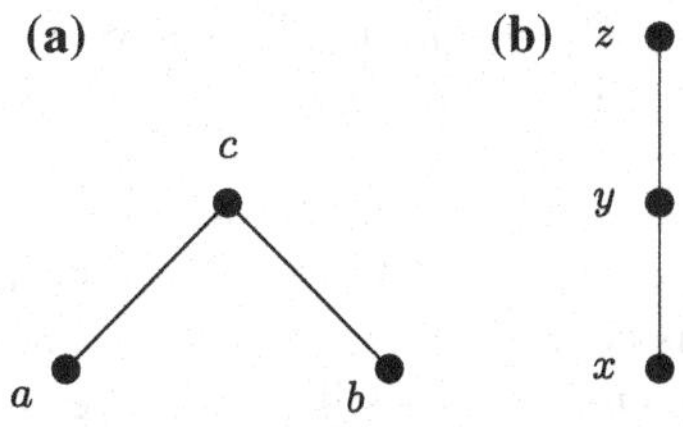

Fig. 11.1 Hasse diagrams. **a** The posets $(P, \leqslant)$, and **b** $(Q, \leqslant)$

Example 11.7 Let S be a set and let $(\mathbf{Seq}(S), \leqslant_{\text{pref}})$ be the poset introduced in Example 2.6. We prove that this is a semilattice by verifying that $\inf\{\mathbf{u}, \mathbf{v}\}$ exists for any sequences $\mathbf{u}, \mathbf{v} \in \mathbf{Seq}(S)$.

Note that any two sequences have at least the null sequence $\boldsymbol{\lambda}$ as a common prefix. If $\mathbf{t}$ and $\mathbf{s}$ are common prefixes of $\mathbf{u}$ and $\mathbf{v}$, then either $\mathbf{t}$ is a prefix of $\mathbf{s}$ or vice-versa. Thus, the finite set of common prefixes of $\mathbf{u}$ and $\mathbf{v}$ is totally ordered by "$\leqslant_{\text{pref}}$" and therefore it has a largest element $\mathbf{z}$. The sequence $\mathbf{z}$ is the longest common prefix of the sequences $\mathbf{u}$ and $\mathbf{v}$. It is clear that $\mathbf{z} = \inf\{\mathbf{u}, \mathbf{v}\}$ in the poset $(\mathbf{Seq}(S), \leqslant_{\text{pref}})$.

We denote the result of the semilattice operation introduced here, which associates with $\mathbf{u}$ and $\mathbf{v}$ their longest common prefix, by $lcp(\mathbf{u}, \mathbf{v})$.

The associativity of this operation can be written as

$$lcp(\mathbf{u}, lcp(\mathbf{v}, \mathbf{w})) = lcp(lcp(\mathbf{u}, \mathbf{v}), \mathbf{w}) \tag{11.1}$$

for all sequences $\mathbf{u}, \mathbf{v}, \mathbf{w} \in \mathbf{Seq}(S)$.

Theorem 11.8 *Let S be a set and let $\boldsymbol{u}, \boldsymbol{v}, \boldsymbol{w}$ be three sequences in* $\mathbf{Seq}(S)$. *Then, at most two of the sequences $lcp(\boldsymbol{u}, \boldsymbol{v})$, $lcp(\boldsymbol{v}, \boldsymbol{w})$, $lcp(\boldsymbol{w}, \boldsymbol{u})$ are distinct. The common value of two of these sequences is a prefix of the third sequence.*

Proof Let $\mathbf{t} = lcp(\mathbf{u}, \mathbf{v})$, $\mathbf{r} = lcp(\mathbf{v}, \mathbf{w})$, and $\mathbf{s} = lcp(\mathbf{w}, \mathbf{u})$. Note that any two of the sequences $\mathbf{t}, \mathbf{r}, \mathbf{s}$ are prefixes of the same sequence. Therefore, they form a chain in the poset $(\mathbf{Seq}(S), \leqslant_{\text{pref}})$.

Suppose, for example, that $\mathbf{t} \leqslant_{\text{pref}} \mathbf{r} \leqslant_{\text{pref}} \mathbf{s}$. Observe that $\mathbf{r}$ is a prefix of $\mathbf{v}$ because it is a prefix of $\mathbf{s}$. Thus, $\mathbf{r}$ is a prefix of both $\mathbf{u}$ and $\mathbf{v}$. Since $\mathbf{t}$ is the longest common prefix of $\mathbf{u}$ and $\mathbf{v}$, it follows that $\mathbf{r}$ is a prefix of $\mathbf{t}$, so $\mathbf{r} = \mathbf{t}$.

The remaining five cases that correspond to the remaining permutation of the sequences $\mathbf{t}$, $\mathbf{r}$, and $\mathbf{s}$ can be treated in a similar manner.

Theorems 11.4 and 11.5 shows that, in principle, a partial order relation on a set S may induce two semilattice structures on S. Traditionally, the semilattice $(S, *)$ has been referred to as the *meet semilattice* , while $(S, \star)$ is called the *join semilattice*, and the operations "$*$" and "$\star$" are denoted by "$\wedge$" and "$\vee$", respectively. This is a notation that we will use from now on.

Definition 11.9 *Let $\mathcal{S}_1 = (S_1, \{\wedge\})$ and $\mathcal{S}_2 = (S_2, \{\wedge\})$ be two semilattices. A* morphism *h from $\mathcal{S}_1$ to $\mathcal{S}_2$ is a function $h : S_1 \longrightarrow S_2$ such that $h(x \wedge y) = h(x) \wedge h(y)$ for $x, y \in S_1$.*

The semilattices $\mathcal{S}_1$ *and* $\mathcal{S}_2$ *are* isomorphic *if there exist two bijective morphisms* $h : S_1 \longrightarrow S_2$ *and* $h' : S_2 \longrightarrow S_1$ *that are inverse to each other.*

A semilattice morphism is a monotonic function between the partially ordered sets $(S_1, \leqslant)$ and $(S_2, \leqslant)$. Indeed, suppose that $x, y \in S_1$ such that $x \leqslant y$, which is equivalent to $x = x \wedge y$. Since h is a morphism, we have $h(x) = h(x) \wedge h(y)$, so $h(x) \leqslant h(y)$. The converse is not true; a monotonic function between the posets $(S_1, \leqslant)$ and $(S_2, \leqslant)$ is not necessarily a semilattice morphism, as the next example shows.

Example 11.10 Let $(P, \{\star\})$ and $(Q, \{\star\})$ be the semilattices defined in Example 11.6. The function $f : P \longrightarrow Q$ given by $f(a) = x$, $f(b) = y$, and $f(c) = z$ is clearly monotonic, and it is even a bijection. However, it fails to be a semilattice morphism because $f(a \star b) = f(c) = z$, while $f(a) \star f(b) = x \star y = y \neq z$.

However, we have the following theorem.

Theorem 11.11 *Let* $\mathcal{S}_1 = (S_1, \{\wedge\})$ *and* $\mathcal{S}_2 = (S_2, \{\wedge\})$ *be two semilattices.* $\mathcal{S}_1$ *and* $\mathcal{S}_2$ *are isomorphic if and only if there exists a bijection* $h : S_1 \longrightarrow S_2$ *such that both* h *and* h^{-1} *are monotonic.*

Proof Suppose that $h : S_1 \longrightarrow S_2$ is a bijection such that both h and h^{-1} are monotonic functions, and let x and y be two elements of S_1. Since $x \wedge y \leqslant x$ and $x \wedge y \leqslant y$, we have $h(x \wedge y) \leqslant h(x)$ and $h(x \wedge y) \leqslant h(y)$, so $h(x \wedge y) \leqslant h(x) \wedge h(y)$. We further prove that $h(x \wedge y)$ is the infimum of $h(x)$ and $h(y)$.

Let $u \in S_2$ such that $u \leqslant h(x) \wedge h(y)$, so $u \leqslant h(x)$ and $u \leqslant h(y)$. Equivalently, we have $h^{-1}(u) \leqslant x$ and $h^{-1}(u) \leqslant y$, which implies $h^{-1}(u) \leqslant x \wedge y$. Therefore, $u \leqslant h(x \wedge y)$, which allows us to conclude that $h(x) \wedge h(y) = \inf\{h(x), h(y)\} = h(x \wedge y)$, so h is indeed a morphism. Similarly, one can prove that h^{-1} is also a morphism, so $\mathcal{S}_1$ and $\mathcal{S}_2$ are isomorphic.

Conversely, if $\mathcal{S}_1$ and $\mathcal{S}_2$ are isomorphic and $h : S_1 \longrightarrow S_2$ and $h' : S_2 \longrightarrow S_1$ are morphisms that are inverse to each other, then they are clearly inverse monotonic mapping.

A structure that combines the properties of join and meet semilattices is introduced next.

Definition 11.12 *A* lattice *is an algebra of type* $(2, 2)$, *that is, an algebra* $\mathcal{L} = (L, \{\wedge, \vee\})$ *such that* $\wedge$ *and* $\vee$ *are both idempotent, commutative, and associative operations and the equalities (known as* absorption laws*)*

$$x \vee (x \wedge y) = x, x \wedge (x \vee y) = x$$

are satisfied for every $x, y \in L$.

Observe that if $(L, \{\wedge, \vee\})$ is a lattice, then both $(L, \{\wedge\})$ and $(L, \{\vee\})$ are semilattices. Thus, by Theorem 11.3, both operations induce partial order relations on L.

Let us denote these operations temporarily by "$\leqslant$" and "$\leqslant'$", respectively. In other words, we have $x \leqslant y$ if $x = x \wedge y$ and $u \leqslant' v$ if $u = u \vee v$.

The absorption laws that link together the operations $\wedge$ and $\vee$ imply that the two partial orders are dual to each other. Indeed, suppose that $x \leqslant y$, that is, $x = x \wedge y$. Then, since $y \vee x = y \vee (y \wedge x) = y$, we have $y \leqslant' x$. We usually use the partial order $\leqslant$ on the lattice $(L, \{\wedge, \vee\})$.

If $(L, \{\wedge, \vee\})$ is a lattice, then for every finite, nonempty subset K of L, $\inf K$ and $\sup K$ exist, as it can be shown by induction on $n = |K|$, where $n \geqslant 1$ (see Exercise 2). Moreover, if $K = \{x_1, \ldots, x_n\}$, then

$$\inf K = x_1 \wedge x_2 \wedge \cdots \wedge x_n,$$
$$\sup K = x_1 \vee x_2 \vee \cdots \vee x_n.$$

Example 11.13 Let S be a set. The algebra $(\mathcal{P}(S), \{\cap, \cup\})$ is a lattice. Also, if $\mathcal{P}_f(S)$ is the set of all finite subsets of S, then $(\mathcal{P}_f(S), \{\cap, \cup\})$ is also a lattice.

Example 11.14 The posets M_5 and N_5 from Example 2.55 are both lattices. Indeed, the operations $\wedge$ and $\vee$ for the first poset is given by the following table.

$(M_5, \wedge)$	0	a	b	c	1	$(M_5, \vee)$	0	a	b	c	1
0	0	0	0	0	0	0	0	a	b	c	1
a	0	a	0	0	a	a	a	a	1	1	1
b	0	0	b	0	b	b	b	1	b	1	1
c	0	0	0	c	c	c	c	1	1	c	1
1	0	a	b	c	1	1	1	1	b	1	1

The similar operations for N_5 are given next.

$(N_5, \wedge)$	0	x	y	z	1	$(N_5, \vee)$	0	x	y	z	1
0	0	0	0	0	0	0	0	x	y	z	1
x	0	x	x	0	x	x	x	x	y	1	1
y	0	x	y	0	y	y	y	y	y	1	1
z	0	0	0	z	z	z	z	1	1	z	1
1	0	x	y	1	1	1	1	1	1	1	1

Example 11.15 The poset of partitions of a finite set $(PART(S), \leqslant)$ introduced in Example 2.3 is a lattice. Indeed, we saw in Sect. 2.5 that for every two partitions π, σ, both $\inf\{\pi, \sigma\}$ and $\sup\{\pi, \sigma\}$ exist.

Example 11.16 Consider the set $\mathbb{N} \times \mathbb{N}$ and the partial order $\preceq$ on this set defined by $(p, q) \preceq (m, n)$ if $p \leqslant m$ and $q \leqslant n$. Then $\inf\{(u, v), (x, y)\} = (\min\{u, x\}, \min\{v, y\})$ and $\sup\{(u, v), (x, y)\} = (\max\{u, x\}, \max\{v, y\})$.

Theorem 11.17 *Let $(L, \{\wedge, \vee\})$ be a lattice. If $x \leqslant y$ and $u \leqslant v$, then $x \wedge u \leqslant y \wedge v$ and $x \vee u \leqslant y \vee v$ (compatibility of the lattice operations with the partial order).*

Proof Note that $x \leqslant y$ is equivalent to $x = x \wedge y$ and to $y = x \vee y$. Similarly, $u \leqslant v$ is equivalent to $u = u \wedge v$ and to $v = u \vee v$. Therefore, we can write

$$(x \wedge u) \wedge (y \wedge v) = (x \wedge y) \wedge (u \wedge v) = x \wedge u,$$

so $x \wedge u \leqslant y \wedge v$. The proof of the second inequality is similar.

Let $(L, \{\wedge, \vee\})$ be a lattice. If the poset $(L, \leqslant)$ has the largest element 1, then we have $1 \wedge x = x \wedge 1 = x$ and $1 \vee x = x \vee 1 = x$. If the poset has the least element 0, then $0 \wedge x = x \wedge 0 = 0$ and $0 \vee x = x \vee 0 = x$. In other words, if a lattice has a largest element 1, then 1 is a unit with respect to the $\wedge$ operation; similarly, if the least element exists, then it plays the role of a unit with respect to $\vee$.

Let K and H be two finite subsets of L, where $(L, \{\wedge, \vee\})$ is a lattice. If $K \subseteq H$, then it is easy to see that $\sup K \leqslant \sup H$ and that $\inf K \geqslant \inf H$. Since $\emptyset \subseteq H$ for every set H, by choosing $H = \{x\}$ for some $x \in L$, it is clear that if a lattice has the least element 0 and the greatest element 1, then we can define $\sup \emptyset = 0$ and $\inf \emptyset = 1$.

Definition 11.18 *A lattice $(L, \{\wedge, \vee\})$ is* bounded *if the poset $(L, \leqslant)$ has the least element and the greatest element* 1.

Note that every finite subset of of a finite lattice (including the empty set) is bounded.

If $\mathfrak{L} = (L, \{\wedge, \vee\})$ is a finite lattice, then $\mathfrak{L}$ is bounded. Indeed, since both $\sup L$ and $\inf L$ exist, it follows that $\sup L$ is the greatest element and $\inf L$ is the least element of $\mathfrak{L}$, respectively.

Definition 11.19 *Let $\mathfrak{L}_1 = (L_1, \wedge, \vee)$ and $\mathfrak{L}_2 = (L_2, \wedge, \vee)$ be two lattices. A* morphism *h from $\mathfrak{L}_1$ to $\mathfrak{L}_2$ is a function $h : L_1 \longrightarrow L_2$ such that $h(x \wedge y) = h(x) \wedge h(y)$ and $h(x \vee y) = h(x) \vee h(y)$ for every $x, y \in L_1$.*

A lattice isomorphism *is a bijective lattice morphism.*

A counterpart of Theorem 11.11 characterizes isomorphic lattices.

Theorem 11.20 *Let $\mathfrak{L}_1 = (L_1, \{\wedge, \vee\})$ and $\mathfrak{L}_2 = (L_2, \{\wedge, \vee\})$ be two lattices. $\mathfrak{L}_1$ and $\mathfrak{L}_2$ are isomorphic if and only if there exists a bijection $h : L_1 \longrightarrow L_2$ such that both h and h^{-1} are monotonic.*

Proof The proof is similar to the proof of Theorem 11.11; we leave the argument to the reader as an exercise.

Definition 11.21 *Let $\mathfrak{L} = (L, \{\wedge, \vee\})$ be a lattice. A* sublattice *of $\mathfrak{L}$ is a subset K of L that is closed with respect to the lattice operations. In other words, for every $x, y \in K$, we have both $x \wedge y \in K$ and $x \vee y \in K$.*

Note that if K is a sublattice of $\mathcal{L}$, then the pair $\mathcal{K} = (K, \{\wedge, \vee\})$ is itself a lattice. We use the term "sublattice" to designate both the set K and the lattice $\mathcal{K}$ when there is no risk of confusion. For example, if K and K' are two sublattices of a lattice $\mathcal{L}$ and $f : K \longrightarrow K'$ is a morphism between the lattices $\mathcal{K} = (K, \{\wedge, \vee\})$ and $\mathcal{K}' = (K', \{\wedge, \vee\})$, we designate f as a morphism between K and K'.

Example 11.22 Let $\mathcal{L} = (L, \{\wedge, \vee\})$ be a lattice and let a and b be a pair of elements of L. The *interval* $[a, b]$ is the set

$$\{x \in L \mid a \leqslant x \leqslant b\}.$$

Clearly, an interval $[a, b]$ is nonempty if and only if $a \leqslant b$. Each such set is a sublattice. Indeed, if $[a, b] = \emptyset$, then $\emptyset$ is clearly a sublattice.

Suppose that $x, y \in [a, b]$; that is, $a \leqslant x \leqslant b$ and $a \leqslant y \leqslant b$. Due to the compatibility of the lattice operations with the partial order, we obtain immediately $a \leqslant x \wedge y \leqslant b$ and $a \leqslant x \vee y \leqslant b$, so $[a, b]$ is a sublattice in all cases.

Example 11.23 Let $[a, b]$ be a nonempty interval of a lattice $\mathcal{L} = (L, \{\wedge, \vee\})$. The function $h : L \longrightarrow [a, b]$ defined by $h(x) = (x \vee a) \wedge b$ is a surjective morphism between $\mathcal{L}$ and the lattice $([a, b], \{\wedge, \vee\})$ because

$$\begin{aligned} h(x \wedge y) &= ((x \wedge y) \vee a) \wedge b = ((x \vee a) \wedge (y \vee a)) \wedge b \\ &= ((x \vee a) \wedge b) \wedge ((y \vee a)) \wedge b)) = h(x) \wedge h(y) \end{aligned}$$

and

$$\begin{aligned} h(x \vee y) &= ((x \vee y) \vee a) \wedge b = ((x \vee a) \vee (y \vee a)) \wedge b \\ &= ((x \vee a) \wedge b) \vee ((y \vee a)) \wedge b)) = h(x) \vee h(y) \end{aligned}$$

for $x, y \in B$.

The elements a and b are invariant under h. Indeed, we have $h(a) = a$ because $h(a) = (a \vee a) \wedge b = a \wedge b = a$ and $h(b) = (b \vee a) \wedge b = b$ by absorption. Moreover, this property is shared by every member of the interval $[a, b]$ because we can write

$$\begin{aligned} h(h(x)) &= h((x \vee a) \wedge b) = h(x \vee a) \wedge h(b) \\ &= (h(x) \vee h(a)) \wedge h(b) = (h(x) \vee a) \wedge b \\ &= (((x \vee a) \wedge b) \vee a) \wedge b \\ &= (x \vee a) \wedge (b \vee a) \wedge b \\ &= (x \vee a) \wedge b = h(x) \end{aligned}$$

for $x \in B$. We refer to h as the *projection* of L on the interval $[a, b]$.

11.3 Special Classes of Lattices

Let $(L, \{\wedge, \vee\})$ be a lattice and let u, v, w be three members of L such that $u \leqslant w$. Since $u \leqslant u \vee v$ and $u \leqslant w$, it follows that

$$u \leqslant (u \vee v) \wedge w. \tag{11.2}$$

Starting from the inequalities $v \wedge w \leqslant v \leqslant u \vee v$ and $v \wedge w \leqslant w$, we have also

$$v \wedge w \leqslant (u \vee v) \wedge w. \tag{11.3}$$

Combining Inequalities (11.2) and (11.3) yields the inequality

$$u \vee (v \wedge w) \leqslant (u \vee v) \wedge w, \tag{11.4}$$

which is satisfied whenever $u \leqslant w$. This inequality is known as the *submodular inequality*.

An important class of lattices is obtained when we replace the submodular inequality (satisfied by every lattice) with an equality, as follows.

Definition 11.24 *A lattice $(L, \{\wedge, \vee\})$ is* modular *if, for every $u, v, w \in L$, $u \leqslant w$ implies*

$$u \vee (v \wedge w) = (u \vee v) \wedge w. \tag{11.5}$$

Observe that if $u = w$, Equality (11.5) holds in every lattice. Therefore, it is sufficient to require that $u < w$ implies $u \vee (v \wedge w) = (u \vee v) \wedge w$ for all $u, v, w \in L$ to ensure modularity.

Example 11.25 The lattice M_5 introduced in Example 11.14 is modular. Indeed, suppose that $x < z$. If $u = 0$ and $w = 1$, it is easy to see that Equality (11.5) is verified. Suppose, for example, that $u = a$ and $w = 1$. Then, $u \vee (v \wedge w) = a \vee (v \wedge 1) = a \vee v$ and $(u \vee v) \wedge w = (a \vee v) \wedge 1 = a \vee v$ for every $v \in \{0, 1, a, b, c\}$. The remaining cases can be analyzed similarly.

On the other hand, the lattice N_5 introduced in the same example is not modular because we have $x < y$, $x \vee (z \wedge y) = x \vee 0 = x$, and $(x \vee z) \wedge y = 1 \wedge y = y \neq x$.

The special role played by N_5 is described next.

Theorem 11.26 *A lattice $\mathcal{L} = (L, \{\wedge, \vee\})$ is modular if and only if it does not contain a sublattice isomorphic to N_5.*

Proof Suppose $\mathcal{L}$ contains a sublattice $K = \{t_0, t_1, t_2, t_3, t_4\}$ isomorphic to N_5, and let $f : K \longrightarrow N_5$ be an isomorphism. Suppose that $f(t_0) = 0$, $f(t_1) = x$, $f(t_2) = y$, $f(t_3) = z$, and $f(t_4) = 1$. Also, let $g : N_5 \longrightarrow K$ be the inverse isomorphism.

Since $x < y$, $g(x) = t_1$, and $g(y) = t_2$, we have $t_1 < t_2$. On the other hand, $t_1 \vee (t_3 \wedge t_2) = g(x) \vee (g(z) \wedge g(y)) = g(x \vee (z \wedge y)) = g(x) = t_1$ and $(t_1 \vee t_3) \wedge t_2 =$

$(g(x) \vee g(z)) \wedge g(y) = g((x \vee z) \wedge y) = g(y) = t_2 \neq t_1$, which shows that $\mathcal{L}$ is not modular.

Conversely, suppose that $\mathcal{L} = (L, \{\wedge, \vee\})$ is not modular. Then, there exist three members of L – u, v, w – such that $u < w$ and $u \vee (v \wedge w) < (u \vee v) \wedge w$ because $\mathcal{L}$ still satisfies the submodular inequality. Observe that the elements $t_0, \ldots, t_4$ given by:

$$t_0 = v \wedge w, t_1 = u \vee (v \wedge w), t_2 = (u \vee v) \wedge w, t_3 = v, t_4 = (u \vee v) \wedge w$$

form a sublattice isomorphic to N_5.

The relationship between intervals of the form $[a \wedge b, a]$ and $[b, a \vee b]$ in modular lattices is shown next.

Theorem 11.27 *Let $\mathcal{L} = (L, \{\wedge, \vee\})$ be a modular lattice and let a and b be two elements. The mappings $\phi : [a \wedge b, a] \longrightarrow [b, a \vee b]$ and $\psi : [b, a \vee b] \longrightarrow [a \wedge b, a]$ defined by $\phi(x) = x \vee b$ and $\psi(y) = y \wedge a$ for $x \in [a \wedge b, b]$ and $y \in [a, a \vee b]$ are inverse monotonic mappings between the sublattices $[a \wedge b, b]$ and $[a, a \vee b]$.*

Proof Note that, for $a \wedge b \leqslant x \leqslant a$, we have $(x \vee b) \wedge a = x \vee (b \wedge a)$ because $\mathcal{L}$ is modular and $x \vee (b \wedge a) = x$ because $a \wedge b \leqslant x$. Thus, $\psi(\phi(x)) = x$ for every $x \in [a \wedge b, a]$.

Similarly, one can prove that $\phi(\psi(y)) = y$ for every $y \in [b, a \vee b]$, which shows that ϕ and ψ are inverse to each other. The monotonicity is immediate.

Corollary 11.28 *Let $\mathcal{L} = (L, \{\wedge, \vee\})$ be a modular lattice and let a and b be two elements such that a and b cover $a \wedge b$. Then $a \vee b$ covers both a and b.*

Proof Since a covers $a \wedge b$, the interval $[a \wedge b, a]$ consists of two elements. Therefore, by Theorem 11.27, the interval $[b, a \vee b]$ also consists of two elements, so $a \vee b$ covers b. A similar argument shows that $a \vee b$ covers a (starting from the fact that b covers $a \wedge b$).

The property of modular lattices described in Corollary 11.28 allows us to introduce a generalization of the class of modular lattices.

Definition 11.29 *A lattice $\mathcal{L} = (L, \{\wedge, \vee\})$ is* semimodular, *if for every $a, b \in L$ such that both cover $a \wedge b$, the same elements are covered by $a \vee b$.*

Clearly, every modular lattice is semimodular. The converse is not true, as the next example shows.

Example 11.30 Let $(PART(S), \{\wedge, \vee\})$ the lattice of partitions of the set $S = \{1, 2, 3, 4\}$, whose Hasse diagram is shown in Fig. 2.6. By Theorem 2.75, a partition σ covers the partition π if and only if there exists a block C of σ that is the union of two blocks B and B' of π, and every block of σ that is distinct of C is a block of π. Thus, it is easy to verify that this lattice is indeed semimodular.

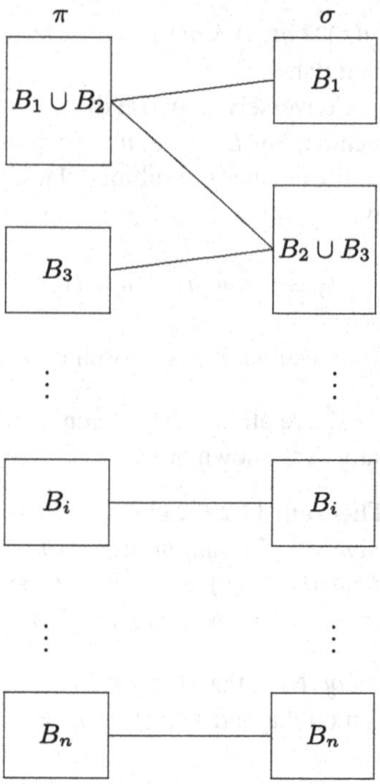

Fig. 11.2 The graph $G_{\pi,\sigma}$ of π and σ

To show that the lattice $(PART(\{1, 2, 3, 4\}), \{\wedge, \vee\})$ is not modular, consider the partitions $\pi_1 = \{12, 3, 4\}$, $\pi_2 = \{123, 4\}$, and $\pi_3 = \{14, 2, 3\}$. It is easy to see that the sublattice $\alpha_S, \pi_1, \pi_2, \pi_3, \omega_S$ is isomorphic to N_5 and therefore the lattice is not modular.

A more general statement follows.

Theorem 11.31 *The partition lattice* $(PART(S), \{\wedge, \vee\})$ *of a nonempty set is semimodular.*

Proof Let $\pi, \sigma \in PART(S)$ be two partitions such that both cover $\pi \wedge \sigma$. By Theorem 2.75, both π and σ are obtained from $\pi \wedge \sigma$ by fusing two blocks of this partition. If $\pi \wedge \sigma = \{B_1, \ldots, B_n\}$, then there exist three blocks of $\pi \wedge \sigma$, B_p, B_q, B_r, such that π is obtained by fusing B_p and B_q, and σ is obtained by fusing B_q and B_r. To simplify the argument we can assume without loss of generality that $p = 1, q = 2$, and $r = 3$.

The graph $G_{\pi,\sigma}$ of the partitions π and σ is given in Fig. 11.2. The blocks of the partition $\pi \vee \sigma$ correspond to the connected components of the graph $G_{\pi,\sigma}$, so $\pi \vee \sigma = \{B_1 \cup B_2 \cup B_3, \ldots, B_n\}$, which covers both π and σ. Thus, $(PART(S), \{\wedge, \vee\})$ is semimodular.

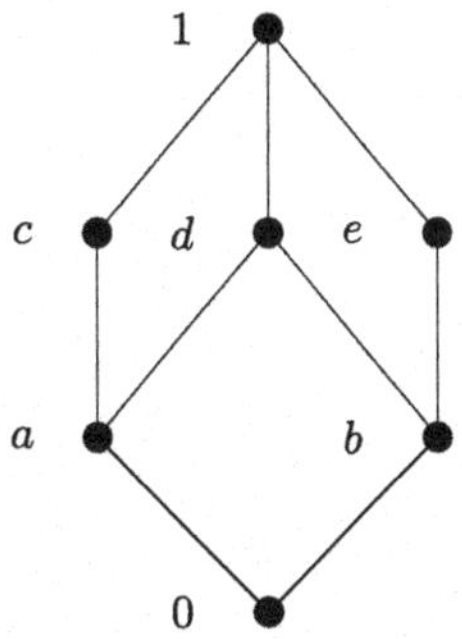

Fig. 11.3 Hasse diagram of lattice $\mathcal{L} = (\{0, a, b, c, d, e, 1\}, \{\wedge, \vee\})$

Example 11.32 Let $\mathcal{L} = (\{0, a, b, c, d, e, 1\}, \{\wedge, \vee\})$ be the lattice whose Hasse diagram is shown in Fig. 11.3. This is a semimodular lattice that is not modular. Indeed, we have $a \leqslant c$ but $(a \vee e) \wedge c = c$, while $a \vee (e \wedge c) = a \neq c$.

Let $\mathcal{L} = (L, \{\wedge, \vee\})$ be a lattice and let x, y, z be three elements of L. We have the inequalities

$$x \wedge (y \vee z) \geq (x \wedge y) \vee (x \wedge z), \tag{11.6}$$

$$x \vee (y \wedge z) \leq (x \vee y) \wedge (x \vee z). \tag{11.7}$$

Indeed, note that $x \geqslant x \wedge y$ and $x \geqslant x \wedge z$, so $x \geqslant (x \wedge y) \vee (x \wedge z)$. Also, $(y \vee z) \geqslant (x \wedge y)$ and $(y \vee z) \geqslant (x \wedge z)$, which implies $(y \vee z) \geqslant (x \wedge y) \vee (x \wedge z)$. Therefore, we conclude that $x \wedge (y \vee z) \geqslant (x \wedge y) \vee (x \wedge z)$. The argument for the second inequality is similar. We refer to Inequalities (11.6) and (11.7) as the *subdistributive inequalities.*

The existence of subdistributive inequalities satisfied by every lattice serves as an introduction to a new class of lattices.

Definition 11.33 *A lattice* $(L, \{\wedge, \vee\})$ *is* distributive *if*

$$x \wedge (y \vee z) = (x \wedge y) \vee (x \wedge z),$$

$$x \vee (y \wedge z) = (x \vee y) \wedge (x \vee z),$$

for $x, y, z \in L$.

It suffices that only one of the equalities of Theorem 11.33 be satisfied to ensure distributivity. Suppose, for example, that $x \wedge (y \vee z) = (x \wedge y) \vee (x \wedge z)$ for $x, y, z \in L$. We have

$$\begin{aligned} x \vee (y \wedge z) &= (x \vee (x \wedge z)) \vee (y \wedge z) \\ &\quad \text{(by absorption)} \\ &= x \vee ((x \wedge z) \vee (y \wedge z)) \end{aligned}$$

$$
\begin{aligned}
&\text{(by the associativity of } \vee)\\
&= x \vee ((z \wedge x) \vee (z \wedge y))\\
&\text{(by the commutativity of } \wedge)\\
&= x \vee (z \wedge (x \vee y))\\
&\text{(by the first distributivity equality)}\\
&= x \vee ((x \vee y) \wedge z)\\
&\text{(by the commutativity of } \wedge)\\
&= (x \wedge (x \vee y)) \vee ((x \vee y) \wedge z)\\
&\text{(by absorption)}\\
&= ((x \vee y) \wedge x) \vee ((x \vee y) \wedge z)\\
&\text{(by the commutativity of } \wedge)\\
&= (x \vee y) \wedge (x \vee z),\\
&\text{(by the first distributivity equality),}
\end{aligned}
$$

which is the second distributivity law. In a similar manner, one could show that the second distributivity law implies the first law.

Theorem 11.34 *Every distributive lattice is modular.*

Proof Let $\mathfrak{L} = (L, \{\wedge, \vee\})$ be a distributive lattice. Suppose that $u \leqslant w$. Applying the distributivity, we can write

$$
\begin{aligned}
u \vee (v \wedge w) &= (u \vee v) \wedge (u \vee w)\\
&= (u \vee v) \wedge w,\\
&\quad \text{(because } u \leqslant w),
\end{aligned}
$$

which shows that $\mathfrak{L}$ is modular.

We saw that the lattice N_5 is not modular and therefore is not distributive. The lattice M_5 is modular (as we have shown in Example 11.25) but not distributive. Indeed, note that $a \vee (b \wedge c) = a \vee 0 = a$ and $(a \vee b) \wedge (a \vee c) = 1 \wedge 1 \neq 0$.

It is easy to see that every sublattice of a distributive lattice is also distributive. Thus, a distributive lattice may not contain sublattices isomorphic to M_5 or N_5. This allows the formulation of a statement for distributive lattices that is similar to Theorem 11.26.

Theorem 11.35 *A lattice $\mathfrak{L} = (L, \{\wedge, \vee\})$ is distributive if and only if it does not contain a sublattice isomorphic to M_5 or N_5.*

Proof The necessity of this condition is clear, so we need to prove only that it is sufficient.

Let $\mathcal{L}$ be a lattice that is not distributive. Then, $\mathcal{L}$ may or may not be modular. If $\mathcal{L}$ is not modular, then by Theorem 11.26 it contains a sublattice isomorphic to N_5. Therefore, we need to consider only the case where $\mathcal{L}$ is modular but not distributive. We show in this case that $\mathcal{L}$ contains a sublattice that is isomorphic to M_5.

The nondistributivity of $\mathcal{L}$ implies the existence of $x, y, z \in L$ such that

$$x \wedge (y \vee z) > (x \wedge y) \vee (x \wedge z), \tag{11.8}$$

$$x \vee (y \wedge z) < (x \vee y) \wedge (x \vee z). \tag{11.9}$$

Let u and v be defined by

$$u = (x \vee y) \wedge (y \vee z) \wedge (z \vee x)$$
$$v = (x \wedge y) \vee (y \wedge z) \vee (z \wedge x).$$

We first prove that $v < u$.

Note that

$$\begin{aligned}
&x \wedge u\\
&= x \wedge ((x \vee y) \wedge (y \vee z) \wedge (z \vee x)) \text{ (by the definition of } u)\\
&= (x \wedge (x \vee y)) \wedge (y \vee z) \wedge (z \vee x) \text{ (by the associativity of } \wedge)\\
&= x \wedge (y \vee z) \wedge (z \vee x) \text{ (by absorption)}\\
&= x \wedge (z \vee x) \wedge (y \vee z) \text{ (by associativity and commutativity of } \wedge)\\
&= x \wedge (x \vee z) \wedge (y \vee z) \text{ (by commutativity of } \vee)\\
&= x \wedge (y \vee z) \text{ (by absorption).}
\end{aligned}$$

Also,

$$\begin{aligned}
&x \vee v\\
&= x \vee ((x \wedge y) \vee (y \wedge z) \vee (z \wedge x)) \quad \text{(by the definition of } v)\\
&= x \vee ((x \wedge y) \vee (x \wedge z) \vee (y \wedge z)) \quad \text{(by associativity and commutativity)}\\
&= ((x \wedge y) \vee (x \wedge z)) \vee (x \wedge (y \wedge z)\\
&\quad \text{(by modularity since} (x \wedge y) \vee (x \wedge z) \leqslant x)\\
&= (x \wedge y) \vee (x \wedge z) \text{ (because } x \wedge y \wedge z \leqslant (x \wedge y) \vee (x \wedge z)).
\end{aligned}$$

Thus, by Inequality (11.8), we have $x \vee v < x \vee u$, which clearly implies $v < u$.

Consider now the projections x_1, y_1, z_1 of x, y, z on the interval $[v, u]$ given by

$$x_1 = (x \wedge u) \vee v, \; y_1 = (y \wedge u) \vee v, \; z_1 = (z \wedge u) \vee v.$$

It is clear that $v \leqslant x_1, y_1, z_1 \leqslant u$. We prove that $\{u, x_1, y_1, z_1, v\}$ is a sublattice isomorphic to M_5 by showing that

$$x_1 \wedge y_1 = y_1 \wedge z_1 = z_1 \wedge x_1 = v \text{ and } x_1 \vee y_1 = y_1 \vee z_1 = z_1 \vee x_1 = u.$$

We have

$$\begin{aligned}
& x_1 \wedge y_1 \\
& = ((x \wedge u) \vee v) \wedge ((y \wedge u) \vee v) \text{ (by the definition of } x_1 \text{ and } y_1) \\
& = ((x \wedge u) \wedge ((y \wedge u) \vee v)) \vee v \text{ (by modularity since } v \leqslant (y \wedge u) \vee v) \\
& = ((x \wedge u) \wedge ((y \vee v) \wedge u)) \vee v \text{ (by modularity since } v \leqslant u) \\
& = ((x \wedge u) \wedge u \wedge (y \vee v)) \vee v \text{ (by associativity and commutativity of } \wedge) \\
& = ((x \wedge u) \wedge (y \vee v)) \vee v \text{ (by absorption)} \\
& = (x \wedge (y \vee z) \wedge (y \vee (x \wedge z))) \vee v \text{ (because } x \wedge u = x \wedge (y \vee z) \text{ and} \\
& \quad y \vee v = y \vee (x \wedge z)) \\
& = (x \wedge (y \vee ((y \vee z) \wedge (x \wedge z)))) \vee v \text{ (by modularity since } y \leqslant y \vee z) \\
& = (x \wedge (y \vee (x \wedge z))) \vee v \text{ (since } x \wedge z \leqslant y \vee z) \\
& = (x \wedge z) \vee (y \wedge z) \vee v \text{ (by modularity since } x \leqslant x \wedge z) \\
& = v \text{ (due to the definition of } v).
\end{aligned}$$

Similar arguments can be used to prove the remaining equalities.

Definition 11.36 *Let $\mathcal{L} = (L, \{\wedge, \vee\})$ be a bounded lattice that has* 0 *as its least element and* 1 *as its largest element.*

The elements x and y are complementary *if $x \wedge y = 0$ and $x \vee y = 1$.*

If x and y are complementary we say that one element is the *complement* of the other. Lattices in which every element has a complement are referred to as *complemented lattices*.

Example 11.37 The lattice N_5 is a complemented lattice. Indeed, x and z are complementary elements and so are y and z. The lattice M_5 is also complemented.

Example 11.38 Let S be a set and let $(\mathcal{P}(S), \cap, \cup)$ be the bounded lattice of its subsets having $\emptyset$ as its first element and S as its last element. Unlike the lattices mentioned in Example 11.37, a set $X \in \mathcal{P}(S)$ has a unique complement $S - X$.

Example 11.39 Let $(\mathbb{N} \cup \{\infty\}, \leqslant)$ be the infinite chain of natural numbers extended by ∞. If $m, n \in \mathbb{N} \cup \{\infty\}$, then $m \wedge n = \min\{m, n\}$ and $m \vee n = \max\{m, n\}$. Clearly, this is a bounded lattice and no two of elements except 0 and ∞ are complementary.

Theorem 11.40 *Let $\mathcal{L} = (L, \{\wedge, \vee\})$ be a bounded, distributive lattice. For every element x, there exists at most one complement.*

Proof Let $x \in L$ and suppose that both r and s are complements of x; that is, $x \wedge r = 0$, $x \vee r = 1$, and $x \wedge s = 0$, $x \vee s = 1$. We can write

$$r = r \wedge 1 = r \wedge (x \vee s) = (r \wedge x) \vee (r \wedge s) = 0 \vee (r \wedge s) = r \wedge s,$$

which implies $r \leqslant s$. Similarly, starting with s, we obtain

$$s = s \wedge 1 = s \wedge (r \vee x) = (s \wedge r) \vee (s \wedge x) = (s \wedge r) \vee 0 = s \wedge r,$$

which implies $s \leqslant r$. Consequently, $s = r$.

Definition 11.41 *A lattice $\mathcal{L} = (L, \{\wedge, \vee\})$ is* relatively complemented *if each interval $[x, y]$ of $\mathcal{L}$ is complemented.*

Example 11.42 We saw that the lattice N_5 is complemented (see Example 11.37). However, this lattice is not relatively complemented.

11.4 Complete Lattices

Definition 11.43 *A* complete lattice *is a poset $(L, \leqslant)$ such that for every subset U of L both* $\sup U$ *and* $\inf U$ *exist.*

Note that if U and V are two subsets of a complete lattice and $U \subseteq V$, then $\sup U \leqslant \sup V$ and $\inf V \leqslant \inf U$. Therefore, for every subset T of L, we have

$$\sup \emptyset \leqslant \sup T \leqslant \sup L \text{ and } \inf L \leqslant \inf T \leqslant \inf \emptyset.$$

If T is a singleton (that is, $T = \{t\}$), then these inequalities amount to

$$\sup \emptyset \leqslant t \leqslant \sup L \text{ and } \inf L \leqslant t \leqslant \inf \emptyset$$

for every $t \in L$. This means that a complete lattice has a least element $0 = \sup \emptyset = \inf L$ and a greatest element $1 = \inf \emptyset = \sup L$.

For a subset U of the complete lattice, we denote $\sup U$ by $\bigvee U$ and $\inf U$ by $\bigwedge U$.

Obviously, every complete lattice is also a lattice since $x \vee y$ and $x \wedge y$ are $\sup\{x, y\}$ and $\inf\{x, y\}$, respectively.

The associative properties of the usual lattices can be extended to complete lattices as follows. Let $(L, \leqslant)$ be a complete lattice and let $\mathcal{C} = \{C_i \mid C_i \subseteq L \text{ for } i \in I\}$ be a collection of subsets of L. Then,

$$\bigvee_{i \in I} \bigvee C_i = \bigvee \bigcup \mathcal{C}, \text{ and } \bigwedge_{i \in I} \bigwedge C_i = \bigwedge \bigcup \mathcal{C}.$$

Theorem 11.44 *Let $(L, \leqslant)$ be a poset such that* $\sup U$ *exists for every subset U of L. Then $(L, \leqslant)$ is a complete lattice.*

Proof It is sufficient to prove that $\inf U$ exists for each subset U of the lattice. By hypothesis, the set U^i of lower bounds of U has a supremum $x = \sup U_i$. Every element of U is an upper bound of U^i, which means that $x \leqslant u$, which implies that x is a lower bound for U. Thus, $x \in U^i \cap (U^i)^s$, which means that $x = \inf U$.

Theorem 11.45 *If $(L, \leqslant)$ is a poset such that $\inf U$ exists for every set U, then $(L, \leqslant)$ is a complete lattice.*

Proof This statement follows by duality from Theorem 11.44.

Example 11.46 Let S be a set. The poset of its subsets $(\mathcal{P}(S), \subseteq)$ is a complete lattice because, for any collection $\mathcal{C}$ of subsets of S, $\sup \mathcal{C} = \bigcup \mathcal{C}$ and $\inf \mathcal{C} = \bigcap \mathcal{C}$.

Example 11.47 Let $\mathcal{C}$ be a closure system on a set S and let $\mathbf{K}$ be the corresponding closure operator. Then, $(\mathcal{C}, \subseteq)$ is a complete lattice because $\inf \mathcal{D} = \bigcap \mathcal{D}$ and $\sup \mathcal{D} = \mathbf{K}\left(\bigcup \mathcal{D}\right)$ for any subcollection $\mathcal{D}$ of $\mathcal{C}$.

It is clear that $\inf \mathcal{D}$ exists and equals $\bigcap \mathcal{D}$ for any subcollection $\mathcal{D}$ of $\mathcal{C}$. We show that $\mathbf{K}\left(\bigcup \mathcal{D}\right)$ equals $\sup \mathcal{D}$. It is clear that $D \subseteq \mathbf{K}\left(\bigcup \mathcal{D}\right)$. Suppose now that E is a subset of $\mathcal{C}$ that is an upper bound for $\mathcal{D}$, that is, $D \subseteq E$ for every $D \in \mathcal{D}$. We have $\bigcup \mathcal{D} \subseteq E$, so $\mathbf{K}\left(\bigcup \mathcal{D}\right) \subseteq \mathbf{K}(E) = E$ because $E \in \mathcal{C}$. Therefore, $\mathbf{K}\left(\bigcup \mathcal{D}\right)$ is the least upper bound of $\mathcal{D}$.

The notion of a lattice morphism is extended to complete lattices.

Definition 11.48 *Let $(L_1, \leqslant)$ and $(L_2, \leqslant)$ be two complete lattices. A function $f : L_1 \longrightarrow L_2$ is a* complete lattice morphism *if $f\left(\bigvee U\right) = \bigvee f(U)$ and $f\left(\bigwedge U\right) = \bigwedge f(U)$ for every subset U of L_1.*

If f is a bijection such that both f and f^{-1} are complete lattice morphisms, then we say that f is a complete lattice isomorphism.

Theorem 11.49 *Every complete lattice is isomorphic to the lattice of closed sets of a closure system.*

Proof Let $(L, \leqslant)$ be a complete lattice and let $I_x = \{t \in L \mid t \leq x\}$ for $x \in L$. We claim that $\mathcal{I} = \{I_x \mid x \in L\}$ is a closure system on L.

Indeed, note that I_1 (where 1 is the largest element of L) coincides with L, so $L \in \mathcal{I}$.

Now let $\{I_x \mid x \in M\}$ be an arbitrary family of sets in $\mathcal{I}$, where M is a subset of L. Note that $\bigcap \{I_x \mid x \in M\} = I_y$, where $y = \inf M$. Thus, $\mathcal{I}$ is a closure system.

It is easy to verify that $f : L \longrightarrow \mathcal{I}$ given by $f(x) = I_x$ is a complete lattice isomorphism.

Definition 11.50 *Let $(S, \leqslant)$ and $(T, \leqslant)$ be two posets. A* Galois connection *between S and T is a pair of mappings (ϕ, ψ), where $\phi : S \longrightarrow T$ and $\psi : T \longrightarrow S$ that satisfy the following conditions.*

(i) *If $s_1 \leqslant s_2$, then $\phi(s_2) \leqslant \phi(s_1)$ for every $s_1, s_2 \in S$.*
(ii) *If $t_1 \leqslant t_2$, then $\phi(t_2) \leqslant \phi(t_1)$ for every $t_1, t_2 \in T$.*
(iii) *$s \leqslant \psi(\phi(s))$ and $t \leqslant \phi(\psi(t))$ for $s \in S$ and $t \in T$.*

Example 11.51 Let X and Y be two sets and let ρ be a relation, $\rho \subseteq X \times Y$. Define $\phi_\rho : \mathcal{P}(X) \longrightarrow \mathcal{P}(Y)$ and $\psi_\rho : \mathcal{P}(Y) \longrightarrow \mathcal{P}(X)$ by

$$\phi_\rho(U) = \{y \in Y \mid (x, y) \in \rho \text{ for all } x \in U\},$$
$$\psi_\rho(V) = \{x \in X \mid (x, y) \in \rho \text{ for all } y \in V\}$$

for $U \in \mathcal{P}(X)$ and $V \in \mathcal{P}(Y)$.

The pair (ϕ_ρ, ψ_ρ) is a Galois connection between the posets $(\mathcal{P}(X), \subseteq)$ and $(\mathcal{P}(Y), \subseteq)$. It is immediate to verify that the first two conditions of Definition 11.50 are satisfied. We discuss here only the third condition of the definition.

To prove that $U \subseteq \psi_\rho(\phi_\rho(U))$, let $u \in U$. We need to show that $(u, y) \in \rho$ for every $y \in \psi_\rho(U)$. By the definition of ψ_ρ, if $y \in \psi_\rho(U)$, we have indeed $(u, y) \in \rho$. The proof of the second inclusion of the third part of the definition is similar.

The pair (ϕ_ρ, ψ_ρ) is referred to as the *polarity generated by the relation* ρ.

Theorem 11.52 *Let $(S, \leqslant)$ and $(T, \leqslant)$ be two posets. A pair of mappings (ϕ, ψ), where $\phi : S \longrightarrow T$ and $\psi : T \longrightarrow S$, is a Galois connection between $(S, \leqslant)$ and $(T, \leqslant)$ if and only if $s \leqslant \psi(t)$ is equivalent to $t \leqslant \phi(s)$.*

Proof Suppose that (ϕ, ψ) is a pair of mappings such that $s \leqslant \psi(t)$ is equivalent to $t \leqslant \phi(s)$. Choosing $t = \phi(s)$, it is clear that $t \leqslant \phi(s)$, so $s \leqslant \psi(t) = \psi(\phi(s))$. Similarly, we can show that $t \leqslant \phi(\psi(t))$, so the pair (ϕ, ψ) satisfies the third condition of Definition 11.50.

Let $s_1, s_2 \in S$ such that $s_1 \leqslant s_2$. Since $s_2 \leqslant \psi(\phi(s_2))$, we have $s_1 \leqslant \psi(\phi(s_2))$, which implies $\phi(s_2) \leqslant \phi(s_1)$. A similar argument can be used to prove that $t_1 \leqslant t_2$ implies $\psi(t_2) \leqslant \psi(t_1)$, so (ϕ, ψ) satisfies the remaining conditions of the definition, and therefore is a Galois connection.

Conversely, let (ϕ, ψ) be a Galois connection. If $s \leqslant \psi(t)$, then $\phi(\psi(t)) \leqslant \phi(s)$. Since $t \leqslant \phi(\psi(t))$, we have $t \leqslant \phi(s)$. The reverse implication can be shown in a similar manner.

The notion of a closure operator, which was discussed in Sect. 1.8, can be generalized to partially ordered sets.

Definition 11.53 *Let $(L, \leqslant)$ be a poset. A mapping $\kappa : L \longrightarrow L$ is a* closure operator *on L if it satisfies the following conditions:*

(i) $u \leqslant \kappa(u)$ (expansiveness),
(ii) $u \leqslant v$ *implies* $\kappa(u) \leqslant \kappa(v)$ (monotonicity), *and*
(iii) $\kappa(\kappa(u)) = \kappa(u)$ (idempotency)

for $u, v \in L$.

Example 11.54 Let $(S, \leqslant)$ and $(T, \leqslant)$ be two posets, and suppose that (ϕ, ψ) is a Galois connection between these posets. Then, $\psi\phi$ is a closure on S and $\phi\psi$ is a closure on T.

By the third part of Definition 11.50, we have $s \leqslant \psi(\phi(s))$, so $\psi\phi$ is expansive. Suppose that $s_1 \leqslant s_2$. This implies $\phi(s_2) \leqslant \phi(s_1)$, which in turn implies $\psi(\phi(s_1)) \leqslant \psi(\phi(s_2))$. Thus, $\psi\phi$ is monotonic.

In exactly the same manner, we can prove that $t \leqslant \phi(\psi(t))$ and that $\phi\psi$ is monotonic.

Since $s \leqslant \psi(\phi(s))$, we have $\phi(\psi(\phi(s))) \leqslant \phi(s)$. On the other hand, choosing $t = \phi(s)$, we have $\phi(s) \leqslant \phi(\psi(\phi(s)))$, so $\phi(s) = \phi(\psi(\phi(s)))$ for every $s \in S$. A similar argument shows that $\psi(t) = \psi(\phi(\psi(t)))$. Therefore we obtain $\psi(\phi(s)) = \psi(\phi(\psi(\phi(s))))$ for every $s \in S$ and $\phi(\psi(t)) = \phi(\psi(\phi(\psi(t))))$, which proves that $\phi\psi$ and $\psi\phi$ are idempotent.

Lemma 11.55 *Let $(L, \leqslant)$ be a complete lattice and let $\kappa : L \longrightarrow L$ be a closure operator. Define the family of κ-closed elements $Q_\kappa = \{x \in L \mid x = \kappa(x)\}$. Then, $1 \in Q_\kappa$, and for each subset D of Q_κ, $\bigwedge D \in Q_\kappa$.*

Proof Since $1 \leqslant \kappa(1) \leqslant 1$, we have $1 \in Q_\kappa$.

Let $D = \{u_i \mid i \in I\}$ be a collection of elements of L such that $u_i = \kappa(u_i)$ for $i \in I$. Since $\bigwedge D \leqslant u_i$, we have $\kappa(\bigwedge D) \leqslant \kappa(u_i) = u_i$ for every $i \in I$. Therefore, $\kappa(\bigwedge D) \leqslant \bigwedge D$, which implies $\kappa(\bigwedge D) = \bigwedge D$. Thus, $\bigwedge D \in Q_\kappa$.

Theorem 11.56 *Let $(L, \leqslant)$ be a complete lattice and let κ be a closure operator on L. Then, $(Q_\kappa, \leqslant)$ is a complete lattice.*

Proof This statement follows from Lemma 11.55 and from Theorem 11.45.

If (ϕ, ψ) is a Galois connection between the posets $(S, \leqslant)$ and $(T, \leqslant)$, then each of the mappings ϕ, ψ is an *adjunct* of the other. The next theorem characterizes those mappings between posets that have an adjunct mapping.

Theorem 11.57 *Let $(S, \leqslant)$ and $(T, \leqslant)$ be two posets and let $\phi : S \longrightarrow T$ be a mapping. There exists a mapping $\psi : T \longrightarrow S$ such that (ϕ, ψ) is a Galois connection between $(S, \leqslant)$ and $(T, \leqslant)$ if and only if for every $t \in T$ there exists $z \in S$ such that $\phi^{-1}(\{v \in T \mid v \leqslant t\}) = \{u \in S \mid u \leqslant z\}$.*

Proof Suppose that the condition of the theorem is satisfied by ϕ. Given $t \in T$, the element $z \in S$ is unique because the equality $\{u \in S \mid u \leqslant z\} = \{u \in S \mid u \leqslant z'\}$ implies $z = z'$. Define the mapping $\psi : T \longrightarrow S$ by $\psi(t) = z$, where z is the element of S whose existence is stipulated by the theorem. Note that $s \leqslant \psi(t)$ is equivalent to $t \leq \phi(s)$, which means that (ϕ, ψ) is a Galois connection according to Theorem 11.52.

The proof of the necessity of the condition of the theorem is immediate.

11.5 Boolean Algebras and Boolean Functions

If $\mathcal{L} = (L, \{\wedge, \vee\})$ is a bounded distributive lattice that is complemented then, by Theorem 11.40, there is a mapping $h : L \longrightarrow L$ such that $h(x)$ is the complement of $x \in L$. This leads to the following definition.

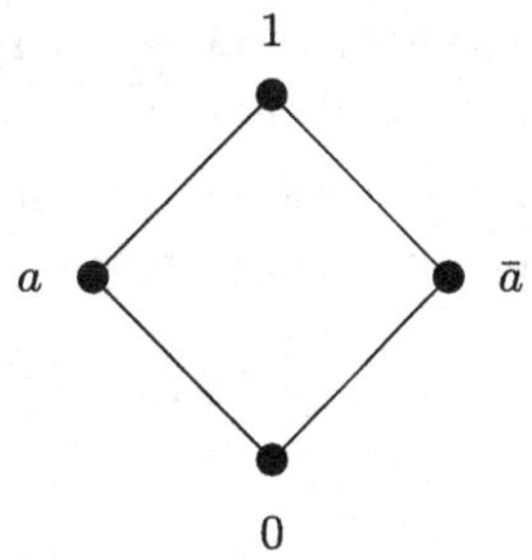

Fig. 11.4 Hasse diagram of the four-element Boolean algebra

Definition 11.58 *A* Boolean lattice *is a bounded distributive lattice that is complemented.*

An equivalent notion that explicitly introduces two zero-ary operations and one unary operation is the notion of *Boolean algebra.*

Definition 11.59 *A* Boolean algebra *is an algebra* $\mathcal{B} = (B, \{\wedge, \vee, \bar{}, 0, 1\})$ *having the type* $(2, 2, 1, 0, 0)$ *that satisfies the following conditions:*

(i) $(B, \{\wedge, \vee\})$ *is a distributive lattice having* 0 *as its least element and* 1 *as its greatest element, and*
(ii) $\bar{}: B \longrightarrow B$ *is a unary operation such that* $\bar{x}$ *is the complement of* x *for* $x \in B$.

Every Boolean algebra has at least two elements, the ones designated by its zero-ary operations.

Example 11.60 The two-element Boolean algebra is the Boolean algebra $\mathcal{B}_2 = (\{0, 1\}, \{\wedge, \vee, \bar{}, 0, 1\})$ defined by:

$$
\begin{aligned}
&0 \wedge 0 = 0, 1 \wedge 1 = 1, 0 \wedge 1 = 1 \wedge 0 = 0,\\
&0 \vee 0 = 0, 1 \vee 1 = 1, 0 \vee 1 = 1 \vee 0 = 1,\\
&\bar{0} = 1, \bar{1} = 0.
\end{aligned}
$$

Example 11.38 can now be recast as introducing a Boolean algebra.

Example 11.61 The set $\mathcal{P}(S)$ of subsets of a set S defines a Boolean algebra $(\mathcal{P}(S), \{\cap, \cup, \bar{}, \emptyset, S\})$, where $\overline{X} = S - X$.

Example 11.62 Let $\mathcal{B}_4 = (\{0, a, \bar{a}, 1\}, \{\wedge, \vee, \bar{}, 0, 1\})$ be the four-element Boolean algebra whose Hasse diagram is given in Fig. 11.4. We leave it to the reader to verify that the poset defined by this diagram is indeed a Boolean algebra.

The existence of the zero-ary operations means that every subalgebra of a Boolean algebra must contain at least 0 and 1.

In a Boolean algebra $\mathcal{B} = (B, \{\wedge, \vee, \bar{}, 0, 1\})$, we have $\bar{\bar{x}} = x$ because of the symmetry of the definition of the complement and because the complement of an element is unique (since $\mathcal{B}$ is a distributive lattice). This property is known as the *involutive property of the complement.*

Theorem 11.63 (DeMorgan Laws) *Let $\mathcal{B} = (B, \{\wedge, \vee, \bar{\ }, 0, 1\})$ be a Boolean algebra. We have $\overline{x \wedge y} = \bar{x} \vee \bar{y}$ and $\overline{x \vee y} = \bar{x} \wedge \bar{y}$ for $x, y \in B$.*

Proof By applying the distributivity, commutativity, and associativity of $\wedge$ and $\vee$ operations, we can write

$$\begin{aligned}(\bar{x} \vee \bar{y}) \wedge (x \wedge y) &= (\bar{x} \wedge (x \wedge y)) \vee (\bar{y} \wedge (x \wedge y)) \\ &= ((\bar{x} \wedge x) \wedge y) \vee ((\bar{y} \wedge y) \wedge x) \\ &= (0 \wedge y) \vee (0 \wedge x) = 0\end{aligned}$$

and

$$\begin{aligned}(\bar{x} \vee \bar{y}) \vee (x \wedge y) &= (\bar{x} \vee \bar{y} \vee x) \wedge (\bar{x} \vee \bar{y} \vee y) \\ &= (1 \vee \bar{y}) \wedge (1 \vee \bar{x}) = 1 \vee 1 = 1\end{aligned}$$

for $x, y \in B$. This shows that $\bar{x} \vee \bar{y}$ is the complement of $x \wedge y$; that is, $\overline{x \wedge y} = \bar{x} \vee \bar{y}$.

The second part of the theorem has a similar argument.

Definition 11.64 *Let $\mathcal{B}_i = (B_i, \{\wedge, \vee, \bar{\ }, 0, 1\})$, $i = 1, 2$, be two Boolean algebras. A* morphism *from $\mathcal{B}_1$ to $\mathcal{B}_2$ is a function $h : B_1 \longrightarrow B_2$ such that $h(x \wedge y) = h(x) \wedge h(y)$, $h(x \vee y) = h(x) \vee h(y)$, and $h(\bar{x}) = \overline{h(x)}$, for $x, y \in B_1$.*

An isomorphism *of Boolean algebras is a morphism that is also a bijection.*

Example 11.65 Let $\mathcal{B} = (B, \{\wedge, \vee, \bar{\ }, 0, 1\})$ be a Boolean algebra and let $c, d \in B$ such that $c \leqslant d$. We can define a Boolean algebra on the interval $[c, d]$ as

$$\mathcal{B}_{[c,d]} = ([c, d], \{\wedge, \vee, \tilde{\ }, c, d\}),$$

where $\wedge, \vee$ are the restrictions of the operations of $\mathcal{B}$ to the set $[c, d]$ and $\tilde{x} = (\bar{x} \vee c) \wedge d$ for $x \in B$.

The projection $h : B \longrightarrow [c, d]$ defined by $h(x) = (x \vee c) \wedge d$ for $x \in B$ is a morphism between $\mathcal{B}$ and $\mathcal{B}_{[c,d]}$. We already saw in Example 11.23 that h is a lattice morphism. Thus, we need to prove only that $\widetilde{h(x)} = h(\bar{x}) = (\bar{x} \vee c) \wedge d$ for $x \in B$. The verification of this equality is left to the reader.

Let $\mathcal{B} = (B, \{\wedge, \vee, \bar{\ }, 0, 1\})$ be a Boolean algebra and let "$\oplus$" be a binary operation on B defined by $a \oplus b = (a \wedge \bar{b}) \vee (\bar{a} \wedge b)$ for $a, b \in B$.

It is easy to verify that

$$\begin{aligned}a \oplus b &= b \oplus a, \\ (a \oplus b) \oplus c &= a \oplus (b \oplus c), \\ a \oplus a &= 0, \\ a \oplus 1 &= \bar{a}, \\ a \wedge (b \oplus c) &= (a \wedge b) \oplus (a \wedge c), \\ a \wedge 1 &= a,\end{aligned}$$

for every $a, b, c \in B$. Thus, the Boolean algebra $\mathcal{B}$ has a related natural structure of a commutative unitary ring $(B, \{0, 1, \oplus, h, \wedge\})$, where the role of the addition or the ring is played by the operation $\oplus$, the additive inverse is given by $h(a) = a$ for $a \in B$, and each element is idempotent.

Example 11.66 For the Boolean algebra $(\mathcal{P}(S), \{\wedge, \vee, \bar{}, \emptyset, S\})$ introduced in Example 11.61, the additive operation of the ring is the symmetric difference of sets

$$U \oplus V = (U - V) \cup (V - U)$$

for $U, V \in \mathcal{P}(S)$. Thus, we obtain a commutative unitary ring structure $(\mathcal{P}, \{\emptyset, S, \oplus, h, \cap\})$, where $h(U) = U$ for $U \in \mathcal{P}(S)$.

A commutative unitary ring in which each element is its own additive inverse and each element is idempotent defines a Boolean algebra, as we show next.

Theorem 11.67 *Let* $\mathcal{I} = (B, \{0, 1, +, h, \cdot, 1\}$ *be a commutative unitary ring such that* $h(b) = b$ *and* $b \wedge b = b$ *for every* $b \in B$. *Define the operations* $\vee, \wedge, \bar{}$ *by*

$$\begin{aligned} a \vee b &= a + b + a \cdot b, \\ a \wedge b &= a \cdot b, \\ \bar{a} &= 1 + a, \end{aligned}$$

for $a \in B$. *Then,* $\mathcal{B} = (B, \{\wedge, \vee, \bar{}, 0, 1\})$ *is a Boolean algebra.*

Proof The operation $\vee$ is commutative because $\mathcal{I}$ is a commutative ring. Observe that

$$\begin{aligned} a \vee (b \vee c) &= a \vee (b + c + bc) \\ &= a + b + c + bc + ab + ac + abc, \\ (a \vee b) \vee c &= a + b + ab + c + ac + bc + abc, \end{aligned}$$

which proves that $\vee$ is also associative. Further, we have $a \vee a = a + a + aa = a$, which proves that $\vee$ is idempotent.

The operation "$\wedge$" is known to be commutative, associative, and idempotent since it coincides with the multiplication of the ring $\mathcal{I}$. To prove the distributivity, note that

$$a \wedge (b \vee c) = a(b + c + bc) = ab + ac + abc$$

and

$$(a \wedge b) \vee (a \wedge c) = ab + ac + (ab)(ac) = ab + ac + abc$$

due to the commutativity and idempotency of multiplication in $\mathcal{I}$. Thus, we have shown that $\mathcal{B} = (B, \{\wedge, \vee, 0, 1\})$ is a distributive lattice having 0 as its least element and 1 as its largest element.

We need to show only that $h(a) = 1 + a$ is the complement of a. This is indeed the case because $a \vee (1+a) = a+1+a+a(1+a) = 1$ and $a \wedge h(a) = a(1+a) = a + a = 0$ for every $a \in B$.

Boolean Functions

Definition 11.68 *Let $\mathcal{B} = (B, \{\wedge, \vee, \bar{}\,, 0, 1\})$ be a Boolean algebra. For $n \in \mathbb{N}$, the set $BF(\mathcal{B}, n)$ of* Boolean functions *of n arguments over $\mathcal{B}$ contains the following functions:*

(i) *For every $b \in B$, the constant function $f_b : B^n \longrightarrow B$ defined by*

$$f_b(x_0, \ldots, x_{n-1}) = b$$

for every $x_0, \ldots, x_{n-1} \in B$ belongs to $BF(\mathcal{B}, n)$.

(ii) *Every* projection function *$p_i^n : B^n \longrightarrow B$ given by $p_i^n(x_0, \ldots, x_{n-1}) = x_i$ for every $x_0, \ldots, x_{n-1} \in B$ belongs to $BF(\mathcal{B}, n)$.*

(iii) *If $f, g \in BF(\mathcal{B}, n)$, then the functions $f \wedge g$, $f \vee g$, and $\bar{f}$ given by*

$$(f \wedge g)(x_0, \ldots, x_{n-1}) = f(x_0, \ldots, x_{n-1}) \wedge g(x_0, \ldots, x_{n-1}),$$
$$(f \vee g)(x_0, \ldots, x_{n-1}) = f(x_0, \ldots, x_{n-1}) \vee g(x_0, \ldots, x_{n-1}),$$

and

$$\bar{f}(x_0, \ldots, x_{n-1}) = \overline{f(x_0, \ldots, x_{n-1})}$$

for every $x_0, \ldots, x_{n-1} \in B$ belong to $BF(\mathcal{B}, n)$.

Definition 11.69 *For $n \in \mathbb{N}$, the set $SBF(\mathcal{B}, n)$ of* simple Boolean functions *of n arguments consists of the following functions:*

(i) *Every projection function $p_i^n : B^n \longrightarrow B$ given by $p_i^n(x_0, \ldots, x_{n-1}) = x_i$ for every $x_0, \ldots, x_{n-1} \in B$.*

(ii) *If $f, g \in SBF(\mathcal{B}, n)$, then the functions $f \wedge g$, $f \vee g$ and $\bar{f}$ given by*

$$(f \wedge g)(x_0, \ldots, x_{n-1}) = f(x_0, \ldots, x_{n-1}) \wedge g(x_0, \ldots, x_{n-1}),$$
$$(f \vee g)(x_0, \ldots, x_{n-1}) = f(x_0, \ldots, x_{n-1}) \vee g(x_0, \ldots, x_{n-1}),$$

and

$$\bar{f}(x_0, \ldots, x_{n-1}) = \overline{f(x_0, \ldots, x_{n-1})}$$

for every $x_0, \ldots, x_{n-1} \in B$ belong to $SBF(\mathcal{B}, n)$.

If $x \in B$ and $a \in \{0, 1\}$, define the function x^a as

$$x^a = \begin{cases} x & \text{if } a = 1 \\ \bar{x} & \text{if } a = 0. \end{cases}$$

Observe that $x = a$ if and only if $x^a = 1$, and $x = \bar{a}$ if and only if $x^a = 0$.

For $\mathbf{A} = (a_1, \ldots, a_n) \in \{0, 1, *\}^n$, define the simple Boolean function $t^{\mathbf{A}} : B^n \longrightarrow B$ as

$$t^{\mathbf{A}}(x_1, \ldots, x_n) = x_{i_1}^{a_{i_1}} \wedge x_{i_2}^{a_{i_2}} \wedge x_{i_p}^{a_{i_p}}$$

for $(x_1, \ldots, x_n) \in B^n$, where $\{a_{i_1}, \ldots, a_{i_p}\} = \{a_i \mid a_i \neq *, 1 \leqslant i \leqslant n\}$. This function is an *n-ary term* for the Boolean algebra $\mathcal{B}$. The set of n-ary terms of a Boolean algebra is denoted by T($\mathcal{B}$, n).

Those components of $\mathbf{A}$ that equal $*$ denote the places of variables that do not appear in the term $t^{\mathbf{A}}$.

Example 11.70 Let $\mathbf{A} = (1, *, 0, 0, *) \in \{0, 1\}^n$. The 5-term $t^{\mathbf{A}}$ is

$$t(x_1, x_2, x_3, x_4, x_5) = x_1 \wedge \bar{x}_3 \wedge \bar{x}_4$$

for every $(x_1, x_2, x_3, x_4, x_5) \in B^5$.

It is easy to see that if $\mathbf{A}, \mathbf{B} \in \{0, 1\}^n$, then

$$t^{\mathbf{B}}(\mathbf{A}) = \begin{cases} 1 & \text{if } \mathbf{A} = \mathbf{B}, \\ 0 & \text{if } \mathbf{A} \neq \mathbf{B}. \end{cases}$$

To simplify the notation, whenever there is no risk of confusion, we omit the symbol "$\wedge$" and denote an application of this operation by a simple juxtaposition of symbols. For example, instead of writing $a \wedge b$, we use the notation ab. For the same reason, we assume that $\wedge$ has higher priority than $\vee$. These assumptions allow us to write $a \vee bc$ instead of $a \vee (b \wedge c)$.

Theorem 11.71 *Let $\mathcal{B} = (B, \{\wedge, \vee, \bar{}\ , 0, 1\})$ be a Boolean algebra. For every $(x_1, \ldots, x_n) \in B^n$, where $n \geqslant 1$, we have*

(i) $t^{\mathbf{A}}(x_1, \ldots, x_n) t^{\mathbf{B}}(x_1, \ldots, x_n) = 0$ *for* $\mathbf{A}, \mathbf{B} \in \{0, 1\}^n$ *and* $\mathbf{A} \neq \mathbf{B}$,
(ii) $\bigvee \{t^{\mathbf{A}}(x_1, \ldots, x_n) \mid \mathbf{A} \in \{0, 1\}^n\} = 1$, *and*
(iii) $\overline{t^{\mathbf{A}}(x_1, \ldots, x_n)} = \bigvee \{t^{\mathbf{B}}(x_1, \ldots, x_n) \mid \mathbf{B} \in \{0, 1\}^n - \{\mathbf{A}\}\}$

for every $(x_1, \ldots, x_n) \in B^n$.

Proof Let $\mathbf{A} = (a_0, \ldots, a_{n-1})$ and $\mathbf{B} = (b_0, \ldots, b_{n-1})$. If $\mathbf{A} \neq \mathbf{B}$, then there exists i such that $0 \leqslant i \leq n-1$ and $a_i \neq b_i$. Therefore, by applying the commutativity and associativity properties of $\wedge$, the expression

$$(t^{\mathbf{A}} t^{\mathbf{B}})(x_1, \ldots, x_n) = x_1^{a_1} \wedge \cdots \wedge x_n^{a_n} \wedge x_1^{b_1} \wedge \cdots \wedge x_n^{b_n}$$

can be written as

$$(t^{\mathbf{A}} t^{\mathbf{B}})(x_1, \ldots, x_n) = \cdots \wedge x_i^{a_i} \wedge x_i^{b_i} \wedge \cdots = \cdots \wedge x_i^1 \wedge x_i^0 \wedge \cdots = 0.$$

The proof of the second part can be done by induction on n. In the base case, $n = 1$, the desired equality amounts to $x_1^0 \vee x_1^1 = 1$, which obviously holds.

Suppose now that the equality holds for n. We have

$$\begin{aligned}
&\bigvee\{(t^{\mathbf{A}}(x_1, \ldots, x_{n+1}) \mid \mathbf{A} \in \{0, 1\}^{n+1}\} \\
&= \bigvee\{(x_1, \ldots, x_n)^{(a_1,\ldots,a_n)} \wedge x_{n+1}^0 \mid (a_1, \ldots, a_n) \in \{0, 1\}^n\} \\
&\quad \bigvee\{(x_1, \ldots, x_n)^{(a_1,\ldots,a_n)} \wedge x_{n+1}^1 \mid (a_1, \ldots, a_n) \in \{0, 1\}^n\} \\
&= \bigvee\{(x_1, \ldots, x_n)^{(a_1,\ldots,a_n)} \mid (a_1, \ldots, a_n) \in \{0, 1\}^n\} \wedge (x_{n+1}^0 \vee x_{n+1}^1) \\
&= \bigvee\{(x_1, \ldots, x_n)^{(a_1,\ldots,a_n)} \mid (a_1, \ldots, a_n) \in \{0, 1\}^n\} \\
&= 1 \text{ (by the inductive hypothesis).}
\end{aligned}$$

Part (iii) of the theorem follows by observing that

$$\overline{t^{\mathbf{A}}(x_1, \ldots, x_n)} = \begin{cases} 1 & \text{if } (x_1, \ldots, x_n) \neq \mathbf{A}, \\ 0 & \text{if } (x_1, \ldots, x_n) = \mathbf{A}. \end{cases}$$

The right-hand member of the equality takes exactly the same values, as can be seen easily.

The set $BF(\mathcal{B}, n)$ is itself a Boolean algebra relative to the operations $\vee$, $\wedge$, and $\bar{\ }$ from Definition 11.68. The least element is the constant function $f_0 : B^n \longrightarrow B$ given by $f_0(x_1, \ldots, x_n) = 0$, and the largest element is the constant function $f_1 : B^n \longrightarrow B$ given by $f_1(x_1, \ldots, x_n) = 0$ for $(x_1, \ldots, x_n) \in B^n$.

A partial order on $BF(\mathcal{B}, n)$ can be introduced by defining $f \leqslant g$ if $f(x_1, \ldots, x_n) \leqslant g(x_1, \ldots, x_n)$ for $x_1, \ldots, x_n \in B$. It is clear that $f \leqslant g$ if and only if $f \vee g = g$ or $f \wedge g = f$.

Theorem 11.72 *Let* $\mathcal{B} = (B, \{\wedge, \vee, \bar{\ }, 0, 1\})$ *be a Boolean algebra,* $\mathbf{A}, \mathbf{B} \in \{0, 1, *\}^n$*, and let* $t^{\mathbf{A}}$ *and* $t^{\mathbf{B}}$ *be the terms in* $\mathrm{T}(\mathcal{B}, \mathrm{n})$ *that correspond to* $\mathbf{A}$ *and* $\mathbf{B}$*, respectively. We have* $t^{\mathbf{A}} \leqslant t^{\mathbf{B}}$ *if and only if* $a_k = *$ *implies* $b_k = *$ *and* $a_k \neq *$ *implies* $a_k = b_k$ *or* $b_k = *$ *for* $1 \leqslant k \leqslant n$.

Proof Suppose that $t^{\mathbf{A}} \leqslant t^{\mathbf{B}}$; that is,

$$x_{i_1}^{a_{i_1}} \wedge x_{i_2}^{a_{i_2}} \wedge x_{i_p}^{a_{i_p}} \leqslant x_{j_1}^{b_{j_1}} \wedge x_{j_2}^{b_{j_2}} \wedge x_{j_q}^{b_{j_q}},$$

for $(x_1, \ldots, x_n) \in B^n$. Here $\{i_1, \ldots, i_p\} = \{i \mid 1 \leqslant i \leqslant n \text{ and } a_i \neq *\}$ and $\{j_1, \ldots, j_q\} = \{j \mid 1 \leqslant j \leqslant n \text{ and } b_j \neq *\}$.

Suppose that $a_k = *$ but $b_k \in \{0, 1\}$. Choose $x_{i_\ell} = a_{i_\ell}$ for $1 \leqslant \ell \leqslant p$ and $x_k = \bar{b}_k$. The remaining components of $(x_1, \ldots, x_k)$ can be chosen arbitrarily. Clearly, $t^{\mathbf{A}}(x_1, \ldots, x_n) = 1$ and $t^{\mathbf{B}}(x_1, \ldots, x_n) = 0$ because $x_k^{b_k} = 0$. This contradicts the inequality $t^{\mathbf{A}} \leqslant t^{\mathbf{B}}$, so we must have $b_k = *$.

Suppose now that $a_k \in \{0, 1\}$ and $b_k \neq *$. This means that x_k occurs in both $t^{\mathbf{A}}$ and $t^{\mathbf{B}}$, so there exists $i_r = j_s = k$ for some r, s, $1 \leqslant r \leqslant p$ and $1 \leqslant s \leqslant q$. Choose as before $x_{i_\ell} = a_{i_\ell}$ for $1 \leqslant \ell \leqslant p$, which implies $t^{\mathbf{A}}(x_1, \ldots, x_n) = 1$, which in turn implies $t^{\mathbf{B}}(x_1, \ldots, x_n) = 1$. This is possible only if $b_k = a_k$, which concludes the argument.

Corollary 11.73 *The minimal elements of the poset* $(\mathrm{T}(\mathcal{B}, \mathrm{n}), \leqslant)$ *are terms that depend on all their arguments, that is, terms of the form*

$$t^{\mathbf{B}}(x_1, \ldots, x_n) = x_1^{b_1} x_2^{b_2} \cdots x_n^{b_n}$$

for $(x_1, \ldots, x_n) \in B^n$.

Proof If $t^{\mathbf{B}}$ is a minimal element of $(\mathrm{T}(\mathcal{B}, \mathrm{n}), \leqslant)$, then $t^{\mathbf{A}} \leqslant t^{\mathbf{B}}$ implies $t^{\mathbf{A}} = t^{\mathbf{B}}$. Suppose that there is k such that $b_k = *$. Then, by defining

$$a_i = \begin{cases} b_i & \text{if } i \neq k, \\ 0 \text{ or } 1 & \text{otherwise,} \end{cases}$$

we would have $t^{\mathbf{A}} < t^{\mathbf{B}}$, which would contradict the minimality of $t^{\mathbf{B}}$.

The minimal terms of the poset $(\mathrm{T}(\mathcal{B}, \mathrm{n}), \leqslant)$, described by Corollary 11.73 are known as n-ary *minterms*.

Definition 11.74 *Let* $\mathcal{B} = (B, \{\wedge, \vee, \bar{\ }, 0, 1\})$ *be a Boolean algebra and let* $f : B^n \longrightarrow B$ *be a Boolean function. A* disjunctive normal form *of* f *is an expression of the form* $\bigvee_{i=1}^{k} t^{\mathbf{A}_i}(x_1, \ldots, x_n) b_{\mathbf{A}_i}$, *where* $\mathbf{A}_i \in \{0, 1, *\}^n$, $\{b_{\mathbf{A}_1}, \ldots, b_{\mathbf{A}_k}\} \subseteq B$, *and*

$$f(x_1, \ldots, x_n) = \bigvee_{i=1}^{k} t^{\mathbf{A}_i}(x_1, \ldots, x_n) b_{\mathbf{A}_i}$$

for $(x_1, \ldots, x_n) \in B^n$.

We can prove the existence of a special disjunctive normal form for every Boolean function, which involves only minterms.

Theorem 11.75 *Let* $\mathcal{B} = (B, \{\wedge, \vee, \bar{\ }, 0, 1\})$ *be a Boolean algebra. A function* $f : B^n \longrightarrow B$ *is a Boolean function if and only if there exists a family* $\{b_{\mathbf{A}} \mid \mathbf{A} \in \{0, 1\}^n\}$ *of elements of* B

$$f(x_1, \ldots, x_n) = \bigvee_{\mathbf{A} \in \{0,1\}^n} t^{\mathbf{A}}(x_1, \ldots, x_n) b_{\mathbf{A}} \tag{11.10}$$

for every $(x_1, \ldots, x_n) \in B^n$.

Proof The sufficiency of this condition is obvious. The necessity is shown by induction on the definition of Boolean functions.

For the base case, we need to consider constant functions and projections. Let $\mathbf{A} = (a_0, \dots, a_{n-1}) \in \{0, 1\}^n$. For a constant function $f_a(x_1, \dots, x_n) = a$ for $(x_1, \dots, x_n)$, we can define $b_{\mathbf{A}} = a$ for every $\mathbf{A} \in \{0, 1\}^n$ because

$$f(x_1, \dots, x_n) = a = \bigvee_{\mathbf{A} \in \{0,1\}^n} t^{\mathbf{A}}(x_1, \dots, x_n)a$$

by the second part of Theorem 11.71.

For a projection $p_i^n : B^n \longrightarrow B$ given by $p_i^n(x_1, \dots, x_n) = x_i$ for $(x_1, \dots, x_n) \in B^n$, let $b_{\mathbf{A}}$ be

$$b_{\mathbf{A}} = \begin{cases} 1 & \text{if } a_i = 1, \\ 0 & \text{otherwise,} \end{cases}$$

for $\mathbf{A} \in \{0, 1\}^n$. We have

$$\begin{aligned}
&\bigvee_{\mathbf{A} \in \{0,1\}^n} t^{\mathbf{A}}(x_1, \dots, x_n)b_{\mathbf{A}} \\
&= \bigvee_{\mathbf{A} \in \{0,1\}^n} t_{(a_1,\dots,a_{i-1},1,a_{i+1},\dots,a_n)}(x_1, \dots, x_n) \\
&= x_i \bigvee_{\mathbf{A} \in \{0,1\}^{n-1}} (x_0, \dots, x_{i-1}, x_{i+1}, \dots, x_{n-1})^{(a_0,\dots,a_{i-1},a_{i+1},\dots,a_{n-1})} \\
&= x_i = p_i^n(x_1, \dots, x_n).
\end{aligned}$$

For the inductive step, suppose that the statement holds for the functions $f, g \in BF(\mathcal{B}, n)$; that is,

$$f(x_1, \dots, x_n) = \bigvee_{\mathbf{A} \in \{0,1\}^n} t^{\mathbf{A}}(x_1, \dots, x_n)b_{\mathbf{A}},$$

$$g(x_1, \dots, x_n) = \bigvee_{\mathbf{A} \in \{0,1\}^n} t^{\mathbf{A}}(x_1, \dots, x_n)c_{\mathbf{A}},$$

for $(x_1, \dots, x_n) \in B^n$. Then, $f \vee g$ is

$$(f \vee g)(x_1, \dots, x_n) = \bigvee_{\mathbf{A} \in \{0, 1\}^n} t^{\mathbf{A}}(x_1, \dots, x_n)(b_{\mathbf{A}} \vee c_{\mathbf{A}})$$

by the associativity, commutativity, and idempotency of "$\vee$".

For $f \wedge g$, we can write

$$
\begin{aligned}
(f \wedge g)(x_1, \ldots, x_n) &= \left(\bigvee_{\mathbf{A} \in \{0,1\}^n} t^{\mathbf{A}}(x_1, \ldots, x_n) b_{\mathbf{A}} \right) \\
&\quad \wedge \left(\bigvee_{\mathbf{A} \in \{0,1\}^n} t^{\mathbf{A}}(x_1, \ldots, x_n) c_{\mathbf{A}} \right) \\
&= \bigvee_{\mathbf{A} \in \{0,1\}^n} t^{\mathbf{A}}(x_1, \ldots, x_n)(b_{\mathbf{A}} \wedge c_{\mathbf{A}})
\end{aligned}
$$

by applying the distributivity properties of the operations $\vee$ and $\wedge$.

For $\overline{f}$, we have

$$
\begin{aligned}
\overline{f(x_1, \ldots, x_n)} &= \bigwedge_{\mathbf{A} \in \{0,1\}^n} \left(\overline{t^{\mathbf{A}}(x_1, \ldots, x_n)} \vee \overline{b_{\mathbf{A}}} \right) \\
&= \bigwedge_{\mathbf{A} \in \{0,1\}^n} \left(\bigvee \{ t^{\mathbf{B}}(x_1, \ldots, x_n) \mid \mathbf{B} \in \{0,1\}^n - \{\mathbf{A}\} \} \vee \overline{b_{\mathbf{A}}} \right) \\
&= \bigwedge_{\mathbf{A} \in \{0,1\}^n} \left(\bigvee \{ t^{\mathbf{B}}(x_1, \ldots, x_n) \mid \mathbf{B} \in \{0,1\}^n - \{\mathbf{A}\} \} \vee \overline{b_{\mathbf{A}}} \right).
\end{aligned}
$$

For $\mathbf{C}, \mathbf{D} \in \{0,1\}^n$ and $(x_1, \ldots, x_n) \in B^n$, define $\phi_{\mathbf{C},\mathbf{D}}(x_1, \ldots, x_n)$ as

$$
\phi_{\mathbf{C},\mathbf{D}}(x_1, \ldots, x_n) = \begin{cases} t^{\mathbf{D}}(x_1, \ldots, x_n) & \text{if } \mathbf{D} \neq \mathbf{C}, \\ \overline{b_{\mathbf{C}}} & \text{if } \mathbf{D} = \mathbf{C}. \end{cases}
$$

Then, we can write

$$
\begin{aligned}
\overline{f(x_1, \ldots, x_n)} &= \bigwedge_{\mathbf{A} \in \{0,1\}^n} \bigvee_{\mathbf{D} \in \{0,1\}^n} \phi_{\mathbf{A},\mathbf{D}}(x_1, \ldots, x_n) \\
&= \bigvee_{\mathbf{D} \in \{0,1\}^n} \bigwedge_{\mathbf{A} \in \{0,1\}^n} \phi_{\mathbf{A},\mathbf{D}}(x_1, \ldots, x_n) \\
&\quad \text{(by the distributivity property)} \\
&= \bigvee_{\mathbf{D} \in \{0,1\}^n} (t^{\mathbf{D}}(x_1, \ldots, x_n) \wedge \overline{b_{\mathbf{D}}}),
\end{aligned}
$$

which concludes the argument.

Equality (11.10) is known as the *standard disjunctive normal form* of the Boolean function f.

Note that by replacing $(x_1, \ldots, x_n)$ by $\mathbf{C} = (c_1, \ldots, c_n) \in \{0,1\}^n$ in Equality (11.10), we obtain $f(c_1, \ldots, c_n) = b_{\mathbf{C}}$, which shows that the elements of the form $b_{\mathbf{A}}$, known as the *standard disjunctive coefficients*, are uniquely determined by the function f. Now, we can rewrite Equality (11.10) as

$$f(x_1, \ldots, x_n) = \bigvee_{\mathbf{A} \in \{0, 1\}^n} t^{\mathbf{A}}(x_1, \ldots, x_n) f(\mathbf{A})$$

for every $(x_1, \ldots, x_n) \in B^n$.

Consider the standard disjunctive normal form of the Boolean function $\overline{f} : B^n \longrightarrow B$ given by

$$\overline{f}(x_1, \ldots, x_n) = \bigvee_{\mathbf{A} \in \{0, 1\}^n} t^{\mathbf{A}}(x_1, \ldots, x_n) \overline{f(\mathbf{A})}.$$

By applying the $\bar{}$ operation in both members, we can write

$$\begin{aligned} f(x_1, \ldots, x_n) &= \bigwedge_{\mathbf{A} \in \{0, 1\}^n} (\overline{t^{\mathbf{A}}}(x_1, \ldots, x_n) \vee f(\mathbf{A})) \\ &= \bigwedge_{(a_1, \ldots, a_n) \in \{0, 1\}^n} \left(\bigvee_{i=1}^{n} x_i^{\bar{a}_i} \vee f(a_1, \ldots, a_n) \right) \\ &= \bigwedge_{(\bar{a}_1, \ldots, \bar{a}_n) \in \{0, 1\}^n} \left(\bigvee_{i=1}^{n} x_i^{a_i} \vee f(\bar{a}_1, \ldots, \bar{a}_n) \right) \\ &= \bigwedge_{(a_1, \ldots, a_n) \in \{0, 1\}^n} \left(\bigvee_{i=1}^{n} x_i^{a_i} \vee f(\bar{a}_1, \ldots, \bar{a}_n) \right). \end{aligned}$$

The last equality is known as the *conjunctive normal form* of the function f.

The existence of the standard disjunctive normal form shows that a Boolean function $f : B^n \longrightarrow B$ is completely determined by its values on n-tuples $\mathbf{A} \in \{0, 1\}^n$. Thus, to fully specify a Boolean function, we can use a table that has 2^n rows, one for each n-tuple $\mathbf{A}$.

Example 11.76 Consider the Boolean function $f : B^3 \longrightarrow B$ given by

x_1	x_2	x_3	$f(x_1, x_2, x_3)$
0	0	0	a
0	0	1	a
0	1	0	a
0	1	1	b
1	0	0	a
1	0	1	b
1	1	0	b
1	1	1	b

Its standard disjunctive normal form is

$$f(x_1, x_2, x_3) = t^{(0,0,0)}(x_1, x_2, x_3)a \vee t^{(0,0,1)}(x_1, x_2, x_3)a \vee t^{(0,1,0)}(x_1, x_2, x_3)a \\ \vee t^{(0,1,1)}(x_1, x_2, x_3)b \vee t^{(1,0,0)}(x_1, x_2, x_3)a \vee t^{(1,0,1)}(x_1, x_2, x_3)b \\ \vee t^{(1,1,0)}(x_1, x_2, x_3)b \vee t^{(1,1,1)}(x_1, x_2, x_3)b.$$

Theorem 11.77 *Let $\mathcal{B} = (B, \{\wedge, \vee, \bar{}, 0, 1\})$ be a Boolean algebra and let $f, g \in BF(\mathcal{B}, n)$. We have $f \leqslant g$ if and only if $f(\mathbf{A}) \leqslant g(\mathbf{A})$ for every $\mathbf{A} \in \{0, 1\}^n$.*

Proof The necessity of the condition is obvious. Suppose that $f(\mathbf{A}) \leqslant g(\mathbf{A})$ for every $\mathbf{A} \in \{0, 1\}^n$. Then, by the monotonicity of the binary operations of the Boolean algebra, we have

$$\begin{aligned} f(x_1, \ldots, x_n) &= \bigvee_{\mathbf{A} \in \{0,1\}^n} t^{\mathbf{A}}(x_1, \ldots, x_n) f(\mathbf{A}) \\ &\leq \bigvee_{\mathbf{A} \in \{0,1\}^n} t^{\mathbf{A}}(x_1, \ldots, x_n) g(\mathbf{A}) = g(x_1, \ldots, x_n) \end{aligned}$$

for $x_1, \ldots, x_n \in B^n$, which gives the desired inequality.

The next theorem (see [1]) is a characterization of simple Boolean functions.

Theorem 11.78 *Let $\mathcal{B} = (B, \{\wedge, \vee, \bar{}, 0, 1\})$ be a Boolean algebra. The following statements that concern a function $f : B^n \longrightarrow B$ are equivalent:*

(i) *f is a simple Boolean function.*
(ii) *f is a Boolean function, and $f(\mathbf{A}) \in \{0, 1\}$ for every $\mathbf{A} \in \{0, 1\}^n$.*
(iii) *$f(x_1, \ldots, x_n) = 0$ for every $(x_1, \ldots, x_n) \in B^n$ or*

$$f(x_1, \ldots, x_n) = \bigvee \{(x_1, \ldots, x_n)^{\mathbf{A}} \mid f(\mathbf{A}) = 1\}.$$

Proof (i) implies (ii): Clearly every simple Boolean function is a Boolean function. The proof that $f(\mathbf{A}) \in \{0, 1\}$ for every $\mathbf{A} \in \{0, 1\}^n$ is by induction on the definition of simple Boolean functions and is left to the reader.

(ii) implies (iii): This implication follows from the existence of the standard disjunctive normal form of Boolean functions.

(iii) implies (i): The constant function $f_0(x_1, \ldots, x_n) = 0$ for $(x_1, \ldots, x_n) \in B^n$ can be written as $f_0(x_1, \ldots, x_n) = x_1 \wedge \bar{x}_1$, so f_0 is a simple Boolean function. It is clear that if $f(x_1, \ldots, x_n) = \bigvee \{(x_1, \ldots, x_n)^{\mathbf{A}} \mid f(\mathbf{A}) = 1\}$, then f is a simple Boolean function.

For a Boolean algebra $\mathcal{B} = (B, \{\wedge, \vee, \bar{}, 0, 1\})$ with $|B| = k$ there exist k^{k^n} functions of the form $f : B^n \longrightarrow B$. The number of Boolean functions can be considerably smaller. Indeed, since a Boolean function f is completely determined by the collection $\{f(\mathbf{A}) \mid A \in \{0, 1\}^n\}$, it follows that the number of Boolean functions in $BF(\mathcal{B}, n)$ is 2^{2^n}. For example, if $\mathcal{B}$ is the four-element Boolean algebra from Example 11.62, there are $4^{4^5} = 2^{2048}$ functions of five arguments defined on the Boolean algebra. However, only 2^{32} of these functions are Boolean functions.

Binary Boolean Functions

Definition 11.79 *Let* $\mathcal{B}_2 = (\{0, 1\}, \{\wedge, \vee, \bar{}\ , 0, 1\})$ *be the two-element Boolean algebra. A* binary Boolean function *is a function* $f : \{0, 1\}^n \longrightarrow \{0, 1\}$.

We saw that in general Boolean algebra there are many functions that are not Boolean. However, in two-element Boolean algebras, any function is a Boolean function, as we show next.

Theorem 11.80 *Every function* $f : \{0, 1\}^n \longrightarrow \{0, 1\}$ *is a binary Boolean function in the two-element Boolean algebra* $\mathcal{B}_2$.

Proof Consider the binary Boolean function $g : \{0, 1\}^n \longrightarrow \{0, 1\}$ defined by $g(x_1, \ldots, x_n) = \bigvee_{\mathbf{A}\in\{0,1\}^n} (x_1, \ldots, x_n)^{\mathbf{A}} f(\mathbf{A})$. It is clear that $g(\mathbf{A}) = f(\mathbf{A})$ for every $\mathbf{A} \in \{0, 1\}^n$, so $g = f$. Thus, $f = \bigvee_{\mathbf{A}\in\{0,1\}^n} (x_1, \ldots, x_n)^{\mathbf{A}} f(\mathbf{A})$, which implies that f is indeed a Boolean function.

Definition 11.81 *An* implicant *of a binary Boolean function* $f : \{0, 1\}^n \longrightarrow \{0, 1\}$ *is a term* $t^{\mathbf{A}} \in \mathrm{T}(\mathcal{B}_2, \mathrm{n})$ *such that* $t^{\mathbf{A}} \leqslant f$.

The rank of an implicant $t^{\mathbf{A}}$ *of* $f : \{0, 1\}^n \longrightarrow \{0, 1\}$ *is the number* $r(t^{\mathbf{A}}) = |\{i \mid 1 \leqslant i \leqslant n, a_i = *\}|$. *Observe that implicants with higher ranks contain fewer literals than implicants with lower rank.*

The set of implicants of rank k of f, $0 \leqslant k \leqslant n$, *is the set* L^k_f *that consists of all implicants of rank k for* f.

The set of implicants of f *is denoted by* $IMPL_f$. *For* $f : \{0, 1\}^n \longrightarrow \{0, 1\}$, *we have* $IMPL_f = \bigcup_{k=0}^{n-1} L^k_f$.

Starting from the standard disjunctive normal form for a function $f : B^n \longrightarrow B$,

$$f(x_1, \ldots, x_n) = \bigvee_{\mathbf{A}\in\{0,1\}^n} t^{\mathbf{A}}(x_1, \ldots, x_n) f(\mathbf{A}),$$

it follows that if $f(\mathbf{A}) = 1$, then the minterm $t^{\mathbf{A}}$ is an implicant of f in L^0_f. Furthermore, each such term is a minimal implicant of f (relative to the partial order introduced on $\mathrm{T}(\mathcal{B}_2, \mathrm{n})$).

In the next definition, we introduce a partial operation on the set $\mathrm{T}(\mathcal{B}_2, \mathrm{n})$.

Definition 11.82 *Let* $\mathbf{A}, \mathbf{B} \in \{0, 1, *\}^n$ *be two n-tuples. Suppose that there exists k,* $1 \leqslant k \leqslant n$ *such that*

1. $a_i = b_i$ *if* $1 \leqslant i \leqslant n$ *and* $i \neq k$;
2. $a_k, b_k \in \{0, 1\}$ *and* $a_k = \bar{b}_k$.

The consensus *of the terms* $t^{\mathbf{A}}$ *and* $t^{\mathbf{B}}$ *is the term* $t^{\mathbf{C}}$, *where* $\mathbf{C} = (c_1, \ldots, c_n)$ *and*

$$c_i = \begin{cases} a_i = b_i & \text{if } i \neq k, \\ * & \text{otherwise}, \end{cases}$$

for $1 \leqslant i \leqslant n$.

The consensus of $t^{\mathbf{A}}$ *and* $t^{\mathbf{C}}$ *is denoted by* $t^{\mathbf{A}} \boxplus t^{\mathbf{B}}$.

Observe that if the consensus $t^{\mathbf{C}}$ of the terms $t^{\mathbf{A}}$ and $t^{\mathbf{A}}$ exists, then $r(t^{\mathbf{C}}) = r(t^{\mathbf{A}}) + 1 = r(t^{\mathbf{B}}) + 1$. Furthermore, it is immediate that $t^{\mathbf{C}} = t^{\mathbf{A}} \vee t^{\mathbf{B}}$ in the Boolean algebra of Boolean functions.

Example 11.83 Let $t^{\mathbf{A}}$ and $t^{\mathbf{B}}$ be the terms

$$t^{\mathbf{A}} = x_1 \wedge \bar{x}_3 \wedge \bar{x}_4 \wedge x_6, t^{\mathbf{B}} = x_1 \wedge x_3 \wedge \bar{x}_4 \wedge x_6,$$

from $\mathrm{T}(\mathcal{B}_2, 6)$. Their consensus is the term

$$\begin{aligned} & t^{\mathbf{A}}(x_1, \ldots, x_6) \vee t^{\mathbf{B}}(x_1, \ldots, x_6) \\ & = (x_1 \wedge \bar{x}_3 \wedge \bar{x}_4 \wedge x_6) \vee (x_1 \wedge x_3 \wedge \bar{x}_4 \wedge x_6) \\ & = x_1 \wedge \bar{x}_4 \wedge x_6. \end{aligned}$$

Theorem 11.84 *Let* $f : \{0, 1\}^n \longrightarrow \{0, 1\}$ *be a Boolean function. If* $t^{\mathbf{A}}$ *and* $t^{\mathbf{B}}$ *are implicants of* f *and their consensus* $t^{\mathbf{C}} = t^{\mathbf{A}} \vee t^{\mathbf{B}}$ *exists, then* $t^{\mathbf{C}}$ *is also an implicant of* f.

Proof The existence of the consensus $t^{\mathbf{C}}$ of $t^{\mathbf{A}}$ and $t^{\mathbf{B}}$ means that there exists k, $1 \leqslant k \leqslant n$ such that $a_i = b_i$ if $1 \leqslant i \leqslant n$ and $a \neq k$, $a_k, b_k \in \{0, 1\}$, and $a_k = \bar{b}_k$.

Since both $t^{\mathbf{A}}$ and $t^{\mathbf{B}}$ are implicants of f, it follows that $t^{\mathbf{A}}(x_1, \ldots, x_n) \leqslant f(x_1, \ldots, x_n)$ and $t^{\mathbf{B}}(x_1, \ldots, x_n) \leqslant f(x_1, \ldots, x_n)$ for every $(x_1, \ldots, x_n) \in \{0, 1\}^n$. Thus,

$$t^{\mathbf{C}}(x_1, \ldots, x_n) = t^{\mathbf{A}}(x_1, \ldots, x_n) \vee t^{\mathbf{B}}(x_1, \ldots, x_n) \leq f(x_1, \ldots, x_n),$$

which means that $t^{\mathbf{C}}$ is an implicant of f.

Definition 11.85 *A* prime implicant *of a function* $f : \{0, 1\}^n \longrightarrow \{0, 1\}$ *is a maximal element of the poset* $(IMPL_f, \leqslant)$.

Theorem 11.86 *For every binary Boolean function* $f : \{0, 1\}^n \longrightarrow \{0, 1\}$, *we have*

$$L_f^{k+1}(\varphi) = \{t^{\mathbf{A}} \vee t^{\mathbf{B}} \mid t^{\mathbf{A}} \vee t^{\mathbf{B}} \in L_f^k \textit{ and } t^{\mathbf{A}} \vee t^{\mathbf{B}} \textit{ exists}\}$$

for $0 \leqslant k \leqslant n - 1$.

Proof We observed already that if $r(t^{\mathbf{A}}) = r(t^{\mathbf{B}}) = k$ and $t^{\mathbf{A}} \vee t^{\mathbf{B}}$ exists, then $r(t^{\mathbf{A}} \vee t^{\mathbf{B}}) = k + 1$. Thus, we have

$$L_f^{k+1}(\varphi) \supseteq \{t^{\mathbf{A}} \vee t^{\mathbf{B}} \mid t^{\mathbf{A}} \vee t^{\mathbf{B}} \in L_f^k \text{ and } t^{\mathbf{A}} \vee t^{\mathbf{B}} \text{ exists}\}$$

for $0 \leqslant k \leqslant n - 1$, and we need to prove only the reverse inclusion.

Let $t^{\mathbf{C}} \in L_f^{k+1}$, where $\mathbf{C} = (c_1, \ldots, c_n)$. There exists ℓ, $1 \leqslant \ell \leqslant n$ such that $c_\ell = *$, so $t^{\mathbf{C}}$ does not depend on x_ℓ. If $t^{\mathbf{C}}(x_1, \ldots, x_n) = x_{i_1}^{c_{i_1}} \cdots x_{i_{n-k-1}}^{c_{i_{n-k-1}}}$, then $\ell \notin \{i_1, \ldots, i_{n-k-1}\}$ and both

$$t^{\mathbf{A}}(x_1, \ldots, x_n) = x_{i_1}^{c_{i_1}} \cdots x_\ell^0 \cdots x_{i_{n-k-1}}^{c_{i_{n-k-1}}} \text{ and}$$
$$t^{\mathbf{A}}(x_1, \ldots, x_n) = x_{i_1}^{c_{i_1}} \cdots x_\ell^1 \cdots x_{i_{n-k-1}}^{c_{i_{n-k-1}}}$$

belong to L_f^k. Clearly, $t^{\mathbf{C}}$ is the consensus of $t^{\mathbf{A}}$ and $t^{\mathbf{B}}$, which yields the reverse inclusion.

Theorem 11.86 suggests that we can generate the posets of all implicants of a binary Boolean function $f : \{0, 1\}^n \longrightarrow \{0, 1\}$ by producing inductively the sets $L_f^0, \ldots, L_f^{n-1}$. The algorithm that implements this idea is the Quine-McCluskey algorithm, discussed next.

Algorithm 11.5.1: Quine-McCluskey Algorithm

Data: A binary Boolean function given in tabular form
Result: The set $IMPL_f$ of all implicants of f

let L_f^0 be the set of minterms for f;
for $0 \leqslant k \leqslant n-2$ **do**
  include in L_f^{k+1} every term that can be obtained as a consensus of two terms from L_f^k
end
return the collection $\bigcup_{k=0}^{n-1} L_f^k$;

The correctness of the algorithm is an immediate consequence of Theorem 11.86.

Example 11.87 Consider the Boolean function $f : \{0, 1\}^3 \longrightarrow \{0, 1\}$ given by

x_1	x_2	x_3	$f(x_1, x_2, x_3)$
0	0	0	0
0	0	1	0
0	1	0	0
0	1	1	1
1	0	0	0
1	0	1	1
1	1	0	1
1	1	1	1

Its standard disjunctive normal form is

$$f(x_1, x_2, x_3) = t^{(0,1,1)}(x_1, x_2, x_3) \vee t^{(1,0,1)}(x_1, x_2, x_3)$$
$$\vee t^{(1,1,0)}(x_1, x_2, x_3) \vee t^{(1,1,1)}(x_1, x_2, x_3),$$

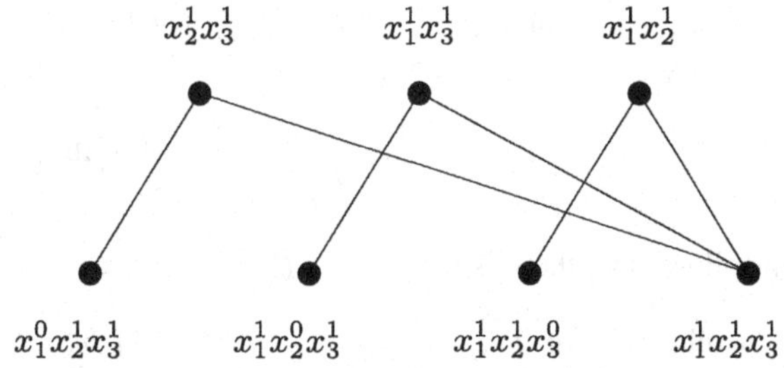

Fig. 11.5 Hasse diagram of $IMPL_f$

so the set L_f^0 consists of the minterms

$$
\begin{aligned}
t^{(0,1,1)}(x_1, x_2, x_3) &= x_1^0 x_2^1 x_3^1, \\
t^{(1,0,1)}(x_1, x_2, x_3) &= x_1^1 x_2^0 x_3^1, \\
t^{(1,1,0)}(x_1, x_2, x_3) &= x_1^1 x_2^1 x_3^0, \\
t^{(1,1,1)}(x_1, x_2, x_3) &= x_1^1 x_2^1 x_3^1,
\end{aligned}
$$

for $(x_1, x_2, x_3) \in \{0, 1\}^3$.

The Hasse diagram of $IMPL_f$ is shown in Fig. 11.5. Clearly, $IMPL_f = L_f^0 \cup L_f^1$ because there is no consensus possible among any two of the implicants from L_f^1.

Definition 11.88 *A nonempty set of terms $T = \{t^{\mathbf{B}_1}, \ldots, t^{\mathbf{B}_m}\}$ of implicants of a binary Boolean function $f : \{0, 1\}^n \longrightarrow \{0, 1\}$ is a* cover *of f if $f(x_1, \ldots, x_n) = \bigvee_{i=1}^m t^{\mathbf{B}_i}(x_1, \ldots, x_n)$.*

T is a minimal cover *of f if T is a cover of f and no proper subset of T is a cover of f.*

The set of all minterms of f is clearly a cover of f. However, other covers may exist for f that contain terms of rank that is higher than 0 and it is important to determine such simpler covers.

Since $(IMPL_f, \leqslant)$ is a finite poset, for every $t^{\mathbf{B}} \in IMPL_f$ there is a prime implicant $t^{\mathbf{A}}$ such that $t^{\mathbf{B}} \leqslant t^{\mathbf{A}}$.

Theorem 11.89 *Let f be a binary Boolean function that is not the constant function f_0. A set of implicants of f, $T = \{t^{\mathbf{B}_1}, \ldots, t^{\mathbf{B}_m}\}$ is a cover of f if and only if for every minterm $t^{\mathbf{A}}$ of f there is an implicant $t^{\mathbf{B}} \in T$ such that $t^{\mathbf{A}} \leqslant t^{\mathbf{B}_i}$.*

Proof Suppose that T satisfies the condition of the theorem. Let $\{\mathbf{A} \in \{0, 1\}^n \mid f(\mathbf{A}) = 1\} = \{\mathbf{A}_1, \ldots, \mathbf{A}_k\}$. Then, since

$$
f(x_1, \ldots, x_n) = \bigvee_{1 \leqslant i \leqslant k} t^{\mathbf{A}}(x_1, \ldots, x_n) \leqslant \bigvee_{1 \leqslant l \leqslant m} t^{\mathbf{B}_l} \leqslant f(x_1, \ldots, x_n),
$$

it is immediate that T is a cover for φ.

Conversely, let T be a cover of f, $T = \{t^{\mathbf{B}_1}, \ldots, t^{\mathbf{B}_m}\}$ and let $t^{\mathbf{A}}$ be a minterm, where $\mathbf{A} = (a_1, \ldots, a_n)$. Since

$$t^{\mathbf{A}}(x_1, \ldots, x_n) \leqslant f(x_1, \ldots, x_n) \leqslant \bigvee \{t^{\mathbf{B}}(x_1, \ldots, x_n) \mid t^{\mathbf{B}} \in T\},$$

it follows that there is $\mathbf{B}$ such that $t^{\mathbf{B}}(a_1, \ldots, a_n) = 1$. This implies $t^{\mathbf{A}} \leqslant t^{\mathbf{B}}$.

Corollary 11.90 *Let $f : \{0, 1\}^n \longrightarrow \{0, 1\}$ be a function that is distinct from the constant function f_0. If $T = \{t^{\mathbf{B}_1}, \ldots, t^{\mathbf{B}_m}\}$ is a cover of f, and $t^{\mathbf{C}}$ is an implicant of f such that $t^{\mathbf{B}_i} < t^{\mathbf{C}}$ for some i, $1 \leqslant i \leqslant m$, then $T' = \{t^{\mathbf{B}_1}, \ldots, t^{\mathbf{B}_{i-1}}, t^{\mathbf{C}}, t^{\mathbf{B}_{i+1}}, \ldots, t^{\mathbf{B}_m}\}$ is a cover of f.*

Proof The statement follows immediately from Theorem 11.89.

We now discuss the Quine-McCluskey systematic construction that starts with the set of prime implicants and the set of minterms of a nonzero Boolean function $f : \{0, 1\}^n \longrightarrow \{0, 1\}$ and yields covers of f that consist of prime implicants.

Let $\mathbf{M}_f = (m_{ij})$ be a $p \times q$-matrix having one row for each prime implicant $t^{\mathbf{B}_1}, \ldots, t^{\mathbf{B}_p}$ and one column for each minterm $\{t^{\mathbf{A}_1}, \ldots, t^{\mathbf{A}_q}\}$ of f. Define

$$m_{ij} = \begin{cases} 1 & \text{if } t^{\mathbf{A}_j} \leqslant t^{\mathbf{A}_i}, \\ 0 & \text{otherwise.} \end{cases}$$

If a column of $\mathbf{M}_f$ contains a single 1, that corresponding prime implicant is referred to as an *essential prime implicant*. Denote by E_f the set of essential prime implicants for f. Clearly, the set E_f must be contained in any cover by prime implicants of f.

Eliminate from $\mathbf{M}$ all essential prime implicants and the columns corresponding to the minterms they dominate.

If the set of rows of $\mathbf{M}_f$ in which a column of a minterm $t^{\mathbf{A}}$ has 1s strictly includes the set of rows in which some other column of a minterm $t^{\mathbf{A}'}$ has 1s, then eliminate column $t^{\mathbf{A}}$. Next, if among the remaining columns several have the same pattern of 1s, then retain only one of them.

Eliminate from $\mathbf{M}_f$ all rows that contain no 1s. The output consists of every minimal set of rows in $\mathbf{M}_f$ such that at least one 1 exists in these rows for every column, to each of which we add the set of essential prime implicants E_f.

Example 11.91 Let $f : \{0, 1\}^4 \longrightarrow \{0, 1\}$ be the binary Boolean function defined by

Starting from the minterms

$$\begin{aligned} t^{\mathbf{A}_1} &= \bar{x}_1\bar{x}_2\bar{x}_3\bar{x}_4, & t^{\mathbf{A}_5} &= \bar{x}_1 x_2 x_3 \bar{x}_4, \\ t^{\mathbf{A}_2} &= \bar{x}_1\bar{x}_2 x_3 \bar{x}_4, & t^{\mathbf{A}_6} &= \bar{x}_1 x_2 x_3 x_4, \\ t^{\mathbf{A}_3} &= \bar{x}_1 x_2 \bar{x}_3 \bar{x}_4, & t^{\mathbf{A}_7} &= x_1 \bar{x}_2 x_3 \bar{x}_4, \\ t^{\mathbf{A}_4} &= \bar{x}_1 x_2 \bar{x}_3 x_4, & t^{\mathbf{A}_8} &= x_1 x_2 x_3 \bar{x}_4, \end{aligned}$$

x_1	x_2	x_3	x_4	$f(x_1, x_2, x_3)$
0	0	0	0	1
0	0	0	1	0
0	0	1	0	1
0	0	1	1	0
0	1	0	0	1
0	1	0	1	1
0	1	1	0	1
0	1	1	1	1
1	0	0	0	0
1	0	0	1	0
1	0	1	0	1
1	0	1	1	0
1	1	0	0	0
1	1	0	1	0
1	1	1	0	1
1	1	1	1	0

we have the following sets of implicants computed by using the Quine-McCluskey algorithm:

$$\begin{aligned}
L_f^0 &= \{\bar{x}_1\bar{x}_2\bar{x}_3\bar{x}_4, \bar{x}_1\bar{x}_2x_3\bar{x}_4, \bar{x}_1x_2\bar{x}_3\bar{x}_4, \bar{x}_1x_2\bar{x}_3x_4,\\
&\quad \bar{x}_1x_2x_3\bar{x}_4, \bar{x}_1x_2x_3x_4, x_1\bar{x}_2x_3\bar{x}_4, x_1x_2x_3\bar{x}_4, \},\\
L_f^1 &= \{\bar{x}_1\bar{x}_2\bar{x}_4, \bar{x}_1\bar{x}_3\bar{x}_4, \bar{x}_1x_3\bar{x}_4, \bar{x}_2x_3\bar{x}_4, \bar{x}_1x_2\bar{x}_3,\\
&\quad \bar{x}_1x_2x_4, \bar{x}_1x_2x_3, x_2x_3\bar{x}_4, x_1x_3\bar{x}_4\},\\
L_f^2 &= \{\bar{x}_1\bar{x}_4, x_3\bar{x}_4, \bar{x}_1x_2\}.
\end{aligned}$$

The prime implicants of f are the terms $t^{\mathbf{B}_1} = \bar{x}_1\bar{x}_4$, $t^{\mathbf{B}_2} = x_3\bar{x}_4$, and $t^{\mathbf{B}_3} = \bar{x}_1x_2$.

The matrix M_f introduced above is a 3×8 matrix:

$$M_f = \begin{pmatrix} 1\,1\,1\,0\,1\,0\,0\,0 \\ 0\,1\,0\,0\,1\,0\,1\,1 \\ 0\,0\,1\,1\,1\,1\,0\,0 \end{pmatrix}.$$

The first, fourth, and the last three columns contain exactly one 1. Thus, all three prime implicants are essential, and they form a unique cover of prime implicants of f.

Definition 11.92 *A partially defined Boolean function (pdBf) on the two-element Boolean algebra is a partial function* $f : \{0, 1\}^n \rightsquigarrow \{0, 1\}$.

A pdBf $f : \{0, 1\}^n \rightsquigarrow \{0, 1\}$ is completely defined by the pair of disjoint sets

$$\begin{aligned}
T_f &= \{\mathbf{A} \in \text{Dom}(f) \mid f(\mathbf{A}) = 1\},\\
F_f &= \{\mathbf{A} \in \text{Dom}(f) \mid f(\mathbf{A}) = 0\}.
\end{aligned}$$

Definition 11.93 *Let $f : \{0, 1\}^n \longrightarrow \{0, 1\}$ be a binary Boolean function and let i be an integer such that $1 \leqslant i \leqslant n$. The function is* i-positive *if*

$$f(x_1, \ldots, x_{i-1}, 0, x_{i+1}, \ldots, x_n) \leq f(x_1, \ldots, x_{i-1}, 1, x_{i+1}, \ldots, x_n)$$

for every $x_1, \ldots, x_{i-1}, x_{i+1}, \ldots, x_n \in \{0, 1\}$.

Similarly, f is i-negative *if*

$$f(x_1, \ldots, x_{i-1}, 0, x_{i+1}, \ldots, x_n) \geq f(x_1, \ldots, x_{i-1}, 1, x_{i+1}, \ldots, x_n)$$

for every $x_1, \ldots, x_{i-1}, x_{i+1}, \ldots, x_n \in \{0, 1\}$.

The function is i-monotonic *if it is either i-positive or i-negative.*

Example 11.94 For every $\mathbf{A} \in \{0, 1, *\}^n$, the term $t^{\mathbf{A}}$ is i-monotonic for $1 \leqslant i \leqslant n$. Indeed, if $a_i \in \{1, *\}$, then $t^{\mathbf{A}}$ is i-positive; if $a_i \in \{0, *\}$, then $t^{\mathbf{A}}$ is i-negative.

Theorem 11.95 *Let $f : \{0, 1\}^n \longrightarrow \{0, 1\}$ be a binary Boolean function and let i be an integer such that $1 \leqslant i \leqslant n$. If f is i-positive, then for every prime implicant $t^{\mathbf{A}}$ of f we have $a_i \in \{1, *\}$, where $\mathbf{A} = (a_1, \ldots, a_n)$.*

If f is i-negative, then $a_i \in \{0, *\}$.

Proof Suppose that f is i-positive. Then,

$$f(x_1, \ldots, x_{i-1}, 0, x_{i+1}, \ldots, x_n) \leq f(x_1, \ldots, x_{i-1}, 1, x_{i+1}, \ldots, x_n)$$

for every $x_1, \ldots, x_{i-1}, x_{i+1}, \ldots, x_n \in \{0, 1\}$.

Suppose that $a_i = 0$, that is,

$$t^{\mathbf{A}}(x_1, \ldots, x_n) = \cdots \wedge \bar{x}_i \wedge \cdots .$$

We claim that this implies the inequality

$$x_1^{a_1} \cdots x_{i-1}^{a_{i-1}} x_{i+1}^{a_{i+1}} \cdots x_n^{a_n} \leqslant f(x_1, \ldots, x_{i-1}, x_i, x_{i+1}, \ldots, x_n)$$

for every $x_1, \ldots, x_n \in \{0, 1\}$. In other words, we have to prove that we have both

$$x_1^{a_1} \cdots x_{i-1}^{a_{i-1}} x_{i+1}^{a_{i+1}} \cdots x_n^{a_n} \leqslant f(x_1, \ldots, x_{i-1}, 0, x_{i+1}, \ldots, x_n)$$

and

$$x_1^{a_1} \cdots x_{i-1}^{a_{i-1}} x_{i+1}^{a_{i+1}} \cdots x_n^{a_n} \leqslant f(x_1, \ldots, x_{i-1}, 1, x_{i+1}, \ldots, x_n).$$

Since f is i-positive, only the proof of the first inequality is necessary. The fact that $t^{\mathbf{A}}$ is an implicant of f means that

$$t^{\mathbf{A}}(x_1, \ldots, x_n) \leqslant f(x_1, \ldots, x_n)$$

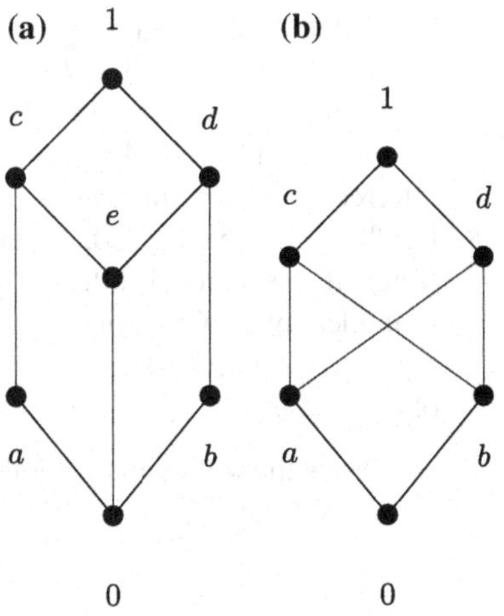

Fig. 11.6 Hasse diagrams of two partially ordered sets

for every $x_1, \ldots, x_n \in \{0, 1\}$. Therefore,

$$\begin{aligned} t^{\mathbf{A}}(x_1, \ldots, x_{i-1}, 0, x_{i+1}, \ldots, x_n) &= x_1^{a_1} \cdots x_{i-1}^{a_{i-1}} x_{i+1}^{a_{i+1}} \cdots x_n^{a_n} \\ &\leqslant f(x_1, \ldots, x_{i-1}, 0, x_{i+1}, \ldots, x_n). \end{aligned}$$

Thus, $t^{\mathbf{B}}(x_1, \ldots, x_n) = x_1^{a_1} \cdots x_{i-1}^{a_{i-1}} x_{i+1}^{a_{i+1}} \cdots x_n^{a_n}$ is also an implicant of f and, since $t^{\mathbf{A}} < t^{\mathbf{B}}$, this contradicts the fact that $t^{\mathbf{A}}$ is a prime implicant of f.

The second part of the theorem can be shown in a similar manner.

Exercises and Supplements

1. Consider the partially ordered sets $(P, \leqslant)$ and $(Q, \leqslant)$ whose Hasse diagrams are given in Figs. 11.6a, b, respectively. Determine which diagram corresponds to a lattice.
2. Prove that if $(L, \{\wedge, \vee\})$ is a lattice, then for every finite, nonempty subset K of L, $\inf K$ and $\sup K$ exist.
3. Prove that every chain is a lattice.
4. Let $\mathcal{L} = (L, \{\wedge, \vee\})$ be a lattice and let x and y be two elements of L. Prove that the least sublattice of $\mathcal{L}$ that contains x and y is $\{x, y, x \wedge y, x \vee y\}$.

Let $\mathcal{L} = (L, \{\wedge, \vee\})$ be a lattice. A nonempty subset I of L is an *ideal* of $\mathcal{L}$ if $x \vee y \in I$ holds if and only if both $x \in I$ and $y \in I$. A *filter* of $\mathcal{L}$ is a nonempty subset F of L such that $x \wedge y \in F$ if and only if $x \in F$ and $y \in F$.

5. Prove that a set K is an ideal of the lattice $\mathcal{L} = (L, \{\wedge, \vee\})$ if and only if $x \in K$ and $y \in K$ imply $x \vee y \in K$, and $x \in K$ and $t \leqslant x$ imply $t \in K$.

6. Prove that a set K is a filter of the lattice $\mathcal{L} = (L, \{\wedge, \vee\})$ if and only if $x \in K$ and $y \in K$ imply $x \wedge y \in I$, and $x \in K$ and $t \geqslant x$ imply $t \in K$.
7. Prove that, for every element x of a lattice $\mathcal{L} = (L, \{\wedge, \vee\})$, the set $I_x = \{t \in L \mid t \leqslant x\}$ is an ideal and the set $F_x = \{t \in L \mid x \leqslant t\}$ is a filter. They are referred to as the *principal ideal* of x and the *principal filter* of x.
8. Let $\mathcal{B} = (B, \{\wedge, \vee, \bar{}\,, 0, 1\})$ be a Boolean algebra. A subset of B is an ideal (filter) if it is an ideal (filter) of the underlying lattice $(B, \{\wedge, \vee\})$. Prove that D is an ideal of $\mathcal{B}$ if and only if $D' = \{\bar{x} \mid x \in D\}$ is a filter.
9. Let $\mathcal{L} = (L, \{\wedge, \vee\})$ be a lattice and let $\mathbf{A} = (a_{ij})$ be an $m \times n$ matrix of elements of L.

   (a) Prove the following generalization of the minimax inequality (see Exercise 23 of Chap. 5):
   $$\bigvee_j \bigwedge_i a_{ij} \leqslant \bigwedge_i \bigvee_j a_{ij}$$
   (b) Suppose that L has the least element 0. Write the inequality that follows from the application of the minimax inequality to the matrix
   $$\mathbf{A} = \begin{pmatrix} 0 & a & b \\ b & 0 & c \\ a & c & 0 \end{pmatrix}.$$

10. Prove the following generalization of Theorem 11.40. In a distributive lattice $\mathcal{L} = (L, \{\wedge, \vee\})$, the equalities $x \vee y = x \vee z$ and $x \wedge y = x \vee z$ imply $y = z$. Conversely, if $x \vee y = x \vee z$ and $x \wedge y = x \vee z$ imply $y = z$ for all $x, y, z \in L$, then L is distributive.
11. Let x, y, z be three elements of a lattice $\mathcal{L} = (L, \wedge, vee)$. Prove that each of the sublattices generated by $x \vee y, y \vee z, z \vee x$ and by $x \wedge y, y \wedge z, z \wedge x$ is distributive.
12. Prove that every lattice having four elements is distributive.
13. Let $\mathcal{L} = (L, \{\wedge, \vee\})$ be a modular lattice. Prove that if $x \leqslant y$ or if $y \leqslant x$, $a \wedge x = a \wedge y$, and $a \vee x = a \vee y$, then $x = y$.
14. Prove that a lattice $\mathcal{L} = (L, \{\wedge, \vee\})$ is modular if and only if
   $$x \wedge (y \vee (x \wedge z)) = (x \wedge y) \vee (x \wedge z)$$
   for every $x, y, z \in L$.
15. Let $\mathcal{S}(\mathbb{R}^n)$ be the collection of subspaces of $\mathbb{R}^n$. Prove that

   (a) for $S, T \in \mathcal{S}(\mathbb{R}^n)$ we have $S = S + T$ if and only if $T \subseteq S$;
   (b) $(\mathcal{S}(\mathbb{R}^n), \cap, +)$ is a lattice having $\{\mathbf{0}\}$ as its least element and S as its greatest element.

16. Prove that if $R, S, T \in \mathcal{S}(\mathbb{R}^n)$ and $R \subseteq T$, then $R + (S \cap T) = (R + S) \cap T$. In other words, $(\mathcal{S}(\mathbb{R}^n), \cap, +)$ is a modular lattice.

Solution: By the submodular inequality we have $R + (S \cap T) \subseteq (R + S) \cap T$. To prove the reverse inclusion let $\mathbf{x} \in (R + S) \cap T$. This implies $\mathbf{x} \in T$ and $\mathbf{x} = \mathbf{y} + \mathbf{z}$, where $\mathbf{y} \in R$ and $\mathbf{z} \in S$. We have $\mathbf{z} = \mathbf{x} - \mathbf{y} \in T$ because $\mathbf{x} \in T$ and $\mathbf{y} \in R \subseteq T$. This implies $\mathbf{z} \in S \cap T$, so $\mathbf{x} \in R + (S \cap T)$. This establishes the modular equality.

17. Let $\mathcal{L} = (L, \{\wedge, \vee\})$ be a lattice that has the least element 0. If the set $\{t \in L \mid x \vee t = 0\}$ has a largest element x^*, then we say that x^* is the *pseudocomplement* of x. If every element of L has a pseudocomplement, then we say that L is a *pseudocomplemented lattice*.

 (a) Prove that any pseudocomplemented lattice has a largest element.
 (b) Prove that if $\mathcal{L}$ is a chain having the least element 0 and the largest element 1, then L is pseudocomplemented.
 (c) Prove that $x \leqslant x^{**}$ and $x^* = x^{***}$ for $x \in L$.
 (d) Prove that $(x \wedge y)^{**} = x^{**} \wedge y^{**}$ for $x, y \in L$.

18. Let $(S, \leqslant)$ be a poset, x be an element of S, and $I_x = \{s \in S \mid s \leqslant x\}$.

 (a) Prove that for every $x \in S$ the set $I_x = \{s \in S \mid s \leqslant x\}$ is an order ideal of $(S, \leqslant)$. This is *the principal order ideal of* x.
 (b) Let $\mathcal{I}_p(S, \leqslant)$ be the collection of principal order ideals of $(S, \leqslant)$ and let $f : S \longrightarrow \mathcal{I}_p(S, \leqslant)$ be the mapping defined by $f(x) = I_x$ for $x \in S$. Prove that f is a monotonic injection.
 (c) Let T be a subset of S. Prove that if $\sup T$ ($\inf T$) exists in $(S, \leqslant)$, then $\sup\{I_x \mid x \in T\}$ in $(\mathcal{I}_p(S, \leqslant), \subseteq)$ is $I_{\sup T}$ ($\inf\{I_x \mid x \in T\}$ in $(\mathcal{I}_p(S, \leqslant), \subseteq)$ is $I_{\inf T}$).
 (d) Prove that the poset of principal order ideals of S, $(\mathcal{I}_p(S, \leqslant), \subseteq)$ is a complete lattice.

19. Let $(L_1, \leqslant)$ and $(L_2, \leqslant)$ be two complete lattices and let $f : L_1 \longrightarrow L_2$ be a monotonic function between these posets. Prove that

$$f\left(\bigvee K\right) \geq \bigvee f(K),$$
$$f\left(\bigwedge K\right) \leq \bigwedge f(K),$$

for every subset K of L_1.

20. Let $(L, \leqslant)$ be a complete lattice and let $f : L \longrightarrow L$ be a monotonic mapping. Prove that there exists $x \in L$ such that $f(x) = x$ (Tarski's fixed-point theorem).
Solution: Let $T = \{x \in L \mid x \leqslant f(x)\}$ and $t = \sup T$. Since $x \leqslant t$, we have $f(x) \leqslant f(t)$ for every $x \in T$, so $x \leqslant f(x) \leqslant f(t)$. This implies $t \leqslant f(t)$, so $t \in T$. Therefore, $f(t) \leqslant f(f(t))$, so $f(t) \in T$, which implies $f(t) \leq t$. This shows that $t = f(t)$.

21. Let S and T be two sets and let $f : S \longrightarrow T$, $g : T \longrightarrow S$ be two injective functions. Define the function $F : \mathcal{P}(S) \longrightarrow \mathcal{P}(S)$ as $F(U) = S - g(T - f(U))$ for every $U \in \mathcal{P}(S)$.

(a) Prove that F is a monotonic mapping between the complete lattices $(\mathcal{P}(S), \subseteq)$ and $(\mathcal{P}(T), \subseteq)$.
(b) Let $U_0 \subseteq S$ be a fixed point of F. Define the function $h : S \longrightarrow T$ by

$$h(x) = \begin{cases} f(x) & \text{if } x \in U_0, \\ y & \text{if } x \notin U_0 \text{ and } g(y) = x. \end{cases}$$

Show that h is well-defined and, moreover, is a bijection. (The existence of a bijection h between S and T when the injections f and g exist is known as the *Schröder-Bernstein theorem.*)

22. Prove that $f : B^n \longrightarrow B$ is a Boolean function if and only if

$$(x_1, \ldots, x_n)^{\mathbf{A}} f(x_1, \ldots, x_n) = (x_1, \ldots, x_n)^{\mathbf{A}} f(\mathbf{A})$$

for every $\mathbf{A} \in \{0, 1\}^n$.
23. Prove that there are $2^{n-k}\binom{n}{k}$ n-ary terms of rank k.
24. A *Horn term* is a term $t^{\mathbf{A}} : B^n \longrightarrow B$ such that $\mathbf{A}$ contains at most one 0. Prove that if the consensus $t^{\mathbf{C}}$ of the Horn $t^{\mathbf{A}}, t^{\mathbf{B}}$ exists, then $t^{\mathbf{C}}$ is a Horn term.
25. Let $\mathcal{B} = (B, \{\wedge, \vee, \bar{\ }, 0, 1\})$ be a Boolean algebra and let $f : B^n \longrightarrow B$ be a Boolean function. Prove that, for every $b \in B$, there exists a Boolean function $f_{(b)} : B^n \longrightarrow B$ such that

$$b \wedge f(x_0, \ldots, x_{n-1}) = f_{(b)}(b \wedge x_0, \ldots, b \wedge x_{n-1})$$

for $(x_0, \ldots, x_{n-1}) \in B^n$.
Hint: The argument is by induction on the definition of Boolean functions.
26. Let S be a set and let π be a partition of S. Prove that the collection of π-saturated subsets of S is a Boolean subalgebra S_π of the Boolean algebra $(\mathcal{P}(S), \{\cap, \cup, \bar{\ }, \emptyset, S\})$.

Let $\mathcal{L} = (L, \{\wedge, \vee\})$ be a lattice that has 0 as its least element and let $A(\mathcal{L})$ be the set of its atoms. Recall that an atom for a poset was introduced in Definition 2.18. $\mathcal{L}$ is an *atomic lattice* if for every $x \in L$ we have $x = \sup\{a \in A(\mathcal{L}) \mid a \leqslant x\}$.

27. Let $\mathcal{B} = (B, \{\wedge, \vee, \bar{\ }, 0, 1\})$ be a Boolean algebra and let $A(\mathcal{B})$ be the set of its atoms. Prove that:

(a) if $a \in A(\mathcal{B})$, then $a \leqslant x \vee y$ if and only if either $a \leqslant x$ or $a \leqslant y$;
(b) $a \leqslant x \wedge y$ if and only if both $a \leqslant x$ and $a \leqslant y$;
(c) if $a \leqslant x \oplus y$, then we have either $a \leqslant x \wedge \bar{y}$ or $a \leqslant \bar{x} \wedge y$.

28. Let $\mathcal{B} = (B, \{\wedge, \vee, \bar{\ }, 0, 1\})$ be a Boolean algebra and let $A(\mathcal{B})$ be the set of its atoms. Define the function $h : B \longrightarrow \mathcal{P}(A(\mathcal{B}))$ as

$$h(x) = \{a \in A(\mathcal{B}) \mid a \leqslant x\}.$$

Prove that

(a) h is a Boolean algebra homomorphism between the Boolean algebras $\mathcal{B}$ and $(\mathcal{P}(A(\mathcal{B})), \{\cap, \cup, \bar{\ }, \emptyset, A(\mathcal{B})\})$;
(b) if $\mathcal{B}$ is an atomic Boolean algebra, then h is injective;
(c) if $\mathcal{B}$ is complete, then h is surjective;
(d) a Boolean algebra is isomorphic to the Boolean algebra of subsets if and only if it is both atomic and complete.

29. Let $\mathcal{C}$ be a finite collection of subsets of a set S. Prove that the subalgebra of $(\mathcal{P}(S), \{\cap, \cup, \bar{\ }, \emptyset, S\})$ generated by $\mathcal{C}$ coincides with the collection of all $\pi_{\mathcal{C}}$-saturated sets, where $\pi_{\mathcal{C}}$ is the partition defined in Supplement 6 of Chap. 1. Further, show that the atoms of this subalgebra are the blocks of the partition $\pi_{\mathcal{C}}$.
30. Let S and T be two sets and let $f : S \longrightarrow T$ be a mapping.

(a) Prove that the function $F : \mathcal{P}(T) \longrightarrow \mathcal{P}(S)$ defined by $F(V) = f^{-1}(V)$ for $V \in \mathcal{P}(T)$ is a Boolean algebra morphism between $(\mathcal{P}(T), \{\cap, \cup, \bar{\ }, \emptyset, T\})$ and $(\mathcal{P}(S), \{\cap, \cup, \bar{\ }, \emptyset, S\})$.
(b) Let $\mathcal{D} = \{D_1, \ldots, D_r\}$ be a finite collection of subsets of T and let $\mathcal{C} = \{F(D) \mid D \in \mathcal{D}\}$ be the corresponding finite collection of subsets of C.

If $\pi_{\mathcal{D}}$ is the partition of T associated to $\mathcal{D}$, then prove that for any block B of this partition, $F(B)$ is either the empty set or a block of the partition $\pi_{\mathcal{C}}$ of S, and each block of the partition $\pi_{\mathcal{C}}$ is of the form $F(B)$. Further, if f is a surjective mapping, then $F(B)$ is always a block of $\pi_{\mathcal{C}}$.
Solution: The first part is a consequence of Theorems 1.63 and 1.65. For the second part, let

$$D_{a_1,\ldots,a_r} = D_1^{a_1} \cap \cdots \cap D_r^{a_r}$$

be an atom of $\pi_{\mathcal{D}}$ for some $(a_1, \ldots, a_r)$. Note that $F(D_{a_1,\ldots,a_r}) = C_1^{a_1} \cap \cdots \cap C_r^{a_r}$, where $C_i = f^{-1}(D_i) = F(D_i)$ for $1 \leqslant i \leqslant r$. If this intersection is nonempty, then it is clearly a block of $\pi_{\mathcal{P}}$.
If f is surjective, the preimage of any nonempty set $f^{-1}(D_{a_1,\ldots,a_r})$ is nonempty and therefore a block of $\pi_{\mathcal{C}}$.

31. Let $F : \{0, 1\}^n \longrightarrow \{0, 1\}^n$ be a function. Prove that there exists a bijection $G : \{0, 1\}^n \longrightarrow \{0, 1\}^n$ such that $G(\mathbf{x}) \wedge \mathbf{x} = F(\mathbf{x}) \wedge \mathbf{x}$ for every $\mathbf{x} \in \{0, 1\}^n$.
Solution: Without loss of generality, we may assume that $F(\mathbf{x}) \leqslant \mathbf{x}$ for $\mathbf{x} \in \{0, 1\}^n$. Thus, we need to prove the existence of G such that $G(\mathbf{x}) \wedge \mathbf{x} = F(\mathbf{x})$. For $\mathbf{x} \subseteq \{0, 1\}^n$ let $K_F(\mathbf{x}) = \{\mathbf{u} \in \{0, 1\}^n \mid \mathbf{u} \wedge \mathbf{x} = F(\mathbf{x})\}$. If $X \subseteq \{0, 1\}^n$, define $K_F(X) = \bigcup\{K(\mathbf{x}) \mid \mathbf{x} \in X\}$. Then, we should have $G(\mathbf{x}) \in K_F(\mathbf{x})$ for each $\mathbf{x} \in \{0, 1\}^n$. To obtain the result it suffices to show that, for every X, we have $|X| \leqslant |K_F(X)|$ because this would imply that there is a bijection G such that $G(\mathbf{x}) \in K_F(\mathbf{x})$.

Note that if $F(\mathbf{x}) = \mathbf{x}$, then $K_F(\mathbf{x}) = \{\mathbf{u} \in \{0,1\}^n \mid \mathbf{u} \geqslant \mathbf{x}\}$, so $\mathbf{x} \in K_F(\mathbf{x})$ for every $\mathbf{x} \in \{0,1\}^n$, which implies $|X| \leqslant |K_F(X)|$.
For $\mathbf{x} = (x_1, \ldots, x_n)$ and $F(\mathbf{x}) = (y_1, y_2, \ldots, y_n)$, define $F_1 : \{0,1\}^n \longrightarrow \{0,1\}^n$ by modifying the first component of $F(\mathbf{x})$ as

$$F_1(\mathbf{x}) = (x_1, y_2, \ldots, y_n).$$

If $\mathbf{u} = (u_1, u_2, \ldots, u_n) \in \{0,1\}^n$, denote by $\mathbf{u}_{[0]}$ and $\mathbf{u}_{[1]}$ the n-tuples $\mathbf{u}_{[0]} = (0, u_2, \ldots, u_n)$ and $\mathbf{u}_{[1]} = (1, u_2, \ldots, u_n)$.

We claim that $|K_{F_1}(X)| \leqslant |K_F(X)|$. To prove this inequality, it suffices to show that $|K_{F_1}(X) \cap \{\mathbf{u}_{[0]}, \mathbf{u}_{[1]}\}| \leqslant |K_F(X) \cap \{\mathbf{u}_{[0]}, \mathbf{u}_{[1]}\}|$ for every $\mathbf{u} \in \{0,1\}^n$.

If $\mathbf{u}_{[0]} \in K_{F_1}(X)$, then $\{\mathbf{u}_{[0]}, \mathbf{u}_{[1]}\} \subseteq K_F(X)$. Indeed, $\mathbf{u}_{[0]} \in K_{F_1}(X)$ implies $\mathbf{u}_{[0]} \wedge \mathbf{x} = F_1(\mathbf{x})$ for some $\mathbf{x} = (x_1, x_2, \ldots, x_n) \in X$, which yields $x_1 = 0$. Since $F(\mathbf{x}) \leqslant \mathbf{x}$, it follows that $(F(\mathbf{x}))_1 = 0$ and $(F(\mathbf{x}))_1 = (F_1(\mathbf{x}))_1$. Thus, $\mathbf{u}_{[1]} \wedge \mathbf{x} = \mathbf{u}_{[0]} \wedge \mathbf{x} = F(\mathbf{x})$, so $\{\mathbf{u}_{[0]}, \mathbf{u}_{[1]} \subseteq K_F(X)\}$. If $\mathbf{u}_{[1]} \in\in F_1(X)$ and $\mathbf{u}_{[0]} \notin F_1(X)$, then $\{\mathbf{u}_{[0]}, \mathbf{u}_{[1]}\} \cap K_F(X) \neq \emptyset$. Under these assumptions, there exists $\mathbf{x} \in X$ such that $\mathbf{u}_{[1]} \wedge \mathbf{x} = F_1(\mathbf{x})$ and $\mathbf{u}_{[0]} \wedge \mathbf{x} \neq F_1(\mathbf{x})$. Note that $(\mathbf{u}_{[0]} \wedge \mathbf{x})_i = (\mathbf{u}_{[1]} \wedge \mathbf{x})_i = (F_1(\mathbf{x}))_i = (F(\mathbf{x}))_i$ for $2 \leqslant i \leqslant n$. Also, we have $(F(\mathbf{x}))_1 = 0 = (\mathbf{u}_{[0]} \wedge \mathbf{x})_1$ or $(F(\mathbf{x}))_1 = 1 = (\mathbf{u}_{[1]} \wedge \mathbf{x})_1$. Thus, either $\mathbf{u}_{[0]} \wedge \mathbf{x} = F(\mathbf{x})$ or $\mathbf{u}_{[1]} \wedge \mathbf{x} = F(\mathbf{x})$.
The treatment applied to the first coordinate can now be repeated for the second component starting from F_1 to produce a function $F_2 : \{0,1\}^n \longrightarrow \{0,1\}^n$ such that $|K_{F_2}(X)| \leq |K_{F_1}(X)|$, etc. After n steps, we reach a function F_n such that $F_n(\mathbf{x}) = \mathbf{x}$. We have $K_{F_n}(\mathbf{x}) = \{\mathbf{u} \in \{0,1\}^n \mid \mathbf{u} \geqslant \mathbf{x}\}$, so $X \subseteq K_{F_n}(X)$, which implies $|X| \leqslant |K_{F_n}(X)| \leqslant |K_F(X)$.

32. Prove that if $\mathcal{L}$ is a relatively complemented lattice of finite length, then $\mathcal{L}$ is an atomic lattice.

Bibliographical Comments

The most known reference for lattice theory is [2]. The books [1, 3] are essential references for Boolean algebra and Boolean functions. A comprehensive reference on pseudo-Boolean functions is the monograph [4]. Supplement 5 appears in a slightly different form in [5].

The modularity of the lattice of subspaces of an n-dimensional linear space was established in [6].

References

1. S. Rudeanu, *Boolean Functions and Equations* (North-Holland, Amsterdam, 1974)
2. G. Birkhoff, *Lattice Theory*, 3rd edn. (American Mathematical Society, Providence, 1973)
3. S. Rudeanu, *Lattice Functions and Equations* (Springer, London, 2001)
4. P.L. Hammer, S. Rudeanu, *Méthodes Booléennes en Recherche Operationnelle* (Dunod, Paris, 1970)
5. D. Pollard, *Empirical Processes: Theory and Applications* (Institute of Mathematical Statistics, Hayward, 1990)
6. G. Birkhoff, J. von Neumann, The logic of quantum mechanics. Ann. Math. **37**, 823–843 (1936)

Chapter 12
Applications to Databases and Data Mining

12.1 Introduction

This chapter presents an introduction to the relational model, which is of paramount importance for data mining. We continue with certain equivalence relations (and partitions) that can be associated to sets of attributes of tables.

An algebraic approach to the notion of entropy and several of its generalizations is also discussed because entropy is used in data mining for evaluating the concentrations of values of certain data features. Generalized measures and data constraints that can be formulated using these measures are included. Finally, we discuss certain types of classifiers, namely, decision trees and perceptrons and the associated learning processes.

12.2 Relational Databases

Relational databases are the mainstay of contemporary databases. The principles of relational databases were developed by Codd in the early 1970s [1, 2], and various extensions have been considered since. In this section, we illustrate applications of several notions introduced earlier to the formalization of database concepts.

The notion of a tabular variable (or relational variable) was introduced by Date in [3]; we also formalize the notion of table of a relational variable. To reflect the implementations of a relational database system we assume that table contents are sequences of tuples (and not just sets of tuples, a simplification often adopted in the literature that is quite distant from reality).

Let $\mathcal{U}$ be a countably infinite injective sequence having pairwise distinct members, $\mathcal{U} = (A_0, A_1, \ldots)$. The components of $\mathcal{U}$ are referred to as *attributes* and denoted, in general, by capital letters from the beginning of the alphabet, $A, B, C, \ldots$ We also consider a collection of sets indexed by the components of $\mathcal{U}$, $\mathcal{D} = \{D_A \mid A \in U\}$.

D. A. Simovici and C. Djeraba, *Mathematical Tools for Data Mining*,
Advanced Information and Knowledge Processing, DOI: 10.1007/978-1-4471-6407-4_12,

The set D_A is referred to as the *domain* of the attribute A and denoted alternatively as $\text{Dom}(A)$. We assume that each set D_A contains at least two elements.

Let H be a finite subset of *set*($\mathfrak{U}$), $H = \{A_{i_1}, \ldots, A_{i_p}\}$. We refer to such a set as a *heading*. In keeping with the tradition of the field of relational databases, we shall denote H as $H = A_{i_1} \cdots A_{i_p}$. For example, instead of writing $H = \{A, B, C, D, E\}$, we shall write $H = ABCDE$.

The *set of tuples on* $H = A_{i_1} \cdots A_{i_p}$ is the set $D_{A_{i_1}} \times \cdots \times D_{A_{i_p}}$ denoted by *tupl*(H). Thus, a *tuple* t *on the heading* $H = A_{i_1} \cdots A_{i_p}$ is a sequence $t = (t_1, \ldots, t_p)$ such that $t_j \in \text{Dom}(A_{i_j})$ for $1 \leqslant j \leqslant p$.

A *tabular variable* is a pair $\tau = (T, H)$, where T is a word over an alphabet to be defined later and H is a heading.

A value of a tabular variable $\tau = (T, H)$ is a triple $\theta = (T, H, \mathbf{r})$, where $\mathbf{r}$ is a sequence on *tupl*(H). We refer to such a triple as a *table of the tabular variable* τ or, a τ*-table*; when the tabular variable is clear from the context or irrelevant, we refer to θ just as a *table*.

The set *set*$(\mathbf{r})$ of tuples that constitute the components of a tuple sequence $\mathbf{r}$ is a p-ary relation on the collection of sets *tupl*(H); this justifies the term "relational" used for the basic database model. If $\theta_1 = (T_1, H, \mathbf{r}_1)$ and $\theta_2 = (T_2, H, \mathbf{r}_2)$ are two tables and *set*$(\mathbf{r}_1) =$ *set*$(\mathbf{r}_2)$ we say that θ_1 and θ_2 are *coextensive* and write $\theta_1 \equiv \theta_2$.

Example 12.1 Consider a tabular variable that is intended to capture the description of a collection of objects,

$$\tau = (\text{OBJECTS, shape length width height color}),$$

where

$$\text{Dom}(shape) = \text{Dom}(color) = \{a, \ldots, z\}^* \text{ and}$$
$$\text{Dom}(length) = \text{Dom}(width) = \text{Dom}(color) = \mathbb{N}.$$

A value of this variable is

$$(\text{OBJECTS, shape length width height color}, \mathbf{r}),$$

where $\mathbf{r}$ consists of the tuples

(cube, 5, 5, 5, red),
(sphere, 3, 3, 3, blue),
(pyramid, 5, 6, 4, blue),
(cube, 2, 2, 2, red),
(sphere, 3, 3, 3, blue),

that belong to *tupl*(shape length width height color). It is convenient to represent this table graphically as

The set *set*$(\mathbf{r})$ of tuples that corresponds to the sequence $\mathbf{r}$ of tuples of the table is

Objects				
Shape	Length	Width	Height	Color
Cube	5	5	5	Red
Sphere	3	3	3	Blue
Pyramid	5	6	4	Blue
Cube	2	2	2	Red
Sphere	3	3	3	Blue

$$set(\mathbf{r}) = \{(\text{cube}, 5, 5, 5, \text{red}), (\text{sphere}, 3, 3, 3, \text{blue}),\\ (\text{pyramid}, 5, 6, 4, \text{blue}), (\text{cube}, 2, 2, 2, \text{red})\}.$$

Note that duplicate tuples do not exist in $set(\mathbf{r})$.

We can now formalize the notion of a relational database.

Definition 12.2 *A* relational database *is a finite, nonempty collection* $\mathcal{D}$ *of tabular variables* $\tau_i = (T_k, H_k)$*, where* $1 \leqslant k \leqslant m$ *such that* $i \neq j$ *implies* $T_i \neq T_j$ *for* $1 \leqslant i, j \leqslant m$.

In other words, a relational database $\mathcal{D}$ is a finite collection of tabular variables that have pairwise distinct names.

Let $\mathcal{D} = \{\tau_1, \ldots, \tau_m\}$ be a relational database. A *state of* $\mathcal{D}$ is a sequence of tables $(\theta_1, \ldots, \theta_m)$ such that θ_i is a table of τ_i for $1 \leqslant i \leqslant m$. The set of states of a relational database $\mathcal{D}$ will be denoted by $\mathcal{S}_{\mathcal{D}}$.

To discuss further applications we need to introduce table projection, an operation on tables that allows us to build new tables by extracting "vertical slices" of the original tables.

Definition 12.3 *Let* $\theta = (T, H, \mathbf{r})$ *be a table, where* $H = A_1 \cdots A_p$ *and* $\mathbf{r} = (t_1, \ldots, t_n)$*, and let* $K = A_{i_1} \cdots A_{i_q}$ *be a subsequence of* $set(H)$.

The projection *of a tuple* $t \in tupl(H)$ *on* K *is the tuple* $t[K] \in tupl(K)$ *defined by* $t[K](j) = t(i_j)$ *for every* j, $1 \leqslant i \leqslant q$.

The projection *of the table* θ *on* K *is the table* $\theta[K] = (T[K], K, \mathbf{r}[K])$*, where* $\mathbf{r}[K]$ *is the sequence* $(t_1[K], \ldots, t_n[K])$.

Observe that, for every tuple $t \in tupl(H)$, we have $t[\emptyset] = \lambda$; also, $t[H] = t$.

Example 12.4 The projection of the table

Objects				
Shape	Length	Width	Height	Color
Cube	5	5	5	Red
Sphere	3	3	3	Blue
Pyramid	5	6	4	Blue
Cube	2	2	2	Red
Sphere	3	3	3	Blue

on the set K = shape color is the table

Objects (shape color)	
Shape	Color
Cube	Red
Sphere	Blue
Pyramid	Blue
Cube	Red
Sphere	Blue

Two simple but important properties of projection are given next.

Theorem 12.5 *Let H be a set of attributes, $u, v \in tupl(H)$, and let K and L be two subsets of H. The following statements hold:*

(i) $u[K][K \cap L] = u[L][K \cap L] = u[K \cap L]$.
(ii) *The equality $u[KL] = v[KL]$ holds if and only if $u[K] = v[K]$ and $u[L] = v[L]$.*

Proof The argument is a straightforward application of Definition 12.3 and is left to the reader. □

Definition 12.6 *Let $\tau = (T, H, \mathbf{t})$ and $\sigma = (S, K, \mathbf{s})$ be two tables and let $t \in set(\mathbf{t})$ and $s \in set(\mathbf{s})$.*

The tuples t and s are joinable *if $t[H \cap K] = s[H \cap K]$.*

If t and s are joinable, their join *is the tuple $u \in tupl(H \cup K)$ given by*

$$u[A] = \begin{cases} t[A] & \text{if } A \in H, \\ s[A] & \text{if } A \in K. \end{cases}$$

The tuple u is denoted by $t \bowtie s$.

If $H \cap K = \emptyset$, then every tuple $t \in set(\mathbf{t})$ is joinable with every tuple $s \in set(\mathbf{s})$. If $H = K$, then t is joinable with s if and only if $t = s$.

Example 12.7 Let $\tau = (T, ABC, \mathbf{r})$ and $\sigma = (S, BCDE, \mathbf{r})$ be the tables

T			
	A	*B*	*C*
t_1	a_1	b_1	c_1
t_2	a_1	b_2	c_2
t_3	a_2	b_1	c_1
t_4	a_3	b_2	c_3
t_5	a_2	b_3	c_1

and

S				
	B	*C*	*D*	*E*
s_1	b_1	c_1	d_1	e_1
s_2	b_1	c_1	d_2	e_2
s_3	b_2	c_2	d_3	e_2
s_4	b_3	c_3	d_2	e_1

It is clear that both t_1 and t_3 are joinable with s_1 and s_2 and that t_2 is joinable with s_3. Their joins are

$$t_1 \bowtie s_1 = (a_1, b_1, c_1, d_1, e_1) \quad t_1 \bowtie s_2 = (a_1, b_1, c_1, d_2, e_2)$$
$$t_3 \bowtie s_1 = (a_2, b_1, c_1, d_1, e_1) \quad t_3 \bowtie s_2 = (a_2, b_1, c_1, d_2, e_2)$$
$$t_2 \bowtie s_3 = (a_1, b_2, c_2, d_3, e_2).$$

Definition 12.8 *Let $\tau = (T, H, \boldsymbol{t})$ and $\sigma = (S, K, \boldsymbol{s})$ be two tables, where $\boldsymbol{t} = (t_1, \ldots, t_m)$ and $\boldsymbol{s} = (s_1, \ldots, s_n)$.*

Their join *is the table $\tau \bowtie \sigma$ given by $\tau \bowtie \sigma = (T \bowtie S, HK, \boldsymbol{w})$, where $\boldsymbol{w}$ is the sequence of tuples obtained by concatenating the sequences obtained by joining each tuple t in* **t** *with each joinable tuple in* **s**.

We denote $set(\boldsymbol{w})$ by $set(\boldsymbol{t}) \bowtie set(\boldsymbol{s})$.

Example 12.9 The join of tables τ and σ defined in Example 12.7 is the table

$T \bowtie S$					
	A	B	C	D	E
$t_1 \bowtie s_1$	a_1	b_1	c_1	d_1	e_1
$t_1 \bowtie s_2$	a_1	b_2	c_2	d_2	e_2
$t_2 \bowtie s_3$	a_2	b_3	c_1	d_3	e_2
$t_3 \bowtie s_1$	a_2	b_1	c_1	d_1	e_1
$t_3 \bowtie s_2$	a_3	b_2	c_3	d_2	e_2

Definition 12.10 *Let $\theta = (T, H, \boldsymbol{r})$ be a table, where $\boldsymbol{r} = (t_1, \ldots, t_n)$. The* indiscernibility relation defined by a set of attributes X, *$X \subseteq set(H)$ is the relation $\epsilon^X \subseteq \{1, \ldots, n\}^2$ given by*

$$\epsilon^X = \{(p, q) \in \{1, \ldots, n\}^2 \mid t_p[X] = t_q[X]\}.$$

It is easy to verify that ϵ^X is an equivalence for every set of attributes X. The partition of $\{1, \ldots, n\}$ that corresponds to this equivalence will be denoted by π^X.

Example 12.11 Consider again the table introduced in Example 12.4.

Objects					
	Shape	Length	Width	Height	Color
t_1	Cube	5	5	5	Red
t_2	Sphere	3	3	3	Blue
t_3	Pyramid	5	6	4	Blue
t_4	Cube	2	2	2	Red
t_5	Sphere	3	3	3	Blue

Several partitions defined by sets of attributes of this table are:

$$\pi^{\text{shape}} = \{\{t_1, t_4\}, \{t_2, t_5\}, \{t_3\}\}$$
$$\pi^{\text{length}} = \{\{t_1, t_3\}, \{t_2, t_5\}, \{t_4\}\},$$

$$\pi^{\text{width}} = \{\{t_1\}, \{t_2, t_5\}, \{t_3\}, \{t_4\}\},$$
$$\pi^{\text{height}} = \{\{t_1\}, \{t_2, t_5\}, \{t_3\}, \{t_4\}\},$$
$$\pi^{\text{color}} = \{\{t_1, t_4\}, \{t_2, t_3, t_5\}\},$$
$$\pi^{\text{shape length}} = \{\{t_1\}, \{t_4\}, \{t_2, t_5\}, \{t_3\}\},$$
$$\pi^{\text{shape color}} = \{\{t_1, t_4\}, \{t_2, t_5\}, \{t_3\}\}.$$

Theorem 12.12 *Let $\theta = (T, H, \mathbf{r})$ be a table and let X and Y be two sets of attributes, $X, Y \subseteq H$. We have $\epsilon^{XY} = \epsilon^X \cap \epsilon^Y$.*

Proof Let $t_p, t_q \in set(\mathbf{r})$ be two tuples such that $(p, q) \in \epsilon^{XY}$. This means that $t_p[XY] = t_q[XY]$. By the second part of Theorem 12.5, this holds if and only if $t_p[X] = t_p[X]$ and $t_p[Y] = t_q[Y]$; that is, if and only if $(p, q) \in \epsilon^X$ and $(p, q) \in \epsilon^Y$. Thus, $\epsilon^{XY} = \epsilon^X \cap \epsilon^Y$. □

Corollary 12.13 *Let $\theta = (T, H, \mathbf{r})$ be a table and let X and Y be two sets of attributes, $X, Y \subseteq H$. We have $\pi^{XY} = \pi^X \wedge \pi^Y$.*

Proof This statement follows immediately from Theorems 12.12 and 10.102. □

Corollary 12.14 *Let $\theta = (T, H, \mathbf{r})$ be a table and let X and Y be two sets of attributes, $X, Y \subseteq H$. If $X \subseteq Y$, we have $\pi^Y \leqslant \pi^X$.*

Proof Since $X \subseteq Y$, we have $XY = Y$, so $\pi^Y = \pi^X \wedge \pi^Y$, which implies $\pi^Y \leqslant \pi^X$. □

Definition 12.15 *A* reduct *of a table $\theta = (T, H, \mathbf{r})$ is a set of attributes L that satisfies the following conditions:*

(i) *$\pi^L = \pi^H$, and*
(ii) *L is a minimal set having the property (i); that is, for every $J \subset L$, we have $\pi^J \geqslant \pi^H$.*

The core *of θ is the intersection of the reducts of the table.*

Example 12.16 Let $\theta = (T, ABCDE, \mathbf{r})$ be the following table:

T					
	A	B	C	D	E
t_1	a_1	b_1	c_1	d_1	e_1
t_2	a_2	b_2	c_2	d_2	e_1
t_3	a_1	b_2	c_2	d_1	e_2
t_4	a_2	b_2	c_1	d_2	e_2
t_5	a_1	b_1	c_1	d_1	e_1
t_6	a_1	b_1	c_1	d_1	e_1
t_7	a_1	b_2	c_2	d_1	e_2

We have $\pi^H = \{\{t_1, t_5, t_6\}, \{t_3, t_7\}, \{t_2\}, \{t_4\}\}$. Note that we have $\pi^{AC} = \pi^H$ and $\pi^{DE} = \pi^H$. On the other hand, we also have

$$\pi^A = \{\{t_1, t_3, t_5, t_6, t_7\}, \{t_2, t_4\}\},$$
$$\pi^C = \{\{t_1, t_4, t_5, t_6\}, \{t_2, t_3, t_7\}\},$$
$$\pi^D = \{\{t_1, t_3, t_5, t_6, t_7\}, \{t_2, t_4\}\},$$
$$\pi^E = \{\{t_1, t_2, t_5, t_6\}, \{t_3, t_4, t_7\}\},$$

which shows that both AC and DE are reducts of this table.

Table reducts are minimal sets of attributes that have the same separating power as the entire set of attributes of the table. Example 12.16 shows that a table may possess several reducts.

Note that no two distinct reducts may be comparable as sets because of the minimality condition. Therefore, each maximal chain of sets in the poset $(\mathcal{P}(H), \subseteq)$ that joins $\emptyset$ to H may include at most one reduct. Thus, the largest number of reducts that a table with n attributes may have is $\binom{n}{\lfloor n/2 \rfloor}$.

Example 12.17 Let θ be the table

	T			
	A	B	C	D
t_1	a_1	b_1	c_1	d_1
t_2	a_1	b_2	c_1	d_2
t_3	a_2	b_1	c_1	d_1
t_4	a_2	b_2	c_1	d_2

It is easy to see that this table has two reducts, AB and AD. Therefore, the core of this table consists of the attribute A.

On the other hand, the core of the two-tuple table

	S			
	A	B	C	D
t_1	a_1	b_1	c_1	d_1
t_2	a_1	b_2	c_1	d_2

is empty because its two reducts, B and D, have no attributes in common.

The following theorem gives a characterization of table reducts.

Theorem 12.18 *Let $\theta = (T, H, \mathbf{r})$ be a table such that $|\mathbf{r}| = n$ and let $\delta: \{1, \ldots, n\}^2 \longrightarrow \mathcal{P}(H)$ be the function defined by*

$$\delta(i, j) = \{A \in set(H) \mid t_i[A] \neq t_j[A]\}$$

for $1 \leqslant i, j \leqslant n$. The set of attributes L is a reduct for θ if and only if $L \cap \delta(i, j) \neq \emptyset$ for every pair $(i, j) \in \{1, \ldots, n\}^2$ such that $\delta(i, j) \neq \emptyset$ and L is minimal with this property.

Proof Suppose that L is a reduct for θ and that (i, j) is a pair such that $\delta(i, j) \neq \emptyset$. The equality $\delta(i, j) \neq \emptyset$ implies that $(i, j) \notin \epsilon^H = \epsilon^L$, so $t_i[L] \neq t_j[L]$. Therefore, $L \cap \delta(i, j) \neq \emptyset$.

Suppose that there is a strict subset G of L such that $G \cap \delta(i, j) \neq \emptyset$ for every pair $(i, j) \in \{1, \ldots, n\}^2$ such that $\delta(i, j) \neq \emptyset$. This implies $\epsilon^G = \epsilon^H$, which contradicts the minimality of the reduct L.

Conversely, suppose that $L \cap \delta(i, j) \neq \emptyset$ for every pair $(i, j) \in \{1, \ldots, n\}^2$ such that $\delta(i, j) \neq \emptyset$ and L is minimal with this property. Since $L \subseteq H$, we have $\epsilon^H \subseteq \epsilon^L$.

Now let (h, k) be a pair in ϵ^L. Since t_h coincides with t_k on every attribute of L, it follows that we must have $\delta(h, k) = \emptyset$, which implies $(h, k) \in \epsilon^H$. Thus, $\epsilon^H = \epsilon^L$. If L is a minimal set satisfying the condition of the theorem, it follows immediately that L is minimal in the collection of sets of attributes that differentiate the tuples of θ, so L is a reduct. □

The notion of a key that is frequently used in databases is related to the notion of a reduct.

Definition 12.19 *A* key *of a table* $\theta = (T, H, \mathbf{r})$ *and* $\mathbf{r} = (t_1, \ldots, t_n)$ *is a set of attributes L that satisfies the following conditions:*

(i) $\pi^L = \alpha_{\{1,\ldots,n\}}$, *and*
(ii) *L is a minimal set having the property (i); that is, for every* $J \subset L$, *we have* $\pi^J \not\leq \alpha_{\{1,\ldots,n\}}$.

A table $\theta = (T, H, \mathbf{r})$ has a key if and only if the sequence $\mathbf{r}$ does not contain duplicate tuples.

12.3 Partitions and Functional Dependencies

Note that if two objects have the same shape, then they have the same color. We also note that the reverse implication is not true because two objects may have the same color without having the same shape. This observation suggests the introduction of a type of constraint that applies to the table contents for every table that is a value of a tabular variable.

Definition 12.20 *Let H be a set of attributes. A* functional dependency *is an ordered pair* (X, Y) *of subsets of H.*

The set of all functional dependencies on a set of attributes H is denoted by $\mathsf{FD}(H)$. If $(X, Y) \in \mathsf{FD}(H)$ we shall write this pair as $X \to Y$ using a well-established convention in database theory.

Definition 12.21 *Let* $\theta = (T, H, \mathbf{r})$ *be a table and let X and Y be two subsets of H. The table* θ satisfies the functional dependency $X \longrightarrow Y$ *if* $u[X] = v[X]$ *implies* $u[Y] = v[Y]$ *for every two tuples* $u, v \in set(\mathbf{r})$.

In other words, a table θ satisfies the functional dependency $X \rightarrow Y$ if and only if $\epsilon^X \subseteq \epsilon^Y$ or, equivalently, $\pi^X \leqslant \pi^Y$.

Example 12.22 Let us consider a tabular variable whose values are intended to store the data reflecting the instructors, students, and musical instruments studied by the students of a community music school. Lessons are scheduled once a week, and each instructor is teaching one instrument:

$$\tau = (\text{SCHEDULE}, \text{student instructor instrument day time room}).$$

Any table θ that is a value of this tabular variable must satisfy functional dependencies that reflect these "business rules" as well as other semantic restrictions:

$$\begin{aligned}
&\text{student instrument} \rightarrow \text{instructor},\\
&\text{instructor} \rightarrow \text{instrument},\\
&\text{student instrument} \rightarrow \text{day time},\\
&\text{room day time} \rightarrow \text{student instructor},\\
&\text{student day time} \rightarrow \text{room},\\
&\text{instructor day time} \rightarrow \text{room}.
\end{aligned}$$

For example, a possible value of this tabular variable is the table:

	Schedule					
	Student	Instructor	Instrument	Day	Time	Room
t_1	Margo	Donna	Piano	Mon	4	A
t_2	Danielle	Igor	Violin	Mon	4	B
t_3	Joshua	Donna	Piano	Mon	5	A
t_4	Ondine	Donna	Piano	Tue	3	A
t_5	Michael	Donna	Piano	Tue	4	A
t_6	Linda	Mary	Flute	Tue	4	B
t_7	Todor	Mary	Flute	Tue	5	A
t_8	Sarah	Emma	Piano	Tue	6	A
t_9	Samuel	Donna	Piano	Tue	6	B
t_{10}	Alex	David	Guitar	Tue	6	C
t_{11}	Dan	Emma	Piano	Wed	3	A
t_{12}	William	Mary	Flute	Wed	4	A
t_{13}	Nora	David	Guitar	Wed	4	B
t_{14}	Amy	Donna	Piano	Wed	5	A
t_{15}	Peter	Igor	Violin	Thr	4	A
t_{16}	Kenneth	David	Guitar	Thr	4	B
t_{17}	Patrick	Donna	Piano	Thr	5	A
t_{18}	Elizabeth	Emma	Piano	Thr	5	B
t_{19}	Helen	Mary	Flute	Thr	5	C
t_{20}	Cris	Mary	Flute	Fri	4	B
t_{21}	Richard	Igor	Violin	Fri	4	C
t_{22}	Yves	Donna	Piano	Fri	5	A
t_{23}	Paul	Emma	Piano	Fri	5	B
t_{24}	Colin	Igor	Violin	Fri	6	C

The reader can easily check that this table satisfies all functional dependencies identified after the definition of the tabular variable.

It is clear that if $X, Y \subseteq H$ and $Y \subseteq X$, any table $\theta = (T, H, \mathbf{r})$ satisfies the functional dependency $X \to Y$.

Definition 12.23 *A functional dependency $X \to Y$ is* trivial *if it is satisfied by every table $\theta = (T, H, \mathbf{r})$ such that $X, Y \in \mathcal{P}(H)$.*

Theorem 12.24 *Let H be a finite set of attributes. A functional dependency $X \to Y \in \mathsf{FD}(H)$ is trivial if and only if $Y \subseteq X$.*

Proof For each attribute $A \in H$, let u_A and v_A be two distinct values in $\text{Dom}(A)$. Suppose that $X \to Y$ is a trivial functional dependency and that Y is not included in X. This means that there exists an attribute $B \in Y - X$. Consider the table $\theta = (T, XY, \mathbf{r})$, where $\mathbf{r} = (t_1, t_2)$, where $t_1[A] = u_A$ for every $A \in XY$, and

$$t_2[A] = \begin{cases} u_A & \text{if } A \neq B, \\ v_B & \text{if } A = B. \end{cases}$$

Since $t_1[X] = t_2[X]$ and $t_1[Y] \neq t_2[Y]$, it follows that θ violates the functional dependency $X \to Y$, which contradicts the fact that $X \to Y$ is trivial.

The sufficiency of the condition is immediate. □

Suppose now that $\theta = (T, H, \mathbf{r})$ satisfies the functional dependencies $X \to Y$ and $Y \to Z$, where X, Y, Z are subsets of H. This means that $\pi^X \leqslant \pi^Y$ and $\pi^Y \leqslant \pi^Z$, which implies $\pi^X \leqslant \pi^Z$. Therefore, θ satisfies the functional dependency $X \to Z$.

If $\theta = (T, H, \mathbf{r})$ satisfies the functional dependency $X \to Y$ and W is a subset of H, then we have $\pi^X \leqslant \pi^Y$. Therefore, we have

$$\pi^{XW} = \pi^X \wedge \pi^W \leqslant \pi^Y \wedge \pi^W = \pi^{YW},$$

which means that θ satisfies the functional dependency $XW \to YW$.

In the database design process, it is necessary to identify functional dependencies satisfied by tables that are values of tabular variables. Thus, for a tabular variable $\tau = (T, H)$, the design of the database entails the construction of the *functional dependency schema* defined as a pair $S = (H, F)$, where $F \subseteq \mathsf{FD}(H)$. Tables that are values of τ are also said to *satisfy the schema* S. The identification of these functional dependencies is based on the meaning of the attributes involved. For example, in a table schema that contains the attributes *ssn* (standing for social security number) and *name*, it is natural to impose the functional dependency *ssn* $\to$ *name*. Every table that satisfies this schema will satisfy this functional dependency.

Suppose that a table satisfies the functional dependencies $A \to B$ and $B \to C$. By a previous observation, the table will also satisfy $A \to C$. Thus, it is not necessary to explicitly stipulate that the table will satisfy $A \to C$. This functional dependency is obtained by applying the rule

$$\frac{A \to B, B \to C}{A \to C},$$

which is an instance of the *transitivity rule*

$$\frac{X \to Y, Y \to Z}{X \to Z} R_{trans},$$

for every $X, Y, Z \in \mathcal{P}(H)$. Here H is the set of attributes of a table. Our previous argument shows that this rule is sound; in other words, if a table satisfies $X \to Y$ and $Y \to Z$, then the table satisfies $X \to Z$.

The previous arguments allow us to identify two more sound rules, the *inclusion rule*

$$\frac{X \subseteq Y}{Y \to X} R_{inc},$$

and the *augmentation rule*

$$\frac{X \to Y}{XW \to YW} R_{aug},$$

for every $X, Y, W \in \mathcal{P}(H)$.

As we saw above, rules are denoted as fractions; the objects that appear in the numerator are known as the *premises* of the rule; the object that appears in the denominator is the *conclusion* of the rule.

The three rules introduced so far (transitivity, augmentation, and inclusion) are known as *Armstrong's rules*.

The previous discussion establishes the *soundness* of the rules R_{inc}, R_{aug}, and R_{tran}. This means that a table that satisfies a set F of functional dependencies will satisfy any functional dependency obtained from F through applications of these rules.

In a certain sense that will be made clear in what follows, these are all the rules we need in order to reason about functional dependencies.

Rules are used to generate in a syntactic manner new functional dependencies starting from existing sets of such dependencies. The process of producing such new functional dependencies is known as a *proof*. This notion is formalized in the next definition.

Definition 12.25 *Let $S = (H, F)$ be a functional dependencies schema. A non-null sequence of functional dependencies:*

$$U_1 \to V_1, \ldots, U_n \to V_n$$

*is an F-*proof *of length n if one of the following conditions is satisfied for every i, $1 \leqslant i \leqslant n$:*

(i) *$U_i \to V_i$ is one of the functional dependencies of F, or*

(ii) $U_i \to V_i$ *is obtained from* 0, 1, *or* 2 *predecessors in the sequence by applying one of Armstrong's rules.*

The last dependency in the sequence $U_n \to V_n$ *is the* target *of the* F*-proof.*

Example 12.26 Suppose that $S = (H, F)$ is a functional dependency schema, where $H = ABCDE$ and F is the set of functional dependencies

$$F = \{A \to C, AB \to D, CD \to E\}.$$

We claim that the sequence

$$A \to C, AB \to BC, AB \to ABC, AB \to D,$$
$$ABC \to CD, AB \to CD, CD \to E, AB \to E$$

is an F-proof of $AB \to E$ for the following reasons:

(i) $A \to C$ belongs to F.
(ii) $AB \to ABC$ is obtained from (i) by applying R_{aug} with $W = AB$.
(iii) $AB \to D$ belongs to F.
(iv) $ABC \to CD$ is obtained from (iii) by applying R_{aug} with $W = C$.
(v) $AB \to CD$ is obtained from (ii) and (iv) by applying R_{tran}.
(vi) $CD \to E$ belongs to F.
(vii) $AB \to E$ is obtained from (v) and (vi) by applying R_{tran}.

The existence of an F-proof that has a functional dependency $U \to V$ as a target is denoted as $F \vdash_{ARM} U \to V$.

Finding an F-proof for a functional dependency can be a daunting task if the number of attributes and functional dependencies is large. Fortunately, there are ways of simplifying this process.

Theorem 12.27 *Let* $S = (H, F)$ *be a functional dependency schema. If* $F \vdash_{ARM} X \to Y$ *and* $F \vdash_{ARM} X \to Z$, *then* $F \vdash_{ARM} X \to YZ$ *for every* $X, Y, Z \subseteq H$.

Proof Let $U_1 \to V_1, \ldots, U_n \to V_n$ and $U'_1 \to V'_1, \ldots, U'_n \to V'_m$ be two F-proofs that have $X \to Y$ and $X \to Z$ as targets. Using the augmentation by X, the first proof generates the F-proof

$$U_1 \to V_1, \ldots, U_n \to V_n = X \to Y, X \to XY.$$

On the other hand, starting from the second proof

$$U'_1 \to V'_1, \ldots, U'_n \to V'_m = X \to Z,$$

by augmenting the last functional dependency by Y, we have the F-proof

$$U'_1 \to V'_1, \ldots, U'_n \to V'_m = X \to Z, XY \to YZ$$

By concatenating the two newly obtained proofs and applying the transitivity property, we have the F-proof

$$U_1 \to V_1, \ldots, U_n \to V_n, X \to XY,$$
$$U'_1 \to V'_1, \ldots, U'_n \to V'_m, XY \to YZ, X \to YZ,$$

which has the desired functional dependency $X \to YZ$ as its target. □

The last theorem shows that we can derive the functional dependency $X \to YZ$ from the functional dependencies $X \to Y$ and $X \to Z$. This fact is interpreted as a "derived rule" known as the *additivity rule* and is denoted by

$$\frac{X \to Y, X \to Z}{X \to YZ} R_{add}.$$

Another derived rule is introduced in the next theorem.

Theorem 12.28 *If* $F \underset{ARM}{\vdash} X \to YZ$, *then* $F \underset{ARM}{\vdash} X \to Y$ *for every* $X, Y, Z \subseteq H$.

Proof Let $U_1 \to V_1, \ldots, U_n \to V_n$ be an F-proof that has $X \to YZ$ as its target. We can add to this proof the functional dependency $YZ \to Y$ obtained by applying R_{inc}. This yields the needed F-proof

$$U_1 \to V_1, \ldots, U_n \to V_n = X \to YZ, YZ \to Y, X \to Y,$$

where the last step was obtained by applying the transitivity rule to the previous two steps. □

Thus, from $X \to YZ$ we can derive the functional dependency $X \to Y$. This derived rule is known as the *projectivity rule* and is denoted by

$$\frac{X \to YZ}{X \to Y} R_{proj}.$$

Note that if F is a set of functional dependencies, $F \subseteq \mathsf{FD}(H)$ and $X \subseteq H$, then it is always possible to find Y such that $F \underset{ARM}{\vdash} X \to Y$. Indeed, it suffices to take $Y = X$ and we can always prove $X \to X$ starting from F because the functional dependency $X \to X$ can be generated by applying R_{inc}.

Let $X \to Y_1, \ldots, X \to Y_p$ be the set of all functional dependencies such that $F \underset{ARM}{\vdash} X \longrightarrow Y$, where $X, Y \subseteq H$. By repeatedly applying the additivity rule, we have $F \underset{ARM}{\vdash} X \longrightarrow Y_1 \cdots Y_p$. The set $Y_1 \cdots Y_p$ is the largest set Y such that $F \underset{ARM}{\vdash} X \longrightarrow Y$. Further, we have $F \underset{ARM}{\vdash} X \longrightarrow V$ if and only if $V \subseteq Y_1 \cdots Y_p$. Indeed, it is clear that if $F \underset{ARM}{\vdash} X \longrightarrow V$, then $V \subseteq Y_1 \cdots Y_p$. Conversely, if

$V \subseteq Y_1 \cdots Y_p$, then $Y_1 \cdots Y_p \rightarrow V$ (by R_{inc}), so $F \underset{ARM}{\vdash} X \rightarrow V$ by R_{tran}. Thus, the set $Y_1 \cdots Y_p$ plays a special role; we will refer to it as the *closure of X under F* and will denote it by $cl_F(X)$.

Theorem 12.29 *Let $S = (H, F)$ be a functional dependency schema. The mapping $\mathsf{cl}_F : \mathcal{P}(H) \longrightarrow \mathcal{P}(H)$ is a closure operator on H.*

Proof We need to show that cl_F satisfies the conditions of Definition 1.171.

Since we have $F \underset{ARM}{\vdash} X \rightarrow X$, it is clear that $X \subseteq cl_F(X)$.

Suppose now that $X, X' \in \mathcal{P}(H)$ and $X' \subseteq X$. Since $F \underset{ARM}{\vdash} X \rightarrow X'$ by R_{inc} and $F \underset{ARM}{\vdash} X' \rightarrow cl_F(X')$, it follows that $F \underset{ARM}{\vdash} X \rightarrow cl_F(X')$. This implies $cl_F(X') \subseteq cl_F(X)$, so cl_F is monotonic.

Finally, note that we have both $F \underset{ARM}{\vdash} X \rightarrow cl_F(X)$ and $F \underset{ARM}{\vdash} cl_F(X) \rightarrow cl_F(cl_F(X))$, which yields $F \underset{ARM}{\vdash} X \rightarrow cl_F(cl_F(X))$ by R_{tran}. This implies $cl_F(cl_F(X)) \subseteq cl_F(X)$. The converse inclusion follows from the fact that $X \subseteq cl_F(X)$ and the monotonicity of cl_F. Thus, $cl_F(cl_F(X)) = cl_F(X)$ for every $X \in \mathcal{P}(H)$. □

The statement that $F \underset{ARM}{\vdash} X \rightarrow Y$ has a syntactic character; it can be shown by constructing an F-proof that has $X \rightarrow Y$ as its target. Actually, a computation of $cl_F(X)$ allows us to decide whether $F \underset{ARM}{\vdash} X \rightarrow Y$ without constructing the F-proof, as shown in the next theorem.

Theorem 12.30 *Let $S = (H, F)$ be a functional dependency schema. We have $F \underset{ARM}{\vdash} X \rightarrow Y$ if and only if $Y \subseteq cl_F(X)$.*

Proof If $Y \subseteq cl_F(X)$, then we have $F \underset{ARM}{\vdash} cl_F(X) \rightarrow Y$ by a single application of R_{inc}. Then, since $F \underset{ARM}{\vdash} X \rightarrow cl_F(X)$, (by the definition of $cl_F(X)$), another application of R_{tran} yields $F \underset{ARM}{\vdash} X \rightarrow Y$.

Conversely, if $F \underset{ARM}{\vdash} X \rightarrow Y$, then $Y \subseteq cl_F(X)$ by the definition of $cl_F(X)$. □

Now we introduce a semantic counterpart of relation $\underset{ARM}{\vdash}$.

Definition 12.31 *Let $S = (H, F)$ be a functional dependency schema. The set F* logically implies *the functional dependency $X \rightarrow Y$ if every table that satisfies all functional dependencies of F also satisfies $X \rightarrow Y$. This is denoted by $F \models X \rightarrow Y$.*

The soundness of Armstrong's rules means that if $F \underset{ARM}{\vdash} X \rightarrow Y$, then $F \models X \rightarrow Y$. It is interesting that the reverse implication also holds. This fact is known as the *completeness* of Armstrong's axioms and will be established next. To this end, we introduce the notion of an *Armstrong table*.

Definition 12.32 *Let $S = (H, F)$ be a functional dependency schema, where $H = A_1 \cdots A_n$, and let $X \in \mathcal{P}(H)$. For each attribute $A \in H$, let u_A and v_A be two distinct values from* $\mathrm{Dom}(A)$. *The* Armstrong table $\theta_{F,X} = (T_{F,X}, H, \mathbf{r}_{F,X})$ *contains a two-row sequence $(\mathbf{r}_{F,X}) = (t_0, t_1)$, where $t_0(A) = u_A$ for $A \in H$ and*

$$t_1(A) = \begin{cases} u_A & \text{if } A \in cl_F(X), \\ v_A & \text{if } A \in H - cl_F(X). \end{cases}$$

Note that the existence of Armstrong relations is assured by our assumption that the domain of every attribute contains at least two values.

Lemma 12.33 *Let F be a set of functional dependencies $F \subseteq \mathsf{FD}(H)$, $H = A_1 \cdots A_n$, and let $X \in \mathcal{P}(H)$. The* Armstrong table $\theta_{F,X} = (T_{F,X}, H, \mathbf{r}_{F,X})$ *satisfies all dependencies that can be proven from F.*

Proof Suppose that $U \to V$ is a functional dependency that can be proven from F (which means that $F \underset{ARM}{\vdash} U \to V$) and that this dependency is violated by $\theta_{F,X}$. Since $\theta_{F,X}$ contains two tuples, this is possible only if these tuples have the same projection on U but distinct projections on V. The definition of $\theta_{F,X}$ allows this only if $U \subseteq cl_F(X)$ and $V \not\subseteq cl_F(X)$. By the definition of $cl_F(X)$ this is possible only if $F \underset{ARM}{\vdash} X \to U$ and $F \underset{ARM}{\not\vdash} X \to V$. This leads to a contradiction because $F \underset{ARM}{\vdash} X \to U$ and $F \underset{ARM}{\vdash} U \to V$ imply $F \underset{ARM}{\vdash} X \to V$ (by R_{tran}). □

Theorem 12.34 (Completeness of Armstrong's Rules) *Let F be a set of functional dependencies, $F \subseteq \mathsf{FD}(H)$, $H = A_1 \cdots A_n$, and let $X, Y \in \mathcal{P}(H)$. If $F \models X \to Y$, then $F \underset{ARM}{\vdash} X \to Y$.*

Proof Suppose that $F \models X \to Y$ but $F \underset{ARM}{\not\vdash} X \to Y$, which means that $Y \not\subseteq cl_F(X)$. The Armstrong table $\theta_{F,X} = (T_{F,X}, H, \mathbf{r}_{F,X})$ satisfies $X \to Y$ because it satisfies all functional dependencies of F. Since $X \subseteq cl_F(X)$, this implies $Y \subseteq cl_F(X)$, which yields a contradiction. □

Corollary 12.35 *Let $S = (H, F)$ be a functional dependency schema and let $X \to Y$ be a functional dependency in $\mathsf{FD}(F)$. The following three statements are equivalent:*

(i) $Y \subseteq cl_F(X)$.
(ii) $F \underset{ARM}{\vdash} X \to Y$.
(iii) $F \models X \to Y$.

Proof (i) is equivalent to (ii) by Theorem 12.30. We have (ii) implies (iii) by the soundness of Armstrong's rules and (iii) implies (ii) by the completeness of these rules. □

Theorem 12.36 *Let $\theta = (T, H, \mathbf{r})$ be a table and let X and Y be two subsets of H. If $Z = H - (XY)$, then*

$$set(\mathbf{r}) \subseteq set(\mathbf{r}[XY]) \bowtie set(\mathbf{r}[XZ]).$$

Proof Let u be a tuple in $set(\mathbf{r})$. Since $u = t \bowtie s$ for some $t \in set(\mathbf{r}[XY])$ and $s \in set(\mathbf{r}[XZ])$, it follows that t and s are joinable (because $u[X] = t[X] = s[X]$) and, therefore, $u \in set(\mathbf{r}[XY]) \bowtie set(\mathbf{r}[XZ])$.

Corollary 12.37 *If θ satisfies the functional dependency $X \longrightarrow Y$ the inclusion of Theorem 12.36 becomes an equality.*

Proof Indeed, let $v \in set(\mathbf{r}[XY]) \bowtie set(\mathbf{r}[XZ])$. There exist two joinable tuples $t \in set(\mathbf{r}[XY])$ and $s \in set(\mathbf{r}[XZ])$ such that $v = t \bowtie s$, which means that $v[XY] = t[XY]$ and $v[XZ] = s[XZ]$. This implies

$$\begin{aligned} v[X] &= t[X] = s[X] = x, \\ v[Y] &= t[Y] = y, \\ v[Z] &= s[Z] = z. \end{aligned}$$

for some x, y, z. The existence of t and s implies the existence of $t_1, s_1 \in set(\mathbf{r})$ such that $t_1 = (x, y, z_1)$ and $s_1 = (x, y_1, z)$. Since θ satisfies the functional dependency $X \longrightarrow Y$ it follows that $y = y_1$, which implies $s = (x, y, z) = v$. Thus, $v \in set(\mathbf{r})$.

12.4 Partition Entropy

The notion of entropy is a probabilistic concept that lies at the foundation of information theory. Our goal is to define entropy in an algebraic setting by introducing the notion of entropy of a partition of a finite set. This approach allows us to take advantage of the partial order that is naturally defined on the set of partitions. Actually, we introduce a generalization of the notion of entropy that has the Gini index and Shannon entropy as special cases.

Let S be a finite set and let $\pi = \{B_1, \ldots, B_m\}$ be a partition of S. The *Shannon entropy of* π is the number

$$\mathcal{H}(S, \pi) = -\sum_{i=1}^{m} \frac{|B_i|}{|S|} \log_2 \frac{|B_i|}{|S|}.$$

The *Gini index of* π is the number

$$gini(S, \pi) = 1 - \sum_{i=1}^{m} \left(\frac{|B_i|}{|S|}\right)^2.$$

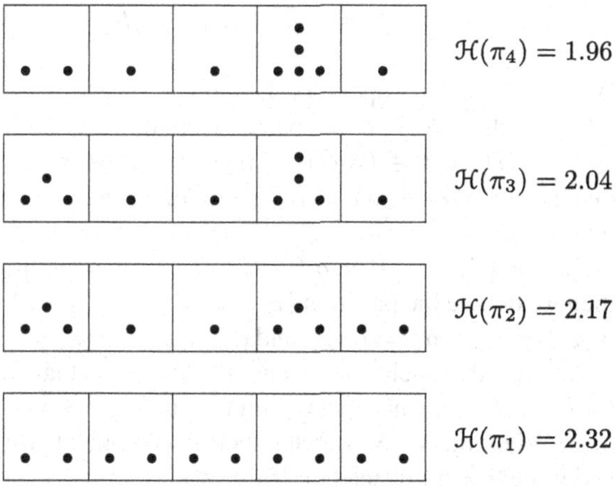

Fig. 12.1 Entropy increasing with partition uniformity

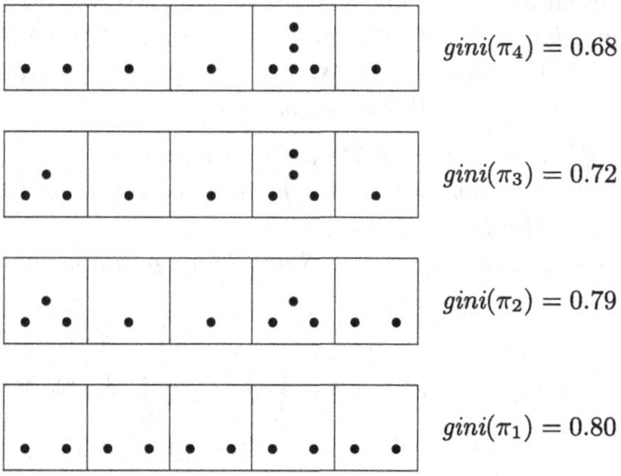

Fig. 12.2 Gini index increasing with partition uniformity

Both numbers can be used to evaluate the uniformity of the distribution of the elements of S in the blocks of π because both values increase with the uniformity of the distribution of the elements of S.

Example 12.38 Let S be a set containing ten elements and let $\pi_1, \pi_2, \pi_3, \pi_4$ be the four partitions shown in Fig. 12.1.

The partition π_1, which is the most uniform (each block containing two elements), has the largest entropy. At the other end of the range, partition π_4 has a strong concentration of elements in its fourth block and the lowest entropy. Similar results involving the Gini index are shown in Fig. 12.2.

If S and T are two disjoint and nonempty sets, $\pi \in PART(S)$ and $\sigma \in PART(T)$, where $\pi = \{B_1, \ldots, B_m\}$, $\sigma = \{C_1, \ldots, C_n\}$, then the partition $\pi + \sigma$ is the partition of $S \cup T$ given by

$$\pi + \sigma = \{B_1, \ldots, B_m, C_1, \ldots, C_n\}.$$

Whenever the "+" operation is defined, then it is easily seen to be associative. In other words, if S, T, U are pairwise disjoint and nonempty sets and $\pi \in PART(S)$, $\sigma \in PART(T)$, $\tau \in PART(U)$, then $\pi + (\sigma + \tau) = (\pi + \sigma) + \tau$. If S and T are disjoint, then $\alpha_S + \alpha_T = \alpha_{S \cup T}$. Also, $\omega_S + \omega_T$ is the partition $\{S, T\}$ of the set $S \cup T$.

If $\pi = \{B_1, \ldots, B_m\}, \sigma = \{C_1, \ldots, C_n\}$ are partitions of two arbitrary sets S, T, then we denote the partition $\{B_i \times C_j \mid 1 \leqslant i \leqslant m, 1 \leqslant j \leqslant n\}$ of $S \times T$ by $\pi \times \sigma$. Note that $\alpha_S \times \alpha_T = \alpha_{S \times T}$ and $\omega_S \times \omega_T = \omega_{S \times T}$.

We introduce below a system of four axioms that define a class of functions of the form $\mathcal{H}_\beta(S, \pi)$, where β is a number such that $\beta > 1$, S is a set, and π is a partition of S. When the set S is understood from context, we shall omit the first argument and write $\mathcal{H}_\beta(\pi)$ instead of $\mathcal{H}_\beta(S, \pi)$.

Definition 12.39 *Let $\beta \in \mathbb{R}$, $\beta \geqslant 1$, and $\Phi : \mathbb{R}^2_{\geqslant 0} \longrightarrow \mathbb{R}_{\geqslant 0}$ be a continuous function such that $\Phi(x, y) = \Phi(y, x)$, and $\Phi(x, 0) = x$ for $x, y \in \mathbb{R}_{\geqslant 0}$.*

A (Φ, β)-system of axioms for a partition entropy $\mathcal{H}_\beta : PART(S) \longrightarrow \mathbb{R}_{\geqslant 0}$ consists of the following axioms:

(P1) *If $\pi, \pi' \in PART(S)$ are such that $\pi \leqslant \pi'$, then $\mathcal{H}_\beta(S, \pi') \leqslant \mathcal{H}_\beta(S, \pi)$.*

(P2) *If S and T are two finite sets such that $|S| \leqslant |T|$, then $\mathcal{H}_\beta(S, \alpha_S) \leqslant \mathcal{H}_\beta(T, \alpha_T)$.*

(P3) *For all disjoint sets S and T and partitions $\pi \in PART(S)$ and $\sigma \in PART(T)$ we have*

$$\mathcal{H}_\beta(S \cup T, \pi + \sigma) = \left(\frac{|S|}{|S| + |T|}\right)^\beta \mathcal{H}_\beta(S, \pi) + \left(\frac{|T|}{|S| + |T|}\right)^\beta \mathcal{H}_\beta(T, \sigma) + \mathcal{H}_\beta(S \cup T, \{S, T\}).$$

(P4) *We have $\mathcal{H}_\beta(S \times T, \pi \times \sigma) = \Phi(\mathcal{H}_\beta(S, \pi), \mathcal{H}_\beta(T, \sigma))$ for $\pi \in PART(S)$ and $\sigma \in PART(T)$.*

Since the range of every function $\mathcal{H}_\beta$ is $\mathbb{R}_{\geqslant 0}$, it follows that $\mathcal{H}_\beta(S, \pi) \geqslant 0$ for any partition $\pi \in PART(S)$.

Lemma 12.40 *For every (Φ, β)-entropy $\mathcal{H}_\beta$ and set S, we have $\mathcal{H}_\beta(S, \omega_S) = 0$.*

Proof Let S and T be two disjoint sets that have the same cardinality, $|S| = |T|$. Since $\omega_S + \omega_T$ is the partition $\{S, T\}$ of the set $S \cup T$, by Axiom **(P3)** we have

$$\mathcal{H}_\beta(S \cup T, \omega_S + \omega_T) = \left(\frac{1}{2}\right)^\beta (\mathcal{H}_\beta(S, \omega_S) + \mathcal{H}_\beta(T, \omega_T)) + \mathcal{H}_\beta(S \cup T, \{S, T\}),$$

which implies $\mathcal{H}_\beta(S, \omega_S) + \mathcal{H}_\beta(T, \omega_T) = 0$. It follows that $\mathcal{H}_\beta(S, \omega_S) = \mathcal{H}_\beta(T, \omega_T) = 0$ because both $\mathcal{H}_\beta(S, \omega_S)$ and $\mathcal{H}_\beta(T, \omega_T))$ are non-negative numbers. □

Lemma 12.41 *Let S and T be two disjoint sets and let $\pi, \pi' \in PART(S \cup T)$ be defined by $\pi = \sigma + \alpha_T$ and $\pi' = \sigma + \omega_T$, where $\sigma \in PART(S)$. Then,*

$$\mathcal{H}_\beta(S \cup T, \pi) = \mathcal{H}_\beta(S \cup T, \pi') + \left(\frac{|T|}{|S| + |T|}\right)^\beta \mathcal{H}_\beta(T, \alpha_T).$$

Proof By Axiom **(P3)**, we can write

$$\begin{aligned}\mathcal{H}_\beta(S \cup T, \pi) = &\left(\frac{|S|}{|S| + |T|}\right)^\beta \mathcal{H}_\beta(S, \sigma) + \left(\frac{|T|}{|S| + |T|}\right)^\beta \mathcal{H}_\beta(T, \alpha_T)\\ &+ \mathcal{H}_\beta(S \cup T, \{S, T\})\end{aligned}$$

and

$$\begin{aligned}\mathcal{H}_\beta(S \cup T, \pi') = &\left(\frac{|S|}{|S| + |T|}\right)^\beta \mathcal{H}_\beta(S, \sigma) + \left(\frac{|T|}{|S| + |T|}\right)^\beta \mathcal{H}_\beta(T, \omega_T)\\ &+ \mathcal{H}_\beta(S \cup T, \{S, T\})\\ = &\left(\frac{|S|}{|S| + |T|}\right)^\beta \mathcal{H}_\beta(S, \sigma) + \mathcal{H}_\beta(S \cup T, \{S, T\})\\ &\text{(by Lemma 12.40).}\end{aligned}$$

The previous equalities yield the equality of the lemma. □

Theorem 12.42 *For every (Φ, β)-entropy and partition $\pi = \{B_1, \ldots, B_m\} \in PART(S)$, we have*

$$\mathcal{H}_\beta(\pi) = \mathcal{H}_\beta(\alpha_S) - \sum_{i=1}^m \left(\frac{|B_i|}{|S|}\right)^\beta \mathcal{H}_\beta(\alpha_{B_i}).$$

Proof Starting from the partition π, consider the following sequence of partitions in $PART(S)$:

$$\begin{aligned}\pi_0 &= \omega_{B_1} + \omega_{B_2} + \omega_{B_3} + \cdots + \omega_{B_m}\\ \pi_1 &= \alpha_{B_1} + \omega_{B_2} + \omega_{B_3} + \cdots + \omega_{B_m}\\ \pi_2 &= \alpha_{B_1} + \alpha_{B_2} + \omega_{B_3} + \cdots + \omega_{B_m}\\ &\vdots\\ \pi_n &= \alpha_{B_1} + \alpha_{B_2} + \alpha_{B_3} + \cdots + \alpha_{B_m}.\end{aligned}$$

Let $\sigma_i = \alpha_{B_1} + \cdots + \alpha_{B_i} + \omega_{B_{i+2}} + \cdots + \omega_{B_m}$. Then, $\pi_i = \sigma_i + \omega_{B_{i+1}}$ and $\pi_{i+1} = \sigma_i + \alpha_{B_{i+1}}$; therefore, by Lemma 12.41, we have

$$\mathcal{H}_\beta(S, \pi_{i+1}) = \mathcal{H}_\beta(S, \pi_i) + \left(\frac{|B_{i+1}|}{|S|}\right)^\beta \mathcal{H}_\beta(B_{i+1}, \alpha_{B_{i+1}})$$

for $0 \leqslant i \leqslant m - 1$.

A repeated application of this equality yields

$$\mathcal{H}_\beta(S, \pi_m) = \mathcal{H}_\beta(S, \pi_0) + \sum_{i=0}^{m-1} \left(\frac{|B_{i+1}|}{|S|}\right)^\beta \mathcal{H}_\beta(B_{i+1}, \alpha_{B_{i+1}}).$$

Since $\pi_0 = \pi$ and $\pi_m = \alpha_S$, we have

$$\mathcal{H}_\beta(S, \pi) = \mathcal{H}_\beta(S, \alpha_S) - \sum_{i=1}^{m} \left(\frac{|B_i|}{|S|}\right)^\beta \mathcal{H}_\beta(B_i, \alpha_{B_i}).$$

□

Note that if S and T are two sets such that $|S| = |T| > 0$, then, by Axiom **(P2)**, we have $\mathcal{H}_\beta(S, \alpha_S) = \mathcal{H}_\beta(T, \alpha_T)$. Therefore, the value of $\mathcal{H}_\beta(S, \alpha_S)$ depends only on the cardinality of S, and there exists a function $\mu : \mathbb{N}_1 \longrightarrow \mathbb{R}_{\geqslant 0}$ such that $\mathcal{H}_\beta(S, \alpha_S) = \mu(|S|)$ for every nonempty set S. Axiom **(P2)** also implies that μ is an increasing function. We will refer to μ as the *kernel* of the (Φ, β)-system of axioms.

Corollary 12.43 *Let $\mathcal{H}_\beta$ be a (Φ, β)-entropy. For the kernel μ defined in accordance with Axiom **(P2)** and every partition $\pi = \{B_1, \ldots, B_m\} \in PART(S)$, we have*

$$\mathcal{H}_\beta(S, \pi) = \mu(|S|) - \sum_{i=1}^{m} \left(\frac{|B_i|}{|S|}\right)^\beta \mu(|B_i|). \tag{12.1}$$

Proof The statement is an immediate consequence of Theorem 12.42. □

Theorem 12.44 *Let $\pi = \{B_1, \ldots, B_m\}$ be a partition of the set S. Define the partition π' obtained by fusing the blocks B_1 and B_2 of π as $\pi' = \{B_1 \cup B_2, B_3, \ldots, B_m\}$ of the same set. Then*

$$\mathcal{H}_\beta(S, \pi) = \mathcal{H}_\beta(S, \pi') + \left(\frac{|B_1 \cup B_2|}{|S|}\right)^\beta \mathcal{H}_\beta(B_1 \cup B_2, \{B_1, B_2\}).$$

Proof A double application of Corollary 12.43 yields

$$\begin{aligned}
&\mathcal{H}_\beta(S, \pi) - \mathcal{H}_\beta(S, \pi)\\
&= \left(\frac{|B_1 \cup B_2|}{|S|}\right)^\beta \mu(|B_1 \cup B_2|) - \left(\frac{|B_1|}{|S|}\right)^\beta \mu(|B_1|) - \left(\frac{|B_2|}{|S|}\right)^\beta \mu(|B_2|)\\
&= \left(\frac{|B_1 \cup B_2|}{|S|}\right)^\beta \left(\mu(|B_1 \cup B_2|) - \left(\frac{|B_1|}{|B_1 \cup B_2|}\right)^\beta \mu(|B_1|)\right.
\end{aligned}$$

$$- \left(\frac{|B_2|}{|B_1 \cup B_2|} \right)^{\beta} \mu(|B_2|) \Bigg)$$
$$= \left(\frac{|B_1 \cup B_2|}{|S|} \right)^{\beta} \mathcal{H}_{\beta}(B_1 \cup B_2, \{B_1, B_2\}),$$

which is the desired equality. □

Theorem 12.44 allows us to extend Axiom **(P3)**:

Corollary 12.45 *Let $B_1, \ldots, B_m$ be m nonempty, disjoint sets and let $\pi_i \in PART(B_i)$ for $1 \leqslant i \leqslant m$. We have*

$$\mathcal{H}_{\beta}(S, \pi_1 + \cdots + \pi_m) = \sum_{i=1}^{m} \left(\frac{|B_i|}{|S|} \right)^{\beta} \mathcal{H}_{\beta}(B_i, \pi_i) + \mathcal{H}_{\beta}(S, \{B_1, \ldots, B_m\}),$$

where $S = B_1 \cup \cdots \cup B_m$.

Proof The argument is by induction on $m \geqslant 2$. The basis step, $m = 2$, is Axiom **(P3)**. Suppose that the statement holds for m, and let $B_1, \ldots, B_m, B_{m+1}$ be $m + 1$ disjoint sets. Further, suppose that $\pi_1, \ldots, \pi_m, \pi_{m+1}$ are partitions of these sets, respectively. Then, $\pi_m + \pi_{m+1}$ is a partition of the set $B_m \cup B_{m+1}$. By the inductive hypothesis, we have

$$\mathcal{H}_{\beta}(S, \pi_1 + \cdots + (\pi_m + \pi_{m+1})) = \sum_{i=1}^{m-1} \left(\frac{|B_i|}{|S|} \right)^{\beta} \mathcal{H}_{\beta}(B_i, \pi_i)$$
$$+ \left(\frac{|B_m| + |B_{m+1}|}{|S|} \right)^{\beta} \mathcal{H}_{\beta}(B_m \cup B_{m+1}, \pi_m + \pi_{m+1})$$
$$+ \mathcal{H}_{\beta}(S, \{B_1, \ldots, (B_m \cup B_{m+1})\}),$$

where $S = B_1 \cup \cdots \cup B_m \cup B_{m+1}$.

Axiom **(P3)** and Theorem 12.44 give the desired equality. □

Theorem 12.46 *Let μ be the kernel of a (Φ, β)-system. If $a, b \in \mathbb{N}_1$, then*

$$\mu(ab) - \mu(a) \cdot b^{1-\beta} = \mu(b).$$

Proof Let $A = \{x_1, \ldots, x_a\}$ and $B = \{y_1, \ldots, y_b\}$ be two nonempty sets. The relation $\omega_A \times \alpha_B$ consists of b blocks of size a: $A \times \{y_1\}, \ldots, A \times \{y_b\}$. By Axiom **(P4)**,

$$A \times B, \mathcal{H}_{\beta}(\omega_A \times \alpha_B)$$
$$= \Phi(A, \mathcal{H}_{\beta}(\omega_A), \mathcal{H}_{\beta}(B, \alpha_B)) = \Phi(0, \mathcal{H}_{\beta}(B, \alpha_B)) = \mathcal{H}_{\beta}(B, \alpha_B) = \mu(b).$$

On the other hand,

$$\begin{aligned}\mathcal{H}_\beta(A \times B, \omega_A \times \alpha_B) &= \mathcal{H}_\beta(A \times B, \alpha_{A\times B}) \\ &\quad - \sum_{i=1}^{b} \left(\frac{1}{b}\right)^{\beta} \mathcal{H}_\beta(A \times \{y_i\}, \alpha_{A\times\{y_i\}}) \\ &= \mu(ab) - \frac{1}{b^\beta} b \cdot \mu(a),\end{aligned}$$

which gives the needed equality. □

An entropy is said to be *non-Shannon* if it is defined by a (Φ, β)-system of axioms such that $\beta > 1$; otherwise (that is if $\beta = 1$), the entropy will be referred to as a *Shannon* entropy. As we shall see, the choice of the parameter β determines the form of the function Φ.

Initially we focus on non-Shannon entropies, that is, on (Φ, β)-entropies, where $\beta > 1$.

Lemma 12.47 *Let $f : (0, \infty) \longrightarrow \mathbb{R}$ be a function such that*

$$f(x + y) = f(x) + f(y)$$

for $x, y \in (0, \infty)$. If there exists an interval on which f is bounded above, then $f(x) = xf(1)$ for $x > 0$.

Proof Let $g(x) = f(x) - xf(1)$. The function g satisfies the functional equation $g(x + y) = g(x) + g(y)$ for $x, y > 0$; in addition, $g(1) = 0$. It is easy to verify, by induction on n that $g(nx) = ng(x)$. Therefore, $g\left(\frac{n}{m}\right) = ng\left(\frac{1}{m}\right)$. On another hand, $0 = g(1) = mg\left(\frac{1}{m}\right)$, so $g\left(\frac{1}{m}\right) = 0$. Consequently, $g(r) = 0$ for every rational number r.

To prove that $g(x) = 0$ for every $x > 0$ suppose that $g(x_0) \neq 0$ for some x_0. If $g(x_0) < 0$ and r is a rational number, since $g(r) = 0$ and $g(r - x_0) + g(x_0) = g(r) = 0$, it follows that $g(r - x_0) > 0$. Thus, we can assume that there exists a number z_0 such that $g(z_0) > 0$.

Let (c, d) be an open interval such that g has an upper bound on (c, d) and let u be a number. Let n be a number such that $ng(z_0) > u$ and let r be a rational number such that $nz_0 + r \in (c, d)$.

If $r > 0$, we have $g(r + nz_0) = g(r) + g(nz_0) = g(nz_0) = ng(z_0)$; if $r < 0$, then $ng(z_0) = g(nz_0) = g(-r + nz_0 + r) = g(-r) + g(nz_0 + r) = g(nz_0 + r)$. In either case, $g(nz_0 + r) = ng(z_0) > u$. Since u is arbitrary, ths contradicts the fact that g was supposed to be bounded on (c, d). Thus, $g(x) = 0$ for every $x > 0$, so $f(x) = xf(1)$. □

Lemma 12.48 *Let $h : (0, \infty) \longrightarrow \mathbb{R}$ be a function such that*

$$h(xy) = h(x) + h(y)$$

for $x, y \in (0, \infty)$. If there exists an interval on which h is bounded above, then there exists $c \in \mathbb{R}$ such $h(t) = c \ln t$ for $t > 0$.

Proof Define the function p as $p(x) = h(e^x)$. The hypothesis implies that there exists an interval of $\mathbb{R}$ such that p is bounded above on this interval. We have

$$p(x + y) = h(e^{x+y}) = h(e^x e^y) = h(e^x) + h(e^y) = p(x) + p(y),$$

for $x, y \in \mathbb{R}$. By Lemma 12.47 we have $p(x) = xp(1)$, so $h(e^x) = xh(e)$, which implies $h(t) = h(e) \ln t = c \log t$, where $c = h(e) \ln 10$.

Theorem 12.49 *Let $t : (0, \infty) \longrightarrow \mathbb{R}$ be a function such that $t(xy) = yt(x) + xt(y)$ for $x, y \in (0, \infty)$. If there is an interval I and a constant k such that $t(x) \leqslant kx$ for $x \in I$, then $t(x) = dx \log x$ for some constant d.*

Proof By hypothesis, we have

$$\frac{t(xy)}{xy} = \frac{t(x)}{x} + \frac{t(y)}{y}$$

for $x, y \in (0, \infty)$. Let $h : (0, \infty) \longrightarrow \mathbb{R}$ be the function defined by $h(x) = \frac{t(x)}{x}$ for $x \in (0, \infty)$. Note that h has is upper bounded when x ranges over I, so $h(x) = c \ln x$ by Lemma 12.48. Therefore, $t(x) = xh(x) = cx \ln x$ for some constant c and $x \in (0, \infty)$.

Theorem 12.50 *Let $\mathcal{H}_\beta$ be a non-Shannon entropy defined by a (Φ, β)-system of axioms and let μ be the kernel of this system of axioms.*

There is a number $k > 0$ such that $\mu(a) = k \cdot (1 - a^{1-\beta})$ for every $a \in \mathbb{N}_1$.

Proof Theorem 12.46 implies $\mu(ab) = \mu(a) \cdot b^{1-\beta} + \mu(b) = \mu(b) \cdot a^{1-\beta} + \mu(a)$ for every $a, b \in \mathbb{N}_1$. Consequently, $\frac{\mu(a)}{1-a^{1-\beta}} = \frac{\mu(b)}{1-b^{1-\beta}} = k$ for every $a, b \in \mathbb{N}_1$, which gives the desired equality. □

Corollary 12.51 *If $\mathcal{H}_\beta$ is a non-Shannon entropy defined by a (Φ, β)-system of axioms and $\pi \in PART(S)$, where $\pi = \{B_1, \ldots, B_m\}$, then there exists a constant $k \in \mathbb{R}$ such that*

$$\mathcal{H}_\beta(S, \pi) = k\left(1 - \sum_{i=1}^{m}\left(\frac{|B_i|}{|S|}\right)^\beta\right). \tag{12.2}$$

Proof By Corollary 12.43 and Theorem 12.50, we have

$$\begin{aligned}\mathcal{H}_\beta(S, \pi) &= \mu(|S|) - \sum_{i=1}^{m}\left(\frac{|B_i|}{|S|}\right)^\beta \mu(|B_i|)\\ &= k\left(1 - \frac{1}{|S|^{\beta-1}}\right) - k\sum_{i=1}^{m}\left(\frac{|B_i|}{|S|}\right)^\beta \cdot \left(1 - \frac{1}{|B_i|^{\beta-1}}\right)\end{aligned}$$

$$= k\left(1 - \frac{1}{|S|^{\beta-1}}\right) - k\sum_{i=1}^{m}\left(\frac{|B_i|}{|S|}\right)^{\beta} + k\sum_{i=1}^{m}\frac{|B_i|}{|S|^{\beta}}$$

$$= k\left(1 - \sum_{i=1}^{m}\left(\frac{|B_i|}{|S|}\right)^{\beta}\right).$$

The last equality follows from the fact that $\sum_{i=1}^{m}|B_i| = |S|$. □

The constant k introduced in Theorem 12.50 is given by

$$k = \lim_{a\to\infty}\mu(a), \tag{12.3}$$

and the range of values assumed by μ is $[0, k]$.

Our axiomatization defines entropies (and therefore the kernel μ) up to the multiplicative constant k and the Equality (12.3) expresses this constant in terms of the limit of $\mu(a)$ when a tends to infinity.

The next theorem shows that the function Φ introduced by Definition 12.39 and used in Axiom **(P4)** is essentially determined by the choices made for β and k.

Theorem 12.52 *Let $\mathcal{H}_\beta$ be the non-Shannon entropy defined by a (Φ, β)-system and let k be as defined by Equality* (12.3)*, where μ is the kernel of the (Φ, β)-system of axioms.*

The function Φ of Axiom **(P4)** *is given by $\Phi(x, y) = x + y - \frac{1}{k}\cdot xy$ for $x, y \in \mathbb{R}_{\geqslant 0}$.*

Proof Let $\pi = \{B_1, \ldots, B_m\} \in PART(S)$ and $\sigma = \{C_1, \ldots, C_n\} \in PART(T)$ be two partitions. Since

$$\sum_{i=1}^{m}\left(\frac{|B_i|}{|S|}\right)^{\beta} = 1 - \frac{1}{k}\mathcal{H}_\beta(\pi) \text{ and } \sum_{j=1}^{n}\left(\frac{|C_j|}{|T|}\right)^{\beta} = 1 - \frac{1}{k}\mathcal{H}_\beta(\sigma),$$

we can write

$$\mathcal{H}_\beta(\pi \times \sigma) = k\left(1 - \sum_{i=1}^{m}\sum_{j=1}^{n}\left(\frac{|B_i||C_j|}{|S||T|}\right)^{\beta}\right)$$

$$= k\left(1 - \left(1 - \frac{1}{k}\mathcal{H}_\beta(\pi)\right)\left(1 - \frac{1}{k}\mathcal{H}_\beta(\sigma)\right)\right)$$

$$= \mathcal{H}_\beta(\pi) + \mathcal{H}_\beta(\sigma) - \frac{1}{k}\mathcal{H}_\beta(\pi)\mathcal{H}_\beta(\sigma).$$

Suppose initially that $\beta > 1$. Observe that the set of rational numbers of the form

$$1 - \sum_{l=1}^{n} r_l^{\beta},$$

where $r_l \in \mathbb{Q}, 0 \leqslant r_l \leqslant 1$ for $1 \leqslant l \leqslant n$ and $\sum_{l=1}^{n} r_l = 1$, for some $n \in \mathbb{N}_1$, is dense in the interval $[0, 1]$. Thus, Formula (12.2) shows that the set of entropy values is dense in the interval $[0, k]$ because the sets $B_1, \ldots, B_m$ are finite but of arbitrarily large cardinalities. Since the set of values of entropies is dense in the interval $[0, k]$, the continuity of Φ implies the desired form of Φ. □

Choosing $k = \frac{1}{1-2^{1-\beta}}$ in Equality (12.2), we obtain the Havrda-Charvat entropy (see [4]):

$$\mathcal{H}_\beta(S, \pi) = \frac{1}{1-2^{1-\beta}} \cdot \left(1 - \sum_{i=1}^{m} \left(\frac{|B_i|}{|S|}\right)^\beta\right).$$

If $\beta = 2$, we obtain $\mathcal{H}_2(\pi)$, which is twice the Gini index,

$$\mathcal{H}_\beta(S, \pi) = 2 \cdot \left(1 - \sum_{i=1}^{m} \left(\frac{|B_i|}{|S|}\right)^2\right).$$

The *Gini index*, $gini(\pi) = 1 - \sum_{i=1}^{m} \left(\frac{|B_i|}{|S|}\right)^2$, is widely used in machine learning and data mining.

The limit case, $\lim_{\beta \to 1} \mathcal{H}_\beta(\pi)$, yields

$$\begin{aligned}
\lim_{\beta \to 1} \mathcal{H}_\beta(S, \pi) &= \lim_{\beta \to 1} \frac{1}{1-2^{1-\beta}} \cdot \left(1 - \sum_{i=1}^{m} \left(\frac{|B_i|}{|S|}\right)^\beta\right) \\
&= \lim_{\beta \to 1} \frac{1}{2^{1-\beta} \ln 2} \cdot \left(- \sum_{i=1}^{m} \left(\frac{|B_i|}{|S|}\right)^\beta \ln \frac{|B_i|}{|S|}\right) \\
&= - \sum_{i=1}^{m} \frac{|B_i|}{|S|} \log_2 \frac{|B_i|}{|S|},
\end{aligned}$$

which is the Shannon entropy of π.

If $\beta = 1$, by Theorem 12.46, we have $\mu(ab) = \mu(a) + \mu(b)$ for $a, b \in \mathbb{N}_1$. If $\eta : \mathbb{N}_1 \longrightarrow \mathbb{R}$ is the function defined by $\eta(a) = a\mu(a)$ for $a \in \mathbb{N}_1$, then η is clearly an increasing function and we have $\eta(ab) = ab\mu(ab) = b\eta(a) + a\eta(b)$ for $a, b \in \mathbb{N}_1$. By Theorem 12.49, there exists a constant $c \in \mathbb{R}$ such that $\eta(a) = ca \log_2 a$ for $a \in \mathbb{N}_1$, so $\mu(a) = c \log_2(a)$. Then, Eq. (12.1) implies:

$$\mathcal{H}_\beta(S, \pi) = c \cdot \sum_{i=1}^{m} \frac{a_i}{a} \log_2 \frac{a_i}{a}$$

for every partition $\pi = \{A_1, \ldots, A_m\}$ of a set A, where $|A_i| = a_i$ for $1 \leqslant i \leqslant m$ and $|A| = a$. This is the expression of Shannon's entropy.

The continuous function Φ is determined as in the previous case. Indeed, if A, B are two sets such that $|A| = a$ and $|B| = b$, then we must have

$$c \cdot \log_2 ab = \mathcal{H}_\beta(A \times B, \alpha_A \times \alpha_B) = \Phi(c \cdot \log_2 a, c \cdot \log_2 b)$$

for any $a, b \in \mathbb{N}_1$ and any $c \in \mathbb{R}$. The continuity of Φ implies $\Phi(x, y) = x + y$.

The β-entropy of α_S is given by

$$\mathcal{H}_\beta(S, \alpha_S) = \frac{1 - |S|^{1-\beta}}{1 - 2^{1-\beta}}. \tag{12.4}$$

The entropies previously introduced generate corresponding conditional entropies.

Let $\pi \in PART(S)$ and let $C \subseteq S$. Denote by π_C the "trace" of π on C given by

$$\pi_C = \{B \cap C | B \in \pi \text{ such that } B \cap C \neq \emptyset\}.$$

Clearly, $\pi_C \in PART(C)$; also, if C is a block of π, then $\pi_C = \omega_C$.

Definition 12.53 *Let* $\pi, \sigma \in PART(S)$ *and let* $\sigma = \{C_1, \ldots, C_n\}$. *The* β-conditional entropy *of the partitions* $\pi, \sigma \in PART(S)$ *is the function* $\mathcal{H}_\beta : PART(S)^2 \longrightarrow \mathbb{R}_{\geqslant 0}$ *defined by*

$$\mathcal{H}_\beta(\pi|\sigma) = \sum_{j=1}^{n} \left(\frac{|C_j|}{|S|}\right)^\beta \mathcal{H}_\beta(\pi_{C_j}).$$

Note that $\mathcal{H}_\beta(\pi|\omega_S) = \mathcal{H}_\beta(\pi)$ and that $\mathcal{H}_\beta(\omega_S|\pi) = \mathcal{H}_\beta(\pi|\alpha_S) = 0$ for every partition $\pi \in PART(S)$.

For $\pi = \{B_1, \ldots, B_m\}$ and $\sigma = \{C_1, \ldots, C_n\}$, the conditional entropy can be written explicitly as

$$\begin{aligned}\mathcal{H}_\beta(\pi|\sigma) &= \sum_{j=1}^{n} \left(\frac{|C_j|}{|S|}\right)^\beta \sum_{i=1}^{m} \frac{1}{1 - 2^{1-\beta}} \left[1 - \left(\frac{|B_i \cap C_j|}{|C_j|}\right)^\beta\right] \\ &= \frac{1}{1 - 2^{1-\beta}} \sum_{j=1}^{n} \left(\left(\frac{|C_j|}{|S|}\right)^\beta - \sum_{i=1}^{m} \left(\frac{|B_i \cap C_j|}{|S|}\right)^\beta\right).\end{aligned} \tag{12.5}$$

For the special case when $\pi = \alpha_S$, we can write

$$\mathcal{H}_\beta(\alpha_S|\sigma) = \sum_{j=1}^{n} \left(\frac{|C_j|}{|S|}\right)^\beta \mathcal{H}_\beta(\alpha_{C_j}) = \frac{1}{1 - 2^{1-\beta}} \left(\sum_{j=1}^{n} \left(\frac{|C_j|}{|S|}\right)^\beta - \frac{1}{|S|^{\beta-1}}\right). \tag{12.6}$$

Theorem 12.54 *Let* S *be a finite set and let* $\pi, \sigma \in PART(S)$. *We have* $\mathcal{H}_\beta(\pi|\sigma) = 0$ *if and only if* $\sigma \leqslant \pi$.

Proof Suppose that $\sigma = \{C_1, \ldots, C_n\}$. If $\sigma \leqslant \pi$, then $\pi_{C_j} = \omega_{C_j}$ for $1 \leqslant j \leqslant n$ and therefore

$$\mathcal{H}_\beta(\pi|\sigma) = \sum_{j=1}^{n} \left(\frac{|C_j|}{|S|}\right)^\beta \mathcal{H}_\beta(\omega_{C_j}) = 0.$$

Conversely, suppose that

$$\mathcal{H}_\beta(\pi|\sigma) = \sum_{j=1}^{n} \left(\frac{|C_j|}{|S|}\right)^\beta \mathcal{H}_\beta(\pi_{C_j}) = 0.$$

This implies $\mathcal{H}_\beta(\pi_{C_j}) = 0$ for $1 \leqslant j \leqslant n$, which means that $\pi_{C_j} = \omega_{C_j}$ for $1 \leqslant j \leqslant n$ by a previous remark. This means that every block C_j of σ is included in a block of π, so $\sigma \leqslant \pi$. □

The next statement is a generalization of a well-known property of Shannon's entropy.

Theorem 12.55 *Let π and σ be two partitions of a finite set S. We have*

$$\mathcal{H}_\beta(\pi \wedge \sigma) = \mathcal{H}_\beta(\pi|\sigma) + \mathcal{H}_\beta(\sigma) = \mathcal{H}_\beta(\sigma|\pi) + \mathcal{H}_\beta(\pi),$$

Proof Let $\pi = \{B_1, \ldots, B_m\}$ and that $\sigma = \{C_1, \ldots, C_n\}$. Observe that

$$\pi \wedge \sigma = \pi_{C_1} + \cdots + \pi_{C_n} = \sigma_{B_1} + \cdots + \sigma_{B_m}.$$

Therefore, by Corollary 12.45, we have

$$\mathcal{H}_\beta(\pi \wedge \sigma) = \sum_{j=1}^{n} \left(\frac{|C_j|}{|S|}\right)^\beta \mathcal{H}_\beta(\pi_{C_j}) + \mathcal{H}_\beta(\sigma),$$

which implies $\mathcal{H}_\beta(\pi \wedge \sigma) = \mathcal{H}_\beta(\pi|\sigma) + \mathcal{H}_\beta(\sigma)$. The second equality has a similar proof. □

Corollary 12.56 *If $\mathcal{H}_\beta(\pi \wedge \sigma) = \mathcal{H}_\beta(\pi)$, then $\pi \leqslant \sigma$.*

Proof Since $\mathcal{H}_\beta(\pi \wedge \sigma) = \mathcal{H}_\beta(\pi)$, Theorem 12.55 implies $\mathcal{H}_\beta(\sigma|\pi) = 0$. By Theorem 12.54, we have $\pi \leqslant \sigma$. □

Lemma 12.57 *Let $w_1, \ldots, w_n$ be n positive numbers such that $\sum_{i=1}^{n} w_i = 1$, $a_1, \ldots, a_n \in [0, 1]$, and let $\beta \geqslant 1$. We have*

$$1 - \left(\sum_{i=1}^{n} w_i a_i\right)^\beta - \left(\sum_{i=1}^{n} w_i (1 - a_i)\right)^\beta \geqslant \sum_{i=1}^{n} w_i^\beta \left(1 - a_i^\beta - (1 - a_i)^\beta\right).$$

Proof It is easy to see that $x^\beta + (1 - x)^\beta \leqslant 1$ for $x \in [0, 1]$. This implies

$$w_i\left(1-a_i^{\beta}-(1-a_i)^{\beta}\right) w_i^{\beta}\left(1-a_i^{\beta}-(1-a_i)^{\beta}\right)$$

because $w_i \in (0, 1)$ and $\beta \geqslant 1$.

By applying Jensen's inequality to the convex function $h(x) = x^{\beta}$ we have

$$\left(\sum_{i=1}^{n} w_i a_i\right)^{\beta} \leqslant \sum_{i=1}^{n} w_i a_i^{\beta}$$
$$\left(\sum_{i=1}^{n} w_i (1-a_i)\right)^{\beta} \leqslant \sum_{i=1}^{n} w_i (1-a_i)^{\beta}.$$

These inequalities allow us to write

$$\begin{aligned}
&1-\left(\sum_{i=1}^{n} w_i a_i\right)^{\beta}-\left(\sum_{i=1}^{n} w_i(1-a_i)\right)^{\beta}\\
&=\sum_{i=1}^{n} w_i-\left(\sum_{i=1}^{n} w_i a_i\right)^{\beta}-\left(\sum_{i=1}^{n} w_i(1-a_i)\right)^{\beta}\\
&\geqslant \sum_{i=1}^{n} w_i-\sum_{i=1}^{n} w_i a_i^{\beta}-\sum_{i=1}^{n} w_i(1-a_i)^{\beta}\\
&=\sum_{i=1}^{n} w_i\left(1-a_i^{\beta}-(1-a_i)^{\beta}\right) \geqslant \sum_{i=1}^{n} w_i^{\beta}\left(1-a_i^{\beta}-(1-a_i)^{\beta}\right),
\end{aligned}$$

which is the desired inequality. □

Theorem 12.58 *Let S be a set, $\pi \in PART(S)$ and let C and D be two disjoint subsets of S. For $\beta \geqslant 1$, we have*

$$|C \cup D|^{\beta}\mathcal{H}_{\beta}(\pi_{C\cup D}) \geqslant |C|^{\beta}\mathcal{H}_{\beta}(\pi_C) + |D|^{\beta}\mathcal{H}_{\beta}(\pi_D).$$

Proof Let $\pi = \{B_1, \ldots, B_n\} \in PART(S)$. Define

$$w_i = \frac{|B_i \cap (C \cup D)|}{|C \cup D|},\; a_i = \frac{|B_i \cap C|}{|B_i \cap (C \cup D)|}$$

for $1 \leqslant i \leqslant n$, so $1 - a_i = \frac{|B_i \cap D|}{|B_i \cap (C \cup D)|}$.

By Lemma 12.57, we have

$$1-\left(\sum_{i=1}^{n} \frac{|B_i \cap C|}{|C \cup D|}\right)^{\beta}-\left(\sum_{i=1}^{n} \frac{|B_i \cap D|}{|C \cup D|}\right)^{\beta}$$

$$\geqslant \sum_{i=1}^{n} \left(\frac{|B_i \cap (C \cup D)|}{|C \cup D|}\right)^{\beta} \left(1 - \left(\frac{|B_i \cap C|}{|B_i \cap (C \cup D)|}\right)^{\beta} - \left(\frac{|B_i \cap D|}{|B_i \cap (C \cup D)|}\right)^{\beta}\right),$$

which is equivalent to

$$|C \cup D|^{\beta} - \sum_{i=1}^{n} |B_i \cap (C \cup D)|^{\beta}$$
$$\geqslant |C|^{\beta} - \sum_{i=1}^{n} |B_i \cap C|^{\beta} + |D|^{\beta} - \sum_{i=1}^{n} |B_i \cap D|^{\beta}.$$

This last inequality leads immediately to the inequality of the theorem. □

The next result shows that the β-conditional entropy is dually monotonic with respect to its first argument and is monotonic with respect to its second argument.

Theorem 12.59 *Let $\pi, \sigma, \sigma' \in PART(S)$, where S is a finite set. If $\sigma \leqslant \sigma'$, then $\mathcal{H}_{\beta}(\sigma|\pi) \geqslant \mathcal{H}_{\beta}(\sigma'|\pi)$ and $\mathcal{H}_{\beta}(\pi|\sigma) \leqslant \mathcal{H}_{\beta}(\pi|\sigma')$.*

Proof Since $\sigma \leqslant \sigma'$, we have $\pi \wedge \sigma \leqslant \pi \wedge \sigma'$, so $\mathcal{H}_{\beta}(\pi \wedge \sigma) \geqslant \mathcal{H}_{\beta}(\pi \wedge \sigma')$. Therefore, $\mathcal{H}_{\beta}(\sigma|\pi) + \mathcal{H}_{\beta}(\pi) \geqslant \mathcal{H}_{\beta}(\sigma'|\pi) + \mathcal{H}_{\beta}(\pi)$, which implies $\mathcal{H}_{\beta}(\sigma|\pi) \geqslant \mathcal{H}_{\beta}(\sigma'|\pi)$.

For the second part of the theorem, it suffices to prove the inequality for partitions σ, σ' such that $\sigma \prec \sigma'$. Without loss of generality we may assume that $\sigma = \{C_1, \ldots, C_{n-2}, C_{n-1}, C_n\}$ and $\sigma' = \{C_1, \ldots, C_{n-2}, C_{n-1} \cup C_n\}$. Thus, we can write

$$\begin{aligned}
&\mathcal{H}_{\beta}(\pi|\sigma') \\
&= \sum_{j=1}^{n-2} \left(\frac{|C_j|}{|S|}\right)^{\beta} \mathcal{H}_{\beta}(\pi_{C_j}) + \left(\frac{|C_{n-1} \cup C_n|}{|S|}\right)^{\beta} \mathcal{H}_{\beta}(\pi_{C_{n-1} \cup C_n}) \\
&\geqslant \left(\frac{|C_j|}{|S|}\right)^{\beta} \mathcal{H}_{\beta}(\pi_{C_j}) + \left(\frac{|C_{n-1}|}{|S|}\right)^{\beta} \mathcal{H}_{\beta}(\pi_{C_{n-1}}) + \left(\frac{|C_n|}{|S|}\right)^{\beta} \mathcal{H}_{\beta}(\pi_{C_n}) \\
&\quad \text{(by Theorem 12.58)} \\
&= \mathcal{H}(\pi|\sigma).
\end{aligned}$$

□

Corollary 12.60 *We have $\mathcal{H}_{\beta}(\pi) \geqslant \mathcal{H}_{\beta}(\pi|\sigma)$ for every $\pi, \sigma \in PART(S)$.*

Proof We noted that $\mathcal{H}_{\beta}(\pi) = \mathcal{H}_{\beta}(\pi|\omega_S)$. Since $\omega_S \geqslant \sigma$, the statement follows from the second part of Theorem 12.59. □

Corollary 12.61 *Let ξ, θ, θ' be three partitions of a finite set S. If $\theta \geqslant \theta'$, then*

$$\mathcal{H}_{\beta}(\xi \wedge \theta) - \mathcal{H}_{\beta}(\theta) \geqslant \mathcal{H}_{\beta}(\xi \wedge \theta') - \mathcal{H}_{\beta}(\theta').$$

Proof By Theorem 12.55, we have

$$\mathcal{H}_\beta(\xi \wedge \theta) - \mathcal{H}_\beta(\xi \wedge \theta') = \mathcal{H}_\beta(\xi|\theta) + \mathcal{H}_\beta(\theta) - \mathcal{H}_\beta(\xi|\theta') - \mathcal{H}_\beta(\theta').$$

The monotonicity of $\mathcal{H}_\beta(|)$ in its second argument means that: $\mathcal{H}_\beta(\xi|\theta) - \mathcal{H}_\beta(\xi|\theta') \geqslant 0$, so $\mathcal{H}_\beta(\xi \wedge \theta) - \mathcal{H}_\beta(\xi \wedge \theta') \geqslant \mathcal{H}_\beta(\theta) - \mathcal{H}_\beta(\theta')$, which implies the desired inequality. □

The behavior of β-conditional entropies with respect to the "addition" of partitions is discussed in the next statement.

Theorem 12.62 *Let S be a finite set and π and θ be two partitions of S, where $\theta = \{D_1, \ldots, D_h\}$. If $\sigma_i \in PART(D_i)$ for $1 \leqslant i \leqslant h$, then*

$$\mathcal{H}_\beta(\pi|\sigma_1 + \cdots + \sigma_h) = \sum_{i=1}^{h} \left(\frac{|D_i|}{|S|}\right)^\beta \mathcal{H}_\beta(\pi_{D_i}|\sigma_i).$$

If $\tau = \{F_1, \ldots, F_k\}$ and $\sigma = \{C_1, \ldots, C_n\}$ are two partitions of S, let $\pi_i \in PART(F_i)$ for $1 \leqslant i \leqslant k$. Then,

$$\mathcal{H}_\beta(\pi_1 + \cdots + \pi_k|\sigma) = \sum_{i=1}^{k} \left(\frac{|F_i|}{|S|}\right)^\beta \mathcal{H}_\beta(\pi_i|\sigma_{F_i}) + \mathcal{H}_\beta(\tau|\sigma).$$

Proof Suppose that $\sigma_i = \{E_i^\ell \mid 1 \leqslant \ell \leqslant p_i\}$. The blocks of the partition $\sigma_1 + \cdots + \sigma_h$ are the sets of the collection $\bigcup_{i=1}^h \{E_i^\ell \mid 1 \leqslant \ell \leqslant p_i\}$. Thus, we have

$$\mathcal{H}_\beta(\pi|\sigma_1 + \cdots + \sigma_h) = \sum_{i=1}^{h} \sum_{\ell=1}^{p_i} \left(\frac{|E_i^\ell|}{|S|}\right)^\beta \mathcal{H}_\beta(\pi_{E_i^\ell}).$$

On the other hand, since $(\pi_{D_i})_{E_i^\ell} = \pi_{E_i^\ell}$, we have

$$\begin{aligned}\sum_{i=1}^{h} \left(\frac{|D_i|}{|S|}\right)^\beta \mathcal{H}_\beta(\pi_{D_i}|\sigma_i) &= \sum_{i=1}^{h} \left(\frac{|D_i|}{|S|}\right)^\beta \sum_{\ell=1}^{p_i} \left(\frac{|E_i^\ell|}{|D_i|}\right)^\beta \mathcal{H}_\beta(\pi_{E_i^\ell}) \\ &= \sum_{i=1}^{h} \sum_{\ell=1}^{p_i} \left(\frac{|E_i^\ell|}{|S|}\right)^\beta \mathcal{H}_\beta(\pi_{E_i^\ell}),\end{aligned}$$

which gives the first equality of the theorem.

To prove the second part, observe that $(\pi_1 + \cdots + \pi_k)_{C_j} = (\pi_1)_{C_j} + \cdots + (\pi_k)_{C_j}$ for every block C_j of σ. Thus, we have

$$\mathcal{H}_\beta(\pi_1 + \cdots + \pi_k|\sigma) = \sum_{j=1}^{n} \left(\frac{|C_j|}{|S|}\right)^\beta \mathcal{H}_\beta((\pi_1)_{C_j} + \cdots + (\pi_k)_{C_j}).$$

By applying Corollary 12.45 to partitions $(\pi_1)_{C_j}, \ldots, (\pi_k)_{C_j}$ of C_j, we can write

$$\mathcal{H}_\beta((\pi_1)_{C_j} + \cdots + (\pi_k)_{C_j}) = \sum_{i=1}^{k} \left(\frac{|F_i \cap C_j|}{|C_j|}\right)^\beta \mathcal{H}_\beta((\pi_i)_{C_j}) + \mathcal{H}_\beta(\tau_{C_j}).$$

Thus,

$$\begin{aligned}
&\mathcal{H}_\beta(\pi_1 + \cdots + \pi_k|\sigma)\\
&= \sum_{j=1}^{n}\sum_{i=1}^{k} \left(\frac{|F_i \cap C_j|}{|S|}\right)^\beta \mathcal{H}_\beta((\pi_i)_{C_j}) + \sum_{j=1}^{n} \left(\frac{|C_j|}{|S|}\right)^\beta \mathcal{H}_\beta(\tau_{C_j})\\
&= \sum_{i=1}^{k} \left(\frac{|F_i|}{|S|}\right)^\beta \sum_{j=1}^{n} \left(\frac{|F_i \cap C_j|}{|F_i|}\right)^\beta \mathcal{H}_\beta((\pi_i)_{F_i \cap C_j}) + \mathcal{H}_\beta(\tau|\sigma)\\
&= \sum_{i=1}^{k} \left(\frac{|F_i|}{|S|}\right)^\beta \mathcal{H}_\beta(\pi_i|\sigma_{F_i}) + \mathcal{H}_\beta(\tau|\sigma),
\end{aligned}$$

which is the desired equality. □

Theorem 12.63 *Let π, σ, τ be three partitions of the finite set S. We have*

$$\mathcal{H}_\beta(\pi|\sigma \wedge \tau) + \mathcal{H}_\beta(\sigma|\tau) = \mathcal{H}_\beta(\pi \wedge \sigma|\tau).$$

Proof By Theorem 12.55, we can write

$$\begin{aligned}
\mathcal{H}_\beta(\pi|\sigma \wedge \tau) &= \mathcal{H}_\beta(\pi \wedge \sigma \wedge \tau) - \mathcal{H}_\beta(\sigma \wedge \tau)\\
\mathcal{H}_\beta(\sigma|\tau) &= \mathcal{H}_\beta(\sigma \wedge \tau) - \mathcal{H}_\beta(\tau).
\end{aligned}$$

By adding these equalities and again applying Theorem 12.55, we obtain the equality of the theorem. □

Corollary 12.64 *Let π, σ, τ be three partitions of the finite set S. Then, we have*

$$\mathcal{H}_\beta(\pi|\sigma) + \mathcal{H}_\beta(\sigma|\tau) \geqslant \mathcal{H}_\beta(\pi|\tau).$$

Proof By Theorem 12.63, the monotonicity of β-conditional entropy in its second argument, and the antimonotonicity of the same in its first argument, we can write

$$\begin{aligned}\mathcal{H}_\beta(\pi|\sigma) + \mathcal{H}_\beta(\sigma|\tau) &\geqslant \mathcal{H}_\beta(\pi|\sigma \wedge \tau) + \mathcal{H}_\beta(\sigma|\tau)\\ &= \mathcal{H}_\beta(\pi \wedge \sigma|\tau)\\ &\geqslant \mathcal{H}_\beta(\pi|\tau),\end{aligned}$$

which is the desired inequality. □

Corollary 12.65 *Let π and σ be two partitions of the finite set S. Then, we have*

$$\mathcal{H}_\beta(\pi \vee \sigma) + \mathcal{H}_\beta(\pi \wedge \sigma) \leqslant \mathcal{H}_\beta(\pi) + \mathcal{H}_\beta(\sigma).$$

Proof By Corollary 12.64, we have $\mathcal{H}_\beta(\pi|\sigma) \leqslant \mathcal{H}_\beta(\pi|\tau) + \mathcal{H}_\beta(\tau|\sigma)$. Then, by Theorem 12.55, we obtain

$$\mathcal{H}_\beta(\pi \wedge \sigma) - \mathcal{H}_\beta(\sigma) \leqslant \mathcal{H}_\beta(\pi \wedge \tau) - \mathcal{H}_\beta(\tau) + \mathcal{H}_\beta(\tau \wedge \sigma) - \mathcal{H}_\beta(\sigma),$$

hence

$$\mathcal{H}_\beta(\tau) + \mathcal{H}_\beta(\pi \wedge \sigma) \leqslant \mathcal{H}_\beta(\pi \wedge \tau) + \mathcal{H}_\beta(\tau \wedge \sigma).$$

Choosing $\tau = \pi \vee \sigma$ implies immediately the inequality of the corollary. □

The property of $\mathcal{H}_\beta$ described in Corollary 12.65 is known as the *submodularity* of the generalized entropy. This result generalizes the modularity of the Gini index proven in [5] and gives an elementary proof of a result shown in [6] concerning Shannon's entropy.

12.5 Generalized Measures and Data Mining

The notion of a *measure* is important for data mining since, in a certain sense, the support count and the support of an item sets are generalized measures.

The notion of *generalized measure* was introduced in [5], where generalizations of measures of great interest to data mining are considered.

We need first to introduce four properties that apply to real-valued functions defined on a lattice.

Definition 12.66 *Let $(L, \{\wedge, \vee\})$ be a lattice. A function $f : L \longrightarrow \mathbb{R}$ is*

- submodular *if $f(u \vee v) + f(u \wedge v) \leqslant f(u) + f(v)$,*
- supramodular *if $f(u \vee v) + f(u \wedge v) \geqslant f(u) + f(v)$,*
- modular *if it is both submodular and supramodular;*
- logarithmic submodular *if $f(u \vee v) \cdot f(u \wedge v) \leqslant f(u) \cdot f(v)$, and*
- logarithmic supramodular *if $f(u \vee v) \cdot f(u \wedge v) \geqslant f(u) \cdot f(v)$,*

for every $u, v \in L$.

Clearly, if f is a submodular or supramodular function then a^f is logarithmic submodular or supramodular, respectively, where a is a fixed positive number.

Generalized measures are real-valued functions defined on the lattice of subsets $(\mathcal{P}(S), \{\cup, \cap\})$ of a set S. The first two properties introduced in Definition 12.66 may be combined with the monotonicity or antimonotonicity properties to define four types of generalized measures.

Definition 12.67 *A* generalized measure *or* g-measure *on a set S is a mapping $m : \mathcal{P}(S) \longrightarrow \mathbb{R}$ that is either monotonic or anti-monotonic and is either submodular or supramodular.*

Example 12.68 Let S be a finite nonempty set of nonnegative numbers, $S = \{x_1, x_2, \ldots, x_n\}$ such that $x_1 \leqslant x_2 \leqslant \cdots \leqslant x_n$. Define the mapping $\max : \mathcal{P}(S) \longrightarrow \mathbb{R}_{\geqslant 0}$ by

$$\max(U) = \begin{cases} \text{the largest element of } U & \text{if } U \neq \emptyset, \\ x_1 & \text{if } U = \emptyset, \end{cases}$$

for $U \in \mathcal{P}(S)$.

Note that the definition of max is formulated to ensure that the function is monotonic; that is, $U \subseteq V$ implies $\max U \leqslant \max V$.

The function max is submodular. Indeed, let U and V be two subsets of S and let $u = \max U$ and $v = \max V$. Without restricting the generality, we may assume that $u \leqslant v$. In this case, it is clear that $\max(U \cup V) = v$ and that $\max(U \cap V) \leqslant \max U$ and $\max(U \cap V) \leqslant \max V$. This implies immediately that max is submodular and therefore that max is a g-measure.

The function min is defined similarly by

$$\min(U) = \begin{cases} \text{the least element of } U & \text{if } U \neq \emptyset, \\ x_n & \text{if } U = \emptyset. \end{cases}$$

The function min is antimonotonic; that is, $U \subseteq V$ implies $\min V \leqslant \min U$. If $U = \emptyset$, then we have $\emptyset \subseteq V$ for every subset V of S and therefore $\min V \leqslant \min \emptyset = x_n$, which is obviously true.

It is easy to show that min is a supramodular function, so it is also a g-measure.

Let $f : \mathcal{P}(S) \longrightarrow \mathbb{R}_{\geqslant 0}$ be a nonnegative function defined on the set of subsets of S. The functions f^{neg} and f^{co} introduced in [5] are defined by

$$\begin{aligned} f^{\text{neg}}(X) &= f(S) + f(\emptyset) - f(X), \\ f^{\text{co}}(X) &= f(S - X), \end{aligned}$$

for $X \in \mathcal{P}(X)$.

Theorem 12.69 *Let $f : \mathcal{P}(S) \longrightarrow \mathbb{R}_{\geqslant 0}$ be a nonnegative function defined on the set of subsets of S. The following statements are equivalent:*

(i) *f is monotonic.*
(ii) *f^{neg} is antimonotonic.*
(iii) *f^{co} is antimonotonic.*
(iv) *$(f^{co})^{neg}$ is monotonic.*

Also, the following statements are equivalent:

(i) *f is submodular.*
(ii) *f^{neg} is supramodular.*
(iii) *f^{co} is submodular.*
(iv) *$(f^{co})^{neg}$ is supramodular.*

We have $(f^{neg})^{neg} = f$ and $(f^{co})^{co} = f$ (the involutive property *of* neg *and* co*).*

Proof The arguments are straightforward and are left to the reader. □

Example 12.70 Let S be a set and let $\mathcal{B}$ be a finite collection of subsets of S. Define the function $\phi : \mathcal{P}(S) \longrightarrow \mathbb{N}$ as

$$\phi(W) = |\{B \in \mathcal{B} \mid W \subseteq B\}|,$$

for $W \in \mathcal{P}(S)$. Starting from ϕ we can provide four examples of g-measures [5].

It is immediate that ϕ is an anti-monotonic function. Moreover, since

$$\begin{aligned}\{B \in \mathcal{B}|X \cap Y \subseteq B\} &\supseteq \{B \in \mathcal{B}|X \subseteq B\} \cup \{B \in \mathcal{B}|Y \subseteq B\},\\ \{B \in \mathcal{B}|X \cup Y \subseteq B\} &= \{B \in \mathcal{B}|X \subseteq B\} \cap \{B \in \mathcal{B}|Y \subseteq B\},\end{aligned}$$

it follows that

$$\begin{aligned}&\phi(X \cap Y) \geqslant |\{B \in \mathcal{B}|X \subseteq B\} \cup \{B \in \mathcal{B}|Y \subseteq B\}|\\ &= |\{B \in \mathcal{B}|X \subseteq B\}| + |\{B \in \mathcal{B}|Y \subseteq B\}| - |\{B \in \mathcal{B}|X \subseteq B\} \cap \{B \in \mathcal{B}|Y \subseteq B\}|\\ &= |\{B \in \mathcal{B}|X \subseteq B\}| + |\{B \in \mathcal{B}|Y \subseteq B\}| - |\{B \in \mathcal{B}|X \cup Y \subseteq B\}|\\ &= \phi(X) + \phi(Y) - \phi(X \cup Y),\end{aligned}$$

so ϕ is supramodular.

Note that $\phi(S) = \begin{cases} 1 & \text{if } S \in \mathcal{B}, \\ 0 & \text{otherwise} \end{cases}$ and $\phi(\emptyset) = |\mathcal{B}|$.

Suppose now that $S \notin \mathcal{B}$. The function ϕ^{neg} is given by

$$\begin{aligned}\phi^{\text{neg}}(X) &= |\mathcal{B}| - \phi(X) = |\mathcal{B}| - |\{B \in \mathcal{B} \mid X \subseteq B\}|\\ &= \{B \in \mathcal{B}|X \not\subseteq B\}.\end{aligned}$$

Thus, the function $\psi : \mathcal{P}(S) \longrightarrow \mathbb{N}$ given by $\psi(X) = \phi^{\text{neg}}(X) = \{B \in \mathcal{B}|X \not\subseteq B\}$ is monotonic and submodular.

The function $\zeta = \phi^{\text{co}}$ given by

$$\zeta(X) = \phi^{\text{co}}(X) = \phi(\overline{X}) = |\{B \in \mathcal{P} \mid \overline{X} \subseteq B\}|$$

is monotonic and supramodular.

Finally, the function $\eta = (\phi^{\text{co}})^{\text{neg}}$ given by

$$\eta(X) = (\phi^{\text{co}})^{\text{neg}}(X) = |\{B \in \mathcal{B} | \overline{X} \not\subseteq B\}|$$

is anti-monotonic and submodular.

We consider next two important examples of g-measures related to database tables and sets of transactions, respectively. Recall that the partition generated by the set of attributes X of a table was denoted by π^X.

Definition 12.71 *Let* $\theta = (T, H, \mathbf{r})$ *be a table and let* X *be a set of attributes,* $X \subseteq H$*. The* β-entropy of X, $\mathsf{H}_\beta(X)$, *is the* β*-entropy of the partition of the set of tuples set*($\mathbf{r}$) *generated by* X*:*

$$\mathsf{H}_\beta(X) = \mathcal{H}_\beta(\pi^X).$$

Example 12.72 We claim that H_β is a monotonic submodular g-measure on the set of attributes of the table on which it is defined.

Indeed, if $X \subseteq Y$, we saw that $\pi^Y \leqslant \pi^X$, so $\mathcal{H}_\beta(\pi^X) \leqslant \mathcal{H}_\beta(\pi^Y)$ by the first axiom of partition entropies. Thus, H_β is monotonic.

To prove the submodularity, we start from the submodularity of the β-entropy on partitions shown in Corollary 12.65. We have

$$\mathcal{H}_\beta(\pi^X \vee \pi^Y) + \mathcal{H}_\beta(\pi^X \wedge \pi^Y) \leqslant \mathcal{H}_\beta(\pi^X) + \mathcal{H}_\beta(\pi^Y);$$

hence

$$\mathcal{H}_\beta(\pi^X \vee \pi^Y) + \mathcal{H}_\beta(\pi^{X \cup Y}) \leqslant \mathcal{H}_\beta(\pi^X) + \mathcal{H}_\beta(\pi^Y)$$

because $\pi^{X \cup Y} = \pi^X \wedge \pi^Y$. Since $X \cap Y \subseteq X$ and $X \cap Y \subseteq Y$ it follows that $\pi^X \leqslant \pi^{X \cap Y}$ and $\pi^Y \leqslant \pi^{X \cap Y}$, so $\pi^X \vee \pi^Y \leqslant \pi^{X \cap Y}$. By Axiom **(P1)**, we have $\mathcal{H}_\beta(\pi^X \vee \pi^Y) \geqslant \mathcal{H}_\beta(\pi^{X \cap Y})$, which implies

$$\mathcal{H}_\beta(\pi^{X \cap Y}) + \mathcal{H}_\beta(\pi^{X \cup Y}) \leqslant \mathcal{H}_\beta(\pi^X) + \mathcal{H}_\beta(\pi^Y),$$

which is the submodularity of the g-measure H_β.

Example 12.73 Let T be a transaction data set over a set of items I as introduced in Definition 13.1. The functions $suppcount_T$ and $supp_T$ introduced in Definition 13.3 are antimonotonic, supramodular g-measures over $\mathcal{P}(I)$. The antimonotonicity of these functions was shown in Theorem 13.6.

Let K and L be two item sets of T. If k is the index of a transaction such that either $K \subseteq T(k)$ or $L \subseteq T(k)$, then it is clear that $K \cap L \subseteq T(k)$. Therefore, we have

$$suppcount_T(K \cap L) \geqslant |\{k \mid K \subseteq T(k)\} \cup \{k \mid L \subseteq T(k)\}|.$$

This allows us to write

$$\begin{aligned}
&suppcount_T(K \cap L)\\
&= |\{k \mid K \cap L \subseteq T(k)\}|\\
&\geqslant |\{k \mid K \subseteq T(k)\} \cup \{k \mid L \subseteq T(k)\}|\\
&= |\{k \mid K \subseteq T(k)\}| + |\{k \mid L \subseteq T(k)\}|\\
&\quad - |\{k \mid K \subseteq T(k)\} \cap \{k \mid K \subseteq T(k)\}|\\
&= suppcount_T(K) + suppcount_T(L) - suppcount_T(K \cup L)
\end{aligned}$$

for every $K, L \in \mathcal{P}(I)$, which proves that $suppcount_T$ is supramodular. The supramodularity of $supp_T$ follows immediately.

The definition of conditional entropy of partitions allows us to extend this concept to attribute sets of tables.

Definition 12.74 *Let $\theta = (T, H, \mathbf{r})$ be a table and let X and Y be two attribute sets of this table. The* β-conditional entropy *of X, Y is defined by*

$$\mathsf{H}_\beta(X|Y) = \mathcal{H}_\beta(\pi^X|\pi^Y).$$

Properties of conditional entropies of partitions can now be easily transferred to conditional entropies of attribute sets. For example, Theorem 12.55 implies

$$\mathsf{H}_\beta(Y|X) = \mathsf{H}_\beta(XY) - \mathsf{H}_\beta(X).$$

12.6 Differential Constraints

Differential constraints have been introduced in [7]. They apply to real-valued functions defined over the set of subsets of a set. Examples of such functions are abundant in databases and data mining. For example, we have introduced the entropy of attribute sets mapping sets of attributes into the set of real numbers and the support count of sets of items mapping such sets into natural numbers. Placing restrictions on such functions help us to better express the semantics of data.

Definition 12.75 *Let $\mathcal{C}$ be a collection of subsets of a set S and let $f : \mathcal{P}(S) \longrightarrow \mathbb{R}$ be a function. The* $\mathcal{C}$-differential *of f is the function $D_f^{\mathcal{C}} : \mathcal{P}(S) \longrightarrow \mathbb{R}$ defined by*

$$D_f^{\mathcal{C}}(X) = \sum_{\mathcal{D} \subseteq \mathcal{C}} (-1)^{|\mathcal{D}|} f\left(X \cup \bigcup \mathcal{D}\right)$$

for $X \in \mathcal{P}(S)$.

The density function of f *is the function* $d_f : \mathcal{P}(S) \longrightarrow R$ *defined by*

$$d_f(X) = \sum_{X \subseteq U \subseteq S} (-1)^{|U|-|X|} f(U) \tag{12.7}$$

for $X \in \mathcal{P}(S)$.

Recall that in Example 3.29 we have shown that the Möbius function of a poset $(\mathcal{P}(S), \subseteq)$ is given by

$$\mu(X, U) = \begin{cases} (-1)^{|U|-|X|} & \text{if } X \subseteq U, \\ 0 & \text{otherwise,} \end{cases}$$

for $X, U \in \mathcal{P}(S)$. Therefore, the density d_f can be written as

$$d_f(X) = \sum_{X \subseteq U \subseteq S} \mu(X, U) f(U)$$

for $X, U \in \mathcal{P}(S)$, which implies

$$f(X) = \sum_{X \subseteq U \subseteq S} d_f(U) \tag{12.8}$$

by the Möbius dual inversion theorem (Theorem 3.28).

The density of f can be expressed also as a $\mathcal{C}$-differential. For $X \in \mathcal{P}(S)$, define the collection $\mathcal{C}_X$ as $\mathcal{C}_X = \{\{y\} | y \notin X\}$. Then, we can write

$$\begin{aligned} D_f^{\mathcal{C}_X}(X) &= \sum_{\mathcal{D} \subseteq \mathcal{C}_X} (-1)^{|\mathcal{D}|} f\left(X \cup \bigcup \mathcal{D}\right) \\ &= \sum_{D \subseteq S-X} (-1)^{|D|} f(X \cup D) \\ &= \sum_{X \subseteq U \subseteq S} (-1)^{|U|-|X|} f(U). \end{aligned}$$

In the last equality, we denoted $U = X \cup D$. Since D is a subset of $S - X$, we have immediately $D = U - X$, which justifies the last equality. This allows us to conclude that $d_f(X) = D_f^{\mathcal{C}_X}(X)$ for $X \in \mathcal{P}(S)$.

Example 12.76 Let S be a finite set and let $f : \mathcal{P}(S) \longrightarrow \mathbb{R}$. If $\mathcal{C} = \emptyset$, we have $D^{\emptyset}(X) = f(X)$. Similarly, if $\mathcal{C}$ is a one-set collection $\mathcal{C} = \{Y\}$, then $D^{\mathcal{C}}(X) = f(X) - f(X \cup Y)$. When $\mathcal{C} = \{Y, Z\}$, we have $D^{\mathcal{C}}(X) = f(X) - f(X \cup Y) - f(X \cup Z) + f(X \cup Y \cup Z)$ for $X \in \mathcal{P}(S)$.

Example 12.77 Let $S = \{a, b, c, d\}$ and let $f : \mathcal{P}(S) \longrightarrow \mathbb{R}$ be a function. For $X = \{a, b\}$, we have $\mathcal{C}_X = \{\{c\}, \{d\}\}$. Thus, $d_f(\{a, b\}) = D_f^{\{c\},\{d\}}(\{a, b\})$. Note that the collections of subsets of S included in the collection $\{\{c\}, \{d\}\}$ are $\emptyset$, $\{\{c\}\}$, $\{\{d\}\}$, and $\{\{c\}, \{d\}\}$. Therefore,

$$D_f^{\{\{c\},\{d\}\}}(\{a, b\}) = f(\{a, b\}) - f(\{a, b, c\}) - f(\{a, b, d\}) + f(\{a, b, c, d\}),$$

which equals $d_f(\{a, b\})$, as computed directly from Equality (12.7).

Definition 12.78 *Let $\mathcal{C}$ be a collection of subsets of a set S. A subset W of S is a* witness set of $\mathcal{C}$ *if $W \subseteq \bigcup \mathcal{C}$ and $X \cap W \neq \emptyset$ for every $X \in \mathcal{C}$.*

The collection of all witness sets for $\mathcal{C}$ is denoted by $\mathcal{W}(\mathcal{C})$.

Observe that $\mathcal{W}(\emptyset) = \{\emptyset\}$.

Example 12.79 Let $S = \{a, b, c, d\}$ and let $\mathcal{C} = \{\{b\}, \{c, d\}\}$ be a collection of subsets of S. The collection of witness sets of $\mathcal{C}$ is

$$\mathcal{W}(\mathcal{C}) = \{\{b, c\}, \{b, d\}, \{b, c, d\}\}.$$

For the collection $\mathcal{D} = \{\{b, c\}, \{b, d\}\}$, we have

$$\mathcal{W}(\mathcal{D}) = \{\{b\}, \{b, c\}, \{b, d\}, \{c, d\}, \{b, c, d\}\}.$$

Definition 12.80 *Let $\mathcal{C}$ be a collection of subsets of S and let X be a subset of S. The* decomposition *of $\mathcal{C}$ relative to X is the collection $\mathcal{L}[X, \mathcal{C}]$ of subsets of S defined as a union of intervals by*

$$\mathcal{L}[X, \mathcal{C}] = \bigcup_{W \in \mathcal{W}(\mathcal{C})} [X, \overline{W}],$$

where $\overline{W} = S - W$.

Example 12.81 The decomposition of the collection $\mathcal{C} = \{\{b\}, \{c, d\}\}$ considered in Example 12.79 relative to the set $X = \{a\}$ is given by

$$\begin{aligned}\mathcal{L}[X, \mathcal{C}] &= [\{a\}, \overline{\{b, c\}}] \cup [\{a\}, \overline{\{b, d\}}] \cup [\{a\}, \overline{\{b, c, d\}}] \\ &= [\{a\}, \{a, d\}] \cup [\{a\}, \{a, c\}] \cup [\{a\}, \{a\}] \\ &= \{\{a\}, \{a, d\}, \{a, c\}\}.\end{aligned}$$

Similarly, we can write for the collection $\mathcal{D} = \{\{b, c\}, \{b, d\}\}$

$$\begin{aligned}&\mathcal{L}[X, \mathcal{D}] \\ &= [\{a\}, \overline{\{b\}}] \cup [\{a\}, \overline{\{b, c\}}] \cup [\{a\}, \overline{\{b, d\}}] \cup [\{a\}, \overline{\{c, d\}}] \cup [\{a\}, \overline{\{b, c, d\}}] \\ &= [\{a\}, \{a, c, d\}] \cup [\{a\}, \{a, d\}] \cup [\{a\}, \{a, c\}] \cup [\{a\}, \{a, b\}] \cup [\{a\}, \{a\}] \\ &= \{\{a\}, \{a, c\}, \{a, d\}, \{a, c, d\}, \{a, b\}\}\end{aligned}$$

Example 12.82 We have $\mathcal{L}[X, \emptyset] = \bigcup_{W \in \mathcal{W}(\emptyset)} [X, \overline{W}] = [X, S]$ because $\mathcal{W}(\emptyset) = \{\emptyset\}$. Consequently, $\mathcal{L}[\emptyset, \emptyset] = \mathcal{P}(S)$ for every set S.

Note that if $X \neq \emptyset$, then $\mathcal{W}(\{X\}) = \mathcal{P}(X) - \{\emptyset\}$. Therefore, $\mathcal{L}[X, \{X\}] = \bigcup_{W \in \mathcal{W}(\mathcal{C})} [X, \overline{W}] = \emptyset$ because there is no set T such that $X \subseteq T \subseteq \overline{W}$.

Theorem 12.83 *Let S be a finite set, $X, Y \in \mathcal{P}(S)$, and let $\mathcal{C}$ be a collection of subsets of S. We have*

$$\mathcal{L}[X, \mathcal{C}] = \mathcal{L}[X, \mathcal{C} \cup \{Y\}] \cup \mathcal{L}[X \cup Y, \mathcal{C}].$$

Proof We begin the proof by showing that $\mathcal{L}[X, \mathcal{C} \cup \{Y\}] \subseteq \mathcal{L}[X, \mathcal{C}]$. Let $U \in \mathcal{L}[X, \mathcal{C} \cup \{Y\}]$. There is a witness set W for $\mathcal{C} \cup \{Y\}$ such that $X \subseteq U \subseteq \overline{W}$.

The set $W' = W \cap \bigcup \mathcal{C}$ is a witness set for $\mathcal{C}$. Indeed, we have $W' \subseteq \bigcup \mathcal{C}$, and for every set $Z \in \mathcal{C}$ we have $W' \cap Z \neq \emptyset$. Since $W' \subseteq W$, we have $\overline{W} \subseteq \overline{W'}$, so $X \subseteq U \subseteq \overline{W'}$. Therefore, $U \in \mathcal{L}[X, \mathcal{C}]$.

Next, we show that $\mathcal{L}[X \cup Y, \mathcal{C}] \subseteq \mathcal{L}[X, \mathcal{C}]$. Let $V \in \mathcal{L}[X \cup Y, \mathcal{C}]$, so $V \in [X \cup Y, \overline{W}]$ for some witness set of $\mathcal{C}$. Since $[X \cup Y, \overline{W}] \subseteq [X, \overline{W}]$, the desired conclusion follows immediately. Thus, we have shown that

$$\mathcal{L}[X, \mathcal{C} \cup \{Y\}] \cup \mathcal{L}[X \cup Y, \mathcal{C}] \subseteq \mathcal{L}[X, \mathcal{C}].$$

To prove the converse inclusion, let $U \in \mathcal{L}[X, \mathcal{C}]$. There is a witness set W of $\mathcal{C}$ such that $X \subseteq U \subseteq \overline{W}$. Depending on the relative positions of the sets U and Y, we can distinguish three cases:

(i) $Y \subseteq U$;
(ii) $Y \not\subseteq U$ and $Y \cap W \neq \emptyset$;
(iii) $Y \not\subseteq U$ and $Y \cap W = \emptyset$.

Note that the first condition of the second case is superfluous because $Y \cap W \neq \emptyset$ implies $Y \not\subseteq U$.

In the first case, we have $U \in \mathcal{L}[X \cup Y, \mathcal{C}]$.

In Case (ii), W is a witness set for $\mathcal{C} \cup \{Y\}$, and therefore $U \in \mathcal{L}[X, \mathcal{C} \cup \{Y\}]$.

Finally, in the third case, define $W_1 = W \cup (Y - U)$. We have $W_1 \subseteq \bigcup \mathcal{C} \cup Y$. Since every member of $\mathcal{C}$ has a nonempty intersection with W_1 and $Y \cap W_1 \neq \emptyset$, it follows that W_1 is a witness set of $\mathcal{C} \cup \{Y\}$. Note that $U \subseteq \overline{W_1}$. Therefore $U \in \mathcal{L}[X, \mathcal{C} \cup \{Y\}]$. □

The connection between differentials and density functions is shown in the next statement.

Theorem 12.84 *Let S be a finite set, $X \in \mathcal{P}(S)$ and let $\mathcal{C}$ be a collection of subsets of S. If $f : \mathcal{P}(S) \longrightarrow \mathbb{R}$, then*

$$D_f^{\mathcal{C}}(X) = \sum \{d_f(U) \mid U \in \mathcal{L}[X, \mathcal{C}]\}.$$

Proof By Definition 12.75, the $\mathcal{C}$-*differential* of f is

$$\begin{aligned}
D_f^{\mathcal{C}}(X) &= \sum_{\mathcal{D}\subseteq\mathcal{C}} (-1)^{|\mathcal{D}|} f\left(X \cup \bigcup \mathcal{D}\right) \\
&= \sum_{\mathcal{D}\subseteq\mathcal{C}} (-1)^{|\mathcal{D}|} \sum_{X\cup\bigcup\mathcal{D}\subseteq U\subseteq S} d_f(U), \\
&\quad \text{(by Equality 12.8)} \\
&= \sum_{X\subseteq U\subseteq S} d_f(U) \sum_{\mathcal{D}\subseteq\{Y\in\mathcal{C}\,|\,Y\subseteq U\}} (-1)^{|\mathcal{D}|} \\
&= \sum \{d_f(U) \mid X \subseteq U \subseteq S \text{ and } \{Y \in \mathcal{D} \mid Y \subseteq U\} = \emptyset\}.
\end{aligned}$$

□

Constraints can be formulated on real-valued functions defined on collection subsets using differentials of functions or density functions. Both types of constraints have been introduced and studied in [5, 7, 8].

Definition 12.85 *Let S be a set, $\mathcal{C}$ a collection of subsets of S, and $f : \mathcal{P}(S) \to \mathbb{R}$ a function.*

The function f satisfies the differential constraint $X \rightarrowtail \mathcal{C}$ *if* $D_f^{\mathcal{C}}(X) = 0$.

The function f satisfies the density constraint $X \rightsquigarrow \mathcal{C}$ *if* $d_f(U) = 0$ *for every* $U \in \mathcal{L}[X, \mathcal{C}]$.

By Theorem 12.84, if f satisfies the density constraint $X \rightsquigarrow \mathcal{C}$, then f satisfies the differential constraint $X \rightarrowtail \mathcal{C}$. If the density of f takes only nonnegative values (or only nonpositive values), then, by the same theorem, the reverse also holds. However, in general, this is not true, as the next example shows. Thus, differential constraints are weaker than density constraints.

Example 12.86 Let $S = \{a\}$ and let $f:\mathcal{P}(S) \to \mathbb{R}$ be defined by $f(\emptyset) = 0$ and $f(\{a\}) = 1$. Observe that $D_f^{\emptyset}(\emptyset) = f(\emptyset) = 0$, so f satisfies the differential constraint $\emptyset \rightarrowtail \emptyset$.

Since $\mathcal{C}_\emptyset = \{\{a\}\}$ and $\mathcal{C}_{\{a\}} = \emptyset$, we have

$$\begin{aligned}
d_f(\emptyset) &= D_f^{\mathcal{C}_\emptyset}(\emptyset) = f(\emptyset) - f(\{a\}) = -1, \\
d_f(\{a\}) &= D_f^{\mathcal{C}_{\{a\}}}(\{a\}) = f(\{a\}) = 1.
\end{aligned}$$

On the other hand, $\mathcal{L}[\emptyset, \emptyset] = \mathcal{P}(S)$ by Example 12.82, and f fails to satisfy the density constraint $\emptyset \rightsquigarrow \emptyset$.

Example 12.87 Consider the β-entropy H_β defined on the set of subsets of the heading of a table $\theta = (T, H, \mathbf{r})$. We saw that H_β is a monotonic, submodular g-measure on $\mathcal{P}(H)$.

We claim that H_β satisfies the differential constraint $X \rightarrowtail \{Y\}$ if and only if the table θ satisfies the functional dependency $X \rightarrow Y$.

By Example 12.76, H_β satisfies the differential constraint $X \rightarrowtail \{Y\}$, that is, $D_f^{\{Y\}}(X) = 0$ if and only if $\mathsf{H}_\beta(X) = \mathsf{H}_\beta(X \cup Y)$. This is equivalent to $\mathcal{H}_\beta(\pi^{XY}) = \mathcal{H}_\beta(\pi^X)$ or to $\mathcal{H}_\beta(\pi^X \wedge \pi^Y) = \mathcal{H}_\beta(\pi^X)$ by Corollary 12.13. This equality implies: $\pi^X \leqslant \pi^Y$ by Corollary 12.56, which shows that θ satisfies the functional dependency $X \rightarrow Y$.

The observation contained in this example generalizes the result of Sayrafi [5] proven for the Gini index and the result contained in [9–11] that involves the Shannon entropy.

Note also that this shows that

$$\mathsf{H}_\beta(Y|X) = -D_f^{\{Y\}}(X)$$

for $X, Y \in \mathcal{P}(H)$.

Example 12.88 Let $S = \{a, b, c\}$. To compute $\mathcal{L}[\{a\}, \{\{b\}\}]$, note that

$$\mathcal{W}(\{\{b\}\}) = \{\{b\}\}.$$

Thus, $\mathcal{L}[\{a\}, \{\{b\}\}] = [\{a\}, \{a, c\}] = \{\{a\}, \{a, c\}\}$. In general, we have for $x, y, z \in S$ that are pairwise distinct

$$\mathcal{L}[\{x\}, \{\{y\}\}] = \{\{x\}, \{x, z\}\}.$$

Consider now a function $f : \mathcal{P}(S) \rightarrow \mathbb{R}$ such that

$$f(X) = \begin{cases} 2 & \text{if } X \in \{\emptyset, \{c\}\}, \\ 1 & \text{otherwise.} \end{cases}$$

We have

$$\begin{aligned} d_f(\{c\}) &= f(\{c\}) - f(\{a, c\}) - f(\{b, c\}) + f(\{a, b, c\}) = 1, \\ d_f(\{a, b, c\}) &= f(\{a, b, c\}) = 1. \end{aligned}$$

For $X \notin \{\{c\}, \{a, b, c\}\}$, we have $d_f(X) = 0$. For example,

$$\begin{aligned} d_f(\{b\}) &= f(\{b\}) - f(\{a, b\}) - f(\{b, c\}) + f(\{a, b, c\}) = 0, \\ d_f(\{b, c\}) &= f(\{b, c\}) - f(\{a, b, c\}) = 0. \end{aligned}$$

This shows that f satisfies the density constraints $\{a\} \rightsquigarrow \{\{b\}\}$ and $\{b\} \rightsquigarrow \{\{c\}\}$ but fails to satisfy $\{c\} \rightsquigarrow \{\{a\}\}$ because $\mathcal{L}[\{c\}, \{\{a\}\}]$ consists of $\{c\}$ and $\{b, c\}$.

12.7 Decision Systems and Decision Trees

Classifiers are algorithms that place objects in certain classes based on characteristic features of those objects. Classifiers are constructed starting from a set of objects known as a *training set*; for each object of the training set the class of the object is known and the classifier can be regarded as a function that maps as well as possible the objects of a training set to their respective classes.

It is desirable that the classifiers constructed from a training set do a reliable job of placing objects that do not belong to the training set in their correct classes. When the classifier works well on the training set but does a poor job of classifying objects outside the training set, we say that the classifier overfits the training set.

Decision systems use tables to formalize the notion of a training set. The features of the objects are represented by the attributes of the table; a special attribute (called a decision attribute) represents the class of the objects.

Definition 12.89 *A* decision system *is a pair* $\mathcal{D} = (\theta, D)$*, where* $\theta = (T, H, \mathbf{r})$ *is a table and* D *is a special attribute of* H *called a* decision attribute. *The attributes of* H *that are distinct from* D *are referred to as* conditional attributes, *and the set of conditional attributes of* H *will be denoted by* H_c.

Clearly, H_c is obtained by removing D from H.

Example 12.90 The decision system considered in this example is based on a data set that is well-known in the machine-learning literature (see [12, 13]). The heading H and domains of the attributes are specified below:

Attribute	Domain
Outlook	{Sunny, Overcast, Rain}
Temperature	{Hot, Mild, Cool}
Humidity	{Normal, High}
Wind	{Weak, Strong}

The decision attribute is PlayTennis; this attribute has the domain {yes, no}.

The sequence $\mathbf{r}$ consists of 14 tuples, $t_1, \ldots, t_{14}$, shown in Table 12.1:

The partitions of the form π^A (where A is an attribute) are

$$\begin{aligned}
\pi^{Outlook} &= \{\{1, 2, 8, 9, 11\}, \{3, 7, 12\}, \{4, 5, 6, 10, 13, 14\}\},\\
\pi^{Temperature} &= \{\{1, 2, 3, 13\}, \{4, 8, 10, 11, 12, 14\}, \{5, 6, 7, 9\}\},\\
\pi^{Humidity} &= \{\{1, 2, 3, 4, 8, 12, 14\}, \{5, 6, 7, 9, 10, 11, 13\}\},\\
\pi^{Wind} &= \{\{1, 3, 4, 5, 8, 9, 10, 13\}, \{2, 6, 7, 11, 12, 14\}\},\\
\pi^{PlayTennis} &= \{\{1, 2, 6, 8, 14\}, \{3, 4, 5, 7, 9, 10, 11, 12, 13\}\}.
\end{aligned}$$

Let $\mathcal{D} = (\theta, D)$ be a decision system, where $\theta = (T, H, \mathbf{r})$ and $\mathbf{r} = (t_1, \ldots, t_n)$. The *decision function of* $\mathcal{D}$ is the function $\delta_{\mathcal{D}} : \{1, \ldots, n\} \longrightarrow \mathrm{Dom}(D)$ given by

Table 12.1 The content of the decision system

	Outlook	Temperature	Humidity	Wind	PlayTennis
1	Sunny	Hot	High	Weak	No
2	Sunny	Hot	High	Strong	No
3	Overcast	Hot	High	Weak	Yes
4	Rain	Mild	High	Weak	Yes
5	Rain	Cool	Normal	Weak	Yes
6	Rain	Cool	Normal	Strong	No
7	Overcast	Cool	Normal	Strong	Yes
8	Sunny	Mild	High	Weak	No
9	Sunny	Cool	Normal	Weak	Yes
10	Rain	Mild	Normal	Weak	Yes
11	Sunny	Mild	Normal	Strong	Yes
12	Overcast	Mild	High	Strong	Yes
13	Rain	Hot	Normal	Weak	Yes
14	Rain	Mild	High	Strong	No

$$\delta_{\mathfrak{D}}(i) = \{d \in \text{Dom } D \mid (i, j) \in \epsilon^{H_c} \text{ and } t_j[D] = d\},$$

where ϵ^{H_c} is the indiscernibility relation defined by the set of conditional attributes of $\mathfrak{D}$. Equivalently, we can write

$$\delta_{\mathfrak{D}}(i) = \{t_j[D] \mid j \in [i]_{\epsilon^{H_c}}\}$$

for $1 \leq i \leq n$.

If $|\delta_{\mathfrak{D}}(i)| = 1$ for every i, $1 \leq i \leq n$, then $\mathfrak{D}$ is a *deterministic (consistent) decision system*; otherwise, $\mathfrak{D}$ is a nondeterministic (inconsistent) decision system. In other words, a decision system $\mathfrak{D}$ is consistent if the values of the components of a tuple that correspond to the conditional attributes determine uniquely the value of the tuple for the decision attribute.

If there exists $d \in \text{Dom}(D)$ such that $\delta_{\mathfrak{D}}(i) = \{d\}$ for every i, $1 \leq i \leq n$, then $\mathfrak{D}$ is a *pure decision system*.

Definition 12.91 *Let* $\mathfrak{D} = (\theta, D)$ *be a decision system, where* $\theta = (T, H, \mathbf{r})$ *and* $|\mathbf{r}| = n$. *The* classification generated by $\mathfrak{D}$ *is the partition* π^D *of the set* $\{1, \ldots, n\}$.

If B_d *is the block of* π^D *that corresponds to the value* d *of* $\text{Dom}(D)$, *we refer to* B_d *as the* d*-decision class of* $\mathfrak{D}$.

Note that the partition π^D contains a block for every element of $\text{Dom}(D)$ that occurs in $set(\mathbf{r}[D])$.

If X is a set of attributes, we denote the functions uap_{ϵ^X} and lap_{ϵ^X} by $\overline{X}$ and $\underline{X}$, respectively.

Example 12.92 The decision classes of the decision system $\mathcal{D}$ of Example 12.90 are

$$B_{no} = \{1, 2, 6, 8, 14\},$$
$$B_{yes} = \{3, 4, 5, 7, 9, 10, 11, 12, 13\}.$$

Definition 12.93 *Let U be a set of conditional attributes of $\mathcal{D} = (\theta, D)$, where $\theta = (T, H, \boldsymbol{r})$ and $|\boldsymbol{r}| = n$. The U-positive region of the decision system $\mathcal{D}$ is the set*

$$POS_U(\mathcal{D}) = \bigcup\{\underline{U}(B_d) \mid d \in set(\boldsymbol{r}[D])\}.$$

The tuples whose indexes occur in $POS_{H_c}(\mathcal{D})$ can be unambiguously placed in the d-decision classes of $\mathcal{D}$.

Example 12.94 For the decision system $\mathcal{D}$ of Example 12.90, we have

$$POS_{Outlook}(\mathcal{D}) = \{3, 7, 12\},$$
$$POS_{Temperature}(\mathcal{D}) = POS_{Humidity}(\mathcal{D}) = POS_{Wind}(\mathcal{D}) = \emptyset.$$

Thus, using the value of a single attribute, we can reach a classification decision only for the tuples t_3, t_7, and t_{12} (based on the *Outlook* attribute).

Next, we attempt to classify tuples using partitions generated by two attributes. We have six such partitions:

$$\pi^{Outlook,Temperature} = \{\{1, 2\}, \{3\}, \{4, 10, 14\}, \{5, 6\}, \{7\}, \{8, 11\}, \{9\}, \{12\}, \{13\}\},$$
$$\pi^{Outlook,Humidity} = \{\{1, 2, 8\}, \{3, 12\}, \{4, 14\}, \{5, 6, 10, 13\}, \{7\}, \{9, 11\}\},$$
$$\pi^{Outlook,Wind} = \{\{1, 8, 9\}, \{2, 11\}, \{3\}, \{4, 5, 10, 13\}, \{6, 14\}, \{7, 12\}\},$$
$$\pi^{Temperature,Humidity} = \{\{1, 2, 3\}, \{4, 8, 12, 14\}, \{5, 6, 7, 9\}, \{10, 11\}, \{13\}\},$$
$$\pi^{Temperature,Wind} = \{\{1, 3, 13\}, \{2\}, \{4, 8, 10\}, \{5, 9\}, \{6, 7\}, \{11, 12, 14\}\},$$
$$\pi^{Humidity,Wind} = \{\{1, 3, 4, 8\}, \{2, 12, 14\}, \{5, 9, 10, 13\}, \{6, 7, 11\}\},$$

and their corresponding positive regions are

$$POS_{Outlook,Temperature}(\mathcal{D}) = \{1, 2, 3, 7, 9, 13\},$$
$$POS_{Outlook,Humidity}(\mathcal{D}) = \{1, 2, 3, 7, 8, 9, 11, 12\},$$
$$POS_{Outlook,Wind}(\mathcal{D}) = \{3, 4, 5, 6, 7, 10, 12, 13, 14\},$$

$$POS_{Temperature,Humidity}(\mathfrak{D}) = \{10, 11, 13\},$$
$$POS_{Temperature,Wind}(\mathfrak{D}) = \{2, 5, 9\},$$
$$POS_{Humidity,Wind}(\mathfrak{D}) = \{5, 9, 10, 13\}.$$

There are four partitions generated by three attributes:

$$\begin{aligned}\pi^{Outlook,Temperature,Humidity} &= \{\{1, 2\}, \{3\}, \{4, 14\}, \{5, 6\},\\ &\quad \{7\}, \{8\}, \{9\}, \{10\}, \{11\}, \{12\}, \{13\}\},\\ \pi^{Outlook,Temperature,Wind} &= \{\{1\}, \{2\}, \{3\}, \{4, 10\}, \{5\}, \{6\},\\ &\quad \{7\}, \{8\}, \{9\}, \{11\}, \{12\}, \{13\}, \{14\}\},\\ \pi^{Outlook,Humidity,Wind} &= \{\{1, 8\}, \{2\}, \{3\}, \{4\}, \{5, 10, 13\},\\ &\quad \{6\}, \{7\}, \{9\}, \{11\}, \{12\}, \{14\}\},\\ \pi^{Temperature,Humidity,Wind} &= \{\{1, 3\}, \{2\}, \{4, 8\}, \{5, 9\},\\ &\quad \{6, 7\}, \{10\}, \{11\}, \{12, 14\}, \{13\}\}.\end{aligned}$$

Their positive regions are

$$POS_{Outlook,Temperature,Humidity}(\mathfrak{D}) = \{1, 2, 3, 7, 8, 9, 10, 11, 12, 13\},$$
$$POS_{Outlook,Temperature,Wind}(\mathfrak{D}) = set(H),$$
$$POS_{Outlook,Humidity,Wind}(\mathfrak{D}) = set(H),$$
$$POS_{Temperature,Humidity,Wind}(\mathfrak{D}) = \{2, 5, 9, 10, 11, 13\}.$$

This computation shows that a classification decision can be reached for every tuple starting from its components on the projection on either the set Outlook, Temperature, Wind or the set Outlook, Humidity, Wind.

Finally, note that $\mathfrak{D}$ is a deterministic system because $\pi^{set(H_c)} = \alpha_{set(H_c)}$.

Theorem 12.95 *Let $\mathfrak{D} = (\theta, D)$ be a decision system, where $\theta = (T, H, \mathbf{r})$ and $|\mathbf{r}| = n$. The following statements are equivalent:*

(i) *$\mathfrak{D}$ is deterministic;*
(ii) $\pi^{H_c} \leq \pi^D$;
(iii) $POS_{H_c}(\mathfrak{D}) = \{1, \ldots, n\}$.

Proof (i) implies (ii): Suppose that $\mathfrak{D}$ is deterministic. Then, if two tuples u and v in $\mathbf{r}$ are such that $u[H_c] = v[H_c]$, then $u[D] = v[D]$. This is equivalent to saying that $\pi^{H_c} \leq \pi^D$.

(ii) implies (iii): This implication follows immediately from Theorem 1.209.

(iii) implies (i): Suppose that $POS_{H_c}(\mathfrak{D}) = \{1, \ldots, n\}$, that is,

$$\bigcup\{\underline{H_c}(B_d) \mid d \in set(\mathbf{r}[D])\} = \{1, \ldots, n\}. \tag{12.9}$$

Note that $\underline{H_c}(B_d) = lap_{\epsilon^{H_c}}(B_d) \subseteq B_d$. Therefore, we have $lap_{\epsilon^{H_c}}(B_d) = B_d$ for every $d \in set(\mathbf{r}[D])$ because if the inclusion were strict for any of the sets B_d, then Equality (12.9) would be violated. Thus, each block B_d is a union of blocks of the partition π^{H_c}. In other words, each equivalence class of ϵ^{H_c} is included in a block B_d, which means that for every $j \in [i]_{\epsilon^{H_c}}$ we have $t_j[D] = d$ and the set $\delta_{\mathcal{D}}(i)$ contains a single element. Thus, $\mathcal{D}$ is deterministic. □

Next, we discuss informally a classification algorithm that makes use of decision systems. This algorithm is recursive and begins with the selection of a conditional attribute (referred to as a *splitting attribute*) by applying a criterion that is specific to the algorithm. The algorithm halts when it applies to a pure decision system (or to a decision system where a minimum percentage of tuples have the same decision value).

The splitting attribute is chosen here as one of the attributes whose positive region has the largest number of elements. This choice is known as a *splitting criterion*. The table of the decision system is split into a number of tables such that each new table is characterized by a value of the splitting attribute. Thus, we obtain a new set of decision systems, and the algorithm is applied recursively to the new decision systems. The process must stop because the tables become smaller with each split; its result is a tree of decision systems known as a *decision tree*.

Example 12.96 As we saw in Example 12.94, a classification decision can be reached immediately for the tuples t_3, t_7, t_{12}, which are characterized by the condition Outlook = 'overcast'.

Define the tables

$$\begin{aligned}\theta_0 &= (\theta \textbf{ where } \text{Outlook} = \text{'sunny'})[K],\\ \theta_1 &= (\theta \textbf{ where } \text{Outlook} = \text{'overcast'})[K],\\ \theta_2 &= (\theta \textbf{ where } \text{Outlook} = \text{'rain'})[K],\end{aligned}$$

where

$$K = \text{Temperature Humidity Wind PlayTennis}$$

is the heading obtained by dropping the Outlook attribute and the decision systems $\mathcal{D}_i = (\theta_i, D)$ for $0 \leq i \leq 3$.

Note that the decision system $\mathcal{D}_1$ is pure because the tuples of θ_1 belong to the same PlayTennis-class as shown in Table 12.2.

For the remaining systems

$$\mathcal{D}_0 = (\theta_0, \text{PlayTennis})\text{and } \mathcal{D}_2 = (\theta_2, \text{PlayTennis}),$$

we have the tables shown in Table 12.3a, b.

The same process is now applied to the decision systems $\mathcal{D}_0$ and $\mathcal{D}_2$. The positive regions are:

Table 12.2 The table θ_1

	Temperature	Humidity	Wind	PlayTennis
3	Hot	High	Weak	Yes
7	Cool	Normal	Strong	Yes
12	Mild	High	Strong	Yes

Table 12.3 The tables θ_0 and θ_2

	Temperature	Humidity	Wind	PlayTennis
(a)				
1	Hot	High	Weak	No
2	Hot	High	Strong	No
8	Mild	High	Weak	No
9	Cool	Normal	Weak	Yes
11	Mild	Normal	Strong	Yes
(b)				
4	Mild	High	Weak	Yes
5	Cool	Normal	Weak	Yes
6	Cool	Normal	Strong	No
10	Mild	Normal	Weak	Yes
13	Hot	Normal	Weak	Yes
14	Mild	High	Strong	No

$$\begin{aligned}
&POS_{Temperature}(\mathcal{D}_0) = \{1, 2, 9\},\\
&POS_{Humidity}(\mathcal{D}_0) = \{1, 2, 8, 9, 11\},\\
&POS_{Wind}(\mathcal{D}_0) = \emptyset,\\
&POS_{Temperature}(\mathcal{D}_2) = \{13\},\\
&POS_{Humidity}(\mathcal{D}_2) = \emptyset,\\
&POS_{Wind}(\mathcal{D}_2) = \{4, 5, 10, 13, 6, 14\}.
\end{aligned}$$

Thus, the splitting attribute for $\mathcal{D}_0$ is *Humidity*; the splitting attribute for $\mathcal{D}_2$ is *Wind*.

The decision system $\mathcal{D}_0$ yields the decision systems

$$\mathcal{D}_{00} = (\theta_{00}, PlayTennis) \text{ and } \mathcal{D}_{01} = (\theta_{01}, PlayTennis),$$

where θ_{00} and θ_{01} are given by

$$\begin{aligned}
\theta_{00} &= (\theta_0 \textbf{ where } \text{Humidity} = \text{`high'})[K_0],\\
\theta_{01} &= (\theta_0 \textbf{ where } \text{Humidity} = \text{`normal'})[K_0],
\end{aligned}$$

where

$$K_0 = \text{ Temperature Wind PlayTennis.}$$

Table 12.4 The tables θ_{00} and θ_{01}

	Temperature	Wind	PlayTennis
(a)			
1	Hot	Weak	No
2	Hot	Strong	No
8	Mild	Weak	No
(b)			
9	Cool	Weak	Yes
11	Mild	Strong	Yes

Table 12.5 The tables θ_{00} and θ_{01}

	Temperature	Humidity	PlayTennis
(a)			
4	Mild	High	Yes
5	Cool	Normal	Yes
10	Mild	Normal	Yes
13	Hot	Normal	Yes
(b)			
6	Cool	Normal	No
14	Mild	High	No

The tables θ_{00} and θ_{01} are shown in Table 12.4a, b, respectively. Note that both decision systems $\mathcal{D}_{00}$ and $\mathcal{D}_{01}$ are pure, so no further action is needed.

The decision system $\mathcal{D}_2$ produces the decision systems

$$\mathcal{D}_{20} = (\theta_{20}, PlayTennis) \text{ and } \mathcal{D}_{21} = (\theta_{21}, PlayTennis),$$

where θ_{20} and θ_{21} are given by

$$\theta_{20} = (\theta_0 \textbf{ where } \text{Wind} = \text{`weak'})[K_2]$$
$$\theta_{21} = (\theta_0 \textbf{ where } \text{Wind} = \text{`strong'})[K_2],$$

where

$$K_2 = \text{Temperature Humidity PlayTennis.}$$

The tables θ_{20} and θ_{21} are shown in Table 12.5a, b, respectively.

Again, both decision systems are pure, so no further splitting is necessary. The decision tree that is obtained from this process is shown in Fig. 12.3.

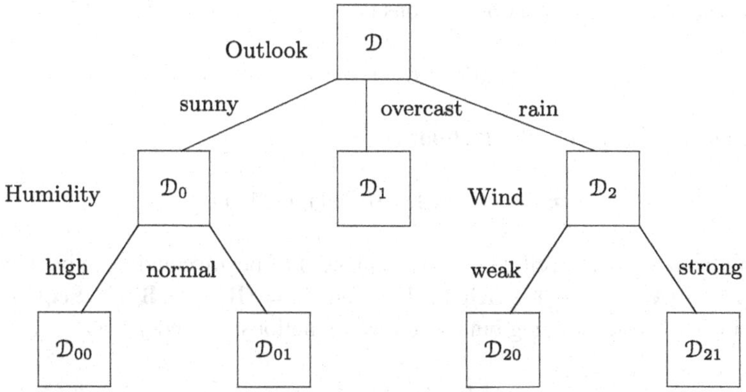

Fig. 12.3 Decision tree

12.8 Logical Data Analysis

Logical data analysis (LDA) is a methodology that aims to discover patterns in data using Boolean methods. LDA techniques were introduced by the Rutcor group (see [14]).

Let $\mathcal{D} = (\theta, C)$ be a pair (also referred to as a *decision system*), where $\theta = (T, H, \mathbf{r})$ is a table having the heading $H = A_1 \cdots A_n C$ and C is the decision attribute. All attributes are binary, that is, we have $\text{Dom}(A_1) = \cdots = \text{Dom}(A_n) = \text{Dom}(C) = \{0, 1\}$. The content of θ represents a sequence of observations that consists of the projections $t[A_1 \cdots A_n]$ of the tuples of $\mathbf{r}$. The component $t[C]$ of t is the class of the observation t. The sequence of *positive observations* is

$$\mathbf{r}^+ = \{t[A_1 \cdots A_n] \text{ in } \mathbf{r} \mid t[C] = 1\};$$

the sequence of *negative observations* is

$$\mathbf{r}^- = \{t[A_1 \cdots A_n] \text{ in } \mathbf{r} \mid t[C] = 0\}.$$

It is clear that the sets *set*($\mathbf{r}^+$) and *set*($\mathbf{r}^-$) are disjoint and thus define a partial Boolean function of n arguments. We refer to τ as an *observation table*.

Example 12.97 Consider the decision system $\mathcal{D} = ((T, A_1A_2A_3C, \mathbf{r}), C)$ given by

T			
A_1	A_2	A_3	C
0	0	1	0
0	1	1	0
1	1	1	0
0	0	0	1
1	0	0	1
1	1	0	1

The sequence of positive observations is

$$\mathbf{r}^+ = ((0, 0, 0), (1, 0, 0), (1, 1, 0)).$$

The sequence of negative observations is

$$\mathbf{r}^- = ((0, 0, 1), (0, 1, 1), (1, 1, 1)).$$

Starting from a pdBf specified by two sequences of positive and negative observations, $\mathbf{r}^+ = (\mathbf{A}_1, \ldots, \mathbf{A}_p) \in \mathbf{Seq}(\{0, 1\}^n)$ and $\mathbf{r}^- = (\mathbf{B}_1, \ldots, \mathbf{B}_q) \in \mathbf{Seq}(\{0, 1\}^n)$, we define two corresponding binary Boolean functions f^+ and f^- as

$$f^+(x_1, \ldots, x_n) = \begin{cases} 1 & \text{if } (x_1, \ldots, x_n) \text{ does not occur in } \mathbf{r}^-, \\ 0 & \text{otherwise,} \end{cases}$$

for $(x_1, \ldots, x_n) \in \{0, 1\}^n$, and

$$f^-(x_1, \ldots, x_n) = \begin{cases} 1 & \text{if } (x_1, \ldots, x_n) \text{ does not occur in } \mathbf{r}^+, \\ 0 & \text{otherwise,} \end{cases}$$

for $(x_1, \ldots, x_n) \in \{0, 1\}^n$. Clearly, we have

$$\begin{aligned} f^+(x_1, \ldots, x_n) &= \bigvee \{(x_1, \ldots, x_n)^{\mathbf{A}} \mid \mathbf{A} \text{ occurs in } \mathbf{r}^-\}, \\ f^-(x_1, \ldots, x_n) &= \bigvee \{(x_1, \ldots, x_n)^{\mathbf{A}} \mid \mathbf{A} \text{ occurs in } \mathbf{r}^+\}. \end{aligned}$$

Definition 12.98 *The* positive (negative) patterns *of a decision system $\mathcal{D} = (\theta, C)$, where $\theta = (T, H, \mathbf{r})$, are the prime implicants of the binary Boolean function f^+ (of the function f^-) that cover at least one minterm $t^{\mathbf{A}}$, where $\mathbf{A}$ is a positive (negative) observation of τ.*

Example 12.99 For the positive and negative observations considered in Example 12.97, the binary Boolean functions f^+ and f^- are given by

$$\begin{aligned} f^+(x_1, x_2, x_3, x) = \bar{x}_1\bar{x}_2\bar{x}_3 \vee \bar{x}_1 x_2 \bar{x}_3 \vee x_1 \bar{x}_2 \bar{x}_3 \\ \vee x_1 \bar{x}_2 x_3 \vee x_1 x_2 \bar{x}_3 \end{aligned}$$

and

$$\begin{aligned} f^-(x_1, x_2, x_3, x) = \bar{x}_1\bar{x}_2 x_3 \vee \bar{x}_1 x_2 \bar{x}_3 \vee \bar{x}_1 x_2 x_3 \\ \vee x_1 \bar{x}_2 x_3 \vee x_1 x_2 x_3 \end{aligned}$$

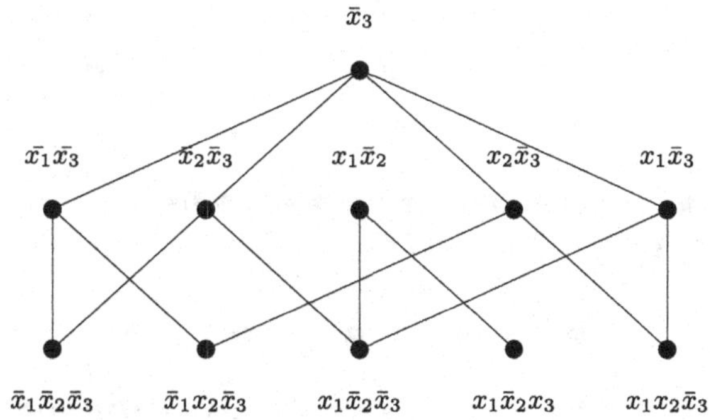

Fig. 12.4 Hasse diagram of $(IMPL_{f^+}, \leqslant)$

for $(x_1, x_2, x_3) \in \{0, 1\}^3$. The Hasse diagram of the poset of implicants $(IMPL_{f^+}, \leqslant)$ is shown in Fig. 12.4. The prime implicants of f^+ are $\bar{x}_3$ and $x_1\bar{x}_2$, and they are both positive patterns. Indeed, $\bar{x}_3$ covers every positive observation, while $x_1\bar{x}_2$ covers the minterm that corresponds to the positive observation $(1, 0, 0)$.

A positive pattern can be regarded as a combination of values taken by a small number of variables that never appeared in a negative observation and did appear in some positive observation. Thus, if a new observation is covered by a positive pattern, this fact can be regarded as an indication that the observation is a positive one.

Next, we discuss algorithms for generating positive patterns (the negative patterns can be found using similar techniques).

Two basic approaches are described for finding positive patterns: a top-down approach and a bottom-up approach (directions are defined relative to the Hasse diagram of the poset $(\mathtt{T}(\mathcal{B}_2, n), \leqslant)$). In the bottom-up approach, we start with the minterms that correspond to positive observations, which are clearly positive patterns of rank 0. If one or more literals are removed from such a pattern, the resulting term may still be a pattern if it does not cover any negative examples. The process consists of a systematic removal of literals from minterms and verifying whether the resulting minterms remain positive patterns until prime patterns are reached.

The top-down approach begins with terms of rank $n - 1$; that is, with patterns that contain one literal. If such a term does not cover any negative observations, then it is a positive pattern; otherwise, literals are added systematically until a positive pattern is obtained.

The number of positive patterns can be huge, which suggests that seeking only patterns whose rank is sufficiently high (and therefore contain few literals) is a good practical compromise.

Example 12.100 The Hasse diagram of the poset $(\mathtt{T}(\mathcal{B}_2, 3), \leqslant)$ is shown in Fig. 12.5. We apply the top-down method to determine the positive patterns of the decision system introduced in Example 12.97.

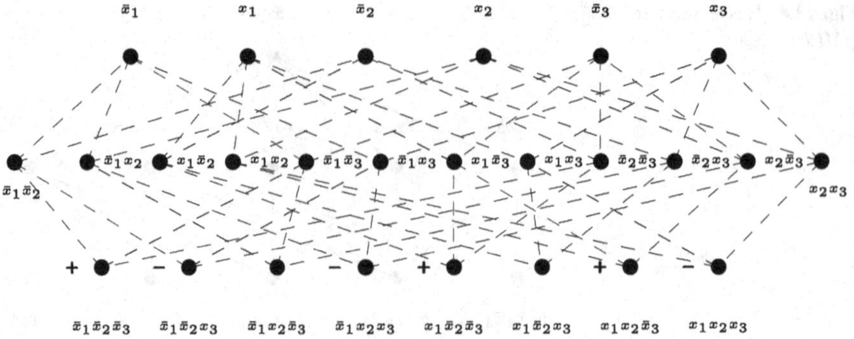

Fig. 12.5 Hasse diagram of the poset $(\mathrm{T}(\mathcal{B}_2, 3), \leqslant)$

We begin with the minterms that correspond to positive observations,

$$t^{(0,0,0)}(x_1, x_2, x_3) = \bar{x}_1\bar{x}_2\bar{x}_3,$$
$$t^{(1,0,0)}(x_1, x_2, x_3) = x_1\bar{x}_2\bar{x}_3,$$
$$t^{(1,1,0)}(x_1, x_2, x_3) = x_1x_2\bar{x}_3.$$

The terms that can be obtained from these terms by discarding one literal are listed below.

Original term	Derived term	Negative patterns covered
$\bar{x}_1\bar{x}_2\bar{x}_3$	$\bar{x}_2\bar{x}_3$	None
$\bar{x}_1\bar{x}_2\bar{x}_3$	$\bar{x}_1\bar{x}_3$	None
$\bar{x}_1\bar{x}_2\bar{x}_3$	$\bar{x}_1\bar{x}_2$	$\bar{x}_1\bar{x}_2x_3$
$x_1\bar{x}_2\bar{x}_3$	$\bar{x}_2\bar{x}_3$	None
$x_1\bar{x}_2\bar{x}_3$	$x_1\bar{x}_3$	None
$x_1\bar{x}_2\bar{x}_3$	$x_1\bar{x}_2$	None
$x_1x_2\bar{x}_3$	$x_2\bar{x}_3$	None
$x_1x_2\bar{x}_3$	$x_1\bar{x}_3$	None
$x_1x_2\bar{x}_3$	x_1x_2	$x_1x_2x_3$

The preceding table yields a list of seven positive patterns of rank 1:

$$\bar{x}_2\bar{x}_3, \bar{x}_1\bar{x}_3, \bar{x}_2\bar{x}_3, x_1\bar{x}_3, x_1\bar{x}_2, x_2\bar{x}_3, x_1\bar{x}_3.$$

By discarding one literal, we have the following patterns of rank 2:

Original term	Derived term	Negative patterns covered
$\bar{x}_2\bar{x}_3$	$\bar{x}_2$	$\bar{x}_1\bar{x}_2x_3$
$\bar{x}_2\bar{x}_3$	$\bar{x}_3$	None
$\bar{x}_1\bar{x}_3$	$\bar{x}_3$	None
$\bar{x}_1\bar{x}_3$	$\bar{x}_1$	$\bar{x}_1\bar{x}_2x_3, \bar{x}_1x_2x_3$
$\bar{x}_2\bar{x}_3$	$\bar{x}_3$	None
$\bar{x}_2\bar{x}_3$	$\bar{x}_2$	$\bar{x}_1\bar{x}_2x_3$
$x_1\bar{x}_3$	$\bar{x}_3$	None
$x_1\bar{x}_3$	x_1	$\bar{x}_1\bar{x}_2x_3, \bar{x}_1x_2x_3$
$x_1\bar{x}_2$	$\bar{x}_2$	$\bar{x}_1\bar{x}_2x_3$
$x_1\bar{x}_2$	x_1	$x_1x_2x_3$
$x_2\bar{x}_3$	$\bar{x}_3$	None
$x_2\bar{x}_3$	x_2	$\bar{x}_1x_2x_3, x_1x_2x_3$
$x_1\bar{x}_3$	$\bar{x}_3$	None
$x_1\bar{x}_3$	x_1	$x_1x_2x_3$

The unique positive pattern of rank 2 is $\bar{x}_3$, which covers all positive patterns of rank 1 with the exception of $x_1\bar{x}_2$. Thus, we retrieve the results obtained in Example 12.99, where we used the Hasse diagram of $(IMPL_{f^+}, \leqslant)$.

A pattern generation algorithm is given in [14]. The algorithm gives preference to high-ranking patterns and attempts to cover every positive observation.

Data binarization is a preparatory process for LAD. Its goal is to allow the application of the Boolean methods developed in the LAD, and there are other computational benefits that follow from the binarization process.

Let $\mathcal{D} = (\theta, C)$ be a decision system where $\theta = (T, H, \mathbf{r})$ is a table having the heading $H = A_1 \cdots A_n C$. We assume that the attributes of H, except C, are nominal or numerical rather than binary. Nominal attributes have discrete domains that do not admit a natural partial order. For example, a *color* attribute having as domain the set {red, white, blue} is a nominal attribute. The domain of C is the set $\{0, 1\}$, and we continue to refer to θ as an *observation table.*

As in Sect. 12.8, the content of θ represents a sequence of observations that consists of the projections $t[A_1 \cdots A_n]$ of the tuples of $\mathbf{r}$, and $t[C]$ of t is the class of the observation t. The sequence of *positive observations* is

$$\mathbf{r}^+ = \{t[A_1 \cdots A_n] \text{ in } \mathbf{r} \mid t[C] = 1\};$$

the sequence of *negative observations* is

$$\mathbf{r}^- = \{t[A_1 \cdots A_n] \text{ in } \mathbf{r} \mid t[C] = 0\}.$$

Data binarization consists of replacing nominal or numerical attributes by binary attributes. The technique that is described here was introduced in [14].

In the case of a nominal attribute B whose domain consists of the values $\{b_1, \ldots, b_p\}$, we introduce p attributes $B_1, \ldots, B_p$. The B-component of a tuple

$t[B]$ will be replaced with p components corresponding to the attributes $B_1, \ldots, B_p$ such that

$$t[B_i] = \begin{cases} 1 & \text{if } t[B] = b_i, \\ 0 & \text{otherwise,} \end{cases}$$

for $1 \leqslant i \leqslant p$.

For a numerical attribute A, we define a set of cut points. Suppose that the set of values that appear under the attribute A is $\{a_1, \ldots, a_k\}$ such that $a_1 < a_2 < \cdots < a_k$. If two consecutive values a_j and a_{j+1} belong to two different classes, a cut point v is defined as $v = \frac{a_j + a_{j+1}}{2}$. The role of cut points is to separate consecutive values of an attribute that belong to different classes.

There is no sense in choosing cut points below $\min_i a_i$ or above $\max_i a_i$ because they could not distinguish between positive or negative observations. Also, there is no reason to choose cut points between consecutive values that correspond to two positive observations or two negative observations. Therefore, we need to consider at most one cut point between any two consecutive values of A that correspond to different classes.

Each cut point v defines a *level binary attribute* A_v. The A_v-component of a tuple t is

$$t[A_v] = \begin{cases} 0 & \text{if } t[A] < v, \\ 1 & \text{if } t[A] \geqslant v. \end{cases}$$

Each pair (v, v') of consecutive cut points defines an *interval binary attribute* $A_{vv'}$, where the $A_{vv'}$-component of t is

$$t[A_{vv'}] = \begin{cases} 0 & \text{if } v \leqslant t[A] < v', \\ 1 & \text{otherwise.} \end{cases}$$

Example 12.101 Consider the decision system $\mathcal{D} = (\theta, PlayTennis)$, where θ is the following table.

Tennis

Outlook	Temperature	Humidity	Wind	PlayTennis
Overcast	90	70	Weak	1
Rain	65	72	Weak	1
Rain	50	60	Weak	1
Overcast	55	55	Strong	1
Rain	89	58	Weak	1
Rain	58	52	Strong	0
Sunny	75	75	Weak	0
Rain	77	77	Strong	0

The attributes *Outlook* and *Wind* are nominal, while the attributes *Temperature* and *Humidity* are numerical.

Since there exist three distinct values (overcast, rain, and sunny) for the attribute *Outlook*, this attribute will be replaced with three binary attributes, O_o, O_r, O_s that correspond to these values. Similarly, the attribute *Wind* will be replaced by two binary attributes, W_w and W_s.

The sequence of values for *Temperature* is shown together with the class of the observations:

50	55	58	65	75	77	89	90
+	+	−	+	−	−	+	+

This requires four cut points placed at the midpoints of intervals determined by consecutive values that belong to distinct classes: 56.5, 61.5, 70, 83.

Similarly, the sequence of values for *Humidity* is

52	55	58	60	70	72	75	77
−	+	+	+	+	+	−	−

In this case, we need two cut points: 53.5 and 73.5. We use a simplified notation for binary attributes shown in the right column of the next table.

Attribute	Simplified notation
O_o	B_1
O_r	B_2
O_s	B_3
$T_{56.5}$	B_4
$T_{61.5}$	B_5
T_{70}	B_6
T_{83}	B_7
$T_{56.5,61.5}$	B_8
$T_{61.5,70}$	B_9
$T_{70,83}$	B_{10}
$H_{53.5}$	B_{11}
$H_{73.5}$	B_{12}
$H_{53.5,73.5}$	B_{13}
W_w	B_{14}
W_s	B_{15}

The binarized table is

	B_1	B_2	B_3	B_4	B_5	B_6	B_7	B_8	B_9	B_{10}	B_{11}	B_{12}	B_{13}	B_{14}	B_{15}	C
t_1	1	0	0	1	1	1	1	0	0	0	1	0	1	1	0	1
t_2	0	1	0	1	1	0	0	0	1	0	1	0	1	1	0	1
t_3	0	1	0	0	0	0	0	0	0	0	1	0	1	1	0	1
t_4	1	0	0	0	0	0	0	0	0	0	1	0	1	0	1	1
t_5	0	1	0	1	1	1	1	0	0	0	1	0	1	1	0	1
t_6	0	1	0	1	0	0	0	1	0	0	0	0	0	0	1	0
t_7	0	0	1	1	1	1	0	0	0	1	1	1	0	1	0	0
t_8	0	1	0	1	1	1	0	0	0	1	1	1	0	0	1	0

The number of binary attributes can be reduced; however, the remaining attributes must allow the differentiation between positive and negative cases.

Definition 12.102 *Let $\mathcal{D} = (\theta, C)$ be a decision system where $\theta = (T, H, \boldsymbol{r})$ is a table having the heading $H = B_1 \cdots B_p C$ that consists of binary attributes.*

A support heading *for $\mathcal{D}$ is a subset $L = B_{i_1} \cdots B_{i_q}$ of $B_1 \cdots B_p$ such that if u occurs in $\boldsymbol{r}^+$ and v occurs in $\boldsymbol{r}^-$, then $u[L] \neq v[L]$. For any two such tuples, define their* difference set $\Delta_{\mathcal{D}}(u, v) = \{i \in H \mid u[B_i] \neq v[B_i]\}$.

A support heading L is irredundant *if no proper subset of L is a support heading of $\mathcal{D}$.*

If L is a support set for a decision system $\mathcal{D} = (\theta, C)$, then L must have a nonempty intersection with each set of the form $\{B_i \mid i \in \Delta_{\mathcal{D}}(u, v)\}$ for each positive example u and each negative example v. Finding an irredundant support heading can be expressed as a discrete optimization problem, as was shown in [14].

Example 12.103 Let $\mathbf{y} = (y_1, \ldots, y_{14})$ be the characteristic sequence of a support heading L; that is,

$$y_i = \begin{cases} 1 & \text{if } B_i \in L, \\ 0 & \text{otherwise.} \end{cases}$$

The decision system introduced in Example 12.101 has five positive examples and three negative examples, so there are 15 difference sets:

$$\begin{aligned}
\Delta_{\mathcal{D}}(t_1, t_6) &= \{1, 2, 5, 6, 7, 8, 11, 13, 14, 15\}, \\
\Delta_{\mathcal{D}}(t_1, t_7) &= \{1, 3, 7, 10, 12, 13\}, \\
\Delta_{\mathcal{D}}(t_1, t_8) &= \{1, 2, 7, 10, 12, 13, 14, 15\}, \\
\Delta_{\mathcal{D}}(t_2, t_6) &= \{5, 8, 9, 11, 13, 14, 15\}, \\
\Delta_{\mathcal{D}}(t_2, t_7) &= \{2, 3, 6, 9, 10, 12, 13\}, \\
\Delta_{\mathcal{D}}(t_2, t_8) &= \{6, 9, 10, 12, 13, 14, 15\}, \\
\Delta_{\mathcal{D}}(t_3, t_6) &= \{4, 8, 11, 13, 14, 15\}, \\
\Delta_{\mathcal{D}}(t_3, t_7) &= \{2, 3, 4, 5, 6, 10, 12, 13\}, \\
\Delta_{\mathcal{D}}(t_3, t_8) &= \{4, 5, 6, 10, 12, 13, 14, 15\}, \\
\Delta_{\mathcal{D}}(t_4, t_6) &= \{1, 2, 4, 8, 11, 13\}, \\
\Delta_{\mathcal{D}}(t_4, t_7) &= \{1, 3, 4, 5, 6, 10, 12, 13, 14, 15\}, \\
\Delta_{\mathcal{D}}(t_4, t_8) &= \{1, 2, 4, 5, 6, 10, 12, 13\}, \\
\Delta_{\mathcal{D}}(t_5, t_6) &= \{5, 6, 7, 8, 11, 13, 14, 15\}, \\
\Delta_{\mathcal{D}}(t_5, t_7) &= \{2, 3, 7, 10, 12, 13\}, \\
\Delta_{\mathcal{D}}(t_5, t_8) &= \{7, 10, 12, 13, 14, 15\}.
\end{aligned}$$

The requirement that a support heading intersect each of these sets leads to 15 inequalities. For example, the requirement that $\Delta_{\mathcal{D}}(t_1, t_6) \cap L \neq \emptyset$ amounts to

$$y_1 + y_2 + y_5 + y_6 + y_7 + y_8 + y_{11} + y_{13} + y_{14} + y_{15} \geqslant 1.$$

Fourteen other similar inequalities can be similarly written.

Note that the set $\{13\}$ intersects all these sets, so it is a minimal support heading.

To find an irredundant support heading for a decision system $\mathcal{D} = (\theta, C)$, where $\theta = (T, H, \mathbf{r})$ is a table having the heading $H = B_1 \cdots B_p C$, we need to minimize the sum $\sum_{i=1}^{p} y_i$ subjected to restrictions of the form:

$$\sum \{y_i \mid i \in \Delta_{\mathcal{D}}(t_i, t_j)\} \geqslant 1$$

for every pair of tuples (t_i, t_j) such that t_i is a positive example and t_j is a negative example.

12.9 Perceptrons

Perceptrons are classifiers that use hyperplanes to separate sets of vectors in $\mathbb{R}^n$. Data analyzed consists of finite sequences of pairs of the form $\begin{pmatrix} \mathbf{x} \\ y \end{pmatrix}$, where $\mathbf{x} \in \mathbb{R}^n$ and $y \in \{-1, 1\}$.

If $\begin{pmatrix} \mathbf{x} \\ 1 \end{pmatrix} \in S$, we say that $\mathbf{x}$ is a *positive example*; otherwise, that is, if $y = -1$, $\mathbf{x}$ is a *negative example*. A sequence

$$S = \left(\begin{pmatrix} \mathbf{x}_1 \\ y_1 \end{pmatrix}, \ldots, \begin{pmatrix} \mathbf{x}_\ell \\ y_\ell \end{pmatrix} \right)$$

is *linearly separable* if there exists a hyperplane $\boldsymbol{w}'\mathbf{x} + b = 0$ such that $\boldsymbol{w}'\mathbf{x}_i + b \geqslant 0$ if $y_i = 1$ and $\boldsymbol{w}'\mathbf{x}_i + b < 0$ if $y_i = -1$.

Example 12.104 Let

$$\mathbf{x}_1 = \begin{pmatrix} 0 \\ 0 \end{pmatrix}, \mathbf{x}_2 = \begin{pmatrix} 0 \\ 1 \end{pmatrix}, \mathbf{x}_3 = \begin{pmatrix} 1 \\ 1 \end{pmatrix}, \mathbf{x}_4 = \begin{pmatrix} 1 \\ 0 \end{pmatrix}.$$

The sequence

$$S_1 = \left(\begin{pmatrix} \mathbf{x}_1 \\ 1 \end{pmatrix}, \begin{pmatrix} \mathbf{x}_2 \\ 1 \end{pmatrix}, \begin{pmatrix} \mathbf{x}_3 \\ -1 \end{pmatrix}, \begin{pmatrix} \mathbf{x}_4 \\ 1 \end{pmatrix} \right)$$

is linearly separable, as shown in Fig. 12.6a. On the other hand the sequence

$$S_2 = \left(\begin{pmatrix} \mathbf{x}_1 \\ 1 \end{pmatrix}, \begin{pmatrix} \mathbf{x}_2 \\ -1 \end{pmatrix}, \begin{pmatrix} \mathbf{x}_3 \\ 1 \end{pmatrix}, \begin{pmatrix} \mathbf{x}_4 \\ -1 \end{pmatrix}, \right)$$

shown in Fig. 12.6b. is not linearly separable.

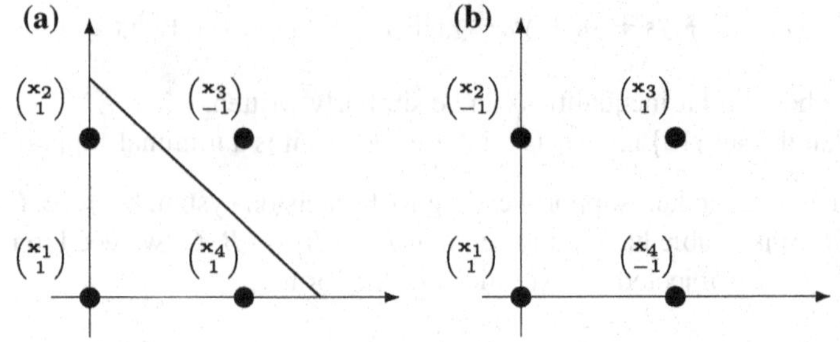

Fig. 12.6 A linearly separable sequence and a sequence that is not linearly separable

Next, we discuss an algorithm that begins with a linearly separable sequences and produces a separating hyperplane. The algorithm is due to Rosenblatt [15] and aims to produce a mathematical device known as a *perceptron* that is described by a vector $\boldsymbol{w} \in \mathbb{R}^n$ called the *weight vector* and a number $b \in \mathbb{R}$ called *bias*. Together, the pair $(\boldsymbol{w}, b)$ determine a hyperplane $\boldsymbol{w}'\mathbf{x} + b = 0$ in $\mathbb{R}^n$.

Let R be the minimum radius of a closed ball centered in $\mathbf{0}$, that is $R = \max\{\| \mathbf{x}_i \| \mid 1 \leqslant i \leqslant \ell\}$.

If $\begin{pmatrix} \mathbf{x}_i \\ y_i \end{pmatrix}$ is a member of the sequence S and H is the target hyperplane $\boldsymbol{w}'\mathbf{x}+b = 0$, where $\| \boldsymbol{w} \| = 1$, define the functional margin of $\begin{pmatrix} \mathbf{x}_i \\ y_i \end{pmatrix}$ as $\gamma_i = y_i(\boldsymbol{w}'\mathbf{x}_i + b)$. Observe that if y_i and $\boldsymbol{w}'\mathbf{x}_i + b$ have the same sign, then $\begin{pmatrix} \mathbf{x}_i \\ y_i \end{pmatrix}$ is classified correctly; otherwise, it is incorrectly classified and we say that a mistake occurred.

A perceptron is constructed starting from the sequence S and from a parameter $\eta \in (0, 1)$ known as a *learning rate*. There are several variants of this algorithm in the literature [15–18]. In Algorithm 12.9.1 we present the variant of [18].

Algorithm 12.9.1: Learning Algorithm for Perceptron

Data: labelled training sequence S and learning rate η
Result: weight vector $\boldsymbol{w}$ and parameter b defining classifier

initialize $\boldsymbol{w} = \mathbf{0}$, $b_0 = 0$, $k = 0$;
$R = \max\{\| \mathbf{x}_i \| \mid 1 \leqslant i \leqslant \ell\}$;
repeat
  for $i = 1$ *to* ℓ **do**
    if $y_i(\boldsymbol{w}_k'\boldsymbol{x}_i + b_k) \leqslant 0$ **then**
      $\boldsymbol{w}_{k+1} = \boldsymbol{w}_k + \eta y_i \mathbf{x}_i$;
      $b_{k+1} = b_k + \eta y_i R^2$;
      $k = k + 1$;
    end
  end
until *no mistakes are made in the for loop*;
return k, $(\boldsymbol{w}_k, b_k)$ where k is the number of mistakes;

Theorem 12.105 *Let* $S = \left(\begin{pmatrix} x_1 \\ y_1 \end{pmatrix}, \ldots, \begin{pmatrix} x_\ell \\ y_\ell \end{pmatrix} \right)$ *be a non-trivial training sequence that is linearly separable, and let* $R = \max\{\parallel x_i \parallel \mid 1 \leqslant i \leqslant \ell\}$*. Suppose there exists an optimal weight vector* w_{opt} *and an optimal bias* b_{opt} *such that*

$$\parallel w_{opt} \parallel = 1 \text{ and } y_i(w'_{opt} x_i + b_{opt}) \geqslant \gamma,$$

for $1 \leqslant i \leqslant \ell$*. The, the number of mistakes made by the algorithm is at most*

$$\left(\frac{2R}{\gamma} \right)^2$$

Proof Let t be the update counter

$$\hat{\mathbf{w}} = \begin{pmatrix} w \\ \frac{b}{R} \end{pmatrix} \text{ and } \hat{\mathbf{x}}_i = \begin{pmatrix} \mathbf{x}_i \\ R \end{pmatrix}$$

for $1 \leqslant i \leqslant \ell$.

The algorithm begins with an augmented vector $\hat{\mathbf{w}}_0 = \mathbf{0}$ and updates it at each mistake.

Let $\hat{\mathbf{w}}_{t-1}$ be the augmented weight vector prior to the tth mistake. The tth update is performed when

$$y_i \hat{\mathbf{w}}'_{t-1} \hat{\mathbf{x}}_i = y_i(w'_{t-1}\mathbf{x}_i + b_{t-1}) \leqslant 0,$$

where $(\mathbf{x}_i, y_i)$ is the example incorrectly classified by

$$\hat{\mathbf{w}}_{t-1} = \begin{pmatrix} w_{t-1} \\ \frac{b_{t-1}}{R} \end{pmatrix}.$$

The update is

$$\begin{aligned} \hat{\mathbf{w}}_t &= \begin{pmatrix} w_t \\ \frac{b_t}{R} \end{pmatrix} = \begin{pmatrix} w_{t-1} + \eta y_i \mathbf{x}_i \\ \frac{b_{t-1} + \eta y_i R^2}{R} \end{pmatrix} \\ &= \begin{pmatrix} w_{t-1} + \eta y_i \mathbf{x}_i \\ \frac{b_{t-1}}{R} + \eta y_i R \end{pmatrix} = \begin{pmatrix} w_{t-1} \\ \frac{b_{t-1}}{R} \end{pmatrix} + \begin{pmatrix} \eta y_i \mathbf{x}_i \\ \eta y_i R \end{pmatrix} \\ &= \hat{\mathbf{w}}_{t-1} + \eta y_i \hat{\mathbf{x}}_i, \end{aligned}$$

where we used the fact that $b_t = b_{t-1} + \eta y_i R^2$.

Since

$$y_i \hat{\mathbf{w}}'_{opt} \hat{\mathbf{x}}_i = y_i \begin{pmatrix} w'_{opt} & \frac{b}{R} \end{pmatrix} \begin{pmatrix} \mathbf{x}_i \\ R \end{pmatrix} = y_i(w'_{opt}\mathbf{x}_i + b) \geqslant \gamma,$$

we have

$$\hat{\mathbf{w}}'_{opt} \hat{\mathbf{w}}_t = \hat{\mathbf{w}}'_{opt} \hat{\mathbf{w}}'_{t-1} + \eta y_i \hat{\mathbf{w}}'_{opt} \hat{\mathbf{x}}_i \geqslant \hat{\mathbf{w}}'_{opt} \hat{\mathbf{w}}_{t-1} + \eta\gamma.$$

By repeated application of the inequality $\hat{\mathbf{w}}'_{opt}\hat{\mathbf{w}}_t \geqslant \eta\gamma$ we obtain

$$\hat{\mathbf{w}}'_{opt}\boldsymbol{w}_t \geqslant t\eta\gamma.$$

Since $\hat{\mathbf{w}}_t = \hat{\mathbf{w}}_{t-1} + \eta y_i \hat{\mathbf{x}}_i$, we have

$$\begin{aligned}
\| \hat{\mathbf{w}}_t \|^2 &= \hat{\mathbf{w}}'_t\hat{\mathbf{w}}_t = (\hat{\mathbf{w}}'_{t-1} + \eta y_i \hat{\mathbf{x}}'_i)(\hat{\mathbf{w}}_{t-1} + \eta y_i \hat{\mathbf{x}}_i) \\
&= \| \hat{\mathbf{w}}_{t-1} \|^2 + 2\eta y_i \hat{\mathbf{w}}'_{t-1}\hat{\mathbf{x}}_i + \eta^2 \| \hat{\mathbf{x}}_i \|^2 \\
&\quad \text{(because } y_i \hat{\mathbf{w}}'_{t-1}\hat{\mathbf{x}}_i \leqslant 0 \text{ when an update occurs)} \\
&\leqslant \| \hat{\mathbf{w}}_{t-1} \|^2 + \eta^2 \| \hat{\mathbf{x}}_i \|^2 \\
&\leqslant \| \hat{\mathbf{w}}_{t-1} \|^2 + \eta^2 (\| \mathbf{x}_i \|^2 + R^2) \\
&\leqslant \| \hat{\mathbf{w}}_{t-1} \|^2 + 2\eta^2 R^2,
\end{aligned}$$

which implies $\| \hat{\mathbf{w}}_t \|^2 \leqslant 2t\eta^2 R^2$. By combining the inequalities

$$\hat{\mathbf{w}}'_{opt}\boldsymbol{w}_t \geqslant t\eta\gamma \text{ and } \| \hat{\mathbf{w}}_t \|^2 \leqslant 2t\eta^2 R^2$$

we have

$$\| \hat{\mathbf{w}}_{opt} \| \sqrt{2t}\eta R \geqslant \| \hat{\mathbf{w}}_{opt} \| \| \hat{\mathbf{w}}_t \| \geqslant \hat{\mathbf{w}}'_{opt}\hat{\mathbf{w}}_t \geqslant t\eta\gamma,$$

which imply

$$t \leqslant 2\left(\frac{R}{\gamma}\right)^2 \| \hat{\mathbf{w}}_{opt} \|^2 \leqslant \left(\frac{2R}{\gamma}\right)^2$$

because $b_{opt} \leqslant R$ for a non-trivial separation of data and hence

$$\| \hat{\mathbf{w}}_{opt} \|^2 \leqslant \| \boldsymbol{w}_{opt} \|^2 + 1 = 2.$$

Exercises and Supplements

1. Let $H = A_1 \cdots A_n$ be a set of n attributes. Prove that $|\mathsf{FD}(H)| = 4^n$ and that there exist $4^n - 3^n$ nontrivial functional dependencies in $\mathsf{FD}(H)$.
2. How many functional dependency schemas can be defined on a set H that contains n attributes?
3. Consider the table

T

A	B	C	D	E
a_1	b_1	c_1	d_1	e_1
a_1	b_1	c_2	d_2	e_2
a_1	b_1	c_2	d_3	e_2

Show several functional dependencies that this table violates.

4. Let $\theta = (T, H, \mathbf{r})$ be a table of a table schema (H, F) such that $\mathbf{r}$ contains no duplicate rows. Prove that the set of attributes K is a key of θ if and only if $cl_F(K) = H$ and for every proper subset L of K we have $cl_F(L) \subset cl_F(K)$. Formulate and prove a similar statement for reducts.
5. Let (S, F) be a functional dependency schema. Prove that if A is an attribute in H that does not occur on the right member of any functional dependency, then A is a member of the core of the schema.
6. Let $S = (ABCDE, \{AB \to D, BD \to AE, C \to B\})$ be a functional dependency schema. Is the functional dependency $ABC \to DE$ a logical consequence of F?
7. Let $S = (A_1, \ldots, A_n, B_1, \ldots, B_n, F)$ be a table schema, where $F = \{A_i \to B_i, B_i \to A_i \mid 1 \leqslant i \leqslant n\}$. Prove that any table of this schema has 2^n reducts.
8. Let S be a finite set. Prove that for every partition $\pi \in PART(S)$ we have $\mathcal{H}_\beta(\alpha_S|\pi) = \mathcal{H}_\beta(\alpha_S) - \mathcal{H}_\beta(\pi)$.
9. Let $\pi = \{B_1, \ldots, B_m\}$ be a partition of the finite set S, where $|S| = n$. Use Jensen's inequality (Theorem 9.35) applied to suitable convex functions to prove the following inequalities:

   (a) for $\beta > 1$, $\frac{1}{m^{\beta-1}} \leqslant \sum_{i=1}^m \left(\frac{|B_i|}{|S|}\right)^\beta$;
   (b) $\log m \geqslant -\sum_{i=1}^m \frac{|B_i|}{|S|} \log \frac{|B_i|}{|S|}$;
   (c) $me^{\frac{1}{m}} \leqslant \sum_{i=1}^m e^{\frac{|B_i|}{|S|}}$.

   Solution: Choose, in all cases $p_1 = \cdots = p_m = \frac{1}{m}$. The needed convex functions are x^β with $\beta > 1$, $x \log x$, and e^x, respectively.
10. Use Supplement 9 to prove that if $\beta > 1$, then $\mathcal{H}_\beta(\pi) \leqslant \frac{1-m^{1-\beta}}{1-2^{1-\beta}}$.
11. Prove that if K is a reduct of a table $\theta = (T, H, \mathbf{r})$, then $\mathsf{H}_\beta(K) = \min\{\mathsf{H}_\beta(L) \mid L \in \mathcal{P}(H)\}$.
12. Let S be a finite set and be a set and let $f : \mathcal{P}(S) \longrightarrow \mathbb{R}$ be a function. Prove that f is a submodular function if and only if for $U, V \subseteq S$ and $x \in S - V$ such that $U \subseteq V$ the function f satisfies the inequality

$$f(U \cup \{x\}) - f(U) \geqslant f(V \cup \{x\}) - f(V) \tag{12.10}$$

   (the diminishing return property of submodular functions).

   Solution: Let f be a submodular function and let $C = U \cup \{x\}$ and $D = V$. Since $C \cup D = V \cup \{x\}$ and $C \cap D = U$, we have

$$\begin{aligned} & f(U \cup \{x\}) - f(U) - f(V \cup \{x\}) + f(V) \\ & = f(C) + f(D) - f(C \cup D) - f(C \cap D) \geqslant 0, \end{aligned}$$

   which proves that f satisfies the condition given above.

Conversely, suppose that Inequality (12.10) is satisfied and let $X = \{x_1, \ldots, x_m\}$ be such that $V \cap X = \emptyset$. By adding up the inequalities

$$f(U \cup \{x_1, \ldots, x_k\}) - f(U \cup \{x_1, \ldots, x_{k-1}\}) \\ \geqslant f(V \cup \{x_1, \ldots, x_k\}) - f(V \cup \{x_1, \ldots, x_{k-1}\})$$

for $1 \leqslant k \leqslant m$, we obtain $f(U \cup X) - f(U) \leqslant f(V \cup X) - f(V)$. Take $U = A \cap B$, $V = B$ and $X = A - B$. Clearly, $U \subseteq V$ and $V \cap X = \emptyset$. By the previous assumption we have $f(A) - f(A \cap B) \leqslant f(A \cup B) - f(B)$, which is the submodular property.

13. Let S be a finite set and be a set and let $f : \mathcal{P}(S) \longrightarrow \mathbb{R}$ be a function. Prove that f is a submodular function if and only if for all $U \in \mathcal{P}(S)$ and $x, y \notin U$, we have
$$f(U \cup \{y\}) - f(U) \geqslant f(U \cup \{x, y\}) - f(U \cup \{x\}).$$
14. Let $\mathbf{a} \in \mathbb{R}^n$ and let $f : \mathcal{P}(\{1, \ldots, n\}) \longrightarrow \mathbb{R}$ be the function defined by $f(S) = \mathbf{a}'\mathbf{c}_S$, where $S \subseteq \{1, \ldots, n\}$ and c_S is the characteristic vector of S. Prove that f is a modular set function.
15. Let $f : \mathcal{P}(\{1, \ldots, n\}) \longrightarrow \mathbb{R}$ be a function such that $f(\emptyset) = 0$. The *Lovász extension of f at* $\mathbf{z}$, where $\mathbf{z}' = (z_1, \ldots, z_n) \in \mathbb{R}^n$ is the function $\hat{f} : \mathbb{R}^n \longrightarrow \mathbb{R}$ given by
$$f(\mathbf{z}) = \sum_{k=1}^{n-1} f(\{j_1, \ldots, j_k\})(z_{j_k} - z_{j_{k+1}}) + f(\{1, \ldots, n\})z_{j_n}$$
where $z_{j_1} \geqslant \cdots \geqslant z_{j_n}$. Prove that

    (a) $\hat{f}(\mathbf{c}_S) = f(S)$, where $S \in \mathcal{P}(\{1, \ldots, n\})$ and $\mathbf{c}_S$ is the characteristic vector of S;
    (b) The function f is submodular if and only if $\hat{f}$ is convex.
16. Let $\mathcal{L} = (L, \{\wedge, \vee\})$ be a lattice and let $f : L \longrightarrow \mathbb{R}$ be an antimonotonic mapping. Prove that the following statements are equivalent:

    (a) f is submodular.
    (b) $f(z) + f(x \wedge y) \leqslant f(x \wedge z) + f(z \wedge y)$ for $x, y, z \in L$.

    Solution: To prove that (i) implies (ii), apply the submodular inequality to $x \wedge z$ and $z \wedge y$. This yields
$$f((x \wedge z) \vee (z \wedge y)) + f(x \wedge y \wedge z) \leqslant f(x \wedge z) + f(z \wedge y).$$
By the subdistributive inequality (11.6) and the anti-monotonicity of f, we have
$$f((x \wedge z) \vee (z \wedge y)) \leqslant f(z \vee (x \wedge y)) \leqslant f(z).$$

Since $f(x \wedge y) \leqslant f(x \wedge y \wedge z)$, the desired inequality follows immediately. The reverse implication follows immediately by replacing z by $x \vee y$ and using the absorption properties of the lattice.

17. Let $\mathcal{L} = (L, \{\wedge, \vee\})$ be a lattice and let $f : L \longrightarrow \mathbb{R}$ be an anti-monotonic mapping. Prove that the following statements are equivalent:

    (a) f is supramodular.
    (b) $f(z) + f(x \vee y) \geqslant f(x \vee z) + f(z \vee y)$ for $x, y, z \in L$.

18. Let S be a set, $\pi \in PART(S)$ and let C and D be two disjoint subsets of S. Prove that if $\beta \geqslant 1$ then

$$\mathcal{H}_\beta(C \cup D, \pi_{C \cup D}) \geqslant \mathcal{H}_\beta(C \cup D, \pi_C + \pi_D) + \mathcal{H}_\beta(C \cup D, \{C, D\}).$$

19. Let $\phi : \mathcal{P}(S) \longrightarrow \mathbb{R}$ be a logarithmic supramodular on $\mathcal{P}(S)$. For a collection $\mathcal{E} \subseteq \mathcal{P}(S)$ define $\phi(\mathcal{E}) = \sum_{E \in \mathcal{E}} \phi(E)$.
    Prove that if $\mathcal{A}, \mathcal{B}$ are two collections of subsets of S, then $\phi(\mathcal{A})\phi(\mathcal{B}) \leqslant \phi(\mathcal{A} \vee \mathcal{B})\phi(\mathcal{A} \wedge \mathcal{B})$.
20. Let $f : L \longrightarrow \mathbb{R}_{\geqslant 0}$ be a real-valued, nonnegative function, where $\mathcal{L} = (L, \{\wedge, \vee\})$ is a lattice. Define the mapping $d : L^2 \longrightarrow \mathbb{R}_{\geqslant 0}$ as $d(x, y) = f(x) + f(y) - 2f(x \vee y)$ for $x, y \in L$. Prove that d is a semimetric on L if and only if f is anti-monotonic and supramodular.

Bibliographical Comments

Extensive presentations of functional dependencies and their role in database design are offered in [19–21].

The identification of functional dependencies satisfied by database tables is a significant topic in data mining [22–24]. Generalized entropy was introduced in [4, 25]. The algebraic axiomatization of partition entropy was done in [26] and various applications of Shannon and generalized entropies in data mining were considered in [27, 28].

Basic references for logical data analysis are [14, 29].

Generalized measures and their differential constraints were studied in [5, 7, 8].

References

1. E.F. Codd, A relational model of data for large shared data banks. Commun. ACM **13**, 377–387 (1970)
2. E.F. Codd, *The Relational Model for Database Management, Version 2* (Addison-Wesley, Reading, 1990)
3. C.J. Date, *An Introduction to Database Systems*, 8th edn. (Addison-Wesley, Boston, 2003)

4. J.H. Havrda, F. Charvat, Quantification methods of classification processes: concepts of structural α-entropy. Kybernetica **3**, 30–35 (1967)
5. B. Sayrafi, A Measure-Theoretic Framework for Constraints and Bounds on Measurements of Data. Ph.D. thesis, Indiana University, 2005
6. E.H. Lieb, M. Ruskai, Proof of the strong subadditivity of quantum-mechanical entropy. J. Math. Phys. **14**, 1938–1941 (1973)
7. B. Sayrafi, D. van Gucht, in *Principles of Database Systems*, ed. by C. Li. Differential Constraints, Baltimore, MD, (ACM, New York, 2005), pp. 348–357
8. B. Sayrafi, D. van Gucht, M. Gyssens, Measures in databases and datamining. Technical Report TR602, Indiana University, 2004
9. F.M. Malvestuto, Statistical treatment of the information content of a database. Inf. Syst. **11**, 211–223 (1986)
10. T.T. Lee, An information-theoretic analysis of relational databases. IEEE Trans. Softw. Eng. **13**, 1049–1061 (1997)
11. M.M. Dalkilic, E.L. Robertson, Information dependencies. Technical Report TR531, Indiana University, 1999
12. T.M. Mitchell, *Machine Learning* (McGraw-Hill, New York, 1997)
13. J.R. Quinlan, *C4.5: Programs for Machine Learning* (Morgan Kaufmann, San Mateo, 1993)
14. E. Boros, P.L. Hammer, T. Ibaraki, A. Kogan, E. Mayoraz, I. Muchnik, An implementation of logical analysis of data. IEEE Trans. Knowl. Data Eng. **12**, 292–306 (2000)
15. F. Rosenblatt, The perceptron: a probabilistic model for information storage and organization in the brain. Psychol. Rev. **65**, 386–407 (1958)
16. A.B.J. Novikoff, On convergence proofs on perceptrons, in *Proceedings of the Symposium on Mathematical Theory of Automata*
17. Y. Freund, R.E. Shapire, Large margin classification using the perceptron algorithm. Mach. Learn. **37**, 277–296 (1999)
18. N. Cristianini, J. Shawe-Taylor, *Support Vector Machines* (Cambridge University, Cambridge, 2000)
19. D. Maier, *The Theory of Relational Databases* (Computer Science Press, Rockville, 1983)
20. J.D. Ullman, *Database and Knowledge-Base Systems* (2 vols.) (Computer Science Press, Rockville, 1988)
21. D.A. Simovici, R.L. Tenney, *Relational Database Systems* (Academic Press, New York, 1995)
22. Y. Huhtala, J. Kärkkäinen, P. Porkka, H. Toivonen, Efficient discovery of functional and approximate dependencies using partitions (extended version). TR C-79, University of Helsinki, Department of Computer Science, Helsinki, Finland, 1997
23. J. Kivinen, H. Mannila, Approximate dependency inference from relations. Theor. Comput. Sci. **149**, 129–149 (1995)
24. D.A. Simovici, D. Cristofor, L. Cristofor, Impurity measures in databases. Acta Informatica **38**, 307–324 (2002)
25. Z. Daróczy, Generalized information functions. Inf. Control **16**, 36–51 (1970)
26. S. Jaroszewicz, D.A. Simovici, On axiomatization of conditional entropy, *Proceedings of the 29th International Symposium for Multiple-Valued Logic*, Freiburg, Germany (IEEE Computer Society, Los Alamitos, 1999), pp. 24–31
27. D. Simovici, S. Jaroszewicz, Generalized conditional entropy and decision trees, in *Proceedings of Extraction et Gestion des connaissances—EGC 2003* (Lavoisier, Paris, 2003), pp. 363–380
28. D.A. Simovici, S. Jaroszewicz, in *Finite Versus Infinite*, ed. by C. Calude, G. Paun. On Information-Theoretical Aspects of Relational Databases (Springer, London, 2000), pp. 301–321
29. E. Boros, P.L. Hammer, T. Ibaraki, A. Kogan, A logical analysis of numerical data. Math. Prog. **79**, 163–190 (1997)

Chapter 13
Frequent Item Sets and Association Rules

13.1 Introduction

Association rules have received lots of attention in data mining due to their many applications in marketing, advertising, inventory control, and many other areas.

A typical supermarket may well have several thousand items on its shelves. Clearly, the number of subsets of the set of items is immense. Even though a purchase by a customer involves a small subset of this set of items, the number of such subsets is very large. For example, even if we assume that no customer has more than five items in his shopping cart, there are $\sum_{i=1}^{5} \binom{10000}{i}$ possible contents of this cart, which corresponds to the subsets having no more than five items of a set that has 10,000 items, and this is indeed a large number!

The supermarket is interested in identifying associations between item sets; for example, it may be interested to know how many of the customers who bought bread and cheese also bought butter. This knowledge is important because if it turns out that many of the customers who bought bread and cheese also bought butter, the supermarket will place butter physically close to bread and cheese in order to stimulate the sales of butter. Of course, such a piece of knowledge is especially interesting when there is a substantial number of customers who buy all three items and a large fraction of those individuals who buy bread and cheese also buy butter.

We will formalize this problem and will explore its algorithmic aspects.

13.2 Frequent Item Sets

Suppose that I is a finite set; we refer to the elements of I as *items*.

Definition 13.1 *A* transaction data set on I *is a function* $T : \{1, \ldots, n\} \longrightarrow \mathcal{P}(I)$. *The set* $T(k)$ *is the* kth transaction of T. *The numbers* $1, \ldots, n$ *are the transaction identifiers (*tids*).*

D. A. Simovici and C. Djeraba, *Mathematical Tools for Data Mining*,

Advanced Information and Knowledge Processing, DOI: 10.1007/978-1-4471-6407-4_13,

An example of a transaction set is the set of items present in the shopping cart of a consumer that completed a purchase in a store.

Example 13.2 The table below describes a transaction data set on the set of over-the-counter medicines in a drugstore.

Trans.	Content
$T(1)$	{Aspirin, Vitamin C}
$T(2)$	{Aspirin, Sudafed}
$T(3)$	{Tylenol}
$T(4)$	{Aspirin, Vitamin C, Sudafed}
$T(5)$	{Tylenol, Cepacol}
$T(6)$	{Aspirin, Cepacol}
$T(7)$	{Aspirin, Vitamin C}

The same data set can be presented as a $0/1$ table:

	Aspirin	Vitamin C	Sudafed	Tylenol	Cepacol
$T(1)$	1	1	0	0	0
$T(2)$	1	0	1	0	0
$T(3)$	0	0	0	1	0
$T(4)$	1	1	1	0	0
$T(5)$	1	0	0	0	1
$T(6)$	1	0	0	0	1
$T(7)$	1	1	0	0	0

The entry in the row $T(k)$ and the column i_j is set to 1 if $i_j \in T(k)$; otherwise, it is set to 0.

Example 13.2 shows that we have the option of two equivalent frameworks for studying frequent item sets: tables or transaction item sets.

Given a transaction data set T on the set I, we would like to determine those subsets of I that occur often enough as values of T.

Definition 13.3 *Let $T : \{1, \ldots, n\} \longrightarrow \mathcal{P}(I)$ be a transaction data set on a set of items I. The* support count *of a subset K of the set of items I in T is the number $suppcount_T(K)$ given by*

$$suppcount_T(K) = |\{k \mid 1 \leqslant k \leqslant n \text{ and } K \subseteq T(k)\}|.$$

The support of an item set K is the number

$$supp_T(K) = \frac{suppcount_T(K)}{n}.$$

Example 13.4 For the transaction data set T considered in Example 13.2, we have

$$suppcount_T(\{Aspirin, VitaminC\}) = 3$$

because $\{Aspirin, VitaminC\}$ is a subset of three of the sets $T(k)$. Therefore, $supp_T(\{Aspirin, VitaminC\}) = \frac{3}{7}$.

Example 13.5 Let $I = \{i_1, i_2, i_3, i_4\}$ be a collection of items. Consider the transaction data set T given by

$$\begin{aligned} T(1) &= \{i_1, i_2\}, \\ T(2) &= \{i_1, i_3\}, \\ T(3) &= \{i_1, i_2, i_4\}, \\ T(4) &= \{i_1, i_3, i_4\}, \\ T(5) &= \{i_1, i_2\}, \\ T(6) &= \{i_3, i_4\}. \end{aligned}$$

Thus, the support count of the item set $\{i_1, i_2\}$ is 3; similarly, the support count of the item set $\{i_1, i_3\}$ is 2. Therefore, $supp_T(\{i_1, i_2\}) = \frac{1}{2}$ and $supp_T(\{i_1, i_3\}) = \frac{1}{3}$.

The following rather straightforward statement is fundamental for the study of frequent item sets.

Theorem 13.6 *Let $T : \{1, \ldots, n\} \longrightarrow \mathcal{P}(I)$ be a transaction data set on a set of items I. If K and K' are two item sets, then $K' \subseteq K$ implies $supp_T(K') \geqslant supp_T(K)$.*

Proof Note that every transaction that contains K also contains K'. The statement follows immediately. □

If we seek those item sets that enjoy a minimum support level relative to a transaction data set T, then it is natural to start the process with the smallest nonempty item sets.

Definition 13.7 *An item set K is* μ-frequent *relative to the transaction data set T if $supp_T(K) \geqslant \mu$.*

We denote by $\mathcal{F}_T^{\mu}$ the collection of all μ-frequent item sets relative to the transaction data set T and by $\mathcal{F}_{T,r}^{\mu}$ the collection of μ-frequent item sets that contain r items for $r \geqslant 1$.

Note that

$$\mathcal{F}_T^{\mu} = \bigcup_{r \geqslant 1} \mathcal{F}_{T,r}^{\mu}.$$

If μ and T are clear from the context, then we may omit either or both adornments from this notation.

Let $I = \{i_1, \ldots, i_n\}$ be an item set that contains n elements.

Denote by $\mathcal{G}_I = (\mathcal{P}(I), E)$ the Rymon tree of $\mathcal{P}(I)$. Recall that the root of the tree is $\emptyset$. A vertex $K = \{i_{p_1}, \ldots, i_{p_k}\}$ with $i_{p_1} < i_{p_2} < \cdots < i_{p_k}$ has $n - i_{p_k}$ children $K \cup \{j\}$, where $i_{p_k} < j \leqslant n$.

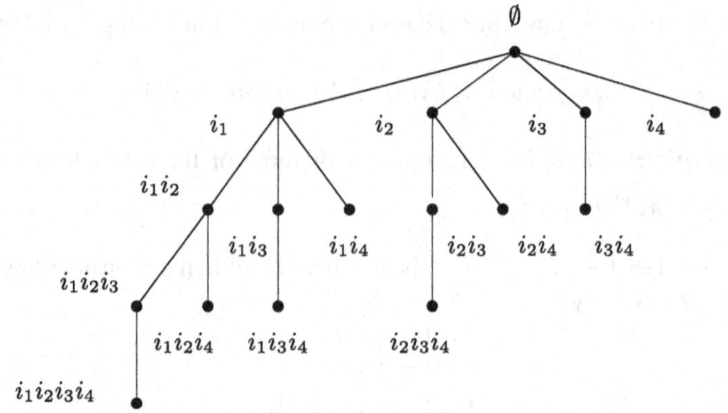

Fig. 13.1 Rymon tree for $\mathcal{P}(\{i_1, i_2, i_3, i_4\})$

Let $\mathcal{S}_r$ be the collection of item sets that have r elements. The next theorem suggests a technique for generating $\mathcal{S}_{r+1}$ starting from $\mathcal{S}_r$.

Theorem 13.8 *Let $\mathcal{G}$ be the Rymon tree of $\mathcal{P}(I)$, where $I = \{i_1, \ldots, i_n\}$. If $W \in \mathcal{S}_{r+1}$, where $r \geqslant 2$, then there exists a unique pair of distinct sets $U, V \in \mathcal{S}_r$ that has a common immediate ancestor $T \in \mathcal{S}_{r-1}$ in $\mathcal{G}$ such that $U \cap V \in \mathcal{S}_{r-1}$ and $W = U \cup V$.*

Proof Let u and v be the two elements of W that have the largest and the second-largest subscripts, respectively. Consider the sets $U = W - \{u\}$ and $V = W - \{v\}$. Both sets belong to $\mathcal{S}_r$. Moreover, $Z = U \cap V$ belongs to $\mathcal{S}_{r-1}$ because it consists of the first $r - 1$ elements of W. Note that both U and V are descendants of Z and that $U \cup V = W$.

The pair (U, V) is unique. Indeed, suppose that W can be obtained in the same manner from another pair of distinct sets $U', V' \in \mathcal{S}_r$ such that U' and V' are immediate descendants of a set $Z' \in \mathcal{S}_{r-1}$. The definition of the Rymon tree $\mathcal{G}_I$ implies that $U' = Z' \cup \{i_m\}$ and $V' = Z' \cup \{i_q\}$, where the letters in Z' are indexed by a number smaller than $\min\{m, q\}$. Then, Z' consists of the first $r - 1$ symbols of W, so $Z' = Z$. If $m < q$, then m is the second-highest index of a symbol in W and q is the highest index of a symbol in W, so $U' = U$ and $V' = V$. □

Example 13.9 Consider the Rymon tree of the collection $\mathcal{P}(\{i_1, i_2, i_3, i_4\})$ shown in Fig. 13.1.

The set $\{i_1, i_3, i_4\}$ is the union of the sets $\{i_1, i_3\}$ and $\{i_1, i_4\}$ that have the common ancestor $\{i_1\}$.

Next we discuss an algorithm that allows us to compute the collection $\mathcal{F}_T^\mu$ of all μ-frequent item sets for a transaction data set T. The algorithm is known as the *Apriori algorithm*.

We begin with the procedure `apriori_gen`, which starts with the collection $\mathcal{F}_{T,k}^\mu$ of frequent item sets for the transaction data set T that contain k elements and generates a collection $\mathcal{C}_{k+1}$ of sets of items that contains $\mathcal{F}^{\mu}_{T,k+1}$, the collection of the frequent item sets that have $k+1$ elements. The justification for this procedure is based on the next statement.

Theorem 13.10 *Let T be a transaction data set on a set of items I and let $k \in \mathbb{N}$ such that $k > 1$.*

If W is a μ-frequent item set and $|W| = k+1$, then there exists a μ-frequent item set Z and two items i_m and i_q such that $|Z| = k-1$, $Z \subseteq W$, $W = Z \cup \{i_m, i_q\}$, and both $Z \cup \{i_m\}$ and $Z \cup \{i_q\}$ are μ-frequent item sets.

Proof If W is an item set such that $|W| = k+1$, then we already know that W is the union of two subsets U and V of I such that $|U| = |V| = k$ and that $Z = U \cap V$ has $k-1$ elements. Since W is a μ-frequent item set and Z, U, V are subsets of W, it follows that each of these sets is also a μ-frequent item set. □

Note that the reciprocal statement of Theorem 13.10 is not true, as the next example shows.

Example 13.11 Let T be the transaction data set introduced in Example 13.5. Note that both $\{i_1, i_2\}$ and $\{i_1, i_3\}$ are $\frac{1}{3}$-frequent item sets; however,

$$supp_T(\{i_1, i_2, i_3\}) = 0,$$

so $\{i_1, i_2, i_3\}$ fails to be a $\frac{1}{3}$-frequent item set.

The procedure apriori_gen mentioned above is Algorithm 13.2.1. This procedure starts with the collection of item sets $\mathcal{F}_{T,k}$ and produces a collection of item sets $\mathcal{C}_{T,k+1}$ that includes the collection of item sets $\mathcal{F}_{T,k+1}$ of frequent item sets having $k+1$ elements.

Note that in apriori_gen no access to the transaction data set is needed.

The *Apriori* Algorithm 13.2.2 operates on "levels". Each level k consists of a collection $\mathcal{C}^{\mu}_{T,k}$ of candidate item sets of μ-frequent item sets. To build the initial collection of candidate item sets $\mathcal{C}^{\mu}_{T,1}$, every single item set is considered for membership in $\mathcal{C}^{\mu}_{T,1}$. The initial set of frequent item sets consists of those singletons that pass the minimal support test. The algorithm alternates

Algorithm 13.2.1: The Procedure apriori_gen

Data: a minimum support μ, the collection $\mathcal{F}^{\mu}_{T,k}$ of frequent item sets having k elements
Result: the set of candidate frequent item sets $\mathcal{C}^{\mu}_{T,k+1}$

```
set j = 1;
C^μ_{T,j+1} = ∅;
for L, M ∈ F^μ_{T,k} such that L ≠ M and L ∩ M ∈ F^μ_{T,k-1} do
    add L ∪ M to C^μ_{T,k+1}
end
remove all sets K in C^μ_{T,k+1} where there is a subset of K containing k elements that does not
belong to F^μ_{T,k};
```

between a candidate generation phase (accomplished by using `apriori_gen`) and an evaluation phase that involves a data set scan and is therefore the most expensive component of the algorithm.

Algorithm 13.2.2: The Apriori Algorithm

Data: transaction data set T and a minimum support μ
Result: the collection $\mathcal{F}^{\mu}_{T}$ of μ-frequent item sets
$\mathcal{C}^{\mu}_{T,1} = \{\{i\} \mid i \in I\}$; set $i = 1$; **while** $\mathcal{C}^{\mu}_{T,i} \neq \emptyset$ **do**
  /* evaluation phase */ $\mathcal{F}^{\mu}_{T,i} = \{L \in \mathcal{C}^{\mu}_{T,i} \mid supp_T(L) \geqslant \mu\}$;
  /* candidate generation */ $\mathcal{C}^{\mu}_{T,i+1} =$ `apriori_gen`$(\mathcal{F}^{\mu}_{T,i})$;
  $i++$;
end
return $\mathcal{F}^{\mu}_{T} = \bigcup_{j<i} \mathcal{F}^{\mu}_{T,j}$;

Example 13.12 Let T be the data set given by

	i_1	i_2	i_3	i_4	i_5
$T(1)$	1	1	0	0	0
$T(2)$	0	1	1	0	0
$T(3)$	1	0	0	0	1
$T(4)$	1	0	0	0	1
$T(5)$	0	1	1	0	1
$T(6)$	1	1	1	1	1
$T(7)$	1	1	1	0	0
$T(8)$	0	1	1	1	1

The support counts of various subsets of $I = \{i_1, \ldots, i_5\}$ are given below:

i_1	i_2	i_3	i_4	i_5
5	6	5	2	5

i_1i_2	i_1i_3	i_1i_4	i_1i_5	i_2i_3	i_2i_4	i_2i_5	i_3i_4	i_3i_5	i_4i_5
3	2	1	3	5	2	3	2	3	2

$i_1i_2i_3$	$i_1i_2i_4$	$i_1i_2i_5$	$i_1i_3i_4$	$i_1i_3i_5$	$i_1i_4i_5$	$i_2i_3i_4$	$i_2i_3i_5$	$i_2i_4i_5$	$i_3i_4i_5$
2	1	1	1	1	1	2	3	2	2

$i_1i_2i_3i_4$	$i_1i_2i_3i_5$	$i_1i_2i_4i_5$	$i_1i_3i_4i_5$	$i_2i_3i_4i_5$
1	1	1	1	2

$i_1i_2i_3i_4i_5$
0

Starting with $\mu = 0.25$ and $\mathcal{F}^{\mu}_{T,0} = \{\emptyset\}$, the Apriori algorithm computes the following sequence of sets:

$$\begin{aligned}
\mathcal{C}^{\mu}_{T,1} &= \{i_1, i_2, i_3, i_4, i_5\},\\
\mathcal{F}^{\mu}_{T,1} &= \{i_1, i_2, i_3, i_4, i_5\},\\
\mathcal{C}^{\mu}_{T,2} &= \{i_1i_2, i_1i_3, i_1i_4, i_1i_5, i_2i_3, i_2i_4, i_2i_5, i_3i_4, i_3i_5, i_4i_5\},\\
\mathcal{F}^{\mu}_{T,2} &= \{i_1i_2, i_1i_3, i_1i_5, i_2i_3, i_2i_4, i_2i_5, i_3i_4, i_3i_5, i_4i_5\},\\
\mathcal{C}^{\mu}_{T,3} &= \{i_1i_2i_3, i_1i_2i_5, i_1i_3i_5, i_2i_3i_4, i_2i_3i_5, i_2i_4i_5, i_3i_4i_5\},\\
\mathcal{F}^{\mu}_{T,3} &= \{i_1i_2i_3, i_2i_3i_4, i_2i_3i_5, i_2i_4i_5, i_3i_4i_5\},\\
\mathcal{C}^{\mu}_{T,4} &= \{i_2i_3i_4i_5\},\\
\mathcal{F}^{\mu}_{T,4} &= \{i_2i_3i_4i_5\},\\
\mathcal{C}^{\mu}_{T,5} &= \emptyset.
\end{aligned}$$

Thus, the algorithm will output the collection

$$\begin{aligned}
\mathcal{F}^{\mu}_{T} &= \bigcup_{i=1}^{4} \mathcal{F}^{\mu}_{T,i}\\
&= \{i_1, i_2, i_3, i_4, i_5, i_1i_2, i_1i_3, i_1i_5, i_2i_3, i_2i_4, i_2i_5, i_3i_4, i_3i_5, i_4i_5,\\
&\quad i_1i_2i_3, i_2i_3i_4, i_2i_3i_5, i_2i_4i_5, i_3i_4i_5, i_2i_3i_4i_5\}.
\end{aligned}$$

13.3 Borders of Collections of Sets

Let $\mathcal{I}$ be a collection of sets such that $\mathcal{I} \subseteq \mathcal{P}(I)$, where I is a set.

Definition 13.13 *The* border of $\mathcal{I}$ *is the collection*

$$BD(\mathcal{I}) = \{L \in \mathcal{P}(I) \mid U \subset L \text{ implies } U \in \mathcal{I} \text{ and } L \subset V \text{ implies } V \notin \mathcal{I}\}.$$

The positive border of $\mathcal{I}$ *is the collection*

$$\begin{aligned}
BD^{+}(\mathcal{I}) &= BD(\mathcal{I}) \cap \mathcal{I}\\
&= \{L \in \mathcal{I} \mid U \subset L \text{ implies } U \in \mathcal{I} \text{ and } L \subset V \text{ implies } V \notin \mathcal{I}\},\}
\end{aligned}$$

while the negative border is

$$\begin{aligned}
BD^{-}(\mathcal{I}) &= BD(\mathcal{I}) - \mathcal{I}\\
&= \{L \in \mathcal{P}(I) - \mathcal{I} \mid U \subset L \text{ implies } U \in \mathcal{I} \text{ and } L \subset V \text{ implies } V \notin \mathcal{I}\}.
\end{aligned}$$

Clearly, we have $BD(\mathcal{I}) = BD^{+}(\mathcal{I}) \cup BD^{-}(\mathcal{I})$.

If $\mathcal{I}$ is a hereditary collection of sets (see Definition 1.14), then the positive and the negative borders of $\mathcal{I}$ are given by

$$BD^+(\mathcal{I}) = \{L \in \mathcal{I} \mid L \subset V \text{ implies } V \notin \mathcal{I}\}$$

and

$$BD^-(\mathcal{I}) = \{L \in \mathcal{P}(I) - \mathcal{I} \mid U \subset L \text{ implies } U \in \mathcal{I}\},$$

respectively. Thus, for a hereditary collection of subsets $\mathcal{I}$, the positive border consists of the maximal subsets of $\mathcal{I}$, while the negative border of $\mathcal{I}$ consists of the minimal subsets of the collection $\mathcal{P}(I) - \mathcal{I}$.

Note that if $\mathcal{I}$ and $\mathcal{I}'$ are two hereditary collections of subsets of I and $BD^+(\mathcal{I}) = BD^+(\mathcal{I}')$, then $\mathcal{I} = \mathcal{I}'$. Indeed, if $K \in \mathcal{I}$, one of the following two cases may occur:

1. If K is not a maximal set of $\mathcal{I}$, then there is a maximal set H of $\mathcal{I}$ such that $K \subset H$. Since $H \in BD^+(\mathcal{I}) = BD^+(\mathcal{I}')$, it follows that $H \in \mathcal{I}'$, hence $K \in \mathcal{I}'$ because $\mathcal{I}'$ is hereditary.
2. If K is a maximal set of $\mathcal{I}$, then $K \in BD^+(\mathcal{I}) = BD^+(\mathcal{I}')$; hence, $K \in \mathcal{I}'$.

In either case $K \in \mathcal{I}'$, so $\mathcal{I} \subseteq \mathcal{I}'$. The reverse inclusion can be proven in a similar way, so $\mathcal{I} = \mathcal{I}'$.

Similarly, we can show that for two hereditary collections $\mathcal{I}, \mathcal{I}'$ of subsets of I, $BD^-(\mathcal{I}) = BD^-(\mathcal{I}')$ implies $\mathcal{I} = \mathcal{I}'$. Indeed, suppose that $K \in \mathcal{I} - \mathcal{I}'$. Since $K \notin \mathcal{I}'$, there exists a minimal subset V of K such that $V \notin \mathcal{I}'$ and each of its subsets is in $\mathcal{I}'$. The set V belongs to the negative border $BD^-(\mathcal{I}')$ and, therefore to $BD^-(\mathcal{I})$. This leads to a contradiction because $K \in \mathcal{I}$ and V is subset of K does not belong to $\mathcal{I}$, thereby contradicting the fact that $\mathcal{I}$ is a hereditary family of sets.

Since no such set K may exist, it follows that $\mathcal{I} \subseteq \mathcal{I}'$. The reverse inclusion can be be shown in the same manner.

Borders of collections of sets play an important role in the study of the Apriori algorithm. Observe, for example, that after computing the collection $\mathcal{F}^{\mu}_{T,3} = \{i_1i_2i_3, i_2i_3i_4, i_2i_3i_5, i_2i_4i_5, i_3i_4i_5\}$ in Example 13.12, the candidate set $\mathcal{C}^{\mu}_{T,4} = \{i_2i_3i_4i_5\}$ is the negative border of $\mathcal{F}^{\mu}_{T,3}$. In general, $\mathcal{C}^{\mu}_{T,i+1}$ is the negative border $BD^-(\mathcal{F}^{\mu}_{T,i})$.

For the same example, the negative and the positive borders of the collection of frequent sets $\mathcal{F}^{\mu}_{T}$ are given by

$$BD^+(\mathcal{F}^{\mu}_{T}) = \{i_1i_5, i_1i_2i_3, i_2i_3i_4i_5\},$$
$$BD^-(\mathcal{F}^{\mu}_{T}) = \{i_1i_4, i_1i_2i_5, i_1i_3i_5\},$$

respectively. Clearly, $BD^+(\mathcal{F}^{\mu}_{T})$ consists of the maximal μ-frequent item sets, while $BD^-(\mathcal{F}^{\mu}_{T})$ consists of the minimal μ-infrequent item sets.

The time complexity of the Apriori algorithm is dominated by the number of accesses to the data set T that is required for computing the support of candidate item sets.

Theorem 13.14 *The Apriori algorithm performs* $|\mathcal{F}_T^\mu| + |BD^-(\mathcal{F}_T^\mu)|$ *support computations.*

Proof The Apriori algorithm selects the μ-frequent item sets from among the candidate item sets, and for each candidate item set it must perform a support computation. A number of $|\mathcal{F}_T^\mu|$ candidate sets turn out to be μ-frequent, so the algorithm will perform $|\mathcal{F}_T^\mu|$ computations for these sets. On the other hand, a candidate set C is not retained as a μ-frequent set if and only if all its subsets are μ-frequent (a requirement of `apriori-gen`) and C itself is not μ-frequent, which means that none of its supersets are μ-frequent. This happens if and only if C belongs to the negative border of $\mathcal{F}_T^\mu$. Thus, the total number of support computations is $|\mathcal{F}_T^\mu| + |BD^-(\mathcal{F}_T^\mu)|$. □

Theorem 13.15 *Let* I *be a set and let* $\mathcal{I}$ *be a hereditary family of subsets of* I. *Consider the collection of sets*

$$\mathcal{E} = \{I - L \mid L \in BD^+(\mathcal{I})\}$$

and the hypergraph $\mathcal{H} = (I, \mathcal{E})$. *Then, the collection of minimal transversals of the hypergraph* $\mathcal{E}$ *equals* $BD^-(\mathcal{I})$, *the negative border of* $\mathcal{I}$.

Proof The following statements concerning a subset X of I are easily seen to be equivalent:

(i) X is a transversal of $\mathcal{H}$.
(ii) $X \cap Y \neq \emptyset$ for every $Y \in \mathcal{E}$.
(iii) $X \cap (I - L) \neq \emptyset$ for every $L \in BD^+(\mathcal{I})$.
(iv) X is not included in any maximal set L of $\mathcal{I}$.
(v) $X \notin \mathcal{I}$.

Thus, X is a transversal of $\mathcal{H}$ if and only if $X \notin \mathcal{I}$. Consequently, X is a minimal transversal of $\mathcal{H}$ if and only if X is a minimal set with the property that $X \notin \mathcal{I}$, which means that $X \in BD^-(\mathcal{I})$. □

13.4 Association Rules

Definition 13.16 *An* association rule *on an item set* I *is a pair of nonempty disjoint item sets* (X, Y).

Note that if $|I| = n$, then there exist $3^n - 2^{n+1} + 1$ association rules on I. Indeed, suppose that the set X contains k elements; there are $\binom{n}{k}$ ways of choosing X. Once X is chosen, Y can be chosen among the remaining $2^{n-k} - 1$ nonempty subsets of $I - X$. In other words, the number of association rules is

$$\sum_{k=1}^{n} \binom{n}{k} (2^{n-k} - 1) = \sum_{k=1}^{n} \binom{n}{k} 2^{n-k} - \sum_{k=1}^{n} \binom{n}{k}.$$

By taking $x = 2$ in the equality

$$(1+x)^n = \sum_{k=0}^{n} \binom{n}{k} x^{n-k},$$

we obtain

$$\sum_{k=1}^{n} \binom{n}{k} 2^{n-k} = 3^n - 2^n.$$

Since $\sum_{k=1}^{n} \binom{n}{k} = 2^n - 1$ the desired equality follows immediately. The number of association rules can be quite considerable even for small values of n. For example, for $n = 10$, we have $3^{10} - 2^{11} + 1 = 57002$ association rules.

An association rule (X, Y) is denoted by $X \Rightarrow Y$. The confidence of $X \Rightarrow Y$ is the number

$$conf_T(X \Rightarrow Y) = \frac{supp_T(XY)}{supp_T(X)}.$$

Definition 13.17 *An association rule holds in a transaction data set T with support μ and confidence c if $supp_T(XY) \geqslant \mu$ and $conf_T(X \Rightarrow Y) \geqslant c$.*

Once a μ-frequent item set Z is identified, we need to examine the support levels of the subsets X of Z to ensure that an association rule of the form $X \Rightarrow Z - X$ has a sufficient level of confidence, $conf_T(X \Rightarrow Z - X) = \frac{\mu}{supp_T(X)}$. Observe that $supp_T(X) \geqslant \mu$ because X is a subset of Z. To obtain a high level of confidence for $X \Rightarrow Z - X$, the support of X must be as small as possible.

Clearly, if $X \Rightarrow Z - X$ does not meet the level of confidence, then it is pointless to look for rules of the form $X' \Rightarrow Z - X'$ among the subsets X' of X.

Example 13.18 Let T be the transaction data set introduced in Example 13.12. We saw that the item set $L = i_2 i_3 i_4 i_5$ has support count equal to 2 and therefore $supp_T(L) = 0.25$. This allows us to obtain the following association rules having three item sets in their antecedent that are subsets of L:

Rule	$suppcount_T(X)$	$conf_T(X \Rightarrow Y)$
$i_2 i_3 i_4 \Rightarrow i_5$	2	1
$i_2 i_3 i_5 \Rightarrow i_4$	3	$\frac{2}{3}$
$i_2 i_4 i_5 \Rightarrow i_3$	2	1
$i_3 i_4 i_5 \Rightarrow i_2$	2	1

Note that $i_2 i_3 i_4 \Rightarrow i_5$, $i_2 i_4 i_5 \Rightarrow i_3$, and $i_3 i_4 i_5 \Rightarrow i_2$ have 100 % confidence. We refer to such rules as *exact association rules*.

The rule $i_2 i_3 i_5 \Rightarrow i_4$ has confidence $\frac{2}{3}$. It is clear that the confidence of rules of the form $U \Rightarrow V$ with $U \subseteq i_2 i_3 i_5$ and $UV = L$ will be lower than $\frac{2}{3}$ since $supp_T(U)$ is at least 3. Indeed, the possible rules of this form are:

Rule	$suppcount_T(X)$	$conf_T(X \Rightarrow Y)$
$i_2i_3 \Rightarrow i_4i_5$	5	$\frac{2}{5}$
$i_2i_5 \Rightarrow i_3i_4$	3	$\frac{2}{3}$
$i_3i_5 \Rightarrow i_2i_4$	3	$\frac{2}{3}$
$i_2 \Rightarrow i_3i_4i_5$	6	$\frac{2}{6}$
$i_3 \Rightarrow i_2i_4i_5$	5	$\frac{2}{5}$
$i_5 \Rightarrow i_2i_3i_4$	5	$\frac{2}{5}$

Obviously, if we seek association rules having a confidence larger than $\frac{2}{3}$, no such rule $U \Rightarrow V$ can be found such that U is a subset of $i_2i_3i_5$.

Suppose, for example, that we seek association rules $U \Rightarrow V$ that have a minimal confidence of 80 %. We need to examine subsets U of the other sets, $i_2i_3i_4$, $i_2i_4i_5$, or $i_3i_4i_5$, which are not subsets of $i_2i_3i_5$ (since the subsets of $i_2i_3i_5$ cannot yield levels of confidence higher than $\frac{2}{3}$). There are five such sets:

Rule	$suppcount_T(X)$	$conf_T(X \Rightarrow Y)$
$i_2i_4 \Rightarrow i_3i_5$	2	1
$i_3i_4 \Rightarrow i_2i_5$	2	1
$i_4i_5 \Rightarrow i_2i_3$	2	1
$i_3i_4 \Rightarrow i_2i_5$	2	1
$i_4 \Rightarrow i_2i_3i_5$	2	1

Indeed, all these sets yield exact rules, that is, rules having 100 % confidence.

Many transaction data sets produce a huge number of frequent item sets and therefore a huge number of association rules, particularly when the levels of support and confidence required are relatively low. Moreover, it is well-known [1] that limiting the analysis of association rules to the support/confidence framework can lead to dubious conclusions. The data mining literature contains many references that attempt to derive interestingness measures for association rules in order to focus data analysis of those rules that may be more relevant [2–7].

13.5 Levelwise Algorithms and Posets

This section focuses on the levelwise algorithms, a powerful and elegant generalization of the Apriori algorithm that was introduced in [8].

Let $(P, \leqslant)$ be a partially ordered set and let Q be a subset of P.

Definition 13.19 *The border of Q is the set*

$$BD(Q) = \{p \in P \mid u < p \textit{ implies } u \in Q \textit{ and } p < v \textit{ implies } v \notin Q\}.$$

The positive border of Q is the set:

$$BD^+(Q) = BD(Q) \cap Q,$$

while the negative border of Q is

$$BD^-(Q) = BD(Q) - Q.$$

Clearly, we have $BD(Q) = BD^+(Q) \cup BD^-(Q)$.

An alternative terminology exists that makes use of the terms *generalization* and *specialization*. If $r, p \in P$ and $r < p$, then we say that r is a *generalization* of p or that p is a *specialization* of r. Thus, the border of a set Q consists of those elements p of P such that all of their generalizations are in Q and none of their specializations is in Q.

Theorem 13.20 *Let $(P, \leqslant)$ be a partially ordered set. If Q and Q' are two disjoint subsets of P, then $BD(Q \cup Q') \subseteq BD(Q) \cup BD(Q')$.*

Proof Let $p \in BD(Q \cup Q')$. Suppose that $u < p$, so $u \in Q \cup Q'$. Since Q and Q' are disjoint, we have either $u \in Q$ or $u \in Q'$. On the other hand, if $p < v$, then $v \notin Q \cup Q'$, so $v \notin Q$ and $v \notin Q'$. Thus, we have $p \in BD(Q) \cup BD(Q')$. □

The notion of a hereditary subset of a poset is an immediate generalization of the notion of a hereditary family of sets.

Definition 13.21 *A subset Q of a poset $(P, \leqslant)$ is said to be hereditary if $p \in Q$ and $r \leqslant p$ imply $r \in Q$.*

Theorem 13.22 *If Q is a hereditary subset of a poset $(P, \leqslant)$, then the positive and the negative borders of Q are given by*

$$BD^+(Q) = \{p \in Q \mid p < v \text{ implies } v \notin Q\}$$

and

$$BD^-(Q) = \{p \in P - Q \mid u < p \text{ implies } u \in Q\},$$

respectively.

Proof Let t be an element of the positive border $BD^+(Q) = BD(Q) \cap Q$. We have $t \in Q$ and $t < v$ implies $v \notin Q$ because $t \in BD(Q)$.

Conversely, suppose that t is an element of Q such that $t < v$ implies $v \notin Q$. Since Q is hereditary, $u < t$ implies $u \in Q$, so $t \in BD(Q)$. Therefore, $t \in BD(Q) \cap Q = BD^+(Q)$.

Now let s be an element of the negative border of Q; that is, $s \in BD(Q) - Q$. We have immediately $s \in P - Q$. If $u < s$, then $u \in Q$, because Q is hereditary. Thus, $BD^-(Q) \subseteq \{p \in P - Q \mid u < p \text{ implies } u \in Q\}$.

Conversely, suppose that $s \in P - Q$ and $u < s$ implies $u \in Q$. If $s < v$, then v cannot belong to Q because this would entail $s \in Q$ due to the hereditary property of Q. Consequently, $s \in BD(Q)$ and so $s \in BD(Q) - Q = BD^{-}(Q)$. □

Theorem 13.22 can be paraphrased by saying that for a hereditary subset Q of P the positive border consists of the maximal elements of Q, while the negative border of Q consists of the minimal elements of $P - Q$.

Note that if Q and Q' are two hereditary subsets of P and $BD^{+}(Q) = BD^{+}(Q')$, then $Q = Q'$. Indeed, if $z \in P$, one of the following two cases may occur:

1. If z is not a maximal element of Q, then there is a maximal element w of Q such that $z < w$. Since $w \in BD^{+}(Q) = BD^{+}(Q')$, it follows that $w \in Q'$; hence $z \in Q'$ because Q' is hereditary.
2. If z is a maximal element of Q, then $z \in BD^{+}(Q) = BD^{+}(Q')$, hence $z \in Q'$.

In either case $z \in Q'$, so $Q \subseteq Q'$. The reverse inclusion can be proven in a similar way, so $Q = Q'$.

Similarly, we can show that for two hereditary collections Q and Q' of subsets of I, $BD^{-}(Q) = BD^{-}(Q')$ implies $Q = Q'$. Indeed, suppose that $z \in Q - Q'$. Since $z \notin Q'$, there exists a minimal element v such that $v \notin Q'$ and each of its lower bounds is in Q'. Since v belongs to the negative border $BD^{-}(Q')$, it follows that $v \in BD^{-}(Q)$. This leads to a contradiction because $z \in Q$ and v (for which we have $v < z$) does not, thereby contradicting the fact that Q is a hereditary subset. Since no such z may exist, it follows that $Q \subseteq Q'$. The reverse inclusion can be shown in the same manner.

Definition 13.23 *Let $\mathcal{D}$ be a relational database, $\mathcal{S}_{\mathcal{D}}$ be the set of states of $\mathcal{D}$, and $(B, \leq, h)$ be a ranked poset referred to as the ranked poset of objects.*

A query is a function $q : \mathcal{S}_{\mathcal{D}} \times B \longrightarrow \{0, 1\}$ such that $D \in \mathcal{S}_{\mathcal{D}}$, $b \leqslant b'$, and $q(D, b') = 1$ imply $q(D, b) = 1$.

Definition 13.23 is meant to capture the framework of the Apriori algorithm for identification of frequent item sets. As was shown in [8], this framework can capture many other situations.

Example 13.24 Let $\mathcal{D}$ be a database that contains a tabular variable (T, H) and let $\theta = (T, H, \rho)$ be the table that is the current value of (T, H) contained by the current state D of $\mathcal{D}$.

The graded poset $(B, \leq, h)$ is $(\mathcal{P}(H), \subseteq, h)$, where $h(X) = |X|$. Given a number μ, the query is defined by

$$q(D, K) = \begin{cases} 1 & \text{if } supp_T(K) \leqslant \mu, \\ 0 & \text{otherwise.} \end{cases}$$

Since $K \subseteq K'$ implies $supp_T(K') \leqslant supp_T(K)$, it follows that q satisfies the condition of Definition 13.23.

Example 13.25 As in Example 13.24, let $\mathcal{D}$ be a database that contains a tabular variable (T, H), and let $\theta = (T, H, \rho)$ be the table that is the current value of (T, H) contained by the current state D of $\mathcal{D}$. The graded poset $(\mathcal{P}(H), \supseteq, g)$ is the dual of the graded poset considered in Example 13.24, where $g(K) = |H| - |K|$. If L is a set of attributes, the function q_L is defined by

$$q_L(D, K) = \begin{cases} 1 & \text{if } K \rightarrow L \text{ holds in } \theta, \\ 0 & \text{otherwise.} \end{cases}$$

Note that if $K' \subseteq K$ and D satisfies the functional dependency $K' \rightarrow L$, then D satisfies $K \rightarrow L$. Thus, q is a query in the sense of Definition 13.23.

Definition 13.26 *The set of interesting objects for the state D of the database and the query q is given by*

$$INT(D, q) = \{b \in B \mid q(D, b) = 1\}.$$

Note that the set of interesting objects is a hereditary set $(B, \leqslant)$. Indeed, if $b \in INT(D, q)$ and $c \leq b$, then $c \in INT(D, q)$ according to Definition 13.23. Thus,

$$BD^+(INT(D, q)) = \{b \in INT(D, q) \mid b < v \text{ implies } v \notin INT(D, q)\},$$
$$BD^-(INT(D, q)) = \{b \in B - INT(D, q) \mid u < b \text{ implies } u \in INT(D, q)\}.$$

In other words, $BD^+(INT(D, q))$ is the set of maximal objects that are interesting, while $BD^-(INT(D, q))$ is the set of minimal objects that are not interesting.

Algorithm 13.5.1, which we discuss next, is a general algorithm that seeks to compute the set of interesting objects for a state of a database. The algorithm is known as the *levelwise algorithm* because it identifies these objects by successively scanning the levels of the graded poset of objects.

If $L_0, L_1, \ldots$ are the levels of the graded poset $(B, \leq, h)$, then the algorithm begins by examining all objects located on the initial level. The set of interesting objects located on the level L_i is denoted by $\mathcal{F}_i$; for each level L_i, the computation of $\mathcal{F}_i$ is preceded by a computation of the set of potentially interesting objects $\mathcal{C}_i$ referred to as the set of *candidate objects*.

The first set of candidate objects, $\mathcal{C}_1$, coincides with the level L_i. Only the interesting objects on this level are retained for the set $\mathcal{F}_1$.

The next set of candidate objects, $\mathcal{C}_{i+1}$, is constructed by examining the level L_{i+1} and keeping those objects b having all their subobjects c in the interesting sets of the previous levels.

Algorithm 13.5.1: General Levelwise Algorithm

Data: a database state D, a graded poset of objects $(B, \leq, h)$, and a query q
Result: the set of interesting objects for D

$\mathcal{C}_1 = L_1$;
$i = 1$;
while $\mathcal{C}_i \neq \emptyset$ **do**
  /* evaluation phase */ $\mathcal{F}_i = \{b \in \mathcal{C}_i \mid q(D, b) = 1\}$;
  /* candidate generation */
  $\mathcal{C}_{i+1} = \{b \in L_{i+1} \mid c < b \text{ implies } c \in \bigcup_{j \leqslant i} \mathcal{F}_j\} - \bigcup_{j \leqslant i} \mathcal{C}_j$;
  $i{+}{+}$;
end
output $\bigcup_{j<i} \mathcal{F}_j$;

Example 13.27 For frequent item sets, we can work in the framework described in Example 13.24. The algorithm, which is essentially the Apriori algorithm described in Sect. 13.2, goes through the **while** loop no more than $k + 1$ times, where

$$k = \max\{|X| \mid X \subseteq H, supp_T(X) > \mu\}.$$

Example 13.28 In Example 13.25, we defined the grading query q_L as

$$q_L(D, K) = \begin{cases} 1 & \text{if } K \to L \text{ holds in } \theta, \\ 0 & \text{otherwise,} \end{cases}$$

for $K \in \mathcal{P}(H)$. The levelwise algorithm allows us to identify those subsets K such that a table $\theta = (T, H, \rho)$ satisfies the functional dependency $K \to L$. The first level consists of all subsets K of H that have $|H| - 1$ attributes. There are, of course, $|H| - 1$ such subsets, and the set $\mathcal{F}_1$ will contain all these sets such that $K \to H$ is satisfied. Successive levels contain sets that have fewer and fewer attributes. Level L_i contains sets that have $|H| - i$ attributes.

The algorithm will go through the **while** loop at most $1 + |H - K|$ times, where K is the smallest set such that $K \to L$ holds.

Observe that the computation of $\mathcal{C}_{i+1}$ in the generic levelwise algorithm

$$\mathcal{C}_{i+1} = \left\{ b \in L_{i+1} \mid c < b \text{ implies } c \in \bigcup_{j \leqslant i} \mathcal{F}_j \right\} - \bigcup_{j \leqslant i} \mathcal{C}_j$$

can be written as

$$\mathcal{C}_{i+1} = BD^{-}\left(\bigcup_{j \leqslant i} \mathcal{F}_j\right) - \bigcup_{j \leqslant i} \mathcal{C}_j.$$

This shows that the set of candidate objects at level L_{i+1} is the negative border of the interesting sets located on the lower level, excluding those objects that have already been evaluated.

The most expensive component of the levelwise algorithm is the evaluation of $q(D, b)$ since this requires a scan of the database state D. Clearly, we need to evaluate this function for each candidate element, so we will require $|\bigcup_{i=1}^{\ell} \mathcal{C}_i|$ evaluations, where ℓ is the number of levels that are scanned. Some of these evaluations will result in including the evaluated object b in the set $\mathcal{F}_i$. Objects that will not be included in $INT(D, q)$ are such that any of their generalizations are in $INT(D, q)$, even though they fail to belong to this set. They belong to $BD^{-}(INT(D, q))$. Thus, the levelwise algorithm performs $|INT(D, q)| + |BD^{-}(INT(D, q))|$ evaluations of $q(D, b)$.

13.6 Lattices and Frequent Item Sets

Galois connections discussed in Sect. 11.4 are useful in the study of frequent item sets. This approach was introduced for the first time in [9].

Let I be a set of items and $T : \{1, \ldots, n\} \longrightarrow \mathcal{P}(I)$ be a transaction data set. Denote by D the set of transaction identifiers $D = \{1, \ldots, n\}$. The functions $items_T : \mathcal{P}(D) \longrightarrow \mathcal{P}(I)$ and $tids_T : \mathcal{P}(I) \longrightarrow \mathcal{P}(D)$ are defined by

$$items_T(E) = \bigcap \{T(k) \mid k \in E\},$$
$$tids_T(H) = \{k \in D \mid H \subseteq T(k)\},$$

for every $E \in \mathcal{P}(D)$ and every $H \in \mathcal{P}(I)$.

Note that $suppcount_T(H) = |tids_T(H)|$ for every $H \in \mathcal{P}(I)$.

Theorem 13.29 *Let $T : \{1, \ldots, n\} \longrightarrow \mathcal{P}(I)$ be a transaction data set. The pair $(items_T, tids_T)$ is a Galois connection between the posets $(\mathcal{P}(D), \subseteq)$ and $(\mathcal{P}(I), \subseteq)$.*

Proof We need to prove that

1. if $E \subseteq E'$, then $items_T(E') \subseteq items_T(E)$,
2. if $H \subseteq H'$, then $tids_T(H') \subseteq tids_T(H)$,
3. $E \subseteq tids_T(items_T(E))$, and
4. $H \subseteq items_T(tids_T(H))$

for every $E, E' \in \mathcal{P}(D)$ and every $H, H' \in \mathcal{P}(I)$.

The first two properties follow immediately from the definitions of the functions $items_T$ and $tids_T$.

To prove Part (iii), let $k \in E$ be a transaction identifier. Then, the item set $T(e)$ includes $items_T(E)$ by the definition of $items_T(E)$. By Part (ii), $tids_T(T(e)) \subseteq tids_T(items_T(E))$. Since $e \in tids_T(T(e))$, it follows that $e \in tids_T(items_T(E))$, so $E \subseteq tids_T(items_T(E))$.

The argument for Part (iv) is similar. □

The theorem can be obtained directly by noting that $(items_T, tids_T)$ is the polarity determined by the relation

$$\rho = \{(k, i) \in D \times I \mid i \in T(k)\}.$$

Corollary 13.30 *Let $T : D \longrightarrow \mathcal{P}(I)$ be a transaction data set and let $\mathbf{K}_i : \mathcal{P}(I) \longrightarrow \mathcal{P}(I)$ and $\mathbf{K}_d : \mathcal{P}(D) \longrightarrow \mathcal{P}(D)$ be defined by $\mathbf{K}_i(H) = items_T(tids_T(H))$ for $H \in \mathcal{P}(I)$ and $\mathbf{K}_d(E) = tids_T(items_T(E))$ for $E \in \mathcal{P}(D)$. Then, $\mathbf{K}_i$ and $\mathbf{K}_d$ are closure operators on I and D, respectively.*

Proof The argument was made in Example 11.54. □

Theorem 13.31 *Let $T : D \longrightarrow \mathcal{P}(I)$ be a transaction data set. We have*

$$\mathbf{K}_i(H_1 \cup H_2) = \mathbf{K}_i(H_1) \cap \mathbf{K}_i(H_2),$$
$$\mathbf{K}_d(E_1 \cup E_2) = \mathbf{K}_d(E_1) \cap \mathbf{K}_d(E_2),$$

for $H_1, H_2 \subseteq I$ and $E_1, E_2 \subseteq D$.

Proof This statement is a direct consequence of the definitions of $\mathbf{K}_i$ and $\mathbf{K}_d$. □

Closed sets of items (that is, sets of items H such that $H = \mathbf{K}_i(H)$) can be characterized as follows.

Theorem 13.32 *Let $T : \{1, \ldots, n\} \longrightarrow \mathcal{P}(I)$ be a transaction data set.*

A set of items H is closed if and only if, for every set $L \in \mathcal{P}(I)$ such that $H \subset L$, we have $supp_T(L) < supp_T(H)$.

Proof Suppose that for every superset L of H we have $supp_T(H) > supp_T(L)$ and that H is not a closed set of items. Therefore, the set $\mathbf{K}_i(H) = items_T(tids_T(H))$ is a superset of H and consequently $suppcount_T(H) > suppcount_T(items_T(tids_T(H)))$. Since

$$suppcount_T(items_T(tids_T(H))) = |tids_T(items_T(tids_T(H)))| = |tids_T(H)|,$$

this leads to a contradiction. Thus, H must be closed.

Conversely, suppose that H is a closed set of items,

$$H = \mathbf{K}_i(H) = items_T(tids_T(H)),$$

and let L be a strict superset of H. Suppose that $supp_T(L) = supp_T(H)$. This means that $|tids_T(L)| = |tids_T(H)|$.

Since $H = items_T(tids_T(H)) \subset L$, it follows that

$$tids_T(L) \subseteq tids_T(items_T(tids_T(H))) = tids_T(H),$$

which implies the equality $tids_T(L) = tids_T(items_T(tids_T(H)))$ because the sets $tids_T(L)$ and $tids_T(H)$ have the same number of elements. Thus, we have $tids_T(L) = tids_T(H)$. In turn, this yields

$$H = items_T(tids_T(H)) = items_T(tids_T(L)) \supseteq L,$$

which contradicts the initial assumption $H \subset L$. □

The importance of determining the closed item sets is based on the equality $suppcount_T(items_T(tids_T(H))) = |tids_T(items_T(tids_T(H)))| = |tids_T(H)|$. Thus, if we have the support counts of the closed sets, we have the support count of every set of items and the number of closed sets can be much smaller than the total number of item sets. An interesting algorithm focused on closed item sets was developed in [10].

Exercises and Supplements

1. Let $I = \{a, b, c, d\}$ be a set of items and let T be a transaction data set defined by

$$T(1) = abc,$$
$$T(2) = abd,$$
$$T(3) = acd,$$
$$T(4) = bcd,$$
$$T(5) = ab.$$

   (a) Find item sets whose support is at least 0.25.
   (b) Find association rules having support at least 0.25 and a confidence at least 0.75.

2. Let $I = i_1 i_2 i_3 i_4 i_5$ be a set of items. Find the 0.6-frequent item sets of the transaction data set T on I defined by

$$\begin{array}{ll} T(1) = i_1, & T(6) = i_1 i_2 i_4, \\ T(2) = i_1 i_2, & T(7) = i_1 i_2 i_5, \\ T(3) = i_1 i_2 i_3, & T(8) = i_2 i_3 i_4, \\ T(4) = i_2 i_3, & T(9) = i_2 i_3 i_5, \\ T(5) = i_2 i_3 i_4, & T(10) = i_3 i_4 i_5. \end{array}$$

   Also, determine all rules whose confidence is at least 0.75.

3. Let T be a transaction data set T on an item set I, $T : \{1, \ldots, n\} \longrightarrow \mathcal{P}(I)$. Define the bit sequence of an item set X as sequence $\mathbf{b}^X = (b_1, \ldots, b_n) \in Seq_n(\{0, 1\})$, where

$$b_i = \begin{cases} 1 & \text{if } X \subseteq T(i), \\ 0 & \text{otherwise,} \end{cases}$$

for $1 \leqslant i \leqslant n$.
For $\mathbf{b} \in \mathbf{Seq}_n(\{0,1\})$, the number $\sqrt{|\{i | 1 \leqslant i \leqslant n, b_i = 1\}|}$ is denoted by $\| \mathbf{b} \|$. The distance between the sequences $\mathbf{b}$ and $\mathbf{c}$ is defined as $\| \mathbf{b} \oplus \mathbf{c} \|$. Prove that:

(a) $\mathbf{b}^{X \cup Y} = \mathbf{b}^X \wedge \mathbf{b}^Y$ for every $X, Y \in \mathcal{P}(I)$.
(b) $\mathbf{b}^{K \oplus L} = \mathbf{b}^L \oplus \mathbf{b}^K$, where $K \oplus L$ is the symmetric difference of the item sets K and L.
(c) $|\sqrt{supp_T(K)} - \sqrt{supp_T(L)}| \leq \frac{d(\mathbf{b}^K, \mathbf{b}^L)}{\sqrt{|T|}}$.

4. For a transaction data set T on an item set $I = \{i_1, \ldots, i_n\}$, $T : \{1, \ldots, n\} \longrightarrow \mathcal{P}(I)$ and a number h, $1 \leqslant h \leqslant n$, define the number $\nu_T(h)$ by

$$\nu_T(h) = 2^{n-1} b_n + \cdots + 2b_2 + b_1,$$

where

$$b_k = \begin{cases} 1 & \text{if } i_k \in T(h), \\ 0 & \text{otherwise,} \end{cases}$$

for $1 \leqslant k \leqslant n$. Prove that $i_k \in T(h)$ if and only if the result of the integer division $\nu_T(h)/k$ is an odd number.

Suppose that the tabular variables of a database $\mathcal{D}$ are $(T_1, H_1), \ldots, (T_p, H_p)$. An *inclusion dependency* is an expression of the form $T_i[K] \subseteq T_j[L]$, where $K \subseteq H_i$ and $L \subseteq H_j$ for some i, j, where $1 \leqslant i, j \leqslant p$ are two sets of attributes having the same cardinality. Denote by $\mathrm{ID}_{\mathcal{D}}$ the set of inclusion dependencies of $\mathcal{D}$.
Let $D \in \mathcal{S}_{\mathcal{D}}$ be a state of the database $\mathcal{D}$, $\phi = T_i[K] \subseteq T_j[L]$ be an inclusion dependency, and $\theta_i = (T_i, H_i, \rho_i)$, $\theta_j = (T_j, H_j, \rho_j)$ be the tables that correspond to the tabular variables (T_i, H_i) and (T_j, H_j) in D. The inclusion dependency ϕ is satisfied in the state D of $\mathcal{D}$ if for every tuple $t \in \rho_i$ there is a tuple $s \in \rho_j$ such that $t[K] = s[L]$.

5. For $\phi = T_i[K] \subseteq T_j[L]$ and $\psi = T_d[K'] \subseteq T_e[L']$, define the relation $\phi \leqslant \psi$ if $d = i$, $e = j$, $K \subseteq K'$, and $H \subseteq H'$. Prove that "$\leq$" is a partial order on $\mathrm{ID}_{\mathcal{D}}$.
6. Prove that the triple $(\mathrm{ID}_{\mathcal{D}}, \leq, h)$ is a graded poset, where $h(T_i[K] \subseteq T_j[L]) = |K|$.
7. Prove that the function $q : \mathcal{S}_{\mathcal{D}} \times \mathrm{ID}_{\mathcal{D}} \longrightarrow \{0, 1\}$ defined by

$$q(D, \phi) = \begin{cases} 1 & \text{if } \phi \text{ is satisfied in } D, \\ 0 & \text{otherwise,} \end{cases}$$

is a query (as in Definition 13.23).

8. Specialize the generic levelwise algorithm to an algorithm that retrieves all inclusion dependencies satisfied by a database state.
Let $T : \{1, \ldots, n\} \longrightarrow \mathcal{P}(D)$ be a transaction data set on an item set D. The contingency matrix of two item sets X and Y is the 2×2-matrix:

$$M_{XY} = \begin{pmatrix} m_{11} & m_{10} \\ m_{01} & m_{00} \end{pmatrix},$$

where

$$\begin{aligned} m_{11} &= |\{k | X \subseteq T(k) \text{ and } Y \subseteq T(k)\}|, \\ m_{10} &= |\{k | X \subseteq T(k) \text{ and } Y \not\subseteq T(k)\}|, \\ m_{01} &= |\{k | X \not\subseteq T(k) \text{ and } Y \subseteq T(k)\}|, \\ m_{00} &= |\{k | X \not\subseteq T(k) \text{ and } Y \not\subseteq T(k)\}|. \end{aligned}$$

Also, let $m_{1.} = m_{11} + m_{10}$ and $m_{.1} = m_{11} + m_{01}$.

9. Let $X \Rightarrow Y$ be an association rule. Prove that

$$conf_T(X \Rightarrow Y) = \frac{m_{11}}{m_{11} + m_{10}}.$$

Which significance has the number m_{10} for $X \Rightarrow Y$?

10. Let $T : \{1, \ldots, n\} \longrightarrow \mathcal{P}(I)$ be a transaction data set on a set of items I and let π be a partition of the set $\{1, \ldots, n\}$ of transaction identifiers, $\pi = \{B_1, \ldots, B_p\}$. Let $n_i = |B_i|$ for $1 \leqslant i \leqslant p$.
A *partitioning* of T is a sequence $T_1, \ldots, T_p$ of transaction data sets on I such that $T_i : \{1, \ldots, n_i\} \longrightarrow \mathcal{P}(I)$ is defined by $T_i(\ell) = T(k_\ell)$, where $B_i = \{k_1, \ldots, k_{n_i}\}$ for $1 \leqslant i \leq p$.

Intuitively, this corresponds to splitting horizontally the table of T into p tables that contain $n_1, \ldots, n_p$ consecutive rows, respectively.

Let K be an item set. Prove that if $supp_T(K) \geqslant \mu$, there exists j, $1 \leqslant j \leqslant p$, such that $supp_{T_j}(K) \geqslant \mu$. Give an example to show that the reverse implication does not hold; in other words, give an example of a transaction data set T, a partitioning $T_1, \ldots, T_p$ of T, and an item set K such that K is μ-frequent in some T_i but not in T.

11. Piatetsky-Shapiro formulated in [2] three principles that a rule interestingness measure R should satisfy:

(a) $R(X \Rightarrow Y) = 0$ if $m_{11} = \frac{m_{1.} m_{.1}}{n}$,
(b) $R(X \rightarrow Y)$ increases with m_{11} when other parameters are fixed, and
(c) $R(X \rightarrow Y)$ decreases with $m_{.1}$ and with $m_{1.}$ when other parameters are fixed.

The *lift* of a rule $X \Rightarrow Y$ is the number $lift(X \Rightarrow Y) = \frac{nm_{11}}{m_{1.}m_{.1}}$. The *PS* measure is $PS(X \rightarrow Y) = m_{11} - \frac{m_{1.}m_{.1}}{n}$. Do *lift* and *PS* satisfy Piatetsky-Shapiro's principles? Give examples of interestingness measures that satisfy these principles.

12. Let I be a set of items and $T : \{1, \ldots, n\} \longrightarrow \mathcal{P}(I)$ be a transaction data set. Recall that in Sect. 13.6 we introduced the function $tids_T : \mathcal{P}(I) \longrightarrow \mathcal{P}(\{1, \ldots, n)$ by $tids_T(H) = \{k \in \{1, \ldots, n\} \mid H \subseteq T(k)\}$ for any item set H.

    (a) Prove that if $L, J \subseteq I$, $J \subseteq L$, and $L - J = \{i_1, \ldots, i_p\}$, then $tids_T(L) = \bigcap_{\ell=1}^{p} tids_T(J \cup \{i_\ell\})$.
    (b) Let F_J^L be the number $F_J^L = |\{(h, J') \mid J' \subseteq T(h) \text{ and } J' \cap L = J|$. Prove that
    $$F_J^L = \left| \bigcup_{k=1}^{p} tids_T(J \cup \{i_k\}) \right| - |tids_T(J)|.$$
    (c) By applying the Inclusion-Exclusion Principle, prove that
    $$suppcount(L) - (-1)^p F_J^L = \sum_{J \subseteq J' \subset L} (-1)^{|L-J'|+1} suppcount(J').$$

13. Let $T : \{1, \ldots, n\} \longrightarrow \mathcal{P}(I)$ be a transaction data set on the set of items I. Prove that if $f = suppcount_T$, then for the density d_f we have $d_f(K) = |\{i \mid T(i) = K\}|$ for every $K \in \mathcal{P}(I)$.
14. Let $T : \{1, \ldots, n\} \longrightarrow \mathcal{P}(I)$ be a transaction data set on the set of items I and let $\mathcal{C}$ be a collection of sets of items. Prove that $D^{\mathcal{C}}_{suppcount}(K) = \sum\{d_{suppcount}(U) \mid U \in \mathcal{L}[K, \mathcal{C}]\}$.
15. Let $T : \{1, \ldots, n\} \longrightarrow \mathcal{P}(I)$ be a transaction data set over a set of items I. Prove that the mapping $d : (\mathcal{P}(I))^2 \longrightarrow R_{\geqslant 0}$ defined by $d(H, K) = suppcount(H) + suppcount(K) - 2suppcount_T(HK)$ for $K, H \subseteq I$ is a semi-metric on the collection of item sets.

Bibliographical Comments

In addition to general data mining references [1, 11], the reader should consult [12], a monograph dedicated to frequent item sets and association rules. Seminal work in this area, in addition to the original paper [13], has been done by Mannila and Toivonen [8] and by Zaki [10]; these references lead to an interesting and rewarding journey through the data mining literature. An alternative method for detecting frequent item sets based on a very interesting condensed representation of the data set was developed by Han et al. [14].

An algorithm that searches the collection of item sets in a depth-first manner with the purpose of discovering maximal frequent item sets was proposed in [15, 16].

Exercises 5–8 are reformulations of results obtained in [8].

References

1. P.N. Tan, M. Steinbach, V. Kumar, *Introduction to Data Mining* (Addison-Wesley, Reading, 2005)
2. G. Piatetsky-Shapiro, in *Knowledge Discovery in Databases*, ed. by G. Piatetsky-Shapiro, W. Frawley. Discovery, analysis and presentation of strong rules, (MIT Press, Cambridge, MA, 1991), pp. 229–248
3. C.C. Aggarwal, P.S. Yu, Mining associations with the collective strength approach. IEEE Trans. Knowl. Data Eng. **13**, 863–873 (2001)
4. R. Bayardo, R. Agrawal, in *Proceedings of the 5th KDD*, ed. by S. Chaudhuri, D. Madigan. Mining the most interesting rules, San Diego, CA, (ACM, New York, 1999), pp. 145–153
5. S. Brin, R. Motwani, C. Silverstein, in *Proceedings of the ACM SIGMOD International Conference on Management of Data*, ed. by J. Pekham. Beyond Market Baskets: Generalizing Association Rules to Correlations. Tucson, AZ, (ACM, New York, 1997), pp. 265–276
6. S. Jaroszewicz, D. Simovici, in *Proceedings of the Tenth ACM SIGKDD International Conference on Knowledge Discovery and Data Mining*, eds. by ed. by W. Kim, R. Kohavi, J. Gehrke, W. DuMouchel. Interestingness of Frequent Item Sets Using Bayesian Networks as Background Knowledge, Seattle, WA, (ACM, New York, 2004), pp. 178–186
7. R. Hilderman, H. Hamilton. Knowledge discovery and interestingness measures: a survey. Technical Report CS 99–04, Department of Computer Science, University of Regina, 1999
8. H. Mannila, H. Toivonen. Levelwise search and borders of theories in knowledge discovery. Technical Report C-1997-8, University of Helsinki, 1997
9. N. Pasquier, Y. Bastide, R. Taouil, L. Lakhal, in *Database Theory - ICDT'99*, ed. by C. Beeri, P. Buneman. Lecture Notes in Computer Science. Discovering frequent closed itemsets for association rules, vol. 1540, Jerusalem, Israel, (Springer-Verlag, Berlin, 1999), pp. 398–416
10. M.J. Zaki, C.J. Hsiao, Efficient algorithms for mining closed itemsets and their lattice structure. IEEE Trans. Knowl. Data Eng. **17**, 462–478 (2005)
11. M. Steinbach, G. Karypis, V. Kumar, in *A Comparison of Document Clustering Techniques*, eds. by M. Grobelnik, D. Mladenic, N. Milic-Freyling. (KDD Workshop on Text Mining, Boston 2000)
12. J.-M. Adamo, *Data Mining for Association Rules and Sequential Patterns* (Springer-Verlag, New York, 2001)
13. R. Agrawal, T. Imielinski, A.N. Swami, in *Proceedings of the 1993 International Conference on Management of Data*, eds. by P. Buneman, S. Jajodia. Mining association rules between sets of items in large databases. Washington, D.C., (ACM, New York, 1993), pp. 207–216
14. J. Han, J. Pei, Y. Yin, in *Proceedings of the ACM-SIGMOD International Conference on Management of Data*, eds. by W. Chen, J.F. Naughton, P.A. Bernstein. Mining Frequent Patterns Without Candidate Generation, Dallas, TX, (ACM, New York, 2000), pp. 1–12
15. R.C. Agarwal, C.C. Aggarwal, V. V. V. Prasad, in *Proceedings of the 6th Conference on Knowledge Discovery in Data*, ed. by R. Bayardo, R. Ramakrishnan, S.J. Stolfo. Depth First Generation of Long Patterns, Boston, MA, (ACM, New York, 2000), pp. 108–118
16. R.C. Agarwal, C.C. Aggarwal, V.V.V. Prasad, A tree projection algorithm for generation of frequent item sets. J. Parallel Distrib. Comput. **61**(3), 350–371 (2001)

Chapter 14
Special Metrics

14.1 Introduction

Clustering and classification, two central data mining activities, require the evaluation of degrees of dissimilarity between data objects. This task is accomplished using a variety of specializations of the notion of dissimilarity, such as tree metrics and ultrametrics, which we introduced in Sect. 1.9.

Substantial attention is paid to ultrametrics due to their importance for clustering algorithms. Various modalities for generating ultrametrics are discussed, starting with hierarchies on sets, equidistant trees, and chains of partitions (or equivalences).

Metrics on several quite distinct data types (beyond metrics on linear spaces discussed in Chap. 5) are considered: subsets of finite sets, partitions of finite sets, and sequences. The chapter concludes with a section dedicated to the application of elementary properties of metrics to searching in metric spaces. Further applications of metrics are presented in subsequent chapters.

14.2 Ultrametrics and Ultrametric Spaces

Recall that an ultrametric on a set S was defined in Sect. 1.9 as a mapping $d : S^2 \longrightarrow \mathbb{R}_{\geqslant 0}$ that has the following properties:

(i) $d(x, y) = 0$ if and only if $x = y$ for $x, y \in S$;
(ii) $d(x, y) = d(y, x)$ for $x, y \in S$;
(iii) $d(x, y) \leqslant \max\{d(x, z), d(z, y)\}$ for $x, y, z \in S$.

As we did for metrics, if property (i) is replaced by the weaker requirement that $d(x, x) = 0$ for $x \in S$, then d is a *quasi-ultrametric* on S.

Example 14.1 Let $\pi = \{B, C\}$ be a two-block partition of a nonempty set S. Define the mapping $d_\pi : S^2 \longrightarrow \mathbb{R}_{\geqslant 0}$ by

D. A. Simovici and C. Djeraba, *Mathematical Tools for Data Mining*,
Advanced Information and Knowledge Processing, DOI: 10.1007/978-1-4471-6407-4_14,

$$d_\pi(x, y) = \begin{cases} 0 & \text{if } \{x, y\} \subseteq B \text{ or } \{x, y\} \subseteq C \\ 1 & \text{otherwise,} \end{cases}$$

for $x, y \in S$. We claim that d_π is a quasi-ultrametric. Indeed, it is clear that $d_\pi(x, x) = 0$ and $d_\pi(x, y) = d_\pi(y, x)$ for $x, y \in S$. Now let x, y, z be three arbitrary elements in S. If $d_\pi(x, y) = 1$, then x and y belong to two distinct blocks of the partition π, say to B and C, respectively. If $z \in B$, then $d_\pi(x, z) = 0$ and $d_\pi(z, y) = 1$; similarly, if $z \in C$, then $d_\pi(x, z) = 1$ and $d_\pi(z, y) = 0$. In either case, the ultrametric inequality is satisfied.

Theorem 14.2 *Let $a_0, a_1, a_2 \in \mathbb{R}$ be three numbers. If $a_i \leqslant \max\{a_j, a_k\}$ for every permutation (i, j, k) of the set $\{0, 1, 2\}$, then two of the numbers are equal and the third is not larger than the two others.*

Proof Suppose that a_i is the least of the numbers a_0, a_1, a_2 and a_j, a_k are the remaining numbers. Since $a_j \leqslant \max\{a_i, a_k\} = a_k$ and $a_k \leqslant \max\{a_i, a_j\} = a_j$, it follows that $a_j = a_k \geqslant a_i$.

Triangles in ultrametric spaces have an interesting property that is given next.

Corollary 14.3 *Let (S, d) be an ultrametric space. For every $x, y, z \in S$, two of the numbers $d(x, y), d(x, z), d(y, z)$ are equal and the third is not larger than the two other equal numbers.*

Proof Since d satisfies the ultrametric inequality, the statement follows immediately from Theorem 14.2.

Theorem 14.4 *Let $B(x, r)$ be a closed sphere in the ultrametric space (S, d). If $z \in B(x, d)$, then $B(x, r) = B(z, r)$. In other words, in an ultrametric space, a closed sphere has all its points as centers.*

Proof Suppose that $z \in B(x, r)$, so $d(x, z) \leqslant r$. Let $y \in B(z, r)$. Since $d(y, x) \leqslant \max\{d(y, z), d(z, x)\} \leqslant r$, we have $y \in B(x, r)$. Conversely, if $y \in B(x, r)$, we have $d(y, z) \leqslant \max\{d(y, x), d(x, z)\} \leqslant r$, hence $y \in B(z, r)$.

Both closed and open spheres in ultrametric spaces are clopen sets as we show next.

Theorem 14.5 *If d is an ultrametric on S, then any closed sphere $B(t, r)$ and any open sphere $C(t, r)$ are clopen sets in the topological ultrametric space $(S, \mathcal{O}_d)$.*

Proof We already know that $B(t, r)$ is closed. To prove that this set is also open if d is an ultrametric, let $s \in B(t, r)$. By Theorem 14.4 s is a center of the sphere. Therefore, $C\left(s, \frac{r}{2}\right) \subseteq B(t, r)$, so $B(t, r)$ is open. We leave the proof that $C(t, r)$ is also closed to the reader.

By Theorem 4.35, the border of a closed sphere or of an open sphere in an ultrametric space is empty.

Theorem 14.6 *Let (S, d) be an ultrametric space, $x, y \in S$, and let $S(x, y) \subseteq$ **Seq**(S) be the set of sequences that start with x and end with y. We have $d(x, y) = \min\{amp_d(\mathbf{s}) \mid \mathbf{s} \in S(x, y)\}$.*

Proof Since d is an ultrametric, we have $d(x, y) \leqslant amp_d(\mathbf{s})$ for any nonnull sequence $\mathbf{s} = (s_1, \ldots, s_n)$ such that $s_1 = x$ and $s_n = y$. Therefore,

$$d(x, y) \leqslant \min\{amp_d(\mathbf{s}) \mid \mathbf{s} \in S(x, y)\}.$$

The equality of the theorem follows from the fact that $(x, y) \in S(x, y)$.

Theorem 14.7 *If two closed spheres $B(x, r)$ and $B(y, r')$ of an ultrametric space have a point in common, then one of the closed spheres is included in the other.*

Proof The statement follows directly from Theorem 14.4.

Theorem 14.4 implies that the entire space S equals the closed sphere $B(x, diam_{S,d})$ for any point $x \in S$.

The next statement gives a method of constructing ultrametrics starting from chains of equivalence relations.

Theorem 14.8 *Let S be a finite set and let $d : S \times S \longrightarrow \mathbb{R}_{\geqslant 0}$ be a function whose range is $\mathsf{Ran}(d) = \{r_1, \ldots, r_m\}$, where $r_1 = 0$ such that $d(x, y) = 0$ if and only if $x = y$. Define the relations $\eta_{r_i} = \{(x, y) \in S \times S \mid d(x, y) \leqslant r_i\}$ for $1 \leqslant i \leqslant m$.*

The function d is an ultrametric on S if and only if the sequence of relations $\eta_{r_1}, \ldots, \eta_{r_m}$ is an increasing chain of equivalences on S such that $\eta_{r_1} = \iota_S$ and $\eta_{r_m} = \theta_S$.

Proof Suppose that d is an ultrametric on S. We have $(x, x) \in \eta_{r_i}$ because $d(x, x) = 0$, so all relations η_{r_i} are reflexive. Also, it is clear that the symmetry of d implies $(x, y) \in \eta_{r_i}$ if and only if $(y, x) \in \eta_{r_i}$, so these relations are symmetric.

The ultrametric inequality is essential for proving the transitivity of the relations η_{r_i}. If $(x, y), (y, z) \in \eta_{r_i}$, then $d(x, y) \leqslant r_i$ and $d(y, z) \leqslant r_i$, which implies $d(x, z) \leqslant \max\{d(x, y), d(y, z)\} \leqslant r_i$. Thus, $(x, z) \in \eta_{r_i}$, which shows that every relation η_{r_i} is transitive and therefore an equivalence.

It is straightforward to see that $\eta_{r_1} \leqslant \eta_{r_2} \leqslant \cdots \leqslant \eta_{r_m}$; that is, this sequence of relations is indeed a chain of equivalences.

Conversely, suppose that $\eta_{r_1}, \ldots, \eta_{r_m}$ is an increasing sequence of equivalences on S such that $\eta_{r_1} = \iota_S$ and $\eta_{r_m} = \theta_S$, where $\eta_{r_i} = \{(x, y) \in S \times S \mid d(x, y) \leqslant r_i\}$ for $1 \leqslant i \leqslant m$ and $r_1 = 0$.

Note that $d(x, y) = 0$ is equivalent to $(x, y) \in \eta_{r_1} = \iota_S$, that is, to $x = y$.

We claim that

$$d(x, y) = \min\{r \mid (x, y) \in \eta_r\}. \tag{14.1}$$

Indeed, since $\eta_{r_m} = \theta_S$, it is clear that there is an equivalence η_{r_i} such that $(x, y) \in \eta_{r_i}$. If $(x, y) \in \eta_{r_i}$, the definition of η_{r_i} implies $d(x, y) \leqslant r_i$, so $d(x, y) \leqslant \min\{r \mid$

$(x, y) \in \eta_r\}$. This inequality can be easily seen to become an equality since $(x, y) \in \eta_{d(x,y)}$. This implies immediately that d is symmetric.

To prove that d satisfies the ultrametric inequality, let x, y, z be three members of the set S. Let $p = \max\{d(x, z), d(z, y)\}$. Since $(x, z) \in \eta_{d(x,z)} \subseteq \eta_p$ and $(z, y) \in \eta_{d(z,y)} \subseteq \eta_p$, it follows that $(x, y) \in \eta_p$, due to the transitivity of the equivalence η_p. Thus, $d(x, y) \leqslant p = \max\{d(x, z), d(z, y)\}$, which proves the triangular inequality for d.

Of course, Theorem 14.8 can be formulated in terms of partitions.

Theorem 14.9 *Let S be a finite set and let $d : S \times S \longrightarrow \mathbb{R}_{\geqslant 0}$ be a function whose range is $\mathsf{Ran}(d) = \{r_1, \ldots, r_m\}$, where $r_1 = 0$ such that $d(x, y) = 0$ if and only if $x = y$. For $u \in S$ and $r \in \mathbb{R}_{\geqslant 0}$, define the set $D_{u,r} = \{x \in S \mid d(u, x) \leqslant r\}$ and let $\pi_{r_i} = \{D(u, r_i) \mid u \in S\}$ for $1 \leqslant i \leqslant m$.*

The function d is an ultrametric on S if and only if the sequence $\pi_{r_1}, \ldots, \pi_{r_m}$ is an increasing sequence of partitions on S such that $\pi_{r_1} = \alpha_S$ and $\pi_{r_m} = \omega_S$.

Proof The argument is entirely similar to the proof of Theorem 14.8 and is omitted.

14.2.1 Hierarchies and Ultrametrics

Definition 14.10 *Let S be a set. A* hierarchy *on S is a collection of sets $\mathcal{H} \subseteq \mathcal{P}(S)$ that satisfies the following conditions:*

(i) *the members of $\mathcal{H}$ are nonempty sets;*
(ii) *$S \in \mathcal{H}$;*
(iii) *for every $x \in S$, we have $\{x\} \in \mathcal{H}$;*
(iv) *if $H, H' \in \mathcal{H}$ and $H \cap H' \neq \emptyset$, then we have either $H \subseteq H'$ or $H' \subseteq H$.*

A standard technique for constructing a hierarchy on a set S starts with a rooted tree $(\mathcal{T}, v_0)$ whose nodes are labeled by subsets of the set S. Let V be the set of vertices of the tree $\mathcal{T}$. The function $\mu : V \longrightarrow \mathcal{P}(S)$, which gives the label $\mu(v)$ of each node $v \in V$, is defined as follows:

(i) the tree $\mathcal{T}$ has $|S|$ leaves, and each leaf v is labeled by a distinct singleton $\mu(v) = \{x\}$ for $x \in S$;
(ii) if an interior vertex v of the tree has the descendants $v_1, v_2, \ldots, v_n$, then $\mu(v) = \bigcup_{i=1}^n \mu(v_i)$.

The set of labels $\mathcal{H}_\mathcal{T}$ of the rooted tree $(\mathcal{T}, v_0)$ forms a hierarchy on S. Indeed, note that each singleton $\{x\}$ is a label of a leaf. An easy argument by induction on the height of the tree shows that every vertex is labeled by the set of labels of the leaves that descend from that vertex. Therefore, the root v_0 of the tree is labeled by S.

Suppose that H, H' are labels of the nodes u, v of $\mathcal{T}$, respectively. If $H \cap H' \neq \emptyset$, then the vertices u, v have a common descendant. In a tree, this can take place only if u is a descendant of v or v is a descendant of u; that is, only if $H \subseteq H'$, or $H' \subseteq H$, respectively. This gives the desired conclusion.

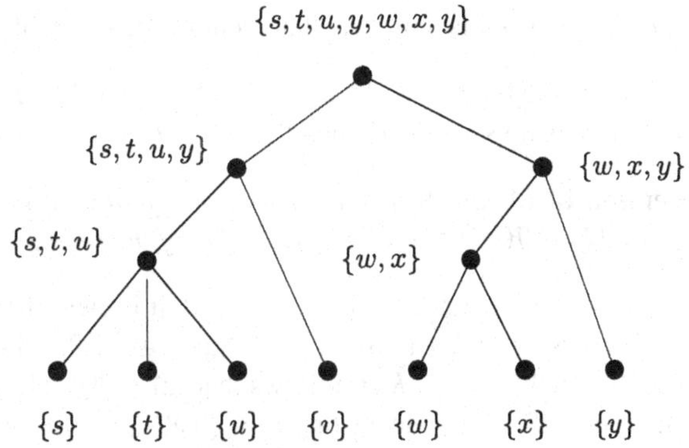

Fig. 14.1 Tree labeled by subsets of S

Example 14.11 Let $S = \{s, t, u, v, w, x, y\}$ and let $\mathcal{T}$ be a tree whose vertices are labeled as shown in Fig. 14.1. It is easy to verify that the family of subsets of S that label the nodes of $\mathcal{T}$,

$$\begin{aligned}\mathcal{H} = \{&\{s\}, \{t\}, \{u\}, \{v\}, \{w\}, \{x\}, \{y\},\\ &\{s, t, u\}, \{w, x\}, \{s, t, u, v\}, \{w, x, y\}, \{s, t, u, v, w, x, y\}\}\end{aligned}$$

is a hierarchy on the set S.

Chains of partitions defined on a set generate hierarchies, as we show next.

Theorem 14.12 *Let S be a set and let $C = (\pi_1, \pi_2, \ldots, \pi_n)$ be an increasing chain of partitions $(PART(S), \leqslant)$ such that $\pi_1 = \alpha_S$ and $\pi_n = \omega_S$. Then, the collection $\mathcal{H}_C = \bigcup_{i=1}^{n} \pi_i$ that consists of the blocks of all partitions in the chain is a hierarchy on S.*

Proof The blocks of any of the partitions are nonempty sets, so $\mathcal{H}_C$ satisfies the first condition of Definition 14.10.

We have $S \in \mathcal{H}_C$ because S is the unique block of $\pi_n = \omega_S$. Also, since all singletons $\{x\}$ are blocks of $\alpha_S = \pi_1$, it follows that $\mathcal{H}_C$ satisfies the second and the third conditions of Definition 14.10. Finally, let H and H' be two sets of $\mathcal{H}_C$ such that $H \cap H' \neq \emptyset$. Because of this condition, it is clear that these two sets cannot be blocks of the same partition. Thus, there exist two partitions π_i and π_j in the chain such that $H \in \pi_i$ and $H' \in \pi_j$. Suppose that $i < j$. Since every block of π_j is a union of blocks of π_i, H' is a union of blocks of π_i and $H \cap H' \neq \emptyset$ means that H is one of these blocks. Thus, $H \subseteq H'$. If $j > i$, we obtain the reverse inclusion. This allows us to conclude that $\mathcal{H}_C$ is indeed a hierarchy.

Theorem 14.12 can be stated in terms of chains of equivalences; we give the following alternative formulation for convenience.

Theorem 14.13 *Let S be a finite set and let $(\rho_1, \ldots, \rho_n)$ be a chain of equivalence relations on S such that $\rho_1 = \iota_S$ and $\rho_n = \theta_S$. Then, the collection of blocks of the equivalence relations ρ_r (that is, the set $\bigcup_{1 \leqslant r \leqslant n} S/\rho_r$) is a hierarchy on S.*

Proof The proof is a mere restatement of the proof of Theorem 14.12.

Define the relation "$\prec$" on a hierarchy $\mathcal{H}$ on S by $H \prec K$ if $H, K \in \mathcal{H}$, $H \subset K$, and there is no set $L \in \mathcal{H}$ such that $H \subset L \subset K$.

Lemma 14.14 *Let $\mathcal{H}$ be a hierarchy on a finite set S and let $L \in \mathcal{H}$. The collection $\mathcal{P}_L = \{H \in \mathcal{H} \mid H \prec L\}$ is a partition of the set L.*

Proof We claim that $L = \bigcup \mathcal{P}_L$. Indeed, it is clear that $\bigcup \mathcal{P}_L \subseteq L$.

Conversely, suppose that $z \in L$ but $z \notin \bigcup \mathcal{P}_L$. Since $\{z\} \in \mathcal{H}$ and there is no $K \in \mathcal{P}_L$ such that $z \in K$, it follows that $\{z\} \in \mathcal{P}_L$, which contradicts the assumption that $z \notin \bigcup \mathcal{P}_L$. This means that $L = \bigcup \mathcal{P}_L$.

Let $K_0, K_1 \in \mathcal{P}_L$ be two distinct sets. These sets are disjoint since otherwise we would have either $K_0 \subset K_1$ or $K_1 \subset K_0$, and this would contradict the definition of $\mathcal{P}_L$.

Theorem 14.15 *Let $\mathcal{H}$ be a hierarchy on a set S. The graph of the relation $\prec$ on $\mathcal{H}$ is a tree whose root is S; its leaves are the singletons $\{x\}$ for every $x \in S$.*

Proof Since $\prec$ is an antisymmetric relation on $\mathcal{H}$, it is clear that the graph $(\mathcal{H}, \prec)$ is acyclic. Moreover, for each set $K \in \mathcal{H}$, there is a unique path that joins K to S, so the graph is indeed a rooted tree.

Definition 14.16 *Let $\mathcal{H}$ be a hierarchy on a set S. A* grading function *for $\mathcal{H}$ is a function $h : \mathcal{H} \longrightarrow \mathbb{R}$ that satisfies the following conditions:*

(i) *$h(\{x\}) = 0$ for every $x \in S$, and*
(ii) *if $H, K \in \mathcal{H}$ and $H \subset K$, then $h(H) < h(K)$.*

If h is a grading function for a hierarchy $\mathcal{H}$, the pair $(\mathcal{H}, h)$ is a graded hierarchy.

Example 14.17 For the hierarchy $\mathcal{H}$ defined in Example 14.11 on the set $S = \{s, t, u, v, w, x, y\}$, the function $h : \mathcal{H} \longrightarrow \mathbb{R}$ given by

$$\begin{aligned}
&h(\{s\}) = h(\{t\}) = h(\{u\}) = h(\{v\}) = h(\{w\}) = h(\{x\}) = h(\{y\}) = 0,\\
&h(\{s, t, u\}) = 3, h(\{w, x\}) = 4, h(\{s, t, u, v\}) = 5, h(\{w, x, y\}) = 6,\\
&h(\{s, t, u, v, w, x, y\}) = 7,
\end{aligned}$$

is a grading function and the pair $(\mathcal{H}, h)$ is a graded hierarchy on S.

Theorem 14.12 can be extended to graded hierarchies.

Theorem 14.18 *Let S be a finite set and let $C = (\pi_1, \pi_2, \ldots, \pi_n)$ be an increasing chain of partitions $(PART(S), \leqslant)$ such that $\pi_1 = \alpha_S$ and $\pi_n = \omega_S$.*

If $f : \{1, \ldots, n\} \longrightarrow \mathbb{R}_{\geqslant 0}$ is a function such that $f(1) = 0$, then the function $h : \mathcal{H}_C \longrightarrow \mathbb{R}_{\geqslant 0}$ given by $h(K) = f\left(\min\{j \mid K \in \pi_j\}\right)$ is a grading function for the hierarchy $\mathcal{H}_C$.

Proof Since $\{x\} \in \pi_1 = \alpha_S$, it follows that $h(\{x\}) = 0$, so h satisfies the first condition of Definition 14.16.

Suppose that $H, K \in \mathcal{H}_C$ and $H \subset K$. If $\ell = \min\{j \mid H \in \pi_j\}$ it is impossible for K to be a block of a partition that precedes π_ℓ. Therefore, $\ell < \min\{j \mid K \in \pi_j\}$, so $h(H) < h(K)$, and $(\mathcal{H}_C, h)$ is indeed a graded hierarchy.

A graded hierarchy defines an ultrametric, as shown next.

Theorem 14.19 *Let $(\mathcal{H}, h)$ be a graded hierarchy on a finite set S. Define the function $d : S^2 \longrightarrow \mathbb{R}$ as $d(x, y) = \min\{h(U) \mid U \in \mathcal{H}$ and $\{x, y\} \subseteq U\}$ for $x, y \in S$. The mapping d is an ultrametric on S.*

Proof Observe that for every $x, y \in S$ there exists a set $H \in \mathcal{H}$ such that $\{x, y\} \subseteq H$ because $S \in \mathcal{H}$.

It is immediate that $d(x, x) = 0$. Conversely, suppose that $d(x, y) = 0$. Then, there exists $H \in \mathcal{H}$ such that $\{x, y\} \subseteq H$ and $h(H) = 0$. If $x \neq y$, then $\{x\} \subset H$, hence $0 = h(\{x\}) < h(H)$, which contradicts the fact that $h(H) = 0$. Thus, $x = y$.

The symmetry of d is immediate.

To prove the ultrametric inequality, let $x, y, z \in S$, and suppose that $d(x, y) = p$, $d(x, z) = q$, and $d(z, y) = r$. There exist $H, K, L \in \mathcal{H}$ such that $\{x, y\} \subseteq H$, $h(H) = p$, $\{x, z\} \subseteq K$, $h(K) = q$, and $\{z, y\} \subseteq L$, $h(L) = r$. Since $K \cap L \neq \emptyset$ (because both sets contain z), we have either $K \subseteq L$ or $L \subseteq K$, so $K \cup L$ equals either K or L and, in either case, $K \cup L \in \mathcal{H}$. Since $\{x, y\} \subseteq K \cup L$, it follows that

$$d(x, y) \leqslant h(K \cup L) = \max\{h(K), H(L)\} = \max\{d(x, z), d(z, y)\},$$

which is the ultrametric inequality.

We refer to the ultrametric d whose existence is shown in Theorem 14.19 as the *ultrametric generated by the graded hierarchy* $(\mathcal{H}, h)$.

Example 14.20 The values of the ultrametric generated by the graded hierarchy $(\mathcal{H}, h)$ on the set S introduced in Example 14.17 are given in the following table:

d	s	t	u	v	w	x	y
s	0	3	3	5	7	7	7
t	3	0	3	5	7	7	7
u	3	3	0	5	7	7	7
v	5	5	5	0	7	7	7
w	7	7	7	7	0	4	6
x	7	7	7	7	4	0	6
y	7	7	7	7	6	6	0

The hierarchy introduced in Theorem 14.13 that is associated with an ultrametric space can be naturally equipped with a grading function, as shown next.

Theorem 14.21 *Let (S, d) be a finite ultrametric space. There exists a graded hierarchy $(\mathcal{H}, h)$ on S such that d is the ultrametric associated to $(\mathcal{H}, h)$.*

Proof Let $\mathcal{H}$ be the collection of equivalence classes of the equivalences $\eta_r = \{(x, y) \in S^2 \mid d(x, y) \leqslant r\}$ defined by the ultrametric d on the finite set S, where the index r takes its values in the range R_d of the ultrametric d. Define $h(E) = \min\{r \in R_d \mid E \in S/\eta_r\}$ for every equivalence class E.

It is clear that $h(\{x\}) = 0$ because $\{x\}$ is an η_0-equivalence class for every $x \in S$.

Let $[x]_t$ be the equivalence class of x relative to the equivalence η_t.

Suppose that E and E' belong to the hierarchy and $E \subset E'$. We have $E = [x]_r$ and $E' = [x]_s$ for some $x \in X$. Since E is strictly included in E', there exists $z \in E' - E$ such that $d(x, z) \leqslant s$ and $d(x, z) > r$. This implies $r < s$. Therefore,

$$h(E) = \min\{r \in R_d \mid E \in S/\eta_r\} \leqslant \min\{s \in R_d \mid E' \in S/\eta_s\} = h(E'),$$

which proves that $(\mathcal{H}, h)$ is a graded hierarchy.

The ultrametric e generated by the graded hierarchy $(\mathcal{H}, h)$ is given by

$$\begin{aligned} e(x, y) &= \min\{h(B) \mid B \in \mathcal{H} \text{ and } \{x, y\} \subseteq B\} \\ &= \min\{r \mid (x, y) \in \eta_r\} = \min\{r \mid d(x, y) \leqslant r\} = d(x, y), \end{aligned}$$

for $x, y \in S$; in other words, we have $e = d$.

Example 14.22 Starting from the ultrametric on the set $S = \{s, t, u, v, w, x, y\}$ defined by the table given in Example 14.20, we obtain the following quotient sets:

Values of r	S/η_r
$[0, 3)$	$\{s\}, \{t\}, \{u\}, \{v\}, \{w\}, \{x\}, \{y\}$
$[3, 4)$	$\{s, t, u\}, \{v\}, \{w\}, \{x\}, \{y\}$
$[4, 5)$	$\{s, t, u\}, \{v\}, \{w, x\}, \{y\}$
$[5, 6)$	$\{s, t, u, v\}, \{w, x\}, \{y\}$
$[6, 7)$	$\{s, t, u, v\}, \{w, x, y\}$
$[7, \infty)$	$\{s, t, u, v, w, x, y\}$

We shall draw the tree of a graded hierarchy $(\mathcal{H}, h)$ using a special representation known as a *dendrogram*. In a dendrogram, an interior vertex K of the tree is represented by a horizontal line drawn at the height $h(K)$. For example, the dendrogram of the graded hierarchy of Example 14.17 is shown in Fig. 14.2.

By Theorem 14.19, the value $d(x, y)$ of the ultrametric d generated by a hierarchy $\mathcal{H}$ is the smallest height of a set of a hierarchy that contains both x and y. This allows us to "read" the value of the ultrametric generated by $\mathcal{H}$ directly from the dendrogram of the hierarchy.

Example 14.23 For the graded hierarchy of Example 14.17, the ultrametric extracted from Fig. 14.2 is clearly the same as the one that was obtained in Example 14.20.

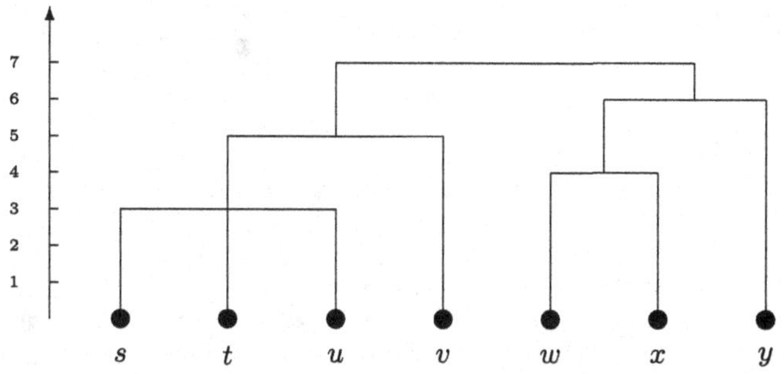

Fig. 14.2 Dendrogram of graded hierarchy of Example 14.17

14.2.2 The Poset of Ultrametrics

Let S be a set. Recall that we denoted the set of dissimilarities by $\mathcal{D}_S$. Define a partial order $\leqslant$ on $\mathcal{D}_S$ by $d \leqslant d'$ if $d(x, y) \leqslant d'(x, y)$ for every $x, y \in S$. It is easy to verify that $(\mathcal{D}_S, \leqslant)$ is a poset.

The set $\mathcal{U}_S$ of ultrametrics on S is a subset of $\mathcal{D}_S$.

Theorem 14.24 *Let d be a dissimilarity on a set S and let U_d be the set of ultrametrics $U_d = \{e \in \mathcal{U}_S \mid e \leqslant d\}$. The set U_d has a largest element in the poset $(\mathcal{D}_S, \leqslant)$.*

Proof The set U_d is nonempty because the zero dissimilarity d_0 given by $d_0(x, y) = 0$ for every $x, y \in S$ is an ultrametric and $d_0 \leqslant d$.

Since the set $\{e(x, y) \mid e \in U_d\}$ has $d(x, y)$ as an upper bound, it is possible to define the mapping $e_1 : S^2 \longrightarrow \mathbb{R}_{\geqslant 0}$ as $e_1(x, y) = \sup\{e(x, y) \mid e \in U_d\}$ for $x, y \in S$. It is clear that $e \leqslant e_1$ for every ultrametric e. We claim that e_1 is an ultrametric on S.

We prove only that e_1 satisfies the ultrametric inequality. Suppose that there exist $x, y, z \in S$ such that e_1 violates the ultrametric inequality; that is,

$$\max\{e_1(x, z), e_1(z, y)\} < e_1(x, y).$$

This is equivalent to

$$\sup\{e(x, y) \mid e \in U_d\} > \max\{\sup\{e(x, z) \mid e \in U_d\}, \sup\{e(z, y) \mid e \in U_d\}\}.$$

Thus, there exists $\hat{e} \in U_d$ such that

$$\hat{e}(x, y) > \sup\{e(x, z) \mid e \in U_d\}, \text{ and } \hat{e}(x, y) > \sup\{e(z, y) \mid e \in U_d\}.$$

In particular, $\hat{e}(x, y) > \hat{e}(x, z)$ and $\hat{e}(x, y) > \hat{e}(z, y)$, which contradicts the fact that $\hat{e}$ is an ultrametric.

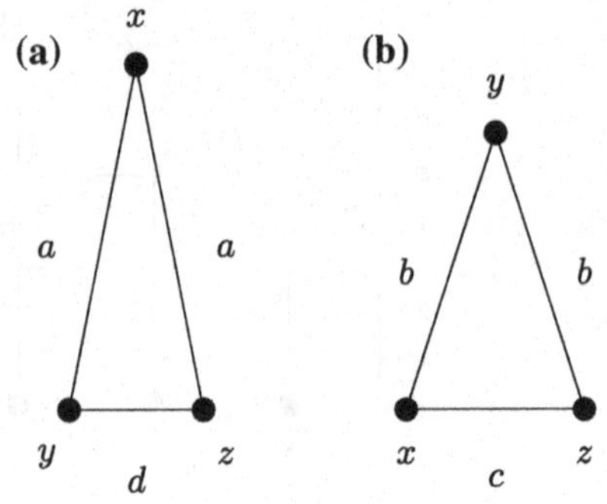

Fig. 14.3 Two ultrametrics on the set $\{x, y, z\}$

The ultrametric defined by Theorem 14.24 is known as the *maximal subdominant ultrametric for the dissimilarity d*.

The situation is not symmetric with respect to the infimum of a set of ultrametrics because, in general, the infimum of a set of ultrametrics is not necessarily an ultrametric.

For example, consider a three-element set $S = \{x, y, z\}$, four distinct nonnegative numbers a, b, c, d such that $a > b > c > d$ and the ultrametrics d and d' defined by the triangles shown in Figs. 14.3a, b, respectively.
The dissimilarity d_0 defined by $d_0(u, v) = \min\{d(u, v), d'(u, v)\}$ for $u, v \in S$ is given by

$$d_0(x, y) = b, d_0(y, z) = d, \text{ and } d_0(x, z) = c,$$

and d_0 is clearly not an ultrametric because the triangle xyz is not isosceles.

In what follows, we give an algorithm for computing the maximal subdominant ultrametric for a dissimilarity defined on a finite set S.

We define inductively an increasing sequence of partitions $\pi_1 \prec \pi_2 \prec \cdots$ and a sequence of dissimilarities $d_1, d_2, \ldots$ on the sets of blocks of $\pi_1, \pi_2, \ldots$, respectively.

For the initial phase, $\pi_1 = \alpha_S$ and $d_1(\{x\}, \{y\}) = d(x, y)$ for $x, y \in S$.

Suppose that d_i is defined on π_i. If $B, C \in \pi_i$ is a pair of blocks such that $d_i(B, C)$ has the smallest value, define the partition π_{i+1} by

$$\pi_{i+1} = (\pi_i - \{B, C\}) \cup \{B \cup C\}.$$

In other words, to obtain π_{i+1}, we replace two of the closest blocks B and C, of π_i (in terms of d_i) with new block $B \cup C$. Clearly, $\pi_i \prec \pi_{i+1}$ in *PART*(S) for $i \geqslant 1$. Note that the collection of blocks of the partitions π_i forms a hierarchy $\mathcal{H}_d$ on the set S. The dissimilarity d_{i+1} is given by

$$d_{i+1}(U, V) = \min\{d(x, y) \mid x \in U, y \in V\} \tag{14.2}$$

for $U, V \in \pi_{i+1}$.

We introduce a grading function h_d on the hierarchy defined by this chain of partitions starting from the dissimilarity d. The definition is done for the blocks of the partitions π_i by induction on i.

For $i = 1$ the blocks of the partition π_1 are singletons; in this case we define $h_d(\{x\}) = 0$ for $x \in S$.

Suppose that h_d is defined on the blocks of π_i, and let D be the block of π_{i+1} that is generated by fusing the blocks B and C of π_i. All other blocks of π_{i+1} coincide with the blocks of π_i. The value of the function h_d for the new block D is given by $h_d(D) = \min\{d(x, y) \mid x \in B, y \in C\}$. It is clear that h_d satisfies the first condition of Definition 14.16.

For a set U of $\mathcal{H}_d$, define $p_U = \min\{i \mid U \in \pi_i\}$ and $q_U = \max\{i \mid U \in \pi_i\}$.

To verify the second condition of Definition 14.16, let $H, K \in \mathcal{H}_d$ such that $H \subset K$. It is clear that $q_H \leqslant p_K$. The construction of the sequence of partitions implies that there are $H_0, H_1 \in \pi_{p_H - 1}$ and $K_0, K_1 \in \pi_{p_K - 1}$ such that $H = H_0 \cup H_1$ and $K = K_0 \cup K_1$. Therefore,

$$\begin{aligned} h_d(H) &= \min\{d(x, y) \mid x \in H_0, y \in H_1\}, \\ h_d(K) &= \min\{d(x, y) \mid x \in K_0, y \in K_1\}. \end{aligned}$$

Since H_0 and H_1 were fused (to produce the partition π_{p_H}) before K_0 and K_1 were (to produce the partition π_{p_K}), it follows that $h_d(H) < h_d(K)$.

By Theorem 14.19, the graded hierarchy $(\mathcal{H}_d, h_d)$ defines an ultrametric; we denote this ultrametric by e and will prove that e is the maximal subdominant ultrametric for d. Recall that e is given by

$$e(x, y) = \min\{h_d(W) \mid \{x, y\} \subseteq W\}$$

and that $h_d(W)$ is the least value of $d(u, v)$ such that $u \in U, v \in V$ if $W \in \pi_{p_W}$ is obtained by fusing the blocks U and V of $\pi_{p_W - 1}$. The definition of $e(x, y)$ implies that we have neither $\{x, y\} \subseteq U$ nor $\{x, y\} \subseteq V$. Thus, we have either $x \in U$ and $y \in V$ or $x \in V$ and $y \in U$. Thus, $e(x, y) \leqslant d(x, y)$.

We now prove that:

$$e(x, y) = \min\{amp_d(\mathbf{s}) \mid \mathbf{s} \in \mathbf{S}(x, y)\}$$

for $x, y \in S$.

Let D be the minimal set in $\mathcal{H}_d$ that includes $\{x, y\}$. Then, $D = B \cup C$, where B and C are two disjoint sets of $\mathcal{H}_d$ such that $x \in B$ and $y \in C$. If $\mathbf{s}$ is a sequence included in D, then there are two consecutive components of $\mathbf{s}$, s_k and s_{k+1}, such that $s_k \in B$ and $s_{k+1} \in C$. This implies

$$\begin{aligned} e(x, y) &= \min\{d(u, v) \mid u \in B, v \in C\} \\ &\leqslant d(s_k, s_{k+1}) \\ &\leqslant amp_d(\mathbf{s}). \end{aligned}$$

If s is not included in D, let s_q and s_{q+1} be two consecutive components of s such that $s_q \in D$ and $s_{q+1} \notin D$. Let E be the smallest set of $\mathcal{H}_d$ that includes $\{s_q, s_{q+1}\}$. We have $D \subseteq E$ (because $s_k \in D \cap E$) and therefore $h_d(D) \leqslant h_d(E)$. If E is obtained as the union of two disjoint sets E' and E'' of $\mathcal{H}_d$ such that $s_k \in E'$ and $s_{k+1} \in E''$, we have $D \subseteq E'$. Consequently,

$$h_d(E) = \min\{d(u, v) \mid u \in E', v \in E''\} \leqslant d(s_k, s_{k+1}),$$

which implies

$$e(x, y) = h_d(D) \leqslant h_d(E) \leqslant d(s_k, s_{k+1}) \leqslant amp_d(s).$$

Therefore, we conclude that $e(x, y) \leqslant amp_d(s)$ for every $s \in S(x, y)$.

We now show that there is a sequence $\mathbf{w} \in S(x, y)$ such that $e(x, y) \geqslant amp_d(\mathbf{w})$, which implies the equality $e(x, y) = amp_d(\mathbf{w})$. To this end, we prove that for every $D \in \pi_k \subseteq \mathcal{H}_d$ there exists $\mathbf{w} \in S(x, y)$ such that $amp_d(\mathbf{w}) \leqslant h_d(D)$. The argument is by induction on k.

For $k = 1$, the statement obviously holds. Suppose that it holds for $1, \ldots, k - 1$, and let $D \in \pi_k$. The set D belongs to π_{k-1} or D is obtained by fusing the blocks B, C of π_{k-1}. In the first case, the statement holds by inductive hypothesis. The second case has several subcases:

(i) If $\{x, y\} \subseteq B$, then by the inductive hypothesis, there exists a sequence $\mathbf{u} \in S(x, y)$ such that $amp_d(\mathbf{u}) \leqslant h_d(B) \leqslant h_d(D) = e(x, y)$.
(ii) The case $\{x, y\} \subseteq C$ is similar to the first case.
(iii) If $x \in B$ and $y \in C$, there exist $u, v \in D$ such that $d(u, v) = h_d(D)$. By the inductive hypothesis, there is a sequence $\mathbf{u} \in S(x, u)$ such that $amp_d(\mathbf{u}) \leqslant h_d(B)$ and there is a sequence $\mathbf{v} \in S(v, y)$ such that $amp_d(\mathbf{v}) \leqslant h_d(C)$. This allows us to consider the sequence $\mathbf{w}$ obtained by concatenating the sequences $\mathbf{u}$, (u, v), $\mathbf{v}$; clearly, $\mathbf{w} \in S(x, y)$ and $amp_d(\mathbf{w}) = \max\{amp_d(\mathbf{u}), d(u, v), amp_d(\mathbf{v})\} \leqslant h_d(D)$.

To complete the argument, we need to show that if e' is another ultrametric such that $e(x, y) \leqslant e'(x, y) \leqslant d(x, y)$, then $e(x, y) = e'(x, y)$ for every $x, y \in S$. By the previous argument, there exists a sequence $s = (s_0, \ldots, s_n) \in S(x, y)$ such that $amp_d(s) = e(x, y)$. Since $e'(x, y) \leqslant d(x, y)$ for every $x, y \in S$, it follows that $e'(x, y) \leqslant amp_d(s) = e(x, y)$. Thus, $e(x, y) = e'(x, y)$ for every $x, y \in S$, which means that $e = e'$. This concludes our argument.

14.3 Tree Metrics

The distance d between two vertices of a connected graph $\mathcal{G} = (V, E)$ introduced in Definition 10.21 is a metric on the set of vertices V. Recall that $d(x, y) = m$ if m is the length of the shortest path that connects x and y.

We have $d(x, y) = 0$ if and only if $x = y$. The symmetry of d is obvious. If $\mathbf{p}$ is a shortest path that connects x to z and $\mathbf{q}$ is a shortest path that connects z to y, then $\mathbf{pq}$ is a path of length $d(x, z) + d(z, y)$ that connects x to y. Therefore, $d(x, y) \leqslant d(x, z) + d(z, y)$.

The notion of distance between the vertices of a connected graph can be generalized to weighted graphs as follows. Let $(\mathcal{G}, w)$ be a weighted graph where $\mathcal{G} = (V, E)$, and let $w : E \longrightarrow \mathbb{R}_{\geqslant 0}$ be a positive weight. Define $d_w(x, y)$ as

$$d_w(x, y) = \min\{w(\mathbf{p}) \mid \mathbf{p} \text{ is a path joining } x \text{ to } y\}$$

for $x, y \in V$. We leave to reader to verify that d_w is indeed a metric.

If $(\mathcal{T}, w)$ is a weighted tree the metric d_w is referred to as a *tree metric*. Since $\mathcal{T}$ is a tree, for any two vertices $u, v \in V$ there is a unique simple path $\mathbf{p} = (v_0, \ldots, v_n)$ joining $u = v_0$ to $v = v_n$. In this case, $d_w(u, v) = \sum_{i=0}^{n-1} w(v_i, v_{i+1})$. Moreover, if $t = v_k$ is a vertex on the path $\mathbf{p}$, then

$$d_w(u, t) + d_w(t, v) = d_w(u, v), \tag{14.3}$$

a property known as the *additivity* of d_w.

We already know that d_w is a metric for arbitrary connected graphs. For trees, we have the additional property given in the next statement.

Theorem 14.25 *If $(\mathcal{T}, w)$ is a weighted tree, then d_w satisfies Buneman's inequality*

$$d_w(x, y) + d_w(u, v) \leqslant \max\{d_w(x, u) + d_w(y, v), d_w(x, v) + d_w(y, u)\}$$

for every four vertices x, y, u, v of the tree $\mathcal{T}$.

Proof Let x, y, u, v be four vertices in $\mathcal{T}$. If $x = u$ and $y = v$, the inequality reduces to an obvious equality. Therefore, we may assume that at least one of the pairs (x, u) and (y, v) consists of distinct vertices.

Suppose that $x = u$. In this case, the inequality amounts to

$$d_w(x, y) + d_w(x, v) \leqslant \max\{d_w(y, v), d_w(x, v) + d_w(y, x)\},$$

which is obviously satisfied. Thus, we may assume that we have both $x \neq u$ and $y \neq v$. Since $\mathcal{T}$ is a tree, there exists a simple path $\mathbf{p}$ that joins x to y and a simple path $\mathbf{q}$ that joins u to v.

Two cases may occur, depending on whether $\mathbf{p}$ and $\mathbf{q}$ have common edges.

Suppose initially that there are no common vertices between $\mathbf{p}$ and $\mathbf{q}$. Let s be a vertex on the path $\mathbf{p}$ and t be a vertex on $\mathbf{q}$ such that $d_w(s, t)$ is the minimal distance between a vertex located on the path $\mathbf{p}$ and one located on $\mathbf{q}$; here $d_w(s, t)$ is the sum of the weights of the edges of the simple path $\mathbf{r}$ that joins s to t.

The path $\mathbf{r}$ has no other vertices in common with $\mathbf{p}$ and $\mathbf{q}$ except s and t, respectively (see Fig. 14.4). We have

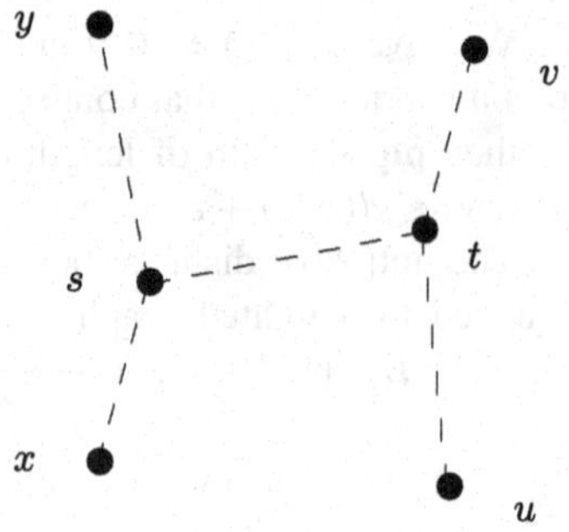

Fig. 14.4 Paths that have no common vertices

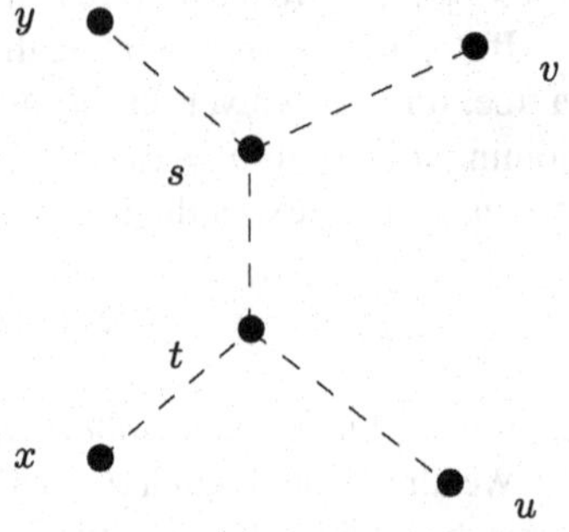

Fig. 14.5 Paths that share vertices

$$\begin{aligned}
d_w(x,u) &= d_w(x,s) + d_w(s,t) + d_w(t,u),\\
d_w(y,v) &= d_w(y,s) + d_w(s,t) + d_w(t,v),\\
d_w(x,v) &= d_w(x,s) + d_w(s,t) + d_w(t,v),\\
d_w(y,u) &= d_w(y,s) + d_w(s,t) + d_w(t,u).
\end{aligned}$$

Thus, $d_w(x,u)+d_w(y,v) = d_w(x,v)+d_w(y,u) = d_w(x,s)+d_w(s,t)+d_w(t,u)+d_w(y,s)+d_w(s,t)+d_w(t,v) = d_w(x,y)+d_w(u,v)+2d_w(s,t)$, which shows that Buneman's inequality is satisfied.

If **p** and **q** have some vertices in common, the configuration of the graph is as shown in Fig. 14.5. In this case, we have

$$\begin{aligned}
d_w(x,y) &= d_w(x,t) + d_w(t,s) + d_w(s,y),\\
d_w(u,v) &= d_w(u,t) + d_w(t,s) + d_w(s,v),\\
d_w(x,u) &= d_w(x,t) + d_w(t,u),\\
d_w(y,v) &= d_w(y,s) + d_w(s,v),\\
d_w(x,v) &= d_w(x,t) + d_w(t,s) + d_w(s,v),\\
d_w(y,u) &= d_w(y,s) + d_w(s,t) + d_w(t,u),
\end{aligned}$$

which implies

$$d_w(x, y) + d_w(u, v) = d_w(x, t) + 2d_w(t, s) + d_w(s, y) + d_w(u, t) + d_w(s, v),$$
$$d_w(x, u) + d_w(y, v) = d_w(x, t) + d_w(t, u) + d_w(y, s) + d_w(s, v),$$
$$d_w(x, v) + d_w(y, u) = d_w(x, t) + 2d_w(t, s) + d_w(s, v) + d_w(y, s) + d_w(t, u).$$

Thus, Buneman's inequality is satisfied in this case, too, because

$$d_w(x, y) + d_w(u, v) = d_w(x, v) + d(y, u) \geqslant d_w(x, u) + d_w(y, v).$$

By Theorem 14.2, Buneman's inequality is equivalent to saying that of the three sums $d(x, y) + d(u, v)$, $d(x, u) + d(y, v)$, and $d(x, v) + d(y, u)$, two are equal and the third is no less than the two others.

Next, we examine the relationships that exist between metrics, tree metrics, and ultrametrics.

Theorem 14.26 *Every tree metric is a metric, and every ultrametric is a tree metric.*

Proof Let S be a nonempty set and let d be a tree metric on S, that is, a dissimilarity that satisfies the inequality $d(x, y)+d(u, v) \leqslant \max\{d(x, u)+d(y, v), d(x, v)+ d(y, u)\}$ for every $x, y, u, v \in S$. Choosing $v = u$, we obtain $d(x, y) \leqslant \max\{d(x, u)+ d(y, u), d(x, u) + d(y, u)\} = d(x, u) + d(u, y)$ for every x, y, u, which shows that d satisfies the triangular inequality.

Suppose now that d is an ultrametric. We need to show that

$$d(x, y) + d(u, v) \leqslant \max\{d(x, u) + d(y, v), d(x, v) + d(y, v)\}$$

for $x, y, u, v \in S$. Several cases are possible depending on which of the elements u, v is the closest to x and y, as the next table shows:

Case	Closest to x	Closest to y	Implications
1	u	u	$d(x, v) = d(u, v), d(y, v) = d(u, v)$
2	u	v	$d(x, v) = d(u, v), d(y, u) = d(u, v)$
3	v	u	$d(x, u) = d(u, v), d(y, v) = d(u, v)$
4	v	v	$d(x, u) = d(u, v), d(y, u) = d(u, v)$

We discuss here only the first two cases; the remaining cases are similar and are left to the reader.

In the first case, by Corollary 14.3, we have $d(x, u) \leqslant d(x, v) = d(u, v)$ and $d(y, u) \leqslant d(y, v) = d(u, v)$. This allows us to write

$$\begin{aligned}
&\max\{d(x, u) + d(y, v), d(x, v) + d(y, v)\}\\
&= \max\{d(x, u) + d(u, v), d(u, v) + d(y, v)\}\\
&= \max\{d(x, u), d(y, v)\} + d(u, v)
\end{aligned}$$

$\geqslant \max\{d(x, u), d(u, y)\} + d(u, v)$
(because u is closer to$y$ than v)
$\geqslant d(x, y) + d(u, v)$
(since d is an ultrametric),

which concludes the argument for the first case.

For the second case, by the same theorem mentioned above, we have $d(x, u) \leqslant d(x, v) = d(u, v)$ and $d(y, v) \leqslant d(y, u) = d(u, v)$. This implies $d(x, u) + d(y, v) \leqslant d(x, v) + d(y, v) = 2d(u, v)$. Thus, it remains to show only that $d(x, y) \leqslant d(u, v)$. Observe that we have $d(x, u) \leqslant d(u, v) = d(u, y)$. Therefore, in the triangle x, y, u, we have $d(x, y) = d(u, y) = d(u, v)$, which concludes the argument in the second case.

Theorem 14.26 implies that for every set S, $\mathcal{U}_S \subseteq \mathcal{T}_S \subseteq \mathcal{M}_S$.

As shown in [1], Buneman's inequality is also a sufficient condition for a graph to be a tree in the following sense.

Theorem 14.27 *A graph $\mathcal{G} = (V, E)$ is a tree if and only if it is connected, contains no triangles, and its graph distance satisfies Buneman's inequality.*

Proof By our previous discussions, the conditions are clearly necessary. We show here that they are sufficient.

Let $\mathbf{p}$ be a cycle of minimal length ℓ. Since $\mathcal{G}$ contains no triangles, it follows that $\ell \geqslant 4$. Therefore, ℓ can be written as $\ell = 4q+r$, where $q \geqslant 1$ and $0 \leqslant r \leqslant 3$. Since $\mathbf{p}$ is a minimal circuit, the distance between its end points is given by the least number of edges of the circuit that separate the points. Therefore, we can select vertices x, u, y, v (in this order) on the cycle such that the distances $d(x, u), d(u, y), d(y, v), d(v, x)$ are all either q or $q+1$ and $d(x, u)+d(u, y)+d(y, v)+d(v, x) = 4q+r$. Then, $2q \leqslant d(x, y) \leqslant 2q+2$ and $2q \leqslant d(u, v) \leqslant 2q+2$, so $4q \leqslant d(x, y)+d(u, v) \leqslant 4q+4$, which prevents d from satisfying the inequality $d(x, y)+d(u, v) \leqslant \max\{d(x, u)+d(y, v), d(x, v)+d(y, u)\}$. This condition shows that $\mathcal{G}$ is acyclic, so it is a tree.

In data mining applied in biology, particularly in reconstruction of phylogenies, it is important to determine the conditions that allow the construction of a weighted tree $(\mathcal{T}, w)$ starting from a metric space (S, d) such that the tree metric induced by $(\mathcal{T}, w)$ coincides with d when restricted to the set S.

Example 14.28 Let $S = \{a, b, c\}$ be a three-element set and let d be a distance defined on S. Suppose that (a, b) are the closest points in S, that is, $d(a, b) \leqslant d(a, c)$ and $d(a, b) \leqslant d(b, c)$.

We shall seek to determine a weighted tree $(\mathcal{T}, w)$ such that the restriction of the metric induced by the tree to the set S coincides with d. To this end, consider the weighted tree shown in Fig. 14.6. The distances between vertices can be expressed as $d(a, b) = m + n$, $d(a, c) = m + p + q$, and $d(b, c) = n + p + q$. It is easy to see that

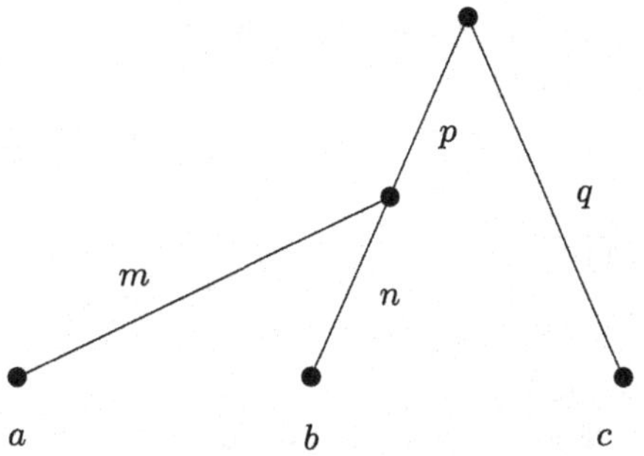

Fig. 14.6 Weighted tree

$$p + q = \frac{d(a, c) + d(b, c) - d(a, b)}{2}.$$

A substitution in the last two equalities yields

$$m = \frac{d(a, c) - d(b, c) + d(a, b)}{2} \geqslant 0, n = \frac{d(b, c) - d(a, c) + d(a, b)}{2} \geqslant 0,$$

which determines the weights of the edges that end in a and b, respectively. For the remaining two edges, one can choose p and q as two arbitrary positive numbers whose sum equals $\frac{d(a,c)+d(b,c)-d(a,b)}{2}$.

Theorem 14.29 *Starting from a tree metric d on a nonempty set S, there exists a weighted tree $(\mathcal{T}, w)$ whose set of vertices contains S and such that the metric induced by this weighted tree on S coincides with d.*

Proof The argument is by induction on $n = |S|$. The basis step, $n = 3$, is immediate.

Suppose that the statement holds for sets with fewer than n elements, and let S be a set with $|S| = n$. Define a function $f : S^3 \longrightarrow \mathbb{R}$ as $f(x, y, z) = d(x, z) + d(y, z) - d(x, y)$. Let $(p, q, r) \in S^3$ be a triple such that $f(p, q, r)$ is maximum. If $x \in S - \{p, q\}$, we have $f(x, q, r) \leqslant f(p, q, r)$ and $f(p, x, r) \leqslant f(p, q, r)$. These inequalities are easily seen to be equivalent to

$$d(x, r) + d(p, q) \leqslant d(x, q) + d(p, r) \text{ and } d(x, r) + d(p, q) \leqslant d(x, p) + d(q, r),$$

respectively. Using Buneman's inequality, we obtain

$$d(x, q) + d(p, r) = d(x, p) + d(q, r). \tag{14.4}$$

Similarly, for any other $y \in S - \{p, q\}$, we have

$$d(y, q) + d(p, r) = d(y, p) + d(q, r),$$

so $d(x, q) + d(y, p) = d(x, p) + d(y, q)$.

Consider now a new object t, $t \notin S$. The distances from t to the objects of S are defined by

$$d(t, p) = \frac{d(p,q) + d(p,r) - d(q,r)}{2},$$

and

$$d(t, x) = d(x, p) - d(t, p), \tag{14.5}$$

where $x \neq p$.

For $x \neq p$, we can write

$$\begin{aligned} d(t,x) &= d(x,p) - d(t,p) \\ &= d(x,p) - \frac{d(p,q) + d(p,r) - d(q,r)}{2} \\ &\quad \text{(by the definition of } d(t,p)\text{)} \\ &= d(x,p) - \frac{d(p,q) + d(p,x) - d(q,x)}{2} \\ &\quad \text{(by Equality (14.4))} \\ &= \frac{d(p,x) - d(p,q) + d(q,x)}{2} \geqslant 0. \end{aligned}$$

Choosing $x = q$ in Equality (14.5), we have

$$\begin{aligned} d(t,q) = d(q,p) - d(t,p) &= d(p,q) - \frac{d(p,q) + d(p,r) - d(q,r)}{2} \\ &= \frac{d(p,q) - d(p,r) + d(q,r)}{2} \geqslant 0, \end{aligned}$$

which shows that the distances of the form $d(t, \cdot)$ are all nonnegative.

Also, we can write for $x \in S - \{p, q\}$

$$\begin{aligned} d(q,t) + d(t,x) &= \frac{d(p,q) - d(p,r) + d(q,r)}{2} + \frac{d(p,x) - d(p,q) + d(q,x)}{2} \\ &= \frac{d(p,q) - d(p,x) + d(q,x)}{2} + \frac{d(p,x) - d(p,q) + d(q,x)}{2} \\ &\quad \text{(by Equality (14.4))} \\ &= d(q,x). \end{aligned}$$

It is not difficult to verify that the expansion of d to $S \cup \{t\}$ using the values defined above satisfies Buneman's inequality.

Consider the metric space $((S - \{p, q\}) \cup \{t\}, d)$ defined over a set with $n - 1$ elements. By inductive hypothesis, there exists a weighted tree $(\mathcal{T}, w)$ such that the metric induced on $(S - \{p, q\}) \cup \{t\}$ coincides with d. Adding two edges (t, p) and (t, q) having the weights $d(t, p)$ and $d(t, q)$, we obtain a tree that generates the distance d on the set S.

Fig. 14.7 An equidistant tree

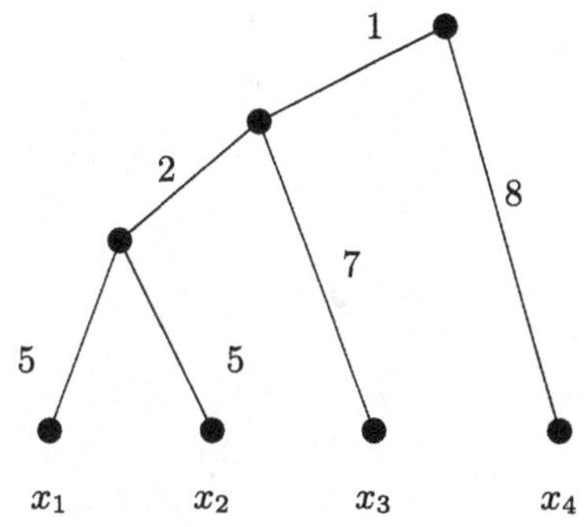

A class of weighted trees that is useful in clustering algorithms and in phylogenetics is introduced next.

Definition 14.30 *An* equidistant tree *is a triple* $(\mathcal{T}, w, v_0)$, *where* $(\mathcal{T}, v_0)$ *is a rooted tree and* w *is a weighting function defined on the set of edges of* $\mathcal{T}$ *such that* $d_w(v_0, v)$ *is the same for every leaf of the rooted tree* $(\mathcal{T}, v_0)$.

In an equidistant tree $(\mathcal{T}, w; v_0)$, for every vertex u there is a number k such that $d_w(u, t) = k$ for every leaf that is joined to v_0 by a path that passes through u. In other words, the equidistant property is inherited by subtrees.

Example 14.31 The tree shown in Fig. 14.7 is an equidistant tree. The distance d_w from the root to each of its four leaves is equal to 8.

Theorem 14.32 *A function* $d : S \times S \longrightarrow \mathbb{R}$ *is an ultrametric if and only if there exists an equidistant tree* $(\mathcal{T}, w; v_0)$ *having* S *as its set of leaves and* d *is the restriction of the tree distance* d_w *to* S.

Proof To prove that the condition is necessary, let $(\mathcal{T}, w; v_0)$ be an equidistant tree and let $x, y, z \in L$ be three leaves of the tree. Suppose that u is the common ancestor of x and y located on the shortest path that joins x to y. Then, $d_w(x, y) = 2d_w(u, x) = 2d_w(u, y)$.

Let v be the common ancestor of y and z located on the shortest path that joins y to z. Since both u and v are ancestors of y, they are located on the path that joins v_0 to y. Two cases may occur:

`Case 1` occurs when $d_w(v_0, v) \leqslant d_w(v_0, u)$ (Fig. 14.8a).

`Case 2` occurs when $d_w(v_0, v) > d_w(v_0, u)$ (Fig. 14.8b).

In the first case, we have $d(u, x) = d(u, y)$ and $d(v, z) = d(v, u) + d(u, x) = d(v, u) + d(u, y)$ because $(\mathcal{T}, w, v_0)$ is equidistant. Therefore, $d_w(x, y) = 2d_w(x, u)$, $d_w(y, z) = 2d_w(u, v) + 2d_w(u, y)$, and $d_w(x, z) = 2d_w(u, v) + 2d_w(u, x)$. Since $d_w(u, y) = d_w(u, x)$, the ultrametric inequality $d_w(x, y) \leqslant \max\{d_w(x, z), d_w(z, y)\}$ follows immediately. The second case is similar and is left to the reader.

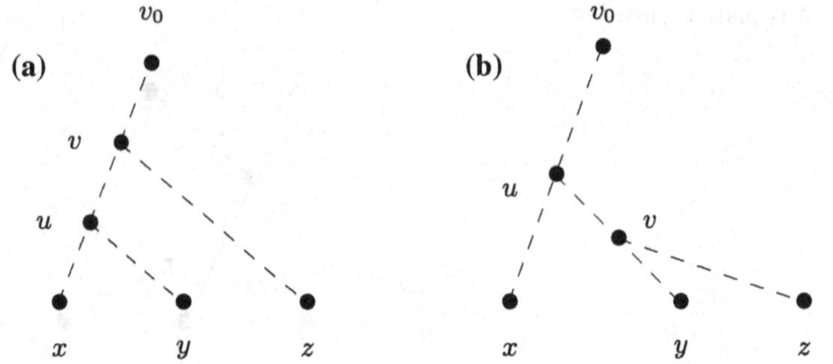

Fig. 14.8 Two equidistant trees

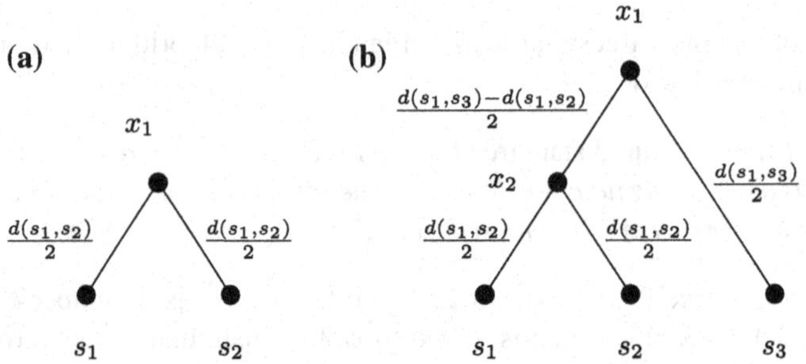

Fig. 14.9 Small equidistant weighted trees

Conversely, let $d : S \times S \longrightarrow \mathbb{R}$ be an ultrametric, where $S = \{s_1, \ldots, s_n\}$. We prove by induction of $n = |S|$ that an equidistant tree can be constructed that satisfies the requirements of the theorem.

For $n = 2$, the simple tree shown in Fig. 14.9a, where $w(x_0, s_1) = w(x_0, s_2) = \frac{d(s_1,s_2)}{2}$ satisfies the requirements of the theorem. For $n = 3$, suppose that $d(s_1, s_2) \leqslant d(s_1, s_3) = d(s_2, s_3)$. The tree shown in Fig. 14.9b is the desired tree for the ultrametric because $d(s_1, s_3) - d(s_1, s_2) \geqslant 0$.

Suppose now that $n \geqslant 4$. Let s_i, s_j be a pair of elements of S such that the distance $d(s_i, s_j)$ is minimal. By the ultrametric property, we have $d(s_k, s_i) = d(s_k, s_j) \geqslant d(s_i, s_j)$ for every $k \in \{1, \ldots, n\} - \{i, j\}$.

Define $S' = S - \{s_i, s_j\} \cup \{s\}$, and let $d' : S' \times S' \longrightarrow \mathbb{R}$ be the mapping given by $d'(s_k, s_l) = d(s_k, s_l)$ if $s_k, s_l \in S$, and $d'(s_k, s) = d(s_k, s_i) = d(s_k, s_j)$. It is easy to see that d' is an ultrametric on the smaller set S', so, by inductive hypothesis, there exists an equidistant weighted tree $(\mathcal{T}', w'; v_0)$ that induces d' on the set of its leaves S'.

Let z be the direct ancestor of s in the tree $\mathcal{T}'$ and let s_m be a neighbor of s. The weighted rooted tree $(\mathcal{T}, w; v_0)$ is obtained from $(\mathcal{T}', w'; v_0)$ by transforming s into an interior node that has the leaves s_i and s_j as immediate descendants, as shown in Fig. 14.10. To make the new tree $\mathcal{T}$ be equidistant, we keep all weights of the edges of $\mathcal{T}'$ in the new tree $\mathcal{T}$ except the weight of the edge (z, s), which is defined now as

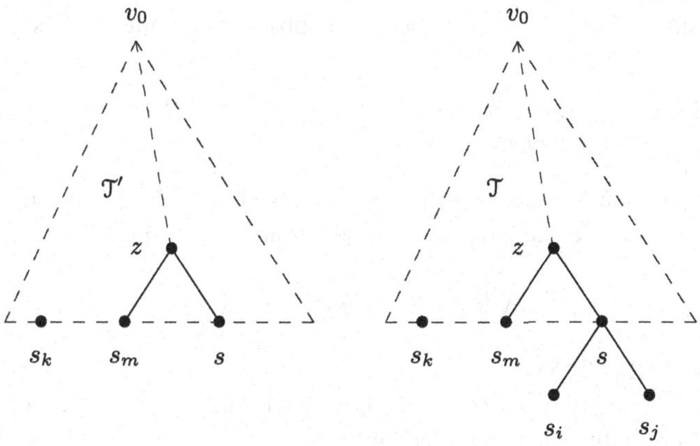

Fig. 14.10 Constructing $(\mathcal{T}, w; v_0)$ starting from $(\mathcal{T}', w'; v_0)$

$$w(z, s) = \frac{d(s_m, s_i) - d(s_i, s_j)}{2}.$$

We also define the weight of the edges (s, s_i) and (s, s_j) as

$$w(s, s_i) = w(s, s_j) = \frac{d(s_i, s_j)}{2}.$$

These definitions imply that $\mathcal{T}$ is an equidistant tree because

$$\begin{aligned} d'(s_m, z) &= \frac{d(s_m, s_i)}{2} \\ &\quad \text{(because of the definition of } d\text{)} \\ &= d'(z, s) \\ &\quad \text{(since } \mathcal{T}' \text{ is equidistant)}, \end{aligned}$$

$$d(z, s_i) = w(z, s) + w(s, s_i) = \frac{d(s_m, s_i) - d(s_i, s_j)}{2} + \frac{d(s_i, s_j)}{2} = \frac{d(s_m, s_i)}{2}.$$

and $d(z, z_m) = d'(z, s_m)$.

14.4 Metrics on Collections of Sets

Dissimilarities between subsets of finite sets have an intrinsic interest for data mining, where comparisons between sets of objects are frequent. Also, metrics defined on subsets can be transferred to metrics between binary sequences using the characteristic sequences of the subsets and thus become an instrument for studying binary data.

A very simple metric on $\mathcal{P}(S)$, the set of subsets of a finite set S is given in the next theorem.

Theorem 14.33 *Let S be a finite set. The mapping $\delta : (\mathcal{P}(S))^2 \longrightarrow \mathbb{R}_{\geqslant 0}$ defined by $\delta(X, Y) = |X \oplus Y|$ is a metric on $\mathcal{P}(S)$.*

Proof The function δ is clearly symmetric and we have $\delta(X, Y) = 0$ if and only if $X = Y$. Therefore, we need to prove only the triangular inequality

$$|X \oplus Y| \leqslant |X \oplus Z| + |Z \oplus Y|$$

for every $X, Y, Z \in \mathcal{P}(S)$.

Since $X \oplus Y = (X \oplus Z) \oplus (Z \oplus Y)$, we have $|X \oplus Y| \leqslant |X \oplus Z| + |Z \oplus Y|$, which is precisely the triangular inequality for δ.

For $U, V \in \mathcal{P}(S)$, we have $0 \leqslant \delta(U, V) \leqslant |S|$, where $\delta(U, V) = |S|$ if and only if $V = S - U$.

Lemma 14.34 *Let $d : S \times S \longrightarrow \mathbb{R}_{\geqslant 0}$ be a metric and let $u \in S$ be an element of the set S. Define the* Steinhaus transform *of d as the mapping $d_u : S \times S \longrightarrow \mathbb{R}_{\geqslant 0}$ given by*

$$d_u(x, y) = \begin{cases} 0 & \text{if } x = y = u \\ \frac{d(x,y)}{d(x,y)+d(x,u)+d(u,y)} & \text{otherwise.} \end{cases}$$

Then, d_u is a metric on S.

Proof It is easy to see that d_u is symmetric and, further, that $d_u(x, y) = 0$ if and only if $x = y$.

To prove the triangular inequality, observe that $a \leqslant a'$ implies

$$\frac{a}{a+k} \leqslant \frac{a'}{a'+k}, \tag{14.6}$$

which holds for every positive numbers a, a', k. Then, we have

$$\begin{aligned} d_u(x, y) &= \frac{d(x, y)}{d(x, y) + d(x, u) + d(u, y)} \\ &\leqslant \frac{d(x, z) + d(z, y)}{d(x, z) + d(z, y) + d(x, u) + d(u, y)} \\ &\quad \text{(by Inequality (14.6))} \\ &= \frac{d(x, z)}{d(x, z) + d(z, y) + d(x, u) + d(u, y)} \\ &\quad + \frac{d(z, y)}{d(x, z) + d(z, y) + d(x, u) + d(u, y)} \\ &\leqslant \frac{d(x, z)}{d(x, z) + d(z, y) + d(z, u)} + \frac{d(z, y)}{d(z, y) + d(z, u) + d(u, y)} \end{aligned}$$

$$= d_u(x, z) + d_u(z, y),$$

which is the desired triangular inequality.

Theorem 14.35 *Let S be a finite set. The function $d : \mathcal{P}(S)^2 \longrightarrow \mathbb{R}_{\geqslant 0}$ defined by $d(X, Y) = \frac{|X \oplus Y|}{|X \cup Y|}$ for $X, Y \in \mathcal{P}(S)$ is a metric on $\mathcal{P}(S)$.*

Proof It is clear that d is symmetric and that $d(X, Y) = 0$ if and only if $X = Y$. So, we need to prove only the triangular inequality. The mapping δ defined by $\delta(X, Y) = |X \oplus Y|$ is a metric on $\mathcal{P}(X)$, as we proved in Theorem 14.33. By Lemma 14.34, the mapping $\delta_\emptyset$ is also a metric on $\mathcal{P}(S)$. We have

$$\delta_\emptyset(X, Y) = \frac{|X \oplus Y|}{|X \oplus Y| + |X \oplus \emptyset| + |\emptyset \oplus Y|}.$$

Since $X \oplus \emptyset = X, \emptyset \oplus Y = Y$, we have

$$|X \oplus Y| + |X \oplus \emptyset| + |\emptyset \oplus Y| = |X \oplus Y| + |X| + |Y| = 2|X \cup Y|,$$

which means that $2\delta_\emptyset(X, Y) = d(X, Y)$ for every $X, Y \in \mathcal{P}(S)$. This implies that d is indeed a metric.

Theorem 14.36 *Let S be a finite set. The function $d : \mathcal{P}(S)^2 \longrightarrow \mathbb{R}_{\geqslant 0}$ defined by $d(X, Y) = \frac{|X \oplus Y|}{|S| - |X \cap Y|}$ for $X, Y \in \mathcal{P}(S)$ is a metric on $\mathcal{P}(S)$.*

Proof We only prove that d satisfies the triangular axiom. The argument begins, as in Theorem 14.35, with the metric δ. Again, by Lemma 14.34, the mapping δ_S is also a metric on $\mathcal{P}(S)$. We have

$$\begin{aligned}
\delta_S(X, Y) &= \frac{|X \oplus Y|}{|X \oplus Y| + |X \oplus S| + |S \oplus Y|} \\
&= \frac{|X \oplus Y|}{|X \oplus Y| + |S - X| + |S - Y|} \\
&= \frac{|X \oplus Y|}{|X \oplus Y| + |S - X| + |S - Y|} \\
&= \frac{|X \oplus Y|}{2(|S| - |X \cap Y|)}
\end{aligned}$$

because $|X \oplus Y| + |S - X| + |S - Y| = 2(|S| - |X \cap Y|)$, as the reader can easily verify. Therefore, $d(X, Y) = 2\delta_S(X, Y)$, which proves that d is indeed a metric.

A general mechanism for defining a metric on $\mathcal{P}(S)$, where S is a finite set, $|S| = n$, can be introduced starting with two functions:

1. a *weight function* $w : S \longrightarrow \mathbb{R}_{\geqslant 0}$ such that $\sum\{w(x) \mid x \in S\} = 1$ and
2. an injective function $\varphi : \mathcal{P}(S) \longrightarrow (S \longrightarrow \mathbb{R})$.

The metric defined by the pair (w, φ) is the function $d_{w,\varphi} : \mathcal{P}(S)^2 \longrightarrow \mathbb{R}_{\geqslant 0}$ defined by

$$d_{w,\varphi}(X, Y) = \left(\sum_{s \in S} w(s) |\varphi(X)(s) - \varphi(Y)(s)|^q \right)^{\frac{1}{q}}$$

for $X, Y \in \mathcal{P}(S)$.

The function w is extended to $\mathcal{P}(S)$ by

$$w(T) = \sum \{ w(x) \mid x \in T \}.$$

Clearly, $w(\emptyset) = 0$ and $w(S) = 1$. Also, if P and Q are two disjoint subsets, we have $w(P \cup Q) = w(P) + w(Q)$.

We refer to both w and its extension to $\mathcal{P}(S)$ as *weight functions*.

The value $\varphi(T)$ of the function φ is itself a function $\varphi(T) : S \longrightarrow \mathbb{R}$, and each subset T of S defines such a distinct function. These notions are used in the next theorem.

Theorem 14.37 *Let S be a set, $w : S \longrightarrow \mathbb{R}_{\geqslant 0}$ be a weight function, and $\varphi : \mathcal{P}(S) \longrightarrow (S \longrightarrow \mathbb{R})$ be an injective function.*

If $w(x) > 0$ for every $x \in S$, then the mapping $d_{w,\varphi} : (\mathcal{P}(S))^2 \longrightarrow \mathbb{R}$ defined by

$$d_{w,\varphi}(U, V) = \left(\sum_{x \in S} w(x) |\varphi(U)(x) - \varphi(V)(x)|^p \right)^{\frac{1}{p}} \tag{14.7}$$

for $U, V \in \mathcal{P}(S)$ is a metric on $\mathcal{P}(S)$.

Proof It is clear that $d_{w,\varphi}(U, U) = 0$. If $d_{w,\varphi}(U, V) = 0$, then $\varphi(U)(x) = \varphi(V)(x)$ because $w(x) > 0$, for every $x \in S$. Thus, $\varphi(U) = \varphi(V)$, which implies $U = V$ due to the injectivity of φ.

The symmetry of $d_{w,\varphi}$ is immediate.

To prove the triangular inequality, we apply Minkowski's inequality. Suppose that $S = \{x_0, \ldots, x_{n-1}\}$, and let $U, V, W \in \mathcal{P}(S)$. Define the numbers

$$a_i = (w(x_i))^{\frac{1}{p}} \varphi_U(x_i), b_i = (w(x_i))^{\frac{1}{p}} \varphi_V(x_i), c_i = (w(x_i))^{\frac{1}{p}} \varphi_W(x_i),$$

for $0 \leqslant i \leqslant n - 1$. Then, by Minkowski's inequality, we have

$$\left(\sum_{i=0}^{n-1} |a_i - b_i|^p \right)^{\frac{1}{p}} \leqslant \left(\sum_{i=0}^{n-1} |a_i - c_i|^p \right)^{\frac{1}{p}} + \left(\sum_{i=0}^{n-1} |c_i - b_i|^p \right)^{\frac{1}{p}},$$

which amounts to the triangular inequality $d_{w,\varphi}(U, V) \leqslant d_{w,\varphi}(U, W) + d_{w,\varphi}(W, V)$. Thus, we may conclude that $d_{w,\varphi}$ is indeed a metric on $\mathcal{P}(S)$.

Example 14.38 Let $w : S \longrightarrow [0, 1]$ be a positive weight function. Define the function φ by

$$\varphi(U)(x) = \begin{cases} \frac{1}{\sqrt{w(U)}} & \text{if } x \in U, \\ 0 & \text{otherwise.} \end{cases}$$

It is easy to see that $\varphi(U) = \varphi(V)$ if and only if $U = V$, so φ is an injective function.

Choosing $p = 2$, the metric defined in Theorem 14.37 becomes

$$d^2_{w,\varphi}(U, V) = \left(\sum_{x \in S} w(x) |\varphi(U)(x) - \varphi(V)(x)|^2 \right)^{\frac{1}{2}}.$$

Suppose initially that neither U nor V are empty. Several cases need to be considered:

1. If $x \in U \cap V$, then

$$|\varphi(U)(x) - \varphi(V)(x)|^2 = \frac{1}{w(U)} + \frac{1}{w(V)} - \frac{2}{\sqrt{w(U)w(V)}}.$$

The total contribution of these elements of S is

$$w(U \cap V) \left(\frac{1}{w(U)} + \frac{1}{w(V)} - \frac{2}{\sqrt{w(U)w(V)}} \right).$$

If $x \in U - V$, then

$$|\varphi(U)(x) - \varphi(V)(x)|^2 = \frac{1}{w(U)}$$

and the total contribution is $w(U - V)\frac{1}{w(U)}$.

2. When $x \in V - U$, then

$$|\varphi(U)(x) - \varphi(V)(x)|^2 = \frac{1}{w(V)}$$

and the total contribution is $w(V - U)\frac{1}{w(V)}$.

3. Finally, if $x \notin U \cup V$, then $|\varphi(U)(x) - \varphi(V)(x)|^2 = 0$.

Thus, we can write

$$\begin{aligned} d^2_{w,\varphi}(U, V) &= w(U \cap V) \left(\frac{1}{w(U)} + \frac{1}{w(V)} - \frac{2}{\sqrt{w(U)w(V)}} \right) \\ &\quad + w(U - V)\frac{1}{w(U)} + w(V - U)\frac{1}{w(V)} \\ &= \frac{w(U \cap V) + w(U - V)}{w(U)} + \frac{w(U \cap V) + w(V - U)}{w(V)} \end{aligned}$$

$$
\begin{aligned}
&- \frac{2w(U \cap V)}{\sqrt{w(U)w(V)}} \\
&= 2\left(1 - \frac{w(U \cap V)}{\sqrt{w(U)w(V)}}\right),
\end{aligned}
$$

where we used the fact that $w(U \cap V) + w(U - V) = w(U)$ and $w(U \cap V) + w(V - U) = w(V)$. Thus,

$$
d_{w,\varphi}(U, V) = \sqrt{2\left(1 - \frac{w(U \cap V)}{\sqrt{w(U)w(V)}}\right)}.
$$

If $U \neq \emptyset$ and $V = \emptyset$, then it is immediate that $d_{w,\varphi}(U, \emptyset) = 1$. Of course, $d_{w,\varphi}(\emptyset, \emptyset) = 0$.

Thus, the mapping $d_{w,\varphi}$ defined by

$$
d_{w,\varphi}(U, V) = \begin{cases} 0 & \text{if } U = V = \emptyset, \\ 1 & \text{if } U \neq \emptyset \text{ and } V = \emptyset, \\ 1 & \text{if } U = \emptyset \text{ and } V \neq \emptyset, \\ \sqrt{2\left(1 - \frac{w(U \cap V)}{\sqrt{w(U)w(V)}}\right)} & \text{if } U \neq \emptyset \text{ and } V \neq \emptyset, \end{cases}
$$

for $U, V \in \mathcal{P}(S)$ is a metric, which is known as the *Ochïai metric* on $\mathcal{P}(S)$.

Example 14.39 Using the same notation as in Example 14.38 for a positive weight function $w : S \longrightarrow [0, 1]$, define the function φ by

$$
\varphi(U)(x) = \begin{cases} \frac{1}{w(U)} & \text{if } x \in U, \\ 0 & \text{otherwise.} \end{cases}
$$

It is easy to see that φ is an injective function.

Suppose that $p = 2$ in Equality (14.7). If $U \neq \emptyset$ and $V \neq \emptyset$, we have the following cases:

1. If $x \in U \cap V$, then

$$
|\varphi(U)(x) - \varphi(V)(x)|^2 = \frac{1}{w(U)^2} + \frac{1}{w(V)^2} - \frac{2}{w(U)w(V)}.
$$

   The total contribution of these elements of S is

$$
w(U \cap V)\left(\frac{1}{w(U)^2} + \frac{1}{w(V)^2} - \frac{2}{w(U)w(V)}\right).
$$

If $x \in U - V$, then

$$|\varphi(U)(x) - \varphi(V)(x)|^2 = \frac{1}{w(U)^2}$$

and the total contribution is $w(U - V)\frac{1}{w(U)^2}$.

2. When $x \in V - U$, then

$$|\varphi(U)(x) - \varphi(V)(x)|^2 = \frac{1}{w(V)^2}$$

and the total contribution is $w(V - U)\frac{1}{w(V)^2}$.

3. Finally, if $x \notin U \cup V$, then $|\varphi(U)(x) - \varphi(V)(x)|^2 = 0$.

Summing up these contributions, we can write

$$\begin{aligned} d^2_{w,\varphi}(U, V) &= \frac{1}{w(U)} + \frac{1}{w(V)} - 2\frac{w(U \cap V)}{w(U)w(V)} \\ &= \frac{w(U) + w(V) - 2w(U \cap V)}{w(U)w(V)} \\ &= \frac{w(U \oplus V)}{w(U)w(V)}. \end{aligned}$$

If $V = \emptyset$, $d_{w,\varphi}(U, \emptyset) = \sqrt{\frac{1}{w(U)}}$; similarly, $d_{w,\varphi}(\emptyset, V) = \sqrt{\frac{1}{w(V)}}$. We proved that the mapping $d_{w,\varphi}$ defined by

$$d_{w,\varphi}(U, V) = \begin{cases} \sqrt{\frac{w(U \oplus V)}{w(U)w(V)}} & \text{if } U \neq \emptyset \text{ and } V \neq \emptyset, \\ \sqrt{\frac{1}{w(U)}} & \text{if } U \neq \emptyset \text{ and } V = \emptyset, \\ \sqrt{\frac{1}{w(V)}} & \text{if } U = \emptyset \text{ and } V \neq \emptyset, \\ 0 & \text{if } U = V = \emptyset, \end{cases}$$

for $U, V \in \mathcal{P}(S)$, is a metric on $\mathcal{P}(S)$ known as the χ^2 *metric*.

14.5 Metrics on Partitions

Metrics on sets of partitions of finite sets are useful in data mining because attributes induce partitions on the sets of tuples of tabular data. Thus, they help us determine interesting relationships between attributes and to use these relationships for classification, feature selection, and other applications. Also, exclusive clusterings can be regarded as partitions of the set of clustered objects, and partition metrics can be used for evaluating clusterings, a point of view presented in [2].

Let S be a finite set and let π and σ be two partitions of S. The equivalence relations ρ_π and ρ_σ are subsets of $S \times S$ that allow us a simple way of defining a metric between π and σ as the relative size of the symmetric difference between the sets of pairs ρ_π and ρ_σ,

$$\delta(\pi, \sigma) = \frac{1}{|S|^2}|\rho_\pi \oplus \rho_\sigma| = \frac{1}{|S|^2|}\left(|\rho_\pi| + |\rho_\sigma| - 2|\rho_\pi \cap \rho_\sigma|\right). \tag{14.8}$$

If $\pi = \{B_1, \ldots, B_m\}$ and $\sigma = \{C_1, \ldots, C_n\}$, then there are $\sum_{i=1}^{m} |B_i|^2$ pairs in ρ_π, $\sum_{j=1}^{n} |C_j|^2$ pairs in ρ_σ, and $\sum_{i=1}^{n} \sum_{j=1}^{n} |B_i \cap C_j|^2$ pairs in $\rho_\pi \cap \rho_\sigma$. Thus, we have

$$\delta(\pi, \sigma) = \frac{1}{|S|^2}\left(\sum_{i=1}^{m} |B_i|^2 + \sum_{j=1}^{n} |C_j|^2 - 2\sum_{i=1}^{n}\sum_{j=1}^{n} |B_i \cap C_j|^2\right). \tag{14.9}$$

The same metric can be linked to a special case of a more general metric related to the notion of partition entropy.

We can now show a central result.

Theorem 14.40 *For every $\beta \geqslant 1$ the mapping $d_\beta : PART(S)^2 \longrightarrow \mathbb{R}_{\geqslant 0}$ defined by*

$$d_\beta(\pi, \sigma) = \mathcal{H}_\beta(\pi|\sigma) + \mathcal{H}_\beta(\sigma|\pi)$$

for $\pi, \sigma \in PART(S)$ is a metric on $PART(S)$.

Proof A double application of Corollary 12.64 yields

$$\mathcal{H}_\beta(\pi|\sigma) + \mathcal{H}_\beta(\sigma|\tau) \geqslant \mathcal{H}_\beta(\pi|\tau),$$
$$\mathcal{H}_\beta(\sigma|\pi) + \mathcal{H}_\beta(\tau|\sigma) \geqslant \mathcal{H}_\beta(\tau|\pi).$$

Adding these inequalities gives

$$d_\beta(\pi, \sigma) + d_\beta(\sigma, \tau) \geqslant d_\beta(\pi, \tau),$$

which is the triangular inequality for d_β.

The symmetry of d_β is obvious, and it is clear that $d_\beta(\pi, \pi) = 0$ for every $\pi \in PART(S)$.

Suppose now that $d_\beta(\pi, \sigma) = 0$. Since the values of β-conditional entropies are nonnegative, this implies $\mathcal{H}_\beta(\pi|\sigma) = \mathcal{H}_\beta(\sigma|\pi) = 0$. By Theorem 12.54, we have both $\sigma \leqslant \pi$ and $\pi \leqslant \sigma$, so $\pi = \sigma$. Thus, d_β is a metric on $PART(S)$.

An explicit expression of the metric between two partitions can now be obtained using the values of conditional entropies given by Equality (12.5),

$$d_\beta(\pi, \sigma) = \frac{1}{(1 - 2^{1-\beta})|S|^\beta} \left(\sum_{i=1}^{m} |B_i|^\beta + \sum_{j=1}^{n} |C_j|^\beta - 2 \cdot \sum_{i=1}^{m} \sum_{j=1}^{n} |B_i \cap C_j|^\beta \right),$$

where $\pi = \{B_1, \ldots, B_m\}$ and $\sigma = \{C_1, \ldots, C_n\}$ are two partitions from $PART(S)$.

In the special case $\beta = 2$, we have

$$d_2(\pi, \sigma) = \frac{2}{|S|^2} \left(\sum_{i=1}^{m} |B_i|^2 + \sum_{j=1}^{n} |C_j|^2 - \sum_{i=1}^{m} \sum_{j=1}^{n} 2|B_i \cap C_j|^2 \right),$$

which implies $d_2(\pi, \sigma) = 2\delta(\pi, \sigma)$, where δ is the distance introduced by using the symmetric difference in Equality (14.9).

It is clear that $d_\beta(\pi, \omega_S) = \mathcal{H}_\beta(\pi)$ and $d_\beta(\pi, \alpha_S) = \mathcal{H}(\alpha_S|\pi)$. Another useful form of d_β can be obtained by applying Theorem 12.55. Since $\mathcal{H}_\beta(\pi|\sigma) = \mathcal{H}_\beta(\pi \wedge \sigma) - \mathcal{H}_\beta(\sigma)$ and $\mathcal{H}_\beta(\sigma|\pi) = \mathcal{H}_\beta(\pi \wedge \sigma) - \mathcal{H}_\beta(\sigma)$, we have

$$d_\beta(\pi, \sigma) = 2\mathcal{H}_\beta(\pi \wedge \sigma) - \mathcal{H}_\beta(\pi) - \mathcal{H}_\beta(\sigma), \tag{14.10}$$

for $\pi, \sigma \in PART(S)$.

The behavior of the metric d_β with respect to partition addition is discussed in the next statement.

Theorem 14.41 *Let S be a finite set and π and θ be two partitions of S, where $\theta = \{D_1, \ldots, D_h\}$. If $\sigma_i \in PART(D_i)$ for $1 \leqslant i \leqslant h$, then we have $\sigma_1 + \cdots + \sigma_h \leqslant \theta$ and*

$$d_\beta(\pi, \sigma_1 + \cdots + \sigma_h) = \sum_{i=1}^{h} \left(\frac{|D_i|}{|S|} \right)^\beta d_\beta(\pi_{D_i}, \sigma_i) + \mathcal{H}_\beta(\theta|\pi).$$

Proof This statement follows directly from Theorem 12.62.

Theorem 14.42 *Let σ and θ be two partitions in $PART(S)$ such that*

$$\theta = \{D_1, \ldots, D_h\}$$

and $\sigma \leqslant \theta$. Then, we have

$$d_\beta(\theta, \sigma) = \sum_{i=1}^{h} \left(\frac{|D_i|}{|S|} \right)^\beta d_\beta(\omega_{D_i}, \sigma_{D_i}).$$

Proof In Theorem 14.41, take $\pi = \theta$ and $\sigma_i = \sigma_{D_i}$ for $1 \leqslant i \leqslant h$. Then, it is clear that $\sigma = \sigma_1 + \cdots + \sigma_h$ and we have

$$d_\beta(\theta, \sigma) = \sum_{i=1}^{h} \left(\frac{|D_i|}{|S|}\right)^\beta d_\beta(\omega_{D_i}, \sigma_{D_i})$$

because $\theta_{D_i} = \omega_{D_i}$ for $1 \leqslant i \leqslant h$.

The next theorem generalizes a result from [2].

Theorem 14.43 *In the metric space* $(PART(S), d_\beta)$, *we have that*

(i) *if* $\sigma \leqslant \pi$, *then* $d_\beta(\pi, \sigma) = \mathcal{H}_\beta(\sigma) - \mathcal{H}_\beta(\pi)$,
(ii) $d_\beta(\alpha_S, \sigma) + d_\beta(\sigma, \omega_S) = d_\beta(\alpha_S, \omega_S)$, *and*
(iii) $d_\beta(\pi, \pi \wedge \sigma) + d_\beta(\pi \wedge \sigma, \sigma) = d_\beta(\pi, \sigma)$

for all partitions $\pi, \sigma \in PART(S)$.

Furthermore, we have $d_\beta(\omega_T, \alpha_T) = \frac{1-|T|^{1-\beta}}{1-2^{1-\beta}}$ *for every subset* T *of* S.

Proof The first three statements of the theorem follow immediately from Equality (14.10); the last part is an application of the definition of d_β.

A generalization of a result obtained in [2] is contained in the next statement, which gives an axiomatization of the metric d_β.

Theorem 14.44 *Let* $d : PART(S)^2 \longrightarrow \mathbb{R}_{\geqslant 0}$ *be a function that satisfies the following conditions:*

(D1) d *is symmetric; that is,* $d(\pi, \sigma) = d(\sigma, \pi)$.
(D2) $d(\alpha_S, \sigma) + d(\sigma, \omega_S) = d(\alpha_S, \omega_S)$.
(D3) $d(\pi, \sigma) = d(\pi, \pi \wedge \sigma) + d(\pi \wedge \sigma, \sigma)$.
(D4) *if* $\sigma, \theta \in PART(S)$ *such that* $\theta = \{D_1, \ldots, D_h\}$ *and* $\sigma \leqslant \theta$, *then we have*

$$d(\theta, \sigma) = \sum_{i=1}^{h} \left(\frac{|D_i|}{|S|}\right)^\beta d(\omega_{D_i}, \sigma_{D_i}).$$

(D5) $d(\omega_T, \alpha_T) = \frac{1-|T|^{1-\beta}}{1-2^{1-\beta}}$ *for every* $T \subseteq S$.

Then, $d = d_\beta$.

Proof Choosing $\sigma = \alpha_S$ in axiom (D4) and using (D5), we can write

$$\begin{aligned} d(\alpha_S, \theta) &= \sum_{i=1}^{h} \left(\frac{|D_i|}{|S|}\right)^\beta d(\omega_{D_i}, \alpha_{D_i}) \\ &= \sum_{i=1}^{h} \left(\frac{|D_i|}{|S|}\right)^\beta \frac{1 - |D_i|^{1-\beta}}{1 - 2^{1-\beta}} = \frac{\sum_{i=1}^{h} |D_i|^\beta - |S|}{(1 - 2^{1-\beta})|S|^\beta}. \end{aligned}$$

From Axioms (D2) and (D5) it follows that

$$d(\theta, \omega_S) = d(\alpha_S, \omega_S) - d(\alpha_S, \theta)$$
$$= \frac{1 - |S|^{1-\beta}}{1 - 2^{1-\beta}} - \frac{\sum_{i=1}^{h} |D_i|^\beta - |S|}{(1 - 2^{1-\beta})|S|^\beta} = \frac{|S|^\beta - \sum_{i=1}^{h} |D_i|^\beta}{(1 - 2^{1-\beta})|S|^\beta}.$$

Now let $\pi, \sigma \in PART(S)$, where $\pi = \{B_1, \ldots, B_m\}$ and $\sigma = \{C_1, \ldots, C_n\}$. Since $\pi \wedge \sigma \leqslant \pi$ and $\sigma_{B_i} = \{C_1 \cap B_i, \ldots, C_n \cap B_i\}$, an application of Axiom (D4) yields:

$$d(\pi, \pi \wedge \sigma) = \sum_{i=1}^{m} \left(\frac{|B_i|}{|S|}\right)^\beta d(\omega_{B_i}, (\pi \wedge \sigma)_{B_i})$$
$$= \sum_{i=1}^{m} \left(\frac{|B_i|}{|S|}\right)^\beta d(\omega_{B_i}, \sigma_{B_i})$$
$$= \sum_{i=1}^{m} \left(\frac{|B_i|}{|S|}\right)^\beta \frac{|B_i|^\beta - \sum_{j=1}^{n} |B_i \cap C_j|^\beta}{(1 - 2^{1-\beta})|B_i|^\beta}$$
$$= \frac{1}{(1 - 2^{1-\beta})|S|^\beta} \left(\sum_{i=1}^{m} |B_i|^\beta - \sum_{j=1}^{n} \sum_{i=1}^{n} |B_i \cap C_j|^\beta\right)$$

because $(\pi \wedge \sigma)_{B_i} = \sigma_{B_i}$.

By Axiom (D1), we obtain the similar equality

$$d(\pi \wedge \sigma, \sigma) = \frac{1}{(1 - 2^{1-\beta})|S|^\beta} \left(\sum_{i=1}^{m} |B_i|^\beta - \sum_{j=1}^{n} \sum_{i=1}^{n} |B_i \cap C_j|^\beta\right),$$

which, by Axiom (D3), implies:

$$d(\pi, \sigma) = \frac{1}{(1 - 2^{1-\beta})|S|^\beta} \left(\sum_{i=1}^{m} |B_i|^\beta + \sum_{j=1}^{n} |C_j|^\beta - 2\sum_{j=1}^{n} \sum_{i=1}^{n} |B_i \cap C_j|^\beta\right);$$

that is, $d(\pi, \sigma) = d_\beta(\pi, \sigma)$.

14.6 Metrics on Sequences

Sequences are the objects of many data mining activities (text mining, biological applications) that require evaluation of the degree to which they are different from each other.

The Hamming distance introduced for sequences in Example 1.192 is not very useful due to its inability to measure anything but the degree of coincidence between symbols that occur in similar position. A much more useful tool is Levenshtein's distance, introduced in [3], using certain operations on sequences.

Recall that we introduced the notion of replacement of an occurrence $(\mathbf{y}, i)$ in a sequence $\mathbf{x}$ in Definition 1.91 on page 25.

Definition 14.45 *Let S be a set and let $\boldsymbol{x} \in \mathbf{Seq}(S)$. The* insertion of $s \in S$ in $\boldsymbol{x}$ at position i *yields the sequence* $\mathsf{i}_{s,i}(\boldsymbol{x}) = \texttt{replace}(\boldsymbol{x}, (\boldsymbol{\lambda}, i), s)$*, where* $0 \leqslant i \leqslant |\boldsymbol{x}|$.

The deletion of the symbol located at position i *yields the sequence* $\mathsf{d}_i(\boldsymbol{x}) = \texttt{replace}(\boldsymbol{x}, (\boldsymbol{x}(i), i), \boldsymbol{\lambda})$*, where* $0 \leqslant i \leqslant |\boldsymbol{x}| - 1$.

The substitution of $s \in S$ at position i by s' *produces the sequence* $\mathsf{s}_{s,i,s'}(\boldsymbol{x}) = \texttt{replace}(\boldsymbol{x}, (s, i), s')$*, where* $0 \leqslant i \leqslant |\boldsymbol{x}| - 1$.

In Definition 14.45, we introduced three types of partial functions on the set of sequences $\mathbf{Seq}(S)$, $\mathsf{i}_{s,i}$, d_i, and $\mathsf{s}_{s,i,s'}$, called *insertion*, *deletion*, and *substitution*, respectively. There partial functions are collectively referred to as *editing functions*. Observe that, in order to have $\mathbf{x} \in \text{Dom}(\mathsf{d}_i)$, we must have $|\mathbf{x}| \geqslant i$.

Definition 14.46 *An* edit transcript *is a sequence* $(f_0, f_1, \ldots, f_{k-1})$ *of edit operations.*

Example 14.47 Let S be the set of small letters of the Latin alphabet, $S = \{a, b, \ldots, z\}$, and let $\mathbf{x} = (m, i, c, k, e, y)$, $\mathbf{y} = (m, o, u, s, e)$. The following sequence of operations transforms $\mathbf{x}$ into $\mathbf{y}$:

Step	Sequence	Operation
0	(m, i, c, k, e, y)	$\mathsf{s}_{i,1,o}$
1	(m, o, c, k, e, y)	$\mathsf{s}_{c,2,u}$
2	(m, o, u, k, e, y)	d_3
3	(m, o, u, e, y)	d_3
4	(m, o, u, y)	$\mathsf{i}_{s,3}$
5	(m, o, u, s, y)	$\mathsf{s}_{y,4,e}$
6	(m, o, u, s, e)	

The edit transcript $(\mathsf{s}_{i,1,o}, \mathsf{s}_{c,2,u}, \mathsf{d}_3, \mathsf{d}_3, \mathsf{i}_{s,3}, \mathsf{s}_{y,4,e})$ has length 6.

If $(f_0, f_1, \ldots, f_{k-1})$ is an edit transcript that transforms a sequence $\mathbf{x}$ into a sequence $\mathbf{y}$, then we have the sequences $\mathbf{z}_0, \mathbf{z}_1, \ldots, \mathbf{z}_k$ such that $\mathbf{z}_0 = \mathbf{x}$, $\mathbf{z}_i \in \text{Dom}(f_i)$, and $f_i(\mathbf{z}_i) = \mathbf{z}_{i+1}$ for $0 \leqslant ik - 1$ and $\mathbf{z}_k = \mathbf{y}$. Moreover, we can write $\mathbf{y} = f_{k-1}(\cdots f_1(f_0(\mathbf{x}))\cdots)$.

Theorem 14.48 *Let* $\ell : \mathbf{Seq}(S) \times \mathbf{Seq}(S) \longrightarrow \mathbb{R}_{\geqslant 0}$ *be a function defined by* $\ell(\boldsymbol{x}, \boldsymbol{y}) = n$ *if n is the length of the shortest edit transcript needed to transform $\boldsymbol{x}$ into $\boldsymbol{y}$. The function ℓ is a metric on* $\mathbf{Seq}(S)$.

Proof It is clear that $\ell(\mathbf{x}, \mathbf{x}) = 0$ and that $\ell(\mathbf{x}, \mathbf{y}) = \ell(\mathbf{y}, \mathbf{x})$ for every $\mathbf{x}, \mathbf{y} \in \mathbf{Seq}(S)$. Observe that the triangular inequality is also satisfied because the sequence of operations that transform $\mathbf{x}$ into $\mathbf{y}$ followed by the sequence of operations that transform

$\mathbf{y}$ into $\mathbf{z}$ will transform $\mathbf{x}$ into $\mathbf{z}$. Since the smallest such number of transformations is $\ell(\mathbf{x}, \mathbf{z})$, it follows that $\ell(\mathbf{x}, \mathbf{z}) \leqslant \ell(\mathbf{x}, \mathbf{y}) + \ell(\mathbf{y}, \mathbf{z})$. This allows us to conclude that ℓ is a metric on $\mathbf{Seq}(S)$.

We refer to ℓ as the *Levenshtein distance* between $\mathbf{x}$ and $\mathbf{y}$.

Recall that we introduced on page 23 the notation $\mathbf{x}_{i,j}$ for the infix $(x_i, \ldots, x_j)$ of a sequence $\mathbf{x} = (x_0, \ldots, x_{n-1})$.

Let $\mathbf{x} = (x_0, \ldots, x_{n-1})$ and $\mathbf{y} = (y_0, \ldots, y_{m-1})$ be two sequences and $l_{ij}(\mathbf{x}, \mathbf{y})$ be the length of a shortest edit transcript needed to transform $\mathbf{x}_{0,i}$ $\mathbf{y}_{0,j}$ for $-1 \leqslant i \leqslant |\mathbf{x}|$ and $-1 \leqslant j \leqslant |\mathbf{y}|$. In other words,

$$l_{ij} = \ell(x_{0,i}, y_{0,j}) \tag{14.11}$$

for $0 \leqslant i \leqslant n-1$ and $0 \leqslant j \leqslant m-1$, where $n = |\mathbf{x}|$ and $m = |\mathbf{y}|$.

When $i = -1$, we have $\mathbf{x}_{0,-1} = \boldsymbol{\lambda}$; similarly, when $j = -1$, $\mathbf{y}_{0,-1} = \boldsymbol{\lambda}$. Therefore, $l_{-1,j} = j$ since we need to insert j elements of S into $\boldsymbol{\lambda}$ to obtain $y_{0,j-1}$ and $l_{i,-1} = i$ for similar reasons.

To obtain an inductive expression of l_{ij}, we distinguish two cases. If $x_i = y_j$, then $l_{i,j} = l_{i-1,j-1}$; otherwise (that is, if $x_i \neq y_i$) we need to choose the edit transcript of minimal length among the following edit transcripts:

(i) the shortest edit transcript that transforms $\mathbf{x}_{0,i}$ into $\mathbf{y}_{0,j-1}$ followed by $\mathsf{i}_{x_i,j}$;
(ii) the shortest edit transcript that transforms $\mathbf{x}_{0,i-1}$ into $\mathbf{y}_{0,j}$ followed by d_i;
(iii) the shortest edit transcript that transforms $\mathbf{x}_{0,i-1}$ into $\mathbf{y}_{0,j-1}$ followed by substitution s_{x_i,i,y_j} if $x_i \neq y_j$.

Therefore,

$$l_{ij} = \min\{l_{i-1,j} + 1, l_{i,j-1} + 1, l_{i-1,j-1} + \delta(i, j)\}, \tag{14.12}$$

where

$$\delta(i, j) = \begin{cases} 0 & \text{if } x_i = y_j \\ 1 & \text{otherwise.} \end{cases}$$

The numbers l_{ij} can be computed using a bidimensional $(m+1) \times (n+1)$ array L. The rows of the array are numbered from -1 to $|\mathbf{x}| - 1$, while the columns are numbered from -1 to $|\mathbf{y}| - 1$. The component L_{ij} of L consists of a pair of the form (l_{ij}, A_{ij}), where A_{ij} is a subset of the set $\{\uparrow, \leftarrow, \nwarrow\}$.

Initially, the first row of L is $L_{-1,j} = (l_{-1,j}, \{\leftarrow\})$ for $-1 \leqslant j \leqslant |\mathbf{y}| - 1$; the first column of L is $L_{i,-1} = (l_{i,-1}, \uparrow)$ for $-1 \leqslant i \leqslant |\mathbf{x}| - 1$.

For each of the numbers $l_{i-1,j} + 1$, $l_{i,j-1} + 1$, or $l_{i-1,j-1} + \delta(i, j)$ that equals l_{ij}, we include in A_{ij} the symbols $\uparrow$, $\rightarrow$, or $\nwarrow$, respectively, pointing to the surrounding cells that help define l_{ij}. This will allow us to extract an edit transcript by following the points backward from $L_{m-1,n-1}$ to the cell $L_{-1,-1}$. Each symbol $\leftarrow$ denotes an insertion of y_j into the current string, each symbol $\uparrow$ as a deletion of x_i from the current string, and each diagonal edge as a match between x_i and y_j or as a substitution of x_i by y_j.

Example 14.49 Consider the sequences $\mathbf{x} = (a, b, a, b, c, a, b, a, c)$ and $\mathbf{y} = (a, b, c, a, a, c)$. The content of $L_{8,5}$ shows that $\ell(\mathbf{x}, \mathbf{y}) = 3$. Following the path

			a	b	c	a	a	c
		-1	0	1	2	3	4	5
	−1	0	$\leftarrow 0$	$\leftarrow 0$	$\leftarrow 0$	$\leftarrow 0$	$\leftarrow 0$	$\leftarrow 0$
a	0	$\uparrow 0$	$\nwarrow 0$	$\leftarrow\nwarrow 1$	$\leftarrow\nwarrow 2$	$\nwarrow 2$	$\leftarrow\nwarrow 3$	$\leftarrow 4$
b	1	$\uparrow 1$	$\uparrow\nwarrow 1$	$\nwarrow 1$	$\leftarrow\nwarrow 2$	$\leftarrow\uparrow\nwarrow 3$	$\nwarrow 3$	$\leftarrow\nwarrow 4$
a	2	$\uparrow 2$	$\nwarrow 1$	$\leftarrow\uparrow\nwarrow 2$	$\nwarrow 2$	$\nwarrow 2$	$\leftarrow\nwarrow 3$	$\leftarrow\nwarrow 4$
b	3	$\uparrow 3$	$\uparrow 2$	$\nwarrow 1$	$\uparrow\nwarrow 3$	$\uparrow\nwarrow 3$	$\nwarrow 3$	$\uparrow\nwarrow 4$
c	4	$\uparrow 4$	$\uparrow 3$	$\uparrow 2$	$\nwarrow 1$	$\uparrow\nwarrow 4$	$\uparrow\nwarrow 4$	$\nwarrow 3$
a	5	$\uparrow 5$	$\uparrow\nwarrow 4$	$\uparrow 3$	$\uparrow 2$	$\nwarrow 1$	$\leftarrow 2$	$\leftarrow 3$
b	6	$\uparrow 6$	$\uparrow 5$	$\uparrow\nwarrow 4$	$\uparrow 3$	$\uparrow 2$	$\nwarrow 2$	$\leftarrow\nwarrow 3$
a	7	$\uparrow 7$	$\uparrow\nwarrow 6$	$\uparrow 5$	$\uparrow 4$	$\uparrow\nwarrow 3$	$\uparrow\nwarrow 3$	$\nwarrow 3$
c	8	$\uparrow 8$	$\uparrow 7$	$\uparrow 6$	$\uparrow\nwarrow 5$	$\uparrow 4$	$\uparrow\nwarrow 4$	$\nwarrow 3$

$L_{8,5}$	$L_{7,4}$	$L_{6,3}$	$L_{5,3}$	$L_{4,2}$	$L_{3,1}$	$L_{2,0}$	$L_{1,0}$	$L_{0,0}$
$\nwarrow 3$	$\uparrow\nwarrow 3$	$\uparrow 2$	$\nwarrow 1$	$\nwarrow 1$	$\nwarrow 1$	$\uparrow 1$	$\uparrow 0$	0

that leads from $L_{8,5}$ to $L_{0,0}$, we obtain the following edit transcript:

Step	Sequence	Operation	Remark
0	$(a, b, a, b, c, a, b, a, c)$		match $x_8 = y_5 = c$
1	$(a, b, a, b, c, a, b, a, c)$		match $x_7 = y_4 = a$
2	(a, b, a, b, c, a, a, c)	d_6	
3	(a, b, a, b, c, a, a, c)		match $x_5 = y_3 = a$
4	(a, b, a, b, c, a, a, c)		match $x_4 = y_2 = c$
5	(a, b, a, b, c, a, a, c)		match $x_3 = y_1 = b$
6	(a, b, a, b, c, a, a, c)		match $x_2 = y_0 = a$
7	(a, a, b, c, a, a, c)	d_1	
8	(a, b, c, a, a, c)	d_0	

The notion of an edit distance can be generalized by introducing costs for the edit functions.

Definition 14.50 *A* cost scheme *is a triple* $(c_i, c_d, c_s) \in \hat{\mathbb{R}}_{\geqslant 0}$, *where the components* c_i, c_d, *and* c_s *are referred to as the costs of an insertion, deletion, and substitution, respectively.*

The *cost of an edit transcript* $\mathbf{t} = (f_0, f_1, \ldots, f_{k-1})$ according to the cost scheme (c_i, c_d, c_s) is $n_i c_i + n_d c_d + n_s c_s$, where n_i, n_d, and n_s are the number of insertions, deletions, and substitutions that occur in $\mathbf{t}$.

When $c_i = c_d = c_s = 1$, the cost of **t** equals the length of **t**, and finding the Levenshtein distance between two strings **x**, **y** can now be seen as determining the length of the shortest editing transcript that transforms **x** into **y** using the cost schema $(1, 1, 1)$. It is interesting to remark that, for any cost schema, the minimal cost of a transcript that transforms **x** into **y** remains an extended metric on $\mathbf{Seq}(S)$. This can be shown using an argument that is similar to the one we used in Theorem 14.48.

Note that a substitution can always be replaced by a deletion followed by an insertion. Therefore, for a cost scheme $(1, 1, \infty)$, the edit transcript of minimal cost will include only insertions and deletions. Similarly, if $c_s = 1$ and $c_i = c_d = \infty$, then the edit transcript will contain only substitutions if the two sequences have equal lengths and the distance between strings will be reduced to the Hamming distance.

The recurrence (14.12) that allowed us to compute the length of the shortest edit transcript is now replaced by a recurrence that allows us to compute the least cost C_{ij} of transforming the prefix $\mathbf{x}_{0,i}$ into $\mathbf{y}_{0,j}$:

$$C_{ij} = \min\{C_{i-1,j} + c_i, l_{i,j-1} + c_d, l_{i-1,j-1} + \delta(i, j)c_s\}. \tag{14.13}$$

The computation of the edit distance using the cost scheme (c_i, c_d, c_s) now proceeds in a tabular manner similar to the one used for computing the length of the shortest edit transcript.

14.7 Searches in Metric Spaces

Searches that seek to identify objects that reside in the proximity of other objects are especially important in data mining, where the keys or the ranges of objects of interest are usually unknown. This type of search is also significant for multimedia databases, where classical, exact searches are often meaningless. For example, querying an image database to find images that contain a sunrise is usually done by providing an example image and then, identifying those images that are similar to the example. The natural framework for executing such searches is provided by metric spaces or, more generally, by dissimilarity spaces [4], and we examine the usefulness of metric properties for efficient searching algorithms. We show how various metric properties benefit the design of searching algorithms.

Starting from a finite collection of members of S, $T \subseteq S$, and a *query object* q, we consider two types of searching problems:

(i) *range queries* that seek to compute the set $B(q, r) \cap T$, for some positive number r, and
(ii) *k-nearest-neighbor queries* that seek to compute a set N_k such that $N_k \subseteq T$, $|N_k| = k$ and for every $x \in N_k$ and $y \in T - N_k$, we have $d(x, q) \leqslant d(y, q)$.

In the case of k-nearest-neighbor queries the set N_k is not uniquely identified because of the ties that may exist. For $k = 1$, we obtain the *nearest-neighbor queries*.

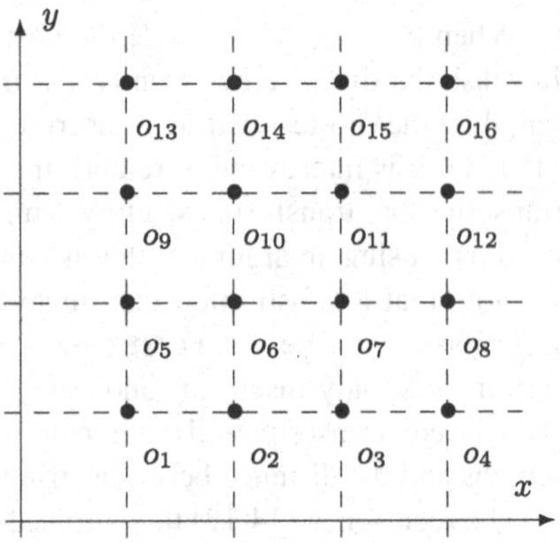

Fig. 14.11 Set of 16 points in $\mathbb{R}^2$

The triangular inequality that is satisfied by every metric plays an essential role in reducing the amount of computation required by proximity queries.

Suppose that we select an element p of S (referred to as a *pivot*) and we compute the set of distances $\{d(p, x) \mid x \in S\}$ before executing any proximity searches. If we need to compute a range query $B(q, r) \cap T$, then by the triangular inequality, we have $d(q, x) \geqslant |d(p, q) - d(p, x)|$. Since the distances $d(p, q)$ and $d(p, x)$ have been already computed, we can exclude from the search all elements x such that $|d(p, q) - d(p, x)| > r$.

The triangular inequality also ensures that the results of the search are plausible. Indeed, it is expected that if both x and y are in the proximity of q, then a certain degree of similarity exists between x and y. This is implied by the triangular inequality that requires $d(x, y) \leqslant d(x, q) + d(q, y)$.

To execute any of these searches, we need to examine the entire collection of objects C unless specialized data structures called *indexes* are prepared in advance.

One of the earliest types of indexes is the *Burkhard-Keller tree* (see [5]), which can be used for metric spaces where the distance is discrete; that is, the range of the distance function is limited to a finite set. To simplify our presentation, we assume that $\mathsf{Ran}(d) = \{0, 1, \ldots, k\}$, where $k \in \mathbb{N}$. The pseudo-code for this construction is given in the Algorithm 14.7.1.

Example 14.51 Consider the collection of points $C = \{o_1, \ldots, o_{16}\}$ in $\mathbb{R}^2$ shown in Fig. 14.11. Starting from their Euclidean distance $d_2(o_i, o_j)$, we construct the discrete distance d as in Exercise 16, namely, we define $d(o_i, o_j) = \lceil d_2(o_i, o_j) \rceil$ for $1 \leqslant i, j \leqslant 16$.

The Manhattan distances $d_1(o_i, o_j)$ are given by the following matrix and we shall use this distance to construct the Burkhard-Keller tree.

$$D = \begin{pmatrix}
0 & 1 & 2 & 3 & 1 & 2 & 3 & 4 & 2 & 3 & 4 & 5 & 3 & 4 & 5 & 6 \\
1 & 0 & 1 & 2 & 2 & 1 & 2 & 3 & 3 & 3 & 3 & 4 & 4 & 3 & 4 & 5 \\
2 & 1 & 0 & 1 & 3 & 2 & 1 & 2 & 4 & 3 & 2 & 3 & 5 & 4 & 3 & 4 \\
3 & 2 & 1 & 0 & 4 & 3 & 2 & 1 & 9 & 4 & 3 & 2 & 6 & 5 & 4 & 3 \\
1 & 2 & 3 & 4 & 0 & 1 & 2 & 3 & 1 & 2 & 3 & 4 & 2 & 3 & 4 & 5 \\
2 & 1 & 2 & 3 & 1 & 0 & 1 & 2 & 2 & 1 & 2 & 3 & 3 & 2 & 3 & 4 \\
3 & 2 & 1 & 2 & 2 & 1 & 0 & 1 & 3 & 2 & 1 & 2 & 4 & 3 & 2 & 3 \\
4 & 2 & 2 & 1 & 3 & 2 & 1 & 0 & 4 & 3 & 2 & 1 & 5 & 4 & 3 & 2 \\
2 & 3 & 4 & 5 & 1 & 2 & 3 & 4 & 0 & 1 & 2 & 3 & 1 & 2 & 3 & 4 \\
3 & 2 & 3 & 4 & 2 & 1 & 2 & 3 & 1 & 0 & 1 & 2 & 2 & 1 & 2 & 3 \\
4 & 3 & 2 & 3 & 3 & 2 & 1 & 2 & 2 & 1 & 0 & 1 & 3 & 2 & 1 & 2 \\
5 & 4 & 3 & 2 & 4 & 3 & 2 & 1 & 3 & 2 & 1 & 0 & 4 & 3 & 2 & 1 \\
3 & 4 & 5 & 6 & 2 & 3 & 4 & 5 & 1 & 2 & 3 & 4 & 0 & 1 & 2 & 3 \\
4 & 3 & 4 & 5 & 3 & 2 & 3 & 4 & 2 & 1 & 2 & 3 & 1 & 0 & 1 & 2 \\
5 & 4 & 3 & 4 & 4 & 3 & 2 & 3 & 3 & 2 & 1 & 2 & 2 & 1 & 0 & 1 \\
6 & 5 & 4 & 3 & 5 & 4 & 3 & 2 & 4 & 3 & 2 & 1 & 3 & 2 & 1 & 0
\end{pmatrix}.$$

Algorithm 14.7.1: Construction of the Burkhard-Keller Tree

Data: a collection of elements C of a metric space (S, d), where
$\mathrm{Ran}(d) = \{0, 1, \ldots, k\}$
Result: a tree $\mathcal{T}_C$ whose nodes are labeled by objects of C
1 **if** $|C| = 1$ **then**
2 $\quad$ return a single-vertex tree whose root is labeled p
3 **else**
4 $\quad$ select randomly an object $p \in C$ to serve as root of $\mathcal{T}_C$;
5 $\quad$ partition C into the sets $C_1, \ldots, C_k$ defined by $C_i = \{o \in C \mid d(o, p) = i\}$ for
$\quad$ $1 \leqslant i \leqslant k$;
6 $\quad$ construct the trees corresponding to $C_{l_0}, \ldots, C_{l_{m-1}}$, which are the nonempty sets among
$\quad$ $C_1, \ldots, C_k$;
7 $\quad$ connect the trees $\mathcal{T}_{C_{l_0}}, \ldots, \mathcal{T}_{C_{l_{m-1}}}$ to p;
8 $\quad$ return $\mathcal{T}_C$
9 **end**

We begin by selecting o_6 as the first pivot. Then, we create trees for the sets

$$C_1 = \{o_2, o_5, o_7, o_{10}\},$$
$$C_2 = \{o_1, o_3, o_8, o_{11}, o_{14}\},$$
$$C_3 = \{o_4, o_{12}, o_{13}, o_{15}\},$$
$$C_4 = \{o_{16}\}.$$

Choose o_7, o_8, and o_{13} as pivots for the sets C_1, C_2, and C_3, respectively. Note that $\mathcal{T}_{C_4}$ is completed because it consists of one vertex. Assuming certain choices of pivots, the construction results in a tree shown in Fig. 14.12.

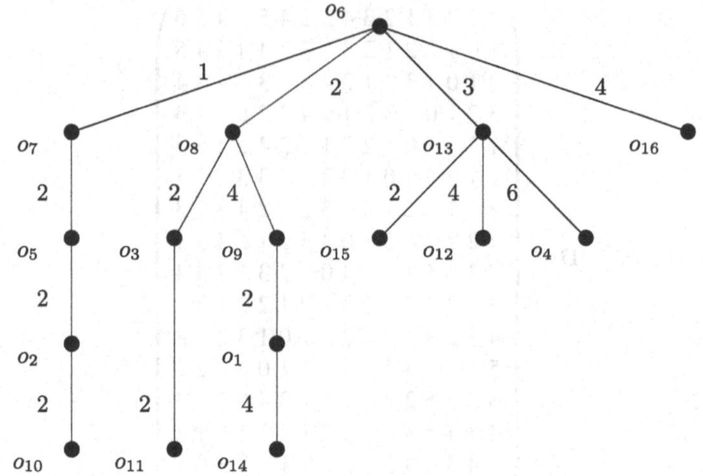

Fig. 14.12 Burkhard-Keller tree

Burkhard-Keller trees can be used for range queries that seek to compute sets of the form $O_{q,r,C} = B(q, r) \cap C$, where d is a discrete metric. In other words, we seek to locate all objects o of C such that $d(q, o) \leqslant r$ (see Algorithm 14.7.2).

By Exercise 62 we have $|d(p, o) - d(p, q)| \leqslant d(q, o) \leqslant r$, where p is the pivot that labels the root of the tree. This implies $d(p, q) - r \leqslant d(p, o) \leqslant d(p, q) + r$, so we need to visit recursively the trees $\mathcal{T}_{C_i}$ where $d(p, q) - r \leqslant i \leqslant d(p, q) + r$.

Algorithm 14.7.2: Searching in Burkhard-Keller Trees

Data: a collection of elements C of a metric space (S, d), a query object q and a radius r
Result: the set $O(q, r, C) = B(q, r) \cap C$

```
O(q, r, C) = ∅;
if d(p, q) ⩽ r then
    O(q, r, C) = O(q, r, C) ∪ {p}
end
compute I = {i | 1 ⩽ i ⩽ k, d(p, q) − r ⩽ i ⩽ d(p, q) + r};
compute ⋃_{i∈I} O(q, r, C_i);
O(q, r, C) = O(q, r, C) ∪ ⋃_{i∈I} O(q, r, C_i);
return O(q, r, C);
```

Example 14.52 To solve the query $B(o_{11}, 1) \cap C$, where C is the collection of objects introduced in Example 14.51, we begin by observing that $d(o_{11}, o_6) = 2$, so the pivot itself does not belong to $O(o_{11}, 1, C)$. The set I in this case is $I = \{1, 2, 3\}$.

We need to execute three recursive calls, namely $O_{o_{11},1,C_1}$, $O_{o_{11},1,C_2}$, and $O_{o_{11},1,C_3}$.

For the set C_1 having the pivot o_7, we have $o_7 \in O(o_{11}, 1, C_1)$ because $d(o_7, o_{11}) = 1$. Thus, $O(o_{11}, 1, C_1)$ is initialized to $\{o_7\}$ and the search proceeds with the set $C_{1,2}$, which consists of the objects o_5, o_2, o_{10} located at distance 2 from the pivot o_7.

Since $d(o_5, o_{11}) = 3$, o_5 does not belong to the result. The set $C_{1,2,2}$ consists of $\{o_2, o_{10}\}$. Choosing o_2 as the pivot, we can exclude it from the result because $d(o_2, o_{11}) = 3$. Finally, the set $C_{1,2,2,2}$ consists of $\{o_{10}\}$ and $d(o_{10}, o_{11}) = 1$. Thus,

$$O(o_{11}, 1, C_{1,2,2,2}) = O(o_{11}, 1, C_{1,2,2}) = O(o_{11}, 1, C_{1,2}) = \{o_{10}\},$$

so $O(o_{11}, 1, C_1) = \{o_7, o_{10}\}$.

Similarly, we have $O(o_{11}, 1, C_2) = o_{11}$ and $O(o_{11}, 1, C_3) = \{o_{12}, o_{15}\}$. The result of the query is $O_{o_{11},1,C} = \{o_7, o_{10}, o_{11}, o_{12}, o_{15}\}$.

Orchard's algorithm [6] aims to solve the nearest-neighbor problem and proceeds as follows (see Algorithm 14.7.3).

Algorithm 14.7.3: Orchard's Algorithm

Data: a finite metric space (S, d) and a query $q \in S$
Result: a member of S that is closest to the query q

```
for w ∈ S do
    establish a list L_w of elements of S in increasing order of their distance to w; select an
    initial candidate c (pre-processing phase)
end
repeat
    compute d(c, q);
    scan L_c until a site s closer to q is found;
    c = s;
until L_c is completely traversed or s is found in L_c such that d(c, s) > 2d(c, q);
return c;
```

Since L_c lists all elements of the space in increasing order of their distance to c, observe that if the scanning of a list L_c is completed without finding an element that is closer to q, then it is clear that p is one of the elements that is closest to q and the algorithm halts. Let s be the first element in the current list L_c such that $d(c, s) > 2d(c, q)$ (if such elements exist at all). Then, none of the previous elements on the list is closer to q than s since otherwise we would not have reached s in L_c. By Exercise 67, (with $k = 2$) we have $d(q, s) > d(c, q)$, so c is still the closest element to q on this list. If z is an element of L_c situated past s, it follows that $d(z, c) \geqslant d(s, c)$ because L_c is arranged in increasing order of the distances to c, so $d(c, z) > 2d(c, q)$, which ensures that z is more distant from q than c. So, in all cases where c is closest to q, the algorithm works correctly.

The preprocessing phase requires an amount of space that grows as $\Theta(n^2)$ with n, and this limits the usefulness of the algorithm to rather small sets of objects.

An alternative algorithm known as the *annulus algorithm*, proposed in [7], allows reduction of the volume of preprocessing space to $\Theta(n)$.

Suppose that the finite metric space (S, d) consists of n objects. The preprocessing phase consists of selecting a pivot o and constructing a list of objects $L = (o_1, o_2, \ldots, o_n)$ such that $d(o, o_1) \leqslant d(o, o_2) \leqslant \cdots \leqslant d(o, o_n)$. Without loss of generality, we may assume that $o = o_1$.

Suppose that u is closer to the query q than v; that is, $d(q, u) \leqslant d(q, v)$. Then we have

$$|d(u, o - d(q, o)| \leqslant d(u, q) \leqslant d(q, v)$$

by Exercise 62, which implies

$$d(q, o) - d(q, v) \leqslant d(u, o) \leqslant d(q, p) + d(q, v).$$

Thus, u is located in an annulus centered in o that contains the query point q and is delimited by two spheres, $B(o, d(q, o) - d(q, v))$ and $B(o, d(q, p) + d(q, v))$.

Algorithm 14.7.4: Annulus algorithm

Data: a finite metric space (S, d) and a query $q \in S$
Result: the member of S that is closest to the query q

select a pivot object o;
establish a list L of elements of S in increasing order of their distances to o (preprocessing phase);
select an initial candidate v;
compute the set of U_v that consists of those u such that $d(q, o) - d(q, v) \leqslant d(u, o) \leqslant d(q, p) + d(q, v)$;
scan U_v for an object w closer to q;
if *such a vector exists* **then**
    replace v by w, recompute U_v and resume scan
else
    output v
end

The advantage of the annulus algorithm over Orchard's algorithm consists of the linear amount of space required for the preprocessing phase. Implementations of these algorithms and performance issues are discussed in detail in [8].

A general algorithm for the nearest-neighbor search in dissimilarity spaces was formulated in [9]. The algorithm uses the notion of *basis for a subset of a dissimilarity space*.

Definition 14.53 *Let (S, d) be a dissimilarity space and let (α, β) be a pair of numbers such that $\alpha \geqslant \beta > 0$. A* basis *at level (α, β) for a subset H, $H \subseteq S$ is a finite set of points $\{z_1, \ldots, z_k\} \subseteq S$ if for any $x, y \in H$ we have*

$$\alpha d(x, y) \geqslant \max\{|d(x, z_i) - d(y, z_i)| \mid 1 \leqslant i \leqslant k\} \geqslant \beta d(x, y).$$

A dissimilarity space is k-dimensional *if there exist* α, β, *and* k *depending only on* (S, d) *such that for any bounded subset* H *of* S, *there are* k *points of* S *that form a basis at level* (α, β) *for* H.

Example 14.54 Consider the metric space $(\mathbb{R}^n, d_2)$, and let H be a bounded subset of $\mathbb{R}^n$. A basis at level $(1, 0.5)$ for H can be formed by the $n+1$ vertices of a sufficiently large n-dimensional simplex S_n that contains H. Indeed, observe that $d_2(\mathbf{x}, \mathbf{y}) \geqslant |d_2(\mathbf{x}, \mathbf{z}_i) - d_2(\mathbf{y}, \mathbf{z}_i)|$ for $1 \leqslant i \leqslant n+1$ by Exercise 62, which shows that the first condition of Definition 14.53 is satisfied.

On the other hand, if the $n+1$ points of the n-dimensional simplex are located at sufficient distances from the points of the set H, then there exists at least one vertex $\mathbf{z}_i$ such that $|d_2(\mathbf{u}, \mathbf{z}_i) - d_2(\mathbf{v}, \mathbf{z}_i)| \geqslant 0.5 d_2(\mathbf{u}, \mathbf{v})$; that is, $\max\{|d_2(\mathbf{u}, \mathbf{z}_i) - d_2(\mathbf{v}, \mathbf{z}_i)| \mid 1 \leqslant i \leqslant k\} \geqslant 0.5 d(\mathbf{u}, \mathbf{v})$. Indeed, let h_i be the distance from $\mathbf{z}_i$ to the line determined by $\mathbf{u}$ and $\mathbf{v}$, and let $\mathbf{w}_i$ be the projection of $\mathbf{z}_i$ on this line (see Fig. 14.13). We discuss here only the case where $\mathbf{w}_i$ is located outside the segment $(\mathbf{u}, \mathbf{v})$. Let $k_i = \min\{d_2(\mathbf{u}, \mathbf{w}_i), d_2(\mathbf{v}, \mathbf{w}_i)\}$.

To satisfy the condition

$$|d_2(\mathbf{u}, \mathbf{z}_i) - d_2(\mathbf{v}, \mathbf{z}_i)| \geqslant \frac{\ell}{2}$$

or the equivalent equality

$$\left| \sqrt{h_i^2 + (\ell + k_i)^2} - \sqrt{h_i^2 + k_i^2} \right| \geqslant \frac{\ell}{2},$$

it suffices to have

$$\ell + 2k_i \geqslant \sqrt{h_i^2 + (\ell + k_i)^2}.$$

This inequality is satisfied if $k_i \geqslant \frac{1}{3}\left(\sqrt{\ell + 4h_i^2} - \ell\right)$. Thus, if $\mathbf{z}_i$ is chosen appropriately, the set $\mathbf{z}_1, \ldots, \mathbf{z}_n, \mathbf{z}_{n+1}$ is a $(1, 0.5)$ basis for H.

Algorithm 14.7.5: Faragó-Linder-Lugosi Algorithm

Data: a finite subset $H = \{x_1, \ldots, x_n\}$ of a dissimilarity space (S, d), an (α, β)-basis $z_1, \ldots, z_k$ for H, and a query $x \in S$

Result: the element of H that is closest to the query x

compute and store all dissimilarities $d(x_i, z_j)$ for $1 \leqslant i \leqslant n$ and $1 \leqslant j \leqslant k$ (preprocessing phase);

$\mathcal{I} = \{x_1, \ldots, x_n\}$;

$\gamma(x_i) = \max_{1 \leqslant j \leqslant k} |d(x_i, z_j) - d(x, z_j)|$ for $1 \leqslant i \leqslant n$;

$t_0 = \min_{1 \leqslant i \leqslant n} \gamma(x_i)$;

delete all points x_i from $\mathcal{I}$ such that $\gamma(x_i) > \frac{\alpha}{\beta} t_0$;

find the nearest neighbor of x in the remaining part of $\mathcal{I}$ by exhaustive search and output $x_{nn} = \arg\min_{1 \leqslant i \leqslant n} \gamma(x_i)$;

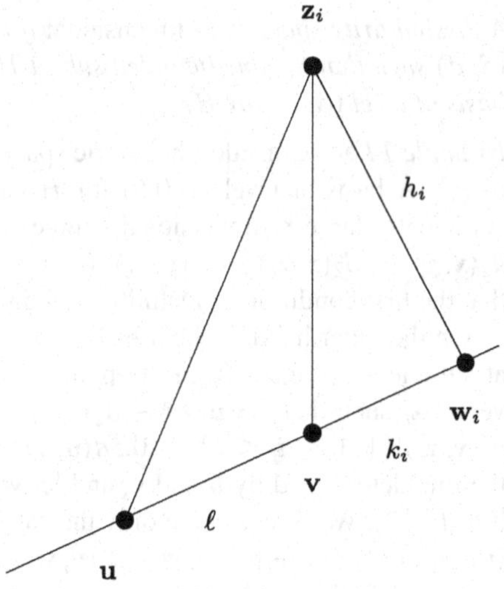

Fig. 14.13 Point of the basis for a set H in $\mathbb{R}^n$

If a tie occurs in the last step of the algorithm, then an arbitrary element is chosen among the remaining elements of $\mathfrak{I}$ that minimize $\gamma(x_i)$.

The first phase of the algorithm is designated as the preprocessing phase because it is independent of the query x and can be executed only once for the data set H and its base. Its time requirement is $O(nk)$.

To prove the correctness of the algorithm, we need to show that if an element of H is deleted from $\mathfrak{I}$, then it is never the nearest neighbor x_{nn} of x. Suppose that the nearest neighbor x_{nn} were removed. This would imply $\gamma(x_{nn}) > \frac{\alpha}{\beta} t_0$ or, equivalently,

$$\frac{1}{\alpha}\gamma(x_{nn}) > \frac{1}{\beta} \min_{1 \leqslant i \leqslant n} \gamma(x_i).$$

Since $\{z_1, \ldots, z_k\}$ is a basis for the set H, we have

$$d(x, x_{nn}) \geqslant \frac{1}{\alpha} \max_{1 \leqslant j \leqslant k} |d(x, z_j) - d(x_{nn}, z_j)| \geqslant \frac{1}{\alpha}\gamma(x_{nn})$$

and

$$\frac{1}{\beta} \min_{1 \leqslant i \leqslant n} \gamma(x_i) \geqslant \min_{1 \leqslant i \leqslant n} d(x, x_i),$$

which implies $d(x, x_{nn}) > \min_{1 \leqslant i \leqslant n} d(x, x_i)$. This contradicts the definition of x_{nn}, so it is indeed impossible to remove x_{nn}. Thus, in the worst case, the algorithm performs n dissimilarity calculations.

Next, we present a unifying model of searching in metric spaces introduced in [4] that fits several algorithms used for proximity searches. The model relates equivalence relations (or partitions) to indexing schemes.

Definition 14.55 *The index defined by the equivalence ρ on the set S is the surjection $I_\rho : S \longrightarrow S/\rho$ defined by $I_\rho(x) = [x]$, where $[x]$ is the equivalence class of x in the quotient set S/ρ.*

A general searching strategy can be applied in the presence of an index and involves two phases:

(i) identify the equivalence classes that contain the answers to the search, and
(ii) exhaustively search the equivalence classes identified in the first phase.

The cost of the first phase is the *internal complexity* of the search, while the cost of the second phase is the *external complexity*.

If $\rho_1, \rho_2 \in EQ(S)$ and $\rho_1 \leqslant \rho_2$, then $|S/\rho_1| \geqslant |S/\rho_2|$. Therefore, the internal complexity of the search involving I_{ρ_1} is larger than the internal complexity of the search involving I_{ρ_2} since we have to search more classes, while the external complexity of the search involving I_{ρ_1} is smaller than the external complexity of the search involving I_{ρ_2} since the classes that need to be exhaustively searched form a smaller set.

Let (S, d) be a metric space and let ρ be an equivalence on the set S. The metric d generates a mapping $\delta_{d,\rho} : (S/\rho)^2 \longrightarrow \mathbb{R}_{\geqslant 0}$ on the quotient set S/ρ, where $\delta_{d,\rho}([x], [y]) = \inf\{d(u, v) \mid u \in [x] \text{ and } v \in [y]\}$. We refer to $\delta_{d,\rho}$ as the *pseudo-distance generated by d and ρ*. It is clear that $\delta_{d,\rho}([x], [y]) \leqslant d(x, y)$, for $x, y \in S$, but $\delta_{d,\rho}$ is not a metric because if fails to satisfy the triangular inequality in general.

If a range query $B(q, r) \cap T = \{y \in T \mid d(q, y) \leqslant r\}$ must be executed we can transfer this query on the quotient set S/ρ (which is typically a smaller set than S) as the range query $B([q], r) \cap \{[t] \mid t \in T\}$. Note that if $y \in B(q, r)$, then $d(q, y) \leqslant r$, so $\delta_{d,\rho}([q], [y]) \leqslant d(q, y) \leqslant r$. This setting allows us to reduce the search of the entire set T to the search of the set of equivalence classes $\{[y] \mid y \in T, \delta_{d,\rho}([q], [y]) \leqslant r\}$.

Since $\delta_{d,\rho}$ is not a metric, it is not possible to reduce the internal complexity of the algorithm. In such cases, a solution is to determine a metric e on the quotient set S/ρ such that $e([x], [y]) \leqslant \delta_{d,\rho}([x], [y])$ for every $[x], [y] \in S/\rho$. If this is feasible, then we can search the quotient space for classes $[y]$ such that $e([q], [y]) \leqslant r$ using the properties of the metric e.

Let $\rho_1, \rho_2 \in EQ(S)$ be two equivalences on S such that $\rho_1 \leqslant \rho_2$. Denote by $[z]_i$ the equivalence class of z relative to the equivalence ρ_i for $i = 1, 2$.

If $\rho_1 \leqslant \rho_2$, then $[z]_1 \subseteq [z]_2$ for every $z \in S$ and therefore

$$\begin{aligned}\delta_{d,\rho_1}([x]_1, [y]_1) &= \inf\{d(u, v) \mid u \in [x]_1 \text{ and } v \in [y]_1\}\\ &\geqslant \inf\{d(u, v) \mid u \in [x]_2 \text{ and } v \in [y]_2\}\\ &= \delta_{d,\rho_2}([x]_2, [y]_2).\end{aligned}$$

Thus, $\delta_{d,\rho_1}([x]_1, [y]_1) \geqslant \delta_{d,\rho_2}([x]_2, [y]_2)$ for every $x, y \in S$, and this implies

$$\{[y]_1 \mid \delta_{d,\rho_1}([q], [y]) \leqslant r\} \subseteq \{[y]_2 \mid \delta_{d,\rho_2}([q], [y]) \leqslant r\},$$

confirming that the external complexity of the indexing algorithm based on ρ_2 is greater than the same complexity for the indexing algorithm based on ρ_1.

Example 14.56 Let (S, d) be a metric space, $p \in S$, and let $\mathbf{r}$ be a sequence of positive real numbers $\mathbf{r} = (r_0, r_1, \ldots, r_{n-1})$ such that $r_0 < r_1 < \ldots < r_{n-1}$. Define the collection of sets $\mathcal{E} = \{E_0, E_1, \ldots, E_n\}$ by $E_0 = \{x \in S \mid d(p, x) < r_0\}$, $E_i = \{x \in S \mid r_{i-1} \leqslant d(p, x) < r_i\}$ for $1 \leqslant i \leqslant n-1$, and $E_n = \{x \in S \mid r_{n-1} \leqslant d(p, x)\}$. The subcollection of $\mathcal{E}$ that consists of nonempty sets is a partition of S denoted by $\pi_{\mathbf{r}}$. Denote the corresponding equivalence relation by $\rho_{\mathbf{r}}$.

If $E_i \neq \emptyset$, then E_i is an equivalence class of $\rho_{\mathbf{r}}$ that can be imagined as a circular ring around p. Then, if $i < j$ and E_i and E_j are equivalence classes, we have $\delta_{d,\rho}(E_i, E_j) > r_{j-1} - r_i$.

Example 14.57 Let (S, d) be a metric space and p be a member of S. Define the equivalence $\rho_p = \{(x, y) \in S \times S \mid d(p, x) = d(p, y)\}$. We have $\delta_{d,\rho}([x], [y]) = |d(p, x) - d(p, y)|$. It is easy to see that the pseudo-distance $\delta_{d,\rho}$ is actually a distance on the quotient set S/ρ.

Exercises and Supplements

1. Let d be a metric on the set S. Prove that the function $d' : S \times S \longrightarrow \mathbb{R}$ given by

$$d'(x, y) = 1 - e^{-kd(x,y)},$$

where k is a positive constant and $x, y \in S$, is also a metric on S. This metric is known as the *Schoenberg transform of d* (see [10]).

If d is an ultrametric, does it follow that d' is an ultrametric?

2. Let S be a set and let $\phi : \mathbf{Seq}(S) \longrightarrow \mathbb{R}_{>0}$ be a function such that $\mathbf{u} \leqslant_{\text{pref}} \mathbf{v}$ implies $\phi(\mathbf{u}) \geqslant \phi(\mathbf{v})$ for $\mathbf{u}, \mathbf{v} \in \mathbf{Seq}(S)$.

(a) Define the mapping $d_\phi : (\mathbf{Seq}(S))^2 \longrightarrow \mathbb{R}_{\geqslant 0}$ by

$$d_\phi(\mathbf{u}, \mathbf{v}) = \begin{cases} 0 & \text{if } \mathbf{u} = \mathbf{v}, \\ \phi(lcp(\mathbf{u}, \mathbf{v})) & \text{otherwise}, \end{cases}$$

for $\mathbf{u}, \mathbf{v} \in \mathbf{Seq}(S)$. Prove that d_ϕ is an ultrametric on $\mathbf{Seq}(S)$.

(b) Consider an extension of the function d_ϕ to the set $\mathbf{Seq}_\infty(S)$ obtained by replacing the sequences $\mathbf{u}, \mathbf{v}$ in the definition of d_ϕ by infinite sequences. Note that this extension is possible because the longest common prefix of two distinct infinite sequences is always a finite sequence. Prove that the extended function is an ultrametric on $\mathbf{Seq}_\infty(S)$.

(c) Give examples of functions ϕ that satisfy the conditions of Part (a) and the associated ultrametrics.

Solution for Part (a): It is clear that $d_\phi(\mathbf{u}, \mathbf{v}) = 0$ if and only if $\mathbf{u} = \mathbf{v}$ and that $d_\phi(\mathbf{u}, \mathbf{v}) = d_\phi(\mathbf{v}, \mathbf{u})$. Thus, we need to prove only the ultrametric inequality. Let $\mathbf{u}, \mathbf{v}, \mathbf{w}$ be three sequences. In Theorem 11.8 we have shown that at most two of the sequences $lcp(\mathbf{u}, \mathbf{v})$, $lcp(\mathbf{v}, \mathbf{w})$, $lcp(\mathbf{w}, \mathbf{u})$ are distinct and that the common value of two of these sequences is a prefix of the third sequence. This is equivalent to the ultrametric inequality for d_ϕ.

3. Let $\mathbf{x}, \mathbf{y} \in \mathbf{Seq}_\infty(\{0, 1\})$ be two infinite binary sequences. Define

$$d : (\mathbf{Seq}_\infty(\{0, 1\}))^2 \longrightarrow \hat{\mathbb{R}}_{\geqslant 0}$$

as $d(\mathbf{x}, \mathbf{y}) = \sum_{i=0}^{\infty} \frac{|x_i - y_i|}{a^i}$, where $a > 1$.
Prove that d is a metric on $\mathbf{Seq}_\infty(\{0, 1\})$ such that $d(\mathbf{x}, \mathbf{y}) = d(\mathbf{x}, \mathbf{z})$ implies $\mathbf{y} = \mathbf{z}$ for all $\mathbf{x}, \mathbf{y}, \mathbf{z} \in \mathbf{Seq}_\infty(\{0, 1\})$.
4. Give an example of a topological metric space where the inclusion $\mathbf{K}(C(x, r)) \subseteq B(x, r)$ can be strict.
Solution: Let $(\mathbf{Seq}_\infty(\{0, 1\}), d_\phi)$ be the ultrametric space, where the ultrametric d_ϕ was introduced in Supplement 2 and let $\phi(\mathbf{u}) = \frac{1}{|\mathbf{u}|}$ for $\mathbf{u} \in \mathbf{Seq}(\{0, 1\})$.
It is clear that $B(\mathbf{x}, 1) = \mathbf{Seq}_\infty(\{0, 1\})$ because $d_\phi(\mathbf{x}, \mathbf{y}) \leqslant 1$ for every $\mathbf{x}, \mathbf{y} \in \mathbf{Seq}_\infty(\{0, 1\})$. On the other hand, $C(\mathbf{x}, 1)$ contains those sequences $\mathbf{y}$ that have a non-null longest common prefix with $\mathbf{x}$; the first symbol of each such sequence is the same as the first symbol s of $\mathbf{x}$.
Since d_ϕ is an ultrametric, the open sphere $C(\mathbf{x}, 1)$ is also closed, so $C(\mathbf{x}, 1) = \mathbf{K}(C(\mathbf{x}, 1))$. Let s' be a symbol in S distinct from s and let $\mathbf{z} = (s', s', \ldots) \in \mathbf{Seq}_\infty(S)$. Note that $d_\phi(\mathbf{x}, \mathbf{z}) = 1$, so $\mathbf{z} \notin C(\mathbf{x}, 1) = \mathbf{K}(C(\mathbf{x}, 1))$. Thus, we have $\mathbf{K}(C(\mathbf{x}, 1) \subset B(\mathbf{x}, 1)$.
5. Consider the ultrametric space $(\mathbf{Seq}_\infty(\{0, 1\}), d_\phi)$, where $\phi(\mathbf{u}) = \frac{1}{2^{|\mathbf{u}|}}$ for $\mathbf{u} \in \mathbf{Seq}(\{0, 1\})$. For $\mathbf{u} \in \mathbf{Seq}(\{0, 1\})$, let $P_{\mathbf{u}} = \{\mathbf{ut} \mid \mathbf{t} \in \mathbf{Seq}_\infty(\{0, 1\})\}$ be the set that consists of all infinite sequences that begin with $\mathbf{u}$. Prove that $\{P_{\mathbf{u}} \mid \mathbf{u} \in \mathbf{Seq}(\{0, 1\})\}$ is a basis for the topological ultrametric space defined above.
6. Prove that $\mathbf{x} = (x_0, x_1, \ldots)$ is a Cauchy sequence in a topological ultrametric space $(S, \mathcal{O}_d)$ if and only if $\lim_{n\to\infty} d(x_n, x_{n+1}) = 0$.
7. Let (S, d) be a finite ultrametric space, where $S = \{x_1, \ldots, x_n\}$ and $n \geqslant 3$. Prove that d takes at most $n - 1$ distinct positive values.

Solution: The argument is by induction on n. The statement is clearly true when $n = 3$. Let $n \geqslant 4$ and and suppose that the statement is true for all $m < n$. Without loss of generality we may assume that $d(x_1, x_2) \leqslant d(x_i, x_j)$ for all i, j such that $1 \leqslant i \neq j \leqslant n$. Then, $d(x_k, x_1) = d(x_k, x_2)$ for all k such that $3 \leqslant k \leqslant n$.
Let $\{a_1, \ldots, a_r\} = \{d(x_k, x_1) \mid 3 \leqslant k \leqslant n\}$, where $0 = a_0 < a_1 < \cdots < a_r$. Define $B_j = \{x_k \mid k \geqslant 3 \text{ and } d(x_k, x_1) = a_j\}$ for $1 \leqslant j \leqslant r$. The collection $B_1, \ldots, B_r$ is a partition of the set $\{x_3, x_4, \ldots, x_n\}$. Let $m_j = |B_j|$; then $m_1 + \cdots + m_r = n - 2$.

If $u \in B_i$ and $v \in B_j$ with $i < j$, then $d(u, x_1) = a_i < a_j = d(v, x_1)$, hence $d(u, v) = a_j$. Therefore, the values of $d(u, v)$ for u, v in distinct blocks of the partition belong to the set $\{a_1, \ldots, a_r\}$. By the induction hypothesis, for each k, $1 \leqslant k \leqslant r$, the restriction of d to B_k can take at most $m_k - 1$ distinct positive values; therefore, d can take at most $1 + r + (m_1 - 1) + \cdots + (m_r - 1) = n - 1$ distinct values on S, which concludes the argument.

8. Let $(S, \mathcal{E})$ be a measurable space and let $m : \mathcal{E} \longrightarrow \hat{\mathbb{R}}_{\geqslant 0}$ be a measure. Prove that the d_m defined by $d_m(U, V) = m(U \oplus V)$ is a semimetric on $\mathcal{E}$.
9. Let $f : L \longrightarrow \mathbb{R}_{\geqslant 0}$ be a real-valued, nonnegative function, where $\mathcal{L} = (L, \{\wedge, \vee\})$ is a lattice. Define the mapping $d : L^2 \longrightarrow \mathbb{R}_{\geqslant 0}$ as $d(x, y) = 2f(x \wedge y) - f(x) - f(y)$ for $x, y \in L$. Prove that d is a semimetric on L if and only if f is anti-monotonic and submodular.
   Hint: Use Supplement 16 of Chap. 12.
10. Let $d : S \times S \longrightarrow \mathbb{R}_{\geqslant 0}$ be a metric on a set S and let k be a number $k \in \mathbb{R}_{\geqslant 0}$. Prove that the function $e : S \times S \longrightarrow \mathbb{R}_{\geqslant 0}$ defined by $\epsilon(x, y) = \min\{d(x, y), k\}$ is a metric on S.
11. Let S be a set and $e : \mathcal{P}(S)^2 \longrightarrow \mathbb{R}_{\geqslant 0}$ be the function defined by $e(X, Y) = |X - Y|$ for $X, Y \in \mathcal{P}(S)$. Prove that e satisfies the triangular axiom but fails to be a dissimilarity.
12. Let S be a set and $e : S^2 \longrightarrow \mathbb{R}$ be a function such that

    - $e(x, y) = 0$ if and only if $x = y$ for $x, y \in S$,
    - $e(x, y) = e(y, x)$ for $x, y \in S$, and
    - $e(x, y) \leqslant e(x, z) + e(z, y)$

    for $x, y, z \in S$. Prove that $e(x, y) \geqslant 0$ for $x, y \in S$.
13. Let S be a set and $f : S^2 \longrightarrow \mathbb{R}$ be a function such that

    - $f(x, y) = 0$ if and only if $x = y$ for $x, y \in S$;
    - $f(x, y) = f(y, x)$ for $x, y \in S$;
    - $f(x, y) \geqslant f(x, z) + f(z, y)$ for $x, y, z \in S$.

    Note that the triangular inequality was replaced with its inverse. Prove that the set S contains at most one element.
14. Let $d : S^2 \longrightarrow \mathbb{R}$ be a function such that $d(x, y) = 0$ if and only if $x = y$ and $d(x, y) \leqslant d(x, z) + d(y, z)$ (note the modification of the triangular axiom), for $x, y, z \in S$. Prove that d is a metric, that is, prove that $d(x, y) \geqslant 0$ and $d(x, y) = d(y, x)$ for all $x, y \in S$.
15. Let $f : \mathbb{R}_{\geqslant 0} \longrightarrow \mathbb{R}_{\geqslant 0}$ be a function that satisfies the following conditions:

    a) $f(x) = 0$ if and only if $x = 0$.
    b) f is monotonic on $\mathbb{R}_{\geqslant 0}$; that is, $x \leqslant y$ implies $f(x) \leqslant f(y)$ for $x, y \in \mathbb{R}_{\geqslant 0}$.
    c) f is subadditive on $\mathbb{R}_{\geqslant 0}$; that is, $f(x + y) \leqslant f(x) + f(y)$ for $x, y \in \mathbb{R}_{\geqslant 0}$.

    Prove that if d is a metric on a set S, then fd is also a metric on S.
16. Let $d : S \times S \longrightarrow \mathbb{R}_{\geqslant 0}$ be a metric on the set S. Prove that the function $e : S \times S \longrightarrow \mathbb{R}_{\geqslant 0}$ defined by $e(x, y) = \lceil d(x, y) \rceil$ for $x, y \in S$ is also be a metric on the set S. Also, prove that if the ceiling function is replaced by the

floor function, then this statement is no longer valid. Note that e is a discretized version of the metric d.

17. Let S be a set and let $c : S^2 \longrightarrow [0, 1]$ be a function such that $c(x, y)+c(y, x) = 1$ and $c(x, y) \leqslant c(x, t) + c(t, y)$ for every $x, y, t \in S$.

    (a) Prove that the relation $\rho_c = \{(x, y) \in S^2 \mid c(x, y) = 1\}$ is a strict partial order on S.

    (b) Let "<" be a strict partial order on a set S. Define the function $e : S^2 \longrightarrow \{0, \frac{1}{2}\}$ by

$$e(x, y) = \begin{cases} 1 & \text{if } x < y, \\ \frac{1}{2} & \text{if } x, y \text{ are incomparable} \\ 0 & \text{if } y < x, \end{cases}$$

    for $x, y \in S$. Prove that $e(x, y)+e(y, x) = 1$ and $c(x, y) \leqslant c(x, t)+c(t, y)$ for every $x, y, t \in S$.

18. Let S be a finite set and let $d : S^2 \longrightarrow \mathbb{R}_{\geqslant 0}$ be a dissimilarity. Prove that there exists $a \in \mathbb{R}_{\geqslant 0}$ such that the dissimilarity d_a defined by

$$d_a(x, y) = \begin{cases} (d(x, y))^a & \text{if } x \neq y \\ 0 & \text{if } x = y, \end{cases}$$

    for $x, y \in S$ satisfies the triangular inequality.
    Hint: Observe that $\lim_{a \to 0} d_a(x, y)$ is a dissimilarity that satisfies the triangular inequality.

19. Let U be a set and let $f : U \longrightarrow S$ be an injective function. Show that if (S, d) is a metric space, then the pair (U, d') is also a metric space, where $d'(u, v) = d(f(u), f(v))$.

20. Let $(S_1, d_1), \ldots, (S_n, d_n)$ be n metric spaces, where $n \geqslant 1$, and let ν be a norm on $\mathbb{R}^n$. Define the mapping $D_\nu : (S_1 \times \cdots \times S_n)^2 \longrightarrow \hat{\mathbb{R}}_{\geqslant 0}$ as $D_\nu(\mathbf{x}, \mathbf{y}) = \nu(d_1(x_1, y_1), \ldots, d_n(x_n, y_n))$ for $\mathbf{x} = (x_1, \ldots, x_n)$ and $\mathbf{y} = (y_1, \ldots, y_n)$.

    (a) Prove that D_ν is a metric on $S_1 \times \cdots \times S_n$.
    We refer to $(S_1 \times \cdots \times S_n, D_\nu)$ as the *ν-product of the metric spaces* $(S_1, d_1), \ldots, (S_n, d_n)$. When ν is the Euclidean norm, we refer to $(S_1 \times \cdots \times S_n, D_\nu)$ simply as the *product of the metric spaces* $(S_1, d_1), \ldots, (S_n, d_n)$.

    (b) Let (S_1, d_1), (S_2, d_2) be two metric spaces. Consider the functions $\delta, \delta' : S_1 \times S_2 \longrightarrow \hat{\mathbb{R}}_{\geqslant 0}$ given by

$$\delta((x, y), (u, v)) = d(x, u) + d(y, v),$$
$$\delta'((x, y), (u, v)) = \max\{d(x, u), d(y, v)\},$$

    for every $(x, y), (u, v) \in S_1 \times S_2$. Prove that both δ and δ' are metrics on the set product $S_1 \times S_2$.

21 Let (S, d) be a finite metric space. Prove that there exists a graph $\mathcal{G} = (S, E)$ such that d is the distance associated with this graph if and only if $d(x, y) \in \mathbb{N}$ and $d(x, y) \geqslant 2$ implies the existence of $z \in S$ such that $d(x, y) = d(x, z) + d(z, y)$ for $x, y \in S$.

22. Prove that every metric defined on a finite set S such that $|S| = 3$ is a tree metric.

23. Let S be a finite set, $d : S \times S \longrightarrow \mathbb{R}_{\geqslant 0}$ be a dissimilarity on S, and s be an element of S. Define the mapping $d_{s,k} : S \times S \longrightarrow \mathbb{R}$ by

$$d_{s,k}(x, y) = \begin{cases} \frac{k+d(x,y)-d(x,s)-d(y,s)}{2} & \text{if } x \neq y \\ 0 & \text{if } x = y, \end{cases}$$

for $x, y \in S$.

(a) Prove that there is $k > 0$ such that $d_{s,k} \geqslant 0$ for every $s \in S$.
(b) Prove that d is a tree metric if and only if there exists k such that $d_{s,k}$ is an ultrametric for all $s \in S$.

24. Let S be a set, π be a partition of S, and a, b be two numbers such that $a < b$. Prove that the mapping $d : S^2 \longrightarrow \mathbb{R}_{\geqslant 0}$ given by

$$d(x, y) = \begin{cases} 0 & \text{if } x = y, \\ a & \text{if } x \neq y \text{ and } x \equiv_{\pi} y, \\ b & \text{if } x \not\equiv_{\pi} y, \end{cases}$$

is an ultrametric on S.

25. Prove the following extension of the statement from Exercise 24.
Let S be a set, $\pi_0 < \pi_1 < \cdots < \pi_{k-1}$ be a chain of partitions on S, where $\pi_0 = \alpha_S$ and $\pi_{k-1} = \omega_S$ and let $0 < a_1 \ldots < a_{k-1} < a_k$ be a chain of positive reals. Prove that the mapping $d : S^2 \longrightarrow \mathbb{R}_{\geqslant 0}$ given by $d(x, y) = 0$ if $x = y$ and $d(x, y) = a_i$ if $i = \min\{p > 0 \mid (x, y) \in \pi_p\}$ is an ultrametric on S.

Solution: It is clear that $d(x, y) = 0$ if and only if $x = y$ and that $d(x, y) = d(y, x)$ for any $x, y \in S$. Thus, we need to show only that d satisfies the ultrametric property.
Suppose that $x, y, z \in S$ are such that $d(x, y) = a_i$ and $d(y, z) = a_j$, where $a_i < a_j$ and $i < j$. The definition of d implies that $x \not\equiv_{\pi_{i-1}} y$, $x \equiv_{\pi_i} y$, and $y \not\equiv_{\pi_{j-1}} z$, $y \equiv_{\pi_j} z$. Since $\pi_i < \pi_j$, it follows that $x \equiv_{\pi_j} z$ by the transitivity of $\equiv_{\pi_j}$. Thus, $d(x, z) \leqslant j = \max\{d(x, y), d(y, z)\}$.

26. Using the Steinhaus transform (Lemma 14.34 on page 673), prove that if the mapping $d : \mathbb{R}^n \times \mathbb{R}^n \longrightarrow \mathbb{R}_{\geqslant 0}$ is defined by

$$d(\mathbf{x}, \mathbf{y}) = \frac{d_2(\mathbf{x}, \mathbf{y})}{d_2(\mathbf{x}, \mathbf{y}) + \parallel \mathbf{x} \parallel + \parallel \mathbf{y} \parallel},$$

for $\mathbf{x}, \mathbf{y} \in \mathbb{R}^n$, then d is a metric on $\mathbb{R}^n$ such that $d(\mathbf{x}, \mathbf{y}) < 1$.

27. Using Exercises 17 and 26 prove that the mapping $e : \mathbb{R}^n \times \mathbb{R}^n \longrightarrow \mathbb{R}_{\geqslant 0}$ given by

$$e(\mathbf{x}, \mathbf{y}) = \frac{d_2(\mathbf{x}, \mathbf{y})}{\| \mathbf{x} \| + \| \mathbf{y} \|}$$

for $\mathbf{x}, \mathbf{y} \in \mathbb{R}^n$ is a metric on $\mathbb{R}^n$.

28. Prove that the following statements that concern a subset U of $(\mathbb{R}^n, d_p)$ are equivalent:

    (a) U is bounded.
    (b) There exists n closed intervals $[a_1, b_1], \ldots, [a_n, b_n]$ such that $U \subseteq [a_1, b_1] \times \cdots \times [a_n, b_n]$.
    (c) There exists a number $k \geqslant 0$ such that $d_p(\mathbf{0}, \mathbf{x}) \leqslant k$ for every $\mathbf{x} \in U$.

29. Let $S = \{\mathbf{x}_1, \ldots, \mathbf{x}_m\}$ be a finite subset of the metric space $(\mathbb{R}^2, d_p)$. Prove that there are at most m pairs of points $(\mathbf{x}_i, \mathbf{x}_j) \in S \times S$ such that $d_p(x_i, x_j) = diam(S)$.
30. Let $\mathbf{x}, \mathbf{y} \in \mathbb{R}^2$. Prove that $\mathbf{z}$ is outside the circle that has the diameter $\overline{\mathbf{x}, \mathbf{y}}$ if and only if $d_2^2(\mathbf{x}, \mathbf{z}) + d_2^2(\mathbf{y}, \mathbf{z}) > d_2^2(\mathbf{x}, \mathbf{y})$.
31. Let $S = \{\mathbf{x}_1, \ldots, \mathbf{x}_m\}$ be a finite subset of the metric space $(\mathbb{R}^2, d_p)$. The *Gabriel graph* of S is the graph $\mathcal{G} = (S, E)$, where $(\mathbf{x}_i, \mathbf{x}_j) \in E$ if and only if $d_2^2(\mathbf{x}_i, \mathbf{x}_k) + d_2^2(\mathbf{x}_j, \mathbf{x}_k) > d_2^2(\mathbf{x}, \mathbf{y})$ for every $k \in \{1, \ldots, n\} - \{i, j\}$.
    Prove that if $\mathbf{x}, \mathbf{y}, \mathbf{z} \in S$ and $(\mathbf{y} - \mathbf{x}) \cdot (\mathbf{z} - \mathbf{x}) < 0$, for some $\mathbf{y} \in S$, then there is no edge $(\mathbf{x}, \mathbf{z})$ in the Gabriel graph of S. Formulate an algorithm to compute the Gabriel graph of S that requires an amount of time that grows as m^2.
32. Let $\mathbf{C} \in \mathbb{R}^{n \times n}$ be a square matrix such that $\mathbf{C}\mathbf{w} = \mathbf{0}$ implies $\mathbf{w} = \mathbf{0}$ for $\mathbf{w} \in \mathbb{R}^{n \times 1}$. Define $d_\mathbf{C} : \mathbb{R}^n \times \mathbb{R}^n \longrightarrow \mathbb{R}$ by $d_\mathbf{C}(\mathbf{x}, \mathbf{y}) = (\mathbf{x} - \mathbf{y})'\mathbf{C}'\mathbf{C}(\mathbf{x} - \mathbf{y})$ for $\mathbf{x}, \mathbf{y} \in \mathbb{R}^{n \times 1}$. Prove that $d_\mathbf{C}$ is a metric on $\mathbb{R}^n$.
33. Let $U_n = \{(x_1, \ldots, x_n) \in \mathbb{R}^n \mid \sum_{i=1}^n x_i^2 = 1\}$ be the set of unit vectors in $\mathbb{R}^n$. Prove that the mapping $d : U_n^2 \longrightarrow \mathbb{R}_{\geqslant 0}$ defined by

$$d(\mathbf{x}, \mathbf{y}) = \arccos \left(\sum_{i=1}^n x_i y_i \right),$$

where $\mathbf{x} = (x_1, \ldots, x_n)$ and $\mathbf{y} = (y_1, \ldots, y_n)$ belong to U_n, is a metric on U_n.

34. Let (S, d) be a finite metric space. Prove that the functions $D, E : \mathcal{P}(S)^2 \longrightarrow \mathbb{R}$ defined by

$$D(U, V) = \max\{d(u, v) \mid u \in U, v \in V\},$$
$$E(U, V) = \frac{1}{|U| \cdot |V|} \sum \{d(u, v) \mid u \in U, v \in V\},$$

for $U, V \in \mathcal{P}(S)$ such that $U \neq V$, and $D(U, U) = E(U, U) = 0$ for every $U \in \mathcal{P}(S)$ are metrics on $\mathcal{P}(S)$.

35. Prove that if we replace max by min in Exercise 14.7.1, then the resulting function $F : \mathcal{P}(S)^2 \longrightarrow \mathbb{R}$ defined by

$$D(U, V) = \min\{d(u, v) \mid u \in U, v \in V\}$$

for $U, V \in \mathcal{P}(S)$ is not a metric on $\mathcal{P}(S)$, in general.
Solution: Let $S = U \cup V \cup W$, where

$$\begin{aligned} U &= \{(0,0), (0,1), (1,0), (1,1,)\}, \\ V &= \{(2,0), (2,1), (2+\ell, 0), (2+\ell, 1)\}, \\ W &= \{(\ell+1, 0), (\ell+1, 1), (\ell+2, 0), (\ell+2, 1)\}. \end{aligned}$$

The metric d is the usual Euclidean metric in $\mathbb{R}^2$. Note that $F(U, V) = F(V, W) = 1$; however, $F(U, W) = \ell + 2$. Thus, if $\ell > 0$, the triangular axiom is violated by F.

36. Let (S, d) be a metric space. Prove that:

(a) $d(x, T) \leqslant d(x, y) + d(y, T)$ for every $x, y \in S$ and $T \in \mathcal{P}(S)$.
(b) If U and V are nonempty subsets of S, then:

$$\inf_{x \in U} d(x, V) = \inf_{x \in V} d(x, U).$$

37. Let S be a finite set and let $\delta : \mathcal{P}(S)^2 \longrightarrow \mathbb{R}_{\geqslant 0}$ defined by

$$\delta(X, Y) = \frac{|X \oplus Y|}{|X| + |Y|}$$

for $X, Y \in \mathcal{P}(S)$. Prove that δ is a dissimilarity but not a metric.
Hint: Consider the set $S = \{x, y\}$ and its subsets $X = \{x\}$ and $Y = \{y\}$. Compare $\delta(X, Y)$ with $\delta(X, S) + \delta(S, Y)$.

38. Let S be a finite set and let π and $\sigma \in PART(S)$. Prove that

$$d_\beta(\pi, \sigma) \leqslant d_\beta(\alpha_S, \omega_S) = \mathcal{H}_\beta(\alpha_S)$$

for every $\beta \geqslant 1$.

39. Let S be a finite set and let π, σ be two partitions on S. Prove that if σ covers the partition π, then there exist $B_i, B_j \in \pi$ such that

$$d_2(\pi, \sigma) = \frac{4 \cdot |B_i| \cdot |B_j|}{|S|^2}.$$

40. Let X be a set of attributes of a table $\theta = (T, H, \mathbf{r})$, $\mathbf{r} = (t_1, \ldots, t_n)$, and let π^X be the partition of $\{1, \ldots, n\}$ defined on page 571. For $\beta \in \mathbb{R}$ such that $\beta > 1$, prove that:

(a) We have $\mathcal{H}_\beta(\pi^{UV}) = \mathcal{H}_\beta(\pi^U|\pi^V) + \mathcal{H}(\pi^V)$.
(b) If $D_\beta : \mathcal{P}(H)^2 \longrightarrow \mathbb{R}_{\geqslant 0}$ is the semimetric defined by $D_\beta(U, V) = d_\beta(\pi^U, \pi^V)$, show that $D_\beta(U, V) = 2\mathcal{H}_\beta(\pi^{UV}) - \mathcal{H}_\beta(\pi^U) - \mathcal{H}_\beta(\pi^V)$.
(c) Prove that if θ satisfies the functional dependency $U \longrightarrow V$, then $D_\beta(U, V) = \mathcal{H}_\beta(\pi^U) - \mathcal{H}_\beta(\pi^V)$.
(d) Prove that $D_\beta(U, V) \leqslant \mathcal{H}_\beta(\pi^{UV}) - \mathcal{H}_\beta(\pi^U \vee \pi^V)$.

41. An attribute A of a table θ is said to be *binary* if $\text{Dom}(A) = \{0, 1\}$. Define the contingency matrix of two binary attributes of a table $\theta = (T, H, \mathbf{r})$ as the 2×2-matrix

$$K_{AB} = \begin{pmatrix} n_{00} & n_{01} \\ n_{10} & n_{11} \end{pmatrix},$$

where $\mathbf{r} = (t_1, \ldots, t_n)$ and $n_{ij} = |\{k \mid t_k[AB] = (i, j)\}|$. Prove that $D_2(U, V) = \frac{4}{n}(n_{00} + n_{11})(n_{10} + n_{01})$, where D_2 is a special case of the semimetric D_β introduced in Exercise 14.7.1.

42. Let $\pi = \{B_1, \ldots, B_m\}$ and $\sigma = \{C_1, \ldots, C_n\}$ be two partitions of a set S. The *Goodman-Kruskal* coefficient of π and σ is the number

$$GK(\pi, \sigma) = 1 - \frac{1}{|S|}\sum_{i=1}^{m} \max_{1 \leqslant j \leqslant n} |B_i \cap C_j|.$$

(a) Prove that $GK(\pi, \sigma) = 0$ if and only if $\pi \leqslant \sigma$.
(b) Prove that the function GK is monotonic in the first argument and dually monotonic in the second argument.
(c) If $\theta, \pi, \sigma \in PART(S)$, then prove that:

$$GK(\pi \wedge \theta, \sigma) + GK(\theta, \pi) \geqslant GK(\theta, \pi \wedge \sigma).$$

(d) Prove that $GK(\theta, \pi) + GK(\pi, \sigma) \geqslant GK(\theta, \sigma)$ for $\theta, \pi, \sigma \in PART(S)$.
(e) Prove that the mapping $d_{GK} : PART(S) \times PART(S) \longrightarrow \mathbb{R}$ given by

$$d_{GK}(\pi, \sigma) = GK(\pi, \sigma) + GK(\sigma, \pi)$$

for $\pi, \sigma \in PART(S)$, is a metric on $PART(S)$.

A longest common subsequence of two sequences $\mathbf{x}$ and $\mathbf{y}$ is a sequence $\mathbf{z}$ that is a subsequence of both $\mathbf{x}$ and $\mathbf{y}$ and is of maximal length. For example, if $\mathbf{x} = (a_1, a_2, a_3, a_4, a_2, a_2)$ and $\mathbf{y} = (a_3, a_2, a_1, a_3, a_2, a_1, a_1, a_2, a_1)$, then both (a_2, a_3, a_2, a_2) and (a_1, a_3, a_2, a_2) are both longest subsequences of $\mathbf{x}$ and $\mathbf{y}$. The length of a longest common subsequence of $\mathbf{x}$ and $\mathbf{y}$ is denoted by $llcs(\mathbf{x}, \mathbf{y})$.

43. Let $\mathbf{x} = (x_0, \ldots, x_{n-1})$ and $\mathbf{y} = (y_0, \ldots, y_{m-1})$ be two sequences. Prove that we have

$$llcs(\mathbf{x}, \mathbf{y}) = \begin{cases} 0 & \text{if } \mathbf{x} = \lambda \text{ or } \mathbf{y} = \lambda, \\ llcs(\mathbf{x}_{0,n-2}, \mathbf{y}_{0,m-2}) + 1 & \text{if } x_{n-1} = y_{m-1}, \\ \max\{llcs(\mathbf{x}_{0,n-2}, \mathbf{y}), llcs(\mathbf{x}, \mathbf{y}_{0,m-2})\}. \end{cases}$$

Based on this equality, formulate a tabular algorithm (similar to the one used to compute Levenshtein's distance) that can be used to compute $llcs(\mathbf{x}, \mathbf{y})$ and all longest common subsequences of $\mathbf{x}$ and $\mathbf{y}$.

44. Let d be the string distance calculated with the cost scheme $(1, 1, \infty)$. Prove that $d(\mathbf{x}, \mathbf{y}) = |\mathbf{x}| + |\mathbf{y}| - 2llcs(\mathbf{x}, \mathbf{y})$.
45. Let d be the string distance calculated with the cost scheme $(\infty, \infty, 1)$. Show that if $\mathbf{x} = (x_0, \ldots, x_{n-1})$ and $\mathbf{y} = (y_0, \ldots, y_{m-1})$, then

$$d(\mathbf{x}, \mathbf{y}) = \begin{cases} \infty & \text{if } |\mathbf{x}| \neq |\mathbf{y}|, \\ |\{i \mid 0 \leqslant i \leqslant n-1, x_i \neq y_i\}| & \text{if } |\mathbf{x}| = |\mathbf{y}|. \end{cases}$$

Let $\pi = \{B_1, \ldots, B_m\}$ and $\sigma = \{C_1, \ldots, C_n\}$ of a finite set $S = \{s_1, \ldots, s_\ell\}$. The *contingency matrix* of the partitions π and σ is the $m \times n$ matrix $Q(\pi, \sigma)$, where $Q(\pi, \sigma)_{ij} = |B_i \cap C_j|$ for $1 \leqslant i \leqslant m$ and $1 \leqslant j \leqslant n$. The element $Q(\pi, \sigma)_{ij}$ will be denoted by q_{ij} for $1 \leqslant i \leqslant m$ and $1 \leqslant j \leqslant n$.
The *marginal totals* of $Q(\pi, \sigma)$ are:

$$q_{\cdot j} = \sum_{i=1}^{m} q_{ij} \text{ for } 1 \leqslant j \leqslant n, \text{ and,}$$

$$q_{i\cdot} = \sum_{j=1}^{n} q_{ij} \text{ for } 1 \leqslant i \leqslant m.$$

Clearly, $|C_j| = q_{\cdot j}$, $|B_i| = q_{i\cdot}$ and $|S| = \sum_{i=1}^{m} q_{i\cdot} = \sum_{j=1}^{n} q_{\cdot j} = |S|$. Also, we have:

$$\sum_{i=1}^{m}\sum_{j=1}^{n} q_{ij} = \sum_{i=1}^{m} q_{i\cdot} = \sum_{j=1}^{n} q_{\cdot j} = \ell.$$

A subset T of S is π-*homogeneous* if there exists a block B_i such that $T \subseteq B_i$. The set of unordered pairs of elements of S was denoted by $\mathcal{P}_2(S)$. If $|S| = \ell$, then it is easy to see that $|\mathcal{P}_2(S)| = \frac{\ell^2-\ell}{2}$ distinct unordered pairs of elements. An unordered pair $\wp = \{s, s'\}$ is said to be of

1. *type 1* if $\wp$ is both π-homogeneous and σ-homogeneous;
2. *type 2* if $\wp$ is neither π-homogeneous nor σ-homogeneous;
3. *type 3* if $\wp$ is not π-homegeneous but is σ-homogeneous;
4. *type 4* if $\wp$ is π-homegeneous but not σ-homogeneous.

The *number of agreements* $agr(\pi, \sigma)$ of the partitions π, σ is the total number of pairs of type 1 and 2; the number of disagreements of these partitions $dagr(\pi, \sigma)$ is the total number of pairs of types 3 and 4. Clearly, we have:

$$agr(\pi, \sigma) + dagr(\pi, \sigma) = \frac{\ell^2 - \ell}{2}.$$

46. Prove that:

(a) the number of pairs of type 1 equals

$$\ell_1 = \sum_{i=1}^{m}\sum_{j=1}^{n} \frac{q_{ij}^2 - q_{ij}}{2} = \frac{1}{2}\left(\sum_{i=1}^{m}\sum_{j=1}^{n} q_{ij}^2 - \ell\right);$$

(b) the number of pairs of type 3 is

$$\ell_3 = \sum_{j=1}^{n} \frac{q_{\cdot j}^2 - q_{\cdot j}}{2} - \sum_{i=1}^{m}\sum_{j=1}^{n} \frac{q_{ij}^2 - q_{ij}}{2} = \frac{1}{2}\left(\sum_{j=1}^{n} q_{\cdot j}^2 - \sum_{i=1}^{m}\sum_{j=1}^{n} q_{ij}^2\right);$$

(c) the number of pairs of type 4 is:

$$\ell_4 = \sum_{i=1}^{m} \frac{q_{i\cdot}^2 - q_{i\cdot}}{2} - \sum_{i=1}^{m}\sum_{j=1}^{n} \frac{q_{ij}^2 - q_{ij}}{2} = \frac{1}{2}\left(\sum_{i=1}^{m} q_{i\cdot}^2 - \sum_{i=1}^{m}\sum_{j=1}^{n} q_{ij}^2\right).$$

(d) the number of pairs of type 2 is:

$$\ell_2 = \frac{\ell^2 - \ell}{2} - \ell_1 - \ell_3 - \ell_4$$
$$= \frac{1}{2}\left(\ell^2 + \sum_{i=1}^{m}\sum_{j=1}^{n} q_{ij}^2 - \sum_{i=1}^{m} q_{i\cdot}^2 - \sum_{j=1}^{n} q_{\cdot j}^2\right).$$

47. Prove that $agr(\pi, \sigma)$ is:

$$agr(\pi, \sigma) = \frac{1}{2}\left(2\sum_{i=1}^{m}\sum_{j=1}^{n} q_{ij}^2 + \ell^2 - \ell - \sum_{i=1}^{m} q_{i\cdot}^2 - \sum_{j=1}^{n} q_{\cdot j}^2\right).$$

48. Prove that the number of disaggrements is:

$$dagr(\pi, \sigma) = \frac{1}{2}\left(\sum_{j=1}^{n} q_{\cdot j}^2 + \sum_{i=1}^{m} q_{i\cdot}^2 - 2\sum_{i=1}^{m}\sum_{j=1}^{n} q_{ij}^2\right).$$

49. The *Rand index of two partitions* $\pi, \sigma \in PART(S)$ is the number:

$$rand(\pi, \sigma) = \frac{agr(\pi, \sigma)}{\binom{|S|}{2}},$$

Prove that:

(a) $dagr(\pi, \sigma) = \frac{|S|^2}{4} \cdot d_2(\pi, \sigma)$;
(b) $rand(\pi, \sigma) = 1 - \frac{|S|^2}{2(|S|^2 - |S|)} d_2(\pi, \sigma)$;
(c) $0 \leqslant rand(\pi, \sigma) \leqslant 1$; moreover, $rand(\pi, \sigma) = 1$ if and only if $\pi = \sigma$.

50. A *shortest common supersequence* of two sequences $\mathbf{x}$ and $\mathbf{y}$ is a sequence of minimum length that contains both $\mathbf{x}$ and $\mathbf{y}$ as subsequences. Prove that the length of a shortest common supersequence of $\mathbf{x}$ and $\mathbf{y}$ equals $|\mathbf{x}| + |\mathbf{y}| - llcs(\mathbf{x}, \mathbf{y})$.
51. Let (S, d) be a finite metric space. A *metric tree for* (S, d) (see [11]) is a binary tree $\mathcal{T}(S, d)$, defined as follows:

- If $|S| = 1$, then $\mathcal{T}(S, d)$ consists of a single node that is sphere $B(s, 0)$, where $S = \{s\}$.
- If $|S| > 1$ create a node v labeled by a sphere $B(s, r) \subseteq S$ and construct the trees $\mathcal{T}(B(s, r), d)$ and $\mathcal{T}(S - B(s, r), d)$. Then, make $\mathcal{T}(B(s, r)$ and $\mathcal{T}(S - B(s, r)$ the direct descendants of v.
  Design an algorithm for retrieving the k-nearest members of S to a query $q \in S$ using an existing metric tree.

The AESA algorithm (an acronym of Approximating and Eliminating Search Algorithm) shown as Algorithm 14.7.6 starts with a finite metric space (S, d), a subset $X = \{x_1, \ldots, x_n\}$ of S, and a query $q \in S$ and produces the nearest neighbors of q in X.
The values of distances between the members of X are precomputed. The algorithm uses the semimetric D_T defined in Exercise 65.
The algorithm partitions the set X into three sets: K, A, E, where K is the set of *known elements* (that is, the set of elements of S for which $d(x, q)$ has been computed), A is the set of *active elements*, and E is the set of *eliminated elements* defined by

$$E = \{x \in X \mid D_K(x, q) > \min\{d(y, q) \mid y \in K\}\}.$$

The algorithm is given next.

Algorithm 14.7.6: The Approximating and Eliminating Search Algorithm

Data: a metric space (S, d), a subset X of S, and a query $q \in S$
Result: the set of nearest neighbors of q in X
compute the matrix of dissimilarities $(d(x_i, x_j))$ of the elements of X (preprocessing phase);
$A = X$; $K = \emptyset$; $E = \emptyset$;
while $(A \neq \emptyset)$ **do**
  $D_K(x, q) = \infty$;
  select $x \in A$ such that $x = \arg\min\{D_K(x, q) \mid x \in A\}$;
  compute $d(x, q)$; $K = K \cup \{x\}$; $A = A - \{x\}$;
  update $r = \min\{d(x, q) \mid x \in K\}$;
  update $D_K(x', q)$ for all $x' \in A$ as
  $D_{K \cup \{x\}}(x', q) = \max\{D_K(x', q), |d(x, q) - d(x, x')|\}$;
  $K = K \cup \{x\}$;
  if $D_K(x', q) > r$ **then**
    $A = A - \{x'\}$;
    $E = E \cup \{x'\}$;
  end
  return the set K
end

52. Prove that the AESA algorithm is indeed computing the set of nearest neighbors of q.
53. Let (S, d) be a metric space and let $X = \{x_1, \ldots, x_n\}$ be a finite subset of S. Suppose that not all distances between the elements of X are known. The *distance graph* of X is a weighted graph $(\mathcal{G}_X, w)$, where $\mathcal{G}_X = (X, E)$. An edge (x, y) exists in the underlying graph $\mathcal{G}_X$ if $d(x, y)$ is known; in this case, $w(x, y) = d(x, y)$.
If $\mathbf{p}$ is a simple path in the graph $(\mathcal{G}_X, w)$ that joins x to y, define $\eta(\mathbf{p}) = w(\hat{e}) - \sum\{w(e) \mid e \text{ is in } \mathbf{p}, e \neq \hat{e}\}$, where $\hat{e}$ is the edge of maximum weight in $\mathbf{p}$. Prove that $d(x, y) \geqslant \eta(\mathbf{p})$.
54. Let *paths*(x, y) be the set of simple paths in $(\mathcal{G}_X, w)$ that joins x to y. Define the *approximate distance map* for X as an $n \times n$ matrix $\mathbf{A} = (a_{ij})$ such that

$$a_{ij} = \max\{\eta(\mathbf{p}) \mid \mathbf{p} \in paths(x_i, x_j)\}$$

for $1 \leqslant i, j \leqslant n$. Also, define the $n \times n$-matrix $\mathbf{M} = (m_{ij})$ as

$$m_{ij} = \min\{w(\mathbf{p}) \mid \mathbf{p} \in paths(x_i, x_j)\}$$

for $1 \leqslant i, j \leqslant n$. Prove that $a_{ij} \leqslant d(x_i, x_j) \leqslant m_{ij}$ for $1 \leqslant i, j \leqslant n$.
55. Let $paths_k(x, y)$ be the set of simple paths in $(\mathcal{G}_X, w)$ that join x to y and do not pass through any of the vertices numbered higher than k, where $k \geqslant 1$. Denote by $a_{ij}^k, m_{ij}^k, b_{ij}^k$ the numbers

$$a_{ij}^k = \max\{\eta(\mathbf{p}) \mid \mathbf{p} \in paths_k(x_i, x_j)\},$$
$$m_{ij}^k = \min\{w(\mathbf{p}) \mid \mathbf{p} \in paths_k(x_i, x_j)\},$$
$$b_{ij}^k = \max\{\eta(\mathbf{p}) \mid \mathbf{p} \in paths_k(x_i, x_k)paths_k(x_k, x_j)\}$$

for $1 \leqslant i, j \leqslant n$. Prove that $b_{ij}^k = \max\{a_{ij}^{k-1} - m_{ij}^{k-1}, a_{jk}^{k-1} - m_{ki}^{k-1}\}$ for $1 \leqslant k \leqslant n$.

Bibliographical Comments

A comprehensive, research-oriented source for metric spaces is [10].

The reader should consult two excellent surveys [4, 12] on searches in metric spaces. Nearest-neighbor searching is comprehensively examined in [11, 13]. The AESA algorithm was introduced in [14, 15]. Weighted graphs of partially defined metric spaces and approximate distance maps are defined and studied in [16, 17], where Exercises 53–55 originate. Finally, we refer the reader to two fundamental references for all things about metrics [10],[18].

Supplement 7 is due to C. Zara [19]; see also [20].

References

1. P. Buneman, A note on metric properties of trees. J. Comb. Theory **17**, 48–50 (1974)
2. M. Meilă, in *International Conference on Machine Learning, Bonn, Germany*, ed. by L. DeRaedt, S. Wrobel. Comparing clusterings: an axiomatic view (ACM, New York, 2005), pp. 577–584
3. V.I. Levenshtein, Binary code capable of correcting deletions, insertions and substitutions. Cybern. Control Theory **163**, 845–848 (1966)
4. E. Chávez, G. Navarro, R. Baeza-Yates, J.L. Marroquín, Searching in metric spaces. ACM Comput. Surv. **33**, 273–321 (2001)
5. W.A. Burkhard, R.M. Keller, Some approaches to best-match file searching. Commun. ACM **16**, 230–236 (1973)
6. M.T. Orchard, A fast nearest-neighbor search algorithm. Int. Conf. Acoust. Speech Signal Process. **4**, 2297–3000 (1991)
7. C.M. Huang, Q. Bi, G.S. Stiles, R.W. Harris, Fast full search equivalent encoding algorithms for image compression using vector quantization. IEEE Trans. Image Process. **1**, 413–416 (1992)
8. K. Zakloutal, M.H. Johnson, R.E. Ladner, in *DIMACS Series in Discrete Mathematics: Data Structures, Near Neighbor Searches, and Methodology: Fifth and Sixth Implementation Challenges*, ed. by M.H. Goldwasser, D.S. Johnson, C.C. McGeogh. Nearest neighbor search for data compression (American Mathematical Society, Providence, 2002), pp. 69–86
9. A. Faragó, T. Linder, G. Lugosi, Fast nearest-neighbor search in dissimilarity spaces. IEEE Trans. Pattern Anal. Mach. Intell. **15**, 957–962 (1993)
10. M.M. Deza, M. Laurent, *Geometry of Cuts and Metrics* (Springer, Heidelberg, 1997)
11. K.L. Clarkson, in *Nearest-Neighbor Methods for Learning and Vision: Theory and Practice*, ed. by G. Shakhnarovich, T. Darrell, P. Indyk. Nearest-neighbor searching and metric space dimensions (MIT Press, Cambridge, 2006), pp. 15–59

12. C. Böhm, S. Berthold, D.A. Keim, Searching in high-dimensional spaces—index structures for improving the performance of multimedia databases. ACM Comput. Surv. **33**, 322–373 (2001)
13. K. L. Clarkson, in *Proceedings of the 29th ACM Symposium on Theory of Computation*. Nearest neighbor queries in metric spaces, El Paso (ACM, New York, 1997), pp. 609–617
14. E. Vidal, An algorithm for finding nearest neighbours in (approximately) average constant time. Pattern Recogn. Lett. **4**, 145–157 (1986)
15. E. Vidal, New formulation and improvements of the nearest neighbours in approximating end eliminating search algorithms (AESA). Pattern Recogn. Lett. **15**, 1–7 (1994)
16. T. L. Wang, D. Shasha, in *Proceedings of the 16th VLDB Conference*, ed. by D. McLeod, R. Sacks-Davis, and H.-J.Schek. Query processing for distance metrics, Brisbane (Morgan Kaufmann, San Francisco, 1990), pp. 602–613
17. D. Shasha, T.L. Wang, New techniques for best-match retrieval. ACM Trans. Inf. Syst. **8**, 140–158 (1990)
18. E. Deza, M.M. Deza, *Dictionary of Distances* (Elsevier, Amsterdam, 2006)
19. C. Zara, Personal communication (2013)
20. A.J. Lemin, V.A. Lemin, On a universal ultrametric space. Topol. Appl. **103**, 339–345 (2000)

Chapter 15
Dimensions of Metric Spaces

15.1 Introduction

Subsets of $\mathbb{R}^n$ may have "intrinsic" dimensions that are much lower than n. Consider, for example, two distinct vectors $\mathbf{a}, \mathbf{b} \in \mathbb{R}^n$ and the line $L = \{\mathbf{a} + t\mathbf{b} \mid t \in \mathbb{R}\}$. Intuitively, L has the intrinsic dimensionality 1; however, L is embedded in $\mathbb{R}^n$ and from this point of view is an n-dimensional object. In this chapter we examine formalisms that lead to the definition of this intrinsic dimensionality.

Difficulties related to the high number of correlated features that occur when data mining techniques are applied to data of high dimensionality are collectively designated as the *dimensionality curse*. In Sect. 15.3 we discuss properties of the $\mathbb{R}^n$ spaces related to the dimensionality curse and we show how the reality of highly dimensional spaces contradicts the common intuition that we acquire through our common experience with lower dimensional spaces. Higher dimensional spaces are approached using analogies with lower dimensional spaces.

15.2 The Euler Functions and the Volume of a Sphere

The functions B and Γ defined by the integrals

$$B(a, b) = \int_0^1 x^{a-1}(1-x)^{b-1}\, dx \text{ and } \Gamma(a) = \int_0^\infty x^{a-1}e^{-x}\, dx,$$

are known as *Euler's integral of the first type* and *Euler's integral of the second type*, respectively. We assume here that a and b are positive numbers to ensure that the integrals are convergent.

Replacing x by $1 - x$ yields the equality

$$B(a, b) = -\int_1^0 (1-x)^{a-1}(x)^{b-1}\, dx = B(b, a),$$

D. A. Simovici and C. Djeraba, *Mathematical Tools for Data Mining*, Advanced Information and Knowledge Processing, DOI: 10.1007/978-1-4471-6407-4_15,

which shows that B is symmetric.

Integrating $B(a, b)$ by parts, we obtain

$$\begin{aligned} B(a,b) &= \int_0^1 x^{a-1}(1-x)^{b-1}\,dx = \int_0^1 (1-x)^{b-1}\,d\frac{x^a}{a} \\ &= \left.\frac{x^a(1-x)^{1-b)}}{a}\right|_0^1 + \frac{b-1}{a}\int_0^1 x^a(1-x)^{b-2}\,dx \\ &= \frac{b-1}{a}\int_0^1 x^{a-1}(1-x)^{b-2}\,dx - \frac{b-1}{a}\int_0^1 x^{a-1}(1-x)^{b-1}\,dx \\ &= \frac{b-1}{a}B(a,b-1) - \frac{b-1}{a}B(a,b), \end{aligned}$$

which yields

$$B(a,b) = \frac{b-1}{a+b-1}B(a,b-1). \tag{15.1}$$

The symmetry of the function B allows us to infer the formula

$$B(a,b) = \frac{a-1}{a+b-1}\cdot B(a-1,b).$$

If $b = n \in \mathbb{N}$, a repeated application of Equality (15.1) allows us to write

$$B(a,n) = \frac{n-1}{a+n-1}\cdot\frac{n-2}{a+n-2}\cdots\frac{1}{a+1}\cdot B(a,1).$$

It is easy to see that $B(a, 1) = \frac{1}{a}$. Thus,

$$B(a,n) = B(n,a) = \frac{1\cdot 2\cdot\cdots\cdot(n-1)}{a\cdot(a+1)\cdot\cdots\cdot(a+n-1)}.$$

If a is also a natural number, $a = m \in \mathbb{N}$, then

$$B(m,n) = \frac{(n-1)!(m-1)!}{(m+n-1)!}.$$

Next, we show the connection between Euler's integral functions:

$$B(a,b) = \frac{\Gamma(a)\Gamma(b)}{\Gamma(a+b)}. \tag{15.2}$$

Replacing x in the integral

$$\Gamma(a) = \int_0^\infty x^{a-1}e^{-x}\,dx$$

by $x = ry$ with $r > 0$ gives $\Gamma(a) = r^a \int_0^\infty y^{a-1}e^{-ry}\,dy$.

Replacing a by $a+b$ and r by $r+1$ yields the equality

$$\Gamma(a+b)(r+1)^{-(a+b)} = \int_0^\infty y^{a+b-1}e^{-(r+1)y}\,dy.$$

By multiplying both sides by r^{a-1} and integrating, we have

$$\Gamma(a+b)\int_0^\infty r^{a-1}(r+1)^{-(a+b)}\,dr = \int_0^\infty r^{a-1}\left(\int_0^\infty y^{a+b-1}e^{-(r+1)y}\,dy\right)dr.$$

By the definition of B, the last equality can be written

$$\Gamma(a+b)B(a,b) = \int_0^\infty r^{a-1}\left(\int_0^\infty y^{a+b-1}e^{-(r+1)y}\,dy\right)dr.$$

By permuting the integrals from the right member (we omit the justification of this manipulation), the last equality can be written as

$$\Gamma(a+b)B(a,b) = \int_0^\infty y^{a+b-1}e^{-y}\left(\int_0^\infty r^{a-1}e^{-ry}\,dr\right)dy.$$

Note that $\int_0^\infty r^{a-1}e^{-ry}\,dr = \frac{\Gamma(a)}{y^a}$. Therefore,

$$\Gamma(a+b)B(a,b) = \int_0^\infty y^{a+b-1}e^{-y}\frac{\Gamma(a)}{y^a}\,dy = \int_0^\infty y^{b-1}e^{-y}\Gamma(a)\,dy = \Gamma(a)\Gamma(b),$$

which is Formula (15.2).

The Γ function is a generalization of the factorial. Starting from the definition of Γ and integrating by parts, we obtain

$$\Gamma(x) = \int_0^\infty x^{a-1}e^{-x}\,dx = \frac{x^a}{a}e^{-x}\Bigg|_0^\infty + \frac{1}{a}\int_0^\infty x^a e^{-x}\,dx = \frac{1}{a}\Gamma(a+1).$$

Thus, $\Gamma(a+1) = a\Gamma(a)$. Since $\Gamma(1) = \int_0^\infty e^{-x}\,dx = 1$, it is easy to see that $\Gamma(n+1) = n!$ for $n \in \mathbb{N}$.

Using an argument from classical analysis it is possible to show that Γ has derivatives of arbitrary order and that we can compute these derivatives by deriving the function under the integral sign. Namely, we can write:

$$\Gamma'(a) = \int_0^\infty x^{a-1}(\ln x)e^{-x}\, dx,$$

and, in general, $\Gamma^{(n)}(a) = \int_0^\infty x^{a-1}(\ln x)^n e^{-x}\, dx$. Thus, $\Gamma^{(2)}(a) > 0$, which shows that the first derivative is increasing.

Since $\Gamma(1) = \Gamma(2) = 1$, there exists $a \in [1, 2]$ such that $\Gamma'(a) = 0$. For $0 < x < a$, we have $\Gamma'(x) \leqslant 0$, so Γ is decreasing. For $x > a$, $\Gamma'(x) \geqslant 0$, so Γ is increasing. It is easy to see that

$$\lim_{x \to 0+} \Gamma(x) = \frac{\Gamma(x+1)}{x} = \infty,$$

and $\lim_{x\to\infty} \Gamma(x) = \infty$.

An integral that is useful for a variety of applications is

$$I = \int_{\mathbb{R}} e^{-\frac{1}{2}t^2}\, dt.$$

We prove that $I = \sqrt{2\pi}$.

We can write

$$I^2 = \int_{\mathbb{R}} e^{-\frac{1}{2}x^2}\, dx \cdot \int_{\mathbb{R}} e^{-\frac{1}{2}y^2}\, dy = \int_{\mathbb{R}^2} e^{-\frac{x^2+y^2}{2}}\, dx\, dy.$$

Changing to polar coordinates by using the transformation $x = \rho\cos\theta$ and $y = \rho\sin\theta$ whose Jacobian is

$$\begin{vmatrix} \frac{\partial x}{\partial \rho} & \frac{\partial x}{\partial \theta} \\ \frac{\partial y}{\partial \rho} & \frac{\partial y}{\partial \theta} \end{vmatrix} = \begin{vmatrix} \cos\theta & -\rho\sin\theta \\ \sin\theta & \rho\cos\theta \end{vmatrix} = \rho,$$

we have

$$I^2 = \int_{\mathbb{R}^2} e^{-\frac{\rho^2}{2}} \rho\, d\rho\, d\theta = \int_0^{2\pi} d\theta \int_0^\infty e^{-\frac{\rho^2}{2}} \rho\, d\rho = 2\pi.$$

Thus, $I = \sqrt{2\pi}$. Since $e^{-\frac{1}{2}t^2}$ is an even function, it follows that

$$\int_0^\infty e^{-\frac{1}{2}t^2}\, dt = \sqrt{\frac{\pi}{2}}.$$

Using this integral, we can compute the value of $\Gamma\left(\frac{1}{2}\right)$. Note that Since $\Gamma\left(\frac{1}{2}\right) = \int_0^\infty \frac{e^{-x}}{\sqrt{x}}\, dx$, by applying the change of variable $x = \frac{t^2}{2}$, we have

$$\Gamma\left(\frac{1}{2}\right) = \sqrt{2}\cdot\int_0^\infty e^{-\frac{1}{2}t^2}\,dt = \sqrt{\pi}. \tag{15.3}$$

The last equality allows us to compute the values of the form $\Gamma\left(\frac{2p+1}{2}\right)$. It is easy to see that

$$\Gamma\left(\frac{2p+1}{2}\right) = \frac{(2p-1)\cdot(2p-3)\cdots 3\cdot 1}{2^p}\sqrt{\pi} = \frac{(2p)!}{p!2^{2p}}\sqrt{\pi}. \tag{15.4}$$

A closed sphere centered in $(0,\ldots,0)$ and having the radius R in $\mathbb{R}^n$ is defined as the set of points:

$$S_n(R) = \left\{(x_1,\ldots,x_n)\in\mathbb{R}^n \mid \sum_{i=1}^n x_i^2 = 1\right\}.$$

The volume of this sphere is denoted by $V_n(R)$.

We approximate the volume of an n-dimensional sphere of radius R as a sequence of $n-1$-dimensional spheres of radius $r(u) = \sqrt{R^2 - u_2}$, where u varies between $-R$ and R. This allows us to write

$$V_{n+1}(R) = \int_{-R}^{R} V_n(r(u))\,du.$$

We seek $V_n(R)$ as a number of the form $V_n(R) = k_n R^n$. Thus, we have

$$\begin{aligned} V_{n+1}(R) &= k_n\int_{-R}^{R}(r(u))^n\,du = k_n\int_{-R}^{R}(R^2-u^2)^{\frac{n}{2}}\,du \\ &= k_n R^n\int_{-R}^{R}\left(1-\left(\frac{u}{R}\right)^2\right)^{\frac{n}{2}}du \\ &= V_n(R)\int_{-R}^{R}\left(1-\left(\frac{u}{R}\right)^2\right)^{\frac{n}{2}}du = RV_n(R)\int_{-1}^{1}(1-x^2)^{\frac{n}{2}}\,dx. \end{aligned}$$

In turn, this yields the recurrence

$$k_{n+1} = k_n\int_{-1}^{1}(1-x^2)^{\frac{n}{2}}\,dx.$$

Note that

$$\int_{-1}^{1}(1-x^2)^{\frac{n}{2}}\,dx = 2\cdot\int_0^1(1-x^2)^{\frac{n}{2}}\,dx$$

because the function $(1-x^2)^{\frac{n}{2}}$ is even. To compute the latest integral, substitute $u = x^2$. We obtain

$$\int_0^1 (1-x^2)^{\frac{n}{2}}\, dx = \frac{1}{2}\int_0^1 u^{-\frac{1}{2}}(1-u)^{\frac{n}{2}}\, du,$$

which equals $\frac{1}{2} \cdot B(\frac{1}{2}, \frac{n}{2}+1)$. Using the Γ function, the integral can be written as

$$\int_0^1 (1-x^2)^{\frac{n}{2}}\, dx = \frac{1}{2} \cdot \frac{\Gamma(\frac{1}{2})\Gamma(\frac{n}{2}+1)}{\Gamma\left(\frac{n}{2}+\frac{3}{2}\right)}.$$

Thus,

$$k_{n+1} = k_n \frac{\Gamma\left(\frac{1}{2}\right)\Gamma\left(\frac{n}{2}+1\right)}{\Gamma\left(\frac{n+1}{2}+1\right)}.$$

Since $k_1 = 2$, this implies

$$k_n = 2\left(\Gamma\left(\frac{1}{2}\right)\right)^{n-1} \frac{\Gamma\left(\frac{1}{2}+1\right)}{\Gamma\left(\frac{n}{2}+1\right)} = \left(\Gamma\left(\frac{1}{2}\right)\right)^{n} \frac{1}{\Gamma\left(\frac{n}{2}+1\right)} = \pi^{\frac{n}{2}} \frac{1}{\Gamma\left(\frac{n}{2}+1\right)}.$$

Thus, the volume of the n-dimensional sphere of radius R equals

$$\frac{\pi^{\frac{n}{2}} R^n}{\Gamma\left(\frac{n}{2}+1\right)}.$$

For $n = 1, 2, 3$, by applying Formula (15.4), we obtain the well-known values $2R$, πR^2, and $\frac{4\pi R^3}{3}$, respectively. For $n = 4$, the volume of the sphere is $\frac{\pi^2 R^4}{2}$.

15.3 The Dimensionality Curse

The term "dimensionality curse," invented by Richard Bellman in [1], is used to describe the difficulties of exhaustively searching a space of high dimensionality for an optimum value of a function defined on such a space. These difficulties stem from the fact that the size of the sets that must be searched increases exponentially with the number of dimensions. Moreover, phenomena that are at variance with the common intuition acquired in two- or three-dimensional spaces become more significant. This section is dedicated to a study of these phenomena.

The dimensionality curse impacts many data mining tasks, including classification and clustering. Thus, it is important to realize the limitations imposed by high-dimensional data on designing data mining algorithms.

Let $Q_n(\ell)$ be an n-dimensional cube in $\mathbb{R}^n$. The volume of this cube is ℓ^n. Consider the n-dimensional closed sphere of radius R that is centered in the center of the cube $Q_n(2R)$ and is tangent to the opposite faces of this cube. We have:

$$\lim_{n\to\infty} \frac{V_n(R)}{2^n R^n} = \frac{\pi^{\frac{n}{2}}}{2^n \Gamma\left(\frac{n}{2}+1\right)} = 0.$$

In other words, as the dimensionality of the space grows, the fraction of the cube volume that is located inside the sphere decreases and tends to become negligible for very large values of n.

It is interesting to compare the volumes of two concentric spheres of radii R and $R(1-\epsilon)$, where $\epsilon \in (0,1)$. The volume located between these spheres relative to the volume of the larger sphere is

$$\frac{V_n(R) - V_n(R(1-\epsilon))}{V_n(R)} = 1 - (1-\epsilon)^n,$$

and we have

$$\lim_{n\to\infty} \frac{V_n(R) - V_n(R(1-\epsilon))}{V_n(R)} = 1.$$

Thus, for large values of n, the volume of the sphere of radius R is concentrated mainly near the surface of this sphere.

Let $Q_n(1)$ be a unit side-length n-dimensional cube, $Q_n(1) = [0,1]^n$, centered in $\mathbf{c}_n = (0.5, \ldots, 0.5) \in R^n$. The d_2-distance between the center of the cube $\mathbf{c}_n$ and any of its vertices is $\sqrt{0.5^2 + \cdots 0.5^2} = 0.5\sqrt{n}$, and this value tends to infinity with the number of dimensions n despite the fact that the volume of the cube remains equal to 1. On the other hand, the distance from the center of the cube to any of its faces remains equal to 0.5. Thus, the n-dimensional cube is exhibits very different properties in different directions; in other words the n-dimensional cube is an anisotropic object.

An interesting property of the unit cube $Q_n(1)$ is observed in [2]. Let $P = (p, \ldots, p) \in \mathbb{R}^n$ be a point located on the main diagonal of $Q_n(1)$ and let K be the subcube of $Q_n(1)$ that includes $(0, \ldots, 0)$ and P and has a side of length p; similarly, let K' be the subcube of $Q_n(1)$ that includes P and $(1, \ldots, 1)$ and has side of length $1-p$. The ratio of the volumes V and V' of the cubes K and K' is

$$r(p) = \left(\frac{p}{1-p}\right)^n.$$

To determine the increase δ of p needed to double the volume of this ratio, we must find δ such that $\frac{r(p+\delta)}{r(p)} = 2$, that is

$$\frac{p(1-p) + \delta(1-p)}{p(1-p) - \delta p} = \sqrt[n]{2}.$$

Equivalently, we have

$$\delta = \frac{p(1-p)(\sqrt[n]{2}-1)}{1-p+p\sqrt[n]{2}}.$$

The first factor $\frac{p(1-p)}{1-p+p\sqrt[n]{2}}$ remains almost constant for large values of n. However, the second factor $\sqrt[n]{2}-1$ tends toward 0, which shows that within large dimensionality smaller and smaller moves of the point p are needed to double the ratio of the volumes of the cubes K and K'. This suggests that the division of $Q_n(1)$ into subcubes is very unstable. If data classifications are attempted based on the location of data vectors in subcubes, this shows in turn the instability of such classification schemes.

Another interesting example of the counterintuitive behavior of spaces of high dimensionality is given in [3]. Now let $Q_n(1)$ be the unit cube centered in the point $\mathbf{c}_n \in \mathbb{R}^n$, where $\mathbf{c}_n = (0.5, \ldots, 0.5)$. For $n = 2$ or $n = 3$, it is easy to see that every sphere that intersects the sides of $Q_2(1)$ or all faces of $Q_3(1)$ must contain the center of the cube $\mathbf{c}_n$. We shall see that, for sufficiently high values of n a sphere that intersects all $(n-1)$-dimensional faces of $Q_n(1)$ does not necessarily contain the center of $Q_n(1)$.

Consider the closed sphere $B(\mathbf{q}_n, r)$, whose center is the point $\mathbf{q}_n = (q, \ldots, q)$, where $q \in [0, 1]$. Clearly, we have $\mathbf{q}_n \in Q_n(1)$ and $d_2(\mathbf{c}_n, \mathbf{q}_n) = \sqrt{n(q^2 - q + 0.25)}$.

If the radius r of the sphere $B(\mathbf{q}_n, r)$ is sufficiently large, then $B(\mathbf{q}_n, r)$ intersects all faces of Q_n. Indeed, the distance from $\mathbf{q}_n$ to an $(n-1)$-dimensional face is no more than $\max\{q, 1-q\}$, which shows that $r \geqslant \max\{q, 1-q\}$ ensures the nonemptiness of all these intersections. Thus, the inequalities

$$n\,(q-0.5)^2 > r^2 > \max\{q^2, (1-q)^2\} \tag{15.5}$$

ensure that $B(\mathbf{q}_n, r)$ intersects every $(n-1)$-dimensional face of Q_n, while leaving $\mathbf{c}_n$ outside $B(\mathbf{q}_n, r)$. This is equivalent to requiring

$$n > \frac{\max\{q^2, (1-q)^2\}}{(q-0.5)^2}.$$

For example, if we choose $q = 0.3$, then $n > \frac{0.7^2}{0.2^2} = 12.25$. Thus, in the case of R^{13}, Inequality (15.5) amounts to $0.52 > r^2 > 0.49$. Choosing $r = \frac{\sqrt{2}}{2}$ gives the sphere with the desired "paradoxical" property.

The examples discussed in this section suggest that precautions and sound arguments are needed when trying to extrapolate familiar properties of two- or three-dimensional spaces to spaces of higher dimensionality.

15.4 Inductive Dimensions of Topological Metric Spaces

We present two variants of the inductive dimensions of topological metric spaces: the *small inductive dimension* $ind(S, \mathcal{O}_d)$ and the *large inductive dimension* $IND(S, \mathcal{O}_d)$. Informally, these dimensions capture the intuitive idea that a sphere $B(\mathbf{x}, r)$ in $\mathbb{R}^{n+1}$ has a border that is n-dimensional. They are defined by inductive definitions, which we present next.

Definition 15.1 *Let $(S, \mathcal{O}_d)$ be a topological metric space. The* large inductive dimension *of $(S, \mathcal{O}_d)$ is a member of the set $\{n \in \mathbb{Z} \mid n \geqslant -1\} \cup \{\infty\}$ defined by:*

1. *if $S = \emptyset$ and $\mathcal{O}_d = \{\emptyset\}$, then $IND(S, \mathcal{O}_d) = -1$;*
2. *$IND(S, \mathcal{O}_d) \leqslant n$ for $n \geqslant 0$ if, for every closed set H and every open set L such that $H \subseteq L$, there exists an open set V such that*

$$H \subseteq V \subseteq L \text{ and } IND(\partial V, \mathcal{O}_d{\restriction}_{\partial V}) \leqslant n - 1;$$

3. *$IND(S, \mathcal{O}_d) = n$ if $IND(S, \mathcal{O}_d) \leqslant n$ and $IND(S, \mathcal{O}_d) \not\leqslant n - 1$;*
4. *if there is no integer $n \geqslant -1$ such that $IND(S, \mathcal{O}_d) = n$, then $IND(S, \mathcal{O}_d) = \infty$.*

Theorem 15.2 *If $IND(S, \mathcal{O}_d) \in \mathbb{Z}$, then $IND(S, \mathcal{O}_d)$ is the smallest integer n such that $n \geqslant -1$, and for every closed set H and every open set L of the topological metric space $(S, \mathcal{O}_d)$ such that $H \subseteq L$, there exists an open set V such that $H \subseteq V \subseteq L$ such that $IND(\partial V, \mathcal{O}_d{\restriction}_{\partial V}) \leqslant n - 1$.*

Proof The statement is an immediate consequence of Definition 15.1.

If we relax the requirement of Definition 15.1 by asserting the existence of the set V only when the closed set H is reduced to an element of S, we obtain the following definition of the small inductive dimension.

Definition 15.3 *Let $(S, \mathcal{O}_d)$ be a topological metric space. The* small inductive dimension *of $(S, \mathcal{O}_d)$ is a member of the set $\{n \in \mathbb{Z} \mid n \geqslant -1\} \cup \{\infty\}$ defined by:*

(i) *if $S = \emptyset$ and $\mathcal{O}_d = \{\emptyset\}$, then $ind(S, \mathcal{O}_d) = -1$;*
(ii) *$ind(S, \mathcal{O}_d) \leqslant n$, where $n \geqslant 0$, if for $x \in S$ and every open set L that contains x, there exists an open set V such that $x \in V \subseteq L$ such that $ind(\partial V, \mathcal{O}_d{\restriction}_{\partial V}) \leqslant n - 1$;*
(iii) *$ind(S, \mathcal{O}_d) = n$ if $IND(S, \mathcal{O}_d) \leqslant n$ and $ind(S, \mathcal{O}_d) \not\leqslant n - 1$;*
(iv) *if there is no integer $n \geqslant -1$ such that $ind(S, \mathcal{O}_d) = n$, then $ind(S, \mathcal{O}_d) = \infty$.*

Theorem 15.4 *If $ind(S, \mathcal{O}_d) \in \mathbb{Z}$, then $ind(S, \mathcal{O}_d)$ is the smallest integer n such that $n \geqslant -1$, and for every $x \in S$ and every open set L that contains x, there is an open set V such that $x \in V \subseteq L$ and $ind(\partial V, \mathcal{O}_d{\restriction}_{\partial V}) \leqslant n - 1$.*

Proof The statement is an immediate consequence of Definition 15.3.

Since $\{x\}$ is a closed set for every $x \in S$, it is clear that, for every topological metric space $(S, \mathcal{O}_d)$, we have $ind(S, \mathcal{O}_d) \leqslant IND(S, \mathcal{O}_d)$.

If there is no risk of confusion, we denote $ind(S, \mathcal{O}_d)$ and $IND(S, \mathcal{O}_d)$ by $ind(S)$ and $IND(S)$, respectively.

Definition 15.5 *A topological metric space* $(S, \mathcal{O}_d)$ *is* zero-dimensional *if* $ind(S, \mathcal{O}_d) = 0$.

Clearly, if $IND(S, \mathcal{O}_d) = 0$, then $(S, \mathcal{O}_d)$ is zero-dimensional.

Theorem 15.6 *Let* $(S, \mathcal{O}_d)$ *be a nonempty topological metric space. The space is zero-dimensional if and only if there exists a basis for* $\mathcal{O}_d$ *that consists of clopen sets.*

Proof Suppose that $ind(S) = 0$. By Definition 15.3, for every $x \in S$ and every open set L, there is an open set V such that $x \in V \subseteq L$ and $ind(\partial V) \leqslant -1$, which implies $ind(\partial V) = -1$ and thus $\partial V = \emptyset$. This shows that V is a clopen set and the collection of all such sets V is the desired basis.

Conversely, if there exists a basis $\mathcal{B}$ for $\mathcal{O}_d$ such that each set in $\mathcal{B}$ is clopen, then for every $x \in S$ and open set L there exists $V \in \mathcal{B}$ such that $\partial V = \emptyset$, $x \in V = \mathbf{K}(V) \subseteq L$. This implies $ind(S) = 0$.

Theorem 15.7 *Let* $(S, \mathcal{O}_d)$ *be a zero-dimensional separable topological metric space. If* H_1 *and* H_2 *are two disjoint closed subsets of* S*, there exists a clopen set* U *such that* $H_1 \subseteq U$ *and* $U \cap H_2 = \emptyset$.

Proof Since $ind(S) = 0$, by Theorem 15.6 there exists a base $\mathcal{B}$ of $(S, \mathcal{O}_d)$ that consists of clopen sets.

Let $x \in S$. If $x \notin H_1$, then x belongs to the open set $S - H_1$, so there exists $U_x \in \mathcal{B}$ such that $x \in U_x \subseteq S - H_1$, which implies $U_x \cap H_1 = \emptyset$.

If $x \notin H_2$, a similar set U_x can be found such that $x \in U_x \cap H_2 = \emptyset$. Since every x is in either of the two previous cases, it follows that $\mathcal{U} = \{U_x \mid x \in S\}$ is an open cover of S and each set U_x is disjoint from H_1 or H_2.

By Theorem 8.32, the separability of $(S, \mathcal{O}_d)$ implies that $\mathcal{U}$ contains a countable subcover, $\{U_{x_1}, U_{x_2}, \ldots\}$. Let $V_1, V_2, \ldots$ be the sequence of clopen sets defined inductively by $V_1 = U_{x_1}$, and $V_n = U_{x_n} - \bigcup_{i=1}^{n-1} V_i$ for $n \geqslant 1$. The sets V_i are pairwise disjoint, $\bigcup_{i \geqslant 0} V_i = S$, and each set V_i is disjoint from H_1 or H_2.

Let $U = \bigcup\{V_i \mid V_i \cap H_2 = \emptyset\}$. The set U is open, $H_1 \subseteq U$, and $U \cap H_2 = \emptyset$. Note that the set U is also closed because $S - U = \bigcup\{V_i \mid V_i \cap H_2 \neq \emptyset\}$ is also open. This means that U is clopen and satisfies the conditions of the theorem.

Theorem 15.7 can be restated by saying that in a zero-dimensional space $(S, \mathcal{O}_d)$, for any two disjoint closed subsets of S, H_1 and H_2, there exist two disjoint clopen subsets U_1 and U_2 such that $H_1 \subseteq U_1$ and $H_2 \subseteq U_2$. This follows from the fact that we can choose $U_1 = U$ and $U_2 = S - U$.

Corollary 15.8 *Let* $(S, \mathcal{O}_d)$ *be a zero-dimensional separable topological metric space. If* H *is a closed set and* L *is an open set such that* $H \subseteq L$*, then there exists a clopen set* U *such that* $H \subseteq U \subseteq L$.

Proof This statement is immediately equivalent to Theorem 15.7.

Corollary 15.9 *Let $(S, \mathcal{O}_d)$ be a separable topological metric space. We have $ind(S) = 0$ if and only if $IND(S) = 0$.*

Proof We saw that $IND(S) = 0$ implies $ind(S) = 0$. Suppose that $ind(S) = 0$. By Corollary 15.9, if H is a closed set and L is an open set such that $H \subseteq L$, then there exists a clopen set U such that $H \subseteq U \subseteq L$. This implies $IND(S) = 0$ by Theorem 15.2.

Example 15.10 If $(S, \mathcal{O}_d)$ is a nonempty topological ultrametric space, then $ind(S) = 0$ because the collection of open spheres is a basis for $\mathcal{O}_d$ that consists of clopen sets (see Theorem 14.5).

Example 15.11 For any nonempty, finite topological metric space $(S, \mathcal{O}_d)$, we have $ind(S) = 0$. Indeed, consider the a open sphere $C(x, \epsilon)$. If we choose $\epsilon < \min\{d(x, y) \mid x, y \in S \text{ and } x \neq y\}$, then each open sphere consists of $\{x\}$ itself and thus is a clopen set.

Example 15.12 Let $\mathbb{Q}$ be the set of rational numbers and let $\mathbb{I} = \mathbb{R} - \mathbb{Q}$ be the set of irrational numbers. Consider the topological metric spaces $(\mathbb{Q}, \mathcal{O}')$ and $(\mathbb{I}, \mathcal{O}'')$, where the topologies $\mathcal{O}'$ and $\mathcal{O}''$ are obtained by restricting the usual metric topology $\mathcal{O}_d$ of $\mathbb{R}$ to $\mathbb{Q}$ and $\mathbb{I}$, respectively. We claim that $ind(\mathbb{Q}, \mathcal{O}') = ind(\mathbb{I}, \mathcal{O}'') = 0$.

Let r be a rational number and let α be an irrational positive number. Consider the set $D(r, \alpha) = \{q \in \mathbb{Q} \mid |r - q| < \alpha\}$. It is easy to see that the collection $\{D(r, \alpha) \mid r \in \mathbb{Q}, \alpha \in \mathbb{I}\}$ is a basis for $(\mathbb{Q}, \mathcal{O}')$. We have

$$\begin{aligned} \partial D(r, \alpha) &= \{q \in \mathbb{Q} \mid |q - r| \leqslant \alpha\} \cap \{q \in \mathbb{Q} \mid |q - r| \geqslant \alpha\} \\ &= \{q \in \mathbb{Q} \mid |q - r| = \alpha\} = \emptyset \end{aligned}$$

because the difference of two rational numbers is a rational number. Therefore, the sets of the form $D(r, \alpha)$ are clopen and $ind(\mathbb{Q}, \mathcal{O}') = 0$.

Let r and p be two rational numbers. Consider the set of irrational numbers

$$E(r, p) = \{\alpha \in \mathbb{I} \mid |r - \alpha| < p\}.$$

We claim that the collection $\mathcal{E} = \{E(r, p) \mid r, p \in \mathbb{Q}\}$ is a basis for $(\mathbb{I}, \mathcal{O}'')$. Indeed, let $\beta \in \mathbb{I}$, and let L be an open set in $\mathcal{O}''$. There exists an open sphere $C(\beta, u)$ such that $u > 0$ and $C(\beta, u) \subseteq L$. Let $r_1, r_2 \in \mathbb{Q}$ be two rational numbers such that $\beta - u < r_1 < \beta < r_2 < \beta + u$. If we define $r = (r_1 + r_2)/2$ and $p = (r_2 - r_1)/2$, then $\beta \in E(r, p) \subseteq C(\beta, u) \subseteq L$, which proves that $\mathcal{E}$ is indeed a basis. We have

$$\begin{aligned} \partial E(r, p) &= \{\alpha \in \mathbb{I} \mid |r - \alpha| \leqslant p\} \cap \{\alpha \in \mathbb{I} \mid |r - \alpha| \geqslant p\} \\ &= \{\alpha \in \mathbb{I} \mid |r - \alpha| = p\} = \emptyset \end{aligned}$$

for reasons similar to the ones given above. The sets in the basis $\mathcal{E}$ are clopen, and therefore $ind(\mathbb{I}, \mathcal{O}'') = 0$.

Example 15.13 We have $ind(\mathbb{R}, \mathcal{O}) = 1$. Indeed, by Theorem 8.53, its topological dimension is not 0 and, on the other hand, it has a basis that consists of spheres $C(x, r)$, that are open intervals of the form $(x - r, x + r)$. Clearly, $\partial(x - r, x + r)$ is the finite set $\{-r, r\}$, which has a small inductive dimension equal to 0. Therefore, $ind(\mathbb{R}, \mathcal{O}) = 1$. It is interesting to observe that this shows that the union of two zero-dimensional sets is not necessarily zero-dimensional because $ind(\mathbb{Q}, \mathcal{O}') = ind(\mathbb{I}, \mathcal{O}'') = 0$, as we saw in Example 15.12.

Theorem 15.14 *Let* $(S, \mathcal{O}_d)$ *be a topological metric space and let* T *be a subset of* S. *We have* $ind(T, \mathcal{O}_d \restriction_T) \leqslant ind(S, \mathcal{O}_d)$.

Proof The statement is immediate if $ind(S, \mathcal{O}_d) = \infty$. The argument for the finite case is by strong induction on $n = \dim(S, \mathcal{O}_d) \geqslant -1$.

For the base case, $n = -1$, the space $(S, \mathcal{O}_d)$ is the empty space $(\emptyset, \{\emptyset\})$, so $T = \emptyset$ and the inequality obviously holds.

Suppose that the statement holds for topological metric spaces of dimension no larger than n. Let $(S, \mathcal{O}_d)$ be a metric topological space such that $ind(S, \mathcal{O}_d) = n + 1$, T be a subset of S, t be an element of T, and L be an open set in $(T, \mathcal{O}_d \restriction_T)$ such that $t \in L$.

There is an open set $L_1 \in \mathcal{O}_d$ such that $L = L_1 \cap T$. Since $ind(S, \mathcal{O}_d) = n + 1$, n is the least integer such that there is an open set $W \subseteq S$ for which

$$t \in W \subseteq L_1 \tag{15.6}$$

and $ind(\partial W) \leqslant n$. The set $V = W \cap T$ is an open set in $(T, \mathcal{O}_d \restriction_T)$, and we have

$$t \in V \subseteq L$$

by intersecting the sets involved in Inclusions (15.6) with T. Theorem 4.31 implies that

$$\partial_T(V) = \partial_T(W \cap T) \subseteq \partial_S(W)$$

and, by the inductive hypothesis, $ind(\partial_T(V)) \leqslant ind(\partial_S(W)) \leqslant n$. Therefore, the small inductive dimension of T cannot be greater than $n + 1$, which is the desired conclusion.

A similar statement involving the large inductive dimension can be shown.

Theorem 15.15 *Let* $(S, \mathcal{O}_d)$ *be a topological metric space and let* T *be a subset of* S. *We have* $IND(T, \mathcal{O}_d \restriction_T) \leqslant IND(S, \mathcal{O}_d)$.

Proof The argument is similar to the one given in the proof of Theorem 15.14.

We denote $ind(T, \mathcal{O}_d \restriction_U)$ by $ind(T)$.

An extension of Theorem 15.6 is given next.

Theorem 15.16 *Let $(S, \mathcal{O}_d)$ be a topological metric space. We have $ind(S) = n$, where $n \geqslant 0$, if and only if n is the smallest integer such that there exists a basis for $\mathcal{O}_d$ such that for every $B \in \mathcal{B}$ we have $ind(\partial B) \leqslant n - 1$.*

Proof The necessity of the condition is immediate from Definition 15.3 because the sets V constitute a basis that satisfies the condition.

To prove that the condition is sufficient, note that from the proof of Theorem 8.25 we obtain the existence of two open disjoint sets V_1 and V_2 such that $\{x\} \subseteq V_1$ and $S - L \subseteq V_2$ because $\{x\}$ and $S - L$ are two disjoint closed sets. This is equivalent to $x \in V_1 \subseteq S - V_2 \subseteq L$ and, because $S - V_2$ is closed, we have $x \in V_1 \subseteq \mathbf{K}(V_1) \subseteq L$. Let B be a set in the basis such that $x \in B \subseteq V_1$. We have $x \in B \subseteq L$ and $ind(\partial B) \leqslant n - 1$; since n is the least number with this property, we have $ind(S) = n$.

Corollary 15.17 *For every separable topological metric space $(S, \mathcal{O}_d)$, we have $ind(S) = n$, where $n \geqslant 0$, if and only if n is the smallest integer such that there exists a countable basis for $\mathcal{O}_d$ such that, for every $B \in \mathcal{B}$, we have $ind(\partial B) \leqslant n - 1$.*

Proof This statement is a consequence of Theorems 15.16 and 4.49.

The inductive dimensions can be alternatively described using the notion of set separation.

Definition 15.18 *Let $(S, \mathcal{O})$ be a topological space, and let X and Y be two disjoint subsets of S. The subset T of S* separates the sets X and Y *if there exists two open, disjoint sets L_1 and L_2 in $(S, \mathcal{O})$ such that $X \subseteq L_1$, $Y \subseteq L_2$, and $T = S - (L_1 \cup L_2)$.*

It is clear that if T separates X and Y, then T must be a closed subset of S.

Observe that the empty set separates the sets X and Y if and only if the space S is the union of two open disjoint sets L_1 and L_2 such that $X \subseteq L_1$ and $Y \subseteq L_2$. Since L_1 is the complement of L_2, both L_1 and L_2 are clopen sets.

Theorem 15.19 *Let $(S, \mathcal{O})$ be a topological space, and let X and Y be two disjoint subsets of S. The set T separates the sets X and Y if and only if the following conditions are satisfied:*

(i) *T is a closed set in $(S, \mathcal{O})$, and*
(ii) *there exist two disjoint sets K_1 and K_2 that are open in the subspace $S - T$ such that $X \subseteq K_1$, $Y \subseteq K_2$, and $S - T = K_1 \cup K_2$.*

Proof Suppose that T separates the sets X and Y. It is clear that we have both $X \subseteq S - T$ and $Y \subseteq S - T$. We already observed that T is closed, and so $S - T$ is open. Therefore, the sets L_1 and L_2 considered in Definition 15.18 are open in $S - T$ and the second condition is also satisfied.

Conversely, suppose that conditions (i) and (ii) are satisfied. Since T is closed, $S - T$ is open. Since K_1 and K_2 are open in $S - T$, they are open in $(S, \mathcal{O})$, so T separates X and Y.

Theorem 15.20 *Let $(S, \mathcal{O})$ be a topological space, H be a closed set, and L be an open set such that $H \subseteq L$. The set T separates the disjoint sets H and $S - L$ if and only if there exists an open set V and a closed set W such that the following conditions are satisfied:*

(i) $H \subseteq V \subseteq W \subseteq S - L$ *and*
(ii) $T = W - V$.

Proof Suppose that T separates H and $S - L$. There are two disjoint open sets L_1 and L_2 such that $H \subseteq L_1$, $S - L \subseteq L_2$, and $T = S - (L_1 \cup L_2)$. This implies $S - L_2 \subseteq L$, and $T = (S - L_1) \cap (S - L_2)$. Let $V = L_1$ and $W = S - L_2$. Since L_1 and L_2 are disjoint, it is clear that $V \subseteq W$. Also, $T = (S - V) \cap W = W - V$.

Conversely, if the conditions of the theorem are satisfied, then T separates H and $S - L$ because V and $S - W$ are the open sets that satisfy the requirements of Definition 15.18.

Using the notion of set separation, we have the following characterization of topological metric spaces having large or small inductive dimension n.

Theorem 15.21 *Let $(S, \mathcal{O}_d)$ be a topological metric space and let $n \in \mathbb{N}$. The following statements hold:*

(i) *$IND(S) = n$ if and only if n is the smallest integer such that for every closed subset H and open set L of S such that $H \subseteq L$ there exists a set W with $IND(W) \leqslant n - 1$ that separates H and L;*
(ii) *$ind(S) = n$ if and only if n is the smallest integer such that for any element x of S and any open set L that contains x there exists a set W with $ind(W) \leqslant n - 1$ that separates $\{x\}$ and L.*

Proof Suppose that $IND(S) = n$. By Definition 15.1, n is the smallest integer such that $n \geqslant -1$, and for every closed set H and every open set L such that $H \subseteq L$, there exists an open set V such that $H \subseteq V \subseteq \mathbf{K}(V) \subseteq L$ such that $IND(\partial V) \leqslant n - 1$. Let $W = \mathbf{K}(V) - V$. It is clear that W separates H and L. Since

$$W = \mathbf{K}(V) - V = \mathbf{K}(V) \cap (S - V) \subseteq \mathbf{K}(V) \cap \mathbf{K}(S - V) = \partial(V),$$

it follows that $IND(W) \leqslant n - 1$.

Conversely, suppose that n is the least integer such that for any closed set H and open set L such that $H \subseteq L$ there exist an open set V and a closed set U such that $H \subseteq V \subseteq U \subseteq L$, $W = U - V$, and $IND(W) \leqslant n - 1$. Clearly, we have $\mathbf{K}(V) \subseteq U$ and therefore

$$H \subseteq V \subseteq \mathbf{K}(V) \subseteq L.$$

Note that

$$\begin{aligned}
\partial(V) &= \mathbf{K}(V) \cap \mathbf{K}(S - V) \\
&= \mathbf{K}(V) \cap (S - V) \\
&\quad \text{because } S - V \text{ is a closed set} \\
&\subseteq U \cap (S - V) = U - V = W,
\end{aligned}$$

which implies $IND(V) \leqslant n - 1$.

The proof of the second part of the theorem is similar.

The next statement shows the possibility of lifting the separation of two closed sets from a subspace to the surrounding space.

Theorem 15.22 *Let $(S, \mathcal{O}_d)$ be a topological metric space, and let H_1 and H_2 be two closed and disjoint subsets of S. Suppose that U_1 and U_2 are two open subsets of S such that $H_1 \subseteq U_1$, $H_2 \subseteq U_2$ and $\mathbf{K}(U_1) \cap \mathbf{K}(U_2) = \emptyset$.*

If $T \subseteq S$ and the set K separates the sets $T \cap \mathbf{K}(U_1)$ and $T \cap \mathbf{K}(U_2)$ in the subspace $(T, \mathcal{O}_d \upharpoonright_T)$, then there exists a subset W of S, that separates H_1 and H_2 in $(S, \mathcal{O}_d)$ such that $W \cap T \subseteq K$.

Proof Since K separates the sets $T \cap \mathbf{K}(U_1)$ and $T \cap \mathbf{K}(U_2)$ in T there are two open, disjoint subsets V_1 and V_2 of T such that $T \cap \mathbf{K}(U_1) \subseteq V_1$, $T \cap \mathbf{K}(U_2) \subseteq V_2$, and $T - K = V_1 \cup V_2$.

We have

$$\begin{aligned} U_1 \cap V_2 &= U_1 \cap (T \cap V_2) \\ &= (U_1 \cap T) \cap V_2 \\ &\subseteq \mathbf{K}(U_1) \cap T \cap V_2 \\ &\subseteq V_1 \cap V_2 = \emptyset, \end{aligned}$$

and therefore $U_1 \cap \mathbf{K}(V_2) = \emptyset$, because U_1 is open (by Theorem 4.9). Therefore, $H_1 \cap \mathbf{K}(V_2) = \emptyset$. Similarly, $H_2 \cap \mathbf{K}(V_1) = \emptyset$. Consequently,

$$\begin{aligned} (H_1 \cup V_1) \cap (\mathbf{K}(H_2 \cup V_2)) &= (H_1 \cup V_1) \cap (\mathbf{K}(H_2) \cup \mathbf{K}(V_2)) \\ &= (H_1 \cup V_1) \cap (H_2 \cup \mathbf{K}(V_2)) = \emptyset \end{aligned}$$

and similarly

$$\mathbf{K}(H_1 \cup V_1) \cap (H_2 \cup V_2) = \emptyset.$$

By Supplement 11 of Chap. 8, there exist two disjoint open sets Z_1 and Z_2 such that $H_1 \cup V_1 \subseteq Z_1$ and $H_2 \cup V_2 \subseteq Z_2$. Then, the set $W = S - (Z_1 \cup Z_2)$ separates H_1 and H_2, and $W \cap T \subseteq T - (Z_1 \cup Z_2) \subseteq T - (V_1 \cup V_2) = K$.

We can now extend Corollary 15.8.

Corollary 15.23 *Let $(S, \mathcal{O}_d)$ be a separable topological metric space and let T be a zero-dimensional subspace of S. For any disjoint closed sets H_1 and H_2 in S, there exist two disjoint open sets L_1 and L_2 such that $H_1 \subseteq L_1$, $H_2 \subseteq L_2$, and $T \cap \partial L_1 = T \cap \partial L_2 = \emptyset$.*

Proof By Theorem 8.28, there are two open sets V_1 and V_2 such that $H_1 \subseteq V_1$, $H_2 \subseteq V_2$, and $\mathbf{K}(V_1) \cap \mathbf{K}(V_2) = \emptyset$. By Theorem 15.7, there exists a clopen subset U of T such that $T \cap \mathbf{K}(V_1) \subseteq U$ and $T \cap \mathbf{K}(V_2) \subseteq T - U$. Therefore, we have

$$T - U \subseteq S - \mathbf{K}(V_1) \subseteq S - V_1 \subseteq S - H_1,$$
$$U \subseteq T - \mathbf{K}(V_2) \subseteq S - V_2 \subseteq S - H_2,$$

which implies that the sets $H_1 \cup U$ and $H_2 \cup (T - U)$ are disjoint.

Let $f, g : S \longrightarrow \mathbb{R}$ be the continuous functions defined by $f(x) = d_{H_1 \cup U}(x)$ and $g(x) = d_{H_2 \cup (T-U)}(x)$ for $x \in S$. The open sets

$$L_1 = \{x \in S \mid f(x) < g(x)\},$$
$$L_2 = \{x \in S \mid f(x) > g(x)\},$$

are clearly disjoint. Note that if $x \in H_1$ we have $f(x) = 0$ and $g(x) > 0$, so $H_1 \subseteq L_1$. Similarly, $H_2 \subseteq L_2$.

Since U is closed in T, we have $f(x) = 0$ and $g(x) > 0$ for every $x \in U$; similarly, since $T - U$ is closed in T, we have $f(x) > 0$ and $g(x) = 0$. Thus, $U \subseteq L_1$ and $T - U \subseteq L_2$, so $T \subseteq L_1 \cup L_2$.

Note that we have the inclusions

$$\begin{aligned}
\partial L_1 &= \mathbf{K}(L_1) \cap \mathbf{K}(S - L_1) \\
&\quad \text{(by the definition of the border)} \\
&= \mathbf{K}(L_1) \cap (S - L_1) \\
&\quad \text{(because } S - L_1 \text{ is a closed set)} \\
&\subseteq \mathbf{K}(S - L_2) \cap S - L_1 \\
&\quad \text{(since } L_1 \subseteq S - L_2\text{)} \\
&= (S - L_2) \cap (S - L_1) = S - (L_1 \cup L_2) \subseteq S - T.
\end{aligned}$$

Similarly, we can show that $\partial L_2 \subseteq S - T$, so $T \cap \partial L_1 = T \cap \partial L_2 = \emptyset$.

Theorem 15.24 *Let T be a zero-dimensional, separable subspace of the topological metric space $(S, \mathcal{O}_d)$. If H_1 and H_2 are disjoint and closed subsets of S, then there exists a set W that separates H_1 and H_2 such that $W \cap T = \emptyset$.*

Proof By Theorem 8.28, there exist two open sets U_1 and U_2 such that $H_1 \subseteq U_1$, $H_2 \subseteq U_2$, and $\mathbf{K}(U_1) \cap \mathbf{K}(U_2) = \emptyset$.

Since T is zero-dimensional, the empty set separates the sets $T \cap \mathbf{K}(U_1)$ and $T \cap \mathbf{K}(U_1)$ in the space T. By Theorem 15.22, there exists a set W of S that separates H_1 and H_2 in $(S, \mathcal{O}_d)$ such that $W \cap T = \emptyset$, as stipulated in the statement.

Theorem 15.25 *Let $(S, \mathcal{O}_d)$ be a nonempty separable topological metric space that is the union of a countable collection of zero-dimensional closed sets $\{H_n \mid n \in \mathbb{N}\}$. Then, $(S, \mathcal{O}_d)$ is zero-dimensional.*

Proof Let $x \in S$ and let L be an open set such that $x \in L$. By Corollary 8.29, two open sets U_0 and W_0 exist such that $x \in U_0$, $L \subseteq W_0$, and $\mathbf{K}(U_0) \cap \mathbf{K}(W_0) = \emptyset$.

We define two increasing sequences of open sets $U_0, U_1, \ldots, U_n, \ldots$ and $W_0, W_1, \ldots, W_n, \ldots$ such that

(i) $\mathbf{K}(U_i) \cap \mathbf{K}(W_i) = \emptyset$ for $i \geqslant 0$;
(ii) $H_i \subseteq U_i \cup W_i$ for $i \geqslant 1$.

Suppose that we have defined the sets U_n and W_n that satisfy the conditions above. Observe that the disjoint sets $H_{n+1} \cap \mathbf{K}(U_n)$ and $H_{n+1} \cap \mathbf{K}(W_n)$ are closed in the subspace H_{n+1}.

Since $\dim(H_{n+1}) = 0$, by Theorem 15.7, there is a clopen set K in H_{n+1} such that $H_{n+1} \cap \mathbf{K}(U_n) \subset K$ and $K \cap (H_{n+1} \cap \mathbf{K}(W_n)) = \emptyset$. Both K and $H_{n+1} - K$ are closed sets in the space S because H_{n+1} is a closed subset of S, which implies that the sets $K \cup \mathbf{K}(U_n)$ and $(H_{n+1} - K) \cup \mathbf{K}(W_n)$ are also closed. Moreover, we have

$$(K \cup \mathbf{K}(U_n)) \cap ((H_{n+1} - K) \cup \mathbf{K}(W_n)) = \emptyset,$$

so there exist two open subsets of S, U_{n+1} and W_{n+1}, such that $K \cup \mathbf{K}(U_n) \subseteq U_{n+1}$, $(H_{n+1} - K) \cup \mathbf{K}(W_n) \subseteq W_{n+1}$, and $\mathbf{K}(U_{n+1}) \cap \mathbf{K}(W_{n+1}) = \emptyset$.

Consider the open sets $U = \bigcup_{n\in\mathbb{N}} U_n$ and $W = \bigcup_{n\in\mathbb{N}} W_n$. It is clear that $U \cap W = \emptyset$ and $U \cup W = S$, so both U and W are clopen. Since $x \in U_0 \subseteq U = S - W \subseteq V$ and $S - V \subseteq W_0 \subseteq W$, it follows that S is zero-dimensional.

It is interesting to contrast this theorem with Example 15.13, where we observed that $ind(\mathbb{R}, \mathcal{O}) = 1$ and $ind(\mathbb{Q}, \mathcal{O}') = ind(\mathbb{I}, \mathcal{O}'') = 0$. This happens, of course, because the subspaces $\mathbb{Q}$ and $\mathbb{I}$ are not closed.

Theorem 15.26 *Let $(S, \mathcal{O}_d)$ be a separable metric space. If X and Y are two subsets of S such that $S = X \cup Y$, $ind(X) \leqslant n - 1$, and $ind(Y) = 0$, then $ind(S) \leqslant n$.*

Proof Suppose that S can be written as $S = X \cup Y$ such that X and Y satisfy the conditions of the theorem. Let $x \in S$ and let L be an open set such that $x \in L$. By applying Theorem 15.24 to the closed sets $\{x\}$ and $S - L$, we obtain the existence of a set W that separates $\{x\}$ and $S - L$ such that $W \cap Y = \emptyset$, which implies $W \subseteq X$. Thus, $ind(W) \leqslant n - 1$, and this yields $ind(S) \leqslant n$.

Theorem 15.27 (The Sum Theorem) *Let $(S, \mathcal{O}_d)$ be a separable topological metric space that is a countable union of closed subspaces, $S = \bigcup_{i\geqslant 1} H_i$, where $ind(H_i) \leqslant n$. Then, $ind(S) \leqslant n$, where $n \geqslant 0$.*

Proof The argument is by strong induction on n. The base case, $n = 0$, was discussed in Theorem 15.25.

Suppose that the statement holds for numbers less than n, and let S be a countable union of closed subspaces of small inductive dimension less than n. By Corollary 15.17, each subspace H_i has a countable basis $\mathcal{B}_i$ such that $ind(\partial_{H_i} B_i) \leqslant n - 1$ for every $B_i \in \mathcal{B}_i$.

Each border $\partial_{H_i} B_i$ is closed in H_i and therefore is closed in S because each H_i is closed. Define $X = \bigcup_{i\geqslant 1} \bigcup\{\partial B_i \mid B_i \in \mathcal{B}_i\}$. By the inductive hypothesis, we have $ind(X) \leqslant n - 1$.

Define the sets $K_i = H_i - X$ for $i \geqslant 1$. The collection $\mathcal{C}_i = \{K_i \cap B \mid B \in \mathcal{B}\}$ consists of sets that are clopen in K_i, and therefore, for each nonempty set K_i, we have $ind(K_i) = 0$. Let $Y = S - X$. Since Y is a countable union of the closed subspaces $K_i = H_i \cap Y$, it follows that $ind(Y) = 0$. By Theorem 15.26, it follows that $ind(S) \leqslant n$.

The next statement complements Theorem 15.26.

Corollary 15.28 *Let $(S, \mathcal{O}_d)$ be a separable metric space. If $ind(S) \leqslant n$, then there exist two subsets X and Y of S such that $S = X \cup Y$, $ind(X) \leqslant n-1$, and $ind(Y) = 0$.*

Proof Let S be such that $ind(S) \leqslant n$ and let $\mathcal{B}$ be a countable basis such that $ind(\partial B) \leqslant n - 1$ for every $B \in \mathcal{B}$. The existence of such a basis is guaranteed by Corollary 15.17. Define $X = \bigcup\{\partial B \mid B \in \mathcal{B}\}$. By the Sum Theorem, we have $ind(X) \leqslant n - 1$. If $Y = S - X$, then $\{Y \cap B \mid B \in B\}$ is a base of Y that consists of clopen sets (in Y), so $ind(Y) \leqslant 0$.

Theorem 15.29 (The Decomposition Theorem) *Let $(S, \mathcal{O}_d)$ be a separable metric space such that $S \neq \emptyset$. We have $ind(S) = n$, where $n \geqslant 0$ if and only if S is the union of $n + 1$ zero-dimensional subspaces.*

Proof This statement follows from Theorem 15.26.

Theorem 15.30 (The Separation Theorem) *Let $(S, \mathcal{O}_d)$ be a separable topological metric space such that $ind(S) \leqslant n$, where $n \geqslant 0$. If H_1 and H_2 are two disjoint closed subsets, then there exist two disjoint open subsets L_1 and L_2 that satisfy the following conditions:*

(i) *$H_1 \subseteq L_1$ and $H_2 \subseteq L_2$;*
(ii) *$ind(\partial L_1) \leqslant n - 1$ and $ind(\partial L_2) \leqslant n - 1$.*

Proof By Theorem 15.26, there exist two subsets X and Y of S such that $S = X \cup Y$, $ind(X) \leqslant n - 1$, and $ind(Y) = 0$. By Corollary 15.23, there exist two disjoint open sets L_1 and L_2 such that $H_1 \subseteq L_1$, $H_2 \subseteq L_2$, and $Y \cap \partial L_1 = Y \cap \partial L_2 = \emptyset$. Therefore, $\partial L_1 \subseteq X$ and $\partial L_2 \subseteq X$, so $ind(\partial L_1) \leqslant n - 1$ and $ind(\partial L_2) \leqslant n - 1$.

The next statement is an extension of Theorem 15.24.

Theorem 15.31 *Let T be a subspace of a separable topological metric space $(S, \mathcal{O}_d)$ such that $ind(T) = k$, where $k \geqslant 0$. If H_1 and H_2 are disjoint closed subsets of S, then there exists a subset U of S that separates H_1 and H_2 such that $ind(T \cap U) \leqslant k - 1$.*

Proof The case $k = 0$ was discussed in Theorem 15.24.

If $k \geqslant 1$, then $T = X \cup Y$, where $ind(X) = k - 1$ and $ind(Y) = 0$. By Theorem 15.24, the closed sets H_1 and H_2 are separated by a set W such that $W \cap Y = \emptyset$, which implies $W \cap T \subseteq X$. Thus, $ind(W \cap T) \leqslant k - 1$.

Theorem 15.32 *Let $(S, \mathcal{O}_d)$ be a separable topological metric space. We have $ind(S) = IND(S)$.*

Proof We observed already that $ind(S) \leqslant IND(S)$ for every topological metric space. Thus, we need to prove only the reverse equality, $IND(S) \leqslant ind(S)$. This clearly holds if $ind(S) = \infty$.

The remaining argument is by induction on $n = ind(S)$. The base case, $n = 0$, was discussed in Corollary 15.9.

Suppose that the inequality holds for numbers less than n. If H_1 and H_2 are two disjoint and closed sets in S, then they can be separated by a subset U of S such that $ind(U) \leqslant n - 1$ by Theorem 15.31. By the inductive hypothesis, $IND(U) \leqslant n - 1$, so $IND(S) \leqslant n$.

15.5 The Covering Dimension

Definition 15.33 *Let $\mathcal{E}$ be a family of subsets of a set S. The* order *of $\mathcal{E}$ is the least number n such that any $n + 2$ sets of $\mathcal{E}$ have an empty intersection.*

The order of $\mathcal{E}$ is denoted by $ord(\mathcal{E})$.

If $ord(\mathcal{E}) = n$, then there exist $n + 1$ sets in $\mathcal{E}$ that have a nonempty intersection. Also, we have $ord(\mathcal{E}) \leqslant |\mathcal{E}| + 1$.

Example 15.34 If $ord(\mathcal{E}) = -1$, this means that any set of $\mathcal{E}$ is empty, so $\mathcal{E} = \{\emptyset\}$.

The order of any partition is 0.

Definition 15.35 *A topological metric space $(S, \mathcal{O}_d)$ has the* covering dimension *n if n is the least number such that $n \geqslant -1$ and every open cover $\mathcal{C}$ of S has a refinement $\mathcal{D}$ that consists of open sets with $ord(\mathcal{D}) = n$. If no such number n exists, then the covering dimension is ∞.*

The covering dimension of $(S, \mathcal{O}_d)$ is denoted by $cov(S, \mathcal{O}_d)$, or just by $cov(S)$, when there is no risk of confusion.

Theorem 15.36 *Let $(S, \mathcal{O}_d)$ be a topological metric space. The following statements are equivalent:*

(i) *$cov(S) \leqslant n$;*
(ii) *for every open cover $\mathcal{L} = \{L_1, \ldots, L_p\}$ of $(S, \mathcal{O}_d)$, there is an open cover $\mathcal{K} = \{K_1, \ldots, K_p\}$ such that $ord(\mathcal{K}) \leqslant n$ and $K_i \subseteq L_i$ for $1 \leqslant i \leqslant p$;*
(iii) *for every open cover $\mathcal{L} = \{L_1, \ldots, L_{n+2}\}$ there exists an open cover $\mathcal{K} = \{K_1, \ldots, K_{n+2}\}$ such that $\bigcap \mathcal{K} = \emptyset$ and $K_i \subseteq L_i$ for $1 \leqslant i \leqslant n + 2$;*
(iv) *for every open cover $\mathcal{L} = \{L_1, \ldots, L_{n+2}\}$ there exists a closed cover $\mathcal{H} = \{H_1, \ldots, H_{n+2}\}$ such that $\bigcap \mathcal{H} = \emptyset$ and $H_i \subseteq L_i$ for $1 \leqslant i \leqslant n + 2$.*

Proof (i) implies (ii): If $cov(S) \leqslant n$, then for the open cover $\mathcal{L} = \{L_1, \ldots, L_p\}$ of $(S, \mathcal{O}_d)$ there exists an open cover $\mathcal{U}$ that is a refinement of $\mathcal{L}$ such that $ord(\mathcal{U}) \leqslant n$. We need to derive from $\mathcal{U}$ another open cover that is also a refinement of $\mathcal{L}$, contains the same number of sets as $\mathcal{L}$, and satisfies the other conditions of (ii).

For $U \in \mathcal{U}$, let i_U be the least number i such that $U \subseteq L_i$. Define the open set $K_i = \bigcup\{U \in \mathcal{U} \mid i_U = i\}$ for $1 \leqslant i \leqslant p$. Observe that $\mathcal{K} = \{K_1, \ldots, K_p\}$ is an open cover.

An arbitrary element $x \in S$ belongs to at most $n+1$ members of the collection $\mathcal{U}$ because $ord(\mathcal{U}) \leqslant n$. Observe that $x \in K_i$ only if $x \in U$ for some $U \in \mathcal{U}$, which implies that x belongs to at most $n+1$ members of $\mathcal{K}$. Thus, $ord(\mathcal{K}) \leqslant n$.

(ii) implies (iii): This implication is immediate.

(iii) implies (iv): Suppose that (iii) holds, so for every open cover $\mathcal{L} = \{L_1, \ldots, L_{n+2}\}$ there exists an open cover $\mathcal{K} = \{K_1, \ldots, K_{n+2}\}$ such that $\bigcap \mathcal{K} = \emptyset$ and $K_i \subseteq L_i$ for $1 \leqslant i \leqslant n+2$. Starting from the open cover $\mathcal{K}$, by Supplement 35(b) of Chap. 4, we obtain the existence of the closed cover $\mathcal{H} = \{H_1, \ldots, H_{n+2}\}$ such that $H_i \subseteq K_i$ for $1 \leqslant n+2$. This implies immediately that $\mathcal{H}$ satisfies the requirements.

(iv) implies (iii): Suppose that (iv) holds, so for every open cover $\mathcal{L} = \{L_1, \ldots, L_{n+2}\}$ there exists a closed cover $\mathcal{H} = \{H_1, \ldots, H_{n+2}\}$ such that $\bigcap \mathcal{H} = \emptyset$ and $H_i \subseteq L_i$ for $1 \leqslant i \leqslant n+2$. By Part (b) of Supplement 36 of Chap. 4, there exists an open cover $\mathcal{K} = \{K_1, \ldots, K_{n+2}\}$ such that $K_i \subseteq L_i$ for $1 \leqslant i \leqslant n+2$ and $\bigcap \mathcal{K} = \emptyset$.

(iii) implies (ii): Suppose that $(S, \mathcal{O})$ satisfies condition (iii), and let $\mathcal{L} = \{L_1, \ldots, L_p\}$ be an open cover of $(S, \mathcal{O}_d)$. If $p \leqslant n+1$, then the desired collection is $\mathcal{L}$ itself. Thus, we may assume that $p \geqslant n+2$.

We need to prove that there exists an open cover $\mathcal{K} = \{K_1, \ldots, K_p\}$ such that $ord(\mathcal{K}) \leqslant n$ and $K_i \subseteq L_i$ for $1 \leqslant i \leqslant p$. This means that we have to show that the intersection on any $n+2$ sets of $\mathcal{K}$ is empty. Without loss of generality, we can prove that the intersection of the first $n+2$ sets of $\mathcal{K}$ is empty.

Consider the open cover $\{L_1, \ldots, L_{n+1}, L_{n+2} \cup \cdots \cup L_p\}$. By (iii), there exists an open cover $\mathcal{Q} = \{Q_1, \ldots, Q_{n+2}\}$ such that $\bigcap \mathcal{Q} = \emptyset$ and $Q_i \subseteq L_i$ for $1 \leqslant i \leqslant n+1$ and $Q_{n+2} \subseteq L_{n+2} \cup \cdots \cup L_p$. For $1 \leqslant i \leqslant p$, define the open sets

$$K_i = \begin{cases} Q_i & \text{if } 1 \leqslant i \leqslant n+1, \\ Q_{n+2} \cap L_i & \text{if } n+2 \leqslant i \leqslant p. \end{cases}$$

The collection $\mathcal{K} = \{K_1, \ldots, K_p\}$ is clearly an open cover, $K_i \subseteq L_i$ for $1 \leqslant i \leqslant p$, and $\bigcap_{i=1}^{n+2} K_i = \emptyset$.

(ii) implies (i): This implication is immediate.

Corollary 15.37 *Let $(S, \mathcal{O}_d)$ be a nonempty topological metric space. The following statements are equivalent:*

(i) *$cov(S) = 0$;*
(ii) *for all open sets L_1 and L_2 such that $L_1 \cup L_2 = S$, there exist two disjoint open sets K_1 and K_2 such that $K_i \subseteq L_i$ for $i \in \{1, 2\}$;*
(iii) *for all open sets L_1 and L_2 such that $L_1 \cup L_2 = S$ there exist two disjoint closed sets H_1 and H_2 such that $H_i \subseteq L_i$ for $i \in \{1, 2\}$.*

Proof This corollary is a special case of Theorem 15.36.

Theorem 15.38 *Let $(S, \mathcal{O}_d)$ be a topological metric space. We have $cov(S) = 0$ if and only if $IND(S) = 0$.*

Proof Suppose that $cov(S) = 0$. Let H_1 and H_2 be two disjoint closed sets. Then $\{S - H_1, S - H_2\}$ is an open cover of S. By Part (ii) of Corollary 15.37, there exist two disjoint open sets K_1 and K_2 such that $K_1 \subseteq S - H_1$ and $K_2 \subseteq S - H_2$. Thus, $K_1 \cup K_2 \subseteq (S - H_1) \cup (S - H_2) = S - (H_1 \cap H_2) = S$, which means that both K_1 and K_2 are clopen. This implies $IND(S) = 0$.

Conversely, suppose that $IND(S) = 0$, so $ind(S) = 0$. Let L_1 and L_2 be two open sets such that $L_1 \cup L_2 = S$. The closed sets $S - L_1$ and $S - L_2$ are disjoint, so by Theorem 15.7 there exists a clopen set U such that $S - L_1 \subseteq U$ (that is, $S - U \subseteq L_1$) and $U \cap (S - L_2) = \emptyset$ (that is, $U \subseteq L_2$). Since the sets $S - U$ and U are also closed, it follows that $cov(S) = 0$ by the last part of Corollary 15.37.

Theorem 15.39 *If $(S, \mathcal{O}_d)$ is a separable topological space, then $cov(S) \leqslant ind(S)$.*

Proof The statement clearly holds if $ind(S) = \infty$. Suppose now that $ind(S) = n$.

By the Decomposition Theorem (Theorem 15.29), S is the union of $n + 1$ zero-dimensional spaces, $S = \bigcup_{i=1}^{n+1} T_i$. If $\mathcal{L} = \{L_1, \ldots, L_m\}$ is a finite open cover of S, then $\mathcal{C} = \{L_1 \cap T_i, \ldots, L_m \cap T_i\}$ is a finite open cover of the subspace T_i. Since $ind(T_i) = 0$, we have $cov(T_i) = 0$ by Theorem 15.38. Therefore, the open cover $\mathcal{C}$ has a finite refinement that consists of disjoint open sets of the form K_{ij} such that $K_{ij} \subseteq L_j$ and $\bigcup_{j=1} K_{ij} \subseteq T_i$. Consequently, the collection $\mathcal{K} = \{B_{ij} \mid 1 \leqslant i \leqslant n + 1, 1 \leqslant j \leqslant m\}$ is a cover of S that refines the collection $\mathcal{L}$. Every subcollection $\mathcal{K}'$ of $\mathcal{K}$ that contains $n + 2$ sets must contain two sets that have the same second index, so any such intersection is empty. This allows us to conclude that $cov(S) \leqslant n = ind(S)$.

15.6 The Cantor Set

We introduce a special subset of the set of real numbers that plays a central role in the dimension theory of metric spaces.

Let $v_n : \{0, 1\}^n \longrightarrow \mathbb{N}$ be the function defined by

$$v_n(b_0, b_1, \ldots, b_{n-1}) = 2^{n-1} b_0 + \cdots + 2b_{n-2} + b_{n-1}$$

for every sequence $(b_0, \ldots, b_n) \in \{0, 1\}^n$. Clearly, $v_n(b_0, \ldots, b_{n-2}, b_{n-1})$ yields the number designated by the binary sequence $(b_0, \ldots, b_{n-2}, b_{n-1})$. For example, $v_3(110) = 2^2 \cdot 1 + 2^1 \cdot 1 + 0 = 6$.

Similarly, let $w_n : \{0, 1, 2\}^n \longrightarrow \mathbb{N}$ be the function defined by

$$w_n(b_0, b_1, \ldots, b_{n-1}) = 3^{n-1} b_0 + \cdots + 3b_{n-2} + b_{n-1}$$

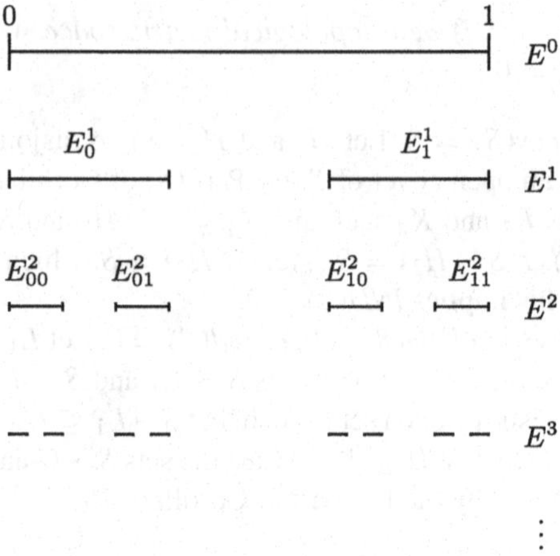

Fig. 15.1 Construction of the Cantor dust

for every sequence $(b_0, \ldots, b_n) \in \{0, 1, 2\}^n$. Then, $w_n(b_0, \ldots, b_{n-2}, b_{n-1})$ is the number designated by the ternary sequence $(b_0, \ldots, b_{n-2}, b_{n-1})$. For example, $w_3(110) = 3^2 \cdot 1 + 3^1 \cdot 1 + 0 = 12$.

Consider a sequence of subsets of $\mathbb{R}$, $E^0, E^1, \ldots$, where $E^0 = [0, 1]$ and E^1 is obtained from E^0 by removing the middle third $(1/3, 2/3)$ of E^0. If the remaining closed intervals are E_0^1 and E_1^1, then E^1 is defined by $E^1 = E_0^1 \cup E_1^1$.

By removing the middle intervals from the sets E_0^1 and E_1^1, four new closed intervals $E_{00}^2, E_{01}^2, E_{10}^2, E_{11}^2$ are created. Let $E^2 = E_{00}^2 \cup E_{01}^2 \cup E_{10}^2 \cup E_{11}^2$.

E^n is constructed from E^{n-1} by removing 2^{n-1} disjoint middle third intervals from E^{n-1} (see Fig. 15.1). Namely, if $E_{i_0 \cdots i_{n-1}}^n$ is an interval of the set E^n, by removing the middle third of this interval, we generate two closed intervals $E_{i_0 \cdots i_{n-1} 0}^{n+1}$ and $E_{i_0 \cdots i_{n-1} 1}^{n+1}$.

In general, E_n is the union of 2^n closed intervals

$$E^n = \bigcup_{i_0, \ldots, i_{n-1}} \{E_{i_0, \ldots, i_{n-1}}^n \mid (i_0, \ldots, i_{n-1}) \in \{0, 1\}^n\},$$

for $n \geqslant 0$.

An argument by induction on $n \in \mathbb{N}$ shows that

$$E_{i_0 \cdots i_{n-1}}^n = \left[\frac{2w_n(i_0, \ldots, i_{n-1})}{3^n}, \frac{2w_n(i_0, \ldots, i_{n-1}) + 1}{3^n} \right].$$

Indeed, the equality above holds for $n = 0$. Suppose that it holds for n, and denote by a and b the endpoints of the interval $E_{i_0 \cdots i_{n-1}}^n$; that is,

$$a = \frac{2w_n(i_0, \ldots, i_{n-1})}{3^n},$$
$$b = \frac{2w_n(i_0, \ldots, i_{n-1}) + 1}{3^n}.$$

By the inductive hypothesis, the points that divide $E^n_{i_0 \cdots i_{n-1}}$ are

$$\frac{2a+b}{3} = \frac{6w_n(i_0, \ldots, i_{n-1}) + 1}{3^{n+1}} = \frac{2w_{n+1}(i_0, \ldots, i_{n-1}, 0) + 1}{3^{n+1}}$$

and

$$\frac{a+2b}{3} = \frac{6w_n(i_0, \ldots, i_{n-1}) + 2}{3^{n+1}} = \frac{2w_{n+1}(i_0, \ldots, i_{n-1}, 1)}{3^{n+1}}.$$

Thus, the remaining left third of $E^n_{i_0 \cdots i_{n-1}}$ is

$$\begin{aligned} E^{n+1}_{i_0 \cdots i_{n-1} 0} &= \left[\frac{2w_n(i_0, \ldots, i_{n-1})}{3^n}, \frac{2w_{n+1}(i_0, \ldots, i_{n-1}, 0) + 1}{3^{n+1}}\right] \\ &= \left[\frac{2w_{n+1}(i_0, \ldots, i_{n-1}, 0)}{3^{n+1}}, \frac{2w_{n+1}(i_0, \ldots, i_{n-1}, 0) + 1}{3^{n+1}}\right], \end{aligned}$$

while the remaining right third is

$$\begin{aligned} E^{n+1}_{i_0 \cdots i_{n-1} 1} &= \left[\frac{2w_{n+1}(i_0, \ldots, i_{n-1}, 1)}{3^{n+1}}, \frac{2w_n(i_0, \ldots, i_{n-1}) + 1}{3^n}\right] \\ &= \left[\frac{2w_{n+1}(i_0, \ldots, i_{n-1}, 1)}{3^{n+1}}, \frac{2w_{n+1}(i_0, \ldots, i_{n-1}, 1) + 1}{3^{n+1}}\right], \end{aligned}$$

which concludes the inductive argument.

Each number x located in the leftmost third $E^1_0 = [0, 1/3]$ of the set $E_0 = [0, 1]$ can be expressed in base 3 as a number of the form $x = 0.0d_2d_3 \cdots$; the number $1/3$, the right extreme of this interval, can be written either as $x = 0.1$ or $x = 0.022 \cdots$. We adopt the second representation which allows us to say that all numbers in the rightmost third $E^1_1 = [2/3, 1]$ of E^0 have the form $0.2d_2d_3 \cdots$ in the base 3.

The argument applies again to the intervals $E^2_{00}, E^2_{01}, E^2_{10}, E^2_{11}$ obtained from the set E^1. Every number x in the interval E^2_{ij} can be written in base 3 as $x = 0.i'j' \cdots$, where $i' = 2i$ and $j' = 2j$.

The Cantor set is the intersection $C = \bigcap \{E^n \mid n \geqslant 0\}$.

Let us evaluate the total length of the intervals of which a set of the form E_n consists. There are 2^n intervals of the form $E^n_{i_0 \cdots i_{n-1}}$, and the length of each of these intervals is $\frac{1}{3^n}$. Therefore, the length of E_n is $(2/3)^n$, so this length tends toward 0 when n tends towards infinity. In this sense, the Cantor set is very sparse. Yet, surprisingly, the Cantor set is equinumerous with the interval $[0, 1]$. To prove this fact, observe that the Cantor set consists of the numbers x that can be expressed as $x = \sum_{n=1}^{\infty} \frac{a_n}{3^n}$, where $a_n \in \{0, 2\}$ for $n \geqslant 1$. For example, $1/4$ is a member of this set since $1/4$ can be expressed in base 3 as $0.020202\cdots$. Define the function $g : C \longrightarrow [0, 1]$ by $g(x) = y$ if $x = 0.a_1a_2\cdots$ (in base 3), where $a_i \in \{0, 2\}$ for $i \geqslant 1$ and $y = 0.b_1b_2\cdots$ (in base 2), where $b_i = a_i/2$ for $i \geqslant 1$. It is easy to see that this is a bijection between C and $[0, 1]$, which shows that these sets are equinumerous.

We now study the behavior of the sets

$$E^n_{i_0\cdots i_{n-1}} = \left[\frac{2w_n(i_0, \ldots, i_{n-1})}{3^n}, \frac{2w_n(i_0, \ldots, i_{n-1}) + 1}{3^n}\right]$$

relative to two mappings $f_0, f_1 : [0, 1] \longrightarrow [0, 1]$ defined by

$$f_0(x) = \frac{x}{3} \text{ and } f_1(x) = \frac{x + 2}{3}$$

for $x \in [0, 1]$.

Note that

$$\begin{aligned} f_0(E^n_{i_0\cdots i_{n-1}}) &= \left[\frac{2w_n(i_0, \ldots, i_{n-1})}{3^{n+1}}, \frac{2w_n(i_0, \ldots, i_{n-1}) + 1}{3^{n+1}}\right] \\ &= \left[\frac{2w_{n+1}(0i_0, \ldots, i_{n-1})}{3^{n+1}}, \frac{2w_{n+1}(0i_0, \ldots, i_{n-1}) + 1}{3^{n+1}}\right] \\ &= E^{n+1}_{0i_0\cdots i_{n-1}}. \end{aligned}$$

Similarly,

$$f_1(E^n_{i_0\cdots i_{n-1}}) = E^{n+1}_{1i_0\cdots i_{n-1}}.$$

Thus, in general, we have $f_i(E^n_{i_0\cdots i_{n-1}}) = E^{n+1}_{ii_0\cdots i_{n-1}}$ for $i \in \{0, 1\}$.

This allows us to conclude that $E^{n+1} = f_0(E^n) \cup f^1(E^n)$ for $n \in \mathbb{N}$. Since both f_0 and f_1 are injective, it follows that

$$\begin{aligned} C = \bigcap_{n\geqslant 1} E^n &= \bigcap_{n\geqslant 0} E^{n+1} \\ &= \bigcap_{n\geqslant 0} [f_0(E^n) \cup f_1(E^n)] \\ &= \left(\bigcap_{n\geqslant 0} f_0(E^n)\right) \cup \left(\bigcap_{n\geqslant 0} f_1(E^n)\right) \\ &= f_0\left(\bigcap_{n\geqslant 0} E^n\right) \cup f_1\left(\bigcap_{n\geqslant 0} E^n\right). \end{aligned}$$

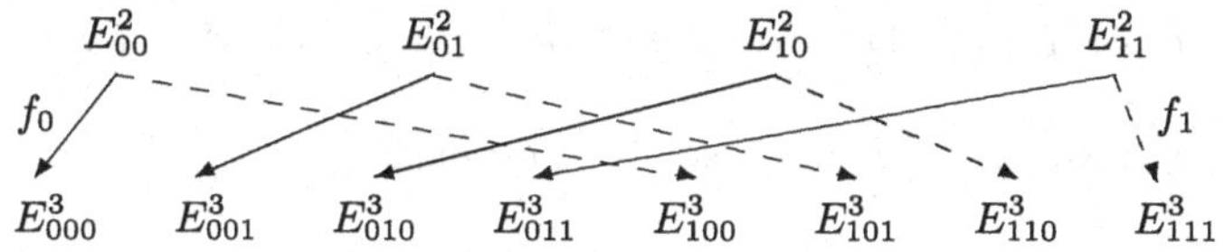

Fig. 15.2 Mapping sets E^2_{ij} into sets E^3_{ijk}

In Fig. 15.2 we show how sets of the form E^2_{ij} are mapped into sets of the form E^3_{ijk} by f_0 (represented by plain arrows) and f_1 (represented by dashed arrows).

Theorem 15.40 *The small inductive dimension of the Cantor set is* 0.

Proof We saw that

$$C = \bigcap_{n\in\mathbb{N}} E^n = \bigcap_{n\in\mathbb{N}} \bigcup_{i_0\cdots i_{n-1}} E^n_{i_0\cdots i_{n-1}}.$$

The sets $C \cap E^n_{i_0\cdots i_{n-1}}$ form a base for the open sets of the subspace C of $(\mathbb{R}, \mathcal{O})$. Note that the length of a closed interval $E^n_{i_0\cdots i_{n-1}}$ is $\frac{1}{3^n}$ and the distance between two distinct intervals $E^n_{i_0\cdots i_{n-1}}$ and $E^n_{j_0\cdots j_{n-1}}$ is at least $\frac{1}{3^n}$. Thus, $C \cap E^n_{i_0\cdots i_{n-1}}$ is closed in C. On the other hand, the same set is also open because

$$C \cap E^n_{i_0\cdots i_{n-1}} = C \cap \left(a - \frac{1}{3^n}, b + \frac{1}{3^n}\right),$$

where

$$a = \frac{2w_n(i_0, \ldots, i_{n-1})}{3^n},$$
$$b = \frac{2w_n(i_0, \ldots, i_{n-1}) + 1}{3^n}.$$

If $x \in C$, then $C \cap E^n_{i_0\cdots i_{n-1}} \subseteq C \cap S(x, r)$ provided that $\frac{1}{3^n} < r$. This shows that C has a basis that consists of clopen sets, so $ind(C) = 0$ by Theorem 15.6.

15.7 The Box-Counting Dimension

The box-counting dimension reflects the variation of the results of measuring a set at a diminishing scale, which allows the observation of progressively smaller details.

Let $(S, \mathcal{O}_d)$ be a topological metric space and let T be a precompact set. For every positive r, there exists an r-net for T; that is, a finite subset N_r of S such that $T \subseteq \bigcup\{C(x, r) \mid x \in N_r\}$ for every $r > 0$. Denote by $n_T(r)$ the *smallest* size of an r-net of T. It is clear that $r < r'$ implies $n_T(r) \geqslant n_T(r')$.

Definition 15.41 *Let $(S, \mathcal{O}_d)$ be a topological metric space and let T be a precompact set.*

The upper box-counting dimension *of T is the number*

$$ubd(T) = \limsup_{r \to 0} \frac{n_T(r)}{\log \frac{1}{r}}.$$

The lower box-counting dimension *of T is the number*

$$lbd(T) = \liminf_{r \to 0} \frac{n_T(r)}{\log \frac{1}{r}}.$$

If $ubd(T) = lbd(T)$, we refer to their common values as the box-counting dimension *of T, denoted by $bd(T)$.*

Example 15.42 Let $T = \{0\} \cup \{\frac{1}{n} \mid n \geqslant 1\}$ be a subset of $\mathbb{R}$. The interval $[0, r]$ contains almost all members of T because if $n \geqslant \lceil \frac{1}{r} \rceil$, we have $\frac{1}{n} \in [0, r]$.

It is easy to verify that

$$\frac{1}{n-1} - \frac{1}{n} > \frac{1}{n} - \frac{1}{n+1},$$

for $n > 1$. Note that $\frac{1}{n-1} - \frac{1}{n} > r$ is equivalent to $n^2 - n - \frac{1}{r} < 0$, and this happens when

$$n < \frac{1 + \sqrt{1 + \frac{4}{r}}}{2}.$$

Thus, each number of the form $\frac{1}{n}$ for

$$n \leqslant n_0 = \left\lfloor \frac{1 + \sqrt{1 + \frac{4}{r}}}{2} \right\rfloor$$

requires a distinct interval of size r to be covered.

The portion of T that is located to the left of $\frac{1}{n_0}$ and ends with the number r has length $\frac{1}{n_0} - r$ and can be covered with no more than $\frac{1}{rn_0} - 1$ intervals of length r. The least number of intervals of length r that are needed has

$$\frac{1}{rn_0} + \left\lfloor \frac{1 + \sqrt{1 + \frac{4}{r}}}{2} \right\rfloor$$

as an upper bound, a number that has the order of magnitude $\Theta(r^{-1/2})$. Thus, $ubd(T) \leqslant \frac{1}{2}$.

The notion of an r-net is related to two other notions, which we introduce next.

Definition 15.43 *Let (S, d) be a metric space, T be a subset of S, and let r be a positive number.*

A collection $\mathcal{C}$ of subsets of S is an r-cover *of T of S if, for every $C \in \mathcal{C}$, $diam_d(C) \leqslant 2r$ and $T \subseteq \bigcup \mathcal{C}$;*

A subset W of T is r-separated *if, for every $x, y \in W$, $x \neq y$ implies $d(x, y) > r$. The cardinality of the* largest *r-separated subset W of T is denoted by $\wp_T(r)$ and will be referred to as the* r-separation number *of T.*

Observe that an r-net for a set T is an r-cover.

Example 15.44 Consider the metric space $([0, 1]^2, d_2)$, where d_2 is the Euclidean metric. Since the area of a circle $B_{d_2}(x, r)$ is πr^2, it follows that for any $2r$-cover $\mathcal{C}$ that consists of circles, we have $\pi \cdot r^2 \cdot |\mathcal{C}| \geqslant 1$. Thus, a cover of this type of $([0, 1]^2, d_2)$ must contain at least $\frac{1}{\pi \cdot r^2}$ circles.

In general, the volume of a sphere $B_{d_2}(x, r)$ in $\mathbb{R}^n$ is

$$\frac{\pi^{\frac{n}{2}}}{\Gamma\left(\frac{n}{2}+1\right)} r^n,$$

which means that in the metric space $([0, 1]^n, d_2)$, a cover by spheres of radius r contains at least

$$\frac{\Gamma\left(\frac{n}{2}+1\right)}{\pi^{\frac{n}{2}} r^n}$$

spheres.

Example 15.45 Let $W = \{w_1, \ldots, w_n\}$ be an r-separated subset of the interval $[0, 1]$, where $w_1 < \cdots < w_n$. We have $1 \geqslant w_n - w_1 \geqslant (n-1)r$, so $n \leqslant \frac{1}{r} + 1$. This implies

$$\wp_{[0,1]}(r) = \left\lfloor \frac{1}{r} + 1 \right\rfloor.$$

Example 15.46 Let $T = \{0\} \cup \{\frac{1}{n} \mid n \in \mathbb{N}_1\}$. We seek to determine an upper bound for $\wp_T(r)$. Note that if $p > n$, then

$$\frac{1}{n} - \frac{1}{p} \geqslant \frac{1}{n} - \frac{1}{n+1}.$$

By the Mean Value Theorem, there exists $c \in (n, n+1)$ such that

$$\frac{1}{n} - \frac{1}{n+1} = \frac{1}{c^2}$$

and therefore

$$\frac{1}{n^2} > \frac{1}{n} - \frac{1}{n+1} > \frac{1}{(n+1)^2}.$$

Let n_1 be the largest number such that $\frac{1}{(n_1+1)^2} \geqslant r$. Then, an r-separated subset of T has at least n_1 elements; thus, the number $\frac{1}{\sqrt{r}}$ is a lower bound for the number of elements of an r-separating set.

Theorem 15.47 *Let T be a subset of a metric space (S, d). The following statements are equivalent:*

(i) *For each $r > 0$, there exists an r-net for T.*
(ii) *For each $r > 0$, every r-separated subset of T is finite.*
(iii) *For each $r > 0$, there exists a finite r-cover of T.*

Proof (i) implies (ii): Let W be an r-separated subset of T and let $N_{\frac{r}{2}}$ be an $\frac{r}{2}$-net for T. By the definition of $\frac{r}{2}$-nets, for each $w \in W$ there exists $t \in N_{\frac{r}{2}}$ such that $d(w, t) < \frac{r}{2}$; that is, $w \in C\left(t, \frac{r}{2}\right)$. Note that each sphere $C(t, \frac{r}{2})$ contains at most one member of w because W is an r-separated subset of T. The finiteness of T implies that W is finite too.

(ii) implies (iii): Let $W = \{w_1, \ldots, w_n\}$ be a maximal finite r-separated subset of T. If $t \in T$, then there exists $w_i \in W$ such that $d(t, w_i) \leqslant r$ since otherwise the maximality of W would be contradicted. Thus, $T \subseteq \bigcup_{i=1}^{n} B(w_i, r)$, each set $B(w_i, r)$ has a diameter of $2r$, and this implies that $\{B(w_i, r) | 1 \leqslant i \leqslant n\}$ is an r-cover of T.

(iii) implies (i): Let $\mathcal{D} = \{D_1, \ldots, D_n\}$ be a finite $\frac{r+\epsilon}{2}$-cover of T, where $\epsilon > 0$. Select $y_i \in D_i$ for $1 \leqslant i \leqslant n$, and define the set $Y = \{y_1, \ldots, y_n\}$. Since the diameter of every set D_i is not larger than $r + \epsilon$, for every $t \in T$ there exists y_i such that $d(t, y_i) \leqslant r + \epsilon$ for every $\epsilon > 0$, so $d(t, y_i) < r$. Therefore, Y is an r-net for T.

The connection between the numbers $n_T(r)$ and $\wp_T(r)$ is discussed next.

Corollary 15.48 *For every precompact set T of a topological metric space $(S, \mathcal{O}_d)$, we have*

$$n_T(r) \leqslant \wp_T(r) \leqslant n_T\left(\frac{r}{2}\right),$$

for every positive r.

Proof The first inequality follows from the proof of the first implication in Theorem 15.47. The second is a consequence of the last two implications of the same theorem.

The open spheres of radius r in the definition of a box-counting dimension of a precompact set T can be replaced with arbitrary sets of diameter $2r$. Indeed, suppose $n_T(r)$ is the smallest number of open spheres of radius r that cover T and $n_T(2r)'$ is the least number of sets of diameter $2r$ that cover T. Since each sphere of radius r has diameter $2r$, we have $n_T(2r)' \leqslant n_T(r)$. Observe that each set of diameter

$2r$ that intersects T is enclosed in an open sphere with radius $2r$ centered in T, so $n_T(2r) \leqslant n_T(2r)'$. The inequalities $n_T(2r) \leqslant n_T(2r)' \leqslant n_T(r)$ imply that the replacement previously mentioned does not affect the value of the box dimension. For example, open spheres can be replaced by closed spheres without affecting the value of the box-counting dimension.

Theorem 15.49 *Let T be a subset of a topological metric space $(S, \mathcal{O}_d)$. We have $ubd(\mathbf{K}(T)) = ubd(T)$ and $lbd(\mathbf{K}(T)) = lbd(T)$.*

Proof Let $\{B(x_1, r), \ldots, B(x_n, r)\}$ be a finite collection of closed spheres such that $T \subseteq \bigcup_{i=1}^n B(x_i, r)$. Clearly, we have $\mathbf{K}(T) \subseteq \bigcup_{i=1}^n B(x_i, r)$. Thus, a finite collection of closed spheres covers T if and only if it covers $\mathbf{K}(T)$. The conclusion follows immediately.

15.8 The Hausdorff-Besicovitch Dimension

The Hausdorff-Besicovitch measure plays a fundamental role in the study of fractals. The best-known definition of fractals was formulated by B. Mandelbrot [4], who is the founder of this area of mathematics, and states that a fractal is a geometrical object whose Hausdorff-Besicovitch dimension is greater than its small inductive dimension. The most famous example is the Cantor set whose small inductive dimension is 0 (by Theorem 15.40) and whose Hausdorff-Besicovitch dimension is $\frac{\ln 2}{\ln 3}$, as we shall prove below.

Recall that a collection $\mathcal{C}$ of subsets of a metric space (S, d) is an *r-cover* of a subset U of S if, for every $C \in \mathcal{C}$, $diam_d(C) \leqslant 2r$ and $U \subseteq \bigcup \mathcal{C}$.

Let (S, d) be a metric space and let $\mathfrak{C}_r(U)$ be the collection of all countable r-covers for a set U. Observe that $r_1 \leqslant r_2$ implies $\mathfrak{C}_{r_1}(U) \subseteq \mathfrak{C}_{r_2}(U)$ for $r_1, r_2 \in \mathbb{R}_{>0}$.

Let s be a positive number. We shall use the outer measure HB_r^s obtained by applying Method I (see Theorem 4.128) to the function $f : \mathcal{C} \longrightarrow \mathbb{R}_{\geqslant 0}$ given by $f(C) = (diam(C))^s$ for $C \in \mathcal{C}$, which is given by

$$HB_r^s(U) = \inf_{\mathcal{C} \in \mathfrak{C}_r(U)} \sum \{(diam(C))^s \mid C \in \mathcal{C}\}.$$

The function $HB_r^s(U)$ is antimonotonic with respect to r. Indeed, if $r_1 \leqslant r_2$, then $\mathfrak{C}_{r_1}(U) \subseteq \mathfrak{C}_{r_2}(U)$, so

$$\inf_{\mathcal{C} \in \mathfrak{C}_{r_2}(U)} \sum \{(diam(C))^s \mid C \in \mathcal{C}\} \leqslant \inf_{\mathcal{C} \in \mathfrak{C}_{r_1}(U)} \sum \{(diam(C))^s \mid C \in \mathcal{C}\},$$

which means that $HB_{r_2}^s(U) \leqslant HB_{r_1}^s(U)$. Because of this, $\lim_{r \to 0} HB_r^s(U)$ exists for every set U, and this justifies the next definition.

Definition 15.50 *The* Hausdorff-Besicovitch outer measure HB^s *is given by*

$$HB^s(U) = \lim_{r \to 0} HB^s_r(U)$$

for every $U \in \mathcal{P}(S)$.

Theorem 15.51 *Let* (S, d) *be a metric space and let* U *be a Borel set in this space. If* s *and* t *are two positive numbers such that* $s < t$ *and* $HB^s(U)$ *is finite, then* $HB^t(U) = 0$. *Further, if* $HB^t(U) > 0$, *then* $HB^s(U) = \infty$.

Proof If $s < t$ and $\mathcal{C}$ is an r-cover of U, then

$$\begin{aligned}\sum\{(diam(C))^t \mid C \in \mathcal{C}\} &= \sum\{(diam(C))^{t-s}(diam(C))^s \mid C \in \mathcal{C}\}\\ &\leqslant r^{t-s}\sum\{(diam(C))^s \mid C \in \mathcal{C}\},\end{aligned}$$

which implies $HB^t_r(U) \leqslant r^{t-s} HB^s_r(U)$. This, in turn, yields

$$HB^t(U) = \lim_{r \to 0} HB^t_r(U) \leqslant \lim_{r \to 0} r^{t-s} HB^s(U).$$

If $HB^s(U)$ is finite, then $HB^t(U) = 0$. On the other hand, if $HB^t(U) > 0$, the last inequality implies $HB^s(U) = \infty$.

Corollary 15.52 *Let* (S, d) *be a metric space and let* U *be a Borel set. There exists a unique* s_0 *such that* $0 \leqslant s_0 \leqslant \infty$ *and*

$$HB^s(U) = \begin{cases} \infty & \text{if } s < s_0, \\ 0 & \text{if } s > s_0. \end{cases}$$

Proof This statement follows immediately from Theorem 15.51 by defining $s_0 = \inf\{s \in \mathbb{R}_{\geqslant 0} \mid HB^s(U) = 0\} = \sup\{s \in \mathbb{R}_{\geqslant 0} \mid HB^s(U) = \infty\}$.

Corollary 15.52 justifies the following definition.

Definition 15.53 *Let* (S, d) *be a metric space and let* U *be a Borel set. The* Hausdorff-Besicovitch dimension of U *is the number*

$$HBdim(U) = \sup\{s \in \mathbb{R}_{\geqslant 0} \mid HB^s(U) = \infty\}.$$

Theorem 15.54 *The Hausdorff-Besicovitch dimension is monotonic; that is,* $U \subseteq U'$ *implies* $HBdim(U) \leqslant HBdim(U')$.

Proof If $U \subseteq U'$, then it is clear that $\mathfrak{C}_r(U') \subseteq \mathfrak{C}_r(U)$. Therefore, we have $HB^s_r(U) \leqslant HB^s_r(U')$, which implies $HB^s(U) \leqslant HB^s(U')$ for every $s \in \mathbb{R}_{\geqslant 0}$. This inequality yields $HBdim(U) \leqslant HBdim(U')$.

Theorem 15.55 *If $\{U_n \mid n \in \mathbb{N}\}$ is a countable family of sets, then*

$$HBdim\left(\bigcup_{n\in\mathbb{N}} U_n\right) = \sup\{HBdim(U_n) \mid n \in \mathbb{N}\}.$$

Proof By Theorem 15.55, we have $HBdim(U_n) \leqslant HBdim\left(\bigcup_{n\in\mathbb{N}} U_n\right)$, so sup
$\{HBdim(U_n) \mid n \in \mathbb{N}\} \leqslant HBdim\left(\bigcup_{n\in\mathbb{N}} U_n\right)$.

If $HBdim(U_n) < t$ for $n \in \mathbb{N}$, then $HB^t(U_n) = 0$, so $HB^t\left(\bigcup_{n\in\mathbb{N}} U_n\right) = 0$ since the Hausdorff-Besicovitch outer measure HB^t is subadditive. Therefore, $HBdim\left(\bigcup_{n\in\mathbb{N}} U_n\right) < t$. This implies $HBdim\left(\bigcup_{n\in\mathbb{N}} U_n\right) \leqslant \sup\{HBdim(U_n) \mid n \in \mathbb{N}\}$, which gives the desired equality.

Example 15.56 If $U = \{u\}$ is a singleton, then $HB^0(\{u\}) = 0$. Thus, $HBdim(\{x\}) = 0$. By Theorem 15.55, we have $HBdim(T) = 0$ for every countable set T.

Example 15.57 Let $f : [0, 1]^2 \longrightarrow \mathbb{R}$ be a function that is continuous and has bounded partial derivatives in the square $[0, 1]^2$ and let S be the surface in $\mathbb{R}^2$ defined by $z = f(x, y)$. Under these conditions, there is a constant k such that $|f(x', y') - f(x, y)| \leqslant k(|x' - x| + |y' - y|)$. We prove that $HBdim(S) = 2$.

Suppose that S is covered by spheres of diameter d_i, $S \subseteq \bigcup\{B(x_i, \frac{d_i}{2}) \mid i \in I\}$. Then, the square $[0, 1]^2$ is covered by disks of diameter d_i and therefore $\sum_{i\in I} \frac{\pi d_i^2}{4} \geqslant 1$, which is equivalent to $\sum_{i\in I} d_i^2 \geqslant \frac{4}{\pi}$. Therefore, $HB^2(S) > 0$, so $HBdim(S) \geqslant 2$. Observe that, in this part of the argument, the regularity of f played no role.

To prove the converse inequality, $HBdim(S) \leqslant 2$, we show that $HB^{2+\epsilon}(S) = 0$ for every $\epsilon > 0$; that is, $\lim_{r\to 0} HB_r^{2+\epsilon}(U) = 0$ for every $\epsilon > 0$.

Divide the square $[0, 1]^2$ into n^2 squares of size $\frac{1}{n}$. Clearly, for any two pairs (x, y) and (x', y') located in the same small square, we have $|f(x', y') - f(x, y)| \leqslant \frac{2k}{n}$, which means that the portion of S located above a small square can be enclosed in a cube of side $\frac{2k}{n}$ and therefore in a sphere of diameter $\frac{2\sqrt{3}k}{n}$. For the covering $\mathcal{C}$ that consists of these n^2 spheres, we have

$$\sum\{(diam(C))^{2+\epsilon} \mid C \in \mathcal{C}\} = n^2\left(\frac{2\sqrt{3}k}{n}\right)^{2+\epsilon} = \frac{2\sqrt{3}k}{n^\epsilon}.$$

If n is chosen such that $\frac{2\sqrt{3}k}{n} < r$, we have $HB_r^{2+\epsilon}(S) \leqslant \frac{2\sqrt{3}k}{n^\epsilon}$. Thus, $\lim_{r\to 0} HB_r^{2+\epsilon}(S) = 0$, so $HBdim(S) \leqslant 2$.

Example 15.58 We saw that the Cantor set C is included in each of the sets E_n that consists of 2^n closed intervals of length $\frac{1}{3^n}$. Thus, we have

$$HB^s_{\frac{1}{3^n}}(C) \leqslant \frac{2^n}{3^{sn}} = \left(\frac{2}{3^s}\right)^n$$

for every $n \geqslant 1$. If $\frac{2}{3^s} < 1$ $\left(that\ is, if\ s > \frac{\ln 2}{\ln 3}\right)$, we have $\lim_{n\to\infty} HB^s_{\frac{1}{3^n}} = 0$. If $s < \frac{\ln 2}{\ln 3}$, then $\lim_{n\to\infty} HB^s_{\frac{1}{3^n}} = \infty$, so $HBdim(C) = \frac{\ln 2}{\ln 3}$.

Theorem 15.59 *Let $(S, \mathcal{O}_d)$ be a topological metric space and let T be a precompact set. We have $HBdim(T) \leqslant lbd(U)$.*

Proof Suppose that T can be covered by $n_T(r)$ sets of diameter r. By the definition of the outer measure $HB^s_r(U)$, we have

$$HB^s_r(U) \leqslant r^s n_T(r).$$

Since $HB^s(U) = \lim_{r\to 0} HB^s_r(U)$, if $HB^s(U) > 1$, then if r is sufficiently small we have $HB^s_r(U) > 1$, so $\log HB^s_r(U) > 0$, which implies $s \log r + \log n_T(r) > 0$. Thus, if r is sufficiently small, $s < \frac{n_T(r)}{\log \frac{1}{r}}$, so $s \leqslant lbd(U)$. This entails $HBdim(U) \leqslant lbd(U)$.

The following statement is known as the *mass distribution principle* (see [5]).

Theorem 15.60 *Let (S, d) be a metric space and let μ be a Carathéodory outer measure on S such that there exist $s, r > 0$ such that $\mu(U) \leqslant c \cdot diam(U)^s$ for all $U \in \mathcal{P}(S)$ with $diam(U) \leqslant r$. Then, $HB^s(W) \leqslant \frac{\mu(W)}{c}$ and $s \leqslant HBdim(W) \leqslant lbd(W)$ for every precompact set $W \in \mathcal{P}(S)$ with $\mu(W) > 0$.*

Proof Let $\{U_i \mid i \in I\}$ be a cover of W. We have

$$0 < \mu(W) \leqslant \mu\left(\bigcup_i U_i\right) \leqslant \sum_{i\in I} \mu(U_i) \leqslant c \sum_i (diam(U_i))^s,$$

so $\sum_i (diam(U_i))^s \geqslant \frac{\mu(W)}{c}$. Therefore, $HB^s_r(W) \geqslant \frac{\mu(W)}{c}$, which implies $HB^s(W) \geqslant \frac{\mu(W)}{c} > 0$. Consequently, $HBdim(W) \geqslant \frac{\mu(W)}{c} > 0$.

15.9 Similarity Dimension

The notions of similarity and contraction between metric spaces were introduced in Definition 8.65.

Definition 15.61 *Let $\mathbf{r} = (r_1, \ldots, r_n)$ be a sequence of numbers such that $r_i \in (0, 1)$ for $1 \leqslant i \leqslant n$ and let (S, d) be a metric space.*

An iterative function system *on (S, d) that realizes a sequence of ratios $\mathbf{r} = (r_1, \ldots, r_n)$ is a sequence of functions $f = (f_1, \ldots, f_n)$, where $f_i : S \longrightarrow S$ is a contraction of ratio r_i for $1 \leqslant i \leqslant n$.*

A subset T of S is an invariant set for the iterative function system $(f_1, \ldots, f_n)$ *if $T = \bigcup_{i=1}^n f_i(T)$.*

Example 15.62 Let $f_0, f_1 : [0, 1] \longrightarrow [0, 1]$ defined by

$$f_0(x) = \frac{x}{3} \text{ and } f_1(x) = \frac{x+2}{3}$$

for $x \in [0, 1]$, which are contractions of ratio $\frac{1}{3}$.

The Cantor set C is an invariant set for the iterative function system $\mathbf{f} = (f_0, f_1)$, as we have shown in Sect. 15.6.

Lemma 15.63 *Let $r_1, \ldots, r_n$ be n numbers such that $r_i \in (0, 1)$ for $1 \leqslant i \leqslant n$ and $n > 1$. There is a unique number d such that*

$$r_1^d + r_2^d + \cdots + r_n^d = 1.$$

Proof Define the function $\phi : \mathbb{R}_{\geqslant 0} \longrightarrow \mathbb{R}_{\geqslant 0}$ by

$$\phi(x) = r_1^x + r_2^x + \cdots + r_n^x$$

for $x > 0$. Note that $\phi(0) = n$, $\lim_{x \to \infty} \phi(x) = 0$, and $\phi'(x) = r_1^x \ln r_1 + r_2^x \ln r_2 + r_n^x \ln r_n < 0$. Since $\phi'(x) < 0$, ϕ is a strictly decreasing function, so there exists a unique d such that $\phi(d) = 1$.

Definition 15.64 *Let $\mathbf{r} = (r_1, \ldots, r_n)$ be a sequence of ratios such that $r_i \in (0, 1)$ for $1 \leqslant i \leqslant n$ and $n > 1$. The* dimension of $\mathbf{r}$ *is the number d, whose existence was proven in Lemma 15.63.*

Observe that if the sequence $\mathbf{r}$ has length 1, $\mathbf{r} = (r_1)$, then $r_1^d = 1$ implies $d = 0$.

Example 15.65 The dimension of the sequence $\mathbf{r} = (\frac{1}{3}, \frac{1}{3})$ is the solution of the equation

$$2 \cdot \left(\frac{1}{3}\right)^d = 1;$$

that is, $d = \frac{\log 2}{\log 3}$.

Lemma 15.66 *Let $(S, \mathcal{O}_d)$ be a complete topological metric space and let $f = (f_1, \ldots, f_n)$ be an iterative function system that realizes a sequence of ratios $\mathbf{r} = (r_1, \ldots, r_n)$. The mapping $F : \mathcal{K}(S, \mathcal{O}_d) \longrightarrow \mathcal{K}(S, \mathcal{O}_d)$ defined on the Hausdorff metric hyperspace $(\mathcal{K}(S, \mathcal{O}_d), \delta)$ by*

$$F(U) = \bigcup_{i=1}^{n} f_i(U)$$

is a contraction.

Proof We begin by observing that F is well-defined. Indeed, since each contraction f_i is continuous and the image of a compact set by a continuous function is compact (by Theorem 4.69), it follows that if U is compact, then $F(U)$ is compact as the union of a finite collection of compact sets.

Next, we prove that $\delta(F(U), F(V)) \leqslant r\delta(U, V)$ for $r = \max_{0 \leqslant i \leqslant n-1} r_i < 1$.

Let $x \in F(U)$. There is $u \in U$ such that $x = f_i(u)$ for some i, $1 \leqslant i \leqslant n$. By the definition of δ, there exists $v \in V$ such that $d(u, v) \leqslant \delta(U, V)$. Since f_i is a contraction, we have $d(u, v) = d(f_i(v), f_i(u)) \leqslant r_i d(u, v) \leqslant rd(u, v) \leqslant r\delta(U, V)$, so $F(U) \subseteq C(F(V), r\delta(U, V))$. Similarly, $F(V) \subseteq C(F(U), r\delta(U, V))$, so $\delta(F(U), F(V)) \leqslant r\delta(U, V)$, which proves that F is a contraction of the Hausdorff metric hyperspace $(\mathcal{K}(S, \mathcal{O}_d), \delta)$.

Theorem 15.67 *Let $(S, \mathcal{O}_d)$ be a complete topological metric space and let $\mathbf{f} = (f_1, \ldots, f_n)$ be an iterative function system that realizes a sequence of ratios $\mathbf{r} = (r_1, \ldots, r_n)$. There exists a unique compact set U that is an invariant set for $\mathbf{f}$.*

Proof By Lemma 15.66, the mapping $F : \mathcal{K}(S, \mathcal{O}_d) \longrightarrow \mathcal{K}(S, \mathcal{O}_d)$ is a contraction. Therefore, by the Banach fixed point theorem (Theorem 8.71), F has a fixed point in $\mathcal{K}(S, \mathcal{O}_d)$, which is an invariant set for $\mathbf{f}$.

The unique compact set that is an invariant for an iterative function system $\mathbf{f}$ is usually referred to as the *attractor* of the system.

Definition 15.68 *Let $(S, \mathcal{O}_d)$ be a topological metric space and let U be an invariant set of an iterative function system $\mathbf{f} = (f_1, \ldots, f_n)$ that realizes a sequence of ratios $\mathbf{r} = (r_1, \ldots, r_n)$.*

The similarity dimension *of the pair $(U, \mathbf{f})$ is the number SIMdim($\mathbf{f}$), which equals the dimension of $\mathbf{r}$.*

In principle, a set may be an invariant set for many iterative function systems.

Example 15.69 Let $p, q \in (0, 1)$ such that $p + q \geqslant 1$ and let $f_0, f_1 : [0, 1] \longrightarrow [0, 1]$ be defined by $f_0(x) = px$ and $f_1(x) = qx + 1 - q$. Both f_0 and f_1 are contractions. The sequence $\mathbf{f} = (f_0, f_1)$ realizes the sequence of ratios $\mathbf{r} = (p, q)$ and we have $(0, 1) = f_0(0, 1) \cup f_1(0, 1)$. The dimension of the pair $(U, \mathbf{r})$ is the number d such that $p^d + q^d = 1$; this number depends on the values of p and q.

Theorem 15.70 *Let $(S, \mathcal{O}_d)$ be a complete topological metric space, $\mathbf{f} = (f_1, \ldots, f_n)$ be an iterative function system that realizes a sequence of ratios $\mathbf{r} = (r_1, \ldots, r_n)$, and U be the attractor of $\mathbf{f}$. Then, we have $HBdim(U) \leqslant$ SIMdim($\mathbf{f}$).*

Proof Suppose that $r_1^d + r_2^d + \cdots + r_n^d = 1$, that is, d is the dimension of $\mathbf{f}$. For a subset T of S denote the set $f_{i_1}(f_{i_2}(\cdots f_{i_p}(T) \cdots))$ by $f_{i_1 i_2 \cdots i_p}(T)$.

If U is the attractor of $\mathbf{f}$, then

$$U = \bigcup \{f_{i_1 i_2 \cdots i_p}(U) \mid (i_1, i_2, \ldots, i_p) \in \mathbf{Seq}_p(\{1, \ldots, n\})\}.$$

This shows that the sets of the form $f_{i_1 i_2 \cdots i_p}(U)$ constitute a cover of U.

Since $f_{i_1}, f_{i_2}, \ldots, f_{i_p}$ are similarities of ratios $r_{i_1}, r_{i_2}, \ldots, r_{i_p}$, respectively, it follows that $diam(f_{i_1 i_2 \cdots i_p}(U)) \leqslant (r_{i_1} r_{i_2} \cdots r_{i_p}) diam(U)$. Thus,

$$\begin{aligned}
&\sum\{(diam(f_{i_1 i_2 \cdots i_p}(U)))^d \mid (i_1, i_2, \ldots, i_p) \in \mathbf{Seq}_p(\{1, \ldots, n\})\} \\
&\leq \sum\{r_{i_1}^d r_{i_2}^d \cdots r_{i_p}^d diam(U)^d \mid (i_1, i_2, \ldots, i_p) \in \mathbf{Seq}_p(\{1, \ldots, n\})\} \\
&= \left(\sum_{i_1} r_{i_1}^d\right)\left(\sum_{i_2} r_{i_2}^d\right) \cdots \left(\sum_{i_p} r_{i_p}^d\right) diam(U)^d \\
&= diam(U)^d.
\end{aligned}$$

For $r \in \mathbb{R}_{>0}$, choose p such that $diam(f_{i_1 i_2 \cdots i_p}(U)) \leqslant (\max r_i)^p diam(U) < r$. This implies $HB_r^d(U) \leqslant diam(U)^d$, so $HB^d(U) = \lim_{r \to 0} HB_r^d(U) \leqslant diam(U)^d$. Thus, we have $HBdim(U) \leqslant d = SIMdim(\mathbf{f})$.

The next statement involves an iterative function system $\mathbf{f} = (f_1, \ldots, f_m)$ acting on a closed subset H of the metric space $(\mathbb{R}^n, d_2)$ such that each contraction f_i satisfies the double inequality

$$b_i d_2(\mathbf{x}, \mathbf{y}) \leqslant d_2(f_i(\mathbf{x}), f_i(\mathbf{y})) \leqslant r_i d_2(\mathbf{x}, \mathbf{y})$$

for $1 \leqslant i \leqslant m$ and $\mathbf{x}, \mathbf{y} \in H$. Note that each of the functions f_i is injective on the set H.

Theorem 15.71 *Let $\mathbf{f} = (f_1, \ldots, f_m)$ be an iterative function system on a closed subset H of $\mathbb{R}^n$ that realizes a sequence of ratios $\mathbf{r} = (r_1, \ldots, r_m)$. Suppose that, for every i, $1 \leqslant i \leqslant m$, there exists $b_i \in (0, 1)$ such that $d_2(f_i(\mathbf{x}), f_i(\mathbf{y})) \geqslant b_i d_2(\mathbf{x}, \mathbf{y})$ for $\mathbf{x}, \mathbf{y} \in H$.*

If U is the nonempty and compact attractor of $\mathbf{f}$ and $\{f_1(U), \ldots, f_m(U)\}$ is a partition of U, then U is a totally disconnected set and $HBdim(U) \geqslant c$, where c is the unique number such that $\sum_{i=1}^n b_i^c = 1$.

Proof Let

$$t = \min\{d_2(f_i(U), f_j(U) \mid 1 \leqslant i, j \leqslant m \text{ and } i \neq j\}.$$

Using the same notation as in the proof of Theorem 15.70, observe that the collection of sets

$$\{f_{i_1 \cdots i_p}(U) \mid (i_1, \ldots, i_p) \in \mathbf{Seq}(\{1, \ldots, m\})\}$$

is a sequential cover of the attractor U. Note also that all sets $f_{i_1 \cdots i_p}(U)$ are compact and therefore closed. Also, since each collection $\{f_{i_1 \cdots i_p}(U) \mid (i_1, \ldots, i_p) \in \mathbf{Seq}_p(\{1, \ldots, m\})$ is a partition of U, it follows that each of these sets is clopen in U. Thus, U is totally disconnected.

Define

$$m(f_{i_1 i_2 \cdots i_p}(U)) = (b_{i_1} b_{i_2} \cdots b_{i_p})^c.$$

Note that

$$\begin{aligned}
\sum_{i=1}^{m} m(f_{i_1 i_2 \cdots i_p i}(U)) &= \sum_{i=1}^{m} (b_{i_1} b_{i_2} \cdots b_{i_p} b_i)^c \\
&= (b_{i_1} b_{i_2} \cdots b_{i_p})^c \sum_{i=1}^{m} b_i^c \\
&= (b_{i_1} b_{i_2} \cdots b_{i_p})^c = m(f_{i_1 i_2 \cdots i_p}(U)) \\
&= m\left(\bigcup_{i=1}^{m} f_{i_1 i_2 \cdots i_p i}(U) \right).
\end{aligned}$$

For $x \in U$, there is a unique sequence $(i_1, i_2, \ldots) \in \mathbf{Seq}_\infty(\{1, \ldots, m\})$ such that $x \in U_{i_1 \cdots i_k}$ for every $k \geqslant 1$. Observe also that

$$U \supseteq U_{i_1} \supseteq U_{i_1 i_2} \supseteq \cdots \supseteq U_{i_1 i_2 \cdots i_k} \supseteq \cdots.$$

Consider the decreasing sequence $1 > b_{i_1} > b_{i_1} b_{i_2} > \cdots > b_{i_1} \cdots b_{i_n} > \cdots$. If $0 < r < t$, let k be the least number j such that $\frac{r}{t} \geqslant b_{i_1} \cdots b_{i_j}$. We have

$$b_{i_1} \cdots b_{i_{k-1}} > \frac{r}{t} \geqslant b_{i_1} \cdots b_{i_k}$$

so $m(U_{i_1 \cdots i_{k-1}}) > \frac{r^c}{t^c} \geqslant m(U_{i_1 \cdots i_k})$.

Let $(i'_1, \ldots, i'_k)$ be a sequence distinct from $(i_1, \ldots, i_k)$. If ℓ is the least integer such that $i'_\ell \neq i_\ell$, then $U_{i'_1 \cdots i'_\ell} \subseteq U_{i'_\ell}$ and $U_{i_1 \cdots i_\ell} \subseteq U_{i_\ell}$. Since U_{i_ℓ} and $U_{i_{\ell'}}$ are disjoint and separated by t, it follows that the sets $U_{i_1 \ldots i_k}$ and $U_{i'_1 \ldots i'_k}$ are disjoint and separated by at least $b_{i_1} \cdots b_{i_\ell} t > r$. Thus, $U \cap B(x, r) \subset U_{i_1 \cdots i_k}$, so

$$m(U \cap B(x, r)) \leqslant m(U_{i_1 \cdots i_k}) = (b_{i_1} \cdots b_{i_k})^c \leqslant \left(\frac{r}{t}\right)^c.$$

If $U \cap W \neq \emptyset$, then $W \subset B(x, r)$ for some $x \in U$ with $r = diam(W)$. Thus, $m(W) \leqslant \frac{diam(W)}{t^c}$, so $HB^c(U) > 0$ and $HBdim(U) \geqslant c$.

Exercises and Supplements

1. Let $Q_n(\ell)$ be an n-dimensional cube in $\mathbb{R}^n$. Prove that:
   (a) there are $\binom{n}{k} \cdot 2^{n-k}$ k-dimensional faces of $Q_n(\ell)$;
   (b) the total number of faces of $Q_n(\ell)$ is $\sum_{k=1}^{n} \binom{n}{k} \cdot 2^{n-k} = 3^n - 1$.

2. Let $\{U_i \mid i \in I\}$ be a collection of pairwise disjoint open subsets of the topological metric space $(\mathbb{R}^n, \mathcal{O}_{d_2})$ such that for each $i \in I$ there exist $\mathbf{x}_i, \mathbf{y}_i \in \mathbb{R}^n$ such that $B(\mathbf{x}_i, ar) \subseteq U_i \subseteq B(\mathbf{y}_i, br)$. Prove that for any $B(\mathbf{u}, r)$, we have $|\{U_i \mid \mathbf{K}(U_i) \cap B(\mathbf{u}, r) \neq \emptyset\}| \leqslant \left(\frac{1+2b}{a}\right)^n$.

   Solution: Suppose that $\mathbf{K}(U_i) \cap B(\mathbf{u}, r) \neq \emptyset$. Then $\mathbf{K}(U_i) \subseteq B(\mathbf{u}, r + 2br)$. Recall that the volume of a sphere of radius r in $\mathbb{R}^n$ is $V_n(r) = \frac{\pi^{\frac{n}{2}} r^n}{\Gamma(\frac{n}{2}+1)}$. If $m = |\{U_i \mid \mathbf{K}(U_i) \cap B(\mathbf{u}, r) \neq \emptyset\}|$, the total volume of the spheres $B(\mathbf{x}_i, ar)$ is smaller than the volume of the sphere $B(\mathbf{u}, (1 + 2b)r)$, and this implies $ma^n \leqslant (1 + 2b)^n$.
3. Let T be a subset of $\mathbb{R}$. Prove that T is zero-dimensional if and only if it contains no interval.
4. Prove that the Cantor set C is totally disconnected.

   Hint: Suppose that $a < b$ and b belongs to the connected components K_a of a. By Example 4.83, this implies $[a, b] \subseteq K_a \subseteq C$, which leads to a contradiction.
5. Prove the following extension of Example 15.42. If $T = \{0\} \cup \{\frac{1}{n^a} \mid n \geqslant 1\}$, then $(\mathbf{T}) = \frac{\mathbf{1}}{\mathbf{1+a}}$.
6. Let $(S, \mathcal{O}_d)$ be a compact topological metric space, $x \in S$, and let H be a closed set in $(S, \mathcal{O})$. Prove that if the sets $\{x\}$ and $\{y\}$ are separated by a closed set K_{xy} with $ind(K_{xy}) \leqslant n - 1$ for every $y \in H$, then $\{x\}$ is separated from H by a closed set K with $ind(K) \leqslant n - 1$.
7. Prove that every zero-dimensional separable topological space $(S, \mathcal{O})$ is homeomorphic to a subspace of the Cantor set.

   Hint: By the separability of $(S, \mathcal{O})$ and by Theorem 15.6, $(S, \mathcal{O})$ has a countable basis $\{B_0, B_1, \ldots, B_n, \ldots\}$ that consists of clopen sets. Consider the mapping $f : S \longrightarrow \mathbf{Seq}_\infty(\{0, 1\})$ defined by $f(x) = (b_0, b_1, \ldots)$, where $b_i = I_{B_i}(x)$ for $i \in \mathbb{N}$.
8. Let T be a subset of $\mathbb{R}^n$. A function $f : T \longrightarrow \mathbb{R}^n$ satisfies the *Hölder condition of exponent* α if there is a constant k such that $|f(\mathbf{x}) - f(\mathbf{y})| \leqslant k|\mathbf{x} - \mathbf{y}|^\alpha$ for $\mathbf{x}, \mathbf{y} \in \mathbb{R}^n$. Prove that
$$HB^{\frac{s}{\alpha}}(f(T)) \leqslant k^{\frac{s}{\alpha}} HB^s(T).$$

   Solution: If $C \subseteq \mathbb{R}^n$ is a set of diameter $diam(C)$, then $f(C)$, the image of C under f, has a diameter no larger than $k(diam(C))^\alpha$. Therefore, if $\mathcal{C}$ is an r-cover of T, then $\{f(T \cap C) \mid C \in \mathcal{C}\}$ is a kr^α-cover of $f(T)$. Therefore,
$$\begin{aligned}\sum\{(diam(T \cap C))^{\frac{s}{\alpha}} \mid C \in \mathcal{C}\} &\leqslant \sum\{(k(diam(C))^\alpha)^{\frac{s}{\alpha}} \mid C \in C\}\\ &= k^{\frac{s}{\alpha}} \sum\{(diam(C))^s \mid C \in \mathcal{C}\},\end{aligned}$$
   which implies $HB_r^{\frac{s}{\alpha}}(T) \leqslant k^{\frac{s}{\alpha}} HB_r^s(T)$. Since $\lim_{r\to 0} kr^\alpha = 0$, we have $HB^{\frac{s}{\alpha}}(f(T)) \leqslant k^{\frac{s}{\alpha}} HB^s(T)$.

9. Consider the ultrametric space $(\mathbf{Seq}_\infty(\{0, 1\}), d_\phi)$ introduced in Supplement 2 of Chap. 14, where $\phi(\mathbf{u}) = 2^{-|\mathbf{u}|}$ for $\mathbf{u} \in \mathbf{Seq}(\{0, 1\})$. Prove that if $(S, \mathcal{O}_d)$ is a separable topological metric space such that $ind(S) = 0$, then S is homeomorphic to a subspace of the topological metric space $(\mathbf{Seq}_\infty\{0, 1\}, \mathcal{O}_{d_\phi})$.
   Solution: Since $ind(S) = 0$, there exists a basis $\mathcal{B}_0$ for $\mathcal{O}_d$ that consists of clopen sets (by Theorem 15.6). Further, since $(S, \mathcal{O}_d)$ is countable, there exists a basis $\mathcal{B} \subseteq \mathcal{B}_0$ that is countable. Let $\mathcal{B} = \{B_0, B_1, \ldots\}$.
   If $\mathbf{s} = (s_0, s_1, \ldots, s_{p-1}) \in \mathbf{Seq}_p(\{0, 1\})$, let $B(\mathbf{s})$ be the clopen set $B(\mathbf{s}) = B_0^{s_0} \cap B_1^{s_1} \cap \cdots \cap B_{p-1}^{s_1}$.
   Define the mapping $h : S \longrightarrow \mathbf{Seq}_\infty(\{0, 1\})$ by $h(x) = (s_0, s_1, \ldots)$, where $s_i = 1$ if $x \in B_i$ and $s_i = 0$ otherwise for $x \in S$. Thus, $\mathbf{s}$ is a prefix of $h(x)$ if and only if $x \in B(\mathbf{s})$. The mapping h is injective. Indeed, suppose that $x \neq y$. Since $S - \{y\}$ is an open set containing x, there exists i with $x \in B_i \subseteq S - \{y\}$, which implies $(h(x))_i = 1$ and $(h(y))_i = 0$, so $h(x) \neq h(y)$. Thus, h is a bijection between S and $h(S)$.
   In Exercise 4 of Chap. 8, we saw that the collection $\{P_\mathbf{u} \mid \mathbf{u} \in \mathbf{Seq}(\{0, 1\})\}$ is a basis for $\mathbf{Seq}_\infty(\{0, 1\})$ and $h^{-1}(P_\mathbf{u}) = B(\mathbf{u})$, so $h^{-1} : h(S) \longrightarrow S$ is continuous.
   Note that $h(U_i) = h(S) \cap \{0, 1\}^{i-1}\mathbf{Seq}_\infty(\{0, 1\})$ is open in $\mathbf{Seq}_\infty(\{0, 1\})$ for every $i \in \mathbb{N}$, so h^{-1} is continuous. Thus, h is a homeomorphism of S into $h(S)$.
10. Let $\mathbf{f} = (f_1, \ldots, f_m)$ be an iterative function system on $\mathbb{R}^n$ that realizes the sequence $(r, \ldots, r)$ with $r \in (0, 1)$ and let H be a nonempty compact set in $\mathbb{R}^n$. Prove that if U is the attractor of $\mathbf{f}$, then

$$\delta(H, U) \leqslant \frac{1}{1-r}\delta\left(H, F(H)\right),$$

   where δ is the metric of the Hausdorff hyperspace of compact subsets and $F(T) = \bigcup_{i=1}^n f_i(T)$ for $T \in \mathcal{P}(S)$.

   Solution: In the proof of Lemma 15.66, we saw that $\delta(F(H), F(U)) \leqslant r\delta(H, U)$. This allows us to write

$$\begin{aligned}\delta(H, U) &\leqslant \delta(H, F(H)) + \delta(F(H), U)\\ &= \delta(H, F(H)) + \delta(F(H), F(U))\\ &\leqslant \delta(H, F(H)) + r\delta(H, U),\end{aligned}$$

   which implies the desired inequality.
11. Consider the contractions $f_0, f_1 : \mathbb{R} \longrightarrow \mathbb{R}$ defined by $f_0(x) = rx$ and $f_1(x) = rx+1-r$ for $x \in \mathbb{R}$, where $r \in (0, 1)$. Find the attractor of the iterative function system $\mathbf{f} = (f_0, f_1)$.
12. Let $(S, \mathcal{O}_d)$ be a compact topological metric space. Prove that $cov(S) \leqslant n$ if and only if for every $\epsilon > 0$ there is an open cover $\mathcal{C}$ of S with $ord(\mathcal{C}) \leqslant n$ and $\sup\{diam(C) \mid C \in \mathcal{C}\} < \epsilon$.

Solution: The set of open spheres $\{C(x, \epsilon/2) \mid x \in S\}$ is an open cover that has a refinement with order not greater than n; the diameter of each of the sets of the refinement is less than ϵ.

Conversely, suppose that for every $\epsilon > 0$ there is an open cover $\mathcal{C}$ of S with $ord(\mathcal{C}) \leqslant n$ and $\sup\{diam(C) \mid C \in \mathcal{C}\} < \epsilon$.

Let $\mathcal{D}$ be a finite open cover of S. By Lebesgue's Lemma (Theorem 8.21) there exists $r > 0$ such that for every subset U of S with $diam(U) < r$ there is a set $L \in \mathcal{D}$ such that $U \subseteq L$.

Let $\mathcal{C}'$ be an open cover of S with order not greater than n and such that $\sup\{diam(C') \mid C' \in \mathcal{C}'\} < \min\{\epsilon, r\}$. Then $\mathcal{C}'$ is a refinement of $\mathcal{D}$, so $cov(S) \leqslant n$.

13. Prove that if $f : \mathbb{R}^n \longrightarrow \mathbb{R}^m$ is an isometry, then $HB^{\frac{s}{\alpha}}(f(T)) = HB^s(T)$.
14. Let F be a finite set of a metric space. Prove that $HB^0(F) = |F|$.
15. A useful variant of the Hausdorff-Besicovitch outer measure can be defined by restricting the r-covers to closed spheres of radius no greater than r. Let $\mathfrak{B}_r(U)$ be the set of all countable covers of a subset U of a metric space (S, d) that consist of closed spheres of radius no greater than r. Since $\mathfrak{B}_r(U) \subseteq \mathfrak{C}_r(U)$, it is clear that $HB^s_r(U) \leqslant HB'^s_r(U)$, where

$$HB'^s_r(U) = \inf_{\mathcal{C} \in \mathfrak{B}_r(U)} \sum\{(diam(C))^s \mid C \in \mathcal{C}\}.$$

Prove that:

(a) $HB'^s_r(U) \leqslant 2^s HB^s_r(U)$,
(b) $HB^s U \leqslant HB'^s U \leqslant 2^s HB^s U$, where $HB'^s(U) = \lim_{r \to 0} HB'^s_r(U)$, and
(c) $HBdim(U) = HBdim'(U)$, where $HBdim'(U) = \sup\{s \in \mathbb{R}_{\geqslant 0} \mid HB'^s(U) = \infty\}$

for every Borel subset U of S.

16. Prove that if U is a subset of $\mathbb{R}^n$ such that $\mathbf{I}(U) \neq \emptyset$, then $HBdim(U) = n$.

Let (S, d) be a metric space, s and r be two numbers in $\mathbb{R}_{>0}$, U be a subset of S, and

$$P^s_r(U) = \sup\left\{\sum_i diam(B_i)^s \mid B_i \in \mathcal{B}_r(U)\right\},$$

where $\mathcal{B}_r(U)$ is the collection of disjoint closed spheres centered in U and having diameter not larger than r. Observe that $\lim_{r \to 0} P^s_r(U)$ exists because $P^s_r(U)$ decreases when r decreases. Let $P^s(U) = \lim_{r \to 0} P^s_r(U)$.

17. Let $(S, \mathcal{O}_d)$ be a topological metric space. Define $PK^s(U)$ as the outer measure obtained by Method I starting from the function P^s,

$$PK^s(U) = \inf_{\mathcal{C} \in \mathfrak{C}_r(U)} \sum\left\{P^s(C) \mid C \in \mathcal{C}\right\},$$

where $\mathfrak{C}_r(U)$ is the collection of all countable r-covers for a set U.

(a) Prove that if U is a Borel set in $(S, \mathcal{O}_d)$, $0 < s < t$, and $PK^s(U)$ is finite, then $PK^t(U) = 0$. Further, prove that if $PK^t(U) > 0$, then $PK^s(U) = \infty$.
(b) The *packing dimension* of U is defined as

$$PKdim(U) = \sup\{s \mid PK^s(U) = \infty\}.$$

Prove that $HBdim(U) \leqslant PKdim(U)$ for any Borel subset U of $\mathbb{R}^n$.

Bibliographical Comments

The first monograph dedicated to dimension theory is the book by Hurewicz and Wallman [6]. A topology source with a substantial presentation of topological dimensions is [7]. The literature dedicated to fractals has several excellent references for dimension theory [5, 8, 9]. Supplement 10 appears in the last reference. Example 15.5 is from [10], where an interesting connection between entropy and the Hausdorff-Besicovitch dimension is discussed.

References

1. R. Bellman, *Adaptive Control Processes: A Guided Tour* (Princeton University Press, Princeton, NJ, 1961)
2. M. Köppen. The curse of dimensionality. 5th Online World Conference on Soft Computing in Industrial Applications (WSC5), 2000
3. C. Böhm, S. Berthold, D.A. Keim, Searching in high-dimensional spaces—index structures for improving the performance of multimedia databases. ACM Comput. Surv. **33**, 322–373 (2001)
4. B. Mandelbrot, *The Fractal Geometry of Nature* (W. H. Freeman, New York, 1982)
5. K. Falconer, *Fractal Geometry* (Wiley, New York, second edition , 2003)
6. W. Hurewicz, H. Wallman, *Dimension Theory* (Princeton University Press, Princeton, 1948)
7. R. Engelking, K. Siekluchi, *Topology - A Geometric Approach* (Heldermann Verlag, Berlin, 1992)
8. G. Edgar, *Measure, Topology, and Fractal Geometry* (Springer, New York, 1990)
9. G. Edgar, *Integral, Probability, and Fractal Measures* (Springer, New York, 1998)
10. P. Billingsley, *Ergodic Theory and Information* (John Wiley, New York, 1965)

Chapter 16
Clustering

16.1 Introduction

Clustering is the process of grouping together objects that are similar. The groups formed by clustering are referred to as *clusters*.

Similarity between objects that belong to a set V is usually measured using either a similarity function $s : V \times V \longrightarrow [0, 1]$, or a definite dissimilarity $d : V \times V \longrightarrow \mathbb{R}_{\geqslant 0}$ (see Sect. 1.9). The similarities between the objects are grouped in a symmetric similarity matrix $S \in \mathbb{R}^{m \times m}$, where $s_{ij} = s(v_i, s_j)$ for $1 \leqslant i, j \leqslant m$, where $m = |V|$.

There are several points of view for examining clustering techniques. We follow here the taxonomy of clustering presented in [1].

Clustering may or may not be *exclusive*, where an exclusive clustering technique yields clusters that are disjoint, while a nonexclusive technique produces overlapping clusters. From an algebraic point of view, an exclusive clustering algorithm generates a partition $\kappa = \{C_1, \ldots, C_k\}$ of the set of objects whose blocks $C_1, \ldots, C_k$ are referred to as *clusters*.

Clustering may be *intrinsic* or *extrinsic*. Intrinsic clustering is an unsupervised activity that is based only on the dissimilarities between the objects to be clustered. Most clustering algorithms fall into this category. Extrinsic clustering relies on information provided by an external source that prescribes, for example, which objects should be clustered together and which should not.

Finally, clustering may be *hierarchical* or *partitional*.

In hierarchical clustering algorithms, a sequence of partitions is constructed. In *hierarchical agglomerative algorithms* this sequence is increasing and it begins with the least partition of the set of objects whose blocks consist of single objects; as the clustering progresses, certain clusters are fused together. As a result, an agglomerative clustering is a chain of partitions on the set of objects that begins with the least partition α_S of the set of objects S and ends with the largest partition ω_S. In a *hierarchical divisive algorithm*, the sequence of partitions is decreasing. Its first member is the one-block partition ω_S, and each partitions is built by subdividing the blocks of the previous partition.

D. A. Simovici and C. Djeraba, *Mathematical Tools for Data Mining*,
Advanced Information and Knowledge Processing, DOI: 10.1007/978-1-4471-6407-4_16,

Partitional clustering creates a partition of the set of objects whose blocks are the clusters such that objects in a cluster are more similar to each other than to objects that belong to different clusters. A typical representative algorithm is the k-means algorithm and its many extensions.

Our presentation is organized around the last dichotomy. We start with a class of hierarchical agglomerative algorithms. This is continued with a discussion of the k-means algorithm, a representative of partitional algorithms. Then, we continue with a discussion of certain limitations of clustering centered around Kleinberg's impossibility theorem. We conclude with an evaluation of clustering quality.

Clustering can be regarded as a special type of classification, where the clusters serve as classes of objects. It is a widely used data mining activity with multiple applications in a variety of scientific activities ranging from biology and astronomy to economics and sociology.

16.2 Hierarchical Clustering

Hierarchical clustering is a recursive process that begins with a metric space of objects (S, d) and results in a chain of partitions of the set of objects. In each of the partitions, similar objects belong to the same block and objects that belong to distinct blocks tend to be dissimilar.

In agglomerative hierarchical clustering, the construction of this chain begins with the unit partition $\pi^1 = \alpha_S$. If the partition constructed at step k is

$$\pi^k = \{U_1^k, \ldots, U_{m_k}^k\},$$

then two distinct blocks U_p^k and U_q^k of this partition are selected using a *selection criterion*. These blocks are fused and a new partition

$$\pi^{k+1} = \{U_1^k, \ldots, U_{p-1}^k, U_{p+1}^k, \ldots, U_{q-1}^k, U_{q+1}^k, \ldots, U_p^k \cup U_q^k\}$$

is formed. Clearly, we have $\pi^k \prec \pi^{k+1}$. The process must end because the poset $(PART(S), \leqslant)$ is of finite height. The algorithm halts when the one-block partition ω_S is reached.

As we saw in Theorem 14.12, the chain of partitions $\pi^1, \pi^2, \ldots$ generates a hierarchy on the set S. Therefore, all tools developed for hierarchies, including the notion of a dendrogram, can be used for hierarchical algorithms.

When data to be clustered are numerical (that is, when $S \subseteq \mathbb{R}^n$), we can define the *centroid* of a nonempty subset U of S as:

$$\mathbf{c}_U = \frac{1}{|U|}\sum\{\mathbf{o} | \mathbf{o} \in U\}.$$

If $\pi = \{U_1, \ldots, U_m\}$ is a partition of S, then the *sum of the squared errors* of π is the number

$$sse(\pi) = \sum_{i=1}^{m} \sum \left\{ d^2(\mathbf{o}, \mathbf{c}_{U_i}) | \mathbf{o} \in U_i \right\}, \tag{16.1}$$

where d is the Euclidean distance in $\mathbb{R}^n$.

If two blocks U and V of a partition π are fused into a new block W to yield a new partition π' that covers π, then the variation of the sum of squared errors is given by

$$\begin{aligned} sse(\pi') - sse(\pi) = & \sum \left\{ d^2(\mathbf{o}, \mathbf{c}_W) | \mathbf{o} \in U \cap V \right\} \\ & - \sum \left\{ d^2(\mathbf{o}, \mathbf{c}_U) | \mathbf{o} \in U \right\} - \sum \left\{ d^2(\mathbf{o}, \mathbf{c}_V) | \mathbf{o} \in V \right\}. \end{aligned}$$

The centroid of the new cluster W is given by

$$\mathbf{c}_W = \frac{1}{|W|} \sum \{\mathbf{o} | \mathbf{o} \in W\} = \frac{|U|}{|W|} \mathbf{c}_U + \frac{|V|}{|W|} \mathbf{c}_V.$$

This allows us to evaluate the increase in the sum of squared errors:

$$\begin{aligned} sse(\pi') - sse(\pi) = & \sum \left\{ d^2(\mathbf{o}, \mathbf{c}_W) \mid \mathbf{o} \in U \cup V \right\} \\ & - \sum \left\{ d^2(\mathbf{o}, \mathbf{c}_U) \mid \mathbf{o} \in U \right\} - \sum \left\{ d^2(\mathbf{o}, \mathbf{c}_V) \mid \mathbf{o} \in V \right\} \\ = & \sum \left\{ d^2(\mathbf{o}, \mathbf{c}_W) - d^2(\mathbf{o}, \mathbf{c}_U) \mid \mathbf{o} \in U \right\} \\ & + \sum \left\{ d^2(\mathbf{o}, \mathbf{c}_W) - d^2(\mathbf{o}, \mathbf{c}_V) \mid \mathbf{o} \in V \right\}. \end{aligned}$$

Observe that:

$$\begin{aligned} & \sum \left\{ d^2(\mathbf{o}, \mathbf{c}_W) - d^2(\mathbf{o}, \mathbf{c}_U) \mid \mathbf{o} \in U \right\} \\ & = \sum_{\mathbf{o} \in U} ((\mathbf{o} - \mathbf{c}_W)(\mathbf{o} - \mathbf{c}_W) - (\mathbf{o} - \mathbf{c}_U)(\mathbf{o} - \mathbf{c}_U)) \\ & = |U|(\mathbf{c}_W^2 - \mathbf{c}_U^2) + 2(\mathbf{c}_U - \mathbf{c}_W) \sum_{\mathbf{o} \in U} \mathbf{o} \\ & = |U|(\mathbf{c}_W^2 - \mathbf{c}_U^2) + 2|U|(\mathbf{c}_U - \mathbf{c}_W)\mathbf{c}_U \\ & = (\mathbf{c}_W - \mathbf{c}_U)\left(|U|(\mathbf{c}_W + \mathbf{c}_U) - 2|U|\mathbf{c}_U\right) \\ & = |U|(\mathbf{c}_W - \mathbf{c}_U)^2. \end{aligned}$$

Using the equality $\mathbf{c}_W - \mathbf{c}_U = \frac{|U|}{|W|}\mathbf{c}_U + \frac{|V|}{|W|}\mathbf{c}_V - \mathbf{c}_U = \frac{|V|}{|W|}(\mathbf{c}_V - \mathbf{c}_U)$, we obtain $\sum \left\{ d^2(\mathbf{o}, \mathbf{c}_W) - d^2(\mathbf{o}, \mathbf{c}_U) \mid \mathbf{o} \in U \right\} = \frac{|U||V|^2}{|W|^2}(\mathbf{c}_V - \mathbf{c}_U)^2$.

Similarly, we have

$$\sum \left\{ d^2(\mathbf{o}, \mathbf{c}_W) - d^2(\mathbf{o}, \mathbf{c}_V) \mid \mathbf{o} \in V \right\} = \frac{|U|^2 |V|}{|W|^2} (\mathbf{c}_V - \mathbf{c}_U)^2 ,$$

so,

$$sse(\pi') - sse(\pi) = \frac{|U||V|}{|W|} (\mathbf{c}_V - \mathbf{c}_U)^2 . \tag{16.2}$$

The dissimilarity between two clusters U and V can be defined using one of the following real-valued, two-argument functions defined on the set of subsets of S:

$$\begin{aligned} sl(U, V) &= \min \{d(u, v) | u \in U, v \in V\} ; \\ cl(U, V) &= \max \{d(u, v) | u \in U, v \in V\} ; \\ gav(U, V) &= \frac{\sum \{d(u, v) | u \in U, v \in V\}}{|U| \cdot |V|} ; \\ cen(U, V) &= (\mathbf{c}_U - \mathbf{c}_V)^2 ; \\ ward(U, V) &= \frac{|U||V|}{|U| + |V|} (\mathbf{c}_V - \mathbf{c}_U)^2 . \end{aligned}$$

The names of the functions *sl*, *cl*, *gav*, and *cen* defined above are acronyms of the terms "single link", "complete link", "group average", and "centroid", respectively. They are linked to variants of the hierarchical clustering algorithms that we discuss in later. Note that in the case of the *ward* function the value equals the increase in the sum of the square errors when the clusters U, V are replaced with their union.

The specific selection criterion for fusing blocks defines the clustering algorithm. All algorithms store the dissimilarities between the current clusters $\pi^k = \{U_1^k, \ldots, U_{m_k}^k\}$ in an $m_k \times m_k$-matrix $D^k = (d_{ij}^k)$, where d_{ij}^k is the dissimilarity between the clusters U_i^k and U_j^k. As new clusters are created by merging two existing clusters, the distance matrix must be adjusted to reflect the dissimilarities between the new cluster and existing clusters.

The general form of the algorithm is shown as Algorithm 16.2.1.

Algorithm 16.2.1: Matrix Agglomerative Clustering

Data: the initial dissimilarity matrix D^1
Result: the cluster hierarchy on the set of objects S, where $|S| = n$

```
k = 1;
initialize clustering: π^1 = α_S;
while π^k contains more than one block do
    merge a pair of two of the closest clusters;
    output new cluster;
    k++;
    compute the dissimilarity matrix D^k;
end
```

To evaluate the space and time complexity of hierarchical clustering, note that the algorithm must handle the matrix of the dissimilarities between objects, and this is a symmetric $n \times n$-matrix having all elements on its main diagonal equal to 0; in other words, the algorithm needs to store $\frac{n(n-1)}{2}$ numbers. To keep track of the clusters, an extra space that does not exceed $n-1$ is required. Thus, the total space required is $O(n^2)$.

The time complexity of agglomerative clustering algorithms has been evaluated in [2].

The computation of the dissimilarity between a new cluster and existing clusters is described next.

Theorem 16.1 *Let U and V be two clusters of the clustering π that are joined into a new cluster W. Then, if $Q \in \pi - \{U, V\}$, we have*

$$
\begin{aligned}
sl(W, Q) &= \frac{1}{2} sl(U, Q) + \frac{1}{2} sl(V, Q) - \frac{1}{2}\left| sl(U, Q) - sl(V, Q)\right|;\\
cl(W, Q) &= \frac{1}{2} cl(U, Q) + \frac{1}{2} cl(V, Q) + \frac{1}{2}\left| cl(U, Q) - cl(V, Q)\right|;\\
gav(W, Q) &= \frac{|U|}{|U| + |V|} gav(U, Q) + \frac{|V|}{|U| + |V|} gav(V, Q);\\
cen(W, Q) &= \frac{|U|}{|U| + |V|} cen(U, Q) + \frac{|V|}{|U| + |V|} cen(V, Q)\\
&\quad - \frac{|U||V|}{(|U| + |V|)^2} cen(U, V);\\
ward(W, Q) &= \frac{|U| + |Q|}{|U| + |V| + |Q|} ward(U, Q) + \frac{|V| + |Q|}{|U| + |V| + |Q|} ward(V, Q)\\
&\quad - \frac{|Q|}{|U| + |V| + |Q|} ward(U, V).
\end{aligned}
$$

Proof The first two equalities follow from the fact that

$$
\begin{aligned}
\min\{a, b\} &= \frac{1}{2}(a + b) - \frac{1}{2}|a - b|,\\
\max\{a, b\} &= \frac{1}{2}(a + b) + \frac{1}{2}|a - b|,
\end{aligned}
$$

for every $a, b \in \mathbb{R}$.

For the third equality, we have

$$
\begin{aligned}
gav(W, Q) &= \frac{\sum\{d(w, q) | w \in W, q \in Q\}}{|W| \cdot |Q|}\\
&= \frac{\sum\{d(u, q) | u \in U, q \in Q\}}{|W| \cdot |Q|} + \frac{\sum\{d(v, q) | v \in V, q \in Q\}}{|W| \cdot |Q|}
\end{aligned}
$$

$$= \frac{|U|}{|W|}\frac{\sum\{d(u,q)|u \in U, q \in Q\}}{|U|\cdot|Q|} + \frac{|V|}{|W|}\frac{\sum\{d(v,q)|v \in V, q \in Q\}}{|V|\cdot|Q|}$$
$$= \frac{|U|}{|U|+|V|}gav(U,Q) + \frac{|V|}{|U|+|V|}gav(V,Q).$$

The equality involving the function *cen* is immediate. The last equality can be easily translated into

$$\frac{|Q||W|}{|Q|+|W|}(\mathbf{c}_Q - \mathbf{c}_W)^2 = \frac{|U|+|Q|}{|U|+|V|+|Q|}\frac{|U||Q|}{|U|+|Q|}(\mathbf{c}_Q - \mathbf{c}_U)^2$$
$$+ \frac{|V|+|Q|}{|U|+|V|+|Q|}\frac{|V||Q|}{|V|+|Q|}(\mathbf{c}_Q - \mathbf{c}_V)^2$$
$$- \frac{|Q|}{|U|+|V|+|Q|}\frac{|U||V|}{|U|+|V|}(\mathbf{c}_V - \mathbf{c}_U)^2,$$

which can be verified replacing $|W| = |U| + |V|$ and $\mathbf{c}_W = \frac{|U|}{|W|}\mathbf{c}_U + \frac{|V|}{|W|}\mathbf{c}_V$. □

The equalities contained by Theorem 16.1 are often presented as a single equality involving several coefficients.

Corollary 16.2 (The Lance-Williams Formula) *Let U and V be two clusters of the clustering π that are joined into a new cluster W. Then, if $Q \in \pi - \{U, V\}$, the dissimilarity between W and Q can be expressed as*

$$d(W,Q) = a_U d(U,Q) + a_V d(V,Q) + b d(U,V) + c|d(U,Q) - d(V,Q)|,$$

where the coefficients a_U, a_V, b, c are given by the following table:

Function	a_U	a_V	b	c
sl	$\frac{1}{2}$	$\frac{1}{2}$	0	$-\frac{1}{2}$
cl	$\frac{1}{2}$	$\frac{1}{2}$	0	$\frac{1}{2}$
gav	$\frac{\lvert U\rvert}{\lvert U\rvert+\lvert V\rvert}$	$\frac{\lvert V\rvert}{\lvert U\rvert+\lvert V\rvert}$	0	0
cen	$\frac{\lvert U\rvert}{\lvert U\rvert+\lvert V\rvert}$	$\frac{\lvert V\rvert}{\lvert U\rvert+\lvert V\rvert}$	$-\frac{\lvert U\rvert\lvert V\rvert}{(\lvert U\rvert+\lvert V\rvert)^2}$	0
ward	$\frac{\lvert U\rvert+\lvert Q\rvert}{\lvert U\rvert+\lvert V\rvert+\lvert Q\rvert}$	$\frac{\lvert V\rvert+\lvert Q\rvert}{\lvert U\rvert+\lvert V\rvert+\lvert Q\rvert}$	$-\frac{\lvert Q\rvert}{\lvert U\rvert+\lvert V\rvert+\lvert Q\rvert}$	0

Proof This statement is an immediate consequence of Theorem 16.1. □

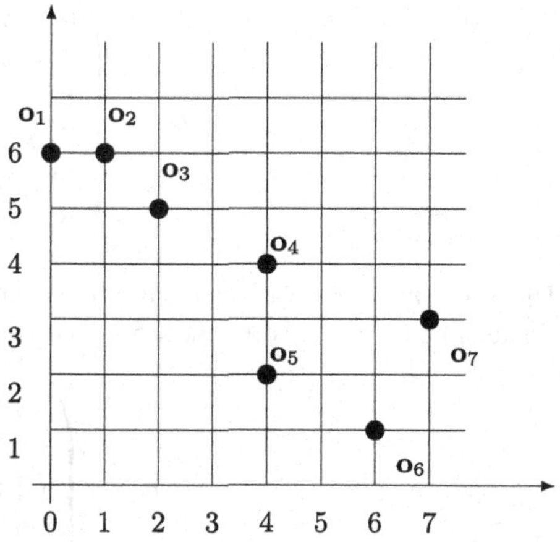

Fig. 16.1 Set of seven points in $\mathbb{R}^2$

The variant of the algorithm that makes use of the function *sl* is known as the *single-link* clustering. It tends to favor elongated clusters.

Example 16.3 We use single-link clustering for the metric space (S, d_1), where $S \subseteq \mathbb{R}^2$ consists of seven objects, $S = \{\mathbf{o}_1, \ldots, \mathbf{o}_7\}$ (see Fig. 16.1).

The distances $d_1(\mathbf{o}_i, \mathbf{o}_j)$ for $1 \leqslant i, j \leqslant 7$ between the objects of S are specified by the 7×7 matrix

$$D^1 = \begin{pmatrix} 0 & 1 & 3 & 6 & 8 & 11 & 10 \\ 1 & 0 & 2 & 5 & 7 & 10 & 9 \\ 3 & 2 & 0 & 3 & 5 & 8 & 7 \\ 6 & 5 & 3 & 0 & 2 & 5 & 4 \\ 8 & 7 & 5 & 2 & 0 & 3 & 4 \\ 11 & 10 & 8 & 5 & 3 & 0 & 3 \\ 10 & 9 & 7 & 4 & 4 & 3 & 0 \end{pmatrix}.$$

We apply the hierarchical clustering algorithm using the single-link variant to the set S. Initially, the clustering consists of singleton sets:

$$\pi^1 = \{\{\mathbf{o}_i\} \mid 1 \leqslant i \leqslant 7\} \{\{\mathbf{o}_1\}, \{\mathbf{o}_2\}, \{\mathbf{o}_3\}, \{\mathbf{o}_4\}, \{\mathbf{o}_5\}, \{\mathbf{o}_6\}, \{\mathbf{o}_7\}\}.$$

Two of the closest clusters are $\{\mathbf{o}_1\}, \{\mathbf{o}_2\}$; these clusters are fused into the cluster $\{\mathbf{o}_1, \mathbf{o}_2\}$, the new partition is

$$\pi^2 = \{\{\mathbf{o}_1, \mathbf{o}_2\}, \ldots, \{\mathbf{o}_7\}\},$$

and the matrix of dissimilarities becomes the 6×6-matrix

$$D^2 = \begin{pmatrix} 0 & 2 & 5 & 7 & 10 & 9 \\ 2 & 0 & 3 & 5 & 8 & 7 \\ 5 & 3 & 0 & 2 & 5 & 4 \\ 7 & 5 & 2 & 0 & 3 & 4 \\ 10 & 8 & 5 & 3 & 0 & 3 \\ 9 & 7 & 4 & 4 & 3 & 0 \end{pmatrix}.$$

The next pair of closest clusters is $\{\mathbf{o}_1, \mathbf{o}_2\}$ and $\{\mathbf{o}_3\}$. These clusters are fused into the cluster $\{\mathbf{o}_1, \mathbf{o}_2, \mathbf{o}_3\}$, and the new 5×5-matrix is:

$$D^3 = \begin{pmatrix} 0 & 3 & 5 & 8 & 7 \\ 3 & 0 & 2 & 5 & 4 \\ 5 & 2 & 0 & 3 & 4 \\ 8 & 5 & 3 & 0 & 3 \\ 7 & 4 & 4 & 3 & 0 \end{pmatrix},$$

which corresponds to the partition

$$\pi^3 = \{\{\mathbf{o}_1, \mathbf{o}_2, \mathbf{o}_3\}, \{\mathbf{o}_4\}, \ldots, \{\mathbf{o}_7\}\}.$$

Next, the closest clusters are $\{\mathbf{o}_4\}$ and $\{\mathbf{o}_5\}$. Fusing these yields the partition

$$\pi^4 = \{\{\mathbf{o}_1, \mathbf{o}_2, \mathbf{o}_3\}, \{\mathbf{o}_4, \mathbf{o}_5\}, \{\mathbf{o}_6\}, \{\mathbf{o}_7\}\}$$

and the 4×4-matrix

$$D^4 = \begin{pmatrix} 0 & 3 & 8 & 7 \\ 3 & 0 & 3 & 4 \\ 8 & 3 & 0 & 3 \\ 7 & 4 & 3 & 0 \end{pmatrix}.$$

We have three choices now since there are three pairs of clusters at distance 3 of each other: $(\{\mathbf{o}_1, \mathbf{o}_2, \mathbf{o}_3\}, \{\mathbf{o}_4, \mathbf{o}_5\})$, $(\{\mathbf{o}_4, \mathbf{o}_5\}, \{\mathbf{o}_6\})$, and $(\{\mathbf{o}_6\}, \{\mathbf{o}_7)$. By choosing to fuse the first pair, we obtain the partition

$$\pi^5 = \{\{\mathbf{o}_1, \mathbf{o}_2, \mathbf{o}_3, \mathbf{o}_4, \mathbf{o}_5\}, \{\mathbf{o}_6\}, \{\mathbf{o}_7\}\},$$

which corresponds to the 3×3-matrix

$$D^5 = \begin{pmatrix} 0 & 3 & 4 \\ 3 & 0 & 3 \\ 4 & 3 & 0 \end{pmatrix}.$$

Observe that the large cluster $\{\mathbf{o}_1, \mathbf{o}_2, \mathbf{o}_3, \mathbf{o}_4, \mathbf{o}_5\}$ formed so far has an elongated shape, which is typical for single-link variant of the algorithm.

Next, we coalesce $\{\mathbf{o}_1, \mathbf{o}_2, \mathbf{o}_3, \mathbf{o}_4, \mathbf{o}_5\}$ and $\{\mathbf{o}_6\}$, which yields

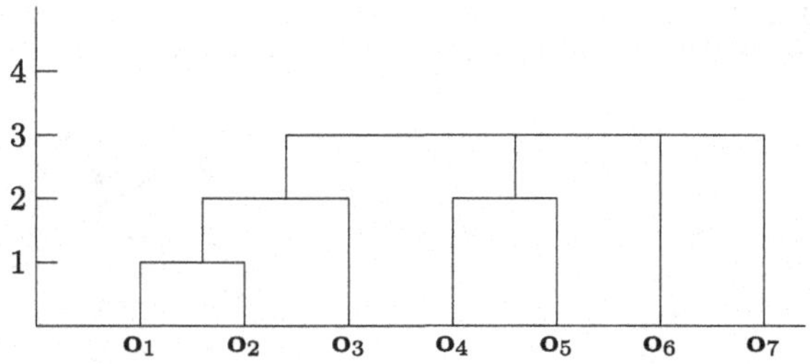

Fig. 16.2 Dendrogram of single-link clustering

$$\pi^6 = \{\{\mathbf{o}_1, \mathbf{o}_2, \mathbf{o}_3, \mathbf{o}_4, \mathbf{o}_5, \mathbf{o}_6\}, \{\mathbf{o}_7\}\}$$

and

$$D^6 = \begin{pmatrix} 0 & 3 \\ 3 & 0 \end{pmatrix}.$$

Finally, we join the last two clusters, and the clustering is completed.

The dendrogram of the hierarchy produced by the algorithm is given in Fig. 16.2.

The variant of the algorithm that uses the function *cl* is known as the *complete-link* clustering. It tends to favor globular clusters.

Example 16.4 Now we apply the complete-link algorithm to the set S considered in Example 16.3. It is easy to see that the initial two partitions and the initial matrix are the same as for the single-link algorithm.

However, after creating the first cluster $\{\mathbf{o}_1, \mathbf{o}_2\}$, the distance matrices begin to differ. The next matrix is

$$D^2 = \begin{pmatrix} 0 & 3 & 6 & 8 & 11 & 10 \\ 3 & 0 & 3 & 5 & 8 & 7 \\ 6 & 3 & 0 & 2 & 5 & 4 \\ 8 & 5 & 2 & 0 & 3 & 4 \\ 11 & 8 & 5 & 3 & 0 & 3 \\ 10 & 7 & 4 & 4 & 3 & 0 \end{pmatrix},$$

which shows that the closest clusters are now $\{\mathbf{o}_4\}$ and $\{\mathbf{o}_5\}$. Thus,

$$\pi^3 = \{\{\mathbf{o}_1, \mathbf{o}_2\}, \{\mathbf{o}_3\}, \{\mathbf{o}_4, \mathbf{o}_5\}, \{\mathbf{o}_6\}, \{\mathbf{o}_7\}\}$$

and the new matrix is

$$D^3 = \begin{pmatrix} 0 & 3 & 8 & 11 & 10 \\ 3 & 0 & 5 & 8 & 7 \\ 8 & 5 & 0 & 5 & 3 \\ 11 & 8 & 5 & 0 & 3 \\ 10 & 7 & 3 & 3 & 0 \end{pmatrix}.$$

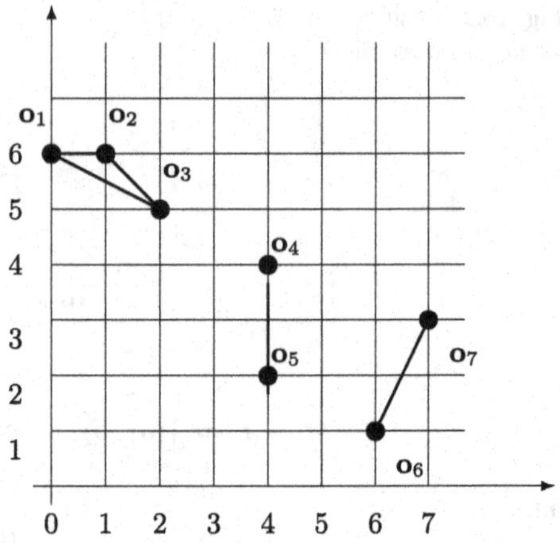

Fig. 16.3 Partial clustering obtained by the complete-link method

Three pairs of clusters correspond to the minimal value 3 in D^3:

$$(\{\mathbf{o}_1, \mathbf{o}_2\}, \{\mathbf{o}_3\}),$$
$$(\{\mathbf{o}_4, \mathbf{o}_5\}, \{\mathbf{o}_3\}),$$
$$(\{\mathbf{o}_6\}, \{\mathbf{o}_7\}).$$

If we merge the last pair, we get the partition

$$\pi^4 = \{\{\mathbf{o}_1, \mathbf{o}_2\}, \{\mathbf{o}_3\}, \{\mathbf{o}_4, \mathbf{o}_5\}, \{\mathbf{o}_6, \mathbf{o}_7\}\}$$

and the matrix

$$D^4 = \begin{pmatrix} 0 & 3 & 8 & 11 \\ 3 & 0 & 5 & 8 \\ 8 & 5 & 0 & 5 \\ 11 & 8 & 5 & 0 \end{pmatrix}.$$

Next, the closest clusters are $\{\mathbf{o}_1, \mathbf{o}_2\}, \{\mathbf{o}_3\}$. Merging those clusters results in the partition $\pi^5 = \{\{\mathbf{o}_1, \mathbf{o}_2, \mathbf{o}_3\}, \{\mathbf{o}_4, \mathbf{o}_5\}, \{\mathbf{o}_6, \mathbf{o}_7\}\}$ and the matrix

$$D^5 = \begin{pmatrix} 0 & 8 & 11 \\ 8 & 0 & 5 \\ 11 & 5 & 0 \end{pmatrix}.$$

The current clustering is shown in Fig. 16.3. Observe that in the case of the clusters obtained by the complete-link method that appear early tend to enclose objects that are closed in the sense of the distance.

Now the closest clusters are $\{\mathbf{o}_4, \mathbf{o}_5\}$ and $\{\mathbf{o}_6, \mathbf{o}_7\}$. By merging those clusters, we obtain the partition $\pi^5 = \{\{\mathbf{o}_1, \mathbf{o}_2, \mathbf{o}_3\}, \{\mathbf{o}_4, \mathbf{o}_5, \mathbf{o}_6, \mathbf{o}_7\}\}$ and the matrix

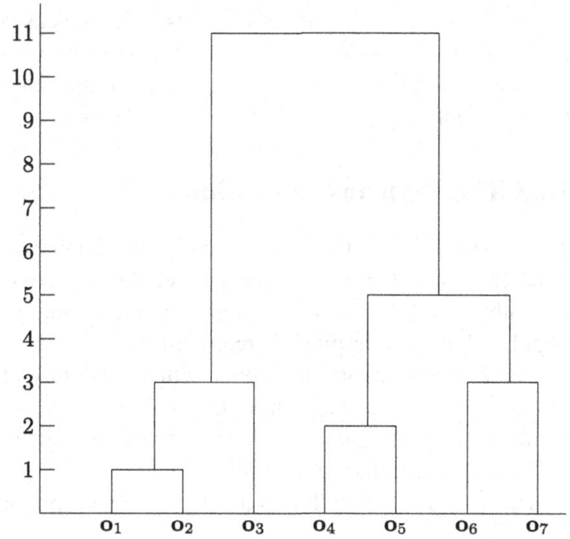

Fig. 16.4 Dendrogram of complete-link clustering

$$D^6 = \begin{pmatrix} 0 & 11 \\ 11 & 0 \end{pmatrix}.$$

The dendrogram of the resulting clustering is given in Fig. 16.4.

The *group average method*, which makes use of the *gav* function generates an intermediate approach between the single-link and the complete-link method. What the methods mentioned so far have in common is the *monotonicity property* expressed by the following statement.

Theorem 16.5 *Let (S, d) be a finite metric space and let $D^1, \ldots, D^m$ be the sequence of matrices constructed by any of the first three hierarchical methods (single, complete, or average link), where $m = |S|$. If μ_i is the smallest entry of the matrix D^i for $1 \leqslant i \leqslant m$, then $\mu_1 \leqslant \mu_2 \leqslant \cdots \leqslant \mu_m$. In other words, the dissimilarity between clusters that are merged at each step is nondecreasing.*

Proof Suppose that the matrix D^{j+1} is obtained from the matrix D^j by merging the clusters C_p and C_q that correspond to the lines p and q and to columns p, q of D^j. This happens because $d_{pq} = d_{qp}$ is one of the minimal elements of the matrix D^j. Then, these lines and columns are replaced with a line and column that corresponds to the new cluster C_r and the dissimilarities between this new cluster and the previous clusters C_i, where $i \neq p, q$. The elements d_{rh}^{j+1} of the new line (and column) are obtained either as $\min\{d_{ph}^j, d_{qh}^j\}$, $\max\{d_{ph}^j, d_{qh}^j\}$, or $\frac{|C_p|}{|C_r|} d_{ph}^j + \frac{|C_q|}{|C_r|} d_{qh}^j$, for the single-link, complete-link, or group average methods, respectively. In any of these cases, it is not possible to obtain a value for d_{rh}^{j+1} that is less than the minimal value of an element of D^j. □

The last two methods captured by the Lance-Williams formula are the centroid method and the Ward method of clustering. As we observed before, Formula (16.2) shows that the dissimilarity of two clusters in the case of Ward's method equals the increase in the sum of the squared errors that results when the clusters are merged. The centroid method adopts the distance between the centroids as the distance between the corresponding clusters. Either method lacks the monotonicity properties.

16.3 The k-Means Algorithm

The k-means algorithm is a partitional algorithm that requires the specification of the number of clusters k as an input. The set of objects to be clustered $S = \{\mathbf{o}^1, \ldots, \mathbf{o}^n\}$ is a subset of $\mathbb{R}^m$. Due to its simplicity and its many implementations it is a very popular algorithm despite this requirement.

The k-means algorithm begins with a randomly chosen collection of k points $\mathbf{c}^1, \ldots, \mathbf{c}^k$ in $\mathbb{R}^m$ called centroids. An initial partition of the set S of objects is computed by assigning each object $\mathbf{o}^i$ to its closest centroid $\mathbf{c}^j$. Let U_j be the set of points assigned to the centroid $\mathbf{c}^j$.

The assignments of objects to centroids are expressed by a matrix (b_{ij}), where

$$b_{ij} = \begin{cases} 1 & \text{if } \mathbf{o}^i \in U_j, \\ 0 & \text{otherwise.} \end{cases}$$

Since each object is assigned to exactly one cluster, we have $\sum_{j=1}^{k} b_{ij} = 1$. Also, $\sum_{i=1}^{n} b_{ij}$ equals the number of objects assigned to the centroid $\mathbf{c}^j$.

After these assignments, expressed by the matrix (b_{ij}), the centroids $\mathbf{c}^j$ must be re-computed using the formula:

$$\mathbf{c}^j = \frac{\sum_{i=1}^{n} b_{ij}\mathbf{o}^i}{\sum_{i=1}^{n} b_{ij}} \tag{16.3}$$

for $1 \leqslant j \leqslant k$.

The sum of squared errors of a partition $\pi = \{U_1, \ldots, U_k\}$ of a set of objects S was defined in Equality (16.1) as

$$sse(\pi) = \sum_{j=1}^{k} \sum_{\mathbf{o} \in U_j} d^2(\mathbf{o}, \mathbf{c}^j),$$

where $\mathbf{c}^j$ is the centroid of U_j for $1 \leqslant j \leqslant k$. The error of such an assignment is the sum of squared errors of the partition $\pi = \{U_1, \ldots, U_k\}$ defined as

$$\begin{aligned} sse(\pi) &= \sum_{i=1}^{n} \sum_{j=1}^{k} b_{ij} \|\mathbf{o}^i - \mathbf{c}^j\|^2 \\ &= \sum_{i=1}^{n} \sum_{j=1}^{k} b_{ij} \sum_{p=1}^{m} \left(o^i_p - c^j_p\right)^2. \end{aligned}$$

The mk necessary conditions for a local minimum of this function,

$$\frac{\partial sse(\pi)}{\partial c_p^j} = \sum_{i=1}^{n} b_{ij}\left(-2\left(o_p^i - c_p^j\right)\right) = 0,$$

for $1 \leqslant p \leqslant m$ and $1 \leqslant j \leqslant k$, can be written as

$$\sum_{i=1}^{n} b_{ij} o_p^i = \sum_{i=1}^{n} b_{ij} c_p^j = c_p^j \sum_{i=1}^{n} b_{ij},$$

or as

$$c_p^j = \frac{\sum_{i=1}^{n} b_{ij} o_p^i}{\sum_{i=1}^{n} b_{ij}}$$

for $1 \leqslant p \leqslant m$. In vectorial form, these conditions amount to

$$\mathbf{c}^j = \frac{\sum_{i=1}^{n} b_{ij} \mathbf{o}^i}{\sum_{i=1}^{n} b_{ij}},$$

which is exactly the formula (16.3) that is used to update the centroids. Thus, the choice of the centroids can be justified by the goal of obtaining local minima of the sum of squared errors of the clusterings.

Since we have new centroids, objects must be reassigned, which means that the values of b_{ij} must be recomputed, which, in turn, affects the values of the centroids, etc.

The halting criterion of the algorithm depends on particular implementations and may involve

(i) performing a certain number of iterations;
(ii) lowering the sum of squared errors $sse(\pi)$ below a certain limit;
(iii) the current partition coinciding with the previous partition.

This variant of the k-means algorithm is known as *Forgy's* Algorithm 16.3.1.

Algorithm 16.3.1: The k-means Forgy's Algorithm

Data: the set of objects to be clustered $S = \{\mathbf{o}^1, \ldots, \mathbf{o}^n\}$ and the number of clusters k
Result: collection of k clusters

```
extract a randomly chosen collection of k vectors c_1, ..., c_k in R^n;
assign each object o^i to the closest centroid c^j;
let π = {U_1, ..., U_k} be the partition defined by c^1, ..., c^k;
recompute the centroids of the clusters U_1, ..., U_k;
while halting criterion is not met do
    compute the new value of the partition π using the current centroids;
    recompute the centroids of the blocks of π;
end
```

The popularity of the k-means algorithm stems from its simplicity and its low time complexity $O(kn\ell)$, where n is the number of objects to be clustered and ℓ is the number of iterations that the algorithm is performing.

Another variant of the k-means algorithm redistributes objects to clusters based on the effect of such a reassignment on the objective function. If $sse(\pi)$ decreases, the object is moved and the two centroids of the affected clusters are recomputed. This variant is carefully analyzed in [3].

16.4 The PAM Algorithm

Another algorithm, named PAM (an acronym of "Partition Around Medoids") developed by Kaufman and Rousseeuw [4], also requires as an input parameter the number k of clusters to be extracted.

The k clusters are determined based on a representative object from each cluster, called the *medoid* of the cluster. The medoid of a cluster is one of the objects that have a most central position in the cluster.

PAM begins with a set of objects S, where $|S| = n$, a dissimilarity $n \times n$ matrix D, and a prescribed number of clusters k. The d_{ij} entry of the matrix D is the dissimilarity $d(o_i, o_j)$ between the objects o_i and o_j.

The algorithm has two phases:

(i) The *building phase* aims to construct a set L of selected objects, $L \subseteq S$. The set of remaining objects is denoted by R; clearly, $R = S - L$. To determine the most centrally located object we compute $Q_i = \sum_{j=1}^{n} d_{ij}$ for $1 \leqslant i \leqslant n$. The most central object o_q is determined by $q = \arg\min_i Q_i$. The set L is initialized as $L = \{o_q\}$.
Suppose now that we have constructed a set L of selected objects and $|L| < k$. We need to add a new selected object to the set L. To do this, we need to examine all objects that have not been included in L so far, that is, all objects in R. The selection is determined by a merit function $M : R \longrightarrow \mathbb{N}$.
To compute the merit $M(o)$ of an object $o \in R$, we scan all objects in R distinct from o. Let $o' \in R - \{o\}$ be such an object. If $d(o, o') < d(L, o')$, then adding o to L could benefit the clustering (from the point of view of o') because $d(L, o')$ will diminish. The potential benefit is $d(o', L) - d(o, o')$. Of course, if $d(o, o') \geqslant d(L, o')$, no such benefit exists (from the point of view of o'). Thus, we compute the merit of o as

$$M(o) = \sum_{o' \in R - \{o\}} \max\{D(L, o') - d(o, o'), 0\}.$$

We add to L the unselected object o that has the largest merit value. The building phase halts when $|L| = k$.

The objects in set L are the potential medoids of the k clusters that we seek to build.

(ii) In a second phase, swapping objects and existing medoids is considered. A cost of a swap is defined with the intention of penalizing swaps that diminish the centrality of the medoids in the clusters. Swapping continues as long as useful swaps (that is, swaps with negative costs) can be found.
The second phase of the algorithm aims to improve the clustering by considering the merit of swaps between selected and unselected objects. So, assume now that o_i is a selected object, $o_i \in L$, and o_h is an unselected object, $o_h \in R = S - L$. We need to determine the cost $C(o_i, o_h)$ of swapping o_i and o_h. Let o_j be an arbitrary unselected object. The contribution c_{ihj} of o_j to the cost of the swap between o_i and o_h is defined as follows:

(a) If $d(o_i, o_j)$ and $d(o_h, o_j)$ are greater than $d(o, o_j)$ for any $o \in L - \{o_i\}$, then $c_{ihj} = 0$.

(b) If $d(o_i, o_j) = d(L, o_j)$, then two cases must be considered depending on the distance $e(o_j)$ from e_j to the second-closest object of S.
(i) If $d(o_h, o_j) < e(o_j)$, then $c_{ihj} = d(o_h, o_j) - d(S, o_j)$.
(ii) If $d(o_h, o_j) \geqslant e(o_j)$, then $c_{ihj} = e(o_j) - d(S, o_j)$.
In either of these two subcases, we have

$$c_{ihj} = \min\{d(o_h, o_j), e_j\} - d(o_i, o_j).$$

(c) If $d(o_i, o_j) > d(L, o_j)$ (that is, o_j is more distant from o_i than from at least one other selected object) and $d(o_h, o_j) < d(L, o_j)$ (which means that o_j is closer to o_h than to any selected object), then $c_{ihj} = d(o_h, o_j) - d(S, o_j)$.

The cost of the swap is $C(o_i, o_h) = \sum_{o_j \in R} c_{ihj}$. The pair that minimizes $C(o_i, o_j)$ is selected. If $C(o_i, o_j) < 0$, then the swap is carried out. All potential swaps are considered.
The algorithm halts when no useful swap exists; that is, no swap with negative cost can be found.

PAM is more robust than Forgy's variant of k-clustering because it minimizes the sum of the dissimilarities instead of the sum of the squared errors.

The pseudocode of the algorithm is given in Algorithm 16.4.1.

Algorithm 16.4.1: The PAM algorithms

Data: a set of objects S, where $|S| = n$, a dissimilarity $n \times n$ matrix D, and a prescribed number of clusters k
Result: a k-clustering of S

```
construct the set L of k medoids;
repeat
    compute the costs C(o_i, o_h) for o_i ∈ L and o_h ∈ R;
    select the pair (o_i, o_h) that corresponds to the minimum m = C(o_i, o_h);
until (m > 0);
```

Note that inside the loop **repeat** $\cdots$ **until** there are $l(n-l)$ pairs of objects to be examined, and for each pair we need to involve $n-l$ non-selected objects. Thus, one execution of the loop requires $O(l(n-l)^2)$, and the total execution may require up to $O\left(\sum_{l=1}^{n-l} l(n-l)^2\right)$, which is $O(n^4)$. Thus, the usefulness of PAM is limited to rather small data set (no more than a few hundred objects).

16.5 The Laplacian Spectrum of a Graph

Spectral clustering is a relatively new clustering technique that applies linear algebra technique to matrices associated to a similarity graph of a set of objects and produces clustering that are often more adequate to clusterings produced via the methods previously discussed. Laplacian matrices of graphs play a central role in spectral clusterings and are the focus of the current section.

Definition 16.6 *Let* $\mathcal{G} = (V, E, w)$ *be a weighted graph. The* Laplacian matrix of $\mathcal{G}$ *is the symmetric matrix* $L_{\mathcal{G}} = D_{\mathcal{G}} - A_{\mathcal{G}}$.

The spectrum of the Laplacian matrix is referred to as the Laplacian spectrum *of the weighted graph.*

Note that the off-diagonal elements of $L_{\mathcal{G}}$ are non-positive numbers. Also, $L_{\mathcal{G}}\mathbf{1}_m = \mathbf{1}'_m A = \mathbf{0}_m$.

The notion of Laplacian can be applied to common, unweighted graphs which can be regarded as weighted graphs such that the weight of every edge is 1.

Example 16.7 Let $\mathcal{G} = (\{v_1, v_2\}, \{(v_1, v_2)\}, w)$ be a two-vertex weighted graph, where $w(v_1, v_2) = a$. The degree matrix $D_{\mathcal{G}}$ and the adjacency matrix are

$$D_{\mathcal{G}} = \begin{pmatrix} a & 0 \\ 0 & a \end{pmatrix}, \text{ and } A_{\mathcal{G}} = \begin{pmatrix} 0 & a \\ a & 0 \end{pmatrix}.$$

Thus, we have

$$L_{\mathcal{G}} = a \cdot \begin{pmatrix} 1 & -1 \\ -1 & 1 \end{pmatrix}.$$

Lemma 16.8 *Let* $\mathcal{G} = (V, E, w)$ *be a weighted graph, where* $V = \{v_1, \ldots, v_m\}$ *and let* $A_{\mathcal{G}} = (w_{ij}) \in \mathbb{R}^{m \times m}$ *be its adjacency matrix. For* $\mathbf{x} \in \mathbb{R}^m$ *we have* $\mathbf{x}' L_{\mathcal{G}} \mathbf{x} = \frac{1}{2} \sum_{i=1}^m \sum_{j=1}^m w_{ij}(x_i - x_j)^2$.

Proof We have

$$\begin{aligned} \mathbf{x}' L_{\mathcal{G}} \mathbf{x} &= \mathbf{x}'(D_{\mathcal{G}} - A_{\mathcal{G}})\mathbf{x} = \mathbf{x}' D_{\mathcal{G}} \mathbf{x} - \mathbf{x}' A_{\mathcal{G}} \mathbf{x} \\ &= \sum_{i=1}^m \mathsf{d}(v_i) x_i^2 - \sum_{i=1}^m \sum_{j=1}^m \omega_{ij} x_i x_j. \end{aligned}$$

Since

$$\sum_{i=1}^{m}\sum_{j=1}^{m} w_{ij}(x_i - x_j)^2 = \sum_{i=1}^{m} x_i^2 \sum_{j=1}^{m} w_{ij} - 2\sum_{i=1}^{m}\sum_{j=1}^{m} w_{ij}x_i x_j + \sum_{j=1}^{m} x_j^2 \sum_{i=1}^{m} w_{ij}$$
$$= 2\sum_{i=1}^{m} x_i^2 \mathsf{d}(v_i) - 2\sum_{i=1}^{m}\sum_{j=1}^{m} w_{ij}x_i x_j,$$

the desired equality is immediate.

The symmetry of $A_{\mathcal{G}}$ implies that the equality of Lemma 16.8 can be written as

$$\mathbf{x}' L_{\mathcal{G}} \mathbf{x} = \sum \{w_{ij}(x_i - x_j)^2 \mid 1 \leqslant i < j \leqslant m\}, \tag{16.4}$$

for every $\mathbf{x} \in \mathbb{R}^m$.

Theorem 16.9 *The Laplacian of a weighted graph $\mathcal{G} = (V, E, w)$ is a singular, symmetric and positive semi-definite matrix that has 0 as its smallest eigenvalue and $\mathbf{1}_m$ as a corresponding eigenvector, where $m = |V|$.*

Proof Let $L_{\mathcal{G}} = D_{\mathcal{G}} - A_{\mathcal{G}}$ be the Laplacian of $\mathcal{G}$. Since both $D_{\mathcal{G}}$ and $A_{\mathcal{G}}$ are symmetric matrices, so is $L_{\mathcal{G}}$.

The positive definiteness of $L_{\mathcal{G}}$ follows immediately from Lemma 16.8. Since the sum of elements of each row of $L_{\mathcal{G}}$ is 0 we have $L_{\mathcal{G}}\mathbf{1}_m = \mathbf{0}_m$, which shows that 0 is an eigenvalue of $L_{\mathcal{G}}$ and $\mathbf{1}_m$ is an eigenvector of this eigenvalue.

Theorem 16.9 implies that all eigenvalues of $L_{\mathcal{G}}$ are real and non-negative, and $L_{\mathcal{G}}$ has a full set of n real and orthogonal eigenvectors. Thus, 0 is the smallest eigenvalue of $L_{\mathcal{G}}$.

There exists an interesting connection between the Laplacian of a graph $\mathcal{G}$ and the incidence matrix of an oriented graph $\mathcal{G}^r$ obtained from $\mathcal{G}$ be applying an orientation to the edges of $\mathcal{G}$, as seen in Definition 10.36.

Theorem 16.10 *Let $\mathcal{G} = (V, E)$ be a graph and let $\mathcal{G}^r = (V, E^r)$ be the directed graph obtained by applying the orientation r. We have $L_{\mathcal{G}} = U_{\mathcal{G}^r} U'_{\mathcal{G}^r}$, where $U_{\mathcal{G}^r}$ is the incidence matrix of $\mathcal{G}^r$.*

Proof Let $V = \{v_1, \ldots, v_m\}$, $E = \{e_1, \ldots, e_n\}$, and let $r : V \times V \longrightarrow \{-1, 0, 1\}$ be an orientation of $\mathcal{G}$. Since

$$(U_{\mathcal{G}^r} U'_{\mathcal{G}^r})_{ij} = \sum_{p=1}^{m} u_{ip} u_{jp},$$

several cases are possible. If $i = j$, each term of the sum $\sum_{p=1}^{n} u_{ip}^2$ that corresponds to an edge e_p equals 1, so the sum equals $\mathsf{d}(v_i)$.

If $i \neq j$ and there exists an edge e_p that joins v_i to v_j, then the only non-zero term of the sum $\sum_{p=1}^{n} u_{ip}u_{jp}$ equals -1. Otherwise, the value of the sum is 0. This justified the equality of the theorem.

Observe that Theorem 16.10 also implies that L_G is positive semi-definite because $\mathbf{x}'L_{\mathcal{G}}\mathbf{x} = \mathbf{x}'U_{\mathcal{G}^r}U'_{\mathcal{G}^r}\mathbf{x} = (U_{\mathcal{G}^r}\mathbf{x}')'U'_{\mathcal{G}^r}\mathbf{x} = \parallel U'_{\mathcal{G}^r}\mathbf{x} \parallel_2^2 \geqslant 0$.

16.5.1 Laplacian Spectra of Special Graphs

Example 16.11 The Laplacian of $\mathcal{K}_m$, the complete graph having m vertices is

$$L_{\mathcal{K}_m} = \begin{pmatrix} m-1 & -1 & \cdots & -1 \\ -1 & m-1 & \cdots & -1 \\ \vdots & \vdots & \cdots & \vdots \\ -1 & -1 & \cdots & m-1 \end{pmatrix}.$$

Note that $L_{\mathcal{K}_m} = -P(1-m)$, where $P(a)$ is the matrix defined in Example 7.8. Its characteristic equation is $\det(\lambda I_m - L_{\mathcal{G}}) = 0$, or $\det(\lambda I_m + P(1-m)) = 0$. Thus, the eigenvalues of $L_{\mathcal{G}}$ are the opposites of the eigenvalues of $P(1-m)$. In other words, $L_{\mathcal{G}}$ has 0 as an simple eigenvalue and m with multiplicity $m-1$. So $\alpha(\mathcal{K}_m) = m$.

For $\mathcal{K}_m$ we have:

$$\mathbf{x}'L_{\mathcal{K}_m}\mathbf{x} = \sum\{(x_i - x_j)^2 \mid 1 \leqslant i < j \leqslant m\}. \tag{16.5}$$

When $\mathbf{x}'\mathbf{1}_m = 0$, we can write

$$\mathbf{x}'L_{\mathcal{K}_m}\mathbf{x} = \mathbf{x}'(mI_m - J_{m,m})\mathbf{x} = m\mathbf{x}'\mathbf{x} = m \parallel \mathbf{x} \parallel_2^2 \tag{16.6}$$

because $\mathbf{x}'J_{m,m} = \mathbf{x}'(\mathbf{1}_m \cdots \mathbf{1}_m) = \mathbf{0}_m$.

Example 16.12 The Laplacian of the star graph shown in Fig. 10.36 is

$$L_{\mathcal{G}} = \begin{pmatrix} q-1 & -1 & \cdots & -1 \\ -1 & 1 & \cdots & 0 \\ \vdots & \vdots & \cdots & \vdots \\ -1 & 0 & \cdots & 1 \end{pmatrix}.$$

Using the determinant $Q_q(q-1-\lambda, 1-\lambda) = (1-\lambda)^{q-2}\lambda(\lambda - q)$ computed in Example 5.145, the eigenvalues of $L_{\mathcal{G}}$ are 0, 1 and q, so $\alpha(\mathcal{G}) = 1$.

Fig. 16.5 Cycle graph $\mathcal{C}_n$

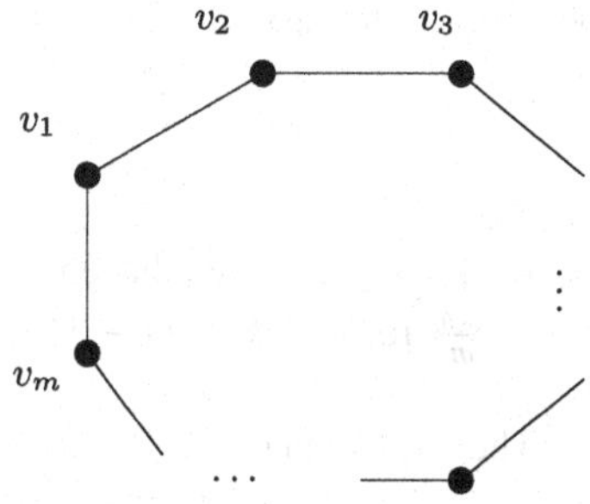

Theorem 16.13 *Let $\mathcal{G} = (V, E)$ be a k-regular graph with $|V| = m$. If the ordinary spectrum of $\mathcal{G}$ is $k = \lambda_1 \geqslant \lambda_2 \geqslant \cdots \geqslant \lambda_m$, then its Laplacian spectrum consists of the numbers $0 = k - \lambda_1 \leqslant k - \lambda_2 \leqslant \cdots \leqslant k - \lambda_m$.*

Proof Since $\mathcal{G}$ is a k-regular graph its Laplacian has the form $L_{\mathcal{G}} = kI_n - A_{\mathcal{G}}$. Therefore, $L_{\mathcal{G}}$ has the eigenvalues $0 = k - \lambda_1 \leqslant k - \lambda_2 \leqslant \cdots \leqslant k - \lambda_m$.

Example 16.14 The graph $\mathcal{C}_m=(\{v_1, \ldots, v_m\}, \{(v_i, v_{i+1}) | 1 \leqslant i \leqslant m\} \cup \{(v_m, v_1)\})$ is a simple cycle shown in Fig. 16.5. The adjacency matrix of $\mathcal{C}_m$ is $A_{\mathcal{C}_m} = A_{\mathcal{D}_m} + A'_{\mathcal{D}_m}$, where $\mathcal{D}_m$ is the directed simple cycle examined in Example 10.156. Recall that $A_{\mathcal{D}_m}$ and $A'_{\mathcal{D}_m}$ have identical spectra.

Suppose that

$$\mathbf{w}_k = \begin{pmatrix} 1 \\ z_k \\ \vdots \\ z_k^{m-1} \end{pmatrix}$$

is an eigenvector that corresponds to the eigenvalue z_k of $\mathcal{D}_m$. We have

$$A'_{\mathcal{D}_m} \begin{pmatrix} 1 \\ z_k \\ \vdots \\ z_k^{m-1} \end{pmatrix} = \begin{pmatrix} z_k^{m-1} \\ 1 \\ \vdots \\ z_k^{m-2} \end{pmatrix} = \frac{1}{z_k} \begin{pmatrix} 1 \\ z_k \\ \vdots \\ z_k^{m-1} \end{pmatrix} = \frac{1}{z_k}\mathbf{w}_k.$$

This shows that the matrices $A_{\mathcal{D}_m}$ and $A'_{\mathcal{D}_m}$ have both identical spectra and an identical set of eigenvectors. However, while $\mathbf{w}_k$ corresponds to the eigenvalue z_k for $A_{\mathcal{D}_m}$, the same $\mathbf{w}_k$ corresponds to the eigenvalue $\frac{1}{z_k}$ for $A'_{\mathcal{D}_m}$. Thus, we have

$$(A_{\mathcal{D}_m} + A'_{\mathcal{D}_m})\mathbf{w}_k = (z_k + z_k^{-1})\mathbf{w}_k,$$

Fig. 16.6 Path graph $\mathcal{P}_m$

which proves that the adjacency matrix of $\mathcal{C}_m$ has the real eigenvalues $z_k + z_k^{-1} = 2\cos\frac{2k\pi}{m}$ for $0 \leqslant k \leqslant m-1$, and the same eigenvectors as the matrices $A_{\mathcal{D}_m}$ and $A'_{\mathcal{D}_m}$.

The Laplacian of $\mathcal{C}_m$ is

$$L_{\mathcal{C}_m} = \begin{pmatrix} 2 & -1 & 0 & 0 & \cdots & -1 \\ -1 & 2 & -1 & 0 & \cdots & 0 \\ 0 & -1 & 2 & 1 & \cdots & 0 \\ \vdots & \vdots & \vdots & \cdots & \vdots & 0 \\ -1 & 0 & 0 & \cdots & -1 & 2 \end{pmatrix}.$$

By Theorem 16.13, $\mathcal{C}$ has the Laplacian spectrum

$$\left\{2 - 2\cos\frac{2k\pi}{m} \,\middle|\, 0 \leqslant k \leqslant m-1\right\}$$

because $\mathcal{C}_m$ is a 2-regular graph.

Example 16.15 Let $\mathcal{P}_m = (\{v_1, \ldots, v_m\}, \{(v_i, v_{i+1}) | 1 \leqslant i \leqslant m-1\}$ be the path graph shown in Fig. 16.6.

The Laplacian matrix of $\mathcal{P}_m$ is the tridiagonal matrix

$$L_{\mathcal{P}_m} = \begin{pmatrix} 1 & -1 & 0 & 0 & \cdots & 0 \\ -1 & 2 & -1 & 0 & \cdots & 0 \\ 0 & -1 & 2 & -1 & \cdots & 0 \\ 0 & 0 & -1 & 2 & \cdots & 0 \\ \vdots & \vdots & \vdots & \vdots & \cdots & \vdots \\ 0 & 0 & 0 & 0 & -1 & 1 \end{pmatrix}$$

To determine the ordinary spectrum of $\mathcal{P}_m$ consider the simple undirected cycle $\mathcal{C}_{2m+2}$. The ordinary spectrum of this graph consists of eigenvalues of the form $z_k + z_k^{-1} = 2\cos\frac{2k\pi}{2m+2} = 2\cos\frac{k\pi}{m+1}$, where $z_k^{2m+2} = 1$. These eigenvalues have the geometric multiplicity 2 and both vectors

$$\mathbf{w}_k = \begin{pmatrix} 1 \\ z_k \\ \vdots \\ z_k^{2m+1} \end{pmatrix} \text{ and } \mathbf{u}_k = \begin{pmatrix} 1 \\ z_k^{-1} \\ \vdots \\ z_k^{-(2m+1)} \end{pmatrix}$$

are eigenvectors that correspond to $2\cos\frac{k\pi}{m+2}$, where $0 \leqslant k \leqslant 2m+1$. Therefore, the vector

$$\mathbf{t}_k = \mathbf{w}_k - \mathbf{u}_k = \begin{pmatrix} 0 \\ z_k - z_k^{-1} \\ \vdots \\ z_k^m - z_k^{-m} \\ 0 \\ z_k^{m+2} - z_k^{-m-2} \\ \vdots \\ z_k^{2m+1} - z_k^{-2m-1} \end{pmatrix} = \begin{pmatrix} 0 \\ z_k - z_k^{-1} \\ \vdots \\ z_k^m - z_k^{-m} \\ 0 \\ -(z_k^m - z_k^{-m}) \\ \vdots \\ (-z_k - z_k^{-1}) \end{pmatrix}$$

is also an eigenvector for $2\cos\frac{k\pi}{m+1}$. We used here the fact that $z_k^{2m+2} = 1$ implies $z_k^{m+1} = z_k^{-m-1}$. The incidence matrix $A_{\mathcal{C}_{2m+2}}$ can be written as

$$A_{\mathcal{C}_{2m+2}} = \begin{pmatrix} A_{\mathcal{P}_{m+1}} & E_{m+1,1} \\ E_{m+2,1} & A_{\mathcal{P}_{m+1}} \end{pmatrix},$$

so the equality $A_{\mathcal{C}_{2m+2}}\mathbf{t}_k = 2\cos\frac{k\pi}{m+1}\mathbf{t}_k$ implies

$$A_{\mathcal{P}_{m+1}} \begin{pmatrix} 0 \\ z_k - z_k^{-1} \\ \vdots \\ z_k^m - z_k^{-m} \end{pmatrix} = 2\cos\frac{k\pi}{m+1}\mathbf{t}_k \begin{pmatrix} 0 \\ z_k - z_k^{-1} \\ \vdots \\ z_k^m - z_k^{-m} \end{pmatrix}.$$

This, in turn, yields

$$A_{\mathcal{P}_m} \begin{pmatrix} z_k - z_k^{-1} \\ \vdots \\ z_k^m - z_k^{-m} \end{pmatrix} = 2\cos\frac{k\pi}{m+1}\mathbf{t}_k \begin{pmatrix} z_k - z_k^{-1} \\ \vdots \\ z_k^m - z_k^{-m} \end{pmatrix}.$$

which shows that $2\cos\frac{k\pi}{m+1}\mathbf{t}_k$ is an ordinary eigenvalue for the path graph $\mathcal{P}_m$. A corresponding eigenvector is

$$\begin{pmatrix} z_k - z_k^{-1} \\ \vdots \\ z_k^m - z_k^{-m} \end{pmatrix}.$$

Next we compute the Laplacian spectrum of $\mathcal{P}_m$. Let z be a complex number such that $z^{2m} = 1$ and let

$$\mathbf{u} = \begin{pmatrix} 1 + z^{2m-1} \\ z + z^{2m-2} \vdots \\ z^j + z^{2m-1-j} \\ \vdots \\ z^{m-2} + z^{m+1} \\ z^{m-1} + z^m \end{pmatrix}.$$

The equalities

$$1 + z^{2m-1} - \left(z + z^{2m-2}\right) = \left(2 - z - z^{-1}\right)\left(1 + z^{2m-1}\right),$$

$$-\left(z^{j-1} + z^{2m-j}\right) + 2\left(z^j + z^{2m-1-j}\right) - \left(z^{j+1} + z^{2m-2-j}\right)$$
$$= \left(2 - z - z^{-1}\right)\left(z^j + z^{2m-1-j}\right),$$

and

$$-\left(z^{m-2} + z^{m+1}\right) + z^{m-1} + z^m = \left(2 - z - z^{-1}\right)\left(z^{m-1} + z^m\right)$$

can be directly verified and they show that $\mathbf{u}$ is an eigenvector of $L_{\mathcal{P}_m}$ that corresponds to the eigenvalue $2 - z - z^{-1}$. If $z_k = \cos\frac{2k\pi}{2m} + i\sin\frac{2k\pi}{2m}$, then the Laplacian spectrum of $\mathcal{P}_m$ consists of numbers of the form

$$2 - 2\cos\frac{2k\pi}{2m} = 4\sin^2\frac{k\pi}{2m},$$

where $0 \leqslant k \leqslant m - 1$.

16.5.2 Graph Connectivity

Theorem 16.16 *Let $\mathcal{G} = (V, E)$ be a graph, where $|V| = m$. The number of connected components of $\mathcal{G}$ equals* $\mathsf{algm}(L_{\mathcal{G}}, 0)$ *and the the characteristic vector of each connected component is an eigenvector of A that corresponds to the eigenvalue* 0.

Proof Let k the number of connected components of $\mathcal{G}$.

When $k = 1$ the graph is connected and this is the case that we examine initially. If $\mathbf{x}$ is an eigenvector that corresponds to the eigenvalue 0 we have $\mathbf{x}'L_{\mathcal{G}}\mathbf{x} = 0$, so by Lemma 16.8, we have $\sum_{i=1}^{m}\sum_{j=1}^{m} a_{ij}(x_i - x_j)^2 = 0$. Thus, $(x_i - x_j)^2 = 0$ for all i, j, $1 \leqslant i, j \leqslant m$ such that $a_{ij} = 1$. This means that the existence of an edge (v_i, v_j) in $\mathcal{G}$ implies $x_i = x_j$. Consequently, the values of the components of $\mathbf{x}$ must be the same and the invariant subspace $S_{L_{\mathcal{G}},0}$ is generated by the vector $\mathbf{1}_m$.

Suppose now that we have k connected components. Without loss of generality we can assume that the vertices of the graph are numbered such that the numbers attributed to the vertices that belong to a connected component are consecutive. In this case, the Laplacian $L_{\mathcal{G}}$ has a block-diagonal form

$$L_{\mathcal{G}} = \begin{pmatrix} L_1 & O & \cdots & O \\ O & L_2 & \cdots & O \\ \vdots & \vdots & \cdots & \vdots \\ O & O & \cdots & L_k \end{pmatrix}.$$

By Corollary 7.32, 0 is an eigenvalue of $L_{\mathcal{G}}$ and of each of the matrices L_j, where $1 \leqslant j \leqslant k$. Furthermore, each L_i is the Laplacian of a connected component of $\mathcal{G}$, so it has 0 as an eigenvalue of multiplicity 1. Therefore, $L_{\mathcal{G}}$ has 0 as an eigenvalue of multiplicity k.

The characteristic vector $\mathbf{c}_p \in \mathbb{R}^m$ of a connected component C_p, where $1 \leqslant p \leqslant k$, is given by

$$(\mathbf{c}_p)_i = \begin{cases} 1 & \text{if } i \text{ is the number of a row that corresponds to } L_i \\ 0 & \text{otherwise.} \end{cases}$$

It is clear that each such vector is an eigenvector of $L_{\mathcal{G}}$.

Corollary 16.17 *If a graph $\mathcal{G} = (V, E)$ is connected then rank $(L_{\mathcal{G}}) = |V| - 1$.*

Proof By Theorem 16.16, if $\mathcal{G}$ is connected, 0 has algebraic multiplicity 1. Thus, by Theorem 7.26, $L_{\mathcal{G}}$ has rank $|V| - 1$.

Definition 16.18 *The* connectivity *of a graph $\mathcal{G}$ is the second smallest eigenvalue $\alpha(\mathcal{G})$ of its Laplacian $L_{\mathcal{G}}$.*

Recall that the smallest eigenvalue of $L_{\mathcal{G}}$ is 0. By Theorem 7.51 we have

$$\alpha(G) = \min_{\mathbf{x}} \left\{ \mathbf{x}^{\mathsf{H}} L_{\mathcal{G}} \mathbf{x} \mid \| \mathbf{x} \|_2 = 1 \text{ and } \mathbf{x}'\mathbf{1} = 0 \right\}. \tag{16.7}$$

Example 16.19 The incidence matrices of the similarity graphs $\mathcal{G}_{0.2}$ and $\mathcal{G}_{0.4}$ introduced in Example 10.10 are $A_{0.2}$:

$$\begin{pmatrix}
0&1&1&0&0&0&0&0&0&0&0&0&0&0\\
1&0&0&1&0&0&0&0&0&0&0&0&0&0\\
1&0&0&1&1&0&0&0&0&0&0&0&0&0\\
0&1&1&0&1&0&1&0&0&0&0&0&0&0\\
0&0&1&1&0&0&1&0&0&0&0&0&0&0\\
0&0&0&0&0&0&0&1&1&1&0&0&0&0\\
0&0&0&1&1&0&0&0&0&0&1&1&0&0\\
0&0&0&0&0&1&0&0&1&0&0&0&0&1\\
0&0&0&0&0&1&0&1&0&1&0&0&0&1\\
0&0&0&0&0&1&0&0&1&0&0&0&0&1\\
0&0&0&0&0&0&1&0&0&0&0&1&1&0\\
0&0&0&0&0&0&1&0&0&0&1&0&1&0\\
0&0&0&0&0&0&0&0&0&0&1&1&0&0\\
0&0&0&0&0&0&0&1&1&1&0&0&0&0
\end{pmatrix}$$

and $A_{0.4}$, given by

$$\begin{pmatrix}
0&0&0&0&0&0&0&0&0&0&0&0&0&0\\
0&0&0&0&0&0&0&0&0&0&0&0&0&0\\
0&0&0&0&1&0&0&0&0&0&0&0&0&0\\
0&0&0&0&1&0&0&0&0&0&0&0&0&0\\
0&0&1&1&0&0&1&0&0&0&0&0&0&0\\
0&0&0&0&0&0&0&0&1&0&0&0&0&0\\
0&0&0&0&1&0&0&0&0&0&1&0&0&0\\
0&0&0&0&0&0&0&0&1&0&0&0&0&0\\
0&0&0&0&0&1&0&1&0&1&0&0&0&1\\
0&0&0&0&0&0&0&0&1&0&0&0&0&0\\
0&0&0&0&0&0&1&0&0&0&0&1&1&0\\
0&0&0&0&0&0&0&0&0&0&1&0&0&0\\
0&0&0&0&0&0&0&0&0&0&1&0&0&0\\
0&0&0&0&0&0&0&0&1&0&0&0&0&0
\end{pmatrix}$$

The spectrum of the Laplacian of the first matrix is 0 (with $\mathsf{algm}(0, L_{0.2}) = 2$), 0.41, 1.56, 2.39, 3.00, 3 (with $\mathsf{algm}(3, L_{0.2}) = 3$), 4, 4.10, 4.77, 5 (with $\mathsf{algm}(5, L_{0.2}) = 2$) and 5.73. Thus, $\alpha(\mathcal{G}_{0.2}) = 0.41$.

For the second graph $\mathcal{G}_{0.4}$ the eigenvalues of the Laplacian are 0 (with $\mathsf{algm}(0, L_{0.2}) = 4$), 0.26, 1 (with $\mathsf{algm}(0, L_{0.2}) = 5$), 1.58, 3.73, 4.41, 5. Thus, the connectivity of $\mathcal{G}_{0.4}$ is 0.26, which is lower than the connectivity of the first graph.

As expected, the multiplicity of 0, the least eigenvalues of the Laplacians equals the number of connected components of these graphs.

Definition 16.20 *The* edge connectivity *of a graph $\mathcal{G} = (V, E)$ is the minimal number of edges whose removal would result is losing connectivity. This number is denoted by $\mathsf{e}(\mathcal{G})$. The least number of vertices whose removal (with the corresponding edges) would result in losing connectivity is the* vertex connectivity *and is denoted by $v(\mathcal{G})$.*

For a complete graph we have $v(\mathcal{K}_m) = m - 1$.

Theorem 16.21 *Let $\mathcal{G} = (V, E)$ be a graph. We have*

$$v(\mathcal{G}) \leqslant e(\mathcal{G}) \leqslant \min_{v \in V} \mathsf{d}(v).$$

Proof Suppose that $\mathcal{G}$ is a connected graph which is not complete.

Let v_0 be a vertex such that $\mathsf{d}(v_0) = \min_{v \in V} \mathsf{d}(v)$. By removing all edges incident with v_0 the graph becomes disconnected, so $e(\mathcal{G}) \leqslant \min_{v \in V} \mathsf{d}(v)$ because $e(\mathcal{G})$ was defined as the minimal number of edges whose removal renders the graph disconnected. By the same definition, there exists a set of edges E_0 with $|E_0| = e(\mathcal{G})$ whose removal partitions the graph into two sets of vertices S and T. Note that any edge of E_0 joins a vertex in S with a vertex in T, because adding back any edge in E_0 restores the connectivity of the graph.

The graph $\mathcal{G}$ can also be disconnected by removing vertices from S or T joined by edges from E_0 and the subgraphs generated by S and T may lose their connectedness before $e(\mathcal{G})$ vertices are removed. Thus, $v(\mathcal{G}) \leqslant e(\mathcal{G})$.

Theorem 16.21 allows us to obtain a counterpart of Menger's Theorem for Digraphs (Theorem 10.146).

Theorem 16.22 (Menger's Theorem for Graphs) *Let $\mathcal{G} = (V, E)$ be a graph and let x and y be two vertices. The maximum number of paths that join x to y whose sets* *of edges are pairwise disjoint equals the minimum number of edges whose removal eliminates all paths from x to y.*

Proof Let E_1 be a set of edges whose removal separates x from y, which means that the removal destroys all paths between x and y. Thus, the number of link-disjoint paths between x and y cannot exceed $|E_1|$. If we choose E_1 to contain the least number of edges the desired conclusion follows.

Theorem 16.23 *Let $\mathcal{G}_i = (V, E_i)$, $i = 1, 2$ be two graphs having the same set of vertices such that $E_1 \cap E_2 = \emptyset$. If $\mathcal{G}_1 \cup \mathcal{G}_2 = (V, E_1 \cup E_2)$, then $\alpha(\mathcal{G}_1 \cup \mathcal{G}_2) \geqslant \alpha(\mathcal{G}_1) + \alpha(\mathcal{G}_2)$.*

Proof It is easy to see that $L_{\mathcal{G}_1 \cup \mathcal{G}_2} = L_{\mathcal{G}_1} + L_{\mathcal{G}_2}$, so by Equality (16.7) we have

$$\begin{aligned}\alpha(\mathcal{G}_1 \cup \mathcal{G}_2) &= \min_{\mathbf{x}}\{\mathbf{x}' L_{\mathcal{G}_1}\mathbf{x} + \mathbf{x}' L_{\mathcal{G}_2}\mathbf{x} \mid \parallel \mathbf{x} \parallel_2 = 1 \text{ and } \mathbf{x}'\mathbf{1} = 0\} \\ &\geqslant \min_{\mathbf{x}}\{\mathbf{x}' L_{\mathcal{G}_1}\mathbf{x} \mid \parallel \mathbf{x} \parallel_2 = 1 \text{ and } \mathbf{x}'\mathbf{1} = 0\} \\ &\quad + \min_{\mathbf{x}}\{\mathbf{x}' L_{\mathcal{G}_2}\mathbf{x} \mid \parallel \mathbf{x} \parallel_2 = 1 \text{ and } \mathbf{x}'\mathbf{1} = 0\} \\ &= \alpha(\mathcal{G}_1) + \alpha(\mathcal{G}_2).\end{aligned}$$

If $\mathcal{G}_1 = (V, E_1)$ and $\mathcal{G}_2 = (V, E_2)$ are two graphs having the same set of vertices we write $\mathcal{G}_1 \subseteq \mathcal{G}_2$ if $E_1 \subseteq E_2$.

Corollary 16.24 *If $\mathcal{G}_1$ and $\mathcal{G}_2$ are two graphs on the same set of vertices, then $\mathcal{G}_1 \subseteq \mathcal{G}_2$ implies $\alpha(\mathcal{G}_1) \leqslant \alpha(\mathcal{G}_2)$.*

Proof This statement is an immediate consequence of Theorem 16.23.

Theorem 16.25 *Let $\mathcal{G}' = (V', E')$ be a graph obtained from the graph $\mathcal{G} = (V, E)$ by removing k vertices and all edges incident to these vertices. Then, $\alpha(\mathcal{G}') \geqslant \alpha(\mathcal{G}) - k$.*

Proof The proof is by induction on $k \geqslant 1$ and the base case $k = 1$ is the only non-trivial part.

For $k = 1$ suppose that $V = \{v_1, \ldots, v_m\}$ and $V' = V - \{v_m\}$. Define the graph $\mathcal{G}_1 = (V' \cup \{v_m\}, E' \cup \{(v_i, v_m) \mid 1 \leqslant i \leqslant m - 1\})$. Clearly $\mathcal{G} \subseteq \mathcal{G}_1$, so $\alpha(\mathcal{G}) \leqslant \alpha(\mathcal{G}_1)$. The Laplacian of $\mathcal{G}_1$ has the form

$$L_{\mathcal{G}_1} = \begin{pmatrix} L_{\mathcal{G}'} + I_{m-1} & -\mathbf{1}_{m-1} \\ -\mathbf{1}'_{m-1} & m-1 \end{pmatrix}.$$

If $\mathbf{t}$ is an eigenvector of $L_{\mathcal{G}'}$, then

$$\mathbf{z} = \begin{pmatrix} \mathbf{t} \\ 0 \end{pmatrix}$$

is an eigenvector of $L_{\mathcal{G}_1}$ and we have

$$L_{\mathcal{G}_1}\begin{pmatrix}\mathbf{t}\\0\end{pmatrix} = (\alpha(\mathcal{G}')+1)\begin{pmatrix}\mathbf{t}\\0\end{pmatrix},$$

which proves that $\alpha(\mathcal{G}')+1$ is a non-zero eigenvalue of $L_{\mathcal{G}_1}$. In other words, we have $\alpha(\mathcal{G}_1) \leqslant \alpha(\mathcal{G}')+1$, which implies $\alpha(\mathcal{G}') \geqslant \alpha(\mathcal{G}) - 1$.

The induction step is immediate.

Theorem 16.26 *Let $\mathcal{G}_1 = (V_1, E_1)$ and $\mathcal{G}_2 = (V_2, E_2)$ be two graphs, where $V_1 = \{v_1, \ldots, v_p\}$ and $V_2 = \{u_1, \ldots, u_q\}$. We have $L_{\mathcal{G}_1 \times \mathcal{G}_2} = L_{\mathcal{G}_1} \oplus L_{\mathcal{G}_2}$.*

Proof If the vertex (v_i, u_j) of $\mathcal{G}_1 \times \mathcal{G}_2$ occupies the ℓth place in the list of vertices we have

$$(D_{\mathcal{G}_1\times\mathcal{G}_2})_{ll} = \mathsf{d}_{\mathcal{G}_1}(v_i) + \mathsf{d}_{\mathcal{G}_2}(u_j),$$

where $i = \left\lceil \frac{\ell}{q} \right\rceil$ and $j = \ell - q\left(\left\lceil \frac{\ell}{q} \right\rceil - 1\right)$. This shows that $D_{\mathcal{G}_1\times\mathcal{G}_2} = D_{\mathcal{G}_1} \oplus D_{\mathcal{G}_2}$, as it can be verified easily from Definition 5.163. Thus,

$$\begin{aligned}
L_{\mathcal{G}_1\times\mathcal{G}_2} &= D_{\mathcal{G}_1\times\mathcal{G}_2} - A_{\mathcal{G}_1\times\mathcal{G}_2} = (D_{\mathcal{G}_1} \oplus D_{\mathcal{G}_2}) - (A_{\mathcal{G}_1} \oplus A_{\mathcal{G}_2})\\
&= (D_{\mathcal{G}_1} \otimes I_q + I_p D_{\mathcal{G}_2}) - (A_{\mathcal{G}_1} \otimes I_q + I_p A_{\mathcal{G}_2})\\
&= (D_{\mathcal{G}_1} - A_{\mathcal{G}_1}) \otimes I_q + I_p \otimes (D_{\mathcal{G}_2} - A_{\mathcal{G}_2}) = L_{\mathcal{G}_1} \oplus L_{\mathcal{G}_2}.
\end{aligned}$$

Theorem 16.27 *Let $\mathcal{G}_1 = (V_1, E_1)$ and $\mathcal{G}_2 = (V_2, E_2)$ be two graphs. We have $\alpha(\mathcal{G}_1 \times \mathcal{G}_2) = \min\{\alpha(\mathcal{G}_1), \alpha(\mathcal{G}_2)\}$.*

Proof By Supplement 68 of Chap. 7 the eigenvalues of the Laplacian $L_{\mathcal{G}_1\times\mathcal{G}_2}$ have the form $\lambda + \mu$, where λ is an eigenvalue of $L_{\mathcal{G}_1}$ and μ is an eigenvalue of $L_{\mathcal{G}_2}$. Therefore, the second smallest eigenvalue of $L_{\mathcal{G}_1\times\mathcal{G}_2}$ is either $\alpha(\mathcal{G}_1) + 0$ or $0 + \alpha(\mathcal{G}_2)$, which implies the desired statement.

Theorem 16.28 *Let $A \in \mathbb{R}^{m\times m}$ be a symmetric, positive semidefinite matrix such that $A\mathbf{1}_m = \mathbf{0}_m$ and $\mathsf{spec}(A) = \{\lambda_1, \ldots, \lambda_m\}$, where $0 = \lambda_1 \leqslant \lambda_2 \leqslant \cdots \leqslant \lambda_m$. The second smallest eigenvalue λ_2 satisfies the inequality*

$$\lambda_2 \leqslant \frac{m}{m-1}\min\{a_{ii} \mid 1 \leqslant i \leqslant m\}.$$

Proof Note that the smallest eigenvalue of A is 0 and $\mathbf{1}_m$ is an eigenvector that corresponds to this eigenvalue. Therefore, by Equality (16.7),

$$\lambda_2 = \min\{\mathbf{x}'A\mathbf{x} \mid \parallel \mathbf{x} \parallel_2 = 1, \mathbf{x}'\mathbf{1}_m = 0\}.$$

Let $B = A - \lambda_2(I_m - \frac{1}{m}J_{m,m})$. Note that if $\mathbf{y} \in \mathbb{R}^m$ we can write $\mathbf{y} = c_1\mathbf{1}_m + c_2\mathbf{x}$, where $\mathbf{x}'\mathbf{1} = 0$ and $\parallel \mathbf{x} \parallel_2 = 1$.

Since $B\mathbf{1}_m = 0$ it follows that $\mathbf{y}'B\mathbf{y} = c_2^2\mathbf{x}'B\mathbf{x} = c_2^2(\mathbf{x}'A\mathbf{x} - \lambda_2) \geqslant 0$. Thus, B is positive semidefinite and the least diagonal entry of B, $\min\{a_{ii} \mid 1 \leqslant i \leqslant m\} - \lambda_2\left(1 - \frac{1}{m}\right)$ is non-negative.

Corollary 16.29 *Let $\mathcal{G} = (V, E)$ be a graph, where $|V| = m$ Then,*

$$\alpha(\mathcal{G}) \leqslant \frac{m}{m-1} \min_{v \in V} \mathsf{d}(v) \leqslant 2 \frac{|E|}{m-1}.$$

Proof The first inequality follows from Theorem 16.28; the second is a consequence of the fact that $m \min_{v \in V} \leqslant \sum_{v \in V} \mathsf{d}(v) = 2\,|E|$.

Let $\mathcal{G} = (V, E)$ be a graph. Its *complement* is the graph $\overline{\mathcal{G}} = (V, \bar{E})$, where $\bar{E} = \{(x, y) \mid x \neq y \text{ and } (x, y) \notin E\}$.

For a graph $\mathcal{G} = (V, E)$ with $|V| = m$ define $\beta(\mathcal{G}) = m - \alpha(\overline{\mathcal{G}})$. The next theorem was obtained by Fiedler in [5].

Theorem 16.30 *For every graph $\mathcal{G} = (V, E)$, where $|V| = m$, the following statements hold:*

(i) $\beta(\mathcal{G}) = \max\{\boldsymbol{x}' L_{\mathcal{G}} \boldsymbol{x} \mid \| \boldsymbol{x} \|_2 = 1 \text{ and } \boldsymbol{x}'\boldsymbol{1}_m = 0\}$;
(ii) $\alpha(\mathcal{G}) \leqslant \beta(\mathcal{G})$ *and the equality holds if and only if $\mathcal{G}$ is a complete graph or a void graph;*
(iii) *if $\mathcal{G}_1, \ldots, \mathcal{G}_k$ are the connected components of $\mathcal{G}$, then* $\beta(\mathcal{G}) = \max\{\beta(\mathcal{G}_i) \mid 1 \leqslant i \leqslant k\}$;
(iv) *if $\mathcal{G}_1 = (V, E_1)$ and $\mathcal{G}_2 = (V, E_2)$ with $E_1 \subseteq E_2$, then $\beta(\mathcal{G}_1) \leqslant \beta(\mathcal{G}_2)$;*
(v) *if $\mathcal{G}_1$ and $\mathcal{G}_2$ have the same set of vertices, then $\beta(\mathcal{G}_1 \cup \mathcal{G}_2) \leqslant \beta(\mathcal{G}_1) + \beta(\mathcal{G}_2) - \alpha(\mathcal{G}_1 \cap \mathcal{G}_2)$;*
(vi) $\frac{m}{m-1} \max_{v \in V} \mathsf{d}(v) \leqslant \beta(\mathcal{G}) \leqslant 2 \max_{v \in V} \mathsf{d}(v)$.

Proof Note that $L_{\mathcal{G}} + L_{\overline{\mathcal{G}}} = mI_m - J_{m,m}$. Also, for $\mathbf{x} \in \mathbb{R}^m$ we have $\| \mathbf{x} \|_2 = \mathbf{x}'\mathbf{x} = 1$ and $\mathbf{x}'\mathbf{1}_m = 0$; therefore, $\mathbf{x}'(mI_m - J_{m,m})\mathbf{x} = m$. This implies

$$\begin{aligned}\max\{\mathbf{x}' L_{\mathcal{G}} \mathbf{x} \mid \| \mathbf{x} \|_2 = 1, \mathbf{x}'\mathbf{1}_m\} &= m - \min\{\mathbf{x}' L_{\overline{\mathcal{G}}} \mathbf{x} \mid \| \mathbf{x} \|_2 = 1, \mathbf{x}'\mathbf{1}_m\} \\ &= n - \alpha(\overline{\mathcal{G}}) = \beta(\mathcal{G}),\end{aligned}$$

which proves Part (i) and implies $\alpha(\mathcal{G}) \leqslant \beta(\mathcal{G})$. If $\alpha(\mathcal{G}) = \beta(\mathcal{G})$, then $\mathbf{x}' L_{\mathcal{G}} \mathbf{x}$ is constant on the set $W = \{\mathbf{x} \mid \| \mathbf{x} \|_2 = 1, \mathbf{x}'\mathbf{1}_m\}$. Note that $\mathbf{x} = \sqrt{\frac{1}{m(m-1)}}(m\mathbf{e}_i - \mathbf{1}_m)$ belongs to W. Since $\mathbf{x}' L_{\mathcal{G}} \mathbf{x} = m^2 (L_{\mathcal{G}})_{ii}$ it follows that all diagonal elements of $L_{\mathcal{G}}$ are constant. Similarly, choosing $\mathbf{x} = \frac{1}{\sqrt{2}}(\mathbf{e}_i - \mathbf{e}_j)$ (with $i \neq j$) it follows that all off-diagonal elements of $L_{\mathcal{G}}$ are 0 (so all are equal to 0 or -1). Thus, $\mathcal{G}$ is either void or a complete graph.

Part (iii) follows immediately from Part (i). Part (iv) follows from $\alpha(\mathcal{G}_1) \leqslant \alpha(\mathcal{G}_2)$.

To prove Part (v) note that

$$\begin{aligned}\beta(\mathcal{G}_1 \cup \mathcal{G}_2) &= m - \alpha\left(\overline{\mathcal{G}_1 \cup \mathcal{G}_2}\right) \\ &= m - \alpha\left(\overline{\mathcal{G}_1 \cup \mathcal{G}_2}\right) = m - \alpha(\overline{\mathcal{G}}_1 \cap \overline{\mathcal{G}}_2)\end{aligned}$$

$$
\begin{aligned}
&= m - \left(\alpha(\overline{\mathcal{G}}_1) - \alpha(\overline{\mathcal{G}}_2) + \alpha(\overline{\mathcal{G}}_1 \cup \overline{\mathcal{G}}_2)\right) \\
&= \beta(\mathcal{G}_1) + \beta(\mathcal{G}_2) - m + \alpha(\overline{\mathcal{G}}_1 \cup \overline{\mathcal{G}}_2) \\
&\leqslant \beta(\mathcal{G}_1) + \beta(\mathcal{G}_2) - \alpha(\mathcal{G}_1 \cap \mathcal{G}_2).
\end{aligned}
$$

To prove the left inequality of (vi), by Corollary 16.29 we have

$$
\alpha(\overline{\mathcal{G}}) \leqslant \frac{m}{m-1} \min_{v \in V} \mathsf{d}_{\overline{\mathcal{G}}}(v),
$$

which is equivalent to

$$
m - \beta(\mathcal{G}) \leqslant \frac{m}{m-1} \left(m - 1 - \max_{v \in V} \mathsf{d}_{\mathcal{G}}(v)\right).
$$

Finally, the right part of the inequality of (vi) is discussed in Supplement 26.

Corollary 16.31 *Let $\mathcal{G} = (V, E)$ be a graph with $|V| = m$. We have $\alpha(\mathcal{G}) \geqslant 2 \min_{v \in V} \mathsf{d}(v) - m + 2$.*

Proof By Part (vi) of Theorem 16.30 we have $\beta(\overline{\mathcal{G}}) \leqslant 2 \max_{v \in V} \mathsf{d}_{\overline{\mathcal{G}}}(v) = 2 \max_{v \in V}(m - 1 - \mathsf{d}_{\mathcal{G}}(v))$. Since $\alpha(\mathcal{G}) = m - \beta(\overline{\mathcal{G}})$ the inequality follows immediately.

Theorem 16.32 (Fiedler's Matrix Theorem) *Let $A \in \mathbb{R}^{m \times m}$ be a symmetric, non-negative and irreducible matrix with eigenvalues $\lambda_1 \geqslant \cdots \geqslant \lambda_m$. Let $\mathbf{v} \in \mathbb{R}^m$ be a vector such that for a fixed $s \in \mathbb{N}$, $s \geqslant 2$, such that $A\mathbf{v} \geqslant \lambda_s \mathbf{v}$. If*

$$
P = \{i \mid 1 \leqslant i \leqslant p \text{ and } v_i \geqslant 0\},
$$

then $P \neq \emptyset$ and the degree of reducibility of the submatrix $A\begin{bmatrix} P \\ P \end{bmatrix}$ does not exceed $s - 2$.

Proof Suppose that $P = \emptyset$, that is $v_i < 0$ for $1 \leqslant i \leqslant m$. In this case the vector $\mathbf{z} = -\mathbf{v}$ has positive components and $A\mathbf{z} \leqslant \lambda_s \mathbf{z}$. This implies that all off-diagonal elements of $\lambda_1 I_m - A$ are non-positive and all its eigenvalues are non-negative. Furthermore, $\lambda_1 I_m - A$ is irreducible because A is irreducible.

Note that the matrix A' is also symmetric, non-negative and irreducible. Therefore, by Perron-Frobenius Theorem, there exists a positive eigenvector that corresponds to λ_1 such that $A'\mathbf{y} = \lambda_1 \mathbf{y}$, or $\mathbf{y}'A = \lambda_1 \mathbf{y}'$. This implies $\mathbf{y}'A\mathbf{v} = \lambda_1 \mathbf{y}'\mathbf{v} > \lambda_s \mathbf{y}'\mathbf{v}$, since λ_1 is a simple value. Thus, $\mathbf{u}'A\mathbf{v} \leqslant \lambda_s \mathbf{u}'\mathbf{z}$, which is a contradiction. Consequently, we have $P \neq \emptyset$.

If $P = \{1, \ldots, n\}$, then $A\begin{bmatrix} P \\ P \end{bmatrix} = A$ and the proof is complete because A is irreducible and, therefore, its degree of reducibility is 0. Therefore, we can assume that $\emptyset \subset P \subset \{1, \ldots, n\}$.

We prove that if we assume that the degree of reducibility of A is at least $s - 1$, then we obtain a contradiction. With this assumption, without loss of generality, we can assume that $P = \{1, \ldots, p\}$, where $p < n$, and that the symmetric matrix A can be written as a block matrix:

$$A = \begin{pmatrix} A_{11} & O & \cdots & O & A_{1\,r+1} \\ O & A_{22} & \cdots & O & A_{2\,r+1} \\ \vdots & \vdots & \cdots & \vdots & \vdots \\ A'_{1\,r+1} & A'_{2\,r+1} & \cdots & A'_{r\,r+1} & A_{r+1\,r+1} \end{pmatrix},$$

where $r \geqslant s$, the matrices $A_{jj} \in \mathbb{R}^{m_j \times m_j}$ are irreducible for $1 \leqslant j \leqslant r$ and $\sum_{j=1}^{r} m_j = p$.

The vector $\mathbf{v}$ can be partitioned in blocks as

$$\mathbf{v} = \begin{pmatrix} \mathbf{v}^{(1)} \\ \vdots \\ \mathbf{v}^{(r)} \\ \mathbf{v}^{(r+1)} \end{pmatrix}.$$

We have $\mathbf{v}^{(j)} \geqslant \mathbf{0}_{m_j}$ for $1 \leqslant j \leqslant r$ and $\mathbf{v}^{(r+1)} < \mathbf{0}_q$, where $q = n - \sum_{j=1}^{r} m_j$. By hypothesis, we have

$$(A_{jj} - \lambda_s I_{m_j})\mathbf{v}^{(j)} > -A_{j\,r+1}\mathbf{v}^{(r+1)} \tag{16.8}$$

for $1 \leqslant j \leqslant r$.

Since the matrix

$$B = \begin{pmatrix} \lambda_s I_{m_1} - A_{11} & O & \cdots & O \\ O & \lambda_s I_{m_2} - A_{22} & \cdots & O \\ \vdots & \vdots & \cdots & \vdots \\ O & O & \cdots & \lambda_s I_{m_r} - A_{rr} \end{pmatrix}$$

is a principal submatrix of $\lambda I - A$, it follows that it has at most $s - 1$ negative eigenvalues. Thus, at least one of the matrices $\lambda_s I_{m_j} - A_{jj}$ has only non-negative eigenvalues values. We can assume that is is the case for $\lambda_s I_{m_1} - A_{11}$. Therefore, $\lambda_s I_{m_1} - A_{11} \in \mathbf{K}_0$ and is irreducible.

Equality (16.8) implies

$$(\lambda_s I_{m_1} - A_{11})\mathbf{v}^{(1)} \leqslant A_{1\,r+1}\mathbf{v}^{(r+1)} \leqslant 0. \tag{16.9}$$

Suppose that $\lambda_s I_{m_1} - A_{11}$ is non-singular. Then, $\lambda_s I_{m_1} - A_{11} \in \mathbf{K}$ and we have $(\lambda_s I_{m_1} - A_{11})^{-1} > O$ by Part (xi) of Theorem 10.125. Equality (16.9) implies

$$\mathbf{v}^{(1)} \leqslant (\lambda_s I_{m_1} - A_{11})^{-1} A_{1\,r+1} \mathbf{v}^{(r+1)} \leqslant \mathbf{0}_{m_1}.$$

Thus, $\mathbf{v}^{(1)} = \mathbf{0}_{m_1}$ and $A_{1\,r+1}\mathbf{v}^{(r+1)} = \mathbf{0}$ which implies $A_{1\,r+1} = O$. This contradicts the irreducibility of A.

Suppose now that $\lambda_s I_{m_1} - A_{11}$ is singular, which means that $\mathbf{u}'(\lambda_s I_{m_1} - A_{11}) = \mathbf{0}'$. Therefore, $\mathbf{u}'(\lambda_s I_{m_1} - A_{11})\mathbf{v}^{(1)} = 0$. Since $(\lambda_s I_{m_1} - A_{11})\mathbf{v}^{(1)} \leqslant \mathbf{0}$, it follows that $(\lambda_s I_{m_1} - A_{11})\mathbf{v}^{(1)} = \mathbf{0}$ by Equality (16.9). Thus, $A_{1,r+1}\mathbf{v}^{(r+1)} = \mathbf{0}$, so $A_{1,r+1} = O$, which is again a contradiction. This concludes the argument.

Corollary 16.33 *Let $A \in \mathbb{R}^{m\times m}$ be a symmetric, non-negative, and irreducible matrix having the eigenvalues $\lambda_1 \geqslant \lambda_2 \geqslant \cdots \geqslant \lambda_m$. Let $s \in \mathbb{N}$ be a natural number such that $s \geqslant 2$ and let v an eigenvector corresponding to λ_s. Then, the set $P = \{i \in \mathbb{N} \mid v_i \geqslant 0\}$ is non-void and the degree of reducibility of the submatrix $A\begin{bmatrix} P \\ P \end{bmatrix}$ does not exceed $s - 2$.*

Proof The corollary is immediate from Theorem 16.32.

Corollary 16.34 *Let $A \in \mathbb{R}^{m\times m}$ be a symmetric, non-negative, and irreducible matrix having the eigenvalues $\lambda_1 \geqslant \lambda_2 \geqslant \cdots \geqslant \lambda_m$.*

If $(\lambda_1, \boldsymbol{u})$ and $(\lambda_2, \boldsymbol{v})$ are eigenpairs of A with $\boldsymbol{u} > \mathbf{0}$, then, for any $\alpha \geqslant 0$, the submatrix $A\begin{bmatrix} P_\alpha \\ P_\alpha \end{bmatrix}$ is irreducible, where $P_\alpha = \{i \in \mathbb{N} \mid v_i + \alpha u_i \geqslant 0\}$.

Proof Note that

$$A(\mathbf{v} + \alpha\mathbf{u}) = \lambda_2\mathbf{v} + \alpha\lambda_1\mathbf{u} \geqslant \lambda_2(\mathbf{v} + \alpha\mathbf{u}),$$

which means that the vector $\mathbf{v} + \alpha\mathbf{u}$ satisfies the condition of Theorem 16.32 for $s = 2$. The statement of the corollary follows immediately from the theorem.

The notion of connectivity can be extended to weighted graphs.

Definition 16.35 *Let $(\mathcal{G}, w)$ be a weighted graph. Its* connectivity *$\alpha(\mathcal{G}, w)$ is the second smallest eigenvalue of its Laplacian $L_{\mathcal{G}}$.*

Theorem 16.36 *Let $(\mathcal{G}, w)$ be a connected weighted graph such that $|V| = m$ and $w(v_i, v_j) > 0$ for every $(v_i, v_j) \in E$.*

The algebraic connectivity $\alpha(\mathcal{G}, w)$ is positive and is equal to the minimum of the function $\phi : \mathbb{R}^m \longrightarrow \mathbb{R}$ defined by

$$\phi(\boldsymbol{x}) = m\frac{\sum\{w_{ij}(x_i - x_j)^2 \mid (v_i, v_j) \in E\}}{\sum\{(x_i - x_j)^2 \mid i < j\}},$$

over all vectors $\boldsymbol{x} \in \mathbb{R}^m$ having distinct components. The corresponding eigenvectors to $\alpha(\mathcal{G}, w)$ are those vectors $\mathbf{y}$ having distinct components for which the minimum of ϕ is attained and for which $\sum_{i=1}^m y_i = 0$.

Proof Since $L_{\mathcal{G}}\mathbf{1} = 0$ and $L_{\mathcal{G}}$ is positive semidefinite, the smallest eigenvalue of $L_{\mathcal{G}}$ is 0. Since $\mathcal{G}$ is connected it follows that $\mathbf{1}$ is the only linearly independent solution of $(L_{\mathcal{G}}\mathbf{x}, \mathbf{x}) = 0$, so 0 is a simple eigenvalue. All other eigenvalues are positive, and all eigenvectors that correspond to these values are orthogonal to $\mathbf{1}$. Thus, by Theorem 7.50, the second smallest eigenvalue $\alpha(\mathcal{G}, w)$ is given by

$$\alpha(\mathcal{G}, w) = \min \left\{ \frac{\sum\{w_{ij}(x_i - x_j)^2 \mid (v_i, v_j) \in E\}}{\sum_{i=1}^{m} x_i^2} \middle| \mathbf{x} \neq \mathbf{0}, \mathbf{x}'\mathbf{1} = 0 \right\}$$

and the minimum is attained for any eigenvector corresponding to $\alpha(\mathcal{G}, w)$.

By the elementary identity

$$m \sum_{i=1}^{m} x_i^2 - \left(\sum_{i=1}^{m} x_i \right)^2 = \sum_{i<j} (x_i - x_j)^2,$$

taking into account that $\mathbf{x}'\mathbf{1} = \sum_{i=1}^{m} x_i = 0$, we have

$$m \sum_{i=1}^{m} x_i^2 = \sum_{i<j} (x_i - x_j)^2,$$

which yields the desired equality. Observe that the value of $\phi(\mathbf{x})$ is invariant with respect to adding a multiple of $\mathbf{1}$ to $\mathbf{x}$.

Corollary 16.37 *Let $(\mathcal{G}, w)$ be a connected weighted graph such that $w(v_i, v_j) > 0$ for every $(v_i, v_j) \in E$. If $|V| = m$ we have the inequality*

$$m \sum \{w_{ij}(x_i - x_j)^2 \mid (v_i, v_j) \in E\} \geqslant \alpha(\mathcal{G}) \sum \{(x_i - x_j)^2 \mid i < j\}$$

for every $\boldsymbol{x} \in \mathbb{R}^n$.

Proof The corollary is a direct consequence of Theorem 16.36.

Definition 16.38 *Let $(\mathcal{G}, w)$ be a weighted graph. A* Fiedler vector *of the weighted graph is an eigenvector that corresponds to the second smallest eigenvalue $\alpha(\mathcal{G}, w)$.*

A Fiedler vector is distinct from $\mathbf{0}$ and is determined up to a non-zero factor. If $\mathbf{y}$ is a Fiedler vector of the weighted graph $((\{v_1, \ldots, v_m\}, E), w)$, its component y_i corresponds to the vertex v_i is the $\mathbf{y}$-*valuation of the vertex* v_i.

Theorem 16.37 (Fiedler's Graph Theorem) *Let $\mathcal{G} = (V, E)$ be a connected graph such that $|V| = m$ and let $\mathbf{y}$ be a Fiedler vector of this graph. The subgraphs determined by the sets $V_r = \{v_i \in V \mid y_i + r \geqslant 0\}$ for $r \geqslant 0$ and $V'_r = \{v_i \in V \mid y_i + r \leqslant 0\}$ for $r \leqslant 0$ are connected.*

Proof The matrix $B = -L_{\mathcal{G}}$ has non-negative off-diagonal entries. Therefore, for sufficently large r, the matrix $B + rI_m$ is non-negative.

Note that $\mathbf{v}$ is an eigenvector of $L_{\mathcal{G}}$ if and only if $L_{\mathcal{G}}\mathbf{v} = \lambda\mathbf{v}$, which is equivalent to $(rI_m - L_{\mathcal{G}})\mathbf{v} = (r - \lambda)\mathbf{v}$. Thus, the matrices $rI_m + B$ and $L_{\mathcal{G}}$ have the same eigenvectors. Moreover, the second smallest eigenvalue of $L_{\mathcal{G}}$ corresponds to the second largest eigenvalue of the matrix $B + rI_m + B$, so a Fiedler vector of $\mathcal{G}$ is also an eigevector of $rI_m + B$ that corresponds to the second largest eigenvalue of this matrix.

By Corollary 16.33, $\mathbf{y} + r\mathbf{1}_m$ has the property that the submatrix of $rI_m + B$ with indices in V_r has reducibilty degree 0, that is, it is irreducible and, therefore, the subgraph determined by V_r is connected. The argument for V_r' is similar.

16.6 Spectral Clustering Algorithms

Spectral clustering algorithms compute clusterings starting from local information represented by a similarity matrix of the object space and using global information in the form of eigenvectors of this matrix [6].

If similarities between objects that belong to distinct clusters is 0 (which is usually not the case), the clustering problem is reduced to the determination of the connected components. In this ideal case, clusterings can be found using Theorem 16.16 and identify clusters as subsets of V whose indicator vectors are eigenvectors that span the eigenspace of 0.

If this is not the case, we could use k eigenvectors that correspond to the k smallest eigenvalues, represent the objects to be clustered as k-tuples of components of these vectors located in the same position and apply a clustering algorithm to these representatives.

Next, we introduce two variants of the Laplacian matrix of a graph.

Definition 16.40 *The* symmetric Laplacian *or the* normalized Laplacian *of a graph $\mathcal{G} = (V, E, w)$ is the matrix $L_{\mathcal{G},sym}$ given by*

$$L_{\mathcal{G},sym} = D_{\mathcal{G}}^{-\frac{1}{2}} L_{\mathcal{G}} D_{\mathcal{G}}^{-\frac{1}{2}} = I - D_{\mathcal{G}}^{-\frac{1}{2}} W_{\mathcal{G}} D_{\mathcal{G}}^{-\frac{1}{2}}.$$

The random walk Laplacian *of $\mathcal{G}$ is the matrix $L_{\mathcal{G},rw}$ defined as*

$$L_{\mathcal{G},rw} = D_{\mathcal{G}}^{-1} L_{\mathcal{G}} = I - D_{\mathcal{G}}^{-1} W_{\mathcal{G}}.$$

Theorem 16.41 *The pair $(\lambda, \mathbf{t})$ is an eigenpair of the symmetric Laplacian $L_{\mathcal{G},sym}$, if and only if $(\lambda, D^{-\frac{1}{2}}\mathbf{t})$ is an eigenpair of the random walk Laplacian $L_{\mathcal{G},rw}$.*

Proof Let $(\lambda, \mathbf{t})$ be an eigenpair of the symmetric Laplacian $L_{\mathcal{G},\mathrm{sym}}$. We have $L_{\mathcal{G},\mathrm{sym}}\mathbf{t} = \lambda\mathbf{t}$, or $D_{\mathcal{G}}^{-\frac{1}{2}} L_{\mathcal{G}} D_{\mathcal{G}}^{-\frac{1}{2}}\mathbf{t} = \lambda\mathbf{t}$, so $L_{\mathcal{G}} D_{\mathcal{G}}^{-\frac{1}{2}} = \lambda D_{\mathcal{G}}^{\frac{1}{2}}\mathbf{t}$. By multiplying this equality to the left by D^{-1} we obtain finally $D^{-1}L_{\mathcal{G},\mathrm{sym}}\left(D_{\mathcal{G}}^{\frac{1}{2}}\mathbf{t}\right) = \lambda\left(D_{\mathcal{G}}^{\frac{1}{2}}\mathbf{t}\right)$, or $L_{\mathcal{G},\mathrm{rw}}\left(D_{\mathcal{G}}^{\frac{1}{2}}\mathbf{t}\right) = \lambda\left(D_{\mathcal{G}}^{\frac{1}{2}}\mathbf{t}\right)$, which proves that $\left(\lambda, D_{\mathcal{G}}^{\frac{1}{2}}\mathbf{t}\right)$ is an eigenpair of $L_{\mathcal{G},\mathrm{rw}}$.

The reverse implication follows by observing that all implications mentioned in the previous argument hold in reverse.

An analogue of Lemma 16.8 can be formulated for the symmetric Laplacian:

Lemma 16.42 *Let $\mathcal{G} = (V, E, w)$ be a weighted graph, where $V = \{v_1, \ldots, v_m\}$. For $\mathbf{x} \in \mathbb{R}^m$ we have $\mathbf{x}'L_{\mathcal{G},sym}\mathbf{x} = \frac{1}{2}\sum_{i=1}^{m}\sum_{j=1}^{m} w_{ij}\left(\frac{x_i}{\sqrt{\mathsf{d}(v_i)}} - \frac{x_j}{\sqrt{\mathsf{d}(v_j)}}\right)^2$.*

Proof By the definition of the symmetric Laplacian we have $\mathbf{x}'L_{\mathcal{G},\mathrm{sym}}\mathbf{x} = \mathbf{x}'D_{\mathcal{G}}^{-\frac{1}{2}}L_{\mathcal{G}} D_{\mathcal{G}}^{-\frac{1}{2}}\mathbf{x}$. Note that

$$D_{\mathcal{G}}^{-\frac{1}{2}}\mathbf{x} = \mathsf{diag}\left(\frac{1}{\sqrt{\mathsf{d}(v_1)}}, \ldots, \frac{1}{\sqrt{\mathsf{d}(v_n)}}\right)\mathbf{x} = \begin{pmatrix} \frac{x_1}{\sqrt{\mathsf{d}(v_1)}} \\ \vdots \\ \frac{x_n}{\sqrt{\mathsf{d}(v_n)}} \end{pmatrix}.$$

An application of Lemma 16.8 yields the desired conclusion.

Theorem 16.43 *The symmetric Laplacian of a weighted graph $\mathcal{G} = (V, E, w)$ is a symmetric positive semi-definite matrix that has the eigenvalues $0 = \lambda_1 \leqslant \lambda_2 \leqslant \cdots \leqslant \lambda_n$. Furthermore, $(0, D^{\frac{1}{2}}\mathbf{1}_n)$ is an eigenpair of $L_{\mathcal{G},sym}$.*

Proof The symmetry of $L_{\mathcal{G},\mathrm{sym}}$ is immediate. Its positive semi-definiteness of follows from Lemma 16.42. Theorem 16.9 implies that all eigenvalues of $L_{\mathcal{G},\mathrm{sym}}$ are real and non-negative. Finally, we have

$$L_{\mathcal{G},\mathrm{sym}}D^{\frac{1}{2}}\mathbf{1}_n = D_{\mathcal{G}}^{-\frac{1}{2}}L_{\mathcal{G},\mathrm{sym}}D_{\mathcal{G}}^{-\frac{1}{2}}D^{\frac{1}{2}}\mathbf{1}_n = \mathbf{0}_n.$$

16.6.1 Spectral Clustering by Cut Ratio

Let $\mathcal{G} = (V, E, s)$ be a similarity graph and let $\kappa = \{C_1, \ldots, n\}$ be a clustering. For each cluster C_j define the vector $\mathbf{q}_j \in \mathbb{R}^n$ of the cluster C_j as

$$(\mathbf{q}_j)_i = \begin{cases} \frac{1}{\sqrt{|C_j|}} & \text{if } v_i \in C_j \\ 0 & \text{otherwise,} \end{cases}$$

where $1 \leqslant i \leqslant m$ and $1 \leqslant j \leqslant k$.

Let $Q = (\mathbf{q}_1 \ \cdots \ \mathbf{q}_k)$. The matrix Q has an orthonormal set of columns. Indeed, since the sets $C_1, \ldots, C_k$ are pairwise disjoint, if follows that the vectors $\mathbf{q}_1, \ldots, \mathbf{q}_k$ are pairwise orthogonal. Furthermore,

$$\| \mathbf{q}_j \|_2^2 = \sum_{i=1}^{m} (\mathbf{q}_j)_i^2 = |C_j| \cdot \frac{1}{|C_j|} = 1,$$

for $1 \leqslant j \leqslant k$, which allows us to conclude that the set of columns Q is orthonormal.

The notion of cut ratio of a partition was introduced in [7].

Definition 16.44 *Let $\kappa = \{C_1, \ldots, C_k\}$ be a partition of a set $V = \{v_1, \ldots, v_m\}$. The* cut ratio *of κ is the number $cutratio(\kappa)$ given by:*

$$cutratio(\kappa) = \sum_{j=1}^{k} \frac{cut(C_j, \bar{C}_j)}{|C_j|}.$$

By Lemma 16.8 we have

$$\begin{aligned}
\mathbf{q}_j' L_{\mathcal{G}} \mathbf{q}_j &= \frac{1}{2} \sum_{i=1}^{m} \sum_{\ell=1}^{m} s_{i\ell} (q_{ij} - q_{\ell j})^2 \\
&= \sum_{v_i \in C_j} \sum_{v_\ell \notin C_j} s_{ij} (q_{ij} - q_{\ell j})^2 + \sum_{v_i \notin C_j} \sum_{v_\ell \in C_j} s_{ij} (q_{ij} - q_{\ell j})^2 \\
&= 2 \sum_{v_i \in C_j} \sum_{v_\ell \notin C_j} s_{ij} (q_{ij} - q_{\ell j})^2.
\end{aligned}$$

Since in each of the terms of the sum we have $q_{ij} = \frac{1}{\sqrt{|C_j|}}$ and $q_{\ell j} = 0$, it follows that

$$s_{ij} (q_{ij} - q_{\ell j})^2 = \frac{s_{ij}}{|C_j|}.$$

Therefore, $\mathbf{q}_j' L_{\mathcal{G}} \mathbf{q}_j = 2 \sum_{v_i \in C_j} \sum_{v_\ell \notin C_j} \frac{s_{ij}}{|C_j|} = 2 \frac{cut(C_j, \bar{C}_j)}{|C_j|}$. Also, we have

$$Q' L_{\mathcal{G}} Q = (\mathbf{q}_1' \ \cdots \ \mathbf{q}_k') L_{\mathcal{G}} \begin{pmatrix} \mathbf{q}_1 \\ \vdots \\ \mathbf{q}_k \end{pmatrix},$$

which implies $\mathbf{q}_j' L_{\mathcal{G}} \mathbf{q}_j = (Q' L_{\mathcal{G}} Q)_{jj}$. Now we can write

$$\sum_{j=1}^{k} \mathbf{q}_j' L_{\mathcal{G}} \mathbf{q}_j = \sum_{j=1}^{k} (Q' L_{\mathcal{G}} Q)_{jj}$$

$$= trace(Q'L_{\mathcal{G}}Q) = 2\sum_{j=1}^{k} \frac{cut(C_j, \bar{C}_j)}{|C_j|} = 2cutratio(\kappa).$$

To minimize $cutratio(\kappa)$ is tantamount to seeking Q such that $trace(Q'L_{\mathcal{G}}Q)$ is minimized subjected to the constraint $Q'Q = I_k$. A practical solution this optimization problem is obtained by relaxation, namely by allowing Q to range over $\mathbb{R}^{m\times k}$. By Ky Fan's Theorem (Theorem 7.60), the minimum is obtained by choosing Q such that its columns consist of the eigenvectors $\mathbf{u}_1, \ldots, \mathbf{u}_k$ of $L_{\mathcal{G}}$ that correspond to the k smallest eigenvalues of the Laplacian $L_{\mathcal{G}}$.

If Q is redefined now as $Q = (\mathbf{u}_1 \ \cdots \ \mathbf{u}_k)$, then the m points to be clustered correspond now to the rows $\mathbf{y}_1, \ldots, \mathbf{y}_m$ of Q.

Example 16.45 Let $\mathcal{G}_{0.2} = (\{\}, E, s)$ be the similarity graph examined in Example 16.9. Recall that its Laplacian spectrum consists of 0 (with $\mathsf{algm}(0, L_{0.2}) = 2$), 0.41, 1.56, 2.39, 3.00, 3 (with $\mathsf{algm}(3, L_{0.2}) = 3$), 4, 4.10, 4.77, 5 (with $\mathsf{algm}(5, L_{0.2}) = 2$) and 5.73. The three eigenvectors that correspond to the smallest eigenvalues, 0 and 0.41 are:

0.33	0.41
0.33	0.37
0.33	0.28
0.33	0.18
0.33	0.13
0	0
0.33	0.12
0	0
0	0
0	0
0.33	0.38
0.33	0.38
0.33	0.48
0	0

Note that the 6th, 8th, 9th, 10th and 14th objects are mapped into the same point in $\mathbb{R}^2$, so they constitute a cluster; the remaining objects form another cluster, which is a reasonable good approximation of the groupings that occur in Fig. 10.6a.

16.6.2 Spectral Clustering by Normalized Cuts

Graph partitioning can be used for identifying groups of vertices such that the similarity between vertices that belong to different groups is low by using a divisive approach. The set of vertices is partitioned into two sets, such that the similarity between the objects of these sets is minimal. This similarity is evaluated using the sum of the weights that join objects that belong to the two subsets identified by the partition algorithm. The algorithm is applied recursively to the subgraphs that result from the partition of the initial graph, until a satisfactory clustering of the vertices is achieved.

Let $\mathcal{G} = (V, E, w)$ be a weighted graph and let S, T be two subsets of V. Define the association between S and T as the number

$$\mathsf{assoc}(S, T) = \sum\{w(u, v) \mid u \in S, v \in T\}.$$

In particular, if $\pi = \{S, T\}$ is a bipartition of V, to $\mathsf{assoc}(S, T)$ is the size of the cut π.

Let $U \subseteq V$, where $V = \{v_1, \ldots, v_m\}$, let $\mathbf{c}_U \in \{0, 1\}^m$ be the characteristic vector of U defined by

$$(\mathbf{c}_U)_i = \begin{cases} 1 & \text{if } v_i \in U \\ 0 & \text{otherwise,} \end{cases}$$

for $1 \leqslant i \leqslant m$.

The *volume* of the set U is the number

$$vol(U) = \sum_{u \in U} \mathsf{d}(u).$$

It is easy to see that $vol(U) = \mathbf{c}'_U D_{\mathcal{G}} \mathbf{c}_U$, where $D_{\mathcal{G}}$ is the diagonal matrix $D = \mathsf{diag}(\mathsf{d}(v_1), \ldots, \mathsf{d}(v_m))$. Also, note that

$$\mathbf{c}'_S D_{\mathcal{G}} \mathbf{c}_T = vol(S \cap T) \tag{16.10}$$

for any set of vertices S, T of $\mathcal{G}$. Therefore, for a bipartition π of V we have $\mathbf{c}'_S D_{\mathcal{G}} \mathbf{c}_T = 0$.

In [8], it is shown that a relaxation of the minimum cut problem in graphs can be solved efficiently.

If S, T are two sets of vertices of the weighted graph $\mathcal{G} = (V, E, w)$, then

$$\mathbf{c}'_S A_{\mathcal{G}} \mathbf{c}_T = \sum\{w(u, v) \mid u \in S, v \in T\} = \mathsf{assoc}(S, T). \tag{16.11}$$

This implies

$$\begin{aligned} \mathbf{c}'_U L_{\mathcal{G}} \mathbf{c}_U &= \mathbf{c}'_U (D_{\mathcal{G}} - A_{\mathcal{G}}) \mathbf{c}_U \\ &= \sum\{\mathsf{d}(v) \mid v \in U\} - \sum\{w(u, v) \mid u \in U, v \in U\}. \end{aligned}$$

The *normalized cut* of a bipartition $\pi = \{S, T\}$ of $\mathcal{G}$ is the number

$$\mathsf{ncut}(\pi) = \frac{cut(S, T)}{\mathsf{assoc}(S, V)} + \frac{cut(S, T)}{\mathsf{assoc}(T, V)}.$$

The *normalized association* of π is

$$\mathsf{nassoc}(\pi) = \frac{\mathsf{assoc}(S,S)}{\mathsf{assoc}(S,V)} + \frac{\mathsf{assoc}(T,T)}{\mathsf{assoc}(T,V)}.$$

It is immediate that $\mathsf{ncut}(\pi) + \mathsf{nassoc}(\pi) = 2$. Thus, seeking a bipartition π minimizing the normalized cut is equivalent to seeking a partition that maximizes the normalized association of π.

For a weighted graph $\mathcal{G} = (V, E, w)$ and a bipartition $\pi = \{S, T\}$ of $V = \{v_1, \ldots, v_m\}$ define $\mathbf{x} \in \{-1, 1\}^m$ as

$$x_i = \begin{cases} 1 & \text{if } v_i \in S, \\ -1 & \text{if } v_i \in T. \end{cases}$$

Observe that

$$\mathbf{c}_S = \frac{1}{2}(\mathbf{1}_m + \mathbf{x}) \text{ and } \mathbf{c}_T = \frac{1}{2}(\mathbf{1}_m - \mathbf{x}). \tag{16.12}$$

We have

$$\mathsf{ncut}(\pi) = \frac{cut(S,T)}{\mathsf{assoc}(S,V)} + \frac{cut(S,T)}{\mathsf{assoc}(T,V)} = \frac{\mathbf{c}_S' L_{\mathcal{G}} \mathbf{c}_S}{k \mathbf{1}_m' D_{\mathcal{G}} \mathbf{1}_m} + \frac{\mathbf{c}_T' L_{\mathcal{G}} \mathbf{c}_T}{(1-k)\mathbf{1}_m' D_{\mathcal{G}} \mathbf{1}},$$

where $k = \frac{\sum_{v \in S} \mathsf{d}(v)}{\sum_{v \in V} \mathsf{d}(v)}$. Taking into account Equalities (16.12) we obtain:

$$4 \cdot \mathsf{ncut}(\pi) = \frac{(\mathbf{1}_m + \mathbf{x})' L_{\mathcal{G}} (\mathbf{1}_m + \mathbf{x})}{k \mathbf{1}_m' D_{\mathcal{G}} \mathbf{1}_m} + \frac{(\mathbf{1}_m - \mathbf{x})' L_{\mathcal{G}} (\mathbf{1}_m - \mathbf{x})}{(1-k)\mathbf{1}_m' D_{\mathcal{G}} \mathbf{1}}.$$

Since $L_{\mathcal{G}} \mathbf{1}_m = \mathbf{0}_m$ and $\mathbf{1}_m' L_{\mathcal{G}} = \mathbf{0}_m'$, the last equality can be written as

$$\begin{aligned} 4 \cdot \mathsf{ncut}(\pi) &= \frac{\mathbf{x}' L_{\mathcal{G}} \mathbf{x}}{k \mathbf{1}_m' D_{\mathcal{G}} \mathbf{1}_m} + \frac{\mathbf{x}' L_{\mathcal{G}} \mathbf{x}}{(1-k)\mathbf{1}_m' D_{\mathcal{G}} \mathbf{1}_m} \\ &= \frac{1}{k(1-k)} \cdot \frac{\mathbf{x}' L_{\mathcal{G}} \mathbf{x}}{\mathbf{1}_m' D_{\mathcal{G}} \mathbf{1}_m} = \frac{vol(V)}{vol(S) vol(T)} \mathbf{x}' L_{\mathcal{G}} \mathbf{x}. \end{aligned}$$

Define the vector $\mathbf{y} \in \mathbb{R}^m$ as

$$\mathbf{y} = 2\left(\mathbf{c}_S - \frac{k}{1-k}\mathbf{c}_T\right) = \frac{1-2k}{1-k}\mathbf{1}_m + \frac{1}{1-k}\mathbf{x}.$$

The value of a component y_i of $\mathbf{y}$ is either 2, when $x_i = 1$, or is $-\frac{2k}{1-k}$, when $x_i = -1$. Also, $\mathbf{y}' D_{\mathcal{G}} \mathbf{1}_m = 0$, when $\{S, T\}$ is a partition of V.

We have

$$\begin{aligned}
\mathbf{y}'D\mathbf{y} &= 4\cdot\left(\mathbf{c}'_S - \frac{k}{1-k}\mathbf{c}'_T\right) D_{\mathcal{G}}\left(\mathbf{c}_S - \frac{k}{1-k}\mathbf{c}_T\right)\\
&= 4\cdot\left(\mathbf{c}'_S D_{\mathcal{G}}\mathbf{c}_S + \frac{k^2}{(1-k)^2}\mathbf{c}'_T D_{\mathcal{G}}\mathbf{c}_T\right) \text{ (because } S\cap T=\emptyset)\\
&= 4\cdot\left(vol(S) + \frac{k^2}{(1-k)^2}vol(T)\right) = 4\cdot\left(vol(S)+\frac{vol(S)^2}{vol(T)}\right)\\
&= 4\frac{vol(S)vol(V)}{vol(T)},
\end{aligned}$$

and

$$\begin{aligned}
\mathbf{y}'L_{\mathcal{G}}\mathbf{y} &= \left(\frac{1-2k}{1-k}\mathbf{1}'_m + \frac{1}{1-k}\mathbf{x}'\right) L_{\mathcal{G}}\left(\frac{1-2k}{1-k}\mathbf{1}_m + \frac{1}{1-k}\mathbf{x}\right)\\
&= \frac{1}{(1-k)^2}\mathbf{x}'L_{\mathcal{G}}\mathbf{x} = \left(\frac{vol(V)}{vol(T)}\right)^2 \mathbf{x}'L_{\mathcal{G}}\mathbf{x}.
\end{aligned}$$

Thus, we can write

$$\frac{\mathbf{y}'L_{\mathcal{G}}\mathbf{y}}{\mathbf{y}'D_{\mathcal{G}}\mathbf{y}} = \frac{\left(\frac{vol(V)}{vol(T)}\right)^2\mathbf{x}'L_{\mathcal{G}}\mathbf{x}}{4\frac{vol(S)vol(V)}{vol(T)}} = \frac{1}{4}\frac{vol(V)}{vol(S)vol(T)}\mathbf{x}'L_{\mathcal{G}}\mathbf{x} = \mathsf{ncut}(\pi).$$

If we define $\mathbf{z} = D_{\mathcal{G}}^{\frac{1}{2}}\mathbf{y}$, the normalized cut can be written as a Rayleigh quotient

$$\mathsf{ncut}(\pi) = \frac{\mathbf{z}'D_{\mathcal{G}}^{-\frac{1}{2}'}L_{\mathcal{G}}D_{\mathcal{G}}^{-\frac{1}{2}}\mathbf{z}}{\mathbf{z}'\mathbf{z}} = \frac{\mathbf{z}'L_{\mathcal{G}}\mathbf{z}}{\mathbf{z}'\mathbf{z}}.$$

We saw that the components of $\mathbf{y}$ range over the set $\left\{2, -\frac{2k}{1-k}\right\}$. If we relax this restriction and we allow these components to range over $\mathbb{R}$, then this relaxation will involve also the vector $\mathbf{z}$. Thus, the relaxation of this problem amounts to finding $\mathbf{z}$ such that the Rayleigh quotient $\frac{\mathbf{z}'L_{\mathcal{G}}\mathbf{z}}{\mathbf{z}'\mathbf{z}}$ is minimal.

Normalized cuts of partitions offer another approach to spectral clustering. As before, let $\kappa = \{C_1, \ldots, C_k\}$ be a partition of a set $V = \{v_1, \ldots, v_n\}$ of n objects into k clusters for which we have a similarity matrix $S \in \mathbb{R}^{n\times n}$. Define the characteristic vector $\mathbf{h}_j$ of C_j as

$$(\mathbf{h}_j)_i = \begin{cases}\frac{1}{\sqrt{vol(C_j)}} & \text{if } v_i \in C_j,\\ 0 & \text{otherwise,}\end{cases}$$

for $1 \leqslant j \leqslant k$ and let $H = (\mathbf{h}_1 \ \cdots \ \mathbf{h}_k)$. We have

$$\mathbf{h}_j' D_{\mathcal{G}} \mathbf{h}_j = \sum_{i=1}^{n} \sum_{\ell=1}^{n} (\mathbf{h}_j)_i d_{i\ell} (\mathbf{h}_j)_\ell.$$

The non-zero terms in this sum are such that $i = \ell$ and $v_i \in C_j$. Thus, $\mathbf{h}_j' D_{\mathcal{G}} \mathbf{h}_j = \frac{1}{vol(C_j)} \sum_{v \in C_j} \mathsf{d}(v) = 1$. On the other hand we have $\mathbf{h}_j' D_{\mathcal{G}} \mathbf{h}_m = 0$ if $j \neq m$, so $H' D_{\mathcal{G}} H = I_k$. A similar computation yields

$$\mathbf{h}_j' A_{\mathcal{G}} \mathbf{h}_j = \sum_{i=1}^{n} \sum_{\ell=1}^{n} (\mathbf{h}_j)_i s_{i\ell} (\mathbf{h}_j)_\ell = \frac{1}{vol(C_j)} \sum_{v_i, v_\ell \in C_j} s(v_i, v_\ell).$$

These computations allow us to write

$$\begin{aligned} \mathbf{h}_j' L_{\mathcal{G}} \mathbf{h}_j &= \mathbf{h}_j'(D_{\mathcal{G}} - A_{\mathcal{G}}) \mathbf{h}_j = I_k - \mathbf{h}_j' A_{\mathcal{G}} \mathbf{h}_j = 1 - \frac{1}{vol(C_j)} \sum_{v_i, v_\ell \in C_j} s(v_i, v_\ell) \\ &= \frac{vol(C_j) - \sum_{v_i, v_\ell \in C_j} s(v_i, v_j)}{vol(C_j)} = \frac{cut(C_j, \bar{C}_j)}{vol(C_j)}. \end{aligned}$$

Therefore,

$$trace(H' L_{\mathcal{G}} H) = \sum_{j=1}^{k} \mathbf{h}_j' L_{\mathcal{G}} \mathbf{h}_j = \sum_{j=1}^{k} \frac{cut(C_j, \bar{C}_j)}{vol(C_j)} = \mathsf{ncut}(\kappa).$$

To minimize the normalized cut we need to minimize $trace(H' L_{\mathcal{G}} H)$ subjected to the constraint $H' D H = I_k$. Let $M = D^{\frac{1}{2}} H$. Then, in terms of the matrix M, the optimization problem amounts to minimizing

$$trace(M' D^{-\frac{1}{2}} L_{\mathcal{G}} D^{-\frac{1}{2}} M) = trace(M' L_{\mathcal{G},\mathrm{sym}} M),$$

subjected to the restriction $M' M = I_k$. By allowing M to range over $\mathbb{R}^{n \times k}$, the optimum can be achieved by $M = (\mathbf{m}_1, \ldots, \mathbf{m}_k)$, where $\mathbf{m}_1, \ldots, \mathbf{m}_k$ are the first k eigenvectors of the symmetric Laplacian $L_{\mathcal{G},\mathrm{sym}}$. By Theorem 16.41 $D^{-\frac{1}{2}} \mathbf{m}_1, \ldots, D^{-\frac{1}{2}} \mathbf{m}_k$ are the first k eigenvectors of the of the random walk Laplacian and these are exactly the columns of the matrix H. So, the optimal value of H is obtained by choosing its columns to be equal to the eigenvectors that correspond to the first k eigenvalues of $L_{\mathcal{G},\mathrm{rw}}$.

Observe also, that an eigenvector of $L_{\mathcal{G},\mathrm{rw}}$ is also an eigenvector of the matrix pencil $(L_{\mathcal{G}}, D_{\mathcal{G}})$ and λ is an eigenvalue of $L_{\mathcal{G},\mathrm{rw}}$ if and only if it belongs to $\mathsf{spec}(L_{\mathcal{G}}, D_{\mathcal{G}})$.

Using the random walk Laplacian, Shi and Malik [8] gave the spectral clustering Algorithm 16.6.1.

Algorithm 16.6.1: Random Walk Lagrangean Spectral Clustering

Data: Similarity matrix $S \in \mathbb{R}^{n \times n}$, number k of clusters
Result: A clustering $\kappa = \{C_1, \ldots, C_k\}$
let W be its weighted adjacency matrix;
compute the random walk Laplacian $L_{\mathcal{G},\mathrm{rw}}$;
compute the first k eigenvectors $\mathbf{v}_1, \ldots, \mathbf{v}_k$ of $L_{\mathcal{G},\mathrm{rw}}$;
let $V = (\mathbf{v}_1, \ldots, \mathbf{v}_k) \in \mathbb{R}^{n \times k}$;
define $\mathbf{y}_1, \ldots, \mathbf{y}_n \in \mathbb{R}^k$ such that $V' = (\mathbf{y}_1 \cdots \mathbf{y}_n)$;
cluster $\{\mathbf{y}_1, \ldots, \mathbf{y}_n\} \subseteq \mathbb{R}^k$ using the k-means algorithm into κ;

An asymmetric variant of Shi and Malik technique is proposed in [9] in order to cluster the most salient points of an image. The set of these points is defined as the *foreground set*; the remaining points form the *background set*.

The starting point, as before, is a similarity matrix S defined on a set of objects $V = \{\mathbf{v}_1, \ldots, \mathbf{v}_m\}$. By Eckhart-Young Theorem (Theorem 7.79), S can be approximated by the rank-1 matrix $P = \sigma_1 \mathbf{u}\mathbf{v}'$ such that $\mathbf{u}$ and $\mathbf{v}$ are unit vectors and $|||A - P|||_2 = \sigma_2$. In addition, since S is symmetric, we have $\mathbf{v} = \mathbf{u}'$, so the approximating matrix is $P = \sigma_1 \mathbf{u}\mathbf{u}'$. Let $\mathbf{p}$ be defined by $\mathbf{p} = \sqrt{\sigma_1}\mathbf{u}$, which allows us to write the approximating matrix as $P = \mathbf{p}\mathbf{p}'$. The foreground set F consists of those objects whose indices belong to the set $\{i \mid p_i > 0\}$; the remaining objects constitute the set of background objects B and they are indexed by the set $\{1, \ldots, m\} - \{i \mid p_i > 0\}$.

The set of foreground objects F can be obtained by minimizing an asymmetric form of Shi and Malik's criterion, namely

$$N(F) = \frac{\mathsf{assoc}(F, B)}{\mathsf{assoc}(F, F)}.$$

Let $\mathbf{c}_F$ be the characteristic of the set F. Then,

$$N(F) = \frac{\sum_{i \in F, j \in B} s_{ij}}{\sum_{i \in F, j \in F} s_{ij}} = \frac{\mathbf{c}_F' S(\mathbf{1}_m - \mathbf{c}_F)}{\mathbf{c}_F' S \mathbf{c}_F} = \frac{\mathbf{c}_F' S \mathbf{1}}{\mathbf{c}_F' S \mathbf{c}_F} - 1.$$

Thus, mimimizing $N(F)$ is equivalent to minimizing $N(F) + 1 = \frac{\mathbf{c}_F' S \mathbf{1}}{\mathbf{c}_F' S \mathbf{c}_F}$.

Let $S = UDU'$ be the singular value decomposition of the symmetric matrix S and let $\mathbf{z} = D^{\frac{1}{2}} U' \mathbf{c}_F$. Now we have

$$N(F) + 1 = \frac{\mathbf{c}_F' UDU' \mathbf{1}}{\mathbf{c}_F' UDU' \mathbf{c}_F}.$$

Define $\mathbf{z} = D^{\frac{1}{2}} U' \mathbf{c}_F$. Using this notation $N(F)$ can be written as

$$N(F) + 1 = \frac{\mathbf{z}' D^{\frac{1}{2}} U' \mathbf{1}}{\mathbf{z}'\mathbf{z}} = \frac{\mathbf{z}'\mathbf{u}}{\mathbf{z}'\mathbf{z}},$$

where $\mathbf{u} = D^{\frac{1}{2}} U' \mathbf{1}$.

Since $\mathbf{c}_F$ is a characteristic vector, we seek a vector with 0/1 components that minimizes $\frac{\mathbf{c}'_F S \mathbf{1}}{\mathbf{c}'_F S \mathbf{c}_F}$ and it is not clear how to find such a vector. Instead, the problem is relaxed and we seek a unit vector $\mathbf{z}$ that minimizes $\mathbf{z}'\mathbf{u}$ (since $\mathbf{z}'\mathbf{z} = 1$). This is done by finding the component u_k of $\mathbf{u}$ with the largest absolute value and choosing $\mathbf{z}$ either as $\mathbf{e}_k$ or $-\mathbf{e}_k$ (with an opposite sign of that of u_k) and, thus obtaining $\mathbf{c}_F = D^{-\frac{1}{2}} U \mathbf{z}$. As the result, the following algorithm for computing the foreground set is obtained:

Algorithm 16.6.2: Perona-Freeman algorithm for the foreground set

Data: Similarity matrix $S \in \mathbb{R}^{n \times n}$
Result: Foreground set of objects

let $S = UDU'$ be the singular decomposition of S ;
compute the vector $\mathbf{u} = D^{\frac{1}{2}} U' \mathbf{1}$;
determine the index k of the maximum entry of $\mathbf{u}$;
define $\mathbf{x}$ as the kth column of U;
obtain F as the set of objects that correspond to non-zero entries of $\mathbf{x}$;

Exercises and Supplements

1. Apply hierarchical clustering to the data set given in Example 16.3 using the average-link method, the centroid method, and the Ward method. Compare the shapes of the clusters that are formed during the aggregation process. Draw the dendrograms of the clusterings.
2. Using a random number generator, produce h sets of points in $\mathbb{R}^n$ normally distributed around h given points in $\mathbb{R}^n$. Use k-means to cluster these points with several values for k and compare the quality of the resulting clusterings.
3. A variant of the k-means clustering introduced in [10] is the *bisecting k-means algorithm* described below.
   The cluster C that is bisected may be the largest cluster or the cluster having the largest *sse*.
   Evaluate the time performance of bisecting k-means compared with the standard k-means and with some variant of a hierarchical clustering.
4. One of the issues that the k-means algorithm must confront is that the number of clusters k must be provided as an input parameter. Using clustering validity, design an algorithm that identifies local maxima of validity (as a function of k) to provide a basis for a good choice of k. See [11] for a solution that applies to image segmentation.

Algorithm 16.6.3: Bisecting k-means algorithm

Data: S the set of objects to be clustered, k the desired number of clusters, and nt, the number of trial bisections

Result: A k-clustering of S

1 $set\_of\_clusters = \{S\}$;
2 **while** $|set\_of\_clusters| < k$ **do**
3   select a cluster C from the $set\_of\_clusters$;
4      $k = 0$;
5   **for** $i = 1$ *to* nt **do**
6   let C_{0i}, C_{1i} be the two clusters obtained from C by bisecting C using standard k-means ($k = 2$);
7   **if** $(i = 1)$ **then**
8              $s = sse(\{C_{0i}, C_{1i}\})$
9   **end**
10   **if** $(sse(\{C_{0i}, C_{1i}\}) \leqslant s)$ **then**
11             $k = i$;
12             $s = sse(\{C_{0i}, C_{1i}\})$
13   **end**
14   **end**
15   add C_{0k}, C_{1k} to $set\_of\_clusters$;
16 **end**

---

5. Let $S = \{\mathbf{x}_1, \ldots, \mathbf{x}_m\} \subseteq \mathbb{R}^n$ be a set of m objects and let $C_1, \ldots, C_k$ be the set of clusters computed by the k-means algorithm at any step. Prove that the convex closure of each cluster C_i, $\mathbf{K}_{\text{conv}}(C_i)$ is included in a polytope P_i that contains $\mathbf{c}_i$ for $1 \leqslant i \leqslant k$.

Solution: Let $\mathbf{c}_1, \ldots, \mathbf{c}_k$ be the centroids of the partition $\{C_1, \ldots, C_k\}$ and let $\mathbf{m}_{ij} = \frac{1}{2}(\mathbf{c}_i + \mathbf{c}_j)$ be the midpoint of the segment $\overline{\mathbf{c}_i \mathbf{c}_j}$. Define the hyperplane H_{ij} as the set of points $\mathbf{x}$ such that $(\mathbf{c}_i - \mathbf{c}_j)'(\mathbf{x} - \mathbf{m}_{ij}) = 0$, that is, the perpendicular bisector of the segment $\overline{\mathbf{c}_i \mathbf{c}_j}$. Equivalently,

$$H_{ij} = \{\mathbf{x} \in \mathbb{R}^m \mid (\mathbf{c}_i - \mathbf{c}_j)'\mathbf{x} = \frac{1}{2}(\mathbf{c}_i - \mathbf{c}_j)'(\mathbf{c}_i + \mathbf{c}_j)\}.$$

The halfspaces determined by H_{ij} are described by the inequalities:

$$H_{ij}^+ : (\mathbf{c}_i - \mathbf{c}_j)'\mathbf{x} \leqslant \frac{1}{2}\left(\| \mathbf{c}_i \|_2^2 - \| \mathbf{c}_j \|_2^2 \right)$$
$$H_{ij}^- : (\mathbf{c}_i - \mathbf{c}_j)'\mathbf{x} \geqslant \frac{1}{2}\left(\| \mathbf{c}_i \|_2^2 - \| \mathbf{c}_j \|_2^2 \right).$$

It is easy to see that $\mathbf{c}_i \in H_{ij}^+$ and $\mathbf{c}_j \in H_{ij}^-$. Moreover, if $d_2(\mathbf{c}_i, \mathbf{x}) < d_2(\mathbf{c}_j, \mathbf{x})$, then $\mathbf{x} \in H_{ij}^+$, and if $d_2(\mathbf{c}_i, \mathbf{x}) > d_2(\mathbf{c}_j, \mathbf{x})$, then $\mathbf{x} \in H_{ij}^-$. Indeed, suppose that $d_2(\mathbf{c}_i, \mathbf{x}) < d_2(\mathbf{c}_j, \mathbf{x})$, which amounts to $\| \mathbf{c}_i - \mathbf{x} \|_2^2 < \| \mathbf{c}_j - \mathbf{x} \|_2^2$. This is equivalent to

$$(\mathbf{c}_i - \mathbf{x})'(\mathbf{c}_i - \mathbf{x}) < (\mathbf{c}_j - \mathbf{x})'(\mathbf{c}_j - \mathbf{x}).$$

The last inequality is equivalent to

$$\| \mathbf{c}_i \|_2^2 - 2\mathbf{c}_i'\mathbf{x} < \| \mathbf{c}_j \|_2^2 - 2\mathbf{c}_j'\mathbf{x},$$

which implies that $\mathbf{x} \in H_{ij}^+$. In other words, $\mathbf{x}$ is located in the same half-space as the closest centroid of the set $\{\mathbf{c}_i, \mathbf{c}_j\}$. Note also that if $d_2(\mathbf{c}_i, \mathbf{x}) = d_2(\mathbf{c}_j, \mathbf{x})$, then $\mathbf{x}$ is located in $H_{ij}^+ \cap H_{ij}^- = H_{ij}$, that is, on the hyperplane shared by P_i and P_j.

Let P_i be the closed polytope defined by

$$P_i = \bigcap \{H_{ij}^+ \mid j \in \{1, \ldots, k\} - \{i\}\}$$

Objects that are closer to $\mathbf{c}_i$ than to any other centroid $\mathbf{c}_j$ are located in the closed polytope P_i. Thus, $C_i \subseteq P_i$ and this implies $\mathbf{K}_{\mathrm{conv}}(C_i) \subseteq P_i$.

6. Let $B \subseteq \mathbb{R}^n$ be a finite subset of $\mathbb{R}^n$. The *clustering feature* of B is a triple $(p, \mathbf{s}, \mathbf{q})$, where $p = |B|$, $\mathbf{s} = \sum\{\mathbf{x} \mid \mathbf{x} \in \mathbb{R}^n\}$, and

$$\mathbf{q} = \left(\sum\{x_1^2 \mid \mathbf{x} \in B\}, \ldots, \sum\{x_n^2 \mid \mathbf{x} \in B\}\right).$$

The center of B is $\bar{\mathbf{x}} = \frac{1}{p}\mathbf{s}$, the *average distance between the center and the members* of B is

$$R_B = \sqrt{\frac{(x - \bar{x})^2}{p}}$$

and the average distance between the members of the clusters is

$$D_B = \sqrt{\frac{\sum\{(\mathbf{u} - \mathbf{v})^2 \mid \mathbf{u}, \mathbf{v} \in B\}}{p(p-1)}}.$$

Prove that $\bar{x}$, R_B, and D_B can be computed starting from the cluster feature.

Let $\mathcal{G} = (V, E)$ be a finite graph. A *graph clustering* [12] is a partition $\kappa = \{C_1, \ldots, C_p\}$ of the set V; the clusters are the subgraphs $\mathcal{G}_{C_i} = (C_i, E_{C_i})$ induced by the blocks of κ. The *intracluster edges* are the edges in $E_\kappa = \bigcup_{i=1}^p E_{C_i}$, while the *intercluster edges* are the edges in $E - E_\kappa$. The set of edges between nodes in C and C' is denoted by $E(C, C')$.

7. The quality of a graph clustering κ can be measured by its *modularity index* $q(\kappa)$ given by

$$q(\kappa) = \sum_{C \in \kappa} \left\{ \frac{|E_\kappa|}{|E|} - \left(\frac{|E(C)| + \sum_{C' \in \kappa} |E(C, C')|}{2|E|} \right)^2 \right\}.$$

Prove that $q(\kappa) = \sum_{C \in \kappa} \left\{ \frac{|E_\kappa|}{|E|} - \left(\frac{\sum_{v \in C} \mathsf{d}(v)}{2|E|} \right)^2 \right\}$ What does it take for a clustering to achieve a high value of the modularity index?

8. Prove that $q(\kappa) \in [-0.5, 1]$ for every clustering of a graph $\mathcal{G}$ and the minimum is achieved when all edges are intercluster edges.
9. Prove that there is always a clustering of a graph $\mathcal{G}$ that has maximum modularity in which each cluster consists of a connected subgraph.
10. Let $\mathcal{G} = (V, E)$ be a bipartite graph with the partition $\pi = \{V_1, V_2\}$ (see Definition 10.89). Prove that $q(\kappa) = -0.5$.

In general, a clustering algorithm starts with a definite dissimilarity on S and generates a partition of S whose blocks are regarded as clusters. If $\mathcal{D}'_S$ is the set of definite dissimilarities S, a *clustering function* on S is a mapping $f : \mathcal{D}'_S \longrightarrow PART(S)$.

Let κ be a partition of S and let $d, d' \in \mathcal{D}'_S$. The definite dissimilarity d' is a *κ-transformation of d* if the following conditions are satisfied:

(i) if $x \equiv_\kappa y$, then $d'(x, y) \leqslant d(x, y)$;
(ii) if $x \not\equiv_\kappa y$, then $d'(x, y) > d(x, y)$.

In other words, d' is a κ-transformation of d if for two objects that belong to the same κ-cluster $d'(x, y)$ is smaller than $d(x, y)$, while for two objects that belong to two distinct clusters $d'(x, y)$ is larger than $d(x, y)$.

The following properties are desirable for a clustering function $f : \mathcal{D}'_S \longrightarrow PART(S)$. The function f is:

(i) *scale-invariant* if, for every $d \in \mathcal{D}'_S$ and every $\alpha > 0$, we have $f(d) = f(\alpha d)$;
(ii) *rich*, if f is surjective;
(iii) *consistent* if, for every $d, d' \in \mathcal{D}'_S$ and $\kappa \in PART(S)$ such that $f(d) = \kappa$ and d' is a κ-transformation of d, we have $f(d') = \kappa$.

A dissimilarity $d \in \mathcal{D}'_S$ is *(a, b)-conformant to a clustering* κ if $x \equiv_\kappa y$ implies $d(x, y) \leqslant a$ and $x \not\equiv_\kappa y$ implies $d(x, y) \geqslant b$. A dissimilarity is *conformant* to a clustering κ if it is (a, b)-conformant to κ for some pair of numbers (a, b).

Let $g : \mathbb{R}_{\geqslant 0} \longrightarrow \mathbb{R}_{\geqslant 0}$ be a continuous, nondecreasing and unbounded function and let $S \subseteq \mathbb{R}^n$ be a finite subset of $\mathbb{R}^n$. For $k \in \mathbb{N}$ and $k \geqslant 2$, define a g_k-clustering function as follows.

Begin by selecting a set T of k points from S such that the function $\Lambda^g_d(T) = \sum_{x \in S} g(d(x, T))$ is minimized. Here $d(x, T) = \min\{d(x, t) | t \in T\}$. Then, define a partition of S into k clusters by assigning each point to the point in T that is the closest and breaking the ties using a fixed (but otherwise arbitrary) order on the set of points. The clustering function defined by g_d, denoted by f^g, maps d to this partition.

The *k-median clustering function*, is obtained by choosing $g(x) = x$ for $x \in \mathbb{R}_{\geqslant 0}$; the *k-means clustering function* is obtained by taking $g(x) = x^2$ for $x \in \mathbb{R}_{\geqslant 0}$.

11. Prove that for $k \geqslant 2$ and for sufficiently large sets of objects, the clustering function g_k introduced above is not consistent.
Solution: Suppose that $\kappa = \{C_1, C_2, \ldots, C_k\}$ is a partition of S and d is a definite dissimilarity on S such that $d(x, y) = r_i$ if $x \neq y$ and $\{x, y\} \subseteq C_i$ for some $1 \leqslant i \leqslant k$ and $d(x, y) = r + a$ if x and y belong to two distinct blocks of κ, where $r = \max\{r_i | 1 \leqslant i \leqslant k\}$ and $a > 0$.
Suppose that T is a set of k members of S. Then, the value of $g(d(x, T))$ is $g(r)$ if the closest member of T is in the same block as x and is $g(r + a)$ otherwise. This means that the smallest value of $\Lambda_d^g(T) = \sum_{x \in C_i} g(d(x, T))$ is obtained when each block C_i contains a member t_i of T for $1 \leqslant i \leqslant k$ and the actual value is $\Lambda_d^g(T) = \sum_{i=1}^k (|C_i| - 1)r^2 = (|S| - k)r^2$.
Consider now a partition $\kappa' = \{C_1', C_1'', C_2, \ldots, C_k\}$, where $C_1 = C_1' \cup C_1''$ so $\kappa' < \kappa$. Choose r' to be a positive number such that $r' < r$ and define the dissimilarity d' on S such that $d'(x, y) = r'$ if $x \neq y$ and $x \equiv_{\kappa'} y$ and $d'(x, y) = d(x, y)$ otherwise. Clearly, d' is a κ-transformation of d. The minimal value for $\Lambda_d^g(T')$ is achieved when T' consists of $k+1$ points, one in each block of κ'; as a result, the value of the clustering function for d' is $\kappa' \neq \kappa$, which shows that no clustering function obtained by this technique is consistent.
12. Prove that if d' is a κ-transformation of d, and d is (a, b)-conformant to κ, then d' is also (a, b)-conformant to κ.
13. Let $\kappa \in PART(S)$ be a partition on S and let f be a clustering function on S. A pair of positive numbers (a, b) is *κ-forcing with respect to f* if, for every $d \in \mathcal{D}_S'$ that is (a, b)-conformant to κ, we have $f(d) = \kappa$.
Let f be a consistent clustering function on a set S. Prove that for any partition $\kappa \in \mathsf{Ran}(f)$ there exist $a, b \in \mathbb{R}_{>0}$ such that the pair (a, b) is κ-forcing.
14. Prove that if f is a scale-invariant and consistent clustering function on a set S, then its range is an antichain in poset $(PART(S), \leqslant)$.

Solution: This statement is equivalent to saying that, for any scale-invariant and consistent clustering function, no two distinct partitions of S that are values of f are comparable.
Suppose that there are two clusterings, κ_0 and κ_1, in the range of a scale-invariant and consistent clustering such that $\kappa_0 < \kappa_1$.
Let (a_i, b_i) be a κ_i-forcing pair for $i = 0, 1$, where $a_0 < b_0$ and $a_1 < b_1$. Let a_2 be a number such that $a_2 \leqslant a_1$ and choose ϵ such that

$$0 < \epsilon < \frac{a_0 a_2}{b_0}.$$

By Supplement 25 of Chap. 14 construct a distance d such that

(a) for any points x, y that belong to the same block of κ_0, $d(x, y) \leqslant \epsilon$;

(b) for points that belong to the same cluster of κ_1 but not to the same cluster of κ_0, $a_2 \leqslant d(x, y) \leqslant a_1$; and
(c) for points that do not belong to the same cluster of κ_1, $d(x, y) \geqslant b_1$.

The distance d is (a_1, b_1)-conformant to κ_1, and so we have $f(d) = \kappa_1$. Take $\alpha = \frac{b_0}{a_2}$, and define $d' = \alpha d$. Since f is scale-invariant, we have $f(d') = f(d) = \kappa_1$. Note that for points x, y that belong to the same cluster of κ_0, we have

$$d'(x, y) \leqslant \frac{\epsilon b_0}{a_2} < a_0,$$

while for points x, y that do not belong to the same cluster of κ_0 we have

$$d'(x, y) \geqslant \frac{a_2 b_0}{a_2} \geqslant b_0.$$

Thus, d' is (a_0, b_0)-conformant to κ_0, and so we must have $f(d') = \kappa_0$. Since $\kappa_0 \neq \kappa_1$, this is a contradiction.

15. Prove that if $|S| \geqslant 2$, there is no clustering function that is scale-invariant, rich and consistent. (This fact is known as the Kleinberg's Impossibility Theorem.)

Solution: If S contains at least two elements, then the poset $(PART(S), \leqslant)$ is not an antichain. Therefore, this statement is a direct consequence of Supplement 14.

16. Prove that for every antichain A of the poset $(PART(S), \leqslant)$, there exists a clustering function f that is scale-invariant and consistent such that $\mathsf{Ran}(f) = A$.

Solution: Suppose that A contains more than one partition. We define $f(d)$ as the first partition $\pi \in A$ (in some arbitrary but fixed order) that minimizes the quantity

$$\Phi_d(\pi) = \sum_{x \equiv_\pi y} d(x, y).$$

Note that $\Phi_{\alpha d} = \alpha \Phi_d$. Therefore, f is scale-invariant.
We need to prove that every partition of A is in the range of f.
For a partition $\rho \in A$, define d such that $d(x, y) < \frac{1}{|S|^3}$ if $x \equiv_\rho y$ and $d(x, y) \geqslant 1$ otherwise. Observe that $\Phi_d(\rho) < 1$. Suppose that $\Phi_d(\theta) < 1$. The definition of d means that

$$\Phi_d(\theta) = \sum_{x \equiv_\theta y} d(x, y) < 1,$$

so for all pairs $(x, y) \in \equiv_\theta$ we have $d(x, y) < \frac{1}{|S|^3}$, which means that $x \equiv_\rho y$. Therefore, we have $\pi < \rho$. Since A is an antichain, it follows that ρ must minimize Φ_d over all partitions of A and, consequently, $f(d) = \rho$.
To verify the consistency of f, suppose that $f(d) = \pi$, and let d' be a π-transformation of d. For $\sigma \in PART(S)$, define $\delta(\sigma)$ as $\Phi_d(\sigma) - \Phi_{d'}(\sigma)$. For $\sigma \in A$, we have

$$\delta(\sigma) = \sum_{x \equiv_\sigma y} (d(x, y) - d'(x, y))$$

$$\leqslant \sum_{\substack{x \equiv_\sigma y \\ \text{and} x \equiv_\pi y}} (d(x, y) - d'(x, y))$$

(only terms corresponding to pairs in the same cluster are nonnegative)

$$\leqslant \delta(\pi)$$

(every term corresponding to a pair in the same cluster is nonnegative).

Consequently,

$$\Phi_d(\sigma) - \Phi_{d'}(\sigma) \leqslant \Phi_d(\pi) - \Phi_{d'}(\pi)$$

or $\Phi_d(\sigma) - \Phi_d(\pi) \leqslant \Phi_{d'}(\sigma) - \Phi_{d'}(\pi)$. Thus, if π minimizes $\Phi_d(\pi)$, then $\Phi_d(\sigma) - \Phi_d(\pi) \geqslant 0$ for every $\sigma \in A$ and therefore $\Phi_{d'}(\sigma) - \Phi_{d'}(\pi) \geqslant 0$, which means that π also minimizes $\Phi_{d'}(\pi)$. This implies $f(d') = \pi$, which shows that f is consistent.

17. Let $O = \{u_1, \ldots, u_n\}$ be a collection of objects, $d : O \times O \longrightarrow \mathbb{R}_{\geqslant 0}$ be a dissimilarity on O, and let $f : O \longrightarrow \{C_1, \ldots, C_k\}$ be a clustering function. Define the functions $a, b : O \longrightarrow \mathbb{R}_{\geqslant 0}$ as

$$a(u_i) = \frac{\sum\{d(u_i, u) \mid f(u) = f(u_i) \text{ and } u \neq u_i\}}{|f(u_i)|},$$
$$b(u_i) = \min\{d(u_i, C) \mid C \neq f(u_i)\},$$

for $u_i \in O$. The *silhouette* of the object u_i for which $|f(u_i)| \geqslant 2$ is the number $sil(u_i)$ given by

$$sil(u_i) = \begin{cases} 1 - \frac{a(u_i)}{b(u_i)} & \text{if } a(u_i) < b(u_i) \\ 0 & \text{if } a(u_i) = b(u_i) \\ \frac{b(u_i)}{a(u_i)} - 1 & \text{if } a(u_i) > b(u_i). \end{cases}$$

(a) Prove that $-1 \leqslant sil(u_i) \leqslant 1$ for $1 \leqslant i \leqslant m$.
(b) Discuss the situations when $sil(u_i)$ is close to 1 or to -1.

Let (S, d) be a finite metric space and let $\mathcal{G}_0, \ldots, \mathcal{G}_k$ be a sequence of graphs, where $k = diam_{S,d}$, where $\mathcal{G}_p = (S, E_p)$ is defined by its set of edges

$$E_p = \{(x, y) \in S \times S \mid d(x, y) \leqslant p\}$$

for $0 \leqslant p \leqslant k$. The graph $\mathcal{G}_0$ is $(S, \emptyset)$, while $\mathcal{G}_k$ is a complete graph on the set S. The number of connected components of the threshold graph $\mathcal{G}_i$ is denoted by c_i for $1 \leqslant i \leqslant k$.

18. Consider Algorithm 16.6.4

Algorithm 16.6.4: Graph-based single-link clustering

Data: a finite metric space (S, d) of diameter k, where $|S| = n$
Result: a hierarchical clustering of S

initialize the threshold graph $\mathcal{G}_0$;
$c = n$; // current number of connected components;
$p = 1$;
while $(c_p > 1)$ **do**
  if $(c_p < c)$ **then**
    output the connected components $\mathcal{G}_p$
  end
  $p{+}{+}$;
end

Prove that two connected components C, C' of $\mathcal{G}_{p-1}$ are fused into a connected component of $\mathcal{G}_p$ if there exists one edge in $\mathcal{G}_p$ that joins these components.

19. Let (S, d) be a finite metric space. Construct a chain of partitions $\pi_1, \pi_2, \ldots$ and a chain of dissimilarities $d_1, d_2, \ldots$, where d_i is defined on the set of blocks of π_i as follows. Define $\pi_1 = \alpha_S$. The partition π_{i+1} is obtained from π_i by fusing the blocks B, C of π such that $d_i(B, C)$ has the smallest value, that is,

$$\pi_{i+1} = (\pi_i - \{B, C\}) \cup \{B \cup C\}.$$

(a) Prove that the sequences of partitions $\pi_1, \pi_2, \ldots$ and a chain of dissimilarities $d_1, d_2, \ldots$ coincides with the sequence of partitions $\pi^1, \pi^2, \ldots$ and the sequence of dissimilarities $d^1, d^2, \ldots$ constructed in the single-link clustering algorithm.
(b) Prove that the value $e(x, y)$ of the subdominant ultrametric that corresponds to d equals the least height $h_d(W)$ of a cluster W such that $\{x, y\} \subseteq W$.

20. Let $\mathcal{G} = (V, E)$ be a graph with $|V| = m$. Prove that the adjacency and the Laplacian matrices of the graph complement $\overline{\mathcal{G}}$ are $A_{\overline{\mathcal{G}}} = J_{m,m} - I_m - A_{\mathcal{G}}$ and $L_{\overline{\mathcal{G}}} = mI_m - J_{m,m} - L_{\mathcal{G}}$, respectively.
21. Let $\mathcal{G} = (V, E, w)$ be a weighted graph, where $V = \{v_1, \ldots, v_m\}$. Prove that if λ is an eigenvalue of $L_{\mathcal{G},\mathrm{sym}}$, then $0 \leqslant \lambda \leqslant 2$.

Solution: Since $\lambda \leqslant \sup\{\mathbf{x}' L_{\mathcal{G},\mathrm{sym}} \mathbf{x} \mid \parallel \mathbf{x} \parallel = 1\}$ it follows that

$$\begin{aligned}
\lambda &\leqslant \sup_{\|\mathbf{x}\|=1} \frac{1}{2} \sum_{i=1}^{m} \sum_{j=1}^{m} w_{ij} \left(\frac{x_i}{\sqrt{\mathsf{d}(v_i)}} - \frac{x_j}{\sqrt{\mathsf{d}(v_j)}} \right)^2 \\
&\text{(by Lemma 16.42)} \\
&\leqslant \sup_{\|\mathbf{x}\|=1} \frac{1}{2} \sum_{i=1}^{m} \sum_{j=1}^{m} 2w_{ij} \left(\frac{x_i^2}{\mathsf{d}(v_i)} + \frac{x_j^2}{\mathsf{d}(v_j)} \right) \\
&\leqslant \sup_{\|\mathbf{x}\|=1} 2 \parallel \mathbf{x} \parallel^2 = 2.
\end{aligned}$$

22. If $\mathcal{G}_1 = (V_1, E_1)$ and $\mathcal{G}_2 = (V_2, E_2)$ are two graphs such that $V_1 \subseteq V_2$ and $E_1 \subseteq E_2$. Prove that $\beta(\mathcal{G}_1) \leqslant \beta(\mathcal{G}_2)$.

    Solution: By Corollary 16.24 we have $\alpha(\overline{\mathcal{G}_2}) \leqslant \alpha(\overline{\mathcal{G}_1})$. Therefore, $\beta(\mathcal{G}_1) = |V_1| - \alpha(\overline{\mathcal{G}_1}) \leqslant |V_2| - \alpha(\overline{\mathcal{G}_2}) = \beta(\mathcal{G}_2)$.
    A set of vertices U in a graph $\mathcal{G} = (V, E)$ is *independent* if no two vertices in U are joined by an edge in E.
23. Let $\mathcal{G} = (V, E)$ be a graph with $|V| = m$. Prove that if $\mathcal{G}$ contains an independent set of vertices U, then $\alpha(\mathcal{G}) \geqslant m - |U|$.

    Solution: Observe that $\overline{\mathcal{G}}$ contains a complete subgraph $\mathcal{K}_p$, where $p = |U|$. Therefore, by Supplement 22, $p \leqslant \beta(\overline{\mathcal{G}}) = m - \alpha(\mathcal{G})$, which produces the desired inequality.
24. Prove that if the Laplacian spectrum of a threshold graph $\mathcal{G} = (V, E)$ is $\lambda_1 \geqslant \lambda_2 \geqslant \cdots \geqslant \lambda_n = 0$, then $\lambda_j = |\{v \in V \mid \mathsf{d}(v) \geqslant j\}|$.

    Hint: The argument is by induction on the length of the sequence of operations used to construct the graph.
25. Let $\mathcal{G}_1 = (V, E_1)$ and $\mathcal{G}_2 = (V, E_2)$ be two graphs having the same set of vertices V and let $\mathcal{G}_1 \cup \mathcal{G}_2 = (V, E_1 \cup E_2)$, $\mathcal{G}_1 \cap \mathcal{G}_2 = (V, E_1 \cap E_2)$. Prove that:

    (a) $\mathsf{d}_{\mathcal{G}_1}(v) + \mathsf{d}_{\mathcal{G}_2}(v) = \mathsf{d}_{\mathcal{G}_1 \cup \mathcal{G}_2}(v) + \mathsf{d}_{\mathcal{G}_1 \cap \mathcal{G}_2}(v)$ for every $v \in V$.
    (b) $L_{\mathcal{G}_1} + L_{\mathcal{G}_2} = L_{\mathcal{G}_1 \cup \mathcal{G}_2} + L_{\mathcal{G}_1 \cap \mathcal{G}_2}$.

26. Let $\mathcal{G} = (V, E)$ be a graph. Prove that $\beta(\mathcal{G}) \leqslant |||L_{\mathcal{G}}|||_1 \leqslant 2 \max_{v \in V} \mathsf{d}(v)$.

    Solution: The first inequality is an immediate consequence of Theorem 7.65. Since $|||L_{\mathcal{G}}|||_1 = \max_j \sum_{i=1}^{m} (L_{\mathcal{G}})_{ij}$, where $|V| = m$, the second inequality follows immediately from the definition of the Laplacian matrix.
27. Let $\kappa = \{V_1, \ldots, V_k\}$ be a partition of the set of vertices of a graph $\mathcal{G} = (V, E)$ such that $|V_p| = m_p$ for $1 \leqslant p \leqslant k$ and $\sum_{p=1}^{k} m_p = m = |V|$. Furthermore, assume that the vertices in $V_1, \ldots, V_k$ are numbered consecutively. Let $B \in \mathbb{R}^{m \times m}$ be a block-diagonal matrix having $J_{m_1,m_1}, \ldots, J_{m_k,m_k}$ as its diagonal blocks. Prove that *trace*$(L_{\mathcal{G}} B) = 2|E'|$, where E' be the set of edges in E having their endpoints in two distinct blocks of κ.

    Solution: Note that the element ℓ_{ij} of the Laplacian matrix $L_{\mathcal{G}}$ can be written as $\ell_{ij} = d_i \delta_{ij} - a_{ij}$, where

$$\delta_{ij} = \begin{cases} 1 & \text{if } i = j, \\ 0 & \text{otherwise,} \end{cases}$$

    $d_i = \mathsf{d}(v_i)$ and a_{ij} is the (i, j)-element of the adjacency matrix $A_{\mathcal{G}}$.

For $trace(L_{\mathcal{G}}B)$ we can write:

$$trace(L_{\mathcal{G}}B) = \sum_{p=1}^{m}(L_{\mathcal{G}}B)_{pp} = \sum_{p=1}^{m}\sum_{i=1}^{m}\ell_{pi}b_{ip} = \sum_{p=1}^{m}\sum_{i=1}^{m}(d_i\delta_{pi} - a_{pi})b_{ip}$$

$$= \sum_{p=1}^{m}\sum_{i=1}^{m}d_i\delta_{pi}b_{ip} - \sum_{p=1}^{m}\sum_{i=1}^{m}a_{pi}b_{ip} = \sum_{p=1}^{m}d_pb_{pp} - \sum_{p=1}^{m}\sum_{i=1}^{m}a_{pi}b_{ip}.$$

Since $b_{pp} = 1$, and

$$b_{ip} = \begin{cases} 1 & \text{if } (v_i, v_p) \in E \text{ and} \\ & v_i, v_p \text{ belong to the same block } V_j, \\ 0 & \text{otherwise,} \end{cases}$$

it follows that $trace(L_{\mathcal{G}}B) = 2|E| - 2(|E| - |E') = 2|E'|$.

28. Using the notations of Supplement 27 assume further that $m_1 \geqslant m_2 \geqslant \cdots \geqslant m_k$. If $\lambda_1 \geqslant \lambda_2 \geqslant \cdots \geqslant \lambda_m$ are the eigenvalues of $L_{\mathcal{G}}$, prove that

$$|E_c| \geq \frac{1}{2}\sum_{p=1}^{k} m_p\lambda_p.$$

Solution: The spectrum of B consists of the numbers $m_1, \ldots, m_k$ (each having algebraic multiplicity 1) and 0 having algebraic multiplicity $\sum_{p=1}^{k} m_p - k$. By Supplement 47 of Chap. 10 we have

$$trace(L_{\mathcal{G}}B) \leqslant \sum_{p=1}^{k} m_p\lambda_p.$$

Taking into account Supplement 27 it follows that $|E'| \leqslant \sum_{p=1}^{k} m_p\lambda_p$.

Bibliographical Comments

Several general introductions in data mining [10, 13] provide excellent references for clustering algorithms. Basic reference books for clustering algorithms are [4, 14]. Recent surveys such as [1, 15] allow the reader to get familiar with current issues in clustering. Cluster features discussed in Exercise 6 were considered in the BIRCH algorithm [16]. Exercises 7–10 contain results obtained in [12]. Theorem 16.39 was obtained in [17].

The result described in Supplement 15 was established in [18].

Supplements 27–28 contain results obtained in [19].

References

1. A.K. Jain, M.N. Murty, P.J. Flynn, Data clustering: a review. ACM Comput. Surv. **31**, 264–323 (1999)
2. T. Kurita, An efficient agglomerative clustering algorithm using a heap. Pattern Recogn. **24**, 205–209 (1991)
3. P. Berkhin, J. Becher, Learning simple relations: theory and applications. ed. by R.L. Grossman, J. Han, V. Kumar, H. Mannila, R. Motwani, in *Proceedings of the 2nd SIAM International Conference on Data Mining*, (Arlington, 2002), pp. 420–436
4. L. Kaufman, P.J. Rousseeuw, *Finding Groups in Data—An Introduction to Cluster Analysis* (Wiley Interscience, New York, 1990)
5. M. Fiedler, Algebraic connectivity of graphs. Czechoslovak Math. J. **23**, 298–305 (1973)
6. B. Nadler, M. Galun, Fundamental limitation of spectral clustering, *Advances in Neural Information Processing Systems*, vol 19 (MIT Press, Cambridge, 2007), pp. 1017–1024
7. L. Hagen, A. Kahng, New spectral methods for ratio cut partitioning and clustering. IEEE Trans. Comput. Aided Des. **11**, 1074–1085 (1992)
8. J. Shi, J. Malik, Normalized cuts and image segmentation. IEEE Trans. Pattern Anal. Mach. Intell. **22**, 888–905 (2000)
9. P. Perona, W. Freeman, A factorization approach to grouping. In *European Conference on Computer Vision* (1998), pp. 655–670
10. M. Steinbach, G. Karypis, V. Kumar. A comparison of document clustering techniques. ed. by M. Grobelnik, D. Mladenic, N. Milic-Freyling. *KDD Workshop on Text Mining*, (Boston, 2000)
11. S. Ray, R. Turi, Determination of number of clusters in k-means clustering in colour image segmentation. In *Proceedings of the 4th International Conference on Advances in Pattern Recognition and Digital Technology*, (Narosa, New Delhi, 1984), pp. 137–143
12. U. Brandes, D. Delling, M. Gaertler, R. Görke, M. Hoefer, Z. Nikolski, D. Wagner, On modularity clustering. IEEE Trans. Knowl. Data Eng. **20**(2), 172–188 (2008)
13. P.N. Tan, M. Steinbach, V. Kumar, *Introduction to Data Mining* (Addison-Wesley, Reading, 2005)
14. A.K. Jain, R.C. Dubes, *Algorithm for Clustering Data* (Prentice Hall, Englewood Cliffs, 1988)
15. P. Berkhin, A survey of clustering data mining techniques, in *Grouping Multidimensional Data—Recent Advances in Clustering*, ed. by J. Kogan, C. Nicholas, M. Teboulle(Springer, Berlin, 2006), pp. 25–72
16. T. Zhang, R. Ramakrishnan, M. Livny, Birch: a new data clustering algorithm and its applications. Data Min. Knowl. Disc. **1**(2), 141–182 (1997)
17. M. Fiedler, A property of eigenvectors of nonnegative symmetric matrices and its applications to graph theory. Czechoslovak Math. J. **25**, 619–633 (1975)
18. J. Kleinberg, An impossibility theorem for clustering, in *Advances in Neural Information Processing Systems*, 15, Vancouver, Canada, 2002, ed. by S. Becker, S. Thrun, K. Obermayer. (MIT Press, Cambridge, 2003), pp. 446–453
19. W.E. Donath, A.J. Hoffman, Lower bounds for the partitioning of graphs. IBM J. Res. Dev. **17**, 420–425 (1973)

Index

D. A. Simovici and C. Djeraba, *Mathematical Tools for Data Mining*, Advanced Information and Knowledge Processing, DOI: 10.1007/978-1-4471-6407-4,

P

Q

T

U

Printed in the United States
By Bookmasters